D1089019

THE HISTORY OF CARTOGRAPHY

VOLUME ONE

EDITORIAL ADVISORS

Luís de Albuquerque   Joseph Needham
J. H. Andrews   David B. Quinn
Józef Babicz   Maria Luisa Righini Bonelli†
Marcel Destombes†   Walter W. Ristow
O. A. W. Dilke   Arthur H. Robinson
L. A. Goldenberg   Avelino Teixeira da Mota†
George Kish   Helen M. Wallis
Cornelis Koeman   Lothar Zögner

†Deceased

The History of Cartography

1
*Cartography in Prehistoric, Ancient, and Medieval Europe
and the Mediterranean*

2
Cartography in the Traditional Asian Societies

3
Cartography in the Age of Renaissance and Discovery

4
Cartography in the Age of Science, Enlightenment, and Expansion

5
Cartography in the Nineteenth Century

6
Cartography in the Twentieth Century

THE HISTORY OF CARTOGRAPHY

VOLUME ONE

Cartography in Prehistoric, Ancient, and Medieval Europe and the Mediterranean

Edited by

J. B. HARLEY

and

DAVID WOODWARD

THE UNIVERSITY OF CHICAGO PRESS · CHICAGO & LONDON

J. B. Harley is professor of geography
at the University of Wisconsin—Milwaukee, formerly
Montefiore Reader in Geography at the University of Exeter.

David Woodward is professor of geography
at the University of Wisconsin—Madison.

The University of Chicago Press, Chicago 60637
The University of Chicago Press, Ltd., London

© 1987 by The University of Chicago
All rights reserved. Published 1987
Printed in the United States of America

96 95 94 93 92 91 90 89 88   5 4 3 2

This work is supported in part by grants from the Division of
Research Programs of the National Endowment for the
Humanities, an independent federal agency

Additional funds were contributed by
The Andrew W. Mellon Foundation
The National Geographic Society
The Hermon Dunlap Smith Center for
the History of Cartography, The Newberry Library
The Johnson Foundation
The Luther I. Replogle Foundation
The British Academy

and the following individuals

Richard B. Arkway   John T. Monckton
Joe C. W. Armstrong   Mr. and Mrs. Kenneth Nebenzahl
Clive A. Burden   Rear Admiral G. S. Ritchie
Gerald F. Fitzgerald   Walter E. Scott
Peter J. Guthorn   Richard H. Sigel
Arthur Holzheimer   Mrs. L. M. C. Smith
Mr. and Mrs. Roderick Webster

Library of Congress Cataloging in Publication Data

The history of cartography.

Bibliography: p.
Includes index.
Contents: v. 1. Cartography in prehistoric, ancient,
and medieval Europe and the Mediterranean.
1. Cartography--History. I. Harley, J. B. (John
Brian) II. Woodward, David, 1942-
GA201.H53 1987   526'.09   86 6995
ISBN 0-226-31633-5 (v. 1)

Contents

LIBRARY
ALMA COLLEGE
ALMA, MICHIGAN

v

Illustrations
With Tables and Appendixes

TABLES

Preface

This *History of Cartography* was born of a belief in the importance of maps, and their underlying cartographic concepts and techniques, in the long-term development of human society and culture. Curiosity about space—no less than about the dimension of time—has reached from the familiar immediate surroundings to the wider space of the earth and its celestial context. On another plane, men and women have explored with the inward eye the shape of sacred space and the realms of fantasy and myth. As visual embodiments of these various conceptions of space, maps have deepened and expanded the consciousness of many societies. They are the primary medium for transmitting ideas and knowledge about space. As enduring works of graphic synthesis, they can play a more important role in history than do their makers. In this sense their significance transcends their artifactual value. As images they evoke complex meanings and responses and thus record more than factual information on particular events and places. Viewed in such a light, as a focus for social and cultural history, the history of cartography can be placed in its proper context, an essential part of a much wider humanistic endeavor. In number and scale, the six volumes of this *History* have been planned accordingly.

The present *History* has had to build on new foundations.[1] As an independent subject, the history of cartography occupies a no-man's-land among several paths of scholarship. History, geography, and bibliography, for instance, are well represented in its literature,[2] but the treatment of maps on their own terms is sketchy. Theoretical studies of the nature and historical importance of maps are relatively few. Even basic definitions have not been clearly formulated. As editors, therefore, we have had to turn first to the concepts carried by terms such as "cartography," "map," and "history of cartography," since it is on such clarifications that the scope and content of the entire work must rest. In this Preface, therefore, we will attempt to convey our understanding of these key words.

In existing histories of cartography the current definitions of "map" and "cartography" seem to have been accepted uncritically. Their subject matter has accordingly been selected on the basis of the perceived func-

tions, areas, or periods of map production rather than on the basis of an objective definition. At most there may be a simple statement that the main area of study is geographical maps. One of the more explicit in this respect was Leo Bagrow, in his *History of Cartography*, who quoted the French mathematician J. L. Lagrange (1779): "A geographical map is a plane figure representing the surface of the earth, or part of it."[3] Although Bagrow considered Lagrange's definition "perfectly adequate" for the purposes of his book,[4] it is clear today that it imposed an undue restriction on the scope of the history of cartography. In recent decades, as cartography has become a more distinct field of study, a broader outlook has emerged. In 1964, for instance, the newly established British Cartographic Society clarified its own terms of reference by adopting a much more catholic definition. The society saw cartography as "the art, science and technology of making maps, together with their study as scientific documents and works of art," and it amplified this by explaining that "in this context maps may be regarded as including all types of maps, plans, charts and sections, three-dimensional models and globes, representing the earth or any heavenly body at any scale."

In particular cartography is concerned with all "stages of evaluation, compilation, design and draughting required to produce a new or revised map document from

---

1. For a fuller discussion see pp. 24–26.
2. The aims of the project are described in J. B. Harley and David Woodward, "The History of Cartography Project: A Note on Its Organization and Assumptions," *Technical Papers*, 43d Annual Meeting, American Congress on Surveying and Mapping, March 1983, 580–89.
3. "Une carte géographique n'est autre chose qu'une figure plane qui représente la surface de la Terre, ou une de ses parties." J. L. Lagrange, "Sur la construction des cartes géographiques," *Nouveaux Mémoires de l'Académie Royale des Sciences et Belles-Lettres* (1779), 161–210, quotation on 161.
4. Leo Bagrow, *History of Cartography*, rev. and enl. R. A. Skelton, trans. D. L. Paisey (Cambridge: Harvard University Press; London: C. A. Watts, 1964), 22; Bagrow does, however, discuss on the same page the etymology of the word "chart" (*Karte*), which can also mean "map." For his textbook, Gerald R. Crone, *Maps and Their Makers: An Introduction to the History of Cartography*, 2d ed. (London: Hutchinson University Library, 1962), xi, also defines the purpose of maps in relation to the "earth's surface."

all forms of basic data. It also includes all stages in the reproduction of maps. It encompasses the study of maps, their historical evolution, methods of cartographic presentation and map use."[5]

Such a definition, when also linked to the concept of a history of communication by maps, enlarges the proper subject matter of the history of cartography, as will be made clear below.[6] It is significant that "all types of maps" were specifically included, as were the technical processes of mapmaking. The present *History* will survey a similarly broad field.

Another conceptual obstacle in the history of cartography has been a confusion over the meaning associated with the word "map" in different periods and cultures. In a sense the subject has become a prisoner of its own etymology. The fundamental problem is that in many ancient languages there was no exclusive word for what we now call a map. In European languages such as English, Polish, Spanish, and Portuguese, for example, the word map derives from the Late Latin word *mappa*, meaning a cloth. In most of the other European languages, the words used for map—French *carte*, Italian *carta*, Russian *karta*—derive from the Late Latin *carta*, which meant any sort of formal document. These distinctly different derivations result in ambiguities that persist to this day, since these words continue to carry more than one meaning.[7] In Russian, for example, the word for picture is *kartina*, and in fact in many early historical societies, those of medieval and Renaissance Europe, for instance, it was common to use words such as "picture" or "description" for what we would today call a map. Thus the apparently simple question, What is a map? raises complex problems of interpretation.[8] The answer varies from one period or culture to another. This issue is particularly acute for maps in early societies, but it also occasions difficulty, if not confusion, with those maps that can be regarded as a type of picture and indeed were often produced as such by painters or artists who were not specialist mapmakers.[9] We have not therefore assumed that the lack of vocabulary is in itself sufficient grounds for dismissing the map as a latecomer to the cultural scene. On the contrary, this volume provides ample evidence that maps existed long before they entered the historical record and before their makers and users called them maps.[10]

We have therefore adopted an entirely new definition of "map," one that is neither too restrictive nor yet so general as to be meaningless. What has eventually emerged is a simple formulation:

Maps are graphic representations that facilitate a spatial understanding of things, concepts, conditions, processes, or events in the human world.

Such a definition reflects the fundamental concern of the *History* both with maps as artifacts and with the way maps store, communicate, and promote spatial understanding. It is also designed to free the subject from some of the more restrictive interpretations of its scope. The words "human world" (in the widest sense of man's cosmographic surroundings) signal that the perspective of the *History* is not confined to those maps of the earth whose description constitutes so much of the existing literature. Our treatment thus naturally extends to ce-

5. *Cartographic Journal* 1 (1964): 17. One of the earlier acts of the International Cartographic Association in 1962 was to agree to set up a commission to study the standardization of technical terms. It was formally established in 1964, with national subcommittees, among which the British subcommittee adopted this definition in its *Glossary of Technical Terms in Cartography*, British National Committee for Geography (London: Royal Society, 1966). In an abbreviated form, omitting the final paragraph, it was incorporated in International Cartographic Association, *Multilingual Dictionary of Technical Terms in Cartography*, ed. E. Meynen (Wiesbaden: Franz Steiner Verlag, 1973), and to this extent at least it came to represent an international consensus about the scope of cartography. A revised edition of the *Dictionary* is in preparation.

6. For a discussion of the development of the concept that the mapping process functions as a formal system of communication, see pp. 33–36, and the references cited there.

7. P. D. A. Harvey, *The History of Topographical Maps: Symbols, Pictures and Surveys* (London: Thames and Hudson, 1980), 10. The Latin word *carta* is from the Greek χάρτη (*chartes*, papyrus). Harvey notes that we find a similar pattern in non-European languages. In most Indian languages the word for map derives from the Arabic *naqshah*, but other meanings attached to it include picture, general description, and even official report. In Chinese, *tu* is no less ambiguous: besides map it can also mean a drawing or diagram of any kind.

8. For a discussion of this problem in a prehistoric context see pp. 60–62. In the early literate societies of Europe and the Mediterranean the problem remains, and it is particularly difficult to resolve in archaic and classical Greek—where the two most common words for a map are *periodos* and *pinax*—as well as in Latin, where *forma* can also mean shape. To some extent the problem still exists. In Italian, for example, owing to the various meanings of *carta*, Osvaldo Baldacci invented the word *geocarta*; he has used the new word in his historical work for the past several years. In particular, it is a key word of Baldacci's journal *Geografia*, founded in 1978 in Rome, in the same institute formerly directed by Roberto Almagià. The invention of *geocarta* is an attempt to specify the content of a *carta* (geo stands for geography) in order to avoid confusion with *carta*, a document on paper. Nevertheless, historians of cartography, as we assert in this preface, do not deal only with geographical maps.

9. Examples recur throughout the volume.

10. Mircea Eliade, *A History of Religious Ideas*, trans. Willard R. Trask (Chicago: University of Chicago Press, 1978), vol. 1, *From the Stone Age to the Eleusinian Mysteries*, 7 and n. 4, points to this problem in general in the history of culture. With maps, analogies can be drawn with other classes of objects that existed—and that are shown to exist in the archaeological record—long before the specific words for them are found in the historical record. This applies, of course, to all prehistoric objects; but from the classical period, for example, itineraries are preserved from Augustus's time onward, yet the word *itinerarium* first occurs in Vegetius, writing after A.D. 383, and we know of no equivalent Latin word or phrase. We owe this example to O. A. W. Dilke.

lestial cartography and to the maps of imagined cosmographies. In implementing this definition we have also sought to avoid criteria specific to particular cultures based on the historical-literate experience. Consequently, discussion in this work is not confined, like Samuel Johnson's definition,[11] to those maps revealing a graticule of latitude and longitude. Nor do we necessarily require that they incorporate the projective, coordinate, and Euclidean geometries currently associated with maps and usually linked with systems of numeration and metrology. Many early maps did not possess these geometries, being topologically structured in relation to networks of routes, drainage systems, coastlines, or boundaries.[12]

Some of these points also apply to the word "cartography." This word is a neologism, coined by Manuel Francisco de Barros e Sousa, Viscount of Santarém, in the mid-nineteenth century in particular reference to the study of early maps.[13] The meaning of the word cartography has changed since Santarém's day. It has broadened to include the art and science of contemporary mapmaking as well as the study of early maps. On the other hand, it has also narrowed to such an extent that it is difficult to relate an interpretation of the scope of cartography, as defined for the *History*, to the realities of cartographic practice in the 1980s. The diversification of mapping techniques in recent decades has led to a tendency to divorce from cartography subjects that are nevertheless crucial to our enterprise. International practice in this respect is extremely varied: in some countries modern cartography is defined to exclude the processes of data collection in mapmaking, such as land and hydrographic surveying, aerial photography, and, most recently, remote sensing.[14] There are, moreover, signs that cartography itself is seeking a still narrower perspective. Suggestions have been made that the subject might be confined to those operations concerned with the design of maps or even, more radically still, solely with philosophical and theoretical foundations.[15] Whatever the merits of such definitions in the context of contemporary practice, they have been firmly rejected for the *History*, even though such a decision greatly increases the variety of topics, size of the literature, and diversity of methodology, and thus the problem of synthesis, particularly for the two volumes concerned with the nineteenth and twentieth centuries.

The meanings thus attached to the words "map" and "cartography" in this *History* have also led us to a specific understanding of the "history of cartography." This term too has frequently been a source of confusion. For example, for some the distinction between "history of cartography" and "historical cartography" still remains unclear.[16] Another problem can be anticipated. It is already clear that in the later volumes of the *History* a

distinction will have to be drawn between the history of cartography defined, on the one hand, as the history of methods of making and using maps and, on the other, as the history of the discipline of cartography in terms of its theoretical foundations, principles, and rules for

11. Samuel Johnson defines a map as "a geographical picture on which lands and seas are delineated according to the longitude and the latitude," in *A Dictionary of the English Language* (London, 1755).

12. The cartographic significance of topology as a branch of mathematics is discussed with historical examples by Naftali Kadmon, "Cartograms and Topology," *Cartographica* 19, nos. 3–4 (1982): 1–17. See also Carl B. Boyer, *A History of Mathematics* (New York: John Wiley, 1968); Klaus Mainzer, *Geschichte der Geometrie* (Mannheim: Bibliographisches Institut, 1980); or Nicolas Bourbaki, *Eléments d'histoire des mathématiques*, new ed. (Paris: Hermann, 1974).

13. The word cartography is derived from the Greek word *chartes* used in Late Greek, meaning a sheet of paper or papyrus, that is, the material on which the map was drawn in later times. See p. 12 for further documentation.

14. In fact some of these activities—surveying, photogrammetry and, in particular, remote sensing—have become increasingly independent, with their own literature and their own international organizations. On the other hand, the definition of cartography adopted by the United Nations is very broad: "Cartography is considered as the science of preparing all types of maps and charts, and includes every operation from original surveys to final printing of copies"; *Modern Cartography: Base Maps for World Needs*, document no. 1949.I.19 (New York: United Nations Department of Social Affairs, 1949), 7. It is noted in *Glossary*, 11 (note 5 above), that British practice excluded land and hydrographic surveying and photogrammetry from the field of cartography; similarly, in Austria and Germany a narrower interpretation is given to cartography: see, for example, Erik Arnberger, "Die Kartographie als Wissenschaft und ihre Beziehungen zur Geographie und Geodäsie," in *Grundsatzfragen der Kartographie* (Vienna: Österreichische Geographische Gesellschaft, 1970), 1–28; Günter Hake, *Der wissenschaftliche Standort der Kartographie*, Wissenschaftliche Arbeiten der Fachrichtung Vermessungswesen der Universität Hannover, no. 100 (Hanover, 1981), 85–89; and F. J. Ormeling, "Einige Aspekte und Tendenzen der modernen Kartographie," *Kartographische Nachrichten* 28 (1978): 90–95. The *Multilingual Dictionary* (note 5) excludes from consideration terms relating more specifically to methods and processes of surveying, photogrammetric compilation, and general printing. Remote sensing and photogrammetry now have their own equivalent dictionary: George A. Rabchevsky, ed., *Multilingual Dictionary of Remote Sensing and Photogrammetry* (Falls Church, Va.: American Society of Photogrammetry, 1983).

15. Arthur H. Robinson and Barbara Bartz Petchenik, *The Nature of Maps: Essays toward Understanding Maps and Mapping* (Chicago: University of Chicago Press, 1976), 19. Phillip C. Muehrcke, *Thematic Cartography*, Commission on College Geography Resource Paper no. 19 (Washington, D.C.: Association of American Geographers, 1972), 1.

16. These are still loosely employed as synonyms by some writers. It is now generally accepted that "historical cartography" is conveniently reserved for the practice of compiling maps in the present from historical data: for a discussion see R. A. Skelton, *Maps: A Historical Survey of Their Study and Collecting* (Chicago: University of Chicago Press, 1972), 62–63; David Woodward, "The Study of the History of Cartography: A Suggested Framework," *American Cartographer* 1, no. 2 (1974): 101–15, esp. 107–8; Michael J. Blakemore and J. B. Harley, *Concepts in the History of Cartography: A Review and Perspective*, Monograph 26, *Cartographica* 17, no. 4 (1980): 5–8.

maps and mapping procedures. Setting aside such complications, the definitions adopted for the *History* are thus not an attempt to cater to every major (still less minor) cartographic event that has taken place but an effort to establish broad criteria to underpin the universal aims of the entire work. These criteria can be precisely spelled out. They involve, first, acceptance of a catholic definition of "map"; second, commitment to a discussion of the manifold technical processes that have contributed to the form and content of individual maps; third, recognition that the primary function of cartography is ultimately related to the historically unique mental ability of map-using peoples to store, articulate, and communicate concepts and facts that have a spatial dimension; and fourth, the belief that, since cartography is nothing if not a perspective on the world, a general history of cartography ought to lay the foundations, at the very least, for a world view of its own growth.[17] Together these four criteria summarize the basic scope of the *History of Cartography*.

The organization of the *History* arises from these principles. In planning the volumes it soon became clear that the choice of appropriate time periods, world regions, and identifiable themes would in itself considerably influence not only the choice of the cartographic events described but also the nature of the theories advanced in their interpretation. The overall framework of the *History* is simultaneously chronological and geographical. It is chronological inasmuch as both the individual volumes and their principal sections are generally organized in terms of broad time periods. It is geographical in the sense that the continents of the Old and New Worlds, the major cultural provinces within them, and specific areas of national interest are also used to structure the narrative.[18] In five of the six volumes, the major chronological divisions reflect those devised by Western historians.[19] Thus this first volume, devoted to cartography in Europe and the Mediterranean down to about 1470, is subdivided into sections for the prehistoric, ancient, and medieval periods. Subsequent volumes deal first with the cartography of Renaissance Europe and then in turn with mapping in the eighteenth, nineteenth, and twentieth centuries. In these volumes the perspective is at first European but increasingly becomes a world view corresponding to the growth of international relationships in cartography. These time periods do not avoid the limitations that beset any attempt at periodization in historical writing: by their very nature they are artificial divisions. Even so, we believe they are indispensable and unavoidable. They do provide a means by which the history of cartography can be related to the wider context of other aspects of historical change.[20] They allow us to view individual events within the long-term processes of their own development, and they will eventually facilitate comparative judgments about the cartography of different ages and societies. Indeed, that such comparative judgments cannot be properly made, and that maps cannot be fully understood historically unless we recognize that they are an integral part of the simultaneous histories of art and of science as well as of the wider realms of political and social activity, emerges constantly from each volume. Accordingly, though it is possible to debate the precise meaning and exact limits of such Western terms as the Renaissance or the Enlightenment—and their relevance to all aspects of cartography will indeed be often questioned—they have been retained to help bridge the gap between the specific subject matter of the *History* and the broader context of social and cultural history necessary for its interpretation.[21]

17. We do this despite several distinguished precedents for confining general history to a largely European perspective. See, for example, the argument advanced for this course of action in Charles Singer et al., eds., *A History of Technology*, 7 vols. (Oxford: Clarendon Press, 1954–78), vol. 1, *From Early Times to Fall of Ancient Empires*, vi. See also J. H. Clapham and Eileen Power, *The Cambridge Economic History of Europe from the Decline of the Roman Empire*, 7 vols. (Cambridge: Cambridge University Press, 1941–78), vol. 1, *The Agrarian Life of the Middle Ages*, v, and vol. 4, *The Economy of Expanding Europe in the Sixteenth and Seventeenth Centuries*, ed. E. E. Rich and C. H. Wilson, xiii–xiv, where the "uncompromisingly European" approach was justified on the "conviction that the world-economy which resulted *was* European in incentive, in organization, and in its preoccupations." Although cartographic history has sometimes been written in terms of these assumptions, we have tried to preserve a balance by allowing Asian developments to be reported on their own terms.

18. It is, however, our intention to try to avoid creating, especially in later volumes, what has been described elsewhere as a mere "collection of separate national histories bound together in the same covers": George Clark, "General Introduction: History and the Modern Historian," in *The New Cambridge Modern History* (Cambridge: Cambridge University Press, 1957–79), vol. 1, *The Renaissance, 1493–1520*, ed. G. R. Potter, xxxv.

19. Volume two deals with the cartography of the Asian societies in their traditional periods.

20. The issues of periodization are discussed in Gordon Leff, *History and Social Theory* (University: University of Alabama Press, 1969), 130–51. See also Fritz Schalk, "Über Epoche und Historie," part of "Studien zur Periodisierung und zum Epochebegriff," by Hans Diller and Fritz Schalk, *Abhandlungen der Akademie der Wissenschaften und der Literatur, Mainz*, Geistes- und Sozialwissenschaftliche Klasse (1972): 150–76. The problems of periodization also feature prominently in Marxist historiography; see, for example, *A Dictionary of Marxist Thought*, ed. Tom Bottomore (Oxford: Blackwell Reference, 1983), 365–68.

21. The more extreme view of Otto Neugebauer, *The Exact Sciences in Antiquity*, 2d ed. (Providence: Brown University Press, 1957), 3, that in "the history of mathematics and astronomy the traditional division of political history into Antiquity and Middle Ages is of no significance," for example, has not been accepted for cartography. Nor for the purposes of the general *History* have we adopted Ulrich Freitag's interesting division of the history of cartography into eras of communication: Ulrich Freitag, "Die Zeitalter und Epochen der Kartengeschichte," *Kartographische Nachrichten* 22 (1972): 184–91; see also Ulrich Freitag, "Zur Periodisierung der Geschichte der Kartogra-

The sequence adopted for the individual volumes has also been designed to mitigate the usual tendency to write cartographic history only as seen through European eyes. As editors, we have been all too conscious of the extent to which a deeply entrenched Eurocentricity has dominated the literature of the subject.[22] To redress this imbalance somewhat, volume 2 has been devoted entirely to cartography in the historical Asian societies. The fundamental links between East and West have long been expounded in the literature of the history of cartography,[23] but the three indigenous spheres of Asian mapping—the Islamic, the South and Southeast Asian, and the East Asian—have received very uneven treatment and have been virtually ignored in the standard histories of cartography. Thus we have particularly welcomed the opportunity to create a cartographic history corresponding to the major civilizations of Asia and structured independently of the chronologies, priorities, and values of mapping in the Western world. In so doing, we explicitly recognize that Asian cartographies, just as much as European, have been fundamental pillars of cartographic development when viewed on a world scale. A single volume cannot, of course, entirely compensate for historical imbalances in the literature, but we believe it is at least a step in the right direction.

The detailed subdivisions of the volumes are also an attempt to do justice to the great richness and variety of map genres in different cultures, to the multiplicity of uses to which they have been put, and to the complexity of the technical and social processes that underlie them. In such chapters, more localized chronologies of mapping structure each narrative, along with regional subdivisions or thematic essays where these reflect distinctive cartographic cultures. Indeed, a principal aim of the *History* is to highlight these map-using cultures. The work as a whole has been designed to emphasize the creative contribution of the mapping undertaken *within* such areas, rather than to be a mere commentary on the content of specific landmark maps that happen to show the particular region irrespective of their context or origin.[24] Such a distinction has not been clearly made in previous histories of cartography. It is important because it lets us spotlight the making and using of maps in their primary historical contexts rather than focusing on changes of representation divorced from cartographic process.[25]

Finally, we would like to comment on one aspect of the organization of the *History* as a whole. From the outset, the *History* was planned as a multiauthor work. In taking this road we were aware that some might see a collaborative venture on this scale as more cumbersome than a work of individual or dual authorship. Thus Toynbee, in an attack on "synthetic histories" (which to him represented the "industrialization of historical

thought"), forcefully expressed his preference for the "works of historical literature . . . created by single minds."[26] It is arguable, however, that in the present stage of the subject's development a satisfactory general history of cartography could only with the greatest difficulty be created by a single mind. If Max Eckert could make this point in 1921, there is a much stronger basis for our concurrence today.[27] A project such as the *History* is achievable only through a division of labor. No single scholar with the necessary breadth of linguistic and methodological skills and subject background (and without the commonly revealed nationalistic bias) has emerged to write it alone. The risks of multiple authorship should be no greater for a general history of cartography than for existing specialist works on science, technology, astronomy, and music, or for the collective social, economic, and political histories that have inspired our present plan.[28]

phie Thailands," in *Kartenhistorisches Colloquium Bayreuth '82: Vorträge und Berichte* (Berlin: Reimer, 1983), 213–27. For a recent discussion of periodization in the history of cartography, see also Pápay Gyula, "A kartográfiatörténet korszakolásának módszertani kédései," *Geodézia és Kartografia* 35, no. 5 (1983): 344–48.

22. See pp. 28–29.

23. Most convincingly by Joseph Needham, *Science and Civilisation in China* Cambridge: Cambridge University Press, 1954–), vol. 3, *Mathematics and the Sciences of the Heavens and the Earth.*

24. This has been described as an approach that can "all too easily become little more than a catalogue, a set of descriptions of one map after another"; Harvey, *Topographical Maps*, 7 (note 7).

25. In the case of North America, for example, no detailed review will be provided on how its different regions were portrayed by mapmakers in Berlin, London, Paris, and elsewhere; for example there will be no chapter detailing the history of the representation of California as an island. Such themes have already been extensively described, and some may here form part of the study of the mapping, or map trades, of those European countries in the appropriate volumes.

26. Arnold J. Toynbee, *A Study of History*, 12 vols. (London: Oxford University Press, 1934–[61]), 1:4–5. See also the arguments set out in E. A. Gutkind, *The International History of City Development*, 8 vols. (New York: Free Press of Glencoe, 1964–72), vol. 1, *Urban Development in Central Europe*, 10–11.

27. Max Eckert, *Die Kartenwissenschaft: Forschungen und Grundlagen zu einer Kartographie als Wissenschaft*, 2 vols. (Berlin: Walter de Gruyter, 1921–25), 1:26. "Als ein grosser Mangel ist in der geographischen Wissenschaft das Fehlen einer Geschichte der Karte und damit einer Geschichte der Kartographie empfunden worden. Sie dürfte bis auf weiteres noch kaum geschreiben werden. Die Zeit scheint noch nicht reif dazu zu sein. Es fehlen noch zu viele Vorarbeiten." (The absence of a history of the map and hence a history of cartography has been perceived as a great shortcoming in the science of geography. It probably won't be written in the foreseeable future. The time doesn't seem to be ripe for it yet. Too many preliminary studies are missing. Translation by Guntram Herb, University of Wisconsin–Madison.)

28. Among the works that have especially influenced our design, the following most closely parallel our own intentions: René Taton, *Histoire générale des sciences*, 3 vols. in 4 parts (Paris: Presses Universitaires de France, 1957–64; English edition, *History of Science*, trans. A. J. Pomerans, 4 vols. (London: Thames and Hudson, 1963–66); Singer et al., *History of Technology* (note 17); and *The New*

Given the real possibility of a team of specialist scholars working toward a common goal, the concept of a general *History of Cartography* at once became feasible. It could be amply justified not only by the historical importance of maps, as already asserted, but also in view of the inadequacy of existing general works.[29] Equally persuasive was the urgent need to integrate an increasingly technical and analytical yet highly fragmented literature on the various types of maps. Even where genres belong together, within the same cartographic culture, they have often been treated separately. For instance, globes and other geographical instruments have been studied as independent artifacts and reported in specialized journals; celestial mapping has often been regarded as a branch of the history of astronomy rather than cartography; the history of hydrographic mapping is being drawn into the history of nautical science; and the history of thematic mapping is written about in specialist journals of the natural or social sciences. This is entirely proper from the viewpoint of those other subjects, but it does not deny that there is also an overwhelming case for reintegrating these genres into a single developmental account of the historical meaning, relevance, and significance of maps in general.

It could be said, perhaps, that the moment is never right for this kind of general synthesis. This *History* will certainly reveal its share of the gaps and imbalances in our existing knowledge. Nevertheless, it could—and should—act as a springboard for future developments in the subject as a whole. A particular aim is that it will be able to contribute, as its assumptions and research priorities are developed in line with the wider currents of ideas in the humanities and social sciences, to a strengthening of interest in the history of cartography.

Since the preliminary planning for the *History of Cartography* began in 1975, we have accumulated more scholarly debts than we can ever properly acknowledge or repay. The research foundations and other bodies that, together with a number of individuals, have given us such generous financial support have already been named separately. Their faith in our enterprise has been crucial in developing the concept of a multivolume history, and it has enabled us to build up a small organization to administer the project as a whole. Through this, we have been able to undertake essential research, to mount seminars and discussions, to carry on bibliographical checking, and to search for illustrations with a thoroughness that would have been impossible without such resources. Similar thanks must go to our own parent institutions—the Universities of Exeter and Wisconsin—who not only have provided us with basic facilities to carry on our work but have also granted generous periods of study leave since 1975 so that we could undertake research, writing, and editing.

The Newberry Library, Chicago, where David Woodward was director of the Hermon Dunlap Smith Center for the History of Cartography when the project was conceived, has continued to be its spiritual home. Its president and librarian, Lawrence W. Towner, has lovingly supported it from the very beginning. The present director of the Hermon Dunlap Smith Center for the History of Cartography at the Newberry Library, David Buisseret, has continued to welcome us to seminars and lectures and has provided accommodation for some of the *History*'s editorial meetings. Likewise, the Newberry map curator, Robert Karrow, has remained a fountain of bibliographical knowledge for the project as a whole. In the matter of bibliographical research, the resources of the American Geographical Society Collection at the University of Wisconsin—Milwaukee have also proved indispensable, and we are grateful to its director, Roman Drazniowsky, and the director of libraries, William Roselle, for so helpfully smoothing the path of our many inquiries.

In a collaborative work of this nature our greatest academic indebtedness has been to our advisers and fellow authors. The members of our Editorial Advisory Board have all played far more than a nominal role, and they have greatly assisted us in the initial planning of the volumes, in the difficult task of recruiting authors, and, lately, in discussing a series of structural changes in the organization of the volumes as the scope has continued to evolve. It is with great sadness that we record that three of our most valued editorial advisers in the early years—Maria Luisa Righini Bonelli, Marcel Destombes, and Avelino Teixeira da Mota—did not live to see the publication of this first volume. On a happier note, though, we have discovered that authors can become firm friends while remaining our sternest critics. In this volume, together with other readers, they have read and commented freely on chapters other than their own, and we have no doubt that—though the final responsibility lies elsewhere—the text has benefited considerably from the advice of Michael Conzen, Catherine Delano Smith, D. R. Dicks, O. A. W. Dilke, P. D. A. Harvey, G. Malcolm Lewis, David Quinn, A. L. F. Rivet, and Arthur H. Robinson. All the authors are thanked for their stoical patience as we have wrestled with a series of editorial changes designed to bring the content of

---

Oxford History of Music, 10 vols. (London: Oxford University Press, 1957–82). Methodologically as well as substantively we also owe a tremendous debt to the Cambridge Histories: *The Cambridge Ancient History*, orig. 12 vols. (Cambridge: Cambridge University Press, 1924–39); *The Cambridge Medieval History*, orig. 8 vols. (Cambridge: Cambridge University Press, 1911–36); *The Cambridge Modern History*, orig. 13 vols. (Cambridge: Cambridge University Press, 1902–10). Among comparable works in progress there may be noted the multivolume *General History of Astronomy* (Cambridge: Cambridge University Press, 1984–).

29. See pp. 24–26.

specific chapters in line with the broader aims of the *History* as a whole. Our personal authors' acknowledgments, like those of the other contributors, are recorded as the first footnote to each chapter.

As the *History* has grown in substance and complexity, supported by its funding, we have also been fortunate to work with staff whose efficiency and loyalty prevented the ship from foundering on the rocks of correspondence and footnotes and, in the early days at least, on the arcane mysteries of the word processor. The main office has been in Madison, and here Maureen Reilly has been a tower of strength since the formal inception of the project. For bibliographical checking and research inquiries we were extremely fortunate to have the services of a historian of science, Elaine Stroud, until June 1984. Since then the bulk of this work has been taken on by Judith Leimer, assisted by Gary Chappell, Matthew Edney, Kevin Kaufman, Chingliang Liang, and Barbara Weisman. In the design and production of the line drawings we wish to acknowledge Onno Brouwer and James Hilliard of the University of Wisconsin Cartographic Laboratory. The Inter-Library Loan Department of the University of Wisconsin Memorial Library has likewise provided essential and efficient support.

In January 1984 we were able to appoint Anne Godlewska as assistant project director, and though her main editorial responsibilities lie in future volumes, we have both benefited enormously from her enthusiasm and from having a fresh mind brought to bear on the final effort of getting volume 1 to the Press.

In Exeter, Judy Gorton and Denise Roberts have between them coped with a voluminous correspondence as well as with typing manuscripts. Among graduate students in the Department of Geography, Michael Turner and Sarah Wilmot have provided much intelligent research assistance. In London, Francis Herbert of the Royal Geographical Society has answered numerous bibliographical queries. In their customary fashion, staff members of the British Library, in both the Map Library and the Department of Manuscripts, have assisted us greatly in our capacity both as authors and as editors.

Even a single book is a partnership between author, editor, and publisher. Given the complexity and length of the present *History*, these relationships have become an especially necessary condition of success. We feel indeed fortunate, therefore, that the University of Chicago Press had sufficient faith to take up the idea of a general history of cartography and a commitment that must have seemed, in its early days in particular, very open-ended. For his initial support and enthusiasm and for piloting our proposal toward a contract, we are especially grateful to Allen Fitchen, now director of the University of Wisconsin Press. Barbara Hanrahan, his successor at the University of Chicago Press, was equally positive and supportive. In the designing of the book and in copyediting, as in all other matters, it has been a pleasure to work with the Press.

Especially when they have families, editors cannot shut themselves away in ivory towers. As the *History* has increasingly encroached on our private lives—as it is discussed at the dinner table and the authors become household names—even our younger children sometimes sense the traumas of editorship. We should both like to thank our families for their forbearance, support, and love while we have been engaged in a seemingly endless task. Without them, especially, we would be neither writing this Preface nor contemplating five more volumes.

1 · The Map and the Development of the History of Cartography

J. B. HARLEY

THE HISTORICAL IMPORTANCE OF THE MAP

The principal concern of the history of cartography is the study of the map in human terms. As mediators between an inner mental world and an outer physical world, maps are fundamental tools helping the human mind make sense of its universe at various scales. Moreover, they are undoubtedly one of the oldest forms of human communication. There has probably always been a mapping impulse in human consciousness, and the mapping experience—involving the cognitive mapping of space—undoubtedly existed long before the physical artifacts we now call maps. For many centuries maps have been employed as literary metaphors and as tools in analogical thinking.[1] There is thus also a wider history of how concepts and facts about space have been communicated, and the history of the map itself—the physical artifact—is but one small part of this general history of communication about space.[2] Mapping—like painting—precedes both written language and systems involving number, and though maps did not become everyday objects in many areas of the world until the European Renaissance, there have been relatively few mapless societies in the world at large. The map is thus both extremely ancient and extremely widespread; maps have impinged upon the life, thought, and imagination of most civilizations that are known through either archaeological or written records.

Any appreciation of the historical importance of maps depends upon a clear conception of their nature, of the factors that have shaped their making and transmission, and of their role within human societies. In these respects the starting assumption is that maps constitute a specialized graphic language, an instrument of communication that has influenced behavioral characteristics and the social life of humanity. Maps have often served as memory banks for spatial data and as mnemonics in societies without printing. Scholars over the centuries have been convinced of the eloquence and expressive power of maps, which can speak across the barriers of ordinary language. A group of American historians has asserted that maps "constitute a common language used by men of different races and tongues to express the relationship of their society . . . to a geographic envi-

ronment."[3] In the *History of Cartography* we have gone further and accepted language as a metaphor for the

I owe a considerable debt to those who have helped me formulate the ideas as well as the substance in this inevitably eclectic essay. Alan R. H. Baker (University of Cambridge) provided, through his theoretical writings, the initial stimulus to search for a deeper understanding of the place of maps in history, while the late R. A. Skelton, by his outstanding example of fertile scholarship, long ago convinced me that the history of cartography constitutes a discrete and important field of study. Among those who contributed material to an earlier draft of this Introduction, I am especially grateful to John Andrews (Trinity College, University of Dublin), Michael J. Blakemore (University of Durham), Christopher Board (London School of Economics and Political Science), Tony Campbell (British Library), Catherine Delano Smith (University of Nottingham), O. A. W. Dilke (University of Leeds), P. D. A. Harvey (University of Durham), Francis Herbert (Royal Geographical Society), Roger J. P. Kain (University of Exeter), Cornelis Koeman (University of Utrecht), Monique Pelletier (Bibliothèque Nationale), David B. Quinn (University of Liverpool), Günter Schilder (University of Utrecht), Gerald R. Tibbetts (Senate House Library, University of London), Sarah Tyacke (British Library), Vladimiro Valerio (University of Naples), and Denis Wood (North Carolina State University), and Lothar Zögner (Staatsbibliothek Preußischer Kulturbesitz).

1. The extent to which the map has become an almost universal metaphor is indicated by the second definition of a map in *Webster's Third New International Dictionary of the English Language* (1976): "something (as a significant outward appearance, a pointed or concise verbal description) that indicates or delineates or reveals by representing or showing with a clarity suggestive of that of a map." For a discussion of the importance of the map analogy in scientific research, see Stephen Toulmin, *The Philosophy of Science: An Introduction* (London: Hutchinson University Library, 1953), esp. chap. 4, "Theories and Maps," 105–39. For a recent example of the sustained use of the map analogy in teaching the history and philosophy of science, see units 1–3 in *Mapping Inquiry* (Milton Keynes: Open University Press, 1981). The present *History* cannot be systematically concerned with the development of these metaphorical uses, although it should be borne in mind that in various societies they may provide some index of how much familiarity and sophistication in handling maps writers assumed among their audience or readers.

2. This wider history would include, for example, the study of spatial representation in architecture, dance, drama, geometry, gesture, landscape and town plans, music, and painting as well as in oral speech and written language. Such a list serves also as a guide to topics that are not systematically considered within the *History* even where they provide examples of communication that was spatial in intention.

3. Frank Freidel, ed., *Harvard Guide to American History*, rev. ed., 2 vols. (Cambridge: Belknap Press of Harvard University Press, 1954), 1:44–47, where the importance of maps in the history of geographic exploration, diplomacy, economic development, social planning, and

way maps have been used in past societies as well as a means of tracing their spread through time and space. We must accept, although our general position is founded in semiology, that precise scientific analogies to the structure of language may be impossible to sustain;[4] but as a general metaphor, helping to fashion an approach to the history of cartography, the concept of a graphic language—and the map as a graphic text—is valid. The significance of maps—and much of their meaning in the past—derives from the fact that people make them to tell other people about the places or space they have experienced. This implies that throughout history maps have been more than just the sum of technical processes or the craftsmanship in their production and more than just a static image of their content frozen in time. Indeed, any history of maps is compounded of a complex series of interactions, involving their use as well as their making. The historical study of maps may therefore require a knowledge of the real world or of whatever is being mapped; a knowledge of its explorers or observers; a knowledge of the mapmaker in the narrower sense as the originator of the artifact; a knowledge of the map itself as a physical object; and a knowledge of the users (or—more likely—the community of map users). The *History of Cartography* is concerned, as far as possible, with the historical process by which the graphic language of maps has been created and used. At once a technical, a cultural, and a social history of mapping, it rejects the view of a historian of discovery who wrote that "cartographical studies do not come within the sphere of social history."[5] On the contrary, it favors an approach that is potentially capable of exploring the behavioral and ideological implications of its subject matter.

A major problem in assessing the importance of maps for the historical study of society is the paradox constituted by the map itself. On the one hand, the map appears at first sight as a relatively simple iconic device. Indeed, much of its universal appeal is that the simpler types of map can be read and interpreted with only a little training. Throughout history—though ways of looking at maps have to be learned even within oral societies—formal literacy has not been a precondition for them to be made or read. An anthropologist has remarked that "the making and reading of two dimensional maps is almost universal among mankind whereas the reading and writing of linear scripts is a special accomplishment associated with a high level of social and technical sophistication."[6] Thus maps have been associated with cultures that differ widely in social or technological development, while modern psychological research has shown that children can derive meaning from maps (and indeed draw them) from an early age.[7] There is an immediacy about the message in a map that makes

it more readily perceived than knowledge encoded in other ways. One of the map's properties is that it can be taken in quickly by the eye, contributing to the potency of cartographic images. It has been said that maps have an "extraordinary authority," even when they are in error, that may be lacking in other forms of images.[8]

---

warfare is especially stressed. See also Carl O. Sauer, "The Education of a Geographer," *Annals of the Association of American Geographers* 46 (1956): 287–99, esp. 289, where he wrote that "the map speaks across the barriers of language; it is sometimes claimed as the language of geography."

4. Arthur H. Robinson and Barbara Bartz Petchenik, *The Nature of Maps: Essays toward Understanding Maps and Mapping* (Chicago: University of Chicago Press, 1976), discuss the analogy at length. It is also rejected as an exact analogy by J. S. Keates, *Understanding Maps* (New York: John Wiley, 1982), 86, although he continues to employ it as a metaphor for the way maps "can be studied as ordered structures." Another recent discussion is C. Grant Head, "The Map as Natural Language: A Paradigm for Understanding," in *New Insights in Cartographic Communication*, ed. Christopher Board, Monograph 31, *Cartographica* 21, no. 1 (1984): 1–32, and Hansgeorg Schlichtmann's "Discussion" of the Head article, ibid., 33–36. Our context in relation to semiology is that maps form a system of signification. This is defined by Roland Barthes, *Elements of Semiology*, trans. Annette Lavers and Colin Smith (New York: Hill and Wang, [1968]), 9: "Semiology . . . aims to take in any system of signs, whatever their substance and limits; images, gestures, musical sounds, objects, and the complex association of all these, which form the content of ritual, convention or public entertainment: these constitute, if not *languages*, at least systems of signification."

5. C. R. Boxer, *The Portuguese Seaborne Empire, 1415–1825* (London: Hutchinson, 1969), 396. It is not suggested that this view is typical; see also note 139 below.

6. Edmund Leach, *Culture and Communication: The Logic by Which Symbols Are Connected: An Introduction to the Use of Structuralist Analysis in Social Anthropology* (Cambridge: Cambridge University Press, 1976), 51.

7. Jean Piaget and Bärbel Inhelder, *The Child's Conception of Space*, trans. F. J. Langdon and J. L. Lunzer (London: Routledge and Kegan Paul, 1956), esp. chap. 14. Although Piaget and his followers have dominated the study of human intelligence for nearly six decades, stimulating more recently a number of critical assessments as well as *exposées raisonées*, the spatial aspect has received very little attention in this voluminous literature. Piaget himself produced only one other directly relevant book: Jean Piaget, Bärbel Inhelder, and Alina Szeminska, *The Child's Conception of Geometry* (New York: Basic Books, 1960). Among recent reassessments of the Piagetian theory in general see Linda S. Siegel and Charles J. Brainerd, *Alternatives to Piaget: Critical Essays on the Theory* (New York: Academic Press, 1978), and Herbert Ginsburg and Sylvia Opper, *Piaget's Theory of Intellectual Development*, 2d ed. (Englewood Cliffs, N.J.: Prentice-Hall, 1979). Piaget's ideas have also been adapted in an attempt to provide a genetic epistemology for the study of images as cultural forms in general; see, for example, Sidney J. Blatt, *Continuity and Change in Art: The Development of Modes of Representation* (Hillsdale, N.J.: Lawrence Erlbaum Associates, 1984). This concept—for example, as manifest in the use of topological versus Euclidean geometry in maps—has not been applied to the history of cartographic representation and does not appear to coincide with the cultural sequences that can be observed empirically in this volume.

8. Arthur H. Robinson, "The Uniqueness of the Map," *American Cartographer* 5 (1978): 5–7. Kenneth E. Boulding, *The Image* (Ann

On the other hand, however simple maps may appear at first sight, on analysis they are almost certainly less than straightforward. Mapmaking is not a simple inborn skill, even among "primitive"peoples, as believed by an earlier generation of scholars. Moreover, maps are two-dimensional combinations of "shapes, sizes, edges, orientation, position, and relations of different masses"[9] that require painstaking interpretation in relation to their original purpose, their modes of production, and the context of their use. Maps created for one purpose may be used for others, and they will articulate subconscious as well as conscious values. Even after exhaustive scrutiny maps may retain many ambiguities, and it would be a mistake to think they constitute an easily readable language. Maps are never completely translatable. Historians cannot be alone in suggesting that they find maps an intractable form of evidence and "slippery witnesses" of the past.[10] In some respects—even after the development of a more sophisticated vocabulary of cartographic signs—maps are no less imprecise than written language. Although a key or legend may be provided, a line, a dot, or a color, for instance, may have had several meanings, both manifest and latent, and it is unwise to assume that identical cartographic signs have similar meanings, or even a common origin, when found in different cultures. Thus even today, despite notable advances in the theory of cartography,[11] maps remain "a complex language . . . whose properties we know very little about."[12] The proper understanding of maps, like that of any other ancient or modern language or like the reading of art or music, is a major challenge, even more so since contemporary cartographers are still trying to decide the "grammar" of current maps so that we can better understand how they are used. As representations of belief and ideology—rooted in particular cultures and institutions—as well as "factual" images of scientific knowledge, maps are increasingly being recognized as touching the subject matter of a wide range of scholarly disciplines. The value of the map as a humanly created document is one of the major themes in the *History of Cartography*.

Making a map, it is often said, involves both art and science. Similarly, if the study of how maps have communicated in the past is starting to reflect the hermeneutic concerns of scholars in many fields, then there are narrower scientific aspects of the history of cartography that are part of the traditional history of science and technology. These latter are the better known. The historical importance of maps has often been indexed to the progress of mapmaking as a scientific and practical skill, and this view is still deeply entrenched in the writings on cartographic history. Gerald R. Crone's words, written in 1953, that "the history of cartography is largely that of the increase in the accuracy with which

. . . elements of distance and direction are determined and . . . the comprehensiveness of the map content"[13] still have a measure of acceptance. Other writers have pointed out that the history of the map relates how many have striven to establish cartography as a precise science;[14] that it is concerned with measuring the "rate of cartographic progress";[15] and that it involves the study of "scientific conquest of the unknown."[16] The contribution of these approaches is that they have secured for the history of cartography an established place in the traditional histories of science and technology.[17] We also accept that a fundamental theme in the *History of Cartography* is the scientific development of mapping, with its related instrumentation and increasing mathematical sophistication.

Taken alone, however, this aspect fails to provide a balanced view of the development of maps in history. It assumes a linear historical progression and, moreover (somewhat anachronistically), assumes that accuracy of measurement and comprehensiveness were as important throughout the past as they have been in the modern period. Thus it is at least arguable that an overemphasis on the scientific frontiers and the revolutions of mapping, on landmarks and innovations, or on the saga of

Arbor: University of Michigan Press, 1956), 65–68, is among the philosophers who have commented on the authority of the map.

9. Wilbur Zelinsky, "The First and Last Frontier of Communication: The Map as Mystery," *Bulletin of the Geography and Map Division, Special Libraries Association* 94 (1973): 2–8, quotation on 7–8.

10. J. A. Williamson, *The Voyages of John and Sebastian Cabot*, Historical Association Pamphlet no. 106 (London: G. Bell, 1937), 7; J. H. Parry, "Old Maps Are Slippery Witnesses," *Harvard Magazine* (Alumni ed.), April 1976, 32–41.

11. See below, pp. 33–34.

12. David Harvey, *Explanation in Geography* (London: Edward Arnold, 1969; New York: St. Martin's Press, 1970), 370; see also Robinson and Petchenik, *Nature of Maps*, chap. 3 (note 4).

13. Gerald R. Crone, *Maps and Their Makers: An Introduction to the History of Cartography*, 1st ed. (London: Hutchinson University Library, 1953), xi. The work has been published in five editions: 1953, 1962, 1966, 1968, and 1978.

14. C. Bricker and R. V. Tooley, *Landmarks of Mapmaking: An Illustrated Survey of Maps and Mapmakers*, preface by Gerald R. Crone (Brussels: Elsevier-Sequoia, 1968), 5.

15. R. A. Skelton, *Maps: A Historical Survey of Their Study and Collecting* (Chicago: University of Chicago Press, 1972), 106.

16. Lloyd A. Brown, *The Story of Maps* (Boston: Little, Brown, 1949; reprinted New York: Dover, 1979), 4.

17. The systematic attention paid by George Sarton, *Introduction to the History of Science*, 3 vols. (Baltimore: Williams and Wilkins, 1927–48), to the annals of geographical knowledge, including maps, sets the standard in this respect. Short sections on cartography are generally included even in concise histories of science; see, for example, Charles Singer, *A Short History of Scientific Ideas to 1900* (Oxford: Clarendon Press, 1959; reprinted 1966). Earlier volumes of Charles Singer et al., eds., *A History of Technology*, 7 vols. (Oxford: Clarendon Press, 1954–78), contain important essays on cartography and navigation, but cartography is dropped in volumes 6–7, *The Twentieth Century, c. 1900 to c. 1950*, ed. Trevor I. Williams.

how the unmappable was finally mapped[18] has distorted the history of cartography: the historical importance of maps must also be related to the social implications of their varied format and subject matter. As Robinson and Petchenik put it, the map is at once "so basic and has such a multiplicity of uses" that "the variety of its occurrences is vast," elaborating further that:

> There are specific maps and general maps, maps for the historian, for the meteorologist, for the sociologist, and so on without limit. Anything that can be spatially conceived can be mapped—and probably has been. Maps range in size from those on billboards or projection screens to postage stamps, and they may be monochrome or multicolored, simple or complex. They need not be flat—a globe is a map; they need not be of earth—there are maps of Mars and the moon; or for that matter, they need not be of any place real—there have been numerous maps made of imaginary "places" such as Utopia and even of the "Territory of Love."[19]

The historical record, it will be seen, yields a still longer list. In particular, accuracy is not regarded as the sole criterion for including maps as objects for serious study within the *History*. Historiography confirms that, for example, in many cultures crude, distorted, plagiarized, ephemeral, oversimplified, and small-scale maps have been neglected. Such scientific chauvinism dictates that they are often dismissed as not maps at all or labeled as mere oddities or cartographic curiosities. Many early maps were imagined evocations of space rather than realistic records of geography. The pages of the *History* have been deliberately opened up to this wide range of maps, including those created for nonscientific or nonpractical purposes that, although not part of the history of cartographic science in the narrowest sense, are nevertheless part of the history of human communication by means of maps.

What is true of types of maps, whether classified by purpose or by form, is also true of ways they are known to have been used since the cartographic record began. This aspect also has a history of an ever-widening diversity. Crone remarked that "a map can be considered from several aspects, as a scientific report, a historical document, a research tool, and an object of art";[20] but since he wrote it has become much clearer that these scientific, historical, and artistic dimensions by no means exhaust the importance of maps in human terms. Far from being purely practical documents—surrogates of space[21] or the mind's miniatures of real distribution[22]—maps have played an important role in stimulating the human imagination to reach for the very meaning of life on earth. An appreciation of the way maps have helped shape human beings' ideas of their relationship to the natural world, and of the way they have acted as doc-

uments in the wider history of ideas, is based on an observation of the frequency—starting in prehistoric times and including nonliterate societies—with which maps have been used as teleological instruments, epitomizing the sacred and mythical space of cosmologies as well as the more tangible landscapes of the real world.[23] There is today a growing awareness of the importance of these other cartographic roles.[24] Recognition of the ideological, religious, and symbolic aspects of maps, particularly when linked with a more traditional appreciation of maps for political and practical purposes, greatly enhances the claim that cartography can be regarded as a graphic language in its own right. Maps, we find, are such a basic and universal form of human communication that the *History* has not had to seek its justification in some esoteric backwater in the development of civilization but finds its purposes in some of the most central aspects of human activities.

If these are bold claims for maps, they are based on a conviction of their considerable, yet only partly understood, relevance to the study of the development of human societies. Maps may indeed be a "sensitive indicator of the changing thought of man, and . . . an excellent mirror of culture and civilisation,"[25] but they are also more than a mere reflection: maps in their own right enter the historical process, to which they are linked by means of reciprocally structured relationships. The development of the map, whether it occurred in one

18. A recent extreme example of this emphasis on the "scientific heroes" of map history is John Noble Wilford, *The Mapmakers* (New York: Alfred A. Knopf; London: Junction Books, 1981); see the review by Denis Wood, *Cartographica* 19, nos. 3–4 (1982): 127–31.

19. Robinson and Petchenik, *Nature of Maps*, 15 (note 4).

20. Crone, *Maps and Their Makers*, 1st ed., ix (note 13).

21. Robinson and Petchenik, *Nature of Maps*, 86 (note 4).

22. The phrase is that of Robert Harbison, *Eccentric Spaces* (New York: Alfred A. Knopf, 1977), chap. 7, "The Mind's Miniatures: Maps," 124–39.

23. For an introduction to examples of such sacred and mythical conceptualizations but surprisingly without explicit reference to the literature of the history of cartography see Yi-Fu Tuan, *Topophilia: A Study of Environmental Perception, Attitudes, and Values* (Englewood Cliffs, N.J.: Prentice-Hall, 1974); also idem, *Space and Place: The Perspective of Experience* (Minneapolis: University of Minnesota Press, 1977). A valuable survey is found in Mircea Eliade, *A History of Religious Ideas*, trans. Willard R. Trask (Chicago: University of Chicago Press, 1978), vol. 1, *From the Stone Age to the Eleusinian Mysteries*, chap. 1.

24. For example, Hermann Kern's exploration of the labyrinth in literature and art contains much that is relevant, not least in its interdisciplinary approach, to the history of world maps in various cultures: *Labirinti: Forme e interpretazione, 5000 anni di presenza di un archetipo manuale e file conduttore* (Milan: Feltrinelli, 1981); German edition, *Labyrinthe: Erscheinungsformen und Deutungen, 5000 Jahre Gegenwart eines Urbilds* (Munich: Prestel-Verlag, 1982).

25. Norman J. W. Thrower, *Maps and Man: An Examination of Cartography in Relation to Culture and Civilization* (Englewood Cliffs, N.J.: Prentice-Hall, 1972), 1.

place or at a number of independent hearths, was clearly a conceptual advance—an important increment to the technology of the intellect[26]—that in some respects may be compared to the emergence of literacy or numeracy. An archaeologist has recently observed that when men moved from cognitive mapping to a "mapping process" that "involves the production of a material 'map' . . . we face a documented advance in intelligent behaviour,"[27] an argument that is stated more comprehensively by Robinson:

> The use of a reduced, substitute space for that of reality, even when both can be seen, is an impressive act in itself; but the really awesome event was the similar representation of distant, out of sight, features. The combination of the reduction of reality and the construction of an analogical space is an attainment in abstract thinking of a very high order indeed, for it enables one to discover structures that would remain unknown if not mapped.[28]

It follows that the spread of the idea of the map from its origins, the growth of formal map knowledge, the adoption of distinctive geometrical structures for maps, the acquisition of maps as tools for practical and intellectual purposes, the gradual and sometimes sudden technical improvement of maps through new techniques, and later the ability to reproduce maps exactly by mechanical means have all been of major significance in the societies where they occurred. The processes of transmission underlying these changes—from their earliest beginnings to the age of mass and now computer cartography—also become a central concern of the history of cartography.

Thus several main threads have been identified that are woven through the history of cartography. They all rest on the axiom that the map is a historical phenomenon of great significance in human terms, with a rich harvest to be gleaned from its systematic study. Maps—like books—can be regarded as agents of change in history.[29] The history of cartography represents more than a technical and practical history of the artifact. It may also be viewed as an aspect of the history of human thought, so that while the study of the techniques that influence the medium of that thought is important, it also considers the social significance of cartographic innovation and the way maps have impinged on the many other facets of human history they touch.

RENAISSANCE TO ENLIGHTENMENT: THE EARLY ANTECEDENTS OF THE HISTORY OF CARTOGRAPHY

Ways of thinking about the history of early maps have often proved tenacious. The ideas and preoccupations of each period may survive as important ingredients in the thought and practices of the succeeding period. Most

of this chapter is devoted to a historiographical essay on the history of cartography. In the Western world[30] this development, albeit with national variations in chronology and precise direction, may be divided into three periods. The first deals with developments to about 1800; the second, with the nineteenth century and the early part of the twentieth (up to ca. 1930); and the third with the past fifty years, which have seen the emergence of a scholarly identity for the subject. Although these three periods can be identified in historiographical terms, older ways of approaching the study of early maps survived as orthodoxies into the recent past.

The scope of this review needs to be carefully defined. It is devoted not to the large literature relating to contemporary maps in each period—which will be dealt with in chronological context in the individual volumes—but to the historical writings of successive generations about the maps of earlier generations, a relatively small body of literature. Yet to those who dismiss the achievements of the pioneers in the history of cartography, the volume of such writings may come as a surprise. Moreover, they are sufficiently diverse in character to extend to at least some of the topics regarded today as lying at the heart of the history of cartography.[31] Disappointment awaits those who expect to en-

26. Jack Goody, ed., *Literacy in Traditional Societies* (Cambridge: Cambridge University Press, 1968), 1–11, assessing the social importance of the acquisition of writing.

27. Colin Renfrew, *Towards an Archaeology of Mind*, Inaugural Lecture, University of Cambridge, 30 November 1982 (Cambridge: Cambridge University Press, 1982), 18–19.

28. Arthur H. Robinson, *Early Thematic Mapping in the History of Cartography* (Chicago: University of Chicago Press, 1982), 1. Others have commented on the intellectual achievement of the map. For example, Michael Polanyi, *The Study of Man*, Lindsay Memorial Lectures (London: Routledge and Kegan Paul, 1959), 24, comments on "the great speculative advantage achieved by storing up knowledge in a handy, condensed form. Maps, graphs, books, formulae, etc., offer wonderful opportunities for reorganizing our knowledge from ever new points of view."

29. In this respect the history of the map is directly comparable to the history of the book as envisaged by Lucien Febvre: Lucien Febvre and Henri-Jean Martin, *L'apparition du livre* (Paris: Editions Albin, 1958); English edition, *The Coming of the Book: The Impact of Printing, 1450–1800*, new ed., ed. Geoffrey Nowell-Smith and David Wootton, trans. David Gerard (London: NLB, 1976). See also Kenneth E. Carpenter, ed., *Books and Society in History: Papers of the Association of College and Research Libraries Rare Books and Manuscripts Preconference, 24–28 June 1980, Boston, Massachusetts* (New York: R. R. Bowker, 1983), and especially Elizabeth L. Eisenstein, *The Printing Press as an Agent of Change: Communications and Cultural Transformations in Early Modern Europe*, 2 vols. (Cambridge: Cambridge University Press, 1979).

30. Excluded here is any systematic discussion of writings on the history of Asian cartography produced within those cultures; these will be dealt with in the appropriate context of volume 2 of the present *History*.

31. The potential of this literature has been illustrated by a series of volumes known as *Acta Cartographica* in which selected articles

counter in such studies the same assumptions, priorities, and techniques that can be brought to bear on a history of maps today: but at the same time, such studies cannot be dismissed as merely old-fashioned or antiquarian.

There are three main reasons why the *History*, as a synthetic work, has taken notice of this classic literature and why these older writings need to be reviewed here. In the first place, many older studies preserve the only record (or reproduction) of maps, or their related sources, that no longer survive. The high mortality of maps, albeit not always "more severe than that of any other class of historical document,"[32] is an endemic condition that historians of cartography—like archaeologists—just have to live with. But recent wars and disasters have added their share of destruction.[33] Similarly, the breaking up and dispersal of once organic atlases[34] or the disappearance of cartographic items into inaccessible private collections accentuate the problem of document survival. For such reasons it is as hard to speculate about the number of maps that may originally have been produced in some early societies as it is to base generalizations on the surviving population of maps where the record is known to be so incomplete.

In the second place, so slow and uneven has been the pace of research in the history of cartography that some so called classic works have endured unchallenged as fundamental references. Research for this volume of the *History* has already proved this especially pertinent, and our remaining debt to such pioneer authorities as Nordenskiöld on portolan charts[35] or Konrad Miller on *mappaemundi*[36] will be clear from the appropriate chapters. Thus the sometimes fashionable view that such writings are no longer serviceable must be rejected as false concerning the substantive content of a general history of cartography, although it may be true in the sense that these have indeed ceased to be methodological exemplars.

Finally, past writings on the history of cartography are, after all, the primary sources for any account of that history's intellectual development. As new directions are being sought in the history of cartography, it is appropriate that there be some retrospection and that the past be scanned for the methodological lessons it may hold. It is pertinent not only to ask, for example, When did the study of the history of cartography begin? but also to recognize that the answers lie in the literature of the past. Similarly, the question, Can we sustain the notion of an emerging subject that we call the history of cartography? calls for a historiographical review. Of course much will depend on what is understood by "the history of cartography," and it begs the question whether such a concept was always separately formulated. This chapter will show how an understanding of what constitutes the history of cartography has altered

over the period during which early maps have been studied. The first to frame systematically such an agenda was R. A. Skelton, in his 1966 series of lectures,[37] and some of the same questions will be reexamined here, since they offer a relevant dimension to the intellectual heritage nurturing the present work.

ANTIQUARIES, COLLECTORS, AND MAPMAKERS AS CHRONICLERS OF THE SUBJECT

While it is difficult to put a precise date to the earliest writings in the history of cartography, it is probably safe to assume that it would have been broadly coeval with the origins of historical writing, particularly in cultures where geographical and historical studies were closely interrelated and served political purposes and where there is known to have been some tradition of preserving and collecting the maps of earlier periods.[38] Such conditions were met, for example, in China during the Han

from nineteenth- and early twentieth-century periodicals have been reproduced since 1967 and which now includes over 450 items from over 100 journals; recent volumes have extended the coverage further into the twentieth century. *Acta Cartographica*, vols. 1–27 (Amsterdam: Theatrum Orbis Terrarum, 1967–81).

32. Skelton, *Maps*, 26 (note 15). While Skelton's contention may be true of some categories of working maps—such as sea charts or wall maps constantly in use—it is doubtful that it is true of all maps, some of which, on the contrary, have a particular capacity for survival, being kept when other records are thrown away. There may, however, also be a bias for fine, "collectible" specimens to be preserved, making the surviving sample less representative of everyday cartography.

33. See, for example, the reports of cartographic destruction referred to in [Anon.], "With Fire and Sword," *Imago Mundi* 4 (1947): 30–31; A. Codazzi, "With Fire and Sword," *Imago Mundi* 5 (1948): 37–38; [Anon.], "With Fire and Sword," *Imago Mundi* 6 (1949): 38; Norbert Fischer, "With Fire and Sword, III," *Imago Mundi* 10 (1953): 56; Marian Łodiński, "With Fire and Sword, VI," *Imago Mundi* 14 (1959): 117; Fr. Grenacher, "With Fire and Sword, VII," *Imago Mundi* 15 (1960): 120. On evidence for the destruction of portolan charts and on Italian losses of maps during World War II, see below, chapter 19, "Portolan Charts from the Late Thirteenth Century to 1500."

34. Carl Christoph Bernoulli, "Ein Karteninkunabelnband der öffentlichen Bibliothek der Universität Basel," *Verhandlungen der Naturforschenden Gesellschaft in Basel* 18 (1906): 58–82, reprinted in *Acta Cartographica* 27 (1981): 358–82, describes the contents of an important sixteenth-century composite Italian atlas. Since the atlas has now been broken up, the article is the only remaining historical record of its original composition; for another example, see Wilhelm Bonacker, "Über die Wertsteigerung von Einzelblättern aus zerfledderten alten Atlaswerken," *Kartographische Nachrichten* 13 (1963): 178–79.

35. A. E. Nordenskiöld, *Periplus: An Essay on the Early History of Charts and Sailing-Directions*, trans. Francis A. Bather (Stockholm: P. A. Norstedt, 1897).

36. Konrad Miller, *Mappaemundi: Die ältesten Weltkarten*, 6 vols. (Stuttgart: J. Roth, 1895–98).

37. Skelton, *Maps* (note 15).

38. Some of these conditions in early cultures are described by Herbert Butterfield, *The Origins of History* (New York: Basic Books, 1981). In China, from about 1000 B.C. onward, there were archivists

and Chin dynasties, when contemporary geographical manuals sometimes reviewed earlier maps and indicated their shortcomings.[39] In much the same way, the practice of successive commentators, including Herodotus, Aristotle, Cleomedes, and Ptolemy, of criticizing earlier maps might be said to reflect an intuitive appreciation for a history of maps as a foundation of the modern geographical science of the Greek and Roman periods.[40] Likewise for medieval Europe, an awareness of the intellectual interest of noncontemporary maps can similarly be traced through some surviving texts containing *mappaemundi.* Skelton has suggested that the European Middle Ages saw "an incipient interest in the evolution of geographical ideas as expressed in maps," supporting this contention with the observation that world maps attributed to different types and periods, both Roman and post-Roman, were juxtaposed in some texts so that "we may discern a rudimentary historical sense applied to comparative cartography."[41] It is probable, however, that these maps were reproduced to illustrate changing cosmological ideas rather than to demonstrate the development of cartographic form or technique. For those seeking the beginnings of the study of the history of cartography, there is little weight to such straws in the wind.

During the European Renaissance, however, especially from the sixteenth century onward, it is possible to trace an increasingly systematic attention to the maps of preceding centuries. The extent to which this represented a genuine historical feeling for maps as independent documents should not be exaggerated, especially in view of the general surge of interest in the fifteenth and sixteenth centuries in classical geographical authors and of the fact that maps from the classical sources were valued as useful contemporary tools as well as vaunted as monuments of antiquity. One of these monuments of antiquity, Ptolemy's *Geography*—the touchstone of the Renaissance in European cartography—reveals to the modern historian of cartography the stages by which the antiquarian study of early maps hived off from the practical and technical development of contemporary mapping. In the fifteenth century, the Ptolemaic maps were initially valued as authoritative maps of the world and its regions, and it was only gradually, though accelerated by the application of printing to cartography, that they were replaced by the *tabulae modernae*,[42] leaving the classical maps as primarily historical objects. The Strasbourg Ptolemy of 1513 was the first to separate modern from ancient maps in a discrete section, reflecting the growth of a general critical sense among mapmakers and readers. This practice was confirmed in 1578 when Mercator reissued the Ptolemaic maps alone, without any modern supplements, as a facsimile of a classical atlas, thereby underlining their purely historical interest.[43]

Other historical maps were also reproduced, continuing a medieval tradition of manuscript copying in the Renaissance,[44] but it was the printed facsimiles of such maps that did most to stimulate their study and widen an appreciation of the cartography of earlier centuries. Notable examples, engraved from medieval manuscript sources, were the Peutinger map (reproduced in both the sixteenth and the eighteenth centuries),[45] Marino Sanudo's medieval tract *Liber secretorum fidelium crucis* (in which the map of Palestine, though secondary to the text, was essential for its interpretation) in the early seventeenth century,[46] and in the eighteenth century, Richard Gough's maps of medieval Britain.[47] By the eighteenth century too, in much of Europe, Renaissance

expressly responsible for "maps and records" (p. 142). In Mediterranean lands, however, scholarship, in the sense of research based on written texts, was relatively late in its development and awaited the growth of archives or libraries such as that established in Alexandria by, probably, the mid-third century B.C.: James T. Shotwell, *The History of History* (New York: Columbia University Press, 1939), 55.

39. Joseph Needham, *Science and Civilisation in China* (Cambridge: Cambridge University Press, 1954–), vol. 3, *Mathematics and the Sciences of the Heavens and the Earth*, 538–39, who cites the *Chin Shu* of Pei Hsiu (A.D. 224–71) in this respect. See also volume 2 of this *History* for a more detailed historiographical review.

40. In this sense one might also include Hipparchus's criticism of Eratosthenes, as recounted by Strabo, and Pliny's remarks on the province of Boetia, southern Spain, in Agrippa; see below, pp. 166–67 and 207–8.

41. Skelton, *Maps*, 64–65 (note 15).

42. The early date (1427) at which Claudius Clavus's map of the north was added to the manuscripts of Ptolemy does, however, suggest an early belief that the classical world picture might not be completely applicable to the fifteenth-century world. A full list of "modern" maps and commentary appears in a chapter on the rediscovery of Ptolemy's *Geography* and its reception in western Europe in volume 3 of the present *History*.

43. Skelton, *Maps*, 66 (note 15).

44. James Nelson Carder, *Art Historical Problems of a Roman Land Surveying Manuscript: The Codex Arcerianus A, Wolfenbüttel* (New York: Garland, 1978), 6. See also the account of the *Notitia Dignitatum* below, pp. 244–45.

45. Annalina Levi and Mario Levi, *Itineraria picta: Contributo allo studio della Tabula Peutingeriana* (Rome: Erma di Bretschneider, 1967), 17–25, for details of editions and earlier studies. In Renaissance Europe the first printed facsimile of an ancient map was probably the engraving of the Peutinger map commissioned by Abraham Ortelius, first published in 1598 and afterward included in several editions of his *Parergon.* See also Ekkehard Weber, ed., *Tabula Peutingeriana: Codex Vindobonensis 324* (Graz: Akademische Druck- und Verlagsanstalt, 1976).

46. Skelton, *Maps*, 69 (note 15); Marino Sanudo, *Liber secretorum fidelium crucis*, vol. 2 of *Gesta Dei per Francos*, ed. Jacques Bongars (Hanover: Heirs of J. Aubrius, 1611); Bongars also published facsimiles of three *mappaemundi.*

47. Richard Gough, *British Topography; or, An Historical Account of What Has Been Done for Illustrating the Topographical Antiquities of Great Britain and Ireland*, 2 vols. (London: T. Payne and J. Nichols, 1780); a list of maps is included in Ronald P. Doig, "A Bibliographical Study of Gough's *British Topography*," *Edinburgh Bibliographical Society Transactions* 4 (1963): 103–36, esp. 105–29.

as well as ancient and medieval maps were being recognized as part of the cartographic past. While the line between what was of contemporary and what was of historical interest was less closely drawn than it is today, a consciously antiquarian market for Renaissance maps was catered to both by keeping in print editions from the sixteenth-century copperplates and by issuing reengraved versions of fifteenth-century items such as Andrea Bianco's world maps (Venice, 1783) or Martin Behaim's globe (Nuremberg, 1730 and 1778).[48] Such examples further confirm that the ability of the print trade to produce facsimiles of early maps, together with the contemporary maps that formed its main stock-in-trade, was already laying the foundations for an intellectual climate that favored the recovery and preservation of the cartographic past as well as its more systematic study.

A second influence on this tendency after the sixteenth century was the widespread growth of map collections. This was an integral part of the general European enthusiasm for collecting in this period, but it must be regarded as a central influence in the historical development of the study of early maps. Although there is evidence of systematic map collecting, often for bureaucratic purposes, in the ancient civilizations of both China and Europe,[49] the growth of map collecting is an especially marked feature in the rise of cartographic consciousness in sixteenth-century Europe. Records show that maps were included in medieval libraries, but they were usually integral to the texts they illustrated,[50] and no separate inventories of maps are known to us from the Middle Ages.[51] Only after the sixteenth century can we begin to trace the emergence of maps and atlases as distinct categories within libraries as a whole or as items displayed as a group for decorative purposes. Thereafter, the formation of such map collections was rapid. It has been documented in many countries of Europe, not only in royal map collections (often the founding elements in today's national map libraries),[52] but also in wall displays such as those in the Vatican or the Palazzo Vecchio in Florence, in the houses of the nobility,[53] or in maps and atlases associated with the libraries and working papers of statesmen, leading churchmen, city dignitaries, merchants, and historians, many of whom were particularly interested in collecting plans of cities and towns. By the eighteenth century maps were increasingly being kept in independently designed sections with their own specialist curators, and this too helped create the conditions under which historical studies could develop.[54]

The truly antiquarian dimension in this map collecting, despite an interest in the recovery and preservation of ancient texts and monuments, is more difficult to isolate or to measure. Manuscript copies of Ptolemy's *Geography* were collected throughout Renaissance Europe;[55] but most map libraries owed their birth and development to the working copies of *contemporary* maps assembled as the political and military tools of statecraft, as raw materials in cartographers' workshops, as records of national exploration and discovery, as the working documents of trade and colonization, as specimens of graphic art, or in the case of astronomical maps, for the practice of astrology. Furthermore, one should not underestimate the practical value accorded to older maps in the sixteenth, seventeenth, and eighteenth centuries, when they were extensively drawn upon not only by cartographers but also by lawyers, politicians, and others as sources of information that, though old, were

48. Skelton, *Maps*, 71 (note 15), gives full references.

49. On China see volume 2 forthcoming; on Europe, pp. 210 and 244.

50. Leo Bagrow, "Old Inventories of Maps," *Imago Mundi* 5 (1948): 18–20. With *mappaemundi* the wider textual context of most surviving copies is especially apparent: Marcel Destombes, ed., *Mappemondes A.D. 1200–1500: Catalogue préparé par la Commission des Cartes Anciennes de l'Union Géographique Internationale* (Amsterdam: N. Israel, 1964).

51. Cornelis Koeman, *Collections of Maps and Atlases in the Netherlands: Their History and Present State* (Leiden: E. J. Brill, 1961), 12.

52. Helen Wallis, "The Royal Map Collections of England," *Publicaciónes do Centro de Estudos de Cartografia Antiga*, Série Separatas, 141 (Coimbra, 1981); Mireille Pastoureau, "Collections et collectionneurs de cartes en France, sous l'ancien-régime" (paper prepared for the Tenth International Conference on the History of Cartography, Dublin 1983).

53. On the Vatican maps in this category see Roberto Almagià, *Monumenta cartographica Vaticana*, 4 vols. (Rome: Biblioteca Apostolica Vaticana, 1944–52), vol. 3, *Le pitture murali della Galleria delle Carte Geografiche*. On the mural collection in Florence, see George Kish, "The Japan on the 'Mural Atlas' of the Palazzo Vecchio, Florence," *Imago Mundi* 8 (1951): 52–54; Giuseppe Caraci, "La prima raccolta moderna di grandi carte murali rappresentanti i 'quattro continenti'," *Atti del XVII Congresso Geografico Italiano, Trieste 1961*, 2 vols. (1962), 2:49–60; Koeman, *Collections*, 19 (note 51), notes that in about 1560 fifty maps decorated the rooms and gallery of Batestein castle in the Netherlands. See also Juergen Schulz, "Maps as Metaphors: Mural Map Cycles of the Italian Renaissance," in *Art and Cartography: Six Historical Essays*, ed. David Woodward (Chicago: University of Chicago Press, 1987).

54. There seems to have been a map room or charts room in the British Museum from its opening in 1759: Helen Wallis, "The Map Collections of the British Museum Library," in *My Head Is a Map: Essays and Memoirs in Honour of R. V. Tooley*, ed. Helen Wallis and Sarah Tyacke (London: Francis Edwards and Carta Press, 1973), 3–20. In France, the Dépôt des Cartes et Plans de la Marine was created in 1720, but its unofficial origin can be dated from Colbert, minister of Louis XIV; it existed in 1682, and its keeper was Charles Pene, editor of *Le Neptune François* (first published Paris: Imprimerie Royal, 1693). In 1720 Duc de Luynes was nominated "directeur du dépôt"; in 1721 Philippe Buache was draftsman in the dépôt. For other examples see Skelton, *Maps*, 26–52 (note 15). Dates of foundation of other map libraries are given in John A. Wolter, Ronald E. Grimm, and David K. Carrington, eds., *World Directory of Map Collections*, International Federation of Library Associations Publication Series no. 31 (Munich: K. G. Saur, 1985).

55. See volume 3 of the present *History*.

not necessarily out of date.[56] It would thus be wrong to regard the early collections as evidence, solely or even primarily, for the contemporary study of the history of cartography. There is no neat compartmentalization of motive in the acquisition of maps. On the contrary, the collecting mentality of the age is epitomized by the "cabinets of curiosities," or *Wunderkammer*, with which maps were occasionally associated, albeit valued as artifacts rather than as documents representing the growth of cartography.[57] The antiquarian element in most map collections of the period before 1800 tends to be incidental in origin, and it has to be most carefully sifted out from its broader context. For example, in England, Burghley's map collection was acquired almost wholly through political circumstances,[58] and the interests of Robert Cotton can be identified as partly political and partly historical.[59] So were those of Pepys,[60] linked to his naval responsibilities, and the driving force behind Ortelius's accumulation of maps—to judge by his correspondence—was as much his interest in maps as historical documents as their usefulness in the compilation of new maps.[61] Isaac Vossius, too, used maps to underpin not one but several of the aspects of his polymathic humanism,[62] and in Louvain the learned librarian Viglius ab Aytta Zuichemus had several motives in bringing together a wide variety of maps relating to the half-century before 1575.[63]

Analyzing the motives for forming such collections also lets us generalize about the intellectual and practical uses of early maps before 1800. The breadth of these applications is truly universal. Cosmography, geography, and chorography, as defined by Ptolemy and others, were regarded as inseparable.[64] Indeed, cosmography depended upon astronomical observation and mapping just as much as geography was based on terrestrial exploration and survey. In both cases, earlier maps were eagerly consulted as authorities and critically examined to see if they were in accord with current belief or observed reality. By the seventeenth century, however, maps—as in the textbooks of that age—were becoming more narrowly associated with geography and terrestrial survey and charting, and indeed they were formally regarded as one of the three methods of geographical representation,[65] although cosmological diagrams and celestial maps, atlases, and globes continued to be part of the general mapping impulse just as much as in the Renaissance.

Against such a background there were several reasons earlier maps were examined. First, it was the practicing map- and chartmakers who seem to have taken the lead in exploring the maps of the past. They did so either in their search for raw materials[66] (in a period when the useful life of an individual map was much longer than it is today) or in an attempt to compare the state of geographical knowledge and science in their own age with that of the past. This latter approach, which is still a focus of much research in the history of cartography,

56. The role of early maps was thus then as now; see J. B. Harley and David Woodward, "Why Cartography Needs Its History," forthcoming.

57. See, for example, Georges Duplessis, "Roger de Gaignières et ses collections iconographiques," *Gazette des Beaux-Arts*, 2d ser., 3 (1870): 468–88. On the collections in general of this period, though he does not mention maps, see Arthur MacGregor, "Collectors and Collections of Rarities in the Sixteenth and Seventeenth Centuries," in *Tradescant's Rarities: Essays on the Foundation of the Ashmolean Museum 1683, with a Catalogue of the Surviving Early Collections*, ed. Arthur MacGregor (Oxford: Clarendon Press, 1983), 70–97.

58. R. A. Skelton and John Summerson, *A Description of Maps and Architectural Drawings in the Collection Made by William Cecil, First Baron Burghley, Now at Hatfield House* (Oxford: Roxburghe Club, 1971). In France, the Duke of Sully, Henry IV's war minister, was likewise an avid collector of maps of strategic areas and cities; David Buisseret, "Les ingénieurs du roi au temps de Henri IV," *Bulletin du Comité des Travaux Historiques et Scientifiques: Section de Géographie* 77 (1964): 13–84, esp. 80, describing him as "obsessed by maps."

59. Kevin Sharpe, *Sir Robert Cotton, 1586–1631: History and Politics in Early Modern England* (Oxford: Oxford University Press, 1979); Skelton, *Maps*, 43 (note 15).

60. Pepys's collection was a "modern" reference collection of maps and charts accumulated as a result of his public office. The historical collection of older maps and atlases dates only from the end of the sixteenth century. He did, however, try to collect all the naval textbooks and atlases he could find in order to compile a historical bibliography, and an antiquarian dimension in his collecting has recently been demonstrated by Sarah Tyacke, *The Map of Rome 1625, Paul Maupin: A Companion to the Facsimile* (London: Nottingham Court Press with Magdalene College, Cambridge, 1982). A short note on Pepys as collector appears in Robert Latham and William Matthews, eds., *The Diary of Samuel Pepys*, 11 vols. (Berkeley: University of California Press, 1970–83), 10:34–36.

61. Skelton, *Maps*, 45 (note 15); John Henry Hessels, ed., *Abrahami Ortelii (geographi antverpiensis) et virorum eruditorum ad eundem . . . Epistulae . . . (1524–1628)*, Ecclesiae Londino-Batavae archivum, vol. 1 (London: Nederlandsche Hervormde Gemeente, 1887).

62. Dirk de Vries, "Atlases and Maps from the Library of Isaac Vossius (1618–1689)," *International Yearbook of Cartography* 21 (1981): 177–93.

63. E. H. Waterbolk, "Viglius of Aytta, Sixteenth Century Map Collector," *Imago Mundi* 29 (1977): 45–48; Antoine De Smet, "Viglius ab Aytta Zuichemus, savant, bibliothécaire et collectionneur de cartes du XVIᵉ siècle," in *The Map Librarian in the Modern World: Essays in Honour of Walter W. Ristow*, ed. Helen Wallis and Lothar Zögner (Munich: K. G. Saur, 1979), 237–50.

64. Numa Broc, *La géographie de la Renaissance (1420–1620)* (Paris: Bibliothèque Nationale, 1980), 61–76. For Ptolemy's definitions of geography and chorography see below, p. 183.

65. The other two were by "tables and divisions" and by "treatises": Nicolas Sanson, *Introduction à la géographie* (Paris, 1682), 6.

66. In this respect the chart collections of the European maritime nations, involving both official bodies and the trading companies, were especially extensive; for just two examples see Avelino Teixeira da Mota, "Some Notes on the Organization of Hydrographical Services in Portugal before the Beginning of the Nineteenth Century," *Imago Mundi* 28 (1976): 51–60, and Günter Schilder, "Organization and Evolution of the Dutch East India Company's Hydrographic Office in the Seventeenth Century," *Imago Mundi* 28 (1976): 61–78.

should be set in the context of the Enlightenment and the belief it had engendered in accuracy of measurement as the sine qua non of cartographic progress. While the trend toward cartographic realism should be neither oversimplified nor exaggerated,[67] by the eighteenth century there was an increasing emphasis in mapping on original survey, on more precise instruments, especially at sea, and on more detailed cartographic representation as an end in itself.[68] Moreover, practicing mapmakers were increasingly distancing themselves from—or were openly critical of—their predecessors' maps.[69]

Second, maps were closely associated with the histories of discovery, and these, often with a strong nationalistic or imperialistic flavor, had become a historical genre of their own from the days of Ramusio, Thevet, Hakluyt, and de Bry onward.[70] In the age of European reconnaissance, in the sixteenth and seventeenth centuries, earlier maps were already being used as historical documents. By the eighteenth century they were widely employed not only as historical records—to establish national precedents in discovery, for instance, or rival territorial claims—but also, at a more academic level, to begin to sift out the false topographies and imaginary islands that so abounded in the geographical record.[71]

Third, early maps were increasingly coming into circulation, owing largely to an emergent antiquarian interest and a preoccupation with cartographic evidence for biblical and classical geography. These influences were important in the development of the study of the history of maps. Among the humanist scholars who collected maps were, for example, Konrad Peutinger, who had developed an interest in Roman geography,[72] and Isaac Vossius, who also cultivated the study of classical geography as well as philosophy. Individual mapmakers engaged in reconstructions of antiquity, a research interest that would today be defined as historical cartography.[73] D'Anville, for example, who had collected some nine to ten thousand sheet maps (of which over five hundred were in manuscript), successfully employed some of these materials in the systematic study of the ancient world.[74]

67. Certainly there is no neat progression toward scientific cartography, and all that can be said in the mapping of large areas, for example, is that the less realistic maps, with many instances of regression, were gradually ousted by more realistic maps. There continued to be much fusion of older and newer map sources, often leavened with academic notions or myths, but these were frequently as important as "true geography" in influencing human action. See, for example, P. J. Marshall and Glyndwr Williams, *The Great Map of Mankind: British Perceptions of the World in the Age of Enlightenment* (London: J. M. Dent, 1982), esp. 9; Percy G. Adams, *Travelers and Travel Liars, 1660–1800* (Berkeley and Los Angeles: University of California Press, 1962); and for a different interpretation, also with implications for maps, John L. Allen, "Lands of Myth, Waters of Wonder: The Place of the Imagination in the History of Geographical Exploration," in *Geographies of the Mind: Essays in Historical Geo-*

sophy in Honor of John Kirtland Wright, ed. David Lowenthal and Martyn J. Bowden (New York: Oxford University Press, 1976), 41–61.

68. Singer, *Short History*, 316–21 (note 17), dealing with the "Measurement of the Earth and Cartography"; Margarita Bowen, *Empiricism and Geographical Thought from Francis Bacon to Alexander von Humboldt* (Cambridge: Cambridge University Press, 1981), on the development of scientific empiricism in the eighteenth century.

69. *The Construction of Maps and Globes* (London: T. Horne, 1717), attributed to John Green, on whose critical work as a map editor see Gerald R. Crone, "John Green: Notes on a Neglected Eighteenth Century Geographer and Cartographer," *Imago Mundi* 6 (1949): 85–91. A similarly critical approach to earlier mapping was adopted by French cartographers, who, besides providing lists of previous mapmakers, sometimes inveighed against their inaccuracies; see Abbé Lenglet Dufresnoy, *Catalogue des meilleures cartes géographiques générales et particulières* (reprinted Amsterdam: Meridian, 1965), a reimpression of *Méthode pour étudier la géographie*, 3d ed., vol. 1, pt. 2 (Paris: Rollin Fils, Debure l'Aîné, 1741–42), where he makes value judgments about seventeenth-century printed maps such as those of Blaeu, Jaillot, Sanson, and de Wit; see also Didier Robert de Vaugondy, *Essai sur l'histoire de la géographie* (Paris: Antoine Boudet, 1755), 243. In France particular criticism was leveled at the working methods of the eighteenth-century "armchair cartographers," such as Guillaume Delisle and Jean Baptiste Bourguignon d'Anville, who combined old and new sources in their maps, while in Germany Eberhard David Hauber and his disciple Friedrich Anton Büsching were also advocating a more critical approach to map compilation. See Ruthardt Oehme, *Eberhard David Hauber (1695–1765): Ein schwäbisches Gelehrtenleben* (Stuttgart: W. Kohlhammer, 1976). Hauber also wrote specifically about the history of maps in his *Versuch einer umständlichen Historie der Land-Charten* (Ulm: D. Bartholomäi, 1724).

70. Febvre and Martin, *Coming of the Book*, 280–82 (note 29); Gerald R. Crone and R. A. Skelton, "English Collections of Voyages and Travels, 1625–1846," in *Richard Hakluyt and His Successors*, 2d ser., 93 (London: Hakluyt Society, 1946); on Thevet's maps see Mireille Pastoureau, *Les atlas français, XVIᵉ–XVIIᵉ siècles: Répertoire bibliographique et étude* (Paris: Bibliothèque Nationale, 1984), 481–95.

71. Skelton, *Maps*, 71 (note 15); Philippe Buache, "Dissertation sur l'île Antillia," in *Mémoires sur l'Amérique et sur l'Afrique donnés au mois d'avril 1752* (n.p., 1752).

72. See above, note 45, for the authorities on Peutinger's researches into classical geography. As well as pursuing the antiquarian interests represented by his unearthing the Peutinger map, Peutinger also acquired in Italy the unfinished copperplate for the Nicholas of Cusa map of Central Europe and took it to Germany, where he arranged for Hans Burgkmair to print from it for him.

73. See Preface, note 16, for a definition of the term "historical cartography" and its misuse.

74. Jean-Baptiste Bourguignon d'Anville, *Considérations générales sur l'étude et les connaissances que demande la composition des ouvrages de géographie* (Paris: Galeries du Louvre, 1777), 5–12. A description of d'Anville's collection is given by Charles Du Bus, "Les collections d'Anville à la Bibliothèque Nationale," *Bulletin du Comité des Travaux Historiques et Scientifiques: Section de Géographie* 41 (1926): 93–145 (see also note 150 below). Among d'Anville's works that most reflect his interest in reconstructing the past with maps are *Dissertation sur l'étendue de l'ancienne Jérusalem et de son temple, et sur les mesures hebraiques de longueur* (Paris: Prault Fils, 1747); *Traité des mesures itinéraires anciennes et modernes* (Paris: Imprimerie Royale, 1769); and *Géographie ancienne abrégée* (Paris: A. Delalain, 1782); see also Juliette Taton, "Jean-Baptiste Bourguignon d'Anville," in *Dictionary of Scientific Biography*, 16 vols., ed. Charles Coulston

It will have been noted that the primary objectives in many of these approaches to old maps were those of the historian or historical geographer. This had not, however, led to a complete neglect of other aspects. The foundation of some of the other standard approaches to the study of early maps, which were yet to gather momentum, can also be traced in the eighteenth century. It is fair to say that there was a general lack of interest in the history of cartography as a continuous process; but against this can be offset the first attempts at a "systematic, if naïve, summary of cartographic history."[75] Especially marked was the development of a biobibliographic approach to early maps. This reflects a sense of responsibility toward the relics of the past as well as the methods of the encyclopedist. The biobibliographical approach had already taken root in the Renaissance, so that by the 1603 edition of Ortelius's *Theatrum orbis terrarum* the number of cartographers listed as constructing his maps had risen to 183, some having started work in the fifteenth century.[76] At the end of the seventeenth century, Jean Mabillon was listing mapmakers among geographers, together with their works, as part of his recommendations for the contents of a well-ordered ecclesiastical library.[77] During the eighteenth century such lists of earlier cartographic works became both more numerous and more extensive. In Venice, Vincenzo Coronelli compiled a "Cronologia de' geografi antichi, e moderni," naming ninety-six geographers and mapmakers from Homer to Ponza.[78] In France, Didier Robert de Vaugondy, in a comparative essay on the history of geography, listed the earlier maps of a number of European countries.[79] In Germany, J. G. Gregorii, who had noted that early maps had "now grown very rare and difficult to come by" and that they were "becoming as desirable as old coins," also provided a chronological list of geographers and mapmakers down to his own time (beginning with Moses, whom he regarded as the first geographer).[80] In England, Richard Gough, whose explicit aim was "to trace the progress of Map-making among us,"[81] similarly listed both national and regional mapmakers in his pioneer work *British Topography.*[82]

Not all the eighteenth-century writing about early maps was cast in this biobibliographical mold, and we also encounter specialist studies of individual works. These include, above all, Ptolemy's *Geography,*[83] the *mappaemundi,*[84] and the portolan charts.[85] Especially significant for the future scope of the history of cartography was the tentative beginning of an extension of theaters of interest away from the classical world and the European nations during the Renaissance to encompass maps from the Asian cultures[86] and, in the travel literature of the period, to include reports of the mapping skills of nonliterate peoples as they were first encountered by Europeans.[87] These signs of a historical con-

sciousness relating to the maps of earlier periods and non-European civilizations should not be overemphasized. The eighteenth-century view of the cartographic past not only was strongly Eurocentric—regarding other cultures as inferior—but also was colored by the contemporary appetite for greater precision in mapping. Even during the Renaissance, geographers such as Hakluyt had become aware of the "olde imperfectly com-

Gillispie (New York: Charles Scribner's Sons, 1970–80), 1:175–76.

75. Skelton, *Maps,* 70–71 (note 15).

76. Leo Bagrow, *A. Ortelii catalogus cartographorum* (Gotha: Justus Perthes, 1928), reprinted in *Acta Cartographica* 27 (1981): 65–357.

77. Jean Mabillon, *Traité des études monastiques* (Paris: Charles Robustel, 1691), 463–66.

78. Vincenzo Coronelli, *Cronologia universale* (Venice, 1707), 522–24; this was designed as an introduction to his projected *Biblioteca universale* of 45 volumes, of which only vols. 1–7 seem to have been published (Venice: Antonio Tivani, 1701–5).

79. Robert de Vaugondy, *Essai,* chap. 5 (note 69), considers German, English, Dutch, Flemish, Spanish, Italian, Swedish, Russian, and French maps.

80. Johann Gottfried Gregorii, *Curieuse Gedancken von den vornehmsten und accuratesten alt- und neuen Land-Charten* (Frankfort and Leipzig: H. P. Ritscheln, 1713), 120, author's translation.

81. Quoted by Gwyn Walters, "Richard Gough's Map Collecting for the British Topography 1780," *Map Collector* 2 (1978): 26–29, quotation on 27, from Gough's correspondence with the Reverend Michael Tyson, a Cambridge antiquary.

82. Gough, *British Topography* (note 47); a preliminary essay by Gough, also listing maps, was published as *Anecdotes of British Topography . . .* (London: W. Richardson and S. Clark, 1768).

83. Georg Martin Raidel, *Commentatio critico-literaria de Claudii Ptolemaei Geographia, eiusque codicibus tam manuscriptis quam typis expressis* (Nuremberg: Typis et sumptibus haeredum Felseckerianorum, 1737); Jean Nicholas Buache, *Mémoire sur la Géographie de Ptolémée et particulièrement sur la description de l'intérieur de l'Afrique* (Paris: Imprimerie Royale, 1789), esp. 119, where he complains that Ptolemy's works, although known for sixteen hundred years, were badly understood.

84. Gough, *British Topography,* 1:60–86 (note 47), dealing with medieval maps relating to Britain.

85. Girolamo Francesco Zanetti, *Dell'origine di alcune arti principali appresso i Veneziani,* 2 vols. (Venice: Stefano Orlandini, 1758), 46–48.

86. This interest arose earlier with Chinese cartography than with maps from the Muslim world. Although parts of al-Idrīsī's text were printed in Arabic and also translated during the Renaissance, the cartographic component in Arab geography was long neglected by Europeans; for the general background to this see Marshall and Williams *Great Map of Mankind,* chap. 1 (note 67). Joachim Lelewel, *Géographie du Moyen Age,* 4 vols. and epilogue (Brussels: Pilliet, 1852–57; reprinted, Amsterdam: Meridian, 1966), was the first European scholar to look at Arab cartography in detail; but he was more interested in reconstructing maps from Arab tables of latitude and longitude than in the Arabs' own cartographic efforts. See also volume 2 of the present *History* on traditional Asian societies.

87. For example, Jonathan Carver, *Travels through the Interior Parts of North-America in the Years 1766, 1767, and 1768* (London, 1778), 252–53; Awnsham Churchill and John Churchill, *A Collection of Voyages and Travels,* 6 vols. (London: J. Walthoe, 1732), 6:165, cite Colonel Henry Norwood who in 1649–50 watched a map being drawn in sand in Virginia.

posed" maps (which he contrasted with "the new lately reformed Mappes, Globes, Spheares").[88] By the eighteenth century Richard Gough could write even more dismissively of the medieval maps he had studied as belonging to "the barbarous Monkish system of Geography," while Samuel Johnson collected "the maps drawn in the rude and barbarous times . . . to know the errors of the ancient geographers."[89] By now, too, commentators were starting to look forward to perfection in a science of mapping,[90] even though this only reinforced a view of the maps of the past as lower on the ladder of progress. Given such attitudes, by 1800 it was still true (as Skelton has written of the seventeenth century) that maps were seldom contemplated and analyzed as artifacts, that little notice was taken of the methods by which they were constructed and drawn, and that the "study of cartographic expression and form as a mode of communication had not yet begun."[91] Indeed, in terms of these particular concepts, now taken for granted, the history of cartography had yet to be born as a subject we would recognize today.

THE HISTORY OF CARTOGRAPHY AS HANDMAIDEN: TRADITIONAL THEMES FROM THE NINETEENTH CENTURY

The nineteenth century was marked by a sharp intensification of interest in the study of early maps as an area of inquiry distinct from contemporary cartography. As for the history of science in general, it can be regarded as a major formative period. A number of yardsticks can be used to measure this development, including the volume of periodical and monographic literature, the tendency to issue facsimiles of early maps, and the number of scholars active in the field. This interest was to continue with an unbroken intellectual lineage into the present century. Its principal driving force, especially after 1850, was the rise and institutionalization of geography,[92] together with the growth of specialist map libraries at a national level and—in Europe and North America especially—the development of a distinctive antiquarian map trade. It will also be suggested that these influences combined to give the study of early maps—and the very tentatively emerging field of the history of cartography—certain biases in aim and method of inquiry that differentiate it sharply from those of the present *History of Cartography*. In particular, the history of cartography was not studied then as an independent subject but remained primarily a handmaiden to the history of geography defined as the history of geographical discovery and exploration. However, when viewed in the proper context, these aims and methods are seen to be an essential part of the scholarly legacy of the subject, and a review clarifies the nature of the territory covered here.

EARLY VIEWS ON THE DEVELOPMENT OF THE SUBJECT

By the mid-nineteenth century there were a few scholars who would have considered themselves historians of cartography, and their attitudes to the development of their subject in their own time are revealing. For instance, Manuel Francisco de Barros e Sousa, Viscount of Santarém, the Portuguese scholar-author of the influential facsimile *Atlas* of early maps first published in 1841—also credited with coining, in 1839, the word "cartography" for the study of early maps[93]—was being un-

88. Richard Hakluyt, *The Principall Navigations Voiages and Discoveries of the English Nation*, a photolithographic facsimile (originally imprinted in London, 1589) with an introduction by David Beers Quinn and R. A. Skelton and with a new index by Alison Quinn (Cambridge: For the Hakluyt Society and the Peabody Museum of Salem at the University Press, 1965), 2.

89. Walters, "Richard Gough," 27 (note 81); Samuel Johnson, *Rambler* 82, Sat., 29 Dec. 1750. See below, chapter 2 for evidence of the persistence of those attitudes. For the similarly expressed view of a French writer of the same period, the Abbé Lebeuf, see chapter 18 below, "Medieval *Mappaemundi*," n. 17.

90. Juan Andrés, *Dell'origine, progressi e stato attuale d'ogni letteratura di Giovanni Andrés*, new ed., 8 vols. (Pisa: Presso Niccolò Capurro, 1829–30), vol. 3, pt. 1, 161. On Andrés see *Dizionario biografico degli italiani* (Rome: Istituto della Enciclopedia Italiana, 1960–), 3:155–57.

91. Skelton, *Maps*, 70 (note 15).

92. Horacio Capel, "Institutionalization of Geography and Strategies of Change," in *Geography, Ideology and Social Concern*, ed. D. R. Stoddart (Oxford: Basil Blackwell; Totowa, N.J.: Barnes and Noble, 1981), 37–69.

93. The word was soon applied to cartography in general in the senses we use it today and was to appear in many European languages by the second half of the nineteenth century. The fullest account of Santarém's career and contribution to the history of cartography is Armando Cortesão, *History of Portuguese Cartography*, 2 vols. (Coimbra: Junta de Investigações do Ultramar-Lisboa, 1969–71), 1:7–26; editions of Santarém's *Atlas composé de cartes des XIVᵉ, XVᵉ, XVIᵉ, et XVIIᵉ siècles* are described in 1:15–22. The circumstances of Santarém's coining "cartography" in a letter to the Luso-Brazilian historian Francisco Adolfo de Varnhagen in 1839—but with Santarém's *Essai sur l'histoire de la cosmographie et de la cartographie pendant le Moyen-Age et sur les progrès de la géographie après les grandes découvertes du XVᵉ siècle*, 3 vols. (Paris: Maulde et Renou, 1849–52) being the first major work to bring the word into its title—are also discussed by Cortesão, *History of Portuguese Cartography*, 1:4–5 (above). It is tempting to find earlier examples. See the Matteo Pagano view of Venice of ca. 1565, described by Juergen Schulz, "The Printed Plans and Panoramic Views of Venice (1486–1797)," *Saggi e Memorie di Storia dell'Arte* 7 (1970): 52, where the word "Cortografia" appears on the map. This is no doubt a misprint for *corografia* (chorography), although it was mistranscribed as "Cartografia" in Giandomenico Romanelli and Susanna Biadene, *Venezia piante e vedute: Catalogo del fondo cartografico a stampa* (Venice: Museo Correr, 1982), 5. Another example concerns a gilded brass surveying instrument of 1557 now in the Museum of History of Science in Florence. This was reported to include the word "Cartographia" in its engraved inscription as described by Cornelis Koeman in "Hoe oud is het woord kartografie?" *Geografisch Tijdschrift* 8 (1974): 230–31. It is now clear, however, that the word does not exist on the instrument: Helen Wallis, "Cartographic Innovation: An Historical Perspective," in *Canadian Institute of Surveying Centennial Convention Proceedings*, 2 vols. (Ottawa: Canadian Institute of Surveying, 1982), 2:50–63.

usually precise when he commented that his subject was "quite new—the study of ancient maps is hardly a little over sixty years old."[94] Some of the evidence he presented does not, as reviewed here, bear him out; but what is of interest is that he was placing the beginning of the subject at the end of the eighteenth century. He observed that it was only after Zurla and Andrés that some scholars began to be interested in the study of medieval maps in a more general manner.[95] In relation to his own special interest in medieval cartography, he listed those authors who had managed to contribute to the historical aspects of geography without mentioning the Middle Ages at all (or who had skipped over the period in two or three pages). Certainly, by the time Santarém was writing there had been an upsurge of interest in the study of early maps as historical documents, and it had been accompanied by an understandable and increasing tendency for some individuals among those most closely involved to consider themselves the true founders of the new subject. Signs of this attitude may be detected in the public debate between Santarém and Jomard. Edme-François Jomard was by then established as head of the map room that had been created in the Bibliothèque Royale in Paris in 1828. The debate focused on the two men's competing claims to have been the first to conceive the idea of publishing an atlas of medieval maps in facsimile,[96] but it was echoed by scholars working in North America who also saw themselves as the first of a new breed of map historians. So Charles P. Daly asserted in 1879 that there was little "specifically devoted to such an inquiry" as the history of cartography.[97] So too J. G. Kohl, the German scholar, claimed in his Smithsonian lecture two years later that "*the history of geographical maps* has scarcely ever been thought of" and that as a "branch of geographical research" it had remained "a perfect blank" until his own day.[98] While it is understandable that in their more enthusiastic moments such scholars were perhaps apt to exaggerate the significance of their own efforts, it is also clear that they had started to develop a sense of destiny and purpose for the study of early maps. This awakening is nowhere better expressed than in the words of Jomard when he wrote that among his tasks was "that of provoking a more complete search [for] all the as yet unknown geographical monuments . . . to make them rise from the dust and come out from the oblivion in which they are buried."[99]

Twentieth-century historians of cartography have so far done little, except perhaps at a national level,[100] to assess this crucial period in the development of their subject. Only two, Armando Cortesão and R. A. Skelton, even attempted a general historiographic treatment. Both laid much stress on the seminal contributions of a number of leading scholars and on the landmarks in publication thus created. Both singled out atlases of

facsimiles of early maps that they saw as holding a preeminent place in the literature of the subject from the mid-nineteenth century onward. The main emphasis of Cortesão's approach was biobibliographic. For him, writing in the 1960s, the systematic study of the history of cartography had started about a century earlier, and its subsequent development was best explained by the contributions of a procession of leading scholars. Thus he listed over ninety individuals in his chapter "Cartography and Its Historians," although he admitted that even so he had "not mentioned all who have, directly or indirectly, contributed to the advancement of this important and enthralling branch of the history of science."[101] Valuable as they are, though, Cortesão's bio-

94. Manuel Francisco de Barros e Sousa, Viscount of Santarém, "Notice sur plusieurs monuments géographiques inédits du Moyen Age et du XVIᵉ siècle qui se trouvent dans quelques bibliothèques de l'Italie, accompagné de notes critiques," *Bulletin de la Société de Géographie*, 3d ser., 7 (1847): 289–317, quotation on 289, author's translation. The article is reprinted in *Acta Cartographica* 14 (1972): 318–46.

95. Viscount of Santarém, *Essai*, 1:XLIV (note 93), designed to serve as an introduction to his *Atlas*. On Placido Maria Zurla, see Cortesão, *History of Portuguese Cartography*, 1:36 (note 93); on Andrés see note 90 above.

96. Jomard's atlas, *Les monuments de la géographie; ou, Recueil d'anciennes cartes européennes et orientales* (Paris: Duprat), appeared serially from 1842 to 1862, the first issue being a facsimile of the Hereford map in six colored plates. Although Cortesão, *History of Portuguese Cartography*, 1:29–32 (note 93), tends to take a patriotic Portuguese line, regarding Santarém as "the creator of the systematic history of cartography" (1:23), he does give the fullest account of the affair. The issue is probably worth reexamination as a chapter in the intellectual development of the history of cartography. Taking the evidence presented by Cortesão, together with that from various writings by Santarém and Jomard, it seems unlikely that by the 1830s the idea of a facsimile atlas of early maps was the private property of any one scholar. Besides Jomard and Santarém, other scholars such as Marie Armand Pascal d'Avezac-Macaya and Joachim Lelewel were engaged in similar projects. For a reference to d'Avezac's scheme, undertaken with Thomas Wright, an Englishman, but dropped in favor of Jomard, see Edme-François Jomard, *Sur la publication des Monuments de la géographie* (Paris, 1847), 6; on Lelewel see Marian Henryk Serejski, *Joachim Lelewel, 1786–1861: Sa vie et son oeuvre* (Warsaw: Zakład Narodowy imienia Ossolińskich, 1961); also Zbigniew Rzepa, "Joachim Lelewel, 1786–1861," *Geographers: Biobibliographical Studies* 4 (1980): 103–12.

97. Charles P. Daly, "On the Early History of Cartography; or, What We Know of Maps and Map-Making, before the Time of Mercator," Annual Address, *Bulletin of the American Geographical Society* 11 (1879): 1–40, quotation on 1.

98. Johann Georg Kohl, "Substance of a Lecture Delivered at the Smithsonian Institution on a Collection of the Charts and Maps of America," *Annual Report of the Board of Regents of the Smithsonian Institution . . . 1856*, (1857), 93–146, quotation on 95.

99. Edme-François Jomard (posthumously published by M. E. Cortambert), *Introduction à l'atlas des Monuments de la géographie* (Paris: Arthus Bertrand, 1879), 6, author's translation.

100. See below, p. 37.

101. Cortesão, *History of Portuguese Cartography*, 1:1–70, quotation on 69 (note 93).

graphies should be regarded as merely the raw material for future study of the intellectual development of the history of cartography. They stop short of the potential insights of modern prosopography in failing to go on to reveal the wider processes at work behind the contributions of individuals, which remain as unconnected episodes in the history of the study of early maps.[102]

In his discussion of the historical study of early maps in the nineteenth century, R. A. Skelton drew heavily on the material quarried by Cortesão. Like Cortesão, Skelton laid stress on the scholarly achievements of the individual pioneers of the subject—notably Jomard, Santarém, Lelewel, Kohl, Nordenskiöld—and in particular on the contribution made through their facsimile atlases. In Skelton's view these atlases had continued, down to his own day, to enable scholars to formulate "the central problems in the comparative study and use of early maps as historical documents."[103] On the other hand, Skelton recognized that the study of early maps did not develop in an institutional vacuum, and he drew attention to the way the emergence of geography as an independent discipline, and in particular its "newly formed societies . . . hungry for work,"[104] provided the essential framework for a rapid expansion of interest in the history of cartography during the nineteenth century. Notwithstanding the major influence exerted by geography on the practice of the history of cartography, not least in reinforcing the tendency to equate it with a study of geographical maps, its significance has been overlooked elsewhere. A more explicit argument for its role is therefore offered below.

THE RISE OF GEOGRAPHY

The anatomy of the growth of geography in the nineteenth century—as a subject in the universities and with a growing professional community—has been traced in several recent studies,[105] but for the history of cartography perhaps the most important single facet of that growth was the foundation of national geographical societies. Beginning with the Société de Géographie de Paris (1821), followed by the Gesellschaft für Erdkunde zu Berlin (1828) and the Royal Geographical Society of London (1830), new societies were established not only in Europe but also in the New World. By 1885, it has been estimated, the world had ninety-four geographical societies with over forty-eight thousand members.[106] Most of these societies published journals, and through their extensive collections of topographic maps and atlases, as well as by serving as outlets for publication, they provided opportunities for research into early cartography. Of course, neither these societies nor their journals regarded the history of cartography as a major preoccupation, nor indeed did the journals hold a monopoly on such writings.[107] Yet even though they failed in general to initiate directly studies in the history of cartography, it is noticeable that in many countries it was the members of these societies who, for whatever reasons, were the most active in studying early maps and that a large proportion of the relevant literature from the nineteenth century is to be found in periodicals sponsored by the national and regional geographical societies.[108] Moreover, just as geography was studied within national and regional frameworks, so too was the history of cartography. Maps were key documents in both a practical and an ideological sense in shaping European nationalism and imperialism, and it is of interest that Santarém's facsimile *Atlas* had its origin in the use of old maps in a sovereignty dispute between Portugal and France over Casamance in Senegal. Even today, related aspects of the politics of knowledge are manifest in writings on the history of cartography, as, for example, in the case of the hotly disputed primacy between the nations of southern Europe in the development of the portolan chart.[109]

Later in the nineteenth century the history of cartography was also to be caught up by international cooperation, designed to promote the study of academic geography, growing out of these national institutions and

102. Lewis Pyenson, " 'Who the Guys Were': Prosopography in the History of Science," *History of Science* 15 (1977): 155–88, has some pointers for directions in the history of cartography.

103. Skelton, *Maps,* 82 (note 15).

104. Skelton, *Maps,* 74 (note 15).

105. Capel, "Institutionalization" (note 92); also Robert E. Dickinson, *The Makers of Modern Geography* (New York: Frederick A. Praeger, 1969), 267–68.

106. H. Wichmann, "Geographische Gesellschaften, Zeitschriften, Kongresse und Ausstellungen," *Geographisches Jahrbuch* 10 (1884): 651–74, esp. 654.

107. In the Netherlands, for example, after 1850 many contributions to the history of cartography were published in the periodicals of the learned literary societies (personal communication from Cornelis Koeman).

108. On the early role of the Société de Géographie in this respect, see Alfred Fierro, *La Société de Géographie, 1821–1946* (Geneva: Librairie Droz, 1983), 4–18.

109. On Santarém see Skelton, *Maps,* 77 (note 15); on portolan charts see chapter 19 of this volume, "Portolan Charts from the Late Thirteenth Century to 1500"; Cortesão, *History of Portuguese Cartography,* 1:59 (note 93), for example, says that Italian historians of the discoveries and of cartography write "with a more or less bitter and spiteful bias against the Portuguese," and, similarly, Samuel E. Morison, *Portuguese Voyages to America in the Fifteenth Century* (Cambridge: Harvard University Press, 1940), represented a violently anti-Portuguese attitude. Something of the flavor of this debate is given by Heinrich Winter, "Catalan Portolan Maps and Their Place in the Total View of Cartographic Development," *Imago Mundi* 11 (1954): 1–12. Indeed, disagreement is particularly rife among such historians in the discussion of early maps, and it has often been expressed in bitter personal as well as nationalistic forms. As serious study has become more professionalized, however, argument, though still lively, is conducted less acrimoniously.

the territorial and commercial ambitions they supported. It was perhaps coincidence that the first International Geographical Congress, in Belgium in 1871, called for specialists interested in *la science de la terre* and in *les sciences cosmographiques*, had decided to honor the mapmakers Mercator and Ortelius by erecting statues in the towns of their birth. But early maps were exhibited at the Antwerp meeting, and the *questions de géographie historique* scheduled to be discussed by the Comité d'Honneur included a number relating to the history of cartography.[110] At subsequent congresses during the nineteenth century, however, very few papers were given on the history of cartography,[111] and abundant interest in the subject within the organization, noted by Skelton, was primarily a feature of the first half of the twentieth century.[112] Even then, in the period from the congress of 1904 to that of 1972, historical subjects represented only some 14 percent of all the recorded cartographic papers and of course a relatively insignificant proportion of all geographical papers presented at the congresses.[113]

THE GROWTH OF MAP LIBRARIES

Another major source of institutional support for the history of cartography in the nineteenth and early twentieth centuries was to develop within the newly established map libraries, especially those of the national libraries. Before about 1790, there had been a slow growth of map collecting by individuals, as already noted, but after this date the expansion and institutionalization of this activity parallel the rise of geography itself.[114] Again it has to be stressed that these libraries were not set up primarily to meet the needs of the history of cartography. On the contrary, their objectives varied widely, ranging from a need to house national map collections, as in the case of the Bibliothèque Nationale, the British Museum, and the Library of Congress, through the often complementary activities of the various geographical societies in collecting maps, to the establishment of map rooms in public and university libraries or in specialist government departments, not least those concerned with military and naval matters or with administration of overseas empires. While allowing for the wider functions of these map libraries, it can also be accepted that they played a crucial role in the development of the history of cartography by acting as repositories for antiquarian charts and maps and by arranging for their cataloging and exhibition. Moreover, they were especially important in helping to cultivate a new attitude toward the preservation and comparative study of early maps. The writings of Jomard, as he sought to justify the role of the *collection géographique spéciale* established within the Bibliothèque Royale in 1828, strongly reflect this philosophic underpinning.[115]

In 1839, for example, he noted that "the comparative study and attentive examination of geographical maps" had "served more than once to resolve political, diplomatic, or historical questions, such as illuminating legal disputes." He insisted that it was not "enough to possess the newest maps"; a map library, he continued, should not "banish . . . the first products of the printing press," for it was "only by comparing the successive product of a science that history can be written, and it is sometimes in the oldest that the solution to a difficulty is found."[116] True to this principle, Jomard tells us that it was

> especially toward the search for the oldest medieval maps, those venerable monuments of primitive geography, that I directed my efforts. In 1828 nobody dreamed of gathering those precious remains and of reuniting them in a national collection. Since, all has been very much changed; these objects are avidly hunted for everywhere; they are gathered up for the

110. *Compte-rendu du Congrès des Sciences Géographiques, Cosmographiques et Commerciales*, 2 vols. (Antwerp: L. Gerrits and Guil. Van Merlen, 1872), ii–xv.

111. The tally is one (Paris, 1875), one (Venice, 1881), three (Paris, 1889), five (London, 1895), and three (Berlin, 1899): for details see George Kish, ed., *Bibliography of International Geographical Congresses, 1871–1976* (Boston: G. K. Hall, 1979).

112. Skelton, *Maps*, 98 (note 15). This interest can again be monitored from the papers relating to early maps presented at successive congresses: five (Rome, 1913), three (Cairo, 1925), eight (Cambridge, 1928), fourteen (Paris, 1931), seven (Warsaw, 1934), twenty-one (Amsterdam, 1938). Thereafter few papers were offered until the London congress of 1964, when there were twenty-three. For details see Kish, *Geographical Congresses*, passim (note 111).

113. John A. Wolter, "The Emerging Discipline of Cartography" (Ph.D. diss., University of Minnesota, 1975), 199–200.

114. See John A. Wolter, "Geographical Libraries and Map Collections," in *Encyclopedia of Library and Information Science*, ed. Allen Kent, Harold Lancour, and Jay E. Daily (New York: Marcel Dekker, 1968–), 9 (1973): 236–66; see also Wolter, "Emerging Discipline," 78–94 (note 113).

115. Edme-François Jomard, *Considérations sur l'objet et les avantages d'une collection spéciale consacrée aux cartes géographiques et aux diverses branches de la géographie* (Paris: E. Duverger, 1831), where he stressed that geographical *maps* have advantages peculiar to themselves (p. 8) and need to be treated separately from books. Within his map library he also proposed to have a special section for maps from the Middle Ages, or European maps up to ca. 1600 (p. 57). For modern studies of Jomard and the establishment of the Département des Cartes et Plans see Charles Du Bus, "Edme-François Jomard et les origines du Cabinet des Cartes (1777–1862)," Union Géographique Internationale, *Comptes rendus du Congrès International de Géographie, Paris 1931* 3 (1934): 638–42; Edmond Pognon, "Les collections du Département des Cartes et Plans de la Bibliothèque Nationale de Paris," in *Map Librarian*, 195–204 (note 63); and Monique Pelletier, "Jomard et le Département des Cartes et Plans," *Bulletin de la Bibliothèque Nationale* 4 (1979): 18–27.

116. Edme-François Jomard, *De l'utilité qu'on peut tirer de l'étude comparative des cartes géographiques* (Paris: Burgogne et Martinet, 1841), 4, author's translation (reprinted from *Bulletin de la Société de Géographie*, 2d ser., 15 [1841]: 184–94).

enrichment of public depositories from which they cannot again come out, and they are becoming an extreme rarity.[117]

It was with aims such as this, the pursuit of "the linking of the history of science and its graphic products,"[118] while simultaneously helping to ensure that maps were "raised to the dignity of historical documents,"[119] that the larger map libraries became crucial institutions for the study of the history of cartography. And it is also clear from the history of other libraries, besides the Département des Cartes et Plans, that some of these collections became more than just repositories for antiquarian maps and plans. Rather, as with the Department of Maps and Charts established in the British Museum in 1867[120] or the Hall of Maps and Charts (later called the Geography and Map Division) founded in the Library of Congress in 1897,[121] they started to function as national and even international clearinghouses for new findings in the history of cartography. Their curators were encouraged to engage in scholarship, and in this way and a number of other ways they have continued to play a central part in the development of the subject.

PRIVATE COLLECTORS AND THE ANTIQUARIAN MAP TRADE

In addition to the institutional infrastructure for the study of early maps provided through the growth of geography and through the emergence of map libraries, private collectors and the antiquarian trade together contributed a distinctive patina to research and writing in the history of cartography. Their influence was narrowing, akin to the role of connoisseurship in traditional art history. It has tended to favor the study of collectible printed maps and perhaps to encourage an excessive delectation of decorative maps as objets d'art. In practice this usually meant European printed maps of the period from the late fifteenth century to the end of the eighteenth, although in the nineteenth century in particular collectors and scholars joined in concentrating their interests upon the flowering of European cartography during the Renaissance.

The specific legacy of wealthy private collectors to the study of the history of cartography was especially felt in western Europe and North America. It has already been highlighted by Skelton, who discerned three main areas of influence.[122] First, the main fruit of some collecting has been the concentration, to the benefit of later generations of scholars, of the major resources of early maps in the so-called treasure house libraries. For example, for North America attention has recently been drawn to those nineteenth-century collectors, such as John Carter Brown, James Lenox, and Edward E. Ayer,

who specialized in collecting early maps of particular regions or topics even though they did not themselves engage in research.[123] Second—and more directly linked to the intellectual development of the subject—is the group Skelton defined as the scholar-collectors, including J. T. Bodel Nijenhuis and Abraham van Stolk, both from the Netherlands, General von Hauslab in Austria, and the explorer Nordenskiöld in Sweden,[124] whose map collections were the raw materials of their writings on the history of cartography. Third, there were the scholar-dealers, for whom the archetype could have been Frederik Muller, the Netherlands collector, publisher, and bibliographer of early maps and atlases. Cornelis Koeman has described Muller as "the innovator and propagandist par excellence of the scientific management of antiquarian bookselling in general and the first promoter of cartography as an historical source in particular,"[125] and it was Muller who laid the foundations for the history of Dutch cartography that was later written by others. Yet though a leader, Muller was by no means

117. Edme-François Jomard, *De la collection géographique créée à la Bibliothèque Royale* (Paris: E. Duverger, 1848), 79, author's translation.

118. Jomard, *Collection géographique*, 13 (note 117).

119. Kohl, "Lecture," 95 (note 98).

120. Wallis, "Map Collections," 17 (note 54).

121. John A. Wolter et al., "A Brief History of the Library of Congress Geography and Map Division, 1897–1978," in *Map Librarian*, 47–105 (note 63).

122. Skelton, *Maps*, 52–61 (note 15).

123. Douglas Marshall, "The Formation of a Nineteenth-Century Map Collection: A. E. Nordenskiöld of Helsinki," *Map Collector* 21 (1982): 14–19. See also the continuing series in *Map Collector* on "America's Treasure House Libraries," including Norman J. W. Thrower, "The Treasures of UCLA's Clark Library," *Map Collector* 14 (1981): 18–23; Carey S. Bliss, "The Map Treasures of the Huntington Library," *Map Collector* 15 (1981): 32–36; Thomas R. Adams, "The Map Treasures of the John Carter Brown Library," *Map Collector* 16 (1981): 2–8; John Parker, "The Map Treasures of the James Ford Bell Library, Minnesota," *Map Collector* 20 (1982): 8–14; Philip Hoehn, "The Cartographic Treasures of the Bancroft Library," *Map Collector* 23 (1983): 28–32; Robert Sidney Martin, "Treasures of the Cartographic Library at the University of Texas at Arlington," *Map Collector* 25 (1983): 14–19; Barbara McCorkle, "Cartographic Treasures of the Yale University Library," *Map Collector* 27 (1984): 8–13; Robert W. Karrow, "The Cartographic Collections of the Newberry Library," *Map Collector* 32 (1985): 10–15.

124. Skelton, *Maps*, 54–55 (note 15); for insight into Nordenskiöld's scholarly collecting see Ann-Mari Mickwitz, "Dear Mr. Nordenskiöld, Your Offer Is Accepted!" in *Map Librarian*, 221–35 (note 63). To Skelton's list might be added the map collections of some of the nineteenth-century academic geographers. In Germany, for example, that of Carl Ritter (some twelve thousand maps) was notable and a factor in the foundation of a national map library: see Lothar Zögner, "Die Carl-Ritter-Ausstellung in Berlin—eine Bestandsaufnahme," in *Carl Ritter—Geltung und Deutung*, ed. Karl Lenz (Berlin: Dietrich Reimer Verlag, 1979), 213–23. Other map collections from geographers were those of Alexander von Humboldt, Carl Wilhelm von Oesfeld, Carl Friedrich von Klöden, and Heinrich Kiepert.

125. Koeman, *Collections*, 93 (note 51).

an isolated figure. As the correspondence over the foundation of Nordenskiöld's collection has revealed, by the late-nineteenth century a complex network of European dealers in early maps was coming into being, including, among others, Muller in Amsterdam, Lissa in Berlin, Kellings in Stockholm, Quaritch and Stevens in London, Perrella in Naples, and Olschki in Venice and Verona.[126] While the caution caveat emptor may sometimes have to be applied to their work, it is also true that such men helped to raise the standard of bibliographical description of maps, as in the succession of notable catalogs by Muller and his successor Anton Mensing (after 1902 assisted by F. C. Wieder), and through their own scholarly works, as in the case of the study by Henry N. Stevens of the printed editions of Ptolemy's *Geography*.[127]

TRADITIONAL PATTERNS OF CARTOGRAPHIC HISTORY

The three influences on the history of cartography just outlined—the institutionalization of geography, the growth of specialist map libraries, and the interplay of scholarly collecting with an expanding antiquarian market for early maps—together determined the nature of most writings on the history of cartography to the middle of the present century. On the face of it, many of the substantive themes and distinctive topics that still head the history of cartography agenda today—and which will feature prominently in these volumes—were present in the literature of the century and a half after 1800. Such themes include, as R. A. Skelton has suggested, medieval cartography and its Roman origins, mathematical cartography and the history of map projections, the expansion of original topographical surveys, changing navigation techniques and their effects on the design of sea charts, the development of cartographic representation, the growth of the map trade, and the application of printing techniques to cartography.[128] Without qualification, however, such a list is misleading and exaggerates the maturity of the subject in the nineteenth century. Some of the topics—the history of map printing, to take a single example—were hardly studied systematically.[129] Moreover, in perhaps the majority of cases, the maps themselves—considered as independent artifacts and images—were still subservient to the pragmatic aims of the institutional or collecting context, so that it is difficult to acknowledge a history of cartography that possessed a sense of scholarly identity at this stage.

The relationship between the history of cartography and the development of geography from the early nineteenth century onward illustrates how interest in early maps was subordinated to problems external to the map itself. In part this was related to the fact that, in the first half of the nineteenth century, geography itself tended

to be regarded not as a subject in its own right, but as an adjunct to history.[130] Geography provided the necessary background for understanding historical events, especially those of classical history. It is not surprising that early maps should have been viewed primarily as historical documents, to be used in reconstructing the geographies of the past, whether of the ancient world, the biblical lands, or the age of the great discoveries when the foundations were laid for the overseas empires of the nineteenth century.

Accepting such an interpretation of the primary motivation for the study of early maps in the nineteenth and early twentieth century means that much of the writing loosely described as belonging to the history of cartography can also be allocated to other branches of knowledge. In particular, some of the so-called landmarks in the subject could equally well be seen as belonging to the historiography of geographical discoveries or to cognate parts of other historical specialisms. Note that Alexander von Humboldt's interest in the first maps of the New World was very largely connected with his study of the discovery and exploration of the Americas,[131] and Santarém was preoccupied with the "immense utility" of ancient maps for the history of geography and the history of discoveries[132] and (with a touch of chauvinism) the establishment of "the priority of the Portuguese discoveries in western Africa, and the services this nation rendered to geographical sciences."[133] The example of Santarém also serves to make the point that the atlases of map facsimiles[134] which, it

---

126. Mickwitz, "Nordenskiöld" (note 124).

127. Henry N. Stevens, *Ptolemy's Geography: A Brief Account of All the Printed Editions down to 1730*, 2d ed. (London: Henry Stevens, Son and Stiles, 1908, reprinted, Amsterdam: Theatrum Orbis Terrarum, [1973]); see also Stevens's *Recollections of James Lenox and the Formation of His Library*, ed., rev., and elucidated by Victor Hugo Paltsits (New York: New York Public Library, 1951).

128. The list is paraphrased from that of Skelton, *Maps*, 90 (note 15).

129. David Woodward, ed., *Five Centuries of Map Printing* (Chicago: University of Chicago Press, 1975), vii, states that "for many aspects of the study of historical map printing methods, the field is untouched."

130. Capel, "Institutionalization," 39–47 (note 92), summarized the evidence for a number of European countries in this period.

131. Alexander von Humboldt, *Examen critique de l'histoire de la géographie du nouveau continent et des progrès de l'astronomie nautique au XVᵉ et XVIᵉ siècles*, 5 vols. (Paris: Gide, 1836–39).

132. Viscount of Santarém, *Essai*, 1:LVI (note 93).

133. Viscount of Santarém, "Notice," 290 (note 94), author's translation.

134. A distinction has to be drawn between the prephotographic "facsimile" atlases, which were fresh engravings of hand-drawn copies imitating the originals (and which may contain erroneous graphic and textual transcriptions), and photographic facsimiles that date from the second half of the nineteenth century onward. For an early criticism of the reliability of the pioneer "facsimiles" of Jomard and Santarém, see Gabriel A. Marcel, *Reproductions de cartes et de globes relatifs à*

is usually assumed, represent the commanding heights of the history of cartography in the nineteenth century and beyond were in fact often designed as much to make early maps accessible for underpinning the history of geographical discovery as to promote the study of these maps as independent historical objects. Such interests are made absolutely plain by Jomard, whose explicit motives in compiling *L'atlas des monuments de la géographie* were "to illuminate successive eras with the progress of science and the principal discoveries."[135] Similarly, the atlas accompanying Lelewel's *Géographie du Moyen Age* assigns early maps to their place within the history of geography, a relationship that is reinforced by the wider context of his other historical works.[136] Toward the end of the nineteenth century, the same objective in the study of early maps is reflected in Nordenskiöld's preface to his *Facsimile-Atlas to the Early History of Cartography*, where he observes that "the era of the great geographical discoveries can hardly be fully intelligible without a comparative study of the maps which were then accessible."[137] Here too, as in his later *Periplus*, concentrating on "the early history of charts and sailing-directions,"[138] Nordenskiöld was esteeming that cartographic history was being written to provide the technical analyses necessary for the student of discovery and exploration.

Although twentieth-century historians of discovery, with notable exceptions, have failed to exploit early maps to the same degree as these pioneers,[139] links between the historical study of cartography and of discovery have remained firmly intact. As late as the 1960s, Cortesão was writing of the difficulty of disentangling the vast bibliographies relating to the histories of discovery, navigation, nautical science, and cartography.[140] It is true that by this date the range of technical studies in the history of cartography was increasing, in an attempt to match some of the new applications of early maps as sources in other historical specialisms. It is also true that an increasing number of articles and monographs no longer made such direct reference to the historical applications of the maps being studied. Yet lurking in the epistemology of many historians of cartography there was always still the feeling that their primary duty was to make accessible—and to interpret—the cartographic documents required by scholars in other fields.

First properly formulated by Jomard and Santarém, this approach found enduring expression in the continuing interest in the publication of facsimile atlases in the twentieth century. This was one of the objectives embraced by the International Geographical Union. Commissions for the reproduction of early maps were appointed in 1908 and in 1913; but though this project was continued by the Cambridge Congress of 1928, the members seem to have done more talking than publishing, and by 1938, as Leo Bagrow noted in a report on that year's congress, the various proposals for the recording and study of old maps "had not yet materialised."[141] At a national level in many countries, and reaping the technical benefits of modern methods of map reproduction such as collotype and offset lithography using fine screens,[142] scholars were more successful in bringing to fruition schemes for the reproduction of early maps relating to their own countries. There was a very real continuity between the pioneer atlases of Jomard, Santarém, and Nordenskiöld and the succeeding generation of *Monumenta cartographica* (to give them their generic title) that were organized on national and regional lines or in relation to particular collections,[143] perhaps culminating with the *Monumenta cartographica Africae et Aegypti* edited by F. C. Wieder for Prince Youssouf Kamal of Egypt.[144] Like their predecessors,

la découverte de l'Amérique du XVIe au XVIIIe siècle avec texte explicatif (Paris: Ernest Leroux, 1893–94), preface, 8, where he makes a plea for "true" facsimiles.

135. Jomard, *Introduction*, 6 (note 99), author's translation.

136. Rzepa, "Lelewel," 106–7 (note 96), for a "Selective and Thematic Bibliography" of works by Joachim Lelewel; see also Michael J. Mikoś, "Joachim Lelewel: Polish Scholar and Map Collector," *Map Collector* 26 (1984): 20–24.

137. A. E. Nordenskiöld, *Facsimile-Atlas to the Early History of Cartography*, trans. Johan Adolf Ekelöf and Clements R. Markham (Stockholm, 1889), preface.

138. Nordenskiöld, *Periplus*, quotation from title page (note 35).

139. Many historians of discovery have treated maps as incidental contributors to what is primarily narrative history, and their indexes may be searched in vain for maps and charts. An exception is John H. Parry, who was a vigorous pragmatic user of early maps as an integral part of his subject, although his caution in generalizing from cartographic evidence has already been referred to (see Parry, note 10 above).

140. Cortesão, *History of Portuguese Cartography*, 1:70 (note 93).

141. [Leo Bagrow], "Sixteenth International Geographical Congress, 1938," *Imago Mundi* 3 (1939): 100–102, quotation on 101. The ambitious scheme for the publication of a *Monumenta cartographica Europea*, set in motion through a subcommission in 1931, was eventually overtaken by World War II. In 1949, at the Lisbon meeting, the Commission pour la Reproduction et la Publication des Cartes Anciennes gave way to the Commission pour la Bibliographie des Cartes Anciennes on the grounds that the former (reproduction) could not be done satisfactorily without the latter (a bibliography), in order to establish priorities in reproduction. See Roberto Almagià, *Rapport au XVIIe Congrès international: Contributions pour un catalogue des cartes manuscrites, 1200–1500*, ed. Marcel Destombes, International Geographical Union, Commission on the Bibliography of Ancient Maps ([Paris], 1952), 5.

142. Walter W. Ristow, "Recent Facsimile Maps and Atlases," *Quarterly Journal of the Library of Congress* 24 (1967): 213–29.

143. They are discussed as a group by Skelton, *Maps*, 93–95 (note 15), and also in Cornelis Koeman, "An Increase in Facsimile Reprints," *Imago Mundi* 18 (1964): 87–88, and relate to countries or regions including Bohemia, Denmark, Iceland and the Faroes, Italy, the Netherlands, Portugal, Sudetenland, and the Ukraine.

144. Youssouf Kamal, *Monumenta cartographica Africae et Aegypti*, 5 vols. in 16 pts. (Cairo: 1926–51). See below, p. 40, where the full contents are itemized.

the "systematic plan and elaborate commentaries" of the later facsimile atlases gave them "the character of monographs or histories."[145] Yet despite the value of the scholarly commentaries that accompanied them (epitomizing the geographical contribution to the study of early maps), these atlases have today declined in relative importance within the subject as a whole. With the development of new ideas about the history of cartography, taking it beyond the study of maps primarily as historical documents, their influence and significance will continue to decline so that they may even come to be regarded as dinosaurs whose time-absorbing nature (and the finance needed to present them in all their glory) is a thing of the past. The extinction of the commission of the International Geographical Union concerned with early maps at the 1964 congress was a partial reflection, among the wider geographical community, of this changing intellectual landscape.[146] In the event, the abandonment of the history of cartography by the International Geographical Union has done nothing to check the flood of facsimile publications of early maps. Indeed, they have become so numerous that they now have their own bibliography,[147] have attracted a growing secondary literature, and continue unabated to attract some of the more ambitious publishing ventures within the history of cartography as a whole.

The specialist map libraries have also made their mark on the literature of the history of cartography. If the geographical societies from the nineteenth century onward did much to consolidate an approach to early maps as records of discovery, simultaneously initiating the facsimile tradition, then it can equally well be said that the principal legacy to the history of cartography of the great research libraries—apart from the opportunities they offered for research—has been the bibliographical one. The output of bibliographical works relating to early maps is truly substantial.[148] They reflect the diversity both of maps and of the types of libraries where they are conserved. It has been suggested that these bibliographies may be divided into two categories: the lists institutions produce describing their own holdings, and cartobibliographies—lists that offer a more exhaustive analysis and description of the variant forms of each map as a means of illuminating its printing history or the interrelationships of manuscript copies.[149]

As already noted, the history of the cataloging of early maps, so vital an adjunct to historians of cartography, can be traced back to primitive listings of maps undertaken in the seventeenth and eighteenth centuries. However, detailed cataloging was to be given particular impetus by the emergence of the independent map libraries during the nineteenth century. For example, in France in the 1820s, a manuscript *Catalogue géographique raisonné* had been compiled of d'Anville's massive collection, which was by then in the Département des Affaires

Etrangères,[150] and Jomard is said to have proposed the publication of a *catalogue raisonné* of the geographical collections of the Bibliothèque Royale before 1839.[151] Such ambitious projects did not, however, rapidly come to fruition. Of the national map collections on the eve of World War I, only the British Museum had issued a complete catalog of its holdings,[152] although the Library of Congress had already published a list of the maps of America in its possession and had completed the first volumes of a list of its geographical atlases.[153] Elsewhere, the most solid progress by this date was the completion of major lists of manuscript maps in the archives of repositories in the Netherlands (the Rijksarchief) and Spain (the Archivo General de Indias) and of the medieval and Renaissance materials in German and Italian collections,[154] together with the map collections of some

145. Skelton, *Maps*, 95 (note 15).

146. In fact, the voting was fifteen for the continuation of the commission, nineteen against, and three abstentions: *International Geographical Union Newsletter* 16 (1965): 6.

147. Walter W. Ristow, *Facsimiles of Rare Historical Maps: A List of Reproductions for Sale by Various Publishers and Distributors* (Washington, D.C.: Library of Congress, 1960 and subsequent editions). This is selective, and Skelton, *Maps*, 107 (note 15), included among future tasks the preparation of a full index of published facsimiles.

148. John A. Wolter, "Research Tools and the Literature of Cartography," *AB Bookman's Yearbook*, pt. 1 (1976): 21–30, esp. 21, noted that out of 1,169 entries classified in *Bibliotheca cartographica* as "bibliographies, collections, documentations" from 1957 to 1972, over 75 percent were bibliographies.

149. Robert W. Karrow, "Cartobibliography," *AB Bookman's Yearbook*, pt. 1 (1976): 43–52, esp. 43.

150. D'Anville's collection was sold to Louis XVI in 1780. In 1782, the year of d'Anville's death and of the transfer of the collection to Versailles, the inventory started by d'Anville was finished by his assistant, Jean Denis Barbié du Bocage. Barbié du Bocage had to make a new classification after the collection was received by the Ministry of Foreign Affairs; after an interruption, the work was continued by his son, Jean Guillaume, who signed, in January 1827, the introduction to the *Catalogue géographique raisonné* reflecting the new classification (four manuscript parts in five volumes), 1826–27. This catalog has remained unpublished.

151. Jean Bernard Marie Alexander Dezos de La Roquette, *Notice sur la vie et les travaux de M. Jomard* (Paris: L. Martinet, 1863), 15–16.

152. British Museum, *Catalogue of the Manuscript Maps, Charts, and Plans, and of the Topographical Drawings in the British Museum*, 3 vols. (London: Trustees of the British Museum, 1844–61), and idem, *Catalogue of the Printed Maps, Plans and Charts in the British Museum*, 2 vols. (London: W. Clowes by order of the Trustees of the British Museum, 1885).

153. Philip Lee Phillips, *A List of Maps of America in the Library of Congress* (Washington, D.C.: Government Printing Office, 1901); reprinted, Burt Franklin Bibliography and Reference Series no. 129 (New York: Burt Franklin, [1967]), and Library of Congress, *A List of Geographical Atlases in the Library of Congress, with Bibliographical Notes*, 8 vols. (Washington, D.C.: Government Printing Office, 1909–74), vols. 1–4 (1909–20), ed. Philip Lee Phillips, supp. vols. 5–8 (1958–74), ed. Clara Egli LeGear.

154. They are listed by Skelton, *Maps*, 86–89 (note 15).

of the military survey organizations.[155] Given the "gigantic and endless labour," as Skelton calls it, required for recording early map resources, these pioneer catalogs remain landmarks. Even today, over a century since the first major institutional catalogs appeared, there are still numerous collections of early maps lacking adequate published catalogs.[156]

The listing of maps—like the listing of books in bibliographical scholarship—has held a particular fascination for students of early maps since the nineteenth century. No doubt, as a handmaiden of connoisseurship, such an interest was partly rooted in the collecting and listing mentality of that age, an enthusiasm that persisted well into the present century.[157] But it was also a response to the inconvenient lack of published catalogs of many collections that has just been noted. In any case, the interest in producing and publishing catalogs was rapidly diversified so that, quite apart from the catalogs of institutional holdings, there is today an equally large number of early map bibliographies organized according to type and form, by period, by geographical area (world, continent, country, province), by function (cadastral maps, road maps, sea charts, celestial maps), by format (atlases, globes, wall maps), or by various combinations of these criteria.[158]

Another way the bibliographical approach came to exert such an influence on the developing character of the subject was through the frequent introduction of lists of maps, often accompanied by mapmakers' biographies, as the central element of a monograph or of the essays introducing facsimile atlases. Thus biobibliography came to be regarded as the heartbeat of the history of cartography. Many of the classic studies in the history of cartography have a major bibliographical dimension: Nordenskiöld included a list of later world maps in his *Facsimile Atlas*, and nautical charts and sailing directions were listed in his *Periplus*. The central organizing principle of Konrad Miller's *Mappaemundi* was a description of all the medieval world maps known to him,[159] and Henry Harrisse, nineteenth-century bibliographer-historian of the discovery of North America, had proposed a "Cartographia Americana" to accompany his *Bibliotheca Americana vetustissima*.[160] Since these men wrote, at the turn of the nineteenth and the twentieth centuries, the growth of bibliographical research has expanded to such an extent that it has become widely characteristic of the literature of the history of cartography, and many new projects today contain a bibliographical element.[161]

The term "cartobibliography" was coined by the English map historian Herbert George Fordham at the beginning of the present century. Fordham was particularly concerned with developing principles of analysis and classification that could be applied specifically to maps

as opposed to nonmap materials.[162] Subsequent research in a number of countries has refined his methods, particularly by adapting concepts first developed within literary bibliography, so that cartobibliography, or what Karrow has defined as "physical cartobibliography,"[163] has now developed from being concerned merely with

155. In Germany, for example, the nineteenth-century published catalogs of this type include: *Katalog über die im Königlichen Bayerischen Haupt Conservatorium der Armee befindlichen Landkarten und Pläne* (Munich, 1832); *Katalog der Bibliothek und Karten-Sammlung des Königlichen Sächsischen Generalstabes* (Dresden, 1878); and *Katalog der Kartensammlung des Königlichen Preußischen Generalstabes* (Berlin, 1893). The last-named collection was integrated into the map collections of the Preußische Staatsbibliothek in 1919; see Lothar Zögner, "Die Kartenabteilung der Staatsbibliothek, Bestände und Aufgaben," *Jahrbuch Preußischer Kulturbesitz* 14 (1977): 121–32.

156. For a recent plea see Monique Pelletier, "L'accès aux collections cartographiques en France," in *Le patrimoine des bibliothèques: Rapport à Monsieur le directeur du livre et de la lecture par une commission de douze membres*, ed. Louis Desgraves and Jean-Luc Gautier, 2 vols. (Paris: Ministère de la Culture, 1982), 2:253–59.

157. It is also closely paralleled in the literature of the history of science. See, especially, the approach of George Sarton, *Introduction to the History of Science* (note 17) in vol. 1, *From Homer to Omar Khayyam*, 39, where he writes, "Bibliography is another essential basis of historical or scientific investigations of any kind. My account is brief, often of Linnaean brevity, but I have attempted to complete each item with a list of the main sources and of many other publications. Thus the reader will have abundant means of controlling every word of my statements and of continuing the study of any topic to any extent." This could also have been a testament of many historians of cartography, and it has justifiably been regarded as a cornerstone of the subject.

158. There is no general published "bibliography of cartobibliographies." For an institutional list of this nature, however, see Annemieke van Slobbe, *Kartobibliografieën in het Geografisch Instituut Utrecht*, Utrechtse Geografische Studies 10 (Utrecht: Geografisch Instituut Rijksuniversiteit Utrecht, 1978). An introduction to the main bibliographical finding aids is provided by Wolter, "Research Tools" (note 148).

159. See below, chapter 18, for a discussion of the development of his ideas into a proper classification.

160. This was to consist of a list and description of all the maps, whether published or in manuscript, relating to the New World and drawn before 1550: Richard W. Stephenson, "The Henry Harrisse Collection of Publications, Papers, and Maps Pertaining to the Early Exploration of America" (paper prepared for the Tenth International Conference on the History of Cartography, Dublin 1983), 10.

161. The continuing need for—and fundamental importance of—primary bibliographical listing for the history of cartography is illustrated by the inclusion of such lists in some of the subsequent chapters in this volume, such as, inter alia, those on the prehistoric period, portolan charts, and local maps and plans of medieval Europe: see pp. 93–97, 449–61, 498–500.

162. Herbert George Fordham, *Studies in Carto-bibliography, British and French, and in the Bibliography of Itineraries and Road-Books* (Oxford: Clarendon Press, 1914; reprinted, London: Dawson, 1969); the essay in the volume "Descriptive Catalogues of Maps: Their Arrangement, and the Details They Should Contain," 92–127, shows the extent to which Fordham was thinking out new principles for cartobibliography.

163. Karrow, "Cartobibliography," 47–50 (note 149).

the simple enumeration of maps to more complex forms of descriptive analysis that can answer questions about the production of maps, about their provenance, and, in particular, about their chronological relationship to each other in genetic sequences created by printing or manuscript copying. Increasingly, such techniques are being used to answer wider questions in the history of cartography, including the historicity of maps as documents, the nature of the map trade in different periods and places, the statistical growth of cartographic output, and the transmission and dissemination of early maps and their images.[164] Some cartobibliographers also attempted to define an elaborate terminology in connection with their subject,[165] but this has now been challenged by recent developments in the description of nontextual material in books in general rather than maps in particular.[166] Despite the poverty of historical interpretations of cartographic change built around bibliographic listings, which have been justifiably criticized,[167] a substantial volume of bibliographical research remains as the most characteristic and valuable legacy of the influence of map librarians on the development of the study of early maps.

As a group, large map libraries have also contributed significantly to the development of the history of cartography by mounting specialist cartographic exhibitions accompanied by published catalogs. Indeed, this has always been regarded as a major function of the curator of rare early maps. Yet it has been neglected by historians of cartography in helping to explain the rise of the study of the history of cartography.[168] For countries rich in older map resources, however, such as Italy or Spain, such exhibitions have often marked the pioneer phase in a historical awareness of early maps.[169] Exhibitions designed to bring early maps to the attention of a wider public or to expose them to scholars as an appetizer for subsequent detailed study have regularly been mounted in the larger map libraries since the nineteenth century. They have been reported in many national geographical journals, and were a regular feature of the meetings of some societies, and, from 1935 onward, of the "Chronicle" section of *Imago Mundi*.[170] The diversity of subject matter presented in the exhibitions themselves and in the accompanying catalogs is considerable, but it is possible to pick out recurrent themes.[171] There have been major exhibitions to mark the occasions of conferences and congresses; exhibitions designed to highlight the cartography of particular cultures, periods, or places; and commemorative exhibitions to mark the anniversaries of the great names in cartography—the national heroes of the world of maps—or of their landmark publications. As in the history of art and in the museum world in general, an exhibition can serve wider intellectual purposes than merely creating a factual rec-

ord of what was displayed. In some cases the published catalogs (the best of which assume the character of illustrated monographs) provide an original synthesis of a specific subject in the history of cartography.[172] Exhibitions can also be innovative, pioneering new conceptual approaches to maps, as with the exhibition *Cartes et figures de la terre*, held in Paris in 1980.[173] In

164. For these wider uses of bibliography in map history see Michael J. Blakemore and J. B. Harley, *Concepts in the History of Cartography: A Review and Perspective*, Monograph 26, *Cartographica* 17, no. 4 (1980): 37–42; see also, for a discussion with many implications for the history of cartography, G. Thomas Tanselle, "From Bibliography to *Histoire Totale*: The History of Books as a Field of Study," *Times Literary Supplement*, 5 June 1981, 647–49 (text of the second Hanes Lecture in the History of the Book, University of North Carolina, 15 April 1981).

165. Especially Coolie Verner, "The Identification and Designation of Variants in the Study of Early Printed Maps," *Imago Mundi* 19 (1965): 100–105, and idem, "Carto-bibliographical Description: The Analysis of Variants in Maps Printed from Copper Plates," *American Cartographer* 1 (1974): 77–87.

166. G. Thomas Tanselle, "The Description of Non-letterpress Material in Books," *Studies in Bibliography* 35 (1982): 1–42.

167. For example, P. D. A. Harvey, *The History of Topographical Maps: Symbols, Pictures and Surveys* (London: Thames and Hudson, 1980), 7.

168. Skelton, *Maps* (note 15), does not refer to exhibitions in his historical survey of the study and collecting of maps.

169. This still holds true; see, for example, for Spain, Biblioteca Nacional, *La historia en los mapas manuscritos de la Biblioteca Nacional*, exhibition catalog (Madrid: Ministerio de Cultura, Dirección General del Libro y Biblioteca, 1984); *Puertos y fortificaciones en América y Filipinas* (Comision de Estudios Historicos de Obras Publicas y Urbanismo, 1985).

170. *Petermanns Geographische Mitteilungen* is a reliable source for the recording of map exhibitions after 1855. From 1892, exhibitions on early cartography were associated in Italy with the Congressi Geografici Nazionali; see Elio Migliorini, *Indice degli Atti dei Congressi Geografici Italiani dal primo al decimo (1892–1927)* (Rome: Presso la Reale Società Geografica Italiana, 1934); and Luigi Cardi, *Indice degli Atti dei Congressi Geografici Italiani dall'undicesimo al ventesimo (1930–1967)* (Naples: Comitato dei Geografi Italiani, 1972). In *Imago Mundi* the listings begin with "Chronik," *Imago Mundi* 1 (1935): 68–73.

171. The list in British Museum, *Catalogue of Printed Maps, Charts and Plans*, 15 vols. (London: Trustees of the British Museum, 1967), 15: col. 787, and in the supp. for 1965–74 (1978), cols. 1347 et seq., has provided the sample for this discussion. Exhibition catalogs are also currently recorded in the annual volumes of *Bibliographia cartographica* (section IV C). Such specialist catalogs, of course, underestimate the extent to which early maps have been exhibited; they often appear in exhibitions of art, science, and culture of a more general nature.

172. Marijke de Vrij, *The World on Paper: A Descriptive Catalogue of Cartographical Material Published in Amsterdam during the Seventeenth Century* (Amsterdam: Theatrum Orbis Terrarum, 1967); Arend Wilhelm Lang, *Das Kartenbild der Renaissance*, Ausstellungskataloge der Herzog August Bibliothek, no. 20 (Wolfenbüttel: Herzog August Bibliothek, 1977); Sarah Tyacke and John Huddy, *Christopher Saxton and Tudor Map-making* (London: British Library, 1980).

173. *Cartes et figures de la terre*, exhibition catalog (Paris: Centre Georges Pompidou, 1980); see also the comparable *Arte e scienza per*

other cases, they can highlight little-known aspects of maps,[174] introduce new map genres, or reveal to scholars the research potential of a specialist map library.[175] Such exhibitions have educational functions in the broadest sense, and they have contributed significantly to an awareness of the historical importance of early maps.

Finally, the private collectors and the antiquarian map trade, identified as one of the three major forces in the early development of the history of cartography, can also be assessed in terms of their historiographic impact. The major effect has been to reinforce tendencies consequent to the rise, already noted, of institutional geography and of the independent map libraries. On the one hand, there are the publishers who cater to the collector and to the antiquarian market by imitating the scholarly emphasis on facsimile publication, either through high-quality reproductions that have an antique appeal or through the medium of international coffee-table books combining lavish illustration of early maps with relatively brief texts.[176] On the other hand, collectors and the trade are served by catalogs with a bibliographical quality, which strengthens the existing domination of such works within the literature. Some of the catalogs of the leading antiquarian map sellers have for so long had the character of carto-bibliographies, contributing new editions and impressions to the printing history of atlases and maps,[177] that they are recognized as part of the research literature of the history of cartography.

A secondary effect of the links with map collecting and the antiquarian trade is the appearance of a popular face to the history of cartography. There is nothing unusual in this; other subjects, including the history of the fine arts in general, as well as archaeology, landscape history, and family history, have likewise breached the confines of their narrow academic circles. In the history of cartography, however, the tendency has been pronounced. Amateur[178] and professional interests are closely intertwined, and the subject has always been open to all comers, so that this impact is becoming clearly visible in recent decades.

First, there has been a spate of books, in several languages, aimed at popularizing antique maps and at stimulating the marketplace by encouraging beginners to discover and identify them, or offering advice about the investment potential of cartographic items or about the care and preservation of a collection. Such works, as already noted, have the effect of focusing attention upon the decorative and more highly collectible maps from before 1800. Yet since the supply of these early maps is clearly finite, there are now signs of a growing awareness of later periods and of the potential fascination (for example) of maps produced by nineteenth- and twentieth-century official mapmaking agencies.[179]

Second, there has been the founding of both regional and national societies for collectors of early maps. A society of cartophiles was established in New England in the 1920s and reestablished in the 1950s,[180] and in Europe Leo Bagrow founded the Circle of Lovers of Russian Antiquities in Berlin in 1927, clearly much concerned with maps.[181] But permanent developments were to occur only after World War II. In 1952, for example, The International Coronelli Society for the Study of Globes and Instruments was founded (and has continued to publish the scholarly journal *Der Globusfreund*),[182]

il disegno del mondo, exhibition catalog, city of Turin (Milan: Electa Editrice, 1983).

174. See, for example, Gillian Hill, *Cartographical Curiosities* (London: British Museum Publications, 1978); *Het aards paradijs: Dierenvoorstellingen in de Nederlanden van de 16de en 17de eeuw*, exhibition catalog (Antwerp: Zoo Antwerpen, 1982); Omar Calabrese, Renato Giovannoli, and Isabella Pezzini, eds., *Hic sunt leones: Geografia fantastica e viaggi straordinari*, catalog of exhibition, Rome (Milan: Electa Editrice, 1983). Recent catalogs concerning the relation between art and cartography include: "Art and Cartography: Two Exhibitions, October 1980–January 1981," *Mapline* special no. 5 (October 1980); Jasper Johns et al., *Four Artists and the Map: Image/Process/Data/Place* (Lawrence: Spencer Museum of Art, University of Kansas, 1981), and a concurrent exhibition in Lawrence, "A Delightful View: Pictures as Maps," 6 April–31 August 1981; "CARTography" was the title of a two-part exhibition at the John Michael Kohler Arts Center, Sheboygan, Wisconsin, from 16 November 1980 to 11 January 1981; part 1, "An Historical Selection of Maps, Globes, and Atlases from the American Geographical Society Collection," and part 2, "Cartographic Images in Contemporary American Art."

175. See, for example, *A la découverte de la terre, dix siècles de cartographie*, trésors du Département des Cartes et Plans de la Bibliothèque Nationale, Paris, May–July 1979, 122, which was a presentation of their main resources; or *The Italians and the Creation of America: An Exhibition at the John Carter Brown Library*, prepared by Samuel J. Hough (Providence: John Carter Brown Library, 1980).

176. A model for this genre could be *Landmarks of Mapmaking*, with maps chosen and displayed by R. V. Tooley, text written by Charles Bricker, and preface by Gerald R. Crone (note 14); recent examples include George Kish, *La carte: Image des civilisations* (Paris: Seuil, 1980); Tony Campbell, *Early Maps* (New York: Abbeville Press, 1981); and its sequel by Robert Putman, *Early Sea Charts* (New York: Abbeville Press, 1983); see also George Sergeant Snyder, *Maps of the Heavens* (New York: Abbeville Press, 1984).

177. Examples would include some of the catalogs issued by Weinreb and Douwma, Francis Edwards, Nico Israel, H. P. Kraus, Frederick Muller, Kenneth Nebenzahl, Leo Olschki, Rosenthal of Munich, and Henry Stevens, Son and Stiles.

178. "Amateur" is not used here in a derogatory sense but serves to indicate the large number of contributions from scholars and others whose main fields lie outside the history of cartography.

179. For example, the Charles Close Society for the study of Ordnance Survey maps is largely devoted to their history. It publishes a newsletter, *Sheetlines*, no. 1– (October 1981–).

180. Erwin Raisz, "The Cartophile Society of New England," *Imago Mundi* 8 (1951): 44–45.

181. See obituary of Leo Bagrow, [R. A. Skelton], "Leo Bagrow: Historian of Cartography and Founder of *Imago Mundi*, 1881–1957," *Imago Mundi* 14 (1959): 4–12, esp. 8.

182. Wilhelm Bonacker, "The First International Symposium of the Coronelli Weltbund der Globusfreunde," *Imago Mundi* 18 (1964): 83–84.

and a Finnish Map Society, Chartarum Amici, was created in 1965.[183] The real spurt in the founding of such societies came in the 1970s. Both the Chicago Map Society (started with the aim of supporting and encouraging the study and preservation of maps and related materials) and the British Columbia Map Society were founded in 1976, and they were soon joined by half a dozen other societies in North America. In England the International Map Collectors' Society was formed in 1980,[184] and in the Netherlands *Caert-Thresoor*, though not published by a society, reflected the informal growth of similar interests.[185]

A third result of the popularization of the history of cartography has been the launching of monograph and periodical publications with the needs of map collectors specifically in mind. First there was *The Map Collectors' Circle*, started in 1963 by R. V. Tooley, which continued publication until 1975.[186] Then in 1977 *The Map Collector* began publication. Today, as still the only magazine of its kind, it aims to encourage a community of interest between dealers, collectors, and scholars and all who share an interest in early maps. It might be regarded as only the latest event in the long development of the symbiotic contacts between these groups that have characterized research and writing in the history of cartography.

THE GROWTH OF A SCHOLARLY IDENTITY

Since the 1930s the history of cartography has been slowly emerging as a subject with its own scholarly identity.[187] The basis of this claim needs support, however, and, as we shall see below, there are signs that the history of cartography is indeed developing a conscious epistemology and sense of purpose in research that increasingly distance it from its nineteenth- and early twentieth-century phase. An understanding of this change can be derived from three main developments within the subject. First, there is the publication of the general histories of cartography, attempts at a synthesis of its subject matter, of which the present work is the latest reflection. Second, there is the influence of *Imago Mundi*, the journal founded by Leo Bagrow in 1935, which has been the only international journal devoted exclusively to the history of cartography. Third, and probably by far the most significant influence, there is the emergence of cartography as an independent academic and practical discipline providing new theoretical frameworks as well as a reinforced raison d'être for the study of cartographic history.

Even accepting that awareness of their subject is increasing among historians of cartography, the relevant criteria for the changes are difficult to isolate. Moreover, the process can neither be dated precisely nor measured in other than the most qualitative terms. It would be hard to discover, for example, from even the recent literature, any sustained conceptual shift[188] in the ways of thinking of cartographic historians. All that can be discerned is much continuity and some slow change. The former is represented by the survival of approaches toward the study of early maps established in the nineteenth century, while the evidence for change can be traced back to the 1930s. But there is no neat line separating two phases of intellectual development, and there are no grounds yet for believing in succeeding eras of an old and a new history of cartography.

Historiographic analysis of the history of cartography is also complicated by the fragmentation of its literature. The framework of three main influences outlined above is valid in general terms, but it fails to take account of the scattering of writings about early maps in the literature of other subjects. No less than any other historical artifact or document, early maps are not the exclusive property of one subject. From an inspection of the first annual bibliography in *Imago Mundi*, relating to the year 1933, Skelton concluded that of the items listed there:

> Some were published (as we might expect) in journals of geography, history, local history, geodesy and survey, hydrography and navigation, the history of science. But articles on early maps appeared in many other less obvious quarters—periodicals devoted to physical science, biology, agriculture, magnetism, economics, political science, art history, oriental studies, the classics, archaeology, printing history, bibliography and library science, archives.[189]

This enumeration can easily be extended to reveal the wider field of the history of cartography. Bibliographies published subsequently in *Imago Mundi* also list material relating to cartographic history in books or peri-

183. *Map Collector* 10 (1980): 32.

184. Like a number of other societies, it publishes a newsletter, *IMCOS, Journal of the International Map Collectors' Society*, vol. 1– (1980–); for an account of one of the North American societies see Noël L. Diaz, "The California Map Society: First Years," *Bulletin of the Society of University Cartographers* 18, no. 2 (1984): 103–5.

185. *Caert-Thresoor*, no. 1– (1982–).

186. *Map Collectors' Circle*, nos. 1–110 (1963–75).

187. Cornelis Koeman, "Sovremenniye issledovaniya v oblasti istoricheskoy kartografii i ikh znacheniye dlya istorii kul'tury i razvitiya kartograficheskikh nauk" (Modern investigations in the field of the history of cartography: Their contribution to cultural history and to the development of the science of cartography), in *Puti razvitiya kartografii* (Paths to the evolution of cartography), a collection of papers on the occasion of Professor K. A. Salishchev's seventieth birthday (Moscow: Izdatel'stvo Moskovskogo Universiteta, 1975), 107–21.

188. The ideas of Thomas S. Kuhn, *The Structure of Scientific Revolutions* (Chicago: University of Chicago Press, 1962), have, however, been applied to the recent growth of cartography as a subject by Wolter, "Emerging Discipline" (note 113).

189. Skelton, *Maps*, 101–2 (note 15).

odicals devoted to the history of astronomy and mathematics, to geology and medicine, to the histories of architecture, cosmology, religion, and numismatics, to literature and philology, to anthropology and sociology, to historical geography, and to urban history and town planning. Moreover, especially for the recent history of cartography, relevant contributions are increasingly to be found in remote sensing and computer journals and in the recently founded cartographic journals noted below.[190] That there are so many diverse contexts for the history of cartography can, of course, be taken as additional evidence for the universality of maps and for the increasing interest in early maps of a diversified circle of readers. But the scattered literature—and the centrifugal tendencies it has encouraged—also highlights the problem of communication which has so far inhibited the development of a coherent subject. Thus, without ignoring the value of the contribution of many subjects to the history of cartography, there is a justification for trying to understand those long-term influences that have operated to draw together the scattered disciplinary threads. Such influences are therefore discussed below as being of particular significance in the growth of the scholarly identity that is also a central concern of this *History of Cartography*.

ATTEMPTS AT SYNTHESIS

Histories of cartography—setting out to provide a synthesis of the whole field as perceived by their authors—have not exerted a dominant role in the development of the subject. If they are regarded as barometers of integration and self-awareness within the history of cartography, then one measure of the immaturity of the subject is that the composite histories seem to have lagged a long way behind developments in specialized research in the past fifty years. Even today, at neither an academic nor a more popular level is there an up-to-date or balanced treatment of the full spread of the subject matter of the history of maps. In some respects this is a negative chapter in the study of the history of cartography, yet at the same time the histories reviewed here have helped to mold an image of the history of cartography still widely accepted in the 1980s.

It is doubtful that the need for a specific history of maps—apart from the study of their content as records of geographical discovery and exploration—would have been perceived in the nineteenth century in the way this need is viewed today. It is not until the present century that the growth of the idea of independent histories of cartography can be detected, and even then its translation into practice was to make little progress until about 1940. In view of his later dominance in the history of cartography, it is worth noting that as early as 1918 Leo

Bagrow had published in Russia a preliminary essay titled "The History of the Geographical Map: Review and Survey of Literature."[191] This work, however, was intended mainly as a bibliography (it listed no fewer than 1,881 items on the history of cartography), and it was heavily biased in its text and illustrations toward Russian examples. As Bagrow stated in his foreword, his work was undertaken in isolation, and he lacked the library resources available to scholars elsewhere. Indeed, at this date the country where one might expect that a history of cartography would have been conceived was Germany. Berlin, it might be said, had assumed the mantle that Paris had worn in the mid-nineteenth century as the center of "the real ferment in cartographic history."[192] By the early twentieth century there had been important developments in cartography in both Austria and Germany.[193] Moreover, it was from Germany that the most outstanding cartographic thinker of the generation—Max Eckert—was to come. Not only did Eckert begin to offer a philosophical basis for the study of cartography,[194] but in his two-volume treatise on cartography, *Die Kartenwissenschaft* (1921–25), he provided both a substantial and a seminal work which was to exert a considerable influence on modern cartography. Maintaining an approach characterized by a strong historical emphasis, Eckert systematically analyzed the character and evolution of different types of maps and established genetic principles for their formal study, so his *Die Kartenwissenschaft* may also be regarded as a contribution to the history of cartography along systematic lines. It did, however, quite deliberately stop short of any attempt at historical synthesis. Eckert's views on this matter are of particular interest in relation to the question why, despite the considerable interest in the subject, no general history of cartography was produced in the Germany of the 1920s. While Eckert acknowledged, as already noted, that a major deficiency in geo-

190. See below, pp. 32–33.

191. Leo Bagrow, *Istoriya geograficheskoy karty: Ocherk i ukazatel' literatury* (The history of the geographical map: Review and survey of literature), *Vestnik arkheologii i istorii, izdavayemyy Arkheologicheskim Istitutom* (Archaeological and historical review published by the Archaeological Institute) (Petrograd, 1918).

192. Skelton, *Maps*, 76 (note 15).

193. See below p. 32 nn. 262, 263.

194. See, for example, Max Eckert, "Die Kartographie als Wissenschaft," *Zeitschrift der Gesellschaft für Erdkunde zu Berlin* (1907): 539–55; idem, "Die wissenschaftliche Kartographie im Universitäts-Unterricht," in *Verhandlungen des Sechszehnten Deutschen Geographentages zu Nürnberg*, ed. Georg Kollm (Berlin: D. Reimer, 1907), 213–27; idem, "On the Nature of Maps and Map Logic," trans. W. Joerg, *Bulletin of the American Geographical Society* 40 (1908): 344–51; see also Wolfgang Scharfe, "Max Eckert's 'Kartenwissenschaft'—The Turning-Point in German Cartography" (paper prepared for the Eleventh International Conference on the History of Cartography, Ottawa, 1985).

graphy was felt to be the lack of a history of carto-graphy, he believed that it was unlikely to be written in the near future because too many preliminary studies were lacking.[195] Eckert's view prevailed for the next two decades. Konstantin Cebrian, a lecturer in a military school in Danzig, projected a multivolume history of cartography, but only the first volume was actually produced.[196] Several shorter works also appeared in the interwar years; but notwithstanding the aspirations as general histories conveyed in their titles, these either were heavily biased toward particular areas of the world[197] or contained a mélange of topics that did not properly reflect the broad sweep of cartographic development through time.[198]

Toward the middle of the century, however, there appeared three short general histories of cartography, and it is these that, in various editions, have held much of the field until the present day and which deserve notice as broadly diagnostic of existing attempts at synthesis.[199] The first text to be completed, in 1943, was Leo Bagrow's *Die Geschichte der Kartographie*, but the material for the reproduction of the illustrations was destroyed by wartime bombing and it was not finally published until 1951 in Berlin.[200] The revised and enlarged English-language edition did not appear until 1964.[201] Meanwhile, and over a decade before this English edition, Lloyd Brown's *The Story of Maps* was published in 1949.[202] This was followed by the third of this group of general histories, Gerald R. Crone's *Maps and Their Makers* (first edition 1953).[203] Whatever their shortcomings as reviewed today, these three books made a substantial contribution to the development of an appreciation of early cartography. They should first be judged in the light of history of cartography studies at the time of their conception and of their own terms of reference as introductory summaries for popular and student use. But even allowing for such considerations, all three point to the poverty of synthetic writing on the history of cartography. As contemporary reviewers sometimes made plain, they all left much to be desired, not so much on the grounds of technical scholarship as in internal balance and in total coverage.

When Bagrow's *Die Geschichte der Kartographie* first appeared, it was understandably praised as a remarkable performance in synthesis.[204] A closer look, however, reveals that its terms of reference were doubly restrictive. In the first place, as Bagrow himself explained, it excluded some matters such as scientific methods of mapmaking, the way material is collected, and the compilation of maps,[205] yet these are aspects now regarded as vital in cartographic history. In the second place, his narrative ended in the eighteenth century, at the point he saw as "where maps ceased to be works of art, the products of individual minds, and where craftsmanship was finally superseded by specialised science and the

195. Max Eckert, *Die Kartenwissenschaft: Forschungen und Grundlagen zu einer Kartographie als Wissenschaft*, 2 vols. (Berlin and Leipzig: Walter de Gruyter, 1921–25), 1:26–27. He stated: "What impedes the construction of the great edifice of a history of cartography is to be found partly in personal qualifications and partly in the subject matter. A historian of the map and its theory must be equally well versed in the history of geography . . . and philology. With this he must combine considerable mathematical knowledge. It is in the fullest mastery of these branches of science that the formula is to be found that will lead to the desired result" (author's translation).

196. Konstantin Cebrian, *Geschichte der Kartographie: Ein Beitrag zur Entwicklung des Kartenbildes und Kartenwesens* (Gotha: Perthes, 1922), vol. 1, *Altertum: Von den ersten Versuchen der Länderabbildungen bis auf Marinos und Ptolemaios (zur Alexandrinischen Schule)*. This project is described by Wilhelm Bonacker, "Eine unvollendet gebliebene Geschichte der Kartographie von Konstantin Cebrian," *Die Erde* 3 (1951–52): 44–57. Cebrian died in World War I; unpublished materials are preserved in the Map Department of the Staatsbibliothek Preußischer Kulturbesitz.

197. For example, Herbert George Fordham, *Maps, Their History, Characteristics and Uses: A Handbook for Teachers*, 2d ed. (Cambridge: Cambridge University Press, 1927), and Arthur L. Humphreys, *Old Decorative Maps and Charts* (London: Halton and Smith; New York: Minton, Balch, 1926), rev. by R. A. Skelton as *Decorative Printed Maps of the 15th to 18th Centuries* (London: Staples Press, 1952).

198. W. W. Jervis, *The World in Maps: A Study in Map Evolution* (London: George Philip, 1936); see Bagrow's scathing review of this work in *Imago Mundi* 2 (1937): 98.

199. There have been similar short or popular histories in competition that will not be specifically reviewed here: Hans Harms, *Künstler des Kartenbildes: Biographien und Porträts* (Oldenburg: E. Völker, 1962); A. Libault, *Histoire de la cartographie* (Paris: Chaix, 1959); Thrower, *Maps and Man* (note 25); and more recently, Georges Grosjean and Rudolf Kinauer, *Kartenkunst und Kartentechnik vom Altertum bis zum Barock* (Bern and Stuttgart: Hallwag, 1970); Kish, *Carte* (note 176); A. G. Hodgkiss, *Understanding Maps: A Systematic History of Their Use and Development* (Folkestone: Dawson, 1981); Wilford, *Mapmakers* (note 18); and Ivan Kupčík, *Alte Landkarten: Von der Antike bis zum Ende des 19. Jahrhunderts* (Hanau am Main: Dausien, 1980), or the French edition: *Cartes géographiques anciennes: Evolution de la représentation cartographique du monde de l'antiquité à la fin du XIXᵉ siècle*, trans. Suzanne Bartošek (Paris: Edition Gründ, 1981).

200. Leo Bagrow, *Die Geschichte der Kartographie* (Berlin: Safari-Verlag, 1951).

201. Leo Bagrow, *The History of Cartography*, rev. and enl. R. A. Skelton, trans. D. L. Paisey (Cambridge: Harvard University Press; London: C. A. Watts, 1964), preface, 5. This version was also translated into German; see *Meister der Kartographie* (Berlin: Safari-Verlag, 1963).

202. Brown, *Story of Maps* (note 16).

203. Crone, *Maps and Their Makers* (note 13).

204. It was reviewed by C. B. Odell, *Annals of the Association of American Geographers* 43 (1953): 69–70; W. Horn, *Petermanns Geographische Mitteilungen* 97 (1953): 222; and A. W. Lang, *Erdkunde* 7 (1953): 311–12. The 1964 English edition was reviewed by George Kish, *Geographical Review* 56 (1966): 312–13, and J. B. Harley, *Geographical Journal* 131 (1965): 147.

205. Bagrow, *History of Cartography*, 22 (note 201). See the criticism by David Woodward, "The Study of the History of Cartography:

machine."[206] There is no doubt that this view of Bagrow's, which sees the subject as having its center of gravity before the nineteenth century, has been influential. It is easy to understand why so many scholars have been attracted by the flowering of cartography in the European Renaissance, but the narrowness of their focus has skewed the total effort in relation to the cartographic record in its entirety. This tendency is manifest in much subsequent writing on the history of cartography, not only in the coffee-table publications that, as already noted, have served collectors as specimen books for the decorative printed maps of the European period from the sixteenth century to the eighteenth, but also in the content (as will be seen) of *Imago Mundi*, edited by Bagrow for many years and the main journal for the history of cartography.[207] In retrospect, this "paleocartographic bias," as it has been aptly described,[208] is an unnecessary and unjustifiable truncation of cartographic history, but it continues to be reflected in the subject at both a research and a popular level.

Nor have the books by Crone and Brown escaped criticism; but of the two, that by Crone is better organized and arguably remains the best summary of the history of cartography published so far.[209] By its second edition (1962), a chapter had even been added dealing with contemporary cartography. But though its terms of reference—treating maps as scientific reports, historical documents, research tools, and objects of art and regarding them as "products of a number of processes and influences"—are unexceptionable,[210] as one of a series of student texts its length was inevitably restricted. It was intended to describe "the main stages of cartographical development to which many countries have contributed in turn,"[211] but the maps of nonliterate peoples are dealt with in a short paragraph, those of Egypt and Mesopotamia are allotted a second paragraph, and there is no treatment at all of West Asian or East Asian cartography.

Lloyd Brown's book was also written in response to the need for a general survey of the history of cartography. "There is no other such chronicle in print," he could state in 1949, "though in the past seventy-five years or so it has been many times reasserted that the world is becoming increasingly aware of and interested in maps."[212] The reasons for delay seemed to him clear enough: the limited biographical material on earlier mapmakers; the high mortality of maps, leading to the destruction of the relevant evidence; and the unwillingness of specialist scholars "to confine the story" to other than "a straight and more or less narrow path."[213] Such problems are still with us, but Brown's solutions to them disappointed his scholarly reviewers. *The Story of Maps* is in effect "a personal, independent narrative" rather than a history of cartography, and it introduces, as one

early reviewer put it, "such a variety of subjects, handled with so little attention to order or precision, that the real theme can only be conjectured."[214] These words ring true today, yet for want of alternatives Brown's book perforce remains on our reading lists, with a reprint edition even welcomed by a new generation of reviewers. These are the reviewers, too, who have heralded as succeeding, where other books have failed, new general works such as John Noble Wilford's *The Mapmakers*, an eminently readable book but one that lacks the scholarly apparatus for a work of reference and fails to fill the vacuum left by Bagrow, Crone, and Brown.[215] These retrospective comments are not made in a critical spirit. But as revealed in the light of the changing needs of the subject, the earlier works' deficiencies as general histories—whether imposed by a conceptual climate different from our own, by length, or by the authors' interpretation of their terms of reference—both justify and make opportune this *History of Cartography*.

LEO BAGROW, *IMAGO MUNDI*, AND THEIR INFLUENCE

A second identifiable contribution to the growing identity of the history of cartography since the mid-1930s has been made by the periodical *Imago Mundi*, founded by Leo Bagrow. Initially designed as a yearbook for the subject, it is now described as the "Journal of the International Society for the History of Cartography." Leo Bagrow (1881–1957) was born Lev Semenovich Bagrov, an émigré from Saint Petersburg first to Berlin and then to Sweden.[216] Through his personality and his scholarly conception of his subject, he came to dominate the his-

---

A Suggested Framework,"*American Cartographer* 1 (1974): 101–15, esp. 102.

206. Bagrow, *History of Cartography*, 22 (note 201).

207. See below, pp. 27–28.

208. Robinson, *Thematic Mapping*, ix (note 28).

209. Reviews included George Kish, *Geographical Review* 45 (1955): 448–49, and E. M. J. Campbell, *Geographical Journal* 120 (1954): 107–8.

210. Crone, *Maps and Their Makers*, 2d ed., ix (note 13).

211. Crone, *Maps and Their Makers*, 2d ed., ix (note 13).

212. Brown, *Story of Maps*, 3 (note 16).

213. Brown, *Story of Maps*, 4 (note 16).

214. Edward Lynam, *Geographical Review* 40 (1950): 496–99, quotation on 496; it was praised by other reviewers such as "F. G." (Frank George), *Geographical Journal* 116 (1950): 109.

215. Alan M. MacEachren, *American Cartographer* 9 (1982): 188–90; Peter Gould, *Annals of the Association of American Geographers* 72 (1982): 433–34; and J. A. Steers, *Geographical Journal* 149 (1983): 102–3; but see also the more reflective review by Denis Wood, *Cartographica* 19, nos. 3–4 (1982): 127–31, setting the work in the context of similar general histories.

216. "Leo Bagrow" (note 181), which included a bibliography; see also Wilhelm Bonacker, "Lev Semenovič Bagrov (1888–1957): Ein Leben für die Geschichte alter Karten," *Petermanns Geographische Mitteilungen* 101 (1957): 308–9.

tory of cartography over three decades, from the 1930s to the 1950s. In the 1930s he had already conceived several massive projects for the synthesis of materials for the history of cartography. These included a catalog of printed maps of the sixteenth century (surviving only as a typewritten list); a series of monographs with facsimiles of early maps; an encyclopedia of early maps (which never emerged from the planning stage during Bagrow's lifetime); a history of cartography (eventually published, as already noted, in 1951); and finally, a periodical devoted to the history of cartography, to be published annually. The first two of these projects perpetuated nineteenth-century traditions, but the last three point to Bagrow's recognition of what would now be called an "identity crisis" within the history of cartography.[217] Against this background he founded his periodical, in an explicit attempt to create a more unified subject.

First published in 1935, *Imago Mundi* was the first journal to be devoted entirely to the history of cartography, and it is still the only scholarly international one. By giving historians of cartography their own forum it has contributed to their sense of self-awareness, and it has become a barometer for the development of the subject in general. It represents Bagrow's most important contribution to the consolidation of the field. Although other scholars shared in its inception and early development, it was Bagrow who remained at the editorial helm until his death. It was later noted that he had "devoted to it his formidable energy, his authority as a scholar and the greater part of his time" and that, almost dictatorially, "with only occasional reference to corresponding editors, he made all decisions on acceptance or rejection of contributions and on the contents and lay-out of each issue; he conducted all correspondence, and compiled such regular items as the Chronicle and Bibliography."[218] Bagrow's anxiety for integration within the history of cartography is reflected in the "Editorial" to volume 2, where he noted that, although the literature of early cartography was increasing very rapidly and though this increase was a reflection of widespread activity in the study and collection of early maps, there was still little coordination. He continued:

> Students in different countries have very inadequate means of knowing what is being done and published in other countries; many rare and important maps in state archives and private collections are little known and have never been described; and librarians, students, collectors and booksellers, despite the assistance which geographical societies willingly give, often have difficulty in dealing with the various problems, bibliographical, historical and scientific, presented by maps in their possession.[219]

This prospectus confirms that *Imago Mundi* was not conceived as a narrowly based academic journal. It was to be an open forum—designed to create "an international center of information"—in which the diverse streams of interest in the study of early maps that had long been touching without interacting could be brought together. Bagrow planned his format to achieve these ends.[220] Each issue was to consist of a number of major articles accompanied by shorter articles and notices, reviews, and an annual bibliography of items published in the history of cartography. It was also to include what Bagrow called a "Chronicle," a summary of relevant events such as conferences, exhibitions, and major publications, and a means of following the migration or destruction of important maps. Bagrow's *Imago Mundi* has now been published almost every year for nearly fifty years except for a break between 1939 and 1947,[221] and it has become part of the history of the subject. It is thus possible to assess the extent to which it has contributed to change in the history of cartography.

Account, however, must first of all be taken of Bagrow's own terms of reference and of his background as a gentleman dealer. Bagrow explicitly defined the scope of *Imago Mundi* as a "review of early cartography." While no precise date was set down, it is clear that he envisaged a period, as in his own *History of Cartography*, that stopped short of the end of the eighteenth century. As Franz Grenacher, who knew him well, remarked, "Bagrow's interest tended towards material which was difficult of access, rare, primitive, or out-of-the-way; he would have preferred . . . to add some pages on Armenian, Abyssinian and Burmese maps, of which he had evidence, rather than deal with the dry, over-commercial or scientifically constructed maps of the 17th and 18th centuries."[222] He was consistent in his prejudice against modern materials, and his personal tastes have tended to reinforce the wider bias in research

217. "Leo Bagrow," 8–9 (note 181). It is another example of the durability of Bagrow's ideas that, some fifty years after his proposal, the scheme for an encyclopedia of early maps has at last been taken up in Vienna in a modified form as the *Lexikon zur Geschichte der Kartographie*, ed. I. Kretschmer, J. Dörflinger, and F. Wawrik, 2 vols. (Vienna, 1986).

218. "Foreword of the Management Committee," *Imago Mundi* 16 (1962): XI, referring to the first thirteen volumes of the journal, those edited by Bagrow.

219. "Editorial," *Imago Mundi* 2 (1937): prelim.

220. Bagrow's format has remained with relatively little modification to the present day.

221. R. A. Skelton, "Historical Notes on *Imago Mundi*," *Imago Mundi* 21 (1967): 109–10, gives details of the publication arrangements and changes of publisher for the series as a whole. It is thus a major reference work for the subject: volumes 1–36 (1935–84) have generated some 4,749 printed pages, comprising 315 major articles, some 260 shorter articles and notices, a Chronicle in 32 of the volumes, 367 reviews, 55 obituaries, and bibliographies relating to the history of cartography which enumerate some 7,000 items.

222. F. Grenacher, review of Bagrow's *Meister der Kartographie* in *Imago Mundi* 18 (1964): 100–101, quotation on 101.

and writing against more recent cartographic history. It is not surprising, therefore, to find that only 4.7 percent of the articles in volumes 1–30 (1935–78) of *Imago Mundi* deal with the period after 1800.[223]

A closer look at the content of the articles published in *Imago Mundi* enables us to explore how far the history of cartography has become genuinely international in outlook and practice by our own day. A clear picture of Eurocentricity remains. Though Bagrow himself was said to be intensely interested in the maps of non-European peoples, the thirty-six issues published up to 1985 are largely filled with European authors writing about European subjects and much less frequently feature non-European authors writing about their indigenous maps. For example, of the papers published in the journal between 1935 and 1978, nearly four-fifths relate to European cartographers and their products.

No more international in scope or authorship was the Chronicle section, to which Bagrow invited foreign scholars to contribute. This attracted a response from at most thirty countries over a forty-year period (1935–75), and every nation with five or more entries was in Europe or North America.[224] Of these thirty countries, eleven make only a single appearance, pointing to their isolation within the subject if not to the low priority the contributors gave to such contacts. Thus it is clear, at least as reflected by the journal *Imago Mundi*, that the history of cartography was, as Skelton remarked, organized predominantly according to national political boundaries[225] and that the main effect of Bagrow's journal was to reinforce existing links, namely those between historians of cartography in Europe and the English-speaking world. No obvious or significant increase of activity in other areas of the world occurred, a situation that still remains, to judge from both the numbers and the distribution of subscribers to the journal.[226]

This Eurocentric tendency is confirmed by two other aspects of the history of cartography that can be monitored from *Imago Mundi*, the origins of the books and articles contained in its annual bibliographies and the languages in which they were published. The bibliographies have appeared in the journal since its inception.[227] They relate to the total literature of the subject either as gathered by their compilers or as supplied by contributors. For the period 1935 to 1983, they contain approximately seven thousand entries.[228] In terms of a trend, there is no sign of any exponential increase or of a take-off even in the past two decades. Nor is the geographical distribution different from that already noted for papers published in *Imago Mundi* itself. Although the total of seventy-three countries contributing items in the bibliographies is more than double that recorded in the *Imago Mundi* Chronicles, Europe and North America are again the outstanding contributors to the

literature of the history of cartography. Ten countries account for not less than 70 percent of the recorded entries in the period, while only two countries from outside these regions (Japan and Argentina) feature in the first twenty.[229]

When it comes to language of publication, the nar-

223. Blakemore and Harley, *Concepts*, 16 (note 164).

224. Countries most frequently submitting entries for the *Imago Mundi* Chronicle are ranked as follows in the period 1935–75: twenty (United States), seventeen (Great Britain), fifteen (Germany), twelve (France), twelve (Netherlands), nine (Italy), nine (Russia), nine (Switzerland), eight (Austria), eight (Belgium), seven (Czechoslovakia), seven (Sweden), six (Poland), and five (Denmark). No Chronicle was included in volumes 7 and 22.

225. Skelton, *Maps*, 95–96 (note 15), where he discusses the implications of this tendency.

226. The 1980 membership statistics, though their geography is partly obscured by sales through booksellers, show that out of over seven hundred subscribers, under fifty relate to countries outside Europe and North America, with the majority of these being recorded in Australia and New Zealand, and Japan. I owe this information to the secretary and treasurer of *Imago Mundi*.

227. No bibliographies appeared in volumes 7 and 15.

228. I am grateful to Francis Herbert of the Royal Geographical Society and Michael Turner of the University of Exeter for assistance with this analysis in notes 229 and 231. These statistics should be used with care. In any one year they will reflect the information provided by foreign contributors; the assiduity and accuracy of the compiler and the criteria used for their selection; and the space available for bibliographies as a matter of editorial policy. Moreover, no indication is given in the original bibliographies of which journals and sources were searched, and the problem of identifying places of publication—especially items in earlier journals—has also been considerable.

229. Totals of entries for the top twenty countries contributing items to the *Imago Mundi* bibliographies, 1935–83, with percentage shares in the total recorded literature, are as follows:

Country	Total Items in *Imago Mundi* Bibliographies	% Share
England	1,055	15.2
United States	937	13.5
Germany	879	12.7
Netherlands	638	9.2
Italy	422	6.1
USSR	363	5.2
Austria	270	3.9
France	240	3.4
Sweden	231	3.3
Poland	196	2.8
Portugal	157	2.3
Belgium	151	2.2
Switzerland	150	2.2
Japan	148	2.1
Canada	126	1.8
Hungary	126	1.8
Czechoslovakia	100	1.4
Spain	93	1.3
Scotland	89	1.3
Argentina	63	0.9
Other	514	7.4
	N = 6,948	

rowness of the geographical base of the history of cartography is again confirmed.[230] Obviously there is a tendency, in the history of cartography as in other subjects, for scholars to publish in one of the main scientific languages, but even so it is clear that English (accounting for 3,048 items or nearly 43.5 percent in the bibliographies) has strengthened its position as the main language of publication in the history of cartography, especially since World War II.[231] This no doubt reflects the strong interest shown in the subject, as already noted, by scholars from the English-speaking world, especially the British Isles, the United States, and a number of Commonwealth countries.[232] The second most important language is German (19.4 percent). The use of German has maintained the same position as in the 1930s, mirroring the importance of Germany in the rise of cartography as an academic and practical subject[233] as well as the continuing interest in the history of cartography in that country and in Austria and Switzerland. In brief, the use of four languages—English, German, French, and Italian—for over three-quarters of the publications enumerated in the bibliographies merely confirms the established and traditional interest of European countries in the history of cartography, a trend already discernible in the nineteenth century. As to its apparent neglect in other parts of the world, this must in part reflect a real lack of interest and opportunity, for whatever reasons, in the study of a subject that is so well entrenched in the western European nations. At the same time, the linguistic spread of writings in the history of cartography can easily be underestimated if based on the *Imago Mundi* bibliographies alone. They are never exhaustive. The smallness of the tally of recorded publications in any one of the "minority" languages—seven in Chinese, one in Greek, and little from the South American countries, for example—very likely reflects that the compilers of the bibliographies (and their contributors) acquired items by chance rather than by systematic searches among relatively inaccessible national and regional publications from distant or less familiar parts of the world.

Whatever the limitations of the data derived from *Imago Mundi* and its contents, a general conclusion may be reached about the role of the journal in the development of the history of cartography. Despite the dedication of Bagrow and his editorial successors, their policies for *Imago Mundi* have done relatively little to widen the history of cartography. Some problems—including that of proper international communication—remain almost as Bagrow diagnosed them fifty years ago. The development of systematic interdisciplinary contacts has not been seriously attempted. Largely untouched by recent academic debates in the humanities and social sciences, the journal has maintained a conservative posi-

tion. It has continued to project a connoisseur's image for the subject: it has stressed the cartography of the period before 1800, and it has generally emphasized the cartographic history of the developed nations of the Western world. Moreover, in being not only the sole specialist journal in the field but also, as it happened, published mainly in English, it has probably done most to consolidate the study of the history of cartography in Europe and North America—precisely the regions where the subject was already well entrenched by 1935—and to stimulate it particularly in the English-speaking nations within those two continents.[234] It is on such narrow intellectual foundations—constricted by *Imago Mundi*[235]—that the scholarly identity of the history of cartography has traditionally been built.

---

230. There is the particular problem with this analysis that some items have been transliterated in the bibliography using a different orthography from that of their original languages.

231. The top fifteen languages recorded in *Imago Mundi* bibliographies, 1935–83, with percentage shares in the total recorded literature, are as follows:

Language	Total Items in *Imago Mundi* Bibliographies	% Share
English	3,048	43.5
German	1,359	19.4
French	504	7.2
Italian	416	5.9
Russian	333	4.7
Dutch	276	3.9
Spanish	198	2.8
Polish	159	2.3
Portuguese	140	2.0
Japanese	123	1.8
Hungarian	120	1.7
Swedish	90	1.3
Czech	89	1.3
Danish	38	0.5
Norwegian	27	0.4
Other	90	1.3
	N = 7,010	

232. It is, however, normal that the country/language of publication of the particular bibliography has most citations; this has been widely observed in the literature of scientific disciplines.

233. See p. 24 and the subsection on "The Rise of Cartography," esp. pp. 32–33.

234. Conversely, comparison with other bibliographies confirms that *Imago Mundi* is least representative of research in the history of cartography in countries such as Germany, France, and Italy: only volume 1 was published in German; in subsequent volumes a few articles have been published in French.

235. This tendency has been further reinforced in the past twenty years by the series of biennial international conferences on the history of cartography partly organized under the aegis of *Imago Mundi*: the conferences have been held in London (1964, in association with the Twentieth Congress of the International Geographical Union), London (1967), Brussels (1969), Edinburgh (1971), Warsaw-Jadwisin (1973), Greenwich (1975), Washington, D.C. (1977), Berlin (1979), Pisa, Florence, and Rome (1981), Dublin (1983), and Ottawa (1985). Many of

THE RISE OF CARTOGRAPHY

The third principal influence on the history of cartography as a scholarly field, and on its definition and scope, has been the growth of cartography as an increasingly independent academic subject and practical activity. The distinction must be made between cartography as the ancient art and science of making maps in a practical sense (and its products) and cartography as an organized method by which maps are studied, investigated, and analyzed.[236] One can argue that, among all factors, it is the latter influence that lies at the root of the changes taking place in the history of cartography today. Cartography influences the history of cartography in two ways. In the first place, the organizations set up to promote cartography have also increased the opportunities for meetings and publication in the history of cartography. In the second place, academic cartography acts as an intellectual leaven, offering a new philosophical basis, alternative theoretical frameworks, and a range of appropriate techniques for the study of early maps.[237] Furthermore, the growing autonomy of cartography is having repercussions on the history of cartography in such a way that the latter has now the opportunity of becoming, among its other scholarly roles, the "discipline" history for an expanding subject and its practitioners.[238]

For an increasing number of historians of cartography, this relatively new relationship with cartography is clearly stimulating. It must be set against the background of a relative decline of interest among geographers in the study of early maps. Since the 1960s, the history of cartography has been losing its niche within academic geography. This is partly a reflection of geographers' attitudes toward cartography as a whole. While it is true that many of today's academic cartographers were trained as geographers, to others cartography has tended to be regarded primarily as a technical service, very useful but clearly lower in the intellectual hierarchy. Signs of impatience with the closeness of the relationship between cartography and geography have been noticeable since the 1930s. Richard Hartshorne, for example, while applauding the association, evidently preferred to see cartography as a discrete specialist subject: "Because it is more essential to geography than in any other science, and has been developed to the highest extent in geography . . . it is both natural and reasonable that it should be most closely associated with our science, but it is no more a branch of geography, logically speaking, than is statistics a branch of economics."[239] The position of the history of cartography was also weakened in the more recent period of major conceptual change and innovation within geography.[240] This left the history of cartography stranded with its concern for old maps and its

old-fashioned (early twentieth century) image. If not consciously then subliminally, it has been relegated to an antiquarian periphery of geography. Despite pleas for the importance of "graphicacy" in geographical education,[241] early maps—like maps in general—do not seem to have been considered humanistic documents in their own right in the paradigm changes in human and historical geography of recent years. Instead, at the very time when geography was discovering cognitive space, it tended to forget conventional maps. A general decline of the perceived importance of maps in geography is widely reported;[242] recent assessments of the development of human geography, mirroring the emphasis of conceptual changes, have not seriously reviewed carto-

the papers, especially from earlier conferences, were published in *Imago Mundi*, which has also carried reports of proceedings in all cases. For a wider discussion of *Imago Mundi* and its role in developing the history of cartography see also J. B. Harley, "*Imago Mundi*: The First Fifty Years and the Next Ten" (paper prepared for the Eleventh International Conference on the History of Cartography, Ottawa, 1985).

236. The distinction is based on Daniel E. Gershenson and Daniel A. Greenberg, "How Old Is Science?" *Columbia University Forum* (1964), 24–27, esp. 27.

237. It is significant that some German cartographers, without being very explicit, regard the history of cartography as an integral part of theoretical cartography: see, for example, Rudi Ogrissek, "Ein Strukturmodell der theoretischen Kartographie für Lehre und Forschung," *Wissenschaftliche Zeitschrift der Technischen Universität Dresden* 29, no. 5 (1980): 1121–26; Ingrid Kretschmer, "The Pressing Problems of Theoretical Cartography," *International Yearbook of Cartography* 13 (1978): 33–40. This follows Eckert's view of the content of scientific cartography expressed in *Die Kartenwissenschaft*.

238. Paul T. Durbin, ed., *A Guide to the Culture of Science, Technology, and Medicine* (New York: Free Press, 1980), 33, discusses the development of histories of scientific disciplines within the literature of the history of science.

239. Richard Hartshorne, *The Nature of Geography: A Critical Survey of Current Thought in the Light of the Past* (Lancaster, Pa.: Association of American Geographers, 1939), 398–99; his *Perspective on the Nature of Geography* (Chicago: Rand McNally for the Association of American Geographers, 1959) contains no further discussion of cartography.

240. The rift with maps was accelerated by a shift to statistical rather than cartographic techniques in geographical analysis: what became a widely held view was set out by William Bunge, *Theoretical Geography*, Lund Studies in Geography, ser. C, General and Mathematical Geography no. 1 (Lund: C. W. K. Gleerup, 1962; 2d ed., 1966), 71, when he concluded that notwithstanding "much pre-commitment to maps" among geographers and "in spite of certain advantages of maps over mathematics, mathematics is the broader and more flexible medium for geography."

241. Most recently by David Boardman, *Graphicacy and Geography Teaching* (London: Croom Helm, 1983), who summarizes the history of the term.

242. Phillip Muehrcke, "Maps in Geography," in *Maps in Modern Geography: Geographical Perspectives on the New Cartography*, ed. Leonard Guelke, Monograph 27, *Cartographica* 18, no. 2 (1981): 1–41. Some statistical data on the falling percentage of cartographic articles in selected geographical journals up to 1968 were given by Wolter, "Emerging Discipline," 206 (note 113).

graphy, let alone the history of cartography.[243] Neither have their counterparts in historical geography[244] taken much notice of the existence of the history of cartography. Even in the literature on cognitive maps, conceptions of space, and environmental images—containing ambitious attempts to reconstruct the geographies of the mind—there has been a failure to relate the manifestations of these internal cognitive processes to the "real" maps that must in an increasing number of cases in modern societies have helped to fashion them.[245] There are very tentative signs that this undervaluation of "real" maps, old and new, which was particularly characteristic of Anglo-American geography, may be coming to an end. At least one American geographer, referring specifically to the history of cartography, has recently written of this "most fundamental part of our discipline."[246]

The decline of interest among some geographers in the study of early maps has been partly counterbalanced by the rise of interest among those who increasingly regard themselves as cartographers. So, whereas during the last century a major formative influence in the history of cartography was the rise of geography, currently it is that of academic cartography. Certainly, for most cartographers, early maps have always been regarded as maps in their own right, and this tends to reinforce the affinity between cartography and the history of cartography. It is not easy to foresee the future relationship of academic cartography to geography and therefore of the history of cartography to geography and other subjects concerned with the management of the environment, but it is certain that the more recent links with academic cartography, which have already led to a rethinking of the nature of early maps (discussed below), will remain influential.

Particular examples of how the growth of an independent cartography provides support for the history of cartography are given by John A. Wolter in his study of the emergence of cartography as a discipline.[247] First, using bibliometric methods, he traces the history of subject bibliographies of cartography back to the nineteenth century.[248] During that century, and for most of the first half of the present century, the cartographic entries were usually an integral part of geographical bibliographies. Even in the most comprehensive of such bibliographies—notably the cartographic sections of the *Geographisches Jahrbuch*,[249] the *Bibliographie géographique internationale*,[250] and the *Research Catalogue of the American Geographical Society*[251]—there is a marked tendency to underrecord the literature of cartography and, consequently, the writings on the history of cartography.[252] Since the middle of the present century, however, the literary output of cartography as a whole has been independently listed, as in Hans-Peter

Kosack and Karl-Heinz Meine, *Die Kartographie*,[253] and later in the *Bibliotheca cartographica*,[254] the *Biblio-*

243. For example, Paul Claval, *Essai sur l'évolution de la géographie humaine*, new ed. (Paris: Belles Lettres, 1976); R. J. Johnston, *Geography and Geographers: Anglo-American Human Geography since 1945*, 2d ed. (London: Edward Arnold, 1983); Preston E. James and Geoffrey J. Martin, *All Possible Worlds: A History of Geographical Ideas*, 2d ed. (New York: John Wiley, 1981). An exception is *Progress in Human Geography*, vol. 1– (1977–), which has maintained a series of "Progress Reports" on cartography.

244. For example, Alan R. H. Baker, ed., *Progress in Historical Geography* (Newton Abbot: David and Charles, 1972), where the references to maps in the various essays concern only their role as landscape evidence or as a means of presenting data.

245. Robert Lloyd, "A Look at Images," *Annals of the Association of American Geographers* 72 (1982): 532–48, for a review of literature, critical of geographical studies of mental maps and images.

246. Peter Gould, *Annals* 72 (1982): 433 (note 215).

247. Wolter, "Emerging Discipline" (note 113). The following three paragraphs are largely based on this thesis. Wolter also considers other measures of growth, including textbooks and manuals written for students of cartography, and (in the context of the United States) provision for the education and training of cartographers. An examination of these last two types of evidence, however, suggests that while the history of cartography has been accepted as a valid research activity within cartography, it plays only a minor part in the training of cartographers.

248. Defined here as bibliographies that list the literature of cartography rather than cartobibliographies, which refer to lists of maps. It is, however, difficult to isolate the two, especially in the nineteenth century; bibliographies often contain references to the publication of items such as maps, atlases, and globes as well as to the literature pertaining to cartography.

249. *Geographisches Jahrbuch* (Gotha: Perthes, 1866–). Cartography is found in particular volumes: a brief subject analysis of the *Jahrbuch* appears in J. K. Wright and E. T. Platt, *Aids to Geographical Research: Bibliographies, Periodicals, Atlases, Gazetteers and Other Reference Books*, 2d ed., American Geographical Society Research Series no. 22 (New York: Columbia University Press for American Geographical Society, 1947), 52–57.

250. *Bibliographie géographique internationale* (Paris: Centre National de la Recherche Scientifique, 1891–), annual.

251. *Research Catalogue of the American Geographical Society*, 15 vols. and map supplement (Boston: G. K. Hall, 1962); it has been updated by *Current Geographical Publications: Additions to the Research Catalogue of the American Geographical Society* (New York: American Geographical Society, 1938–78; Milwaukee: American Geographical Society Collection, 1978–).

252. Wolter, "Emerging Discipline," 138–39 (note 113); see also 204–6 for an analysis of the cartographic content of selected geographical journals.

253. Hans-Peter Kosack and Karl-Heinz Meine, *Die Kartographie, 1943–1954: Eine bibliographische Übersicht*, Kartographische Schriftenreihe, vol. 4 (Lahr/Schwarzwald: Astra Verlag, 1955).

254. *Bibliotheca cartographica: Bibliographie des kartographischen Schrifttums; Bibliography of Cartographic Literature; Bibliographie de la littérature cartographique* (Bonn-Bad Godesberg: Institut für Landeskunde and Deutsche Gesellschaft für Kartographie, 1957–72); its title was changed to *Bibliographia cartographica: Internationale Dokumentation des kartographischen Schrifttums; International Documentation of Cartographical Literature* with the 1975 issue (renumbered 1–). For a bibliographical note and statistics on its contents, see Lothar Zögner, "25 Jahre 'Bibliographia cartographica,'" *Zeitschrift für Bibliothekswesen und Bibliographie* 29 (1982): 153–56.

graphy of Cartography,[255] and the *Referativnyĭ zhurnal: Geografiia*.[256] The appearance of these bibliographies— and the rate of growth of the literature they portray— can be taken as a measure of the increasing independence of the field of cartography. The point is that all such bibliographies recognized and thus helped to demarcate—and to stimulate—the history of cartography as a distinct subject area within cartography. For example, in *Die Kartographie* monographs dealing with historical topics are listed separately, and of the approximately 5,000 entries in this work, a total of 354 (7 percent) relate to the history of cartography. Similarly, an analysis of the citation structure of *Bibliotheca cartographica* and *Bibliographia cartographica* from 1957 to 1981—containing in all some 43,314 entries—reveals that the history of cartography (accorded its own place in the classification) was the third most important subject category in that period, with a total of 6,298 entries (14.5 percent).[257] These rather dry facts illustrate the new forces working for the history of cartography. Moreover, the attention paid to the history of cartography in these international bibliographies has had its counterpart at the national level. New journals of cartography have listed or reviewed the history of cartography literature, and abstracting journals dealing with cartography have now also recognized the history of cartography as a discrete subject area.[258]

A second example can be given of how the rise of cartography has benefited the history of cartography. It concerns the foundation of new cartographic societies and their associated specialist journals, which have provided a wide range of new outlets for the history of cartography. In comparison with the foundation of new geographical societies in the second half of the nineteenth century,[259] the establishment of societies devoted exclusively to cartography gained momentum much more slowly.[260] Kartografiska Sällskapet was the first modern cartographic society, founded in Stockholm in 1908, but its periodical *Globen* did not begin publication until 1922.[261] Before World War II there were also several attempts to establish cartographic societies and journals in Austria[262] and Germany.[263] These attempts reflected the interest in a science of mapping in those two countries, but not until after 1950 did the more general takeoff occur. The Deutsche Gesellschaft für Kartographie was founded in 1950 and began publishing the *Kartographische Nachrichten* in 1951.[264] This soon gained a reputation as a leading scholarly journal for cartography. In 1958 the *Bulletin* of the Comité Français de Cartographie and the Dutch journal *Kartografie* were first issued. By 1972 there were twenty-six cartographic societies and forty-three cartographic journals; by 1980 the number of journals had risen to sixty-seven.[265]

The importance of such cartographic organizations

and their periodicals for the history of cartography has not always been recognized. It is clear, however, that most new societies of cartography included the advancement of research into the history of cartography among their objectives.[266] Their journals reflect this interest,

255. In 1897, Philip Lee Phillips began collecting entries for the *Bibliography*. What had been collected was inserted as a preface to *A List of Maps of America in the Library of Congress* in 1901. Additions continued to be made, although the effort lagged for some years and was then renewed. Over several years, all of the entries (from the early nineteenth century to 1971) were compiled on twenty-nine reels of sixteen-millimeter microfilm and were finally published: United States Library of Congress, Geography and Map Division, *The Bibliography of Cartography*, 5 vols. (Boston: G. K. Hall, 1973), with subsequent supplements.

256. *Referativnyĭ zhurnal: Geografiia* (Moscow: Institut Nauchnoĭ Informatsii, Akademiia Nauk SSSR, 1956–), monthly.

257. Zögner, "25 Jahre 'Bibliographia cartographica,'" 155 (note 254). This total of over six thousand entries for a shorter time period confirms a shortfall in the *Imago Mundi* bibliographies analyzed above (note 229).

258. *Geo Abstracts*, sec. G, "Remote Sensing, Photogrammetry and Cartography," has had since 1979 a separate heading for the historical aspects of cartography.

259. In some cases, of course, the journals of these societies were of seminal importance in promoting the systematic study of cartography in the nineteenth century. In the German-speaking world, especially, *Petermanns Geographische Mitteilungen*—the earliest European geographical journal—and the *Ergänzungshefte* were of overwhelming importance in this respect: for further evidence see Wolter, "Emerging Discipline," 156–59 (note 113).

260. These societies were listed and discussed by Wilhelm Bonacker, "Kartographische Gesellschaften: Vorläufer und Wegbereiter der internationalen kartographischen Vereinigung," *Geographisches Taschenbuch* (1960–61), supp., 58–77; T. A. Stanchul, "Natsional'nye kartograficheskye obshchestva mira" (National cartographic societies of the world), *Doklady Otdeleniy i Komissiy* 10 (1969): 89–99 (Geograficheskogo obshchestva SSSR, Leningrad).

261. It remains the oldest periodical devoted to cartography that is still being published.

262. In Austria, the *Kartographische und Schulgeographische Zeitschrift* was published from 1912 to 1922; *Die Landkarte: Fachbücherei für Jederman in Länderaufnahme und Kartenwesen* was more short-lived (1925–27), as was the *Kartographische Mitteilungen* (1930–32). For further details see Wolter, "Emerging Discipline," 165–68 (note 113).

263. Early attempts at publishing a regular cartographic journal in Germany were also unsuccessful: the *Deutsche Kartographische Gesellschaft* existed from 1937 to 1949, but not until 1941 did it publish a *Jahrbuch der Kartographie*, which ceased publication in the following year: Wolter, "Emerging Discipline," 168–70 (note 113).

264. *Kartographische Nachrichten* 25, no. 3 (1975), was a special issue: "1950–1975: 25 Jahre Deutsche Gesellschaft für Kartographie."

265. Wolter, "Emerging Discipline," 171 with a list of serials on 303–5 (note 113). In 1980, John D. Stephens listed sixty-seven cartographic serials in six categories (including some categories not included by Wolter in his listing, i.e., bibliographic serials); see John D. Stephens, "Current Cartographic Serials: An Annotated International List," *American Cartographer* 7 (1980): 123–38.

266. *American Cartographer*, for example, established in 1974, despite its strong technical emphasis, included "the history of mapmaking" among its terms of reference: Robert D. Reckert, "A Message from the President of ACSM," *American Cartographer* 1 (1974): 4.

and a few societies have even established special interest groups to promote the history of cartography.[267] Indeed, for at least one of the new journals—the *Canadian Cartographer* (now *Cartographica*)—the history of cartography seems to have been the primary interest in the years 1964 to 1972, with 30 percent of its articles devoted to this subject. Elsewhere, less space was given to articles on the history of cartography: in the (British) *Cartographic Journal* only 16 percent (1964–72); in *Surveying and Mapping* 11 percent (of cartographic articles 1944–72); in *Kartographische Nachrichten* 11 percent (1952–82); in the (Australian) *Cartographer* 3 percent (1954–69); and in the *International Yearbook of Cartography* a mere 2 percent (1961–72).[268]

The existence of the history of cartography was also acknowledged at an international level. In 1972 the International Cartographic Association formally extended its activities to the history of mapmaking when it established the "Working Group on the History of Cartography," its terms of reference being the investigation of cartographic techniques and map production before 1900. In 1976 it was given commission status and a brief to prepare a "historical glossary of cartographic innovations and their diffusion."[269] In this way—but in many countries in cartography rather than in geography—the void left by the phasing out of a commission for ancient maps from the International Geographical Union has been filled.

There is more to the relationship between the new cartography and the history of cartography than the infrastructural matters of bibliographies, societies, journals, and international organizations. Of even greater potential importance to the history of cartography has been the intellectual infusion from a rethinking of concepts and from techniques in cartography. The development of the idea of cartography as being quintessentially concerned with communication—while not the only major concept to have attracted attention in recent years—is nevertheless the one that most nearly offers a set of general principles for the humanistic study of early maps. That these ideas have only slowly filtered into the history of cartography partly reflects the generally belated appearance[270] of a search for theoretical frameworks in cartography itself. As Robinson and Petchenik observe,

> During most of the long history of cartography, cartographers have been chiefly concerned with technical problems: acquiring and perfecting geographic data, devising ways of symbolizing it, and inventing methods of mechanically preparing and duplicating the physical map. Remarkably little concern was ever expressed about how a map actually accomplished what it was supposed to do—communicate. . . . there were thousands of maps made with little or no

thought given to the images evoked in the minds of those who looked at them.[271]

If this was still true in cartography in the 1970s, then it was doubly true of the history of cartography at the same time. In vain are the journals searched for papers concerned explicitly with the nature of maps, as opposed to accounts of mapmakers or descriptions and evaluations[272] of the content of maps. Even when an interest in theory finally began to enliven the history of cartography—mainly in the 1960s—it was first directed at the problems of assessing map content as documentary record rather than at illuminating their study as artifacts or images on their own terms.[273] One can go so far as to suggest that the eventual awareness of early maps as maps in the history of cartography derives mainly from cartography. It is probably too early to predict whether modern cartographic thinking will produce a lasting change of direction, but three signs of a shift in interest are beginning to permeate the history of

267. For example, the Deutsche Gesellschaft für Kartographie set up such an interest group in 1954, the Canadian Cartographic Association in 1976. There is also a working group on the history of cartography of the Nederlandsche Vereniging voor Kartografie (NVK) and a Commission on the History of Cartography in the National Committee of Cartographers of the USSR. In France there is no special group on the history of cartography, but a commission on cartographic documentation was created in 1980 by the Comité Français de Cartographie.

268. Wolter, "Emerging Discipline," 187–98 (note 113), using the *Bibliotheca cartographica* classification system; I am grateful to Francis Herbert for the statistics relating to *Kartographische Nachrichten*.

269. International Cartographic Association, *Cartographical Innovations: An International Handbook of Mapping Terms to 1900*, ed. Helen Wallis and Arthur H. Robinson (Tring, Hertfordshire: Map Collector Publications, forthcoming); International Cartographic Association, *Map-making to 1900: An Historical Glossary of Cartographic Innovations and Their Diffusion*, ed. Helen Wallis (London: Royal Society, 1976). See also Helen Wallis, "Working Group on the History of Cartography," *International Geographical Union Bulletin* 25, no. 2 (1974): 62–64; Henry W. Castner, "Formation of the I.C.A. Working Group on the History of Cartography," *Proceedings of the Eighth Annual Conference of the Association of Canadian Map Libraries* (1974): 73–76.

270. There were a few exceptions: see above, on Max Eckert, pp. 24–25.

271. Robinson and Petchenik, *Nature of Maps*, vii–viii (note 4).

272. The critical appraisal of earlier maps, already visible in the seventeenth- and eighteenth-century literature, did, however, gather strength in the nineteenth century: a notable example is Gregorius Mees, *Historische atlas van Noord-Nederland van de XVI eeuw tot op heden* (Rotterdam: Verbruggen en Van Duym, 1865), where the introduction consists of a critical examination of atlases published in Europe since the seventeenth century; Mees, incidentally, was also the first Dutchman to use the word "cartographie" in print (personal communication to author from Cornelis Koeman).

273. R. A. Skelton, *Looking at an Early Map* (Lawrence: University of Kansas Libraries, 1965); see also the Conference on the History of Cartography, London, September 1967, which took as its theme "Early Maps as Historical Evidence." A selection of papers, some of them methodological, were published in *Imago Mundi* 22 (1968).

cartography: the greater concern with the meaning of the words "map" and "cartography" already commented upon;[274] a greater emphasis on maps as artifacts and on the technical processes by which they are produced; and finally, an initiation of communication approaches to the study of early maps. These last two points are taken up here.

By the 1960s greater emphasis was being placed in cartography on the rapidly changing technical processes by which maps are produced, while in the history of cartography a similar interest in maps as artifacts was growing. In cartography, however, this emphasis on technical processes was soon challenged. A body of literature based on empirical research in psychophysics sought to explain the responses of the map reader to various map elements as an aid to effective map design, and this contributed to a number of seminal papers published in the 1960s.[275] These papers anticipated the developing theories of mapping as a cognitive science that involves communication from mapmaker to map user. By the 1970s these new theories were firmly rooted in the subject,[276] thus stressing the nature of cartography as a process rather than maps as a product. This also led to the modified definitions of "map" and "cartography" already noted. By 1974 cartography was seen as becoming "a science . . . allied in part with the science of graphic communication";[277] by 1976 it could be positively asserted that cartography was the science of communicating information between individuals by the use of maps;[278] and by 1981 it was described as "a formal system for the communication of spatial information."[279] Theoretical cartographers were dismantling their early information flow models, crudely derived from engineering, and seeking to refine their concepts through semiology.[280] They looked for parallels between language and cartography[281] and explored the cognitive dimension in cartographic communication.[282]

For the past two decades this revitalized cartography has increasingly been a major source of new ideas for the study of early maps. From the 1960s onward, the two major preoccupations of the cartographers—the technical aspects of mapmaking and the study of how maps communicated their information—were both reflected in writings on the history of cartography. We can detect a number of theoretical statements designed to reconcile the more traditional study of maps as historical documents with the intensified interest in their characteristics as physical products resulting from human workmanship. Historians of cartography were now exhorted to train their emphasis more on the artifactual nature of the map and less on its content. Skelton recognized the dichotomy in research in 1966 when he clarified the distinction between form and content in the study of early maps. The form of the map artifact, he said, represented "the mind, eye, and hand of the contemporary mapmaker" and the content of the map "the geographical data presented in it."[283] But Skelton's approach to early maps was by both apprenticeship and inclination that of a historian of their content, and a study of form and content were for him aspects of research that would "mutually control and support each other."[284] Others saw it differently. By the 1970s, some felt urgently that the study of design and technique should be given greater emphasis in the history of maps. Thus F. A. Shibanov, a specialist in early Russian maps,

274. See above, Preface, pp. xv–xviii.

275. Barbara Bartz Petchenik, "A Map Maker's Perspective on Map Design Research, 1950–1980," in *Graphic Communication and Design in Contemporary Cartography*, ed. D. R. Fraser Taylor, Progress in Contemporary Cartography, vol. 2 (New York: John Wiley, 1983), 37–68. By 1960 Arthur H. Robinson was envisaging the primary process of cartography as "the conceptual planning and designing of the map as a medium for communication or research": Arthur H. Robinson, *Elements of Cartography*, 2d ed. (New York: John Wiley, 1960), v. Another important paper was Christopher Board, "Maps as Models," in *Models in Geography*, ed. Richard J. Chorley and Peter Haggett (London: Methuen, 1967), 671–725; and Jacques Bertin's *Sémiologie graphique: Les diagrammes, les réseaux, les cartes* (Paris: Gauthier-Villars, 1967), attempted to codify a body of theory for cartography derived from semiotics. Bertin's book was published in English as *Semiology of Graphics: Diagrams, Networks, Maps*, ed. Howard Wainer, trans. William J. Berg (Madison: University of Wisconsin Press, 1983).

276. See the collection of essays in Leonard Guelke, ed., *The Nature of Cartographic Communication*, Monograph 19, *Cartographica* (1977); but a good guide through the literature of the period is Christopher Board, "Cartographic Communication," in *Maps in Modern Geography*, 42–78 (note 242). See also Lech Ratajski, "The Main Characteristics of Cartographic Communication as a Part of Theoretical Cartography," *International Yearbook of Cartography* 18 (1978): 21–32.

277. Joel L. Morrison, "Changing Philosophical-Technical Aspects of Thematic Cartography," *American Cartographer* 1 (1974): 5–14, quotation on 12.

278. Joel L. Morrison, "The Science of Cartography and Its Essential Processes," *International Yearbook of Cartography* 16 (1976): 84–97.

279. M. J. Blakemore, "Cartography," in *The Dictionary of Human Geography*, ed. R. J. Johnston (Oxford: Blackwell Reference, 1981), 29–33, quotation on 29.

280. Bertin, *Sémiologie graphique* (note 275), was probably the first to attempt to work out a "grammar" of graphic symbols applied to cartography; Ulrich Freitag, "Semiotik und Kartographie: Über die Anwendung kybernetischer Disziplinen in der theoretischen Kartographie," *Kartographische Nachrichten* 21 (1971): 171–82; Hansgeorg Schlichtmann, "Codes in Map Communication," *Canadian Cartographer* 16 (1979): 81–97; idem, "Characteristic Traits of the Semiotic System 'Map Symbolism,'" *Cartographic Journal* 22 (1985): 23–30.

281. Christopher Board, "Maps and Mapping," *Progress in Human Geography* 1 (1977): 288–95; Head, "Natural Language" (note 4).

282. Barbara Bartz Petchenik, "Cognition in Cartography," in *Nature of Cartographic Communication*, 117–28 (note 276); Ratajski, "Characteristics of Cartographic Communication," 24–26 (note 276).

283. Skelton, *Maps*, 63 (note 15).

284. Skelton, *Maps*, 63 (note 15).

suggested cogently that what was "of importance for the history of cartography is not what has been represented on a map but how it has been portrayed cartographically."[285] This line of argument had already been taken even further by David Woodward when he set out to show that the study of early maps as a product of cartographic skill and practice had, with certain notable exceptions, remained a major gap within the history of cartography.[286] Woodward approached the problem by classifying stages in the production process according to the resultant cartographic form, summarizing these in terms of a simple matrix and concluding: "The study of the form of maps is that part of the field which we might call the technical history of cartography and is usually attempted by those historians of cartography with a background in cartography. In short, it is the cartographer's view of his craft."[287] These words amount to a statement regarding the cartographer's place in the history of cartography, and they were accepted as such by historically minded cartographers. They were thus also a sign of the coming of age of a larger technical component in the history of cartography. A general analogy could be that this trend belatedly matches the rise of the history of technology as distinct from the history of science in the period since World War II. As another practicing cartographer expressed it, "chronological map knowledge," involving "the history of cartographical technics and technology," ought to be set to increase its relative share of the subject.[288]

So far, only the harbingers rather than the substance of a change in the balance of the history of cartography can be detected.[289] At the very least, however, the emergence of cartography as an independent discipline had the effect of recruiting for the history of cartography a new group of scholars, with technical training and a different intellectual outlook, who were attracted to research in their own specialist fields. An example has been the increased attention paid to the history of thematic mapping, progressively related to the growing importance of this subject in cartography as a whole.[290] Yet these trends must be kept in proportion: the history of cartography has clearly not identified entirely with cartography. For many practicing cartographers, historical studies have inevitably remained a sideline to their contemporary researches, and this tendency has weakened the impact of their contribution on the history of cartography. Systematic studies of form are only just beginning to complement a continuing and proper concern for the content of early maps as historical documents.

An interest in early maps as a means of communication in the past shows a similar process of gradual colonization. Although such models became well established in cartography from the late 1960s onward, they were only slowly taken up in the history of cartography.

The idea that maps represent a form of graphic language is not new. Almost as soon as mapmakers had become aware of the special nature of their craft and had recorded its practice in written treatises, they seem also to have grasped the nature of the communicative properties of maps. For example, Leonard Digges, in his *Pantometria* of 1571, referred to the advantages not only of exactness but also of "dispatch" in the reading of maps, although it was left to John Green, writing in the eighteenth century, to restate the well-established belief that "a Draught shews at once what many Words can't express."[291] But if such a view was often echoed—and had wide acceptance among historians of cartography trained as geographers—it was a truth implicitly understood and conveyed in their writings rather than one that had been fully developed in their research. Statements such as one to the effect that the signs on early maps represented a "cartographical alphabet,"[292] another that studies of early maps should be concerned with the language or vocabulary of mapmakers,[293] or that historians of cartography might focus on the "expressive terms by which [a map] makes its communication"[294] can easily be found in the literature.

285. F. A. Shibanov, "The Essence and Content of the History of Cartography and the Results of Fifty Years of Work by Soviet Scholars," in *Essays on the History of Russian Cartography, 16th to 19th Centuries*, ed. and trans. James R. Gibson, introduction by Henry W. Castner, Monograph 13, *Cartographica* (1975), 141–45, quotation on 142.

286. Woodward, "Suggested Framework" (note 205); see also David Woodward, "The Form of Maps: An Introductory Framework," *AB Bookman's Yearbook*, pt. 1 (1976), 11–20. Woodward's exceptions to this tendency "to slight or ignore the processes by which maps were made" ("Suggested Framework," 109 and n. 17) were Brown, *Story of Maps* (note 16), and François de Dainville, *Le langage des géographes* (Paris: A. et J. Picard, 1964).

287. Woodward, "Suggested Framework," 107 (note 205).

288. Lech Ratajski, "The Research Structure of Theoretical Cartography," *International Yearbook of Cartography* 13 (1973): 217–28.

289. See Blakemore and Harley, *Concepts*, 48–50 (note 164), for examples of the imbalances in the historical study of such cartographic processes.

290. This connection is synthesized in—and epitomized by—Robinson, *Thematic Mapping* (note 28).

291. Leonard Digges, *A Geometrical Practise, Named Pantometria* (London: Henrie Bynneman, 1571), preface; [John Green], *The Construction of Maps and Globes* (note 69), quoted in J. B. Harley, "The Evaluation of Early Maps: Towards a Methodology," *Imago Mundi* 22 (1968): 62–74, quotation on 62.

292. E. M. J. Campbell, "The Beginnings of the Characteristic Sheet to English Maps," pt. 2 of "Landmarks in British Cartography," *Geographical Journal* 128 (1962): 411–15, quotation on 414.

293. De Dainville, *Langage des géographes*, x (note 286); it should be pointed out, however, that de Dainville was not interested in maps for their own sake in this work—or in the history of cartography—but used maps as documents in the service of history.

294. Skelton, *Maps*, 101 (note 15); Skelton's later writings in particular are full of suggestive pointers that reveal his understanding of the potential of the analogy between maps and language.

The theoretical basis, however, was never formally set out, nor was there an interchange with developments in other subjects, such as art history, literature, or social anthropology, where these concepts had been more thoroughly exploited.

Not until the early 1970s can we detect the first deliberate historical adaptations of ideas derived from the cartographer's concern with theories of communication. In 1972, for example, Freitag suggested dividing the history of cartography into eras and epochs corresponding to Marshall McLuhan's eras of communication, starting with the "chirographic or manuscript era" and going on to the eras of "typographic or printed maps" and of "telegraphic (or screened) maps."[295] By the mid-1970s the theme of maps as a means of communication was increasingly being identified in the history of cartography. Woodward had reviewed communication models as part of his "framework" for the subject;[296] Wallis had stressed the place of communication in the study of the history of thematic cartography;[297] at the level of documented research strategies, Andrews was writing about "medium and message" in connection with early Ordnance Survey maps of Dublin City[298] and Lewis had modeled the "message images" transmitted through selected maps of the Great Plains in the eighteenth century;[299] and in 1975 Harley had proposed a systematic documentation using historical evidence for the "user segment" of the communication model of Robinson and Petchenik.[300] By the end of the decade a similar approach to the history of maps was being developed independently by scholars in other disciplines. Some research by art historians on early maps, for example, not only has adopted an iconographic strategy, strongly influenced by the concept of art as language, but has also attempted to make more explicit its assumptions about art (broadly defined to include some types of prints and maps) as a graphic language.[301] Such developments are forcing historians of cartography to consider the contemporary meaning of maps and their social significance as well as their qualities as artifacts or historical documents.[302] In another example, a historian of science wrote about the emergence of "a visual language," in the sense of maps and diagrams, for geology, while historians of the book can now envisage their subject in general in terms of "the communications circuit."[303] For a formalization of an interest in the properties of maps as communicators of knowledge about space, the history of cartography is perhaps primarily indebted to the rise of academic cartography in the past two decades.

RECENT DEVELOPMENTS IN THE HISTORY OF
CARTOGRAPHY

The literature reviewed above may be taken to suggest that a changed scholarly identity for the history of car-

tography had already taken shape by the end of the 1970s. It has to be stressed, however, that the history of cartography cannot be defined as an academic subject by criteria such as the number of university departments or established chairs devoted to its pursuit. In Portugal, where the Junta de Investigações do Ultramar made provision in the period 1958–60 for the study of early cartography in Lisbon and Coimbra, a formal status has emerged, albeit on a small scale.[304] And in the Netherlands, in 1968, a chair of cartography was established in the University of Utrecht, which also formally incor-

---

295. Ulrich Freitag, "Die Zeitalter und Epochen der Kartengeschichte," *Kartographische Nachrichten* 22 (1972): 184–91. He drew on the ideas in Marshall McLuhan, *Understanding Media: The Extensions of Man*, 2d ed. (New York: New American Library, 1964), esp. 145–46.

296. Woodward, "Suggested Framework," 103–5 (note 205).

297. Helen Wallis, "Maps as a Medium of Scientific Communication," in *Studia z dziejów geografii i kartografii: Etudes d'histoire de la géographie et de la cartographie*, ed. Józef Babicz, Monografie z Dziejów Nauki i Techniki, vol. 87 (Warsaw: Zakład Narodowy Imienia Ossolińskich Wydawnictwo Polskiej Akademii Nauk, 1973), 251–62.

298. J. H. Andrews, "Medium and Message in Early Six-Inch Irish Ordnance Maps: The Case of Dublin City," *Irish Geography* 6 (1969–73): 579–93.

299. G. Malcolm Lewis, "The Recognition and Delimitation of the Northern Interior Grasslands during the Eighteenth Century," in *Images of the Plains: The Role of Human Nature in Settlement*, ed. Brian W. Blouet and Merlin P. Lawson (Lincoln: University of Nebraska Press, 1975), 23–44; idem, "Changing National Perspectives and the Mapping of the Great Lakes between 1775 and 1795," *Cartographica* 17, no. 3 (1980): 1–31.

300. J. B. Harley, "The Map User in Eighteenth-Century North America: Some Preliminary Observations," in *The Settlement of Canada: Origins and Transfer*, ed. Brian S. Osborne, Proceedings of the 1975 British-Canadian Symposium on Historical Geography (Kingston, Ont.: Queen's University, 1976), 47–69.

301. Michael Twyman, "A Schema for the Study of Graphic Language," in *Processing of Visible Language*, ed. Paul A. Kolers, Merald E. Wrolstad, and Herman Bouma (New York: Plenum Press, 1979), 1:117–50.

302. Juergen Schulz, "Jacopo de' Barbari's View of Venice: Map Making, City Views, and Moralized Geography before the Year 1500," *Art Bulletin* 60 (1978): 425–74; J. B. Harley, "Meaning and Ambiguity in Tudor Cartography," in *English Map-Making, 1500–1650*, ed. Sarah Tyacke (London: British Library, 1983), 22–45, for an example of an iconographic-linguistic approach aimed at uncovering the contemporary meaning of a group of early maps; also J. B. Harley, "The Iconology of Early Maps," in *Imago et mensura mundi: Atti del IX Congresso Internazionale di Storia della Cartografia*, 2 vols., ed. Carla Clivio Marzoli (Rome: Enciclopedia Italiana, 1985), 1:29–38.

303. Martin J. S. Rudwick, "The Emergence of a Visual Language for Geological Science, 1760–1840," *History of Science* 14 (1976): 149–95; Robert Darnton, "What Is the History of Books?" in *Books and Society in History*, 3–26 (note 29); a graphic model of the "communications circuit" appears on p. 6.

304. See "Portugal" in the Chronicle section of *Imago Mundi* 17 (1963): 105–6, and *Imago Mundi* 24 (1970): 147–48.

porated the history of cartography.[305] But such institutional support for the history of cartography is still relatively fragile, and its growth has to be measured in terms of the activity of individuals rather than permanent endowments. Outside the universities, the only important development has been the establishment (in 1970) of the Hermon Dunlap Smith Center for the History of Cartography at the Newberry Library, Chicago. Created as a research institute, it was designed as much to promote the subject as to exploit that library's rich holdings of early maps. It remains, so far, the only permanent center of its kind.[306]

Some compensation for the lack of formal institutional support is available in the growing self-awareness discernible among those who regard themselves as, first and foremost, historians of cartography. This self-awareness is providing its own support. It could be said that lines of communication now exist for the emergence of an "invisible college" of historians of cartography.[307] These contacts already operate both through national groups and through international links and meetings. What can also be seen is the way the groups, increasingly conscious of the identity of the history of cartography, are beginning to advance the intellectual development of their subject and to exploit its past achievements and its potential to this end. Steps in the process of subject building already noted include the development of special interest groups in the national cartographic societies, the continuing series of international conferences, and the establishment of an International Cartographic Association Commission for the History of Cartography. An additional supporting influence is the regular publication of an international directory of research.[308] Although only forty-four countries are represented in the 1985 edition (compared with seventy-three countries recorded in the *Imago Mundi* bibliographies), even this geographical spread points to an increased flow of ideas across the national boundaries within which the history of cartography has been traditionally constrained.

Taken singly, many of these developments may seem no more than a taste of a different future for the subject; but in recent years they have been supported by a number of writings of an explicitly methodological nature, concerned either with stocktaking at a national level or with criticism of the aims and purposes of the history of cartography of a more general nature. Most convincing is the extent to which this critique is not exclusive to one or two countries but can be traced in most countries where there is an established tradition of research in the history of cartography. As already noted, there is nothing particularly new in the practice of bibliographical stocktaking, but over the past two decades, for example, there have appeared Baldacci's review of studies by Italian scholars;[309] Buczek's bibliographical essay on

the history of cartography in Poland and that of Koeman on the Netherlands;[310] Ruggles's account of the history of cartography in Canada;[311] and Scharfe's description of the state of the art in Germany.[312] In addition, there are the detailed Chronicle entries relating to the United States published in *Imago Mundi*.[313]

Most significant, from the point of view of intellectual change, however, is the parallel tendency toward introspection and self-criticism among historians of cartography. Looking no further than Britain, for example, one finds that as early as 1962 Crone had pointed to an antiquarian and bibliographical bias in the history of cartography,[314] though it was left to Skelton to mount a more systematic critique in 1966. It was Skelton's clearly enunciated view that the subject, as he surveyed it, was loosely defined and lacked philosophical and

305. In 1981 the chair was split into a chair of cartography and a personal professorship in the history of cartography.

306. David Woodward, *The Hermon Dunlap Smith Center for the History of Cartography: The First Decade* (Chicago: Newberry Library, 1980). Other unsuccessful proposals were made in the early 1960s: see G. Jacoby, "Über die Gründung einer internationalen Zentralstelle für die Geschichte der Kartographie," *Kartographische Nachrichten* 12 (1962): 27–28; Wilhelm Bonacker, "Stellungsnahme zu dem Plan einer internationalen Zentralstelle für Geschichte der Kartographie," *Kartographische Nachrichten* 12 (1962): 147–50. Jacoby's main objective was to create an international archive of photographic negatives of all old or rare maps, together with appropriate information and reference material.

307. Diana Crane, *Invisible Colleges: Diffusion of Knowledge in Scientific Communities* (Chicago: University of Chicago Press, 1972).

308. Elizabeth Clutton, ed. and comp., *International Directory of Current Research in the History of Cartography and in Carto-bibliography*, no. 5 (Norwich: Geo Books, 1985).

309. Osvaldo Baldacci, "Storia della cartografia," in *Un sessantennio di ricerca geografica italiana*, Memorie della Società Geografica Italiana, vol. 26 (Rome: Società Geografica Italiana, 1964), 507–52.

310. Karol Buczek, *History of Polish Cartography from the 15th to the 18th Century*, 2d ed., trans. Andrzej Potocki (Amsterdam: Meridian, 1982), 7–15; Cornelis Koeman, *Geschiedenis van de kartografie van Nederland: Zes eeuwen land- en zeekaarten en stadsplattegronden* (Alphen aan den Rijn: Canaletto, 1983); chap. 2 is concerned with "Biografieën van Nederlandse schrijvers over kartografie," 6–13.

311. Richard I. Ruggles, "Research on the History of Cartography and Historical Cartography of Canada, Retrospect and Prospect," *Canadian Surveyor* 31 (1977): 25–33.

312. Wolfgang Scharfe, "Geschichte der Kartographie—heute?" in *Festschrift für Georg Jensch aus Anlaß seines 65. Geburtstages*, ed. F. Bader et al., Abhandlungen des 1. Geographischen Instituts der Freien Universität Berlin, 20 (Berlin: Reimer, 1974), 383–98.

313. For example, Walter W. Ristow in the Chronicle section, *Imago Mundi* 17 (1963): 106–14; idem, *Imago Mundi* 20 (1966): 90–94; and other issues up to *Imago Mundi* 29 (1977), when a new arrangement for the Chronicle, cutting across national divisions—and designed to foster internationalism—was introduced.

314. G. R. Crone, "Early Cartographic Activity in Britain," pt. 1 of "Landmarks in British Cartography," 406–10 (note 292); referring to Crone's observation, similar tendencies were noted in the United States by Walter W. Ristow in the Chronicle section of *Imago Mundi* 17 (note 313).

methodological direction. In particular, he said, it needed "a firm general base, secure lines of communication, and an accepted methodology."[315] Recently, however, some of these ideas have been developed. In England, Blakemore and Harley reviewed them critically in the context of recent Anglo-American writings on the history of cartography.[316] In the United States, Woodward had already concluded in 1974 that the collective picture in the history of cartography was "that of a body of literature lacking consistency in terminology, approach, and general purpose,"[317] to which Denis Wood added his support, inveighing more stridently against what he sees as the dominant "collecting mentality" of many historians of cartography.[318]

The new critical spirit is by no means confined to Great Britain and North America. Those European countries in which there are strong traditions of research in the history of cartography are also adding to the methodological debate. In the Netherlands Koeman, promoting the idea of the wider relevance of the history of cartography, has examined "modern investigations" in the field in terms of their contribution to cultural history and to the development of cartography.[319] In Italy, where discussion centers on the dynamism of the subject, Elio Manzi rejected the notion of decline in the history of cartography as practiced in that country, demonstrating its vigor by enumerating 136 items in a recent review paper;[320] but Gaetano Ferro's answer was that these were mainly local in scope, were fragmented, and were undertaken without an awareness of unifying concepts.[321] Vladimiro Valerio has also injected a systematic note of criticism into the study of the history of cartography by Italian scholars.[322] In Poland, historians of cartography have likewise examined the situation and needs of their subject,[323] and in Switzerland Eduard Imhof, writing in 1964, was one of the earliest scholars to complain of the extensive gaps in historical cartographic research, referring in particular to the emphasis he saw being given to biobibliographical studies at the expense of technical analyses of the map artifact.[324] In Germany too, Ruthardt Oehme had already remarked in 1971 that 'early cartography is now looked on mainly as a hobby and it receives little consideration for study or research in German universities."[325] Since he wrote, an awareness of the history of cartography in Germany has been raised by the activities in the Deutsche Gesellschaft für Kartographie of a working group devoted to its study, and its potential and the need for change have been recognized in a recent review by Scharfe.[326] In France there has been relatively little interest in theoretical matters by the few practicing historians of cartography, but Philippe Pinchemel, a geographer, has sought to clarify the relation between the history of geography and the history of cartography, noting that historians of cartography have only rarely been aware of epistemological issues.[327] Finally, in Russia, where the history of cartography has attracted substantial scholarly attention,[328] there has also been published a systematic review, "The Use of Old Maps in Geographical and Historical Investigations."[329] This, as its title suggests, is primarily concerned with early maps as sources for physical and human geography, but it serves to reemphasize the wider role of a history of maps in historical research in general.

315. Skelton, *Maps*, 92 (note 15).

316. Blakemore and Harley, *Concepts* (note 164).

317. Woodward, "Suggested Framework," 102 (note 205).

318. Denis Wood, review of *The History of Topographical Maps: Symbols, Pictures and Surveys* by P. D. A. Harvey in *Cartographica* 17, no. 3 (1980): 130–33.

319. Cornelis Koeman, "Moderne onderzoekingen op het gebied van de historische kartografie," *Bulletin van de Vakgroep Kartografie* 2 (1975): 3–24.

320. Elio Manzi, "La storia della cartografia," in *La ricerca geografica in Italia, 1960–1980* (Milan: Ask Edizioni, 1980), 327–36.

321. Gaetano Ferro, "Geografia storica, storia delle esplorazioni e della cartografia" (Introduzione), in *Ricerca geografica*, 317–18 (note 320). Italian scholars have recently launched a bulletin, *Cartostorie: Notiziario di Storia della Cartografia e Cartografia Storica* (Genoa), no. 1– (1984–).

322. Vladimiro Valerio, "A Mathematical Contribution to the Study of Old Maps," in *Imago et mensura mundi: Atti del IX Congresso Internazionale di Storia della Cartografia*, 2 vols., ed. Carla Clivio Marzoli (Rome: Enciclopedia Italiana, 1985), 2:497–504; idem, "Sulla struttura geometrica di alcune carte di Giovanni Antonio Rizzi Zannoni (1736–1814)," published as offprint only; idem, "La cartografia Napoletana tra il secolo XVIII e il XIX: Questioni di storia e di metodo," *Napoli Nobilissima* 20 (1980): 171–79; idem, "Per una diversa storia della cartografia," *Rassegna ANIAI* 3, no. 4 (1980): 16–19 (periodical of the Associazione Nazionale Ingegneri e Architetti d'Italia).

323. Zbigniew Rzepa, "Stan i potrzeby badań nad historia Kartografii w Polsce (I Ogólnopolska Konferencja Historyków Kartografii)," *Kwartalnik Historii Nauki i Techniki* 21 (1976): 377–81.

324. Eduard Imhof, "Beiträge zur Geschichte der topographischen Kartographie," *International Yearbook of Cartography* 4 (1964): 129–53, quotation on 130.

325. Ruthardt Oehme, "German Federal Republic," in Chronicle, *Imago Mundi* 25 (1971): 93–95, quotation on 93.

326. Wolfgang Scharfe, "Die Geschichte der Kartographie im Wandel," *International Yearbook of Cartography* 21 (1981): 168–76.

327. Philippe Pinchemel, "Géographie et cartographie, réflexions historiques et épistémologiques," *Bulletin de l'Association de Géographes Français* 463 (1979): 239–47. This interest arose from a report presented to the Centre National de la Recherche Scientifique in 1978; in France the history of cartography is often subsumed under the history of geography, an association that is reflected in recent writings; see, for example, Broc, *Géographie de la Renaissance* (note 64).

328. Shibanov, "Essence and Content," 143 (note 285), reports that in an unpublished bibliography he had compiled there were some 550 studies representing the work of Soviet scholars in the history of cartography in the period 1917–62.

329. L. A. Goldenberg, ed., *Ispol'zovaniye starykh kart v geograficheskikh i istoricheskikh issledovaniyakh* (The use of old maps in geographical and historical investigations) (Moscow: Moskovskiy Filial Geograficheskogo Obschestva SSSR [Moscow Branch, Geographical Society of the USSR], 1980). For a complete listing of the contents see *Imago Mundi* 35 (1983): 131–32.

Such studies may form only a small percentage of the total new literature of the history of cartography, but they do reflect a heightened consciousness of its place in the humanities. They reflect, too, an awareness of an academic subject that has to be understood in terms of its own problems and potential. By 1980 the history of cartography was at a crossroads. The divergence was not only between its historical associations with geography and map librarianship and its newer, enhanced role within an increasingly independent cartography. It was also between its traditional work in the interpretation of the content of early maps as documents and its more recently clarified aims to study maps as artifacts in their own right and as a graphic language that has functioned as a force for change in history.

BIBLIOGRAPHY
CHAPTER 1 THE MAP AND THE DEVELOPMENT OF THE HISTORY OF CARTOGRAPHY

This bibliography also includes a selection of major reference works relevant to the whole volume.

Acta cartographica. Vols. 1–27. Reprints of monographs and articles published since 1801. 3 vols. per year. Amsterdam: Theatrum Orbis Terrarum, 1967–81.

Almagià, Roberto. *Monumenta Italiae cartographica.* Florence: Istituto Geografico Militare, 1929.

———.*Monumenta cartographica Vaticana.* Vol. 1, *Planisferi, carte nautiche e affini dal secolo XIV al XVII esistenti nella Biblioteca Apostolica Vaticana.* Vol. 2, *Carte geografiche a stampa di particolare pregio o rarità dei secoli XVI e XVII.* Rome: Biblioteca Apostolica Vaticana, 1944, 1948.

Arentzen, Jörg-Geerd. *Imago Mundi Cartographica: Studien zur Bildlichkeit mittelalterlicher Welt- und Ökumenekarten unter besonderer Berücksichtigung des Zusammenwirkens von Text und Bild.* Münstersche Mittelalter-Schriften 53. Munich: Wilhelm Fink, 1984.

Aujac, Germaine. *La géographie dans le monde antique.* Paris: Presses Universitaires, 1975.

Bagrow, Leo. *Die Geschichte der Kartographie.* Berlin: Safari-Verlag, 1951. English edition, *History of Cartography.* Revised and enlarged by R. A. Skelton. Translated by D. L. Paisey. Cambridge: Harvard University Press; London: C. A. Watts, 1964; republished and enlarged Chicago: Precedent Publishing, 1985. German edition, *Meister der Kartographie.* Berlin: Safari-Verlag, 1963.

Barthes, Roland. *Elements of Semiology.* Translated by Annette Lavers and Colin Smith. New York: Hill and Wang, [1968].

Beazley, Charles Raymond. *The Dawn of Modern Geography: A History of Exploration and Geographical Science from the Conversion of the Roman Empire to A.D. 900.* 3 vols. London: J. Murray, 1897–1906.

Bertin, Jacques. *Sémiologie graphique: Les diagrammes, les réseaux, les cartes.* Paris: Gauthier-Villars, 1967. English edition, *Semiology of Graphics: Diagrams, Networks, Maps.* Edited by Howard Wainer. Translated by William J. Berg. Madison: University of Wisconsin Press, 1983.

Bibliographia cartographica: Internationale Dokumentation des kartographischen Schrifttums: International Documentation of Cartographical Literature. Munich: Staatsbibliothek Preußischer Kulturbesitz und Deutsche Gesellschaft für Kartographie, 1974–. Preceded by *Bibliotheca cartographica* (1957–72) and Hans Peter Kosack and Karl-Heinz Meine, *Die Kartographie, 1943–1954: Eine bibliographische Übersicht.* Lahr-Schwartzwald: Astra, 1955.

Bibliographie cartographique. Centre National de la Recherche Scientifique. Paris. Annual, 1936–.

Bibliothèque Nationale. *Choix de documents géographiques conservés à la Bibliothèque.* Paris: Maisonneuve, 1883.

Blakemore, Michael J., and J. B. Harley. *Concepts in the History of Cartography: A Review and Perspective.* Monograph 26. *Cartographica* 17, no. 4 (1980).

Bonacker, Wilhelm. *Kartenmacher aller Länder und Zeiten.* Stuttgart: Anton Hiersemann, 1966.

Bowen, Margarita. *Empiricism and Geographical Thought from Francis Bacon to Alexander von Humboldt.* Cambridge: Cambridge University Press, 1981.

Bricker, C., and R. V. Tooley. *Landmarks of Mapmaking: An Illustrated Survey of Maps and Mapmakers.* Brussels: Elsevier-Sequoia, 1968.

British Museum. *Catalogue of Printed Maps, Charts and Plans.* 15 vols and suppls. London: Trustees of the British Museum, 1967.

Brown, Lloyd A. *The Story of Maps.* Boston: Little, Brown, 1949; reprinted New York: Dover, 1979.

Bunbury, Edward Herbert. *A History of Ancient Geography among the Greeks and Romans from the Earliest Ages till the Fall of the Roman Empire.* 2d ed., 2 vols., 1883; republished with a new introduction by W. H. Stahl, New York: Dover, 1959.

Capel, Horacio. "Institutionalization of Geography and Strategies of Change." In *Geography, Ideology and Social Concern,* ed. David R. Stoddart, 37–69. Oxford: Basil Blackwell; Totowa, N.J.: Barnes and Nobel, 1981.

Carpenter, Kenneth E., ed. *Books and Society in History: Papers of the Association of College and Research Libraries Rare Books and Manuscripts Preconference, 24–28 June 1980, Boston, Massachusetts.* New York: R. R. Bowker, 1983.

Cartes et figures de la terre. Exhibition catalog. Paris: Centre Georges Pompidou, 1980.

Cebrian, Konstantin. *Geschichte der Kartographie: Ein Beitrag zur Entwicklung des Kartenbildes und Kartenwesens.* Vol. 1, *Altertum: Von den ersten Versuchen der Länderabbildungen bis auf Marinos und Ptolemaios (zur Alexandrinischen Schule).* Gotha: Justes Perthes, 1922.

Clutton, Elizabeth, ed. and comp. *International Directory of Current Research in the History of Cartography and in Carto-bibliography.* No. 5. Norwich: Geo Books, 1985.

Cortesão, Armando. *History of Portuguese Cartography.* 2 vols. Coimbra: Junta de Investigações do Ultramar-Lisboa, 1969–71.

Cortesão, Armando, and Avelino Teixeira da Mota. *Portugaliae monumenta cartographica*. 6 vols. Lisbon, 1960.

Crone, Gerald R. *Maps and Their Makers: An Introduction to the History of Cartography*. 1st ed. London: Hutchinson University Library, 1953. 5th ed. Folkestone, Kent: Dawson; Hamden, Conn.: Archon Books, 1978.

Daly, Charles P. "On the Early History of Cartography; or, What We Know of Maps and Mapmaking, before the Time of Mercator." Annual Address. *Bulletin of the American Geographical Society* 11 (1879): 1–40.

Destombes, Marcel. *Mappemondes A.D. 1200–1500: Catalogue préparé par la Commission des Cartes Anciennes de l'Union Géographique Internationale*. Amsterdam: N. Israel, 1964.

Dicks, D. R. *Early Greek Astronomy to Aristotle*. Ithaca: Cornell University Press, 1970.

Dilke, O. A. W. *Greek and Roman Maps*. London: Thames and Hudson, 1985.

Duhem, Pierre. *Le système du monde: Histoire des doctrines cosmologiques de Platon à Copernic*. 10 vols. Paris: Hermann, 1913–59.

Eckert, Max. *Die Kartenwissenschaft: Forschungen und Grundlagen zu einer Kartographie als Wissenschaft*. 2 vols. Berlin and Leipzig: Walter de Gruyter, 1921–25.

Febvre, Lucien, and Henri-Jean Martin. *L'apparition du livre*. Paris: Editions Albin, 1958. English edition, *The Coming of the Book: The Impact of Printing, 1450–1800*. New edition. Edited by Geoffrey Nowell-Smith and David Wootton. Translated by David Gerard. London: NLB, 1976.

Fiorini, Matteo. *Le projezioni delle carte geografiche*. Bologna: Zanichelli, 1881.

————. *Sfere terrestre e celeste di autore italiano oppure fatte o conservate in Italia*. Rome: Società Geografica Italiana, 1899.

Fischer, Theobald. *Sammlung mittelalterlicher Welt- und Seekarten italienischen Ursprungs und aus italienischen Bibliotheken und Archiven*. Venice: F. Ongania, 1886; reprinted Amsterdam: Meridian, 1961.

Gibson, James R., ed. and trans. *Essays on the History of Russian Cartography, 16th to 19th Centuries*. Introduction by Henry W. Castner. Monograph 13. *Cartographica* (1975).

Der Globusfreund. Coronelli Weltbund der Globusfreunde. Vienna. 3 nos. per year, 1952–.

Goldenberg, L. A., ed. *Ispol'zovaniye starykh kart v geograficheskikh i istoricheskikh issledovaniyakh* (The use of old maps in geographical and historical investigations). Moscow: Moskovskiy Filial Geograficheskogo Obschestva SSSR (Moscow Branch, Geographical Society of the USSR), 1980. For a complete listing of the contents see *Imago Mundi* 35 (1983): 131–32.

Gough, Richard. *British Topography; or, An Historical Account of What Has Been Done for Illustrating the Topographical Antiquities of Great Britain and Ireland*. 2 vols. London: T. Payne, and J. Nichols, 1780.

Guelke, Leonard. *Maps in Modern Geography: Geographical Perspectives on the New Cartography*. Monograph 27. *Cartographica* 18, no. 2 (1981).

————, ed. *The Nature of Cartographic Communication*. Monograph 19. *Cartographica* (1977).

Harley, J. B. "The Evaluation of Early Maps: Towards a Methodology." *Imago Mundi* 22 (1968): 62–74.

Harvey, P. D. A. *The History of Topographical Maps: Symbols, Pictures and Surveys*. London: Thames and Hudson, 1980.

Heidel, William Arthur. *The Frame of the Ancient Greek Maps*. New York: American Geographical Society, 1937.

Imago Mundi. Journal of the International Society for the History of Cartography. Berlin, Amsterdam, Stockholm, London. Annual, 1935–.

International Cartographic Association. *Multilingual Dictionary of Technical Terms in Cartography*. Edited by E. Meynen. Wiesbaden: Franz Steiner Verlag, 1973.

————. *Map-Making to 1900: An Historical Glossary of Cartographic Innovations and Their Diffusion*. Edited by Helen Wallis. London: Royal Society, 1976.

————. *Cartographical Innovations: An International Handbook of Mapping Terms to 1900*. Edited by Helen Wallis and Arthur H. Robinson. Tring, Hertfordshire: Map Collector Publications, 1986.

International Cartographic Association (British National Committee for Geography Subcommittee). *Glossary of Technical Terms in Cartography*. London: Royal Society, 1966.

Janni, Pietro. *La mappa e il periplo: Cartografia antica e spazio odologico*. Università di Macerata, Pubblicazioni della Facoltà di Lettere e Filosofia, 19. Rome: Bretschneider, 1984.

Jomard, Edme-François. *Considérations sur l'objet et les avantages d'une collection spéciale consacrée aux cartes géographiques et aux diverses branches de la géographie*. Paris: E. Duverger, 1831.

————. *Les monuments de la géographie; ou, Recueil d'anciennes cartes européennes et orientales* (Atlas). Paris: Duprat, etc., 1842–62.

————. *Introduction à l'atlas des Monuments de la géographie*. Posthumously published by M. E. Cortambert. Paris: Arthus Bertrand, 1879.

Kamal, Youssouf. *Monumenta cartographica Africae et Aegypti*. 5 vols. in 16 pts. Cairo, 1926–51. The individual titles to the volumes are as follows: Vol. 1, *Epoque avant Ptolemée*, 1–107 (1926). Vol. 2, parts 1–3, *Ptolemée et époque gréco-romaine*, 108–233 (1928), 234–360 (1932), and 361–480 (1932). Vol. 2, part 4, *Atlas antiquus* and index (1933). Vol. 3, parts 1–5, *Epoque arabe*, 481–582 (1930), 583–691 (1932), 692–824 (1933), 825–945 (1934), and 946–1072 (1935). Vol. 4, parts 1–4, *Epoque des portulans, suivie par l'époque des découvertes*, 1073–1177 (1936), 1178–1290 (1937), 1291–1383 (1938), and 1384–1484 (1939). Vol. 5, parts 1–2, *Additamenta: Naissance et évolution de la cartographie moderne*, 1485–1653 (1951).

Keates, J. S. *Understanding Maps*. New York: John Wiley, 1982.

Kish, George, ed. *Bibliography of International Geographical Congresses, 1871–1976*. Boston: G. K. Hall, 1979.

Koeman, Cornelis. *Collections of Maps and Atlases in the Netherlands: Their History and Present State*. Leiden: E. J. Brill, 1961.

————. "Moderne onderzoekingen op het gebied van de historische kartografie." *Bulletin van de Vakgroep Kartografie* 2 (1975): 3–24 (Utrecht, Geografisch Institut der Rijksuniversiteit).

Kohl, Johann Georg. "Substance of a Lecture Delivered at the Smithsonian Institution on a Collection of the Charts and Maps of America." *Annual Report of the Board of Regents of the Smithsonian Institution . . . 1856* (1857), 93–146.

Kretschmer, I., J. Dörflinger, and F. Wawrik. *Lexikon zur Geschichte der Kartographie.* 2 vols. Vienna, 1986.

Leithäuser, Joachim G. *Mappae mundi: Die geistige Eroberung der Welt.* Berlin: Safari-Verlag, 1958.

Lelewel, Joachim. *Géographie du Moyen Age.* 4 vols. and epilogue. Brussels: Pilliet, 1852–57; reprinted Amsterdam: Meridian, 1966.

Library of Congress. *A List of Geographical Atlases in the Library of Congress, with Bibliographical Notes.* 8 vols. Vols. 1–4 edited by Phillip Lee Phillips; suppl. vols. 5–8 (1958–74) edited by Clara Egli LeGear. Washington, D.C.: Government Printing Office, 1909–74.

Library of Congress, Geography and Map Division. *The Bibliography of Cartography.* 5 vols. Boston: G. K. Hall, 1973.

Lloyd, G. E. R. *Early Greek Science: Thales to Aristotle.* New York: W. W. Norton, 1970.

Map Collector. Tring, Hertfordshire. Quarterly, 1977–.

Map Collectors' Circle. Map Collectors' Series, nos. 1–110. London, 1963–75.

Migne, J. P., ed. *Patrologiæ cursus completus.* 221 vols. and suppls. Paris, 1844–64; suppls. 1958–.

Miller, Konrad. *Mappaemundi: Die ältesten Weltkarten.* 6 vols. Stuttgart: J. Roth, 1895–98. Vol. 1, *Die Weltkarte des Beatus* (1895). Vol. 2, *Atlas von 16 Lichtdruck-Tafeln* (1895). Vol. 3, *Die kleineren Weltkarten* (1895). Vol. 4, *Die Herefordkarte* (1896). Vol. 5, *Die Ebstorfkarte* (1896). Vol. 6, *Rekonstruierte Karten* (1898).

——. *Itineraria Romana.* Stuttgart: Strecker und Schröder, 1916.

Muller, Frederik and Co. *Remarkable Maps of the XV*th*, XVI*th *and XVII*th *Centuries Reproduced in Their Original Size.* Amsterdam: F. Muller, 1894–98.

Müller, Karl, ed. *Geographi Graeci minores.* 2 vols. and tabulae. Paris: Firmin-Didot, 1855–56.

Murdoch, John Emery. *Antiquity and the Middle Ages.* Album of Science. New York: Charles Scribner's Sons, 1984.

Needham, Joseph. *Science and Civilisation in China.* Especially vol. 3, *Mathematics and the Sciences of the Heavens and the Earth.* Cambridge: Cambridge University Press, 1959.

Neugebauer, Otto. *The Exact Sciences in Antiquity.* 2d ed. Providence: Brown University Press, 1957.

——. *A History of Ancient Mathematical Astronomy.* New York: Springer-Verlag, 1975.

Nordenskiöld, A. E. *Facsimile-Atlas to the Early History of Cartography.* Translated by Johan Adolf Ekelöf and Clements R. Markham. Stockholm, 1889.

——. *Periplus: An Essay on the Early History of Charts and Sailing-Directions.* Translated by Francis A. Bather. Stockholm: P. A. Norstedt, 1897.

North, Robert. *A History of Biblical Map Making.* Beihefte zum Tübinger Atlas des Vorderen Orients, B32. Wiesbaden: Reichert, 1979.

Paassen, Christaan van. *The Classical Tradition of Geography.* Groningen: Wolters, 1957.

Pauly, August, Georg Wissowa, et al., eds. *Paulys Realencyclopädie der classischen Altertumswissenschaft.* Stuttgart: J. B. Metzler, 1894–.

Pelletier, Monique. "Jomard et le Département des Cartes et Plans." *Bulletin de la Bibliothèque Nationale* 4 (1979): 18–27.

Piaget, Jean, and Bärbel Inhelder. *The Child's Conception of Space.* Translated by F. J. Langdon and J. L. Lunzer. London: Routledge and Kegan Paul, 1956.

Ristow, Walter W. *Guide to the History of Cartography: An Annotated List of References on the History of Maps and Mapmaking.* Washington, D.C.: Library of Congress, 1973.

Robinson, Arthur H., and Barbara Bartz Petchenik. *The Nature of Maps: Essays toward Understanding Maps and Mapping.* Chicago: University of Chicago Press, 1976.

Santarém, Manuel Francisco de Barros e Sousa, Viscount of. *Atlas composé de mappemondes, de portulans et de cartes hydrographiques et historiques depuis le VI*e *jusqu'au XVII*e *siècle.* Paris, 1849. Reprint, Amsterdam: R. Muller, 1985.

——. *Essai sur l'histoire de la cosmographie et de la cartographie pendant le Moyen-Age et sur les progrès de la géographie après les grandes découvertes du XV*e *siècle.* 3 vols. Paris: Maulde et Renou, 1849–52.

Sarton, George. *Introduction to the History of Science.* 3 vols. Baltimore: Williams and Wilkins, 1927–48.

Scharfe, Wolfgang. "Die Geschichte der Kartographie im Wandel." *International Yearbook of Cartography* 21 (1981): 168–76.

Shirley, Rodney W. *The Mapping of the World: Early Printed World Maps 1472–1700.* London: Holland Press, 1983.

Singer, Charles, et al. *A History of Technology.* 7 vols. Oxford: Clarendon Press, 1954–78.

[Skelton, R. A.] "Leo Bagrow: Historian of Cartography and Founder of *Imago Mundi* 1881–1957." *Imago Mundi* 14 (1959): 4–12.

Skelton, R. A. *Looking at an Early Map.* Lawrence: University of Kansas Libraries, 1965.

——. *Maps: A Historical Survey of Their Study and Collecting.* Chicago: University of Chicago Press, 1972.

Skelton, R. A., and P. D. A. Harvey, eds. *Local Maps and Plans from Medieval England.* Oxford: Clarendon Press, 1986.

Stahl, William Harris. *Roman Science: Origins, Development, and Influence to the later Middle Ages.* Madison: University of Wisconsin Press, 1962.

Strayer, Joseph, ed. *Dictionary of the Middle Ages.* New York: Scribner's Sons, 1982–.

Taton, René. *Histoire générale des sciences.* 3 vols. in 4 pts. Paris: Presses Universitaires de France, 1957–64. English edition, *History of Science.* 4 vols. Translated by A. J. Pomerans. London: Thames and Hudson, 1963–66.

Taylor, Eva G. R. *The Haven-Finding Art: A History of Navigation from Odysseus to Captain Cook.* London: Hollis and Carter, 1956.

Terrae Incognitae. Annals of the Society for the History of Discoveries. Amsterdam. Annual, 1969–.

Thomson, J. Oliver. *History of Ancient Geography.* Cambridge: Cambridge University Press, 1948; reprinted New York: Biblo and Tannen, 1965.

Thrower, Norman J. W. *Maps and Man: An Examination of Cartography in Relation to Culture and Civilization.* Englewood Cliffs, N.J.: Prentice-Hall, 1972.

Tooley, Ronald V. *Tooley's Dictionary of Mapmakers.* New York: Alan Liss, 1979 and supplement 1985.

Tozer, H. F. *A History of Ancient Geography.* 1897. 2d ed. reprinted New York: Biblo and Tannen, 1964.

Uzielli, Gustavo, and Pietro Amat di San Filippo. *Mappamondi, carte nautiche, portolani ed altri monumenti cartografici specialmente italiani dei secoli XIII–XVII.* 2d ed., 2 vols. Studi Biografici e Bibliografici sulla Storia della Geografia in Italia. Rome: Società Geografica Italiana, 1882; reprinted Amsterdam: Meridian, 1967.

Wallis, Helen. "The Map Collections of the British Museum Library." In *My Head Is a Map: Essays and Memoirs in Honour of R. V. Tooley,* ed. Helen Wallis and Sarah Tyacke, 3–20. London: Francis Edwards and Carta Press, 1973.

Wallis, Helen, and Lothar Zögner, eds. *The Map Librarian in the Modern World: Essays in Honour of Walter W. Ristow.* Munich: K. G. Saur, 1979.

Weiss, Roberto. *The Renaissance Discovery of Classical Antiquity.* Oxford: Blackwell, 1969.

Wieder, Frederik Caspar. *Monumenta cartographica: Reproductions of Unique and Rare Maps, Plans and Views in the Actual Size of the Originals: Accompanied by Cartographical Monographs.* 5 vols. The Hague: M. Nijhoff, 1925–33.

Wolter, John A. "Geographical Libraries and Map Collections." In *Encyclopedia of Library and Information Science,* ed. Allen Kent, Harold Lancour, and Jay E. Daily, 9:236–66. New York: Marcel Dekker, 1968–.

Wolter, John A., Ronald E. Grimm, and David K. Carrington, eds. *World Directory of Map Collections.* International Federation of Library Associations Publication Series no. 31. Munich: K. G. Saur, 1985.

Woodward, David. "The Study of the History of Cartography: A Suggested Framework." *American Cartographer* 1, no. 2 (1974): 101–15.

———. *The Hermon Dunlap Smith Center for the History of Cartography: The First Decade.* Chicago: Newberry Library, 1980.

Wright, John Kirtland. *The Geographical Lore of the Time of the Crusades: A Study in the History of Medieval Science and Tradition in Western Europe.* American Geographical Society Research Series no. 15. New York: American Geographical Society, 1925; republished with additions, New York: Dover, 1965.

PART ONE

CARTOGRAPHY IN PREHISTORIC EUROPE
AND THE MEDITERRANEAN

2 · Prehistoric Maps and the History of Cartography: An Introduction

CATHERINE DELANO SMITH

The study of prehistoric mapping in Europe and its borderlands, as in other continents, requires a new beginning. In the past scholars have been handicapped not only by a severe shortage of evidence but also by misguided attitudes toward the intellectual capacity of early man. In addition, they have failed to consider either the diagnostic characteristics of prehistoric maps or the principles that should be developed for their identification and study. Accounts of the origins of mapping have tended to be confused and contradictory, and any new study must necessarily adopt a critical viewpoint. It seems obvious that the origins of European cartography must be sought in the period before that of the earliest recorded maps in the historic societies and that if examples of maps have survived from the prehistoric period they will be found in the archaeological material.

Richard Andree seems to have been the first to focus specifically on the origins of mapping,[1] but it was not until the middle of the twentieth century that the real problem was diagnosed. In 1949 Lloyd Brown had remarked that "map making is perhaps the oldest variety of primitive art . . . as old as man's first tracings on the walls of caves and in the sands."[2] Yet it was not until 1951 that Leo Bagrow belatedly drew attention to the fact that, notwithstanding these prehistoric origins, actual information about early maps is hard to come by and that early maps had been known for a much shorter time than many other products of civilization.[3]

Surveys of the origins of mapping can be counted on the fingers of one hand. The first of three pioneering works is Andree's monograph, which, despite its promising title, "Die Anfänge der Kartographie" (The beginnings of cartography), is a straightforward account of mapping by "primitive people." It does not include any discussion of the relation between such mapping and the earliest development of the idea of the map or of spatial skills in the prehistoric period, although these were obviously well developed by the time of the earliest historical maps.[4] Andree's paper, which set the tone for much of the subsequent literature, starts with a comment on the way many "primitive people," lacking the benefit of the magnetic compass, are nevertheless able to produce maps of surprising exactitude and accuracy. At-

tention is drawn to the two conditions present among "primitive peoples" that account for their cartographic abilities: first, an unparalleled sense of direction, related to their knowledge of the terrain; second, their technical skill in drawing. The main discussion concerns examples of "picture maps" (Kartenbilder), starting with Ainu sand maps and Eskimo maps and finishing with early Chinese and Japanese maps. Although the paper was later incorporated verbatim in one of his major works, Ethnographische Parallelen und Vergleiche, which also contained an informed chapter on petroglyphs from all over the world, Andree still did not link such images, most of which are prehistoric in date, with the origins of mapping concepts.[5]

The second of the pioneering works, Wolfgang Dröber's "Kartographie bei den Naturvölkern" (Mapmaking among primitive peoples), appeared at the beginning of the present century.[6] Dröber's title provides a more honest description of its preoccupation with examples of "primitive maps" rather than the origins of mapping. Dröber was obviously indebted to Andree[7] and, in particular, took up Andree's comments on the basic skills

1. Richard Andree, "Die Anfänge der Kartographie," *Globus: Illustrierte Zeitschrift für Länder* 31 (1877): 24–27, 37–43.

2. Lloyd A. Brown, *The Story of Maps* (Boston: Little, Brown, 1949; reprinted New York: Dover, 1979), 32; five years previously, David Greenhood, "The First Graphic Art," *Newsletter of the American Institute of Graphic Arts* 78 (1944): 1, had said that "cartography is not only the oldest of the graphic arts but also the most composite of them."

3. Leo Bagrow, *Die Geschichte der Kartographie* (Berlin: Safari-Verlag, 1951), 14. The translation is from page 25 of his *History of Cartography*, rev. and enl. by R. A. Skelton, trans. D. L. Paisey (Cambridge: Harvard University Press; London: C. A. Watts, 1964).

4. Dating from about 3000 B.C.; see p. 57.

5. Richard Andree, *Ethnographische Parallelen und Vergleiche* (Stuttgart: Julius Maier, 1878), 197–221; idem, "Anfänge der Kartographie," with figures ("Petroglyphen") (note 1).

6. Wolfgang Dröber, "Kartographie bei den Naturvölkern" (Mapmaking among primitive peoples) (Diss., Erlangen University, 1903; reprinted Amsterdam: Meridian, 1964); summarized under the same title in *Deutsche Geographische Blätter* 27 (1904): 29–46.

7. In addition to Andree's *Ethnographische Parallelen* (note 5), Dröber frequently cites his *Geographie des Welthandels*, 2 vols. (Stuttgart, 1857–72).

of "primitive peoples," adding to the list one other condition—their sharp eyesight.

Finally came Bruno Adler's Russian essay, "Karty pervobytnykh narodov" (Maps of primitive peoples).[8] Still

8. Bruno F. Adler, "Karty pervobytnykh narodov" (Maps of primitive peoples), *Izvestiya Imperatorskogo Obshchestva Lyubiteley Yestestvoznaniya, Antropologii i Etnografii: Trudy Geograficheskogo Otdeleniya* 119, no. 2 (1910). This has never been translated from the Russian, and insofar as it is known at all to historians of cartography, it is probably through H. de Hutorowicz's brief synopsis "Maps of Primitive Peoples," *Bulletin of the American Geographical Society* 43, no. 9 (1911): 669–79. A better idea of the wide-ranging scope of Adler's work may be derived from its contents, as tabulated here using Adler's headings:

Chapter 1
  1. "Orientation" in humans
  2. [Navigational] markers
  3. Drawing
Chapter 2
  1. Maps of primitive peoples
    A. Chukchi
    B. Eskimos
    C. Koryaks
    D. Yukagirs
    E. Yenesei
    F. Samoyeds
    G. Yuraks
    H. Dolgane
    I. Tungusii (Yenesei valley)
    J. Yakuts
    K. Russian peasants of Turukhansk Kray
    L. Ostyaks
    M. Gilyaks
    N. Ainu
    O. Karagas and Sayoti (?)
    P. Mongols and Buryats
    Q. Indians of North America
    R. Indians of South America
    S. Natives of Africa
    T. Ancient Ethiopian (?) map
    U. Australians
    V. Oceanians
    W. Maps of prehistoric peoples
Chapter 3. Maps of semicultured and cultured peoples of antiquity and a comparison of these with the maps of primitive peoples.
    A. Mexicans and Incas
    B. Assyro-Babylonians
    C. Ancient Jews
    D. Ancient Persians
    E. Ancient Indians
    F. Ancient Chinese
    G. Japanese and Koreans
    H. Ancient Egyptians
    I. Ancient Greeks
    J. Ancient Romans
    K. Ancient Arabs
    L. Maps of the Middle Ages
    M. An ancient Russian map
    N. Maps of Russian missionaries among the Yakuts
Chapter 4. Comparison of maps of primitive peoples with maps of literate peoples.
    A. Orientation according to the points of the horizon
    B. The compass

the only substantial work on the subject, it failed to become a seminal text. This may be attributed in part to the language barrier, but it was not a theoretical work, nor did Adler speculate in it about the origins of mapping. What it does contain is an important corpus of "primitive maps" gathered during the decade before its publication. These came from contemporary expeditions, especially those into Siberia; from a library and museum search throughout Europe; and from contributions sent in by American scientific institutions.[9] It also contains, in the wide range of Adler's survey, germs of inspiration that could have stimulated further research (the section on maps and religion, for instance), but these have been left dormant. Notwithstanding all this promise, even Adler had very little to say under his section headed "Maps of Prehistoric Peoples."[10]

It was here, in a largely undeveloped state, that the matter of prehistoric cartography rested for the most part until the 1980s. In the interval, only Leo Bagrow made any contribution to the subject, and even he devoted relatively little space either to prehistoric maps or

    C. Auxiliary lines on a map
    D. Observance of accuracy of distances and areas
    E. *Nomina geographica*
Chapter 5. Materials, instruments, techniques, coloring of maps, etc., of primitive peoples.
    A. Material
      a. Maps on sand, snow, etc.
      b. Relief maps
      c. Maps on stone
      d. Maps on bark and birch bark
      e. Maps on animal hides, cloth, and paper
      f. Maps on chance objects
      g. Stick maps
    B. Map-drawing instruments
    C. Map techniques
    D. The coloring of maps
    E. Geographical landscape portrayed on maps
      a. Rivers
      b. Relief of earth's surface
      c. Vegetation
      d. Anthropogeographical features on the map
      e. The animal world
Chapter 6
    A. Chief types of maps of primitive peoples
    B. The maps of primitive peoples as an educational aid
    C. Atlases of primitive peoples
    D. Maps in religion
    E. Capabilities of primitive peoples in cartography
Findings and conclusions
(Translated by Alexis Gibson, London.)

9. Hutorowicz, "Maps," 669 (note 8), said this added up to fifty-five maps from Asia, forty from Australia and Oceania, fifteen from America, three from Africa, and two from the East Indies.

10. Adler, "Karty" (note 8), cols. 217 (3 lines only) and 218–20, thus taking up only three columns (one and a half pages) to dispose of the full range of his examples; Hutorowicz, "Maps," 675 (note 8), however, said Adler gave "many pages" to a discussion of recently discovered maps. See pp. 64–66 for Adler's comments on the Kesslerloch artifacts.

to the origins of mapping.[11] The various synoptic texts on the history of cartography that appeared later—for example, those of Herbert George Fordham, Lloyd A. Brown, Gerald R. Crone, and Norman J. W. Thrower—were equally brief.[12] All these paid lip service, usually in the opening paragraph, to what they saw as early man's "almost instinctive" ability to draw, though they neither supported such claims nor demonstrated their significance in connection with the origins of mapping. All started their histories of the map with the Babylonians and Egyptians, at the earliest, or with the maps of the classical period. All ignored the prehistoric period.

Thus the first confusion in the bulk of the literature on the earliest maps derives from a lack of proper attention to the distinction between prehistoric cartography and the "primitive" cartography associated with indigenous cultures in the historical period. Another basic aspect of the neglect of prehistoric cartography follows from that and is the second source of confusion, namely the almost exclusive use of anthropological sources by Andree, Dröber, Adler, and Bagrow. Archaeological evidence, unless encountered in the course of ethnographic studies in the New World,[13] was ignored, and European and Old World cartographic prehistory, to say nothing of that in other areas of the world, went largely unacknowledged.[14] The consequence of this bias was that early historians of cartography were distracted from searching the archaeological evidence for the first signs of cartographic activity. Instead, they concentrated on the regional distribution of largely contemporary indigenous maps. Had these authors made a clear distinction between prehistoric and historical indigenous, and had they appreciated the interdependence of interpretations of these two categories, their research might have substantially contributed to the study of the origins of mapping. Only Bagrow recognized the potential of such an approach, pointing out that "we must therefore look at the primitive tribes of today, whose cartographic art has stopped at a certain point in its development [and where] we may find evidence . . . by analogy for what happened in the Mediterranean world in earlier times."[15] Thus, for Bagrow, in the absence of contextual evidence from the prehistoric period itself, the major line of approach to prehistoric cartography would have to be through the maps of historical indigenous cultures. Nevertheless, this would be only a means to an end.

A third source of confusion arose from yet another blurred distinction, the lack of differentiation between the well-documented wayfinding and navigational skills of many indigenous peoples and the practice of making maps within these early societies.[16] Moreover, this whole discussion was clouded by the general acceptance of a Darwinian viewpoint, which stresses an irreversible evolutionary sequence from primitive to advanced, savage to civilized, and simple to complex in thought and behavior.[17] Adler quoted Schurtz's condescending admission that some "rude and awkward attempts" at mapmaking may have been made in prehistoric times; Brown was led to see cartography as evolving "slowly and painfully" from obscure origins; while Fordham's choice of the word "savages" blocked further argument.[18] Their writings thus implied a contradiction. On the one hand was the claim regarding the antiquity of the art they described and on the other was the incapacity of the prehistoric "savages" to produce it. Refuge was taken in the word "instinct." As late as 1953, Crone could accept that "primitive peoples of the present day . . .

11. This treatment can be traced back to his first major publication, Leo Bagrow, *Istoriya geograficheskoy karty: Ocherk i ukazatel' literatury* (The history of the geographical map: Review and survey of literature), *Vestnik arkheologii i istorii, izdavayemyy Arkheologicheskim Istitutom* (Archaeological and historical review, published by the Archaeological Institute) (Petrograd, 1918), where what he had to say about prehistoric maps took one page, "primitive" maps took another, and by page 3 he was discussing the clay tablet maps from Babylonia. This balance was maintained in his 1951 text *Geschichte* (note 3) and in his *Meister der Kartographie* (Berlin: Safari-Verlag, 1963), which is identical in content to the English version of 1964, *History of Cartography* (note 3). For details of his comments on European prehistoric maps see below, pp. 65–66 n.61, 72–73 n.90, 85.

12. Herbert George Fordham, *Maps: Their History, Characteristics and Uses: A Handbook for Teachers*, 2d ed. (Cambridge: Cambridge University Press, 1927); Brown, *Story of Maps* (note 2); Gerald R. Crone, *Maps and Their Makers: An Introduction to the History of Cartography*, 1st ed. (London: Hutchinson, 1953; 5th ed., Folkestone: Dawson; Hamden, Conn: Archon Books, 1978); and Norman J. W. Thrower, *Maps and Man: An Examination of Cartography in Relation to Culture and Civilization* (Englewood Cliffs, N.J.: Prentice-Hall, 1972).

13. For example, Alexander von Humboldt, *Views of Nature*, trans. E. C. Otté and H. G. Bohn (London: Bell and Daldy, 1872), said that the petroglyphs he found in the vicinity of the Orinoco could not possibly have been carved by the existing "naked, wandering savages . . . who occupy the lowest place in the scale of humanity" (p. 147) and concluded that they attest the area was "once the seat of a higher civilisation" (p. 20).

14. The exception being Adler's reference to the bone plaques from Schaffhausen, Switzerland, in "Karty," col. 218 (note 8), which was taken up by Bagrow, *Istoriya*, 2 (note 11), *Geschichte*, 16 (note 3), and *History of Cartography*, 26 (note 3).

15. Bagrow, *Geschichte*, 14 (note 3), and *History of Cartography*, 25 (note 3).

16. This distinction was pointed out by Michael J. Blakemore, "From Way-finding to Map-making: The Spatial Information Fields of Aboriginal Peoples," *Progress in Human Geography* 5, no. 1 (1981): 1–24, esp. 1.

17. On Darwinism in the history of cartography see Michael J. Blakemore and J. B. Harley, *Concepts in the History of Cartography: A Review and Perspective*, Monograph 26, *Cartographica* 17, no. 4 (1980): 17–23.

18. Adler, "Karty," col. 220, n. 2 (note 8), refers to Heinrich Schurtz, *Istoriya pervobytnoy kul'tury* (History of primitive cultures) (Moscow, 1923), 657, translated from the German *Urgeschichte der Kultur* (Leipzig and Vienna, 1900); Brown, *Story of Maps*, 12 (note 2); Fordham, *Maps*, 1 ff. (note 12).

have an almost instinctive ability to produce rough but quite accurate sketches."[19] Moreover, he conjectured, similar abilities would be found at the origins of mapmaking in the Middle East and around the shores of the eastern Mediterranean. Suggestions such as these ignored anthropological evidence. It is well known that indigenous peoples, far from relying on instinct, have developed elaborate and exacting, usually ritualistic, mechanisms to ensure the dissemination of the most valued knowledge within their society and its transmission from one generation to another. Such cartographic skills as these peoples have are not instinctive but are as much acquired and learned as those of members of modern societies.

The fourth confusion characterizing the literature concerns the relative importance and distribution of maps in prehistoric and indigenous societies. It is perfectly fair to point out, as did Dröber, that not all these peoples are equally "good" at cartography,[20] but further qualification is needed. Not all prehistoric and indigenous peoples choose to be interested in graphic forms of expression or communication.[21] It is also necessary to consider the influence of different physical environments on the mapping stimulus. Thus, it can hardly be considered fortuitous that the stick charts of the Marshall Islanders, which are still given prominence in virtually every text or paper touching upon the subject of indigenous mapping, come from Oceania, or the Eskimos' carved maps from the frozen North; they both meet the demands of a highly specialized way of life involving regular navigation in extensive areas of undifferentiated terrain. Land-based tribes, at least those not living in the deserts, need no such artifices and have not normally produced them for their own use. Too much emphasis has been placed on these familiar and well-worn aspects of nonliterate cartography and too little on the nature of prehistoric maps and the origins of cartography.

The final confusion in the literature concerns the narrow interpretation of the function of both prehistoric and indigenous mapping. The tendency has been to assume that both these categories exclusively served what was perceived as a basic need, that of wayfinding. Until very recently, there was no real attempt by historians of cartography to understand indigenous societies on their own terms. Thus Fordham, in a tantalizing but abortive section on cartographic ideas, selected direction and distance as the crucial concepts in the genesis of maps.[22] For him, early maps were never more than route maps, in due course embellished with collateral information to give rise to the topographical map. Such an interpretation ignores well-known anthropological facts. The acknowledged skills of indigenous peoples at navigating without artificial aids, including maps, and the paramount importance to them of memorizing all knowledge, were glossed over.[23] Also ignored was potential insight into the function of prehistoric maps to be gained through prehistorians' and anthropologists' studies of rock art in the prehistoric and historical periods. These studies suggest that prehistoric maps may have been produced in a religious context, that matters of belief governed their execution, and that their function would have been abstract and symbolic rather than exclusively practical wayfinding and recording.[24]

Taking all these points together, we see that historians of cartography are on unfamiliar ground when it comes to a study of the origins of European cartography. They are faced with a new set of concepts and the need for a new approach. There are already signs of a change of attitude in the literature of the history of cartography. In 1980 P. D. A. Harvey's *History of Topographical Maps: Symbols, Pictures and Surveys* was published.[25] In that year, too, Michael Blakemore and J. B. Harley warned of the "ever present danger . . . that we will apply our own standards unthinkingly to those of the cartography of the past."[26] In the following year, Michael Blakemore went on to question why aboriginal (indigenous) peoples should draw maps at all when their directional skills were so developed,[27] and in 1982 an attempt was made to look again at the prehistoric maps in European rock art.[28] It is now time to reconsider the evidence for early maps and for the origins of cartography in a new light.

Taking the broadest view of graphic forms of spatial representation, evidence of early maps can be sought in many different types of art, artifacts, and cultural activities. It has been associated with a wide range of geographically scattered and temporally distributed cultures. Examples of prehistoric maps and maps made by indigenous peoples of the historical period have been reported in the literature of diverse subject disciplines

19. Crone, *Maps and Their Makers*, 15 (note 12).
20. Dröber, "Kartographie," 78 (note 6).
21. Robert Thornton, "Modelling of Spatial Relations in a Boundary-Marking Ritual of the Iraqw of Tanzania," *Man*, n.s., 17 (1982): 528–45.
22. Fordham, *Maps*, 1–2 (note 12), is followed by Crone, *Maps and Their Makers*, i (note 12), among others.
23. A point noted by Bagrow in *Geschichte*, 14 (note 3), and *History of Cartography*, 25 (note 3). See also Frances A. Yates, *The Art of Memory* (London: Routledge and Kegan Paul, 1966).
24. Mircea Eliade, *A History of Religious Ideas*, trans. Willard R. Trask (Chicago: University of Chicago Press, 1978), vol. 1, *From the Stone Age to the Eleusinian Mysteries*, chap. 1.
25. P. D. A. Harvey, *The History of Topographical Maps: Symbols, Pictures and Surveys* (London: Thames and Hudson, 1980).
26. Blakemore and Harley, *Concepts*, 22 (note 17).
27. Blakemore, "Way-finding" (note 16).
28. Catherine Delano Smith, "The Emergence of 'Maps' in European Rock Art: A Prehistoric Preoccupation with Place," *Imago Mundi* 34 (1982): 9–25.

and preserved in map, museum, and archival collections as well as—in the case of rock art—in the field.

In this *History*, a working distinction is drawn between the maps associated with prehistoric and with indigenous societies within the historical period in both Old and New World contexts. The basis of the distinction involves the nature of the evidence. The primary source material for all prehistoric periods is by definition exclusively archaeological. For indigenous mapping, it is primarily anthropological and historical and only secondarily archaeological. The two classes of evidence are not, of course, mutually exclusive, and anthropological findings are crucial in illuminating the archaeological record of the prehistoric period.[29] Adopting this criterion, a more or less clear line, based on the appearance of writing in a culture, can be drawn, often but not universally, to mark the separation between the prehistoric and the historical eras. The present volume deals with the prehistoric period of only part of the Old World. The focus is on Europe, although the sweep is broadened to take in the adjacent parts of southwestern Asia and northern Africa. In these regions the prehistoric period ended approximately in the third millennium B.C., at the time of the appearance of Babylonian pictographs (about 3100 B.C.) and cuneiform writing (after 2700 B.C.) and Egyptian hieroglyphs (about 3000 B.C.), followed by Cretan pictographs (2000 B.C.).[30] It closed slightly later in China (ca. 2000 B.C.). In Southeast Asia and in Japan the arrival of writing and the dawn of the historical period was later still, being scarcely perceptible until the first century A.D. Thus, discussion of Asian prehistoric cartography is deferred until the second volume of the *History*, where it will take its place as a prologue to the great cartographic achievements of that part of the Old World. In the New World, and in many peripheral regions of the Old World, the prehistoric period continued—generally speaking—until the arrival of European voyagers, explorers, and settlers in the fif-

teenth century or later. Apart from some notable exceptions, such as Mayan pictographs and the use of an Old Javanese-based script in the Philippines before the arrival of Magellan in 1521, literacy came to these areas only with European conquest. It is appropriate, therefore, to delay discussion of these maps—both prehistoric and historical, of the Americas, Africa (south of the Sahara), Australasia, and Oceania—until they can be included in a discrete section in volume 4 devoted to the major period of European contact with many of those societies.

Although this division may seem unfamiliar to those accustomed to seeing all prehistoric and indigenous mapping treated as a prologue to the history of cartography proper, it is amply justified. The aim is to be able to describe both the qualitative individuality and the chronological sequence of the main contexts for such mapping in Europe, Asia, and the New World. For the New World, treating indigenous cartography in the context of Old World colonialism maintains the fuller historical perspective as well as the narrative arrangement of the *History* as a whole. Likewise, the following discussion of the origins of cartography, which precedes the survey of the prehistoric cartography of Europe (including Russia west of the Urals), the Middle East, and North Africa (with the Sahara), serves to bring into sharper focus man's earliest involvement in what is now recognized as cartography.

---

29. See below, chap. 4, "Cartography in the Prehistoric Period in the Old World: Europe, the Middle East, and North Africa," pp. 54–101.

30. Information on the different writing systems, their origins, and date of appearance for the present discussion is derived mainly from David Diringer, *The Alphabet: A Key to the History of Mankind*, 3d ed. rev. (London: Hutchinson, 1968), with reference also to Hans Jensen, *Symbol and Script: An Account of Man's Efforts to Write*, 3d ed. rev. and enl. (London: George Allen and Unwin, 1970). The relationship of the different forms of early writing in the Middle East is summarized in *The Times Atlas of World History*, ed. Geoffrey Barraclough (Maplewood: Hammond, 1979), 52–53.

3 · The Origins of Cartography

G. MALCOLM LEWIS

One can argue that man's need to make maps arose during a fairly early stage in the coevolution of brain and culture. While gene mutations created new potentialities, culture would have bestowed advantages on those individuals and groups who could best perform specific mental and mechanical activities.[1] At a certain stage it would have become advantageous for man to structure information about the spatial aspects of his world and to communicate it to others. Unlike temporally structured information such as narratives, which can be transmitted—as speech or music—in a sequential mode, spatial information would not have been easy to transmit by the earliest of man's communication systems. Speech and music were ephemeral as well as sequential, and so were gesture and dance, though those could be two- or at best three-dimensional. However, once graphic forms of communication were developed in the Upper Paleolithic (some forty thousand years ago), they had the advantage of being both more permanent and two- or three-dimensional. Thus it would have been from these graphic forms that the means of expressing and communicating information about the world in spatially structured images first emerged. Although the advantages of such a means of communication would have been accruing from the start, for a long time mapmaking was almost certainly an unconscious, barely differentiable form of graphic expression. Indeed, this is still its status in certain indigenous societies, though in other parts of the world it began to emerge as a distinctive practical art some three thousand or more years ago. Mapmaking appears to have remained undifferentiated in those cultures in which cognitive development, even in adults, terminated at the preoperational stage, which is distinguished by the topological structuring of space. Those societies in which adults first began to manifest operative modes of cognition were the ones that first began to formalize projective and Euclidean geometries, and it was within these that cartography first emerged as a distinctive practical art.

The capacity to transmit information about spatial relationships between phenomena and events and to receive such information in message form was already well developed in many animals long before the emergence of *Homo sapiens*, though their message systems were genetically predetermined and thus unmodifiable either by mental reflection or by group interaction. Since these animals have evolved far less rapidly than man during the past forty thousand years, we can assume that their means of communication were much the same then as now. Studies of animal behavior have revealed examples of mapping procedures. Most involve scent marking of the environment and require the receiver to be in the area.[2] In certain respects such scent marking of territory can be likened to the way man employs markers to indicate boundaries where no maps exist. There are also a few known animal systems for communicating spatially structured information about the environment to receivers outside it, but these are ephemeral, lacking the relative permanence of artifacts. The best known is the round and waggle dance performed by honeybees on returning to the hive, by means of which they indicate to other hive members the direction and distance at which nectar has been found.[3]

Although this example might be dismissed as exceptional, "it is quite possible that we have yet to learn about the specialized languages of many organisms," as John Bonner says, since "each case is rather like cracking a code, and few people have the gift."[4] All the animal systems so far deciphered for transmitting a "map" to others of the species are genetically inherited. In consequence, they are unadaptable and are transmitted in

1. Charles J. Lumsden and Edward O. Wilson, *Promethean Fire: Reflections on the Origin of Mind* (Cambridge: Harvard University Press, 1983), 1–21.

2. For example, wolves in northeastern Minnesota cover their 100 to 300 square kilometer ranges approximately once every three weeks, leaving scent marks at regular intervals along well-established routes and in greater concentration at route junctions and near the edges of their territories. Roger Peters, "Mental Maps in Wolf Territoriality," in *The Behavior and Ecology of Wolves: Proceedings of the Symposium on the Behavior and Ecology of Wolves Held on 23–24 May 1975 in Wilmington, N.C.*, ed. Erich Klinghammer (New York and London: Garland STPM Press, 1979), 122–25.

3. Karl von Frisch, *The Dance Language and Orientation of Bees*, trans. Leigh E. Chadwick (Cambridge: Belknap Press of Harvard University Press, 1967).

4. John T. Bonner, *The Evolution of Culture in Animals* (Princeton: Princeton University Press, 1980), 129.

forms that, though remembered (at least by higher animals), are otherwise unstorable. It is in these aspects of spatial consciousness and the ability to communicate it that *Homo sapiens* is different.

Like all animals, but far more so than most, early *Homo sapiens*, of forty thousand or more years ago, was mobile. People moved in an essentially two-dimensional space for a variety of reasons, either searching out or avoiding a diverse range of objects, conditions, processes, and events. Consciousness of the world involved monitoring it for novelty—for both unanticipated events in time and unexpected objects and conditions in space, which might constitute hazards or, alternatively, afford opportunities. In either case they compelled attention. More than in other primates and far more than in other animals, the well developed eyesight of *Homo sapiens* provided the necessary sensory basis for developing a spatial mental schema against which to relate these hazards or opportunities. In contrast to the forest habitats of most primates, the grassland habitat of *Homo sapiens* afforded a more extensive visual world. Survival involved developing strategies for achieving at the same time prospect through vision and refuge through self-concealment.[5] Not surprisingly, therefore, "spatialization" was probably the "first and most primitive aspect of consciousness," so much so that attributes of space such as distance, location, networks, and area continue to pervade many other areas of human thought and language.[6]

Unlike modern scientific awareness, with its search for order and regularity, the awareness of early *Homo sapiens* focused on irregularities in the world and on uncertainties rather than certainties.[7] Consciousness would have constituted "a form of re-presentation of the current perceptual input on a mental screen," thus maintaining a continuous state of alertness for the unanticipated and unexpected.[8] However, survival and success were not dependent only on consciousness and on response in individuals. They also depended on cooperation between individuals and within the society and on the ability to communicate between individuals and within the group, to store and transmit information, and to decode it in message form. Hence the development of the several forms of language—including those for communicating spatial information—which ensured the emergence of society and the handing on of its accumulated culture to later generations.

As early as 400,000 B.P., *Homo erectus* (i.e., Peking man) was capable of group pursuit and a degree of coordinated action in capturing and slaughtering large animals. These activities involved sporadic forays, systematic searches, and occasional migrations away from established territories (as distinct from the cyclical migrations of many other species). Such abilities were in part the cause and in part the consequence of intellectual and social developments. Success in hunting was to increase further with growth in the ability to adapt behavior to particular circumstances and to communicate and collaborate with others.

Both involved tremendous increases in intelligence and learning ability. *Homo sapiens* developed four important mental capacities that may also be regarded as necessary conditions for the eventual acquisition of mapping skills. First, there was the ability to delay an instinctive response in favor of a pause for exploration; second, the facility of storing acquired information; third, the ability to abstract and generalize; and fourth, the capacity to carry out the required responses to information thus processed. Collaborative effort in hunting, in particular, involved coding information and a capacity to transmit it rapidly and effectively between individuals. Language (gestural and graphic as well as spoken) was the enabling device that ensured this. Unlike the "here and now" language of the other higher primates, human language began to bind "events in space and time within a web of logical relations governed by grammar and metaphor."[9] Wittgenstein's proposition that "the limits of my language mean the limits of my world" remains valid.[10] One could go further and say that the origins of language and the growth of spatial consciousness in man are closely interrelated. The cognitive schema that underlay primitive speech must have had a strong spatial component. Not all messages were spatial in content or manifestation, but many would have been, and these

5. Jay Appleton, *The Experience of Landscape* (New York: John Wiley, 1975), 73.

6. Julian Jaynes, *The Origins of Consciousness in the Breakdown of the Bicameral Mind* (Boston: Houghton Mifflin, 1976), 59–61. Time is, and long has been, described in the terminology of space. Unconsciously, and without giving rise to confusion, we talk, albeit metaphorically, of the *distant* past, *points* in time, and the *way* ahead. Our lives *meander*, *diverge* along different *paths* from those of others, and have *turning points*. We *locate* problems and have different *areas* of interest. Our minds have their *regions* and *frontiers*, and our lives are *circumscribed*. Much less frequently, we reverse the metaphorical process by describing aspects of space in terms of time. Journeys take *minutes*, *hours*, or *days*. Yet for most people, and perhaps from the beginning of human consciousness, "what fails to exist *now* has seemed less real than what merely fails to exist *here*": Alan Robert Lacey, *A Dictionary of Philosophy* (London: Routledge and Kegan Paul, 1976), 204. Hence we *chart* our progress to date, *plan* our careers, and *map* out our lives.

7. Interestingly, forty thousand years or so later, the idea that in order to constitute a message, information must contain a degree of surprise for the receiver has been used by mathematicians in defining it as a precisely measurable commodity.

8. John Hurrell Crook, *The Evolution of Human Consciousness* (Oxford: Clarendon Press, 1980), 35.

9. Crook, *Evolution*, 148 (note 8).

10. Ludwig Wittgenstein, *Tractatus Logico-Philosophicus*, trans. D. F. Pears and B. F. McGuinness (London: Routledge and Kegan Paul, 1961), para. 5.6.

would have helped to provide the structural as well as the functional foundations of language. It has been argued that these foundations helped to promote

> the ability to construct with ease sequences of representations of routes and location. . . . Once hominids had developed names (or other symbols) for places, individuals, and actions, cognitive maps and strategies would provide a basis for production and comprehension of sequences of these symbols. . . . Shared network-like or hierarchical structures, when externalized by sequences of vocalizations or gestures, may thus have provided the structural foundations of language. . . . In this way, cognitive maps may have been a major factor in the intellectual evolution of hominids . . . cognitive maps provided the structure necessary to form complex sequences of utterances. Names and plans for their combination then allowed the transmission of symbolic information not only from individual to individual, but also from generation to generation.[11]

A related way forward is through modern studies of spatial cognition in humans. This has been well researched, and the spatial consciousness of modern indigenous peoples can be used to help unravel prehistoric mapping and hence the origins of cartography. For instance, researchers such as Christopher Hallpike, following the Piagetian school of developmental psychology, have identified a list of spatial concepts dominating aboriginal spatial thought.[12] This is composed of opposites such as inner and outer, center and periphery, left and right, high and low, closed and open, and symmetrical and asymmetrical order. "Boundary" is another important spatial concept. Orderings are "basically topological, as opposed to Euclidean or projective, and are associated with concrete physical features of the natural environment."[13] Here too we have evidence of the cognitive maps that underlay the emergence of maps in material form.

It is in the development of language in its broader sense that the origins of mapping are to be found. Crucial to this development would have been the emergence of teaching beyond the level of mere imitation and of communication systems capable of expressing relationships. Of the latter, aural systems (speech and music) were ephemeral and limited to the temporal dimension.[14] They were therefore least effective as means of communicating spatial messages. Of the visual systems for communication, gesture and dance, though also ephemeral, were themselves spatially three-dimensional forms and therefore would have been more effective in conveying a "map" to members of the group who were present and within range at the time of transmission. Drawings, models, pictographs, and notations were, potentially at least, three-dimensional but had the addi-

tional advantage of combining immediacy with a greater degree of permanence. It was from this visual group of systems for communicating that cartography, along with other graphic images, eventually emerged as a specialist form of language.

Neither the sequence of emergence nor the relative rates of development of these human systems of communication is recorded. A possible key stage, however, linking mental maps to their specialized expression as graphic representation, may be found in the use of gesture. Gesture, says Gordon Hewes, probably "reached the limits of its capacity to cope with cultural phenomena by the end of the Lower Paleolithic" but "gained a new lease of life in the Upper Paleolithic and thereafter, with the birth of drawing, painting and sculpture."[15] Gesture and ephemeral graphics are still used in bridging the gap between different linguistic groups, and they are sometimes preferred or used as an adjunct to speech, especially as a means of communicating locative messages. The literatures of anthropology and of European exploration and discovery from the fifteenth century onward are rich in examples of the way gesture was used in communicating with native peoples, many of whom were still following an essentially Upper Paleolithic way of life.[16] Gesture is frequently described as having been used to solicit or to communicate information about terra incognita. In such cases both European interrogators and native respondents tended to use sketch maps, and occasionally dance, in conjunction with gesture.

A link between gesture and simple mapping is also to be found in pictography. Unlike syllabic alphabetic writing, pictography was not unilinear and was readily adaptable to represent the spatial distribution of things and events.[17] Most early peoples used some form of

---

11. Roger Peters, "Communication, Cognitive Mapping, and Strategy in Wolves and Hominids," in *Wolf and Man: Evolution in Parallel*, ed. Roberta L. Hall and Henry S. Sharp (New York and London: Academic Press, 1978), 95–107, esp. 106.

12. Jean Piaget and Bärbel Inhelder, *The Child's Conception of Space*, trans. F. J. Langdon and J. L. Lunzer (London: Routledge and Kegan Paul, 1956). Christopher R. Hallpike, *The Foundations of Primitive Thought* (New York: Oxford University Press; Oxford: Clarendon Press, 1979), 285. See also James M. Blaut, George S. McCleary, and America S. Blaut, "Environmental Mapping in Young Children," *Environment and Behavior* 2 (1970): 335–49.

13. Hallpike, *Foundations*, 285 (note 12).

14. The spatial and temporal dimensions of messages are discussed in Abraham Moles, *Information Theory and Esthetic Perception*, trans. Joel E. Cohen (Urbana: University of Illinois Press, 1966), 7–9.

15. Gordon W. Hewes, "Primate Communication and the Gestural Origin of Language," *Current Anthropology* 14, nos. 1–2 (1973): 5–24, quotation on 11.

16. Hewes, "Primate Communication," 11, especially n. 7 (note 15).

17. In written Chinese, the character for map (and diagram) is itself a highly stylized map. This suggests that mapping and maps had

pictography, with signs derived in part from the objects being represented and in part from related gestures. Moreover, in surviving maps made by indigenous peoples of historical times, gesture is frequently an important part of the iconography. For example, a hand with the index finger outstretched is used to indicate direction. In other cases, a line of hoofprints or human footprints is used to show both the route and the direction of movement along it.[18]

Such modern analogies are clearly suggestive, but they are not conclusive indicators of the way permanent material maps might have originated. The researcher is brought up against the barrier that the evidence of gesture and ephemeral graphics—by its very nature—has not survived from the Upper Paleolithic. Thus it is in the more permanent art forms—especially in the rock art and mobiliary art of Upper Paleolithic societies in the midlatitude belt of Eurasia—that one might expect to find the earliest evidence of maps. However, just as in ethology one has to be cautious about translating animal signals into human language, so with prehistoric art forms one has to be careful before ascribing specific meaning or function to patterns, textures, symbols, or colors. Furthermore, the mapping of topographical information per se was almost certainly not of practical importance (in the modern sense) to early man. Mapping may, however, have served to achieve what in modern behavioral therapy is known as desensitization: lessening fear by the repeated representation of what is feared.[19] Representing supposedly dangerous terrae incognitae in map form as an extension of familiar territory may well have served to lessen fear of the peripheral world. Similarly, from the Upper Paleolithic onward, man was greatly concerned with his fate after death, and cosmological maps may well have lessened fear of the afterlife. Since early cosmology and religion were also associated with a rather more empirical astronomy, it is reasonable to suppose, too, that celestial maps may have been developed early. For historians of cartography, the difficulty lies not so much in accommodating such ideas as in finding unambiguous evidence to support them and thereby being able to move away from speculation and assumption to firmer intellectual ground.

---

emerged as distinctive activities and products before the final development of writing, which in China is generally supposed to have attained essentially its present form by 2800 B.P. Joseph Needham, *Science and Civilisation in China* (Cambridge: Cambridge University Press, 1954–), vol. 3, *Mathematics and the Sciences of the Heavens and the Earth* (1959), 498.

18. Footprints also feature, in pairs or singly, among the Scandinavian petroglyphs of Bronze Age date, but it is possible that they have a quite different meaning here; see H. R. Ellis Davidson, *Pagan Scandinavia* (London: Thames and Hudson, 1967), 54–55. In the Mixtec picture writing of ancient southern Mexico, human footprints or a band containing footprints usually signified a road: see Mary Elizabeth Smith, *Picture Writing from Ancient Southern Mexico: Mixtec Place Signs and Maps* (Norman: University of Oklahoma Press, 1973), 32–33. Australian aborigines distinguish on their pictographic maps between the tracks of men and those of different types of animals: see Norman B. Tindale, *Aboriginal Tribes of Australia: Their Terrain, Environmental Controls, Distribution, Limits and Proper Names* (Berkeley, Los Angeles, and London: University of California Press, 1974), fig. 33.

19. Julian Jaynes, "The Evolution of Language in the Late Pleistocene," *Annals of the New York Academy of Sciences* 280 (1976): 322.

4 · Cartography in the Prehistoric Period in the Old World: Europe, the Middle East, and North Africa

Catherine Delano Smith

Prehistoric Maps and Historians of Cartography

As was made clear in the Introduction to this section on prehistoric maps, historians of cartography have had little to say on prehistoric cartography in the Old World. Neither Richard Andree nor Wolfgang Dröber said anything at all.[1] In 1910 Bruno F. Adler discussed two decorated bone plaques that a German antiquarian, Fritz Rödiger, had suggested were maps, but he omitted both from his corpus.[2] In 1917 Leo Bagrow followed Adler in referring to Rödiger and in citing, for European prehistoric maps, the work of only three writers (Rödiger, Kurt Taubner, and Amtsgerichtsrath Westedt)[3] among the 1,881 bibliographical items in his *Istoriya geograficheskoy karty: Ocherk i ukazatel' literatury* (The history of the geographical map: Review and survey of literature).[4] Modern authors have scarcely improved on this: three topographical maps from the prehistoric period were published in the 1960s by Walter Blumer,[5] though only two of these are included by P. D. A. Harvey,[6] and one other has been described from the Middle East.[7] Thus, when research for this chapter was started, the number of topographical maps from the prehistoric pe-

I am grateful to the many who have helped me with material for this study, including Emmanuel Anati (Centro Camuno di Studi Preistorici, Capo di Ponte, Italy); Ernst Burgstaller (Gesellschaft für Vor- und Frühgeschichte, Austria); John M. Coles (University of Cambridge); Ronald W. B. Morris; Gerald L'E. Turner and Anthony V. Simcock (both of the Museum of the History of Science, University of Oxford); Andrew Sherratt (Ashmolean Museum, University of Oxford); and Franz Wawrik (Österreichische Nationalbibliothek, Vienna). Several ideas originated in, or are due to, discussions with G. Malcolm Lewis (University of Sheffield), and I owe him a particular debt for his generous advice and helpful comments. I am also indebted to the British Academy for support from the Small Grants Research Fund (1981–82).

1. Richard Andree, "Die Anfänge der Kartographie," *Globus: Illustrierte Zeitschrift für Länder* 31 (1877): 24–27, 37–43. Wolfgang Dröber, "Kartographie bei den Naturvölkern" (Diss., Erlangen University, 1903; reprinted Amsterdam: Meridian, 1964); summarized under the same title in *Deutsche Geographische Blätter* 27 (1904): 29–46. The Old World is here defined to include Europe (with Russia west of the Urals), the Middle East (to the Tigris), and North Africa (with the Sahara).

2. Fritz Rödiger, "Vorgeschichtliche Kartenzeichnungen in der Schweiz," *Zeitschrift für Ethnologie* 23 (1891): Verhandlungen 237–42. Adler misspelled Rödiger as Rödinger, an error perpetuated by Leo Bagrow in both *Die Geschichte der Kartographie* (Berlin: Safari-Verlag, 1951), 16, and *History of Cartography*, rev. and enl. R. A. Skelton, trans. D. L. Paisey (Cambridge: Harvard University Press; London: C. A. Watts, 1964), 26. In addition, Adler misspelled Taubner as Tauber: see Bruno F. Adler, "Karty pervobytnykh narodov" (Maps of primitive peoples), *Izvestiya Imperatorskogo Obshchestva Lyubiteley Yestestvoznaniya, Antropologii i Etnografii: Trudy Geograficheskogo Otdeleniya* 119, no. 2 (1910): 218. See also the summary review by H. de Hutorowicz, "Maps of Primitive Peoples," *Bulletin of the American Geographical Society* 43, no. 9 (1911): 669–79. This omission meant that Adler had not one map example from Europe to set against the 115 gathered from the rest of the world; namely, 55 maps from Asia, 15 from America, 3 from Africa, 40 from Australia and Oceania, and 2 from the East Indies. The description of Adler's corpus comes from de Hutorowicz, "Maps," 669, and is also cited by Norman J. W. Thrower, *Maps and Man: An Examination of Cartography in Relation to Culture and Civilization* (Englewood Cliffs, N.J.: Prentice-Hall, 1972), 5 n. 7.

3. Rödiger, "Kartenzeichnungen," 237–42 (note 2). Kurt Taubner, "Zur Landkartenstein-Theorie," *Zeitschrift für Ethnologie* 23 (1891): Verhandlungen 251–57. Amtsgerichtsrath Westedt, "Steinkammer mit Näpfchenstein bei Bunsoh, Kirchspiel Albersdorf, Kreis Süderdithmarschen," *Zeitschrift für Ethnologie* 16 (1884): Verhandlungen 247–49.

4. Leo Bagrow, *Istoriya geograficheskoy karty: Ocherk i ukazatel' literatury* (The history of the geographical map: Review and survey of literature), *Vestnik arkheologii i istorii, izdavayemyy Arkheologicheskim Istitutom* (Archaeological and historical review, published by the Archaeological Institute) (Petrograd, 1918). The relevant part of Bagrow's text was incorporated into his *Geschichte* (note 2), but very few of the original references reappear. Bagrow's *History of Cartography* (note 2), the revised and enlarged version of *Geschichte*, was translated and published in German as *Meister der Kartographie* (Berlin: Safari-Verlag, 1963). In all of these works, Bagrow discussed the Maikop vase: see note 90 below.

5. Map 43, map 45, and map 47 in appendix 4.1. Walter Blumer, "The Oldest Known Plan of an Inhabited Site Dating from the Bronze Age, about the Middle of the Second Millennium B.C.," *Imago Mundi* 18 (1964): 9–11 (Bedolina); idem, "Felsgravuren aus prähistorischer Zeit in einem oberitalienischen Alpental ältester bekannter Ortsplan, Mitte des zweiten Jahrtausends v. Chr.," *Die Alpen*, 1967, no. 2 (all three).

6. Seradina and Bedolina; P. D. A. Harvey, *The History of Topographical Maps: Symbols, Pictures and Surveys* (London: Thames and Hudson, 1980), figs. 20 and 21.

7. Map 54 in appendix 4.1. James Mellaart, "Excavations at Çatal Hüyük, 1963: Third Preliminary Report," *Anatolian Studies* 14 (1964): 39–119.

riod in the Old World referred to in recent histories of cartography totaled four.

After a reassessment of the evidence from the prehistoric period in the light of new criteria, over fifty maps or spatial representations from this period have been selected for consideration by historians of cartography and itemized in appendix 4.1—List of Prehistoric Maps. This list attempts to summarize what seem to be maps in the prehistoric source material. It is neither complete nor definitive and some items may prove controversial. It has been compiled with the caution such research demands and which has been lacking in the literature. It does, however, resist the recent tendency to dismiss magic and religious belief as irrelevant to an understanding of indigenous or prehistoric art. The pendulum of opinion has probably swung too far in its reaction to the nineteenth- and twentieth-century antiquaries who oversimplified their role. Scholarly research into such matters as the nature of the primitive mind, the importance of symbolism in primitive cultures, the early history of religion, and the meaning and context of rock art has done much to advance a more balanced and rational assessment of the surviving evidence. The expanded length of the list reflects these considerations.

The present approach is based on three general principles. First, an open mind is needed regarding the range of potential source material. Second, any maps found in these sources cannot be studied apart from other forms of contemporary art or in isolation from the total context in which this art was produced, even if this means relying not only on the archaeological record but also on anthropological parallels. And finally, a new theoretical framework may have to be created for what is in effect a new subject.

THE SOURCE MATERIAL AND ITS INTERPRETATION

All the major forms of prehistoric art are of potential interest to the historian of cartography (fig. 4.1). However, by far the most important are the two classes of rock (or parietal) art: paintings (pictographs) and carvings (petroglyphs). Mobiliary art—art on unfixed surfaces such as pebbles or slates or on bone or metal artifacts, decoration on pottery, even sculptures or relief models—can also contain much of cartographic interest. Rock art is found in daylight situations (rock shelters and overhangs that often were inhabited) as well as in underground caverns and deep recesses that would have been reached only with extreme difficulty in prehistoric times. The art is composed of both naturalistic and nonnaturalistic representations. Animals (mainly bison, mammoth, and horse, but occasionally birds and fish) and human figures make up most of the first category.

A variety of what appear to us as geometric and abstract markings forms the second. Much of the literature emphasizes the naturalistic images, especially those (such as the bison and mammoth from Lascaux and other caves of the Dordogne region and the Cantabrian Pyrenees) famous for their beauty of line and execution. This has resulted in a biased impression of their numerical importance. Recent work is balancing this by showing that the same caves also contain vast numbers of nonnaturalistic markings.[8] The suggestion, however, that the abstract or geometric figures may be later in date than the naturalistic figures is probably little more than speculation.[9]

While it is very difficult to place individual figures into chronological sequence, much less assign precise dates, prehistoric art can be described in the broadest of terms as dating either from the Upper Paleolithic and the Mesolithic, periods of hunter-gatherer-fisher populations, or from the post-Paleolithic period of agricultural populations (fig. 4.2). The Upper Paleolithic dates, in Europe, from about 40,000 B.C. to about 10,000 B.C. Where Upper Paleolithic cultural characteristics are found at a later date (as in northern Africa), the term Epipaleolithic is used. The first datable art in the world comes from Europe near the start of the Upper Paleolithic.[10] It is already highly accomplished, and this must imply that the graphic and sculpting skills involved were by no means in their infancy even at this date. Given the total length of the Upper Paleolithic period—some thirty thousand years—its style of art as well as of life is remarkably homogeneous. In contrast, the economic and social characteristics of the post-Paleolithic era are exceedingly diverse, possibly a reflection of the environmental changes that accompanied the gradual disappearance of the ice sheets from Europe, although this was not matched by major changes in art. Prehistorians have long recognized three major cultural subdivisions: the Neolithic (with its transitional terminal phase, the Chalcolithic or Copper Age); the Bronze Age; and the

8. For example, cave decoration at Niaux (Tarascon-sur-Ariège) includes 2–3 human figures and 114 animal figures but also no fewer than 136 "tectiform" signs of various styles and nearly as many circular signs in addition to numerous other geometric or abstract markings: Antonio Beltran-Martínez, René Gailli, and Romain Robert, *La Cueva de Niaux*, Monografías Arqueologicas 16 (Saragossa: Talleres Editoriales, 1973), 227–46.

9. Magín Berenguer, *Prehistoric Man and His Art: The Caves of Ribadesella*, trans. Michael Heron (London: Souvenir Press, 1973), 79 ff. But see Mircea Eliade, *A History of Religious Ideas*, trans. Willard R. Trask (Chicago: University of Chicago Press, 1978), vol. 1, *From the Stone Age to the Eleusinian Mysteries*, chap. 1.

10. Peter J. Ucko and Andrée Rosenfeld, *Palaeolithic Cave Art* (New York: McGraw-Hill; London: Weidenfeld and Nicolson, 1967), 66; Desmond Collins and John Onians, "The Origins of Art," *Art History* 1 (1978): 1–25.

56   *Cartography in Prehistoric Europe and the Mediterranean*

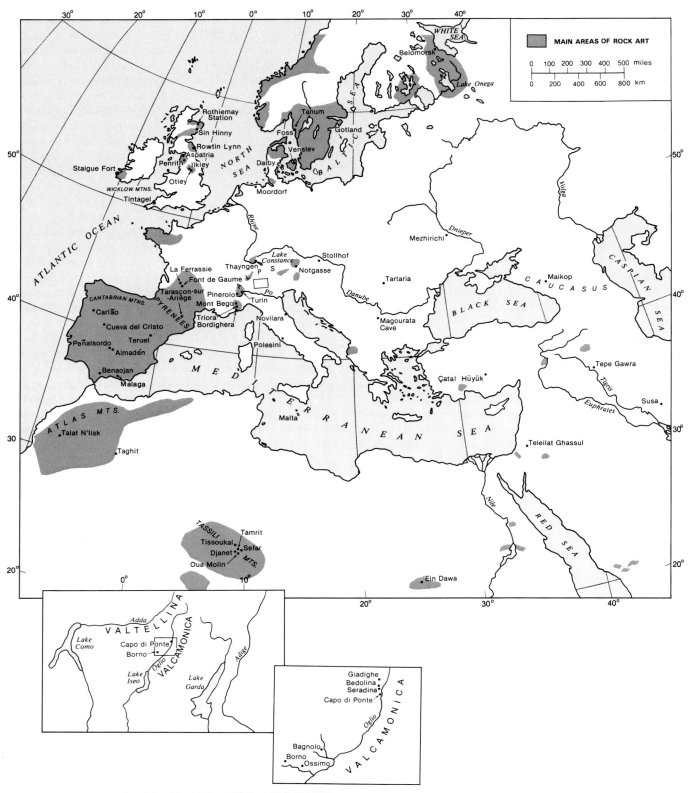

FIG. 4.1. PRINCIPAL AREAS AND SITES ASSOCIATED
WITH PREHISTORIC ROCK ART IN THE OLD WORLD.

Iron Age. Each of these cultural periods started earlier in southern and eastern regions (first Mesopotamia and Egypt, then Asia Minor, Greece, and southern Italy) than in the western Mediterranean or northern Europe. Throughout the Upper Paleolithic, the Scandinavian area lay under an ice sheet and was uninhabited. The main periods of post-Paleolithic rock art coincide with the Neolithic and Bronze ages in southern Europe and with the Bronze and early Iron ages in Scandinavian Europe. The end of the prehistoric period, readily identified by the appearance of writing, likewise varied regionally. In the Middle East the appearance of writing and the rise of the great civilizations of Mesopotamia starts from about 3000 B.C., and the same is true in Egypt. Along the northern and southern shores of the western Mediterranean, however, the prehistoric period lasted well into the final millennium B.C. Northern France and Britain remained prehistoric until the arrival of the Romans. In Scandinavia the Iron Age is generally accepted as continuing until the eighth or ninth century A.D.

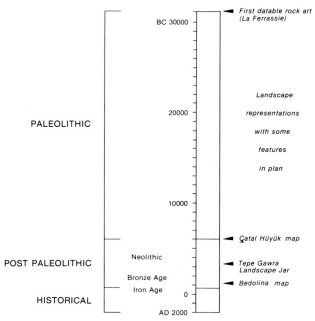

FIG. 4.2. THE PREHISTORIC AND HISTORICAL TIME SCALE. The period of European and Middle Eastern rock and mobiliary art is compared with the historical period. Maps usually described as the "earliest"—such as those on Mesopotamian clay tablets—come from the historical period. The dates of two of the better-known prehistoric maps are indicated.

Notwithstanding these archaeological distinctions based on material culture, the essential characteristics of rock art were maintained throughout the entire prehistoric period. There are detailed differences, such as a change of content according to region or period (differ-

ent animals or objects are depicted) and a change of location (post-Paleolithic rock art tends to be in the open air, being found on exposed rock surfaces and cliff faces and even, in some areas, within sight of contemporary farmland). But one of the crucial features of prehistoric art of any date is the way certain surfaces were used over and over again while neighboring rocks, to our eyes as suitable and as attractive, remain pristine. This characteristic is held to emphasize the sacredness of specific sites. The general distribution of rock art reinforces this conjecture. In part it must reflect accident of discovery or intensity of search (as in Valcamonica, Italy, or in southern Sweden). On the other hand, there are strongly marked clusters of sites within even the best-searched areas (around Mont Bégo in the Ligurian Alps, for example). It is this that has led many to postulate the sacredness of certain localities, and even of specific topographical features, as a factor in the distribution of rock art. Similarly arresting is the absence of composition in European rock art.[11] This makes all the more outstanding those assemblages in rock or mobiliary art where order or regularity is discernible.

Nearly all evidence for human activity in the prehistoric period has been acquired through archaeological investigation. But perhaps surprisingly, considering the extraordinary wealth and detail of information generally available for the Old World, the archaeological record is far from adequate when it comes to the interpretation of prehistoric art and hence its cartographic component. Archaeological information is unevenly spread geographically, through time, and by topic. More often than not, little direct and unambiguous evidence is available for reconstructing a behavioral picture of the people whose beliefs and values account for the different forms of the art. One reason for the ambiguities is that the most basic archaeological criterion—an absolute date

11. As compared with southern Africa, for example, where "narrative compositions are far more common and much more explicit: people are clearly depicted dancing, fighting, hunting or performing . . . ritual activities": J. David Lewis-Williams, *The Rock Art of Southern Africa* (Cambridge: Cambridge University Press, 1983), 11. For the debate over just how much order there may be in European rock art, see Ann Sieveking, *The Cave Artists* (London: Thames and Hudson, 1979), 208–9. For instance, André Leroi-Gourhan has suggested that there is a significant grouping of what he sees as male/female animal types and sexual signs: see his *Art of Prehistoric Man in Western Europe*, trans. Norbert Guterman (London: Thames and Hudson, 1968), and his *The Dawn of European Art: An Introduction to Palaeolithic Cave Painting*, trans. Sara Champion (Cambridge: Cambridge University Press, 1982), but this is not supported by statistical analysis: John Parkington, "Symbolism in Palaeolithic Cave Art," *South African Archaeological Bulletin* 24, pt. 1, no. 93 (1969): 3–13. It has even been suggested that some cave decoration represents local hunter territory, particularly when the natural irregularities of the cave wall are included in the composition: Anne Eastham and Michael Eastham, "The Wall Art of the Franco-Cantabrian Deep Caves," *Art History* 2 (1979): 365–85.

for each picture—cannot be satisfied. Secure dating for a rock art figure is available only if it is found within a properly stratified and datable context, as, for example, at Çatal Hüyük.[12] Prehistorians often attempt to identify the different production stages of a rock art palimpsest by reference to variations in technique and in style of drawing or to degree and nature of subsequent patination. Only slightly safer ground is provided by comparing the depiction of an object with an identical one dated by excavation. But while relative chronologies may perhaps be established in this manner, absolute dates arrived at by such methods have to be regarded with skepticism. This means, for the historian of cartography, that it is difficult to associate a map example with a specific prehistoric culture. Limited knowledge about the art itself means that an advance in understanding its meaning is also held back.

Archaeological investigation has established two important general points, both relating to the purpose of rock art. First, excavations have demonstrated that rock art was associated with belief and religion. The picture map from Çatal Hüyük, for instance, like the other wall paintings from this remarkable Neolithic site in central Turkey, was excavated from a room whose contents and internal arrangements show it was a shrine or some sort of holy room.[13] Second, both at Çatal Hüyük and elsewhere, excavation has also shown that this art was a "product of the moment," created for, or during, ritual and not at all intended to last beyond that event.[14] Although very little rock art is associated with so informative an archaeological context, concordance of ethnographic and archaeological evidence supports such conclusions.

The historian of cartography who is concerned with maps from the prehistoric period in the Old World faces not only conceptual and methodological problems familiar to scholars working on the 'primitive' maps of still-extant New World peoples[15] but also a number of additional problems that limit from the outset any hopes for direct interpretation of the evidence. The greatest conceptual problem involves the researcher's attitude to indigenous and prehistoric cultures. The modern mind is blinkered by its own literacy to the extent that "the effects of oral states of consciousness are bizarre to the literate mind."[16] It is difficult to imagine what a primarily oral culture is like. This may help explain why its products have been all too easily dismissed as irrational, quaint, or (in the pejorative sense) primitive. Another recent revelation is that oral persons tend not to recognize or to have a discrete category for abstract shapes.[17] They see a circle, for instance, as the object they know it represents, so that one circle is described as a plate, another as the moon, and so on. But apart from these general matters, the researcher into European

rock art is at a disadvantage. Suitable ethnic survivors in the Old World are lacking, and the time span between the prehistoric period and the present, or even the recent past, is far too long to allow inference from tradition, myth, or legend.[18]

Hence non-European ethnographic parallels, where they are well documented, can and must be used to provide insight into the context of prehistoric rock art. They are needed as a guide both to its function and to the meaning of its content. The first lesson to be learned

12. See note 7 above. Problems of independent dating are illustrated by Ronald I. Dorn and David S. Whitley, "Chronometric and Relative Age Determination of Petroglyphs in the Western United States," *Annals of the Association of American Geographers* 74 (1984): 308–22.

13. Mellaart, "Excavations," 53–55 (note 7), and James Mellaart, *Çatal Hüyük: A Neolithic Town in Anatolia* (London: Thames and Hudson, 1967), 77. Subsequent excavation revealed an unusual and distinctive feature in this particular shrine: a burial. Moreover, the dead woman was decorated in nonlocal style, prompting speculation about her association with the subject of the wall painting and the eruption of Hasan Dağ (Mellaart, personal communication).

14. Indeed, there is evidence of its having been destroyed after its period of utility: at Çatal Hüyük the walls were regularly replastered and sometimes repainted. Diane Kirkbride, "Umm Dabaghiyah 1974: A Fourth Preliminary Report," *Iraq* 37 (1975): 3–10, esp. 7, and J. B. Hennessy, "Preliminary Report on a First Season of Excavations at Teleilat Ghassul," *Levant* 1 (1969):1–24, also report replastering and repainting, in one case at Teleilat Ghassul up to twenty times. Elsewhere, engraved slate plaques have been found deliberately broken: Gerhard Bosinski, "Magdalenian Anthropomorphic Figures at Gönnersdorf (Western Germany), *Bolletino del Centro Camuno di Studi Preistorici* 5 (1970): 57–97, esp. 67. Also relevant is "the fact that some decorated stones [in Megalithic burial architecture] were never meant to be seen again": Glyn Daniel, review in *Antiquity* 55 (1981): 235, of Elizabeth Shee Twohig, *The Megalithic Art of Western Europe* (Oxford: Clarendon Press, 1981).

15. Discussed by Michael Blakemore, "From Way-finding to Map-making: The Spatial Information Fields of Aboriginal Peoples," *Progress in Human Geography* 5, no. 1 (1981): 1–24.

16. Walter J. Ong, *Orality and Literacy: The Technologizing of the Word* (London and New York: Methuen, 1982), 30.

17. A. R. Luriya, *Cognitive Development: Its Cultural and Social Foundations*, ed. Michael Cole, trans. Martin Lopez-Morillas and Lynn Solotaroff (Cambridge: Harvard University Press, 1976), 32–39. Reported by Ong, *Orality and Literacy*, 50–51 (note 16).

18. The link undoubtedly exists, though it is much complicated by the substitution of characters and events as individual myths pass from one culture to another. See Stephen Toulmin and June Goodfield, *The Discovery of Time* (London: Hutchinson, 1965), 23 ff., and Peter Munz, *When the Golden Bough Breaks: Structuralism or Typology?* (London and Boston: Routledge and Kegan Paul, 1973); Claude Lévi-Strauss, *Structural Anthropology*, trans. Claire Jacobson and Brooke Grundfest Schoepf (New York: Anchor Books, 1967), chap. 11. The usefulness of ethnographic evidence is illustrated by the way certain signs in Australian aboriginal art, which *appear* to have no topographical significance, may be explained by the artists or users as topographical in meaning; signs for hills used by the Walbiri, for example. Nancy D. Munn, "Visual Categories: An Approach to the Study of Representational Systems," *American Anthropologist* 68, no. 4 (1966): 936–50; reprinted in *Art and Aesthetics in Primitive Societies*, ed. Carol F. Jopling (New York: E. P. Dutton, 1971), 335–55.

from the ethnographic evidence, as already noted, is that the maps cannot, in the first instance at least, be studied in isolation, any more than can the rock art corpus as a whole be divorced from its social context. To reach the ideas expressed in the art, it is essential to distill the vital concepts from the ethnography and then show how they are transformed into graphic representation.[19] This can be done only by looking at generalized and repeated features, and not, as has long been the tendency, by selecting the immediately attractive pictures and attempting to match them to specific myths or practices.[20] The second lesson is that rock art is not about the mundane practicalities of daily life. Contrary to views held earlier this century on the importance of "sympathetic magic," it is possible to demonstrate from ethnography that the artists were not concerned with the provision of food.[21] It is also clear that such maps as there may be in rock art are less likely to have been created, as are modern maps, for wayfinding or as a device for the storage of information.[22] Ethnography shows that permanent directional aids are not normally needed within small indigenous, land-based societies,[23] though they might be needed by peoples who must navigate extensive areas of undifferentiated terrain, oceanic or snowbound territories,[24] or by those living in a community which has so outgrown its territory that there are members to whom it is no longer all intimately familiar.[25] What ethnography does show is that the primary aspects of human spatial consciousness may be transferred to the ground (as in settlement planning) or used in the creation of imagined worlds (the cosmos).[26] It has been shown, too, that initiation rites contain the secrets of a society's symbolic knowledge[27] and that it is information about the cosmological world, rather than profane and practical familiarity with the local territory, that is transmitted through those rites.[28] In fact, ideas about the "other world" and the nature of the passage from one part of the cosmos to another are found to be of fundamental importance in indigenous societies. That this was also the case in prehistoric times cannot be doubted, given, for example, the presence in prehistoric art of cosmological symbols such as ladders and trees as well as "guides to the beyond" in the form of labyrinthine designs. Finally, and importantly, ethnography reveals the way the art is composed of "crystallised metaphors"[29] and that it is as resonant with symbolic meaning as any of the more ephemeral gestures or rites of the society that produced it.

The role of image, symbol, and symbolism in oral societies is now well documented. Indeed, it is recognition of the "importance of symbolism in archaic thinking and . . . the fundamental part it plays in the life of any and every primitive society"[30] that distinguishes modern scholars from those of the nineteenth century

in these studies. An appreciation of symbolism closes the gap between prehistoric rock art in general and prehistoric cartography in particular. Maps, like rock art, are executed to convey "a message . . . encoded in visual form."[31] The difficulty is that this code needs to be broken before the message can be reached. It is well known that signs and symbols carry messages particular to a social group, or to individuals within that group (the initiated), and that the meaning of each sign has to be learned. Cross-cultural equivalents do exist, but an apparently familiar symbol with a wide distribution may have not only a wide range of meanings but also mean-

19. Lewis-Williams, *Rock Art*, 37 (note 11).

20. Lewis-Williams, *Rock Art*, 37 (note 11).

21. Lewis-Williams, *Rock Art*, 19 (note 11). See also Lewis-Williams, "Testing the Trance Explanation of Southern African Rock Art: Depictions of Felines," *Bollettino del Centro Camuno di Studi Preistorici* 22 (1985): 47–62.

22. The last phrase is Thrower's, *Maps and Man*, 1 (note 2).

23. See, for example, R. A. Gould, *Living Archaeology* (Cambridge: Cambridge University Press, 1980), 84; David Lewis, "Observations on Route Finding and Spatial Orientation among the Aboriginal Peoples of the Western Desert Region of Central Australia," *Oceania* 46, no. 4 (1976): 249–82, esp. 271. However, there are indications that there may have been a greater demand for navigational aids in small land-based societies than might have been expected. For example, the long journey of the Hopi Indians to fetch salt described by Leo W. Simmons, ed., *Sun Chief: The Autobiography of a Hopi Indian* (New Haven: Yale University Press, 1942), 232–45. I owe this last point to Herbert C. Woodhouse.

24. See discussion in Christopher R. Hallpike, *The Foundations of Primitive Thought* (New York: Oxford University Press; Oxford: Clarendon Press, 1979), 301–13.

25. As in the case of large-scale societies, supported by cultivation and living in permanent settlements. The nature of the agricultural routine and implied division of labor may also mean—as it has in recent times—that few inhabitants of the settlement visit all parts of its territory and that most would not be intimately familiar with all local places: see Catherine Delano Smith, *Western Mediterranean Europe: A Historical Geography of Italy, Spain and Southern France since the Neolithic* (London: Academic Press, 1979), 27–29. On the other hand, it is arguable that such knowledge of the total territory is not needed in the normal pattern of life: Hugh Brody has demonstrated that individual Indian hunters had their own hunting and gathering areas within the same reserve and respected each other's, which must have remained relatively unfamiliar if not wholly unknown: *Maps and Dreams* (New York: Pantheon Books, 1982).

26. Jean Piaget and Bärbel Inhelder, *The Child's Conception of Space*, trans. F. J. Langdon and J. L. Lunzer (London: Routledge and Kegan Paul, 1956); Hallpike, *Foundations*, 285–96 (note 24).

27. For example, Fredrik Barth, *Ritual and Knowledge among the Baktaman of New Guinea* (New Haven: Yale University Press, 1975).

28. Arnold van Gennep, *The Rites of Passage*, trans. Monika B. Vizedom and Gabrielle L. Caffee (London: Routledge and Kegan Paul, 1960), viii, for instance. The original, written in French in 1909, *Les rites de passage: Etude systématique des rites* (Paris: E. Nourry, 1909), was long ignored but is still not superseded.

29. Lewis-Williams, *Rock Art*, 44 (note 11).

30. Mircea Eliade, *Images and Symbols: Studies in Religious Symbolism*, trans. Philip Mairet (London: Harvill Press, 1961), 9.

31. Blakemore, "Way-finding," 3 (note 15).

ings that are total opposites.[32] So even where meanings become relatively fixed (and are therefore available to us), as in the case of pictographic, ideographic, or hieroglyphic writing, it is unwise to transpose meanings from one spatial or temporal context to another.[33]

In the relatively closed world of a small indigenous or traditional society, the messages conveyed by signs and symbols are readily learned. There are many constantly recurring regularities,[34] experiences shared by all and therefore recognized by all. It is this homogeneity of experience that makes signs and symbols an effective and economical form of communication, at least within that particular society and its initiates. To understand these signs and symbols, the historian of cartography must learn to see the same world in the same way as their creators.[35] On a modern map, signs are similarly used to convey the maximum amount of information to the user.[36] However, the nature of this primary information is generally elucidated by a written explanation or key in the course of verbal instruction. On a different level, hidden, symbolic, or coded messages are discovered when the historian of cartography has learned to understand not just the overt content of the map itself but its total context.[37] In the absence of a key or other guide indicating even the primary meanings of the signs employed in prehistoric art, the need to come to terms with the total context of that art is all the more urgent.

A final complication is the matter of style. One difficulty is knowing whether the artist is portraying the object in profile or in plan.[38] Another is that, in rock art no less than in art in general, some artists attempt to economize in representation to the extent that they produce highly stylized figures. These can look like abstract or geometric signs even though the intent is an iconic representation. Henri Breuil, the pioneer authority on European rock art, reproduced a set of figures from the Paleolithic cave of Calapata (Teruel) showing the evolution of the portrayal of a stag.[39] The figures ranged from lively iconic representations to a stylized motif looking like a coarse comb with missing or deformed teeth. Such stylization is also the basis of alphabetic characters.[40] The point is, of course, that "the better an act is understood . . . the more formal and cursory may be the movement that represents it. . . . It becomes an act of *reference* rather than of representation."[41]

Faced with such problems, and in the absence of a title, key, or known context such as can identify a modern map, the historian of cartography has to develop a way of identifying prehistoric maps. Hitherto such prehistoric maps as have entered the literature have been identified by spontaneous recognition ("it looks like a map"). But this is a highly conditioned, optical reaction based on experience of maps from the historical period. In dealing with the enigmatic images and signs of the

rock and mobiliary art of the prehistoric period, it is necessary to construct first principles by asking, What is a "map"? What are the essential visual characteristics of a cartographic image that distinguish it from other

32. Hallpike, *Foundations*, 149–52 (note 24), gives the color white as an example of a symbol with both cross-cultural equivalences and also contradictory meanings. It almost universally symbolizes purity and goodness, but it can also mean disease, destruction, and punishment.

33. Joseph Needham suggests that the Chinese character for a mountain "was once an actual drawing of a mountain with three peaks," while that for fields shows "enclosed and divided spaces": *Science and Civilisation in China* (Cambridge: Cambridge University Press, 1954–), vol. 3, *Mathematics and the Sciences of the Heavens and the Earth*, 497. See also Ulrich Freitag, "Peuples sans cartes," in *Cartes et figures de la terre*, exhibition catalog (Paris: Centre Georges Pompidou, 1980), 61–63. Small visual differences may substantially alter the meaning: in the domestic decoration of the Mesakin of Nuba (Sudan) a row of colored triangles signifies mountains; uncolored, it means female breasts; and two lines enclosing the row make it a nonrepresentational design: Ian Hodder, *Symbols in Action: Ethnoarchaeological Studies of Material Culture* (Cambridge: Cambridge University Press, 1982), 171. These difficulties of interpretation are no doubt what Blakemore had in mind when writing of the "insularity of symbology" in "Way-finding," 20 (note 15).

34. Hallpike, *Foundations*, 167 (note 24). See also Roger William Brown, *Words and Things* (New York: Free Press, 1958), 59–60. Brown suggests that something as familiar as a stick figure ⅄ is a *learned* sign; a child's natural tendency is to draw circular forms, representing the rounded, fleshed body, not the invisible skeleton represented by this type of sign.

35. Brown, *Words and Things*, 59 (note 34).

36. François de Dainville, *Le langage des géographes* (Paris: A. et J. Picard, 1964), 324.

37. Michael J. Blakemore and J. B. Harley, *Concepts in the History of Cartography: A Review and Perspective*, Monograph 26, *Cartographica* 17, no. 4 (1980): esp. 76–86. J. B. Harley, "Meaning and Ambiguity in Tudor Cartography," in *English Map-making, 1500–1650*, ed. Sarah Tyacke (London: British Library, 1983), 22–45.

38. For example, there is little to distinguish the image of a stylized animal shown in comblike profile (see note 39 below) from that of a sheep pen drawn in plan, rather more carefully done but similarly stylized. The latter can be found, for instance, on maps drawn by Antonio di Michele for the Dogana della Mene delle Pecore (a grazier institution) in 1687 (Archivio di Stato, Foggia), one of which is reproduced in Delano Smith, *Western Mediterranean*, 247, pl. 10 (note 25). Essential reading on the problems associated with style is contained in many papers in Peter J. Ucko, ed., *Form in Indigenous Art: Schematisation in the Art of Aboriginal Australia and Prehistoric Europe*, Australian Institute of Aboriginal Studies, Prehistory and Material Culture Series no. 13 (London: Gerald Duckworth, 1977). See also Jan B. Deregowski, *Distortion in Art: The Eye and the Mind* (London: Routledge and Kegan Paul, 1984), and, for a critique of Deregowski's earlier writings, Robert Layton, "Naturalism and Cultural Relativity in Art," in *Indigenous Art*, 34–43 (above).

39. Henri Breuil, "The Palaeolithic Age," in *Larousse Encyclopedia of Prehistoric and Ancient Art*, ed. René Huyghe (London: Paul Hamlyn, 1962), 30–39, esp. 37.

40. S. H. Hooke, "Recording and Writing," in *A History of Technology*, ed. Charles Singer et al., 7 vols. (Oxford: Clarendon Press, 1954–78), vol. 1, *From Early Times to Fall of Ancient Empires*, 744–73.

41. Susanne K. Langer, *Philosophy in a New Key: A Study in the Symbolism of Reason, Rite, and Art*, 3d ed. (Cambridge: Harvard University Press, 1957), 156, her italics.

motifs, ensuring its recognition even where other diagnostics, such as the key or known context, are missing? At some stage, we must also answer the question, What were such maps for? Modern preconceptions about the function of maps, biasing our interpretation of their content or appearance, have to be set aside.[42]

What appears to be spontaneous recognition of a map in fact involves three assumptions: that the artist's intent was indeed to portray the relationship of objects in space; that all the constituent images are contemporaneous in execution; and that they are cartographically appropriate. In the context of prehistoric art, it is difficult to prove that all three conditions are met. The first has to be taken largely for granted, although it is the most basic, once the contemporaneity of the constituent images is assured. Thus, to use an early historical example as a model, the gesticulating stick men and their animals on the Rajum Hani' stone (fig. 4.3) are assumed to have been intentionally placed inside the enclosure, a point confirmed in this case by the accompanying inscription.[43] The demonstration of the second condition, that of contemporaneity, is closely associated with the first and is a vital step in the interpretation of a prehistoric map. Assemblages of images in prehistoric rock art in Europe are outstandingly disordered, lacking any suggestion of deliberate composition.[44] Images are commonly found superimposed,[45] drawn at all angles, or even upside down, and only very exceptionally is there a frame other than the natural edge of the stone or the undecorated portion of the cliff face. It is thus usually difficult to be convinced that the rock art assemblage was originally both intended and executed as an entire composition and that it has not survived merely as a palimpsest or as the result of accidental juxtaposition of individual images that could have been executed at long intervals. For maps drawn in plan, a way out of this problem can be suggested: only where it is reasonably clear that the engraved or painted lines connect neatly with each other, are neither superimposed nor isolated, and are identical in technique and style, should it be assumed that a composition was intended and that the individual images are constituents of a larger whole and are contemporaneous. For picture maps, the only check available is that of stylistic and technical similarity.

The third condition, the cartographic appropriateness of each constituent image of a prehistoric map, presents a different order of problem. A modern topographic map is composed largely of familiar signs, the meaning of which is reinforced by the accompanying key or has been made clear by an alternative form of explanation. Otherwise there would be no way of being certain about the meaning of a sign: any image can be used to stand for any object. It is usual—and sensible—to maintain some degree of correspondence between the image selected

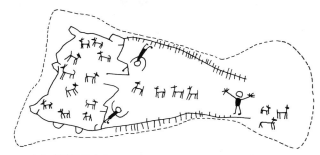

FIG. 4.3. THE RAJUM HANI' STONE. The interpretation of this Jordanian inscription from the early centuries A.D. as a part-plan, part-profile representation of a livestock enclosure is confirmed by the text on the reverse of the stone.
Size of the original: 50 × 95 cm. After G. Lankester Harding, "The Cairn of Hani'," *Annual of the Department of Antiquities of Jordan* 2 (1953): 8–56, fig. 5, no. 73.

42. These problems, of identification and of function of prehistoric maps, are also aired in Catherine Delano Smith, "The Origins of Cartography, an Archaeological Problem: Maps in Prehistoric Rock Art," in *Papers in Italian Archaeology IV*, pt. 2, *Prehistory*, ed. Caroline Malone and Simon Stoddart, British Archaeological Reports, International Series 244 (Oxford: British Archaeological Reports, 1985), 205–19, and her "Archaeology and Maps in Prehistoric Art: The Way Forward?" *Bollettino del Centro Camuno di Studi Preistorici* 23 (1986): forthcoming.

43. The Rajum Hani' stone serves well as a model, since the carved representation is accompanied by an explanatory text. The text reads: "By Mani'at, and he built for Hani'. And he drew a picture of the pen [or, enclosure] and the animals pasturing by themselves": G. Lankester Harding, "The Cairn of Hani'," *Annual of the Department of Antiquities of Jordan* 2 (1953): 8–56, and plates; Harding's translation (with brackets), p. 30. The vertical strokes along the two extended arms of the enclosure have been interpreted as its having been pallisaded and "presumably . . . made of branches of the desert trees and shrubs": "Desert Kites," *Antiquity* 28 (1954): 165–67, quotation on 165. Emmanuel Anati, *L'arte rupestre del Negev e del Sinai* (Milan: Jaca Book, 1979), has an aerial photograph of a livestock enclosure on p. 12 and a reproduction of the Wadi Ramliyeh engraving of similar design but prehistoric in date on p. 57.

44. See note 11 above.

45. There seem to be two quite different situations relating to the superimposition of figures in rock art. On the one hand, the preexisting figure is seen either as destroyed by or irrelevant to the addition of a new one. For example: "This haphazard placing of motifs leads one to deduce that the motifs were inherently significant for the carvers but that their placing on the stone or their relation to one another was unimportant": Elizabeth Shee, "Recent Work on Irish Passage Graves Art," *Bollettino del Centro Camuno di Studi Preistorici* 8 (1972): 199–224, quotation on 218. North American Indian parallels show that individual sites were associated with specific ideas or needs and that the markings made there were part of the current "ritual," prayer, or wish (e.g., for pregnancy), no notice being taken of previous marks: Dale W. Ritter and Eric W. Ritter, "Medicine Men and Spirit Animals in Rock Art of Western North America," in *Acts of the International Symposium on Rock Art: Lectures at Hankø 6–12 August, 1972*, ed. Sverre Marstrander (Oslo: Universitetsforlaget, 1978), 97–125. On the other hand, there may be situations where "superimpositioning was . . . a deliberate way of linking paintings according to certain conventions": Lewis-Williams, *Rock Art*, 40–41, 55, 61 (note 11).

and the object it is intended to represent or symbolize (partly as insurance against forgetting its meaning). So it is reasonable to assume, in the case of prehistoric art, that the naturalistic figures such as those for animals and houses are iconic or pictorial representations, at least at the first level of meaning.[46] Those most likely to be commonplace on a topographic map (a house, for instance, rather than a weapon) can be selected from those less likely to have cartographic significance. Another guideline is the frequency of occurrence, within a single composition, of the individual images. Examination of a modern map shows that it is composed of a range of images, most, if not all, of which occur frequently. This should also be the case with the prehistoric map.

By applying the three diagnostic criteria together—composition, appropriateness of the images, and their frequency within the composition—to prehistoric rock art, some headway has been made in identifying prehistoric maps portraying the landscape from above, as is demonstrated later in this chapter. However, those parts of the Old World examined in this volume have not as yet been found to be rich in examples meeting these criteria.

The arguments for regarding compositions such as those of Mont Bégo or Valcamonica as examples of plan topographic maps are beguiling (fig. 4.4). In the final analysis, however, the matter rests on the intention of the artist who so painstakingly hammered the hard rock surfaces into a complex association of signs and symbols but left no key. An acceptably complete substitute for a key, for at least one of these maps, has yet to be found.

The second type of map—the picture map—is common in prehistoric art. It is characterized by having some images in plan and some in elevation or profile. But while some of the constituent images represent relatively permanent landscape features (mountains, huts, or rivers, for example), others are anthropomorphs or animals. As a whole, such compositions appear to be scenarios in which the spatial layout and the landscape features are of secondary consequence to the event being depicted. This type of map has its counterpart in the historical period, in some of the earliest surviving fragments of classical cartography such as the Mycenean fresco at Thera,[47] the documents of the Roman *agrimensores*,[48] or the European battle plans of the sixteenth and seventeenth centuries. Looking at prehistoric rock and mobiliary art in Europe and its adjacent regions, it is seen that the idea of such picture maps, in which hybridization of plan and profile features is found, dates back to the Upper Paleolithic period. These protocartographic images are, so far, the earliest surviving graphics to reveal, unambiguously, thinking that is manifestly cartographic and a number of examples are listed in appendix 4.2.

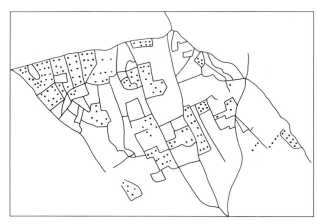

FIG. 4.4. ELEMENTS IN A MODERN TOPOGRAPHICAL MAP. This tracing of field boundaries, paths, roads, streams, and trees is employed to suggest that the visual characteristics of modern maps may be similar to those of comparable prehistoric maps. Shown here are the landscape features which it has been argued are also represented on the Bedolina rock (fig. 4.28), with which it should be compared.
After Istituto Geografico Militare, sheet 164 I NE, 1964, Manfredonia, Italy.

There is no doubt that by the beginning of the Upper Paleolithic man possessed both the cognitive capacity and the manipulative skills to translate mental spatial images into permanently visible images. It is possible to identify alternative modes of cartographic expression in the rock art record, ranging from the supermundane to the real world, for instance, and including perceptions of landscape from sometimes a low, sometimes a high, and occasionally, a vertical angle. An obvious suggestion is that such variations of topographical perspective are linked to terrain. It is tempting to argue that people living in mountainous regions, or in lowlands overlooked by hills, would have found it easier to depict the landscape from above, in plan, but there is as yet far too little evidence to advance this hypothesis.[49] A less contentious suggestion is that the degree of abstract or cognitive interpretation required is greater for depicting large

46. The three levels of meaning in the subject matter of works of art were first defined by Erwin Panofsky, *Studies in Iconology: Humanistic Themes in the Art of the Renaissance* (Oxford: Oxford University Press, 1939), 5–8, and have been applied to maps by J. B. Harley: see Blakemore and Harley, *Concepts*, 76–86 (note 37), and Harley, "Tudor Cartography" (note 37).
47. See p. 132 below.
48. James Nelson Carder, *Art Historical Problems of a Roman Land Surveying Manuscript: The Codex Arcerianus A, Wolfenbüttel* (New York: Garland Publishing, 1978); O. A. W. Dilke, *The Roman Land Surveyors: An Introduction to the Agrimensores* (Newton Abbot: David and Charles, 1971).
49. The concentration of late Neolithic and Bronze Age maps in plan in the southern Alps, for instance, may be no more than a temporary bias of discovery.

areas than it is for portraying small, local areas or for representing stellar patterns on stone or skin or in the sand.[50] Prehistoric maps of very large areas have not been found in the European evidence, despite the views of some antiquarian writers such as Taubner, who thought he discerned a map of the whole of southern Scotland and northern England on a stone at Aspatria (Cumberland).[51]

On the question of the function of topographical maps (as opposed to cosmological or celestial maps) from the prehistoric period, it can be affirmed only that their function would not have been identical to that of maps created in a modern society. By making critical and careful use of ethnographic material to illuminate a study of one of the world's richest storehouses of rock art (that of southern Africa), at least one archaeologist has been able to demonstrate what has long been generally accepted, namely that this art is "an astonishing expression" of those ideas which most seriously moved the primitive artists' minds and filled them with religious feeling.[52] Assuming some sort of link between the object depicted and the artist's intention, any maps or maplike representations found in prehistoric art should be seen as symbolizing abstract attributes, or emotions, associated with the landscape or landscape features portrayed. In this sense they could be relics of fossilized prayers rather than records of the existence or distribution of such landscape features. A review of the rock art evidence leads to the firm conclusion that while some, at least, of the maps of prehistoric and indigenous societies look exactly like those of so-called advanced societies, they would have served a quite different purpose. This point was missed by most antiquarian recorders of prehistoric maps. The discussion now turns to these early observers.

RECOGNITION OF TOPOGRAPHICAL MAPS BY ANTIQUARIANS IN EUROPEAN ROCK ART

The study of European rock art can be documented back to the seventeenth century, but its modern discovery dates mainly from the middle of the nineteenth century.[53] In the case of the Middle East and North Africa, it is even later; many of the most important discoveries in the Sahara, for instance, belong to the period after the First World War.[54] A more significant difference lies in the nature of its discovery. In Europe, the antiquarians of the nineteenth century began to build up a considerable literature on the incidence, form, and interpretation of the newly discovered prehistoric art, but there is no such antiquarian or indigenous literature in either the Middle East or North Africa. In these regions the discovery of rock art, together with that of the ancient settlements and prehistoric sites, was effected by outsiders, namely Europeans. This means that from the start the bulk of reporting and comment in these regions came from the pens of visiting scholars or professional travelers.[55]

The new popular "scientific" climate of nineteenth-century Europe had the effect of directing attention to many aspects of the environment. Not only naturalists

50. Indeed, some historians of cartography deny that simply copying the distribution of stars in a night sky, "without reference to geographic mapping," constitutes cartography at all: for instance, P. D. A. Harvey in "Cartographic Commentary," *Cartographica* 19, no. 1 (1982): 67–69, quotation on 68; see pp. 84–85.

51. Taubner, "Landkartenstein-Theorie" (note 3). No prehistoric maps of very large areas have been identified with certainty in the Old World.

52. Lewis-Williams, *Rock Art*, 66 (note 11).

53. The first publication of rock art seems to have come from Scandinavia. In 1627 a schoolteacher from Kristiania sent his copies of carvings to Ole Worm, but they were not published until 1784, by Peter Frederik Suhm, *Samlinger til den Danske historie* (Copenhagen: A. H. Godishes, 1779–84), vol. 2, no. 3: 215–16, and folded illustration. See P. V. Glob, *Helleristninger i Danmark* (Rock carvings in Denmark), Jysk Arkaeologisk Selskabs Skrifter, vol. 7 (Copenhagen: Gyldendal, 1969), 286 (English summary). Another early commentator on "hieroglyphs," as he called them, was Dimitrie Cantemir (born 1673), prince of Moldavia. Some examples of his notes and sketches were published as fragments from his collected writings, *Operele principelui Demetriu Cantemiru*, 8 vols. (Bucharest, 1872–1901), vol. 7, app. 3. Although one of his sketches is of an early site, none of the petroglyphs appears anything but historical in date. I owe this reference to Dennis Reinhartz (University of Texas at Arlington).

54. For instance, although F. Fourneau had reported in 1894 on the existence of carvings in the Tassili, and Chudeau had shown in 1905 that many carvings could be found in just one small locality, it was 1933 before the Sahara was revealed to be as rich in rock paintings as it was already known to be in rock carvings. Letter from F. Fourneau, *Comptes Rendus des Séances de l'Académie des Inscriptions et Belles-Lettres*, 4th ser., 22 (1894): 98–99. Letters to E. T. Hamy, "Exploration de M. R. Chudeau dans le Sahara," *Géographie: Bulletin de la Société de Géographie* 13 (1906): 304–8. Chudeau was to find over five hundred engravings within a two-kilometer stretch near Ahaygar. Henri Lhote refers to Lieutenant Brenan's police operation in the Oued Djeret in 1933, which initiated the discovery and study of the paintings: *The Search for the Tassili Frescoes*, trans. Alan Houghton Brodrick (London: Hutchinson, 1959), 10. Much of Saharan Africa was explored late. In 1923 and 1927, Douglas Newbold searched for rock carvings in the Libyan desert in areas not previously visited by Europeans or, in some cases, even by the Arabs: "Rock-Pictures and Archaeology in the Libyan Desert," *Antiquity* 2, no. 7 (1928): 261–91.

55. One of the earliest reports of petroglyphs in North Africa came from the explorer Heinrich Barth, who had started his African travels in 1850 from Tripoli. His sketches of engraved animal figures and figures of humans with animal heads from a "desolate valley" in the Fezzan are now lost, but the area was identified and the pictures recopied nearly a century later: Leo Frobenius and Douglas C. Fox, *Prehistoric Rock Pictures in Europe and Africa* (New York: Museum of Modern Art, 1937), 38–41. Other early mentions came from archaeologists, but they were too excited by the prospects of excavation at the great sites to pay any attention to rock art in the deserts, except for Hans Alexander Winkler, *Rock Drawings of Southern Upper Egypt*, 2 vols., Egyptian Exploration Society (London: Oxford University Press, 1938).

and archaeologists but also medical men, clerics, and classicists began to notice and to discuss the meanings of the artificial markings they noticed on certain rock surfaces. For some the markings had cartographic meaning. For example, rocks decorated with cup-and-ring marks found at Staigue Fort and in other parts of Kerry (Ireland) were examined in 1851 by the Very Reverend Charles Graves, president of the Royal Irish Academy, and in 1852 by the Reverend William Greenwell. Heartened by the correlation he thought he noticed between the distribution of the marks on the rocks and the forts on the ground, Graves made his discoveries public only in 1860 and maintained his original conjecture that these carvings were primitive maps, representing the disposition of the neighboring forts.[56] Greenwell appears to have suggested that the cup-and-ring marks on the newly discovered rocks near Rowtin Lynn (Routing Linn, Old Bewick, Northumberland) were plans of the forts themselves, showing their multiple ramparts, scatter of huts within, and single entrance with trackway (fig. 4.5).[57] However, in his presidential address to the Berwickshire Naturalists' Club in July 1853, George Tate disagreed with such interpretations of cup-and-ring marks, pointing out that "their wide distribution, and, notwithstanding differences in detail, their family resemblance, prove that they had a common origin, and indicate a symbolical meaning."[58] The temptation to match local carvings to local features, however, has proved difficult to resist. Even today there remains both in Britain and in Europe a stratum of popular interpretation characterized by its imaginativeness, overfancifulness, and total lack of reference to the wider academic issues involved.

On the European mainland, the situation was very much the same. In Germany, in the last decades of the century, the august Anthropological Society of Berlin was the forum for a spate of incautious enthusiasm concerning rock markings from all over the world and their possible cartographic meaning. Because of the wide circulation of the society's proceedings, (*Zeitschrift für Ethnologie*), some of these contributions received greater attention and a more extensive diffusion than they seem to have merited, at least today. The Russian historian of cartography Bruno Adler was certainly a close follower of the *Zeitschrift*, and it would have been in this way that he encountered the views of the irrepressible but wholly unscientific Fritz Rödiger and of Kurt Taubner. Rödiger, an agriculturalist from Solothurn (Switzerland), was attracted by the patterns he thought he could see in the partly artificial and partly natural markings on cliff faces and on newly excavated prehistoric artifacts in Germany and in Switzerland. By matching these markings with such modern topographic maps as he had in hand, he convinced at least himself of the astounding skill of prehistoric cartographers who mapped trade routes, settlements, major natural features, and even property boundaries.[59] Rödiger's imagination was prolific, but both Adler and Bagrow confined their remarks to the engraved patterns on two shaped bone fragments recovered from the Kesslerloch cave at Thayngen (near Schaffhausen) in Switzerland and on a similarly shaped piece of lignite from the same exca-

56. Charles Graves, "On a Previously Undescribed Class of Monuments," *Transactions of the Royal Irish Academy* 24, pt. 8 (1867): 421–31. Graves discounted any astronomical significance in view of the absence of recognizable signs for key elements such as the sun or the moon (p. 429). The terms "cups," "cup-and-ring marks," "cups and rings" refer to a range of circular and concentric sculpted figures that are probably the most common form of prehistoric petroglyphs throughout the world (for example, see fig. 4.5). They are made up of a basic vocabulary of four key motifs, according to Ronald W. B. Morris, "The Prehistoric Petroglyphs of Scotland," *Bollettino del Centro Camuno di Studi Preistorici* 10 (1973): 159–68, esp. 159. In southern Scotland, 535 sites have cup marks only; 295 sites have cups and rings; 29 sites have rings and grooves; and 15 sites have rings or spirals (p. 161). James Young Simpson (Queen Victoria's physician) made an admirably objective analysis of the various forms: *Archaic Sculpturings of Cups, Circles, etc. upon Stones and Rocks in Scotland, England, and Other Countries* (Edinburgh: Edmonston and Douglas, 1867). They dominate British rock art too. In general terms, British rock art is thought to date from the early Bronze Age (before about 2000 B.C.): see Colin Burgess, *The Age of Stonehenge* (London: J. M. Dent, 1981), 347. Cup-and-ring marks in Sweden are also described by Arthur G. Nordén, *Östergötlands Bronsålder* (Linköping: Henric Carlssons Bokhandels Förlag, 1925), 155.

57. William Greenwell's paper to the Newcastle meeting of the Archaeological Institute in July 1852 was excluded from the "two ponderous volumes professing to be a record of its proceedings": George Tate, *The Ancient British Sculptured Rocks of Northumberland and the Eastern Borders, with Notices of the Remains Associated with These Sculptures* (Alnwick: H. H. Blair, 1865), 3–4. Apparently the paper was lost: Simpson, *Archaic Sculpturings*, 52 (note 56). By 1859 J. Gardner Wilkinson, vice-president of the British Archaeological Association, had retracted his first opinion that neither the cups and rings he himself had seen at Penrith (Cumberland) and on Dartmoor nor those on the Rowtin Lynn stone "related to the circular camps, and certain dispositions connected with them": J. Gardner Wilkinson, "The Rock-Basins of Dartmoor, and Some British Remains in England," *Journal of the British Archaeological Association* 16 (1860): 101–32, quotation on 119.

58. George Tate, address to members at the anniversary meeting held at Embleton, 7 September 1853, *Proceedings of the Berwickshire Naturalists' Club* 3, no. 4 (1854): 125–41, esp. 130. As Evan Hadingham points out, neither Graves nor Greenwell (nor Tate, it should be added) was to know that over two thousand years separated the builders of the forts from the carvers of the rocks: Evan Hadingham, *Ancient Carvings in Britain: A Mystery* (London: Garnstone Press, 1974), 43–44; idem, *Circles and Standing Stones: An Illustrated Exploration of Megalith Mysteries of Early Britain* (Garden City, N.Y.: Anchor Press/Doubleday, 1975), 136–37.

59. Fritz Rödiger, "Vorgeschichtliche Zeichensteine, als Marchsteine, Meilenzeiger (Leuksteine), Wegweiser (Waranden), Pläne und Landkarten," *Zeitschrift für Ethnologie* 22 (1890): Verhandlungen 504–16; idem, "Kartenzeichnungen," 237–42 (note 2); idem, "Erläuterungen und beweisende Vergleiche zur Steinkarten-Theorie," *Zeitschrift für Ethnologie* 23 (1891): Verhandlungen 719–24.

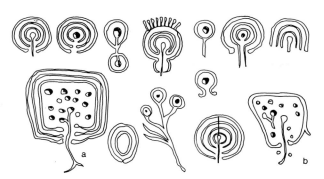

FIG. 4.5. CUP-AND-RING MARKS FROM NORTHUMBERLAND. Figures such as these have been extensively cited in the antiquarian literature as having a cartographic purpose. Two markings, *a* and *b*, were initially seen as being plans of neighboring forts, even though the shape of the supposed camps did not correspond to the rock markings; nor were the markings of the same period as the earthworks.
After George Tate, *The Ancient British Sculptured Rocks of Northumberland and the Eastern Borders, with Notices of the Remains Associated with These Sculptures* (Alnwick: H. H. Blair, 1865), 7.

FIG. 4.6. KESSLERLOCH BONE PLAQUES. Rödiger's drawing of the two decorated bone plaques from Kesslerloch cave, Switzerland (*a* and *b*); *c* shows his interpretation of *a* as a map of the surrounding district, redrawn with named localities.
After Fritz Rödiger, "Vorgeschichtliche Kartenzeichnungen in der Schweiz," *Zeitschrift für Ethnologie* 23 (1891): Verhandlungen 237–42, figs. 8, 6, and 9 respectively.

vation (figs. 4.6 and 4.7).[60] Adler was not entirely unsympathetic to Rödiger's suggestions that each of these represented a prehistoric map of the local area, but Bagrow was skeptical from the outset.[61] Neither discussed

FIG. 4.7. KESSLERLOCH LIGNITE ORNAMENT. The lignite ornament from Kesslerloch cave, worked in the same manner as the two bone plaques in figure 4.6. For Rödiger's interpretation of this as a topographical map of the area between Lake Constance and Schaffhausen, see "Vorgeschichtliche Zeichensteine, als Marchsteine, Meilenzeiger (Leuksteine), Wegweiser (Waranden), Pläne und Landkarten," *Zeitschrift für Ethnologie* 22 (1890): Verhandlungen 504–16.
From Conrad Merk, *Excavations at the Kesslerloch Near Thayngen, Switzerland, a Cave of the Reindeer Period*, trans. John Edward Lee (London: Longmans, Green, 1876), pl. IX, no. 50.

60. The excavator, a local schoolmaster with an established archaeological interest, made no interpretative comments on these designs: Conrad Merk, *Excavations at the Kesslerloch Near Thayngen, Switzerland, a Cave of the Reindeer Period*, trans. John Edward Lee (London: Longmans, Green, 1876). Neither of the items in question was among those later discovered to have been faked: see Merk, *Excavations*, in the preface by Lee (above); see also Robert Munro, *Archaeology and False Antiquities* (London: Methuen, 1905), 55–56. It may be of interest to compare these with the pierced and decorated tablets from Tartaria (Romania), which possibly date from as early as the fifth millennium B.C. See Sarunas Milisauskas, *European Prehistory* (London: Academic Press, 1978), 129–31.

61. Adler agreed that there is a resemblance between the patterns on the Schaffhausen (Kesslerloch) artifacts and the appearance of a map but suggested that this was accidental. On the other hand, he did not dismiss the idea that "primitive man, with his acute powers of observation, hearing and smell which helped him orientate himself and with his proven ability to draw on rock and bone" would have been capable of such an exercise: Adler, "Karty," 218 (note 2), translation by John P. Cole (University of Nottingham). After two of Rödiger's contributions and Taubner's, the president of the Anthropological Society, R. L. C. Virchow, was driven to advising that the study of rock and stone drawings "offers the imagination such easy opportunities, that it is a little difficult to allow for the supposition that those drawings should everywhere have a topographical significance . . . it is with pictures of people as it is with clouds, that a stimulated imagination can see therein all sorts of animal and human shapes.

Taubner's interpretations, though each cited his paper.[62] Taubner, admitting that he had been influenced by A. Ernst,[63] declared that cup marks were topographical representations and that a double circle could represent isolated humps. He went on to describe the Bunsoh stone (Holstein) as a topographical representation of the local area, a suggestion that is not without its proponents even today.[64] Taubner also introduced the idea that stone maps could represent not just the immediate vicinity but much larger regions. By matching the distribution of the cup-and-ring marks and divided circles on the side stone of a cist grave at Aspatria (Cumberland) with a map of Britain taken from a school atlas, he interpreted the pattern as a map of northern England and southern Scotland, complete with settlements such as Carlisle.

One of the fundamental weaknesses of such antiquarian interpretations is the unsystematic approach and lack of discussion of the whole archaeological context and other related points. The underlying assumption is that it is sufficient to look for a simple match between the pattern on the rocks and one in the landscape without questioning such matters as contemporaneity, scale, or appropriate geometry. What fits is included; what does not fit is conveniently disregarded, and the vital fact that prehistoric, like indigenous, maps could only have been constructed according to principles of topological geometry (not Euclidean) remains unappreciated.

A notable exception to such weaknesses was the work of a most remarkable Englishman, Clarence M. Bicknell. Bicknell, born at Herne in Kent and a clergyman in the East End of London before renouncing holy orders, moved to the Italian Riviera for health reasons.[65] He spent his time there botanizing and sketching. Exploring the Maritime Alps inland from Bordighera, he came across the rock carvings below the peak of Mont Bégo (in those days on the Italian side of the frontier) and eventually devoted twelve summers from the end of the century to his death in 1918 to discovering, copying, and commenting on some fourteen thousand individual carved figures—seven thousand from Val Fontanalba and most of the rest from Val Meraviglie. Bicknell's intellectual strength lay in his taxonomic approach, and he classified all these figures into eight subject classes:

1. Horned figures
2. Ploughs
3. Weapons and instruments
4. Men
5. Huts and properties
6. Skins
7. Geometrical forms
8. Miscellaneous indeterminable forms.[66]

It is the fifth group (huts and properties) that Bicknell referred to as maps or "topographical figures" in his

writings. His texts, published from 1897 onward, remain the standard works for the region.[67] There have been additional discoveries, bringing the total number of figures to an estimated one hundred thousand,[68] and

---

May this warning not go unheeded! But may it not be so received as to discourage any further investigation": *Zeitschrift für Ethnologie* 23 (1891): Verhandlungen 258. Bagrow wrote in 1917: "There are grounds for suggesting that prehistoric man was already attempting to represent a locality known to him in order to help a departing traveler orientate himself in unknown territory. Among the finds from the Schaffhausen cave were two bone plaques, covered with a network of marks in which Rödiger tried, by means of a comparison with [modern] maps of the given locality, to discern a map made by the ancient inhabitants. Some scholars see in rock art an attempt to give directions about a place, i.e., a prototype map, but all this remains an unclarified question and the cartography of prehistoric man remains in grave doubt": *Istoriya*, 2 (note 4), translation by John P. Cole (University of Nottingham).

62. Taubner, "Landkartenstein-Theorie" (note 3). In his text Adler refers to Westedt, "who drew attention to the presence of similar [petroglyphic] elements on a stone in Holstein": "Karty," 218 (note 2), translation by John P. Cole (University of Nottingham). But in Westedt's article, "Steinkammer," 247–49 (note 3), there are no interpretations for the markings, let alone the suggestion that they form a map. Taubner, on the other hand, does see this Bunsoh stone as a map, and it looks as though Adler was mistaken in his reference to Westedt instead of Taubner.

63. A. Ernst, "Petroglyphen aus Venezuela," *Zeitschrift für Ethnologie* 21 (1889): Verhandlungen 650–55.

64. Paul Volquart Molt, *Die ersten Karten auf Stein und Fels vor 4000 Jahren in Schleswig-Holstein und Niedersachsen* (Lübeck: Weiland, 1979), 43–92. But see note 121 below.

65. An outline of Bicknell's life and the context of his work is given in Carlo Conti, *Corpus delle incisioni rupestri di Monte Bego: I*, Collezione di Monografie Preistoriche ed Archeologiche 6 (Bordighera: Istituto Internazionale di Studi Liguri, 1972), 6–8. Enzo Bernardini, *Le Alpi Marittime e le meraviglie del Monte Bego* (Genoa: SAGEP Editrice, 1979), 144.

66. Clarence M. Bicknell, *A Guide to the Prehistoric Rock Engravings in the Italian Maritime Alps* (Bordighera: G. Bessone, 1913), 39.

67. Clarence M. Bicknell's first paper was "Le figure incise sulle rocce di Val Fontanalba," *Atti della Società Ligustica di Scienze Naturali e Geografiche* 8 (1897): 391–411, pls. XI–XIII. His major works, besides *Guide* (note 66), were *The Prehistoric Rock Engravings in the Italian Maritime Alps* (Bordighera: P. Gibelli, 1902) and *Further Explorations in the Regions of the Prehistoric Rock Engravings in the Italian Maritime Alps* (Bordighera: P. Gibelli, 1903). For a complete list of his writings see Henry de Lumley, Marie-Elisabeth Fonvielle, and Jean Abelanet, "Vallée des Merveilles," *Union International des Sciences Préhistoriques et Protohistoriques, IXe Congrès, Nice 1976*, Livret-Guide de l'Excursion C1 (Nice: University of Nice), 178. The originals of Bicknell's tracings and notes are now in the University of Genoa (Institute of Geology).

68. Henry de Lumley, Marie-Elisabeth Fonvielle, and Jean Abelanet, "Les gravures rupestres de l'Âge du Bronze dans la région du Mont Bégo (Tende, Alpes-Maritimes)," in *Les civilisations néolithiques et protohistoriques de la France: La préhistoire française*, ed. Jean Guiliane (Paris: Centre National de la Recherche Scientifique, 1976), 2:222–36, esp. 223. Bernardini, *Alpi*, 127 (note 65), says that about 250,000 rock carvings are known.

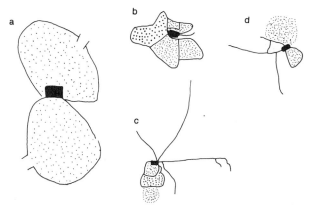

FIG. 4.8. "TOPOGRAPHICAL FIGURES" FROM MONT BÉGO. Bicknell identified these as representations of "huts and properties" or "huts with enclosures" as seen from above. After Clarence M. Bicknell, *Further Explorations in the Regions of the Prehistoric Rock Engravings in the Italian Maritime Alps* (Bordighera: P. Gibelli, 1903), pl. I-13 (*a*), and Clarence M. Bicknell, *A Guide to the Prehistoric Rock Engravings in the Italian Maritime Alps* (Bordighera: G. Bessone, 1913), pls. XVIII-43, XXXIV-12, and XXXII-41 (*b–d*, respectively).

some alternative classifications,[69] but there has been nothing so far to match Bicknell's balanced and systematic studies, nor has there been a definitive analysis of any one of the categories. Most ignored of all have been the so-called topographical figures (in Bicknell's group 5), which have been either misunderstood[70] or simply omitted from discussion in recent literature.[71]

There is little excuse for bypassing Bicknell's remarkably homogeneous category "huts and properties" or "huts with enclosures" (fig. 4.8). The key to his interpretation was simple empiricism. On his many journeys up and down the valleys to Mont Bégo, he repeatedly observed the striking likeness of the carved combinations of solid rectangles, subcircular forms, pecked surfaces, and irregularly interconnecting lines to features in the landscape when these are viewed from above—seen in plan, that is, from a vantage point high up the mountainside. Thus he interpreted the "rectangular figure with semicircle or other sort of closed line joining it" as "signifying huts or sheds with a piece of ground enclosed by a wall"[72] and the interconnecting lines as paths. He also suggested that the variety of enclosures containing stippling, made by hammering with a single blow or repeatedly, arranged with obvious regularity or randomly or left blank, could imply different categories of land use (see fig. 4.20, for example). He concluded, cautiously, that about 194 of the groups of rock-carved figures in the Fontanalba valley and another 15 in Val Meraviglie might be representations of either a hut with a path or huts with enclosed plots.[73]

Not all modern archaeologists are willing to accept Bicknell's interpretation of the "topographical figures." A common objection is that many of these appear to have been "distorted" to fit the outlines of the rocks they were carved on and thus could not be "accurate" representations of some real layout. But this is to ignore the key property of topology, which is the preservation

69. Conti, *Corpus*, 29–32 (note 65), has twelve classes:
1. Human figures of religious nature.
2. Superhuman figures or figures in sacred apparel.
3. Farmers.
4. Fighters or men making sacrifices.
5. Figures of sinister appearance.
6. Horned figures.
7. Weapons.
8. Plows, harrows, and sickles.
9. Recticular figures.
10. Ideographic representations.
11. Signs probably indicating numbers and primitive alphabetic characters.
12. Figures of unknown significance.

70. Recent objections have been on the grounds that they are unlikely to have been topographical figures because there are not, nor can there ever have been, cultivation and permanent settlement at these altitudes (2,000–2,750 m above sea level): see André Blain and Yves Paquier, "Les gravures rupestres de la Vallée des Merveilles," *Bollettino del Centro Camuno di Studi Preistorici* 13–14 (1976): 109–19, and Bernardini, *Alpi*, 171 (note 65), who talks similarly in terms of the *"vocazione pastorale"* of the land. Such objections are irrelevant; drawings are not necessarily made with the subject in sight, nor did Bicknell suggest that there ever had been cultivation at these altitudes. On the contrary, he stressed that "it was not among the wilderness of glaciated rocks or boulders at an elevation of 2,100 m and more that they ploughed. There the land has never been cultivated. . . . But years ago, Val Casterino and the lower parts of the Miniera valley may well have been tilled as they are now, and terraces long since abandoned are still to be discerned far up the steep mountain sides. Here . . . people who stood on the terraces might have looked down at the ploughing in the flat land of the valley, or on other terraces beneath them, and seen the operation from above as it seems to be depicted on the rocks of the higher regions": Bicknell, *Prehistoric Rock Engravings*, 38–39 (note 67). Blain and Paquier seem confused (p. 109) over the distinction between rural settlement types (isolated steading, hamlet, village, etc.) and the social and economic structure or organization associated with each type. A topographical map by definition depicts only the former, the formal aspects of the landscape. Recent mining, as well as grazing, has been responsible for much deforestation. Though there are still some larches in Val Fontanalba, in the seventeenth century Pietro Gioffredo reported thick larch forests: *Corografia delle Alpi Marittime*, 2 books (1824); republished with his *Storia delle Alpi Marittime* in *Monumenta historia patriae*, vol. 3, *Scriptorium I* (Genoa: Augustae Taurinorum, 1840), 47. A. Issel, "Le rupi scolpite nelle alte valli delle Alpi Marittime," *Bollettino di Paletnologia Italiana* 17 (1901): 217–59, simply disagrees with Bicknell's interpretation, holding instead that the so-called topographical figures are not plans but "conventional signs of individuals or tribes."

71. See, for example, de Lumley, Fonvielle, and Abelanet, "Merveilles" (note 67); idem, "Gravures rupestres" (note 68); and Conti, *Corpus* (note 65).

72. Bicknell, *Guide*, 53 (note 66).

73. Bicknell, *Guide*, 53, 56 (note 66).

of contiguity but not shape, and to assess the prehistoric figures according to the then unformulated principles of Euclidean geometry (which stress the properties of distance, direction, and angle that preserve shape and underlie the modern concept of scale). Many of Bicknell's suggested topographical figures do in fact satisfy the cartographic criteria presented here for ichnographic or plan maps, and for this reason (and in the total absence of realistic alternative interpretations) these have been included in the list in appendix 4.1.

The notion that the prehistoric rock artists may have been making graphic representations of parts of the earth's surface is not the only cartographic suggestion to have been made in the nineteenth century and preserved in the antiquarian literature. The apparently randomly distributed cup marks on natural surfaces or on prehistoric monoliths were seen by some observers as representations of the major constellations, while others raised issues of cosmological import. These views are discussed in the sections dealing with celestial and cosmological maps later in this chapter. It must be stressed, however, that of all the theories from the early literature put forward to explain the purpose or original meaning of the rock art figures and motifs, those relating to maps represent but a tiny proportion. Out of no fewer than the 104 such explanations recently amassed by Ronald Morris for the British Isles, all of which "have been put forward in all seriousness from time to time by archaeologists and others," only seven concern maps or plans in any way.[74] Moreover, most of these relate to cup-and-ring markings, probably the most ambiguous of all rock art motifs.

The Classification of Prehistoric Maps from Europe, the Middle East, and North Africa in the Prehistoric Period

The prehistoric material considered here as of cartographic interest has come from a variety of sources. The antiquarian literature, apart from Bicknell's writings, has yielded little that is worth further examination. A number of references to examples of prehistoric art that have already been interpreted as maps can be gleaned from modern archaeological literature, however. Other examples have been described as landscape representations, and these too are part of the history of mapping. The total corpus is thus derived almost wholly from published sources. The examples are discussed under three main headings: topographical maps, celestial maps, and cosmological maps. Nothing has been found that convincingly suggests representations of the sea. The topographical examples, however, fall into two basic categories, picture maps and plan maps; the latter

are further subdivided into simple maps, complex maps, and maps in relief.

TOPOGRAPHICAL MAPS
Picture Maps and Their Antecedents

Four picture maps—as already defined—have been identified in Old World rock and mobiliary art.[75] But just as interesting in the history of cartographic ideas are a number of pictures or small compositions that contain certain landscape features depicted in plan. Some of these plan figures are very simple indeed. Probably the oldest are those from Iberian or French cave paintings, thought to date from the Upper Paleolithic. From the Los Buitres cave (Peñalsordo, Badajoz), for example, comes a composition consisting of a subcircular outline with an external fringe of rays and two sets of markings inside that could represent, in highly stylized form, anthropomorphic figures (fig. 4.9).[76] Other compositions, such as those from Font de Gaume in Dordogne, have similar outlines but lack the internal images, though there may be other markings. These have entered the literature as representations of a "delimited area (hut?)"[77] or "game enclosures" (fig. 4.10).[78] Similar

74. Ronald W. B. Morris, *The Prehistoric Rock Art of Galloway and the Isle of Man* (Poole: Blandford Press, 1979), 15–28. Summarizing these under the headings used in this essay, with Morris's reference number in parentheses, they are:

Topographical maps:	Maps of the countryside (58)
	Building plans (59)
	Plans for megalithic structures (83)
Celestial maps:	Star maps (60)
	Early astronomers' night memoranda (93)
Cosmological maps:	Plans for laying out mazes (84)
	Field plowing plans (85)

The last two both concern labyrinth designs, and since this sign has universal association with death and the afterlife, it has been classed here as cosmological. Morris ascribed each explanation what he calls a "plausible ranking." According to this, the explanations above are to be rejected out of hand, a conclusion with which we do not hesitate to agree. Only explanation 93, that night watchers might have found it useful to have a tactile reference plan of certain constellations handy for use in the dark, is given modest credence by Morris.

75. See p. 62 above for definition.

76. Figure 2 in appendix 4.2. Henri Breuil, *Les peintures rupestres schématiques de la Péninsule Ibérique*, 4 vols., Fondation Singer-Polignac (Paris: Imprimerie de Lagny, 1933), vol. 2, *Bassin du Guadiana*, 58–59 and fig. 16. Maria Ornella Acanfora, *Pittura dell'età preistorica* (Milan: Società Editrice Libraria, 1960), 263.

77. Figure 1 in appendix 4.2. Acanfora, *Pittura*, 262 (note 76). This, from Nuestra Señora del Castillo, Almadén, was first published by Breuil, *Bassin*, pl. VIII (note 76).

78. Two come from the cave of La Pileta, Malaga, and a third from Font de Gaume (Dordogne). All appear under this heading in Johannes Maringer, *The Gods of Prehistoric Man*, trans. Mary Ilford, 2d ed. (London: Weidenfeld and Nicolson, 1960), 95. See also Lya Dams, *L'art paléolithique de la caverne de la Pileta* (Graz: Akademische

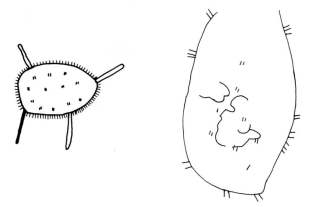

FIG. 4.9. PICTURE WITH POSSIBLE MAP ELEMENTS FROM PEÑALSORDO, BADAJOZ, SPAIN. This has been interpreted as a representation of two figures within a hut or enclosure. If this is the case, then both elements of a picture map (the features being shown both in plan and in profile) are present.
Size of the original: 12 × 10 cm. After Henri Breuil, *Les peintures rupestres schématiques de la Péninsule Ibérique*, 4 vols., Fondation Singer-Polignac (Paris: Imprimerie de Lagny, 1933), vol. 2, *Bassin du Guadiana*, fig. 16f.

FIG. 4.10. ROCK PAINTINGS FROM LA PILETA, MALAGA, SPAIN. Although described as "game enclosures," this is so tenuous an identification that they are excluded as examples of prehistoric maps, though they may suggest the use of a circle to depict in plan the enclosing element of a landscape feature such as a field or a hut.
Width of the originals: 40 cm each. After Henri Breuil, Hugo Obermaier, and W. Verner, *La Pileta à Benaojan (Malaga)* (Monaco, Impr. artistique Vᵛᵉ A. Chêne, 1915). The figure on the left, described by Breuil, Obermaier, and Verner as "tortoise-like," is illustrated in pl. V and also in pl. X (23). That on the right, one of three similar ones, is shown in pl. X (27). The figure here is taken from Johannes Maringer, *The Gods of Prehistoric Man*, trans. Mary Ilford, 2d ed. (London: Weidenfeld and Nicolson, 1960), fig. 21. See also Lya Dams, *L'art paléolithique de la caverne de la Pileta* (Graz: Akademische Druck, 1978), fig. 91 (23-VI and 26-III).

compositions of Neolithic or later date are known from the central Sahara. Among the rock paintings in the Tassili Mountains of southern Algeria are a number that have been interpreted as hut scenes.[79] Each hut is represented in plan by a broad, more or less circular band. The human figures within and just outside are in various postures, and the intention of the artist seems to have been to use the plan outline of the hut as a device for permitting a simultaneous view of both exterior and interior activities (fig. 4.11). Probably much later in date is the example that shows a camellike profile in a circular outline (fig. 4.12).[80] Such examples seem to reflect some essential cartographic concepts, for example, by depicting some landscape features in plan and portraying all features in more or less correct spatial relationships.

There are some later compositions, a similar mixture of picture and plan, in southern Europe, though none is as yet known farther north. One Bronze Age petroglyph from Valcamonica comes from side 4 of the stone found at Borno. This includes what has been described as a composition with ibex running toward the river (fig.

Druck, 1978). For a critique of Breuil's interpretations (from which these are derived) see M. Lorblanchet, "From Naturalism to Abstraction in European Prehistoric Rock Art," in *Indigenous Art*, 44–56 (note 38). Closed shapes, such as circles and rectangles, and the topological concepts of inclusion and separateness are among the primary spatial concepts, and it is not surprising to find them in the earliest drawings: Piaget and Inhelder, *Child's Conception of Space*, 44–79 (note 26).

79. Figures 3–7 in appendix 4.2. Henri Breuil, *Les roches peintes du Tassili-n-Ajjer* (Paris: Arts et Métiers Graphiques, 1954), 33 and fig. 65 (this work is an extract from *Actes du IIᵉ Congrès Panafricain de Préhistoire, Alger 1952*). Another painting (fig. 66) features three smaller circles made with a single line, and it is interesting to speculate whether these two were intended, given the context, to represent huts—that is, were signs for huts. If so, a further speculation is how often, elsewhere or in other periods such as the Upper Paleolithic, circular signs were used as hut signs or settlement signs.

80. Figure 9 in appendix 4.2. Leo Frobenius, *Ekade Ektab: Die Felsbilder Fezzans* (Leipzig: O. Harrassowitz, 1937). Lhote, *Tassili Frescoes*, 202–3 (note 54); Henri Lhote suggests that camel pictures belong to the historical period because this animal was not introduced into North Africa until about the first century A.D.: *Les gravures rupestres du Sud-Oranais*, Mémoires du Centre de Recherches Anthropologiques Préhistoriques et Ethnographiques 16 (Paris: Arts et Métiers Graphiques, 1970), 171. But Michael M. Ripinsky suggests

Çatal Hüyük (Turkey). In addition there is the Great Disk from Talat N'Iisk (Morocco).

The interpretation of the Landscape Jar has already proved controversial. Ten of the twelve painted panels that make up its decoration contain linear and geometric patterns. One of the other two contains what A. J. Tobler concluded was not just a landscape painting but "a kind of map. . . . probably the oldest map yet discovered."[83]

FIG. 4.11. PICTURE WITH POSSIBLE MAP ELEMENTS FROM I-N-ETEN, TASSILI MOUNTAINS, ALGERIA. The circular bands seem to represent (in plan) a hut that contains human figures (shown in profile). Although both groups of figures appear on the same panel and are described by Breuil as two separate groups, they are illustrated as if forming a single group. Hence, the illustration here shows the huts as a single group while in appendix 4.2 they are listed separately (nos. 3 and 4).
Diameter of the upper circle: 25 cm. After Henri Breuil, *Les roches peintes du Tassili-n-Ajjer* (Paris: Arts et Métiers Graphiques, 1954), fig. 65.

FIG. 4.12. PICTURE WITH POSSIBLE MAP ELEMENTS FROM THE TASSILI MOUNTAINS, ALGERIA. This seems to portray a camel within an enclosure.
After Leo Frobenius, *Ekade Ektab: Die Felsbilder Fezzans* (Leipzig: O. Harrassowitz, 1937), fig. 10.

4.13).[81] Much less easily interpretable is the line drawing on a fragment of mammoth tusk excavated in 1966 from a site in Mezhirichi (Ukraine), which lies on the Ros River. The fragment has been dated, like the site, to the Upper Paleolithic. Though most of the markings are narrow bands or simple lines, four shapes along a central strip have been interpreted as profile representations of dwellings on the banks of a river shown in plan (fig. 4.14).[82] Single transverse lines in the river are thought to indicate fishing nets or seines and the domed structures are said to be identical in shape with the excavated Paleolithic huts at the site, which were constructed largely from mammoth bones.

Four more complicated—and arguably more interesting—compositions merit discussion in this section. Three of them have already been described as possible maps: the Landscape Jar from Tepe Gawra (Iraq), the silver vase from Maikop (USSR), and the wall painting from

that the camel was domesticated in the Old World not later than the fourth millennium B.C. and that predynastic Egyptians were acquainted with it: "The Camel in Ancient Arabia," *Antiquity* 49, no. 196 (1975): 295–98.

81. Map 42 in appendix 4.1. Emmanuel Anati, *Camonica Valley*, trans. Linda Asher (New York: Alfred A. Knopf, 1961), 102; idem, *Il masso di Borno* (Brescia: Camuna, 1966), where the same drawing is reproduced (figs. 16 and 17) and described as probably forming a scene. The double line to which the animals are advancing seems to indicate a river (p. 34). On the other hand, figure 15 is a photograph of this side of the stone and shows that the double line continues as a single line, forming a closed subrectangular or roughly circular form. It may be that some of the Mont Bégo figures (e.g., those with herds of oxen or plow teams in enclosures) should be included in this category of picture maps.

82. Personal communication from B. P. Polevoy. Ivan Grigorévich Pidoplichko, *Pozdnepaleoliticheskye zhilishcha iz kostey mamonta na Ukraine* (Late Paleolithic dwellings of mammoth bone in the Ukraine) (Kiev: Izdatelstvo "Naukova Dumka," 1969); idem, *Mezhiricheskye zhilishcha iz kostey mamonta* (Mezhirichi dwellings of mammoth bone) (Kiev: Izdatelstvo "Naukova Dumka," 1976).

83. Map 52 in appendix 4.1. Arthur J. Tobler, *Excavations at Tepe Gawra: Joint Expedition of the Baghdad School and the University Museum to Mesopotamia*, 2 vols. (Philadelphia: University of Pennsylvania Press, 1950), 2:150–51, pl. LXXCIIb. William Harris Stahl also sees the Tepe Gawra vase painting as an example of what he calls the "alternation between planimetric views and vertical projections" taking place in the Near East in Neolithic times: "Cosmology and Cartography," part of "Representation of the Earth's Surface as an Artistic Motif," in *Encyclopedia of World Art* (New York: McGraw-Hill, 1960), 3:cols. 851–54, quotation at 853.

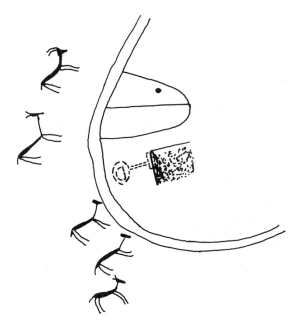

FIG. 4.13. PICTURE WITH POSSIBLE MAP ELEMENTS: SIDE 4 OF THE BORNO STONE FROM VALCAMONICA. This is thought to show deerlike animals (in profile) running toward a river, on the other side of which is a subdivided enclosure (both river and enclosure being in plan).
Size of the original: 70 × 84 cm. After Emmanuel Anati, *Camonica Valley*, trans. Linda Asher (New York: Alfred A. Knopf, 1961; reprinted London: Jonathan Cape, 1964), 102.

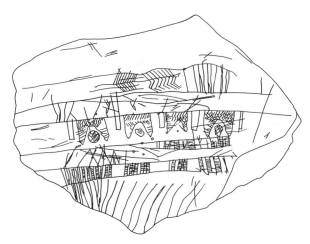

FIG. 4.14. PALEOLITHIC ENGRAVING ON MAMMOTH BONE. Found at Mezhirichi (Ukraine); the four domed features are thought to represent riverside dwellings. The engraving is oriented here as in the original publication; it is difficult, however, to see a close resemblance between the features scratched onto the bone and the excavated mammoth bone dwellings whichever way the drawing is oriented.
After Ivan Grigorévich Pidoplichko, *Pozdnepaleoliticheskie zhilishcha iz kostey mamonta na Ukraine* (Late Paleolithic dwellings of mammoth bone in the Ukraine) (Kiev: Izdatelstvo "Naukova Dumka," 1969), fig. 58.

According to Tobler, the painting shows a hunting scene in a broad valley, the latter flanked by mountains (indicated by the two rows of triangles) and containing the tortuous course of a river with its tributaries (figs. 4.15 and 4.16). He also suggested that the artist must have had some real landscape in mind. However, not all agree either with this interpretation or with his interpretation of the ten geometric panels as representations of different types of terrain such as rolling plains, mountains, deserts, and marshes.[84] Beatrice Goff, for instance, considers the scene to be a highly schematized and not uncommon form of decoration that was a means of giving expression to "deep-seated feelings of aggression" rather than a representation or a picture of a familiar landscape.[85]

Very much more difficult to interpret is the decoration of the Great Disk at Talat N'Iisk in the Atlas Mountains of Morocco.[86] This rock painting, with a diameter of about 100 cm, is by far the largest of any in the area (fig. 4.17). It is its internal decoration, however, rather than its sheer size, that attracts attention, although there does not appear to be any discussion of this aspect in the literature. On the one hand, it is quite unlike anything else in the same district. On the other hand, it has a striking resemblance to the landscape panel on the jar

from Tepe Gawra (and an even closer resemblance to the Babylonian clay tablet from Nuzi).[87] While smaller circles in the district contain either a formless scribble or a simple internal rim pattern, the internal features of the Great Disk seem to have been carefully arranged. Moreover, they could be interpreted as representing a broad valley between two mountain ranges with a major river in the middle, flanked by tributaries or relic channels and abandoned meanders[88] and by two dots, perhaps representing sites or settlements. The schematic nature of the landscape representation (if that is what it is), together with the absence of human or animal figures, distinguishes the Talat N'Iisk disk from other prehistoric picture maps such as the Çatal Hüyük wall painting or the Tepe Gawra Landscape Jar. Its interpretation as an attempt to depict a landscape remains highly subjective and speculative. Nevertheless, in order to draw attention to the existence of such graphic representations and their

84. Tobler, *Tepe Gawra*, 150 (note 83).
85. Beatrice Laura Goff, *Symbols of Prehistoric Mesopotamia* (New Haven: Yale University Press, 1963), 29.
86. Map 57 in appendix 4.1. Jean Malhomme, *Corpus des gravures rupestres du Grand Atlas*, fascs. 13 and 14 (Rabat: Service des Antiquités du Maroc, 1959–61), pt. 1, 91, pl. 4. Paule Marie Grand, *Arte preistorica* (Milan: Parnaso, 1967), fig. 65.
87. See chapter 6, "Cartography in the Ancient Near East," p. 113 and fig. 6.11.
88. In a pattern familiar to anyone who has seen present-day Mediterranean valleys from the air or on aerial photographs.

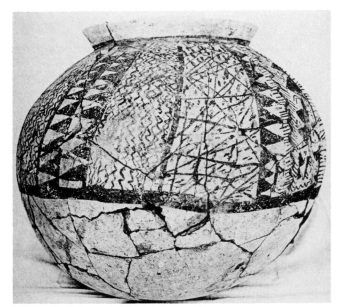

FIG. 4.15. THE TEPE GAWRA (IRAQ) LANDSCAPE JAR. The reconstituted jar is shown with the panel that gives the jar its name on the right (see fig. 4.16).
Diameter of the original: 70 cm. From Arthur J. Tobler, *Excavations at Tepe Gawra: Joint Expedition of the Baghdad School and the University Museum to Mesopotamia*, 2 vols. (Philadelphia: University of Pennsylvania Press, 1950), vol. 2, pl. LXXVIIIa. By permission of the Iraq Museum, Baghdad.

FIG. 4.16. PICTURE MAP ON THE TEPE GAWRA (IRAQ) LANDSCAPE JAR. One of twelve decorated panels, this is by far the most complex and unusual. The juxtaposition of animal figures (absent from other panels), parallel lines of triangles (commonly found on pottery representing mountains), and the sinuous herringbone pattern down the middle (interpreted as a river with its tributaries) led one excavator to suggest the panel portrayed a landscape or even a map of a specific area. From Arthur J. Tobler, *Excavations at Tepe Gawra: Joint Expedition of the Baghdad School and the University Museum to Mesopotamia*, 2 vols. (Philadelphia: University of Pennsylvania Press, 1950), vol. 2, pl. LXXVIIIb. By permission of the Iraq Museum, Baghdad.

potential interest in the history of cartography, it is classed here as a possible example of an early picture map (appendix 4.1).

The silver Maikop vase, with its engraved decoration, was found in the course of excavations of late Neolithic or Chalcolithic tombs in the North Caucasus in 1895 (fig. 4.18).[89] Most prominent are the naturalistically represented large quadrupeds (lions, bulls, horses, goats, and such), together with landscape features. These latter Rostovtzeff considered to be an entirely separate and distinct scheme of ornamentation and "a first timid attempt to subordinate landscape to figures."[90] Two rivers are thought to be shown flowing from the mountains and meeting in a sea or lake. There are also palm trees, waterfowl, a small bear, and some sort of water plant. The rivers are shown in plan, shaded by wavy lines, and the mountains in profile, albeit in varied rather than strictly conventional outlines. In Russia this representation was long considered the earliest geographical

89. Mstislav Farmakovsky, "Arkhaicheskiy period v Rossii: Pamyatniki grecheskogo arkhaicheskogo i drevnego vostochnogo iskusstva, naidënnye v grecheskikh koloniyakh po severnomu beregu Chërnogo morya v kurganakh Skifii i na Kavkaze" (The archaic period in Russia: Relics of Greek archaic and ancient Eastern art found in the Greek colonies along the northern coast of the Black Sea in the barrows

of Scythia and in the Caucasus), *Materialy po Arkheologii Rossii, Izdavayemye Imperatorskoy Arkheologicheskoy Komissiyey* 34 (1914): 15–78, esp. 59.

90. Map 51 in appendix 4.1. Mikhail I. Rostovtzeff, *Iranians and Greeks in South Russia* (Oxford: Clarendon Press, 1922), 22–25, pl. III (1–2), and fig. 2, quotation on 25. Rostovtzeff devoted several paragraphs to a discussion of the decoration, comparing it with Babylonian and Egyptian landscape portrayal, though there he considered landscape subordinate to the figures whereas on the Maikop vase landscape and most of the animals are merely juxtaposed. He also decided that it contains a "survival of prehistoric motives" as well as novelties. A drawing of the vase can be found in Stuart Piggott, *Ancient Europe from the Beginnings of Agriculture to Classical Antiquity* (Edinburgh: Edinburgh University Press, 1965), fig. 37, as well as in

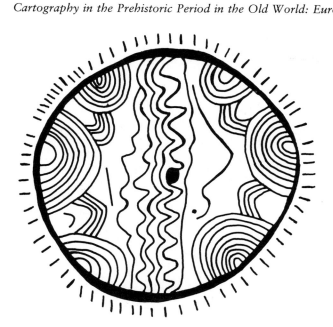

FIG. 4.17. PICTURE MAP: THE "GREAT DISK" FROM TALAT N'IISK, MOROCCO. Outstanding, in the local context, for its size and for the orderliness of its internal decoration, this rock painting could be interpreted as showing, in plan, an arrangement of parallel ranges of mountains and a braided river similar to that of the Landscape Jar of Tepe Gawra (fig. 4.16).
Diameter of the original: approximately 1 m. After Jean Malhomme, *Corpus des gravures rupestres du Grand Atlas*, fascs. 13 and 14 (Rabat: Service des Antiquités du Maroc, 1959–61), pl. 4.

FIG. 4.18. PICTURE MAP ON THE SILVER VASE FROM MAIKOP, RUSSIA. Dating from ca. 3000 B.C., this representation shows two rivers flowing from a range of mountains identified by some as the Caucasus.
Height of the original: 10–12 cm. After Mikhail I. Rostovtzeff, *Iranians and Greeks in South Russia* (Oxford: Clarendon Press, 1922), fig. 2.

map, the mountains being identified with the Caucasus.[91]

Better known to historians of cartography is the oldest of these examples of picture maps, the wall painting from Çatal Hüyük at Konya in west-central Turkey.[92] Only one of a large number of wall paintings discovered from this partially excavated Neolithic site, it was found in 1963 and has been given a date of 6200 ± 97 B.C. (fig. 4.19). In the present context it is unique in several respects: it has been dated relatively precisely; it has a well-documented archaeological context; and it appears to be the only "urban plan" from the prehistoric period in the Old World. Like many, though by no means all, of the other wall paintings at Çatal Hüyük it comes from a shrine, a common item of domestic architecture at the site, as the excavator, James Mellaart, has stressed: "out of 139 living rooms excavated . . . not less than forty . . . appear to have served Neolithic religion."[93] The painting is on two walls that had been regularly replastered and repainted, a point which underlines the contention already made, that it was the context of painting or the act of painting (or both) that was of prime importance rather than the durability of the image itself.[94] The painting itself is nearly three meters long and consists of eighty or so closely packed rectangles, each with a dot or small circle in the angles and a hollow or blank interior. It would be difficult to see in this rectangular pattern anything of cartographic relevance were it not for the extraordinary resemblance of the rectangles in the wall painting to those drawn by the archaeologists as part of their excavation plan. It was this that inspired Mellaart's interpretation that the painting "is a representation of a neolithic town, probably Çatal Hüyük itself, the houses of which rise in exactly the same manner as is shown in the painting."[95] Behind the houses is the profile of a "strange double-peaked object," which Mellaart suggests is identifiable with the two cones of the volcano Hasan Dağ—possibly in eruption—the source of obsidian, one of Çatal Hüyük's most valued commodities and the basis of its wealth.

Plan Maps

The difficulties involved in the unambiguous identification of topographical maps in plan in rock art and the

Bagrow's *Istoriya*, 4 (note 4), *Geschichte*, fig. 97 (note 2), and both *History of Cartography* (note 2) and *Meister* (note 4), fig. 74. Bagrow accepted that the representation could be of the northern Caucasus and suggested that these artistic renderings are "proto-types" of maps and plans; *Istoriya*, 4.

91. K. A. Salishchev, *Osnovy kartovedeniya: Chast' istoricheskaya i kartograficheskiye materialy* (Moscow: Geodezizdat, 1948), 118–19.

92. Map 54 in appendix 4.1.

93. Mellaart, *Çatal Hüyük*, 77 (note 13).

94. See note 14 above.

95. Mellaart, "Excavations," 55 (note 7).

FIG. 4.19. PICTURE MAP: THE NEOLITHIC WALL PAINTING FROM ÇATAL HÜYÜK, TURKEY. This wall painting was identified as a portrayal, in plan, of the former settlement by its similarity to the layout of the excavated houses uncovered by archaeologists. Behind the settlement is a representation of the mountain Hasan Dağ in profile with its volcano erupting.

Length of the original: approximately 3 m. After the copy by Grace Huxtable in James Mellaart, "Excavations at Çatal Hüyük, 1963: Third Preliminary Report," *Anatolian Studies* 14 (1964): 39–119, pl. VI.

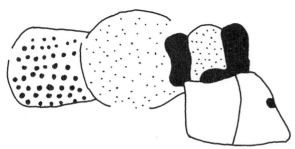

FIG. 4.20. PETROGLYPH MAP FROM VAL FONTANALBA, MONT BEGO. This is typical of the maplike images found as petroglyphs in this valley. With more than six but fewer than eighteen topographical signs, it qualifies as a simple, but not a complex, map as defined in the text.
After Clarence M. Bicknell, *A Guide to the Prehistoric Rock Engravings in the Italian Maritime Alps* (Bordighera: G. Bessone, 1913), pl. XVIII-39.

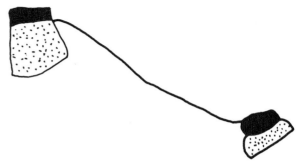

FIG. 4.21. PETROGLYPH MAP FROM VAL FONTANALBA, MONT BEGO. With more than six topographical signs incorporated into its design, this qualifies as a simple map as defined in the text.
After Clarence M. Bicknell, *A Guide to the Prehistoric Rock Engravings in the Italian Maritime Alps* (Bordighera: G. Bessone, 1913), pl. XVIII-32.

three main criteria for their diagnosis—composition, appropriateness of image, and the frequency with which individual images occur within a single composition—have already been discussed (pp. 61–62 above). A more precise threshold is needed, however, to exclude assemblages that are too fragmentary or too ill defined to be worthy of serious attention from the corpus (appendix 4.1). A minimum of six cartographic signs is suggested as this threshold. The usefulness of such a threshold can be demonstrated by reference to three examples (figs. 4.20, 4.21, 4.22). According to our new restrictive definition, only two of these qualify as maps. Figure 4.20 (map 6 in appendix 4.1) is the most clearly cartographic. It not only fulfills all three of the diagnostic criteria but also encompasses a total of ten signs: two

(at least) hut signs; five enclosures (or four enclosures, one with a path across it); and three land-use signs (two forms of stippling and unstippled areas). Figure 4.21 (map 4 in appendix 4.1), with two hut signs, two enclosure signs, one path line, and one or two forms of stippling for land use, making a total of six or seven cartographic signs, just qualifies. Figure 4.22, however, has only four signs (one hut, one enclosure, one path, one land-use sign) and therefore fails to qualify.

Rock art maps or plans identified according to these criteria can be further differentiated, once again based on the number of cartographic elements present, into *simple* and *complex* plans or maps. It has already been suggested that simple maps should contain a minimum of six signs. Complex topographical maps should em

FIG. 4.22. NONCARTOGRAPHIC PETROGLYPH FROM VAL FONTANALBA, MONT BEGO. Unlike the petroglyphs shown in figures 4.20 and 4.21, this does not qualify as a simple map as defined in the text because it contains only four elements.
After Clarence M. Bicknell, *A Guide to the Prehistoric Rock Engravings in the Italian Maritime Alps* (Bordighera: G. Bessone, 1913), pl. XXXII-43.

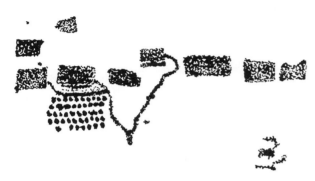

FIG. 4.23. SIMPLE TOPOGRAPHICAL MAP FROM SERADINA, ITALY. Often cited as an example of a prehistoric map, this seems to show an orderly layout of buildings with interconnecting paths and a field or orchard.
Size of the original: 45 × 90 cm. Drawn from a photograph kindly supplied by Centro Camuno di Studi Preistorici, Capo di Ponte, Brescia 25044, Italy.

body at least three times as many signs (i.e., a minimum of eighteen). Compare the examples in figures 4.20 and 4.21 with those in figures 4.26 and 4.27. It is important to note that the quantitative distinction might also suggest a different order of technical effort involved in the production of complex maps. This, in turn, could imply underlying conceptual differences, such as those associated with the purpose or function of the map, the amassing and ordering of the data, and the planning of the map's execution.

Simple Maps The category of simple topographical maps accounts for by far the largest proportion (90 percent) of the present corpus of topographical maps in plan. Interestingly, most examples come from a compact area covering scarcely a dozen square kilometers around Mont Bégo in the Ligurian Alps (dating from as early as the late Neolithic), the remainder coming from elsewhere in the Alps, notably Valcamonica (mostly from the middle or late Bronze Age, 1900–1200 B.C.).[96] The distribution of those at Mont Bégo is highly localized even within that area. As already noted, all but fifteen

of Bicknell's two hundred or so topographical figures come from a single valley, Val Fontanalba, on the northern side of the peak.[97] On the other hand, it is important to stress that even within Val Fontanalba the topographical figures constitute a very minor part (between 2 and 4 percent) of all the petroglyphic subject matter.

Bicknell suggested that 195 of the petroglyphs he studied at Mont Bégo might be interpreted as "topographical figures." A number of these have been discounted in the context of the present discussion as too small, incomplete, or ambiguous to qualify as maps. Of the five examples in this group coming from Valcamonica and discussed here, only one, from Seradina (Capo di Ponte), has already entered the literature of the history of cartography.[98] It seems to depict an orderly layout of buildings with interconnecting paths and at least one (unenclosed) field (fig. 4.23). Another example that is to be found close by, on the riverside rock at Ponte San Rocco, is a much less orderly arrangement of what are thought to be buildings and path signs (fig. 4.25).[99] The topographical representations on side 2 of the Borno stone (thought to be late Neolithic or Chalcolithic in date) were first described before the stone had been completely excavated.[100] The writer Raffaelo Battaglia was already familiar with the larger topographical compositions of Bedolina and Giadighe (see below) and thought he could discern on the Borno stone similar representations of "cultivated fields, fruit groves and paths seen from above."[101] There is little interconnection between any of the markings on the stone, and even the one group of possibly cartographic signs fails to qualify as a map for this reason.[102]

96. Bronze Age maps are still being found by Professor E. Anati and his assistants. Not yet published is a large group of figures (covering about four square meters), thought to be of Bronze Age date, on rock 23 at Foppe di Nadro. This was found in 1982, and I am grateful to Professor Anati and to Tizziana Cittadina for allowing me to see this in the process of recording and for subsequent details.

97. See note 73. Concerning the concentration in the immediate vicinity of Mont Bégo, Bicknell remarked on the awesomeness of this peak, especially under certain weather conditions, and suggested it might have been a "Holy Place," a view accepted by M. C. Burkitt. Bicknell, *Prehistoric Rock Engravings*, 64–65 (note 67). M. C. Burkitt, "Rock Carvings in the Italian Alps," *Antiquity* 3, no. 10 (1929): 155–64.

98. Map 45 in appendix 4.1. Harvey, *Topographical Maps*, 45, fig. 20 (note 6). Additional examples of "maps" have recently been reported by Ausilio Priuli, *Incisioni rupestri della Val Camonica* (Ivrea: Priuli and Verlucca, 1985), including a second one from Seradina (fig. 33) similar to that described here. Few, if any, it would seem, would meet the suggested cartographic criteria.

99. Map 44 in appendix 4.1.

100. Map 42 in appendix 4.1.

101. Raffaello Battaglia and Maria Ornella Acanfora, "Il masso inciso di Borno in Valcamonica," *Bollettino di Paletnologia Italiana* 64 (1954): 225–55, esp. 237.

102. Anati, *Borno*, 20 (note 81), refers to the "plans of cultivated fields, paths, walls, tree plantations."

FIG. 4.24. PHOTOGRAPH OF CAPO DI PONTE, VALCA-MONICA. The winding Oglio River is shown with the hillside rock art sites near Capo di Ponte identified.

Photograph kindly supplied by Ausilio Priuli.

The rest of Italy has little to offer, so far, in the way of possible examples of prehistoric maps, even from other Alpine areas such as Valtellina and Lake Garda, also rich in petroglyphs. Sometimes a topographical motif, such as a field-type rectangle, is found in isolation.[103] A different problem is presented by one of the images on the stela of Novilara (seventh–sixth century B.C.). It has been suggested that this could represent a river with a town in the middle of its course,[104] but despite the proximity of a ship on the same stone, this must remain conjectural. Still less has come from elsewhere in Europe. The rock art of Karelia and the shores of Lake Onega and the White Sea is close to that of Scandinavia in both subject matter and style. Like the Scandinavian art, it has not so far been found to have much in the way of figures of potential cartographic interest, though B. P. Polevoy describes some of the drawings discovered at Zalavruga (on the Vyg River, south of Belomorsk) in 1963–64 as "somewhat reminiscent of geographical maps, with representations of routes, as well as boats,

animals and skier-hunters."[105] In central Norway isolated rectangular motifs, complete with stippled infilling, have been documented as "perhaps pictures of fields."[106]

103. Bric del Selvatico (Lanzo Valley, Turin), for instance. Roberto Roggero, "Recenti scoperte di incisioni rupestri nelle Valli di Lanzo (Torino)," in *Symposium International d'Art Préhistorique Valcamonica, 23–28 Septembre 1968*, Union Internationale des Sciences Préhistoriques et Protohistoriques (Capo di Ponte: Edizioni del Centro, 1970), 125–32.

104. I am grateful to O. A. W. Dilke for drawing this feature to my attention and to Antonio Brancati (director of the Museo Archeologico Oliveriano of Pesaro) for supplying relevant literature. Most archaeological commentators on the stone refer to the "double S" feature only as of "uncertain significance": Gabriele Baldelli, *Novilara: Le necropoli dell'età del ferro*, exhibition catalog (Pesaro: Museo Archeologico Oliveriano, Comune di Pesaro, IV Circoscrizione, n.d.), 28.

105. Written communication, 1982. The drawings are illustrated by Yury A. Savvateyev, *Risunki na skalakh* (Rock drawings) (Petrozavodsk: Karelskoye Knizhnoye Izdelstvo, 1967).

106. Sverre Marstrander, "A Newly Discovered Rock-Carving of Bronze Age Type in Central Norway," in *Symposium International*, 261–72 (note 103).

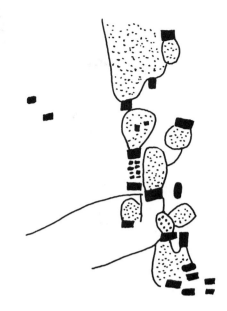

FIG. 4.26. THE "SKIN HILL VILLAGE" MAP FROM VAL FONTANALBA, MONT BEGO. This is one of the most complex assemblages in the area.
Size of the original: 97 × 36 cm. After Clarence M. Bicknell, *A Guide to the Prehistoric Rock Engravings in the Italian Maritime Alps* (Bordighera: G. Bessone, 1913), pl. XLIII-4.

FIG. 4.25. SIMPLE TOPOGRAPHICAL MAP FROM PONTE SAN ROCCO, ITALY. Although confused by the anthropomorphic figures, this seems to show a loosely grouped scatter of buildings linked by paths.
Size of the original: 90 × 45 cm. Drawn from a photograph kindly supplied by Centro Camuno di Studi Preistorici, Capo di Ponte, Brescia 25044, Italy.

Scarcely more promising is a single small composition from Finntörp (near Tanum), Sweden.[107] Comprising an empty rectangle, a number of attached lines, and some scattered dots, it is reminiscent of the style of the Bedolina map in Valcamonica but fails to meet the cartographic criteria. Nor has anything that qualifies as a simple map yet been reported from the Middle East or North Africa. There has been a suggestion that an "inadequately explained feature" associated with the painted star fresco at Teleilat Ghassul in Jordan might represent the plan of a building[108] (see plate 1 and p. 88 below), but this has not received general acceptance.

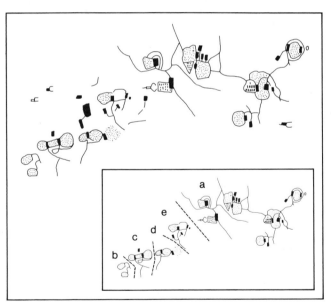

FIG. 4.27. THE "MONTE BEGO VILLAGE." This cannot be accepted in the light of the cartographic criteria as having been intended as a single composition, and the four smaller groups (*b–e*) are classed as simple maps.
Size of the original: 1.40 × 2.40 m. After Clarence M. Bicknell, *A Guide to the Prehistoric Rock Engravings in the Italian Maritime Alps* (Bordighera: G. Bessone, 1913), pl. XLV-1; Bicknell's reproduction shows the long axis vertically on the rock, not horizontally as here.

---

107. I am indebted to John M. Coles, University of Cambridge, for bringing this to my attention and for supplying a photograph.

108. Carolyn Elliott, "The Religious Beliefs of the Ghassulians, c. 4000–3100 B.C.," *Palestine Exploration Quarterly*, January–June 1977, 3–25, though the grounds for her rejection (its early date) are not acceptable to us.

Complex Maps Four prehistoric maps, all petroglyphs and all from the Alps, are described in appendix 4.1 as complex topographical maps. Two come from Mont Bégo in the Ligurian Alps. Both have shared the fate of all Clarence Bicknell's topographical figures; that is, they have been ignored. In what Bicknell christened the "Skin Hill Village" map[109] there are at least nineteen or twenty hut signs; seven complete enclosures and three or four half-completed ones; nine path signs; and at least two types of stippling or land-use signs (fig. 4.26). All are interlinked. The "Monte Bégo Village" map[110] presents more of a problem, since it is clear Bicknell assumed that all five separate groups of topographical images composed a single assemblage. Applying the criterion of linkage, however, it is arguable that the assemblage is composed of a single large composition and four small ones (fig. 4.27). Even on its own, though, the former qualifies as a complex topographical map, containing fifteen hut signs, eleven enclosures, at least twenty interconnecting path signs, and three land-use signs (two types of stippling and some enclosed but unstippled areas). The four small groups have been listed as simple topographical maps. Half a dozen isolated infilled rectangles, possible hut signs, have to be ignored.

In contrast to the "topographical figures" from Mont Bégo, at least one prehistoric map from Valcamonica early received attention in the literature of the history of cartography as the oldest known map. This is the assemblage at Bedolina (Capo di Ponte),[111] which for nearly two decades was the only Old World prehistoric map apparently known to historians of cartography (fig. 4.28). Even so, thirty years had to pass between its first announcement at an archaeological conference in London (and its publication in the proceedings two years later)[112] and its appearance in *Imago Mundi* in 1964.[113] Known as Bedolina 1 (there are a number of fragmentary or incomplete figures of the "topographic" type in the vicinity), the petroglyph occupies most of an ice-polished,[114] undulating rock that projects—like so many others in the district—from the now terraced mountainside. It overlooks the broad, flat-bottomed valley where, some forty meters below, the Oglio River winds its way to the Po. The assemblage covers nearly all of the rock surface exposed today and measures 4.16 by 2.3 meters. Until now, the only detailed study has been a technical and stylistic one, aimed at identifying which figures belonged to the different phases of engraving.[115] This showed that only 134 out of the 183 separate figures engraved on the rock could be considered part of the "map." These come from the second of the four stages of engraving (phase B). The house pictures are later additions, probably of Iron Age date, and should not be regarded as part of the main composition. The range of the topographical images is similar to those at Mont

Bégo, though slightly different in style, and their interpretation as topographical signs was clearly inspired by the work of Clarence Bicknell. One odd feature about the Bedolina composition is that it contains no clear-cut signs, parallel to those of Mont Bégo or even elsewhere in Valcamonica, that can be interpreted as house signs. Despite this, it has been suggested by more than one archaeologist that the Bedolina map was produced as an accurate representation of part of the cultivated landscape on the valley floor during the Bronze Age.[116]

The second petroglyphic composition in Valcamonica that qualifies as a complex topographical map is found on the same hillside, slightly upstream. It is also higher up the mountainside. There is no room for cultivation in the immediate vicinity, nor could there ever have been: the mountain slope falls steeply down to the now cultivated valley floor and the meandering Oglio River over a hundred meters below. Though known also as Plaz d'Ort, the first published reference names the locality Giadighe,[117] the name retained here. Walter Blumer included a photograph of this assemblage in a discussion of Bedolina and Seradina but made no comment on it (fig. 4.29).[118]

Comparison with the Bedolina map reveals some major differences. The close network of "fields" makes the Giadighe representation a more compact composition.

109. Map 35 in appendix 4.1.

110. Map 36 in appendix 4.1.

111. Map 43 in appendix 4.1.

112. Raffaello Battaglia, "Incisioni rupestri di Valcamonica," in *Proceedings of the First International Congress of Prehistoric and Protohistoric Sciences, London, August 1–6, 1932* (London: Oxford University Press, 1934): 234–37.

113. Blumer, "Oldest Known Plan" (note 5).

114. Petroglyphs are usually described as appearing "as new" on discovery, although various degrees of patination are said to be discernible and are useful as an aid to dating. The main reason for this pristine appearance is the hardness of the rock, inevitably the finest grained, closest textured, and most resistant in the district. In some areas, however, this petrological resistance to weathering has been enhanced—it has been suggested—by the way the rock surface has been polished and smoothed by the movement of glacier ice, leaving few irregularities to catch surface water.

115. Miguel Beltrán Lloris, "Los grabados rupestres de Bedolina (Valcamonica)," *Bollettino del Centro Camuno di Studi Preistorici* 8 (1972): 121–58.

116. As has been suggested by Anati, *Camonica Valley*, 104–8 (note 81). Apart from artificial terraces built between the rocks on the lower slopes, there is no room for cultivation except on the valley floor or on deposition cones at the debouchment of tributary streams. It has been suggested by Priuli, *Incisioni rupestri*, 24 (note 98), that maps were executed on rocks, the undulations of which reflected those of the area depicted and that the Bedolina map might portray the zone of Castelliere del Dos dell'Archa. However, it does not seem wise to attempt to infer the Bronze Age landscape from the petroglyphic evidence without further archaeological evidence.

117. Raffaello Battaglia, "Ricerche etnografiche sui petroglifi della Cerchia Alpina," *Studi Etruschi* 8 (1934): 11–48, pls. I–XXII.

118. Blumer, "Felsgravuren" (note 5).

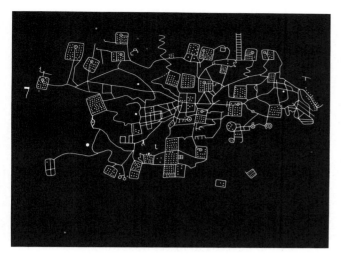

FIG. 4.28. COMPOSITE PETROGLYPH MAP FROM BEDOLINA, VALCAMONICA. The earlier figures and later additions have been removed to reveal a complex topographical map.
Size of the original: 2.30 × 4.16 m. After Miguel Beltràn Lloris, "Los grabados rupestres de Bedolina (Valcamonica)," *Bollettino del Centro Camuno di Studi Preistorici* 8 (1972): 121–58, fig. 48.

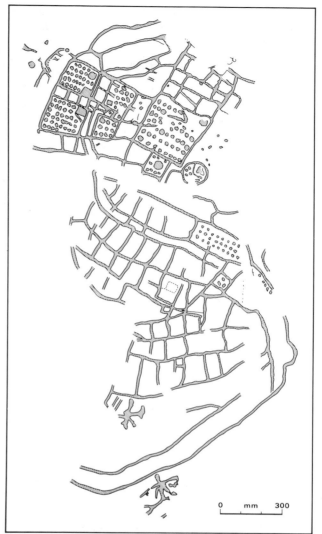

FIG. 4.29. PETROGLYPH MAP FROM GIADIGHE, VAL-CAMONICA. The rock is damaged by fissures and erosion but viewed in the field, the sweep of the broadly spaced double lines that are thought to represent a river is even more striking. There are two anthropomorphic figures toward the bottom.
Size of the original: 2.59 × 1.25 m. Author's field drawing; see also Ausilio Priuli, *Incisioni rupestri della Val Camonica* (Ivrea: Priuli and Verlucca, 1985), fig. 25.

The absence of the points and circles, thought to indicate springs, and of the long, often zigzagging path signs of the Bedolina petroglyph makes the Giadighe figure the more homogeneous of the two. Though one or two lines are executed in a different technique, the only intrusive images (near the bottom of the rock) resemble the anthropomorphic figures common elsewhere in Valcamonica. Only a small proportion of the Giadighe fields are stippled, either by intention or simply because the composition is incomplete, and the stippling is formed of relatively large, regularly spaced, hammered or punched points. Some rectangular depressions could be interpreted as representations of buildings; if this is the case, they would indicate homesteads situated within the enclosures. Although natural fissures in the rock and subsequent erosion have led to discontinuities in the pattern, a particularly striking feature is a double line boldly sweeping in an S-shaped curve across the entire composition from top to bottom. It has been suggested that this represents the meandering river Oglio. Battaglia's interpretation of the Giadighe petroglyph as a map has not so far been challenged. He described it as the valley of the Oglio, "with its enclosed fields and fruitgroves among which the broad ribbon of the river meanders."[119]

No other European region has as yet produced comparable compositions in either rock or mobiliary art,

119. Battaglia, "Incisioni rupestri," 236 (note 112); Battaglia, "Ricerche etnografiche," 44–45 (note 117). See also Priuli, *Incisioni rupestri*, 26 and figs. 24 and 25 (note 98). The modern landscape, with its traditional features of *cultura promiscua* (intercropping), fits both the Bedolina and the Giadighe compositions though Priuli seems to think the stippling represents woodland and that this woodland was part of a "rotation cycle" that allowed, say, a fifteen-year period of soil recuperation (p. 24), though he offers nothing to support such an interpretation. It is possible that *cultura promiscua*, a typical Mediterranean farming system, was already established in Valcamonica in the second millennium B.C.

and nothing similar is known from the Middle East or North Africa. The four examples from Mont Bégo and Valcamonica are outstanding in terms of their cohesiveness and the appropriateness of their signs. They are accepted here as possible examples of prehistoric maps in accordance with the suggested cartographic criteria. On the same basis, other proposed examples have been discarded, notably the Clapier rock (Pinerolo, Italy) (fig. 4.30)[120] and several decorated stones in northern Germany.[121]

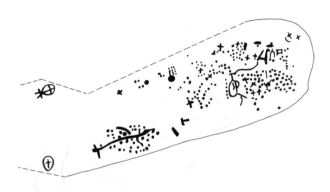

FIG. 4.30. THE CLAPIER ROCK, ITALY. A noncartographic figure, this consists of a collection of cups, rings, and other engravings.
Size of the original: 2.5 × 8.5 m. After Cesare Giulio Borgna, "La mappa litica di rocio Clapier," *L'Universo* 49, no. 6 (1969): 1023–42, pl. following 1042.

The topographical maps, including the simple ones, seem to demonstrate a concept of graphic representation distinct from that represented in picture maps, namely the depiction of all features in plan without apparent regard to the difficulty of interpretation by the uninitiated. This new viewpoint must have constituted a cartographic step every bit as significant in the context of the later prehistoric period as was the reintroduction of the ichnographic city plan in the sixteenth century A.D. The precise cause and context of this modification is, however, far from clear. It may have been related to a change in the original purpose or function of these prehistoric representations of space. It is arguable that what appears to be a new (or perhaps increased) incidence of maps drawn in plan reflects prehistoric man's recognition that depiction in plan provided a more effective means of recording a spatial distribution than did a pictorial map. Although the latter continued to be produced, the apparent proliferation of plan maps in these Alpine districts may suggest a new interest in this sort of factual record. While it is tempting to attribute this change to a more "modern" approach to mapping, it would be rash to impose this interpretation on all complex plan maps. The two examples from Mont Bégo are

high in the mountains, relatively inaccessible, and distant from the homesteads and arable fields seemingly depicted. In the absence of substantial indications to the contrary, it would be more appropriate to attribute these maps to the primarily symbolic purpose behind much prehistoric rock art, in which artistic or visual significance was subordinate to a now unknown abstract context or message. In contrast, the two examples from Valcamonica have a significantly different context, since each overlooks what would have been, even then, a cultivated valley and a route across the Alps. They are also slightly different stylistically, although the fact that they are the largest known plan maps from prehistoric Europe may reflect no more than the availability of large ice-polished rock surfaces. When the arguments and examples are weighed, it remains doubtful that even these two examples can in fact be seen as marking the introduction of the use of maps as factual records in the prehistoric era, although such a transition had taken place by early historical times.

Relief Maps Although nothing has been found in the prehistoric period similar to the three-dimensional representations of topography suggested for some of the coins of the classical period,[122] the archaeology of the post-Paleolithic period is well endowed with finds of small clay models of buildings. These would have been used either as votive offerings or as funerary urns. There are also a number of painted or bas-relief portrayals of buildings and fortifications. These are all wholly profile views; consequently, notwithstanding their accuracy as

120. Cesare Giulio Borgna, "La mappa litica di rocio Clapier," *L'Universo* 49, no. 6 (1969): 1023–42. The Clapier rock (see fig. 4.30) is an extensive exposure (6 × 2 m) high on a mountainside, covered with cup marks (and a few crosses and disjointed lines) in no discernible pattern or order. Borgna sees it as a map of the environs showing the distribution of features of interest to the "seminomadic shepherds" of ancient times, such as springs, natural shelters, and pasturage. The basis for this interpretation is the apparent match of selected marks on the rock with major landscape features, such as neighboring mountain peaks, particularly those that also have rock carvings.

121. Molt, *Karten* (note 64), is an ingenious, sometimes thought-provoking, and painstaking study of several decorated rocks and stones in Schleswig-Holstein and Lower Saxony, including the Bunsoh stone. One interesting suggestion is that the constellations thought to be portrayed on this stone in various arrangements of the cup marks were used to represent individual landscape features of prehistoric times on the rock map (in the way, though Molt himself does not say so, North American Indians used totems on their maps). The chief weaknesses of Molt's approach are the arbitrariness of selection of those cup marks and other marks that appear to fit the supposed pattern from a much larger number present on the stones and the assumption that modern conceptions of mapmaking and surveying (notably the idea of scale and Euclidean geometry) can be applied to prehistoric times (Molt is a retired surveyor). The stones discussed, besides the Bunsoh and Hoisdorf stones, are the megaliths of Plumbohm and Waldhusen.

122. See p. 158 and fig. 9.7 below.

a record of the architecture of specific buildings,[123] they are beyond the scope of this work. What have been said to be plan models from Malta, however, are in a class of their own. One of the two published examples is a sculptured limestone block that was found in a temple built at Tarxien late in the Neolithic period (fig. 4.31). It has been interpreted by some archaeologists as a detailed plan representation of a building with rectangular living spaces.[124] The second, less well known, is a terracotta model found at Hagar Qim. Two fragments survive of a larger original. They are sufficient to show that, when completed, they consisted of a modeled foundation slab on which the lower courses of what have been interpreted as the walls and jambs of a five-apsed temple rested.[125] The archaeological significance of the Hagar Qim fragment rests on the fact that when it was made no buildings of that particular form existed in Malta; its significance for historians of cartography is that it may perhaps be regarded as having been created as a demonstration model or as a three-dimensional reference plan for the actual process of construction.[126]

CELESTIAL MAPS

The idea that there may have been advanced astronomical science in later prehistoric times was strongly held in at least two European countries in the nineteenth century and the early part of the twentieth. Both France and Scotland are rich in those geometric and abstract motifs that are the most difficult to understand and the most vulnerable to fanciful interpretations, namely cup-and-ring marks. They also are rich in concentrations of megalithic monuments and stone alignments, which again have given rise to much controversy about their astronomical or other significance.[127] One of the earliest

proponents of prehistoric astronomical theories in Scotland was George Moore; one of the most persistent was Ludovic MacLellan Mann.[128] But it was George Browne

FIG. 4.31. THE LIMESTONE SCULPTURE FROM TARXIEN, MALTA. This is thought by some archaeologists to be part of a plan representation of a building.
Size of the original: 28 × 28 × 22 × 2 (base) cm, with 2 cm of relief. By permission of the National Museum of Archaeology, Valletta, Malta.

123. For example, Jean Capart, *Primitive Art in Egypt*, trans. A. S. Griffith (London: H. Grevel, 1905), 202, fig. 160, "Clay Model of a Fortified Enclosure." A bronze model from Toprakkale (Turkey) is illustrated by Seton Lloyd, *Early Highland Peoples of Anatolia*, Library of the Early Civilizations (London: Thames and Hudson, 1967), figs. 118–19.

124. Map 49 in appendix 4.1. Themistocles Zammit, *Prehistoric Malta: The Tarxien Temples* (London: Oxford University Press, 1930), 88 and pl. 24 (4). David H. Trump, "I primi architetti: I costruttori dei templi Maltesi" (Rome: Giorgio Bretschneider, 1979), 2113–24 and plates (extract from φιλίας χάριν, *Miscellanea in Onore di Eugenio Manni*).

125. Map 50 in appendix 4.1. Trump, "Primi architetti," 2122 (note 124).

126. Trump, "Primi architetti," 2122 (note 124). It was also once suggested that this apsoidal shape is to be seen "repeated in the symbols carved in the stone altar and sacred slab . . . as well as in the forms of the seven statuettes . . . discovered at Hhagiar Kim, and in the numerous perforations which cover the greater portion of the stones of this building": P. Furse, "On the Prehistoric Monuments in the Islands of Malta and Gozo," *International Congress of Prehistoric Archaeology, Transactions of the Third Session, Norwich 1868* (1869), 407–16, quotation on 412.

127. That there may have been an astronomical motivation behind the construction of a range of ceremonial, burial, and other sites is less contentious than the suggestion that these were based on precise solar and lunar observations, involving a basic knowledge of applied mathematics and surveying, championed by Alexander Thom: "Astronomical Significance of Prehistoric Monuments in Western Europe," in *The Place of Astronomy in the Ancient World*, ed. F. R. Hodson, a joint symposium of the Royal Society and the British Academy (London: Oxford University Press, 1974), 149–56; see also Douglas C. Heggie, *Megalithic Science: Ancient Mathematics and Astronomy in Northwest Europe* (London: Thames and Hudson, 1981); Douglas C. Heggie, ed., *Archaeoastronomy in the Old World* (Cambridge: Cambridge University Press, 1982); James Cornell, *The First Stargazers: An Introduction to the Origins of Astronomy* (New York: Scribner, 1981); Christopher Chippindale, *Stonehenge Complete* (London: Thames and Hudson, 1983). Most authorities are willing to see a secondary symbolic or ritual significance in the sites, while few would doubt that the contemplation of the universe and of celestial phenomena is a practice as old as man himself. A preference for certain art forms may thus have been generated (disk, sphere, etc.): Eugenio Battisti in "Astronomy and Astrology," in *Encyclopedia of World Art*, 2:40 (note 83).

128. George Moore, *Ancient Pillar Stones of Scotland: Their Significance and Bearing on Ethnology* (Edinburgh: Edmonstone and Douglas, 1865). Ludovic MacLellan Mann, *Archaic Sculpturings: Notes on Art, Philosophy, and Religion in Britain 200 B.C. to 900 A.D.* (Edinburgh: William Hodge, 1915); idem, *Earliest Glasgow: A Temple of the Moon* (Glasgow: Mann, 1938).

who, in 1921, attempted to demonstrate in detail how the prehistoric astronomers used cup marks to represent individual constellations on rock and stone. According to Browne, the Sin Hinny stone in Aberdeenshire was an "instructional chart on which the magician could teach his apprentice, instead of teaching him by pointing with his finger to the stars in the sky, with no assurance that the apprentice was looking at the right star,"[129] and to this end he identified the Great Bear, Little Bear, and Corona among the cup marks and hollows on the stone. Similarly, the scatter of 107 cup marks on the Rothiemay stone (also in Aberdeenshire) is said to contain a representation of the Great Bear with accompanying stars—but, curiously, only when the cup-mark pattern is seen as being a mirror image. In France a similar tradition of relating rock carvings to astronomical positions and of recognizing constellations in the cup marks on stones led to publications such as Marcel Baudouin's (1926).[130] To Baudouin, the channels that run outward from some cup-and-ring marks could have been intended as markers of important astronomical axes, while footprint-shaped hollows were made to indicate solar lines. Among the star maps discussed by Baudouin is one that could be the earliest of all, a representation of the Great Bear in a group of seven hollows (out of a total of eighteen) scooped out of a stone excavated from Aurignacian deposits at La Ferrassie.[131]

Enthusiasts like Browne and Baudouin were content to find single constellations in the stone markings. Others, notably Gudmund Schütte, who was well aware of the importance of what he called mythical astronomy in Scandinavia,[132] sought to show that not only individual constellations were portrayed on the rocks but whole portions of the night sky as it would have been seen in the particular locality at a certain time of year. In 1920 Schütte produced a well-illustrated article in which he claimed to have identified at least three star "maps" among the rock carvings of Bohuslän (as illustrated by Baltzer) and in the cup marks of standing stones at Venslev (fig. 4.32) and at Dalby (fig. 4.33) in Denmark.[133] He recounted how it suddenly struck him, as he put it, that one of Baltzer's illustrations of petro-

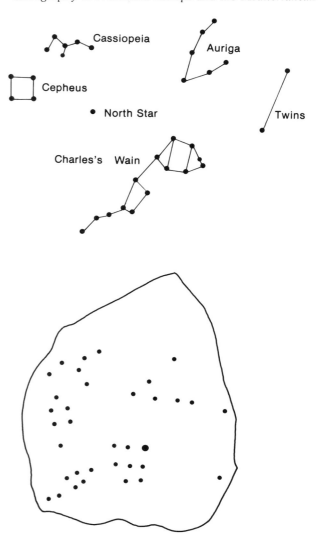

FIG. 4.32. CUP MARKS ON STONES AT VENSLEV, DENMARK. These possibly represent constellations.
After Gudmund Schütte, "Primaeval Astronomy in Scandinavia," *Scottish Geographical Magazine* 36, no. 4 (1920): 244–54, figs. 12 and 13.

129. George Forrest Browne, *On Some Antiquities in the Neighbourhood of Dunecht House Aberdeenshire* (Cambridge: Cambridge University Press, 1921), 159.

130. Marcel Baudouin, *La préhistoire par les étoiles* (Paris: N. Maloine, 1926).

131. Baudouin, *Etoiles*, xv (note 130). For an excellent photograph of the stone, see S. Giedion, *The Eternal Present: The Beginnings of Architecture*, Bollingen Series 35, vol. 6, 2 pts. (New York: Bollingen Foundation, 1962), pt. 1, 137, fig. 78.

132. Gudmund Schütte was the author of a book on home mythology, *Hjemligt Hedenskab: I Almenfattelig Fremstilling* (Copenhagen: Gyldendal, 1919), favorably reviewed in the *Scottish Geographical Magazine* 36, no. 2 (1920): 139–41.

133. Gudmund Schütte, "Primaeval Astronomy in Scandinavia," *Scottish Geographical Magazine* 36, no. 4 (1920): 244–54. Not without interest is the fact that on 5 February 1921 the French journal *La Nature*, no. 2444, 81–83, published an article "L'astronomie préhistorique en Scandinavie," which, though shorter and with fewer illustrations than the *Scottish Geographical Magazine* paper, is otherwise an obvious translation. The author of this paper, however, was given as Dr. M. Schönfeld. Whoever Dr. Schönfeld may or may not have been, Dr. Schütte is a bona fide author, responsible not only for the book on mythology but for several other articles (such as two on Ptolemy's atlas in the *Scottish Geographical Magazine*, vols. 30 and 31). Curiously, though, Browne, *On Some Antiquities*, 162–63 (note 129), referred in detail to the French paper when he might have been expected to have had easier access to Schütte's *Scottish Geographical Magazine* article.

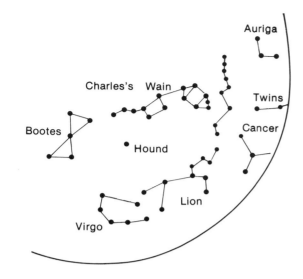

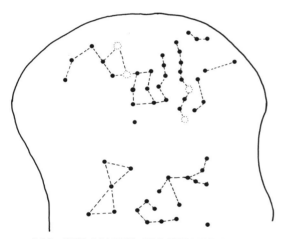

FIG. 4.33. CUP MARKS ON STONES AT DALBY, DEN-MARK. As in figure 4.32, these might represent constellations. After Gudmund Schütte, "Primaeval Astronomy in Scandinavia," *Scottish Geographical Magazine* 36, no. 4 (1920): 244–54, figs. 14 and 15.

glyphs from Tanum "contained an obvious representation of Charles's Wain (Ursa Major) and the Milky Way in fairly correct juxtaposition."[134] Looking more closely he also discerned, to his satisfaction, signs of the zodiac among the other figures on the rock—Cancer, Little (Lesser) Dog, Bull, Foal beside Pegasus, and Capricorn—and he accompanied this interpretation with a figure showing the main constellations visible from Bohuslän on 19 October. One weakness in such interpretations is the inexactitude of the match between the number and positioning of cup marks present on a stone and those needed to make up the constellation thought to be represented. For the Dalby stone, for

example, where there are fifty-six cup marks for fifty-one stars, Schütte admits that the relationship between the two groups of constellations (Charles's Wain and Lynx; Lion, Virgo, and Bootes) is not as correct as it should be.

Other would-be interpreters of cup marks have shown less concern for such matters and are much less specific. Maringer reproduces, apparently the wrong way up, a group of points, lines, and rayed figures from the rock shelter of Pala Pinta (Carlão, Portugal), describing it simply as a "starry sky and ritual axe."[135] Acanfora also reproduces this, together with a similar figure from Cueva del Christo.[136] The idea of a reversed star map also seems widespread. In 1929 one of the excavators of the Tarxien temple reported that the representation of five holes in a floor slab "has suggested to some the image of a constellation, that of the Southern Cross, for example, which at the time the temples were in use was easily seen in our hemisphere."[137] Recently it has been pointed out that it is a mirror image.[138] Also from Malta comes the "star stone" from a small late Neolithic building at Tal Qadi (fig. 4.34). The exact original position of this fragment is unknown. As regards both motifs and the rather crude nature of the engraving, it is different from anything else on the island. The slab is divided by radial lines into five segments, within which there are what appear to be symmetrical arrangements of "star" motifs, short straight lines, and—in one segment on its own—a crescent. The consensus among archaeologists is that it "may have had some religious significance or . . . some astronomical purpose"[139] and that it may have

134. Schütte, "Primaeval Astronomy," 246 (note 133). Charles's Wain is also known as the Plow or the Great Bear; the Milky Way is traditionally the "realm of the dead."

135. Maringer, *Gods*, 169 (note 78). First publication was J. R. dos Santos Júnior, "O abrigo pre-histórico da 'Pala Pinta,'" *Trabalhos da Sociedade Portuguesa de Antropologia e Etnologia* 6 (1933): 33–43.

136. Acanfora, *Pittura*, 260 (note 76). An excellent summary, with bibliography, of star representations in post-Paleolithic art and of archaeological evidence of interest in astronomy and astrology in the prehistoric period is given by Salvatore Puglisi in "Astronomy and Astrology," in *Encyclopedia of World Art*, 2:42–43 (note 83).

137. Themistocles Zammit, *The Neolithic Temples of Hal-Tarxien-Malta*, 3d ed. (Valletta: Empire Press, 1929), 13. The Tarxien temple is dated about 2300 B.C.

138. George Agius and Frank Ventura, "Investigation into the Possible Astronomical Alignments of the Copper Age Temples in Malta," *Archaeoastronomy* 4 (1981): 10–21, esp. 16.

139. Michael Ridley, *The Megalithic Art of the Maltese Islands* (Poole, Dorsetshire: Dolphin Press, 1976), 67. The stone was first described by Luigi M. Ugolini, *Malta: Origini della civiltà mediterranea* (Rome: Libreria dello Stato, 1934), 138 and fig. 79. I am grateful to Gerald L'E. Turner, Museum of the History of Science, University of Oxford, for introducing me to this stone; to Anthony V. Simcock, also of the museum, for the following reference: "The Moon and the Megaliths," *Times Literary Supplement*, 4 June 1971, 633–35; and to David H. Trump, University of Cambridge, for his comments. See also Alexander Marshack, *The Roots of Civilization: The Cognitive*

been used as "some kind of astronomical chart."[140] The word chart here is used in the sense of a calendar; there is no suggestion that the markings represent individual constellations in correct spatial order. Finally, mention should be made of two finds from the Upper Paleolithic. One, recently reported from the USSR, is of a fossilized tortoise said to have the constellation of the Northern Hemisphere (Great Bear and North Star) inscribed on its shell.[141] The other is of a comparatively well-known pebble, excavated in 1956 from the Italian cave of Polesini, marked with the outline of a wolf and a number of pockmarks or carved dots, the latter of which have been interpreted as various constellations of the summer sky as they would have been seen some fifteen to twenty-five thousand years ago.[142]

FIG. 4.34. THE "STAR STONE" FROM TAL QADI, MALTA. The provenance of this loose stone fragment is unknown. It may have served either a religious or an astronomical purpose but, despite the view of some writers, it cannot be considered as a map.
Size of the original: 24 × 29 cm. By permission of the National Museum of Archaeology, Valletta, Malta.

Today there are two schools of thought concerning the place of celestial maps in the history of cartography. According to one, they have no place. Such "sky maps" are "simply pictures of a part of the environment exactly as viewed by the observer, just like pictures of trees or animals," lacking the "highly sophisticated idea of representing landscape as though viewed vertically from every point."[143] According to the other, something is to be gained from a full appreciation of the wealth and, above all, diversity of human experience in spatial representation. All societies in the past have been fascinated by the terrestrial, celestial, and cosmological dimensions

of space.[144] Yet it is arguable that interest was focused first on celestial and cosmological representations—distant and uncertain aspects of life—rather than on those of the known and the familiar local topography. In any case, if societies' achievements have varied at different times in the past, so have their aims, and it is only reasonable, insofar as it is possible, that the former should be judged according to the latter.

Apart from the polemic, there is a practical objection to most celestial "maps" encountered in the literature, and this concerns their definition. The identification of a single or a random group of constellations from large numbers of sometimes variously formed markings on a rock may or may not be a personal or an accidental matter, but in any case such figures cannot generally qualify as celestial *maps*. The requirements for regarding a petroglyph or a rock painting as a sky map are stringent

Beginnings of Man's First Art, Symbol and Notation (London: Weidenfeld and Nicolson, 1972), 344–47, for possible Paleolithic parallels.

140. Ridley, *Megalithic Art*, 32 (note 139).

141. The *Daily Telegraph*, 19 August 1980, reported that "Tass news agency said deep holes in the shell of the tortoise, which symbolized the northern hemisphere in Asian mythology, represented the biggest stars forming the Ursa Major constellation, and the widest hole designated the North Star." However, doubts have been cast on the authenticity of the fossilized tortoise. It is worth noting that the tortoise plays a major role in the cosmological mythology of several cultures, having, according to Chinese legend, the magic square (the four cardinal points and the center of the world) inscribed on its back; see A. Haudricourt and J. Needham, "Ancient Chinese Science," in *History of Science*, 4 vols., ed. Rene Taton, trans. A. J. Pomerans (London: Thames and Hudson, 1963–66), vol. 1, *Ancient and Medieval Science from the Beginnings to 1450*, 161–77, esp. 173 and fig. 22. According to some Indian beliefs, it represents the form of the cosmos itself. Joseph Schwartzberg discusses this further in volume 2 of the present *History*.

142. The interpretation of the Polesini pebble as a star chart is made by Ivan Lee, "Polesini: Upper Palaeolithic Astronomy," *Archaeology 83: The Pro-Am Newsletter* 2 (1983). Thanks are due to Ivan Lee for bringing it to our attention. According to Lee, constellations such as Serpens, Ophiuchus, Scorpio, Lyra, Libra, Aquila, Delphinus, and Saggita can be identified. Earlier commentators on the pebble confined their attention to the wolf outline and to marks on the periphery of the stone suggestive of some sort of notation. See Arturo Mario Radmilli, "The *Movable Art* of the Grotta Polesini," *Antiquity and Survival*, no. 6 (1956): 465–73; Alexander Marshack, "Polesini: A Reexamination of the Engraved Upper Palaeolithic Mobiliary Materials of Italy by a New Methodology," *Rivista di Scienze Preistorici* 24 (1969): 219–81, esp. 272–76; Evan Hadingham, *Secrets of the Ice Age: The World of the Cave Artists* (New York: Walker, 1979) who mentions and illustrates the pebble on p. 254 but whose critique on interpretations of Paleolithic art in general is essential reading; and Martin Brennan, *The Stars and the Stones: Ancient Art and Astronomy in Ireland* (London: Thames and Hudson, 1983), 152.

143. Harvey in "Cartographic Commentary," quotation on 68–69 (note 50).

144. See, for example, Robert David Sack, *Conceptions of Space in Social Thought: A Geographic Perspective* (Minneapolis: University of Minnesota Press; London: Macmillan, 1980).

but straightforward: first, each individual set of markings must correlate in form to make a distinct astronomical entity; and second, the relationship between each of the composite figures must correspond with the relationships between the astronomical entities.[145] These astronomical relationships are observable today and can be calculated for the past. In this respect, given the relatively unchanging nature of the night sky, it should be a simple matter to identify a true star map, particularly in comparison with the problems already encountered of identifying a map of unknown regional topography.

The case for maintaining a category for prehistoric celestial maps rests on solid foundations. The two most important bases are the substantial bodies of ethnographic and of traditional evidence for the importance of stars in routine life among indigenous peoples and, in the latter case, among our European forebears. The use of stars for something as specialized as navigation seems to have been highly developed only in areas of extensive undifferentiated terrain (snow, ocean, and desert).[146] Notwithstanding the use of stars for navigation in the Mediterranean, of much greater significance in the Old World has been the practice of referring to the seasonal appearance of certain constellations to determine the time for the agricultural tasks upon which all livelihood ultimately depended. It is worth noting that the astronomical knowledge needed for these purposes was minimal; it was sufficient to know and observe only a few stars or constellations with relative accuracy.[147] Nilsson pointed out, for example, that the Pleiades was the single most important group of stars among the indigenous people he had studied, owing to the ease with which it can be recognized. Consistent with this is the evidence from classical European literature. Hesiod, for instance, advised timing the whole agricultural year on the movements of no more than four constellations (Sirius, Pleiades, Orion, and Arcturus) and on the two solstices.[148]

As far as the present corpus (appendix 4.1) is concerned, therefore, there is little of substance to include under the heading of prehistoric celestial maps. The view has been adopted that in this period the representation of a single constellation, as opposed to the total celestial sphere, does not constitute a celestial map. As a result, most of the suggested astronomical examples do not qualify as *maps*. Only the Dalby and Venslev stones are included in appendix 4.1, somewhat doubtfully. A point to note is how few of the constellations mentioned in the literature seem to be those relevant to an agricultural population.[149] It is too early, however, to close the lists entirely. As has recently been pointed out, "The weight of the evidence for prehistoric astronomy is cumulative and depends on the apparently repetitive occurrence of indications of the same set of observed phenomena."[150]

The realm of conjecture in prehistoric cartography has already been proved vast, but the case for prehistoric celestial maps should not be judged until more evidence is forthcoming.

COSMOLOGICAL MAPS

In contrast to the fate of celestial maps, historians of cartography have been much more aware of cosmological maps. They usually start with reference to the Babylonians, who are credited with making the earliest recorded attempt at a reasoned conception of the universe.[151] The idea that prehistoric peoples also may have been interested in their cosmos has tended to be rejected as being beyond the intellectual capacity of such "primitive" groups. Bagrow's words, still current in Skelton's edition of his work, enshrine this attitude: "As a rule . . . the maps of primitive peoples are restricted to very small areas. . . . their maps are concrete. . . . they cannot portray the world, or even visualise it in their minds. They have no world maps, for their own locality dominates their thought."[152] It is interesting, therefore, to encounter comments even in the nineteenth century that have bearing not only on prehistoric religion but also—albeit perhaps unintentionally—on prehistoric cosmology. For instance, early in the 1800s the Reverend William Proctor passed on to George Tate his views on the original functions of the decorated rocks they had discovered in Northumberland and on the meaning of the cup-and-ring marks that constituted most of that decoration: "The prevailing figure of the circle . . . may have been designed to symbolise the immortality of the soul. Or the central dot may indicate the individual deceased, the surroundings have reference to his family or temporal circumstances, and the tract from the centre

145. Dorothy Mayer, "Miller's Hypothesis: Some California and Nevada Evidence," *Archaeoastronomy: Supplement to the Journal for the History of Astronomy*, no. 1, suppl. to vol. 10 (1979): 51–74, esp. 52.

146. Hallpike, *Foundations*, 302–3 (note 24).

147. Martin Persson Nilsson, *Primitive Time-Reckoning* (Lund: C. W. K. Gleerup, 1920), 129.

148. Martin Litchfield West, *Hesiod, Works and Days: Edited with Prolegomena and Commentary* (Oxford: Clarendon Press, 1976). The Roman agronomists advised in similar terms.

149. Circumpolar stars are less useful in computing the agricultural cycle, since they tend to remain in view all year round, only changing their position, and they are therefore little utilized by indigenous peoples: Hallpike, *Foundations*, 296–97 (note 24).

150. Richard J. C. Atkinson, review of A. Thom and A. S. Thom, *Megalithic Remains in Britain and Brittany* (Oxford: Clarendon Press, 1978), in *Archaeoastronomy: Supplement to the Journal for the History of Astronomy*, no. 1, suppl. to vol. 10 (1979): 99–102, quotation on 101.

151. Ronald V. Tooley, *Maps and Map-makers*, 6th ed. (London: B. T. Batsford, 1978), 3.

152. Bagrow, *History of Cartography*, 26 (note 2).

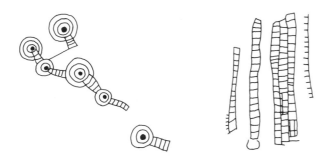

FIG. 4.35. CELESTIAL LADDERS. These examples are taken from stones at Ilkley, Yorkshire, (*left*), and Mont Bégo (*right*). Length of the originals: 23–30 cm (Ilkley), 82–119 cm (Mont Bégo). After Robert Collyer and J. Horsfall Turner, *Ilkley: Ancient and Modern* (Otley: W. Walker, 1885), lxxxvii–lxxxviii; and Clarence M. Bicknell, *A Guide to the Prehistoric Rock Engravings in the Italian Maritime Alps* (Bordighera: G. Bessone, 1913), pl. XXI.

through them may indicate his exit from this round world and its employments."[153] The last idea is essentially cartographic. Moreover, it corresponds closely to the many ethnic and traditional views of the labyrinth motif, which the more complex rings resemble, strongly supported by the modern research discussed below. Nathan Heywood also touched on early cosmological and religious beliefs when, writing in 1888 about the rocks at Ilkley (Yorkshire) (fig. 4.35), he suggested that the ladder motifs "may have been intended as emblematical of some mysterious connection of the earth with the heavens or planets. . . . the cups and rings represented planets, and the circles added to give . . . the appearance of being in motion."[154] Again, the idea of the ladder as a link between earth and heaven has wide currency; it is present in the Babylonian cosmos, for instance.

The approach to the recognition of prehistoric cosmological maps has to be different from the approach to either topographical or celestial maps. It is one thing to postulate that all the cosmic elements should be shown in their correct relative positions but another to be certain not only what these positions were but also what the elements themselves were. The starting point is clear enough, since it is generally accepted that the ancient cosmological beliefs of the Old World were themselves derived from the prehistoric period and were, at the dawn of literacy, in the process of transformation from one already ancient form (mythical) to another form (philosophical).[155] Modern philosophers tend to agree that the Neolithic period would have been the main period of their initial formulation.[156] This was the time when man underwent that "great transformation which endowed him with the gifts of creation and organization"; when the geometric idea of space was formulated; and when the cosmos came to be perceived of after the

human pattern.[157] This is not to deny Paleolithic man a cosmological interest, for which many find evidence in his art,[158] so much as to emphasize the degree to which his "cosmic anguish," the source of religious art if not religion and art themselves,[159] may have been accentuated by these same transformative economic developments. Yi-Fu Tuan is not the first to have observed that fear is most highly, if not exclusively, developed not in the indigenous gathering-hunting people of the world today but among those dependent on agriculture, whose livelihood is far more exposed to man's evil to man and vulnerable to the calamities of nature.[160] Moreover, ethnographic research has shown the widespread and profound interest of indigenous peoples in cosmology; it has also revealed the role of dreams and even of dancing in primitive metaphysics. Dream-laden, induced, or hallucinatory sleep is known to result in petroglyphs and pictographs, their content inspired by the dream.[161] Eth-

153. Cited by Tate, *Sculptured Rocks*, 42 (note 57).

154. Nathan Heywood, "The Cup and Ring Stones on the Panorama Rocks, Near Rombald's Moor, Ilkley, Yorkshire," *Transactions of the Lancashire and Cheshire Antiquarian Society* 6 (1888): 127–28 and figs.

155. G. E. R. Lloyd, "Greek Cosmologies," in *Ancient Cosmologies*, ed. Carmen Blacker and Michael Loewe (London: George Allen and Unwin, 1975), 198–224, esp. 198–200.

156. W. G. Lambert, "The Cosmology of Sumer and Babylon," in *Ancient Cosmologies*, 42–65, esp. 46 (note 155). Juan Eduardo Cirlot, *A Dictionary of Symbols*, trans. Jack Sage (London: Routledge and Kegan Paul, 1971), xvi. Goff, *Prehistoric Mesopotamia*, 169 (note 85), sees the world view of the men of prehistoric Mesopotamia as "an inconsistent, inchoate collection of beliefs," in contrast to most scholars, who view them, on the basis of later Sumerian myths, as ordered. This is, however, a difference of opinion over the nature, not the existence, of prehistoric beliefs in this part of the Old World.

157. Cirlot, *Dictionary*, xvi–xix (note 156) referring to Marius Schneider, *El origen musical de los animales-símbolos en la mitología y la escultura antiguas*, monograph 1 (Barcelona: Instituto Español de Musicología, 1946), and to René Berthelot, *La pensée de l'Asie et l'astrobiologie* (Paris: Payot, 1949).

158. For instance, Giedion, *Eternal Present* (note 131); Marshack, *Roots* (note 139); Gerald S. Hawkins, *Mindsteps to the Cosmos* (New York: Harper and Row, 1983).

159. Giedion, *Eternal Present*, 1:2 (note 131), referring to Wilhem Worringer, *Abstraction and Empathy: A Contribution to the Psychology of Style*, trans. Michael Bullock (London: Routledge and Kegan Paul, 1953), 15.

160. Yi-Fu Tuan, *Landscapes of Fear* (Oxford: Basil Blackwell, 1979), 53; Sieveking, *Cave Artists*, 55 (note 11), quoting James Woodburn, "An Introduction to the Hadza Ecology," in *Man the Hunter*, ed. Richard B. Lee and Irven DeVore (Chicago: Aldine, 1968), 49–55. Woodburn observed that hunting and gathering peoples can be unconcerned to the point of fecklessness as regards their food supply. More generally, in prehistoric art, see Marshack, *Roots* (note 139).

161. David Coxhead and Susan Hiller reproduce pictographs made by Orissans (India) according to instructions received in dreams: *Dreams: Visions of the Night* (New York: Avon Books; London: Thames and Hudson, 1976), 82–83. J. David Lewis-Williams shows that some of the rock paintings of South African bushmen "probably depict the hallucinations of trance performers": "Ethnography and

nologists stress that to the primitive mind (as in Carl Jung's view) dreams are another level of reality, not mere imagination, and in this context *maps* are essential because they "show the way and minimise the risk of becoming lost."[162] For instance, the abstract patterns on an Australian aborigine shaman's drum map his cosmic journey through the center of the three worlds in which he believes[163] just as the tambourines of indigenous peoples in Siberian Asia were decorated with representations of their three worlds, as will be shown in a later volume.[164]

Ancient cosmologies reveal two basic views of the universe.[165] There are the "flat earth" cosmologies, in which the universe is seen as made up of separate layers (heaven, earth, underworld) that are in some way linked—by pillars (in the Egyptian mode) or by a staircase (Babylon) for example—and there are the spherical cosmologies of the Hindus and of Roman and medieval Europe. Either view may include a central or pivotal feature (the *axis mundi*) such as a mountain—the primeval hill of the Egyptians, Mount Meru of the Hindus—or the Tree of Life (Scandinavia). Some of these cosmological features have been discerned in prehistoric art. The Tree of Life, for instance, symbolizing the cosmic life force, is a common motif on Mesopotamian and Egyptian pottery and in Malta, where it also covers a ceiling in the Neolithic temple of Hal Saflien.[166] Ernst Burgstaller sees the Tree of Life as standing for the cosmos itself and suggests that this is the meaning of several treelike motifs in European rock art. He gives as an example a petroglyph from Notgasse (Austria) (fig. 4.36).[167] At least one treelike sign is to be found among the cup-and-ring marks and other rock carvings on the moors at Otley (Yorkshire),[168] and some are found among the ship carvings in Scandinavia.[169] Another petroglyphic motif thought to represent the relationship of the earth to the cosmos or to the sun is formed by a combination of a rectangle and a circle; an example is

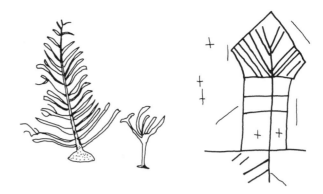

FIG. 4.36. TREE OF LIFE REPRESENTATIONS. These are from Lökeberg, Sweden (*left*) and Notgasse, Austria (*right*). After Oscar Montelius, "Sur les sculptures de rochers de la Suède," in *Congrès International d'Anthropologie et d'Archéologie Préhistoriques, compte rendu de la 7e Session, Stockholm 1874* (Stockholm: P. A. Norstedt, 1876), 453–74, fig. 24; and Ernst Burgstaller, *Felsbilder in Österreich*, Schriftenreihe des Institutes für Landeskunde von Oberösterreich 21 (Linz, 1972), pl. LVII.

found in the cave in the Kienbach Gorge (Austria)[170] and another on the stela from Bagnolo (Valcamonica, Italy) (fig. 4.37).[171] Herbert Kuhn saw the quartered circle as a representation of the cosmos.[172]

Another very common geometric motif in Old World rock art is the labyrinth, a figure widely accepted as having cosmological connotations throughout the world. This mazelike figure is seen variously as difficult

Iconography: Aspects of Southern San Thought and Art," *Man, the Journal of the Royal Anthropological Institute*, n. s., 15, no. 3 (1980): 467–82. See also Klaus F. Wellmann, "Rock Art, Shamans, Phosphenes and Hallucinogens in North America," *Bollettino del Centro Camuno di Studi Preistorici* 18 (1981): 89–103. On dance, see Maria-Gabriele Wosien, *Sacred Dance: Encounter with the Gods* (New York: Avon Books; London: Thames and Hudson, 1974).

162. Coxhead and Hiller, *Dreams* (note 161), excerpt from caption to pl. 19, illustrating the cosmological maps of the Chukchi of Siberia.

163. For the Australian examples, see Coxhead and Hiller, *Dreams*, 94 (note 161), and especially Nancy D. Munn, "The Spatial Presentation of Cosmic Order in Walbiri Iconography," in *Primitive Art and Society*, ed. Anthony Forge (London: Oxford University Press, 1973), 193–220.

164. This will be discussed in volume 4 of the present *History*.

165. Usefully summarized in *Ancient Cosmologies*, ed. Blacker and Loewe, a series of lectures delivered at the University of Cambridge in 1972 (note 155).

166. Ridley, *Megalithic Art*, 63 (note 139).

167. Ernst Burgstaller, "Felsbilder in den Alpenländern Österreichs," in *Symposium International*, 143–47, pl. 62 (note 103). See also fig. I (from Toten Gebirge) in Burgstaller, "Zur Zeitstellung der Österreichischen Felsbilder," in *International Symposium on Rock Art*, 238–46 (note 45). I am grateful to Professor Burgstaller for an informative correspondence on the subject of Austrian rock art and its possible cartographic aspects.

168. "Tree of Life" stone, Low Snowdon. E. T. Cowling, "Cup and Ring Markings to the North of Otley," *Yorkshire Archaeological Journal* 33, pt. 131 (1937): 290–97.

169. As at Lökeberg (Foss, Sweden), illustrated by Oscar Montelius, "Sur les sculptures de rochers de la Suède," in *Congrès International d'Anthropologie et d'Archéologie Préhistoriques, compte rendu de la 7e Session, Stockholm 1874* (Stockholm: P. A. Norstedt, 1876), 453–74 and fig. 24. Also on Viking memorial stones: H. R. Ellis Davidson, "Scandinavian Cosmology," in *Ancient Cosmologies*, 175–197, esp. 175–76 (note 155).

170. Ernst Burgstaller, personal communication, 31 December 1980.

171. Emmanuel Anati, *La stele di Bagnolo presso Malegno*, 2d ed. (Brescia: Camuna, 1965); idem, *Evolution and Style in Camunian Rock Art*, trans. Larryn Diamond (Capo di Ponte: Edizioni del Centro, 1976), fig. 76.

172. Herbert Kühn, *Wenn Steine reden: Die Sprache der Felsbilder* (Wiesbaden: F. A. Brockhaus, 1966), cited by Molt, *Karten*, 57 (note 64). It is more widely seen as a sun disk; see, for example, Glob, *Helleristninger i Danmark*, 56–84 (note 53).

FIG. 4.37. COSMOGRAPHICAL SIGN ON THE BAGNOLO STONE. Originating from near Malegno, Valcamonica, the combination of a rectangle and a circle is thought to represent the relationship of the earth to either the cosmos as a whole or the sun in particular.
Size of the original: approximately 30 × 40 cm. After Emmanuel Anati, *La stele di Bagnolo presso Malegno*, 2d ed. (Brescia: Camuna, 1965), 17, and also 20–21.

to get into or difficult to get out of.[173] In the latter case it may also represent a trap for the uninitiated.[174] But the essential idea is that the figure is connected with the passage of the human soul after death to the afterlife or from one world to another. In current New Hebridean belief the labyrinth is explicitly described as a map embodying "The Path" to the other world: release from the impasse comes to him who knows the way, having committed it to memory during life.[175] Karl Kerenyi has also considered the meaning of spirals and labyrinths, seeing both as symbols of death and concluding they are to be regarded as maps of the underworld, in which the ingoing movement of the spiral indicates death, the outgoing rebirth.[176] It is in this context that the labyrinth of Tintagel (Cornwall) is best interpreted, not as the plowing plan of adjoining fields as Ackroyd Gibson suggests.[177]

All these—except perhaps when labyrinths are representations of structures on the ground—are cosmological signs rather than cosmological maps. Several suggestions have been made regarding prehistoric cosmological maps. One of the earliest, in the view of some, would be the eight-rayed star fresco from Teleilat Ghassul (Jordan, dated to the middle of the fourth millennium B.C.) (plate 1).[178] The archaeologists who discovered it avoided interpretation, but Unger heralded it as a painting of the universe and as a picture representing the cosmos.[179] He based this interpretation on the Babylonian world map, reconstructed from the drawing and cuneiform text of the sixth-century B.C. clay tablet that shows a central earth (with several named localities), the encompassing Bitter River, and the seven ray-

like islands. Although George Kish has recently adopted this cosmological interpretation without comment, there are those who remain doubtful.[180] Apart from the inexactitude of the match between the fresco star and the Babylonian model, and the great difference in dates, the besetting difficulty (common to any interpretation of the meaning of prehistoric art) is that while a design may incorporate certain symbols, this does not necessarily mean that the artist who painted it intended a particular symbolic interpretation.[181] The same problem could be

173. Cirlot, *Dictionary*, 173 (note 156).

174. So, in Scotland, a labyrinth or "tangle threid" may be drawn in piped clay on domestic thresholds as a "no entry" sign, excluding unlucky influences: Janet Bord, *Mazes and Labyrinths of the World* (London: Latimer New Dimensions, 1976), 11. For an excellently illustrated and comprehensive survey of the labyrinth design, see Hermann Kern, *Labirinthi: Forme e interpretazione, 5000 anni di presenza di un archetipo manuale e file conduttore* (Milan: Feltrinelli, 1981); German edition, *Labyrinthe: Erscheinungsformen und Deutungen, 5000 Jahre Gegenwart eines Urbilds* (Munich: Prestel-Verlag, 1982).

175. John W. Layard, *Stone Men of Malekula* (London: Chatto and Windus, 1942), 222, 650–51, quoting A. Bernard Deacon, "Geometrical Drawings from Malekula and the Other Islands of the New Hebrides," *Journal of the Royal Anthropological Institute of Great Britain and Ireland*, n.s., 64 (1934): 129–75. The idea of providing guidance for the journeying dead is widespread; see, for example, Wilhelm Bonacker, "The Egyptian *Book of the Two Ways*," *Imago Mundi* 7 (1950): 5–17; also below p. 120 and pl. 2.

176. Karl Kerényi, *Labyrinth-Studien: Labyrinthos als Linienreflex einer mythologischen Idee*, 2d ed. (Zurich: Rhein-Verlag, 1950), 11–12. See also Jill Purce, *The Mystic Spiral, Journey of the Soul* (London: Thames and Hudson, 1974), which illustrates "the first known spiral in the history of art," a Paleolithic talisman from Siberia (figs. 13 and 14), and a Greek votive object (2800–2000 B.C.) decorated as a mandala with the spiral that was "the symbol which unified the Neolithic world." Purce explains that the "central seven windings represent the six directions and the still centre" (fig. 42).

177. Personal communication from Ronald W. B. Morris. The discovery of two petroglyphic labyrinths in Rocky Valley, Tintagel (Cornwall), was reported by Ackroyd Gibson in "Rock-Carvings Which Link Tintagel with Knossos: Bronze-Age Mazes Discovered in North Cornwall," *Illustrated London News* 224, pt. 1 (9 January 1954): 46–47. He also pointed out the symbolic meanings of such figures and the existence of another British example, the Hollywood stone from the Wicklow Mountains (now in the National Museum of Ireland, Dublin). See also G. N. Russell, "Secrets of the Labyrinth," *Irish Times*, 16 December 1964, 10. I owe this reference to Ronald W. B. Morris, who also drew my attention to the interpretation of the Tintagel figure as a field-plowing plan.

178. Map 53 in appendix 4.1.

179. Alexis Mallon, Robert Koeppel, and René Neuville, *Teleilāt Ghassūl*, 2 vols. (Rome: Institut Biblique Pontifical, 1934–40), 1:135–40 and frontispiece (in color); Eckhard Unger, "From the Cosmos Picture to the World Map," *Imago Mundi* 2 (1937): 1–7, esp. 6; idem, "Ancient Babylonian Maps and Plans," *Antiquity* 9 (1935): 311–12; and William Harris Stahl, "By Their Maps You Shall Know Them," *Archaeology* 8 (1955): 146–55.

180. George Kish, *La carte: Image des civilisations* (Paris: Seuil, 1980), 189, pl. 8. But see chapter 6 below, "Cartography in the Ancient Near East," esp. pp. 111–13.

181. Goff stresses this important point: see *Prehistoric Mesopotamia*, 9 (note 85).

FIG. 4.38. COSMOLOGICAL MAP ON A PREDYNASTIC BOWL FROM EGYPT. The course of the sun from east to west is shown, along with the enclosing primeval ocean and the mountains of the East and the West. The bowl dates from the Amratian period, mid-fourth millennium B.C..

From S. Giedion, *The Eternal Present: The Beginnings of Architecture*, Bollingen Series 35, vol. 6, 2 pts. (New York: Bollingen Foundation, 1962), fig. 69. By permission of the Egyptian Museum, Cairo.

relevant in the case of the painted decoration on an oval pottery dish from predynastic Egypt (fig. 4.38). Giedion sees this as portraying in abstract form the course of the sun from east to west, the enclosing primeval ocean, and the two central mountains of East and West, and indeed there is widespread acceptance of such an idea in the interpretation of decorated pottery from the Middle East and sites such as Susa.[182] From farther west, in the Sahara, comes an intriguing rock-painted figure (fig. 4.39). This was published by Frobenius under the caption "Goblin in house."[183] However, it is arguable that a more appropriate interpretation would be that the house figure is a variation of the labyrinth—or even cosmological—motif, especially since it is wholly unlike the hut figures in the domestic scenes from the same area, already described. The unusual feature is the presence of a central double rectangle, enclosing what may be an anthropomorphic figure, and this lends credence to such an interpretation.

In eastern Europe, the paintings of the Magourata cave (Bulgaria) have been known since the eighteenth century. They may have been executed as early as the early Bronze Age (before about 2000 B.C.). Among them is a group which includes a "solar" figure (two rayed

concentric circles) below which are two motifs (parallel lines and a checkerboard pattern) that Anati has suggested might be symbols of water and of fields (fig. 4.40). The whole composition makes, for him, a representation of the sky and the earth as members or parts of an entity in which the various aspects of nature are synthesized.[184] Rather more promising, having less ambiguous symbols and spatial relationships, are the decorations on at least one of the two stone stelae from northern Italy that have been suggested as cosmological representations. The stelae also touch on a widespread and long-lived tradition

---

182. Map 56 in appendix 4.1. Giedion, *Eternal Present*, 2:129, fig. 69 (note 131) though other authors disagree; see below, p. 117. For interpretations of pottery from Susa, see Robert Klein, *Form and Meaning: Essays on the Renaissance and Modern Art*, trans. Madeline Jay and Leon Wieseltier (New York: Viking Press, 1970), 146. An early attempt to interpret prehistoric pottery decoration in terms of Babylonian cosmology was made by W. Gaerte, "Kosmische Vorstellungen im Bilde prähistorischer Zeit: Erdberg, Himmelsberg, Erdnabel und Weltenströme," *Anthropos* 9 (1914): 956–79.

183. Map 55 in appendix 4.1. Frobenius, *Ekade*, 23, fig. 11 (note 80).

184. Map 1 in appendix 4.1. Emmanuel Anati, "Magourata Cave," *Archaeology* 22 (1969): 92–100, quotation on 100. See also Anati, "Magourata Cave, Bulgaria," *Bollettino del Centro Camuno di Studi Preistorici* 6 (1971): 83–107.

of decorated memorial stones that goes from, possibly, the megalithic period to the Viking Age in northern Europe. The first of these stelae (measuring 53 × 28 × 3.5 cm) was found during World War II at Triora in the Ligurian Alps and is presumed prehistoric in date. It was not discussed in an archaeological review until 1956, when Acanfora described the incised decoration as a figurative composition arranged in two registers with a rayed sunlike figure in the upper, the two parts being separated by a band of decoration.[185] In 1973 Emmanuel Anati reinforced this interpretation by recognizing Acanfora's decorative band as a level in itself

(suggesting too that it may include a hut figure) and summarizing it as showing three registers, symbolizing sky, earth, and the underworld, which together were intended to represent a conception of the universe (fig. 4.41).[186] The stela from Ossimo (Brescia), which has been given a late Neolithic or Chalcolithic date, is less convincing in comparison. It has none of the motifs that can be regarded as conventional cosmological symbols, such as those found on the Triora stone. Instead, the decoration of the upper register is composed of a densely

FIG. 4.39. LABYRINTHLIKE ROCK PAINTING FROM NORTH AFRICA. The labyrinth design is known to be associated, on a world scale and throughout history, with death and the route to the afterlife.
After Leo Frobenius, *Ekade Ektab: Die Felsbilder Fezzans* (Leipzig: O. Harrassowitz, 1937), fig. 11.

FIG. 4.40. COSMOLOGICAL PAINTING FROM THE MAGOURATA CAVE, BULGARIA. Thought to have cosmological significance, the sun may indicate the celestial level, the two parallel lines the earthly level, and the patterned line the netherworld.
After Emmanuel Anati, "Magourata Cave, Bulgaria," *Bollettino del Centro Camuno di Studi Preistorici* 6 (1971): 83–107, figs. 59 and 60.

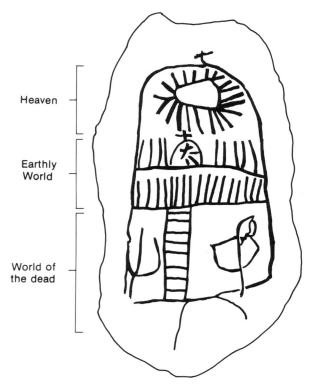

FIG. 4.41. THE TRIORA STELA. The three registers suggested by Emmanuel Anati, "La stele di Triora (Liguria)," *Bollettino del Centro Camuno di Studi Preistorici* 10 (1973): 101–27, are shown. They divide this conception of the universe into sky, earth, and underworld.
Size of the original: 53 × 28 cm. After Maria Ornella Acanfora, "Lastra di pietra figurata da Triora," *Rivista di Studi Liguri* 21 (1955): 44–50, fig. 2c (with additions).

185. Maria Ornella Acanfora, "Singolare figurazione su pietra scoperta a Triora (Liguria)," in *Studi in onore di Aristide Calderini e Roberto Paribeni,* 3 vols. (Milan: Casa Editrice Ceschina, 1956), 3:115–27, esp. 119. Notice of the find had been given by Acanfora the previous year: "Lastra di pietra figurata da Triora," *Rivista di Studi Liguri* 21 (1955): 44–50.

186. Emmanuel Anati, "La stele di Triora (Liguria)," *Bollettino del Centro Camuno di Studi Preistorici* 10 (1973): 101–27, esp. 121. However, Alessandro Bausani, "Interpretazione paleo-astronomica della stele di Triora," *Bollettino del Centro Camuno di Studi Preistorici* 10 (1973): 127–34, esp. 133, considers it a star map.

packed series of hoops; that of the middle includes eight "spectacle-spiral pendants," and the lowest is undecorated. Even so, the Ossimo stela has been described as "an extremely interesting synthesis of a cosmological concept."[187]

From the other side of the Alps and a slightly earlier period comes a different type of artifact, an embossed gold sheet in the form of a disk measuring a few centimeters across. Several such disks have been found, from central Europe to the British Isles. They are usually associated by archaeologists with the solar cult, no significance being attached to the nature of the stylized, rather formal, decoration. Unger, however, was attracted by the decoration on a disk from Moordorf (Germany) (fig. 4.42). Inspired again by his Babylonian model, Unger recognized the same elements which, as he put it, "have been identified . . . as sun-discs, but which, to my mind, are definitely representations of the universe," portraying the central earth, the Bitter River, the mountainous belt of heaven and, finally, the heavenly ocean with the islands.[188] None of the other disks Unger cited has so complete a "cosmological" decoration and he saw these as possibly representing "foreshortened views of the world . . . in which various elements such as the belt of the heavens and the 'islands' are omitted."[189] One of these from the Chalcolithic (ca. 3000 B.C.) comes from Stollhof (Austria). This disk has a rim decoration of three rows of points and three centrally placed bosses as the main elements of its design.[190]

Certain prehistoric traditions lingered on, or were revitalized, in the first millennium A.D. of northern Europe, notably in the art and symbolism of the Scandinavian memorial stones and the associated mythology. The memorial stones are characteristic from the sixth to the eleventh century, though there are several changes in their basic shape and decoration over this period.[191] Throughout, however, there is a tendency for each stone to be divided into two more or less clearly defined registers, in the manner recalling the prehistoric Ligurian stelae. In the early period of the painted stones of Gotland (Sweden), the usual motif on the upper part of the stone is a disk, generally thought to represent the sun or in some way to be linked with a sun or sky deity.[192] Another common motif is the World Tree. By the eighth century the two registers are more clearly differentiated, and the entire decoration is to be made up of mythological scenes apparently depicting the death of the individual, his journey to the afterlife, and his arrival in Valhalla, with the ship the almost inevitable symbol of that cosmic journey.[193] The strong evidence of parallelism in these and other Viking decorations has suggested to several authorities that the sculptors were working in a fixed tradition.[194] Finally, mention may be made of a quite different dimension of cosmological representa-

FIG. 4.42. COSMOLOGICAL MAP: THE GOLD DISK FROM MOORDORF. This disk, thought by some to be a cosmological map, was found near Aurich, West Germany. A central continent is surrounded by concentric rings showing, in turn, a first ocean (marked by lines), another continent (with its mountains), and a second ocean in which are set thirty-two islands (represented by triangles).
Diameter of the original: 15 cm. By permission of the Niedersächsisches Landesmuseum, Hanover.

187. Emmanuel Anati, "La stele di Ossimo," *Bollettino del Centro Camuno di Studi Preistorici* 8 (1972): 51–119, esp. 117 (English summary).

188. Unger, "Cosmos Picture," 5 (note 179).

189. Unger, "Cosmos Picture," n. 19 (note 179).

190. Max Ebert, *Reallexikon der Vorgeschichte* (Berlin: Walter de Gruyter, 1928), 12:442, pl. 110.

191. In view of the prehistoric terms of reference for this chapter, stones with runic inscriptions have been excluded from consideration.

192. H. R. Ellis Davidson, *Pagan Scandinavia* (London: Thames and Hudson, 1967), discusses the continuity of rock decoration traditions from the mid-Neolithic period and the Bronze Age into Viking times and summarizes the stylistic changes of the Viking Age memorial stones. See also Edward O. G. Turville-Petre, *Myth and Religion of the North: The Religion of Ancient Scandinavia* (London: Weidenfeld and Nicolson, 1964), 3–6.

193. Sverre Linquist, *Gotlands Bildsteine*, 2 vols. (Stockholm: Wahlström och Widstrand, 1941–42); David McKenzie Wilson and Ole Klindt-Jensen, *Viking Art* (London: George Allen and Unwin, 1966), 79–82, pl. xxvi, and fig. 42, discuss the compositions on the stones from Tjängvide and Ardre (Gotland).

194. Davidson, *Pagan Scandinavia*, 127 (note 192), and William Gershom Collingwood, *Northumbrian Crosses of the Pre-Norman Age* (London: Faber and Gwyer, 1927), 65.

tion, that manifested in the laying out of ancient Ireland as four great provinces and a center (Tara) to constitute the state as an ordered cosmos.[195]

CONCLUSION

It is obvious from the foregoing that the only evidence we have for the mapmaking inclinations and talents of the inhabitants of Europe and adjacent parts of the Middle East and North Africa during the prehistoric period is the markings and designs on relatively indestructible materials. It is probable, given the prevalence of such activity in historical times among indigenous peoples, that additional cartographic representations were made by prehistoric man on more ephemeral materials such as sand, hide, bark, and the dust of cave floors. All surviving evidence, however, suggests that cartographic depictions in prehistoric rock art constitute a very minor portion of the total sum of that art. Even in Valcamonica, relatively rich in rock art and well searched, the "topographical figures" number a mere half dozen out of a rough total of 180,000 recorded figures from seventy-six sites.[196] The very rarity of cartographic depictions provokes interest in the motivation behind their production. Although some questions will always remain unanswered, there can be no doubt that prehistoric rock and mobiliary art as a whole constitutes a major testimony of early man's expression of himself and his world view.[197] It is reasonable to expect some evidence in this art of the society's spatial consciousness. But when it comes to drawing up the balance sheet of evidence for prehistoric maps, we must admit that the evidence is tenuous and certainly inconclusive. The historian of cartography, looking for maps in the art of prehistoric Europe and its adjacent regions, is in exactly the same position as any other scholar seeking to interpret the content, functions, and meanings of that art. Inferences have to be made about states of mind separated from the present not only by millennia but also—where ethnography is called into service to help illuminate the prehistoric evidence—by the geographical distance and different cultural contexts of other continents.

Despite all these difficulties, a number of statements can be made with confidence. There is, for example, clear evidence in the prehistoric art of Europe that maps—permanent graphic images epitomizing the spatial distribution of objects and events[198]—were being made as early as the Upper Paleolithic. The same evidence shows, too, that the quintessentially cartographic concept of representation in plan was already in use in that period. Moreover, there is sufficient evidence for the use of cartographic signs from at least the post-Paleolithic period. Two of the basic map styles of the historical period—the picture map (perspective view) and the plan (ichno-

graphic view)—also have their prehistoric counterparts. The importance to prehistoric man of his cosmological ideas is reflected in the cartographic record. Less clear, however, is the evidence for celestial mapping. The paucity of evidence of clearly defined representations of constellations in rock art, which should be so easily recognized, seems strange in view of the association of celestial features with religious or cosmological beliefs, though it is understandable if stars were used only for practical matters such as navigation or as the agricultural calendar. What is certainly different is the place and prominence of maps in prehistoric times as compared with historical times, an aspect associated with much wider issues of the social organization, values, and philosophies of two very different types of cultures, the oral and the literate.

It is perhaps fitting to end by being aware of the remaining problems. What is urgently needed as conclusive evidence for the identification of map images such as those of Mont Bégo and Valcamonica, or even the Landscape Jar of Tepe Gawra or the Great Disk of Talat N'Iisk, is a reconstruction of the real-world localities to which at least some maps may refer and the identification of the contemporary mentality. The primary task of recovering the contemporary local landscape is obviously an archaeological one, but there are also archaeologists who would not shirk an attempt to uncover the human reasoning of the times.[199] Then there are problems concerning the difference of function between prehistoric (and historical indigenous) maps and historical topographical maps and the dividing line between them. Should individual examples such as Bedolina or Giadighe be regarded as prototypes of those of the historical period, serving a clearly defined documentary purpose, or as still part of the prehistoric type, having a primarily symbolic function? One point is clear: there is no neat evolution from one type to another, either from prehistoric to historical contexts or even within the historical period. The *mappaemundi* of medieval Europe, for example, may be much closer in concept and purpose to the majority of prehistoric maps than are the estate plans on the clay tablets of protohistoric Babylonia. Whatever the outstanding problems, which are not to be under-

195. Alwyn Rees and Brinley Rees, *Celtic Heritage: Ancient Tradition in Ireland and Wales* (London: Thames and Hudson, 1961), esp. 147–49. I owe this reference to Anthony V. Simcock, Museum of the History of Science, University of Oxford.

196. Emmanuel Anati, "Art with a Message That's Loud and Clear," *Times Higher Educational Supplement*, 12 August 1983, 9.

197. Anati, "Art with a Message" (note 196).

198. See p. xvi above.

199. Colin Renfrew, *Towards an Archaeology of Mind*, Inaugural Lecture, University of Cambridge, 30 November 1982 (Cambridge: Cambridge University Press, 1982), 24–27.

PLATE 1. THE STAR FRESCO FROM TELEILAT GHASSUL, JORDAN. According to some interpretations, this represents a cosmological map, with the known world at the center surrounded by a first ocean, a second world and second ocean, with the eight points perhaps symbolizing the islands of the world beyond and the celestial ocean. The rectangular feature (bottom right) has been suggested as part of a plan drawing of a temple, but again this is highly speculative.
Diameter of the original: 1.84 m. Courtesy of George Kish, University of Michigan, Ann Arbor.

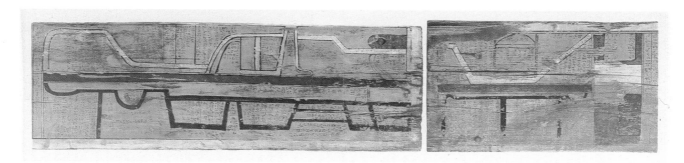

PLATE 2. MAP IN THE *BOOK OF THE TWO WAYS*. An example of a topographical composition, probably intended as a passport to the afterlife, found on many coffin bases from al-Bersha, Middle Egypt, ca. 2000 B.C.
Size of the original: 28 × 63 cm. Photograph courtesy of the American Geographical Society Collection, University of Wisconsin, Milwaukee, from Youssouf Kamal, *Monumenta cartographica Africae et Aegypti*, 5 vols. in 16 pts. (Cairo: 1926–51), 1:6. By permission of the Egyptian Museum, Cairo (coffin 28,083).

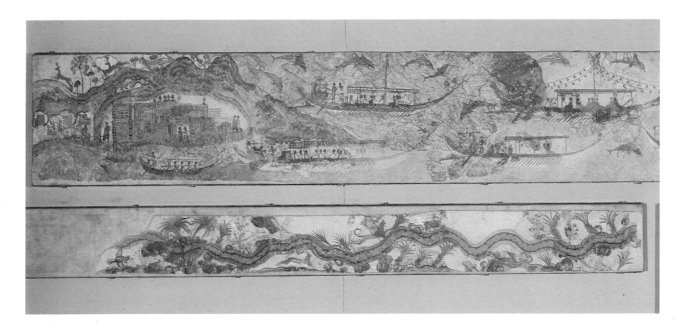

PLATE 3. THE THERA FRESCO. These fragments of a San-
torin fresco, datable to ca. 1500 B.C., contain a number of
cartographic scenes. They also suggest the incipient development of color conventions: the rivers are in blue, but are outlined in gold; the shape of the mountains is also indicated by
a double blue line. The drawings themselves are executed in
plan, in elevation, or from an oblique perspective. The overall
effect is of striking relief, with the different places very clearly
distinguished, so that the fresco is not dissimilar to some of
the many other picture maps that characterize the cartography
of ancient and medieval Europe. There are three frescoes. The
longest (split into two here) contains the story of a fleet: it
departs from a seashore town at the left (*upper section*), and
arrives at its home port at the right (*lower section*). The other
sections show a river in plan and a fragmentary view of war-
riors, flocks, and women.
Lengths of the originals: 3.5 m (river fresco) and 4 m (fresco
of ships). By permission of the National Archaeological Museum, Athens.

PLATE 4. FRESCO FROM THE BOSCOREALE VILLA, NEAR POMPEII. This detail clearly shows a globe drawn in approximate perspective. The object has also been referred to as a sundial.

Size of the original detail: 61 × 39.7 cm. By permission of the Metropolitan Museum of Art, New York (Rogers Fund, 1903 [03.14.2]).

PLATE 5. THE PEUTINGER MAP: ROME. The Peutinger map, dated to the twelfth or early thirteenth century, derives ultimately from a fourth-century archetype, suggested by vignettes such as that of Rome in this segment, in which the city is personified as an enthroned goddess holding a globe, a spear, and a shield.

Size of the original: 33 × 59.3 cm. By permission of the Österreichische Nationalbibliothek, Vienna (Codex Vindobonensis 324, segment IV).

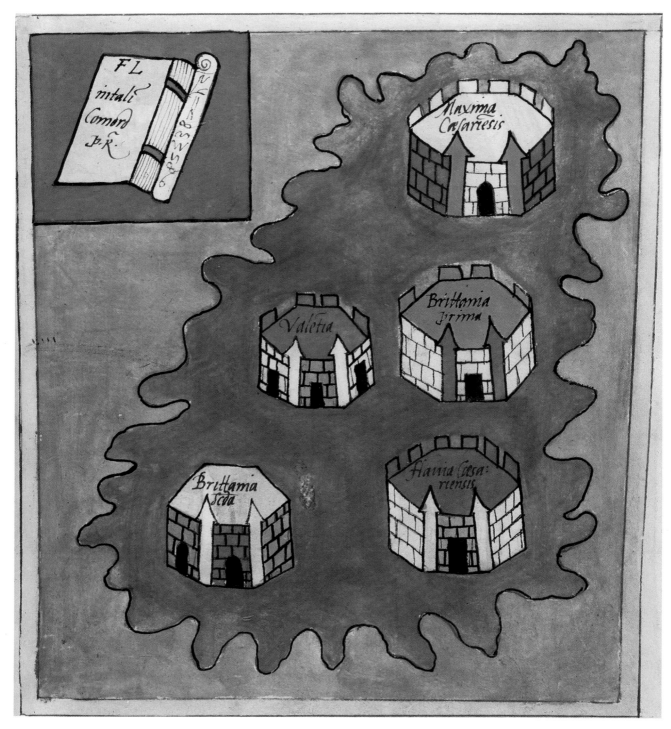

PLATE 6. THE *NOTITIA DIGNITATUM*: BRITAIN. Five provinces are arranged incorrectly in this sixteenth-century copy, at several removes, of a fourth-century original. For example, Maxima Caesariensis, which had London as its capital, is placed not in the southeast but to the northeast near Lincoln.

Size of the original: 31 × 24 cm. By permission of the Bayerische Staatsbibliothek, Munich (Clm. 10291, fol. 212r).

PLATE 7. THE MADABA MOSAIC MAP. Fragment of a sixth-century mosaic now preserved in a church in Madaba, Jordan.

Size of the map as preserved: 5 × 10.5 m. Photograph courtesy of Fr. Michele Piccarillo, Studium Biblicum Franciscanum, Jerusalem.

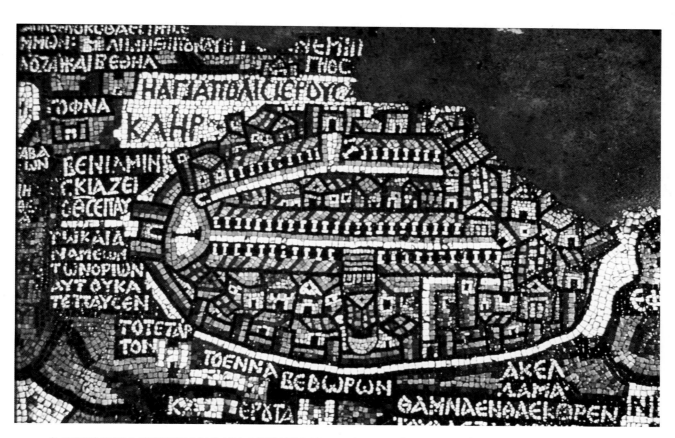

PLATE 8. JERUSALEM ON THE MADABA MOSAIC MAP. The depiction of some churches and other structures is sufficiently realistic for modern scholars to identify them.

Photograph courtesy of Thames and Hudson. By permission of the Department of Antiquities, Jordan.

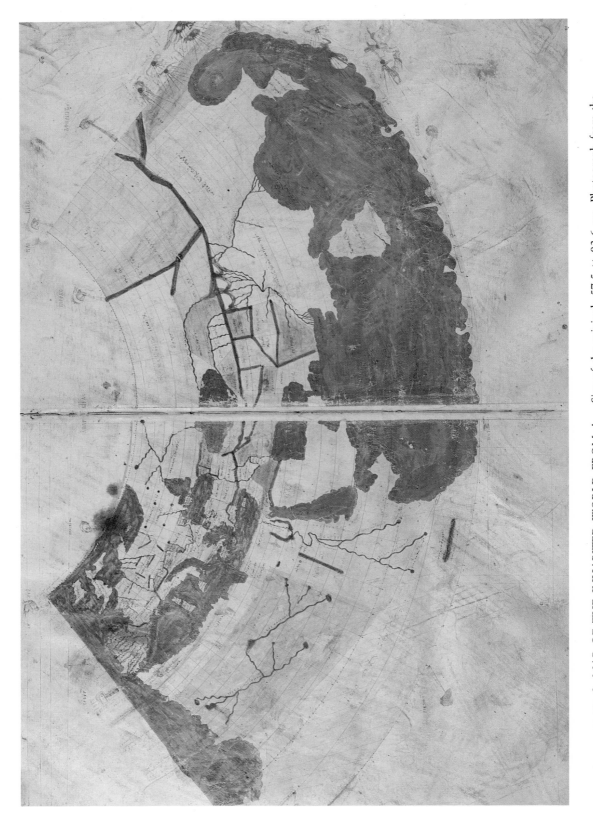

PLATE 9. MAP OF THE INHABITED WORLD FROM A THIRTEENTH-CENTURY BYZANTINE MANUSCRIPT OF PTOLEMY'S *GEOGRAPHY*. Drawn on Ptolemy's first projection, the map is followed in this recension by the twenty-six regional maps. The codex is one of the earliest extant to contain Ptolemaic maps.

Size of the original: 57.5 × 83.6 cm. Photograph from the Biblioteca Apostolica Vaticana, Rome (Urbinas Graecus 82, fols. 60v–61r).

PLATE 11. ORB IN THE LAST JUDGMENT. The tripartite globe or orb is frequently found beneath Christ's feet in medieval representations of the Last Judgment, symbolizing the end of the world.
Size of the original vignette: 12 × 9.8 cm. By permission of the Pierpont Morgan Library, New York (MS. 385, fol. 42v).

PLATE 10. EMPEROR CHARLES IV WITH ORB. This example, from a fourteenth-century armorial, depicts a common theme in medieval art—both sacred and secular—in which Christ or a sovereign is shown with a diagrammatic, tripartite globe, or orb, signifying the rule of its holder over the world. Size of the original detail: 13.6 × 6.5 cm. Copyright Bibliothèque Royale Albert Iᵉʳ, Brussels (MS. 15.652–56, fol. 26r).

PLATE 12. THE THREE SONS OF NOAH. From a fifteenth-century manuscript of Jean Mansel's *La fleur des histoires*, this clearly shows the ark on Mount Ararat and the division of the world between the three sons of Noah: Shem in Asia, Ham in Africa, and Japheth in Europe.
Size of the original: 30 × 22 cm. Copyright Bibliothèque Royale Albert Iᵉʳ, Brussels (MS. 9231, fol. 281v).

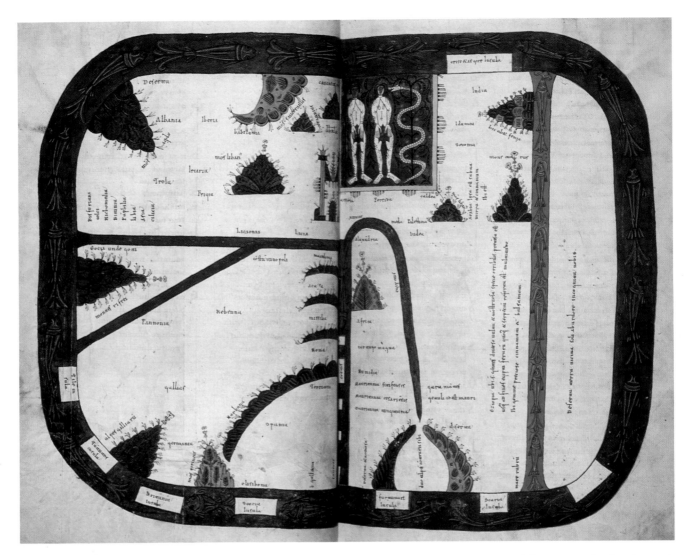

PLATE 13. THE BEATUS MAP FROM THE SILOS APOC-
ALYPSE. Dated to 1109, this map represents a tradition of
rectangular maps that can be traced back to a now-lost prototype of 776–86 in the *Commentary on the Apocalypse of
Saint John* of Beatus of Liebana. Displaying a Spanish-Arabic
style, the main characteristic of this map is the fourth continent, which Beatus considered inhabited.

Size of the original: 32 × 43 cm. By permission of the British
Library, London (Add. MS. 11695, fols. 39v–40r).

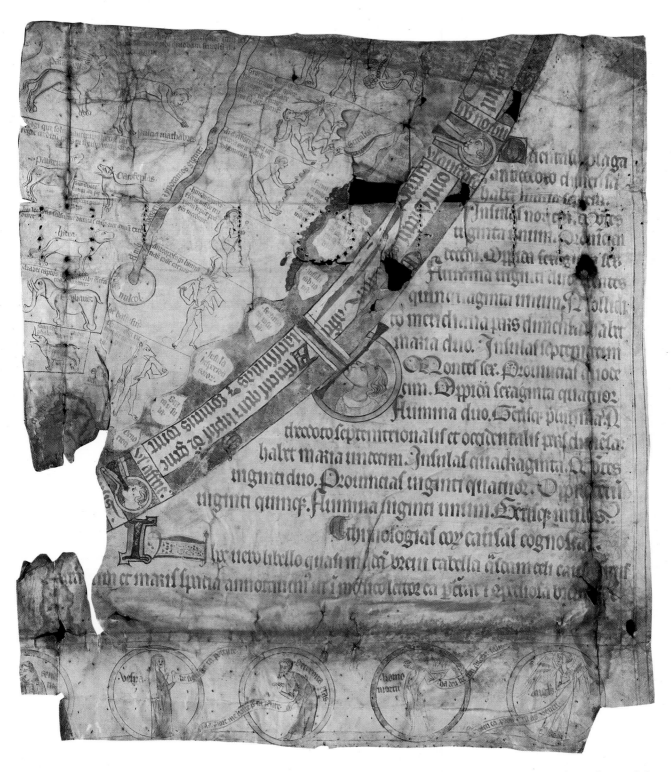

PLATE 14. THE DUCHY OF CORNWALL *MAPPAMUNDI*. This recently discovered fragment is from the lower right corner of a 1.57 m diameter *mappamundi* that has been carbon dated between 1150 and 1220. From a preliminary reading of the legends, the fragment bears similarity to both the Hereford and Ebstorf maps. It shows the area of West Africa.

Size of the fragment: 61 × 53 cm. From the archives of the Duchy of Cornwall, by permission of His Royal Highness the Prince of Wales.

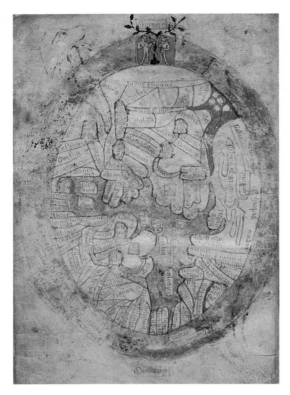

PLATE 15. HIGDEN'S *MAPPAMUNDI*: OVAL TYPE, MID-FOURTEENTH CENTURY. Perhaps following Hugh of Saint Victor's instructions for drawing a world map in the shape of Noah's ark, the oval maps of Higden represent the earliest of three types. Although it has been claimed that this manuscript is in Higden's own hand, most authorities recognize the British Library version (see fig. 18.67) as closer to the original archetype. From Ranulf Higden, *Polychronicon*.

Size of the original: 26.4 × 17.4 cm. By permission of The Huntington Library, San Marino, California (HM 132, fol. 4v).

PLATE 16. VESCONTE'S *MAPPAMUNDI*, 1321. At the beginning of the thirteenth century, *mappaemundi* began to incorporate the content and style of portolan charts. The world maps of Pietro Vesconte, drawn for Marino Sanudo's work promoting a crusade, represent the beginning of this trend. Not only is the Mediterranean Sea derived directly from such charts, but Vesconte also extended a network of rhumb lines over the land. From Marino Sanudo, *Liber secretorum fidelium crucis* 1306–21.

Diameter of the original: 35 cm. By permission of the British Library, London (Add. MS. 27376*, fols. 187v–188r).

PLATE 17. WESTERN EUROPE IN THE CATALAN ATLAS. Forming a segment of the traditional circular *mappamundi*, this late fourteenth-century world map was constructed on twelve panels, with the Mediterranean based on the outlines of the portolan charts.

Size of the original segment: 65 × 50 cm. Photograph from the Bibliothèque Nationale, Paris (MS. Esp. 30, pls. 5v–6r).

PLATE 18. THE FRA MAURO MAP. Representing the culmination of medieval cartography on the eve of the Renaissance, this map is a compendium of geographical sources, including the Portuguese explorations in Africa, Ptolemy's *Geography*, the Marco Polo narratives, and the portolan charts.

The surviving map is a copy—made at the request of the Venetian Signoria—of a map commissioned by Afonso V of Portugal in 1459.

Size of the original: 1.96 × 1.93 m. By permission of the Biblioteca Nazionale Marciana, Venice.

PLATE 19. MAPPAMUNDI OF PIRRUS DE NOHA. From an early fifteenth-century incipit by Pirrus de Noha of the *De cosmographia* of Pomponius Mela. Size of the original: 18 × 27 cm. Photograph from the Biblioteca Apostolica Vaticana, Rome. (Archivio di San Pietro H. 31, fol. 8r).

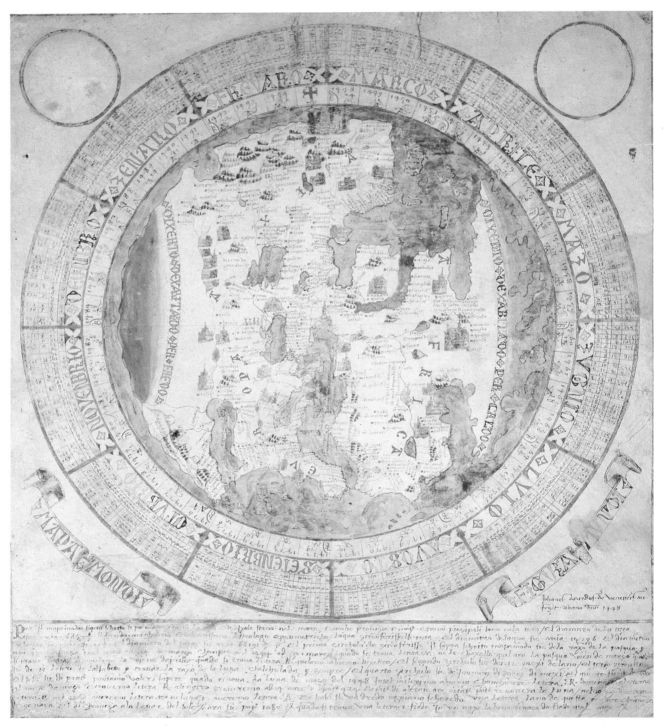

PLATE 20. *MAPPAMUNDI* OF GIOVANNI LEARDO, 1448.
Sharing many characteristics with the other two surviving
world maps of Leardo, among the more striking features are
the surrounding Easter calendar and the strongly colored uninhabited north polar and equatorial torrid zones.

Size of the original: 34.7 × 31.2 cm. By permission of the
Biblioteca Civica Bertoliana, Vicenza (598A).

PLATE 21. THE WORLD MAP OF ANDREAS WALSPER-GER. This 1448 map, which has extensive text explaining the cartographer's intentions, distinguishes between Christian (red) and Islamic (black) cities.
Diameter of the original: 42.5 cm. Photograph from the Biblioteca Apostolica Vaticana, Rome (Pal. Lat. 1362b).

PLATE 22. THE "ANGLO-SAXON" MAP. The heavy gray and bright orange colors on this tenth-century world map depart considerably from the usual blues, greens, and reds on the *mappaemundi*.
Size of the original: 21 × 17 cm. By permission of the British Library (Cotton MS. Tiberius B.V., fol. 56v.).

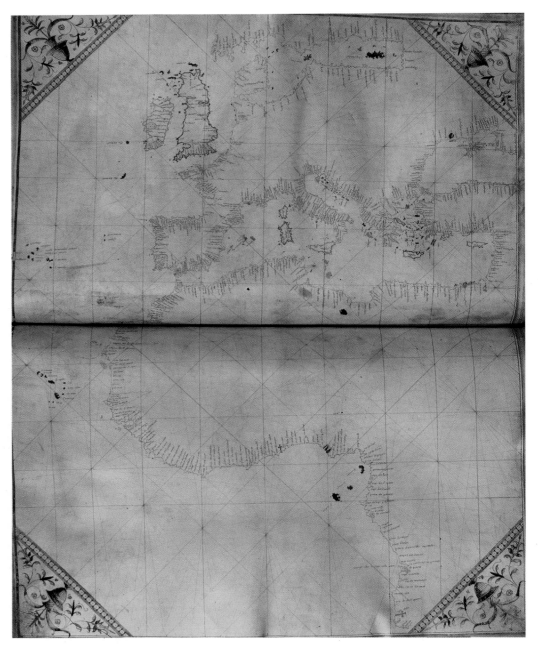

PLATE 23. AN EXTENDED "NORMAL-PORTOLANO."
This example, showing the standard areas of the Mediterranean and Black seas as well as western Africa, is from the fifteenth-century Cornaro atlas. Whereas most portolan charts used sixteen equidistant points to define the intersections of the rhumb lines, this chart has twenty-four.

Size of the original: 53.3 × 40.6 cm. By permission of the British Library, London (Egerton MS. 73, fols. 36–36ᵃv).

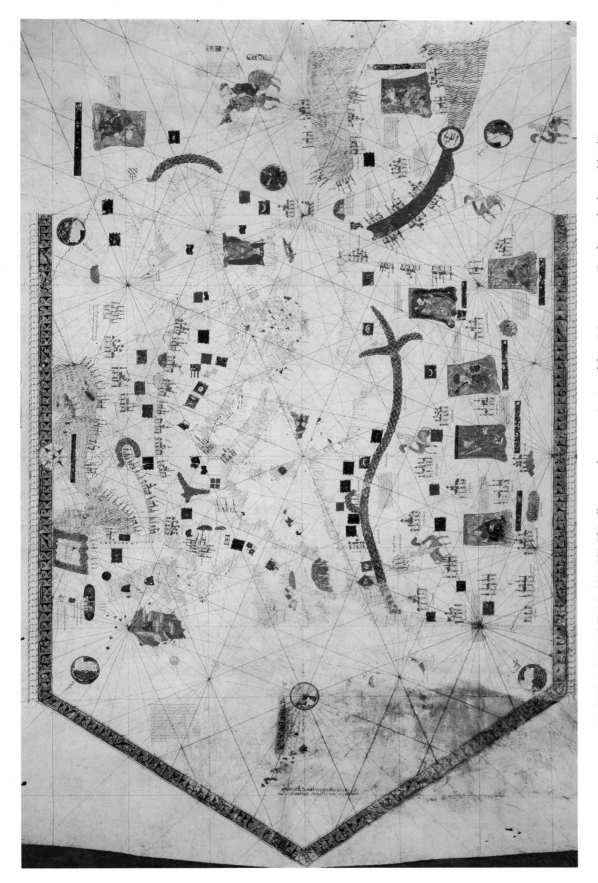

PLATE 24. THE 1439 VALSECA CHART. This illustrates the different color conventions used on portolan charts: the three colors of the rhumb lines (black or brown, green, and red); the rubrication of significant places; and the coloring of islands, such as Rhodes (white or silver cross on a red field), and of certain river deltas. More ornate Catalan-style charts, like this one, added their own elaborate conventions.

Size of the original: 75 × 115 cm. By permission of the Diputación de Barcelona, Museo Marítimo, Barcelona (inv. no. 3236).

PLATE 25. A CONTEMPORARY DERIVATIVE OF A POR-
TOLAN CHART. This map of the Black Sea takes its coastal
outline and names from a portolan chart, but it omits the
navigational rhumb lines. It is from a manuscript island book,
the *Insularum illustratum*, by Henricus Martellus Germanus,
who worked in Florence ca. 1480–96.
By permission of the Biblioteek der Rijksuniversiteit, Leiden
(Codex Voss. Lat. F 23, fols. 75v–76r).

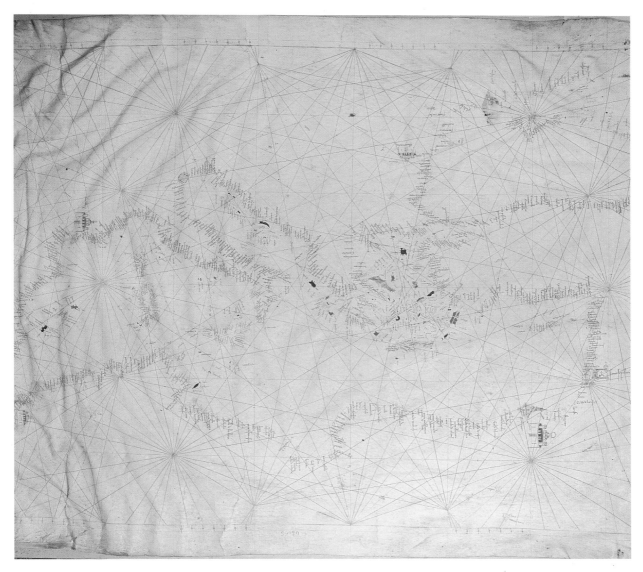

PLATE 26. A CATALAN CHART IN THE ITALIAN STYLE. This unsigned and undated chart emphasizes the difficulty of using only stylistic characteristics to distinguish between the Italian and Catalan portolan charts. Although this example is drawn in the austere fashion associated with Italian work, analysis of its place-names and the presence of town symbols indicate that it was probably produced in Majorca in the late fourteenth century.

Size of the original detail: 63 × 68 cm. By permission of the Biblioteca Nazionale Marciana, Venice (It. IV, 1912).

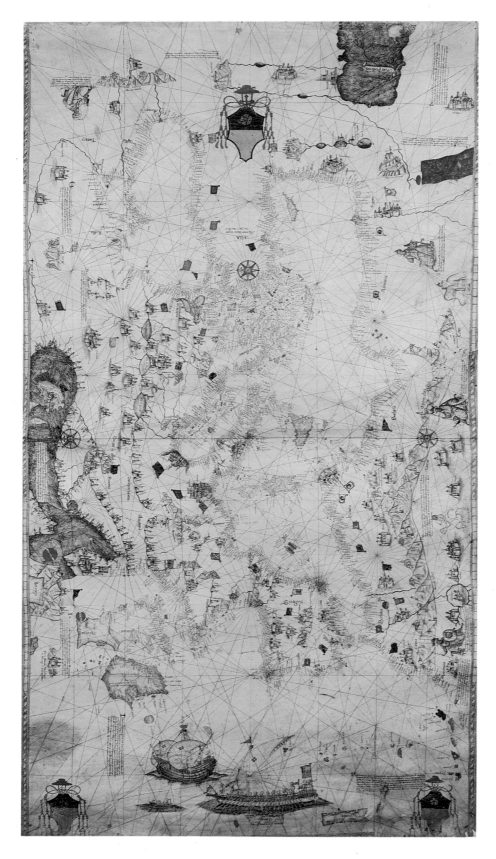

PLATE 27. AN ITALIAN CHART IN THE CATALAN STYLE.
Made in 1482 by Grazioso Benincasa, this chart reverses the
situation in Plate 26. Despite its internal detail and decoration
it was in fact drawn in Bologna by the most prolific of the
fifteenth-century Italian chartmakers. The repeated coats of
arms beneath a cardinal's hat are those of Raffaello Riario,
for whom the chart was made.
Size of the original: 71 × 127.5 cm. By permission of the
Biblioteca Universitaria, Bologna (Rot. 3).

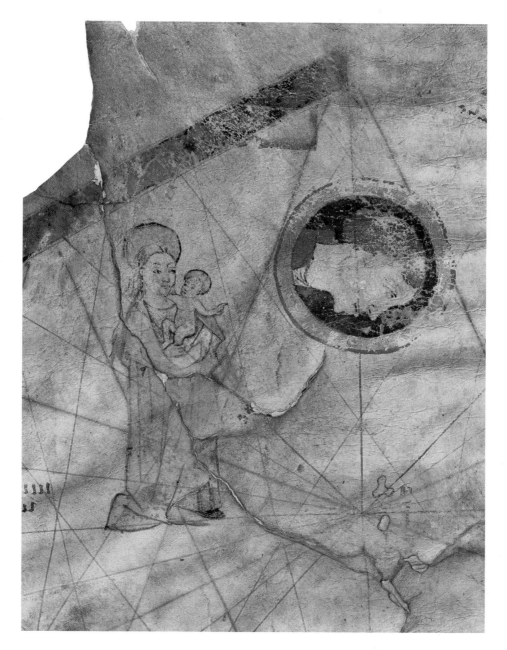

PLATE 28. REPRESENTATION OF THE MADONNA AND CHILD. This particular example is from the neck of the 1464 Petrus Roselli chart. Other charts bear cornerpieces of various saints, in a practice that seems to have been a Venetian hallmark.

Height of the original figure: 7 cm. By permission of the Germanisches Nationalmuseum, Nuremberg (Codex La. 4017).

PLATE 29. CITY FLAGS. The practice of placing flags above cities, as on this chart from an atlas of 1321 attributed to Pietro Vesconte, is less useful for dating than it might appear. The flags are sometimes imprecisely positioned and may be inappropriate for the place concerned. For example, Christian flags were often shown flying above cities many years after their conquest by the Ottoman Turks.

Size of each original: 22.5 × 29.3 cm. Photograph from the Bibliotheca Apostolica Vaticana, Rome (Vat. Lat. 2972).

PLATE 30. THE CARTE PISANE. Probably dating from the end of the thirteenth century, this portolan chart is accepted as the oldest extant example. Pisa is the city from which it supposedly emerged in the nineteenth century; its authorship is generally, though not universally, considered to be Genoese. Among the chart's noteworthy features are the twin rhumb line networks, with centers near Sardinia and the coast of Asia Minor. Outside the two circles, which are inked in here but would be left hidden on later charts, some areas are covered by a grid whose purpose remains unclear.
Size of the original: 50 × 104 cm. Photograph from the Bibliothèque Nationale, Paris (Rés. Ge. B 1118).

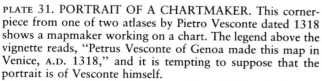

PLATE 31. PORTRAIT OF A CHARTMAKER. This cornerpiece from one of two atlases by Pietro Vesconte dated 1318 shows a mapmaker working on a chart. The legend above the vignette reads, "Petrus Vesconte of Genoa made this map in Venice, A.D. 1318," and it is tempting to suppose that the portrait is of Vesconte himself.
By permission of the Civico Museo Correr, Venice (Collezione Correr, Port. 28, fol. 2).

PLATE 32. THE WHEEL DIAGRAM FROM THE CATALAN ATLAS. This is the most splendid of the lunar calendars found in conjunction with a portolan chart. Moving outward from a symbolic representation of the earth, its concentric rings illustrate, in turn, the other elements, the planets and their astrological qualities, the signs of the zodiac, the moon's stations and (against a deep blue background) its phases, then the six bands of the lunar calendar, followed by further as-

trological texts and figures. The final ring explains the nineteen-year sequence of golden numbers used in conjunction with the lunar calendar. This great wheel diagram is rounded off by cornerpiece female figures representing the seasons, starting upper right with spring and moving counterclockwise.
Size of the original segment: 65 × 50 cm. Photograph from the Bibliothèque Nationale, Paris (MS. Esp. 30).

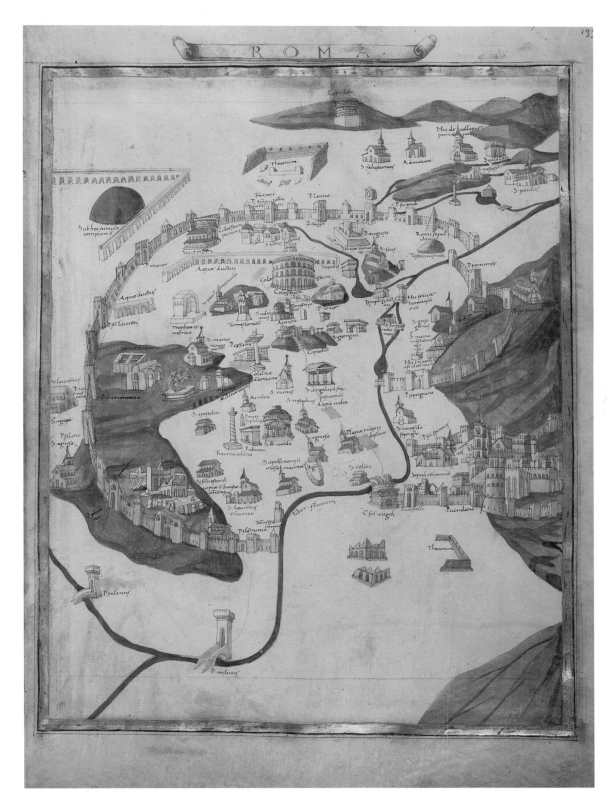

PLATE 33. ROME FROM AN UNDATED MANUSCRIPT OF PTOLEMY'S *GEOGRAPHY*. This is one of many plans of Italian and Near Eastern cities to emerge from the workshop of Pietro del Massaio, a Florentine artist of the late fifteenth century.

Size of the original: 56.8 × 42.1 cm. Photograph from the Bibliothèque Nationale, Paris (MS. Lat. 4802, fol. 133r).

PLATE 34. MAP OF THE DISTRICT AROUND VERONA. Although Lake Garda and the Adige valley may not be drawn to scale on this regional map of the mid-fifteenth century, the idea of a uniform ground scale does seem to have been applied to the detailed representation of Verona. See also fig. 20.13.

Size of the original: 305 × 223 cm. Photograph courtesy of Thames and Hudson. By permission of the Archivio di Stato, Venice.

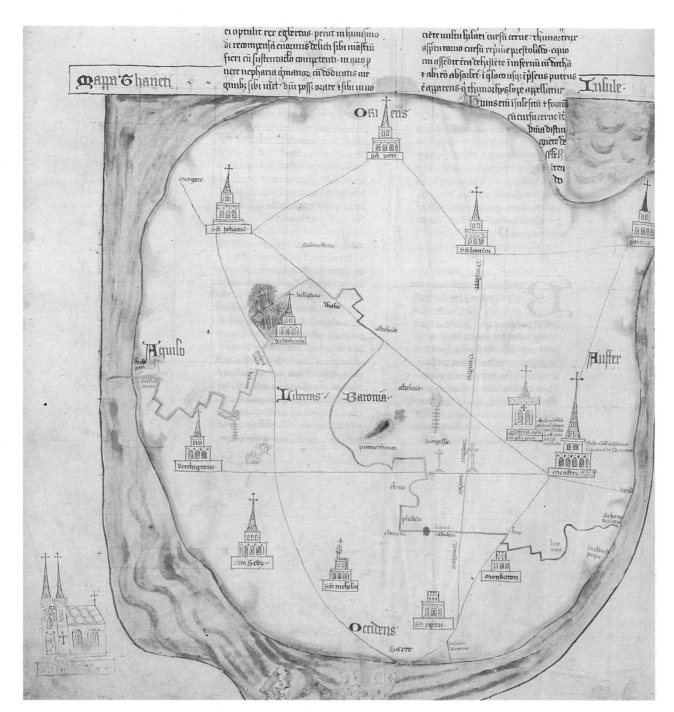

PLATE 35. PLAN OF THE ISLE OF THANET, KENT. Drawn at about the same time as the sketched plan of Clenchwarton (fig. 20.20), ca. 1400, this map represents the other extreme of the stylistic continuum: a carefully drawn and colored artistic work.

Size of the original: 39 × 37.5 cm. Found in Thomas of Elmham's *Historia Abbatiae S. Augustine*. By permission of the Master and Fellows of Trinity Hall, Cambridge (MS. 1, fol. 42v).

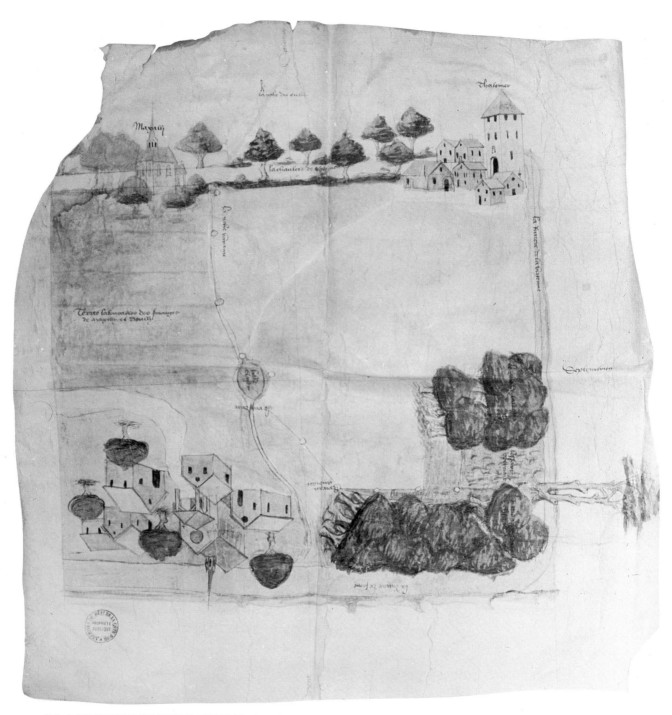

PLATE 36. A PORTION OF THE BOUNDARY OF THE DU-CHY OF BURGUNDY, 1460. The boundary passes through the fields that separate the three villages of Talmay, Maxilly, and Heuilley (Côte-d'Or). The artist has given the map three separate horizons that are labeled in turn: north (to the right), east, and west. At the eastern extreme, beyond Heuilley, is the river Saône. This map was possibly produced as a result of the 1444 boundary dispute between Duke Philip the Good and King Charles VII of France.

Size of the original: 56 × 62 cm. By permission of the Archives Départementales de la Côte-d'Or, Dijon (B 263).

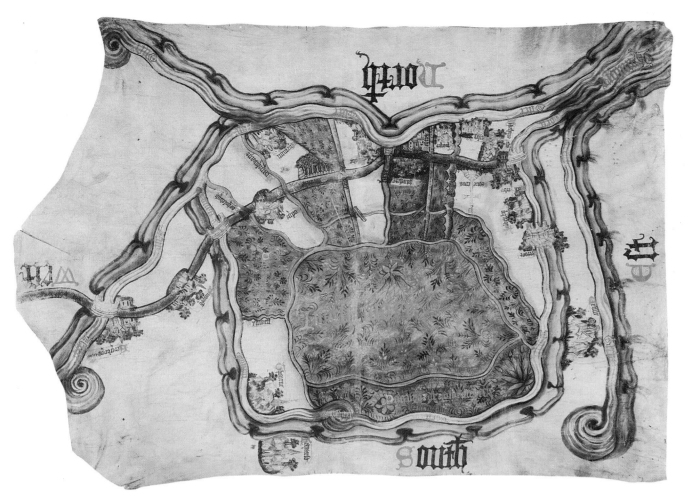

PLATE 37. MAP OF INCLESMOOR, YORKSHIRE. One of two later fifteenth-century copies of a map produced during a dispute between the duchy of Lancaster and Saint Mary's Abbey, York, 1405–8, over the rights to pasture and peat on an area south of the river Humber.

Size of the original: 60 × 74 cm. Crown copyright, by permission of the Public Record Office, Kew (MPC 56, ex DL 31/61).

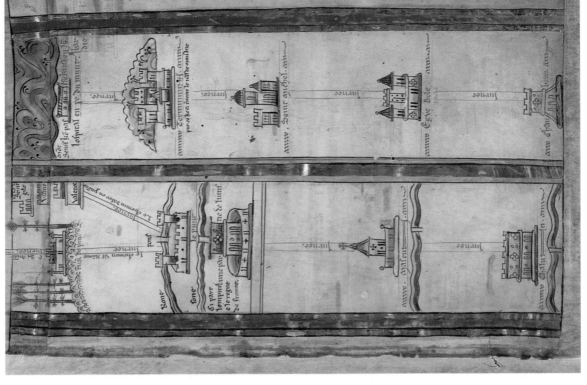

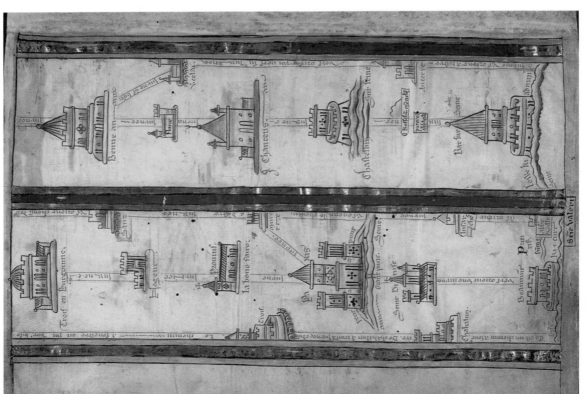

PLATE 38. ITINERARY MAP BY MATTHEW PARIS. This shows two sections of a mid-thirteenth-century itinerary of the route to the Holy Land. The verso depicts Bar-sur-Seine (bottom right) to Troyes (top left); the recto is Tour de Pin (top left) to Chambéry (bottom right). Staging points are depicted, sometimes realistically, by thumbnail sketches set on vertical lines. Intermediate distances are marked with the journey time in days.

Size of each original: 34.8 × 25.2 cm. By permission of the British Library, London (Royal MS. 14.C.vii, fols. 2v–3r).

PLATE 39. GREAT BRITAIN BY MATTHEW PARIS. This famous map, known in four versions, should be read as an itinerary map with its central axis running from Newcastle upon Tyne to Dover in a straight line via the Abbey of Saint Albans (Paris's own monastery).

Size of the original: 33 × 22.9 cm. By permission of the British Library, London (Cotton MS. Claudius D.vi, fol. 12v).

PLATE 40. THE GOUGH MAP, CA. 1360. Deriving its name from its inclusion in the map collection of Richard Gough, the eighteenth-century English antiquary, this map of Great Britain shows five roads radiating from London with branches and crossroads. It is much more detailed than the Matthew Paris maps and, in the positioning of towns, rivers, and coastlines, even beyond the routes themselves, significantly more accurate. Size of the original: 56 × 118 cm. By permission of the Bodleian Library, Oxford (MS Gough Gen. Top. 16).

6 · Cartography in the Ancient Near East

A. R. MILLARD

Under the term "ancient Near East" fall the modern states of Iraq, Syria, Lebanon, Jordan, and Israel. Turkey, Saudi Arabia, the Gulf States, Yemen, and Iran may also be included. The eras embraced begin with the first urban settlements (ca. 5000 B.C.) and continue until the defeat of Darius III by Alexander the Great, who officially introduced Hellenism to the area (330 B.C.). There are few examples of maps as they have been defined in the literature of the history of cartography, but those that remain are important in helping to build a picture of the geographical knowledge available, and of related achievements.

BABYLONIAN GEOGRAPHICAL KNOWLEDGE

Babylonia was open to travelers from all directions. The courses of the Tigris and Euphrates rivers offered major routes to and from the north and the northwest, and the Persian Gulf allowed contact by sea along the coasts of Arabia and east to India (fig. 6.1). It is no surprise, therefore, to find the urban culture which the Sumerians developed during the fourth millennium B.C. spreading far afield through trade and conquest. Recent excavations have revealed a large settlement on the middle Euphrates (Khabuba Kabira) where buildings and pottery have characteristically southern Mesopotamian styles. There is increasing evidence, too, of Sumerian influence eastward into Iran. Arguably the greatest achievement of this culture was the invention of writing, with the development of the cuneiform script, commonly written on clay tablets.

Although there is nothing that qualifies as an unambiguous attempt at mapping in this area during the fourth millennium B.C., the scribal activities and traditions beginning then created the circumstances in which geographical knowledge could be stored and maps could be produced. The extant examples of ancient knowledge and its application have been discovered by chance; new discoveries may add significantly to what is currently available.

The Sumerian scribes compiled long lists of words by category, for reference and teaching, and among these were lists of towns, mountains, and rivers. Good ex-amples of these lists have been unearthed in Babylonia, at Abū Salābīkh near Nippur, and at the northern Syrian settlement of Ebla, the scene of important discoveries by Italian archaeologists, lying fifty-five kilometers south of Aleppo. The scribes who wrote these tablets were working between 2500 and 2200 B.C., but their lists were drawn from earlier sources that reached back as far as the beginning of the third millennium. Besides the names of places in Babylonia, names of Syrian towns appear in the lists from Ebla, including Ugarit (Ra's Shamrah) on the Mediterranean coast.[1] This is one indication of the level Babylonian geographical knowledge had reached at an early date. In support of that may be cited historical sources, contemporary and traditional, for military campaigns by King Sargon of Akkad and his grandson Naram-Sin into northern Syria and even Anatolia in the century 2330–2230 B.C. Place-name lists continued as an element of scribal lore throughout the history of the cuneiform script. In a revised form, they became part of a standard compendium of lexical information that was copied repeatedly with minor variations and explanatory additions. Regrettably, the manuscripts of the second and first millennia B.C. are incomplete, and their total purview remains unknown. As part of a standard, traditional compilation, however, they do not reflect contemporary information.[2]

The marches of armies to distant goals, and the ventures of traders in search of precious metals and stones, timber, and other products, were the obvious means by which the scribes learned about their own and foreign lands. That they knew much more than the lists of place-names reveal is clear from the evidence of links with Iranian towns and the centers of the Indus Valley culture at Mohenjo Daro and Harappa, links formed at least in part by the sea route through the Persian Gulf. Various

1. Robert D. Biggs, "The Ebla Tablets: An Interim Perspective," *Biblical Archaeologist* 43, no. 2 (1980): 76–86, esp. 84; Giovanni Pettinato, "L'atlante geografico del Vicino Oriente antico attestato ad Ebla e ad Abū Salābīkh," *Orientalia*, n.s., 47 (1978): 50–73.

2. Benno Landsberger, *Materialien zum Sumerischen Lexikon: Vokabulare und Formularbücher* (Rome: Pontifical Biblical Institute Press, 1937–), vol. 11, *The Series HAR-ra = ḫubullu: Tablets XX–XXIV*, ed. Erica Reiner and Miguel Civil (1974). (Series title after 1970: *Materials for the Sumerian Lexicon*.)

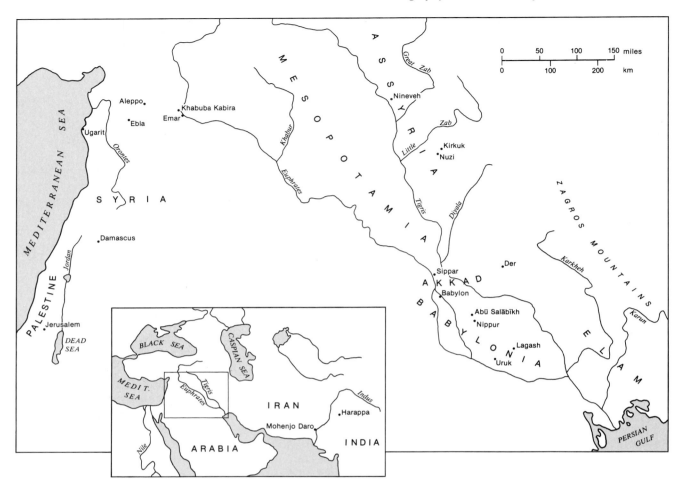

FIG. 6.1. PRINCIPAL PLACES ASSOCIATED WITH MAPS
IN THE ANCIENT NEAR EAST.

objects from those places have been found at several
Babylonian sites, mostly in levels of the middle and later
part of the third millennium B.C.[3]

Occasionally details of such journeys are preserved in
business and administrative records. The most useful
have the form of itineraries, naming the places visited,
some with a note of the time taken to travel from one
to another. The longest example describes a route from
southern Babylonia to Emar (Meskene) on the middle
Euphrates. It appears to have been a military expedition,
although its purpose is unclear; the number of nights
spent at each place is carefully recorded.[4] Other itineraries concern the routes from Assyria to central Anatolia
in the nineteenth century B.C. and the marches of Assyrian armies in the early first millennium B.C.[5] In their
annals the Assyrian kings often included references to
the terrain they crossed, and sometimes to local vegetation and other features. Pictorial records of some of
their campaigns, in which artists attempted to represent
local features, can be seen in some of the bas-reliefs that
decorated the walls of palaces in Nineveh and neighboring cities. In addition, treaties and other documents
might define boundaries, naming towns, villages, or natural features that marked them.[6] For purposes of control or taxation the towns in a territory or kingdom were
also listed.[7]

3. C. C. Lamberg-Karlovsky, "Trade Mechanisms in Indus-Mesopotamian Interrelations," *Journal of the American Oriental Society*
92 (1972): 222–29.

4. William W. Hallo, "The Road to Emar," *Journal of Cuneiform
Studies* 18 (1964): 57–88.

5. Dietz Otto Edzard, "Itinerare," in *Reallexikon der Assyriologie
und vorderasiatischen Archäologie*, ed. Erich Ebeling and Bruno
Meissner (Berlin: Walter de Gruyter, 1932–), 5:216–20.

6. Jean Nougayrol, *Le palais royal d'Ugarit, IV: Textes accadiens
des Archives Sud (Archives Internationales)*, Mission de Ras Shamra,
9 (Paris: Imprimerie Nationale, 1956), 48–52, 63–70; Mervyn E. J.
Richardson, "Hebrew Toponyms," *Tyndale Bulletin* 20 (1969): 95–
104, esp. 97–101.

7. Fritz Rudolf Kraus, "Provinzen des neusumerischen Reiches von
Ur," *Zeitschrift für Assyriologie und vorderasiatische Archäologie*,
n.s., 17 (1955): 45–75.

BABYLONIAN MENSURATION AND CALCULATION

The Babylonians developed means for measuring distances on the basis of the time taken to travel, the main unit being the *bēru*, "double hour," of about ten kilometers. For shorter lengths the cubit (*ammatu*) of about fifty centimeters was used, and this could be divided into "fingers" (*ubānu*), usually thirty to the cubit, but in the late period only twenty-four. The statues of Gudea (see below, "Babylonian Plans") depict graduated rulers, and we may assume knotted cords were the means for measuring longer distances. A goddess is said to carry the rope of cubit and reed measures.[8] Babylonian measurements could be very exact, and the evidence of various mathematical problem texts suggests surveys and plans could be done accurately. Mathematical tables and the problem texts reveal an extensive knowledge of square and cube roots, reciprocal numbers, solutions for quadratic and other equations, and means of calculating areas of rectangular, circular, and irregular figures and the volumes of prisms and cylinders. The Pythagorean theorem was understood both in practice and in theory in the seventeenth century B.C., a millennium before Pythagoras himself was born. Central to Babylonian calculation was the sexagesimal system, in which units of sixty form the base (so $1 + 20 = 80$, $2 + 10 = 130$, etc.). Late in Babylonian history this led to the division of the circumference of the circle into 360 parts.

BABYLONIAN PLANS

Besides their normal habit of writing on clay tablets, the Babylonian scribes also used the tablets as surfaces for drawing. From the days of Sargon of Akkad (ca. 2300 B.C.) until the middle of the first millennium B.C., these drawings included plans of property, land, houses, and temples. Incised lines indicated walls, streets, rivers, and canals, occasionally with wavy lines to denote water. Some of the plans are no more than sketches, perhaps school exercises, but others are carefully drawn, with the walls of buildings of even width and the measurements of the rooms marked precisely in cubits.[9] The most famous is the plan on a statue of Gudea, prince of Lagash (Telloh), ca. 2141–2122 B.C. (figs. 6.2 and 6.3). The seated figure holds on his knees a tablet engraved with the plan of an elaborate enclosure wall, probably for a

FIG. 6.2. THE GUDEA STATUE, CA. 2141–2122 B.C. This statue depicts Gudea, prince of Lagash, with the temple plan illustrated in figure 6.3 on his lap.
Height of the original: 93 cm. By permission of the Musée du Louvre, Paris.

FIG. 6.3. THE TABLET ON THE GUDEA STATUE. A plan of an enclosure wall for a temple or other large building is shown. Note the graduated ruler at the top edge which provided an indication of scale.
Size of the tablet: 12 × 24 cm. By permission of the Musée du Louvre, Paris.

---

8. Ignace J. Gelb et al., eds., *The Assyrian Dictionary* (Chicago: Oriental Institute, 1968), vol. 1, pt. 2, 448.

9. Ernst Heinrich and Ursula Seidl, "Grundrißzeichnungen aus dem alten Orient," *Mitteilungen der Deutschen Orient-Gesellschaft zu Berlin* 98 (1967): 24–45. For a gridded plan of a large building, probably a royal palace, see British Museum, *Cuneiform Texts from Babylonian Tablets, etc., in the British Museum* (London: British Museum, 1906), pt. 22, pl. 50, BM 68841 + 68843 + 68845 and 68840 + 68842.

temple. Beside it lie a stylus and a ruler with graduated divisions, badly damaged. Another statue of the same prince bears a blank tablet with a complete ruler. Since the statues are not life-size (respectively ninety-three and eighty-six centimeters high), it is difficult to discover the exact values of the units of length. Both statues are in the Musée du Louvre, Paris. Most of the plans depict single buildings, but a few show more: the shape of a town, or a part of a town. One fragment marks a temple and adjacent streets, thought to be in Babylon (fig. 6.4),[10] another shows part of the city of Uruk (Erech) and a building inside it (fig. 6.5).[11] A piece of a tablet in the British Museum has part of a town on one side, with a river, a gate, and an intervening suburb. On the other side are measurements that appear to relate to parts of a suburb of Babylon (fig. 6.6).[12] Still incomplete, but far more impressive, is a tablet incised with a plan of Nippur, the religious center of the Sumerians in Babylonia (fig. 6.7). Drawn about 1500 B.C., it marks the principal temple, a park and another enclosure, the river Euphrates, a canal to one side of the city, and another canal running through the center. A wall surrounds the city,

FIG. 6.4. FRAGMENT OF A CITY MAP, PROBABLY BABYLON. This cuneiform tablet probably shows the great temple of Marduk in Babylon, and the adjacent street is probably the sacred procession road that led up to the temple.
Size of the original: 7.5 × 4.5 cm. By permission of the Trustees of the British Museum, London (BM 73319); see also British Museum, *Cuneiform Texts from Babylonian Tablets, Etc., in the British Museum*, pt. 22 (London: British Museum, 1906), pl. 49.

pierced by seven gates which, like all the other features, have their names written beside them. As on some of the house plans, measurements are given for several structures, apparently in units of twelve cubits (about six meters). Scrutiny of the map beside modern surveys of Nippur has led to the claim that it was drawn to scale. At present this is difficult to verify in detail because excavations have not uncovered sufficient remains of the town shown in the plan. How much of the terrain around Nippur was included cannot now be known because of damage to the tablet, nor is there any statement of the plan's purpose, although repair of the city's defenses is suggested.[13]

A plan of a temple drawn, perhaps, in the sixth century B.C. is unique in marking individual bricks of the walls. Here the precise measurements suggest the plan may be a scale drawing. By calculating from the standard brick size of the time, a scale close to 1:66⅔ has been deduced, which Heinrich and Seidl claim to be a common scale in use by architects of the period.[14] Other plans of temples or houses may also follow a scale, but there is no indication of it on the drawings, and some are clearly not in proportion to the measurements given.

Property transactions, sales or disputes, or estimates of yield were probably the reasons for the plans of fields drawn on tablets. Often a single plot of land is delineated, with measurements written along the sides. A few plans set out the relationships of adjacent plots and watercourses (vital to agriculture in southern Babylonia). Examples of this type date from the third millennium B.C. onward. A particularly complex example from Nippur, belonging to the same age as the town plan (about 1500 B.C.), displays the situation of several fields and canals around the hairpin bend of a water-

10. British Museum, *Cuneiform Texts*, pt. 22, pl. 49, BM 73319 (note 9).

11. H. J. Lenzen, Adam Falkenstein, and W. Ludwig, eds., *Vorläufiger Bericht über die von dem Deutschen Archäologischen Institut und der Deutschen Orient-Gesellschaft aus Mitteln der Deutschen Forschungsgemeinschaft unternommenen Ausgrabungen in Uruk-Warka*, Abhandlungen der Deutschen Orient-Gesellschaft, Winter 1953–54, Winter 1954–55 (Berlin: Gebr. Mann, 1956), 42, pl. 23c.

12. British Museum, *Cuneiform Texts*, pt. 22, pl. 49, BM 35385 (note 9); Eckhard Unger, *Babylon, die heilige Stadt nach der Beschreibung der Babylonier* (Berlin: Walter de Gruyter, 1931), 252–53.

13. Samuel Noah Kramer and Inez Bernhardt, "Der Stadtplan von Nippur, der älteste Stadtplan der Welt," *Wissenschaftliche Zeitschrift: Gesellschafts- und Sprachwissenschaftliche Reihe* 19 (1970): 727–30; Samuel Noah Kramer, *From the Tablets of Sumer* (Indian Hills, Colo.: Falcon's Wing Press, 1956), 271–75; idem, *History Begins at Sumer*, 3d ed. (Philadelphia: University of Pennsylvania Press, 1981), 375–79; McGuire Gibson, "Nippur 1975: A Summary Report," *Sumer* 34 (1978): 114–21, esp. 118–20.

14. See Heinrich and Seidl, "Grundrißzeichnungen," 42 (note 9).

FIG. 6.5. FRAGMENT OF A CITY MAP OF URUK. One of the finds resulting from the Uruk-Warka excavations, 1953–55. Unfortunately, no other fragments of this map were found. Size of the original: 8.1 × 11.2 cm. By permission of the Deutsches Archäologisches Institut, Abteilung Baghdad.

course (fig. 6.8).[15] In an area where routes commonly followed rivers or canals, well-defined passes, or coastlines, maps of larger coverage may have been less necessary, but a few tablets do have wider range and broader significance.

BABYLONIAN SMALL-SCALE MAPS

From time to time there were attempts to depict relationships between more widely separated places. A diagrammatic map of the late second millennium B.C., from Nippur, shows nine settlements with canals and a road between them, without noting any distances.[16] On a fragment of a tablet in the British Museum, belonging to the mid-first millennium B.C., a rectangle marks the city of Sippar, parallel lines above it mark the river Euphrates, and parallel lines below mark canals following a sinuous course (fig. 6.9).[17]

The British Museum has long exhibited the famous "Babylonian World Map," drawn about 600 B.C. (fig. 6.10). In the text accompanying the map various legendary beasts are named which were reputed to live in regions beyond the ocean that encircled the Babylonian world. A few ancient heroes reached those places, and the badly damaged text appears to describe conditions in them, one being the region "where the sun is not seen." The map is really a diagram to show the relation of these places to the world of the Babylonians. Each place is drawn as a triangle rising beyond the circle of the salty ocean. There may have been eight originally. Each is marked as being at a certain distance, probably from the next one. Enclosed by the circle of the salt sea lies an oblong marked "Babylon" with two parallel lines

running to it from mountains at the edge of the enclosure, and running on to a marsh marked by two parallel lines near the bottom of the circle. The marsh is the swamp of lower Iraq, its identity secured by the name Bit Yakin at its left end, known to be a tribal territory covering marshland. A trumpet-shaped arm of the sea curves around the right end of the marsh so that its neck

FIG. 6.6. FRAGMENT OF A CITY PLAN, POSSIBLY TÛBU. This cuneiform fragment shows the course of a canal or river, flowing outside the city wall, with one of the city gates, the Gate of Shamash, below.
Largest dimensions of the original: 10.5 × 7.5 cm. By permission of the Trustees of the British Museum, London (BM 35385); see also British Museum, *Cuneiform Texts from Babylonian Tablets, Etc., in the British Museum*, pt. 22 (London: British Museum, 1906).

15. Stephen H. Langdon, "An Ancient Babylonian Map," *Museum Journal* 7 (1916): 263–68; Jacob J. Finkelstein, "Mesopotamia," *Journal of Near Eastern Studies* 21 (1962): 73–92, esp. 80 ff. For a description of seventy late Babylonian field plans in the British Museum, see Karen Rhea Nemet-Nejat, *Late Babylonian Field Plans in the British Museum*, Studia Pohl: Series Maior 11 (Rome: Biblical Institute Press, 1982). See also Wolfgang Röllig, "Landkarten," in *Reallexikon*, 6:464–67 (note 5).

16. Albert Tobias Clay, "Topographical Map of Nippur," *Transactions of the Department of Archaeology, University of Pennsylvania Free Museum of Science and Art* 1, no. 3 (1905): 223–25.

17. British Museum, *Cuneiform Texts*, pt. 22, pl. 49, BM 50644 (note 9).

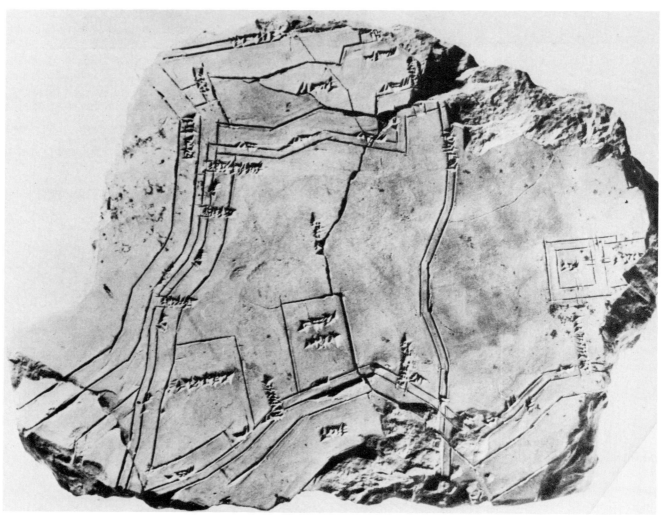

FIG. 6.7. PLAN OF NIPPUR, CA. 1500 B.C. Possibly the earliest town plan drawn to scale, this shows the temple of Enlil in its enclosure on the right edge, city walls, canals, storehouses, and a park.

Size of the original: 18 × 21 cm. By permission of the Hilprecht Collection, Friedrich-Schiller-Universität, Jena.

touches the lines from Babylon. Despite the absence of a name, it is clear that the parallel lines running to and from Babylon represent the river Euphrates. To the right of Babylon an oval marks Assyria, and above it is apparently Urartu (Armenia). Several other cities are marked by small circles; one near the trumpet-shaped sea, named "Fort of the god," is probably Der (Badrah) at the foot of the Zagros Mountains. The name Khabban to the upper left appears to denote an area of Elam southeast of the Zagros, geographically out of place (it might be another town of the same name otherwise unknown).[18] Obviously this is not so much a topographical map as an attempt to illustrate ideas expressed in the accompanying text, greatest attention being paid to the remote regions. The Babylonians evidently viewed the earth as flat, in common with other ancient peoples.[19]

Their references to the "four quarters" relate to the directions of the winds and should not be taken as implying that they thought it was square. (The same is true for Isa. 11:12, which mentions the "four corners of the earth.") There is no reason to suppose, as some have,[20] that the creatures described in the text accompanying this Babylonian world map were intended as zodiacal

18. British Museum, *Cuneiform Texts*, pt. 22, pl. 48, BM 92687 (note 9); Unger, *Babylon*, 254–58 (note 12); A. Leo Oppenheim, "Man and Nature in Mesopotamian Civilization," in *Dictionary of Scientific Biography*, 16 vols., ed. Charles Coulston Gillispie (New York: Charles Scribner's Sons, 1970–80), 15:634–66, esp. 637–38.

19. Wilfred G. Lambert, "The Cosmology of Sumer and Babylon," in *Ancient Cosmologies*, ed. Carmen Blacker and Michael Loewe (London: George Allen and Unwin, 1975), 42–65, esp. 47–48.

20. Eckhard Unger, "From the Cosmos Picture to the World Map," *Imago Mundi* 2 (1937): 1–7, esp. 1–5.

FIG. 6.9. MAP OF SIPPAR AND ITS SURROUNDINGS, FIRST MILLENNIUM B.C. The rectangle marks the city, with the Euphrates above and canals below.
Size of the original: 8 × 9 cm. By permission of the Trustees of the British Museum, London (BM 50644).

FIG. 6.8. PLAN OF FIELDS FROM NIPPUR, CA. 1500 B.C. The fields, belonging to royal and religious estates, are situated on both sides of a hairpin bend in a watercourse, separated by irrigation channels.
Size of original: 13 × 11 cm. By permission of The University Museum, University of Pennsylvania, Philadelphia (CBS 13885).

in any way. At the end is part of the title that might be translated "[These are the drawings] of the four regions (or 'edges') of all [the world]."

Equally significant for the history of cartography is a clay tablet 7.6 × 6.8 centimeters unearthed at Yorghan Tepe near Kirkuk in 1930–31 (fig. 6.11). With it were other tablets from the time of the dynasty of Akkad, and there is no doubt this one belongs to the same date, about 2300 B.C. At that time the place was known as Gasur; a thousand years later it was Nuzi. The surface of the tablet bears a map of a district bounded by two ranges of hills and bisected by a watercourse. Inscriptions identify some features and places. In the center the area of a plot of land is specified as 354 *iku* (about twelve hectares), and its owner is named—Azala. None of the names of other places can be understood except the one in the bottom left corner. This is Mashkan-dur-ibla, a place mentioned in the later texts from Nuzi as Dur-ubla.[21] By the name, the map is identified as of a region near Yorghan Tepe, although the exact location is unknown. Whether the map shows a stream running down a valley to join another or running from that to divide

into three, and whether they are rivers or canals, cannot be determined. The shaded area at the left side, to or from which the channels run, was named, but the writing is illegible. Groups of overlapping semicircles mark ranges of hills, a convention used by artists then and in later times. Finally, the scribe oriented his map, writing "west" at the bottom, "east" at the top, and "north" at the left.[22]

With this, the oldest known example of orientation, and the possibility of a scale drawing in the Nippur map, Babylonian cartographers of the third and second millennia B.C. may be held to have practiced two essential principles of geographical mapmaking. Written itineraries and surveys testify to their awareness of greater

21. A. R. Millard, "Strays from a 'Nuzi' Archive," in *Studies on the Civilization and Culture of Nuzi and the Hurrians*, ed. Martha A. Morrison and David I. Owen (Winona Lake, Ind.: Eisenbrauns, 1981), 433–41, esp. 438 and n. 5 (contributed by Karl-Heinz Deller). There is no justification for linking this place with the famous Ebla or Ibla of northern Syria, as suggested by Nadezhda Freedman, "The Nuzi Ebla," *Biblical Archaeologist* 40, no. 1 (1977): 32–33, 44.

22. Harvard University, Semitic Museum, *Excavations at Nuzi*, 8 vols. (Cambridge: Harvard University Press, 1929–62), vol. 3, Theophile James Meek, *Old Akkadian, Sumerian, and Cappadocian Texts from Nuzi*, XVII ff., pl. 1; idem, "The Akkadian and Cappadocian Texts from Nuzi," *Bulletin of the American Schools of Oriental Research* 48 (December 1932): 2–5.

FIG. 6.10. THE BABYLONIAN WORLD MAP, CA. 600 B.C.
This map shows the relationship between the legendary regions
beyond the ocean and the Babylonian world. The parallel lines
running to and from Babylon (the elongated rectangle) represent the Euphrates, while the circular band represents the
salt sea.
Largest dimensions of the original: 12.5 × 8 cm. By permission
of the Trustees of the British Museum, London (BM 92687).

FIG. 6.11. CLAY TABLET MAP EXCAVATED AT YORGAN
TEPE. This is a cast of the earliest known example, ca. 2300
B.C., of a topographical map in which the cardinal directions
are clearly marked.
Size of the original: 6.8 × 7.6 cm. By permission of the Semitic
Museum, Harvard University, Cambridge (acc. no. SMN
4172); see also Theophile James Meek, *Old Akkadian, Su-
merian, and Cappadocian Texts from Nuzi*, vol. 3 of Harvard
University, Semitic Museum, *Excavations at Nuzi*, 8 vols.
(Cambridge: Harvard University Press, 1929–62), tablet 1.

CELESTIAL GEOGRAPHY

From early times the Babylonians observed how the
heavenly bodies moved or did not move, and in the
second and first millennia B.C. they noted this in writing.
Their basic aims were calendrical and astrological, yet
they went on to make accurate records which are still
of value to scientists.

The practical problems of regulating the calendar pro-
voked Babylonian sky watchers to calculate when the
new moon should appear on the western horizon, so
that they could inaugurate a new month by theory when
weather conditions prevented a sighting. Eventually,
probably in the Persian period (fifth century B.C.), mathematical predictions were generated to give tables of the
moon's position throughout the year. From these it was
possible to compute when a month should be inserted
in the lunar calendar to keep it in step with the solar
year (seven times in nineteen years).

The fixed stars were classed in three parallel bands
called "roads," named after the major gods, Enlil, Anu,
and Ea. Through the central "road of Anu" ran the

distances and spatial relationships, and it may be that
the difficulty of drawing on a flat surface of damp clay
and the limited size of the clay tablets (they are seldom
more than twenty centimeters square) were obstacles to
more extensive mapping. Even allowing for the accidents
of survival, mapmaking cannot have been common
among the scribes of ancient Babylonia. Beside the thousands of administrative and legal documents in cuneiform, the number of plans of houses, properties, and
towns is small, counted in dozens rather than hundreds,
and the number of maps is limited to the few just described. Recently a fragment of a clay tablet originating
in the sixth century B.C. and preserved in the Louvre has
been made known (fig. 6.12). It shows a mountainous
region, the mountains being marked by small squares,
with a road running through it, a river, and a canal with
its secondary streams.[23]

---

23. D. Arnaud, *Naissance de l'écriture*, ed. Béatrice André-Leicknam
and Christiane Ziegler (Paris: Editions de la Réunion des Musées
Nationaux, 1982), 243, no. 189.

FIG. 6.12. THE LOUVRE TABLET MAP. A sixth-century B.C. fragment showing mountains (small squares) with a road, a river, and a canal with secondary streams.
Size of the original: 12 × 7.5 × 2.9 cm. By permission of the Musée du Louvre, Paris (AO 7795).

equator. This concept is described but not specifically illustrated. Other tablets provide computations of the distances between the stars. Related to the scheme of "roads" is a group of texts now labeled "astrolabes" or "planispheres." The earliest known example was written in the twelfth century B.C. Some of these show three concentric circles divided radially into twelve segments, each marked for a month of the year. A star is named in each division, with numbers which increase and decrease in a linear zigzag fashion, a concept basic to later calculations about periods of visibility. These texts are believed to relate to the length of the day as well as to the positions of the stars. Some have linear diagrams of the constellations, making a kind of schematized celestial

map. Other tablets list distances between the heavenly bodies in "double hours," a process somewhat similar to an itinerary.[24]

CARTOGRAPHIC KNOWLEDGE IN SYRIA AND PALESTINE

Where Babylonian cultural and scribal influences were strong, the possibility exists that similar plans and maps were drawn. This was true for most of Syria and, to a lesser extent, most of Palestine during the second millennium B.C. and, as the Ebla texts show, in the previous millennium also. To date, however, no examples of cartography from those ages have come to light in the Levant. As in Babylonia, there are written records which could provide the basis for constructing diagrammatic maps. To the cuneiform texts can be added itineraries in the Old Testament (e.g., Numbers 33), following basically the same form: "They set out from A and camped at B."[25] From the Old Testament, too, come the detailed delineations of the borders of Israel's Promised Land (Num. 34:2–12): "To the east you shall draw a line from Hazar-enan to Shepham; it shall run down from Shepham to Riblah east of Ain, continuing until it strikes the ridge east of the sea of Kinnereth. The frontier shall then run down to the Jordan and its limit shall be the Dead Sea. The land defined by these frontiers shall be your land" (Num. 34:10–12). Similar are the specifications of each tribe's territories by various topographical indicators (Joshua 15–19). A larger horizon is provided by the "Table of Nations" in Genesis 10, which arranges the peoples of the known world mostly on a framework of kinship but with some geographical references.[26] Ancient Israelite scribes, and their colleagues trained in Phoenician and Aramaic, used papyrus as their writing material for all but monumental or ephemeral documents, after the Egyptian fashion, and so their products can hardly be expected to have survived in the damp soil unless through special circumstances of preservation.

24. Ernst F. Weidner, *Handbuch der babylonischen Astronomie, der babylonische Fixsternhimmel* (Leipzig: Hinrichs, 1915; reprinted Leipzig: Zentralantiquariat, 1976); B. L. van der Waerden, "Mathematics and Astronomy in Mesopotamia," in *Dictionary of Scientific Biography*, 15:667–80, esp. 672–76 (note 18).

25. Graham I. Davies, "The Wilderness Itineraries: A Comparative Study," *Tyndale Bulletin* 25 (1974): 46–81; idem, *The Way of the Wilderness: A Geographical Study of the Wilderness Itineraries in the Old Testament* (Cambridge: Cambridge University Press, 1979).

26. Donald J. Wiseman, ed., *Peoples of Old Testament Times* (Oxford: Clarendon Press, 1973), xvi–xviii.

BIBLIOGRAPHY
CHAPTER 6 CARTOGRAPHY IN THE ANCIENT
NEAR EAST

Davies, Graham I. *The Way of the Wilderness: A Geographical Study of the Wilderness Itineraries in the Old Testament.* Cambridge: Cambridge University Press, 1979.

Donald, Trevor. "A Sumerian Plan in the John Rylands Library." *Journal of Semitic Studies* 7 (1962): 184–90.

Edzard, Dietz Otto. "Itinerare." In *Reallexikon der Assyriologie und vorderasiatischen Archäologie*, ed. Erich Ebeling and Bruno Meissner, 5:216–20. Berlin: Walter de Gruyter, 1932–.

Heinrich, Ernst, and Ursula Seidl. "Grundrißzeichnungen aus dem alten Orient." *Mitteilungen der Deutschen Orient-Gesellschaft zu Berlin* 98 (1967): 24–45.

Meek, Theophile James. "The Orientation of Babylonian Maps." *Antiquity* 10 (1936): 223–26.

Nemet-Nejat, Karen Rhea. *Late Babylonian Field Plans in the British Museum.* Studia Pohl: Series Maior 11. Rome: Biblical Institute Press, 1982.

Neugebauer, Otto. *The Exact Sciences in Antiquity.* 2d ed. Providence: Brown University Press, 1957.

North, Robert. *A History of Biblical Map Making.* Beihefte zum Tübinger Atlas des Vorderen Orients, B32. Wiesbaden: Reichert, 1979.

Röllig, Wolfgang. "Landkarten." In *Reallexikon der Assyriologie und vorderasiatischen Archäologie*, ed. Erich Ebeling and Bruno Meissner, 6:464–67. Berlin: Walter de Gruyter, 1932–.

Unger, Eckhard. *Babylon, die heilige Stadt nach der Beschreibung der Babylonier.* Berlin: Walter de Gruyter, 1931.

———. "Ancient Babylonian Maps and Plans." *Antiquity* 9 (1935): 311–22.

Waerden, B. L. van der. "Mathematics and Astronomy in Mesopotamia." In *Dictionary of Scientific Biography*, 16 vols., ed. Charles Coulston Gillispie, 15:667–80. New York: Charles Scribner's Sons, 1970–80.

Weidner, Ernst F. *Handbuch der babylonischen Astronomie, der babylonische Fixsternhimmel.* Leipzig: Hinrichs, 1915; reprinted Leipzig: Zentralantiquariat, 1976.

———. "Fixsterne." In *Reallexikon der Assyriologie und vorderasiatischen Archäologie*, ed. Erich Ebeling and Bruno Meissner, 3:72–82. Berlin: Walter de Gruyter, 1932–.

7 · Egyptian Cartography

A. F. SHORE

Although the so-called Turin map of a gold-bearing region, dating from about 1150 B.C., remains the only map of topographical interest from ancient Egypt, the term map is also commonly applied to representations of cosmological and mythical concepts, such as that of the imaginary land over which the deceased could pass to the afterlife. These are found on a small number of painted wooden coffins from about 2000 B.C. and were first published in scholarly form in 1903.[1] The character of Egyptian drawing produced "picture maps" of a type also found in other contexts in the ancient and medieval worlds, notably battle scenes on temple walls, genre scenes of daily life on walls of tomb chapels, and depictions of cosmological and mythological concepts. Apart from the Turin map of the gold region, no secular map survives except for a very limited number of building plans and cadastral maps. The paucity of material of this type, considering the long span of the ancient civilization of Egypt, makes it difficult to draw firm conclusions concerning the contribution and achievement of the ancient Egyptians to the origins and long-term development of cartography.

About the year 3100 B.C. the land of Egypt, from the Delta south to what is now the first cataract above Aswan, was united under the authority of a single ruler to whom later tradition gave the name Menes. For nearly three thousand years thereafter Egypt was ruled by kings, who in the early second century B.C. were divided by the Egyptian priest Manetho (fl. 280 B.C.) into thirty dynasties. These have been adopted by scholars as the basis of Egyptian chronology. The main historical periods with approximate dates are found in figure 7.1. Later compilers referred to the short period of the Second Persian Period as the Thirty-first Dynasty. This period of Persian occupation was followed by the rule of the Macedonian and Ptolemaic kings. On the deaths of Antony and Cleopatra VII in 30 B.C., Egypt was incorporated as a province of the Roman Empire.

The history of human settlement in the Nile valley may be traced back, unbroken, for over a millennium before the unification (predynastic period). The earliest datable drawing occurs on the decorated pottery of the Amratian (Negada I) period. However, none of this decoration can be unambiguously interpreted as topographical drawings or primitive maps.[2] A rudimentary topography may be depicted on the decorated pottery of the succeeding Gerzean period (Negada II) of this predynastic age (fig. 7.2). Nile boats are shown. Above and below are symbols that can be interpreted as trees and marsh birds. The desert is depicted beyond, with schematic hill formations and antelopes.

TOPOGRAPHICAL DRAWING AND RELIGIOUS CARTOGRAPHY

The unification of Upper and Lower Egypt, about 3100 B.C., initiating the dynastic or pharaonic period, coincided with the appearance of writing and the increased availability of copper for tools, which resulted in the great stone funerary monuments of the Old Kingdom. It is to this period that the first Egyptian maps can be traced. A characteristic style of drawing and composition evolved in the decoration of these funerary monuments in which walls were divided up into a series of separate horizontal strips known as registers, each with its own base line. Objects were drawn from multiple perspectives—in plan, in profile, or in a combination of both—and as isolated images against a flat background.[3] Insofar as topography is reproduced, the mode of representation resembles a bird's-eye view and superficially suggests a picture map by the very character of the drawing. Images were conventionally rendered and placed in

1. Hans Schack-Schackenburg, ed., *Das Buch von den zwei Wegen des seligen Toten (Zweiwegebuch): Texte aus der Pyramidenzeit nach einem im Berliner Museum bewahrten Sargboden des mittleren Reiches* (Leipzig: J. C. Hinrich, 1903). See below, note 8, for recent studies. For the Turin map see note 12 and figure 7.7. For a concise general survey of maps in ancient Egypt see Rold Gundlach, "Landkarte," in *Lexikon der Ägyptologie*, ed. Wolfgang Helck and Eberhard Otto (Wiesbaden: O. Harrassowitz, 1975–), 3:cols. 922–23; Robert North, *A History of Biblical Map Making*, Beihefte zum Tübinger Atlas des Vorderen Orients, B32 (Wiesbaden: Reichert, 1979), 23–29.

2. For an example of a topographic interpretation of Amratian pottery see chapter 4 above, p. 89 and figure 4.38.

3. On the nature of Egyptian drawing, see Heinrich Schäfer, *Principles of Egyptian Art*, ed. with epilogue by Emma Brunner-Traut, ed. and trans. with introduction by John Baines (Oxford: Clarendon Press, 1974). "Maps" are commented upon on page 160.

the manner of the standard signs of more modern maps, that is, so as to indicate the presence of a feature, not its individuality. Landscape generally was only sketchily indicated. The country was distinguished from the town by the presence of a tree or a clump of papyrus. In much

the same manner, no true picture is given of the aspect of the Nile valley by later panoramic depiction.[4]

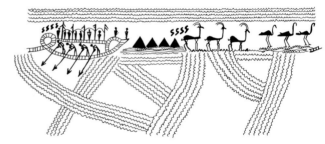

FIG. 7.2. RUDIMENTARY TOPOGRAPHIC DESIGNS ON EGYPTIAN DECORATED POTTERY. Dating from the predynastic Gerzean (Negada II) period, ca. 3700–3100 B.C., Nile boats are shown in the midst of the desert with hills represented schematically.
After redrawing in W. M. F. Petrie, *Prehistoric Egypt, Illustrated by over 1,000 Objects in University College, London* (London: British School of Archaeology in Egypt, 1917), pl. XXI.

The drawings of gardens found in painted tombs of the New Kingdom at Thebes (ca. 1400 B.C.), however, are maplike diagrams, without perspective, that depict regularly laid out paths lined by date palms and sycamores, a rectangular or T-shaped sheet of water on the central axis, lotus plants, fish and birds, and walled orchards with trellised vines (fig. 7.3).[5] The overall impression is that of a formal garden; the balance and for-

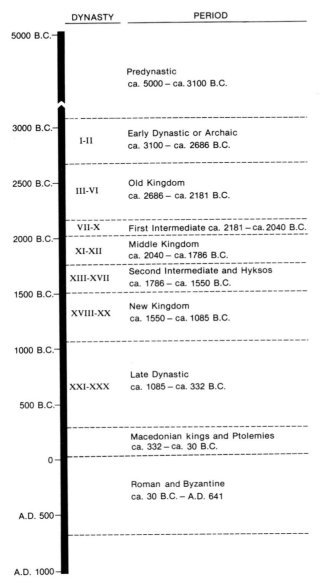

DYNASTY	PERIOD
	Predynastic ca. 5000 – ca. 3100 B.C.
I-II	Early Dynastic or Archaic ca. 3100 – ca. 2686 B.C.
III-VI	Old Kingdom ca. 2686 – ca. 2181 B.C.
VII-X	First Intermediate ca. 2181 – ca. 2040 B.C.
XI-XII	Middle Kingdom ca. 2040 – ca. 1786 B.C.
XIII-XVII	Second Intermediate and Hyksos ca. 1786 – ca. 1550 B.C.
XVIII-XX	New Kingdom ca. 1550 – ca. 1085 B.C.
XXI-XXX	Late Dynastic ca. 1085 – ca. 332 B.C.
	Macedonian kings and Ptolemies ca. 332 – ca. 30 B.C.
	Roman and Byzantine ca. 30 B.C. – A.D. 641

(Scale: 5000 B.C., 3000 B.C., 2500 B.C., 2000 B.C., 1500 B.C., 1000 B.C., 500 B.C., 0, A.D. 500, A.D. 1000)

FIG. 7.1. THE MAIN DYNASTIES AND PERIODS OF EGYPTIAN HISTORY. Ancient Egyptian and Greek sources ascribe the unification of that part of the Nile Valley that we know as Egypt to the conqueror Menes. He was the first of a long line of kings whose reigns were arranged into thirty dynasties by Manetho, an Egyptian priest writing in Greek in the second century B.C. With the advent of Alexander the Great in 333 B.C. Egypt came under the rule of the Macedonian kings and the Ptolemies until it was incorporated within the Roman Empire on the deaths of Antony and Cleopatra VII in 30 B.C. It later formed part of the Eastern Roman and Byzantine empires until it became part of the Islamic world in A.D. 641.

4. On landscape, see Joachim Selim Karig, "Die Landschaftsdarstellung in den Privatgräbern des Alten Reiches" (Ph.D. diss., University of Göttingen, 1962); Helmut Pitsch, "Landschaft (-Beschreibung und -Darstellung)," in *Lexikon*, 3:cols. 923–28 (note 1). A panorama from the ancient world, not from Egypt but possibly of Alexandrine origin, has survived in the Nilotic landscape known as the Palestrina (Praeneste) or Barberini mosaic (p. 246 n. 75), a large composition depicting a peopled landscape, with upper rocky terrain and a lower marshy prospect, forming a sort of picture map of the Nile from source to sea. The most recent studies of the mosaic are Giorgio Gullini, *I mosaici di Palestrina*, Supplemento di Archeologia Classica 1 (Rome: Archeologia Classica, 1956); Helen Whitehouse, *The Dal Pozzo Copies of the Palestrina Mosaic*, British Archaeological Reports, Supplementary Series 12 (Oxford: British Archaeological Reports, 1976); and Angela Steinmeyer-Schareika, *Das Nilmosaik von Palestrina und eine Ptolemäische Expedition nach Äthiopien*, Halbelts Dissertationsdrucke, Reihe Klassische Archäologie 10 (Bonn: Halbelt, 1978).

5. For surveys of gardens in the New Kingdom see Marie-Francine Moens, "The Ancient Egyptian Garden in the New Kingdom: A Study of Representations," *Orientalia Lovaniensia Periodica* 15 (1984): 11–53; Luise Klebs, *Die Reliefs und Malereien des Neuen Reiches* (Heidelberg: C. Winter, 1934), 22–33; Alexander Badawy, *Le dessin architectural chez les anciens Egyptiens* (Cairo: Imprimerie Nationale, 1948), 247–60; idem, *A History of Egyptian Architecture: The Empire (the New Kingdom)* (Berkeley: University of California Press, 1968), 488–99; Leslie Mesnick Gallery, "The Garden of Ancient Egypt," in *Immortal Egypt*, ed. Denise Schmandt-Besserat (Malibu: Undena Publications, 1978), 43–49; Dieter Wildung, "Garten," in *Lexikon*, 2:cols. 376–78, and Wolfgang Helck, "Gartenanlage, -bau," in *Lexikon*, 2:cols. 378–80 (note 1).

FIG. 7.3. PLAN OF AN EGYPTIAN GARDEN. Red and black ink on a wooden tablet surfaced with plaster, excavated at Thebes, and dating from the XVIII Dynasty.
Size of the original: 32.5 × 23.5 cm. By permission of the Metropolitan Museum of Art, New York (gift of N. de Garis Davies, 1914 [14.108]).

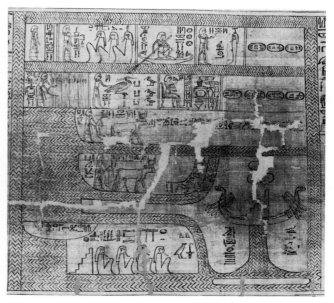

FIG. 7.4. SEKHET-HETEPET, OR THE "FIELDS OF PEACE." Vignette accompanying spell 110 from the *Book of the Dead*, taken from the "Papyrus of Nebseny," depicting the ideal plot of land to be worked in the mythical realm of Osiris, ca. 1400 B.C..
Size of the original: 31.5 × 33 cm. By permission of the Trustees of the British Museum, London (BM 9900, sheet 18).

mality, however, may be misleading and might result from the draftsman's desire for a harmonious composition that may or may not have reflected the real world. A common vignette (illustration) in the collection of spells known as the *Book of the Dead* and inscribed on papyrus about 1400 B.C. is that of an ideal plot of land to be worked by the deceased, set in the mythical realm of Osiris (fig. 7.4). The area depicted is rectangular and cut by canals, and its cartographic appearance is enhanced by the use of color.[6]

There is evidence of an increased interest in the representation of landscape in the New Kingdom, for elements in the great narrative scenes of royal exploits, executed in relief on the walls of temples of the Nineteenth and Twentieth dynasties, are laid out in sequences of maplike images. The progress, for example, of Sethos I (1318–1304 B.C.) past watering stations and frontier forts along the desert highway to Canaan is

drawn in a cartographic manner. The fortress of Kadesh, situated in the valley of the Orontes where the river is joined by a tributary, is conventionally rendered as a fort upon a hilly slope. But in depictions of the attack on the same town by his successor, Ramesses II, preserved in a number of versions, the draftsmen attempted to give more faithful renderings of the topography through the depiction of the river. It encircles the town and continues as a winding strip of water separating the two armies as successive incidents of the battle involving the enemy's crossing of the river are portrayed.[7]

---

6. For a bibliography of publications see Paul Barguet, *Le livre des morts des anciens Egyptiens* (Paris: Editions du Cerf, 1967), 28–30. See also Jean Leclant, "Earu-Gefilde," in *Lexikon*, 1:cols. 1156–60 (note 1). The *Book of the Dead* is the name given in modern times to copies of a heterogeneous collection of funerary spells written down on papyrus and deposited in tombs from the Eighteenth Dynasty down to the Roman period. The number of spells (also called chapters) contained in individual copies and their order vary from copy to copy. The spells derive from similar collections of earlier times, the Coffin Texts of the Middle Kingdom and the Pyramid Texts of the Old Kingdom.

7. For these narrative scenes see Helene J. Kantor, "Narrative in Egyptian Art," *American Journal of Archaeology* 61 (1957): 44–54; G. A. Gaballa, *Narrative in Egyptian Art* (Mainz: Philipp von Zabern, 1976), 99–129; William Stevenson Smith, *Interconnections in the Ancient Near-East: A Study of the Relationships between the Arts of Egypt, the Aegean, and Western Asia* (New Haven: Yale University Press, 1965), 168–79.

FIG. 7.5. COSMOGRAPHICAL MAP: THE LAND OF EGYPT WITH THE GODDESS NUT. South is at the top in this cosmographic representation found on the cover of a stone sarcophagus from Saqqara. It dates from the Thirtieth Dynasty, ca. 350 B.C.
Diameter of the interior circle: 72 cm. By permission of the Metropolitan Museum of Art, New York (gift of Edward S. Harkness, 1914 [14.7.1]).

The most notable example of the cartographic nature of Egyptian topographical drawing is to be found with certain spells, usually copied on the floor of a series of painted coffins of Early Middle Kingdom date (ca. 2000 B.C.) from the single site of al-Bersha in Middle Egypt. These are accompanied by an illustration of a rectangular area with two routes depicted by broad, sinuous bands of color: the upper one is blue and represents a passage by water, the lower one is black and depicts an overland route that takes up about one-third of the composition (plate 2). The presence of color enhances the cartographic quality of the topographical depiction of this mythical land, so that this assemblage of spells forms a sort of compendium of mythical maps, known to modern scholars as *The Book of the Two Ways*.[8]

Funerary spells of this character were of necessity obscurantist. The texts give no clear explanation of the landscape they accompany. They contain allusions to

myths of which we have no complete or connected account. The three surviving versions of *The Book of the Two Ways*, two long and one short, constitute a conflation of differing and conflicting earlier expositions. It is scarcely possible to reconstruct a systematic narrative of the deceased's passage from this text. It would be a mistake to suppose that the topographical drawing accompanying the spells, which occurs in one group of coffins only, was intended to provide a guide by which the deceased might choose his path after death and find his way to a desired goal. The composition may be loosely compared to a passport or *laissez-passer* papers. Knowledge of the spells accompanying the vignette would guarantee safe passage through this mythical and mysterious land, populated by deities and spirits both friendly and hostile, through which the soul would be ever traveling as it left and returned to the body preserved in the coffin. The two routes are not, it seems, to be considered alternative ones. Though they are described as the ways of Osiris, they seem also to be depictions of the paths of the day and night journeys of the sun-god Re, the two paths uniting to form the circuit of the sun as conceived by the Egyptians.

That Egyptian cartography was often more pictorial than planimetric is confirmed by other illustrations of religious geography. Pictures of the structure of the universe as seen by the ancient Egyptians lack even the diagrammatic, maplike quality of their depictions of the imaginary route to, and terrain of, the afterlife. They occur only within religious and magical texts, mostly in

8. Spells 1029–1130, Adriaan de Buck, *The Egyptian Coffin Texts VII*, Oriental Institute Publications, vol. 87 (Chicago: University of Chicago Press, 1961), hieroglyphic texts 252–471, plans 1–15; translation with discussion, Leonard H. Lesko, *The Ancient Egyptian Book of Two Ways*, University of California Near Eastern Studies Publications, vol. 17 (Berkeley: University of California Press, 1972); Alexandre Piankoff, *The Wandering of the Soul*, completed and prepared for publication by Helen Jacquet-Gordon, Egyptian Religious Texts and Representations, Bollingen Series 40, vol. 6 (Princeton: Princeton University Press, 1974); Raymond O. Faulkner, ed. and trans., *The Ancient Egyptian Coffin Texts*, 3 vols. (Warminster: Aris and Phillips, 1978), 3:127–69. For discussion of the interpretation of the text, see also Paul Barguet, "Essai d'interprétation du Livre des deux chemins," *Revue d'Egyptologie* 21 (1969): 7–17; Wilhelm Bonacker, "The Egyptian *Book of the Two Ways*," *Imago Mundi* 7 (1950): 5–17; Hermann Grapow, "Zweiwegebuch und Totenbuch," *Zeitschrift für Ägyptische Sprache und Altertumskunde* 46 (1909): 77–81; Hermann Kees, *Totenglauben und Jenseitsvorstellungen der alten Ägypter* (Berlin: Akademie-Verlag, 1956), 287–302; Leonard H. Lesko, "Some Observations on the Composition of the *Book of Two Ways*," *Journal of the American Oriental Society* 91 (1971): 30–43; Jan Bergman, "Zum Zwei-Wege-Motiv: Religionsgeschichtliche und exegetische Bemerkungen," *Svensk Exegetisk Årsbok* 41–42 (1976–77): 27–56, esp. 51–54; Hellmut Brunner, "Die Unterweltsbücher in den ägyptischen Königsgräbern," in *Leben und Tod in den Religionen: Symbol und Wirklichkeit*, ed. Gunter Stephenson (Darmstadt: Wissenschaftliche Buchgesellschaft, 1980), 215–28. See also Ursula Rössler-Köhler, "Jenseitsvorstellungen," in *Lexikon*, 3:cols. 252–67 (note 1).

the context of a man's journeyings after death, often in the retinue of the sun-god, and are replete with mythical figures in human or animal form. In the New Kingdom the star-studded universe is sometimes depicted with an arched figure of the goddess Nut representing the sky. She is shown held aloft, over the recumbent figure of the earth-god Geb, by the figure of the god Shu in human form representing the space between earth and sky. On the cover of a stone sarcophagus dating to the Thirtieth Dynasty (ca. 350 B.C.), a depiction of the land of Egypt and regions surrounding it is drawn in circular form, probably as a result of foreign influence, below and between the arms and legs of the arched figure of Nut. An inner circular band is occupied by various standards associated with the ancient territorial divisions of Egypt (nomes), at this time still of great religious significance. The exterior ring depicts various peoples and symbols representing Egypt's neighbors. To the left and to the right on the circumference of the outer ring are respectively the goddess of the east and the goddess of the west, the upper part of the diagram therefore representing the south (fig. 7.5).[9]

Astronomical ceilings, the earliest known being that in the tomb of Senmut, the minister of Hatshepsut (ca. 1470 B.C.), depict decans (diagrams of stars for calculating the passage of the hours at night), constellations, and planets. Representations of heavenly bodies with figures such as a standing pregnant hippopotamus or a crocodile tend to destroy any resemblance to a modern map. Only in the case of the depiction of decans is a more chartlike rendering found.[10] Although in the later period there was cross-fertilization of Babylonian and Greek ideas and astronomical observation was necessary for astrology, the same mythical figures were retained. A circular picture of the sky, dating to the end of the Ptolemaic period, which reproduces planets and constellations with some degree of accuracy in their relation to one another, occurs on the ceiling of the chapel of Osiris on the roof of the temple of Dendera, now in the Louvre. It depicts a synthetic image of the sky in which traditional Egyptian constellations and decans are mingled with the twelve signs of the zodiac, imported from Babylonia but Egyptianized in their forms (fig. 7.6).[11]

TURIN MAP OF THE GOLD MINES

In contrast with the number of texts of a religious or funerary character to be found on temple and tomb walls or on papyri from burials, survival of administrative and business documents is relatively rare. Except for the Ramesside period (Nineteenth to Twentieth dynasties), little more than isolated pieces have been preserved from before the sudden increase in everyday documents in both Egyptian and Greek from the Greco-Roman period. One of the best known of all maps from the ancient world belongs to the Ramesside period and is now preserved in the Museo Egizio, Turin. Originally part of the collections of Bernardino Drovetti formed before 1824, the map was first identified by Samuel Birch in 1852 as an ancient plan of gold mines which he located in Nubia. The circumstances of the find are not known, but the

FIG. 7.6. CELESTIAL MAP OF THE PLANETS, CONSTELLATIONS, AND ZODIAC. An example of a late Egyptian astronomical depiction carved on the ceiling of the chapel of Osiris, temple of Dendera, from the end of the Ptolemaic period (first century B.C.).
Size of the original: 2.55 × 2.53 m. By permission of the Musée du Louvre, Paris.

9. C. L. Ranson, "A Late Egyptian Sarcophagus," *Bulletin of the Metropolitan Museum of Art* 9 (1914): 112–20. For a second fragmentary example, see J. J. Clère, "Fragments d'une nouvelle représentation égyptienne du monde," *Mitteilungen des Deutschen Archäologischen Instituts, Abteilung Kairo* 16 (1958): 30–46. On Egyptian representations of the universe see Heinrich Schäfer, *Ägyptische und heutige Kunst und Weltgebäude der alten Ägypter: Zwei Aufsätze* (Berlin: Walter de Gruyter, 1928), 83–128.

10. These texts and representations dealing with decans, star clocks, zodiacs, and planets are assembled and discussed in Otto Neugebauer and Richard A. Parker, eds. and trans., *Egyptian Astronomical Texts*, 3 vols. (Providence and London: Lund Humphries for Brown University Press, 1960–69); for a concise account, see Richard A. Parker, "Ancient Egyptian Astronomy," *Philosophical Transactions of the Royal Society of London*, ser. A, 276 (1974): 51–65.

11. Neugebauer and Parker, *Egyptian Astronomical Texts*, 3:72–74, pl. 35 (note 10).

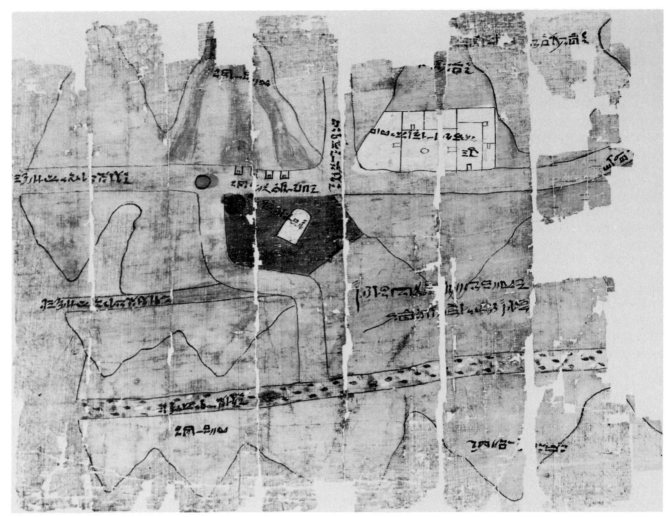

FIG. 7.7. MAP OF GOLD MINES FROM THE TURIN PA-
PYRUS. Dating from the Ramesside Period, this section shows
the location of gold-bearing mountains, gold-working settlements, and roads between the Nile and the coast of the Red
Sea.

Height of the papyrus: 41 cm. By permission of the Soprintendenza per le Antichità Egizie, Turin.

carelessness of Drovetti's agents resulted in fragmentation of the original papyrus and loss of some parts.[12]

The extant papyrus consists of two principal sections,
earlier thought to belong to two different documents.
The more important section is a fragment, measuring
approximately forty centimeters high, generally called
the "map of the gold mines" (fig. 7.7). It depicts two
broad roads, running parallel to each other through
pinkish red mountainous regions. They are drawn horizontally across the papyrus, the lower with indications
of a rocky bed or sparse vegetation, characteristic of the
larger dried-up watercourses or wadis that form the natural routes across the eastern desert from the Nile to the
Red Sea. Legends written in hieratic, the cursive
everyday hand of the time, explain where these routes
to the left are leading. A broad, winding crossway wadi

12. First published as the plan of a royal tomb, Richard Lepsius,
*Auswahl der wichtigsten Urkunden des ægyptischen Alterthums:
Theils zum erstenmale, theils nach den Denkmælern berichtigt* (Leipzig: Wigand, 1842), pl. XXII. See also Samuel Birch, "Upon an Historical Tablet of Ramses II., 19th Dynasty, relating to the Gold Mines
of Æthiopia," *Archaeologia* 34 (1852): 357–91 esp. 382–83; François
J. Chabas, *Les inscriptions des mines d'or* (Chalon-sur-Saône: Dejus-
sieu, 1862), also published in *Bibliothèque Egyptologique* 10 (1902):
183–230; A. H. Gardiner, "The Map of the Gold Mines in a Ramesside
Papyrus at Turin," *Cairo Scientific Journal* 8, no. 89 (1914): 41–46;
G. W. Murray, "The Gold-Mine of the Turin Papyrus," *Bulletin de
l'Institut d'Egypte* 24 (1941–42): 81–86; and the contribution by
G. W. Murray in John Ball, *Egypt in the Classical Geographers* (Cairo:
Government Press, Bulâq, 1942), 180–82 and pls. VII–VIII (in color).
The most recent and fullest discussion of the papyrus is Georges
Goyon, "Le papyrus de Turin dit 'Des mines d'or' et le Wadi Ham-
mamat," *Annales du Service des Antiquités de l'Egypte* 49 (1949):
337–92. For representations of the gold mine area in color photo-
graphs, see Ernesto Scamuzzi, *Museo Egizio di Torino* (Torino: Fratelli
Pozzo, 1964), pl. 88, and Georges Posener et al., *A Dictionary of
Egyptian Civilization*, trans. Alix Macfarland (London: Methuen,
1962), 112.

connects the two routes, from which an alternative route is indicated and labeled, also leading to the left. Running vertically from the upper route is yet another road with hieratic text that gives its destination. The significance of the area painted red is explained by another legend that reads, "the mountains where gold is worked: they are colored in red." The Egyptian term used here for red, *dšr*, is that most generally employed for all shades of red, the color used to depict red granite, sandstone, and the tawny hue of the desert. The term "mountains of gold" is repeated elsewhere in the area colored red, as well as apparently the phrase "mountains of silver and gold." In places the red area is brought to a point and given a distinctive name such as "the peak" or "the peak on which Amun is." The intention was apparently to render the basic outlines of mountains laid down flat on either side of the valley route rather than to delineate precisely and accurately the area of auriferous rocks.

There are other distinctive features outlined, colored, and labeled in hieratic. Near the junction of the cross valley with the upper route a circular, dark-colored image is marked, with a second partially overlapping design in a darker black line. The figure is probably intended to represent a well, though no text identifies it. A little below and to the right of the design is another, more oblong in shape, colored green with the zigzag lines by which the ancient Egyptians conventionally represented water. Within the design there are traces of a hieratic group, apparently to be read as "cistern," "waterplace," or the like. In the same central section of the map a round-topped stela is also indicated in white, with a legend dating it to the reign of Sethos I of the Nineteenth Dynasty. The feature is presumably to be identified with one of the rock-cut stelae executed by that king, depicting Amun or another deity, preserved on the mountain face flanking the wadi.

There are also two man-made features on the upper side of the upper route. One is clearly a large building containing several courts or rooms with connecting doors, described as the "resting place" or "abode" (*ḫnw*) of "Amun of the pure mountain." There are also three small rectangular forms labeled "the houses of the gold-working settlement."

The second section of the papyrus comprises a number of fragments for which the final placement, based on careful study of the fibers of the papyrus, has yet to be made. Its principal feature is the continuation of the wide, winding route of the wadi interspersed with stones. This constitutes the lower route of the other section (fig. 7.8). In contrast with the gold-mine section, the area on each side of the road is colored black, and the legend indicates that in this area the stone known to the ancient Egyptians as *bekhen* is to be found. This black or dark

green stone, generally called schist by Egyptian archaeologists, is more properly identified as graywacke. The surviving fragments give no indication of precise locations comparable to those found on the section depicting the gold-mining region and its settlements.

The Turin papyrus fragments were long considered the earliest surviving topographical map from Egypt to have come to light. The papyrus clearly has a character distinct from the cosmological drawings of the universe or of the routes to or depiction of the afterlife found within the formal context of religious art. The draftsman has distributed distinctive features in accordance with the reality of a particular area, adding clarity by the use of legends and contrasting colors. The texts indicate that the area depicted must be along the natural route from Coptos (Qift) on the Nile through the eastern desert via Wādī al-Ḥammāmāt to the port of Quseir on the Red Sea. This route was used in ancient times in the course of expeditions to the Red Sea for trading voyages south to the land known to the Egyptians as Punt (Pwenet). The central area, between Bīr al-Ḥammāmāt and Bīr Umm Fawākhir, was visited as a source of ornamental stone and of gold, and it is rich in rock tablets recording quarrying expeditions and in archaeological evidence of ancient gold mining. More precise location rests on the interpretation of the orientation of the map. This requires the resolution of questions concerning the placement of fragments in the second section and the identification of the places to which the roads to the left of the viewer are said to lead. In descriptions of property in the later period the points of the compass are given in the order south, north, east, west, suggesting that Egyptians oriented themselves facing south, with north behind them, the west to their right and the east to their left. It would be natural, then, for them to designate the top of the papyrus as south. Such a view seems to be supported by the legend designating the upper route of the gold map leading off to the left as "the road that leads to the *ym*," that is, to "the [Red] sea," taking *ym* in its most common meaning. The route marked as leading off from the cross valley to the left is likewise described as "another road that leads to the *ym*." The placement of the second section to the right of the map of the gold region seems correct, since it would then constitute the beginning of a papyrus roll, which would normally suffer greater damage. The map would then show on the right (that is, the west) the darker "schist" area of the main part of Wādī al-Ḥammāmāt, with the gold mines of the region of Bīr Umm Fawākhir some twenty-five kilometers to the east. A more recent comparison of the features shown on the map with the ground matches the various features specifically mentioned in the gold map with the central area of Wādī al-Ḥammāmāt and with the upper part of the papyrus

FIG. 7.8. THE SECOND SECTION
OF THE TURIN PAPYRUS. As at
present mounted in the Museo Egizio,
this shows the fragments other than
those in figure 7.7.
Height of the papyrus: 41 cm. By permission of the Soprintendenza per le
Antichità Egizie, Turin.

constituting the north.[13] If this placement were correct and the fragments of the second portion were to be placed to the right, it would require the *ym* to which the road now leads westward, that is, back to the Nile, to be taken in some sense other than Red Sea. It would likewise place the area of *bekhen* stone to the east of the location of the main quarry inscriptions in Wādī al-Ḥammāmāt.

The difficulties in matching features depicted and labeled on the papyrus with those on the ground are compounded by the absence of any indications of scale. The map seems to be a freehand drawing. The only indication of its purpose seems to be given in the series of hieratic notations written on those areas left blank above and below the route and the black areas depicted on the fragments of the second section. In contrast with the hieratic texts on the gold map identifying geographical features, these texts refer to the transport of a statue. A text of five lines, of which the first four lack their beginnings, seems to reflect a situation in which a king sent an expedition to the Wādī al-Ḥammāmāt to bring a statue back to Thebes. It was, we are informed, deposited in a workshop beside the mortuary temple of Ramesses II (Ramesseum) on the west bank of the Nile at Thebes and subsequently taken, half-worked, to the Valley of the Kings in a regnal year 6. Such a docket must have been written at Thebes, the papyrus obviously having been at some time in the possession of one of the scribes attached to the work gang responsible for constructing and decorating the royal tombs in the Valley of the Kings. Jottings on the back of the papyrus include a reference to the statue of Ramesses IV of the Twentieth Dynasty, suggesting that year 6 should refer to the reign of that king.[14] The purpose of the map is still obscure. Annotations on the second portion of the papyrus suggest that the document was

drawn up in connection with work on the extraction and transport of stone, ultimately destined perhaps for a royal tomb in the Valley of the Kings. Some of these notes seem to give measurements of blocks; one seems to provide measurements of actual distances separating points on the map. The papyrus may be the result of calculations of distances for logistical purposes. To judge from instructions contained in a model letter copied by a pupil as part of his scribal training (instructions that seem to refer to the same general area as the Turin map), calculations of distance are the kind of work a scribe might be expected to do.[15] What is unusual is that a rough sketch map is included. Surveying rarely resulted in graphic maps, and in this respect ancient Egypt is very similar to medieval Europe until well into the fourteenth and fifteenth centuries.

LAND SURVEY, CADASTRAL MAPS, AND BUILDING PLANS

There is little direct record of the way surveying was practiced, although one passage from Herodotus is used as early evidence for the Egyptian practice of surveying:

> Sesostris was also the king, the priests went on to say, who was responsible for the division and distri-

13. Goyon, "Papyrus de Turin" (note 12). In light of the annotation in demotic script to Papyrus Lille 1 (see below, note 26), it seems more likely that south is represented at the top of the papyrus.

14. Jaroslav Černý, *A Community of Workmen at Thebes in the Ramesside Period* (Cairo: Institut Français d'Archéologie Orientale, 1973), 61–62, 66–67.

15. Papyrus Anastasi VI (London, British Museum, Pap. BM 10245), lines 67–74, text by Alan H. Gardiner, *Late-Egyptian Miscellanies*, Bibliotheca Aegyptiaca 7 (Brussels: Edition de la Fondation Egyptologique Reine Elisabeth, 1937), 77; Ricardo A. Caminos, *Late-Egyptian Miscellanies*, Brown Egyptological Studies 1 (London: Oxford University Press, 1954), 296–98 nn. 70–71.

bution of land into individual square holdings, of equal size, among the Egyptians and for making the plots his source of revenue by fixing the amount of tax to be paid yearly on each holding. Should the river encroach upon any holding, its owner might approach the king and report what had happened. The king would send men to inspect and measure the loss of cultivated land in order that from then on some of the tax proportionate to the report of the loss might be remitted. I attribute the invention of geometry to this cause and from Egypt it spread to Greece.[16]

Wood and stone measuring rods have survived, but these are votive offerings rather than actual implements. Ropes knotted at regular intervals—which presumably were used for measurement for tax purposes—are depicted in agricultural scenes showing fields of standing corn from the New Kingdom tombs at Thebes. There are also statues of high officials from the same period that were intended to be placed in temples or tombs. Sitting back on their heels with coiled measuring ropes resting on their knees, in an activity known as "stretching the cord," they symbolized the role of surveyor in the construction of temples. The rope terminates in a ram's head in honor of the god Khnum-Shu.[17] The basic linear measure was a cubit of differing standards. A unit of one hundred cubits squared, approximately two-thirds of an acre, constituted the basic measure of area, in Egyptian *stꜣt*, which corresponds to the *aroura* of Greek documents.

From later textual material relating to foundation ceremonies in the construction of temples, we know of two more implements that permitted great accuracy in orienting buildings. In Egyptian texts they are called *merkhet*, literally "instrument of knowing," and *bay*, "palm rib." An example of each, purchased in Cairo and dating

perhaps to about 600 B.C., was identified by Ludwig Borchardt in 1899.[18] The *merkhet*, a type of plumb-line sighter, was aligned with an object by means of the *bay*, a palm rib with a V-shaped slot cut at the wider end.

16. Herodotus *History* 2.109, author's translation. On the passage see Henry Lyons, "Two Notes on Land-Measurement in Egypt," *Journal of Egyptian Archaeology* 12 (1926): 242–44. See also the commentary of A. R. Lloyd in *Herodotus, Book II: Commentary 99–182*, Etudes Préliminaires aux Religions Orientales dans l'Empire Romain 43 (Leiden: E. J. Brill, forthcoming). Strabo, *Geography* 17.3 also says geometry was invented to meet the need for the annual remeasurement of holdings; see *The Geography of Strabo*, 8 vols., ed. and trans. Horace Leonard Jones, Loeb Classical Library (Cambridge: Harvard University Press; London: William Heinemann, 1917–32). Similarly Diodorus Siculus, 1.81.2; see *Diodorus of Sicily*, 12 vols. trans. C. H. Oldfather, Loeb Classical Library (Cambridge: Harvard University Press; London: William Heinemann, 1933–67). On ancient Egyptian measurements and measuring see Adelheid Schlott-Schwab, *Die Ausmaße Ägyptens nach altägyptischen Texten* (Wiesbaden: O. Harrassowitz, 1981). For a concise discussion of surveying and surveying implements see Somers Clarke and Reginald Engelbach, *Ancient Egyptian Masonry: The Building Craft* (London: Oxford University Press, 1930), 64–68; O. A. W. Dilke, *The Roman Land Surveyors: An Introduction to the Agrimensores* (Newton Abbot: David and Charles, 1971), 19–30; S. P. Vleeming, "Demotic Measures of Length and Surface, chiefly of the Ptolemaic Period," in P. W. Pestman et al., *Textes et études de papyrologie grecque, démotique et copte*, Papyrologica Lugduno-Batava 23 (Leiden: E. J. Brill, 1985), 208–29.

17. For scenes of the measuring of the fields see Suzanne Berger, "A Note on Some Scenes of Land-Measurement," *Journal of Egyptian Archaeology* 20 (1934): 54–56. For the type of statue, see Jacques Vandier, *Manuel d'archéologie égyptienne*, 6 vols. (Paris: A. et J. Picard, 1952–78), vol. 3, *Les grandes époques: La statuaire* (1958), 476–77. For Khnum-Shu see Paul Barguet, "Khnoum-Chou, patron des arpenteurs," *Chronique d'Egypte* 28 (1953): 223–27.

18. Berlin 14084, 14085: Ludwig Borchardt, "Ein altägyptisches astronomisches Instrument," *Zeitschrift für Ägyptische Sprache und Altertumskunde* 37 (1899): 10–17; Staatliche Museen, Preußischer Kulturbesitz, *Ägyptisches Museum, Berlin* (Berlin: Staatliche Museen, 1967), 54, with photographic illustration. The method by which the

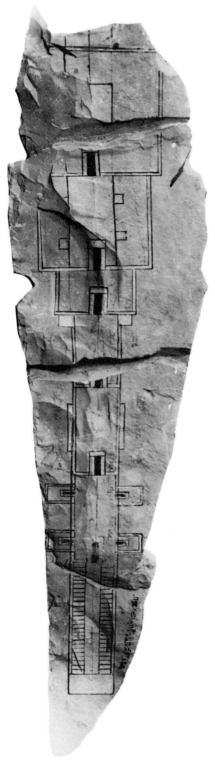

A limestone ostracon in the Cairo Museum, of obscure purport, dating perhaps to the Nineteenth Dynasty, notes distances between tombs in the Valley of the Kings and mentions significant features, which apparently include a willow tree and some form of water.[19] No sketch accompanies this text. However, among a number of documents recording measurements of royal tombs in the Valley of the Kings there are two with plans. One, an ostracon in Cairo, is probably a working plan of Tomb 6 (Ramesses IX); its hieratic legends have faded considerably (fig. 7.9).[20] The other is a more elaborate colored plan carefully drawn on papyrus and preserved in the Museo Egizio, Turin (fig. 7.10).[21] Around the design of the rock-cut tomb the surface of the papyrus is colored brownish, with alternate broken lines of red and black enclosed within a red outline depicting desert hills. The plan of the tomb is neatly executed in parallel thin black lines, as if indicating the walls of a building rather than the sides of rock-cut chambers. A series of rooms and chambers are depicted in plan, but the yellow colored doors are in elevation. Hieratic legends describe the stage of the work and dimensions of the chambers. The drawing is not to scale, and the plan gives only a rough approximation of the real shape and proportions of the rooms. The plan and measurements correspond closely to the tomb of Ramesses IV. The finished nature of the document, including a drawing of the shrines that would have surrounded the sarcophagus, suggests that the plan was a final one drawn up immediately before the burial. Like the map of the gold mining and quarrying area, its purpose cannot be precisely defined,

Egyptians oriented buildings is most authoritatively discussed in Zbyněk Žába, *L'orientation astronomique dans l'ancienne Egypte et la précession de l'axe du monde*, Archiv Orientálni, suppl. 2 (Prague: Editions de l'Académie Tchécoslovaque des Sciences, 1953). See also I. E. S. Edwards, *The Pyramids of Egypt*, new and rev. ed. (Harmondsworth: Viking, 1986), 154–61, and Günther Vittmann, "Orientierung (von Gebäuden)," in *Lexikon*, 4:cols. 607–9 (note 1).

19. Elizabeth Thomas, "Cairo Ostracon J. 72460," in *Studies in Honor of George R. Hughes*, Studies in Ancient Oriental Civilization no. 39 (Chicago: Oriental Institute of the University of Chicago, 1976), 209–16.

20. Georges Daressy, *Ostraca*, Catalogue Général des Antiquités Egyptiennes du Musée du Caire, vol. 1 (Cairo: Institut Français d'Archéologie Orientale, 1901), 35, pl. XXXII (no. 25184); William H. Peck, *Drawings from Ancient Egypt* (London: Thames and Hudson, 1978), no. 130.

21. Howard Carter and Alan H. Gardiner, "The Tomb of Ramesses IV and the Turin Plan of a Royal Tomb," *Journal of Egyptian Archaeology* 4 (1917): 130–58; Peck, *Drawings*, no. 129 (note 20); Clarke and Engelbach, *Ancient Egyptian Masonry*, 48–51 (note 16); Jaroslav Černý, *The Valley of the Kings* (Cairo: Institut Français d'Archéologie Orientale, 1973), 23–34. For sketches in connection with work on nonroyal tombs, see William C. Hayes, *Ostraka and Name Stones from the Tomb of Sen-Mūt (no. 71) at Thebes*, Publications of the Metropolitan Museum of Art Egyptian Expedition, vol. 15 (New York: Metropolitan Museum of Art, 1942), no. 31 and pl. VII.

FIG. 7.9. PLAN OF EGYPTIAN TOMB FROM THE VALLEY OF THE KINGS. This is possibly a working plan on an ostracon of the tomb of Ramesses IX.
Size of the original: 83.5 × 14 cm. By permission of the Egyptian Museum, Cairo (Ostracon 25,184).

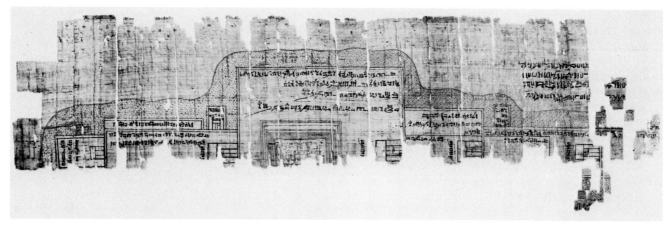

FIG. 7.10. PAPYRUS PLAN OF EGYPTIAN TOMB. Although the plan is not drawn to scale, the measurements in the legends correspond closely to those of the tomb of Ramesses IV.

Size of the original: 31.1 × 104.8 cm. By permission of the Soprintendenza per le Antichità Egizie, Turin.

though it is probable that such drawings of tomb plans are records of the progress of work following inspections.[22] Whatever other land maps might be found, this careful plan of the tomb of Ramesses IV suggests that the Egyptian draftsman, familiar though he was with the use of a strict canon of proportion in the drawing, for instance, of the human figure, did not attempt to convey distance by the deliberate use of scale, being content with a more or less accurate freehand drawing, with precise measurements, if required, written in.

A number of architectural drawings or working plans of buildings and houses are known. One, a sketch on a pottery fragment from Dīr al-Baḥrī, depicts a shrine and enclosure wall with measurements not to scale. It has a text that, if completely preserved, would have allowed the determination of the plan's orientation (fig. 7.11).[23] At present there are no maps of fields or properties known before the Ptolemaic period.[24] Fragments from Gebelein, with notations in Greek and demotic, come from a map of Pathyris and its environs, drawn up perhaps for official administrative purposes or as a

FIG. 7.11. ARCHITECTURAL DRAWING FROM DĪR AL-BAḤRĪ. A shrine and an enclosure wall are shown on this fragment of pottery dating from the twelfth or thirteenth century B.C.
Size of the original: 9.5 × 9.8 cm. By permission of the Trustees of the British Museum, London (BM 41228).

22. See the report of a scribe engaged upon drawing a plan, Černý, *A Community of Workmen*, 12 (note 14).
23. S. R. K. Glanville, "Working Plan for a Shrine," *Journal of Egyptian Archaeology* 16 (1930): 237–39. For house plans see, for example, Clarke and Engelbach, *Ancient Egyptian Masonry*, 51–52 (note 16), and Norman de Garis Davies, "An Architect's Plan from Thebes," *Journal of Egyptian Archaeology* 4 (1917): 194–99. Another example, from the vicinity of the tomb of Hepusonb, in Lydia Collins, "The Private Tombs of Thebes: Excavations by Sir Robert Mond, 1905 and 1906," *Journal of Egyptian Archaeology* 62 (1976): 18–40, esp. 36, is apparently one copied by Černý in the Cairo Museum. For an example from the Meroitic period (ca. 200 B.C.–A.D. 200) see Jean Jacquet, "Remarques sur l'architecture domestique à l'époque méroïtique: Documents recueillis sur les fouilles d'Ash-Shaukan," in *Aufsätze zum 70. Geburtstag von Herbert Ricke*, ed. Abdel Moneim Abubakr et al., Beiträge zur Ägyptischen Bauforschung und Altertumskunde, no. 12 (Wiesbaden: F. Steiner, 1971), 121–31, esp. 130 and pls. 19–20.
24. Sketches of fields are included among diagrams accompanying arithmetical problems concerning the calculation of areas of land, for example, in the Rhind Mathematical papyrus, copied ca. 1600 B.C., T. Eric Peet, *The Rhind Mathematical Papyrus* (Liverpool: University Press of Liverpool, 1923); A. B. Chace, *The Rhind Mathematical Papyrus*, 2 vols. (Oberlin, Ohio: Mathematical Association of America, 1927–29). W. W. Struve, *Mathematischer Papyrus des Staatlichen Museums der Schönen Künste in Moskau* (Berlin: J. Springer, 1930). Similar diagrams are found in demotic in Richard A. Parker, *Demotic Mathematical Papyri* (Providence: Brown University Press, 1972).

diagram of a plot of land forming the matter of some private transaction.[25] Little can be gleaned from it other than that the river or canal is shown in blue. Extensive archives from the region date from the second century to the earlier part of the first century B.C. Two earlier examples of diagrams accompanying memorandums are found among the papers concerned with the management of the large estate of Apollonius, who was head of the civil administration in the Faiyum under Ptolemy II. One, recovered from a cartonnage case of a mummy in the necropolis of Ghoran in the Faiyum, is dated 259 B.C. (fig. 7.12).[26] It shows a schematic plan of a plot intersected by canals and dikes. The orientation of the diagram is given in both Greek (the language of the memorandum) and demotic. The Greek text gives the west as the top of the diagram. In order to read the compass points as drafted by the demotic writer, however, the papyrus must be turned clockwise through 90 degrees so that south is at the top. The second diagram is to be found on another papyrus from the extensive archive of Zenon, who managed the estate on behalf of Apollonius. It shows the course of a canal and the position of a palisade between the house of a certain Artemidorus to the north and the temple of Poremanres to the south (fig. 7.13).[27] The palisade was designed to protect pigs and other animals against flooding.

Such diagrams are very rare, though a large number of private documents concerning the sale, lease, and mortgage of land and buildings survive from Greco-Roman Egypt. No sketch maps accompany public cadastral surveys. Two surveys from the pharaonic period, Papyrus Wilbour and Papyrus Reinhart, likewise lack sketches. Similarly, no map accompanies a "town register" of the west of Thebes between two named locations. This register, dating to the end of the Ramesside period and preserved on the verso of a papyrus concerned with the investigation of tomb robberies, is a list of 182 houses ordered from north to south, in some seven narrow columns of hieratic.[28] Given the importance of the assessment of land for taxation, and of the delimitation of boundaries in a country whose agricultural land is subject to annual flooding, we might expect that field maps would have existed.[29] Their absence might be explained as a matter of chance survival of papyrus or as part of the general lack of administrative documents before the Greco-Roman period. In light of the little evidence of the use of maps for administrative purposes, however, it is perhaps more likely that here, as in other aspects of their culture, the ancient Egyptians

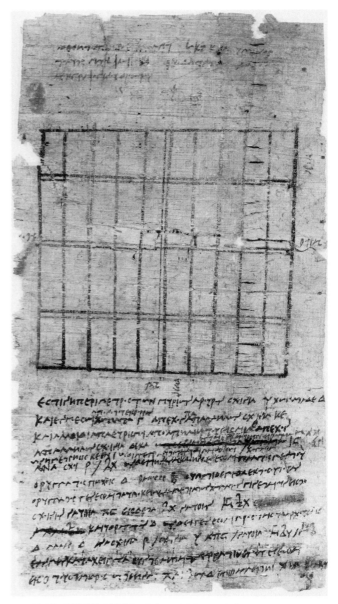

FIG. 7.12. SCHEMATIC PLAN OF DIKES AND CANALS IN THE FAIYUM. A papyrus document dated 259 B.C. from the estate of Apollonius, head of Ptolemy II's civil administration. Size of the papyrus: 61 × 31 cm. Institut de Papyrologie de la Sorbonne, Paris (Papyrus Lille 1). Photograph by Samie Guilbert.

25. Cairo, Egyptian Museum, Pap. dem. Cairo 31163, Wilhelm Spiegelberg, *Die demotischen Denkmäler*, 2 vols. (Leipzig: W. Drugulin, 1904–8), vol. 2, *Die demotischen Papyrus*, 261–63, pl. cv; Greek text, Friedrich Preisigke, *Sammelbuch griechischer Urkunden aus Ägypten* (Strasburg: K. J. Trübner, 1915), vol. 1, no. 4474.

26. Paris, Institut Papyrologique, Papyrus Lille 1, most recently edited with full commentary by P. W. Pestman, *Greek and Demotic Texts from the Zenon Archive*, Papyrologica Lugduno-Batava 20 (Leiden: E. J. Brill, 1980), 253–65 and pl. XXIX.

27. Campbell Cowan Edgar, *Zenon Papyri in the University of Michigan Collection*, Michigan Papyri vol. 1 (Ann Arbor: University of Michigan Press, 1931), 162 (no. 84) and pl. VI.

28. London, British Museum, Pap. BM 10068v; T. Eric Peet, *The Great Tomb-Robberies of the Twentieth Egyptian Dynasty* (Oxford: Clarendon Press, 1930), 83–87, 93 ff.

29. See note 16 above.

showed no great predilection for change or development once a certain level of achievement had been attained. Their principles of drawing incorporated all that was necessary for map drawing, and they possessed the means and the bureaucracy for measuring, calculating, and registering areas. Nevertheless, just as they were reluctant to adopt a system of full alphabetic writing long after such a system had been invented elsewhere, what might seem to us a natural progression in representation was not exploited further, given the circumstances of their agricultural life and economy. It was left to others to develop the potential of mapmaking.

Should more material, from a wider range of dates, be discovered, we may well find that the achievement of the ancient Egyptians in the sphere of cartography was greater than the chance survival of our present material suggests. Such is the preserving quality of parts of Egypt that we may expect new documents relating to the civil administration of the country to come to light. This would certainly help in interpreting such major finds as the Turin papyrus of the gold mines. A definitive edition of both sections of this unique map is clearly a major desideratum.

FIG. 7.13. PLAN OF CANAL AND PALISADE. Associated with a memorandum to Zenon, Apollonius's estate manager (see fig. 7.12), this undated document was found at Philadelphia, Egypt.
Size of the papyrus: 25.5 × 22 cm. By permission of the Department of Rare Books and Special Collections, University of Michigan Library (P. Mich. Inv. 3110).

BIBLIOGRAPHY
CHAPTER 7 EGYPTIAN CARTOGRAPHY

Goyon, Georges. "Le papyrus de Turin dit 'Des mines d'or' et le Wadi Hammamat." *Annales du Service des Antiquités de l'Egypte* 49 (1949): 337–92.

Gundlach, Rold. "Landkarte." In *Lexikon der Ägyptologie,* ed. Wolfgang Helck and Eberhard Otto, 3:col. 922–23. Wiesbaden: O. Harrassowitz, 1975–.

Lesko, Leonard H. *The Ancient Egyptian Book of Two Ways.* University of California Near Eastern Studies Publications, vol. 17. Berkeley: University of California Press, 1972.

Neugebauer, Otto, and Richard A. Parker, eds. and trans. *Egyptian Astronomical Texts.* 3 vols. Providence and London: Lund Humphries for Brown University Press, 1960–69.

North, Robert. *A History of Biblical Map Making.* Beihefte zum Tübinger Atlas des Vorderen Orients, B32. Wiesbaden: Reichert, 1979.

Peck, William H. *Drawings from Ancient Egypt.* London: Thames and Hudson, 1978.

Piankoff, Alexandre. *The Wandering of the Soul.* Completed and prepared for publication by Helen Jacquet-Gordon. Egyptian Religious Texts and Representations, Bollingen Series 40, vol. 6. Princeton: Princeton University Press, 1974.

Žába, Zbyněk. *L'orientation astronomique dans l'ancienne Egypte et la précession de l'axe du monde.* Archiv Orientálni, suppl. 2. Prague: Editions de l'Académie Tchécoslovaque des Sciences, 1953.

8 · The Foundations of Theoretical Cartography in Archaic and Classical Greece

PREPARED BY THE EDITORS FROM MATERIALS SUPPLIED BY GERMAINE AUJAC

Greek civilization started in the Minoan-Mycenaean age (2100–1100 B.C.) and arguably continued to the fall of the empires of Byzantium and Trebizond in the fifteenth century. Within this span of some three thousand years, the main achievements in Greek cartography took place from about the sixth century B.C. to the culminating work of Ptolemy in the second century A.D. This seminal era can be conveniently divided into several periods around which the following chapters are shaped: the archaic and classical period (to the fourth century B.C.), the Hellenistic period (fourth and third centuries B.C.), the early Greco-Roman period (second century B.C. to the second century A.D.), and the age of Ptolemy (second century A.D.).[1]

It has often been remarked that the Greek contribution to cartography lay in the speculative and theoretical realms rather than in the practical realm, and nowhere is this truer than in the earliest period down to the end of the classical era. Large-scale terrestrial mapping, in particular, lacked a firm empirical tradition of survey and firsthand observation. Even at the end of the period, the geographical outlines of the known world or *oikoumene* were only sketchily delineated. Astronomical mapping, while clearly based on direct observation and developed for practical astrological and calendrical purposes, relied more on abstract geometry than on the systematic art of measuring.

Moreover, for the historian of cartography, the early period poses particular problems as much through the scanty nature of the evidence as through the difficulty of its interpretation. No cartographic artifacts clearly define a beginning to the period. The links, for example, with the earlier Babylonian and Egyptian cartography described in the preceding chapters can be only tentatively established, and the extent to which the early Greeks were influenced by such knowledge remains a matter for conjecture. While there is some circumstantial evidence for both the transmission and the reception of important mathematical concepts relevant to cartography—and even for the descent of the basic design of the world map—direct documentary proof for such connections is lacking.[2]

Likewise, it is not always realized that the vast majority of our knowledge about Greek cartography in this early period is known from second- or third-hand accounts. We have no original texts of Anaximander, Pythagoras, or Eratosthenes—all pillars of the development of Greek cartographic thought. In particular, there are relatively few surviving artifacts in the form of graphic representations that may be considered maps. Our cartographic knowledge must therefore be gleaned largely from literary descriptions, often couched in poetic language and difficult to interpret. In addition, many other ancient texts alluding to maps are further distorted by being written centuries after the period they record; they too must be viewed with caution because they are similarly interpretative as well as descriptive.[3] Despite the apparent continuity of some aspects of cartographic thought and practice, we must extrapolate over large gaps to arrive at our conclusions. In the account that follows, therefore, a largely empirical approach is adopted, so that the maximum amount of information about the maps, collected under the names of individual

1. For general works on this period, see G. E. R. Lloyd, *Early Greek Science: Thales to Aristotle* (New York: W. W. Norton, 1970); Armando Cortesão, *History of Portuguese Cartography*, 2 vols. (Coimbra: Junta de Investigações do Ultramar-Lisboa, 1969–71), vol. 1, chap. 2; Edward Herbert Bunbury, *A History of Ancient Geography among the Greeks and Romans from the Earliest Ages till the Fall of the Roman Empire*, 2d ed., 2 vols. (1883; republished with a new introduction by W. H. Stahl, New York: Dover, 1959); J. Oliver Thomson, *History of Ancient Geography* (Cambridge: Cambridge University Press, 1948; reprinted New York: Biblo and Tannen, 1965); H. F. Tozer, *A History of Ancient Geography*, 2d ed. (1897; reprinted New York: Biblo and Tannen, 1964); D. R. Dicks, *Early Greek Astronomy to Aristotle* (Ithaca: Cornell University Press, 1970); Otto Neugebauer, *The Exact Sciences in Antiquity*, 2d ed. (Providence: Brown University Press, 1957); idem, *A History of Ancient Mathematical Astronomy* (New York: Springer-Verlag, 1975); G. S. Kirk, J. E. Raven, and M. Schofield, *The Presocratic Philosophers*, 2d ed. (Cambridge: Cambridge University Press, 1983); W. K. C. Guthrie, *A History of Greek Philosophy*, 6 vols. (Cambridge: Cambridge University Press, 1962–81).

2. Otto Neugebauer, "Survival of Babylonian Methods in the Exact Sciences of Antiquity and the Middle Ages," *Proceedings of the American Philosophical Society* 107 (1963): 528–35.

3. Lloyd, *Early Greek Science*, 10 (note 1).

authors, can be extracted in chronological order from what are often the fragments of lost works.[4]

The earliest literary reference for cartography in early Greece is difficult to interpret. Its context is the description of the shield of Achilles in the *Iliad* of Homer, thought by modern scholars to have been written in the eighth century B.C.[5] Since both Strabo (ca. 64/63 B.C.–A.D. 21) and the Stoics claimed Homer was the founder and father of a geographical science generally understood as involving both maps and treatises, it is tempting to start a history of Greek theoretical cartography with Homer's description of this mythical shield. If this interpretation is valid, then it must also be accepted that Homer was describing a cosmological map. Although from the Hellenistic period onward the original meaning of the term geography was a description of the earth, *gē*, written or drawn (mapping and geographical descriptions were thus inseparable in the Greek world), it is equally clear that Greek mapmaking included not only the representation of the earth on a plane or globe, but also delineations of the whole universe. The shield in Homer's poem, made for Achilles by Hephaestus, god of fire and metallurgy, was evidently such a map of the universe as conceived by the early Greeks and articulated by the poet.

Despite the literary form of the poem, it gives us a clear picture of the various processes in the creation of this great work with its manifestly cartographic symbolism. We are told how Hephaestus forged a huge shield laminated with five layers of metal and with a three-layered metal rim. The five plates that made up the shield consisted of a gold one in the middle, a tin one on each side of this, and finally two of bronze. On the front bronze plate we are told that he fashioned his designs in a concentric pattern; a possible arrangement is suggested in figure 8.1.[6] The scenes of the earth and heavens in the center, two cities (one at peace and one at war), agricultural activity and pastoral life, and "the Ocean, that vast and mighty river" around the edge of the hard shield denote his intention of presenting a synthesis of the inhabited world as an island surrounded by water. Hephaestus depicted the universe in miniature on Achilles' shield, and Homer, in his poetry, only provides a commentary on this pictorial representation. As with the Thera fresco (discussed below), which is roughly contemporaneous with the subject of Homer's poem, the juxtaposition on the shield of scenes and actions that in reality could not occur at the same time shows the artist's desire to portray a syncretism of human activity.

In light of the archaeological discoveries of cultures that certainly influenced Homer's poetry, the content of Achilles' shield seems less extraordinary.[7] Homer was

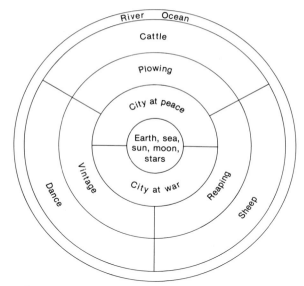

FIG. 8.1. RECONSTRUCTION OF THE SHIELD OF ACHILLES FROM HOMER'S *ILIAD*.
After Malcolm M. Willcock, *A Companion to the Iliad* (Chicago: University of Chicago Press, 1976), 210.

4. For most of the fragments, see H. Diels and W. Kranz, eds., *Die Fragmente der Vorsokratiker*, 6th ed., 3 vols. (Berlin: Weidmann, 1951–52), and an English translation of the fragments from Diels and Kranz in Kathleen Freeman, *Ancilla to the Pre-Socratic Philosophers* (Cambridge: Harvard University Press, 1948).

5. P. R. Hardie, "Imago Mundi: Cosmological and Ideological Aspects of the Shield of Achilles," *Journal of Hellenic Studies* 105 (1985): 11–31; Germaine Aujac, "De quelques représentations de l'espace géographique dans l'Antiquité," *Bulletin du Comité des Travaux Historiques et Scientifiques: Section de Géographie* 84 (1979): 27–38, esp. 27–28. The description of Achilles' shield in the *Iliad* is found in book 18, lines 480–610. For a modern translation and full commentary, see Richmond Lattimore, ed. and trans., *The Iliad of Homer* (Chicago: University of Chicago Press, 1951), 388–91, 411, and based on this translation, Malcolm M. Willcock, *A Companion to the Iliad* (Chicago: University of Chicago Press, 1976), 209–14.

6. The constellations are described thus: "He made the earth upon it, and the sky, and the sea's water, and the tireless sun, and the moon waxing into her fullness, and on it all the constellations that festoon the heavens, the Pleiades and the Hyades and the strength of Orion and the Bear, whom men give also the name of the Wagon, who turns about in a fixed place and looks at Orion and she alone is never plunged in the wash of the Ocean [never falls below the horizon]." Homer *Iliad* 18.483–89; translation by Lattimore, *Iliad*, p. 388 (note 5).

The description of the Ocean Sea and Okeanos, the god of those waters, is as follows: "Not powerful Acheloios matches his strength against Zeus, not the enormous strength of Ocean with his deep-running waters, Ocean, from whom all rivers are and the entire sea and all springs and all deep wells have their waters of him, yet even Ocean is afraid of the lightning of great Zeus and the dangerous thunderbolt when it breaks from the sky crashing." Homer *Iliad* 21.194–99; translation by Lattimore, *Iliad*, p. 423 (note 5).

7. Round shields and vases with comparable ornamentation in concentric bands have been found from this period. See Willcock, *Companion*, 209 (note 5).

writing at a time not much earlier than the first manifestations of what is considered the beginning of Greek science. His poem may be interpreted as the poetic expression of macrocosmic/microcosmic beliefs, held by a society seeking to reconcile a general view of the universe with man's activity within it. Hephaestus, the divine smith, is chosen to give a complete image of the cosmos—earth, sea, and sky together with scenes of human life. The main constellations—Orion, the Hyades, the Pleiades, and the Great Bear—are described, suggesting that a tradition had already developed of using these groupings of stars to identify different parts of the sky. The shield includes a representation of the sun and moon shining simultaneously, again in an attempt to integrate a general knowledge of the sky into one depiction. Even in this poetic form we can glimpse the use of a map, almost as a heuristic device, to bring some order into concept and observation and to codify the early Greeks' reflections on the nature and constitution of their world.

At the same time, we should be clear that the map on Achilles' shield was not intended to communicate a literal view of geographical knowledge of the world as known to the early Greeks. The scenes from rural and urban life are arranged on the surface of the shield in no apparent geographical order. They simply present a generalized and metaphorical view of human activity and of the profound interdependence of human beings in spite of the variety and specialties of their pursuits. This human unity is emphasized by the ocean encircling the whole shield, rendering the world an island. Homer depicts no maritime activity in his social microcosm: the ocean seems to be no more than a geometric framework for the knowable inhabited world, a framework W. A. Heidel considers to be the essential feature of all maps from ancient Greece.[8]

So detailed is Homer's description that, though clearly an imaginary map, Achilles' shield represents a useful glimpse of the early history of efforts to map the world. Probably much of it is conventional, and much also is fanciful. Indeed, it was the subject of ridicule by later writers. Strabo summarized the view:

> Some men, having believed in these stories themselves and also in the wide learning of the poet, have actually turned the poetry of Homer to their use as a basis of scientific investigations. . . . Other men, however, have greeted all attempts of that sort with such ferocity that they not only have cast out the poet . . . from the whole field of scientific knowledge of this kind, but also have supposed to be madmen all who have taken in hand such a task as that.[9]

But the description no doubt reflects elements present in real maps of the time, many of which were widely used later on. Stars are named and grouped into constellations; the limits of the known world are fixed by means of the ocean, real or imaginary, that encircles the inhabited world; and there is an attempt to give pride of place to human activity in this world scene.

In comparison with Homer's poem, the earliest known graphic representation of cartographic significance to have survived from the Greek world is the Thera fresco, fragments of which were discovered in 1971 in the course of archaeological excavation in the House of the Admiral at Akrotiri, Santorini, formerly Thera (plate 3).[10] Rather than depicting the cosmos portrayed on Achilles' shield, it relates to a local area that has been thought to be situated in northern Crete. It probably dates from late Minoan times, the period of the occupation of Thera, about 1500 B.C. The fresco has a picturelike quality and can be reconstructed in detail from the surviving fragments. While its dominant purpose was no doubt decorative, it includes features that have been interpreted as parts of a map, including a coastline, a harbor, a seaside village, a mountain with cattle and wild animals, and a winding river with plants and animals on its banks. Ships and fish are shown in the sea. But besides these geographical features, episodes are also included from what may be the historical past of that society. There are processions of notables going up the hillside, boats in attacking positions along the shore, and battles being fought inland; and there is the departure of the navy and its subsequent triumphal entry into its home port amid general rejoicing. As in Egyptian narrative drawings, events are depicted as occurring simultaneously that are in fact successive in time.

CIRCULAR MAPS AND THE FLAT EARTH: ANAXIMANDER AND HIS SUCCESSORS IN THE SIXTH CENTURY B.C.

With the emergence of Greek science in the sixth century B.C., the context for descriptions of the world changed. It is of course difficult to say how far the greater frequency of allusions to maps in Greek society by this time is due to a fuller survival of literary texts as opposed to real changes and technical advances in the theory and

8. William Arthur Heidel, *The Frame of the Ancient Greek Maps* (New York: American Geographical Society, 1937).

9. Strabo *Geography* 3.4.4; see *The Geography of Strabo*, 8 vols., ed. and trans. Horace Leonard Jones, Loeb Classical Library (Cambridge: Harvard University Press; London: William Heinemann, 1917–32).

10. Peter Warren, "The Miniature Fresco from the West House at Akrotiri, Thera, and Its Aegean Setting," *Journal of Hellenic Studies* 99 (1979): 115–29, and Lajos Stegena, "Minoische kartenähnliche Fresken bei Acrotiri, Insel Thera (Santorini)," *Kartographische Nachrichten* 34 (1984): 141–43. The fresco was first published by Spyridon Marinatos, *Excavations at Thera VI (1972 Season)*, Bibliothēkē tēs en Athēnais Archaiologikēs Hetaireias 64 (Athens: Archailogikē Hetaireia, 1974).

practice of mapmaking. Yet despite the fact that our conclusions must still rest on literary sources (often at several removes from the practices they describe) rather than on map artifacts of the period, there are strong grounds for believing that for the first time natural phi-losophers were asking more systematic questions about the world in general and trying to give naturalistic rather than supernatural explanations for the phenomena they observed. Thus it may be that the Milesian natural phi-losophers were the first Greeks to attempt to map the

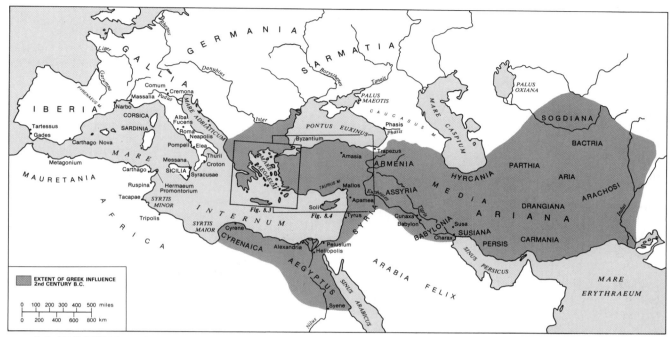

FIG. 8.2. PRINCIPAL PLACES ASSOCIATED WITH MAPS IN THE GREEK WORLD.

FIG. 8.3. THE AEGEAN. Detail from the reference map in figure 8.2.

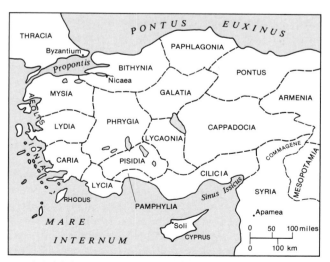

FIG. 8.4 ASIA MINOR FROM THE THIRD CENTURY B.C. Detail from the reference map in figure 8.2.

earth and sky according to recognizable scientific principles.

As viewed through the later Greek authors—who tended to adopt a heroic rather than a contextual view of the development of knowledge—much of the credit for these innovations was given to Anaximander (ca. 610–546 B.C.), who had been a disciple of Thales at Miletus, a city in Asia Minor (see figs. 8.2–8.4). Thales (ca. 624–547 B.C.), one of the Seven Sages of Greece, was considered by later commentators to be an excellent astronomer.[11] It was said he could predict eclipses and calculate the length of the solar year and the lunar month so as to fix the interval between solstices and equinoxes.[12] According to one legend, Thales was so preoccupied with the heavens that he ignored what was beneath his feet, and fell into a well while looking at the stars.[13]

Anaximander, who was also known as a fine astronomer, was particularly interested in the technical aspect of this science. He is alleged to have invented the gnomon and introduced it into Sparta as part of a sundial.[14] In fact, as Herodotus suggests, he may only have borrowed the idea for this instrument from the Babylonians.[15] Whether Anaximander taught that the earth was spherical or cylindrical has also been a point of contention among classical and modern authors—the indirect evidence on his cosmology is contradictory.[16] In any case, according to Diogenes Laertius, the third-century A.D. compiler from whom we derive much of our biographical information about ancient Greek philosophers, Anaximander "was the first to draw the outline of land and sea and also to have constructed a globe."[17] Similarly, Agathemerus, the author of a third-century A.D. geographical treatise and a source of many otherwise lost works, claims that Anaximander was the first "to venture to draw the inhabited world on a map [*pinaki*],"[18] and Strabo calls him the author who "published the first geographical map [*geographikon pinaka*]."[19] It is clear that Anaximander was the first recorded of that long line of Greek craftsmen-philosophers who tried to express concepts in graphic form. The construction of spheres and the drawing of maps were to become characteristic products of the mechanical mind of the Greeks, and their regular occurrence reveals perhaps a more practical side than has traditionally been presented.

It is not certain that Anaximander wrote a commentary on his map or on the construction of his sphere. Hecataeus (fl. 500 B.C.), historian, statesman, and native of Miletus, is thought to be the author of the first *Circuit of the Earth* (*Periodos gēs*).[20] It was divided into two parts: one concerns Europe and the other Asia and Libya (Africa). Hecataeus's treatise is believed to have improved greatly on Anaximander's map; Agathemerus

considered it excellent and even preferred it to the later one of Hellanicus of Lesbos.[21] Figure 8.5 is a reconstruction of Hecataeus's world view.

The materials used for these early maps were probably substantial. The word *pinax*, as defined by later authors, could mean a wooden panel used for writing inscriptions

11. The Seven Sages were statesmen, tyrants, and others who lived between 620 and 550 B.C., each of whom was recognized for a wise maxim. Thales is consistently included among the seven. See John Warrington, *Everyman's Classical Dictionary* (London: J. M. Dent; New York: E. P. Dutton, 1961).

12. Since our evidence of Thales' astronomy is indirect, it must be regarded with caution. Neugebauer, *Exact Sciences in Antiquity*, 142 (note 1), argues that if Thales did predict the solar eclipse of 584 B.C., it was not done on a scientific basis, since the Babylonian theory on which it was supposedly based did not exist in 600 B.C. Legends about Thales are sometimes contradictory, emphasizing either his theoretical or his practical abilities.

13. The story about the well is found in Plato *Theaetetus* 174A; see James Longrigg, "Thales," in *Dictionary of Scientific Biography*, 16 vols., ed. Charles Coulston Gillispie (New York: Charles Scribner's Sons, 1970–80), 13:297, especially n. 7.

14. Diogenes Laertius *Lives of Famous Philosophers* 2.1; see *Vitae philosophorum*, ed. Herbert S. Long, 2 vols. (Oxford: Clarendon Press, 1964); or, for an English translation, *Lives of Eminent Philosophers*, 2 vols., trans. R. D. Hicks, Loeb Classical Library (Cambridge: Harvard University Press; London: William Heinemann, 1925–38).

15. Herodotus *History* 2.109; see *The History of Herodotus*, 2 vols., trans. George Rawlinson (London: J. M. Dent; New York: E. P. Dutton, 1910). See also Herodotus *Histoires*, 10 vols., ed. P. E. Legrand (Paris: Belles Lettres, 1932–39). The actual level of Anaximander's scientific knowledge was probably far less than the secondary and tertiary sources suggest; see D. R. Dicks, "Solstices, Equinoxes, and the Presocratics," *Journal of Hellenic Studies* 86 (1966): 26–40.

16. Dicks, *Early Greek Astronomy*, 45–46 and n. 50 (note 1).

17. Diogenes Laertius *Lives* 2.1 (note 14). See also William Arthur Heidel, "Anaximander's Book: The Earliest Known Geographical Treatise," *Proceedings of the American Academy of Arts and Sciences* 56 (1921): 237–88.

18. Agathemerus *Geographiae informatio* 1.1, in *Geographi Graeci minores*, ed. Karl Müller, 2 vols. and tabulae (Paris: Firmin-Didot, 1855–56), 2:471–87, esp. 471, translation by O. A. W. Dilke; the Greek words rendered here as "on a map" are ἐν πίνακι. The two most common words for a map, (*gēs*) *periodos* and *pinax*, can have other meanings, respectively "circuit of the earth" and "painting." As a result, modern writers have tended to be somewhat cautious in their assessment of Greek cartography, and a proportion of the material presented here is not to be found in published accounts; yet it should be seriously and scientifically considered.

19. γεωγραφικὸν πίνακα. Strabo *Geography* 1.1.11 (note 9), translated by O. A. W. Dilke. See also Strabo, *Géographie*, ed. François Lasserre, Germaine Aujac et al. (Paris: Belles Lettres, 1966–).

20. περίοδος γῆς. The title of Hecataeus's work is sometimes given simply as *Periodos* or *Periegesis*. Most of the extant fragments are from Stephanus of Byzantium, and these are largely lists of place-names. From the fragments in Strabo and Herodotus, however, it is evident that the original work was more extensive. See D. R. Dicks, "Hecataeus of Miletus," in *Dictionary of Scientific Biography*, 6:212–13 (note 13), and Tozer, *History of Ancient Geography*, 70–74 (note 1).

21. Agathemerus *Geographiae informatio* 1.1 (note 18). Hellanicus (ca. 480–400 B.C.), a contemporary of Herodotus, was more a historian than a geographer.

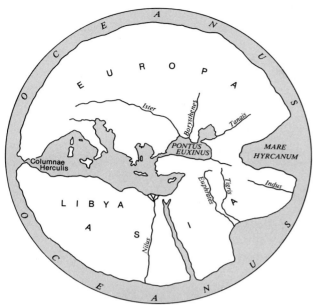

FIG. 8.5. RECONSTRUCTION OF THE WORLD ACCORD-
ING TO HECATAEUS.
After Edward Herbert Bunbury, *A History of Ancient Geography among the Greeks and Romans from the Earliest Ages
till the Fall of the Roman Empire*, 2d ed., 2 vols. (1883; republished with a new introduction by W. H. Stahl, New York:
Dover, 1959), vol. 1, map facing p. 148.

or painting portraits, landscapes, or maps.[22] Herodotus,
on the other hand, speaks of a bronze tablet (*pinax*) with
an engraving of the circuit (*periodos*) of the whole earth
with all the rivers and seas that Aristagoras of Miletus
took with him when he went to Greece about 500 B.C.
in search of allies against the Persians.[23] Herodotus's
reference is important in showing that maps could be
engraved on portable bronze tablets, that general maps
of the inhabited world were frequently made in Ionia,
and that they were more informative than the simple
geometric plans such as the Babylonian clay tablet of
the same era. Aristagoras had in fact been able to show
on that map the regions to be crossed on the way from
Ionia to Persia, comprising Lydia, Phrygia, Cappadocia,
Cilicia stretching to the sea opposite Cyprus, Armenia,
Matiena, and Cissia with the town of Susa. All these
places were inscribed on the "circuit of the earth" engraved on the tablet. The map Aristagoras carried was
probably originally derived from Anaximander's map,
so much admired in antiquity. But we may presume that
it also drew on road measurements compiled by the
Persians for their imperial highways.[24]

We have almost no details of Anaximander's map, but
it is traditionally accepted that "ancient maps" (which
are probably those from Ionia) were circular, with
Greece in the middle and Delphi at the center.[25] Herod-

otus confirms the regularity of the form of these maps:
"For my part, I cannot but laugh when I see numbers
of persons drawing maps of the world without having
any reason to guide them; making, as they do, the ocean-
stream to run all round the earth, and the earth itself to
be an exact circle, as if described by a pair of compasses,
with Europe and Asia just of the same size."[26] It is significant that Herodotus refers here to *periodoi gēs* (circuits of the earth), probably similar to that of Hecataeus.
These works were supposed to be illustrated with diagrams or were accompanied by maps engraved on
bronze or painted on wood.[27]

Aristotle ridiculed his contemporaries who, in their
"circuits of the earth," drew the inhabited world as circular, which he said was illogical.[28] In the first century
B.C. Geminus, the Stoic philosopher and pupil of Posidonius, complained of the artificiality of circular maps
still in use and warned against accepting relative distances in maps of this sort.[29] His use of the word *geographia* in reference to maps illustrates the double meaning of the word. Thus the simple circular maps continued
to be in use long after it was known that the inhabited
world was greater in length (west to east) than in breadth
(south to north).

THE IMPACT OF NEW THEORIES ON CARTOGRAPHY FROM THE SIXTH TO THE FOURTH CENTURY B.C.: PYTHAGORAS, HERODOTUS, AND DEMOCRITUS

Although the tradition of world maps drawn as flat
disks, reflecting a theory that the earth was also a plane

22. πίνακι. These wooden panels were used for public exhibitions,
inserted into the walls of monuments or in porticoes where they were
prominently displayed.
23. Herodotus *History* 5.49 (note 15). Here the word paraphrased
as "circuit," *periodos*, literally means "a going round" and may therefore suggest a roughly circular shape for the map engraved on the
tablet.
24. The description of the Persian Royal Road is in Herodotus
History 5.52–54 (note 15). See also Robert James Forbes, *Notes on
the History of Ancient Roads and Their Construction*, Archaeologisch-
Historische Bijdragen 3 (Amsterdam: North-Holland, 1934), 70–84.
25. At Apollo's oracle in Delphi there was an omphalos (navel), a
stone that symbolized the center of the world. The origin of the notion
of the centrality of Delphi (from Greek mythology) and a general
discussion of the omphalos at Delphi are found in *Oxford Classical
Dictionary*, 2d ed., s.v. "omphalos." See Agathemerus *Geographiae
informatio* 1.2 (note 18).
26. Herodotus *History* 4.36 (note 15).
27. See note 22.
28. Aristotle *Meteorologica* 2.5.362b.13; see *Meteorologica*, trans.
H. D. P. Lee, Loeb Classical Library (Cambridge: Harvard University
Press; London: William Heinemann, 1952).
29. Geminus *Introduction to Phenomena* 16.4.5; see *Introduction
aux phénomènes*, ed. and trans. Germaine Aujac (Paris: Belles Lettres,
1975).

surface, had been entrenched since the time of Homer, the sources indicate that the concept of the heavens and earth as spherical, eventually leading to cartographic representation in the form of celestial and terrestrial globes, came much later. It is very doubtful that the theory of sphericity of the earth can be dated earlier than Pythagoras, a native of Samos who moved to Croton in southern Italy about 530 B.C. The statement by Diogenes Laetius that Anaximander constructed a celestial sphere is unsubstantiated.[30]

The observation that fixed stars seemed to turn around a fixed point (later to be identified as the celestial pole) in regular procession led to the concept of a spherical sky rotating on an axis whose extremities were the celestial poles.[31] Recognition of the spherical nature of the heavens in turn may have led to the supposition that the earth too was a sphere. This concept appears to have been first diffused and taught in the southern Italian cities of Magna Graecia by the Pythagoreans; the first description of a spherical earth has been attributed sometimes to Pythagoras himself (fl. 530 B.C.) and sometimes to Parmenides, a native of Elea (Velia) in southern Italy (fl. ca. 480 B.C.). It was first proposed as a simple hypothesis, not verified scientifically but justified theologically. In the eyes of the Pythagoreans, the geometric perfection of the circle and the sphere was sufficient reason for adopting these ideas. They imagined all parts of the cosmos to be spheres (the stars, the sky in which they were fixed, the terrestrial globe) and all the movements in the sky to be circular (the rotation of fixed stars, the combined circular motions for the movements of the planets). These theories did not, however, have an immediate or dramatic impact on cartography. Since the representation of a sphere on a single plane is a circle, it is probable that the hypothesis of a spherical earth could reinforce, by an understandable misinterpretation of the figure, the idea of a flat, circular inhabited world and perpetuate this kind of representation.

The teachings of Pythagoras (who left no writings) are known only from what was said by his disciples or his successors, who tended to attribute to him all the ideas of the later school. As for Parmenides, he was the author of a philosophical poem, Concerning Nature, of which only fragments remain. Posidonius, who four centuries later described the process leading to the division of the sky and the earth into five zones, considered Parmenides the originator of this division also, and he saw the division itself as the direct result of the hypothesis of the spherical nature of the sky and the earth.[32]

There are no documents to prove whether the Pythagoreans in general and Parmenides in particular, other than producing simple geometric diagrams, put their hypotheses into material representations in the form of globes. But it must be remembered that the making of

mechanical spheres or sphairopoiia flourished in the third century B.C. in this general region, especially Sicily, reinforced by the inventive genius of Archimedes, and that this may represent the continuation of a longer tradition.[33]

It was not until the fifth century B.C. that the traditional Homeric disk-shaped view of the world was systematically challenged by Herodotus (ca. 489–425 B.C.). A native of Halicarnassus (Bodrum) in Caria, but living in Thurii in southern Italy after 444, Herodotus was a friend of Pericles and Anaxagoras and had denounced, as we have seen, the traditional circular maps he viewed as so misleading. According to him, it had not been proven that the inhabited world was surrounded by water on all sides. It was clear to him that Africa was surrounded by sea except on the side where it adjoined Asia, because the Phoenicians sent by Necos (Necho), king of Egypt 609–594 B.C., had been able to go around it by boat in three years.[34] Asia was inhabited only as far as India, and farther to the east there was only a desert about which nothing was known.[35] Similarly for Europe, no one knew "whether any sea girds it round either on the north or on the east."[36] Thus Herodotus refused, in the name of scientific caution, to make a general map of the inhabited world when the outlines were so uncertain. He attacked the theoretical cartographers who based their ideas on geometry alone and seems to have urged a return to empirical cartography founded on exploration and travel. Theory, in his view, should give way to experience.

Another objection Herodotus made to the maps of his day was the way they divided the inhabited world into continents: "I am astonished that men should ever have divided Libya, Asia, and Europe as they have, for they are exceedingly unequal. Europe extends the entire length of the other two, and for breadth will not even (as I think) bear to be compared to them."[37] Herodotus would thus have given the general map of the inhabited world, had he been willing to draw it, a form similar to the T-O maps of the late classical period and Middle

30. Kirk, Raven, and Schofield, Presocratic Philosophers, 104 (note 1).

31. The position of the celestial pole relative to the stars has changed since that time because of the precession of the equinoxes. Our Pole Star, at the end of the tail of the Little Bear, was twelve degrees distant from the pole in Hipparchus's time. See The Geographical Fragments of Hipparchus, ed. D. R. Dicks (London: Athlone Press, 1960), 170.

32. Strabo Geography 2.2.1–2 (note 9).

33. Sphairopoiia means the making of a sphere; it was considered a branch of mechanics that studied the rotation of the sphere. See Hans Joachim Mette, Sphairopoiia: Untersuchungen zur Kosmologie des Krates von Pergamon (Munich: Beck, 1936).

34. Herodotus History 4.42 (note 15).

35. Herodotus History 4.40 (note 15).

36. Herodotus History 4.45 (note 15).

37. Herodotus History 4.42 (note 15).

Ages,[38] except that Europe (and not Asia), would have taken up the transverse part, while Asia and Libya would have been on each side of the vertical line.[39] Yet despite his awareness of the deficiencies of contemporary "geometric" maps, whether originating in Ionia or elsewhere, and perhaps because of his failure to express his ideas in graphic form, Herodotus was never considered a geographer—still less a mapmaker—by his successors.

This was far from the case with his contemporary Democritus (ca. 460 to ca. 370 B.C.), widely acknowledged for his formulation of the concept that the inhabited world was oblong and that the world map could be better accommodated in an oval rather than a circular frame. Born at Abdera in Thrace, Democritus was a great traveler with an inquiring mind. A philosopher and atomist like his master Leucippus, he studied with the Babylonian magi, the Egyptian priests, and even the Indian gymnosophists, at least according to the tradition.[40] He was a prolific writer, but his *Cosmology* (considered a work of physics), *Uranography*, *Geography*, and *Polography* (these three considered mathematical works, the last being perhaps a description of the pole) are all now lost.

The observational work of Democritus is known from the fragments of his calendar preserved in Geminus's *Isagoge* and Ptolemy's *Phaseis*, which gave the dates of the heliacal risings and settings of the chief constellations (the Pleiades, Lyra, Eagle, and Orion) and weather prognostications connected with these.[41] The description and drawing of these constellations was perhaps the main subject of his *Uranography*. In geography and cartography, however, Democritus can be assessed only through the testimony of his successors rather than the substance of his works. Strabo puts him immediately after the Ionians Anaximander and Hecataeus on his list of those who had most served geography and mentions him together with Eudoxus of Cnidus, Dicaearchus, and Ephorus.[42] He considered all four the most distinguished predecessors of Eratosthenes. It is likely that Democritus provided a map, or at least a plan, showing the shape he ascribed to the world in his *Geography*. As already noted, it is probable that this was oblong, its length one and a half times its breadth.[43] This proportion was accepted 150 years later by Dicaearchus. Democritus can thus claim a place in the history of cartography—as among the geographers of the Greek world—on the basis of this new idea of an oval rather than a circular inhabited world, one that by the third century B.C. was to be incorporated in the design of the world map.

While not directly concerned with geography or with the description of Greek maps of the time, Plato (ca. 429–347 B.C.) alluded in his writings to matters broadly associated with cartography in both the *Phaedo* and the *Republic* (both ca. 380 B.C.). In the *Phaedo*, Socrates is made to comment on the shape of the earth:

> Now there are many wondrous regions in the earth, and the earth itself is of neither the nature nor the size supposed by those who usually describe it, as someone has convinced me. . . . I've been convinced that if it is round and in the centre of the heaven, it needs neither air nor any other such force to prevent its falling, but the uniformity of the heaven in every direction with itself is enough to support it, together with the equilibrium of the earth itself.[44]

Whether the word περιφερής (*peripheres*, translated as "round") means circular or spherical has been the subject of a controversy not wholly understandable in view of Plato's obvious spherical analogy of the earth as a ball in a later passage.[45]

He then reveals his view of the earth's size: "And next, that it is of vast size, and that we who dwell between the Phasis River and the Pillars of Heracles inhabit only a small part of it, living around the sea like ants or frogs around a marsh, and that there are many others living elsewhere in many such places."[46] There then follows the passage where he likens the earth to a leather ball made up of twelve pentagonal pieces. This is an allusion to the Pythagorean theory of the dodecahedron, considered in classical times especially significant as the solid most nearly approaching a sphere.[47] In this, Plato also emphasizes the variety of colors of the earth when viewed from above:

> First of all the true earth, if one views it from above, is said to look like those twelve-piece leather balls,

38. The T-O maps were circular maps (hence the O), divided geometrically into three parts by two lines (hence the T). See Marcel Destombes, ed., *Mappemondes A.D. 1200–1500: Catalogue préparé par la Commission des Cartes Anciennes de l'Union Géographique Internationale* (Amsterdam: N. Israel, 1964); see also below, pp. 296–97 and 301.

39. As illustrated in a manuscript of Bede's *De natura rerum*, Bayerische Staatsbibliothek, Munich (Clm. 210, fol. 132v), and in figures 18.38 and 18.55 below.

40. For a general review of the various traditions surrounding Democritus, see G. B. Kerferd, "Democritus," in *Dictionary of Scientific Biography*, 4:30–35 (note 13).

41. Dicks, *Early Greek Astronomy*, 84–85 (note 1).

42. Strabo *Geography* 1.1.1 (note 9).

43. Agathemerus *Geographiae informatio* 1.2 (note 18).

44. Plato *Phaedo* 108e–109a; see the translation by David Gallop (Oxford: Clarendon Press, 1975). Both the *Phaedo* and the *Republic* date to Plato's middle period when he was in close contact with the Pythagorean Archytas, who was called by Horace the "measurer of land and sea"; see Tozer, *History of Ancient Geography*, 169 (note 1).

45. On the controversy, see page 223 of the Gallop translation (note 44). For the reference to the earth as a ball, see below, note 48.

46. Plato *Phaedo* 109b (note 44).

47. Plato, *Phaedo*, ed. John Burnet (Oxford: Clarendon Press, 1911), 131 (110b6).

variegated, a patchwork of colours, of which our colours here are, as it were, samples that painters use. There the whole earth is of such colours, indeed of colours far brighter still and purer than these: one portion is purple, marvellous for its beauty, another is golden, and all that is white is whiter than chalk or snow; and the earth is composed of the other colours likewise, indeed of colours more numerous and beautiful than any we have seen. Even its very hollows, full as they are of water and air, give an appearance of colour, gleaming among the variety of the other colours, so that its general appearance is of one continuous multi-coloured surface.[48]

In the *Republic*, Plato briefly describes the skills of the navigator. He was illustrating the need for government to be in the hands of skilled "pilots" (philosophers). We can perhaps interpret this as confirmation that the art of navigation was fully understood by his readership: "The true pilot must give his attention to the time of the year, the seasons, the sky, the winds, the stars, and all that pertains to his art if he is to be a true ruler of a ship."[49] More directly cartographic in its allusion is Plato's description of a model of the universe within a passage known as the myth of Er. Er is depicted as a Pamphylian warrior who returned from the dead to describe the afterlife. Plato believed in a geocentric universe with the fixed stars on a sphere or band at the outside, and the orbits of the sun, moon, and planets between the earth and the stars. In his description of it, he used a spindle (the Spindle of Necessity) and whorl to symbolize, somewhat imperfectly, its workings.[50] The rims of the whorl—illustrated in figure 8.6—are intended to represent, from the outside in, the fixed stars and the orbits of Saturn, Jupiter, Mars, Mercury, Venus, the sun, and the moon.

Despite the theoretical nature of much of Greek cartography in the fifth and fourth centuries B.C., and though it was mainly the subject of debate among the philosophers rather than the object of much practical mapmaking, it does seem likely that the Greeks' awareness of the place of maps in their society grew in this period. There are even a few fragments of evidence to suggest that a knowledge of maps may have filtered into the experience of ordinary citizens. Three examples show the role maps or plans played in everyday life. Most remarkable, perhaps, is that in a fifth-century comedy by Aristophanes, *The Clouds*, we encounter a stage map that, just as surely as the many cartographic allusions in Shakespeare, suggests that the audience was familiar with the form and content of maps. Strepsiades, an old farmer compelled by war to take up residence in Athens, is intrigued with the paraphernalia of philosophy and questions a student:

STREPSIADES (pointing to a chart): "In the name of heaven, what's *that*?"
STUDENT: That's for astronomy.
STREPSIADES (pointing to surveying instruments): And what are those?
STUDENT: They're for geometry.
STREPSIADES: Geometry? And what's that good for?
STUDENT: Surveying, of course.
STREPSIADES: Surveying what? Lots?
STUDENT: No, the whole world.
STREPSIADES: What a clever gadget! And as patriotic as it is useful.

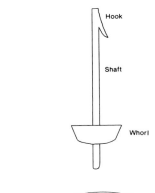

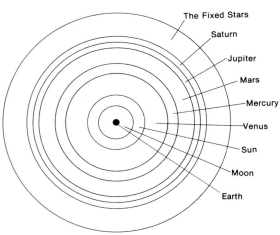

FIG. 8.6. RECONSTRUCTION OF PLATO'S SPINDLE OF NECESSITY. Plato used a spindle as an analogy for the universe, which he believed to be geocentric.
After Plato, *The Republic of Plato*, 2 vols., ed. James Adam (Cambridge: Cambridge University Press, 1902), book 10, figs. iii and iv.

48. Plato *Phaedo* 110b–d (note 44).
49. Plato *Republic* 6.4; see *Plato's Republic*, 2 vols., trans. Paul Shorey, Loeb Classical Library (Cambridge: Harvard University Press; London: William Heinemann, 1935–37).
50. Plato *Republic* 10.14 (note 49). H. D. P. Lee's translation, *The Republic* (London: Penguin Books, 1955), has a description and diagram of the "Spindle of Necessity" on pages 402–5. The diagram is taken from *The Republic of Plato*, ed. James Adam, 2 vols. (Cambridge: Cambridge University Press, 1902), book 10, figs. iii and iv.

STUDENT (pointing to a map): Now then, over here we have a map of the entire world. You see there? That's Athens.

STREPSIADES: *That,* Athens? Don't be ridiculous. Why, I can't see even a single lawcourt in session.

STUDENT: Nonetheless, it's quite true. It really is Athens.

STREPSIADES: Then where are my neighbors of Kikynna?

STUDENT: Here they are. And you see this island squeezed along the coast? That's Euboia.

STREPSIADES: I know that place well enough. Perikles squeezed it dry. But where's Sparta?

STUDENT: Sparta? Right over here.

STREPSIADES: That's MUCH TOO CLOSE! You'd be well advised to move it further away.

STUDENT: But that's utterly impossible.

STREPSIADES: You'll be sorry you didn't, by god.[51]

The passage demonstrates that large-scale cadastral maps and maps of the world were known to an audience of fifth-century Athens, and that the power of the map as a metaphor was realized (Strepsiades thinks he can lessen the threat from Sparta by moving it farther away on the map). The map is thus employed to focus attention on the geography of contemporary issues, and it has also become a vehicle for social criticism of that particular society.

A briefer allusion—this time to the value of maps as propaganda—is found in Plutarch's life of Nicias, in which Alcibiades, the notorious Greek statesman and general of the fifth century, is seeking to persuade the Athenians to undertake an expedition against Sicily:

> Before the assembly had met at all, Alcibiades had already corrupted the multitude and got them into his power by means of his sanguine promises, so that the youth in their training-schools and the old men in their work-shops and lounging-places would sit in clusters drawing maps of Sicily, charts of the sea about it, and plans of the harbours and districts of the island which look towards Libya.[52]

A story in Aelian of Socrates and his rich pupil Alcibiades shows that any Athenian could consult a world map. Seeing Alcibiades blinded by wealth and boasting of his big estates, Socrates took him to a place in the city (Athens) where a world map [*pinakion,* diminutive of *pinax*] was set up. He told Alcibiades to look for Attica; and when he had found it, he told him to look carefully at his own fields. Alcibiades replied: "But they are not drawn in anywhere." Socrates: "Why then, you are boasting of fields which are not even a part of the earth."[53]

Other roles were defined in less ambitious terms. At Thorikos, Attica, on the edge of the horizontal rock face

immediately above the adit to mine 3, is what seems to be a small incised plan of the mine (fig. 8.7).[54] The latter was explored to a distance of 120 meters in 1982 by members of the Belgian Archaeological Mission, and the part explored is said to correspond to the diagram. It may date to the fourth century B.C. While this isolated example is hardly impressive, this and the other glimpses of the practical uses of maps perhaps indicate that some caution should be exercised when defining Greek cartography as a largely theoretical pursuit.

FIG. 8.7. MINE DIAGRAM FROM THORIKOS, ATTICA. Perhaps dating to the fourth century B.C., this seems to be a plan of the mine in front of which it was found incised in the rock.
Length of the original: 35 cm. By permission of the Mission Archéologique Belge en Grèce, Ghent.

Another indication of the Greek bent in practical drawing is afforded by the discovery of detailed architectural plans for parts of Greek buildings. It was thought until recently that no such plans had survived,[55] but a considerable set of incised drawings from the temple of Apollo at Didyma, south of Miletus, has recently

51. Aristophanes *The Clouds* 200–217; see *The Clouds, by Aristophanes,* trans. William Arrowsmith (New York: New American Library, 1962), 30–32.

52. Plutarch *Nicias* 7.1–2 and *Alcibiades* 17.2–3, both in *Plutarch's Lives,* 11 vols., trans. Bernadotte Perrin, Loeb Classical Library, (Cambridge: Harvard University Press; London: William Heinemann, 1914–26).

53. Claudius Aelianus (Aelian) *Varia historia* 3.28, translated by O. A. W. Dilke; see the edition edited by Mervin R. Dilts (Leipzig: Teubner, 1974); cf. Christian Jacob, "Lectures antiques de la carte," *Etudes françaises* 21, no. 2 (1985): 21–46, esp. 42–44.

54. H. F. Mussche, *Thorikos: Eine Führung durch die Ausgrabungen* (Ghent and Nuremberg: Comité des Fouilles Belges en Grèce, 1978), 44 and 48, fig. 53. We may compare another, undated inscription the first words of which may be translated "Boundary of house and shop" and that ends thus: ⊓, in H. W. Catling, "Archaeology in Greece, 1979–80," *Archaeological Reports 1979–80,* no. 26 (1980): 12, col. 2.

55. J. J. Coulton, *Ancient Greek Architects at Work: Problems of Structure and Design* (Ithaca: Cornell University Press, 1977), 53.

been analyzed.[56] This very large temple, planned after 334 B.C., had much work carried out on it about 250 B.C., but it was never completed. The incisions concerned were in fact known earlier but were dismissed by guides as builders' doodles.[57] Perhaps they were ignored for so long because, though one can easily see late Christian incisions on the stylobate, the earlier drawings are in a dark passage near a tunnel.

They include straight lines up to 10 meters long and circles up to 4.5 meters in diameter, originally filled with red chalk. The incisions, mostly full size, represent measurements of the temple and its *naiskos* (interior miniature temple) or their parts. A comparison between the plan of a column base and the actual base shows that the correspondence was very exact, that at two points the drawing was corrected, and that the fluting was not inserted on the plan, evidently because it was carved in situ.[58] The full-scale plans for large architectural members were incised horizontally, not vertically, since only the former gave sufficient length on a suitable flat surface. But in one case we find a column, about 18 meters high, drawn upright but with an unusual type of scale. Half the width only is shown, but at full scale, while the height is at $\frac{1}{16}$ size, or one digit (finger's breadth; Greek *daktylos*, 1.85 centimeters) to one Greek foot (29.6 centimeters). Each foot of height was represented by parallel lines one digit apart. The object was to show a regular Greek feature, entasis or gradual curve on the column, whose diameter is indicated by the end of the line at each foot of height. In this case, to draw the curve, first a straight line was incised between the top and the bottom of the column shaft on the plan, then an arc of a circle was drawn with this line as chord and with a radius thought to be about 3.2 meters.[59] To the left of this diagram is a semicircle showing a half-section of a column, and overlapping both of these is the top quarter of a column drawn horizontally with the entasis shown. On other drawings the only major discrepancy discovered is between the plan of the *naiskos* and its actual dimensions. As to the drawing of the foundations, thin lines were found to have been engraved on the surface of successive layers of them. Although such lines do not reveal great mathematical skill, unlike the others, nevertheless they constitute a closer parallel with cartographic plans than do architectural elevations.

Similar incisions, though far fewer, have been found at two other Greek buildings in Asia Minor—the temple of Athena at Priene and the temple of Artemis at Sardis. They imply a far more systematic concern with scaled drawings in classical Greece than had hitherto been supposed.

THEORY INTO PRACTICE: NEW CELESTIAL GLOBES AND MAPS IN THE FOURTH CENTURY B.C.

The ideas that had been expressed in largely theoretical terms in the fifth century B.C. began to be modified empirically in the fourth century. Significant advances were made both in celestial mapping—especially in the construction of celestial globes—and to a lesser extent in the terrestrial mapping of the inhabited world. Once again, however, the original sources reflect a bias toward the scientific achievements of individuals (which has, moreover, been perpetuated by many modern commentators on the cartographic significance of the texts).[60] In the present state of our knowledge, therefore, the cartographic history of this period still has to be understood in terms of these individuals and their works rather than of the wider social and intellectual milieu in which their ideas were rooted.

Eudoxus of Cnidus (ca. 408–355 B.C.),[61] whom Strabo placed in his long line of philosophers from Homer to Posidonius,[62] apparently initiated great progress in the mapping of both sky and earth. There is some controversy about the sources of Eudoxus's inspiration. He attended lectures of Plato—although apparently not his school—and is said to have spent more than a year in Egypt, some of it studying with the priests at Heliopolis.[63] In any event, Eudoxus of Cnidus is famous for his theory of geocentric and homocentric spheres (twenty-six concentric spheres centered on the earth), which was designed to explain the motion of the planets. His greatest cartographic achievement, however, was that he was the first to draw the stars on a globe representing the sky seen from the outside looking in, rather

56. Lothar Haselberger, "The Construction Plans for the Temple of Apollo at Didyma," *Scientific American*, December 1985, 126–32; idem, "Werkzeichnungen am jüngeren Didymeion," *Mitteilungen des Deutschen Archäologischen Instituts, Abteilung Istanbul* 30 (1980): 191–215.

57. Information from Mrs. J. Lidbrooke of the *Geographical Magazine*.

58. Haselberger, "Temple of Apollo," 128B (note 56).

59. Haselberger, "Temple of Apollo," 131 (note 56), gives 3.2 meters, but this may be an underestimate. The curve is not, as often, parabolic; but care is evidently taken, by substituting a straight line for part of the arc, that the diameter at no point exceeds the diameter at the base.

60. For example, Cortesão, *History of Portuguese Cartography*, 1:74–76 (note 1).

61. Some scholars prefer ca. 400–347 B.C. For example, G. L. Huxley, "Eudoxus of Cnidus," in *Dictionary of Scientific Biography*, 4:465–67, esp. 465 (note 13).

62. Strabo *Geography* 1.1.1 (note 9).

63. Huxley, "Eudoxus," 466 (note 61). But see Heidel, *Frame of Maps*, 100–101 (note 8), and Dicks, *Geographical Fragments*, 13 (note 31), for words of caution on the classical attribution of knowledge to the Egyptian priests.

than as seen by an observer on the earth, together with the positions of the main celestial circles: the equator, the tropics, the arctic (ever-visible) circles, the ecliptic and zodiac, and the colures (fig. 8.8).[64] See also appendix 8.1.

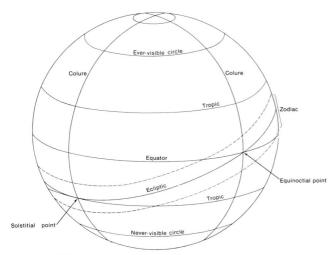

FIG. 8.8. THE CELESTIAL CIRCLES. The arctic circle is constructed for an observer's location at 37°N (see also fig. 10.6).

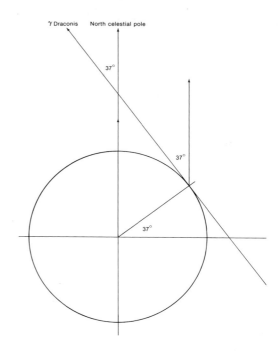

FIG. 8.9. RECONSTRUCTION OF LATITUDE OF OBSERVER. The latitude of intended users of Eudoxus's globe is based on the observation of γ Draconis as it touches the horizon of the observer.

Eudoxus wrote two works, *Phaenomena* and *The Mirror*, to accompany his celestial globe and to help in its interpretation, but both are lost. Fortunately a verse rendering of the *Phaenomena* has survived. It was written by the poet Aratus of Soli (ca. 315–240/239 B.C.) at the request of the Macedonian king Antigonus Gonatas (ca. 320–239 B.C.), a keen patron at his court of scholars, poets, and historians.[65] Aratus undertook this during his stay at Pella, adding to the text of Eudoxus a short prelude in honor of Zeus that has a strong flavor of Stoicism. That the poem accurately reflects the original text is decisively proved in the *Commentary* by Hipparchus.[66]

From Aratus's description of the constellation Draco (the Dragon), it appears that the observer is situated at about 37°N. The southernmost star, γ Draconis, the head of the Dragon, is seen to touch the horizon as the constellation appears to revolve around the heavens.[67] Since the declination of this star at that time was +53°, the angular distance of this star from the pole—37°—is equal to the latitude of the observer, in this case 37°N, the latitude of Athens according to Hipparchus (fig. 8.9).

In his versification of Eudoxus's *Phaenomena*, Aratus not only describes the geometry of the heavens with respect to the visible constellations but also compares the celestial circles to belts that could be linked together on a physical model.[68] Aratus presents the human figures and animals of the constellations in action and in motion

64. Aratus, *Phaenomena* in *Callimachus: Hymns and Epigrams; Lycophron; Aratus*, trans. A. W. Mair and G. R. Mair, Loeb Classical Library (Cambridge: Harvard University Press; London: William Heinemann, 1955), 185–299. See also Aratus, *Phaenomena*, ed. Jean Martin, Biblioteca di Studi Superiori: Filologia Greca 25 (Florence: Nuova Italia, 1956).

65. See G. R. Mair's introduction to his translation of Aratus's *Phaenomena* in *Callimachus: Hymns and Epigrams; Lycophron; Aratus*, 185–89 (note 64).

66. Hipparchus (ca. 190–post 126 B.C.) wrote a commentary on Eudoxus's and Aratus's *Phaenomena*; see *In Arati et Eudoxi Phaenomena commentariorum libri tres*, ed. C. Manitius (Leipzig: Teubner, 1894).

67. Aratus, *Phaenomena*, Mair edition, 211 (note 64). For the declination of γ Draconis see U. Baehr, *Tafeln zur Behandlung chronologischer Probleme*, Veröffentlichungen des Astronomischen Rechen-Instituts zu Heidelberg no. 3 (Karlsruhe: G. Braun, 1955), 58. For Hipparchus's commentary, see *In Arati et Eudoxi Phaenomena* 1.3.12 (note 66), and Dicks, *Geographical Fragments*, 134 (note 31).

68. Aratus, *Phaenomena*, Mair edition, 249 (note 64). A general discussion of celestial globes in antiquity appears in Edward Luther Stevenson, *Terrestrial and Celestial Globes: Their History and Construction, Including a Consideration of Their Value as Aids in the Study of Geography and Astronomy*, 2 vols., Publications of the Hispanic Society of America, no. 86 (New Haven: Yale University Press, 1921; reprinted New York and London: Johnson Reprint Corporation, 1971), 1:14–25. Stevenson's book is based largely on Matteo Fiorini, "Le sfere cosmografiche e specialmente le sfere terrestri," *Bollettino della Società Geografica Italiana* 30 (1893): 862–88, 31 (1894):

like living creatures, and their positions on the globe—and indeed their detailed outlines—are not pure fantasy: they provide a means of identifying the stars and personifying the sky.[69] A close knowledge of the constellations lies behind the poetic language of Aratus. The detailed description of the Dragon, for example, suggests that the stars had been precisely observed in the sky; the constellations are set in relation to one another, and their alignments are clearly explained.[70] We must conclude that the celestial globe of Eudoxus, accompanied by the treatise that described it, was an authentic instrument. It probably helped to conventionalize the figures of the constellations, which have been only slightly modified since Eudoxus's time, and it also gave the Greeks a taste for a mechanical interpretation of the universe.

Although Eudoxus's globe no longer exists, our understanding of its contribution to the development of celestial cartography, as transmitted to later cultures by the poem of Aratus, is enhanced through what is almost certainly a direct descendant. This is the celestial globe resting on the shoulders of a sculptured figure preserved in Naples, known as the Farnese Atlas (figs. 8.10 and 8.11). Although the actual statue dates from the late second century A.D., the style of execution of both the statue and the constellations shows that it is a copy of a Hellenistic original. Forty-three constellations are shown in the form of bas-relief figures (humans, animals, and objects) derived from the Aratus poem; their iconography has not substantially changed since. No individual stars are indicated. The southern sky—invisible from the Mediterranean world—is hidden by the supporting Atlas. At the same time, a cavity in the globe has obliterated the most northerly constellations. The globe was designed to be viewed from the exterior, so that the figures all face in toward the center. As in the Aratus poem, there is also a series of circles: the equator, the two arctic (ever-visible and never-visible) circles, and the two colures. Three oblique parallel circles represent the ecliptic and the zodiac, equally divided into twelve dodecatemories or signs. The intended positions of the arctic circles on this globe, and hence the inferred latitude at which the globe was designed to be used, have been the subject of controversy. The positions of the circles have been precisely measured, but at issue are the intended precision of a statue of this type, the positions of the certain stars (especially Draconis and Canopus) relative to the arctic circles, and the location of constellations in the zodiac relative to the solstitial and equinoctial points. A critical review of existing studies of the globe, together with its detailed reproduction and careful analysis, is urgently needed to resolve these questions.[71]

FIG. 8.10. THE FARNESE ATLAS. Originally belonging to the Farnese family, this is the best surviving example of a genre of statues depicting the mythical figure.
Height of the original: 1.9 m. By permission of the Museo Archeologico Nazionale, Naples.

121–32, 271–81, 331–49, 415–35, and Fiorini's *Sfere terrestri e celesti di autore italiano oppure fatte o conservate in Italia* (Rome: Società Geografica Italiana, 1899). A detailed analysis of Aratus's *Phaenomena* is given in Manfred Erren, *Die Phainomena des Aratos von Soloi: Untersuchungen zum Sach- und Sinnverständnis* (Wiesbaden: Franz Steiner, 1967), 159–200.

69. Dicks, *Early Greek Astronomy*, 158 ff. (note 1).

70. Aratus, *Phaenomena*, Mair edition, 211 (note 64).

71. Costanza Gialanella and Vladimiro Valerio, "Atlas Farnèse," in *Cartes et figures de la terre*, exhibition catalog (Paris: Centre Georges Pompidou, 1980), 84, cite several measurements based on an unpublished photogrammetric study by Valerio, as follows: The tropics are marked at 25°30' on either side of the equator, the band of the zodiac is 13°30' wide, and the arctic circles are 58° distant from the equator,

FIG. 8.11. DETAIL OF THE FARNESE ATLAS. The globe is probably an imitation of Eudoxus's globe and could have been used as an illustration of Aratus's poem.
Diameter of the globe: ca. 64 cm. By permission of the Museo Archeologico Nazionale, Naples.

It is possible that the globe carried by the Farnese Atlas was consciously sculpted as an illustration of Aratus's poem. The cartographic concepts developed by Eudoxus certainly appear to have had wide influence in the classical world: the globe was frequently copied, as was the poem itself, which was translated several times into Latin verse.[72] Both poem and sculpture must have helped to codify the iconography of the various constellations; the legends explaining their presence in the sky served as a mnemonic device, as did the solid graphic representation of the globe.

Eudoxus also wrote a *Circuit of the Earth* (*Periodos gēs*), now lost except for small fragments.[73] He was considered an authority, according to Strabo, in figures (σχημάτων) and in climata (κλιμάτων; latitudes);[74] this praise of his mathematical training and his astronomical skill was no doubt fully justified. The clue to his cartographic contribution lies in the word rendered as "figures," which suggests that his text was accompanied by outline maps of a geometric nature. As a result of these

deliberations, and by modifying the estimate of Democritus, Eudoxus was led to consider the length of the inhabited world to be double its breadth.[75] It is perhaps a measure of his influence that this proportion was adopted in most of the maps in the ancient world that succeeded him.[76]

That a new cartographic image of the inhabited world was being adopted in some quarters by the fourth century B.C. is also suggested by the evidence we have for the map of the historian Ephorus (ca. 405–330 B.C.). His only known contribution is the compilation of a map to illustrate a theoretical geography of the world's peoples, but though he was a contemporary of Eudoxus, the exact nature of his map's construction and content remains partly conjectural. Ephorus was born in Cyme in the Aeolis, and he became a disciple of Isocrates (436–338 B.C.) and an accomplished writer. It is clear that he discussed many geographical questions in his *History* in thirty books, but these are now lost. Once again our knowledge of his cartographic ideas is filtered through the texts of later writers, in this case Strabo (ca. 64/63 B.C. to A.D. 21 or later) and, much later, in the writings of Cosmas Indicopleustes, a Nestorian Christian author of the sixth century A.D.[77]

indicating to them that the globe was designed for an observer at 32°N (the latitude of Alexandria). Other scholars believe that the tropics are intended to be 24°, and the arctic circles 54°, from the equator, pointing out that the star Canopus—at the tip of the rudder (steering oar) of the ship *Argo*—just touches the never-visible circle. This would suggest that the globe was intended for an observer at 36°. For a summary of earlier views of the globe, see Fiorini, *Sfere terrestri e celesti,* 9–25 (note 68).

72. The poem was translated by Cicero, Germanicus, and Avienius: Cicero, *Les Aratea,* ed. and trans. Victor Buescu (Bucharest, 1941; reprinted Hildesheim: Georg Olms, 1966), and *Aratea: Fragments poétiques,* ed. and trans. Jean Soubiran (Paris: Belles Lettres, 1972); Germanicus, *Les Phénomènes d'Aratos,* ed. André Le Boeffle (Paris: Belles Lettres, 1975); *The Aratus Ascribed to Germanicus Caesar,* ed. D. B. Gain (London: Athlone Press, 1976); and Avienius, *Les Phénomènes d'Aratos,* ed. and trans. Jean Soubiran (Paris: Belles Lettres, 1981.

73. Eudoxus *Die Fragmente,* ed. F. Lasserre (Berlin: Walter de Gruyter, 1966).

74. Strabo *Geography* 9.1.2 (note 9).

75. Agathemerus *Geographiae informatio* 1.2 (note 18).

76. This is shown by Geminus, writing in the first century B.C.: "The length of the inhabited world is just about twice its breadth; those who compose geographies according to scale make their drawings on oblong tablets"; translation by O. A. W. Dilke from Geminus *Introduction,* 16.3–4 (note 29).

77. Wanda Wolska, *La topographie chrétienne de Cosmas Indicopleustès: Théologie et science au VIᵉ siècle,* Bibliothèque Byzantine, Etudes 3 (Paris: Presses Universitaires de France, 1962), and Cosmas Indicopleustes *Topographie chrétienne,* ed. Wanda Wolska-Conus in

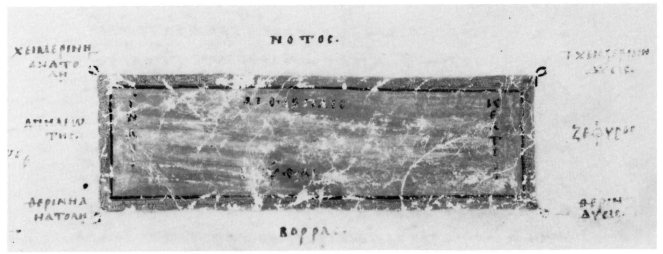

FIG. 8.12. COSMAS'S SCHEMATIC REPRESENTATION OF THE EARTH. An eighth/ninth-century version of a map drawn by Cosmas Indicopleustes (sixth century A.D.) according to the principles explained by Ephorus (ca. 405–330 B.C.).

Photograph from the Biblioteca Apostolica Vaticana, Rome (Vat. Gr. 699, fol. 19r).

It is thus from Strabo's *Geography* that we begin to glimpse the possible map of the world developed by Ephorus. In book 4 of Ephorus's *History*, the part dealing with Europe, says Strabo, we find the opinions of the ancients concerning Ethiopia: "Ephorus, too, discloses the ancient belief in regard to Ethiopia, for in his treatise *On Europe* he says that if we divide the regions of the heavens and of the earth into four parts, the Indians will occupy that part from which Apeliotes blows, the Ethiopians the part from which Notus blows, the Celts the part on the west, and the Scythians the part from which the north wind blows."[78] Ephorus had apparently added that Ethiopia and Scythia were the largest areas, because the Ethiopians seemed to extend from the winter rising to the winter setting, and the Scythians occupied the area from the summer rising to the summer setting.[79]

Cosmas Indicopleustes also quoted the passage of Ephorus from book 4 in full, adding a very interesting detail: Ephorus had stated his opinion "with the help of the enclosed drawings."[80] Indeed, the manuscript of Cosmas is illustrated by a rectangle showing the earth according to the principle explained above (fig. 8.12): the south is in the top part of the length of the rectangle, showing the Ethiopians; the north is in the lower part of its length, showing the Scythians; to the right on its breadth are Zephyrus and the Celts; and to the left are Apeliotes and the Indians. In this map it is clear that the center of such a rectangle—with the positions of summer and winter sunrise and sunset at its corners—must be Greece or the Aegean.

In summary, then, Ephorus included peoples peripheral to the known world in his theoretical geography.

Their distance from the Aegean, or the climate in which they were thought to live, had rendered them mysterious and almost mythical. The map supplied by Ephorus, as reconstructed by Cosmas (although we cannot be quite sure how faithfully), similarly portrays the remote parts of the inhabited world. It is little more than a geometric sketch revealing a general ignorance of these regions. And it also brings home to us to what extent the Mediterranean basin had long remained the best-known part of the inhabited world and the most exactly drawn: distant lands were only vaguely delineated and were inserted into world maps by guesswork.

The culmination of the classical Greek period—at least in terms of a contemporary synthesis of cartography—is seen in the works of Aristotle (384–322 B.C.), teacher of Alexander the Great and founder of the Peripatetic School. Although Aristotle is rarely considered a geographer or a cartographic thinker, he had very definite ideas about the shape of the earth and the outline of the inhabited world. His teaching was ultimately significant for the development of cartography insofar as it not only rationalized the arguments for the sphericity of the earth, but also certainly encouraged the enlargement of the knowledge of the *oikoumene*, particularly

Sources Chrétiennes, nos. 141 (1968), 159 (1970), and 197 (1973). See also pp. 261–63.

78. Strabo *Geography* 1.2.28 (note 9).

79. With mainland Greece or Rhodes as the traditional place of observation, the equinoctial rising and setting of the sun are due east and due west; its summer rising is ENE; its summer setting, WNW; its winter rising ESE; its winter setting, WSW.

80. Cosmas *Topographie chrétienne* 2.80 (note 77), translation by O. A. W. Dilke.

through the Asian journeys and conquests of Alexander the Great.

Aristotle had no doubt at all that the earth was spherical.[81] He proved it by observations that we might make today: the shadow of the earth on the moon in eclipses of the moon is invariably circular, and one sees the celestial pole rising more and more above the horizon as one goes from south to north. Overarching is the idea that it is in the nature of earth and water to move to the center of the universe, since they are the heavy elements (as opposed to air and fire, which are light). He thus saw the natural shape of the earth as spherical.[82]

He also described the system of five zones on the earth that had earlier been introduced by Parmenides,[83] comparing each inhabitable zone on the sphere to a drum and ridiculing his contemporaries who held the earth to be circular: "For there are two habitable sectors of the earth's surface, one, in which we live, towards the upper pole, the other towards the other, that is the south pole. These sectors are drum-shaped—for lines running from the center of the earth cut out this shaped figure on the surface"[84] (fig. 8.13). He goes on to say: "The way in which present maps of the world are drawn is therefore absurd. For they represent the inhabited earth as circular, which is impossible both on factual and theoretical grounds."[85] He believed this view was theoretically impossible because of the geometry of the sphere and empirically impossible because of the proportions between the length of the inhabited world (from the Straits of Gibraltar, or Pillars of Hercules, to India) and its breadth (from Ethiopia to the Sea of Azov, ancient Palus Maeotis), which were more than 5:3. The breadth of the inhabited world, he believed, could not be extended by exploration because of climatic conditions, excessive heat or cold, while between India and the Straits of Gibraltar "it is the ocean which severs the habitable land and prevents it from forming a continuous belt around the globe."[86]

In the next chapter of the *Meteorologica*, which also helps us to understand in summary form some of the principles underlying the construction of the maps of that age, Aristotle uses a diagram (fig. 8.14) to show the relative positions of the winds: "The treatment of their position must be followed with the help of a diagram. For the sake of clarity, we have drawn the circle of the horizon; that is why our figure is round. And it must be supposed to represent the section of the earth's surface in which we live; for the other section could be divided in a similar way."[87] The circle represents the northern temperate zone; the circular horizon has a center where the observer stands, probably in Greece or the Aegean, as in the case of Ephorus's rectangular map. On its circumference are marked the points of the compass as he envisaged them: the equinoctial rising and setting (east

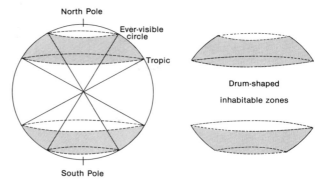

FIG. 8.13. ARISTOTLE'S CONCEPT OF THE POSITION AND SHAPE OF THE INHABITED WORLD. Reconstruction showing the five zones and the corresponding "drums."
After Aristotle, *Meteorologica*, ed. H. D. P. Lee, Loeb Classical Library (Cambridge: Harvard University Press; London: William Heinemann, 1952), 181.

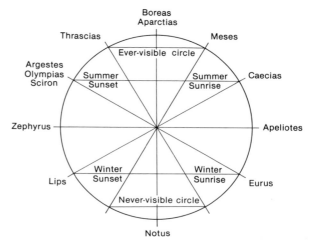

FIG. 8.14. ARISTOTLE'S SYSTEM OF THE WINDS. Reconstruction from the *Meteorologica* 2.6, showing the position of the winds at the summer and winter sunrise and sunset and the ever- and never-visible circles for an observer in Greece or the Aegean.
After Aristotle, *Meteorologica*, ed. H. D. P. Lee, Loeb Classical Library (Cambridge: Harvard University Press; London: William Heinemann, 1952), 187.

81. References are made to Aristotle's *Meteorologica* (note 28), and *De caelo*; see *On the Heavens*, trans. W. K. C. Guthrie, Loeb Classical Library (Cambridge: Harvard University Press; London: William Heinemann, 1939). See also Thomson, *History of Ancient Geography*, 118–21 (note 1).
82. Aristotle *On the Heavens* 2.14 (note 81).
83. According to Dicks, *Geographical Fragments*, 23 (note 31), there is some doubt about Parmenides' division of the earth into zones.
84. Aristotle *Meteorologica* 2.5.362a.33 (note 28).
85. Aristotle *Meteorologica* 2.5.362b (note 28).
86. Aristotle *Meteorologica* 2.5.362b (note 28).
87. Aristotle *Meteorologica* 2.6.363a (note 28). Modern editions of Aristotle reconstruct this diagram from his text rather than from a Madrid manuscript of the twelfth century, for which see Charles Graux

and west), the summer and winter risings and settings (about ENE, WNW, ESE, WSW), north and south. The winds are named according to the directions from which they blow and are diametrically opposed. But Aristotle adds two more winds (about NNE and NNW) that have no named opposites; the chord between them almost corresponds to the "ever-visible circle." Such a schema for the winds, with many modifications, was later to find its way into the paraphernalia of the navigator and cartographer even into the Renaissance period, as chapters 18 and 19 demonstrate.

It is clear that Aristotle's teaching was based on a geocentric hypothesis: the sphere of the fixed stars turns in regular motion on an axis going through the earth, which serves as the center of the celestial sphere. He maintained this hypothesis against others already formulated, which he claimed gave less satisfactory results.[88] Like Eudoxus, he subscribed to the theory of homocentric spheres to explain the apparently irregular movements of the planets, yet his attribution of a physical reality to Eudoxus's general system was to raise as many questions as it answered in the attempt to understand the celestial mechanism.

APPENDIX 8.1
DEFINITIONS OF SOME BASIC TERMS RELATING TO THE CELESTIAL SPHERE

The celestial equator is the celestial great circle whose plane is perpendicular to the axis of the earth; it is the path that the sun seems to describe in its diurnal revolution at the equinoxes. The tropics are two circles parallel to the equator and tangent to the ecliptic; they are the apparent path of the sun in its diurnal rotation on solstitial days.

The zodiac was—for the ancients—an oblique band twelve degrees wide in which the planets appear to move; the ecliptic, the median circle of the zodiac, is the great oblique circle that the sun seems to describe in its annual motion. The colures are two great circles drawn through the poles and the equinoctial or solstitial points; they are perpendicular to one another (see fig. 8.8).

and Albert Martin, "Figures tirées d'un manuscrit des *Météorologiques* d'Aristote," *Revue de Philologie, de Littérature et d'Histoire Anciennes*, n.s., 24 (1900): 5–18. For the connection between this reconstruction and the Pesaro anemoscope see p. 248 and n. 81.

88. Aristotle *On the Heavens* 2.13.293a ff. (note 81). The Pythagoreans had imagined a cosmic system with fire in the center; the earth and the sun moved around it. But the geocentric hypothesis allowed men to study the terrestrial globe geometrically, by reference to the celestial sphere. See Germaine Aujac, "Le géocentrisme en Grèce ancienne?" in *Avant, avec, après Copernic: La représentation de l'univers et ses conséquences épistémologiques*, Centre International de Synthèse, 31ᵉ semaine de synthèse, 1–7 June 1973 (Paris: A. Blanchard, 1975), 19–28.

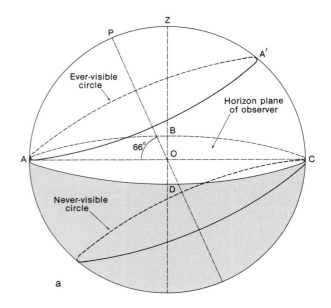

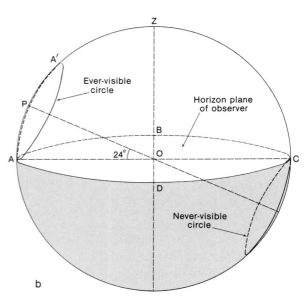

FIG. 8.15. THE "EVER-VISIBLE CIRCLE" AT 66°N (*a*) AND AT 24°N (*b*). The terrestrial globe is an infinitesimal point, O, which also corresponds to the position of the observer. Z is the observer's zenith, ABCD the observer's horizon. The shaded portion of the sky is invisible to the observer. As the stars on the celestial sphere rotate about the celestial pole, P, those above the circle AA′ do not intercept the horizon and thus do not appear to rise or set. In (*a*) the circle AA′ is the ever-visible circle for an observer on the Arctic Circle, latitude 66°N. In (*b*), the circle AA′ is the ever-visible circle for an observer at latitude 24°N (see also fig. 10.6).
After Geminus, *Introduction aux phénomènes*, ed. and trans. Germaine Aujac (Paris: Belles Lettres, 1975).

The arctic circle in the ancient Greek sense is different from our Arctic Circle and varies according to the latitude of the place of observation. It is defined as the limit of the ever-visible stars for a particular latitude. By corollary, the antarctic circle for any given latitude is the limit of never-visible stars. It can be shown that the angular distance of both these circles from the pole is equal to the angular distance of the terrestrial equator from the place of observation (i.e., the latitude). At the terrestrial equator, the ever-visible or arctic circle has no radius; thus all stars rise and set for an observer there. At the North Pole the ever-visible circle is coincident with the observer's horizon; one-half only of the celestial sphere is thus always visible. To an observer on the Arctic Circle (66°N), the arctic circle coincides with the summer tropic of the celestial sphere (fig. 8.15a). To an observer at the tropic (24°N) the arctic circle is 24° distant from the celestial pole and coincides with the polar circle on the celestial sphere (fig. 8.15b).[1]

---

1. On the arctic circles, see *The Geographical Fragments of Hipparchus*, ed. D. R. Dicks (London: Athlone Press, 1960), 165–66.

BIBLIOGRAPHY
CHAPTER 8 THE FOUNDATIONS OF THEORETICAL CARTOGRAPHY IN ARCHAIC AND CLASSICAL GREECE

Aratus. *Phaenomena.* In *Callimachus: Hymns and Epigrams; Lycophron; Aratus,* trans. A. W. Mair and G. R. Mair. Loeb Classical Library. Cambridge: Harvard University Press; London: William Heinemann, 1955.

———. *Phaenomena.* Edited by Jean Martin. Biblioteca di studi superiori: Filologia greca 25. Florence: La Nuova Italia, 1956.

Aristotle. *De caelo.* Edited by D. J. Allan. Oxford: Clarendon Press, 1936.

———. *Meteorologica.* Translated by H. D. P. Lee. Loeb Classical Library. Cambridge: Harvard University Press; London: William Heinemann, 1952.

Brancati, Antonio. *Carte e conoscenze geografiche degli antichi.* Florence: 3M Italia and La nuova Italia, 1972. Detailed notes for use with a set of color transparencies.

Bunbury, Edward Herbert. *A History of Ancient Geography among the Greeks and Romans from the Earliest Ages till the Fall of the Roman Empire.* 2d ed., 2 vols., 1883; republished with a new introduction by W. H. Stahl, New York: Dover, 1959.

Cary, M., and E. H. Warmington. *The Ancient Explorers.* London: Methuen, 1929.

Cebrian, Konstantin. *Geschichte der Kartographie: Ein Beitrag zur Entwicklung des Kartenbildes und Kartenwesens.* Gotha: Perthes, 1922.

Dicks, D. R. *Early Greek Astronomy to Aristotle.* Ithaca: Cornell University Press, 1970.

Diogenes Laertius. *Lives of Eminent Philosophers.* 2 vols. Translated by R. D. Hicks. Loeb Classical Library. Cambridge: Harvard University Press; London: William Heinemann, 1925–38.

———. *Vitae philosophorum.* 2 vols. Edited by Herbert S. Long. Oxford: Clarendon Press, 1964.

Heidel, William Arthur. *The Frame of the Ancient Greek Maps.* New York: American Geographical Society, 1937.

Herodotus. *The History of Herodotus.* 2 vols. Translated by George Rawlinson. London: J. M. Dent; New York: E. P. Dutton, 1910.

———. *Histoires.* 10 vols. Edited by P. E. Legrand. Paris: Belles Lettres, 1932–39.

Kish, George, ed. *A Source Book in Geography.* Cambridge: Harvard University Press, 1978.

Lloyd, G. E. R. *Early Greek Science: Thales to Aristotle.* New York: W. W. Norton, 1970.

Mette, Hans Joachim. *Sphairopoiia: Untersuchungen zur Kosmologie des Krates von Pergamon.* Munich: Beck, 1936.

Müller, Karl, ed. *Geographi Graeci minores.* 2 vols. and tabulae. Paris: Firmin-Didot, 1855–56.

Neugebauer, Otto. *The Exact Sciences in Antiquity.* 2d ed. Providence: Brown University Press, 1957.

Paassen, Christaan van. *The Classical Tradition of Geography.* Groningen: Wolters, 1957.

Thomson, J. Oliver. *History of Ancient Geography.* Cambridge: Cambridge University Press, 1948; reprinted New York: Biblo and Tannen, 1965.

Tozer, H. F. *A History of Ancient Geography.* 1897. 2d ed.; reprinted New York: Biblo and Tannen, 1964.

Warmington, E. H. *Greek Geography.* London: Dent, 1934.

Wolska, Wanda. *La topographie chrétienne de Cosmas Indicopleustès: Théologie et science au VIᵉ siècle.* Bibliothèque Byzantine, Etudes 3. Paris: Presses Universitaires de France, 1962.

9 · The Growth of an Empirical Cartography in Hellenistic Greece

PREPARED BY THE EDITORS FROM MATERIALS SUPPLIED BY GERMAINE AUJAC

There is no complete break between the development of cartography in classical and in Hellenistic Greece. In contrast to many periods in the ancient and medieval world, we are able to reconstruct throughout the Greek period—and indeed into the Roman—a continuum in cartographic thought and practice. Certainly the achievements of the third century B.C. in Alexandria had been prepared for and made possible by the scientific progress of the fourth century. Eudoxus, as we have seen, had already formulated the geocentric hypothesis in mathematical models; and he had also translated his concepts into celestial globes that may be regarded as anticipating the *sphairopoiia*.[1] By the beginning of the Hellenistic period there had been developed not only the various celestial globes, but also systems of concentric spheres, together with maps of the inhabited world that fostered a scientific curiosity about fundamental cartographic questions. The relative smallness of the inhabited world, for example, later to be proved by Eratosthenes, had already been dimly envisaged. It had been the subject of comment by Plato,[2] while Aristotle had quoted a figure for the circumference of the earth from "the mathematicians" at four hundred thousand stades.[3] He does not explain how he arrived at this figure, which may have been Eudoxus's estimate. Aristotle also believed that only the ocean prevented a passage around the world westward from the Straits of Gibraltar to India.

In spite of these speculations, however, Greek cartography might have remained largely the province of philosophy had it not been for a vigorous and parallel growth of empirical knowledge. Indeed, one of the salient trends in the history of Hellenistic cartography is the growing tendency to relate theories and mathematical models to newly acquired facts about the world—especially those gathered in the course of Greek exploration or embodied in direct observations such as those recorded by Eratosthenes in his scientific measurement of the circumference of the earth. Despite a continuing lack of surviving maps and original texts throughout the period—which continues to limit our understanding of the changing form and content of cartography—it can be shown that by its end a markedly different cartographic image of the inhabited world had emerged.

That such a change should occur is due both to political and military factors and to cultural developments within Greek society as a whole. With respect to the latter, we can see how Greek cartography started to be influenced by a new infrastructure for learning that had a profound effect on the growth of formalized knowledge in general. Of particular importance for the history of the map was the growth of Alexandria as a major center of learning, far surpassing in this respect the Macedonian court at Pella. It was at Alexandria that Euclid's famous school of geometry flourished in the reign of Ptolemy II Philadelphus (285–246 B.C.). And it was at Alexandria that this Ptolemy, son of Ptolemy I Soter, a companion of Alexander, had founded the library, soon to become famous throughout the Mediterranean world. The library not only accumulated the greatest collection of books available anywhere in the Hellenistic period but, together with the museum, likewise founded by Ptolemy II, also constituted a meeting place for the scholars of three continents. Demetrius of Phalerum (b. ca. 350 B.C.), Athenian statesman, writer, and disciple of Aristotle, had been asked to start the library, which was endowed with many scientific works

---

1. See chapter 8, "The Foundations of Theoretical Cartography in Archaic and Classical Greece," p. 136 and n. 33.

2. Plato (ca. 429–347 B.C.) was conscious of the relative smallness of the inhabited world on the surface of the globe; see above, p. 137.

3. Aristotle *On the Heavens* 2.14.298a.15 ff; see *On the Heavens*, trans. W. K. C. Guthrie, Loeb Classical Library (Cambridge: Harvard University Press; London: William Heinemann, 1939) and *The Geographical Fragments of Hipparchus*, ed. D. R. Dicks (London: Athlone Press, 1960), 24.

The stade, στάδιον, in origin the distance covered by a plow in a single draft, consisted of 600 Greek feet; but the length of a foot was subject to some local variation in the Greek world. Some authors take the modern equivalent of a stade to be 185 meters or 607 feet; see Jacob Skop, "The Stade of the Ancient Greeks," *Surveying and Mapping* 10 (1950): 50–55, and Dicks, *Geographical Fragments*, 42–46 (above). Other authors disagree and cite 148–58 meters; see Irene Fischer, "Another Look at Eratosthenes' and Posidonius' Determinations of the Earth's Circumference," *Quarterly Journal of the Royal Astronomical Society* 16 (1975): 152–67, and Dennis Rawlins, "The Eratosthenes-Strabo Nile Map," *Archive for History of Exact Sciences* 26 (1982): 211–19. In view of this controversy, the authors and editors have deliberately avoided using modern equivalents for the stade throughout the *History*.

and quickly increased in size. The librarians not only brought together existing texts, they corrected them for publication, listed them in descriptive catalogs, and tried to keep them up to date.[4] Thus Alexandria became a clearinghouse for cartographic and geographical knowledge; it was a center where this could be codified and evaluated and where, we may assume, new maps as well as texts could be produced in parallel with the growth of empirical knowledge.

EXPLORATION AND DISCOVERY IN THE REFORM OF THE WORLD MAP

The other great factor underlying the increasing realism of maps of the inhabited world in the Hellenistic period was the expansion of the Greek world through conquest and discovery, with a consequent acquisition of new geographical knowledge. In this process of strengthening the empirical content of maps the conquests of Alexander the Great, king of Macedon (356–323 B.C.), were especially crucial in providing the Greeks with a far more detailed knowledge of the East than previously had been possible. There is no doubt that Alexander had been influenced by existing geographical lore, some of it expressed in map form and was to contribute substantially to Greek understanding of this Eastern world. In taking up the plan originally conceived by his father Philip to organize an attack against Persia, he was certainly not unprepared. He had been taught by Aristotle and had learned from him that the inhabited world, from the Straits of Gibraltar to India, was relatively small and probably bounded by the ocean. This was clear to the geographers of the period, just as was the fact that the western part of the known world was bounded by the Atlantic Ocean. This belief explains why Alexander not only wanted to explore the whole of the inhabited world east of the Aegean, but also had a firm hope of reaching the Eastern Ocean, the existence of which had not so far been witnessed by the Greeks.

He thus instructed his secretaries to prepare a brief for the journey in great detail and to gather all available information about the countries he would cross. Xenophon had recently described Asia Minor, and Ctesias of Cnidus had described Persia and India.[5] No doubt Alexander collected all kinds of maps, general sketches of the inhabited world or regional maps indicating the main roads through the country, especially Persia, where the roads were very well organized, with posting stations at regular intervals.[6]

It is also clear that the whole expedition was planned with the deliberate aim of expanding existing geographical knowledge. Alexander took a large group of scholars with him—zoologists, doctors, historians, and surveyors—to compile a complete account of all interesting phenomena observed on the way and to verify all information that had been furnished by others. So Eumenes of Cardia (ca. 362–316 B.C.) was entrusted with keeping the daily report of the expedition and wrote *Ephemerides*, used by later historians but now lost. The so-called *bematistai*,[7] Baito and Diognetus, had to keep the record of every distance between halting places and to describe the geographical features—fauna, flora, nature of the soil, and landscape—of each country. The importance of the *bematistai* to the history of mapping is that their *Itinerary* was probably illustrated by sketches or local maps. The whole expedition, in fact, became a most important primary source of new cartographic data.[8] Its topographical notes were drawn on by such later geographers as Isidorus of Charax (fl. ca. A.D. 25).

The information collected by Alexander's expedition was of course greatly influenced by the vagaries of its progress; in no sense can it be regarded as systematic or even reconnaissance mapping—in the modern sense—of the areas traversed. In fact Alexander was prevented from carrying out his original plan by his soldiers, who refused to cross the Indus River and continue east to the external ocean. But he decided to explore at least the Indian (Southern) Ocean, so he sailed down the Indus to the sea. There he split up his troops into three contingents: one of them, under Nearchus, was to sail toward Babylon through the Indian Ocean and the Persian Gulf; another contingent, under Alexander himself, was to go by land along the coast, to support the fleet if necessary; the rest of the army was to return to Babylon by a route farther north.[9] So at least some of the country between the Aegean Sea, the Taurus Mountains, the Indus River, and the Indian Ocean was explored as a result of Alexander's expedition.

---

4. This perhaps explains the complete loss of scientific works of previous periods: they were considered out of date and were replaced by more recent manuscripts. For a description of the library, see Edward Alexander Parson, *The Alexandrian Library: Glory of the Hellenic World* (Amsterdam, London, New York: Elsevier Press, 1952).

5. Xenophon (ca. 430–354 B.C.) had been in charge of the army of mercenaries after Cyrus's death in Cunaxa; he led it through Asia Minor to the Black Sea. His *Anabasis* is the story of this expedition. Ctesias of Cnidus, who was a physician at Artaxerxes' court (he was with him at Cunaxa), wrote a history of Persia, *Persica*, in twenty-three volumes and a history of India, *Indica*, now both lost.

6. Herodotus *History* 5.52–54; see *The History of Herodotus*, trans. George Rawlinson, 2 vols. (London: J. M. Dent; New York: E. P. Dutton, 1910).

7. The *bematistai* had to measure (*bematizein*) the progress of the army every day, but the word itself is not found in this context.

8. For details on Alexander's expeditions, see William Woodthorpe Tarn, *Alexander the Great*, 2 vols. (Cambridge: Cambridge University Press; New York: Macmillan, 1948).

9. Nearchus's periplus is recorded in Arrian's *Indica*, and the whole expedition in Arrian's *Anabasis*.

Later geographers used the accounts of Alexander's journeys extensively to make maps of Asia and to fill in the outline of the inhabited world. The ambition of Eratosthenes to draw a general map of the *oikoumene* based on new discoveries was also partly inspired by Alexander's exploration.[10]

Among the contemporaries of Alexander was Pytheas, a navigator and astronomer from Massalia (Marseilles), who as a private citizen embarked upon an exploration of the oceanic coasts of western Europe. In his treatise *On the Ocean*, Pytheas relates his journey and provides geographical and astronomical information about the countries he observed. This treatise is now lost and is known to us only in fragments through comments made by several later writers whom Strabo quotes. Some of these writers, among whom was Polybius, regarded Pytheas as a liar, a view shared by Strabo himself.[11] It was already well known by Pytheas's time that the continental interior of Europe just north of the Black Sea was extremely cold. Given this, the reports by Pytheas that high latitudes of the Atlantic seaboard were habitable must have been indeed hard to believe.

It is difficult to reconstruct from the fragmentary evidence exactly where Pytheas traveled. While modern scholars agree that the voyage took place, they are not in agreement on its extent, particularly north of the British Isles.[12] Neither are they agreed on exactly when it took place, although the evidence seems to point toward a date between 325 and 320 B.C.[13] It seems, though, that having left Massalia, Pytheas put into Gades (Cádiz), then followed the coasts of Iberia and France to Brittany, crossing to Cornwall and sailing north along the west coast of England and Scotland to the Orkney Islands. From there, some authors believe, he made an Arctic voyage to Thule (probably Iceland) after which he penetrated the Baltic.[14] The confirmation of the sources of tin (in the ancient Cassiterides or Tin Islands) and amber (in the Baltic) was of primary interest to him, together with new trade routes for these commodities.[15]

For the first part of his voyage, from Massalia to Tartessus, Pytheas may have used a version of an ancient Greek geographical description of these coasts known as the *Massaliote Periplus*—it was almost certainly compiled in Massalia—dating probably from about 500 B.C. We know of this periplus through the *Ora Maritima* of the Roman antiquary Rufius Festus Avienius some nine hundred years later. Avienius's work can be traced back to a second-century B.C. version associated with a person known as Pseudo-Scymnus and—from the archaic toponymy and omission of authors later than Thucydides—further back to the fifth century B.C.[16]

While Pytheas's reports of the northern lands branded him as a liar, his skill in mathematics was more widely acknowledged. Eratosthenes and Timaeus both respected his contribution to the world map, although their views are muted by Strabo's prejudiced account.[17] Pytheas was in fact a skilled astronomer who succeeded in establishing the exact position of the celestial pole, a point in the sky not marked by a star, but which, along with three faint stars, makes up a rectangle.[18] He also accurately fixed the latitude of Marseilles, indicating that at midday on the date of the summer solstice "the ratio of the index [gnomon] of the sun-dial to the shadow . . . is that of one hundred and twenty to forty-two minus one-fifth."[19] We can calculate from this the difference in latitude between Marseilles and the summer tropic as 19°12′, compared with its currently measured value of about 19°50′ (fig. 9.1).[20]

10. Strabo *Geography* 1.2.1; see *The Geography of Strabo*, 8 vols., ed. and trans. Horace Leonard Jones, Loeb Classical Library (Cambridge: Harvard University Press; London: William Heinemann, 1917–32). See also Strabo, *Géographie*, ed. Germaine Aujac et al. (Paris: Belles Lettres, 1966–).

11. Strabo *Geography* 1.4.3, 2.4.1, 4.2.1 (note 10).

12. The standard edition of Pytheas is *Fragmenta*, ed. Hans Joachim Mette (Berlin: Walter de Gruyter, 1952). More recent work includes C. F. C. Hawkes, *Pytheas: Europe and the Greek Explorers*, Eighth J. L. Myres Memorial Lecture (Oxford: Blackwell, 1977), which is an excellent modern summary of the issues surrounding Pytheas's voyage with a helpful bibliography, and the articles of Roger Dion, especially his, "Où Pythéas voulait-il aller?" in *Mélanges d'archéologie et d'histoire offerts à André Piganiol*, 3 vols., ed. Raymond Chevallier (Paris: SEVPEN, 1966), 3:1315–36.

13. Hawkes, *Pytheas*, 44 (note 12) dates it 325 B.C.

14. For various views, see J. Oliver Thomson, *History of Ancient Geography* (Cambridge: Cambridge University Press, 1948; reprinted New York: Biblo and Tannen, 1965), 149, and Hawkes, *Pytheas*, 33–39 and map 9 (note 12).

15. Hawkes, *Pytheas*, 1–2 (note 12).

16. Hawkes, *Pytheas*, 17–22 and map 6 (note 12). Editions of Avienius's *Ora Maritima* include *Avieni Ora Maritima (Periplus Massiliensis saec. VI. a.C.)*, ed. Adolf Schulten, Fontes Hispaniae Antiquae, fasc. 1 (Barcelona: A. Bosch, 1922), and his companion book *Tartessos: Ein Beitrag zur ältesten Geschichte des Westens* (Hamburg: L. Friederichsen, 1922); *Ora Maritima*, ed. Edward Adolf Sonnenschein (New York: Macmillan, 1929); *Ora Maritima*, ed. André Berthelot (Paris: H. Champion, 1934); and *Ora Maritima, or a Description of the Seacoast from Brittany Round to Massilia*, ed. J. P. Murphy (Chicago: Ares, 1977). For the transmission of this periplus, Hawkes agrees with the views of Schulten, *Avieni Ora Maritima* and *Tartessos*. For references to commentaries on Schulten's work, see Hawkes, *Pytheas*, 20 n. 52 (note 12). On the influence of Carthaginian knowledge of coasts west of the Straits of Gibraltar transmitted to the Greeks at this time, see Jacques Ramin, *Le Périple d'Hannon/The Periplus of Hanno*, British Archaeological Reports, Supplementary Series 3 (Oxford: British Archaeological Reports, 1976).

17. See footnote 11 above, for example, and Strabo *Geography* 3.4.4, 4.2.1, 7.3.1 (note 10).

18. Hipparchus, *In Arati et Eudoxi Phaenomena commentariorum libri tres*, ed. C. Manitius (Leipzig: Teubner, 1894), 1.4.1. See also Dicks, *Geographical Fragments*, 171 (note 3). Hawkes, *Pytheas*, 44–45 (note 12) misinterprets Dicks in thinking that the Pole Star (α Ursae Minoris) was one of the four stars in the rectangle.

19. Strabo *Geography* 2.5.41 (note 10).

20. It is important to realize that Pytheas would not have arrived at the true value for the latitude of Marseilles using the method outlined

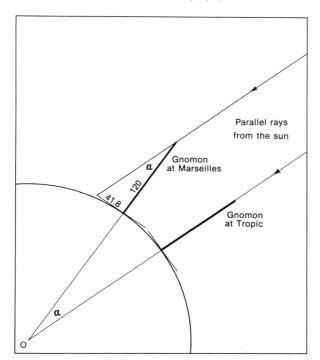

FIG. 9.1. PYTHEAS'S OBSERVATION OF THE LATITUDE OF MARSEILLES. In this diagram, O is the center of the earth and α is 19°12′.

put it—"the [celestial] arctic circle coincides with the [celestial] summer tropic."[23] He did not hesitate to put Thule on the northern boundary of the temperate zone, that is, on our Arctic Circle.

It would appear from what is known about Pytheas's journeys and interests that he may have undertaken his voyage to the northern seas partly in order to verify what geometry (or experiments with three-dimensional models) had taught him. The result was that his observations served not merely to extend geographical knowledge about the places he had visited, but also to lay the foundation for the scientific use of parallels of latitude in the compilation of maps. But it must also be admitted, as Hawkes has suggested, that there may have been wider economic and political motives for the voyage. Detailed information on the inhabitants and resources of the northern lands would have been of interest, for example, to a campaigner of the stature of Alexander. Alexander's death in 323 B.C. would have prevented this information from being fully exploited[24] and, instead, the work became largely the target of derision for later commentators.[25]

As exemplified by the journeys of Alexander and Pytheas, the combination of theoretical knowledge with

His route took him through a wide latitudinal range, and, observing how celestial phenomena and the length of day varied as he moved northward, Pytheas seems to have been the first author to relate systematically the latitude of a place to the length of its longest day, or to the height of the sun at the winter solstice.[21] Building on such observations, he also became the first to use parallels of latitude, drawn on the earth's sphere, to indicate all the places where identical astronomical phenomena could be observed. It is probable, however, that Pytheas indicated the height of the sun at the winter solstice for various latitudes not through observation, but by calculation with the help of geometry. If at least one of the results given by calculation could be empirically checked, then all results of the same series would be reliable. The very short summer nights Pytheas encountered in northern Europe made him confident about the link between latitude and the length of solstitial days. Hence the length of the solstitial day in a place became the usual way of indicating its latitude. The geometry of the sphere had also taught Pytheas that there existed on the earth a parallel of latitude where, at the summer solstice, the day lasted twenty-four hours and the sun did not disappear under the horizon. He located the island of Thule on this particular circle,[22] where—as he

in figure 9.1, since the trigonometric methods necessary to convert the gnomon relationship to degrees were not available until Hipparchus. Some authors have added the value of 19°12′ to the obliquity of the ecliptic as then calculated (24°), to arrive at a figure of 43°12′ for the latitude of Marseilles; but this ignores fully half a degree in the difference between the ancient and present values in the obliquity of the ecliptic. See Dicks, *Geographical Fragments*, 178–79, 188 (note 3).

21. Hipparchus is reported as noticing, for instance, that in the latitude of Celtica and the mouth of the Borysthenes (Dnieper River) the sun rose to only nine cubits (eighteen degrees, the astronomical cubit being two degrees): Strabo *Geography* 2.1.18 (note 10); cf. Dicks, *Geographical Fragments*, 188 (note 3). At that latitude, the longest day had sixteen equinoctial hours: Strabo *Geography* 2.5.42 (note 10). Strabo also reports the following observations relating the length of the longest day to the maximum height of the sun in cubits at the winter solstice:

17 hours	6 cubits	(= 12°)
18 hours	4 cubits	(= 8°)
19 hours	3 cubits	(= 6°)

Strabo *Geography* 2.1.18 and elsewhere (note 10).

22. For the Greeks of this time, since the tropics were reckoned as being roughly 24° distant from the equator, the Arctic Circle (in the modern sense) lay on the parallel 66°N. See Thomson, *History of Ancient Geography*, 153 (note 14). Oenopides of Chio was the first to mention one-fifteenth of a circle (or 24°) for the obliquity of the ecliptic; it was some time before it was applied to cartography.

23. Strabo *Geography* 2.5.8 (note 10). Translation by O. A. W. Dilke.

24. Hawkes, *Pytheas*, 44 (note 12) mentions Alexander's supposed plans discovered at his death to conquer Carthage and to explore the western part of the inhabited world, by land and by sea, to the Straits of Gibraltar (Pillars of Hercules) and the Atlantic (Western) Ocean. Other authors, including Tarn, *Alexander the Great*, 2:376 (note 8), regard these "plans" as later inventions to glorify his reputation further.

25. Hawkes, *Pytheas*, 45 (note 12).

direct observation and the fruits of extensive travel gradually provided new data for the compilation of world maps. While we can assume a priori that such a linkage was crucial to the development of Hellenistic cartography, there is no hard evidence, as in so many other aspects of its history, that allows us to reconstruct the technical processes and physical qualities of the maps themselves. Not even the improved maps that resulted from these processes have survived, and the literary references to their existence (enabling a partial reconstruction of their content) can even in their entirety refer only to a tiny fraction of the number of maps once made and once in circulation. In this case too, our generalizations are founded on the chance survival of references made by individual authors to maps.

First in this category, and roughly contemporaneous with both Alexander and Pytheas, is the map undertaken by Dicaearchus of Messana (Messina) (fl. ca. 326–296 B.C.). A pupil of Aristotle and a contemporary of Theophrastus (ca. 370–288/285 B.C.), Dicaearchus is acknowledged both by ancient writers and by modern historians of cartography and geography to have made a significant contribution.[26] Strabo puts him, with Democritus, Eudoxus, and Ephorus, among philosophers of the second age who were responsible for considerable advances in geographical science.[27] We know that he spent most of his life in the Peloponnese, especially at Sparta, and wrote various works on politics, literature, history, and philosophy.

In his *Circuit of the Earth* (*Periodos gēs*), now lost, Dicaearchus included a map and a description of the inhabited world. Like Democritus, he thought that the known inhabited world was half again as long as it was broad, a proportion of three to two.[28] Strabo, following Polybius, criticizes some distances supplied by Dicaearchus, such as the ten thousand stades from the Peloponnese to the Straits of Gibraltar, or the estimate of over ten thousand stades from the Peloponnese to the head of the Adriatic Sea.[29] Strabo, questioning these figures, criticizes Dicaearchus for having underestimated the length of the inhabited world and overestimated its breadth.

The main cartographic innovation pioneered by Dicaearchus seems to have been the insertion on a map, possibly for the first time, of two lines representing a parallel and a meridian to divide the known world.[30] According to Agathemerus, the parallel drawn by Dicaearchus, albeit somewhat imperfectly, extended eastward from the Straits of Gibraltar. It passed through Sardinia, Sicily, Caria, Lycia, Pamphylia, Cilicia, and along the Taurus range as far as Mount Himaeus (the Himalayas) (fig. 9.2).[31] Various authors have stated that Dicaearchus applied the term *diaphragma* to this arrangement in the sense of a division of the inhabited

world into two parts north and south of this line.[32] It represented an attempt to give his map an east-west coordinate axis crossed by a perpendicular meridian passing approximately through Rhodes. As we shall see, Eratosthenes, working a century later, took up the idea and developed it much further.

Another step toward geographical reality reflected in Dicaearchus's map was that he sketched in the eastward extension of the Taurus Mountains along a parallel, unlike earlier terrestrial maps in which the eastern part of the chain deviated considerably to the north.[33] Eratosthenes, although wanting to make a complete revision of these early geographical maps, was to follow his idea that the Taurus Mountains stretched in a straight course on the parallel of Athens.[34]

Half a century or so later than Dicaearchus, Timosthenes of Rhodes (fl. 270 B.C.) showed the same willingness to modify the early maps rather than to copy

26. See, for example, Armando Cortesão, *History of Portuguese Cartography*, 2 vols. (Coimbra: Junta de Investigações do Ultramar-Lisboa, 1969–71), 1:76–77, and Thomson, *History of Ancient Geography*, 153–54 (note 14).

27. Strabo *Geography* 1.1.1 (note 10). Homer, Anaximander, and Hecataeus are given as representatives of the first age, and the third age comprises Eratosthenes, Polybius, and Posidonius.

28. Agathemerus *Geographiae informatio* 1.2, in *Geographi Graeci minores*, ed. Karl Müller, 2 vols. and tabulae (Paris: Firmin-Didot, 1855–56), 2:471–87, esp. 471.

29. Strabo *Geography* 2.4.2 (note 10). A criticism by John of Lydia, that Dicaearchus made the Nile "flow uphill" from the Atlantic, may be due to a slight misunderstanding. John of Lydia *Liber de mensibus* 4; see the edition by Richard Wünsch (Leipzig, 1898), 147. For an explanation of the controversy concerning the modern value of the stade, see note 3.

30. Despite the view of Aubrey Diller, "Dicaearchus of Messina," in *Dictionary of Scientific Biography*, 16 vols., ed. Charles Coulston Gillispie (New York: Charles Scribner's Sons, 1970–80), 4:81–82, there is no hard evidence for either Dicaearchus or Eudoxus as measuring "a long arc north from Syene (Aswan) and observing the zenith points at the ends." He is referring, no doubt, to the passage by Cleomedes concerning the distance between Syene and Lysimachia (near the Hellespont, or Dardanelles): Cleomedes *De motu circulari* 1.8.44–43; see *De motu circulari corporum caelestium libri duo*, ed. H. Ziegler (Leipzig: Teubner, 1891). But William Arthur Heidel, *The Frame of the Ancient Greek Maps* (New York: American Geographical Society, 1937), 113–19, summarizes the question thoroughly, pointing out that Cleomedes "gives no intimation to whom we should credit this estimate" (p. 115).

31. Agathemerus *Geographiae informatio* 1.5 (note 28).

32. Cortesão, *History of Portuguese Cartography*, 1:77 (note 26); Thomson, *History of Ancient Geography*, 134 (note 14). Bunbury, as Cortesão points out, finds no evidence that Dicaearchus used the term; see Edward Herbert Bunbury, *A History of Ancient Geography among the Greeks and Romans from the Earliest Ages till the Fall of the Roman Empire*, 2d ed., 2 vols. (1883; republished with a new introduction by W. H. Stahl, New York: Dover, 1959), 2:628 n. 6.

33. Strabo *Geography* 2.1.2 (note 10).

34. Strabo *Geography* 2.1.1–2 (note 10).

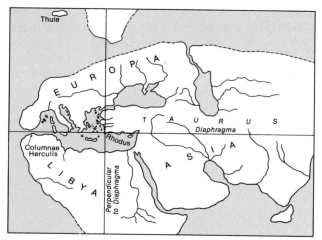

FIG. 9.2. RECONSTRUCTED WORLD MAP OF DICAEAR-CHUS, THIRD CENTURY B.C. The inhabited world is divided by the diaphragm (approximately 36°N) and a perpendicular to this at Rhodes.
After Armando Cortesão, *History of Portuguese Cartography*, 2 vols. (Coimbra: Junta de Investigações do Ultramar-Lisboa, 1969–71), vol. 1, fig. 16.

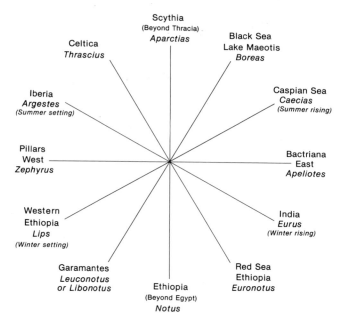

FIG. 9.3. TIMOSTHENES' SYSTEM OF THE WINDS. The relationship between winds, regions, and peoples as proposed by Admiral Timosthenes of Rhodes (third century B.C.).

them slavishly. Timosthenes, as an admiral of Ptolemy II Philadelphus, certainly traveled widely beyond his own island, and he is well known in the history of geography. He wrote a treatise *On Harbors* in ten books (lost), which was used and criticized by Eratosthenes, Hipparchus, and Strabo.[35] Timosthenes, who was later considered an authority on winds (i.e., on directions or rhumbs), added two winds or directions to the ten already mentioned in Aristotle's *Meteorologica*[36] and obtained twelve directions, at regular intervals, based on what would later be recognized as the twelve points of the compass.

According to Agathemerus, Timosthenes used these twelve directions to locate remote peoples or countries of the inhabited world (fig. 9.3).[37] Rather like Ephorus before him, he drew a kind of schematic map of nations, probably imitating the diagram of the winds in Aristotle's *Meteorologica*. The positioning of the Straits of Gibraltar and Bactria on the east-west line and of Scythia beyond Thrace and Ethiopia beyond Egypt on the north-south line suggests that the wind directions were drawn with Rhodes as a center. Thus, between north and east lie the Black Sea and the Sea of Azov (NNE), then the Caspian Sea (ENE); between east and south, India (ESE) and the Red Sea and Ethiopia (SSE); between south and west, the Garamantes (north-central Africa) (SSW) and western Ethiopia (northwestern Africa) (WSW); between west and north, Iberia (WNW) and Celtica (NNW).

Timosthenes probably also drew more detailed maps to illustrate his treatise. Strabo accuses him of being totally ignorant of Iberia, France, Germany, and Britain, and even of Italy, the Adriatic, and the Black Sea, pointing out at least two gross mistakes in his work.[38] First, he mentioned forty islands instead of twenty in the channel between Lesbos and Asia Minor.[39] Second, he put Metagonium (Melilla) opposite (i.e., on the meridian of) Massalia (Marseilles) when in Strabo's opinion it was on the same meridian as Nova Carthago (Cartagena),[40] which is closer to, if not actually, the true location. Timosthenes' treatise and accompanying map, though possibly useful to sailors, seem to have lacked scientific accuracy. But his idea of taking Rhodes as the center of the map was generally adopted by his successors.

THE MEASUREMENT OF THE EARTH AND THE WORLD MAP BY ERATOSTHENES

Few would dispute that in both a theoretical and a practical sense Hellenistic cartography reached its apogee in

35. Strabo *Geography* 9.3.10 (note 10).
36. Aristotle *Meteorologica* 2.6.363a.21 ff.; see *Meteorologica*, trans. H. D. P. Lee, Loeb Classical Library (Cambridge: Harvard University Press; London: William Heinemann, 1952), and pp. 145–46 above.
37. Agathemerus *Geographiae informatio* 2.7 (note 28).
38. Strabo *Geography* 2.1.41 (note 10).
39. Strabo *Geography* 13.2.5 (note 10).
40. Strabo *Geography* 17.3.6 (note 10).

the work of the polymath Eratosthenes (ca. 275–194 B.C.). His was a lasting contribution to the development of mapping, and with some justification he has been variously assigned a founding role in geography, cartography, and geodesy.[41] Although we are acquainted with his contribution only through later writers rather than through his original texts,[42] it is absolutely clear that in two scientific endeavors he surpassed both his predecessors and his contemporaries. The first of these was his measurement of the circumference of the earth, which was methodologically simple but brilliant.[43] The second was his construction of a world map based on both parallels and meridians, which was of seminal importance not only in the subsequent development of map projections but also in the eventual scientific and practical use of maps. Such a cartographic invention was equally applicable in chorographical or regional mapping and in geographical or world mapping, so that its key significance for the history of the map needs to be fully described.

Eratosthenes was born a Greek, in Cyrene (North Africa); going to Athens as a young man, he took lessons at one time from the Stoics and at another from the Academicians, among whom he was particularly influenced by Arcesilas of Pitane (Candarli), who had been a disciple of the mathematician Autolycus. In the work of Eratosthenes, as in that of his predecessors, the importance of his mastery of the geometry of the sphere and of the geocentric hypothesis cannot be overemphasized as providing the point of departure—as well as the theoretical framework—for the development of his cartographic ideas.[44] Eratosthenes' scientific distinction later attracted the attention of Ptolemy III Euergetes, king of Egypt 246–221 B.C. The king asked him to come to Alexandria as tutor to his son Philopator (born ca. 245 B.C.) and to take over the direction of the library when Apollonius left for Rhodes after adverse criticism of his poem *Argonautica*. At Alexandria, Eratosthenes was to compose two works on geographical subjects: one, *Measurement of the Earth*, explained the method used to find the circumference of the earth; the other, entitled *Geographica*, in three books, gave instructions for making a map of the inhabited world. Both works are lost, but Strabo, who begins his own work by a criticism of the *Geographica*, affords us fairly clear knowledge of its contents, and Cleomedes of the second century A.D. gives a brief summary of the *Measurement of the Earth*.[45]

From Cleomedes we learn that the method Eratosthenes used to evaluate the circumference of the earth was based on the geometry of the sphere.[46] According to the geocentric hypothesis, by which the earth was reduced to a point,[47] the sun's rays are parallel when falling on any point of the earth. It was known that

Syene (Aswan) in Egypt was situated under the tropic; at midday on the summer solstice there was no shadow,

41. Cortesão, *History of Portuguese Cartography*, 1:78–79 (note 26); D. R. Dicks, "Eratosthenes," in *Dictionary of Scientific Biography*, 4:388–93 (note 30).

42. R. M. Bentham, "The Fragments of Eratosthenes of Cyrene" (typescript for Ph.D. thesis, University of London, 1948—author died before thesis was submitted); see also *Die geographischen Fragmente des Eratosthenes*, ed. Hugo Berger (Leipzig: Teubner, 1898). We know of Eratosthenes' *Geographica* mostly through Strabo.

43. Gerald R. Crone, *Maps and Their Makers: An Introduction to the History of Cartography*, 5th ed. (Folkestone, Kent: Dawson; Hamden, Conn.: Archon Books, 1978), 3.

44. In this respect Eratosthenes was heir to a continuous tradition of mathematical learning that can be traced back at least as far as Eudoxus, from whose day onward it is likely that treatises entitled *Sphaerica* had existed. Autolycus of Pitane (fl. 310 B.C.) was clearly a link in this chain of writers influencing Eratosthenes. Although Autolycus's textbook *On the Sphere in Motion*, composed about 330 B.C., was not an original work, it was a competent summary of a number of basic theorems concerning celestial phenomena for given places of observation, and it explained clearly the geometric relationship between the sky and the earth and the need for astronomical knowledge to define the position on the earth of any place of observation. See Autolycus of Pitane, *La sphère en mouvement*, ed. and trans. Germaine Aujac, Jean-Pierre Brunet, and Robert Nadal (Paris: Belles Lettres, 1979). It is also likely that the writings of Euclid (fl. Alexandria ca. 300 B.C.) were known to Eratosthenes. The *Elements* had been completed about 300 B.C., but Euclid was also the author of a small treatise entitled *Phaenomena*, which applied specifically to the celestial sphere the conclusions Autolycus drew for rotating spheres in general. After establishing the geometry of the rotating celestial sphere, Euclid examined the rising and setting of stars as a means of measuring time at night; to do this he had to analyze the relationship between the observer's horizon and the ecliptic on the celestial sphere, which is different for each parallel on the earth. A brief summary of this work is given by Pierre Chiron, "Les Phénomènes d'Euclide," in *L'astronomie dans l'antiquité classique*, Actes du Colloque tenu à l'Université de Toulouse–Le Mirail, 21–23 Octobre 1977 (Paris: Belles Lettres, 1979), 83–89. For early spherical astronomy, see Otto Neugebauer, *A History of Ancient Mathematical Astronomy* (New York: Springer-Verlag, 1975), 748–67.

45. Cleomedes *De motu circulari* (note 30). The original Greek title is Κυκλικὴ Θεωρία τῶν Μετεώρων.

46. Cleomedes *De motu circulari* 1.10 (note 30). An English translation of book 1, chap. 10, appears in Cortesão, *History of Portuguese Cartography*, 1:141–43 (note 26).

47. In the geocentric hypothesis, as explained in Euclid's *Phaenomena*, the sky of the fixed stars was compared to a sphere rotating around one diameter called the world axis. In the middle, the earth was reduced to a point that acted as center to the sphere; the fixed stars moved along parallel circles (being on a rotating sphere, they were all circles of the sphere perpendicular to the axis of rotation). The greatest of these parallel circles Euclid recognized as the celestial equator. But two other great circles were important: the oblique circle of the ecliptic (called the "zodiac" by Euclid; see Chiron, "Phénomènes d'Euclide," 85 [note 44]), and the circle of the visible horizon (the astronomical horizon dividing the visible celestial hemisphere from the invisible one), which remained motionless during the apparent motion of the celestial sphere (fig. 8.8). Euclid *Phaenomena* 1 and prop. 1, see Euclid, *Opera omnia*, 9 vols., ed. J. L. Heiberg and H. Menge (Leipzig: Teubner, 1883–1916), vol. 8, *Phaenomena et scripta musica* [and] *Fragmenta* (1916).

the sun being exactly at the zenith.[48] Supposing Alexandria to be on the same meridian as Syene (the difference is only 3°), Eratosthenes measured the angle between the direction of the sun and the vertical in Alexandria at midday on the summer solstice. This angle, one-fiftieth of a circle, was equal to the angle at the earth's center subtended by the arc of the meridian defined by Syene and Alexandria. Estimating the distance between the two towns at roughly 5,000 stades, Eratosthenes calculated the total circumference as 250,000 stades (fig. 9.4). He later extended the value to 252,000 so as to make it divisible by sixty.[49]

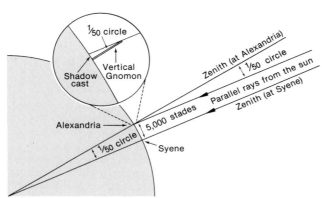

FIG. 9.4. ERATOSTHENES' MEASUREMENT OF THE EARTH. Eratosthenes worked with four assumptions: that Syene was on the tropic (at the summer solstice, the sun was thus directly overhead); that both it and Alexandria were on the same meridian; that the distance between them was 5,000 stades; and that the sun's rays were parallel. He knew that the difference in latitude between Alexandria and Syene was equivalent to the angle between the sun's rays and the zenith at Alexandria. From the lengths of a vertical stick (gnomon) and its shadow, he calculated this angle to be one-fiftieth of a circle. Thus the earth's circumference was estimated at 250,000 stades.
After John Campbell, *Introductory Cartography* (Englewood Cliffs, N.J.: Prentice-Hall, 1984), fig. 1.7.

Eratosthenes' method for calculating the circumference of the earth was sound, but its reliability depended on the accuracy of his base measurements and other assumptions. The angular distance between the two cities is quite accurate (7°12′ instead of the actual 7°7′), but Syene is not directly on the Tropic of Cancer but about 35′ to the north (using the modern figure for the obliquity of the ecliptic), and Alexandria and Syene are not on the same meridian. Furthermore, the distance between Alexandria and Syene is given in stades, the value of which has sparked considerable debate[50] quite apart from the question of the empirical source of the distance thus recorded.[51] Regardless of the actual value

for the stade that Eratosthenes used or of the distance he arrived at—he knew that the distance between the two cities was a very rough estimate, as was his evaluation of the terrestrial circumference—the importance of his calculation lies in its influence. It is probable that after he had measured the circumference of the earth, Eratosthenes henceforth first established any distance in latitude by astronomical means, or by reference to the geometry of the sphere (the distance between equator and tropic being fixed at four-sixtieths of the great circle, for instance), and then evaluated this distance in stades. Thus the distance between equator and tropic, which had never been measured by surveyors, was said to be 16,800 stades.

The availability of knowledge of this estimate of the earth's circumference had three outstanding consequences. First, it was now possible to work out through geometry the length of every parallel circle on the earth. The parallel of Athens, for example, was "less than two hundred thousand stadia in circuit."[52] Second, differences of latitude, found by gnomonic methods and expressed in fractions of the circle, could easily be converted into stades. Third, it was now also possible to define the size of the inhabited world and its position on the surface of the terrestrial globe.

This third issue—the size and location of the inhabited world—was of intense and continuing interest to the Greeks, and having devised a method to answer this question, Eratosthenes was to return to its exposition in his *Geographica*. This work in three books is known to us mainly through Strabo. It was intended to provide a review and solution of all known problems involved in drawing a map of the earth (*gē-graphein*) or, more precisely, a map of the inhabited world on the surface of the terrestrial globe.[53] Starting from the theoretical premise that the earth is spherical, albeit with "certain irregularities of surface,"[54] Eratosthenes located the inhabited world completely in the Northern Hemisphere occupying the northern half of the distance between the Tropic of Cancer and the equator and the entire distance

48. Strabo *Geography* 17.1.48 (note 10).

49. Eratosthenes divided the earth into sixtieths; the use of 360° comes with Hipparchus.

50. On the matter of the modern value of the stade, see note 3 above.

51. Cortesão, *History of Portuguese Cartography*, 1:82 (note 26), speculates that Egyptian cadastral surveys may have been available to Eratosthenes in his calculation of the distance between the two points of observation.

52. Eratosthenes, in Strabo *Geography* 1.4.6 (note 10).

53. Eratosthenes was the first author to attempt this. His work began with a short history of geographical science from the time of Homer and the first mapmakers.

54. Strabo *Geography* 1.3.3 (note 10).

between that tropic and the polar circle. He calculated its width from north to south along the meridian that runs through Meroë, Alexandria, and Rhodes, resulting in a distance of 38,000 stades. Strabo described the overall shape of the *oikoumene* as somewhat like a chlamys, a Macedonian cloak perhaps resembling the shape in fig. 9.5.[55] Its length from west to east, however, he determined in accordance with an established concept, that its length was more than double the known breadth.

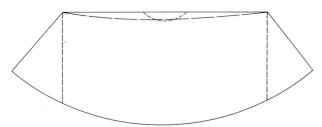

FIG. 9.5. THE CHLAMYS. This is the possible form of a common style of Macedonian cloak used by Strabo to illustrate the shape of the *oikoumene*. The top could be either straight or slightly curved.
Reconstructed from the description in *The Geography of Strabo*, 8 vols., ed. and trans. Horace Leonard Jones, Loeb Classical Library (Cambridge: Harvard University Press; London: William Heinemann, 1917–32), 2.5.6 and p. 435 n. 3.

Eratosthenes first described the distance from the capes of India to the extremities of Iberia as roughly 74,000 stades. Then (according to Strabo) Eratosthenes added 2,000 more stades to both west and east to keep the breadth from being more than half the length.[56] The total length thus became 78,000 stades.

The determination of the length of the inhabited world from India to Iberia was reckoned along the parallel of Athens. Eratosthenes believed this was less than 200,000 stades in circuit, "so that, if the immensity of the Atlantic Sea did not prevent, we could sail from Iberia to India along one and the same parallel over the remainder of the circle, that is, the remainder when you have subtracted the aforesaid distance, which is more than a third of the whole circle."[57] In fact, using a value for the circumference of the earth of 252,000 stades, 78,000 stades on the parallel in question is approximately equivalent to 138° of longitude, which is roughly the distance between the western coast of Spain and Korea rather than India.

It is not surprising that for many centuries to come values representing latitude were always much more reliable than those for longitude. Familiar with the geometry of the sphere, the Greeks were fairly well equipped to derive latitudes from direct observations of the sun and stars. In this respect, straightforward calculations could be undertaken to test the information of travelers. For longitudes the results were much less re-

liable, since it was necessary to observe an eclipse of the moon or other celestial body simultaneously from different places to obtain exact distances between them. Instead, the Greeks had to accept distances given by the itineraries without being able to verify them astronomically.

According to Strabo, it was in the third book of *Geographica* that Eratosthenes explained how to draw a map of the world:

> Eratosthenes, in establishing the map of the inhabited world, divides it into two parts by a line drawn from west to east, parallel to the equatorial line; and as ends of this line he takes, on the west, the Pillars of Heracles [Straits of Gibraltar], on the east, the capes and most remote peaks of the mountain-chain that forms the northern boundary of India. He draws the line from the Pillars through the Strait of Sicily [Straits of Messina] and also through the southern capes both of the Peloponnesus and of Attica, and as far as Rhodes and the Gulf of Issus [Gulf of Iskenderun, Turkey]; . . . then the line is produced in an approximately straight course along the whole Taurus Range as far as India, for the Taurus stretches in a straight course with the sea that begins at the Pillars, and divides all Asia lengthwise into two parts, thus making one part of it northern, the other southern; so that in like manner both the Taurus and the Sea from the Pillars up to the Taurus lie on the parallel of Athens.[58]

We can see from this passage that Eratosthenes had adopted the idea of the *diaphragma* (if not the term) introduced by Dicaearchus to divide the known world by means of a line parallel to the equator, drawn from west to east, beginning at the Straits of Gibraltar and running through Athens and Rhodes to India. It is also clear from other passages in Strabo that Eratosthenes drew a central perpendicular meridian through Rhodes, for he lists the places through which this passes and the distances between them.[59] Eratosthenes used very rough estimates and round numbers. The south-north distances (in stades) he provided between the following regions or towns were:

Between	Stades
Cinnamon country and Meroë	3,400
Meroë and Alexandria	10,000
Alexandria and Hellespont	about 8,100
Hellespont and river Borysthenes	5,000
River Borysthenes and parallel of Thule	about 11,500
Total	38,000

(Strabo *Geography* 1.4.2).

55. Strabo *Geography* 2.5.6 (note 10).
56. Strabo *Geography* 1.4.5 (note 10).
57. Eratosthenes, in Strabo *Geography* 1.4.6 (note 10).
58. Strabo *Geography* 2.1.1 (note 10).
59. Strabo *Geography* 2.5.42 (note 10).

It must be remembered, however, that these figures are not consistently given by Eratosthenes (the cinnamon-producing country, for instance, was often located 3,000 stades south of Meroë); but here Eratosthenes wanted to put the southern limit of the inhabited world halfway between the equator and the tropic.

The intermediate east-west distances used by Eratosthenes between the following regions or towns were:

Between	Stades
Far eastern capes and eastern India	3,000
Eastern India and river Indus	16,000
River Indus and Caspian Gates	14,000
Caspian Gates and river Euphrates	10,000
River Euphrates and river Nile	5,000
River Nile and Carthage	15,000
Carthage and Pillars of Hercules	8,000
Pillars and Hercules and far western capes	3,000
Total	74,000

(Strabo *Geography* 1.4.5).

The next stage in drawing the map suggested by Eratosthenes was to subdivide the northern and southern halves of the map into smaller sections called *sphragides*, literally "seals" but meaning irregular quadrilaterals similar to the shape of document seals. The first two of those in the southern division, with their boundaries, are shown in figure 9.6.[60] Strabo describes only the first three sections of the southern half of the map, but it is enough to show us how Eratosthenes proceeded. It is clear that he had tried to identify the geometric figure characterizing each country, so as to be able to measure the sides (or the diagonals) of each figure and then to insert each in the right position, like the pieces of a jigsaw puzzle. He had to make the boundaries and size of each section fit the general outlines and size of the inhabited world. This was fairly easy with India, the shape of which was rather clear; it was much more difficult with other countries less well known and not bounded by the sea, a range of mountains, or a river.

The northern half of the map, at least as far as Europe was concerned, Eratosthenes divided on the basis of the three promontories projecting southward into the Mediterranean and enclosing both the Adriatic and the Tyrrhenian seas: the Peloponnesian (Greece), the Italian (Italy), and the Ligurian (Corsica and Sardinia).[61] But Hipparchus and Strabo sharply criticized Eratosthenes for overgeneralization, pointing out, for example, that the Peloponnesian promontory was actually made up of a number of smaller capes.

Reconstructions of Eratosthenes' map from Strabo's text, such as those by Bunbury and Cortesão, may be misleading.[62] Eratosthenes' drawing of the central parallel and meridian is not in doubt. Furthermore, he could certainly have drawn other parallels and meridians

where stade distances of places from the reference lines through Athens and Rhodes were the same or almost the same.[63] But Strabo does not say that these were actually shown on Eratosthenes' map, nor can we infer from the evidence in Strabo, as Cortesão does, that "he prepared the ground for the cartographic projection, as developed by Hipparchus, Marinus, and Ptolemy."[64] The apparent precision of these reconstructed maps should therefore be evaluated with care.

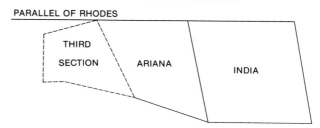

FIG. 9.6. RECONSTRUCTION OF ERATOSTHENES' *SPHRAGIDES*. The seal-like subdivisions for South Asia are part of a system that covered the *oikoumene* described by Strabo in his *Geography*. The dashed lines are the boundaries between geographical areas that Eratosthenes was unable to define properly.
Developed from the description in *The Geography of Strabo*, 8 vols., ed. and trans. Horace Leonard Jones, Loeb Classical Library (Cambridge: Harvard University Press; London: William Heinemann, 1917–32), 2.1.22–23.

THE DISSEMINATION OF CARTOGRAPHIC KNOWLEDGE

On first inspection, the sources for the development of cartography in Hellenistic Greece strongly convey the impression that its knowledge and practice were confined to relatively few in an educated elite. Certainly the names associated with the history of mapping are largely drawn from a handful of outstanding thinkers traditionally associated with the history of Greek science in general. From other sources, however, albeit fragmentary, a broader picture can be drawn in which both the theories underlying mapping and the maps themselves were more widely experienced by the educated class and among the citizens of the major towns. Three-dimensional models of the universe as well as globes and maps were used in schools and sometimes displayed in public places; and considering the maps engraved as emblems for the face of coins, it could even be said that the cartographic image was being popularized.

60. Strabo *Geography* 2.1.22–23 (note 10).

61. Strabo *Geography* 2.1.40 (note 10).

62. Bunbury, *History of Ancient Geography*, facing p. 660 (note 32), and Cortesão, *History of Portuguese Cartography*, 1:83–84 and fig. 19 (note 26).

63. Dicks, *Geographical Fragments*, 159 (note 3).

64. Cortesão, *History of Portuguese Cartography*, 1:84 (note 26).

An interesting pointer to the popularity of maps in Athens at the beginning of the third century B.C. is provided by Theophrastus, disciple and successor of Aristotle and a contemporary of Dicaearchus. He requests in his will "that the small portico adjoining the shrine of the Muses shall be rebuilt no worse than before, and that the panels (*pinakas*) showing maps of the earth (*gēs periodoi*) shall be put up in the lower cloister."[65] This will, transcribed verbatim by Diogenes Laertius, suggests that the custom of drawing maps on wooden panels and showing them in public or semipublic places for information was well established.[66] The maps described were existing fixtures, easily movable like other paintings; they could be displayed with other types of pictures, in the company of which they became a familiar type of image.

Nor was this the sole example of such public cartographic displays. Although the allusion is literary and refers to an earlier period, a passage in the epic of Apollonius Rhodius (fl. ca. 267–260 B.C.) extends the practice outside Athens, claiming that the Colchians, on the southeast coast of the Black Sea, were originally colonists from Egypt. "They preserve," he says, "the engravings of their fathers on pillars, on which are marked all the ways and the limits of the sea and land as you journey on all sides round."[67] The word rendered as "engravings" means "scratchings" in Homer—from whom Apollonius often drew his vocabulary—so the poet is implying roughly incised lines.

A different medium for the dissemination of miniature map images is found in the Ionian coins (fig. 9.7) probably struck by Memnon of Rhodes, who acted as a Persian general in Ephesus until the arrival of Alexander in 334 B.C. In this series of Rhodian-weight tetradrachms, the obverse type is the figure of the Persian king, running or kneeling right. The reverse is a rectangular incuse with irregular raised areas, recognizable as a map depicting the physical relief of the hinterland of Ephesus. It is described thus by Johnston:

> The feature most clearly recognisable is the central loop, with the Tmolus range in the north and the Messogis range in the south, divided by the valley of the Caÿster (now the Küçük Menderes) running towards the sea to the West. Also running east-west are the rivers Hermus (the modern Gediz) to the north of the Tmolus range and the Maeander (Büyük Menderes) to the south of the Messogis range. The tributaries of the Maeander, the Harpasus (Ak) and the Morsynas (Vandalas), divide the southern mountain block into three ridges, visible in the lower part of the reverse.

And Johnston adds:

> If such an accurate and detailed map could be conceived of as a coin type, the maps for ordinary use

must have been the products of a highly developed technique. . . . The whole conception is remarkably close to that of a modern plastic relief map.[68]

FIG. 9.7. IONIAN COIN MAP. This design is found on the reverse of many Ionian silver tetradrachms, one of which is shown, ca. 330 B.C. It depicts an area in Asia Minor near Ephesus.
Diameter of the original: 2.3 cm. By permission of the Trustees of the British Museum, London (BMC Ionia, 323 no. 1).

Although other Greek coins with maps or plans are not of this period, it is convenient to group them here.[69] The map images, of course, serve no practical purpose, but they have a symbolic or propaganda value. Thus the people of Messana chose to give thanks, by a map emblem struck on their coins, for a natural feature that had helped their city develop. This was the sickle-shaped sandspit that protected their harbor and was said to have given the city its original name of Zankle.[70] Some coins of Messana show this; on a few of them are found interior rectangular projections that have been thought to represent harbor buildings.

65. Diogenes Laertius *Lives of Famous Philosophers* 5.51; translation by O. A. W. Dilke. See Diogenes Laertius, *Vitae philosophorum*, 2 vols., ed. Herbert S. Long (Oxford: Clarendon Press, 1964); for an English translation see *Lives of Eminent Philosophers*, 2 vols., trans. R. D. Hicks, Loeb Classical Library (Cambridge: Harvard University Press; London: William Heinemann, 1925–38). See also H. B. Gottschalk, "Notes on the Wills of the Peripatetic Scholarchs," *Hermes* 100 (1972): 314–42; John Patrick Lynch, *Aristotle's School: A Study of a Greek Educational Institution* (Berkeley: University of California Press, 1972), 9 ff. and map, p. 217.

66. Désiré Raoul Rochette, *Peintures antiques inédites, précédées de recherches sur l'emploi de la peinture dans la décoration des édifices sacrés et publics, chez les Grecs et chez les Romains* (Paris: Imprimerie Royale, 1836).

67. Apollonius Rhodius *Argonautica* 4.279–81, translation by O. A. W. Dilke; see *Argonautique*, 3 vols., ed. Francis Vian, trans. Emile Delage (Paris: Belles Lettres, 1974–81), vol. 3.

68. A. E. M. Johnston, "The Earliest Preserved Greek Map: A New Ionian Coin Type," *Journal of Hellenic Studies* 87 (1967): 86–94, quotation on 91.

69. For coins of Cnossos, however, see p. 251 below.

70. George Francis Hill, *Coins of Ancient Sicily* (Westminster: A. Constable, 1903), 38–39 and pl. I.2; Charles Theodore Seltman, *Greek Coins: A History of Metallic Currency and Coinage down to the Fall of the Hellenistic Kingdoms* (London: Methuen, 1933; revised 1955), pl. 14.5 (upper).

A more doubtful example is on a coin of Phocaea (Foca), north of Smyrna (Izmir) in Ionia; this has a drawing of the common seal (the animal; Greek *phoke*, origin of their city name) on the obverse.[71] On the reverse, however, is what could be a city plan, showing three-quarters of a square, with a possible harbor and river. Other coins of Phocaea have a complete square, although the layout differs somewhat from Livy's later description of the town plan.[72]

Another factor contributing to this wider understanding of maps was, as already noted, that the early Greek astronomers and philosophers had a well-developed mechanical flair. Rather than confining their geometric hypotheses solely to diagrams drawn on flat surfaces, they expressed them in three-dimensional models, whether globes or mechanical representations of the workings of the celestial system. In particular, the study of *sphairopoiia* (σφαιροποιία)—a branch of mechanics the object of which was to represent celestial rotations[73]—while serving as a method of research into the laws of cosmic motion through the construction of models of the cosmos, also helped bridge the gap between purely theoretical speculation and its wider understanding in a more tangible form. Pappus, a distinguished Alexandrian mathematician (fl. A.D. 320), defines experts in mechanics as "those who know *sphairopoiia*, a technique used to construct representations of the moving sky through the regular and circular motion of water."[74]

One of these experts was Archimedes (287–212 B.C.), a contemporary and friend of Eratosthenes, exerted an influence on astronomical and geographical thinking, and hence on the maps and models that represented these theories. Born in Syracuse, Archimedes was the son of Pheidias, an astronomer. He visited Alexandria, where he mixed in the intellectual circle of astronomers and philosophers such as Conon of Samos (fl. 245 B.C.) and his pupil Dositheus of Pelusium (fl. 230 B.C.),[75] and where he also met Eratosthenes, with whom he later conducted a scientific correspondence. Archimedes was a favorite at the court of Hieron II (ca. 306–215 B.C.), ruler of Syracuse, becoming famous for the various machines he invented, especially those used to repulse Roman assailants at the siege of Syracuse.[76]

Among numerous other works, Archimedes composed a treatise (lost) on *sphairopoiia*, which led to a greatly improved representation of the universe. His project was indeed ambitious—an attempt to build a model of the "terrestrial system" based on the geocentric hypothesis and to make it work as in reality. The design was such that the celestial sphere, the planets, and the earth were all parts of one intricate mechanism, which could be set in motion so as to simulate the apparent rotation of the stellar sphere and the various motions of the main planets. From this work, as well as from other contemporary and later sources, it becomes clear that in the time of Archimedes it was quite usual to make all kinds of models imitating the various movements in the universe. Furthermore, celestial globes, like the ones introduced by Eudoxus and made familiar by Aratus, were exhibited frequently in schools or in public places. And armillary spheres such as that reconstructed in figure 8.8, in which the sphere of the fixed stars was reduced to its main circles (equator, tropics, ever-visible and never-visible circles, zodiac, colures) and the earth could be envisioned in the center, helped demonstrate to a wider audience the effects of latitude on celestial phenomena.

Archimedes also constructed a number of globes. At least two of them were taken to Rome by Marcus Claudius Marcellus after the fall of Syracuse in 212 B.C.; they are described enthusiastically (but rather vaguely) by Cicero.[77] One of them, apparently of striking appearance, was well known to the people of Rome and had been put in the temple of Virtue. It was probably a solid celestial globe, showing the whole series of constellations, or rather the characters and animals representing them. These kinds of globes were usually elegantly drawn and then brightly colored. A second globe, which

71. William Smith, *Dictionary of Greek and Roman Geography*, 2 vols. (London: J. Murray, 1870), 2:603; S. W. Grose, *Fitzwilliam Museum: Catalogue of the McClean Collection of Greek Coins* (Cambridge: Cambridge University Press, 1929), 3:143 and pl. 285.22. There is, however, some resemblance to the plan given by F. Sartiaux, "Recherches sur le site de l'ancienne Phocée," *Comptes Rendus des Séances de l'Académie des Inscriptions et Belles-Lettres* (1914): 6–18, esp. 6–7.

72. Livy [*History of Rome*] 37.31.7–10; see *Livy*, 14 vols., trans. B. O. Foster et al., Loeb Classical Library (Cambridge: Harvard University Press; London: William Heinemann, 1919–59).

73. The word σφαιροποιία is mentioned by Geminus, in Proclus *Commentary on the Elements of Euclid*; see *In primum Euclidis Elementorum librum commentarii*, ed. G. Friedlein (Leipzig: Teubner, 1873), p. 41, l. 16, and also in Geminus *Introduction to Phenomena* 12.23, 12.27, 14.9, 16.19, 16.27, and 16.29; see *Introduction aux phénomènes*, ed. and trans. Germaine Aujac (Paris: Belles Lettres, 1975).

74. Pappus of Alexandria *Synagoge* (Collection) 8.2, translation by O. A. W. Dilke; see *Collectionis quae supersunt*, 3 vols., ed. F. Hultsch (Berlin: Weidmann, 1867–78), vol. 3.

75. Both were astronomers: Conon of Samos identified a new constellation that he named the Lock of Berenice; Dositheus constructed an astronomical calendar. See Ivor Bulmer-Thomas, "Conon of Samos," and D. R. Dicks, "Dositheus," in *Dictionary of Scientific Biography*, 3:391 and 4:171–72, respectively (note 30).

76. Archimedes' inventions are discussed in Aage Gerhardt Drachmann, *The Mechanical Technology of Greek and Roman Antiquity: A Study of Literary Sources*, Acta Historica Scientiarum Naturalium et Medicinalium, 17 (Copenhagen: Munksgaard, 1961).

77. Cicero *The Republic* 1.14; Edward Luther Stevenson, *Terrestrial and Celestial Globes: Their History and Construction, Including a Consideration of Their Value as Aids in the Study of Geography and Astronomy*, 2 vols., Publications of the Hispanic Society of America, no. 86 (New Haven: Yale University Press, 1921; reprinted New York and London: Johnson Reprint Corporation, 1971), 1:15–17.

was held to be the masterpiece of Archimedes, at first glance did not seem extraordinary. Marcellus kept it in his own house. It was not a celestial globe, but a kind of planetarium. While it lacked the beautiful drawing and attractive colors that probably gave the globe in the temple of Virtue its charm, it was a major contribution to mechanical science, modeling the motions of the sun, the moon, and five planets. Marcus Fadius Gallus, when presenting it to Cicero, declared: "What is admirable in the invention of Archimedes is this: he caused a single rotation to put into constant action bodies moving in various and unequal orbits, at different speeds." And Cicero goes on: "When Gallus put this sphere in motion, the moon placed itself under the sun after as many rotations as days are needed for doing so in the sky itself; so that an eclipse of the sun took place in the artificial sphere, as in the sky. And the moon also went into the shadow thrown by the earth when the light of the sun came from exactly the opposite direction."[78] The construction of this planetarium made Archimedes famous among educated Romans, who were impressed by the high level of knowledge and skill it had required.[79] Later

on, some people—to Cicero's indignation—even expressed the view that the model was better than reality: "that Archimedes, in his imitation of the rotations of the celestial sphere, was a better constructor than Nature itself . . . whereas (Cicero asserts) natural movements are very much better planned than imitated ones."[80]

It is clear that the construction of such models—like their representation in two dimensions—could raise fundamental questions, going beyond the purely astronomical and geographical and expressing concern about man's place in the universe in relation to its very purpose and plan.

78. Cicero *The Republic* 1.14.22, translation by O. A. W. Dilke; see *La République*, 2 vols., trans. Esther Bréguet (Paris: Belles Lettres, 1980), vol. 1.

79. Cf. also Cicero *Tusculan Disputations* 1.25.63; see *Tusculan Disputations*, trans. J. E. King, Loeb Classical Library (Cambridge: Harvard University Press; London: William Heinemann, 1933).

80. Cicero *De natura deorum* 2.35.88, translation by O. A. W. Dilke; see also *De natura deorum* [*and*] *Academica*, trans. H. Rackham, Loeb Classical Library (Cambridge: Harvard University Press; London: William Heinemann, 1924).

BIBLIOGRAPHY
CHAPTER 9 THE GROWTH OF AN EMPIRICAL CARTOGRAPHY IN HELLENISTIC GREECE

Aristotle. *Meteorologica*. Translated by H. D. P. Lee. Loeb Classical Library. Cambridge: Harvard University Press; London: William Heinemann, 1952.

Aujac, Germaine. *La géographie dans le monde antique*. Paris: Presses Universitaires, 1975.

Autolycus of Pitane. *La sphère en mouvement*. Edited and translated by Germaine Aujac, Jean-Pierre Brunet, and Robert Nadal. Paris: Belles Lettres, 1979.

Broche, G. E. *Pythéas le Massaliote*. Paris: Société Française d'Imprimerie, 1936.

Dicks, D. R., ed. *The Geographical Fragments of Hipparchus*. London: Athlone Press, 1960.

Diogenes Laertius. *Lives of Eminent Philosophers*. 2 vols. Translated by R. D. Hicks. Loeb Classical Library. Cambridge: Harvard University Press; London: William Heinemann, 1925–38.

———. *Vitae philosophorum*. 2 vols. Edited by Herbert S. Long. Oxford: Clarendon Press, 1964.

Dubois, M. *Examen de la Géographie de Strabon*. Paris: Imprimerie Nationale, 1891.

Eratosthenes. *Die geographischen Fragmente des Eratosthenes*. Edited by Hugo Berger. Leipzig: Teubner, 1898.

Hawkes, C. F. C. *Pytheas: Europe and the Greek Explorers*. Eighth J. L. Myres Memorial Lecture. Oxford: Blackwell, 1977.

Müller, Karl, ed. *Geographi Graeci minores*. 2 vols. and tabulae. Paris: Firmin-Didot, 1855–56.

Pytheas of Massilia. *Fragmenta*. Edited by Hans Joachim Mette. Berlin: Walter de Gruyter, 1952.

Thalamas, Amédée. *La Géographie d'Eratosthène*. Versailles: Imprimerie Ch. Barbier, 1921.

for it was well known that there was no star at the celestial pole, as Pytheas had correctly noted: the pole was a point in the sky, close to three stars with which it completed a rectangle.[15] A second criticism related to Eudoxus's failure to have observed the general principle regularly used in drawing constellations: "All constellations should be drawn from the observer's point of view, as if they were facing us, unless they are in profile."[16] This rule was a basic one, since in Greece stars were distinguished only by the place they occupied on the figure representing the constellation. For instance, the star we call α Orionis or Betelgeuse was indicated as "the star on the right shoulder of Orion," and our β Orionis or Rigel was called "the star on the left foot of Orion." If Orion was drawn facing the wrong way, the image of the constellation would also be reversed. Thus when Eudoxus, and Aratus after him, declared that the throat and right temple of the Dragon were in line with the Bear's tail,[17] Hipparchus corrected them, saying that the left (not the right) temple of the Dragon was in line with the stars in question. He rejected indignantly the explanation of one of Aratus's commentators who had supposed the head of the Dragon to be turned toward the outside, instead of toward the inside, of the universe.[18] So Hipparchus was emphatic: the figures representing the constellations, when drawn on a solid globe, had to be represented in rear view, looking inward; when drawn on a flat map, they had to be represented in front view, looking at the map reader.

In addition to enunciating these principles, Hipparchus determined a precise location on the globe for most of the stars. Thus, when discussing the couplet in Aratus's poem in which the head of the Dragon is described as brushing the horizon,[19] Hipparchus declared that the star at the extremity of its mouth (μ Draconis) was $34\frac{3}{5}°$ distant from the pole, its southern eye (β Draconis) 35° from it, and its southern temple (γ Draconis) 37°. Thus, anyone in Athens (latitude 37°N) could observe the head of the Dragon turning around entirely inside the ever-visible portion of the sphere, with only the left temple on the ever-visible circle.[20] This example confirms not only that the outlines of the various constellations were drawn on Hipparchus's sphere, but also that the stars thus located were marked as precisely as possible.

The latitudinal position of a star was indicated, in Hipparchus's *Commentary*, by its distance from the pole. For the longitude, its position was noted in relation to the signs of the zodiac, that is, by the degree of the zodiacal sign that is on the same meridian circle as the star, or what is sometimes defined as the polar longitude.[21]

At the end of his *Commentary*, Hipparchus enumerated the main stars situated on twenty-four semicircles constructed from these principles, going from one pole to the other, each separated from the next by a distance of one equinoctial hour.[22] The first is the semicircle going through the summer solstitial point, on which one can see the star at the extreme end of the Dog's tail (η Canis Majoris).[23] Each subsequent interval of one equinoctial hour was equivalent to fifteen degrees of longitude. It is very likely that these twenty-four semicircles, together with a corresponding number of parallel circles, were drawn on Hipparchus's sphere as a graticule. They made it easier for a globe maker to mark in the stars and for the student to find the position of each star. The celestial globe had become a scientific tool that could be used to calculate the time at night or to find how long an eclipse of the moon lasted.[24]

Following these procedures, it is known that Hipparchus produced a catalog listing at least 850 stars.[25] This catalog was certainly illustrated by stellar globes of various kinds, both artistically drawn and scientifically accurate. It is also quite possible, at least for portions of

---

15. Hipparchus *In Arati et Eudoxi Phaenomena* 1.4.1 (note 14).

16. Hipparchus *In Arati et Eudoxi Phaenomena* 1.4.5 (note 14), translation by O. A. W. Dilke.

17. Aratus *Phaenomena* 59–60, in *Callimachus: Hymns and Epigrams; Lycophron; Aratus*, trans. A. W. Mair and G. R. Mair, Loeb Classical Library (Cambridge: Harvard University Press; London: William Heinemann, 1955). Also see Aratus, *Phaenomena*, ed. Jean Martin, Biblioteca di Studi Superiori: Filologia Greca 25 (Florence: Nuova Italia, 1956).

18. Hipparchus *In Arati et Eudoxi Phaenomena* 1.4.5 (note 14).

19. Aratus *Phaenomena* 60–62 (note 17).

20. Hipparchus *In Arati et Eudoxi Phaenomena* 1.4.8 (note 14).

21. Neugebauer, *History of Ancient Mathematical Astronomy*, 277–80 (note 13). In astronomy, a zodiacal sign is either one of the twelve constellations on the zodiac or a segment of the ecliptic circle, thirty degrees in length, named after the corresponding constellation. Here the second meaning is used.

22. An equinoctial hour is the twenty-fourth part of one day. The ancients in ordinary life used "temporary" or "unequal" hours, each hour being the twelfth part of the daytime; these varied according to the season.

23. Hipparchus *In Arati et Eudoxi Phaenomena* 3.5.2 (note 14).

24. Hipparchus *In Arati et Eudoxi Phaenomena* 3.5.1 (note 14).

25. The star catalog has not been preserved, but information about it has been gleaned from Ptolemy's references to it and from Hipparchus's *In Arati et Eudoxi Phaenomena* (note 14). The value of 850 for the number of stars recorded is found in F. Boll, "Die Sternkataloge des Hipparch und des Ptolemaios," *Biblioteca Mathematica*, 3d ser., 2 (1901): 185–95. For a summary of the complexities of the reconstruction of Hipparchus's star catalog, see Neugebauer, *History of Ancient Mathematical Astronomy*, 280–88 (note 13). On Hipparchus's use of coordinates in the catalog, Toomer, "Hipparchus," 217 (note 12) points out that there is no evidence that he assigned coordinates to all—or, indeed, to any—of the large number of stars he listed. On the other hand, in his commentary on Aratus, Hipparchus used a mixture of ecliptic and equatorial coordinates to indicate the positions of stars. See Toomer, "Hipparchus," 217–18 (note 12).

the sky or for individual constellations, that Hipparchus tried to represent the sky on a flat map, with the constellations drawn in front view. A surviving contemporary artifact may demonstrate that such a possibility is not out of the question: it is a bas-relief dated 98 B.C., illustrating the royal horoscope of Antiochus of Commagene (northern Syria). It shows a lion, one of the zodiacal signs, carved in front view, its body outlined with stars, and three planets above it (fig. 10.4).[26] This individual case suggests that other single constellations were also drawn for specific, presumably astrological purposes.[27]

FIG. 10.4. THE LION OF COMMAGENE. This depicts the horoscope of Antiochus I (born 98 B.C.), king of Commagene. The original bas-relief was found on his tomb at Nimroud-Dagh in Turkey.
Size of the original: 1.75 × 2.4 m. After A. Bouché-Leclercq, *L'astrologie grecque* (Paris; E. Leroux, 1899; reprinted Brussels: Culture et Civilisation, 1963), 439 and fig. 41.

Equally interesting from a cartographic viewpoint is Hipparchus's concern with the improvement of geographical maps. This was dependent upon, and closely related to, his studies of the celestial globe. In this case, however, our knowledge of his treatise in three books (now lost), *Against Eratosthenes*, is derived at second hand through Strabo.[28] In this work he criticized Eratosthenes' map as being drawn without sufficient knowledge of the exact position of the different countries. In particular he objected that the Taurus range, following Dicaearchus, still appeared along the central parallel of the map:

> Since we cannot tell either the relation of the longest day to the shortest, or of gnomon to shadow, along the mountain-range that runs from Cilicia on to India, neither can we say whether the direction of the mountains lies on a parallel line, but we must leave the line uncorrected, keeping it aslant as the early maps give it.[29]

Some authorities have emphasized that Hipparchus's criticism of Eratosthenes was "sometimes erroneous and unfair."[30] But one should remember that as a mathematician and astronomer Hipparchus conceived of the value of cartography primarily in those terms. In particular, he emphasized the need for astronomical observations to locate exactly any place on the earth, pointing out the inadvisability of drawing a general map of the inhabited world before such observations had been made for every country. Thus Strabo reports:

> Hipparchus, in his treatise *Against Eratosthenes*, correctly shows that it is impossible for any man, whether layman or scholar, to attain to the requisite knowledge of geography without a determination of the heavenly bodies and of the eclipses which have been observed. For instance, it is impossible to determine whether Alexandria in Egypt is north or south of Babylon, or how far north or south of Babylon it is, without investigation by means of the "climata." In like manner, we cannot accurately fix points that lie at varying distances from us, whether to the East or the West, except by a comparison of eclipses of the sun and the moon.[31]

Hipparchus's *climata* thus appear to be a systematic method of locating towns or countries in their correct latitudinal positions. Strabo reports that he recorded the different celestial phenomena for the regions in the inhabited world between the equator and the North Pole.[32] While he accepted Eratosthenes' meridian through Meroë, and the mouth of the Borysthenes, he faced the same problem as Eratosthenes of estimating distances along it. Figure 10.5 reconstructs that portion of the meridian and also compares the longitudes of the key places along the meridian with their actual values.[33] The

26. A. Bouché-Leclercq, *L'astrologie grecque* (Paris: E. Leroux, 1899; reprinted Brussels: Culture et Civilisation, 1963), 439.

27. For examples of single constellations used on medals and for calendrical purposes, see Georg Thiele, *Antike Himmelsbilder, mit Forschungen zu Hipparchos, Aratos und seinen Fortsetzern und Beiträgen zur Kunstgeschichte des Sternhimmels* (Berlin: Weidmann, 1898), 64–75.

28. Dicks, *Geographical Fragments* (note 14).

29. Hipparchus, in Strabo *Geography* 2.1.11 (note 2), translation adapted from H. L. Jones edition.

30. Armando Cortesão, *History of Portuguese Cartography*, 2 vols. (Coimbra: Junta de Investigações do Ultramar-Lisboa, 1969–71), 1:85; Thomson, *History of Ancient Geography*, 321 (note 6). Strabo defends Eratosthenes against many of Hipparchus's criticisms but also points out some of Eratosthenes' "errors" in his *Geography* 2.1.4–41 (note 2).

31. Strabo *Geography* 1.1.12 (note 2), translation adapted from H. L. Jones edition.

32. Strabo *Geography* 2.5.34 (note 2).

33. Dicks, *Geographical Fragments*, 160–64 (note 14), discusses the nature of Hipparchus's table of latitude and longitude.

celestial phenomena related to latitude were the distance from the pole of the arctic (ever-visible) circle for each latitude (see note 5), the stars situated inside the arctic circle or on its circumference, the relation between the height of the gnomon and its shadow on solstitial and equinoctial days, and the length of the solstitial day. All these phenomena could be either calculated or observed on a globe or an armillary sphere.

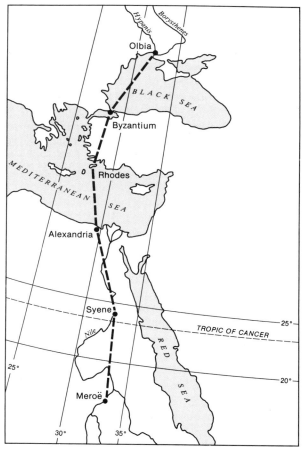

FIG. 10.5. THE MERIDIAN OF ALEXANDRIA. If Strabo's account of Hipparchus's *Against Eratosthenes* is correct, Hipparchus believed the mouth of the Borysthenes, Alexandria, and Meroë to be on the same meridian. This diagram shows its relation to the actual meridian. From *The Geographical Fragments of Hipparchus*, ed. D. R. Dicks (London: Athlone Press, 1960), p. 147, fig. 3.

Hipparchus's contribution to the history of map projections is more controversial. While some authors have claimed that his description of the *oikoumene* as a trapezium indicates an attempt to improve on the rectangular system of Eratosthenes, there is no direct evidence that this is not simply a reference to the chlamys-shaped inhabited world, as Dicks has pointed out.[34] Conversely, although some authors have questioned whether Hip-

parchus was aware of the stereographic projection (for terrestrial or celestial use) or the astrolabe, the weight of opinion now points to his invention of both.[35]

MAPS AND GLOBES IN EDUCATION

In addition to Hipparchus's fundamental challenge to preexisting theories underlying the construction of maps and globes, some treatises were compiled in the first century B.C. that concerned not so much the creation of new knowledge as the wider functions of education. New schools and libraries had been founded as Rome extended its empire over the inhabited world.[36] In Greece and elsewhere, scholars appear in the record who were concentrating on digests of the new discoveries—both theoretical and geographical—for the benefit of their Roman students. While the distinction is less sharply drawn than in the modern world, some of the treatises, it may be said, were pitched at the level of textbooks. Much more succinct than the encyclopedic works of Polybius or Strabo, they presented compendiums of astronomical and geographical knowledge that help to illuminate our understanding of the diffusion of a knowledge of maps in Greek and Roman society.

The nature of such textbooks is indicated by the writings of Theodosius of Bithynia (ca. 150–70 B.C.). Of his known treatises, his *Spherics* (in three books) is the earliest extant textbook of this kind, and his work *On Inhabitable Places* describes the celestial phenomena visible at various latitudes.[37] Both treatises are purely geometric: the first deals implicitly with the celestial sphere and its various circles; the second refers explicitly to the

34. Dicks, *Geographical Fragments*, 148, 206 (note 14).

35. Dicks, *Geographical Fragments*, 207 (note 14). Toomer, "Hipparchus," 219 (note 12), believes the "balance of probability is that Hipparchus used (and perhaps invented) stereographic projection. If that is so, there is no reason to deny his invention of the plane astrolabe." Neugebauer, *History of Ancient Mathematical Astronomy*, 858, 868–79 (note 13), reviews the sources for the theory of stereographic projection and for its possible use in the construction of the astrolabe.

36. Elmer D. Johnson, *A History of Libraries in the Western World* (New York and London: Scarecrow Press, 1965), describes the new schools and libraries that emerged as Rome extended its empire.

37. Theodosius *On Inhabitable Places*; see *De habitationibus liber: De diebus et noctibus libri duo*, ed. Rudolf Fecht, Abhandlungen der Gesellschaft der Wissenschaften zu Göttingen, Philologisch-Historische Klasse, n.s., vol. 19, 4 (Berlin: Weidmann, 1927); and Theodosius's *Sphaerica*; see *Sphaerica*, ed. J. L. Heiberg, Abhandlungen der Gesellschaft der Wissenschaften zu Göttingen, Philologisch-Historische Klasse, n.s., vol. 19, 3 (Berlin: Weidmann, 1927). The name on this translation of *Sphaerica*, "Theodosius Tripolites," has caused some commentators to think that Theodosius was from Tripoli, despite the corrigendum on page XVI. Neugebauer discusses the "textbook" style of Theodosius's works as well as the basic geometry of his spherical astronomy in *History of Ancient Mathematical Astronomy*, 748–51, 757–67 (note 13).

earth at the center of the celestial sphere. The diagrams illustrating the demonstrations show a small circle (theoretically infinitely small) tangent to a diameter of the celestial sphere, itself drawn as a circle (fig. 10.6). The plane ABCD is meant to represent the astronomical horizon, and the small circle represents the terrestrial globe. Through geometry Theodosius demonstrated phenomena already known to Pytheas: that under the polar circle (then regarded as 66°N latitude) the longest day lasted twenty-four hours;[38] that at the pole the solar year was composed of one day six months long and one night six months long; and so forth. The diagram of Theodosius thus apparently represented the armillary sphere, with rings (Latin *armillae* [bracelets]) indicating the main celestial circles and a small globe in the center fixed on the axis of rotation representing the earth. Such models were commonly used in Greece.[39]

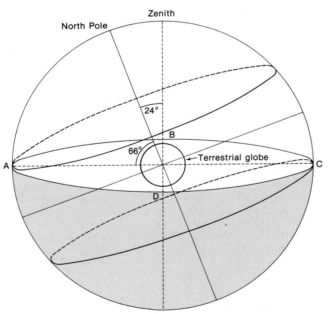

FIG. 10.6. THEODOSIUS'S FIGURE OF A CELESTIAL SPHERE. Theodosius showed that if an observer's ever-visible circle was coincident with the summer tropic, then the observer is situated at the Arctic Circle and has a twenty-four-hour day (see fig. 8.8). The inner circle, representing the terrestrial globe, is theoretically infinitely small. The observer's horizon plane is ABCD.
Developed from the description in Theodosius, *De habitationibus liber: De diebus et noctibus libri duo*, ed. Rudolf Fecht, Abhandlungen der Gesellschaft der Wissenschaften zu Göttingen, Philologisch-Historische Klasse, n.s., vol. 19, 4 (Berlin: Weidmann, 1927).

The better-known contemporary of Theodosius, Posidonius (ca. 135–51/50 B.C.), is generally associated with his measurement of the circumference of the earth. By some scholars, who view the history of mapping as mainly concerned with the diagnosis of increasing accuracy, this measurement has been "deemed disastrous in the history of geography."[40] Depending on the value of the stade we adopt, it may be true that Posidonius, seeking to improve on Eratosthenes, underestimated the size of the earth, and this measurement, copied by Ptolemy, was thereafter transmitted to Renaissance Europe.

But Posidonius clearly did more than measure the earth: such was his reputation as an educator that Strabo described him as "one of the most learned philosophers in our time."[41] He was born in Apamea in Syria; after traveling widely in the western Mediterranean countries and visiting Rome on several occasions, he established himself in Rhodes, where he opened a school. This was patronized by distinguished visitors, including Pompey, the Roman general and statesman, and Cicero, from whom some of our knowledge of Posidonius derives. It was also at Rhodes that he constructed a planetarium in the style of Archimedes, intended for teaching students the laws of the cosmos. Cicero describes "the orrery recently constructed by our friend Posidonius, which at each revolution reproduces the same motions of the sun, the moon and the five planets that take place in the heavens every twenty-four hours."[42]

Besides demonstrating his mechanical skill in this way, Posidonius was engaged in reassessing some of the theories about the earth current in his day. Indeed, in this respect his writing was to serve as an important conceptual link between cartography in the ancient and medieval worlds. In his treatise *The Ocean* (now lost but known to us through Strabo), for example, he discussed the problem of terrestrial zones, which is relevant to an understanding of the zonal *mappaemundi* of the Middle Ages.[43]

38. In fact Theodosius establishes that, in places 66° distant from the equator, the sun does not disappear below the horizon on the solstitial day. But by reason of crepuscular light it is daytime for about a month in those latitudes.

39. Strabo *Geography* 12.3.11 (note 2) confirms that globes existed in the area where Theodosius flourished. On the tradition of sphere construction (*sphairopoiia*) see p. 136 and n. 33 above.

40. Cortesão, *History of Portuguese Cartography*, 1:88 (note 30); Lloyd A. Brown, *The Story of Maps* (Boston: Little, Brown, 1949; reprinted New York: Dover, 1979), 30–32; and see E. H. Warmington, "Posidonius," in *Dictionary of Scientific Biography*, 11:104 (note 12).

41. Strabo *Geography* 16.2.10 (note 2), translation by O. A. W. Dilke.

42. Cicero *De natura deorum* 2.34.88, in *De natura deorum [and] Academica*, trans. H. Rackham, Loeb Classical Library (Cambridge: Harvard University Press; London: William Heinemann, 1924).

43. See below, chapter 18, "Medieval *Mappaemundi*."

In this text, inspired by the work of Pytheas, Posidonius began by criticizing the usual division of the earth into five zones—one uninhabited (torrid) zone, two inhabitable (temperate) zones, and two uninhabited (frigid) zones—for he considered the limits between them to be uncertain and inaccurate.[44] Instead of employing the traditional terms for the zones, based on temperature or habitability, he proposed terms based on clearly defined astronomical criteria. He divided the terrestrial globe by means of the tropics and the polar circles and named the zones as in table 10.1.[45]

TABLE 10.1 Posidonius's Terrestrial Zones

Zone	Area	Meaning
Amphiskian	Between the tropics (one zone)	Where the shadow of the gnomon is directed alternately to the north and to the south
Heteroskian	Between each tropic and each polar circle (two zones)	Where the shadow of the gnomon is directed either to the north or to the south, depending on the hemisphere
Periskian	From each polar circle to each pole (two zones)	Where the shadow of the gnomon makes a complete rotation

Note: The Greek *skia* means shadow.

At the same time, Posidonius appreciated that if he altered the criteria for the division, so as to take temperature distribution more fully into account, the earth could be divided into seven zones. These he identified as the two frigid zones around the poles, the two temperate zones in their usual places, two narrow, extremely arid zones along the terrestrial tropics having the sun directly overhead for about half a month each year, and finally the equatorial zone, more temperate and better watered than the two tropical ones.[46] At one point Posidonius also proposed dividing the inhabited world not into continents, as was usual in his day, but by means of circles parallel to the equator, indicating variations in fauna, flora, and climate. However, Strabo commented unfavorably on this innovative idea. It was, he said, "a mere matter of argument, with no useful end in view."[47]

Equally revisionist was Posidonius's challenge to Eratosthenes' measurement of the circumference of the earth. Our knowledge of his methods, as with the earlier reasoning of Eratosthenes, is derived from Cleomedes,[48] and it becomes clear that it was based on assumptions that were sometimes false.[49] These assumptions included the belief that Rhodes and Alexandria lay on the same

meridian and that the distance between the two places was 5,000 stades. Then Posidonius (according to Cleomedes), noting that Canopus (α Carinae) was seen just on the horizon at Rhodes but rose as far as a quarter of a zodiacal sign (7½°) above the horizon at Alexandria, concluded that the center angle intercepting the Rhodes-Alexandria arc of meridian was one-forty-eighth of the total circle or 7½° (the arc is actually 5°14'). Thus, he argued, the total length of the meridian was forty-eight times the distance between Rhodes and Alexandria, and, assuming this latter to be 5,000 stades, this gave a figure of 240,000 stades for the circumference of the earth.[50]

This, however, is only a partial history of the confusion attached to the measurement. As a teacher interested in promoting discussion, Posidonius seems to have criticized his own assumptions,[51] in particular the estimate of 5,000 stades for the distance from Rhodes to Alexandria. Evidently, at some point in his calculations he employed an alternative value of 3,750 stades, derived from a careful estimate Eratosthenes made "by means of shadow-catching sundials."[52] When applied to the 1:48 ratio, this gave a correspondingly smaller length for the circumference of the earth of 180,000 stades.[53]

What is important for the history of cartography is that it was this measurement—whether directly or

44. According to Strabo *Geography* 2.2.2 (note 2), Posidonius criticized Parmenides' division of the earth into five zones because he represented the torrid zone as almost double its real breadth, and he criticized Aristotle for calling the region between the tropics "torrid" and the regions between the tropics and the arctic circles "temperate." Posidonius disagreed with both Parmenides and Aristotle and asked "how one could determine the limits of the temperate zones, which are non-variable, by means of the 'arctic circles,' which are neither visible among all men nor the same everywhere."

45. Germaine Aujac, "Poseidonios et les zones terrestres: Les raisons d'un échec," *Bulletin de l'Association Guillaume Budé* (1976): 74–78. Neugebauer, *History of Ancient Mathematical Astronomy*, 736–46 (note 13), discusses shadow tables.

46. Strabo *Geography* 2.2.3 (note 2).

47. Strabo *Geography* 2.3.7 (note 2).

48. See chapter 9 above, pp. 154–55.

49. See Cortesão, *History of Portuguese Cartography*, 1:86–88 (note 30), for a fuller discussion with a diagram of the methods Posidonius employed.

50. Cleomedes *De motu circulari* 1.10; see *De motu circulari corporum caelestium libri duo*, ed. H. Ziegler (Leipzig: Teubner, 1891). An English translation of book 1, chap. 10, appears in Cortesão, *History of Portuguese Cartography*, 1:141–43 (note 30).

51. Strabo *Geography* 2.2.2 (note 2). C. M. Taisbak, "Posidonius Vindicated at All Costs? Modern Scholarship versus the Stoic Earth Measurer," *Centaurus* 18 (1973–74): 253–69, attributes more precise scientific theory to Posidonius than seems warranted by the information at our disposal.

52. Strabo *Geography* 2.5.24 (note 2). Neugebauer, *History of Ancient Mathematical Astronomy*, 653 (note 13), dismisses this because sundials cannot measure distance directly.

53. See Neugebauer, *History of Ancient Mathematical Astronomy*, 652–54 (note 13), on Posidonius's measurements.

through an intermediary—that was later adopted by both Marinus of Tyre and Ptolemy. Its main effect was greatly to exaggerate the portion of the globe occupied by the inhabited world, so that the length from the Straits of Gibraltar to India, along the parallel of Rhodes, came to be considered half of the entire parallel around the earth.[54] And such was the authority of Ptolemy that this misconception was carried forward by geographers, cosmographers, and cartographers into the sixteenth century. Historians of discovery have noted that it long colored the perception of that age as to the size of the unknown portion of the world.[55]

A final example from the first century B.C. of the extent to which globes and maps were used in education is provided by Geminus of Rhodes (fl. ca. 70 B.C.). His elementary textbook on astronomy and mathematical geography—*Introduction to Phaenomena*[56]—has survived. It points the historian of cartography to some definite evidence for the regular use of celestial and terrestrial globes in the schools of that day.[57]

To explain celestial phenomena to his students, Geminus uses various types of three-dimensional models: a celestial sphere representing all the constellations, a simpler theoretical celestial sphere containing only the main celestial circles, an armillary sphere, and a planetary model consisting of a series of concentric spheres. He has the most to say about the characteristics of the first category, the celestial spheres. On the solid stellar globe, he tells us, "are designed all the constellations. Now it must not be supposed that all the stars lie upon one surface; some of them are higher, others lower, but as our vision extends only to a certain uniform height, the difference in altitude is imperceptible to us."[58] Such globes depicted five parallel circles (the equator, the two tropics, the two arctic circles), the two colures, and the three oblique circles representing the zodiac.[59] The horizon and the meridian of the place of observation could not, however, be drawn upon it, Geminus explained, because these circles were motionless and did not rotate with the globe itself. It was left to the user of such a globe, he added, to imagine the position of the horizon from the stand on which the globe rested.[60]

Geminus also specified that all stellar globes, at least those used for teaching, should be constructed for the latitude of Rhodes, that is, 36°N,[61] so that the polar axis makes an angle of 36° with the plane of the horizon. Figure 10.7 shows the distances between the parallel circles.[62] The zodiacal zone is twelve degrees wide. Its median circle, called the ecliptic, touches the summer tropic at the first point of Cancer and the winter tropic at the first point of Capricorn. And here it divides the equator into two equal parts at the first point of Aries and at the first point of Libra.[63] Curiously enough, the Milky Way, the only circle visible in the night sky, is not drawn on celestial globes. Geminus explained its absence on the grounds that it is not of the same width throughout.[64] Thus he believed that the Milky Way could not be treated similarly to the other celestial circles.[65]

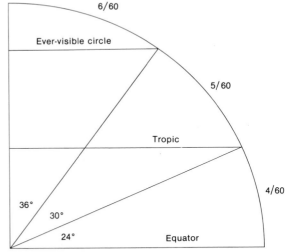

FIG. 10.7. DISTANCES BETWEEN THE PARALLEL CIRCLES ACCORDING TO GEMINUS. The distances between the parallel circles would be:

Fraction	Degrees	Area
4/60 of the circle	24°	From the celestial equator to the tropic
5/60 of the circle	30°	From the tropic to the arctic circle
6/60 of the circle	36°	From the arctic circle to the pole

54. See chapter 11 below on Ptolemy's dimensions for the inhabited world.

55. Most famously in Columbus's belief in the proximity of the Indies when sailing westward; see also chapter 18, p. 354.

56. Geminus *Introduction* (note 1); see also *Elementa astronomiae*, ed. C. Manitius (Leipzig: Teubner, 1898).

57. Germaine Aujac, "Une illustration de la sphéropée: L'*Introduction aux phénomènes* de Géminos," *Der Globusfreund* 18–20 (1970): 21–26; and Aujac, "La sphéropée ou la mécanique au service de la découverte du monde," *Revue d'Histoire des Sciences* 23 (1970): 93–107.

58. Geminus *Introduction* 1.23 (note 1), translation by O. A. W. Dilke.

59. Geminus *Introduction* 5.50–51 (note 1).

60. Geminus *Introduction* 5.62–65 (note 1).

61. Geminus *Introduction* 5.48 (note 1). This means that the ever-visible (or arctic) circle is 36° distant from the pole.

62. This division into zones is reminiscent of Eratosthenes; see D. R. Dicks, "Eratosthenes," in *Dictionary of Scientific Biography*, 4:388–93, esp. 391 (note 12), and Geminus *Introduction* 5.46 (note 1).

63. Geminus *Introduction* 5.51–53 (note 1).

64. Geminus *Introduction* 5.68–69 (note 1).

65. Geminus *Introduction* 5.11 (note 1).

The terrestrial globe was also frequently used for teaching. It was a replica of the celestial globe, similarly constructed according to the latitude of Rhodes, 36°N, so that the arctic circle was fixed at 36° distance from the pole.[66] It also showed the same division into zones by tropics and arctic circles drawn at the same relative distances as the corresponding circles on the celestial globe. Hence, with an earth 252,000 stades in circumference and 84,000 stades in diameter (Geminus took $\pi = 3$), the circle was divided into sixty parts of 4,200 stades each. The frigid zones occupied six-sixtieths or 25,200 stades, the temperate zone five-sixtieths or 21,000 stades, and the torrid zone in the northern hemisphere a breadth of four-sixtieths or 16,800 stades. The same was true of the southern hemisphere. On the semicircle from pole to pole, therefore, the total was 126,000 stades.[67]

To such three-dimensional globes the educated Greeks and Romans were also adding a knowledge of two-dimensional maps. Geminus tells us that both circular and oblong maps were regularly drawn and reiterates the well-established view of earlier writers that oblong maps were more reliable, pointing out that, since the length of the inhabited world was twice its width, a circular map of the same area would distort distances.[68] It must be recognized, therefore, that notwithstanding the public's wider exposure to globes and maps by the first century B.C., there was no standard image of the world that had been disseminated by the Greek writers as a whole and then generally accepted. Indeed, we see from Geminus's comments that alongside the mathematically constructed maps founded on the measurements of Eratosthenes and his revisers there had survived the much older pre-Hellenistic view of the earth mapped in circular form as if representing a flat disk.

This should not be interpreted as an anomaly when set against the record of scientific cartography. Throughout the classical period, as when Geminus wrote, it is clear that maps and globes were not regarded solely as the media of astronomical or geographical instruction seeking to convey a literal view of reality; they were also frequently used as symbols to convey other meanings. As well as appearing in some treatises, they are illustrated in paintings, incorporated in other objets d'art, and, as earlier with the poem of Aratus, described in verse. Globes, in particular, as striking artifacts, were used in symbolic representations, and a general knowledge of their nature may often have depended on their use in this way. Celestial globes, sometimes associated with sundials, can be seen on various Roman sarcophagi as attributes of the Parcae (Fates);[69] and their popularity is also attested by their inclusion in several paintings roughly contemporary with Geminus. A painting in the Casa dei Vettii, Pompeii, represents the Muse Urania

pointing at a celestial globe. On this globe, which rests on a small cubic pedestal, are drawn meridians, parallels, and the oblique zodiacal circle; the polar axis makes an angle with the horizontal plane of the stand.[70] Similarly, another painting in the Casa dell'Argentaria in Pompeii shows Apollo holding in his left hand a celestial globe on which are clearly drawn two great circles, the equator and the ecliptic.[71] From the position of the equator, we may infer that the axis of the poles is inclined to the horizontal plane on which Apollo is standing.

A celestial globe, probably intended also to be a sundial, is found on a fresco acquired by the Metropolitan Museum of Art in 1903 from a villa at Boscoreale near Pompeii, dating from about A.D. 50. It exhibits the celestial circles in lifelike perspective (plate 4).[72] Or again, in a different medium but conveying a similar message for our interpretation of the place of maps in Roman society, we encounter a cartographic motif within the famous "philosopher" mosaic from Torre Annunziata near Pompeii. The design here, like the similar mosaic in the Villa Albani in Rome, was probably inspired by a Hellenistic painting and, along with a sundial, it shows one half of a globe, with meridians and parallels somewhat roughly drawn and wholly inaccurate, emerging from a box supported by four feet.[73] The upper part of the box defines the horizon, dividing the visible hemisphere from the invisible one.

Poetry—sometimes illustrated by maps—also continued to be used as a way of memorizing and popularizing the knowledge or meaning seen in cartographic images. Such literary sources do, however, give the impression that the educated class largely preferred to ignore new discoveries, and earlier Hellenistic concepts of geography persisted long after they had ceased to reflect up-to-date knowledge. A late example is provided by Dionysius, born in Alexandria and called "Periegetes" after the title of his poem.[74] A contemporary of Marinus and Ptolemy, he composed a description in verse of the inhabited world (A.D. 124) that was long used as a school textbook. He presented the *oikoumene* as an island,

66. Geminus *Introduction* 16.12 (note 1); armillary spheres were also constructed to this latitude.

67. Geminus *Introduction* 16.7–8 (note 1).

68. Geminus *Introduction* 16.4–5 (note 1).

69. Otto J. Brendel, *Symbolism of the Sphere: A Contribution to the History of Earlier Greek Philosophy* (Leiden: E. J. Brill, 1977), 70–85 and pls. XXIII, XXVI, XXVIII–XXX.

70. Brendel, *Symbolism of the Sphere*, pl. IX (note 69).

71. Brendel, *Symbolism of the Sphere*, pl. XVII (note 69).

72. Stevenson, *Terrestrial and Celestial Globes*, 1:21 (note 6).

73. Brendel, *Symbolism of the Sphere*, pls. I–VI, VII (note 69).

74. Dionysius *Periegesis*, in *Geographi Graeci minores*, 2 vols. and tabulae, ed. Karl Müller (Paris: Firmin-Didot, 1855–56), 2:103–76. It is followed by Avienius's translation into Latin, 177–89, and by Eustathius's *Commentary*, 201–407.

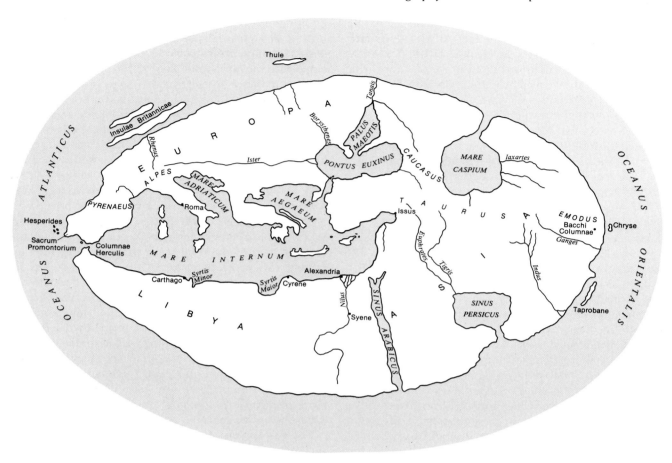

FIG. 10.8. THE RECONSTRUCTED WORLD OF DIONY-
SIUS PERIEGETES, A.D. 124. After Edward Herbert Bunbury, *A History of Ancient Geography among the Greeks and
Romans from the Earliest Ages till the Fall of the Roman*
Empire, 2d ed., 2 vols., (1883; republished with a new introduction by W. H. Stahl, New York: Dover, 1959), vol. 2, map
facing p. 490.

sling-shaped, entirely north of the equator, extending
from Thule to Libya (fig. 10.8). He did not mention
either Agisymba or the promontory of Prasum.[75] He
limited the inhabited world eastward by the river
Ganges, taking into account the Seres (Chinese and Ti-
betans), but locating them much less far east than Marinus.

Dionysius's poem, like Aratus's *Phaenomena*, was a
success partly because it summarized, and made easier
to remember, traditional teachings since Eratosthenes.
It was first translated into Latin by Rufius Festus Avienius (fourth century A.D.), who had also translated Aratus's *Phaenomena*, then by Priscian the grammarian,
who taught in Constantinople in the sixth century A.D.
Cassiodorus (also sixth century), who had founded a
convent in southern Italy, asked the young monks to
learn geography and cosmography through Dionysius's
map and Ptolemy's work.[76] Subsequently, in the twelfth
century, Eustathius, later archbishop of Thessalonica,

who had already commented on the *Iliad* and the
Odyssey, composed a detailed commentary on this poem
that remained in regular use during the whole of the
Middle Ages.

The poem was originally supplied with maps, probably drawn on the models of Eratosthenes' or Strabo's
maps. Various annotations preserved in the margins of
the existing manuscripts refer to maps illustrating the
poem: some of them point out that such and such a
place is lacking on the map or that the outlines of such
and such a country do not agree with Dionysius's de-

75. Agisymba is a kind of generic name referring to central Africa
and the land of the Ethiopians. Cape Prasum is somewhere near Zanzibar, south of Rhapta (possibly Cape Delgado).

76. Cassiodorus *Institutiones divinarum et saecularium litterarum*
1.25; see *Institutiones*, ed. R. A. B. Mynors (Oxford: Clarendon Press,
1937), or for an English translation, Cassiodorus Senator, *An Introduction to Divine and Human Readings*, ed. and trans. Leslie Webber
Jones (New York: Columbia University Press, 1946).

scription. These seem to provide evidence that such map-makers continued to copy their models uncritically and rarely tried to adapt the map to the written description to be illustrated.

In the case of Dionysius, both maps and poem were behind their time, even at the date of their composition; but they reflect the ordinary level of knowledge. His description of the British Isles may be rendered:

Two islands are there, British, off the Rhine,
By Ocean's northern shores; for there the Rhine
Sends out its furthest eddies to the sea.
Enormous is their size: no other isles
Equal the British isles in magnitude.[77]

Such a poor description, and the lack of revision elsewhere, suggests too close a reliance on Eratosthenes.

In all these examples, despite the fragmentary nature of our sources, the wider educative power of the map can be glimpsed. Globes and maps, in particular as displayed in public or incorporated into the composition of paintings, poetry, mosaics, and sculpture, communicated an image of the world with Greece and Rome at its center but with the possibility of other lands and zones—indeed the universe—lying beyond and interrelated in one coherent system.

The Map of the Inhabited World Recommended by Strabo

The contribution of Strabo (ca. 64/63 B.C. to A.D. 21 or later), a native of Amasia (Amasya) in Pontus and a scholar of great stature as philosopher, historian, and geographer,[78] sums up the main themes of this chapter. As a Greek, he epitomizes the continuing importance of the Greek intellectual heritage—and contemporary practice—to the development of cartography in the early Roman world. As the reviser of Eratosthenes, he also illustrates the continuous way later generations had built on the cartographic concepts first clearly set out in the Hellenistic age.

Strabo is said to have visited Rome as a young man (he was an exact contemporary of Augustus), and he pretended to have traveled widely to bring together an enormous amount of geographical knowledge. It is generally accepted, however, that he must have compiled much of this information in the great library at Alexandria, where he had access to many earlier texts now lost. All his writings were firmly set in, if not direct extensions of, the work of his predecessors. Thus his *Historical Memoirs* in forty-seven books, now lost, was a continuation of Polybius. Fortunately, his *Geography* in seventeen books, composed toward the end of his life, has come down to us intact (apart from the seventh book, which survives only as an epitome).[79] The work

is of key importance to our whole knowledge of the history of Greek cartography as well as to the history of science in general.[80] It has already been shown that many of the earlier treatises that touch upon maps are known to us only through Strabo, while the interest of his commentary on these writers is in its critical handling of their theories, albeit he sometimes fails to advance truth by this process.

In many ways the most interesting passages relating to cartography in Strabo's *Geography* are those that, although they contain no maps, give an account, for the first time in a surviving text, of how a description of the known world should be compiled. His motives for writing such a geography (so he tells us) were that he felt impelled to describe the inhabited world because of the considerable strides in geographical knowledge that had been made through the numerous campaigns of the Romans and Parthians.[81] The world map had to be adjusted to take account of these facts, and thus Strabo almost certainly proceeded by taking Eratosthenes' map—and the criticism of it by Polybius, Crates, Hipparchus, and Posidonius—as the basis for his own work.[82]

In this task of compilation Strabo seems to have worked systematically. The first stage was to locate the portion of the terrestrial globe that was known to be inhabited. Strabo reasoned that it lay in a northern quadrant of a globe, in a quadrilateral bounded by the frigid zone, the equator, and two meridians on the sides.[83] In this design Strabo had been influenced not only by Eratosthenes' measurement of the earth but also by the concept of the four inhabited worlds, known and unknown, expounded by Crates, to whom he refers explicitly.[84] Thus far Strabo had relied on theoretical argument derived from his authorities. But he also adduced good empirical grounds for this cartographic reasoning. He continued:

77. Dionysius *Periegesis* 565–69 (note 74), translation by O. A. W. Dilke.

78. E. H. Warmington, "Strabo," in *Dictionary of Scientific Biography*, 13:83–86 (note 12); Germaine Aujac, *Strabon et la science de son temps* (Paris: Belles Lettres, 1966).

79. Strabo *Geography* (note 2).

80. Thus George Sarton writes of the value of this "first attempt to write a geographical encyclopaedia, including mathematical, physical, political and historical geography" in his *Introduction to the History of Science*, 3 vols. (Baltimore: Williams and Wilkins, 1927–48), 1:227, and Cortesão states that as a source it was "second to none in the history of geography and cartography" of this period; see his *History of Portuguese Cartography*, 1:89 (note 30).

81. Strabo *Geography* 1.2.1 (note 2).

82. Cortesão, *History of Portuguese Cartography*, 1:90 n. 3 (note 30).

83. Strabo locates the frigid zone, or arctic circle, at 54° distance from the equator. The so-called quadrilateral, bounded by half of this arctic circle, half of the equator, and segments of two meridians, is a spherical quadrilateral, a portion of a sphere.

84. Strabo *Geography* 2.5.10 (note 2).

But if anyone disbelieves the evidence of reason, it would make no difference, from the point of view of the geographer, whether we make the inhabited world an island, or merely admit what experience has taught us, namely, that it is possible to sail round the inhabited world on both sides, from the east as well as from the west, with the exception of a few intermediate stretches. And, as to these stretches, it makes no difference whether they are bounded by sea or by uninhabited land; for the geographer undertakes to describe the known parts of the inhabited world, but he leaves out of consideration the unknown parts of it—just as he does what is outside of it. And it will suffice to fill out and complete the outline of what we term "the island" by joining with a straight line the extreme points reached on the coasting-voyages made on both sides of the inhabited world.[85]

Despite the extension of the geographical horizons of the inhabited world since the time of Eratosthenes, Strabo's *oikoumene* was smaller. Although Pytheas, Eratosthenes, and perhaps Posidonius had fixed its northern limit on the parallel through Thule (66°N), Strabo, like Polybius, refused to believe that human life was possible so far north, and he blamed Pytheas for having misled so many people by his claim that the "summer tropic" becomes the "arctic circle" at the island of Thule.[86] Again following Polybius, Strabo thus chose as the northern limit of the map and of the inhabited world the parallel through Ierne (Ireland), "which island not only lies beyond Britain but is such a wretched place to live in on account of the cold that the regions on beyond are regarded as uninhabitable."[87] This parallel (54°N) is the projection of the celestial arctic circle constructed for the latitude of Rhodes (36°N); it coincides with the one mentioned by Geminus as the northern limit of the temperate zone. The southern limit of habitable land, for Strabo as for Eratosthenes, is the parallel through the "Cinnamon-producing country" (now in Ethiopia) at about 12°N. He estimated the latitudinal extent of the inhabited world as less than 30,000 stades (compared with Eratosthenes' 38,000 stades) and reduced its length to 70,000 stades instead of Eratosthenes' 78,000.

In order to avoid the deformational problems of flat maps, Strabo stated that he preferred to construct his map on a globe large enough to show all the required detail.[88] He recommended that it be at least ten feet (approximately three meters) in diameter and mentions Crates in this regard. On the other hand, if a globe of this size could not be constructed, Strabo was familiar from Eratosthenes with the transformation necessary to draw it on a plane surface. For a graticule, Strabo adopted the straightforward rectangular network of parallels and meridians. He defended his projection on the ground that it would make only a slight difference if the circles on the earth were represented by straight lines,

"for our imagination can easily transfer to the globular and spherical surface the figure or magnitude seen by the eye on a plane surface."[89] The dimensions of this flat map were also to be generous. Strabo envisaged that it would be at least seven feet long and presumably three feet wide, which would suit the length of the inhabited world (70,000 by 30,000 stades), one foot being equivalent to 10,000 stades (fig. 10.9).

As with all Greek world maps, the great impediment to study for the historian of cartography is that we have only these verbal descriptions, not the images themselves.[90] Nevertheless, apart from the reduced size of the inhabited world, the map Strabo envisaged was similar in its overall shape to that drawn by Eratosthenes.[91] In describing its detailed geography, however, Strabo did not employ, at least overtly, Eratosthenes' division of the world into irregular quadrilaterals or *sphragides*, but he often used geometric figures or comparisons to everyday objects to describe the general outline of a country. For instance, he says that the province of Gallia Narbonensis presents the shape of a parallelogram;[92] that the rivers Garumna (Garonne) and Liger (Loire) are parallel to the Pyrenaeus (Pyrenees), forming with the ocean and the Cemmenus Mountains (Cévennes) two parallelograms;[93] that Britain is triangular;[94] that Italy has been shaped sometimes like a triangle, sometimes like a quadrilateral;[95] that Sicily is indeed triangular,[96] though one side is convex and the two others slightly concave. Similarly, Strabo compares the shape of Iberia to an oxhide,[97] the Peloponnese to a plane-tree leaf,[98] and the northern part of Asia, east of the Caspian, to a kitchen knife with the straight side along the Taurus

85. Strabo *Geography* 2.5.5 (note 2).

86. Strabo *Geography* 2.5.8 (note 2).

87. Strabo *Geography* 2.1.13 (note 2).

88. Strabo *Geography* 2.5.10 (note 2).

89. Strabo *Geography* 2.5.10 and 2.5.16 (note 2), where Strabo says the parallels and meridians allow one to correlate regions that are parallel.

90. Désiré Raoul Rochette, *Peintures antiques inédites, précédées de recherches sur l'emploi de la peinture dans la décoration des édifices sacrés et publics, chez les Grecs et chez les Romains* (Paris: Imprimerie Royale, 1836), suggests that geographical maps, like other pictures, were painted with encaustic, first on easels, and later fixed in their proper place under a portico or in a picture gallery. This again provides further evidence for the wider dissemination of maps in the Greek and Roman world: see above, p. 158.

91. Aujac, *Strabon*, 213 (note 78), speculates on whether Strabo drew a map to accompany the text and on whether he had the map of Eratosthenes in front of him as a model.

92. Strabo *Geography* 4.1.3 (note 2).

93. Strabo *Geography* 4.2.1 (note 2).

94. Strabo *Geography* 4.5.1 (note 2).

95. Strabo *Geography* 5.1.2 (note 2).

96. Strabo *Geography* 6.2.1 (note 2).

97. Strabo *Geography* 2.5.27 (note 2).

98. Strabo *Geography* 8.2.1 (note 2).

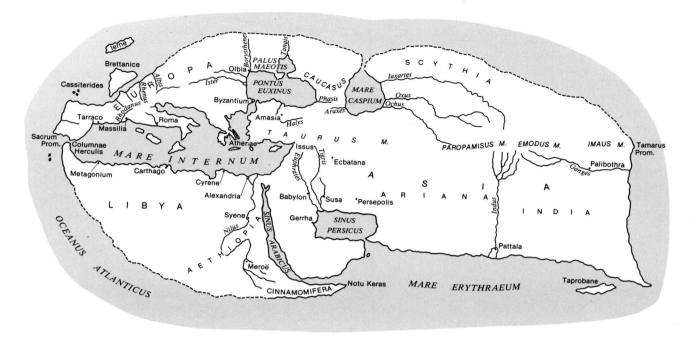

FIG. 10.9. THE SHAPE OF THE INHABITED WORLD RE-CONSTRUCTED FROM STRABO. After Edward Herbert Bunbury, *A History of Ancient Geography among the Greeks and Romans from the Earliest Ages till the Fall of the Roman Empire*, 2d ed., 2 vols. (1883; republished with a new introduction by W. H. Stahl, New York: Dover, 1959), vol. 2, map facing p. 238.

range and the curved side along the northern coastline.[99] India, with two adjacent sides (south and east) much longer than the two others, he described as rhomboidal;[100] Mesopotamia, between the Tigris and Euphrates rivers, he saw as being like a boat drawn in profile, with the deck on the Tigris side and the keel near the Euphrates.[101] Strabo repeats that the river Nile was described by Eratosthenes as a reversed N,[102] and that its mouth was named after the Greek capital letter Δ, delta.[103]

It is not clear how we should interpret these familiar graphic similes Strabo employed to describe to his readers the land areas and other features on the world map. But they do suggest that he was writing with a map in front of him. In some cases, where alternative descriptions are provided, he may have been attempting to collate the outlines of more than one map. It is also probable that students were expected to consult the text of the *Geography* with the help of maps, so that the shapes thus enumerated may have served as a simple mnemonic. Yet if such suggestions must remain speculative, there can be little doubt that by the early Roman period world maps and globes drawn by Greek scholars were encouraging a distinctively geographical way of thinking about the world. And it is likely, among the educated group at least, that an increasingly standard image of the inhabited world had come to be more widely accepted through the use of these maps.

99. Strabo *Geography* 11.11.7 (note 2).
100. Strabo *Geography* 15.1.11 (note 2).
101. Strabo *Geography* 16.1.22 (note 2).
102. Strabo *Geography* 17.1.2 (note 2).
103. Strabo *Geography* 17.1.4 (note 2).

BIBLIOGRAPHY
CHAPTER 10 GREEK CARTOGRAPHY IN THE EARLY ROMAN WORLD

Aujac, Germaine. *Strabon et la science de son temps*. Paris: Belles Lettres, 1966.
———. "Une illustration de la sphéropée: L'*Introduction aux phénomènes* de Géminos." *Der Globusfreund* 18–20 (1970): 21–26.

Brendel, Otto J. *Symbolism of the Sphere: A Contribution to the History of Earlier Greek Philosophy*. Leiden: E. J. Brill, 1977.
Dubois, M. *Examen de la Géographie de Strabon*. Paris: Imprimerie Nationale, 1891.
Fiorini, Matteo. *Sfere terrestri e celesti di autore italiano oppure fatte o conservate in Italia*. Rome: Società Geografica Italiana, 1899.

Geminus. *Elementa astronomiae*. Edited by C. Manitius. Leipzig: Teubner, 1898.

————. *Introduction aux phénomènes*. Edited and translated by Germaine Aujac. Paris: Belles Lettres, 1975.

Hipparchus. *In Arati et Eudoxi Phaenomena commentariorum libri tres*. Edited by C. Manitius. Leipzig: Teubner, 1894.

Mette, Hans Joachim. *Sphairopoiia: Untersuchungen zur Kosmologie des Krates von Pergamon*. Munich: Beck, 1936.

Neugebauer, Otto. *A History of Ancient Mathematical Astronomy*. New York: Springer-Verlag, 1975.

Prontera, Francesco. "Prima di Strabone: Materiali per uno studio della geografia antica come genere letterario." In *Strabone: Contributi allo studio della personalità e dell'opera*. Università degli Studi, Perugia. Rimini: Maggioli, 1984.

Stevenson, Edward Luther. *Terrestrial and Celestial Globes: Their History and Construction, Including a Consideration of Their Value as Aids in the Study of Geography and Astronomy*. 2 vols. Publications of the Hispanic Society of America, no. 86. New Haven: Yale University Press, 1921; reprinted New York and London: Johnson Reprint Corporation, 1971.

Strabo. *The Geography of Strabo*. 8 vols. Edited and translated by Horace Leonard Jones. Loeb Classical Library. Cambridge: Harvard University Press; London: William Heinemann, 1917–32.

————. *Géographie*. Edited by François Lasserre, Germaine Aujac, et al. Paris: Belles Lettres, 1966–.

Walbank, Frank William. *A Historical Commentary on Polybius*. 3 vols. Oxford: Clarendon Press, 1957–79.

11 · The Culmination of Greek Cartography in Ptolemy

O. A. W. DILKE
WITH ADDITIONAL MATERIAL SUPPLIED BY THE EDITORS

By the time of Marinus of Tyre (fl. A.D. 100) and Claudius Ptolemy (ca. A.D. 90–168), Greek and Roman influences in cartography had been fused to a considerable extent into one tradition. There is a case, accordingly, for treating them as a history of one already unified stream of thought and practice. Here, however, though we accept that such a unity exists, the discussion is focused on the cartographic contributions of Marinus and Ptolemy, both writing in Greek within the institutions of Roman society. Both men owed much to Roman sources of information and to the extension of geographical knowledge under the growing empire: yet equally, in the case of Ptolemy especially, they represent a culmination as well as a final synthesis of the scientific tradition in Greek cartography that has been traced through a succession of writers in the previous three chapters.

The remarkable influence of Ptolemy on the development of European, Arabic, and ultimately world cartography can hardly be denied.[1] Through both the *Mathematical Syntaxis* (a treatise on mathematics and astronomy in thirteen books, hereafter called the *Almagest*)[2] and the *Geography* (in eight books), it can be said that Ptolemy tended to dominate both astronomy and geography—and hence their cartographic manifestations—for over fourteen centuries. It is true that during the period from the second century A.D. to the early fifteenth century Ptolemy's geographical writings exerted relatively little influence on Western cartography, though they were known to Arab astronomers and geographers.[3] The *Almagest*, although translated into Latin by Gerard of Cremona in the twelfth century, appears to have had little direct influence on the development of cartography. With translation of the text of the *Geography* into Latin in the early fifteenth century, however, the influence of Ptolemy was to structure European cartography directly for over a century. In the history of the transmission of cartographic ideas it is indeed his work, straddling the European Middle Ages, that provides the strongest link in the chain between the knowledge of mapping in the ancient and early modern worlds.

Notwithstanding his immense importance in the study of the history of cartography, Ptolemy remains in many respects a complicated figure to assess. Many questions about his work remain unanswered. Little is known about Ptolemy the man, and neither his birthplace nor his dates have been positively established.[4] Moreover, in relation to the cartographic component in his writings, we must remember that no manuscript earlier than the twelfth century A.D. has come down to us, and there is no adequate modern translation and critical edition of the *Geography*.[5] Perhaps most serious of all for the student of mapping, however, is the whole debate about the true authorship and provenance of the general and regional maps that accompany the several versions of the Byzantine manuscripts (pp. 268–74 below). Al-

---

1. Some existing authorities tend to eulogize Ptolemy excessively by viewing his work out of context of the sources available to him. See, for example, Lloyd A. Brown, *The Story of Maps* (Boston: Little, Brown, 1949; reprinted New York: Dover, 1979), chap. 3, esp. 79–80. Others criticize his "egregious errors": see R. R. Newton, *The Crime of Claudius Ptolemy* (Baltimore: Johns Hopkins University Press, 1977); cf. note 31 below.

2. Ptolemy, *Ptolemy's Almagest*, trans. G. J. Toomer (London: Duckworth, 1984); Toomer translates the Greek title as "mathematical systematic treatise." The Arabic title given as al-mjsty (consonantal skeleton only) comes from a Greek form, μεγίστη, "the greatest [treatise]" see Toomer *Almagest*, 2. All translations from the *Almagest* appearing in this text are taken from Toomer's edition.

3. See volume 2 of the present *History*.

4. G. J. Toomer, "Ptolemy," in *Dictionary of Scientific Biography*, 16 vols., ed. Charles Coulston Gillispie (New York: Charles Scribner's Sons, 1970–80), 11:186–206, esp. 186–87.

5. Editions of Ptolemy's *Geography* include: *Claudii Ptolemaei Geographia*, 3 vols., ed. C. F. A. Nobbe (Leipzig: C. Tauchnitz, 1843–45), reprinted in one volume with an introduction by Aubrey Diller (Hildesheim: Georg Olms, 1966); *Claudii Ptolemaei Geographia*, 2 vols. and tabulae, ed. Karl Müller (Paris: Firmin-Didot, 1883–1901); *Claudii Ptolemaei Geographiae Codex Urbinas Graecus 82*, 2 vols. in 4, ed. Joseph Fischer, Codices e Vaticanis Selecti quam Simillime Expressi, vol. 19 (Leiden: E. J. Brill; Leipzig: O. Harrassowitz, 1932); and *Geography of Claudius Ptolemy*, trans. Edward Luther Stevenson (New York: New York Public Library, 1932). Because of the complexity and often technical nature of the *Geography*, editions vary substantially in coverage and quality and therefore no single edition was selected for use in the *History*. The Nobbe and Müller editions were consulted for making translations; Stevenson, the only complete English edition, is in many respects inadequate and was not used. Unless otherwise indicated, the translator for all quotations from the *Geography* is James Lowe, Ph.D. candidate (1985) at the University of Wisconsin-Madison. When appropriate or necessary, specific editions are cited in the footnote.

though Bagrow, Crone, and other authors claim it cannot be established whether maps were drawn in connection with the *Geography* in the second century A.D., a rereading of the early Greek version may demonstrate that such maps existed.[6] There is as yet no general agreement on this question, and it illustrates how the whole subject of Ptolemy's place in cartographic development—over the long period of his influence—must be handled with caution. Here we try to bypass the wide swaths of speculation in the earlier literature and to concentrate on reconstructing Ptolemy's work directly from the textual evidence. In particular, we will examine Ptolemy's review of the mapping of Marinus of Tyre, the instructions Ptolemy provides for drawing celestial globes and terrestrial maps, and the likely content of his own maps (if they existed) as inferred from the evidence of the coordinate tables and the maps in the Greek manuscripts of the thirteenth and fourteenth centuries.

UPDATING THE WORLD MAP: PTOLEMY'S CRITICISM OF MARINUS OF TYRE

As the Roman world continued to expand its territorial influence during the first century A.D., and as previous gains were consolidated into the administration of the empire, there could have been some pressure on scholars and administrators to update those maps that were used by the bureaucracy or displayed in public places. The flow of new geographical knowledge can be traced to both military and commercial enterprises. The fleet of Gnaeus Iulius Agricola (A.D. 40–93) had sailed around the British Isles, and it was claimed that the island of Thule had been seen in the distance;[7] it was in fact Mainland, the largest of the Shetland Isles. The campaigns against Germany or Dacia, and the Roman explorations into central Africa or to the sources of the Nile in Ethiopia, had likewise made areas familiar that had previously been considered far outside the inhabited world. Or yet again, by the age of Ptolemy, Chinese merchants were exporting silk to Rome and to other parts of Europe, either by land through Asia, or by sea through the Indian Ocean and the Persian Gulf or the Red Sea.[8] The potential sources for the mapmaker were thus greatly enriched. In the Roman world, just as much as in the Age of Great Discoveries, in a society that was already familiar with maps (see chap. 12 below), this new information in turn created incentives for revising maps to accord with the new knowledge of reality.

It is in this general context of an expanding world that we can place the specific attempts of Marinus of Tyre to modify existing maps from new discoveries. Little is known about Marinus, but the busy Phoenician port from which he originated, maintaining extensive

commercial contacts across the known world, suggests some of the channels by which new knowledge may have reached him.[9] Ptolemy devoted a great deal of space in the *Geography* to a thorough criticism of Marinus's work, describing him as the "latest," in the sense of the most recent, "of the contemporary geographers"[10] and later drawing extensively on his materials in compiling his own *Geography*. The importance of Marinus in the process of updating the world map, although he was not the first to attempt this task, lay in his critical approach to existing maps, even those he had compiled himself, which he revised as new information became available. As Ptolemy suggested in book 1 of the *Geography*, much of Marinus's working life was devoted to this task: "He obviously included many accounts in addition to those already known before his time. He also considered worthy of correction those accounts which both he (the first time) and others had carelessly trusted; this we can see from his editions (of which there are many) of the correction of the map."[11] This passage should not be read to imply that Ptolemy was uncritical of Marinus, and subsequent chapters are devoted to correcting, or to making more intelligible, the text that confronted him. Yet if Marinus was sometimes obscure, Ptolemy revealed himself to be a true cartographer, almost in the modern definition of that word, by focusing primarily on the techniques by which maps were compiled rather than solely on their geographical content. As a result, there emerges from Ptolemy's critique of Marinus a clear perception of three major cartographic problems confronting the mapmaker of that age.

The first of these problems, in a long lineage of Greek work, concerned the size and position of the inhabited world. For these calculations Marinus had adopted, largely uncritically, 180,000 stades as the value for the circumference of the earth. He simply said that "one

6. See below, pp. 189–90; Gerald R. Crone, *Maps and Their Makers: An Introduction to the History of Cartography*, 5th ed. (Folkestone, Kent: Dawson; Hamden, Conn.: Archon Books, 1978), 3; Leo Bagrow, *History of Cartography*, rev. and enl. R. A. Skelton, trans. D. L. Paisey (Cambridge: Harvard University Press; London: C. A. Watts, 1964), 34–37.

7. Tacitus *Agricola* 10; see *The Agricola [and] The Germania*, trans. Harold Mattingly, rev. S. A. Handford (Baltimore: Penguin Books, 1970).

8. John Ferguson, "China and Rome," in *Aufstieg und Niedergang der römischen Welt*, ed. Hildegard Temporini and Wolfgang Haase (Berlin: Walter de Gruyter, 1972–), 2.9.2 (1978): 581–603; Manfred G. Raschke, "New Studies in Roman Commerce with the East," in *Aufstieg und Niedergang* (above), 2.9.2 (1978): 604–1361, and map appendix following.

9. No other Greek writer apart from Ptolemy mentions Marinus of Tyre, nor does any Latin writer; for an Arabic reference see volume 2 of the present *History*.

10. Ptolemy *Geography* 1.6.1 (note 5).

11. Ptolemy *Geography* 1.6.1 (note 5).

part [i.e., degree] contains just about 500 stades,"[12] thus making his measurement the same as the smaller of those estimates ascribed to Posidonius.

According to Marinus, the north-south width of the inhabited world extended from the parallel through Thule, at 63°N, to the parallel through the country of the Ethiopians, named Agisymba, and the promontory of Prasum.[13] This southerly parallel, which Marinus said was below the winter tropic, is in fact the southern tropic, at 24°S. As for the island of Thule, which Pytheas and Eratosthenes located at 66°N, on the polar circle, Marinus did not explain (or at least Ptolemy does not tell us) why he moved it to 63°N. Marinus therefore attributed 87° or 43,500 stades to the latitudinal breadth of the known world. He estimated its length at fifteen hours of longitude[14] between two meridians, that is, 225° or 90,000 stades along the Rhodian parallel, 36°N, on which he had assumed one degree of longitude was about 400 stades.

So the inhabited world according to Marinus occupied well over a quarter of the terrestrial globe. His map also differed greatly from earlier maps in two respects. First, it was drawn on both hemispheres, even though most of it lay in the Northern Hemisphere. Second, the extent of the ocean between the extreme east and west edges of the inhabited world was considerably reduced: it was depicted as 135° longitude or 54,000 stades along the Rhodian parallel, as compared with 225° longitude or 90,000 stades from Spain to China by land.

It is true that Marinus or his source made some astronomical observations, quoted by Ptolemy in *Geography* 1.7.4 and following sections; but Ptolemy dismisses these as inconclusive. Marinus's method was simply to employ the various records of travelers and merchants, by converting into stades the number of days necessary to go by land or sea from one place to another. When the number of stades seemed excessive to him, he arbitrarily reduced it to suit his conceptions. However, Marinus was the first geographer to extend the known world significantly by including in his map the eastern part of Asia, "from the Stone Tower to Sera, the capital of the Seres, a journey of seven months."[15] Likewise, he integrated the part of Africa lying south of the Garamantes (a people living in the Sahara) into the world view, writing about Agisymba, far toward the south and beyond the equator.

A second cartographic problem in which Ptolemy shared an interest with Marinus—and indeed may have built on the foundations he had provided—was that of map projections. Apparently Marinus never completed the final revision of his map of the world, for, as Ptolemy puts it, "he himself says, even in the last edition he has not come to the point of revision in which he corrects the *climata* and the hours."[16] Yet even if this exercise

in mathematical geography did not reach its final cartographic expression, Marinus had nevertheless made a careful study of the problem of representing a portion of the globe on a plane. Like Eratosthenes and Strabo, he adopted a rectangular projection in which the parallels and meridians were all drawn as straight parallel lines, the meridians being perpendicular to the parallels. But unlike Eratosthenes, who had selected only a few parallels and a few meridians at irregular distances, Marinus seems to have used a complete network of parallels and meridians at regular distances from one another (fig. 11.1). In this system all the parallels are the same length: Marinus gave them the length of the parallel through Rhodes. According to Ptolemy, "he kept only the parallel passing through Rhodes proportional to the meridian according to the approximate 4:5 ratio. . . . he had no concern for any of the others with respect to their proportionality or spherical shape."[17] As a result, the distances on the equator fell short by one-fifth of their correct measurement, and the distances on the parallel through Thule were increased by four-fifths.[18] Indeed, Ptolemy stated, "Marinus devoted considerable attention to this and generally found fault with all the systems of the plane-maps; nevertheless he used a system of representation especially unsuitable for keeping distances proportional."[19] Its overall effect was to make Marinus's

---

12. Ptolemy *Geography* 1.7.1 (note 5, Müller edition).

13. As has been previously mentioned, Agisymba refers to central Africa and Cape Prasum is somewhere near Zanzibar, south of Rhapta (possibly Cape Delgado).

14. One hour is fifteen degrees of longitude; but one degree of longitude is equal to 500 stades on the equator (if the circumference of the earth is taken as 180,000 stades) and only 400 stades on the Rhodian parallel.

15. Ptolemy *Geography* 1.11.3 (Müller edition), 1.11.4 (Nobbe edition) (note 5). Sera is the capital of the silk country, China. There is some discrepancy, even within Ptolemy, as to the exact location of the Stone Tower; see J. Oliver Thomson, *History of Ancient Geography* (Cambridge: Cambridge University Press, 1948; reprinted New York: Biblo and Tannen, 1965), 307–9.

16. Ptolemy *Geography* 1.17.1 (note 5, Müller edition).

17. Ptolemy *Geography* 1.20 (note 5). Müller, in his edition, correctly explains ἐπιτέταρτος here as 4:5. The literal sense (cf. ἐπίτριτος, ἐπίπεμπτος) is "a quarter in addition," which could mean 1¼:1 or, as here, 1:1¼. *A Greek-English Lexicon*, 2 vols., comp. Henry George Liddell and Robert Scott, rev. and augmented Henry Stuart Jones (Oxford: Clarendon Press, 1940), translates the adjective wrongly as "ratio of 4:3." (However the Supplement to these volumes, published in 1968, amends the statement to read "ratio of 5:4"; p. 61.) E. L. Stevenson, drawing on a Renaissance Latin translation, brings in a nonexistent character Epitecartus. What Ptolemy means is that Marinus treated the whole world, for the sake of simplification, as if it were like the area around Rhodes, where a degree of longitude was taken as 400 stades, of latitude as 500.

18. Ptolemy *Geography* 1.20.7 (note 5); see also Armando Cortesão, *History of Portuguese Cartography*, 2 vols. (Coimbra: Junta de Investigações do Ultramar-Lisboa, 1969–71), 1:98 n. 52.

19. Ptolemy *Geography* 1.20.3 (note 5).

maps of the inhabited world misleading "and, in many cases, they [the editors following Marinus] go far astray from the general consensus because of the inconvenient and disjointed nature of the directions, as any experienced person can see."[20]

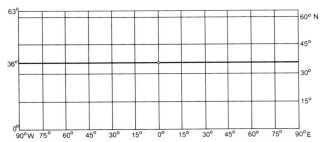

FIG. 11.1. MARINUS'S PROJECTION RECONSTRUCTED FROM PTOLEMY'S DESCRIPTION. Marinus appears to have used a complete network of parallels and meridians in which all the parallels were the same length as that of Rhodes, thereby introducing considerable deformation.
After Otto Neugebauer, *A History of Ancient Mathematical Astronomy* (New York: Springer-Verlag, 1975), fig. 68.

A third and final problem in Marinus's maps related to errors accumulated in the compilation of geographical detail from written commentaries. Ptolemy had discovered that, as a result of uncritical copying, some of the commentaries could not be satisfactorily collated with the content of the maps. He explained that "the constant transfer [of data] from earlier to later models brings about gradual change that usually culminates in a vast discrepancy" and added that many of those working with Marinus's map did not use the latest edition.[21] It seems that mapmakers in Ptolemy's day—and not only Marinus in this respect—usually worked in isolation, incorporating at random the modifications required in their maps as a result of the growth of geographical knowledge.

A combination of such inaccuracies led Ptolemy to reject Marinus's work as a cartographer. As already noted, he believed the information in many of Marinus's maps was neither coherent nor practical. For example, in one work Marinus might correct the latitudes only, in another the longitudes; but the places taken into account were not the same in both works, so that it was difficult to find a place with both sets of coordinates correctly rendered. Consequently Ptolemy—although he evidently made extensive use of Marinus's material—regarded the drawing of a map according to Marinus's commentaries as a hopeless undertaking.

PTOLEMY'S INSTRUCTIONS FOR MAPMAKING

Ptolemy's most crucial legacy to the long-term development of cartography is the instructions he codified as

to how maps of various types should be drawn. These instructions are scattered in various texts, but if brought together they may be said to constitute a technical manual of some sophistication for would-be mapmakers. Moreover, in association with such theoretical guidance, Ptolemy also compiled at length the empirical substance for the content of such maps. As is well known, these survive in the form of coordinate lists of both celestial and terrestrial positions, so that if Ptolemy stopped short of drawing maps or having them drawn for him, which now appears unlikely,[22] he at least left sufficient materials for their construction by others. Indeed, his work was so unambiguously cartographic in its intention that the absence of graphic records would do nothing to diminish its interest for the history of cartography. To make a simple analogy to modern cartographic data bases, we might say that Ptolemy transmitted his cartographic knowledge in digital rather than graphic form, leaving his successors to recreate the images he so clearly envisaged as the end product of the mapping process.

While it is generally believed that Ptolemy was born in Upper Egypt and subsequently lived in Alexandria, he is known to us mainly through his various writings surviving first in a number of Byzantine recensions.[23] The traditional literature of the history of cartography—with its emphasis on geographical maps—has tended to overlook the fact that Ptolemy was a polymath, ranging over topics as diverse as astronomy, mathematics, physics, optics, harmonics, chronology, and geography. As Cortesão has pointed out, however, quite a few of these works contain material relevant to his interests in mapmaking. For example, Ptolemy's *Analemma* deals with the theory of the gnomon and orthographic projection; *Planisphaerium* examines stereographic projection; and *Tetrabiblos*, the large treatise on astrology, also has references to geography.[24] It is, however, through the *Almagest* and through the *Geography*, which in certain manuscripts of the Byzantine recensions contain world or regional maps or both, that his influence on cartographic development was largely transmitted.

So interrelated are the concepts and facts in these last two works that in the history of cartography they have to be considered together. In the *Almagest* Ptolemy taught how to draw a celestial globe; in the *Geography*, how to draw the map of the inhabited world on a globe (said to be simple and similar to the mapping of the celestial sphere) or on a plane surface. In both works he

20. Ptolemy *Geography* 1.18.3 (note 5).

21. Ptolemy *Geography* 1.18.3 (Müller edition), 1.18.2–3 (Nobbe edition) (note 5).

22. See below, pp. 189–90.

23. See below, pp. 268–72.

24. Cortesão, *History of Portuguese Cartography*, 1:92–93 (note 18), and Toomer, "Ptolemy" (note 4).

presented a complete series of coordinates. Since these constitute the only sets of coordinates to survive from classical antiquity, they can justly be said, despite certain imperfections, to mark a critical datum line in the development of celestial and terrestrial cartography.

THE CELESTIAL GLOBE IN THE *ALMAGEST*

Ptolemy, who made astronomical observations in Alexandria between A.D. 127 and 141, was in the first instance an astronomer. His major purpose was to gather as much information as possible and to organize it into an exhaustive synthesis that could be used as an essential tool by every student in this field. So he started by composing a treatise in which he studied all the problems concerning the motion of the celestial bodies, and the relation between the motionless earth and the moving sphere of the sky.

A systematic star catalog, following certain technical rules, is required to make a celestial map or globe.[25] Ptolemy's catalog was derived from that of Hipparchus, and like Hipparchus, according to Toomer, Ptolemy describes the stars "as if they were drawn on the *inside* of a globe, as seen by an observer at the centre of that globe, and facing towards him."[26] All the known stars in the heavens were grouped into constellations: twenty-one constellations lay north of the zodiac, twelve on the zodiac, and fifteen south of it; so forty-eight constellations containing 1,022 stars were listed in the whole catalog. For each star, Ptolemy indicated the longitude and latitude in relation to the ecliptic rather than the equator, so that the positions of the stars would not change owing to the precession of the equinoxes: the latitudes do not vary, and one need only add the values of the precession for any time to find the longitudes.

In a few cases Ptolemy had to amend the naming of the positions of the stars within the constellations to make them relate more closely to the outlines of human figures or animals usually sketched to represent them:

Futhermore, the descriptions which we have applied to the individual stars as parts of the constellation are not in every case the same as those of our predecessors (just as their descriptions differ from their predecessors'): in many cases our descriptions are different because they seemed to be more natural and to give a better proportioned outline to the figures described. Thus, for instance, those stars which Hipparchus places "on the shoulders of Virgo" we describe as "on her sides" since their distance from the stars in her head appears greater than their distance from the stars in her hands, and that situation fits [a location] "on her side," but is totally inappropriate to [a location] "on her shoulders."[27]

The names Ptolemy gave to the stars were thus slightly different from the forms found in Hipparchus. But as Ptolemy explained: "One has a ready means of identifying those stars which are described differently [by others]; this can be done immediately simply by comparing the recorded positions."[28] Their relative positions were, in fact, to remain unaltered on his celestial globe, and it is only the names that were subject to variation.

Having listed all the stars that he wanted to take into account, Ptolemy explained in great detail how to make a solid sphere as an image of the sky. It was advisable to select a dark globe, its color symbolizing the night sky and allowing the stars to be seen clearly. Two points, diametrically opposed, would indicate the poles of the ecliptic. Two great circles would then be drawn, one of them passing through these poles, the other, perpendicular to it, representing the zodiac (one of the points of intersection is selected as the starting point for graduating the ecliptic into 360 degrees).

Nor did Ptolemy neglect to provide practical, mechanical instructions for the globe maker.[29] It would be convenient, he suggested, to attach two semicircles to the globe (Rings A and B in fig. 11.2). so that the relationship between equatorial coordinates and ecliptic coordinates could be demonstrated.

When the exact place of the star has been located, Ptolemy continued, it should be marked by a yellow point or, for some stars, the colors noted in the star catalog, of a size appropriate to the brightness or the magnitude of the star. As for the figures of the constellations, they should be dimly sketched schematically, hardly visible against the dark background of the sphere, so that they do not conceal the stars.[30]

Thus Ptolemy's celestial globe differed greatly from those described by Eudoxus or Aratus or carried by the Farnese Atlas. In those earlier periods, astronomers preferred to group stars into constellations so as to be able to name and identify them, hence the emphasis on the outlines of the constellations. By Ptolemy's day, however, the identification of stars had become less dependent on the constellations, for they could be located—

25. Ptolemy *Almagest* 7.5–8.1 (note 2), contains the tabular layout of the constellations.

26. Toomer, *Ptolemy's Almagest*, introduction, p. 15 (note 2).

27. Ptolemy *Almagest* 7.4 (note 2).

28. Ptolemy *Almagest* 7.4 (note 2).

29. For a full account of the instruments Ptolemy describes, see D. R. Dicks, "Ancient Astronomical Instruments," *Journal of the British Astronomical Association* 64 (1954): 77–85.

30. Ptolemy *Almagest* 8.3 (note 2). For a detailed technical interpretation of some of the Greek at this point, see the translation and notes in Toomer, *Ptolemy's Almagest*, 404 n. 179 and 405 nn. 180, 181 (note 2).

on the sphere as in the catalog—by giving the coordinate positions of each individual star.[31]

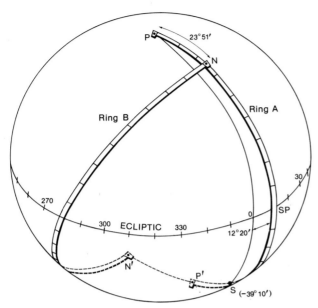

FIG. 11.2. PTOLEMY'S INSTRUCTIONS FOR CONSTRUCTING A STAR GLOBE. The *Almagest* contains explicit instructions for constructing a celestial globe from ecliptic coordinates. Ring A, on the axis PP′, is fixed at the solstitial point, 12°20′E of the meridian of Sirius, which represents its position at that time (first year of the principate of Antonius Pius, A.D. 137). This fixes the axis of the sidereal coordinate system with Sirius as the reference star. Equatorial coordinates could be mechanically converted from the ecliptic coordinates with the help of ring B, which rotates freely around the axis NN′, which is 23°51′ distant from the axis PP′ to allow for the obliquity of the ecliptic.
After Otto Neugebauer, *A History of Ancient Mathematical Astronomy* (New York: Springer-Verlag, 1975), fig. 79.

THE *CLIMATA* IN THE *ALMAGEST*

In book 2 of the *Almagest*, Ptolemy turned to a standard problem in mathematical geography: establishing the position of the inhabited world on the terrestrial globe, and its relation to the celestial sphere, together with the distribution of the *climata*. First of all, he asserted that "our part of the inhabited world is approximately bounded by one of the two northern quarters."[32] Then he decided to compute systematically the celestial phenomena relating to several parallels of the northern hemisphere, noting for each one data such as the height of the pole above the horizon, the ratios of gnomon to shadow on solstitial and equinoctial days, and the length of the longest day. All these values were obtained by calculation (not by observation), as Ptolemy explained in later chapters.

Ptolemy's tables were almost certainly inspired by the table of *climata*[33] drawn up by Hipparchus, in which similar information could have been found. But whereas Hipparchus dealt with parallels spaced at one degree (or 700 stades) apart, Ptolemy used in his calculations a difference of one-quarter of an hour (or sometimes half an hour or even an hour) in the length of the longest day from one parallel to the next. This implies that the parallels employed may not have been equidistant, and hence Ptolemy was obliged to increase the difference of time for the northern parallels.

Curiously enough, Ptolemy mentioned the traditional hypothesis of a probable inhabited world along the terrestrial equator, on the grounds that these latitudes enjoy a milder climate than the areas near the tropics. Yet at the same time he accepted that "what these inhabited regions are we have no reliable grounds for saying. For up to now they are unexplored by men from our part of the inhabited world, and what people say about them must be considered guesswork rather than report. In any case, such, in sum, are the characteristics of the parallel beneath the equator."[34] From the equator to the polar circle, Ptolemy listed thirty-three parallels: the equator is first, followed by the parallel with a 12¼-hour longest day, at 4¼°N, which was assigned to Taprobane (Sri Lanka); the last in the series, with a 24-hour longest day, at approximately 66⅙°N, was not related to a known country. The last parallel to relate to the known inhabited world was the 21-hour parallel, at 64½°N, in the location of unknown Scythian tribes; it was next to the parallel through Thule (day of 20 hours, at 63°N). Beyond the 24-hour parallel, Ptolemy referred to the parallels representing the longest day of one to six months at monthly intervals, the last, of course, being under the pole. It may be noted that the countries or towns related to each parallel were either the traditional ones, for the central part of the map, or, for the southern or northern parts, places that are difficult to identify. In any case, such places were used only for reference in relation to the *climata*.

---

31. Various scholars, and lately Newton, *Crime of Claudius Ptolemy* (note 1), have accused Ptolemy of not having made the observations he claims to have made and of having falsified records obtained by his predecessors in order to bolster his own theories. But in introducing his star catalog, Ptolemy does not conceal his debt to his predecessors, or at any rate to Hipparchus, even if he claims to have verified their observations; no doubt he used celestial globes, quite usual in his time, to choose coherent sets of coordinates. His originality consists in having preferred ecliptic coordinates to equatorial ones and in having provided complete sets of coordinates so that anyone should be able to draw constellations on a globe.

32. Ptolemy *Almagest* 2.1 (note 2).

33. Ernst Honigmann, *Die sieben Klimata und die* πόλεις ἐπίσημοι (Heidelberg: Winter, 1929).

34. Ptolemy *Almagest* 2.6 (note 2).

In other chapters of book 2, Ptolemy selected only a few *climata* to draw up tables of astronomical phenomena according to latitude. In book 2, chapter 8, eleven *climata* are enumerated, from the equator (12-hour) to the Tanais (Don) River (17-hour), with a regular increase of half an hour in the length of the longest day. Later in book 2, chapter 12, he reduced to seven the number of selected *climata*, and these indeed reappear frequently in the *mappaemundi* of the later Middle Ages:
—the 13-hour parallel, through Meroë
—the 13½-hour parallel, through Syene
—the 14-hour parallel, through Lower Egypt
—the 14½-hour parallel, through Rhodes
—the 15-hour parallel, through the Hellespont
—the 15½-hour parallel, through the central part of the Black Sea
—the 16-hour parallel, through the river Borysthenes.

In the commentary on the last (astronomical) table, cited above from the *Almagest*, Ptolemy announced his project of composing a *Geographike hyphegesis* (Manual of geography), now usually known as the *Geography*:

> Now that the treatment of the angles [between ecliptic and principal circles] has been methodically discussed, the only remaining topic in the foundations [of the rest of the treatise] is to determine the coordinates in latitude and longitude of the cities in each province which deserve note, in order to calculate the [astronomical] phenomena for those cities. However, the discussion of this subject belongs to a separate, geographical treatise, so we shall expose it to view by itself [in such a treatise], in which we shall use the accounts of those who have elaborated this field to the extent which is possible. We shall [there] list for each of the cities its distance in degrees of that meridian from the meridian through Alexandria, to the east or west, measured along the equator (for that [Alexandria] is the meridian for which we establish the times of the positions [of the heavenly bodies]).[35]

Ptolemy was to wait some twenty years before executing his project.

THE *GEOGRAPHY*

As with the *Almagest*, there is no doubt that the *Geography* was deliberately planned as a manual for mapmakers. In its opening paragraph Ptolemy explains its scope by defining "geography" as "a graphic representation of the whole known part of the world, along with the things occurring in it."[36] Later he explains the difference between geography and chorography thus: "The aim of chorography is a consideration of the parts, as would be the case for someone depicting [i.e., painting or drawing] just the ear or eye; but the aim of geography is a consideration of the whole, as it is for those (to use

the same analogy) who depict the entire head."[37]

In drawing up his catalog of stars, Ptolemy had simply gathered all the available information and arranged it into a systematic table of coordinates enabling anyone to make a celestial globe. In like manner in his *Geography*, he collected information from his predecessors, especially from the most immediate, Marinus of Tyre,[38] and arranged it within a systematic table of coordinates. Thus, Ptolemy believed, it would be easy for anyone to draw a map of the inhabited world, or regional maps with the main towns and characteristic features of the countries.

In outlining his aim to provide mapmakers with an appropriate tool in handy form, Ptolemy seems to have been fully aware of deficiencies in some of his information. This is shown in his declaration: "But as for [the degrees of latitude and longitude of] places not visited in this manner, it is advisable, because of the scarcity and uncertainty of the accounts, to base the reckoning more completely upon the proximity of reliably known positions or configurations, so that none of the things inserted to fill up the whole world may have an undefined place."[39] It is clear that Ptolemy believed it was preferable for mapmakers to locate as many places as possible in the known world, even where the authority for this location was shaky, perhaps recognizing intuitively that only thus would such maps eventually be challenged and become more complete.

The contents of the *Geography* are as follows:

Book 1 Introduction, including map projections and criticism of Marinus.
Book 2 Ireland, Britain, the Iberian Peninsula, Gaul, Germany, the upper Danube provinces, Dalmatia.
Book 3 Italy and adjacent islands, Sarmatia in Europe, the lower Danube provinces, Greece and adjacent areas.
Book 4 North Africa (west-east), Egypt, interior Libya (Africa), Ethiopia.
Book 5 Asia Minor, Armenia, Cyprus, Syria, Palestine, Arabia Petraea, Mesopotamia, Arabia Deserta, Babylonia.
Book 6 The former Persian empire apart from areas already covered (west-east); the Sacae and Scythia bordering on that empire.
Book 7 India, the Sinae, Taprobane, and adjacent areas. Summary of world map. Description of armillary sphere including the map of the inhabited earth. Summary of regional sections.
Book 8 Brief survey of the twenty-six regional maps.

35. Ptolemy *Almagest* 2.13 (note 2).
36. Ptolemy *Geography* 1.1.1 (note 5, Müller edition).
37. Ptolemy *Geography* 1.1.2 (note 5).
38. Ptolemy *Geography* 1.6–7 (note 5).
39. Ptolemy *Geography* 2.1.2 (note 5).

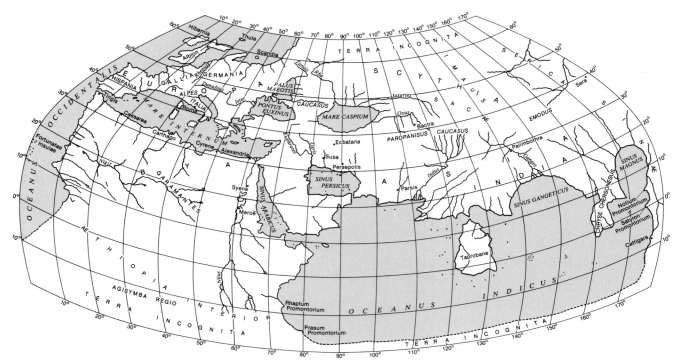

FIG. 11.3. A RECONSTRUCTION OF THE WORLD OF
CLAUDIUS PTOLEMY. After Edward Herbert Bunbury, *A
History of Ancient Geography among the Greeks and Romans*

from the Earliest Ages till the Fall of the Roman Empire, 2d
ed., 2 vols. (1883; republished with a new introduction by
W. H. Stahl, New York: Dover, 1959), map facing p. 578.

In the case of this survey of the regional maps in book
8, standard information is given for each map. In turn,
Ptolemy indicates the proportion of one degree on the
central parallel of the map in question to one degree of
meridian; he describes roughly the outlines of the map;
and then he locates the main towns by a pair of coordinates. But he expresses latitude in terms of length of
the longest day and longitude in number of hours east
or west of the Alexandrian meridian.[40]

SIZE AND DIMENSIONS OF THE INHABITED WORLD IN
THE *GEOGRAPHY*

Ptolemy criticized Marinus for having extended the inhabited world too far. Accepting the Thule parallel
(63°N) as the northern limit, he rejected the southern
tropic (24°S) as the southern limit and located the region
of Agisymba and the promontory of Prasum, the farthest
known countries, on the parallel opposite the one
through Meroë (which one may call Anti-Meroë), at 16°
25′S or about 8,200 stades south of the equator (fig.
11.3). Thus the whole latitudinal extent of the inhabited
world was reduced to 79°25′ or nearly 40,000 stades
(Ptolemy used the value he had borrowed from Marinus
of 180,000 stades for the circumference of the earth).

Similarly, the length of the inhabited world was reduced from Marinus's fifteen-hour longitude to twelve
hours or 180°, from the Fortunate Isles (the Canaries)
in the extreme west to Sera and Cattigara in the extreme
east.[41] Ptolemy claimed to have accomplished this reduction by examining and comparing land and sea journeys, but it is likely that he relied more on guesswork
than on sound calculation. So he fixed the length of the
inhabited world at 72,000 stades, calculated along the
36°N parallel passing through Rhodes, on which one
degree of longitude was reckoned to be 400 stades.[42]
For the distance between the Fortunate Isles and the
Euphrates, Ptolemy indicated 72° or 28,800 stades;[43]
from the Euphrates to Sera or Cattigara, 105°15′ or
42,100 stades via the Stone Tower.[44]

---

40. In the *Almagest* 2.13 a prime meridian of Alexandria is proposed, but when Ptolemy came to write the *Geography,* the extreme
westerly meridian of the Fortunate Isles was preferred—at least in
books 2–7—to allow all longitudes to be expressed as east of this line.
A remnant of the earlier system is found in the *Geography,* book 8.
See Toomer, *Ptolemy's Almagest,* 130, n. 109 (note 2).
41. Ptolemy *Geography* 1.11 (note 5). Cattigara, south of Sera, may
be somewhere near the modern city of Hanoi, although other theories
have been advanced; see pp. 198–99 below.
42. Ptolemy *Geography* 1.11.1, 1.12.10 (note 5).
43. Ptolemy *Geography* 1.12 (note 5).
44. Ptolemy *Geography* 1.12.9 (note 5).

By such calculations in his *Geography*, Ptolemy thus accepted that the inhabited world extended south of the equator in latitude; in longitude, as proposed by Posidonius, it now constituted one-half of the Northern Hemisphere.[45] On the whole, in spite of his criticism of Marinus's map, Ptolemy adopted most of the information transmitted by him, modifying the sections necessary to fit his own concept of the size of the inhabited world: "But were we to find nothing lacking from his last arrangement, it would have been sufficient for us to construct a map of the inhabited world just from these commentaries, and waste no time on anything else."[46]

MAP PROJECTIONS

Since it was usual to draw the map of the inhabited world on a plane surface, Ptolemy examined different types of map projection and their ability to maintain the characteristics of a sphere. With hindsight it may be said that this was perhaps his most vital contribution to the long-term development of the mathematical basis of mapmaking. Ptolemy, revealing a clear conceptual insight into the fundamental problem of map projections, writing of globes and of flat maps respectively, says:

> Each of these conceptual systems would have its advantages. The first system, which locates the map on a sphere, obviously preserves the likeness of the world's shape and obviates the need for any manipulation of it; on the other hand, it hardly provides the size necessary for containing most of the things that must be set in place, nor can it let the entire map be seen from one vantage point: instead, one must move either one's own eyes or the sphere in order to view the rest.
>
> The second system, representation on a flat surface, avoids the aforementioned shortcomings altogether. But it lacks some sort of method for preserving the likeness of the spherical shape whereby it might make the distances recorded on its flat surface as proportional as possible to the true distances.[47]

The *Geography* contains a detailed exposition of four systems of map projection: (1) a projection with straight and perpendicular parallels and meridians, like that of Marinus; (2) a projection with straight converging meridians and curved parallels; (3) a projection with curved converging meridians and curved parallels; and (4) a special projection of the globe as viewed from a distance.[48] This whole subject has generated a substantial literature since the nineteenth century, much of it mathematically confusing, with an emphasis on the modification of the Ptolemaic map projections during the European Renaissance.[49] Ptolemy's explanations certainly need to be handled with caution, and there is the danger, as Keuning argues, of defining his methods too literally in terms more appropriate to the formality of modern projections.[50]

Marinus's Projection

Marinus had selected for his world map what could be defined (in modern terms) as a rectangular projection, represented in a graticule by straight parallel meridians and straight parallels orthogonal to the meridians, forming a grid of rectangles. The scale along the parallel of Rhodes (36°N) and along all meridians was assumed to be constant. Marinus had also assumed the length of the parallel of Rhodes—the central parallel of the inhabited world—to be approximately four-fifths the length of the equator (and thus also of any meridian great circle); Ptolemy modified this slightly by expressing the proportion of the length of a degree of longitude on the central parallel to a degree of latitude on the meridian as being 93:115, which very closely approximates $\cos 36° = 0.809$.

This method of projection was to be used for some versions of the regional or provincial maps that were attached to the later texts of the *Geography*. But Ptolemy rejected the system for the world map on the grounds that the various parallels appear in its construction to be of the same length, causing severe deformation away from the central parallel. He calculated the proportion of the length of the parallel of Thule to the length of the equator, for example, as 52:115 ($\cos 63° = 0.454$), yet it is represented by a line of the same length as the equator on the Marinus projection.[51]

Ptolemy's First Projection

To overcome this disadvantage, Ptolemy devised a system of projection, usually called his first projection, in which the meridians were to be drawn as straight lines from a theoretical point (not the North Pole) and the parallels as arcs of a circle with the same point as center. This, in fact, was the projection usually employed for

45. Strabo *Geography* 2.3.6; see *The Geography of Strabo*, 8 vols., trans. Horace Leonard Jones, Loeb Classical Library (Cambridge: Harvard University Press; London: William Heinemann, 1917–32). Also see Strabo, *Géographie*, ed. François Lasserre, Germaine Aujac, et al. (Paris: Belles Lettres, 1966–).

46. Ptolemy *Geography* 1.6.2 (note 5).

47. Ptolemy *Geography* 1.20.1–2 (note 5).

48. In identifying these four systems we follow Otto Neugebauer, *A History of Ancient Mathematical Astronomy* (New York: Springer-Verlag, 1975), 879–959. See also Cortesão, *History of Portuguese Cartography*, 1:97–109 (note 18), who gives a full summary with extensive quotations of Ptolemy's ideas from the *Geography*.

49. These projections will be dealt with fully in the present *History*, volume 3.

50. See Johannes Keuning, "The History of Geographical Map Projections until 1600," *Imago Mundi* 12 (1955): 1–24, who writes (p. 9), "The Ptolemaic projections appear very like conical projections, though they are not. In antiquity there was no question of projecting on a cone or on a cylinder."

51. Ptolemy *Geography* 1.20 (note 5).

constructing the map of the inhabited world associated with the later manuscripts. Its advantage over Marinus's projection was not only that it maintained constant scale along the central parallel (Rhodes) and the meridians (as had Marinus's) but that the proportion of the length of the parallel of Thule to the length of the equator was also correct. This scale could not, of course, be the same as along the parallel of Rhodes, but since this represented the traditional central parallel, and so many distances were known along it, Ptolemy scaled the whole map to it:

> Since it is impossible for all of the parallels to keep the proportion that there is in a sphere, it will be quite sufficient to observe this proportion in the parallel circle running through Thule and the equinoctial, in order that the sides of our map that represent latitude may be proportionate to the true and natural sides of the earth.
>
> The parallel passing through Rhodes must be inserted because on this parallel very many proofs of distances have been registered, and inserted in right relation to the circumference of the greatest circle, following in this Marinus, who gave the ratio for the equal circumferences of the equator (and the meridians) to the parallel of Rhodes as 5:4. By thus doing, we shall ensure that the longitude of our earth, which is the better known, will be in right proportion to the latitude.[52]

The frame of the map—following the traditional proportions of the inhabited world—would have to be rectangular in shape, the center of the circles representing the parallels lying outside this framework for the map (fig. 11.4). Within it would have to be drawn thirty-six plus one meridians one-third of an hour of longitude (5°) apart. For the section of the map south of the equator he advised drawing one parallel only in addition to the parallel of Anti-Meroë (see above, p. 184): the parallel passing through Rhapta promontory and Cattigara, at a half-hour distance from the equator, which would be the same length as the parallel opposite at 8°25′N (see table below).

To mark the localities that were to be placed on the map, Ptolemy continued, the mapmaker should take a narrow ruler, equal in length to the radius of the circle used to draw the equator. He should attach it to the point taken as the center of the curved parallels, so that it could be made to coincide with any given meridian. Then, using the graduations in latitude inscribed on the ruler and the graduations in longitude inscribed on the equator, he should quite easily be able to mark the towns or geographical features in their true places.[53] We can see from these details that even in the event Ptolemy may not have illustrated his projections either in an actual map or in the form of a diagram of the meridians and

Parallel		Length of Longest Day	Degrees
		North of Equator	
1		12 hr 15 min	4°15′N
2		12 hr 30 min	8°25′N
3		12 hr 45 min	12°30′N
4	Meroë	13 hr	16°25′N
5		13 hr 15 min	20°15′N
6	Syene	13 hr 30 min	23°50′N
7		13 hr 45 min	27°10′N
8		14 hr	30°20′N
9		14 hr 15 min	33°20′N
10	Rhodes	14 hr 30 min	36°N
11		14 hr 45 min	38°35′N
12		15 hr	40°55′N
13		15 hr 15 min	43°05′N
14		15 hr 30 min	45°N
15		16 hr	48°30′N
16		16 hr 30 min	51°30′N
17		17 hr	54°N
18		17 hr 30 min	56°10′N
19		18 hr	58°N
20		18 hr 30 min	61°N
21	Thule	19 hr	63°N
		South of Equator	
Rhapta		12 hr 30 min	8°25′S
Anti-Meroë		13 hr	16°25′S

parallels, his instructions for future mapmakers were nonetheless quite explicit. At the same time, it is hard to imagine how such precise instructions could have been compiled without resorting to graphic experiments.

Ptolemy's Second Projection

Despite its improvements over Marinus's projection, Ptolemy's simple first projection was not without its drawbacks. First, the north and south portions of the meridians form acute angles at the equator; second, the proportions of the parallels between Thule and the equator are not the same as on the sphere. So Ptolemy proposed a further projection—often known as his second projection—to alleviate these problems.[54] It was to be

52. Ptolemy *Geography* 1.21.2 (note 5), translated by O. A. W. Dilke. The translation in Stevenson, *Geography of Claudius Ptolemy* (note 5), following Latin versions, gives a completely incorrect rendering. For another example of Stevenson's unacceptable translation, see note 103 below.

53. Ptolemy *Geography* 1.24.7 (note 5).

54. Ptolemy *Geography* 1.24.9–20 (note 5). Marie Armand Pascal d'Avezac-Macaya, in his pioneer study, *Coup d'oeil historique sur la projection des cartes de géographie* (Paris: E. Martinet, 1863), gave it the name "homeotheric projection"; it may also be regarded as the ancestor of the Bonne projection. D'Avezac-Macaya's study was originally published as "Coup d'oeil historique sur la projection des cartes de géographie," *Bulletin de la Société de Géographie*, 5th ser., 5 (1863): 257–361, 438–85.

constructed with curved parallels and meridians (fig. 11.5). According to Ptolemy, its aim was to give the lines representing the meridians the appearance they have on the sphere when viewed by an observer looking directly at the center of the map.[55]

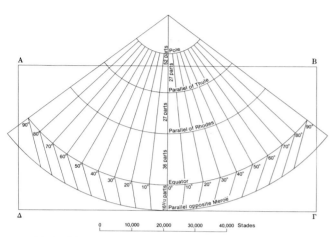

FIG. 11.4. PTOLEMY'S FIRST PROJECTION. The frame of the inhabited world (ΑΒΓΔ) is shown superimposed upon a conic graticule with straight converging meridians and parallels as arcs of circles. Although Ptolemy explained that it was easier to construct and use than his second projection (see fig. 11.5), it did not reflect the spherical shape of the earth as effectively, and only two parallels (as well as all the meridians) maintained their true lengths.
After Erich Polaschek, "Ptolemaios als Geograph," in *Paulys Realencyclopädie der classischen Altertumswissenschaft*, ed. August Pauly, Georg Wissowa, et al. (Stuttgart: J. B. Metzler, 1894–), suppl. 10 (1965): 680–833, fig. 4.

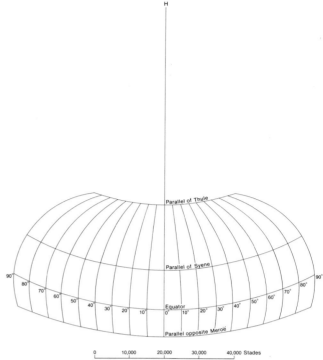

FIG. 11.5. PTOLEMY'S SECOND PROJECTION. Constructed with curved meridians and parallels, this was designed to alleviate some of the problems associated with Ptolemy's first projection (see fig. 11.4). It was especially popular with later editors of the *Geography* in the Renaissance.
After Erich Polaschek, "Ptolemaios als Geograph," in *Paulys Realencyclopädie der classischen Altertumswissenschaft*, ed. August Pauly, Georg Wissowa, et al. (Stuttgart: J. B. Metzler, 1894–), suppl. 10 (1965): 680–833, fig. 5.

The central parallel of the map was designed to run through Syene, at 23°50′ north of the equator. Syene was approximately midway between the parallels of Thule (63°N) and Anti-Meroë (16°25′S). From a center outside the rectangular panel on which the map was to be drawn (H in fig. 11.5), Ptolemy advised that it would be convenient to plot the arcs of the circles representing the main parallels: Thule, Syene, and Anti-Meroë.[56] Second, the thirty-six meridians would be drawn as circular arcs, eighteen on each side of the straight central meridian at five-degree intervals (every one-third hour). Circular meridians were possible, since only three parallels were chosen along which to preserve the true proportions of distances. It was left to later commentators to discover that if the arcs were not circular, all of the parallels in such a projection—not just three—could be drawn preserving their true lengths. Apparently the first person to employ this procedure was Henricus Martellus Germanus on his manuscript world map of about 1490,

now at Yale University. In 1514 Johannes Werner, along with his new translation of book 1 of the *Geography*, added a theoretical discussion of Ptolemy's second projection.[57]

Ptolemy's exposition of his second projection ends on a remarkably pragmatic note. Although he believed it offered a better theoretical solution, the task of drawing the map was rendered more difficult. The curved meridians in particular meant that the geographical details could no longer be plotted, as with the first projection, by the straightforward use of a ruler. Ptolemy thus re-

---

55. Ptolemy *Geography* 1.24.9 (note 5).

56. The parallels would thus be drawn in approximately the following proportions: 5 the equator, 2.25 the Thule parallel, 4.35 the Syene parallel, and 4.4 the Anti-Meroë parallel.

57. Ptolemy, *Geographia*, ed. and trans. Johannes Werner (Nuremburg, 1514). See also the discussion by Neugebauer, *History of Ancient Mathematical Astronomy*, 885–88 (note 48), who does not, however, mention the Martellus map.

tained both methods "for the sake of those who will have recourse to the handier method because of indolence."[58] The recognition that mapmakers sometimes preferred the easier road was prophetic: most of the early scholars attempting to draw a world map from Ptolemy's instructions seem to have preferred the first projection.

Ptolemy's Third Projection

Ptolemy's so-called third projection arises out of his description of the armillary sphere.[59] He mentions that several of his predecessors had attempted to give this demonstration, but inconclusively. It does not seem to have been used in practical map drawing, nor does it appear (unlike the first and second projections) to have influenced the subsequent development of that subject. As Ptolemy describes it: "It is reasonable to add here how the hemisphere in which the inhabited world lies could be represented on a plane surface, with the hemisphere itself being surrounded by an armillary sphere."[60] The aim was to give a plane representation corresponding in some measure to the visual impression of the terrestrial globe in such a way that all the inhabited world could be seen unencumbered by the rings of the armillary sphere. The drawing of its construction in book 7 of later manuscripts of the *Geography* does not entirely clarify the complex exposition in the text.[61] Yet Ptolemy visualized the eye of the viewer situated outside the rings of the imaginary armillary sphere at such a distance that the ring representing the celestial summer tropic would just clear the parallel of Thule on the globe, and the ring representing the celestial equator would just clear the most southerly parallel of the inhabited world (Anti-Meroë). The position of the viewer is thus represented in figure 11.6 by the intersection of the extension of the lines YF and TB. The viewing axis is on a horizontal plane passing through Syene. Figure 11.7 illustrates the concept from the observer's position. In both these diagrams we can see that, in order that the ring on the armillary sphere representing the ecliptic should not obscure the inhabited part of the world, the southern part of the ecliptic should be adjusted on the viewer's side. In his example, in order that all these conditions be met, Ptolemy assumed that the radius of the solid globe was 90 parts, that the ratio of the radius of the armillary sphere to that of the globe must be 4:3, and

58. Ptolemy *Geography* 1.24.22 (Müller edition), 1.24.29 (Nobbe edition) (note 5).

59. Ptolemy *Geography* 7.6.1 (note 5).

60. Ptolemy *Geography* 7.6.1 (note 5, Nobbe edition). Otto Neugebauer, "Ptolemy's *Geography*, Book VII, Chapters 6 and 7," *Isis* 50 (1959): 22–29 gives the best translation and commentary.

61. Ptolemy *Geography* 7.6 and 7.7 (note 5).

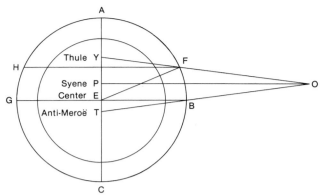

FIG. 11.6. PTOLEMY'S THIRD PROJECTION. This diagram shows the position of the viewer (O) in relation to the position of the globe (center E) within the surrounding armillary ring (ABCG). It thus demonstrates how the inhabited world between the parallels of Thule (Y) and Anti-Meroë (T) can be seen through the armillary rings representing the equator (BG) and the summer tropic (FH). Whereas the armillary sphere itself is seen in perspective, the *oikoumene* is not. Instead, latitudes are preserved along the central meridian. This explains why the line EF does not coincide with PO at the earth's surface, as might otherwise be expected.
After Otto Neugebauer, *A History of Ancient Mathematical Astronomy* (New York: Springer-Verlag, 1975), fig. 78.

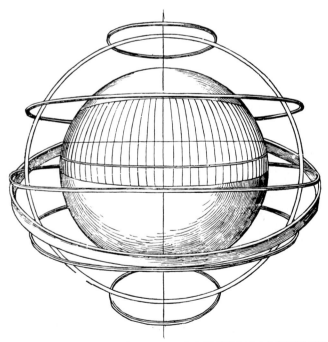

FIG. 11.7. PTOLEMY'S THIRD PROJECTION AS SEEN BY THE OBSERVER. The armillary rings (including the ecliptic) are positioned so as not to impede the full view of the inhabited world.
After Ptolemy, *Geographicae enarrationis libri octo* (Vienne: Gaspar Trechsel, 1541).

that the eye must be placed on the axis described above. The result, once these details had been resolved, would be a projection for the inhabited world in which the parallel and the meridian of Syene were straight lines and all the other parallels and meridians curved, with their concave side toward the central parallel and meridian.[62]

In conclusion, we may emphasize the great significance of Ptolemy's study of map projections for the further development of cartography. Even if he or his contemporaries did not construct maps according to these principles, which is unlikely, and although his instructions lay dormant for centuries, it was largely through the *Geography* that the Greek contribution to the scientific construction of maps was transmitted first to Arab and Byzantine mapmakers and then to the cartographic workshops of Renaissance Europe. Edgerton suggests that book 7, sections 6 and 7, of the *Geography* may also have had an influence on the development of perspective theory in the Renaissance.[63] The best testimony to Ptolemy's instructions is that they were widely followed. All the regional maps in the Greek codices and early Latin translations were drawn on the rectangular projection employed by Marinus and described by Ptolemy. It proved convenient for this purpose, and it was accurate enough so long as the proportion of one degree longitude on the central parallel of the map to one degree of latitude on the meridian was observed (e.g., for Italy 3:4, or for Britain 11:20). Yet one consequence of its adoption was that a map of the whole world could not be obtained by fitting together the regional maps, and the mapmakers were forced to turn to Ptolemy's other projections to represent the inhabited world as a whole. As already noted, they seem to have preferred Ptolemy's first projection. It was used for the world maps in most of the early codices and in the most of the first printed editions of the *Geography*. Ptolemy's second projection was more rarely used[64]—as, for example, in Codex Seragliensis 57, Sultan's Library, Istanbul, of the late thirteenth century—but the history of its dissemination and modification in the Renaissance again points to the long-term, albeit much delayed, importance of his ideas.

Through his discussion of the size and location of the inhabited world, as well as through his projections, Ptolemy also codified for posterity an image of its outlines and general arrangement. Ptolemy's map, as reconstructed or seen in the later manuscripts, depicts the inhabited world as no longer an island in the ocean (see fig. 15.5). It was limited eastward by an unknown land occupying the territory of the East Asian peoples; southward by an equally unknown land surrounding the Indian Sea and the part of Ethiopia south of Libya called Agisymba; westward by an unknown land circling the Ethiopian Gulf in Libya and by the Western Ocean surrounding the western parts of Libya and Europe; and northward by the contiguous ocean, surrounding the British Isles and the northern parts of Europe, and by the unknown land stretching along northern Asia, Sarmatia, Scythia, and the silk land.[65] Inside the inhabited world there were two enclosed seas, the Caspian[66] (or Hyrcanian) and the Indian (with its various gulfs, the Arabian, Persian, Gangetic), and one sea open to the ocean, the Mediterranean.

PTOLEMY AS A MAPMAKER: THE TABLES OF COORDINATES

A final aspect of Ptolemy's instructions for mapmaking concerns the relationship between the tables of coordinates in the *Geography*—which clearly form the raw material for compiling geographical maps—and the geographical maps that first appear in the Byzantine manuscripts of the *Geography*. This may be regarded as a major enigma of Ptolemaic scholarship. It has long been debated whether Ptolemy himself or a contemporary drew maps for the *Geography*, whether they were added after his time under the Roman Empire, or whether those we have go back only to Byzantine times.[67] To Joseph Fischer, for example, that the Biblioteca Apostolica Vaticana's Codex Urbinas Graecus 82 (late thirteenth century) has maps corresponding very closely to the text is an argument that it must depend on a cartographic archetype of the Roman Empire.[68] But if we take as a parallel the maps in the *Corpus Agrimensorum*, we find that the further removed they are from the original the more corrupt they become. Hence Leo Bagrow doubts whether any extant Ptolemaic maps go back to an archetype earlier than about the twelfth century.[69]

The lack of a careful critical edition of the *Geography* is clearly an impediment to our understanding of this

62. Neugebauer, *History of Ancient Mathematical Astronomy*, 883, 889 (note 48); idem, "Ptolemy's *Geography*," 25–29 (note 60); Cortesão, *History of Portuguese Cartography*, 1:106 (note 18).

63. Samuel Y. Edgerton, "Florentine Interest in Ptolemaic Cartography as Background for Renaissance Painting, Architecture, and the Discovery of America," *Journal of the Society of Architectural Historians* 33 (1974): 275–92. See also Neugebauer, *History of Ancient Mathematical Astronomy*, 890 (note 48).

64. Neugebauer, *History of Ancient Mathematical Astronomy*, 885 (note 48).

65. Ptolemy *Geography* 7.5.2 (note 5).

66. For the Caspian as an enclosed sea in Ptolemy, see p. 198.

67. For a list of the authors on either side of the debate, see Erich Polaschek, "Ptolemy's *Geography* in a New Light," *Imago Mundi* 14 (1959): 17–37.

68. Fischer, *Urbinas Graecus 82* (note 5).

69. See Leo Bagrow, "The Origin of Ptolemy's Geographia," *Geografiska Annaler* 27 (1945): 318–87.

question. Whereas in books 1 and 2 Ptolemy speaks of mapping only in the future tense, in 8.2.1 he says "we have had maps drawn up," ἐποιησάμεθα, specifying the twenty-six regional maps as given below. He mentions in this connection not degrees of latitude and longitude, but lengths of daylight and distances from Alexandria. If the passage is genuine, which seems likely, and if this use of the future tense in books 1 and 2 can be taken literally, this suggests, as Polaschek thought, that Ptolemy revised the earlier part of his work but not the later part; that he decided to be content with those regional maps he had already commissioned and to leave to others the compilation of any based on the more exact system of his coordinates, working from degrees and minutes. This is certainly true of book 1 and of book 2, section 1 of the *Geography*. Book 8 was probably composed at a different time from these. The concluding sentence, "These things being settled beforehand we can now attend to that which remains," may indeed apply to the future drawing of maps, but this does not rule out the possibility that some maps or the projections for them had previously been completed. Neither does it tell us whether Ptolemy or others finished the task in his lifetime. In any case, even if there were any, no maps have survived that can positively be ascribed to Ptolemy's period. Marcianus (fourth/fifth century A.D.) may, in bringing the eastern section of the map of the inhabited world up to date, have appended a map of it based on Ptolemy's coordinates. Similarly, Agathodaimon of Alexandria, the technician who drew a map of the world from the *Geography* discussed more fully below (pp. 271–72), may not have drawn any of the regional maps. But unfortunately the information about Agathodaimon does not settle the disputed question of whether Ptolemy's text was in his lifetime accompanied by maps; though to us the preposition ἐκ, "from," in the context that he drew maps *from* Ptolemy's *Geography*,[70] suggests that it was not. There is thus no positive proof that any of the extant maps attached to the later Greek recensions of the *Geography* were copied from maps circulating during the time of the Roman Empire. All features, including such refinements as cartographic signs for tribal areas, may have been reconstructed by Byzantine scholars (p. 268 below) from the text of the *Geography*.

In any event, all that can be said with certainty is that Ptolemy provided raw material for future mapmakers to work. In relation to the content of the geographical maps—as opposed to their mathematical construction—the most important part of the *Geography* was that containing the tables of coordinates (books 2–7). For each country a certain number of towns or places were selected, the positions of which were precisely, if not always accurately, defined by latitude and longitude.

Ptolemy began with the western parts of the inhabited world, Europe first, then Africa, and last Asia. The coordinates were expressed in degrees of longitude east of the meridian drawn through the Fortunate Isles (the prime meridian) and in degrees of latitude north or south of the equator (Ptolemy puts the longitude first, since he expected the mapmaker to draw the map from left to right). He planned twenty-six regional maps: ten for Europe, four for Africa, and twelve for Asia. But the tables of coordinates were meant to be used for both regional and world maps.[71]

CARTOGRAPHIC INSIGHTS FROM PTOLEMY'S TOPOGRAPHY: THE COORDINATES AND THE REGIONAL MAPS

It is not the aim of the present *History* to reconstruct the content of particular maps.[72] In the case of Ptolemy's *Geography* it would be entirely inappropriate to attempt to summarize the massive literature that has sought to reconstruct, often fancifully, his topography for different parts of the known world.[73] Yet though this subject has been of primary concern to classical historians, it also can be harnessed—through the selective assessment of the topography in the coordinates and in the maps—to throw light on broader cartographic questions. A knowledge of the pattern of mapping in the Byzantine manuscripts, for example, may at least enable us to visualize the sort of maps Ptolemy may have had in front of him (such as those of Marinus) even if we do not accept that he drew such maps himself. Moreover, a comparative examination of coordinates and maps may reveal how individual mapmakers might have worked from Ptolemy's instructions and raw materials (or indeed suggest what a modern scholar could achieve by a rigorous re-

70. Not "of," as given by Bagrow, "Origin," 350 (note 69).

71. It is obvious that in establishing these sets of coordinates Ptolemy either relied on previous sources or, more probably, read them off from a map of the inhabited world graduated with latitude and longitude. The coordinates he gives are coherent on the whole and allow anyone to draw a map; but they are largely inaccurate and suggest that he did not verify either his own observations or those made by others. For instance, he stated that the southern coast of Sardinia was at 36°N, probably relying on Dicaearchus's estimates, and did not correct the latitude of Byzantium, erroneously fixed at 43°N by Hipparchus. In this respect Ptolemy's *Geography* may perhaps be regarded as a useful tool for *mechanikoi* (draftsmen) in drawing maps to unspecified accuracy rather than a fully scientific treatise.

72. See Preface, pp. xv–xxi, esp. p. xix, of the present *History*.

73. See each year's issue of *L'Année Philologique: Bibliographie Critique et Analytique de l'Antiquité Gréco-Latine* (1928–), pt. 1, "Auteurs et Textes," s.v. "Ptolemaeus"; William Harris Stahl, *Ptolemy's Geography: A Select Bibliography* (New York: New York Public Library, 1953).

construction along these lines).[74] In any case, only through painstaking topographical research does it become possible to identify Ptolemy's sources, to assess their reliability, to weigh up his skill in reconciling their often conflicting evidence, and not least to be able to decipher the cartographic image of the known world held by the Greek and Roman map users of Ptolemy's day and later that of the scholars of the Byzantine Empire.

This section will therefore illustrate, by means of brief regional examples, the potential of such topographical research. It will be approached first from the evidence of the coordinates and second, from that of the maps in the Greek manuscripts. While both sources are closely interrelated, albeit in ways that depend on our view of their respective origins, they can nevertheless sometimes throw a different light on cartographic questions.

PTOLEMY'S TOPOGRAPHY IN THE COORDINATES AND MAPS: GENERAL CONSIDERATIONS

There are three sources for Ptolemy's gazetteer: the Κανὼν ἐπισήμων πόλεων (*Canon of Significant Places*);[75] books 2–7 of the *Geography*; and book 8 of the *Geography*. Of these only the second gives all the names; the first and third give only a selection of places regarded by Ptolemy as significant for one reason or another. In addition, there are some criticisms of Marinus of Tyre in book 1 of the *Geography* that have a topographical bearing. For topographical details the maps in the Latin text can be mostly ignored, since they are based on extant Greek ones.

In books 2–7 the manuscripts all have coordinates (except that a few lack them toward the end). A fair proportion also have maps. Where coordinate tables and maps disagree in their toponymic detail, the former are usually more reliable, particularly if they are the same in a number of authoritative manuscripts. There are exceptions, however, and in a few cases the names on the maps in the Codex Urbinas Graecus 82 are more correct than the texts, which may mean that the scribe also had another manuscript at hand. For example, whereas on the map of Thrace Byzantium is so called, as it is in Ptolemy's text, on the general map of Europe the name "Konstantinupolis" is found. Here, as elsewhere, the explanation may be that the name was not copied from any ancient manuscript but was inserted by or for Maximus Planudes when maps were being prepared from the text of Ptolemy.

In interpreting the place-names in either the tables of coordinates or the maps, it is also helpful to bear in mind Ptolemy's emphasis in compiling the *Geography* as well as his range as a linguist. He was more interested in establishing latitude and longitude than in place-

names; and he does not seem to have known Latin well. In some cases we cannot be sure that a mistake originated with him, though one may well imagine that he wrote Alpha Bucens for Alba Fucens (Italy), alpha being more familiar than Latin *alba*, "white"; Fucens may have meant "connected with materials for dyeing," but Bucens is meaningless.[76] The most glaring mistake in Latinity concerns a place in Germany that he calls Siatutanda. Tacitus has the phrase, referring to a German tribe, *ad sua tutanda*, "to protect their possessions."[77] Ptolemy has corrupted the last two words of this to Siatutanda and incorporated it into the German section as a place-name.[78]

A further element in reconstructing Ptolemy's topography and its underlying compilation process is that in some cases he may have employed signs rather than toponyms to locate geographical features. This may be regarded as a step in the eventual codification of the cartographic signs employed on maps, which was to become much more general in the Renaissance;[79] in particular, in Codex Urbinas Graecus 82 and the Ptolemaic manuscripts in the Biblioteca Medicea Laurenziana, Florence (see table 15.1), small signs were used to denote tribal territories or subdivisions of provinces.[80] Some of these are astronomical signs; others appear to be invented specially for the purpose. In Ireland, Germany, and the Danube provinces there are none, since the system is inapplicable; in Aquitania and Gallia Lugdunensis, where we would expect signs, there are none. In Britain, Spain, and the other Gallic provinces the signs do not agree in the two manuscripts, whereas in Italy they nearly all do. Codex Urbinas Graecus 82 has signs in Liburnia and Dalmatia; the Laurenziana's Plut. 28.49 has not.

---

74. This type of experimental approach is seldom considered by historians of cartography: it could, however, throw much light on the problems of the Byzantine mapmakers confronted by Ptolemy's manuscripts without maps.

75. Honigmann, *Sieben Klimata* (note 33); Erich Polaschek, "Ptolemaios als Geograph," in *Paulys Realencyclopädie der classischen Altertumswissenschaft*, ed. August Pauly, Georg Wissowa, et al. (Stuttgart: J. B. Metzler, 1894–), suppl. 10 (1965): cols. 680–833, esp. 681–92.

76. Ptolemy *Geography* 3.1.50 (Müller edition), 3.1.57 (Nobbe edition) (note 5).

77. Tacitus *Annals* 4.73; see *The Annals of Tacitus*, trans. Donald R. Dudley (New York: New American Library, 1966).

78. Ptolemy *Geography* 2.11.27 (note 5). The connection between Ptolemy and Tacitus, although not absolutely certain, is generally agreed upon.

79. See Catherine Delano Smith's discussion of cartographic signs in volume 3 of the present *History*.

80. Ptolemy, *Die Geographie des Ptolemaeus: Galliae, Germania, Raetia, Noricum, Pannoniae, Illyricum, Italia*, ed. Otto Cuntz (Berlin: Weidmann, 1923), 18–19.

TABLE 11.1 Selected Greek Manuscripts of Ptolemy's *Geography*

Repository and Collection Number	Date	Maps
Rome, Biblioteca Apostolica Vaticana, Vat. Gr. 191	12th–13th century	None extant, but see pp. 268–69; coordinates omitted from 5.13.16[a]
Copenhagen, Universitetsbiblioteket, Fragmentum Fabricianum Graecum 23	13th century	Fragmentary; originally world and 26 regional
Rome, Biblioteca Apostolica Vaticana, Urbinas Graecus 82	13th century	World and 26 regional[b]
Istanbul, Sultan's Library, Seragliensis 57	13th century	World and 26 regional (poorly preserved)[c]
Rome, Biblioteca Apostolica Vaticana, Vat. Gr. 177	13th century	No extant maps
Florence, Biblioteca Medicea Laurenziana, Plut. 28.49	14th century	Originally world, 1 Europe, 2 Asia, 1 Africa, 63 regional (65 maps extant; see pp. 270–71)
Paris, Bibliothèque Nationale, Gr. Supp. 119	14th century	No extant maps
Rome, Biblioteca Apostolica Vaticana, Vat. Gr. 178	14th century	No extant maps
London, British Library, Burney Gr. 111	14th–15th century	Maps derived from Florence, Plut. 28.49
Oxford, Bodleian Library, 3376 (46)-Qu. Catal. i (Greek), Cod. Seld. 41	14th century	No extant maps
Rome, Biblioteca Apostolica Vaticana, Pal. Gr. 388	Early 15th century	World and 63 regional
Florence, Biblioteca Medicea Laurenziana, Plut. 28.9 (and related manuscript 28.38)	15th century	No extant maps
Venice, Biblioteca Nazionale Marciana, Gr. 516	15th century	Originally world and 26 regional (world map, 2 maps, and 2 half maps missing)
Rome, Biblioteca Apostolica Vaticana, Pal. Gr. 314	Late 15th century	No extant maps; written by Michael Apostolios in Crete

[a]Alexander Turyn, *Codices Graeci Vaticani saeculis XIII et XIV scripti*, Codices e Vaticanis Selecti quam Simillime Expressi, vol. 28 (Rome: Biblioteca Apostolica Vaticana, 1964).

[b]Ptolemy *Claudii Ptolemaei Geographiae Codex Urbinas Graecus 82*, 2 vols. in 4, ed. Joseph Fischer, Codices e Vaticanis Selecti quam Simillime Expressi, vol. 19 (Leiden: E. J. Brill; Leipzig: O. Harrasso-witz, 1932); Aubrey Diller, "The Greek Codices of Palla Strozzi and Guarino Veronese," *Journal of the Warburg and Courtauld Institutes* 24 (1961): 313–21, esp. 316.

[c]Aubrey Diller, "The Oldest Manuscripts of Ptolemaic Maps," *Transactions of the American Philological Association* 71 (1940): 62–67, pls. 1–3.

If all such factors are taken together, the most useful Greek manuscripts for studying topographical details and place-names in Ptolemy are set out in table 11.1.[81]

PTOLEMY'S COORDINATES: THE EXAMPLES OF THE
BRITISH ISLES AND ITALY

Ptolemy's coordinates provided data for mapmaking and were probably modified from Marinus's map. Although anyone reconstructing a regional map from them could not, for example, tell where the coastline was intended to go, the basic pattern would always have been roughly the same. A glance at a map drawn to illustrate the *Geography* will show deformation in parts of the British Isles and Italy, among other regions. This deformation, and the misplacing of certain towns, would become apparent at the first attempt to draw the map. The stress below will be laid on the textual side of Ptolemy's regional chapters, by using the British Isles and Italy as examples not only of his compilation methods, but also of the extent to which some manuscripts were corrupted—

or even underwent very tentative attempts to correct them—at the hands of their subsequent copyists.[82]

Ptolemy's name for Britain is Aluion, that is, Albion,[83] which by his time was an outmoded Greek name. The Romans always knew it as Britannia, and Ptolemy himself calls it the Prettanic island of Aluion. Since he lists European countries roughly from the west, he starts with

81. The principal sources for the table are: Lauri O. T. Tudeer, "On the Origin of the Maps Attached to Ptolemy's Geography," *Journal of Hellenic Studies* 37 (1917): 62–76; Cuntz, *Ptolemaeus* (note 80); Fischer, *Urbinas Graecus 82* (note 5); Paul Schnabel, *Text und Karten des Ptolemäus*, Quellen und Forschungen zur Geschichte der Geographie und Völkerkunde 2 (Leipzig: K. F. Koehlers Antiquarium, 1938); Bagrow, "Origin" (note 69); Polaschek, "Ptolemaios," cols. 680 ff. (note 75); idem, "Ptolemy's Geography" (note 67); Nobbe, *Claudii Ptolemaei Geographia* (note 5).

82. A. L. F. Rivet, "Some Aspects of Ptolemy's Geography of Britain," in *Littérature gréco-romaine et géographie historique: Mélanges offerts à Roger Dion*, ed. Raymond Chevallier, Caesarodunum 9 bis (Paris: A. et J. Picard, 1974), 55–81; A. L. F. Rivet and Colin Smith, *The Place-Names of Roman Britain* (Princeton: Princeton University Press, 1979).

83. Rivet and Smith, *Place-Names*, 247–48 (note 82).

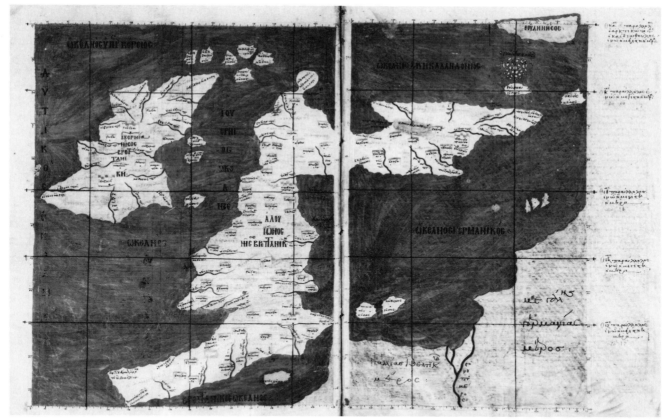

FIG. 11.8. BRITAIN ACCORDING TO PTOLEMY. From a Byzantine manuscript of Ptolemy's *Geography* (late thirteenth century) (see also figs. 11.11 and 15.5).

Size of the original: 41.8 × 57.5 cm. Photograph from the Biblioteca Apostolica Vaticana, Rome (Urbinas Graecus 82, fols. 63v–64r).

Ireland (Iuernia). Four oceans surrounding Britain are named the British, for the Channel; the German, for the North Sea; the Irish; and the Duekaledonios,[84] off northern Scotland. The two others mentioned are the Western, off the west coast of Ireland; and the Hyperborean,[85] off the north coast of Ireland (fig. 11.8).

Like the recensions of Ptolemy, so also the Ordnance Survey map[86] and a recent masterly treatment of place-names[87] give only one latitude and one longitude for each natural feature or town. There is, however, some evidence that the variants in place-names found in manuscripts only in certain cases point to corruption, and that in others there has been a deliberate attempt at slight modification. Thus in the distorted promontory of the Novantae (Mull of Galloway) it makes a difference whether we give the latitude of the river Abravannos (identifed as the Water of Luce) as 61°, with most manuscripts, or as 60°15′, with manuscripts of the fifteenth-century Laurenziana group, that manuscript itself here being corrupt (fig. 11.9). This was probably intended as a deliberate correction by one or the other, since for the

neighboring Iena estuary (either by the Water of Fleet or the river Cree) the fifteenth-century Laurenziana group also has a slight difference of latitude, 60°20′ instead of 60°30′. Sometimes both latitude and longitude are different; thus Tamare on the river Tamar has, according to Codex Graecus 191 (Biblioteca Apostolica

---

84. The expected adjective would be Kaledonios: *due* is a prefix of doubtful meaning; Ammianus's Dicalydones may have no connection. Perhaps Duekaledonios is a corruption of Deukalioneios, from Deucalion—son of Minos. He was one of the Argonauts, who according to one tradition sailed around Ireland. He was also at the Calydonian boar hunt, and this Aetolian adjective could have become confused with Caledonian.

85. "Hyperborean" is usually associated with mountains and tribes of northern continental Europe. To Ptolemy it must have meant "beyond northern lands" and can have had no connection with tribes who traded with the Mediterranean, as some scholars have postulated for northern Europe.

86. Ordnance Survey, *Map of Roman Britain*, 4th ed. (Southampton: Ordnance Survey, 1978), 15.

87. Rivet and Smith, *Place-Names* (note 82). See also O. A. W. Dilke, *Greek and Roman Maps* (London: Thames and Hudson, 1985), 190–92.

Vaticana) and the fifteenth-century Laurenziana group, longitude 15°30' east of the Canaries, latitude 52°40', whereas according to other manuscripts the figures are 15° and 52°15', respectively. Obviously this affects the search for what is in this case an unidentified place.[88]

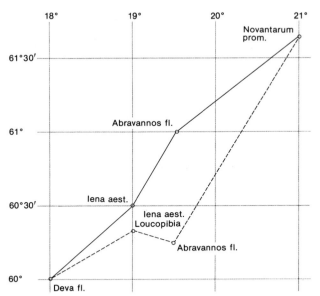

FIG. 11.9. THE MULL OF GALLOWAY IN DIFFERENT VERSIONS OF PTOLEMY. The solid line represents the coastline of Galloway in most manuscripts of the *Geography*, while the dashed line follows the manuscripts of the fifteenth-century Laurenziana recension. Iena aest. (estuary) and Loucopibia are given the same coordinates in some manuscripts. The ratio of meridian to parallel spacing is 11:20.

Ptolemy's two main misunderstandings with respect to the British Isles were the orientation of Scotland and the respective latitudes of Ireland and Great Britain. On the latter point, whereas mainland Britain actually lies between 49°55' and 58°35'N, Ptolemy made it extend from 51°30' to 61°40'; Ireland, which lies between 51°30' and 55°20', was made by Ptolemy to extend from 57° to 61°30'. This means that it was conceived as running parallel to Scotland to an unjustified distance.

The "dogleg" appearance of northern Britain is a Ptolemaic feature that becomes most conspicuous with mapping and that persisted into the Renaissance. A number of factors may have contributed to it. First, Ptolemy was clearly convinced that all to the north of 63°N in this region was terra incognita. At that latitude he had to accommodate Thule, which to him was the Shetland Islands. Since he knew that Britain extended in some direction for 4,000 to 4,500 stades, he could not give coordinates that would have made this distance a straight line without infringing his 63° rule. Second, Eratosthenes, or some other Greek author reflected in Dio-

dorus Siculus, had thought of Britain as an obtuse-angled triangle of the type in figure 11.10. Third, the odd feature of the Mull of Galloway as the most northerly point on mainland Scotland requires some explanation. Rivet's is that the island of Epidion (18°30'E of the Canaries, 62°N) and the promontory of Epidion (Mull of Kintyre) were adjacent in Ptolemy's source map, but that Ptolemy, to account for the difficulties mentioned above, gave coordinates rotating northern Britain clockwise by about one-seventh of 360 degrees.[89] Finally, Ptolemy's second projection, when used for world cartography, could have caused much distortion in the areas of the *oikoumene* farthest to the northwest and northeast.

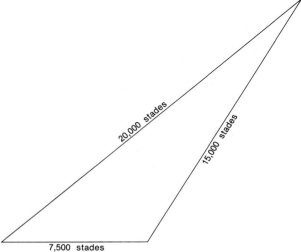

FIG. 11.10. BRITAIN REPRESENTED AS AN OBTUSE-ANGLED TRIANGLE. The belief that Britain was of such a shape, based on measurements recorded in Diodorus Siculus, derived in turn from Pytheas or Eratosthenes, might have influenced Ptolemy's own distinctive version of the island. It appears that Diodorus included the numerous indentations of the west coast, resulting in a longer total westward measurement, thereby accounting for the eastward shift of Scotland.

Some of the mistakes in Ptolemy's Britain—like the many others throughout the *Geography*—must have been present in Marinus's map and not corrected by Ptolemy. Viewed with hindsight, as compared with many nearer provinces that are well portrayed on the whole, it must be admitted to have many defects. Rooted both in his sources and in his compilation decisions, these reflect a variety of components of error, probably

88. Rivet and Smith, *Place-Names*, 464 (note 82), suggest Launceston.

89. A. L. F. Rivet, "Ptolemy's Geography and the Flavian Invasion of Scotland," in *Studien zu den Militärgrenzen Roms, II*, Vorträge des 10. Internationalen Limeskongresses in der Germania Inferior (Cologne: Rheinland-Verlag in Kommission bei Rudolf Habelt, 1977), 45–64.

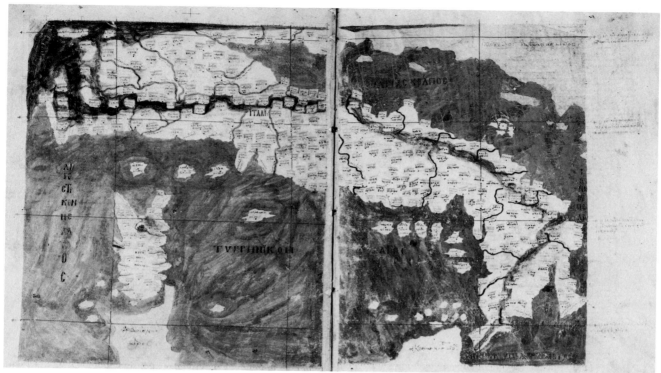

FIG. 11.11. THE PTOLEMAIC MAP OF ITALY.

Size of the original: 41.8 × 57.5 cm. Photograph from the Biblioteca Apostolica Vaticana, Rome (Urbinas Graecus 82, fols. 71v–72r).

including his unwillingness to discard early cartographic attempts such as those of Eratosthenes, insufficient revision of Marinus, and incomplete updating.

Ptolemy's coordinates for Italy[90] are not as good as one might expect for such a well-known area (fig. 11.11). It seems inevitable that anyone attempting to draw a map from them would have problems of orientation over the whole peninsula. Northern and central Italy are bound to be portrayed with a largely west-east rather than northwest-southeast orientation. Although this applies throughout those areas, it can best be illustrated from the Po valley. If we plot the towns along the Via Aemilia (Ptolemy does not give coordinates for roads), we find that many of them lie in a west-east line (fig. 11.12). The result is that the south of Italy appears in too north-south an orientation. This feature applies south of a line Naples–Benevento–Monte Gargano, so that the peninsula, from Ptolemy's coordinates, presents an unwarranted bend. The reason for the Po valley orientation could be that the towns on the Via Aemilia were linked to centuriation schemes, and if Ptolemy looked at centuriation maps he might have concluded that they had north at the top, whereas in fact they tend to follow the orientation of the road (fig. 11.13). For the peninsula in general he may have followed his know-

ledge of a version of Agrippa's map (pp. 207–9 below), which, being designed to be displayed on a colonnade, may well, since it had north or south at the top, have had more room to spread east-west. Also, he was basing his longitudes on Posidonius's measurements, which gave a greater relative width to each degree than did those of Eratosthenes.

The effect of such decisions on Ptolemy's coordinates—and on the maps drawn from them—is also borne out in a number of other examples. The Gulf of Taranto is more subject than other areas of Italy to three differing recensions of manuscript coordinates;[91] but in all of these it is too long and narrow. The coast from the river Var (now in France) to the river Arno is far too straight; the north coast of the Adriatic has inaccuracies; Lake Larius (Lake Como) is located, as a source of the Po, far from Comum (Como); and several important towns are considerably misplaced. A plotting of "significant places" shows three attempts by different scribes at some

---

90. O. A. W. Dilke and Margaret S. Dilke, "Italy in Ptolemy's Manual of Geography," in *Imago et mensura mundi: Atti del IX Congresso Internazionale di Storia della Cartografia*, 2 vols., ed. Carla Clivio Marzoli (Rome: Enciclopedia Italiana, 1985), 2:353–60.

91. Polaschek, "Ptolemaios," plan opposite col. 728, with key in cols. 715–16 (note 75).

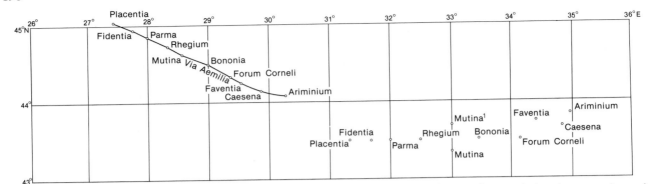

FIG. 11.12. COMPARISON OF PTOLEMAIC AND MODERN COORDINATES OF TOWNS ON THE VIA AEMILIA, NORTHERN ITALY. The shift in orientation may possibly be due to centuriation based upon the Via Aemilia, which was generally oriented perpendicular to the main trend of the road. Ptolemy's coordinates from book 3 of the *Geography* (*right*) assume a west-east trend for the road, perhaps following information derived from such centuriation, in comparison with the modern figures (*left*). The location of Mutina[1] further north is from the Urbinas and Laur. 28.49 manuscripts. The modern equivalents of the towns are, from west to east: Piacenza; Fidenza; Parma; Reggio nell'Emilia; Modena; Bologna; Imola; Faenza; Cesena; Rimini. Longitudes are east of the Canaries.

FIG. 11.13. CENTURIATION IN THE PO VALLEY AROUND PARMA AND REGGIO EMILIA. Centuriation of this area, which was oriented approximately perpendicular to the trend of the Via Aemilia, might have misled Ptolemy into making the error shown in figure 11.12.
After R. Chevallier, "Sur les traces des arpenteurs romains," *Caesarodunum,* suppl. 2 (Orléans-Tours, 1967).

coordinate values: in the *Canon of Significant Places*; in book 3, chapter 1; and in book 8. Five pairs of towns in Italy are given coinciding coordinates in all or some manuscripts.[92] The major islands also suffer from inaccuracies: Corsica is given two northern promontories instead of one; Sardinia is placed on the same latitude as Sicily; and the western part of the north coast of Sicily is made to bend to the southwest.

PTOLEMY'S MAPS: SOME REGIONAL EXAMPLES

The maps included in the Byzantine manuscripts, albeit much delayed in their execution, represent the logical end product of the cartographic processes set in motion by Ptolemy's work. Although it has always been possible to draw regional maps from the coordinates of the text, as the examples of the British Isles and Italy have demonstrated, only with the help of Ptolemy's regional maps does the full cartographic pattern readily appear. These maps can also be subjected to topographical scrutiny; they can be used in assessing the adequacy of Ptolemy's cartographic raw materials (and of his instructions to his successors) and also can enable us to understand the images they may have helped to form in the minds of their contemporary users. The theme of the reliability of the maps will be illustrated from three regions—northern Europe, North Africa and Egypt, and Asia—that supplement the treatment of the British Isles and Italy derived from the coordinates.

First, Ptolemy's knowledge of the areas to the north of continental Europe reflected some of the correct and incorrect concepts of the North current in Greco-Roman literature and earlier cartography.[93] All to the north of $64\frac{1}{2}°$N was terra incognita, and there was a lack of information about much to the south of this. Like other classical writers, Ptolemy had no idea that Norway and Sweden form part of the continent.[94] The north coast of Germany and Poland was for him almost completely straight, just near the fifty-sixth parallel. East of Jutland and in the Baltic north of that coast are one large and a number of small Skandia islands.[95] The central island of the small Skandia islands has a longitude of 41°30' and a latitude of 58°. The large Skandia island is offshore from the Vistula estuary, having a longitude of 43° to 46° and a latitude of 57°40' to 58°30'. Ptolemy lists seven tribes that inhabit this island; manuscripts and printed editions do not always include these owing to lack of space. Since there is no reliable topographical distinction among them, his research must have been based on etymological and historical sources, including the record of tribal movements over the centuries.[96] The most recognizable name is Phinnoi, variously located by scholars in Finmark, Lapland, and Finland. A tribe whose name appears in the manuscripts as Daukiones

or Dankiones may be the ancestors of the Danes. The Khaideinoi have been equated with the Heinnin, and the Goutai with the Gutar. It must be stressed that the Ptolemaic concept of the area north of Jutland was of ocean only very sparsely dotted with islands. As mentioned, Thule was for Ptolemy the Shetland Islands, not part of Scandinavia; and for any suggestions that Renaissance additions to the Ptolemaic cartography of Scandinavia may rest on a much earlier tradition, there is no evidence whatever.[97]

In the more northerly latitudes Ptolemy's coast of the Baltic correctly takes a bend northward. But at the end of the enumeration of Baltic tribes he lists numerous other tribes in what was evidently intended as a southerly direction. As a result, the Byzantine reconstructions fit them in rather close together on a north-south line,[98]

---

92. Polaschek, "Ptolemaios," cols. 719–20 (note 75).

93. Cuntz, *Ptolemaeus* (note 80); D. (S. D. F.) Detlefsen, *Die Entdeckung des germanischen Nordens im Altertum*, Quellen und Forschungen zur Alten Geschichte und Geographie 8 (Berlin: Weidmann, 1904); Gudmund Schütte, *Ptolemy's Maps of Northern Europe: A Reconstruction of the Prototypes* (Copenhagen: Royal Danish Geographical Society, 1917); E. Stechow, "Zur Entdeckung der Ostsee durch die Römer," *Forschungen und Fortschritte* 24 (1948): 240–41; Joseph Gusten Algot Svennung, *Scandinavien bei Plinius und Ptolemaios* (Uppsala: Almqvist och Wiksell, 1974); O. A. W. Dilke, "Geographical Perceptions of the North in Pomponius Mela and Ptolemy," in *Exploring the Arctic*, ed. Louis Rey (Fairbanks: University of Alaska Press, Comité Arctique International, and Arctic Institute of North America, 1984), 347–51. For a critical review of Svennung's book see T. Pekkanen in *Gnomon* 49 (1977): 362–66.

94. Svennung, *Scandinavien* (note 93).

95. No name etymologically connected with the Baltic occurs in Ptolemy. For these names in other ancient writers see Joseph Gusten Algot Svennung, *Belt und Baltisch: Ostseeische Namenstudien mit besonderer Rücksicht auf Adam von Bremen* (Uppsala: Lundequistska Bokhandeln, 1953).

96. Schütte, *Ptolemy's Maps* (note 93); Svennung, *Scandinavien* (note 93).

97. Charles H. Hapgood, *Maps of the Ancient Sea Kings: Evidence of Advanced Civilization in the Ice Age*, rev. ed. (New York: E. P. Dutton, 1979), 124–40, reaches what seems to be an absurd conclusion by comparing a fifteenth-century Vatican manuscript, designed to amplify Ptolemy, with the Zeno map of 1380 (according to Hapgood, see point four below): "If the original source of the Ptolemy map came from the end of the ice age, that of the Zeno map may have originated much earlier" (p. 40). However, first of all, there is not, as Hapgood maintains, the slightest evidence that Ptolemy or this "new Ptolemy" even remotely touched upon the cartography of the Arctic or Antarctic. Second, it is not true, as he claims, that "some authorities have considered that they [Ptolemaic maps] were reconstructed from the tables [coordinates of latitude and longitude] . . . in the 15th century" (p. 133). Any such reconstruction must have been much earlier; see p. 268. Third, because certain parts of a Ptolemaic map contain place-names and others do not, they are not necessarily by two different hands. Fourth, the Zeno map is now thought to date not from 1380, as Nicolò Zeno the younger claimed, but from much later.

98. Müller, *Claudii Ptolemaei Geographia*, tabulae, Europae Tab. VIII (note 5).

whereas presumably Ptolemy had planned them to be more spaced out. In general, the tribes of Sarmatia in Europe, largely corresponding to the European regions of the USSR, seem to be unduly increased in number by duplication of names. Some of these pairs are exactly the same, others are similar. An attempt has been made to show how Ptolemy could have used two regional groupings with different orientations.[99] The explanation could be either that one or both sets of tribes were incorrectly placed by his sources or that these sources took insufficient account of tribal movements.

A second example, concerning the Ptolemaic depiction of North Africa and Egypt, takes us into a realm where, theoretically at least, he was much better informed—in part—by firsthand experience. Ptolemy's coastline of North Africa, granted it was based on an inaccurate latitude for much of his native Egypt (see below), is close enough to reality except in the Tunisian section.[100] Here, among other inaccuracies, the coast from Cape Bon to Monastir was made to run roughly east-southeast instead of south, and from Monastir to Gabès there is a similar deformation. This again, like the Po valley as revealed in the coordinates, could have arisen from the fact that the predominant orientation of centuriated land in Tunisia is at about 45° from the cardinal points.[101] Coastal voyages, however, had to avoid the treacherous sandbanks of Syrtis Major (Gulf of Sidra) and Syrtis Minor (Gulf of Gabès) so that they tended to be far straighter than the coastal outline would suggest; and this too may have had an influence on mapping.

Since Ptolemy spent most of his working life in Alexandria, we should expect him to be an expert on the topography of Egypt. Certainly there is much that is reasonably accurate in the more populated parts. But the Mediterranean coastline starts with a minor inaccuracy, the plotting of his native Alexandria at a latitude of 31° instead of 31°20′, and the coast west of this was kept too close to this latitude. The result was that the border between Egypt and Cyrenaica appeared as 31°20′ instead of 32°40′. The upper Nile was less accurately plotted than the lower, and the Gulf of Suez was made too wide.

In the third example, that of Asia, we can detect the same reliance on a patchwork of older and often imperfect sources. Again, this should be construed not as a criticism of Ptolemy, but rather as a reflection of the geographical lore of the Roman period in which he worked. The nearer parts of Asia presented Ptolemy with relatively little problem.[102] The maps err on the shapes of the Persian Gulf and the Caspian Sea. But whereas earlier writers had tended to make the Caspian flow out northward into the Scythian Ocean, Ptolemy insisted that it was landlocked. At this point, however, Stevenson's English translation has Ptolemy say, most im-

probably, "The Hyrcanian sea, called also the Caspian, is surrounded on all sides by land and has the shape of an island." What he actually meant was "just like the opposite of an island."[103]

Farther east, as the sources became diluted, India was made too small, Taprobane (Sri Lanka) too big. This followed the approach of previous writers, as reflected in the account of the elder Pliny. Some comparison can also be made between Ptolemy's topography and that of Marcianus of Heraclea.[104]

The Ptolemaic outline of Southeast Asian coasts has given rise to considerable discussion. East of the Sinus Gangeticus (Bay of Bengal) the Ptolemaic world maps show the Golden Chersonnese, corresponding roughly to the Malay Peninsula, though on a reduced scale.[105] This is followed to the northeast by what Ptolemy called the Great Gulf, Μέγας Κόλπος (Sinus Magnus). Since this was associated by him with the Chinese, the usual explanation is that it refers to the Gulf of Tonkin. By this explanation Cattigara, the chief town of the area, which is on the coast, would be somewhere in the region of Hanoi. This may represent the farthest point to which sea traders from the West penetrated up to the mid-second century A.D. Marinus's texts evidently contained the itinerary of a Greek called Alexander who had sailed to Cattigara, which he described as being an innumerable number of days from Zabai (perhaps in Kampu-

99. Schütte, *Ptolemy's Maps* (note 93).

100. O. A. W. Dilke, "Mapping of the North African Coast in Classical Antiquity," in *Proceedings of the Second International Congress of Studies on Cultures of the Western Mediterranean* (Paris: Association Internationale d'Etude des Civilisations Méditerranéennes, 1978), 154–60.

101. W. Barthel, "Römische Limitation in der Provinz Africa," *Bonner Jahrbücher* 120 (1911): 39–126; Institut Géographique National, *Atlas des centuriations romaines de Tunisie* (Paris: Institut Géographique National, 1954).

102. F. J. Carmody, *L'Anatolie des géographes classiques: Etude philologique* (Berkeley: Carmody, 1976).

103. Ptolemy *Geography* 7.5.4 (note 5), νήσῳ κατὰ τὸ ἀντικείμενον παραπλησίως, translated by O. A. W. Dilke. Another quite incorrect translation of Stevenson's is at book 1, chap. 20, where, in translating Ptolemy's approval of the latitude of 36° given to Rhodes by Marinus, he makes Ptolemy add: "In this he follows almost exactly the method of Epitecartus." No such person existed, and the Greek text here has an adjective ἐπιτέταρτος, which refers to the proportion 4:5 (see above, p. 179 n.17).

104. For details see Ptolemy, *La Géographie de Ptolémée: L'Inde (VII.1–4)*, ed. Louis Renou (Paris: Champion, 1925); Ptolemy, *Ancient India as Described by Ptolemy*, ed. and trans. John Watson McCrindle (Calcutta: Thacker, Spink, 1885; reprinted Chuckervertty, Chatterjee, 1927).

105. For a masterly interpretation of this question, see Paul Wheatley, *The Golden Khersonese: Studies in the Historical Geography of the Malay Peninsula before A.D. 1500* (Kuala Lumpur: University of Malaya Press, 1961).

chea).[106] But in Ptolemy's concept of the world the sharp turn southward, culminating in unknown land to the south of the Indian Ocean, by no means tallies with the Gulf of Tonkin. For this reason a theory advanced in South America by D. E. Ibarra Grasso, bizarre as it may seem at first, cannot entirely be ruled out.[107] According to him, Ptolemy's Great Gulf is really the Pacific, and Ptolemy's area around Cattigara is actually on the west coast of South America, giving the following equivalents:

Cattigara	=	Trujillo, Peru
Rhabana	=	Tumbes, Peru
Promontory of the Satyrs	=	Aguja Point, Peru

He also maintains that the confusion over distance arose because, whereas Marinus had made the *oikoumene* extend very far east, Ptolemy had decided not to extend it beyond 180° east of the Canaries. One may reply to the theory: first, there is no firm evidence that ancient Europeans reached South America; second, if Ptolemy felt that he had reached the end of the space allotted by himself for the *oikoumene*, he was more likely to distort the orientation, as with Scotland, than to squeeze a vast area, supposedly so represented in Marinus, into a very narrow space; third, the connection with the Chinese should not be ignored; and finally, it is not always safe to argue, as Ibarra Grasso does, from early printed editions of Ptolemy.[108] Admittedly, fifteenth- and sixteenth-century navigators, including Christopher Columbus, used maps that had different interpretations of the Sinus Magnus; but there is no reason to think that ancient navigators or cartographers shared a similar perception of the world.

---

106. It is very likely that Marcianus influenced the readings of coordinates for the Asiatic part of Ptolemy's *Geography*: see Polaschek, "Ptolemy's *Geography*," 35–37 (note 67), who also equates the explorer Alexander supposedly in Marinus's texts with Alexander Polyhistor, born ca. 105 B.C. at Miletus.

107. Dick Edgar Ibarra Grasso, *La representación de América en mapas romanos de tiempos de Cristo* (Buenos Aires: Ediciones Ibarra Grasso, 1970); Paul Gallez, "Walsperger and His Knowledge of the Patagonian Giants, 1448," *Imago Mundi* 33 (1981): 91–93; Ptolemy *Geography* 1.14.1 (note 5); Hans Treidler, "Ζάβαι," in *Realencyclopädie*, 2d ser., 9 (1967): cols. 2197–220 (note 75); André Berthelot, *L'Asie ancienne centrale et sud-orientale d'après Ptolémée* (Paris: Payot, 1930).

108. Ibarra Grasso, *Representación de América* (note 107).

BIBLIOGRAPHY
CHAPTER 11 THE CULMINATION OF GREEK CARTOGRAPHY IN PTOLEMY

Bagrow, Leo. "The Origin of Ptolemy's Geographia." *Geografiska Annaler* 27 (1945) 318–87.

Berthelot, André. *L'Asie ancienne centrale et sud-orientale d'après Ptolémée.* Paris: Payot, 1930.

————. "La côte océanique de Gaule d'après Ptolémée." *Revue des Etudes Anciennes* 35 (1933): 293–302.

————. "La carte de Gaule de Ptolémée." *Revue des Etudes Anciennes* 35 (1933): 425–35.

Carmody, F. J. *L'Espagne de Ptolémée.* Berkeley: Carmody, 1973.

————. *L'Anatolie des géographes classiques: Etude philologique.* Berkeley: Carmody, 1976.

Dilke, O. A. W. "Mapping of the North African Coast in Classical Antiquity." In *Proceedings of the Second International Congress of Studies on Cultures of the Western Mediterranean,* 154–60. Paris: Association Internationale d'Etude des Civilisations Méditerranéennes, 1978.

————. *Greek and Roman Maps.* London: Thames and Hudson, 1985.

Honigmann, Ernst. *Die sieben Klimata und die πόλεις ἐπίσημοι.* Heidelberg: Winter, 1929.

Hövermann, Jürgen. "Das geographische Praktikum des Claudius Ptolemaeus (um 150 p.C.n.) und das geographische Weltbild der Antike." *Abhandlungen der Braunschweigischen Wissenschaftlichen Gesellschaft* 31 (1980): 83–103.

Humbach, Helmut. "Historisch-geographische Noten zum sechsten Buch der Geographie des Ptolemaios." *Jahrbuch des Römisch-Germanischen Zentralmuseums Mainz* 19 (1972): 89–98.

Kubitschek, Wilhelm. "Studien zur Geographie des Ptolemäus. I: Die Ländergrenzen." *Sitzungsberichte der Akademie der Wissenschaften in Wien,* Philosophisch-historische Klasse, 215.5 (1935).

Neugebauer, Otto. "Ptolemy's *Geography,* Book VII, Chapters 6 and 7." *Isis* 50 (1959): 22–29.

————. *A History of Ancient Mathematical Astronomy.* New York: Springer-Verlag, 1975.

Newton, R. R. *The Crime of Claudius Ptolemy.* Baltimore: Johns Hopkins University Press, 1977.

Polaschek, Erich. "Noricum in Ptolemaios' Geographie." In *Festschrift für Rudolf Egger,* 2 vols., 2:247–56. Klagenfurt: Verlag des Geschichtsvereins für Kärnten, 1953.

————. "Ptolemy's *Geography* in a New Light." *Imago Mundi* 14 (1959): 17–37.

————. "Ptolemaios als Geograph." In *Paulys Realencyclopädie der classischen Altertumswissenschaft,* ed. August Pauly, George Wissowa, et al., suppl. 10 (1965): cols. 680–833. Stuttgart: J. B. Metzler, 1894–.

Ptolemy. *Claudii Ptolemaei Geographia.* 3 vols. Edited by C.F. A. Nobbe. Leipzig: C. Tauchnitz, 1843–45; reprinted in one volume with an introduction by Aubrey Diller, Hildesheim: Georg Olms, 1966.

————. *Claudii Ptolemaei Geographia.* 2 vols. and tabulae. Edited by Karl Müller. Paris: Firmin-Didot, 1883–1901.

————. *Ancient India as Described by Ptolemy.* Edited and translated by John Watson McCrindle. Calcutta: Thacker, Spink, 1885; reprinted Chuckervertty, Chatterjee, 1927.

————. *Opera omnia.* Edited by J. L. Heiberg. Leipzig: Teubner, 1898–1907; revised, 1961. (Does not include the *Geography.*)

————. *Die Geographie des Ptolemaeus: Galliae, Germania, Raetia, Noricum, Pannoniae, Illyricum, Italia.* Edited by Otto Cuntz. Berlin: Weidmann, 1923.

————. *Claudii Ptolemaei Geographiae Codex Urbinas Graecus 82.* 2 vols. in 4. Edited by Joseph Fischer. Codices e Vaticanis Selecti quam Simillime Expressi, vol. 19. Leiden: E. J. Brill; Leipzig: O. Harrassowitz, 1932.

————. *Ptolemy's Almagest.* Translated by G. J. Toomer. London: Duckworth, 1984.

Richmond, Ian A. "Ptolemaic Scotland." *Proceedings of the Society of Antiquaries of Scotland* 56 (1921–22): 288–301.

Rivet, A. L. F. "Some Aspects of Ptolemy's Geography of Britain." In *Littérature gréco-romaine et géographie historique: Mélanges offerts à Roger Dion,* ed. Raymond Chevallier, 55–81. Caesarodunum 9 bis. Paris: A. et J. Picard, 1974.

————. "Ptolemy's Geography and the Flavian Invasion of Scotland." In *Studien zu den Militärgrenzen Roms, II,* 45–64. Vorträge des 10. Internationalen Limeskongresses in der Germania Inferior. Cologne: Rheinland-Verlag in Kommission bei Rudolf Habelt, 1977.

Rivet, A. L. F., and Colin Smith. *The Place-Names of Roman Britain.* Princeton: Princeton University Press, 1979.

Roscher, Albrecht. *Ptolemaeus und die Handelsstraßen in Central-Africa: Ein Beitrag zur Erklärung der ältesten und erhaltenen Weltkarte.* Gotha: J. Perthes, 1857; reprinted Amsterdam: Meridian, [1971].

Rosenkranz, Bernhard. "Zu einigen Flußnamen des nordwestlichen Sarmatiens bei Ptolemäus." *Beiträge zur Namenforschung* 4 (1953): 284–87.

Schmitt, P. "Recherches des règles de construction de la cartographie de Ptolémée." In *Colloque international sur la cartographie archéologique et historique, Paris, 1970.* Edited by Raymond Chevallier, 27–61. Tours: Centre de Recherches A. Piganiol, 1972.

————. *Le Maroc d'après la Géographie de Claude Ptolémée.* 2 vols. Tours: Centre de Recherches A. Piganiol, 1973.

Schnabel, Paul. *Text und Karten des Ptolemäus.* Quellen und Forschungen zur Geschichte der Geographie und Völkerkunde 2. Leipzig: K. F. Koehlers Antiquarium, 1938.

Schütte, Gudmund. *Ptolemy's Maps of Northern Europe: A Reconstruction of the Prototypes.* Copenhagen: Royal Danish Geographical Society, 1917.

Stahl, William Harris. *Ptolemy's Geography: A Select Bibliography.* New York: New York Public Library, 1953.

Svennung, Joseph Gusten Algot. *Scandinavien bei Plinius und Ptolemaios.* Uppsala: Almqvist och Wiksell, 1974.

Temporini, Hildegard, and Wolfgang Haase, eds. *Aufstieg und Niedergang der römischen Welt.* Berlin: Walter de Gruyter, 1972–.

Thouvenot, R. "La côte atlantique de la Libye d'après la Géographie de Ptolémée." In *Hommages à la mémoire de Jérome Carcopino,* ed. R. Thomasson, 267–75. Paris: Belles Lettres, 1977.

Tierney, James J. "Ptolemy's Map of Scotland." *Journal of Hellenic Studies* 79 (1959): 132–48.

Tudeer, Lauri O. T. "On the Origin of the Maps Attached to Ptolemy's Geography." *Journal of Hellenic Studies* 37 (1917): 62–76.

————. "Studies in the Geography of Ptolemy." *Suomalaisen tiedeakatemian toimituksia: Annales Academiæ Scientiarum Fennicæ,* ser. B, 21.4 (1927).

Vidal de la Blache, Paul. *Les voies de commerce dans la Géographie de Ptolémée.* Paris: Imprimerie Nationale, 1896.

12 · Maps in the Service of the State: Roman Cartography to the End of the Augustan Era

O. A. W. DILKE

Whereas the Greeks, particularly in Ionia in the early period and at Alexandria in the Hellenistic age, made unparalleled strides in the theory of cosmology and geography, the Romans were concerned with practical applications. This contrast is sometimes exaggerated, yet it can hardly be avoided as a generalization when seeking to understand the overall pattern of cartographic development in the classical world as well as its legacy for the Middle Ages and beyond. Roman writers did not attempt to make original contributions to subjects such as the construction of map projections or the distribution of the *climata*; most cartographic allusions in the literary texts—as well as the surviving maps—are connected with everyday purposes. Whether used for traveling, for trade, for planning campaigns, for establishing colonies, for allocating and subdividing land, for engineering purposes, or as tools of the law, education, and propaganda to legitimize Roman territorial expansion, maps ultimately were related to the same overall organizational ends. In cartography, as in other aspects of material culture, ideas first nurtured in Greek society were taken over and adapted to the service of the Roman state.

It is not only by chance, therefore, that it is in three particular applications of mapping—road organization, land survey for centuriation, and town planning—that Roman maps, or descendants of them, have survived. In the first two Rome was preeminent both in accuracy and in output. Such extant maps, while more numerous than those surviving from ancient Greece, are nevertheless only a tiny fraction of the numbers that were originally produced in the Roman period. The value of the media used—many were cast in metal or painted or carved on stone—contributed directly to their demise. The metals were melted down, and the stones were reused for other purposes in the less organized way of life that followed the fall of Rome. But despite the many gaps in our knowledge arising from such factors, Roman mapping is sufficiently distinctive—in both its impulses and its products—to be treated as a series of discrete chapters in the cartographic history of the classical period.

ETRUSCAN BEGINNINGS

While it is traditionally accepted that Rome owed most of its cartographic knowledge to Greek influences, we cannot rule out the possibility that it may have received independent ideas and practices from Etruscan concepts of cosmology and orientation and even of land division and survey.[1] It was disputed in antiquity, and it is still disputed, whether the Etruscans came to central Italy from Asia Minor or were indigenous. From their homeland in Etruria they expanded southward, and also northeastward into the Po valley (fig. 12.1). By about 500 B.C. they had developed a considerable empire, but it declined owing to Gallic invasions, internal disputes, and the expulsion of their Tarquin dynasty from Rome. The Etruscans were a literate people, well versed in Greek mythology, though their non-Indo-European language is still only partly understood.[2] Artistic and religious, with a respect for divination and a great concern for the afterlife, they had also a practical side that left its imprint on town planning, drainage, tunneling, and administration. The best example of Etruscan planning is Marzabotto, in the northern Apennines, where the rectangular grid pattern of streets is reminiscent of Greek colonies in the West. One may conjecture that their interest in cosmology was reflected in the planning of towns and temples, and certainly the architectural and building skills required in these developments—as with other early societies[3]—may have required simple instruments and measurement, similar to those used in mapping, even where there is no evidence that maps were drawn as part of the design process.

Likewise it may be noted that a number of aspects of Etruscan culture required accuracy in orientation.[4] There may have been a different usage according to whether sky or earth was involved. The elder Pliny writes: "The Etruscans . . . divided the heaven into 16

1. The standard work is Massimo Pallottino, *The Etruscans*, ed. David Ridgway, trans. J. Cremona (London: Allen Lane, 1975).

2. Giuliano Bonfante and Larissa Bonfante, *The Etruscan Language: An Introduction* (Manchester: Manchester University Press, 1984).

3. This will be discussed in relation to south Asia in volume 2 of the present *History*.

4. Carl Olof Thulin, *Die etruskische Disciplin* . . . 3 pts. (1906–9; reprinted Darmstadt: Wissenschaftliche Buchgesellschaft, 1968), pt. 2, *Die Haruspicin*, index s.v. "Orientierung"; Pallottino, *Etruscans*, 145 and fig. 5 (note 1).

parts: (1) from north to east, (2) to south, (3) to west, (4) the remainder from west to north. These parts they then subdivided into four each, and called the eight eastern subdivisions 'left,' the eight on the opposite side 'right.' "[5] This implies that when, for example, a soothsayer was divining from lightning, he would face south. Evidently lightning on the left was considered lucky by the Etruscans, especially if it was visible in front of the soothsayer as well as behind him. It is possible that this division into sixteen parts led later to sixteen-point wind roses as against twelve-point ones.

before the rise of Rome. Against this, literary statements have to be treated with caution: in the case of the extant Etruscan temples, many face to the south, rather than to the west as was apparently held by Varro.

There is one exception, albeit extremely difficult to interpret, to this complete lack of map artifacts surviving from Etruscan culture. This is the Bronze Liver of Piacenza, an unusual religious relic known for over a century, which incorporates a maplike image on part of its surface. This bronze representation, 12.6 centimeters long, of a sheep's liver was found in 1877 between Set-

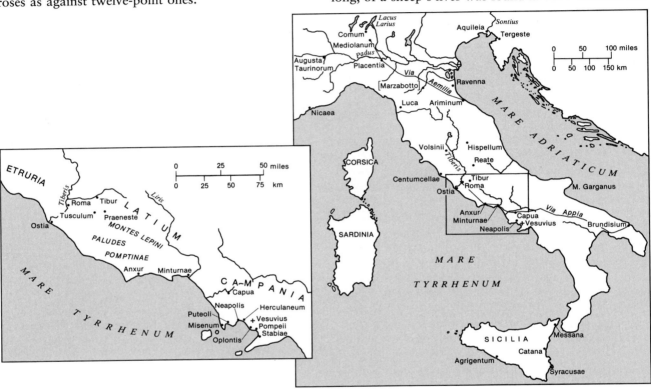

FIG. 12.1. PRINCIPAL PLACES ASSOCIATED WITH MAPS IN ANCIENT ITALY AND SICILY.

When we come to land division, however, Etruscan orientation seems to have been west-facing. Frontinus says: "The origin of centuriation, as Varro observed, is in the Etruscan lore, because their soothsayers divided the earth into two parts, calling that to the north 'right' and that to the south 'left.' They reckoned from east to west, because the sun and moon face that way (*eo spectant*); just as some architects have written that temples should correctly face west."[6] If we accept that some form of centuriation—a system of surveying land in rectangular parcels—was already being practiced in the Etruscan period,[7] then a corollary might be that the land survey methods so characteristic of Roman society, and giving rise to the *agrimensores*, also had their beginnings

tima and Gossolengo and is in the Museo Civico, Piacenza (figs. 12.2 and 12.3).[8] There are no grounds for doubting its genuineness; it dates from about the third century B.C. and has been called a map by some scholars

5. Pliny *Natural History* 2.55.143, author's translation; for an English edition see Pliny *Natural History*, 10 vols., trans. H. Rackham et al., Loeb Classical Library (Cambridge: Harvard University Press; London: William Heinemann, 1940–63).

6. Frontinus *De limitibus* (On centuriation), in *Corpus Agrimensorum Romanorum*, ed. Carl Olof Thulin (Leipzig, 1913; reprinted Stuttgart: Teubner, 1971), 10–15, quotation on 10–11, author's translation; O. A. W. Dilke, "Varro and the Origins of Centuriation," in *Atti del Congresso Internazionale di Studi Varroniani* (Rieti: Centro di Studi Varroniani, 1976), 353–58.

7. O. A. W. Dilke, *The Roman Land Surveyors: An Introduction to the Agrimensores* (Newton Abbot: David and Charles, 1971), 33–34, and for a definition of centuriation, 15–16.

8. G. Körte, "Die Bronzeleber von Piacenza," *Mitteilungen des Kaiserlich Deutschen Archäologischen Instituts, Römische Abteilung* 20

FIG. 12.2. THE BRONZE LIVER OF PIACENZA. A religious relic of the third century B.C., this representation of a sheep's liver has a maplike image on part of its surface. This artifact can be most easily explained as a form of cosmological map.

Size of the original: 7.6 × 12.6 cm. By permission of the Museo Civico, Piacenza.

in this century. Its top side consists of protuberances modeled schematically on those of a sheep's liver and of a flat section divided into boxes representing zones. Each of these has its Etruscan inscription with the name of a deity.[9] The convex underside is divided into two sections, inscribed only with the Etruscan words for sun and moon. The liver was clearly associated with the *disciplina Etrusca*, the art whereby their soothsayers divined the will of the gods by inspecting entrails. A parallel has been drawn with a Chaldean terra-cotta liver in the Budge collection of the British Museum.[10]

On the right of the Piacenza liver's upper side is a pyramid representing the *processus pyramidalis* of the liver. Unlike the left half, the flat part of which has a radial subdivision, under the pyramid are roughly rectangular boxes, that some scholars have associated with the *cardines* and *decumani* of centuriation,[11] considered by Varro to have Etruscan origins.

In other respects the segmented images suggest a cosmological representation of part of the heavens. Martianus Capella, the fifth-century A.D. author of the encylopedic work *De nuptiis Philologiae et Mercurii*, tells us that the ancients divided the soothsayer's view of the

heavens into favorable and unfavorable areas, each represented by particular deities, whose Roman equivalents he gives.[12] Many of those on the Piacenza liver can indeed be equated, such as Tin to Jupiter, Uni to Juno, Fufluns to Liber, Maris to Mars. But attempts to make Martianus's scheme tally with that of the liver have not

(1905): 348–77; Thulin, *Die etruskische Disciplin* (note 4); Massimo Pallottino, *Saggi di Antichità*, 3 vols. (Rome: G. Bretschneider, 1979), 2:779–90.

9. Etruscan lettering, based on early Greek, is quite legible; it is written from right to left. Nevertheless, in Otto-Wilhelm von Vacano, *The Etruscans in the Ancient World*, trans. Sheila Ann Ogilvie (London: Edward Arnold, 1960), 22, and in Tony Amodeo, *Mapline* 14 (July 1979), the Piacenza liver is illustrated in such a way that most of the inscriptions are upside down. James Welland, *The Search for the Etruscans* (London: Nelson, 1973), 146–47, reverses the diagram so that the writing looks as though it were from left to right.

10. London, British Museum, Budge 89-4-26. 238.

11. See Pallottino, *Etruscans*, 164 (note 1), for reference to *cardo* and *decumanus* in Etruscan sacred space.

12. Martianus Capella *De nuptiis Philologiae et Mercurii* 1.45 ff.; see *The Marriage of Philology and Mercury*, trans. William Harris Stahl and Richard Johnson with E. L. Burge (New York: Columbia University Press, 1977); vol. 2 of the series Martianus Capella and the Seven Liberal Arts.

FIG. 12.3. THE BRONZE LIVER OF PIACENZA, SIDE VIEW.

Height of the original: 6 cm. By permission of the Museo Civico, Piacenza.

been entirely satisfactory: it suits Martianus's text best to postulate north at the top, whereas the convex side with "sun" and "moon" suggests west at the top.[13] One must say, therefore, that the liver has a not entirely agreed upon place in the history of religious cartography. But if it cannot be ignored in any attempt to understand Etruscan ways of representing space, it also points to the cosmological use of mapping, whatever its practical foundation, as the dominant motive behind such a representation.

GEOGRAPHICAL AND CADASTRAL MAPS FROM THE REPUBLICAN PERIOD

Rome, thought by the ancients to have been founded in 753 B.C.,[14] developed as a pastoral community on a salt route from the mouth of the Tiber to the hinterland. For two and a half centuries it was ruled by kings, several of whom had strong Etruscan links; the potential influence on Rome of Etruscan concepts of cosmology and orientation has already been discussed. During the early years of the Republican period that followed, Rome also came into contact with the Greek maritime trading colonies of southern Italy, and the long struggles of Rome against Carthage in the third and second centuries B.C. vastly widened her horizons.

The nature of Roman territorial expansion exerted a strong influence on the type of maps it was eventually to generate. Unlike the Greeks—whose cartographic expertise came mainly from theoretical formulations, celestial observations, and maritime explorations—the Romans first expanded by land, and we may conjecture that their earliest rudimentary plans may have been of the small maritime defense colonies set up with land allocations in the fourth century B.C. or of main roads such as the Via Appia, leading in the first place (312 B.C.) to Capua. These roads were provided with milestones, and at a later stage guides to a number of them were compiled in the form of itineraries.

Not until the second century B.C. do we hear of the first two Roman maps. One was concerned with the annexation of Sardinia from the Carthaginian empire; the other was a land survey map relating to Campania and arising from the appropriation and redistribution

13. Pallottino, *Saggi di Antichità*, 2:779–90 (note 8), considers that the axis of the liver was not north-south but perhaps NNW-SSE, corresponding to the orientation of a particular temple.

14. The date may be relevant to the interpretation of the time scales of the history of cartography, since the system of chronology based on years B.C. is comparatively recent. Since 753 B.C. was year 1, conversion is effected by subtracting from 754; e.g., 100 B.C. is A.U.C. 654 (*ab urbe condita*, from the foundation of the city). This chronology has not been adopted here.

of lands after Capua had sided with Hannibal. These earliest allusions in textual sources immediately identify the distinctive nature of Roman mapping. It was both political and practical in character and, above all, was concerned in various ways either with geographical expansion or with the organization and exploitation of settled lands thus brought under political control.

GEOGRAPHICAL MAPS

The earliest mention of a Roman map is attributed to Tiberius Sempronius Gracchus, father of the Gracchi.[15] We are told that in 174 B.C., after his victory in Sardinia, he dedicated to Jupiter, in the temple of Mater Matuta in Rome, a tablet (*tabula*) consisting of a map (*forma*) of Sardinia. On this was an inscription, perhaps originally in Saturnian verse, and pictures of the battles the general had fought. Matuta, whose temple was in the Forum Boarium, was originally a dawn goddess but came to be associated with the deified Greek heroine Leucothea, regarded as a goddess of seas and harbors. The religious connotation is important. By displaying images symbolizing the conquered land, the Romans wished to propitiate appropriate deities. In the same way, personifications of large rivers in enemy countries were carried in triumphal processions, so that Jupiter and other deities might favor Roman military exploits. One should not think of such maps as containing conventional cartographic detail; they were pictorial maps, perhaps with brief sentences recording the victories.[16] We learn that in 164 B.C. a Greek *topographos* (topographical writer or landscape illustrator), Demetrius of Alexandria, was living in Rome. This suggests that there were already a few specialized artists skilled in executing such maps, which, despite the comparatively early date, contained a strong element of propaganda.

Maps may also have been used by writers in helping to compile the histories of the same period, though we have no proof. For example, Cato the Elder (234–149 B.C.) wrote his *Origines*, a work on the origins of Italian cities and tribes, now lost except for fragments, between 168 and 149 B.C. In one fragment he says that the length of Lake Larius (Lake Como) is sixty Roman miles;[17] but this hardly justifies the statement that he must have had a map available.[18]

A more convincing example comes to us from the late Republic, during which period Varro's encyclopedic interests over a long life (116–27 B.C.)[19] may lead us to guess he was very familiar with maps. He sets the scene of his *De re rustica* at the temple of Tellus (Mother Earth)[20] and gives his speakers names associated with the land. They are said to be *spectantes . . . pictam Italiam*, literally "looking at Italy painted" on a wall of this temple or of its portico. This must surely be a map of Italy, not a painting of a personification of it. After

they have seated themselves, the philosopher among them goes on: "The world was divided by Eratosthenes in an essentially natural way, towards north and towards south. There is no doubt that the northern part is healthier than the southern and likewise more fertile."[21] He then compares Italy with Asia Minor and discusses the regions of Italy from the point of view of farming. Again one assumes he was pointing to the map and that such maps were used—as the Greek sources have also indicated—as a regular aid to teaching.[22]

An even more potent force in the development of Roman cartography was, however, geopolitical. There can be little doubt that by the late Republican period Roman rulers and their advisers had come to recognize the value of geographical maps in both administration and propaganda. In particular, it is in this light that the truly imperial scheme initiated by Julius Caesar (100–44 B.C.)—to undertake a survey of the known world—can be interpreted. Even if this was not accompanied by maps and was not completed until the Augustan era, its raw materials were drawn upon for Agrippa's world map.

Caesar's project is known to us from three late sources: first, the *Cosmographia Iulii Caesaris*; second, an anonymous *Cosmographia*;[23] and third, the Hereford world map in which Caesar or Augustus, enthroned, is shown delivering a mandate for the survey of the world (fig. 12.4; see also below, pp. 207 and 309). In 44 B.C., we are told in the *Cosmographia Iulii Caesaris*, four geographers were appointed to measure the four quarters of the earth; if we may believe the ancient sources, they

15. Livy [*History of Rome*] 41.28.8–10, in *Livy*, 14 vols., trans. B. O. Foster et al., Loeb Classical Library (Cambridge: Harvard University Press; London: William Heinemann, 1919–59).

16. Roger Ling, "Studius and the Beginnings of Roman Landscape Painting," *Journal of Roman Studies* 67 (1977): 1–16, esp. 14 and nn. 54–55.

17. Cato the Elder *Origines* 2, fr. 7; see *Originum reliquiae* in M. *Catonis praeter librum de re rustica quae extant*, ed. H. Jordan (Leipzig: Teubner, 1860), 10, and Servius *Commentary on Virgil's Georgics* 2.159, in vol. 3 of *Servii Grammatici qui feruntur in Vergilii carmina commentarii*, 3 vols., ed. Georg Thilo and Hermann Hagen (Leipzig: Teubner, 1881–1902; reprinted Hildesheim: Georg Olms, 1961.

18. See the comment by Jacques Heurgon in his edition of Varro's *De re rustica: Economie rurale: Livre premier* (Paris: Belles Lettres, 1978), 102.

19. Varro was a prolific author and editor, but only two works have survived substantially: *Rerum rusticarum libri III* and *De lingua Latina libri XXV* (books 5–10 preserved in full).

20. Varro *De re rustica* 1.2.1 (note 18).

21. Varro *De re rustica* 1.2.3 (note 18). author's translation.

22. See pp. 254–56.

23. *Cosmographia Iulii Caesaris* and *Cosmographia*, both in *Geographi Latini minores*, ed. Alexander Riese (Heilbronn, 1878; reprinted Hildesheim: Georg Olms, 1964), 21–23 and 71–103, respectively. For a translation of the former see O. A. W. Dilke, *Greek and Roman Maps* (London: Thames and Hudson, 1985), 183.

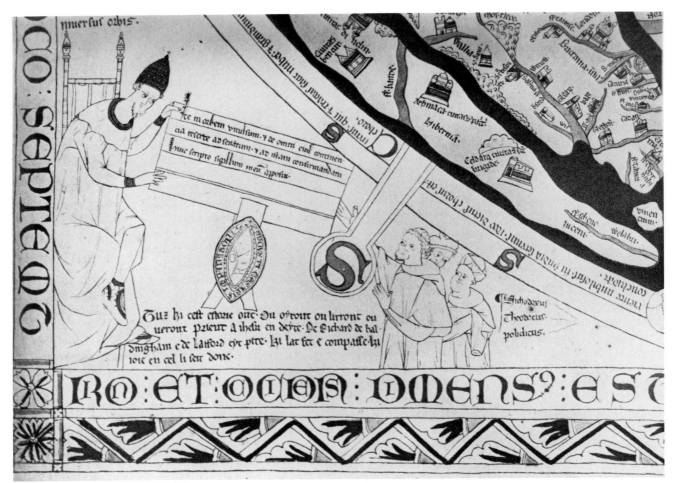

FIG. 12.4. CAESAR'S EDICT ON THE HEREFORD WORLD MAP. Augustus Caesar—whose seal is on the order—is seen ordering a survey of the whole world in this detail from a thirteenth-century *mappamundi*, but the three geographers to whom he is seen entrusting this order belong to the tradition of Julius Caesar's survey.

Size of the original detail: 26.8 × 33 cm. From a facsimile, by permission of the Dean and Chapter of Hereford Cathedral.

each took from twenty-one and a half to thirty-two years to complete their work. The names of the geographers are all Greek, and they were probably freedmen.[24] The periods of work and years of completion do not tally, but the discrepancies are not immediately apparent in the texts, since the year of completion for each was expressed not in years after the foundation of Rome but in names of consuls, the customary method in classical times. The delimitation of the four regions can, however, be guessed at from the anonymous *Cosmographia*: the East is all to the east of Asia Minor; the West is all Europe except Greece, Macedonia, and Thrace; the North contains these three regions (Greece perhaps because it was conquered by Macedonia) and Asia Minor; and the South is Africa. The definition of *provincia* (province) is inconsistent with legal status at any period, and the term is made to extend outside the empire. Per-

haps the geographers' names and work periods are derived from documents of the Julian period,[25] together with the fourfold division, though Caesar's definitions, if they are his, are somewhat different from those of Varro, who divided the inhabited world into two, Europe and Asia-Africa.

One of the few surviving geographical maps of the pre-Augustan period, a recent chance find,[26] may have been associated with Caesar's Gallic campaigns. A block

24. *Cosmographia Iulii Caesaris* 1 (note 23).

25. One manuscript of the *Cosmographia Iulii Caesaris*, Rome, Biblioteca Apostolica Vaticana, Vat. Pal. 73, actually associates that treatise with the measuring of the world instituted by Julius Caesar; and the texts of other manuscripts of the Julian period also imply such a link.

26. Pierre Camus, *Le pas des légions* (Paris: Diffusion Frankelve, 1974), front cover, and for plan of fort, 62.

of local sandstone, with maximum length and width fifty-six by forty-seven centimeters and thickness on average fourteen centimeters, was found in 1976 near the center of the Roman camp of Mauchamp, between Juvincourt and the river Aisne, and is now at Brie Comte Robert (fig. 12.5). It has apparently been worked with a chisel on the sides and will stand up with north roughly at the top. If it is a map of Gaul, as the finder claimed, the western coastline is recognizable, while the other sides could be thought to follow the frontiers of Gaul. Three holes made in a line could represent the Gallic religious centers of Puy de Dôme, Autun, and Grand.

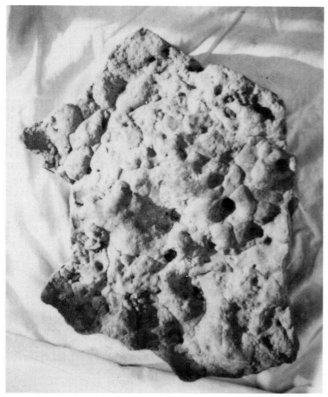

FIG. 12.5. THE STONE "MAP OF GAUL" FROM THE ROMAN CAMP AT MAUCHAMP, FRANCE. Claimed to be a map of Roman Gaul, and perhaps associated with Caesar's Gallic campaigns, the western coastline of the province is clearly recognizable. The three prominent, equally spaced holes—approximately in a straight line from southwest to northeast—may represent the Gallic religious centers of Puy de Dôme, Autun, and Grand.
Size of the original: 56 × 47 cm. By permission of the heirs of the late Pierre Camus, Brie Comte Robert, France.

AGRIPPA'S WORLD MAP

The surveys and measurements ordered by Caesar, but according to our source put in hand only in the year of his death, were thus largely carried out during the reign (27 B.C. to A.D. 14) of his successor, the emperor Augustus. Presumably they took the form largely of itineraries measured along the road network of the empire; and they were used, as mentioned, in the compilation of Agrippa's map. Although this world map, like so many other maps from the classical period, is known to us only from literary texts, some authorities claim it is the most important map in Roman cartography.[27] On the basis of statements by a number of ancient and medieval writers, it is believed to be the prototype for a succession of later world maps such as the thirteenth-century Hereford *mappamundi*.[28]

Political developments under Augustus favored the compilation of such a map. The establishment of the principate, or system of supreme control under an emperor, which effectively took place in 27 B.C., had important results for literary and scientific output. Augustus, often through his ministers, saw to it that as much of this as possible was aligned with his propaganda extolling Rome, the empire, and its leader. As part of the settlement of veterans after the long period of civil wars, Augustus set up colonies, settlements with smallholdings, numbering twenty-eight in Italy and over eighty in the provinces, though some of each were refounded rather than founded in the true sense. To encourage trade inside and outside the Roman Empire and to promote the establishment of colonies, he was clearly keen that a world map should be drawn up and publicly displayed. The man to whom this task was allotted was well connected. M. Vipsanius Agrippa (b. ca. 63 B.C.) had been the admiral of Octavian (later Augustus) at the battle of Actium, 31 B.C., at which Antony and Cleopatra were routed; and after the establishment of the principate he became Augustus's right-hand man, marrying as his third wife Augustus's daughter Julia and being expected to succeed him as emperor.

The world map of Agrippa was set up in the colonnade named after him, Porticus Vipsania, in what is now the Via del Corso area of Rome. This colonnade is also sometimes referred to as the Porticus of the Argonauts. When Agrippa died the work was completed first by Agrippa's sister, Vipsania Polla, then by Augustus.[29] The elder Pliny (A.D. 23/24–79) expresses surprise that the

27. Armando Cortesão, *History of Portuguese Cartography*, 2 vols. (Coimbra: Junta de Investigações do Ultramar-Lisboa, 1969–71), 1:148. See also James J. Tierney, "The Map of Agrippa," *Proceedings of the Royal Irish Academy* 63, sec. C, no. 4 (1963): 151–66. Dilke, *Greek and Roman Maps*, pp. 41–54 (note 23).

28. Gerald R. Crone, *Maps and Their Makers: An Introduction to the History of Cartography*, 5th ed. (Folkestone: Dawson; Hamden, Conn.: Archon Books, 1978). See also chap. 20 below.

29. Dio Cassius *Roman History* 55.8.4 says that it was not yet completed in 7 B.C.; see *Dio's Roman History*, 9 vols., trans. Earnest Cary, Loeb Classical Library (Cambridge: Harvard University Press; London: William Heinemann, 1914–27), vol. 6.

map gave the length of the southern Spanish province of Baetica as 475 Roman miles and its breadth as 258.[30] That, he says, was correct only earlier, when its boundaries extended as far as Carthago Nova (Cartagena). He adds: "Who would believe that Agrippa, who was very careful and took great pains over this work, should, when he was going to set up the map to be looked at by the citizens of Rome, have made this mistake, and together with him the deified Augustus? For it was Augustus who, when Agrippa's sister had begun building the portico, carried it out from the intention and notes (*commentarii*) of M. Agrippa."[31] This comment ascribes to the emperor rather more personal attention to the map than seems likely. He was happy for his name to be linked with a work of potential propaganda, but apart from his association with the large obelisk sundial with a grid on the gnomonic projection recently excavated in Rome,[32] had not the same interest in scientific research as had Julius Caesar.

The dimensions of the map are not known, but it must have been rectangular, not circular.[33] It is thought that its height may have been between two and three meters, and its width greater. Like the later Ptolemy and Peutinger maps, it very likely had north at the top.[34] Whether the map was carved in or painted on marble is disputed.

It is thought from several passages of Pliny's *Natural History*, in addition to the mention of notes, that Agrippa's map was accompanied by a commentary. Detlefsen argued against such a supposition, but Gisinger supported it.[35] In the *Natural History* we are told: "Agrippa calculates this same distance [the length of the Mediterranean], from the Straits of Gibraltar to the Gulf of Iskenderun, as 3,440 miles, but I am inclined to think there is a mistake in this, since he also gives the distance from the Straits of Messina to Alexandria as 1,350 miles."[36] The word "also" in the final clause of this passage suggests a second Agrippa source, and the verb *existimavit* (thought), more suited to a text than to a map, is used twice elsewhere of Agrippa. Moreover, much of the information on the west coast of Africa attributed to Agrippa concerns animals and is hardly suitable for inclusion in an official map.[37] Another such informative passage credits him with writing that the whole Caspian coast from the river Casus consists of very high cliffs, which prevent landing for 425 miles.[38] Since Agrippa died while composing the map, however, the commentary may have been incomplete, and relevant extracts may have been incorporated in the map.

The chief writer who quotes Agrippa as a source is the elder Pliny. He lists him among his sources both for the geography of the Roman Empire and for areas outside it, sometimes quoting him by name, sometimes not. He implies that by his time (the *Natural History* was

completed in A.D. 77) Agrippa's map may well have been out of date in certain respects. For example, just after Agrippa's measurement of Media, Parthia, and Persis, he mentions the town of Charax in Mesopotamia.[39] This is said to have been originally 1¼ miles from the Persian Gulf, on which it was placed by Agrippa; by the time Juba was writing, it was said to be 50 miles inland; by Pliny's time it was 120 miles inland.

Two later treatises clearly indebted to Agrippa are the *Divisio orbis terrarum* and the *Dimensuratio provinciarum*;[40] their measurements sometimes agree, sometimes disagree with those given by Pliny. Similarly, the Irish writer Dicuil (fl. 814–25), although he gives as his authorities the elder Pliny and the mapmakers of Theodosius II, also mentions the "Chorographia" of Augustus, that is, Agrippa's map, as the first to give the threefold division of the world into Europe, Asia, and Africa.[41] From the evidence of these treatises Detlefsen calculated that there were twenty-four regions of the

30. Pliny *Natural History* 3.1.16 ff. (note 5).

31. Pliny *Natural History* 3.2.17 (note 5), author's translation.

32. Pliny *Natural History* 36.14.71–72 (note 5). The sundial excavated was a post-Augustan restoration using Augustan material. Edmund Buchner, "Horologium Solarium Augusti: Vorbericht über die Ausgrabungen 1979/80," *Mitteilungen des Deutschen Archäologischen Instituts, Römische Abteilung* 87 (1980): 355–73, and subsequent reports in this periodical; idem, *Die Sonnenuhr des Augustus: Nachdruck aus RM 1976 und 1980 und Nachtrag über die Ausgrabung 1980/1981* (Mainz: von Zabern, 1982). See review by Andrew Wallace-Hadrill in *Journal of Roman Studies* 75 (1985): 246–47.

33. Konrad Miller, *Mappaemundi: Die ältesten Weltkarten*, 6 vols. (Stuttgart: Roth, 1895–98), 6:145–47, imagined it as being circular, but that would hardly have been suited to a colonnade.

34. It could well, however, have been south, as supposed by F. Gisinger, "Geographie," in *Paulys Realencyclopädie der classischen Altertumswissenschaft*, ed. August Pauly, Georg Wissowa, et al. (Stuttgart: J. B. Metzler, 1894–), suppl. 4 (1924): cols. 521–685, esp. col. 645. This was a classical as well as nonclassical (e.g., Arabic) orientation for world maps. See Ferdinando Castagnoli, "L'orientamento nella cartografia greca e romana," *Rendiconti della Pontificia Accademia Romana di Archeologia* 48 (1975–76): 59–69.

35. D. (S. D. F.) Detlefsen, *Ursprung, Einrichtung und Bedeutung der Erdkarte Agrippas*, Quellen und Forschungen zur Alten Geschichte und Geographie 13 (Berlin: Weidmann, 1906); Gisinger, "Geographie," col. 646 (note 34). The best analysis is by A. Klotz, "Die geographischen Commentarii des Agrippa und ihre Überreste," *Klio* 24 (1931): 38–58, 386–466.

36. Pliny *Natural History* 6.38.207 (note 5), author's translation.

37. Pliny *Natural History* 5.1.9–10 (note 5).

38. Pliny *Natural History* 6.15.39 (note 5).

39. Pliny *Natural History* 6.31.137–38 (note 5). This was the home of the geographer Isidorus (not to be confused with Isidore of Seville).

40. *Divisio orbis terrarum* and *Dimensuratio provinciarum*, both in Riese, *Geographi Latini minores*, 15–20 and 9–14, respectively (note 23).

41. Dicuil *De mensura orbis terrae* (On the measurement of the earth) 1.2; see *Liber de mensura orbis terrae*, ed. and trans. James J. Tierney, Scriptores Latini Hiberniae, no. 6 (Dublin: Dublin Institute for Advanced Studies, 1967).

TABLE 12.1 Ancient Measurements for Sicily in Roman Miles

Source	East Coast	South Coast	North Coast	Circumference
Agrippa (in Pliny, *Natural History*, 3.86)	—	—	—	618[a]
Strabo 6.2.1, sea distance	159	over 165[b]	263	—
Strabo 6.2.1, land distance	168	—	235[c]	—
Pliny, *Natural History* 3.87, land distance	186[d]	200	142[e]	—

[a]Detlefsen proposed emending to 528. For manuscript readings see Pliny, *Naturalis historiae libri XXXVII*, ed. Carolus Mayhoff (Leipzig: Teubner, 1906), vol. 1. The Loeb text is not reliable on these.

[b]Four stages add up to 165 miles, but there should be a fifth to reckon in. Strabo also has figures in stades from Posidonius.

[c]By Via Valeria.

[d]Emended by some editors to 176.

[e]Emended by some editors to 242.

world on Agrippa's map.[42] When Strabo quotes as his source "the chorographer," with entries not in stades but in Roman miles, it has usually been thought that he too refers to Agrippa. This certainly suits the time when Strabo was in Rome, but the measurements do not always agree. Table 12.1, giving in Roman miles the varying estimates of coastal lengths for Sicily, illustrates these discrepancies. Since Ptolemy distorts the shape of Sicily, we may imagine that Agrippa did so in like fashion. Pliny's figure of 200 miles for the south coast of Sicily corresponds to the figure of 1,600 stades given by Agathemerus; and he need not have turned to Agrippa in this case. But there is always the possibility that Strabo was drawing on the measurements instituted by Julius Caesar. The complicated relationship between these various writers and the maps they used seems to confirm that Agrippa's map inaugurated a new line of world maps. But so many are the anomalies in our evidence for the stemma that we must also conclude that many of the links in the chain of its transmission are now lost to us forever.

Either Agrippa's map or his notes or both had measurements of length and breadth, in Roman miles, for each region. Such measurements are reported by the elder Pliny, writing some eighty years later, and in eighteen cases he specifically cites them as coming from Agrippa—more probably from the map than from the notes. It is evident that some of these measurements had been very roughly rounded off. Thus Britain is cited as being 800 Roman miles long and 300 wide; Ireland, 600 by 300 miles. When, on the other hand, we come to Gaul, which from Caesar's Gallic Wars onward was much better known, we find Gallia Narbonensis given as 370 by 248 miles, the rest as 420 by 313.[43] In the case of India, the measurement 3,300 by 1,300 miles foreshadows the figures we may deduce from Ptolemy's coordinates—much too short from north to south. It is feasible to reconstruct the possible appearance of Agrippa's world map, and this is being attempted.[44] But there must always be serious doubts about the accuracy of such reconstructions, since the data are extremely frag-

mentary. Agrippa clearly did not record latitudes and longitudes, and we cannot always trust the figures preserved in the Pliny manuscripts, while later sources are far less reliable. Sea distances are particularly unreliable, since near coasts they may have been measured either from cape to cape or following the indentations of coastlines.

Despite its obvious failings, Agrippa's world map represented new work of a practical Roman type, which must often have been based on data from the extensive network of Roman roads. It lasted, was probably revised, and may have been available in copies erected elsewhere. The practice of erecting a world map on a city colonnade seems to have persisted right down to the late empire. Much later, for example, the rhetorician Eumenius, born about A.D. 264, wrote of the school at Augustodunum (Autun) in Gaul, the rebuilding of which he was lavishly subsidizing after war damage: "Let the boys and girls see on the colonnades all the lands and all the seas . . . the points where rivers rise and where they have their mouths, and the extent of bays."[45] Finally, in assessing the map's influence on cartographic practice in the Roman world, it may be noted that revisions to the "monuments of the ancients" carried out as late as the fifth century A.D. on the orders of Theodosius II are also thought to have been based upon Agrippa's map (see p. 259).

CADASTRAL MAPS

The first Roman land survey map recorded can be dated to 170–165 B.C. The historian Granius Licinianus tells us that Publius Cornelius Lentulus, consul suffectus 162 B.C., when he was urban praetor, had been authorized by the Senate to reclaim state land in Campania, the

42. Detlefsen, *Ursprung der Erdkarte Agrippas*, 21–22 (note 35).
43. The figure 420 is far too small.
44. By John H. Bounds, Sam Houston State University, Texas.
45. Eumenius *Oratio pro instaurandis scholis* 20.2, author's translation; see 9(4) in *XII [Duodecim] Panegyrici Latini*, ed. R. A. B. Mynors (Oxford: Clarendon University Press, 1964), 242.

whole of which had been occupied by private individuals.[46] He therefore bought up for the state 50,000 *iugera* (12,600 hectares) of land and had a map of this area incised on bronze and affixed to the Atrium Libertatis (Hall of Freedom) in Rome. Granius adds that Sulla later (82–79 B.C.) "corrupted" this map, that is, presumably had it considerably changed for political ends. The Atrium Libertatis[47] was a building near the Curia (senate house) particularly associated with slaves and freedmen. This map, since it resulted from a cadastral survey, is likely, except in its material, to have been similar to the cadastral maps of the Roman colony of Orange (Arausio), see below pp. 220–25. The centuriation of Campania is in places well preserved on the land,[48] reflecting mainly later allocations such as those under Gracchus's sons and under Julius Caesar. Inscribed centuriation stones have been found showing that, at least in the allocations by the Gracchi, the survey was carried out carefully.

The state's acceptance of the value of maps was also reflected in the archival provision made for their preservation. In 78 B.C., during the late republic, a Roman state record office, the *tabularium*, was constructed on the slopes of the Capitoline, in what are now the substructures of the Palazzo dei Conservatori. It was this building that, among other things, served as the repository of land surveyors' bronze maps, one copy of which was kept centrally and one locally. Whether these were destroyed, like the tablets in the temple of Jupiter, in the civil war of A.D. 69 is uncertain.[49] At least none has survived. The maps seem in some cases to have been accompanied by a ledger called *liber aeris*, which literally means "book of bronze": it denoted not the substance of the ledger but the fact that it was a commentary on a bronze map. An interesting advantage of bronze as a material was that when revision was necessary, requiring an extension of the map, a new piece could be hammered on to the side.[50] Since centuriation was widespread in Italy and many of the other areas under Roman rule, the number of plans could have been immense.

In the cadastral maps recording centuriation, just as in geographical maps, the Roman rulers saw a vital tool of government, in this case underpinning an orderly system of land registration. It may be noted that Julius Caesar was, according to the *Corpus Agrimensorum*, the founder of the fully organized system of Roman land surveying.[51] This system came to be far more developed in Augustus's principate.

Throughout the Augustan period, maps were thus becoming more and more essential to Rome's highly law-based type of society. A formula of the second half of the second century B.C. emphasizes the legally binding character of land maps. The agrarian law of 111 B.C., referring to a law of the Gracchan period, mentions that

any land in Italy that a land commissioner had given, assigned, abandoned, or entered should be entered on *formae* or *tabulae*.[52] Here we may presume that *formae* (also called *formae publicae*) refer to maps, *tabulae* to land registers. From whenever such maps were legally recognized, they would be turned to as evidence of areas held by private individuals or corporate bodies.

Many of the same considerations apply in the case of aqueduct plans. A highly organized water supply, with legal involvement, must have given surveyors plenty of scope for preparing plans; and fragmentary evidence for this is available from the first century B.C. Roman aqueducts served primarily for town water supplies, but if the supply was sufficient they could also be used for irrigation. For this purpose landowners could apply for a time allocation, as we learn from Frontinus and the Digest (Roman treatise on jurisprudence). Outlets called *calices*, literally "cups," were inserted at an appropriate angle so that water could be turned on and off at the correct times.[53] To explain the timing of the allocations, an inscribed plan would be drawn. One such refers to the Aqua Crabra near Croce del Tuscolo (Tusculum), east of Frascati (fig. 12.6).[54] The inscription adjacent to the plan lays down the number of hours a day when landowners, including C. Julius Caesar (presumably the dictator), could draw water. Another inscription, now

46. Granius Licinianus [*Handbook of Roman History*] 28; see *Grani Liciniani quae supersunt*, ed. Michael Flemisch (Leipzig: Teubner, 1904), 9–10. There is a useful contribution on this section in Robert K. Sherk, "Roman Geographical Exploration and Military Maps," in *Aufstieg und Niedergang der römischen Welt*, ed. Hildegard Temporini (Berlin: Walter de Gruyter, 1972–), 2.1 (1974): 534–62, esp. 558–59. But he is probably incorrect in including this among military maps: Lentulus's office was civilian.

47. Samuel Ball Platner, *A Topographical Dictionary of Ancient Rome*, rev. Thomas Ashby (London: Oxford University Press, 1929), 56–57. It was later used for the first public library in Rome, set up by C. Asinius Pollio.

48. Julius Beloch, *Campanien*, 2d ed. (Breslau, 1890; reprinted Rome: Erma di Bretschneider, 1964), pl. 12; Ferdinando Castagnoli, *Le ricerche sui resti della centuriazione* (Rome: Edizioni di Storia e Letteratura, 1958), 13–14; Dilke, *Roman Land Surveyors*, 144 (note 7).

49. Suetonius *The Deified Vespasian* 8.5; book 8 of *De vita Caesarum* (The lives of the Caesars), in *Suetonius*, 2 vols., trans. J. C. Rolfe, Loeb Classical Library (Cambridge: Harvard University Press; London: William Heinemann, 1913–14).

50. Hyginus Gromaticus *Constitutio limitum* (On setting up centuriation), in *Corpus Agrimensorum*, 131–71, esp. 167, lines 3–5 (note 6).

51. Dilke, *Roman Land Surveyors*, 37 (note 7).

52. *Corpus Inscriptionum Latinarum* (Berlin: Georg Reimer, 1862–), 1.2.1 (1918), 455–64, no. 585; C. G. Bruns, *Fontes iuris Romani antiqui*, 7th ed. (Tübingen: Mohr, 1909), no. 11, para. 7.

53. J. G. Landels, *Engineering in the Ancient World* (London: Chatto and Windus, 1978), 47.

54. *Corpus Inscriptionum Latinarum*, 6.1 (1876): 274, no. 1261 (note 52).

lost, from Tibur (Tivoli) had three lines left blank for what was obviously a similar plan.[55]

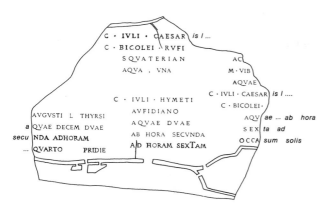

FIG. 12.6. THE INSCRIPTION ON THE TUSCULUM AQUEDUCT. An example of an annotated plan designed to specify the times at which landowners might tap water from the aqueducts shown at top and bottom.
After the *Corpus Inscriptionum Latinarum* (Berlin: Georg Reimer, 1862–), 6.1 (1876), no. 1261.

Such plans were also used for engineering purposes. From the middle of the second century A.D. comes a long aqueduct inscription from Saldae (Bejaia) in North Africa, written with poor grammar in the first person by a retired legionary, Nonius Datus. His specialty was that of *librator*, leveler; and he explains how, after having been robbed by brigands, he managed to save a tunnel, serving a watercourse, where the digging from two directions had failed to meet. He explains: "I carried out the work following the map I had given to the procurator Petronius Celer."[56] It is clear from this inscription that though Nonius Datus had a very sketchy idea of Latin grammar, he was able to draw an aqueduct plan (*forma*) and to work out from it, perhaps by Heron's method (pp. 230–32), the correct drilling directions for the tunnel.

---

55. *Corpus Inscriptionum Latinarum*, 8.1 (1881): 448, no. 4440 (note 52).

56. Author's translation of Nonius Datus; inscription to be found in *Corpus Inscriptionum Latinarum*, 8.1 (1881): 323, no. 2728 (note 52).

BIBLIOGRAPHY
CHAPTER 12 MAPS IN THE SERVICE OF THE STATE: ROMAN CARTOGRAPHY TO THE END OF THE AUGUSTAN ERA

Bunbury, Edward Herbert. *A History of Ancient Geography among the Greeks and Romans from the Earliest Ages till the Fall of the Roman Empire*, 2d ed., 2 vols. 1883; republished with a new introduction by W. H. Stahl, New York: Dover, 1959.

Castagnoli, Ferdinando. *Le ricerche sui resti della centuriazione*. Rome: Edizioni di Storia e Letteratura, 1958.

Detlefsen, D. (S. D. F.). *Ursprung, Einrichtung und Bedeutung der Erdkarte Agrippas*. Quellen und Forschungen zur Alten Geschichte und Geographie, 13. Berlin: Weidmann, 1906.

Dilke, O. A. W. "Illustrations from Roman Surveyors' Manuals." *Imago Mundi* 21 (1967): 9–29.

―――. "Varro and the Origins of Centuriation." In *Atti del Congresso Internazionale di Studi Varroniani*, 353–58. Rieti: Centro di Studi Varroniani, 1976.

―――. *Greek and Roman Maps*. London: Thames and Hudson, 1985.

Gisinger, F. "Geographie." In *Paulys Realencyclopädie der classischen Altertumswissenschaft*, ed. August Pauly, Georg Wissowa, et al., suppl. 4 (1924): cols. 521–685. Stuttgart: J. B. Metzler, 1894–.

Klotz, A. "Die geographischen Commentarii des Agrippa und ihre Überreste." *Klio* 24 (1931): 38–58, 386–466.

Körte, G. "Die Bronzeleber von Piacenza." *Mitteilungen des Kaiserlich Deutschen Archäologischen Instituts, Römische Abteilung* 20 (1905): 348–77.

Miller, Konrad. *Mappaemundi: Die ältesten Weltkarten*. 6 vols. Stuttgart: J. Roth, 1895–98.

Pallottino, Massimo. *The Etruscans*. Edited by David Ridgway. Translated by J. Cremona. London: Allen Lane, 1975.

Sherk, Robert K. "Roman Geographical Exploration and Military Maps." In *Aufstieg und Niedergang der römischen Welt*, ed. Hildegard Temporini, 2.1 (1974): 534–62. Berlin: Walter de Gruyter, 1972–.

Thulin, Carl Olof, ed. *Corpus Agrimensorum Romanorum*. Leipzig, 1913; reprinted Stuttgart: Teubner, 1971

Uhden, Richard. "Die Weltkarte des Martianus Capella." *Petermanns Mitteilungen* 76 (1930): 126.

13 · Roman Large-Scale Mapping in the Early Empire

O. A. W. DILKE

We have already emphasized that in the period of the early empire[1] the Greek contribution to the theory and practice of small-scale mapping, culminating in the work of Ptolemy, largely overshadowed that of Rome. A different view must be taken of the history of large-scale mapping. Here we can trace an analogous culmination of the Roman bent for practical cartography. The foundations for a land surveying profession, as already noted, had been laid in the reign of Augustus. Its expansion had been occasioned by the vast program of colonization carried out by the triumvirs and then by Augustus himself after the civil wars. Hyginus Gromaticus, author of a surveying treatise in the *Corpus Agrimensorum*, tells us that Augustus ordered that the coordinates of surveys be inscribed on the corners of "centuries" and that he fixed the width of main, intermediate, and subsidiary roads within centuriated areas (fig. 13.1).

The early empire was marked by a further expansion, codification, and upgrading of the role of the land surveyors (*agrimensores*) working for the Roman state as well as of that of surveyors on other applications of large-scale mapping in the towns and in engineering projects. As Roman influence spread, more public domains were to be divided up and more colonies founded. Centuriation, the division of land into centuries by surveying, was, as earlier, being applied particularly to colonies consisting of settlements of smallholders with allocations of land. That the demands on the land surveyor were truly immense can be appreciated from the size of some of the new schemes for centuriation. The most important to be initiated under Tiberius—which, we may presume, led to mapping—was the centuriation of a vast area of Roman North Africa corresponding to the mid-south of Tunisia. Some idea of the dimensions may be obtained from surviving centuriation stones,[2] which extend as far as the Chott el Fedjedj, the Bled Segui, and Graïba. There were at least 140 centuries to the right of the *decumanus maximus* and at least 280 beyond the *kardo maximus*, which places the farthest centuriation stone, in all probability, nearly 200 kilometers beyond the *kardo*. No remains of survey maps have been found from this scheme.

A further stimulus to large-scale surveying and mapping practice in the early empire was given by the land reforms undertaken by the Flavians. In particular, a new outlook both on administration and on cartography came with the accession of Vespasian (T. Flavius Vespasianus, emperor A.D. 69–79). Born in the hilly country north of Reate (Rieti), a man of varied and successful military experience, including the conquest of southern Britain, he overcame his rivals in the fierce civil wars of A.D. 69. The treasury had been depleted under Nero, and Vespasian was anxious to build up its assets. Frontinus, who was a prominent senator throughout the Flavian period (A.D. 69–96), stresses the enrichment of the treasury by selling to colonies lands known as *subseciva*. These were of two types, either areas remaining between square or rectangular centuries of allocated land and the outer boundary of the land in question, or unallocated portions of centuries.

A similar tightening of the land regulations by Vespasian resulted in the only official Roman survey maps that have come down to us, the cadasters of Arausio (Orange). It is also thought that there was in the Flavian period a predecessor of the official plan of Rome, the *Forma Urbis Romae*,[3] and the collection of surveyors' manuals known as the *Corpus Agrimensorum*, although it contains manuals and extracts from quite different periods, may owe its real origin to treatises composed in the first century A.D. If we bring these strands together, it can be shown that the early empire is the key period in our understanding of the history of classical large-scale cartography.

---

1. Defined broadly to include the period from the emperor Tiberius (A.D. 14–37) to the emperor Caracalla (A.D. 211–17).

2. *Corpus Inscriptionum Latinarum* (Berlin: Georg Reimer, 1862–), vol. 8 suppl., pt. 4 (1916): nos. 22786 a–m, 22789.

3. Gianfilippo Carettoni et al., *La pianta marmorea di Roma antica: Forma Urbis Romae*, 2 vols. (Rome: Comune di Roma, 1960).

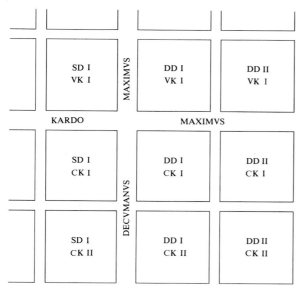

FIG. 13.1. THE METHOD OF NUMBERING CENTURIES. SD = *sinistra decumani* (to left of *decumanus maximus*), DD = *dextra decumani* (to right of *decumanus maximus*), VK = *ultra kardinem* (beyond *kardo maximus*), CK or KK = *citra kardinem* (near side of *kardo maximus*). These abbreviations were carved on the boundary stones identifying the areas. After O. A. W. Dilke, *The Roman Land Surveyors: An Introduction to the Agrimensores* (Newton Abbot: David and Charles, 1971), 92.

SURVEY METHODS OF THE *AGRIMENSORES*[4]

The principal Roman surveying instrument was called the *groma*.[5] It was used in military as well as civilian surveying, and its origin has been traced back to Egyptian practice.[6] Since the *Corpus Agrimensorum* does not give a picture of the *groma*, we have to rely on other sources such as the tombstone of a Roman surveyor, Lucius Aebutius Faustus, dating to the first century B.C., with a relief depicting his profession (fig. 13.2). Below the inscription is a dismantled *groma* or *stella*, with a staff 73 centimeters long and crossarms 35 centimeters. The representation is only schematic, however, and is difficult to interpret. The only example of what is believed to be an actual *groma* was found in 1912 during the excavations at Pompeii. Its metal parts were discovered in the workshop of a surveyor called Verus (fig. 13.3).[7]

The *groma*, a forerunner of the later surveyor's cross,[8] fulfilled the vital need in centuriation for an instrument to lay out long lines at right angles to each other. The cross was placed on a bracket, rather than directly on the staff, to avoid obstruction when sighting from one plumb line to another. The method of operation was for

FIG. 13.2. INSCRIPTION ON A ROMAN SURVEYOR'S TOMBSTONE. Dating from the first century B.C., this tombstone indicates the profession of Lucius Aebutius Faustus by the schematic diagram, below the inscription, of a dismantled *groma*. Ivrea, Museo Civico.
From Hermann Schöne, "Das Visirinstrument der Römischen Feldmesser," *Jahrbuch des Kaiserlich Deutschen Archäologischen Instituts* 16 (1901): 127–32, pl. 2.

4. Carl Olof Thulin, ed., *Corpus Agrimensorum Romanorum* (Leipzig, 1913; reprinted Stuttgart: Teubner, 1971), only one volume published; Friedrich Blume et al., eds., *Die Schriften der römischen Feldmesser*, 2 vols. (Berlin: Georg Reimer, 1848–52; reprinted Hildesheim: G. Olms, 1967); O. A. W. Dilke, *The Roman Land Surveyors: An Introduction to the Agrimensores* (Newton Abbot: David and Charles, 1971).

5. Dilke, *Roman Land Surveyors*, 66–70 (note 4), gives a full description of this instrument and also refers to the Grecian star (*stella*), which was hand-held.

6. Edmond R. Kiely, *Surveying Instruments: Their History* (New York: Teachers College, Columbia University, 1947; reprinted Columbus: Carben Surveying Reprints, 1979), 13–14.

7. For a fuller description see Dilke, *Roman Land Surveyors*, 69–70 (note 4).

8. See P. D. A. Harvey, "Local and Regional Cartography in Medieval Europe," chapter 20 of this volume, pp. 464–501.

the surveyor to plant the *groma* in the ground, keeping the center of the cross one bracket length away from the required center of survey. He then turned it until it faced the required direction, which he had ascertained be-

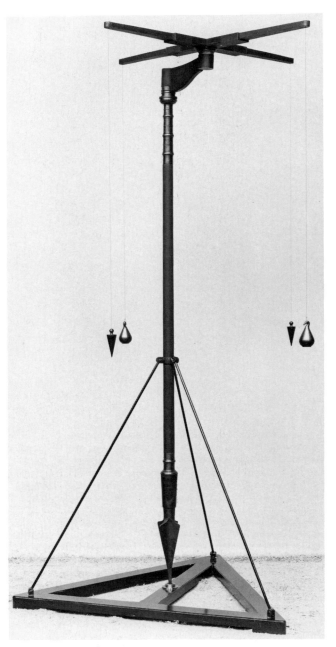

FIG. 13.3. RECONSTRUCTION OF A *GROMA*. This is a replica of the only original *groma* in existence (the original is in the Archaeology Museum, Naples). This Roman surveying instrument for laying out right angles was preserved in the Pompeii workshop of a surveyor named Verus. The triangular support and base are of modern design.
Conjectured height of the original: 2.06 m. By permission of the Trustees of the Science Museum, London.

forehand, finding south either by means of a portable sundial (figs. 13.4 and 13.5) or by observing shadows.[9] Sighting was done by looking from one plumb line to its opposite number, the plumb bobs being grouped into pairs to avoid confusion. Sights could be set onto a second *groma*, positioned first perhaps one *actus* (35.48 m) away, then a similar distance from the first and second *gromae* at right angles. The square would then be completed and cross-checks taken. The *groma* thus had only a limited use: it enabled straight lines, squares, and rectangles to be surveyed. But these were exactly what the *agrimensor* normally required, and more complicated equipment was unnecessary on a straightforward survey.[10]

Assuming these basic instruments were used, we can also reconstruct how the survey proceeded in the field. The usual method of recording allocations was for the land surveyor to divide up the apportioned land, draw lots for the landholdings, and take the settlers to their lands. He also had to make a map (*forma*) of any land he had so divided and compile a register of it. A field notebook must have been used, and one can only surmise that draft maps and notes were executed on wax tablets or on papyrus (later on parchment or vellum): probably size was the determining factor. The longer-term record in bronze would be prepared by the appropriate craftsmen.

The unit of measurement was the *actus* of 120 Roman feet, in origin the distance that oxen plowed before the plowman turned them. The standard Roman foot was 29.57 centimeters, so the standard *actus* was 35.48 meters, but variations are encountered. The measurements of area were:

One square *actus (actus quadratus)* = 14,400 square feet (Roman) = 0.126 hectare.

Two square *actus* = one *iugerum* = 28,800 square feet (Roman) = 0.252 hectare.

9. The portable sundial may have been invented or perfected during Hadrian's principate (A.D. 117–38) and was designed to operate at any latitude. Although discovered specimens date only from later periods, the instrument is depicted on medallions from Claudiopolis (Bithynium) and from Milan, each having on its obverse a representation of Antinous, a Bithynian youth who was a favorite of the emperor Hadrian: the first-mentioned is in a private collection; copy in the Landesmuseum, Trier. See Edmund Buchner, "Römische Medaillons als Sonnenuhren," *Chiron* 6 (1976): 329–48. For the Milan medallion see Giorgio Nicodemi, *Catalogo delle raccolte numismatiche*, 2 vols. (Milan: Bestetti, 1938–40), vol. 2, *Le monete dell'Impero romano da Adriano ad Elio Cesare*, pl. XVI, 3849.

10. That the Roman land surveyor also had other surveying instruments available to him as well as instruments for drawing maps is, however, made clear by the contents of Verus's workshop in Pompeii. Besides the portable sundial, these include the endpieces of a measuring rod, a folding ruler, and bronze compasses: see Dilke, *Roman Land Surveyors*, 73–81 (note 4).

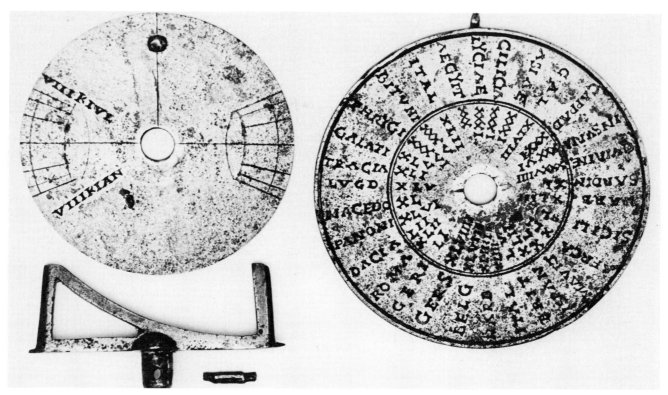

FIG. 13.4. A PORTABLE SUNDIAL, DISASSEMBLED. This shows the individual parts of a bronze instrument said to have been found near Bratislava. It dates from after A.D. 120.

By permission of the Museum of the History of Science, Oxford (R. 40).

Two *iugera* = one *heredium* = 0.504 hectare (little used in practice).

One hundred *heredia* = one *centuria* = about 50.4 hectares (fig. 13.6).

This refers to the regular century of 200 *iugera*, but various other sizes, both larger and smaller, are found. These regular squares had sides of 20 *actus*, properly speaking 709.68 meters, though variants from 703 meters to 714 meters are found. Between each pair of centuries was a *limes*, literally "balk," for which appropriate width was provided; in one direction each of these constituted a *kardo*, and at right angles to it was a *decumanus*.

The system of coordinates used for naming centuries started from the two main roads of a centuriated area, the *kardo maximus* and the *decumanus maximus*. It was assumed that the surveyor counted the centuries from the center, distinguishing (a) *citra kardinem*, CK or KK, "this side of the *kardo*," from *ultra kardinem*, VK, "beyond the *kardo*," (b) *sinistra decumani*, SD, "to the left of the *decumanus*," from *dextra decumani*, DD, "to the right of the *decumanus*." These were followed by the figure denoting the number of the particular century from the intersecting main roads.

FIG. 13.5. A PORTABLE SUNDIAL, ASSEMBLED. The instrument is seen here in its actual form. It could have been used to lay out a centuriation scheme toward the south. Diameter of the original: 6 cm. By permission of the Museum of the History of Science, Oxford (R. 40).

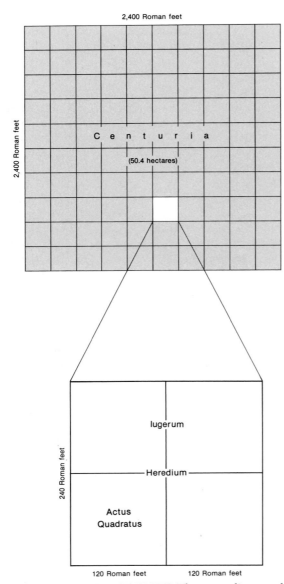

FIG. 13.6. ROMAN AREAL UNITS. The upper diagram shows a regular century of 200 *iugera*, here subdivided into 100 *heredia* (though the *heredium* was not used in calculations). The lower diagram shows that 4 square *actus* = 2 *iugera* = 1 *heredium*.

Three passages of Hyginus Gromaticus give details about the procedure for survey mapping. In one he comments:

> When we have ended all centuries with inscribed stones, we shall surround parts assigned to the state, even if they are centuriated, with a private boundary and shall enter them on the map appropriately, as "public woods" or "common pasture" or both. We shall fill the whole extent with the inscription, so that on the map of the area a more scattered arrangement

of lettering may show greater width. We shall likewise bound excepted or granted farms, giving them inscriptions as with public places. We shall similarly show granted farms, for example, "farm of Seius granted to Lucius Manilius son of Seius." In Augustus's allocation of land, excepted farms have a different status from granted farms.[11]

Besides indicating the central importance of the maps in the process of land allocation and registration, Hyginus Gromaticus shows a cartographic awareness that is almost modern when he implies that the spacing of lettering can also reflect the extent of the land represented. Later he goes on to elaborate how the maps were to be used to record various legal categories of land:

> We shall write both on the maps and on the bronze tablets [*tabulae aeris*, i.e., tablets attached to bronze maps] all mapping indications, "given and assigned," "granted," "excepted," "restored, exchanged for own property," "restored to previous owner," and any other abbreviations in common use, to remain on the map. We shall take to the emperor's record office the mapping registers [*libri aeris*] and the plan of the whole surveyed area drawn in lines according to its particular boundary system, adding the names of the immediate neighbors. If any property, either in the immediate neighborhood or elsewhere, has been given to the colony, we shall enter it in the register of assets. Anything else of surveying interest will have to be held not only by the colony but by the emperor's record office, signed by the founder. This is how we shall allocate undeveloped land in the provinces. But if a borough has its status changed to that of colony, we shall examine local conditions.[12]

It is clear that both the maps, when they were transferred to bronze, and the registers that were to accompany them were prepared in duplicate. One copy of each document was kept locally within the colony; the other was preserved in a central record office in Rome. Roman practice thus anticipated modern cadastral surveys. In the great age of European and American cadastral surveys in the eighteenth and nineteenth centuries,[13] for example, maps provided local and central governments with a similar record of the allocation of land and served to register private and public property. Such surveys, moreover, as in the Roman model, often contained a written and a graphic part and indeed sometimes expressed their indebtedness to their classical lineage.

---

11. Hyginus Gromaticus *Constitutio limitum* (On setting up centuriation), in *Corpus Agrimensorum*, 131–71 (note 4), quotation on 159–60, author's translation.

12. Hyginus Gromaticus *Constitutio limitum*, 165–67 (note 11), author's translation.

13. Such surveys will be dealt with in volumes 4 and 5 of the present *History*.

THE *CORPUS AGRIMENSORUM*

The main textual channel through which Roman ideas about land surveying and mapping were to be transmitted to later societies—as well as a principal source for reconstructing the techniques described above—is known as the *Corpus Agrimensorum*.[14] These documents consist of an extant collection of short works in Latin of quite varied dates that have been recognized as the primary written records of Roman land surveying. One or two items are dated from the republic, three or four from the early empire, and quite a number from the late empire, with even later additions in some manuscripts. The *Corpus* may have been compiled in the middle or late fourth century A.D.[15] and revised from time to time.

Of the writers contained in the *Corpus*, only the first is well known. Sextus Julius Frontinus was governor of Britain from A.D. 74 to 77,[16] during which time he played a prominent part in gaining control for the Romans of southern and central Wales. His works that have fully survived are the *Stratagems* and the *Water Supply of Rome*. The works under his name in the *Corpus Agrimensorum—On the Status of Land, On Land Disputes*, and *On Centuriation*—are clearly only parts of the originals.[17] Balbus, whose short treatise on measurements is preserved, seems to have served in an expedition of Domitian's against northern tribes and to have been writing in the early second century A.D. The two pseudo-Hygini seem to be different writers in the latter part of this period, each presumably claiming the author of his treatise was C. Julius Hyginus, the librarian of Augustus. In addition to these two, there is a work on camp measurement by another pseudo-Hyginus.[18]

The collection is only very roughly based on chronological order: Agennius Urbicus, a late commentator on Frontinus, very reasonably appears immediately after Frontinus, but some items are quite displaced in order. Of the treatises attributable to the early empire, those of Frontinus and the pseudo-Hygini are illustrated; that of Balbus contains geometric drawings. Whether the drawings preserved reflect the authors' illustrations is difficult to say. Frontinus, Balbus, and the earliest pseudo-Hyginus do not refer to drawings, whereas Hyginus Gromaticus uses the adverb *sic* to draw attention to them. The collection was obviously meant to help with the teaching of land survey, and for this purpose the illustrations interspersed with the text represent a great advance in educational method.[19] This was probably helped by the gradual changeover from the use of papyrus to parchment or vellum as the regular writing material.

The chief manuscripts containing illustrations are the Arcerianus A (hereafter abbreviated as A) in the Herzog August Bibliothek, Wolfenbüttel, now dated to about A.D. 500,[20] and the ninth-century Pal. Lat. 1564 (hereafter abbreviated as P) in the Biblioteca Apostolica Vaticana, Rome. The miniatures in each of these are mostly very well preserved, and those of the Wolfenbüttel manuscript have recently been studied from the point of view of art history.[21] In some cases A and P substantially agree in their diagrams, in others they show appreciable differences. On the whole, where they differ, A presents a more accurate interpretation.

The subjects of the illustrations include surveying techniques, material relating to centuriation and boundary stones, and maps of towns and surrounding lands, together with diagrams of legal definitions and other theoretical matters.[22] In terms of convention, too, illustrations suggest that a distinctive style of large-scale mapping was developing during the early empire. The simplest diagrams are monochrome, light brown. In the more elaborate illustrations, however, colors were employed with some measure of consistency. Roads are usually depicted red or brown, sometimes green. Water is blue or bluish green. Buildings are mostly pale brown, yellow, or gray; the predominant color for roofs is red. Mountains are usually mauve or, if wooded, green; sometimes they are brown.

The maps of colonies with centuriated land that appear in the *Corpus* are teaching maps, different from the large-scale survey maps the land surveyors had to produce (figs. 13.7–13.9). Many of the colonies, including five from Hyginus Gromaticus, can be certainly or with some probability identified from either text or map or both; others are called Colonia Iulia, which without

14. In addition to references in note 4, see James Nelson Carder, *Art Historical Problems of a Roman Survey Manuscript: The Codex Arcerianus A, Wolfenbüttel* (New York: Garland, 1978); Hans Butzmann, ed., *Corpus Agrimensorum: Codex Arcerianus A der Herzog-August-Bibliothek zu Wolfenbüttel*, Codices Graeci et Latini 22 (Leiden: A. W. Sijthoff, 1970).

15. This date would suit the *Libri Coloniarum*, lists of colonies with names of founders and other data, which are incorporated into the *Corpus Agrimensorum*; see Blume et al., *Schriften der römischen Feldmesser*, 1:209–62 (note 4).

16. See Anthony R. Birley, *The Fasti of Roman Britain* (Oxford: Clarendon Press, 1981), 71, 77–78.

17. For the probable scope and titles, see Carl Olof Thulin, "Kritisches zu Iulius Frontinus," *Eranos* 11 (1911): 131–44.

18. The best edition is Antonino Grillone, ed., *De metatione castrorum liber* (Leipzig: Teubner, 1977).

19. A parallel can be drawn with the illustration of architectural manuals: Vitruvius *De architectura* 1.1.4; see *On Architecture*, 2 vols., trans. Frank Granger, Loeb Classical Library (Cambridge: Harvard University Press; London: William Heinemann, 1931–34).

20. Carder, *Art Historical Problems*, 205 ff. (note 14). Others give the date as early sixth century, such as Carl Nordenfalk, *Die spätantiken Zierbuchstaben* (Stockholm: published by the author, 1970).

21. Carder, *Art Historical Problems* (note 14).

22. For a fuller classification see O. A. W. Dilke, "Illustrations from Roman Surveyors' Manuals," *Imago Mundi* 21 (1967): 9–29.

further definition could refer to any early imperial foundations; others again contain geographical features such as mountains or sea but are not identifiable. The most clearly recognizable (fig. 13.7) is Colonia Anxurnas, the Volscian Anxur, later Tarracina, where the text says: "In some colonies they set up the *decumanus maximus* in such a way that it contained the trunk road crossing the colony, as at Anxur in Campania. The *decumanus maximus* can be seen along the Via Appia; the cultivable land has been centuriated; the remainder consists of rugged rocks, bounded as unsurveyed land by natural landmarks."[23] The Roman colony at Tarracina was founded on the coast in 329 B.C.; each of the three hundred settlers received a very small allotment of land: two *iugera*. As the text shows, the centuriation was here centered on the Via Appia, at least after that road was built in 312 B.C.[24] *Paludes* in the illustration refers to the Pomptine (Pontine) Marshes, which the Via Appia crossed but which were only partially drained in antiquity. The mountains are the Monti Lepini; between them and the sea, which is also shown on the miniature, the road crossed the plain known as La Valle, where proof of centuriation is in fact available only on the opposite side from that shown in the miniature. A second map of a colony in Hyginus Gromaticus (fig. 13.8) is Minturnae (Minturno Scavi), founded in 295 B.C. This shows the colony on both sides of the river Liris (Garigliano), though in fact the walls lay only on the right bank of the river, to the left of the miniature map. Also featured are the Vescini Mountains, the Augustan new assignation, a bronze statue, and the sea. As with the illustration for Tarracina, this map assumes a viewpoint above the sea looking toward the mountains. A third map shows Hispellum (Spello), founded as a colony probably about 30 B.C. (fig. 13.9). The words *flumen finitimum* refer to a river on the Umbrian plain that separated the territory of Hispellum, the walls of which encircle a hill overlooking it, from that of a neighboring settlement. The river is not easily identifiable;[25] such maps are not always precise depictions of particular locations. They were not derived directly from surveys but were designed for teaching, so their planimetric accuracy may vary considerably.

To illustrate legal definitions, both picture maps and ground plans were used to represent smaller areas. Some of these are of centuriation, and others of features such as farms and common pasture. An example of a picture map depicting a farm is the miniature accompanying a definition in Frontinus (fig. 13.10). The text says: "*Ager arcifinius*, which is unsurveyed, is bounded in accordance with ancient practice by rivers, ditches, mountains, roads, rows of trees, watersheds, and any areas able to be claimed from previous occupation."[26] As many of these features as could easily be shown are

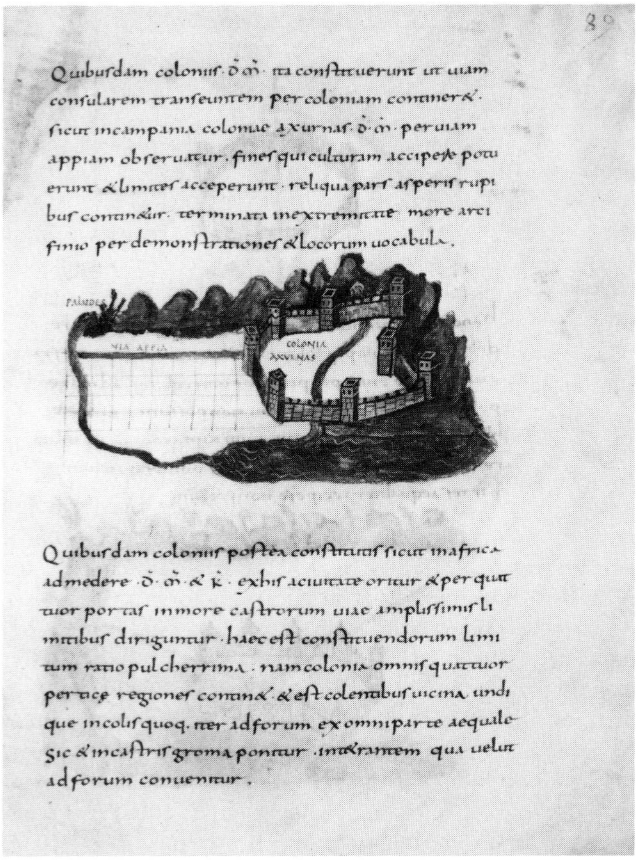

FIG. 13.7. MINIATURE OF THE CENTURIATION AROUND TARRACINA, ITALY. This, like figures 13.8 and 13.9, is a picture map illustrating the geographical relation between the positions of Roman colonies and the centuriation schemes that were associated with them. The colony of Tarracina, known to the Romans as Anxur-Tarracina, was established in Campania in 329 B.C. Only the cultivable land was surveyed, with a centuriation based upon the Via Appia.
Size of the original: 28 × 19.6 cm. Photograph from the Biblioteca Apostolica Vaticana, Rome (Pal. Lat. 1564, fol. 89r).

23. Hyginus Gromaticus, *Constitutio limitum*, 144 (note 11), author's translation.

24. Focke Tannen Hinrichs, *Die Geschichte der gromatischen Institutionen* (Wiesbaden: Franz Steiner, 1974), doubts whether the still-discernible centuriation goes back as early as the fourth century B.C. See O. A. W. Dilke and Margaret S. Dilke, "Terracina and the Pomptine Marshes," *Greece and Rome*, n.s., 8 (1961): 172–78; Gérard Chouquer et al., "Cadastres, occupation du sol et paysages agraires antiques," *Annales: Economies, Sociétés, Civilizations*, 37, nos. 5–6 (1982): 847–82, esp. fig. 7c.

25. According to Adolf Schulten, "Römische Flurkarten," *Hermes* 33 (1898): 534–65, this could have been the river Ose; see O. A. W. Dilke, "Maps in the Treatises of Roman Land Surveyors," *Geographical Journal* 127 (1961): 417–26.

26. Frontinus *De agrorum qualitate*, in *Corpus Agrimensorum*, 1–3 (note 4), quotation on 2, author's translation; Carder, *Art Historical Problems*, 44–46 (note 14).

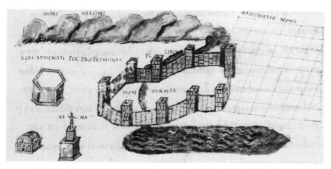

FIG. 13.8. MINIATURE OF THE CENTURIATION AROUND MINTURNAE, ITALY. Founded on the borders of Latium and Campania in 295 B.C., this colony, the modern Minturno Scavi, is depicted here as seen from above the sea, looking toward the Vescini mountains.
Size of the original detail: 8 × 14 cm. Photograph from the Biblioteca Apostolica Vaticana, Rome (Pal. Lat. 1564, fol. 88r).

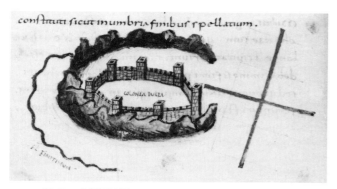

FIG. 13.9. MINIATURE OF THE CENTURIATION AROUND HISPELLUM, ITALY. A schematic representation of the boundary of the territory of the colony of Spello, in Umbria, founded ca. 30 B.C., with that of the neighboring settlement. The boundary follows the river on the left.
Size of the original detail: 10.5 × 14.4 cm. Photograph from the Biblioteca Apostolica Vaticana, Rome (Pal. Lat. 1564, fol. 88v).

incorporated into the map, together with some extra features such as two temples presumably serving as boundaries. The squares on the right of the map were probably intended to illustrate the areas of previous occupation. An example of the ground plan used to define the status of land, a simple painted miniature (fig. 13.11), also illustrates Frontinus's text: "There is also property vested in pastureland, relating to farms, but held in common; so these common pastures are called *communia* in many places in Italy, while in some provinces they are called *pro indiviso*."[27]

Finally, among the treatises attributable to the early empire, a large percentage of the illustrations are devoted to centuriation diagrams, some of which are of carto-

graphic interest, others less so. Thus rectangles of 40 by 20 *actus*, such as are found in Orange Cadaster A, are depicted as in figure 13.12,[28] where both *kardo maximus* and *decumanus maximus* are labeled KM: one should be DM. In another line drawing (fig. 13.13),[29] two adjacent centuriation systems are shown, one perhaps oriented according to the cardinal points, one not.[30] It is, however, in a late and corrupt manuscript that the nearest approach to the Orange cadasters is found, albeit only with specimen names and holdings (fig. 13.14); thus, "The farm of Seius Agerius has ten *iugera* freehold," "Sextilius's farm has thirty *iugera*," "Vennius's farm has fifty *iugera*, noted in register," and so on.[31]

There can be little doubt that these texts and diagrams, particularly those dealing with centuriation, had an effect both on the adoption of similar schemes in more recent times and on the discovery of centuriation systems. In 1833 the Danish sea captain C. T. Falbe, who had read about Roman land surveying, noticed that squares around Carthage had sides of 2,400 Roman feet; and similar work was done in the Po valley in 1846 by E. N. Legnazzi and in the Trieste area in 1848 by P. Kandler.[32] Ever since then the discoveries of centuriation have been so extensive that we can appreciate what an important part the *agrimensores* played in the planning and mapping of the countryside.

Teaching maps for surveyors were not always designed as early as the treatises they were supposed to illuminate. The illustrations accompanying the treatise

27. Frontinus *De controversiis*, in *Corpus Agrimensorum*, 4–10 (note 4), quotation on 6, author's translation; see also fig. 18 (manuscript A).

28. Thulin, *Corpus Agrimensorum*, fig. 78 (manuscript A) (note 4).

29. Thulin, *Corpus Agrimensorum*, fig. 74 (manuscript A) (note 4).

30. Such conflicts of orientation are frequently discovered from aerial photography, a good example being Enfida, Tunisia. But in such cases, unless there is archaeological or other evidence, it is usually impossible to tell whether adjacent systems are contemporary: John Bradford, *Ancient Landscapes* (London: Bell, 1957); Gérard Chouquer and Françoise Favory, *Contribution à la recherche des cadastres antiques*, Annales Littéraires de l'Université de Besançon 236 (Paris, 1980); Institut Géographique National, *Atlas des centuriations romaines de Tunisie* (Paris: Institut Géographique National, 1954). See also Monique Clavel-Lévêque, ed., *Cadastres et espace rural: Approches et réalités antiques* (Paris: Centre National de la Recherche Scientifique, 1983). For a bibliography on centuriation, see O. A. W. Dilke, "Archaeological and Epigraphic Evidence of Roman Land Surveys," in *Aufstieg und Niedergang der römischen Welt*, ed. Hildegard Temporini (Berlin: Walter de Gruyter, 1972–), 2.1 (1974): 564–92, esp. 585–92.

31. Author's translation from illustration in the sixteenth-century Jena manuscript of Frontinus; see Thulin, *Corpus Agrimensorum*, fig. 17 (note 4).

32. C. T. Falbe, *Recherches sur l'emplacement de Carthage* (Paris: Imprimerie Royale, 1833); Pietro Kandler, *Indicazioni per riconoscere le cose storiche del Litorale* (Trieste, 1855); cf. G. Ramilli, *Gli agri centuriati di Padova e di Pola nell'interpretazione di Pietro Kandler* (Trieste: Società Istriana di Archeologia e Storia Patria, 1973).

FIG. 13.10. *AGER ARCIFINIUS.* A miniature picture map of a farm, this is designed to illustrate the accompanying text's definition of unsurveyed farmland in the *Corpus Agrimensorum Romanorum.*

Size of the original map detail: 4.5 × 21 cm. By permission of the Herzog August Bibliothek, Wolfenbüttel (Codex Guelf. 36.23 Aug. 2°, fol. 18r).

of Hyginus Gromaticus, *Constitutio limitum* (On setting up centuriation), include two regional maps[33] that appear in two different forms: one in the Arcerianus A (A), one in Pal. Lat. 1564 (P). Whereas the A map is in each case simple, with little wording and with straight lines, the P map is very complex, with much wording and with winding as well as straight lines. The winding lines designate both roads and rivers, though Roman roads in the areas indicated did not wind in this way. In one case what looks like a single region turns out to portray places widely separated, in Campania, Latium, and the Po valley.[34] The other map[35] does not depict scattered areas, but it is poorly oriented and it uses the word *ut,* meaning "for example," one farm being said to have been made over by P. Scipio, the famous general of perhaps six hundred years earlier. These, then, are likely to have been theoretical maps used for teaching. As such they are probably to be dated to the fourth or fifth century A.D.

THE CADASTERS OF ARAUSIO

Whereas no survey maps in bronze have survived, we do possess substantial fragments of the cadastral maps carved in stone found at Orange (the Roman Arausio) in the Rhône valley. They originally consisted of several rows of tablets that must have been fixed to a wall for permanent display. The term "cadaster" in this sense means a large-scale land survey carried out for taxation purposes.[36] Unfortunately, since the collapse of a floor

at the Orange museum in 1962 caused much damage, not all the fragments survive. But the important surviving fragments, found mostly in 1949–51, are now well displayed on walls, and the lost pieces were carefully documented by Piganiol.[37]

The inscribed stones fall into three categories. First, there is an inscription of A.D. 77 explaining the purpose of the emperor Vespasian's edict; second, there are fragments of three cadasters, now known as A, B, and C; and third, there are several inscribed stones from the Orange public record office (*tabularium*).

The inscription in the first category may be rendered thus when letters have been restored and abbreviations completed:

> The emperor Vespasian, in the eighth year of his tribunician power [i.e., A.D. 77], so as to restore the state lands the emperor Augustus had given to soldiers of the second legion Gallica, but which for some

33. Dilke, "Treatises," 417–26 (note 25).
34. Thulin, *Corpus Agrimensorum,* fig. 136a (note 4); Blume et al., *Schriften der römischen Feldmesser,* vol. 1, fig. 197a (manuscript P) (note 4).
35. Thulin, *Corpus Agrimensorum,* fig. 135a (note 4); Blume et al., *Schriften der römischen Feldmesser,* vol. 1, fig. 196b (manuscript P) (note 4).
36. Its derivation is probably not from medieval Latin *capitastrum* but from Byzantine Greek κατὰ στίχον, "line by line" (of a ledger).
37. André Piganiol, *Les documents cadastraux de la colonie romaine d'Orange, Gallia,* suppl. 16 (Paris: Centre National de la Recherche Scientifique, 1962); cf. Dilke, *Roman Land Surveyors,* 159–77 (note 4).

years had been occupied by private individuals, ordered a survey map to be set up, with a record on each century of the annual rental. This was carried out by . . . Ummidius Bassus, proconsul of the province of Gallia Narbonensis.[38]

The territory involved is that of the colony founded at Arausio in or immediately before 35 B.C. The colony was planned for veterans of the second legion Gallica, replaced in 35 B.C. by the second legion Augusta, under the title colonia Iulia firma Secundanorum. The amount of land received by each veteran (other than centurions) may have been 33⅓ iugera—8.4 hectares or 20.8 acres.

38. Author's translation. The inscription is recorded in Piganiol, *Documents cadastraux d'Orange*, fig. 11 and pl. 3 (note 37).

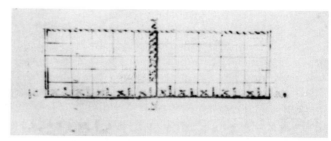

FIG. 13.12. MINIATURE FROM THE *CORPUS AGRIMEN-SORUM*. A depiction of a centuriation scheme, 40 × 20 *actus*. The *Corpus* records that cadasters of such dimensions were executed at two unidentified places in Spain near Mérida. Size of the original: 2.4 × 7 cm. By permission of the Herzog August Bibliothek, Wolfenbüttel (Codex Guelf. 36.23 Aug. 2°, fol. 45r).

FIG. 13.11. A GROUND PLAN FROM THE *CORPUS AGRIMENSORUM*. This miniature was designed to illustrate the status of land, in this case the pasture shared by two centuriated areas.
Size of the original: 5.2 × 14.1 cm. By permission of the Herzog August Bibliothek, Wolfenbüttel (Codex Guelf. 36.23 Aug. 2°, fol. 21r).

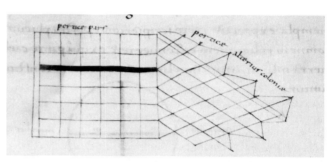

FIG. 13.13. ADJACENT CENTURIATION SCHEMES. This vignette from the *Corpus Agrimensorum* shows two neighboring centuriation schemes with different orientations.
Size of the original: 6 × 11 cm. Photograph from the Biblioteca Apostolica Vaticana, Rome (Pal. Lat. 1564, fol. 84v).

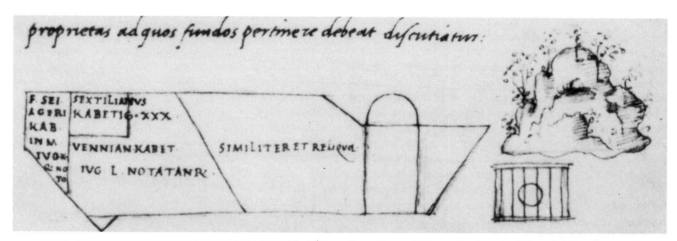

FIG. 13.14. THE MAPPING OF SMALLHOLDINGS. This paradigmatic miniature is the only one from any of the manuscripts of the *Corpus Agrimensorum* that resembles the results of an actual centuriation map.

From James Nelson Carder, *Art Historical Problems of a Roman Land Surveying Manuscript: The Codex Arcerianus A, Wolfenbüttel* (New York: Garland, 1978), ill. 22.
By permission of the Universitätsbibliothek, Friedrich-Schiller-Universität, Jena (MS. Prov. f. 156 [Apographon Jenense], fol. 77r).

Occupation of state land (*ager publicus*) was an offense, since it could, among other things, deprive the state of revenue. After the devastation of the civil war of A.D. 69, Vespasian was keen to build up Rome's financial resources, and this was an easy way.

Although the cadasters are in some respects similar to diagrams in the *Corpus Agrimensorum*, they are strictly confined to centuriated land. It has been shown that the illustrations of the *Corpus* represented topographical features other than the lines of centuriation and, moreover, that they often extended beyond the areas that had been centuriated, to show other landscapes in pictorial form. Such a style is not found in these cadasters, nor was it used on the bronze maps of surveyors, as far as we know. At Arausio, the cadasters list financial requirements and legal status; in the *Corpus*, the main concerns are surveying and legal status.

The Roman colony of Arausio was founded in the territory of the Tricastini,[39] and in Cadaster B, originally the most extensive of the three cadasters, the phrase TRIC RED, *Tricastinis reddita* (restored to the Tricastini), occurs. Although it may well be true that some of the poorest land was among that restored to the local tribe, the move may have been connected with a change in the status of its capital (St-Paul-Trois-Châteaux) which, at some time before A.D. 97, was elevated to colonia Flavia Tricastinorum.

The considerable detail required could be recorded only on a large-scale plan of this type, in this case about 1:6,000 overall, with the main intersecting roads of the centuriated area, the *kardo maximus* and *decumanus maximus*, given double width. A number of roads not in alignment with the centuriation are duly noted, as are also rivers and islands, though without names. The proportions of squares and rectangles are not always correct; some centuries appear as rectangles rather than squares.

The orientation of Cadasters B and C can be established, but that of Cadaster A, which is very fragmentary, has given rise to discussion. Piganiol held that north was at the top, but both Oliver and Salviat maintain that in this case south is at the top and that the cadasters were erected on three walls of the record office in such a way that Surveys A, B, and C would have south, west, and north at the top of the diagram, for easy consultation, though each would have been constructed by a surveyor initially facing west.[40] Whether we can go further with Salviat and claim that all three cadasters were virtually contemporary is very doubtful. If we look at the evidence from aerial photography, we find that there is only one centuriation scheme consisting of rectangles rather than squares.[41] Since Cadaster A is of rectangles (whereas B and C represent squares), it is only logical to equate it with this area roughly between Orange and Carpentras

and not, as Salviat claims, with an area farther south. The rectangles between Orange and Carpentras are not quite on the same orientation as the squares farther north, which again suggests that the three cadasters were not simultaneous. As with so much else in Roman mapping, one has the problem that much material survives only in its final, revised form; if accepted as it stands, it may telescope what happened over a considerable period of history.[42]

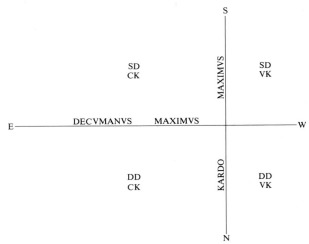

FIG. 13.15. THE PROBABLE LAYOUT OF ORANGE CADASTER A. Unlike the other two cadasters at Orange, this seems to have consisted of rectangles of sides 40 × 20 *actus*, as depicted in figure 13.12.

After F. Salviat, "Orientation, extension et chronologie des plans cadastraux d'Orange," *Revue Archéologique de Narbonnaise* 10 (1977): 107–18, esp. 111.

39. Claude Boisse, *Le Tricastin des origines à la chute de l'Empire romain* (Valence: Sorepi, 1968).

40. J. H. Oliver, "North, South, East, West at Arausio and Elsewhere," in *Mélanges d'archéologie et d'histoire offerts à André Piganiol*, 3 vols., ed. Raymond Chevallier (Paris: SEVPEN, 1966), 2:1075–79; F. Salviat, "Orientation, extension et chronologie des plans cadastraux d'Orange," *Revue Archéologique de Narbonnaise* 10 (1977): 107–18.

41. M. Guy in Raymond Chevallier, "Sur les traces des arpenteurs romains," *Caesarodunum*, suppl. 2 (Orléans-Tours, 1967), 16.

42. Thus, on Cadaster A, annual rents payable by landholders are cited in *denarii* and *asses*. The number of *asses* to the *denarius* was changed in the 140s B.C. from ten to sixteen. At Orange the old decimal terms were used in addition to the *as*; ten *libellae* or twenty *singulae* or forty *teruncii* went to a *denarius*. Abbreviations were as follows:

Ӿ	1 *denarius*
S	½ *denarius* (*semis*)
—	1 *libella*
=	2 *libellae*
≡	3 *libellae*
∠	1 *singula*
T	1 *teruncius*
ЯІ	1 *as*

Thus eleven *asses* are expressed as S-TAI, that is, ½ + ⅒ + 1/40 + 1/16 = 11/16 *denarius*.

FIG. 13.16. FRAGMENT 7 OF ORANGE CADASTER A. This fragment shows not only landownership, but also a river with an island between two roads that run across the grain of the centuriation. Among other abbreviations used on this and other fragments are: EXTR, *ex tributario*—withdrawn from tribute-paying status (tribute was paid there only by the Gauls); REL COL, *reliqua coloniae*—remaining to the colony, that is, lands not allocated to veterans but rented by the community;

RP, *rei publicae*—state lands (occurs only in Cadaster A); SVBS, *subseciva*—either areas remaining between surveyed centuries and the outer boundary of the colony's territory or land within a century that was unsuitable for allocation or for which landholders could not be found.

Size of the original: 29.5 × 36 cm. By permission of the Musée Municipal d'Orange, Vaucluse (fragment 7 of Cadaster A).

Cadaster A has been reconstructed by Salviat as in figure 13.15, the lines indicating the approximate extension in each direction from the center of the survey. The centuries are rectangular, and since the largest preserved total area of allocations is 330 *iugera*, we may confidently surmise that each rectangle measured 400 *iugera*, unlike those in Cadasters B and C. Those of A clearly consisted of 40 *actus* east-west by 20 *actus* north-south, dimensions known from the *Corpus Agrimensorum* to have existed in Spain. Fragment 7 Piganiol (fig. 13.16) is of interest from the topographic point of view. It shows, near the intersection of the *kardo maximus* (given an exaggerated width) and the *decumanus maximus*, a braided river with a road on each side, these roads being at a totally different orientation from the centuriation scheme. This must have been a tributary of

the Rhône not very far from its junction with the main river.

Cadaster B, when discovered, was by far the best represented of the three cadasters, and when complete it must have been over 5.5 meters high and 7 meters long. It is in squares of the standard measurement of 200 *iugera*, representing 20 by 20 *actus*. On figure 13.17, which shows the layout, the numbers indicate the minimum number of centuries in each of the four directions. Thus the north-south extension of the surveyed area occupied at least 63 centuries, that is, over 44 kilometers. It must have reached somewhere near Montélimar to the north; but centuriation does not seem to have extended to the west of the Rhône, and in the northeast not all the land was suitable for allocation. Although such an area is smaller than some in Tunisia, it is much

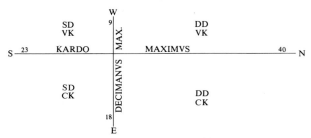

FIG. 13.17. THE PROBABLE LAYOUT OF ORANGE CA-
DASTER B. The figures indicate the minimum number of centuries in each direction, each century being a square of 20 ×
20 *actus.*
After O. A. W. Dilke, *The Roman Land Surveyors: An Introduction to the Agrimensores* (Newton Abbot: David and
Charles, 1971), 166.

larger than the norm in Europe except for parts of the
Po valley. What we cannot do is determine directly the
minimum area of the whole of this cadastral survey from
these measurements, since some centuries may never
have been incorporated. Piganiol's attempt to pinpoint
the survey was inexact,[43] since in the Gorge de Donzère
area north of Orange it would assume an ancient course
of the Rhône too far east in view of the rock formation.
This difficulty can be eliminated by a slight shift of orientation and a slight diminution of the space allocated
to each century. The widest river in Cadaster B occupies
only about 20 percent of the side of a century, that is,
about 140 meters, whereas the present width of the
Rhône is considerably greater; nevertheless, there is reason to believe that the Rhône, as well as some of its
tributaries, is shown.[44] Lands restored to the Tricastini
are in the largest quarter of the survey, that to the northeast (DD CK), which was not a particularly fertile area
(fig. 13.18).

Cadaster C, on a scale of about 1:6,000, is not well
preserved, but it seems to have covered an area south
of Orange. Its western portion included islands in the
Rhône called insulae Furianae, extending at least 5 kilometers north-south and 2 kilometers east-west (fig.
13.19). Changes since Roman times in the course of the
Rhône south of Orange have made it impossible to identify the areas shown with any certainty. The interest of
this western section is that it is the most extensive large-
scale Roman map of a river with islands. It is surprising
that such an area, only partially cultivable and obviously
liable to change of river course, should have been centuriated with the same system of coordinates as the rest;
but this reflects the thoroughness and completeness of
these surveys.

The chronology of the three cadasters is disputed.
Piganiol held that Cadaster A was of the date of Ves-

FIG. 13.18. REASSEMBLED PLAQUE III J FROM ORANGE
CADASTER B. The area is in the northeastern quarter of the
survey (DD CK on fig. 13.17) and shows lands restored to the
native Tricastini.
Size of the original: 121 × 58 cm. By permission of the Musée
Municipal d'Orange, Vaucluse.

pasian's edict and that the other two were later, B preceding C. This seems more likely than the view already
mentioned that the three are contemporary. Cadaster A
is the only one in which state lands are listed; if these,
as happened in certain other colonies, were presented
by Domitian to the local authority, that would account
for their nonappearance in the other cadasters. Domitian
may also have been responsible for the elevation to "colonia" of the capital of the Tricastini, and this suggests
a connection with the comparatively high percentage of
lands restored to the Tricastini in Cadaster B.

43. O. A. W. Dilke, "The Arausio Cadasters," in *Akten des VI.
internationalen Kongresses für griechische und lateinische Epigraphik,
München, 1972,* Vestigia 17 (Munich: C. H. Beck'sche Verlagsbuchhandlung, 1973), 455–57.

44. See diagrams, reprinted from the article cited in the previous
note, in O. A. W. Dilke, *Gli agrimensori di Roma antica* (Bologna:
Edagricole, 1979), 82, fig. 46 bis.

FIG. 13.19. ORANGE CADASTER C. Assembly of the insulae Furianae. These islands in the Rhône, now washed away, were in the western portion of the cadaster.
Size of the original: 109 × 54 cm. By permission of the Musée Municipal d'Orange, Vaucluse.

LARGE-SCALE TOWN MAPS: THE *FORMA URBIS ROMAE*

In view of the importance of cities in Roman life—as well as the skills possessed by the *agrimensores* in large-scale surveying—it is not surprising that by the date of the early empire large-scale plans of towns are also encountered. It has been suggested that such plans, which were constructed upon larger scales than those of the centuriated areas, may have been drawn by architects rather than by land surveyors;[45] the word *forma*, it may

FIG. 13.20. FRAGMENT OF A ROMAN PLAN. This urban property plan, one of the few that have survived, shows private dwellings with the names of the owners, which include women. Antiquarium Comunale, Rome.
Size of the original: 12 × 13 cm. Photograph courtesy of Thames and Hudson.

be noted, could be applied both to a surveyor's map and to an architect's plan of a house.

For both urban and rural properties a few detailed plans have survived that may have originated in (or have been based upon) architectural drawings of some kind. Some plans may have served as a record of town holdings in the same manner as the Orange Cadasters. A fragment of a plan in the Antiquarium Comunale, Rome,[46] has the names of owners of private property inscribed, and as at Orange, these include women's names (fig. 13.20). A plan of baths, however, of uncertain date and now in the Palazzo dei Conservatori, Rome, confines itself to the details of buildings. When it was discovered in 1872 on the Via di Porta San Lorenzo, the figures were thought to refer to military units; in fact they are measurements in Roman feet, and the scale is fairly consistent (fig. 13.21).[47]

The chief specimen of a country estate plan is the "Urbino plan," found on the Via Labicana, Rome, and now in the ducal palace at Urbino (fig. 13.22).[48] It has been thought to be an estate with a funeral monument

45. Dilke, *Roman Land Surveyors*, 112 (note 4).
46. Carettoni et al., *Pianta marmorea*, 207, no. 2 and pl. Q, fig. 47 (note 3); *Corpus Inscriptionum Latinarum*, 6.4.1 (1894): 2897, no. 29846 (note 2); P. D. A. Harvey, *The History of Topographical Maps: Symbols, Pictures and Surveys* (London: Thames and Hudson, 1980), 130, fig. 75.
47. Carettoni et al., *Pianta marmorea*, 209, no. 7 (note 3); *Corpus Inscriptionum Latinarum*, 6.4.1 (1894): 2897, no. 29845 (note 2).
48. For more information on the "Urbino plan" see Carettoni et al., *Pianta marmorea*, 207–8, no. 3, and pl. Q, fig. 51 (note 3); *Corpus Inscriptionum Latinarum*, 6.4.1 (1894): 2897, no. 29847 (note 2).

and adjoining gardens. The sides of the property along the private road are given the measurements of 546 and 524½ Roman feet. It features a public road, with a measurement of 1,683 Roman feet, that led to the property. There is also a series of miniature large-scale maps of farms or villa estates in the *Casae litterarum* (fig. 13.23),[49] a Late Latin treatise included in manuscripts of the *Corpus Agrimensorum*. The word *casa* does not here mean cottage, as in classical Latin, but denotes a farm or villa estate.[50] The exact purpose of these miniature maps is disputed, but there is an interesting possible indication of scale. The title of one set contains the words *in pede V fac pede uno (pedem unum)*. If the V really stands for V̄, that is, 5,000, we have a scale of 1:5,000.[51]

From such detailed plans of individual sites, it was a logical step to large-scale representations of entire towns. While it is not clear how and to what extent such maps were used in town planning, they had become an established part of Roman large-scale mapping by the early empire. Besides the most famous example—the *Forma Urbis Romae*, to be described below—it is likely that other ancient settlements had their own plans of this type. One detailed town plan clearly recognizable as such from the Roman world is a fragment from Isola Sacra (near Ostia) preserved in the museum of ancient Ostia (fig. 13.24).[52] This has no inscription except numerals, which may denote measurements in Roman feet. Walls are represented by double lines, whereas in the *Forma Urbis Romae* they are single lines.

The *Forma Urbis Romae* can be studied as an exceptional artifact not only because of its size and large scale, but also because the many surviving fragments have allowed a detailed reconstruction to be made.[53] Engraved on marble, it may be regarded as an official plan of Rome, covering the exact area of the city's limits at the date when it was constructed. Its original size was 13.03 meters high by up to 18.3 meters wide, and a reproduction at this size, reflecting knowledge of it in 1959 and purposely not updated, is to be seen in a courtyard of the Musei Capitolini. The extant fragments are in the Palazzo Braschi, on Corso Vittorio Emanuele, where they have recently been studied afresh.[54] The date of completion of the plan is between A.D. 203 and 208: the two latest buildings to be included are the Septizodium (an ornamental gateway built in A.D. 203) and a building inscribed SEVERI ET ANTONINI AVGG NN, which presumably means "the unnamed building of the emperors Severus and Antoninus [Caracalla]." Whether it was an original composition or a revision is uncertain; if the latter, it could have been a revision of a city plan of Vespasian (A.D. 69–79) and Titus (A.D. 79–81), who had Rome surveyed in A.D. 74. We do not know whether the work of their surveyors (*mensores*) was displayed in

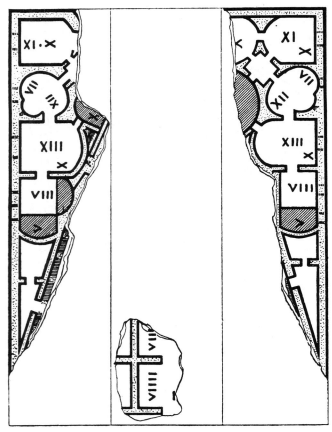

FIG. 13.21. PLAN OF ROMAN BATHS. Fragments of a multicolored mosaic of a thermal bath in the Via di Porta San Lorenzo, Rome.
Size of the original: 1.62 × 1.15 m. From Gianfilippo Carettoni et al., *La pianta marmorea di Roma antica: Forma Urbis Romae*, 2 vols. (Rome: Comune di Roma, 1960), 209. By permission of the Musei Comunali di Roma.

49. Åke Josephson, *Casae litterarum: Studien zum Corpus Agrimensorum Romanorum* (Uppsala: Almqvist och Wiksell, 1950); Blume et al., *Schriften der römischen Feldmesser*, vol. 1, figs. 254 ff. (note 4). Of the two principal manuscripts, Wolfenbüttel Arcerianus A has only the Latin alphabet; both Latin and Greek are included in a Berlin fragment.

50. Adolf Schulten, "Fundus," in *Dizionario epigrafico di antichità romane*, ed. E. de Ruggiero (Rome, 1895–), 3:347.

51. It is a frequent misconception, recently repeated in the popular media, that the Romans could express thousands numerically only by repetition of M: on the contrary, they regularly placed a bar over a numeral or numerals to multiply by one thousand. In mileages this convention is sometimes misunderstood by later writers, who multiply the thousands by the remainder of the figures, obtaining an entirely wrong answer.

52. Carettoni et al., *Pianta marmorea*, 208, no. 5 (note 3); Harvey, *Topographical Maps*, 130, fig. 76 (note 46).

53. Carettoni et al., *Pianta marmorea* (note 3).

54. Emilio Rodríguez Almeida, *Forma Urbis Marmorea: Aggiornamento generale 1980* (Rome: Edizioni Quasar, 1981).

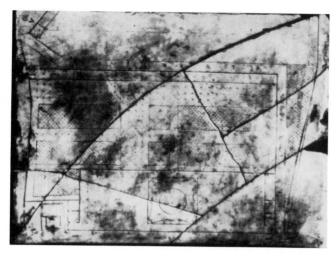

FIG. 13.22. THE "URBINO PLAN." An example of a plan of a Roman country estate. Of uncertain date, it features a *harundinetum* for the cultivation of reeds, a *fossa* (ditch), and measurements along both the public and private (*via privata*) roads.
Size of the original: 74 × 83.2 cm. From Gianfilippo Carettoni et al., *La pianta marmorea di Roma antica: Forma Urbis Romae*, 2 vols. (Rome: Comune di Roma, 1960), pl. Q, fig. 51. By permission of the Musei Comunali di Roma.

public in the form of a plan. In A.D. 191, however, there was a fire in Vespasian's Temple of Peace, and when Septimius Severus restored it, the *Forma Urbis Romae* was affixed to the outer wall of a library attached to that temple (fig. 13.25). This wall still exists outside the Church of Saints Cosmas and Damian, and holes that served to fix the plan can be seen on it. These dowel holes have enabled archaeologists to deduce the original positions of many fragments.

The first fragments were found in 1562, and one published in 1590 was recognizable as a fairly complete plan of the Ludus Magnus, a gladiatorial school founded by Domitian near the Colosseum (fig. 13.26). Other fragments found in the Renaissance but since lost are recorded in Vat. Lat. 3439 (Rome, Biblioteca Apostolica Vaticana). Since 1874, when the *Forma Urbis Romae* was published by H. Jordan, the number of fragments discovered has more than doubled.

Clearly the map first had to be drawn and inscribed on less permanent material by surveyors (*mensores*), presumably under the direction of a senior architect. Unlike land surveyors and camp measurers, urban surveyors have left no manuals. But we have some of their instruments (an inscribed set square, for example, was found recently on the Watling Street site in Canterbury),[55] and we can guess that they started with the most prominent

buildings, since an examination of scales, as mentioned below, has shown special treatment of many of these. For the final version the assistance of *lapicidae*, stone cutters, was required, both to engrave the map and wording on the marble and to affix the marble to the wall.

One of the main advantages of a detailed map of Rome was to improve the efficiency of the city's administration. Augustus had divided Rome into fourteen districts, each subdivided into *vici*. These districts were administered by annually elected magistrates, with officials and public slaves under them. The officials included *vigiles*, municipal police who also acted as a fire service. Another important part of the administration, though not one subordinated to the districts, was the water supply. As an example of possible map use, if the *vigiles* could see from the map the location of the nearest aqueducts and *castella* (local reservoirs), they would be able to fight a fire more easily. For this purpose copies of relevant portions of the *Forma Urbis Romae* may have been made on papyrus or on wax tablets. The long persistence of the Augustan system of district administration may be seen from two late extant topographical manuals, the *Curiosum Urbis regionum XIV* and the *Notitia regionum XIV*, datable to A.D. 354 and 375 respectively.

The orientation of the *Forma Urbis Romae* is approximately to the southeast (about 43° east of south, with variations between 36° and 50°). Such a layout may have been inherited from a previous map, since the first Augustan district of Rome was in the southern part of the city, while the mention in Tacitus of the gardens of Sallust as being in the left part of the city would, if it refers to a plan, suit either east or south at the top.[56] The average scale is approximately 1:240 or 1:250, but there is a tendency to make important buildings bigger than they should be, so that the scale variation in the extant parts is from 1:189 to 1:413. Yet this plan, when allowance is made for some mistakes in outline of build-

55. Hugh Chapman, "A Roman Mitre and Try Square from Canterbury," *Antiquaries Journal* 59 (1979): 403–7; O. A. W. Dilke, "Ground Survey and Measurement in Roman Towns," in *Roman Urban Topography in Britain and the Western Empire*, ed. Francis Grew and Brian Hobley, Council for British Archaeology Research Report no. 59 (London: Council for British Archaeology, 1985), 6–13, esp. 9 and fig. 13.

56. Tacitus *Histories* 3.82; see *The Histories*, trans. Clifford H. Moore, Loeb Classical Library (London: William Heinemann; Cambridge: Harvard University Press, 1925–37 and later editions). Ferdinando Castagnoli, "L'orientamento nella cartografia greca e romana," *Rendiconti della Pontificia Accademia Romana di Archeologia* 48 (1975–76): 59–69, shows that south was a more frequent orientation in the Roman world than is generally supposed. Left and right probably do not refer to the riverbanks. True, the Horti Sallustiani were on what we should call the left bank; but so was 90 percent of the ancient city.

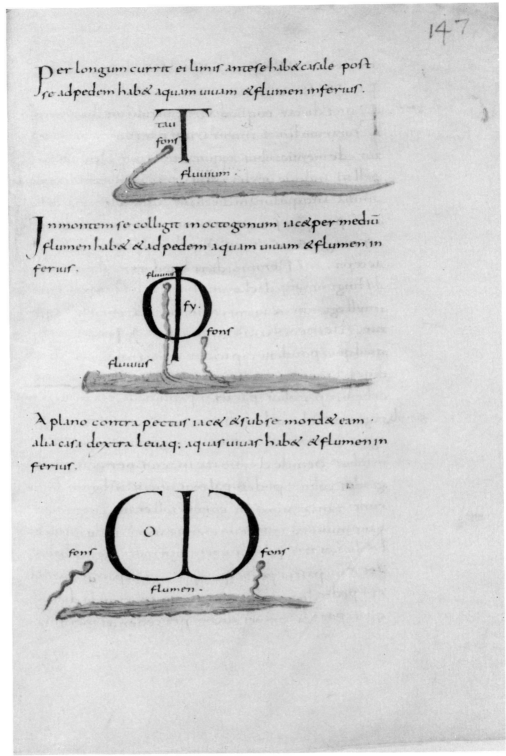

FIG. 13.23. VIGNETTES FROM THE *CASAE LITTERA-RUM.* An example of a genre in a Late Latin treatise, consisting of maps of farms (as letters) in different topographical situations. The descriptive text above the lowest miniature can be translated: "ω is situated away from level ground against the breast of the hill, with another villa set into the hill beneath. It has fresh springs to the left and right, and a river below." Size of the original: 28 × 19.6 cm. Photograph from the Bibliotheca Apostolica Vaticana, Rome (Pal. Lat. 1564, fol. 147r).

ings, represents a degree of accuracy not surpassed in plans of Rome for over fifteen hundred years.[57]

Distortion of scale is caused not only by the enlargement of important buildings, but also by representation of certain features in elevation or sign rather than in plan.[58] Chief among these are the arches carrying aqueducts, which had they not been shown in elevation would have been much more difficult to recognize (fig.

FIG. 13.24. THE ISOLA SACRA FRAGMENT. This is clearly part of a Roman town plan, with double lines for walls. The only inscriptions are numerals that may denote measurements. Museo Ostiense, near Rome.
Size of the original: 14 × 17 cm. Photograph courtesy of Thames and Hudson.

13.27). These may perhaps be reckoned among the cartographic signs used; others include stairs, noticeable in that they are sometimes fully represented, sometimes shown as a triangle with steps inside, and sometimes depicted simply by two lines forming an acute angle and resembling such a triangle. Indeed, as Harvey has pointed out, the conventionalization of much of the detail, while at the same time preserving the accuracy of its position, is one of the most remarkable features of the map. He writes:

> Walls are mostly shown by single lines, though on some important buildings the lines are drawn double and the intervening space hollowed out. . . . Single lines also mark certain boundaries. Dots mark columns or probably sometimes trees, and both dots within rectangles and small rectangles alone, either in outline or hollowed out, presumably mark columns standing on square bases. Larger rectangles and circles are used for bases of statues, altars in front of temples, public fountains and so on, and a wide circle sometimes marks the precincts of a building. . . . All the principal buildings are named, and it is likely that the lettering was coloured—perhaps other parts of the plan as well.[59]

57. Amato Pietro Frutaz, *Le piante di Roma*, 3 vols. (Rome: Istituto di Studi Romani, 1962); O. A. W. Dilke and Margaret S. Dilke, "The Eternal City Surveyed," *Geographical Magazine* 47 (1975): 744–50. Recent research suggests that the shape of a building in the Campus Martius area is more incorrect than most on the *Forma Urbis Romae*.

58. Thomas Ashby, *The Aqueducts of Ancient Rome* (Oxford: Clarendon Press, 1935).

59. Harvey, *Topographical Maps*, 128 (note 46).

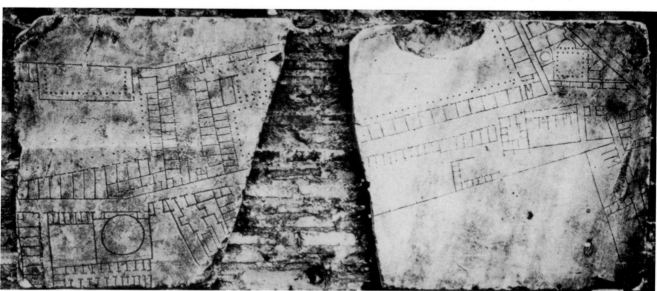

FIG. 13.25. *FORMA URBIS ROMAE*. Reconstructed position of two fragments of the plan on the original wall. The fragments cover an area of the Via Portuense with the south bank of the Tiber at top left.

Height of original fragments: 74 cm. From Gianfilippo Carettoni et al., *La pianta marmorea di Roma antica: Forma Urbis Romae*, 2 vols. (Rome: Comune di Roma, 1960), pl. N, fig. 37. By permission of the Musei Comunali di Roma.

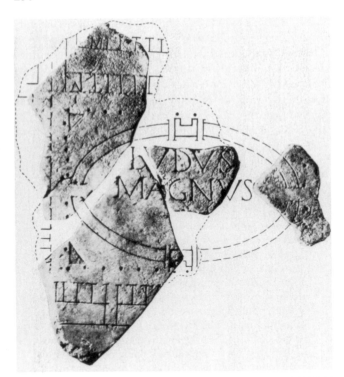

FIG. 13.26. THE LUDUS MAGNUS ON THE *FORMA URBIS ROMAE*. This reconstitution of fragments, first published in 1590, shows a gladiatorial school near the Colosseum.
From A. M. Colini and L. Cozza, *Ludus Magnus* (Rome: Comune di Roma, 1962), 8, fig. 8. By permission of the Musei Comunali di Roma.

The whole map thus represents a sophisticated cartographic achievement that in both concept and execution differs little from the town plans of today. It remains the most impressive surviving monument to the work of Roman surveying and mapping.

If the *Forma Urbis Romae* was exceptional in the early empire, it is clear that not all town plans either were drawn to such a large scale or served practical purposes in connection with property or civic administration. Some were decorative or symbolic in purpose. In this category is a maplike mosaic found in Rome's port of Ostia. This was the principal center for overseas trade, particularly imports to Rome such as grain, oil, and animals for the amphitheater.[60] Among a number of mosaics, one of the most striking representations, though a simple one, is a mosaic of a Roman site virtually in plan, to be seen at ancient Ostia, near the road from the entrance to the museum (fig. 13.28).[61] This is in the meeting place of the *cisiarii*, whose light vehicles were usually drawn by mules, and has been tentatively dated to about A.D. 120.[62] Here the mosaic has as its centerpiece a small, square, walled town (unrealistic, if it is intended to be Ostia), with a tower at each corner. These

FIG. 13.27. AQUEDUCT ARCHES ON THE *FORMA URBIS ROMAE*. The stylized depiction of aqueducts in elevation is typical of cartographic signs used on the *Forma Urbis Romae*. By permission of the Musei Comunali di Roma.

towers are shown not in plan but in elevation, each lying directly opposite the center of the town and supported by giants known as Telamones. The carts, whose mules have satirical names like Podagrosus ("gouty"), are shown with their drivers outside the town.

ENGINEERING PLANS FOR TUNNELS AND AQUEDUCTS

The final category of large-scale surveys and mapping from the early empire is associated with engineering works of various kinds. The construction of tunnels and aqueducts may, in particular, have involved the use of large-scale plans. Tunnels in antiquity could be dug either for roads or for aqueducts. If they were dug from both sides simultaneously, the result might be a near miss, as happened with the Siloam tunnel[63] or that mentioned by Nonius Datus. To avoid this, the Greek mathematician Heron (Hero) of Alexandria, who was evidently writing at some time around A.D. 62,[64] shows by

60. J. H. D'Arms and E. C. Kopff, eds., *The Seaborne Commerce of Ancient Rome: Studies in Archaeology and History*, Memoirs of the American Academy in Rome, 36 (Rome: American Academy in Rome, 1980).

61. For later mosaics see pp. 263–66 below.

62. Giovanni Becatti, ed., *Mosaici e pavimenti marmorei*, 2 pts. (1961); both are vol. 4 of *Scavi di Ostia* (Rome: Istituto Poligrafico dello Stato, 1953–), pt. 1, 42–44 and pt. 2, pl. CVIII.

63. Siloam inscription from Jerusalem of the reign of King Hezekiah, Istanbul Archaeological Museum.

64. Heron is associated with a recollection of the eclipse of A.D. 62; see note 66 below and Otto Neugebauer, "Über eine Methode zur Distanzbestimmung Alexandria-Rom bei Heron," *Historisk-Filologiske Meddelelser udgivne af det Kongelige Danske Videnskabernes Selskab* 26 (1938–39), nos. 2 and 7, esp. no. 2, pp. 21–24. Other dates have been suggested; see, for example, Dimitrios Sakalis, "Die Datierung Herons von Alexandrien," inaugural dissertation (University of Cologne, 1972), who gives A.D. 90–150, pointing out that Proclus (fifth century A.D.) lists him after Menelaus, who was writing about A.D. 98.

FIG. 13.28. THE CISIARII MOSAIC, OSTIA. Dated to about A.D. 120, this mosaic shows a decorative plan of a small walled town with towers at each corner supported by giants. Museo Ostiense, near Rome.

Size of the original: 8.7 × 8.7 m. From Giovanni Becatti, ed., *Mosaici e pavimenti marmorei*, 2 pts. (1961); both are vol. 4 of *Scavi di Ostia* (Rome: Istituto Poligrafico dello Stato, 1953–), pt. 2, pl. CVIII (no. 64).

a plan the method he advocates (fig. 13.29).[65] Keeping the gradient uniform, he draws lines along one side of the hill, then at right angles as many times as is necessary, to the appropriate point on the other side of the hill. In theory at least, the gradient could be worked out by means of his dioptra, which was used for angle observation either in surveying or in astronomical work. Yet although this instrument was most ingenious and elaborately constructed, it was obviously rather heavy and complicated for extensive field survey and is never mentioned in the *Corpus Agrimensorum*.[66]

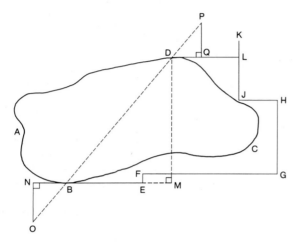

FIG. 13.29. HERON'S TECHNIQUE FOR TUNNELING THROUGH A HILLSIDE. The instructions specify that a tunnel is to be dug from both sides through a hill (in plan) ABCD. From B draw a random line BE, a perpendicular EF, and so on around the hill to K. Construct a line from D to be perpendicular to JK at L. Measure lengths around the hill. Calculate the lengths of BM and DM, being sides of the right-angled triangle BDM. Produce EB to a random point N and construct O such that BN:NO = BM:MD. Likewise, Q and P are located such that DQ:QP = BM:MD. The direction of the tunnel follows sightings along OB and PD.
After Heron of Alexandria, *Opera quae supersunt omnia* (Leipzig: Teubner, 1899–1914), vol. 3, *Rationes dimetiendi et commentatio dioptrica*, ed. Hermann Schöne (1903), 238.

The surveying treatise on the aqueducts of Rome is by Frontinus, who was appointed curator of waterways by the emperor Nerva in A.D. 97. His work, as it has come down to us, is not illustrated, but in one passage he says: "My enthusiasm did not stop at inspecting individual details. I also took care to make maps of the aqueducts. From these it is clear where the valleys are and how big, where rivers are crossed, where channels made on mountain sides demand more care in constantly inspecting and repairing the conduit. This has given me the advantage that I am able to put the situation immediately under review and discuss it as if I were on the spot."[67] Such a description seems to imply something approaching a relief map, far more detailed than the aqueducts on the *Forma Urbis Romae*. Whereas the latter are shown pictorially, Frontinus had a working map, possibly with something like transverse profiles.

---

65. Heron *On the Dioptra*, in his *Opera quae supersunt omnia*, 5 vols. (Leipzig: Teubner, 1899–1914), vol. 3, *Rationes dimetiendi et commentatio dioptrica*, ed. and trans. Hermann Schöne (1903), 238, fig. 95. Schöne also provides a reconstruction of the *dioptra*.

66. For a description and reconstruction of Heron's *dioptra* and his other leveling instruments, see Kiely, *Surveying Instruments*, 20–27 (note 6). Heron also devised a method of distance measurement with his *dioptra*, but again this does not seem to have been taken up in practice. As described in his *On the Dioptra* 35 (note 65), the method involved the use of the *dioptra* together with lunar eclipse observations (a theory already known to Hipparchus), so that the distance between two places could be established. Those he chose were Alexandria and Rome, and his method is known as *analemma*, literally "high rise." The construction of the dioptra is explained by Otto Neugebauer, *A History of Ancient Mathematical Astronomy* (New York: Springer-Verlag, 1975), 845 ff., 1379, fig. 23, summarizing his article cited in note 64. Unfortunately, Heron did not have data in hand for the two locations; he seems to have had Alexandrian data for the eclipse of A.D. 62. His estimate of the time difference, two hours, is wide of the mark; the correct time difference between Alexandria and Rome is one hour, ten minutes.

67. Frontinus *De aquis urbis Romae* (Water supply of Rome) 1.17, author's translation, see Frontinus, *The Stratagems [and] The Aqueducts of Rome*, rev. Charles E. Bennett, based on trans. by Clemens Herschel, ed. Mary B. McElwain, Loeb Classical Library (Cambridge: Harvard University Press; London: William Heinemann, 1925).

BIBLIOGRAPHY
CHAPTER 13 ROMAN LARGE-SCALE MAPPING IN THE EARLY EMPIRE

Blume, Friedrich, et al., eds. *Die Schriften der römischen Feldmesser.* 2 vols. Berlin: Georg Reimer, 1848–52; reprinted Hildesheim: Georg Olms, 1967.

Butzmann, Hans, ed. *Corpus Agrimensorum: Codex Arcerianus A der Herzog-August-Bibliothek zu Wolfenbüttel.* Codices Graeci et Latini 22. Leiden: A. W. Sijthoff, 1970.

Carder, James Nelson. *Art Historical Problems of a Roman Land Surveying Manuscript: The Codex Arcerianus A, Wolfenbüttel.* New York: Garland, 1978.

Carettoni, Gianfilippo, et al. *La pianta marmorea di Roma antica: Forma Urbis Romae.* 2 vols. Rome: Comune di Roma, 1960.

Chevallier, Raymond. "Sur les traces des arpenteurs romains." *Caesarodunum,* suppl. 2. Orléans-Tours, 1967.

——, ed. *Colloque international sur la cartographie archéologique et historique, Paris, 1970.* Tours: Centre de Recherches A. Piganiol, 1972.

Clavel-Lévêque, Monique, ed., *Cadastres et espace rural: Approches et réalités antiques.* Paris: Centre National de la Recherche Scientifique, 1983.

Dilke, O. A. W. "Maps in the Treatises of Roman Land Surveyors." *Geographical Journal* 127 (1961): 417–26.

——. *The Roman Land Surveyors: An Introduction to the Agrimensores.* Newton Abbot: David and Charles, 1971.

——. "The Arausio Cadasters." In *Akten des VI. internationalen Kongresses für griechische und lateinische Epigraphik, München, 1972, Vestigia* 17, 455–57. Munich: C. H. Beck'sche Verlagsbuchhandlung, 1973.

——. *Greek and Roman Maps.* London: Thames and Hudson, 1985.

Frutaz, Amato Pietro. *Le piante di Roma.* 3 vols. Rome: Istituto di Studi Romani, 1962.

Hinrichs, Focke Tannen. *Die Geschichte der gromatischen Institutionen.* Wiesbaden: Franz Steiner, 1974.

Josephson, Åke. *Casae litterarum: Studien zum Corpus Agrimensorum Romanorum.* Uppsala: Almqvist och Wiksell, 1950.

Miller, Konrad. *Die Erdmessung im Altertum und ihr Schicksal.* Stuttgart: Strecker und Schröder, 1919.

Neugebauer, Otto. "Über eine Methode zur Distanzbestimmung Alexandria-Rom bei Heron." *Historisk-Filologiske Meddelelser udgivne af det Kongelige Danske Videnskabernes Selskab* 26 (1938–39), nos. 2 and 7.

Oliver, J. H. "North, South, East, West at Arausio and Elsewhere." In *Mélanges d'archéologie et d'histoire offerts à André Piganiol,* 3 vols., ed. Raymond Chevallier, 2:1075–79. Paris: SEVPEN, 1966.

Piganiol, André. *Les documents cadastraux de la colonie romaine d'Orange. Gallia,* suppl. 16. Paris: Centre National de la Recherche Scientifique, 1962.

Rodríguez Almeida, Emilio. *Forma Urbis marmorea: Aggiornamento generale 1980.* Rome: Edizioni Quasar, 1981.

Sakalis, Dimitrios. "Die Datierung Herons von Alexandrien." Inaugural dissertation. University of Cologne, 1972.

Salviat, F. "Orientation, extension et chronologie des plans cadastraux d'Orange." *Revue Archéologique de Narbonnaise* 10 (1977): 107–18.

Thulin, Carl Olof, ed. *Corpus Agrimensorum Romanorum.* Leipzig, 1913; reprinted Stuttgart: Teubner, 1971.

14 · Itineraries and Geographical Maps in the Early and Late Roman Empires

O. A. W. DILKE

Whereas Greek knowledge about the theory of map-making—as well as the map image of the known world—tended to accumulate and to build on previous writers, Roman efforts in both small- and large-scale mapping tended to be diluted over time. The record is also extremely fragmentary and drawn out over a period of some five hundred years. In this chapter, we discuss itineraries and small-scale geographical maps in both the early and the late empire and conclude with a review of the use of maps in the Roman period as a whole.

The relative decline in the theoretical aspects of Roman cartography cannot be disputed. Although emperors such as Hadrian and Caracalla (M. Aurelius Severus Antoninus) were great philhellenes (and Hadrian's principate—A.D. 117–38—coincided with much of Ptolemy's working life), they do not seem to have encouraged Roman scholars to build on the foundations of Greek geography and astronomy. Latin writers such as Pomponius Mela and the elder Pliny did little to modify Hellenistic concepts of the inhabited world, and in comparison with Greek writers such as Hipparchus or Strabo, the status of maps within their work is relatively ambiguous.

During the late empire, scholars became even further removed from the sources of Greek geographical culture and from the cartographical knowledge it had contained. It is true that mathematics continued to flourish at Alexandria—where Pappus, in the early fourth century, not only commented on Ptolemy's *Almagest* and *Planisphaerium*, but also wrote a *Chorography of the Oikoumene* (now lost) based on Ptolemy's *Geography*[1]—but this did not lead to a revision of the maps. Following the Antonine and Severan dynasties (A.D. 138–235) there was a period of rapidly changing emperors, and, apart from legal writings, the arts and the sciences—including the knowledge that related to maps—cannot be said to have flourished much. If we take accuracy in geographical mapping as a yardstick, standards were declining when compared with the high point of Greek influence. For example, while the Peutinger map was of fourth-century origins but indebted to a first-century A.D. map, the earlier map may have positioned towns and roads with a closer resemblance to reality. Since no ancient handbook of Roman roads, illustrated or not, has survived, this is difficult to ascertain. The bureaucrats' maps attached to the *Notitia Dignitatum*, a directory of officeholders and administrators, follows textual rather than topographical order, sometimes with confused results. Even the maps in the *Corpus Agrimensorum* tended to deteriorate with repeated copying, particularly in their jumbled nomenclature.

The late empire is thus often dismissed as of little consequence by historians of cartography. It is given short shrift in the standard authorities; only one or two surviving maps such as the Dura Europos shield or the Peutinger map are described.[2] However, once we start to evaluate a broader spectrum of evidence it becomes clear that the idea of the map was not only kept alive in western and northern Europe, but also transmitted to the eastern empire after the foundation of Constantinople on the site of Byzantium in A.D. 330. Set against an apparent lack of scientific progress in mapmaking, sound evidence for the widespread use of geographical maps is revealed in literary sources, found in images on coins or incorporated into mosaics, and even seen in the decoration of lamps. As much as in Hellenistic Greece, such maps continued to have meaning in Roman society.

ITINERARIES AND THE PEUTINGER MAP

It is widely accepted that measured itineraries are of fundamental importance in the construction and development of geographical maps and marine charts. This was true of many societies that developed such maps,[3] and the Roman world is no exception, though a clear

---

1. Ivor Bulmer-Thomas, "Pappus of Alexandria," in *Dictionary of Scientific Biography*, 16 vols., ed. Charles Coulston Gillispie (New York: Charles Scribner's Sons, 1970–80), 10:293–304.

2. Armando Cortesão, *History of Portuguese Cartography*, 2 vols. (Coimbra: Junta de Investigações do Ultramar-Lisboa, 1969–71), 1:148–50; Leo Bagrow, *History of Cartography*, rev. and enl. R. A. Skelton, trans. D. L. Paisey (Cambridge: Harvard University Press; London: C. A. Watts, 1964), 37–38; Gerald R. Crone, *Maps and Their Makers: An Introduction to the History of Cartography*, 5th ed. (Folkestone: Dawson; Hamden, Conn.: Archon Books, 1978), 3–4.

3. P. D. A. Harvey, *The History of Topographical Maps: Symbols, Pictures and Surveys* (London: Thames and Hudson, 1980), 135–52.

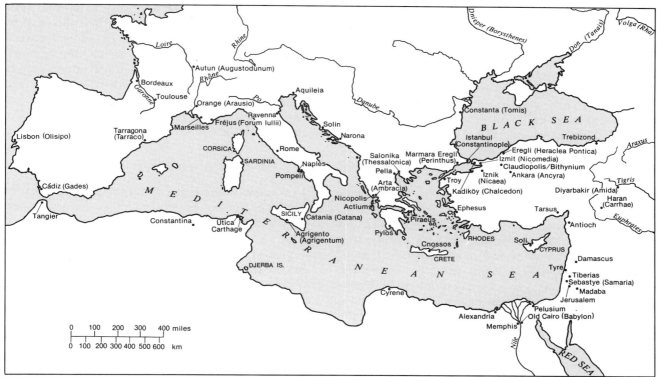

FIG. 14.1. PRINCIPAL PLACES ASSOCIATED WITH ITIN-
ERARIES AND GEOGRAPHICAL MAPS FROM THE
ROMAN EMPIRE. See also figure 12.1.

distinction must be drawn between written itineraries and itinerary maps. In the Roman period the former were more common, being used for both military and civil purposes—and together with the portable sundial providing a principal aid to the well-informed traveler. The earliest surviving Roman itineraries are the Vicarello goblets, which give a list of stages from Cádiz to Rome via the Po valley, with the mileage between successive stages.[4]

The best-preserved examples are the Antonine itinerary, the Bordeaux itinerary, and the Ravenna cosmography that is associated with the Byzantine empire. The first two of these are simply lists of places along routes, giving the distances between them, but because of their close relationship to geographical mapping, all three will be examined alongside the Peutinger map—which remains as the sole surviving example of the itinerary map from the Roman period, unless we assign the Dura Europos shield to this category.

THE ANTONINE ITINERARY

The Antonine itinerary, the most important of the ancient list-type itineraries to be preserved (as opposed to the map or "painted itinerary"), is in two parts: land and sea.[5] The full titles of these are *Itinerarium pro-*

vinciarum Antonini Augusti and *Imperatoris Antonini Augusti itinerarium maritimum.* These titles make it clear that the journeys mentioned were, in origin at least, either planned for or completed by an emperor of the Antonine dynasty; and there is general agreement that this emperor was Caracalla. Since the longest single journey is overland from Rome to Egypt via the Bosporus, it seems only reasonable to link this with such a journey undertaken by Caracalla in A.D. 214–15.[6] A long prestigious journey by an emperor would require careful planning by civil servants, with provision for supplies, changing of horses, and so on, at appropriate staging posts. Every contingency had to be foreseen, and local

4. Jacques Heurgon, "La date des gobelets de Vicarello," *Revue des Etudes Anciennes* 54 (1952): 39–50; Raymond Chevallier, *Les voies romaines* (Paris: Armand Colin, 1972), 46–49, or for an English translation, *Roman Roads*, trans. N. H. Field (London: Batsford, 1978), 47–50. O. A. W. Dilke, *Greek and Roman Maps* (London: Thames and Hudson, 1985), 122–24.

5. Otto Cuntz, ed., *Itineraria Romana* (Leipzig: Teubner, 1929), vol. 1, Itineraria Antonini Augusti et Burdigalense; Konrad Miller, *Itineraria Romana* (Stuttgart: Strecker und Schröder, 1916), LV ff. and regional sections; Dilke, *Greek and Roman Maps*, 125–28 (note 4).

6. D. van Berchem, "L'annone militaire dans l'empire romain au IIIe siècle," *Bulletin de la Société Nationale des Antiquaires de France* 80 (1937): 117–202.

representatives complained that they often had to make such provision at points where in fact the emperor never stopped at all.[7] Nevertheless, the existence in the Antonine itinerary of forms of place-names later than Caracalla's reign, such as Diocletianopolis for Pella and Heraclea for Perinthus (Marmara Eregli), suggests that routes were reused, with or without amendment, over a long period (fig. 14.1). An example of addition is in Sicily, where between Catana (Catania) and Agrigentum (Agrigento) two routes are given, the second including the phrase *mansionibus nunc institutis* (by the staging posts now set up). The date of the final version of the Antonine itinerary may have been between A.D. 280 and 290.

The organization that planned such journeys was the *cursus publicus*,[8] set up by Augustus for transporting officials and their families and for carrying official mail. Hence the *cursus publicus* had its own lists, and in some cases straightforward journeys may well have been copied directly from these. But the Antonine itinerary cannot simply have been a version of those lists, because of the numerous omissions, duplications, and extremely roundabout routes. Thus the Peloponnese, Crete, and Cyprus are unrepresented, and considerable parts of Gaul, the Balkans, and Asia Minor are thinly covered. A good example of a circuitous route is the second journey in Britain, iter II, which reaches Richborough from Birrens via Carlisle, York, Chester, and London.[9] Such a route must have been tailor-made for a particular journey, stopping at the legionary fortresses of York and Chester among other places.

The method in the Antonine itinerary was to list the starting and finishing points of each journey and the total distance in Roman miles (in Gaul, leagues, as mentioned below). Then the individual stages were listed, with the mileage for each. The totals sometimes correspond to the added individual mileages, sometimes do not; in the latter case especially, one or more of the figures may well be corrupt.

The Antonine itinerary begins at Tangier and covers most of the provinces of the empire rather unsystematically. The British section, last before the sea routes, is self-contained and consists of fifteen journeys, some coinciding in the same or the opposite direction.[10] Except in cases where they are clearly corrupt, the mileages are fairly reliable. It has been shown, however, that distances from a settlement sometimes start from the center, sometimes from the outskirts.[11] Since Colchester is in one place called Camulodunum, in another Colonia (it was one of the four colonies of Roman Britain), one may suspect that the routes were not all contemporary.

An interpretation of the Antonine itinerary routes in northern Gaul encounters two difficulties. One is that distances in Gaul are sometimes reckoned in Roman

miles, sometimes in leagues, sometimes in both (1½ Roman miles = 1 *leuga*).[12] The other is that there are in some cases considerable differences of mileage between the same two places according to which journey is followed. More research is needed that will not only study the recorded distances on the modern map, but take account of archaeological and epigraphic evidence, together with such geographical factors as alterations in sea level or in the course of riverbeds.[13]

ITINERARIES FROM THE LATE EMPIRE

The vast extent of the empire, with its expansion of the bureaucracy, encouraged the production of many itineraries, which, since roads with milestones continued to be kept up, provided acceptable accuracy.[14] As barbarians pressed in from north and east, military requirements became more important then ever. Vegetius, the civil servant whose military manual dates from about A.D. 383–95 but draws on much older material, writes of the ideal general:

> In the first place, a commander should have itineraries of all the war zones very fully written out, so that he may thoroughly acquaint himself with the intervening terrain, as regards not only distance but standard of

7. Dio Cassius *Roman History* 78.9.3 and 78.9.6–7; see vol. 9 of *Dio's Roman History*, 9 vols., trans. Earnest Cary, Loeb Classical Library (Cambridge: Harvard University Press; London: William Heinemann, 1914–27).

8. References in Konrat Ziegler and Walther Sontheimer, eds., *Der kleine Pauly*, 5 vols. (Stuttgart: Alfred Druckenmüller, 1964–75), s.v. "cursus publicus," and in *Oxford Classical Dictionary*, 2d ed., s.v. "postal service (Roman)."

9. A. L. F. Rivet and Colin Smith, *The Place-Names of Roman Britain* (Princeton: Princeton University Press, 1979), 157–60, fig. 12.

10. Rivet and Smith, *Place-Names*, 148–82 (note 9).

11. Warwick Rodwell, "Milestones, Civic Territories and the Antonine Itinerary," *Britannia* 6 (1975): 76–101.

12. Gallic leagues were officially recognized by Septimius Severus about A.D. 202.

13. Francis J. Carmody, *La Gaule des itinéraires romains* (Berkeley: Carmody, 1977), contents himself with criticizing Miller's early identification of Vetera, Germania Inferior, with Rheinberg and himself proposing Alpen. But in fact there has been no dispute for many years that Vetera was quite near the previous station, Colonia Traiana (Xanten). There may have been two successive sites, one three kilometers southeast of Xanten, on the Fürstenberg north of Birten, the other on land that in Roman times was on the opposite bank of the Rhine, east-northeast of Vetera I. See H. von Petrikovits, "Vetera," in *Paulys Realencyclopädie der classischen Altertumswissenschaft*, ed. August Pauly, Georg Wissowa, et al. (Stuttgart: J. B. Metzler, 1894–), 2d ser., 8 (1958): cols. 1801–34, esp. 1813–14.

14. The emperor's name on a milestone often indicates substantial repair or improvement to a road rather than a new one. For milestones see the series *Itinera Romana*, ed. Gerald Sicilia Verbrugghe, Ingemar König, and Gerold Walser, 3 vols. (Bern: Kümmerly und Frey, 1967–76); Jeffrey P. Sedgley, *The Roman Milestones of Britain: Their Petrography and Probable Origin*, British Archaeological Reports no. 18 (Oxford: British Archaeological Reports, 1975).

roads, and may study reliable descriptions of short-cuts, deviations, mountains, and rivers. In fact, we are assured that the more careful commanders had, for provinces in which there was an emergency, itineraries that were not merely annotated but even drawn out in color [*picta*], so that the commander who was setting out could choose his route not only with a mental map but with a constructed map to examine.[15]

Itineraries were also used by pilgrims and by soldiers rejoining their legions; they were expected to take good care of them, not to leave the route, and to stop at the *mansiones* (staging posts) indicated.[16]

The official recognition of the Christian church in A.D. 313 affected cartography as it did other branches of science; one direct result was that pilgrimages to Christian shrines created a new use for geographical itineraries. The principal itinerary of the late empire is that of the Bordeaux-Jerusalem pilgrimage, A.D. 333, of which the best manuscript is the Pithoeanus, now Par. Lat. 4808, of the ninth century (no maps).[17] Distances are recorded in leagues (2.22 km) as far as Toulouse, then in Roman miles (1.48 km). Another itinerary attached to this records a journey from the Holy Land to Chalcedon (Kadiköy), in Asia Minor opposite Constantinople, and back via Nicomedia (Izmit), Ancyra (Ankara), Tarsus, and Tyre. A third goes from Heraclea Pontica (Eregli) via Macedonia, Albania, and the east coast of Italy. In addition to this major document, there are fragments of itineraries from monumental inscriptions of various periods from several territories under the Roman Empire.[18]

A second category of written itineraries relates to journeys made by sea. It has already been noted that part of the Antonine itinerary consisted of an *itinerarium maritimum*, and in view of theories about an association between these periploi and the development of portolan charts,[19] such itineraries have been widely discussed in the literature of the history of cartography. An anonymous and incomplete Greek periplus of about the third or fourth century A.D. is known as the *Stadiasmus maris magni*.[20] It records distances in stades between harbors and watering facilities around most of the eastern Mediterranean, covering the North African coast as far west as Utica. Thus a fair amount of detail is given in the entry for even such a small area as Djerba Island, Tunisia. Rhodes is particularly well covered, with sea distances to twenty-seven harbors of the eastern Mediterranean and Aegean. But some areas, such as the Levant, are very poorly covered. Müller questionably attributed the whole work to a much later period.[21]

It is likely that writers in some ports specialized in the production of these aids to the mariner, and such was Marcianus of Heraclea Pontica, a Greek writer of periploi, who is thought to have been a contemporary of

Synesius of Cyrene, about 370–413. Among his sources he mentions the *Geography* of "the most divine and wise Ptolemy," whose coordinates he clearly edited.[22] He accepts the size of the earth according to Ptolemy, not Eratosthenes' measurement of the circumference of the earth. The surviving parts of his *Periplus maris exteri*[23] cover the southern coasts of Asia (which he may well have illustrated by a map based on Ptolemy's coordinates) and the coasts of the less familiar parts of Europe; some, such as the Iberian Peninsula, are covered in

15. Vegetius *De re militari* (Military Institutions of the Romans) 3.6, author's translation. No annotated itineraries are extant, but from Vegetius's previous words they must have commented among other things on quality of road surface. The words "there was an emergency" render the past tense *gerebatur*, which, however, is a conjecture for the present *geritur* by Carl Lang in his edition of Vegetius, *Epitoma rei militaris*, 2d ed. (Leipzig: Teubner, 1885), 75. Pascal Arnaud (criticism at a seminar 12 December 1983 on "La Tabula Peutingeriana et le Corpus Agrimensorum" at Centre Jean Bérard, Institut Français de Naples) takes *picta* differently as referring to itineraries with paintings.

16. Saint Ambrose (bishop of Milan) *Expositio in psalmum 118* 5.2; see *Opera, pars quinta (V): Expositio psalmi CXVIII*, ed. Michael Petschenig in *Corpus Scriptorum Ecclesiasticorum Latinorum*, vol. 62 (Vienna: F. Tempsky; Leipzig: G. Freytag, 1913), 82–83.

17. Cuntz, *Itineraria Romana* (note 5); P. Geyer and Otto Cuntz, "Itinerarium Burdigalense," in *Itineraria et alia geographica*, in *Corpus Christianorum*, Series Latina, vols. 175 and 176 (1965), 175:XVIII–26; Henri Leclercq, "Itinéraires," in *Dictionnaire d'archéologie chrétienne et de liturgie*, 15 vols., ed. Fernand Cabrol and Henri Leclercq (Paris: Letouzey et Ané, 1907–53), 7.2 (1927): cols. 1841–1922, esp. 1853–58; Aubrey Stewart, trans., *Itinerary from Bordeaux to Jerusalem: "The Bordeaux Pilgrim,"* Palestine Pilgrims Text Society, vol. 1, no. 2 (London: Palestine Exploration Fund, 1896), 15 ff., abstracted by George Kish, ed., *A Source Book in Geography* (Cambridge: Harvard University Press, 1978), 156–58.

18. Wilhelm Kubitschek, "Itinerarien," in *Realencyclopädie* 9 (1916): cols. 2308–63, esp. 2314 ff. (note 13); Miller, *Itineraria Romana*, LIV ff. (note 5); Leclercq, "Itinéraires," cols. 1841 ff. (note 17); Annalina Levi and Mario Levi, *Itineraria picta: Contributo allo studio della Tabula Peutingeriana* (Rome: Erma di Bretschneider, 1967), 27–28, n. 29. Two of these are on pilasters, one on a column, others on stone or terra-cotta tablets. An inscription from Solin, Yugoslavia, the town near Diocletian's palace at Split, lists four roads leading from there. Of three Gallic inscriptions, one at Autun lists part of the road from there to Rome; one from Valence refers to the road to Vienna (Vienne, France); and one in Luxembourg covers part of the area between there and Mainz. A columbarium fragment at Vigna Codini is thought to refer to the main road from Cilicia or Syria to Rome.

19. See chapter 19 in this volume, "Portolan Charts from the Late Thirteenth Century to 1500," pp. 371–436. The latest writer on periploi, Pietro Janni, *La mappa e il periplo: Cartografia antica e spazio odologico*, Università di Macerata, Pubblicazioni della Facoltà di Lettere e Filosofia 19 (Rome: Bretschneider, 1984), rightly stresses the lack of maps in periploi.

20. See Karl Müller, ed., *Geographi Graeci minores*, 2 vols. and tabulae (Paris: Firmin-Didot, 1855–56), 1:427–514.

21. Müller, *Geographi Graeci minores*, 1:cxxiii–cxxviii (note 20).

22. Marcianus *Periplus maris exteri* in Müller, *Geographi Graeci minores*, 1:515–62, quotation on 516 (note 20), author's translation.

23. See note 22 above.

greater detail than others. For southern Asia he gives fuller measurements as far as Gedrosia (Pakistan west of the Indus), then in less detail.

THE PEUTINGER MAP

The road map known as the Peutinger map,[24] Codex Vindobonensis 324, was originally a long, narrow parchment roll, 6.75 meters long but only 34 centimeters wide. It is in the Nationalbibliothek, Vienna, and has been divided into sections for preservation. Its date of transcription is twelfth or early thirteenth century, but it has long been recognized as a copy of an ancient map. In his will of 1508, the humanist Konrad Celtes of Vienna left to Konrad Peutinger (in whose hands it had been since the previous year) what he called *Itinerarium Antonini*. This was not justified as a title: it is indeed a road map, but not connected with the Antonine emperors and different from the Antonine itinerary described above. It was first published in 1598 by Markus Welser, a relative of the Peutingers, and since 1618 it has generally been known as the *Tabula Peutingeriana* or translations of that phrase.[25] The original roll at the time of its transcription in the early Middle Ages was of eleven sheets, but as such it was incomplete, since much of Britain, Spain, and the western part of North Africa were already missing at the time of copying; there may also have been an introductory sheet forming part of an earlier prototype version. It was evidently not, as was once thought, the work of the Dominican monk Konrad of Colmar, who in 1265 quite independently produced a *mappamundi* that he says he copied onto twelve parchment pages; the paleography suggests an earlier date. The second sheet of the Peutinger map was treated as if it had been the first, with spellings of truncated names containing false initial capitals (for example, Ridumo for what was originally Moriduno). Hence a total of twelve sheets extant at the time of copying can be accounted for only by assuming that, when the copyist mentioned this number, he was including a title sheet.[26]

The Peutinger map was primarily drawn to show main roads, totaling some 70,000 Roman miles (104,000 km), and to depict features such as staging posts, spas, distances between stages, large rivers, and forests (represented as groups of trees). It is not a military map, though it could have been used for military purposes, but the words of Vegetius quoted above (pp. 236–37) give an indication of its possible function. They suggest that, whether or not the term *itinerarium pictum* (painted itinerary), was in current use, it is a convenient phrase for this unique map.[27] The distances are normally recorded in Roman miles, but for Gaul they are in leagues, for Persian lands in parasangs, and for India evidently in Indian miles.

The proportions of the Peutinger map are such that distances east-west are represented at a much larger scale than distances north-south; for example, Rome looks as though it were nearer to Carthage than Naples is to Pompeii. The archetype may well have been on a papyrus roll, designed for carrying round in a *capsa* (roll box). As such, its width would be severely limited, whereas its length would not. In the extant map a north-south road tends to appear at only a slightly different angle from an east-west one, and distances are calculated not by the map's scale but by adding up the mileages of successive staging posts.

The date of the archetype is likely to have been between A.D. 335 and 366. Such dating is suggested by the three personifications placed on Rome, Constantinople (labeled Constantinopolis, not Byzantium), and Antioch; and it fits in well enough with biblical references on the map. Sometime after the foundation of Constantinople in A.D. 330 as a new Rome on the site of Byzantium, Antioch was recognized as the important bastion against the Parthians. But the suggestion that this fourth-century archetype was based on a much earlier map would account for the inclusion of Herculaneum, Oplontis, and Pompeii, which had been destroyed in the eruption of Vesuvius in A.D. 79 and not rebuilt, except for parts of Pompeii. It is also perhaps easier, on this supposition, to see why certain roads are omitted, such as the major routes through the Parthian empire mentioned in the *Mansiones Parthicae* (Parthian stations) of Isidorus of Charax. This work is believed to have been compiled in the late first century A.D.[28]

---

24. Ekkehard Weber, ed., *Tabula Peutingeriana: Codex Vindobonensis 324* (Graz: Akademische Druck- und Verlagsanstalt, 1976); Konrad Miller, *Die Peutingersche Tafel* (Stuttgart: F. A. Brockhaus, 1962); Levi and Levi, *Itineraria picta* (note 18); Luciano Bosio, *La Tabula Peutingeriana: Una descrizione pittorica del mondo antico*, I Monumenti dell'Arte Classica, vol. 2 (Rimini: Maggioli, 1983). Dilke, *Greek and Roman Maps*, 113–20 (note 4).

25. *Tabula* is one Latin word for a map, but *forma* is more common. Inexplicably, the word *tabula* has been translated "table" rather than "picture" or "map" in popular usage. It is now time to call it the "Peutinger map" to avoid any misconception that the original image was somehow carved on a table or was like a statistical table. The alternative naming of the Peutinger map as the "world map of Castorius" has met with very little support. Castorius, a geographical writer of the fourth century A.D., is several times mentioned as a source in the Ravenna cosmography; but there is no evidence to link him directly with the Peutinger map.

26. During this century preservation has been a major problem, particularly since the green coloring used on the parchment has resulted in deterioration of the sea portions of the map. Photographing maps through the glass covers used for preservation has produced inaccurate colors in some reproductions.

27. Hence the title of Levi and Levi, *Itineraria picta* (note 18). For an alternative suggestion, see end of note 15.

28. For a discussion of the date and the significance of Isidorus of Charax, see Sheldon Arthur Nodelman, "A Preliminary History of

Around the personification of Rome—a female figure on a throne holding a globe, a spear, and a shield—are twelve main roads, each with its name attached, a practice not adopted elsewhere (plate 5). The Tiber is correctly shown with 90 percent of the city on its left bank. But owing to the personification the city surround is formally shown as a circle, enlarged in proportion to the very narrow width of the Italian peninsula. The Via Triumphalis is indicated as leading to a church of Saint Peter; the words *ad scm [sanctum] Petrum* are given in large minuscules on the medieval copy. Ostia is shown with a harbor occupying about one-third of a circle, in a fashion similar to that of miniatures in the early manuscripts of Virgil's *Aeneid*.[29] Constantinople is represented by a helmeted female figure seated on a throne and holding in her left hand a spear and a shield (fig. 14.2). Nearby is a high column (rather than a lighthouse)[30] surmounted by the statue of a warrior, presumably Constantine the Great. Antioch has a similar female personification, perhaps originating in a statue of the Tyche (fortune) of the city, together with arches of an aqueduct or possibly of a bridge. Nearby is the park of Daphne, dedicated to Apollo and other gods and famous for its natural beauty and as a leisure center (fig. 14.3). Even though the temple of Apollo was burned down in 362, there were many other temples, so that this is not necessarily a guide to the dating. It has been claimed that in A.D. 365–66 all three personified cities were important, since the pretender Procopius had his seat of power in Constantinople, Valentinian I in Rome, and his brother Valens in Antioch.[31] But in fact, although Valens set out for Antioch, he was diverted to fight Procopius and he cannot be correctly associated with that last-named city.[32]

Throughout the map, mountains are marked in pale brown and principal rivers in green. Names of countries and of some tribes are recorded. Apart from the personifications, cartographic signs include representations of harbors, altars, granaries, spas, and settlements. A unique sign is that for a tunnel, used for the Crypta Neapolitana, near Pozzuoli. Harbors, if indicated, are given the arcuate shape mentioned in connection with Ostia. The sign for a spa is an ideogram of a roughly square building with an internal courtyard, often with a gabled tower at each end of the near side. There are fifty-two such buildings represented, of which twenty-eight are at places specifically called Aquae; in some other cases there is reason to think that a place so denoted had prominent baths.[33] There are also in the Peutinger map places with cartographic signs for granaries, denoted as rectangular roofed buildings. One such is Centumcellae (Civitavecchia), which had a corn-importing harbor of some size. Variants of a two-gabled building were used to depict some settlements, but most

were distinguished by no more than a name (fig. 14.4). Attempts to differentiate between types of settlements on the map and to establish criteria for the attribution of signs have not been entirely successful. Certain important cities are shown with walls: Aquileia, Ravenna, Thessalonica (Salonika), Nicaea (Iznik), Nicomedia, and Ancyra. But why should the triple-gable sign appear only at Forum Iulii (Fréjus), Augusta Taurinorum (Turin), Luca (Lucca), Narona (on the Neretva River), and Tomis (Constanta)?[34] It is interesting to see that, just as there is one personification in the West and two in the East, so two cities of the second rank, symbolically given walls, are in the West and four in the East. Important cities like Carthage, Ephesus, and Alexandria are not shown with a distinctive sign.

The road network is thought to have been based (at least within the empire) on information held by the *cursus publicus*, responsible for organizing the official transport system set up by Augustus.[35] This system, extended under the late empire to troop movements, relied very largely on staging posts at more or less regular intervals; couriers traveled an average of fifty Roman miles (74 km) a day.

The part of the British section of the Peutinger map that survives is so fragmentary that it covers only a limited area of the southeast, not even including London, and an even smaller area around Exeter.[36] Colchester, surprisingly, is given no cartographic sign. The most northerly place extant in Britain appears as "Ad Taum";

Characene," *Berytus* 13 (1960): 83–121, esp. 107–8, and Fergus Millar, "Emperors, Frontiers and Foreign Relations, 31 B.C. to A.D. 378," *Britannia* 13 (1982): 1–23, esp. 16.

29. For example, Rome, Biblioteca Apostolica Vaticana, Vat. Lat. 3225, of about A.D. 420; the importance of its miniatures for the history of cartography has been recognized.

30. The latter is the interpretation of Levi and Levi, *Itineraria picta*, 153–54 (note 18).

31. Levi and Levi, *Itineraria picta*, 65 ff. (note 18).

32. Glanville Downey, *A History of Antioch in Syria: From Seleucus to the Arab Conquest* (Princeton: Princeton University Press, 1961), 399–400.

33. The closest parallel to the use of this convention is in a work of doubtful authenticity, the so-called Bellori picture, found on the Esquiline in 1668 but now lost. Which settlement was intended to be represented on that *veduta prospettiva* is uncertain; it may have been Pozzuoli. See P. S. Bartoli's drawing in Giovanni Pietro Bellori, *Ichnographia veteris Romae* (Rome: Chalcographia R.C.A., 1764), 1. *Ichnographia* is a word Vitruvius used for the drawing of a ground plan in *De architectura* 1.2.2; see *On Architecture*, 2 vols., trans. Frank Granger, Loeb Classical Library (Cambridge: Harvard University Press; London: William Heinemann, 1931–34).

34. Levi and Levi, *Itineraria picta*, 92–93 (note 18), compare this with an incised glass beaker depicting seaside houses with triple gables in the Bay of Naples (New York, Metropolitan Museum): R. W. Smith, "The Significance of Roman Glass," *Metropolitan Museum Bulletin* 8 (1949): 56.

35. See above, p. 236 and n. 8.

36. Rivet and Smith, *Place-Names*, 149–50 (note 9).

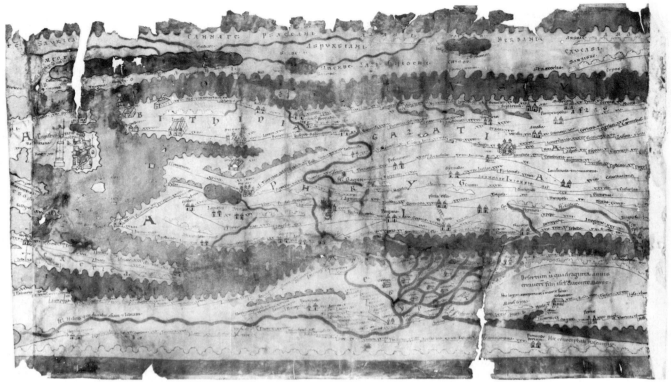

FIG. 14.2. THE PEUTINGER MAP: WESTERN ASIA MINOR AND EGYPT. The elongated deformation of the map is shown by the ribbonlike representation of (*top to bottom*) the Gulf of Azov and the Black, Aegean, Mediterranean, and Red seas. Constantinople is named as such and not as Byzantium, which confirms the pre-fifth-century date of the archetype. Its personification is in the form of a female warrior, enthroned with a shield and spear. Close by are a column and a statue, presumably of Constantine the Great.

Size of the original: 33 × 56.3 cm. By permission of the Österreichische Nationalbibliothek, Vienna (Codex Vindobonensis 324, segment VIII).

but it is very far removed from the river Tay. This name, however, really consists of the ends of [Ven]ta [Icenor]um (Caistor Saint Edmund, Norwich), and the only unusual feature is *ad*, which may have belonged to an adjacent name.

One of the important features of the map is that it records so many small places. This can be well illustrated by a name in Italy otherwise recorded only (in corrupt form) in the Ravenna cosmography. On the Gulf of Naples, marked as being six Roman miles from Herculaneum and three miles each from Pompeii and Stabiae (Castellammare di Stabia), is shown a large building with the name Oplont⟨i⟩s. Until recently scholars could not place this name, like a number of others. But since 1964 a large palace, which probably belonged to Nero's empress Poppaea, has been excavated at Torre Annunziata, and it seems to authenticate the detail on the map.[37] Or

FIG. 14.4. TOWN SIGNS ON THE PEUTINGER MAP. Most towns are shown only by a name, but the more prominent are represented by signs.
After Annalina Levi and Mario Levi, *Itineraria picta: Contributo allo studio della Tabula Peutingeriana* (Rome: Erma di Bretschneider, 1967), 197–201.

37. Alfonso de Franciscis, "La villa romana di Oplontis," in *Neue Forschungen in Pompeji und den anderen vom Vesuvausbruch 79 n. Chr. verschütteten Städten*, ed. Bernard Andreae and Helmut Kyrieleis

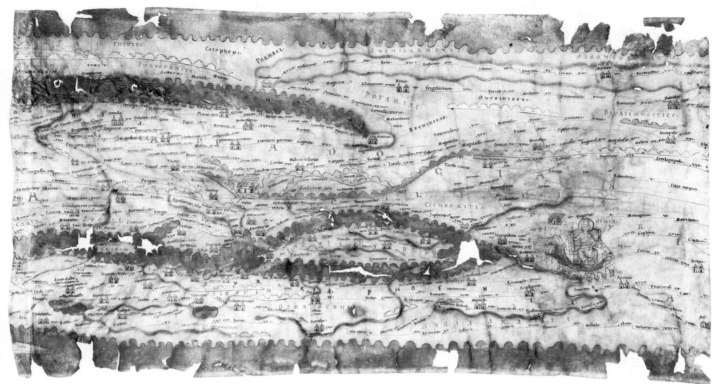

FIG. 14.3. THE PEUTINGER MAP: THE EASTERN MEDI-
TERRANEAN. The north-south axis of Syria and the Holy
Land is here shown parallel to Asia Minor and Cyprus. The
prominent city to the right of Cyprus is Antioch, the third of
the cities on the map personified as a seated figure with spear
and shield.

Size of the original: 33 × 61.4 cm. By permission of the
Österreichische Nationalbibliothek, Vienna (Codex Vindobonensis 324, segment IX).

again, a much earlier discovery near Aquileia in 1830
appears to correspond to an entry on the Peutinger map.
A large bathing establishment, mentioned also by the
elder Pliny, was discovered on the lower reaches of the
river Isonzo.[38] This is probably the place given the cartographic sign for a spa, with the words Fonte Timavi
(spring of the river Timavus). Its fresh waters by the sea
were regarded as an unusual phenomenon and obviously
worth mapping.

Owing to the shape of the map, the Nile could not
be represented as a long river if it were made to flow
northward throughout its course. Instead it is made to
rise in the mountains of Cyrenaica and to flow "eastward" to a point just above the delta. The delta itself is
shown in less compressed form from south to north than
most parts of the Peutinger map (see fig. 14.2). The
distributaries of the Nile are shown to have many islands, three of them marked with temples of Serapis,
three with temples of Isis, while the roads are somewhat
discontinuous. On the Sinai desert we find the words
*desertum ubi quadraginta annis erraverunt filii Israelis
ducente Moyse* (the desert where the children of Israel

wandered for forty years guided by Moses), and there
are other biblical references. There is also an area in
central Asia labeled *Hic Alexander responsum accepit
usq[ue] quo Alexander* (Here Alexander was given the
oracular reply: "How far, Alexander?"). Perhaps these
ampler descriptions, whether Christian or pagan, were
added on otherwise empty space about the fifth or sixth
century A.D. In several areas research is in progress combining fieldwork with study of the Peutinger map and
of the history of place-names. One such is the area between the Gulf of Aqaba and Damascus.[39] A question
that emerges is the extent to which we can argue from
silence: Does the absence of an important road on the

(Recklinghausen: Aurel Bongers, 1975), 9–38; Carlo Melandrino,
Oplontis (Naples: Loffredo, 1977).

38. Luciano Bosio, *La "Tabula Peutingeriana": Una carta stradale
romana del IV secolo* (Florence: 3M Italia and Nuova Italia, 1972),
16 (text accompanying filmstrip).

39. Dr. D. L. Kennedy of the University of Sheffield is researching
on behalf of the Aerial Photographic Archive for Archaeology in the
Middle East, and Professor G. W. Bowersock gave a talk on the subject
at the University of London in 1980.

Peutinger map suggest that the mapmaker, perhaps of the fourth century A.D., was relying for less familiar areas on an earlier map, of the first or second century, made before such a road was built?

THE LATIN GEOGRAPHICAL MANUSCRIPTS AND THEIR MAPS

Many of the geographical manuscripts of Roman origin are of less cartographic interest than their Greek counterparts. Whatever the reasons for this, no continuous tradition of writing in Latin about these subjects took root. It is often difficult to say if a Roman author composed with a map in front of him or indeed whether a map was drawn at all to illustrate a particular text. And in other Latin manuscripts with maps, such as the *Notitia Dignitatum*, the maps suggest that the compilers either did not have access to such standard maps as those of Agrippa or Ptolemy or lacked the cartographic knowledge to exploit these sources properly.

LATIN GEOGRAPHICAL WRITERS IN THE EARLY EMPIRE

This tendency to ignore maps, even when Greek influence was at its height, is probably shown in the work of Pomponius Mela (fl. A.D. 37–42), one of the few Latin geographical writers from the early empire whose text has come down to us. Mela was born in southern Spain. His *Chorographia*, in three books, written under Gaius or Claudius, is a brief world geography, but there is no evidence that it ever contained maps.[40] Mela's world is surrounded by seas and divided into two hemispheres, Asia in the eastern, Europe and Africa in the western. From north to south, as in Eratosthenes' poem *Hermes* and Virgil's *Georgics*, it is divided into five zones, two cold, two temperate, and one hot.

In much the same way, there is relatively little of explicitly cartographic interest in the geographical compendium of the elder Pliny (A.D. 23/24–79), a native of Como, who held important offices under Vespasian but who as admiral of the fleet at Misenum perished in the eruption of Vesuvius. His *Natural History* in thirty-seven books was completed in A.D. 77.[41] It is an encyclopedia based on one hundred major and a large number of subsidiary Greek and Latin authors. For each section he lists the major writers he has followed (often quite closely, as we can tell from extant works), and much research has been devoted to these sources. Pliny's information includes both useful up-to-date material and old travelers' tales. Of Latin writers, he mostly quotes Cornelius Nepos (author of—besides biographies—a lost geographical work) and Agrippa. He incorporates information from inscriptions, statistical data, and lists of tribes and places in provinces.[42] These lists are sometimes in geographical order (for Italy he follows Au-

gustus's division into eleven regions), sometimes alphabetical, though as often in antiquity only the first letter is necessarily in strict order.[43] Only book 2, mainly concerned with the universe (which Pliny calls a constantly revolving perfect sphere),[44] and books 3–6, covering the geography of the *oikoumene*, contain materials of potential cartographic interest.[45]

40. See Pomponius Mela, *De chorographia*, ed. Gunnar Ranstrand, Studia Graeca et Latina Gothoburgensia 28 (Gothenburg: Acta Universitatis Gothoburgensis; distribution Stockholm: Almqvist och Wiksell, 1971), and a newer edition, Pomponius Mela, *De chorographia libri tres*, ed. Piergiorgio Parroni, Storia e Letteratura, vol. 160 (Rome: Edizioni di Storia e Letteratura, 1984). Also see Nicolaus Sallmann, "De Pomponio Mela et Plinio Maiore in Africa describenda discrepantibus," in *Africa et Roma: Acta omnium gentium ac nationum Conventus Latinis litteris linguaeque fovendis*, ed. G. Farenga Ussani (Rome: Erma di Bretschneider, 1979), 164–73.

41. Pliny, *Naturalis historia*, 6 vols. in 3, ed. D. (S. D. F.) Detlefsen (Berlin: Weidmann, 1866–82; reprinted Hildesheim: Georg Olms, 1982); Pliny, *Natural History*, 10 vols., trans. H. Rackham et al., Loeb Classical Library (Cambridge: Harvard University Press; London: William Heinemann, 1940–63). For a French translation with notes of book 5, part 1, sections 1–46 (on North Africa), see Pliny, *Histoire naturelle*, ed. and trans. Jehan Desanges (Paris: Belles Lettres, 1980); of book 6, part 2 (on central and east Asia), see Pliny, *Histoire naturelle: L'Asie centrale et orientale, L'Inde*, ed. Jacques André and Jean Filliozat (Paris: Belles Lettres, 1980); and of book 7, Pliny, *Histoire naturelle, livre VII*, ed. and trans. Robert Schilling (Paris: Belles Lettres, 1977).

42. These include the Trophée des Alpes at La Turbie, quite wrongly located in the Loeb translation by Rackham, *Natural History* 2:100, note g (note 41).

43. The eleven regions are conveniently listed in the *Oxford Classical Dictionary*, 2d ed., s.v. "Italy," sec. 6.

44. Pliny *Natural History* 2.64.160 (note 41). After the section dealing with earth and water (2.66.166), he discusses, in no particular order, maritime exploration: the Sea of Azov (Palus Maeotis), he says, certainly exists, but is it a gulf of the ocean or an overflow from it? Africa had been circumnavigated by Hanno and several others (2.67.167 ff., esp. 169). He then discusses sundials and hours of daylight throughout the known world (2.77.186 ff.). There is also a section on earthquakes (2.81.191) and changes in the coastline; thus at the harbors of Ambracia and Piraeus the sea receded by ten and five miles respectively (2.87.201); no date is given, and in the case of Piraeus, at least, the measurement is inconsistent with known topography. A number of islands or mountains are known to have appeared or disappeared, and towns have disappeared; of these he quotes examples (2.89.202 ff.). Later in the second book he records sea depths, underground rivers, and other phenomena (2.105.224 ff.). The book concludes with some overall measurements: India to Gades is 8,568 or 9,818 miles according to Artemidorus and Isidorus (of Charax), respectively (2.112.242 ff.).

45. The material is organized thus:
Book 3 Western and central Europe bordering on the Mediterranean and the northern Atlantic.
Book 4 Greece and adjacent areas; the Black Sea and adjacent European areas; northern Europe.
Book 5 Africa bordering on the Atlantic and Mediterranean, including all of Egypt; Asia bordering on the Mediterranean and Aegean.
Book 6 The Black Sea and adjacent Asiatic areas; other parts of Asia; Ethiopia and the upper Nile valley.
As is evident from this summary, the Black Sea itself is treated twice.

From these books it is clear that Pliny was a user of maps rather than a contributor to the theory of their construction or compilation. Unfortunately, in giving his sources he does not distinguish between maps and written geographies. Thus we have to infer his use of maps by indirect means such as the cited measurements between places or his descriptions of countries and cities in terms of their shapes. With respect to the latter, for example, Italy is not as today described as boot-shaped but is said to be like an elongated oak leaf, bending to the left at the top and "ending in the shape of an Amazon's shield"[46] (following Strabo and Eratosthenes); the Peloponnese has the shape of a plane tree leaf; and so on.[47]

There are two passages where the context may be considered fully cartographic. The first is where he criticizes the length and breadth of Baetica, southern Spain, as given in Agrippa's map and approved by Augustus.[48] The second is where he says: "There are a number of segments of the earth which we Romans have called circles, while the Greeks have called them parallels."[49] He then gives what we may define as seven *climata* (he does not use the word) from south to north, with lengths of gnomon shadows, numbers of hours in longest days, and principal countries or cities. These range from southern India and the two provinces of Mauretania (the latter are in fact at a very different latitude from southern India) to the north of the Black Sea and Aquitania. Pliny attributes these to earlier Greek theory, but he also gives three additional zones to the north, which he says later Greek geographers have added.

Other evidence is not quite so conclusive; it may depend on our interpretation of *posuere* (they have placed). Pliny may well be thinking of placing on a map, but we cannot be sure. Thus in one passage he writes: "Then there projects into the sea a promontory with a vast horn, which some have called Artabrum [from Cape Roca to Lisbon]. . . . The distance from here to the Pyrenees is given by quite a number as 1,250 miles, and they record there a non-existent tribe of Artabres: they have placed [*posuere*] in this area, by a change of lettering, the Arrotrebae whom I mentioned before the Celtic Promontory [Finisterre]."[50] Similarly in another passage, "Others have placed [*posuere*] the Gedrusi and Sires over a stretch of 138 miles, then the fish-eating Oritae, who do not speak the Indian language, for 200 miles."[51] We may also feel that Pliny is more likely to have been looking at a map than a book when he compares the Arabian peninsula to Italy not merely for being surrounded by two seas but for having what he calls the same orientation.[52]

LATIN GEOGRAPHICAL WRITERS IN THE LATE EMPIRE

During the late empire, Latin geographical writing seems to have been confined to relatively few channels of trans-

mission. Indeed, although there are other manuscripts with maps, only three main writers—Avienius, Macrobius, and Julius Honorius—can be noted under the present heading.[53] Only the second of these can definitely be said to have written a work containing maps. This poverty in the Roman tradition may partly reflect the fact that some of the more important texts containing cartographic knowledge continued to be written in Greek rather than in Latin. This is shown by the work of Agathemerus, whose brief prose manual summarizes Greek mapmaking up to the first century B.C.[54] Rufus Festus Avienius,[55] a senator from Volsinii in Etruria, wrote two geographical works in Latin verse about A.D. 380–400. The first is a general work in hexameters, *Descriptio orbis terrae*, intended as a revision of the Greek work of Dionysius Periegetes. The second, *Ora maritima*, contains only 703 iambs but is thought to be incomplete.[56]

The commentary by Ambrosius Theodosius Macrobius (fl. A.D. 399–422) on Cicero's *Somnium Scipionis* (The dream of Scipio, incorporated in his *Republic*) contains cosmology and Macrobius's impression of the appearance of the world.[57] Whereas he accepts Eratosthenes' sphericity and measurement, he turns to Crates

46. Pliny *Natural History* 3.5.43 (note 41). Shaped like a crescent but with two curves on the inner side and a promontory between them.

47. See chapter 10, "Greek Cartography in the Early Roman World," esp. pp. 174–75.

48. Pliny *Natural History* 3.1.16–3.2.17 (note 41); see also above, pp. 207–8.

49. Pliny *Natural History* 6.39.211–20, esp. 212 (note 41), author's translation.

50. Pliny *Natural History* 4.21.113–4.22.14 (note 41), author's translation.

51. Pliny *Natural History* 6.25.95 (note 41), author's translation.

52. Pliny *Natural History* 6.32.143 (note 41).

53. There are also several extant geographical manuscripts in Latin without maps. For example, Vibius Sequester, *De fluminibus, fontibus, lacubus, nemoribus, paludibus, montibus, gentibus per litteras libellus*, ed. Remus Gelsomino (Leipzig: Teubner, 1967), is an alphabetical gazetteer listing, under separate headings, rivers, springs, lakes, groves, marshes, mountains, and peoples. Similarly, two accounts are provided of the boundaries of provinces and of countries outside the empire, based on Agrippa's map or a revision: *Dimensuratio provinciarum* and *Divisio orbis terrarum*, both in *Geographi Latini minores*, ed. Alexander Riese (Heilbronn, 1878; reprinted Hildesheim: Georg Olms, 1964), 9–14 and 15–23 respectively.

54. Agathemerus *Geographica antiqua*, ed. Jacobus Gronovius (Leiden: J. Luchtmans, 1700); and *Geographiae informatio*, in Müller, *Geographi Graeci minores* 2:471–87 (note 20).

55. Avienius, not Avienus, is the correct form: Alan Cameron, "Macrobius, Avienus, and Avianus," *Classical Quarterly*, n.s., 17 (1967): 385–99, esp. 392 ff.

56. See Avienius, *Rufi Festi Avieni Carmina*, ed. Alfred Holder (Innsbruck, 1887; reprinted Hildesheim: Georg Olms, 1965), chap. 3 and chap. 4, which contains both works. For further references on the *Ora Maritima*, see chapter 9, p. 150 and n. 16.

57. Macrobius, *Commentary on the Dream of Scipio*, ed. and trans. W. H. Stahl (New York: Columbia University Press, 1952).

of Mallos for the concept of ocean and land masses: "Separating us from the people of the southern hemisphere, Ocean flows along the whole extent of the equator; again, as its streams branch out at the extremities of both regions, it forms two islands on the upper face of the earth and two on the underside."[58] Later he adds: "The accompanying diagram will lay everything before our eyes" and "from our diagram we shall also understand Cicero's statement that our quarter is narrow at the top and broad at the sides."[59] Macrobius's map of the world was circular (2.5.13), with north at the top and with cold, temperate, and hot regions, though perhaps its equatorial ocean was narrower in latitude than is shown in manuscripts and printed editions. The manuscript maps accompanying the commentary had a strong influence on zone maps of the Middle Ages.

The *Cosmographia* of Julius Honorius is an inaccurate compilation of about the fifth century; only excerpts survive.[60] His list starting *Seres oppidum, Theriodes oppidum* . . . totally confuses settlements, tribes, and even rivers, calling them all towns.[61] He selects a few places in northern Italy, partly in topographical order, partly not, before suddenly switching to Dalmatia. One of the most incorrect entries may be rendered: "The river Chrysorroas rises in the plains of Syria and flows through Syria, Antioch, and Palestine and the remaining cities of Syria. Its mouth is in the Aegean, where the island of Cyprus is. It runs for 830 miles."[62]

THE *NOTITIA DIGNITATUM*

The full title of this work may be translated "directory of officeholders and administration, both civil and military."[63] There are primary manuscripts at Cambridge, Frankfort, Munich, Oxford, and Paris.[64] Some of these are known, and all are thought, to have been copied from a codex Spirensis (that is, of Speyer cathedral), which was written in the tenth century but disappeared in the sixteenth century. They are all illustrated, and the Munich manuscript has two sets of illustrations.[65]

The main divisions in the work are between the eastern and the western empires and between civil and military officials. The official list was kept by the head of the civil service in the West, though it is disputed whether the extant work is governmental or an amateur's copy. Its date is between 395 and 413, and it may have been revised even later.

Illustrations consist of insignia of officials, personifications of provinces, picture maps, and miscellaneous items. Many maps are such as only a bureaucrat unfamiliar with the areas could have produced; and it is considered likely that there has been much change or addition.[66] Thus the map of Britain under the heading "vicarius Britanniarum" has five provinces arranged 1, 2, 2, from north (top) to south, with "Britannia Prima"

in the southeast (plate 6).[67] But we know from the metrical epitaph of a governor that Corinium (Cirencester) was the capital of Britannia Prima; and it is likely that Maxima Caesariensis had the most important settlement, London, as its capital, yet this province is placed on the map somewhere near Lincoln. Probably, if the civil servant who compiled this list or its official counterpart had gone to the maps department, he could have been put right by the *comes formarum* (director of maps), who was under the Rome city prefect and who provided the only record from the Roman world of an official working for what must have been a civil service maps and plans department.[68]

The *comes Italiae* had as part of his command the region of Italy near the Alps, but this is illustrated in the *Notitia Dignitatum* only by a walled hilltop settlement. Isauria, western Cilicia, is likewise mountainous; it is given something approaching a perspective map, with south at the top. Mount Taurus, complete with a wild animal's hindquarters, is in the center, and Tarsus and the Mediterranean are in the background. Mesopotamia has the Tigris and Euphrates correctly placed, and also Carrhae, but there is a great deal of confusion. Thus Amida (Diyarbakir) and Constantina are each in-

58. Macrobius *Commentary* 2.9.5 (note 57); see also Kish, *Source Book in Geography*, 140–42 (note 17).

59. Macrobius *Commentary* 2.9.7–8 (note 57).

60. Julius Honorius *Excerpta eius sphaerae vel continentia*; see *Iulii Honorii Cosmographia*, in Riese, *Geographi Latini minores*, 24–55 (note 53).

61. Julius Honorius *Excerpta* 6 ff. (note 60).

62. Author's translation of Julius Honorius *Excerpta* 10 (note 60).

63. Otto Seeck, ed., *Notitia Dignitatum* (Berlin: Weidmann, 1876; reprinted Frankfort: Minerva, 1962); R. Goodburn and P. Bartholomew, eds., *Aspects of the Notitia Dignitatum*, British Archaeological Reports, Supplementary Series 15 (Oxford: British Archaeological Reports, 1976); Erich Polaschek, "Notitia Dignitatum," in *Realencyclopädie*, 17.1 (1936): cols. 1077–116 (note 13).

64. This is the manuscript called Londiniensis by I. G. Maier, "The Giessen, Parma and Piacenza Codices of the 'Notitia Dignitatum' with Some Related Texts," *Latomus* 27 (1968): 96–141; it was presented to the Fitzwilliam Museum by Professor F. Wormald. Important manuscripts are: Oxford, Bodleian Library, MS. Canon. Misc. 378, A.D. 1436; Paris, Bibliothèque Nationale, MS. Lat. 9661, fifteenth century; Munich, Bayerische Staatsbibliothek, Clm. 10291, 1542–50; Cambridge, Fitzwilliam Museum, A.D. 1427; and Frankfort, Stadts- und Universitätsbibliothek, Lat. qu. 76, fifteenth century. There are many other manuscripts, but without maps, the two labeled Tridentinus-Vindobonensis being not at Vienna but at Trento.

65. In 1548 Count Ottheinrich requested the loan of the *alt Exemplar* of the codex, but this was denied. Ottheinrich was clearly interested in seeing the original illustrations that accompanied the text and would not accept an untraced copy of them. In 1550 he received permission for his artist to make tracings—now lost—of the original illustrations on *geoldrenckt pappeir* (oiled paper). See I. G. Maier, "The Barberinus and Munich Codices of the *Notitia Dignitatum Omnium*," *Latomus* 28 (1969): 960–1035, esp. 995–99 and 1024–30.

66. Polaschek, "Notitia Dignitatum" (note 63).

67. Seeck, *Notitia Dignitatum*, 171 (note 63).

68. Seeck, *Notitia Dignitatum*, 113 (note 63).

cluded twice, because each has two entries in the list of military units.

Lower Egypt, sphere of the *comes limitis Aegypti*, has the Nile as a single water channel on its map (fig. 14.5). This is correct for the area around Memphis, but not for Pelusium; and Thamudeni should be east of the Red Sea. Upper Egypt, area of the Dux Thebaidos, is particularly muddled, with Syene (Aswan) shown incorrectly on a tributary and not as near the frontier as it should be. Coptos is shown in two different places, because both a legion and a cavalry unit are listed there under separate headings. The map follows neither topographical nor bureaucratic order and even, unlike the text, misspells and misplaces the famous temple of Philae. These shortcomings do not mean that the *Notitia Dignitatum* was a useless document, but in fact the useful information would have had to be gathered from the text rather than from the maps.[69]

MAPS AS DECORATIVE AND SYMBOLIC IMAGES

From the Roman period, as from the classical world in general, the cartographic record extends to maps found as images on diverse artifacts and in other than textual sources. These maps have often gone unnoticed in the literature, but they add a further dimension to an understanding of cartography in the classical period. Examples include the well-known map on the Dura Europos shield, the incised and inscribed stone named as the Pesaro wind rose map, and the map images found on coins, in frescoes, as part of the design of mosaics, and even on Roman lamps. While such representations are perhaps peripheral to a reconstruction of the history of scientific mapmaking, they help in assessing the dissemination (and extent of understanding) of maps in the Roman period. Like the maps described in poems, they suggest one way the idea of the map (if not the formal knowledge underpinning the construction of maps) was kept alive in a period that lacked scientific innovation in the accepted sense. Their context, however, is usually an archaeological one, and the chance circumstances of their survival and discovery, often unrelated to the documentary record, have led us to organize the material typologically in terms of the artifacts carrying the map images.

COINS WITH PLANS OF ROME'S HARBOR

As earlier in the Greek period, there are Roman examples where maps were used emblematically to face coins.[70] About three kilometers north of ancient Ostia at the mouth of the Tiber, Claudius and Trajan constructed new harbors. These were commemorated on coins and medallions, some representations being in pictorial form,

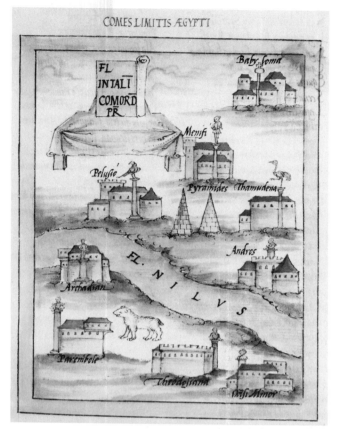

FIG. 14.5. THE *NOTITIA DIGNITATUM*: LOWER EGYPT. The area controlled by the *comes limitis Aegypti* is schematically represented, complete with its major features such as the Nile, Babylon, Memphis, and the Pyramids.
Size of the original: 31 × 24 cm. By permission of the Bayerische Staatsbibliothek, Munich (Clm. 10291, fol. 113r).

some in ground plan. The Claudian harbor appears in plan on a coin of Nero, about A.D. 64 (fig. 14.6).[71] Its two jetties are shown as arcs of a circle, part of the right jetty perhaps being shown on arches. The harbor is on an island, and at its mouth is a lighthouse surmounted by the statue of an emperor; ships are shown entering, leaving, or standing in the harbor. Trajan's harbor, in a hexagonal basin with sides of 358 meters, appears in plan on a coin of Trajan, about A.D. 113 or after.[72] On the right are shown warehouses, on the left what is

---

69. For example, for Britain see Rivet and Smith, *Place-Names*, 216–25 (note 9). We can plot the forts in northern England under the Dux Britanniarum, with the title of the army unit stationed at each; we can reconstruct with reasonable probability the names of all the Saxon Shore forts; and we can read of a *gynaeceum* (imperial textile factory) at Venta, but the question is whether it was at Venta Icenorum (Caistor Saint Edmund) or at Venta Belgarum (Winchester).

70. See pp. 158 and 164 above.

71. Russell Meiggs, *Roman Ostia* (Oxford: Clarendon Press, 1960), pl. XVIIIa.

72. Meiggs, *Roman Ostia*, pl. XVIIIb (note 71).

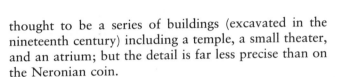

FIG. 14.6. SESTERCE OF NERO SHOWING OSTIA HARBOR. Issued ca. A.D. 64, this coin shows the Claudian harbor of Ostia. Two jetties are represented by arcs of circles, together with a lighthouse surmounted by a statue, Roman ships, and Neptune.

Diameter of the original: 3.5 cm. By permission of the Trustees of the British Museum, London (BMC Emp.I, Nero 132).

thought to be a series of buildings (excavated in the nineteenth century) including a temple, a small theater, and an atrium; but the detail is far less precise than on the Neronian coin.

MAPS IN MOSAICS

Although the mosaic map of Byzantine age at Madaba has appeared in many popular works on the history of cartography, it is by no means the earliest map to occur in this medium. An example dating to the early empire is found in Ostia, associated with the many mosaic floors that have been preserved there in situ.[73] They are in the "Forum of the Corporations," next to the theater, which had a temple of Ceres in the middle and a series of offices belonging to trading corporations around the sides. From the names of foreign ports within the mosaics it is possible to reconstruct many of Rome's trade routes. Only one of these mosaics is in map form, but unfortunately it has no inscription (fig. 14.7).[74] It shows a river spanned by a pontoon bridge, with three branches that could be either tributaries or distributaries. The bridge is supported by three vessels, and on each side of it is a gateway surmounted by military trophies. The Nile delta seems the most probable location. Another office in the same forum was occupied by an Alexandrian corporation, and corn imports from Egypt were considerable. If these two offices are linked, the likely interpretation is that the three principal ancient distributaries of the Nile are depicted, giving a more simplified version of the delta than is to be seen in the Madaba mosaic. Since the pontoon bridge is above the lowest point on the undivided river, one may conjecture that it was between Memphis, the dynastic capital, which was still of some importance under the Roman Empire, and Babylon (Old Cairo). The military trophies would be appropriate to Babylon, which was fortified as a legionary camp under Augustus, while Memphis was the center for the export of wild animals from Egypt to Rome.[75]

---

73. Giovanni Becatti, ed., *Mosaici e pavimenti marmorei*; 2 pts. (1961); both are vol. 4 of *Scavi di Ostia* (Rome: Istituto Poligrafico dello Stato, 1953–).

74. Foro delle Corporazioni, Statio 27: Becatti, *Mosaici*, 74, no. 108 and pl. CLXXXIV (note 73).

75. The mosaic at the temple of Fortuna Primigenia at Praeneste (Palestrina), central Italy, mentioned in chapter 7, note 4 above, also gives a pictorial representation of scenes around the Nile. It has been given various dates but is perhaps of the second century A.D. The inscriptions, in Greek capitals, specify typical Egyptian animals. The attempt to show that it is a map in oblique perspective does not appeal to most art historians, and it must remain in the sometimes debatable borderland between the completely pictorial and the partly cartographic. See references in Wilhelm Kubitschek, "Karten," in *Realencyclopädie*, 10 (1919): cols. 2022–2149, esp. 2023 (note 13); Levi and Levi, *Itineraria picta*, 44 n. 65 (note 18). A black-and-white reproduction is found in Moses Hadas, *Imperial Rome* (Alexandria, Va.: Time-Life Books, 1979), 70–71. For the Alexandrian tradition of landscape painting, which may have influenced the Palestrina mosaic and others, see Roger Ling, "Studius and the Beginnings of Roman Landscape Painting," *Journal of Roman Studies* 67 (1977): 1–16, with bibliography in note on p. 1, and p. 14 n. 53.

FIG. 14.7. THE OSTIA RIVER MOSAIC. Preserved in the Forum of the Corporations in Ostia, this mosaic without an inscription depicts a river, with three tributaries or distributaries, and a pontoon bridge with a military gateway at each end. Of three possible locations—the Nile delta, the lower Tiber, and the Rhône delta—the first seems most likely. Museo Ostiense, near Rome.

Size of the original: 7 × 3.5 m. From Giovanni Becatti, ed., *Mosaici e pavimenti marmorei*, 2 pts. (1961); both are vol. 4 of *Scavi di Ostia* (Rome: Istituto Poligrafico dello Stato, 1953–), pt. 2, pl. CLXXXIV (no. 108).

What have been recognized as maplike designs, in addition to those of mazes described below, also occur occasionally in mosaics dated to the late empire. Mosaics with views of estates, one category of such images, can hardly be regarded even as picture maps: the most conspicuous is the fourth-century mosaic from Carthage, now in the Bardo Museum, Tunis, showing the estate of Julius.[76] But one temple scene, also from Carthage, is represented partly in plan. This is a late fourth- or fifth-century mosaic known as the Offering of the Crane.[77] In the center is a shrine facade with Apollo and Diana between two columns; at their feet is the sacrificial crane, and below it a series of almost concentric squares clearly representing the ground plan of this same shrine.

Some mosaics also contain celestial maps in the form of zodiac diagrams. The earliest Palestine zodiac mosaic, for example, thought to be of the third century A.D., is in the synagogue at Hammath, south of Tiberias.[78] It represents the twelve signs of the zodiac surrounding the sun, with personifications of the four seasons occupying the corners. The inspiration of this Hebrew mosaic is Greek, with the sun as Phoebus Apollo driving a four-horse chariot, Virgo as a veiled Persephone carrying a torch, and Libra as a king, clearly Minos or Rhadamanthys, holding a scepter and balance. Other similar mosaics are of very much later date, when Palestine was part of the Byzantine empire.

THE PESARO WIND ROSE MAP

In some cases an unusual physical object carries a map, so that there is a temptation to describe it as "unique" in artifactual terms although its image may also be familiar from other sources. Such an object is the "Boscovich" anemoscope,[79] a cylindrical block of Luna marble 55.3 centimeters in diameter and 6.8 centimeters in width, on which is engraved a wind rose map (fig. 14.8).[80] It was found near the Via Appia at Rome, outside the Porta Capena, in 1759. The name is inexact, since the astronomer R. G. Boscovich only helped the owner, P. M. Paciaudi, in his researches. Now in the Oliveriano Museum at Pesaro, it is thought to date from about A.D. 200. It is inscribed *Eutropius feci* (I, Eutropius [a Greek name], made it).

To construct his wind rose map, Eutropius engraved a meridian, divided this equally into six, contrary to Aristotle,[81] and from the dividing points drew five lines at right angles to the meridian. These are labeled, in descending order: TOTVS INFRA TERRA⟨M⟩ (the Antarctic Circle); BRVMALIS (the Tropic of Capricorn); AEQVINOCTIALIS (the equator); SOLI⟨S⟩TITIALIS (the Tropic of Cancer); and TOTVS SVPRA TERRA⟨M⟩ (the Arctic Circle). At each end of the meridian and of each line are small holes intended for bronze pegs. From these twelve holes,

lines are drawn to the center, where a large depression presumably served to insert a metal base for the flag of the wind rose. On the rim, opposite the holes, are the names of the twelve winds.

The scheme probably arose from a reading of the passage from Aristotle mentioned above (pp. 145–46) without the benefit of a diagram.[82] Aparcias (Aparctias, Septentrio) occupies a far larger sector than neighboring Boreas; the two are identical in Aristotle. This is not important, since in antiquity the number, names, and directions of winds on a wind rose varied.[83] Neverthe-

76. Very frequently reproduced: references in Levi and Levi, *Itineraria picta*, 68 n. 3 (note 18); Katherine M. D. Dunbabin, *The Mosaics of Roman North Africa: Studies in Iconography and Patronage* (Oxford: Clarendon Press, 1978), 119–21, pl. 109.

77. Dunbabin, *Mosaics*, 57–58, pls. 35–37 (note 76).

78. Michael Avi-Yonah, *Ancient Mosaics* (London: Cassell, 1975), 51–53. For the difference in approach between Christian and Jewish mosaicists see Dunbabin, *Mosaics*, 232–33 and n. 174 (note 76), and for the zodiac in North African mosaics see her Index IV, s.v. "Zodiac."

79. Italo Zicàri, "L'anemoscopio Boscovich del Museo Oliveriano di Pesaro," *Studia Oliveriana* 2 (1954): 69–75; Robert Böker, "Windrosen," in *Realencyclopädie*, 2d ser., 8.2 (1958): cols. 2325–81, esp. 2358–60 (note 13); Antonio Brancati, *La biblioteca e i musei Oliveriani di Pesaro* (Pesaro: Banca Popolare Pesarese, 1976), 210–11; Dilke, *Greek and Roman Maps*, 110–11 (note 4).

80. Wind rose maps are also found in textual sources. A wind rose, for example, is preserved in the oldest manuscript of Vitruvius *De architectura* (note 33), in which it is the only illustration (1.6.4 ff.). In the text he says it is "so delineated that it is clear where winds come from": London, British Library, Harl. MS. 2767; see O. A. W. Dilke, *Roman Books and Their Impact* (Leeds: Elmete Press, 1977), 26–27, quotation on 26. A possible parallel to the Boscovich anemoscope is a small Greek world map to be found in manuscripts of the scholia to Ptolemy's *Handy Tables*; see Otto Neugebauer, "A Greek World Map," in *Le monde grec: Hommages à Claire Préaux*, ed. Jean Bingen, Guy Cambier, and Georges Nachtergael (Brussels: Université de Bruxelles, 1975), 312–17 and pl. III.2. The manuscripts date from the thirteenth century onward, but the map clearly originated under the Roman Empire, probably in Egypt, since it names certain southerly places such as Syene (Aswan) and Hiera Sykaminos (above Syene on the Nile). These are places from which astronomical observations could be made. Ptolemy's instructions for making tables were later worked up by Theon of Alexandria (fl. A.D. 364). Unlike the anemoscope, this world map, which is in the form of a rough sketch, does not use lines linking the wind points to the center. It may go back to an archetype earlier than Theon.

81. But a miniature in a manuscript of Aristotle's *Meteorologica* at the Royal Library in Madrid makes the spacing of the zones rather even by attaching lines to small semicircles inscribed within the perimeter: Charles Graux and Albert Martin, "Figures tirées d'un manuscrit des *Météorologiques* d'Aristote," *Revue de Philologie, de Littérature et d'Histoire Anciennes*, n.s., 24 (1900): 5–18.

82. In modern reconstructions of Aristotle's diagram, four out of five of Eutropius's horizontal lines are visible, though not equally spaced. Aristotle himself may have inserted only the diameters. Aristotle *Meteorologica* 2.6; see *Meteorologica*, trans. H. D. P. Lee, Loeb Classical Library (Cambridge: Harvard University Press; London: William Heinemann, 1952).

83. Vitruvius *De architectura* 1.6.4 ff. (note 33), discussed four, eight, and sixteen winds, the last corresponding to Etruscan usage. In

less, the anemoscope must have been intended as a meteorological device, partly to help the traveler who, as he set out from Rome on the Via Appia, would be facing south as the map does. The flag would show the name, origin, and direction of the wind.

FIG. 14.8. THE PESARO WIND-ROSE MAP. A diagram of winds engraved on a cylindrical marble block, probably dating from about A.D. 200. There is a central hole for a pole supporting a pennant and small holes near the rim for wooden pegs indicating the winds.
Diameter of the original: 55 cm. By permission of the Museo Archeologico Oliveriano, Pesaro (inv. 3.302).

THE DURA EUROPOS SHIELD

At Dura Europos on the Euphrates a parchment fragment, now housed in the Bibliothèque Nationale, Paris, was excavated in 1923.[84] On the sheet of parchment, which had covered a soldier's shield, had been painted a rough map of the Black Sea and surrounding areas, the extant part showing the west and north coasts (fig. 14.9). Although the wording is in Greek, the shield must have belonged to a soldier in the Roman army. It is thought to date to shortly before A.D. 260, when the Romans withdrew from Dura Europos. The measurements of the portion discovered are forty-five (originally about sixty-five) by eighteen centimeters, and it has roughly west-southwest at the top.

On the left, in blue, is the Euxine (Black) Sea, which has two large ships and four heads appearing out of the water that may be those of sailors from other vessels. The shore is indicated by a pale curved line, with no promontories or indentations. Along the coastal route

are shown staging points, each having as a cartographic sign a building with courses of pale green stonework. After each place-name was added the number of Roman miles from the previous staging point for one traveling northward and then eastward. That coastal route ran from Byzantium via Tomis to the mouth of the Danube and beyond. On two distributaries of the Danube are written the names Istros and Danubis, whereas properly speaking Istros was the name given by the Greeks to the whole of the lower Danube.

After an illegible entry the stages shown, with mileages where preserved, are: Odessos, Bybona (Byzone), Kallatis, Tomea (Tomis) 33, river Histros 40, river Danubios, Tyra 84, Borysthenes (Olbia is meant), Chersonesos, Trap . . . , and Arta. Chersonesos is the Tauric Chersonese, the Crimea (fig. 14.10). But the following two have been misunderstood. Trap does indeed stand for Trapezus, but this is not Trebizond, which is on the southern shore of the Black Sea, but the "Table Mountain" of antiquity, the Krymskie Gory. Finally, Arta is not Artaxata, capital of Armenia, which is nowhere near the Black Sea, but the Latin word *arta* (narrows) transliterated into Greek.[85] This must obviously refer to the Straits of Kerch, where the chief ancient settlement was Panticapaeum (Kerch). We may therefore consider this a "painted itinerary" (if that and not a picture is what Vegetius had in mind by his phrase *itineraria picta*) taking the soldier to Panticapaeum, ruled by a native prince but with a Roman garrison.

The shield may be regarded then, like the Peutinger map, as an *itinerarium pictum*. Unlike the latter, it is reasonably orthomorphic, but somewhat oversimplified. It has some semblance of reliability on the northwest coast of the Black Sea, and even eastward of that it is more reliable than has been thought. From its ornamental character and the use of Greek rather than Latin, it is likely to have been an unofficial composition.

other cases, in addition to the cardinal points, summer and winter sunrises and sunsets often appear as directions, though these, as Strabo observes, vary according to latitude: Strabo *Geography* 1.2.20–21; see *The Geography of Strabo*, 8 vols., ed. and trans. Horace Leonard Jones, Loeb Classical Library (Cambridge: Harvard University Press; London: William Heinemann, 1917–32). See also Strabo, *Géographie*, ed. François Lasserre, Germaine Aujac, et al. (Paris: Belles Lettres, 1966–). But for mainland Greece the summer and winter sunrises, together with the summer and winter sunsets, make an angle of about 30° with the east-west axis; and it is likely that Aristotle situated the winds at 30° from the north-south axis: Aristotle *Meteorologica* 2.6 (note 82).

84. Bibliothèque Nationale, Gr. Supp. 1354, no. 5. Franz Cumont, "Fragment de bouclier portant une liste d'étapes," *Syria* 6 (1925): 1–15 and pl. I; idem, *Fouilles de Doura Europos (1922–1923)*, text and atlas (Paris: P. Geuthner, 1926), 323–37, pls. CIX, CX.

85. Richard Uhden, "Bemerkungen zu dem römischen Kartenfragment von Dura Europos," *Hermes* 67 (1932): 117–25. The suggested interpretation of *arta* given here is the author's.

FIG. 14.9. THE MAP ON THE DURA EUROPOS SHIELD. Dating to shortly before the Roman withdrawal from Dura Europos in A.D. 260, this design was drawn upon parchment found covering a Roman soldier's shield. It shows the coastal route from Byzantium to the mouth of the Danube and beyond, complete with the mileages between staging posts.

Size of the original: 18 × ca. 65 cm. Photograph from the Bibliothèque Nationale, Paris (Gr. Suppl. 1354, no. 5).

PLANS ON LAMPS

An unusual series of designs on Roman lamps, incorporating rudimentary plans, has been discovered in Palestine, probably datable to the early fourth century A.D.[86] One from Samaria (Sebastye) shows what appears to be a Roman fort with rooms around the four sides and with intersecting central roads (fig. 14.11). The correctness of the identification as a military diagram is shown by the emblem of two crossed swords on similar lamps from the same area. A fragmentary lamp from the Ophel (near Jerusalem's Old City) appears to depict two L-shaped Roman road stations, and other less certainly identifiable plans exist.[87]

CLASSICAL PLANS OF MAZES

The maps associated with the artifacts described above show that they were adapted as motifs in a variety of historical contexts. Such maps should be thought of as different from, rather than historically less interesting than, the formal maps of the textual sources. They reflect alternative ways of thinking about maps and confirm their more general acceptance within Roman society. At the same time, this evidence is important in showing how some maps were recurrent motifs in the classical world. This is illustrated by a common category of clas-

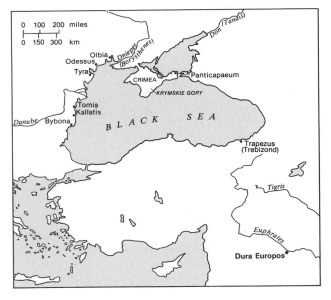

FIG. 14.10. PRINCIPAL PLACE-NAMES FROM THE DURA EUROPOS SHIELD.

86. Mordechai Gichon, "The Plan of a Roman Camp Depicted upon a Lamp from Samaria," *Palestine Exploration Quarterly* 104 (1972): 38–58, who also illustrates the lamp from the Ophel.

87. Robert Alexander Stewart Macalister, *The Excavation of Gezer, 1902–1905 and 1907–1909*, Palestine Exploration Fund, 3 vols. (London: John Murray, 1911–12), 3:pl. LXXII, 18.9: perhaps this depicts a barracks block.

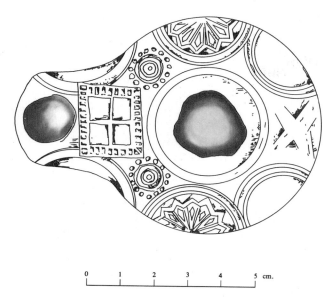

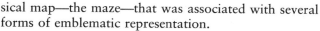

FIG. 14.11. PLAN ON A ROMAN LAMP. Found at Samaria (Sebastye), in Palestine, and dating from the early fourth century A.D.. The illustration is a view from above.
Size of the original: 6.5 × 8.9 × 3.2 cm. After Mordechai Gichon, "The Plan of a Roman Camp Depicted upon a Lamp from Samaria," *Palestine Exploration Quarterly* 104 (1972): 38–58, esp. 39 (top).

FIG. 14.12. LABYRINTH DESIGNS ON CRETAN COINS. These silver coins show both a square maze (ca. 80 B.C.) and a rarer circular one (300–280 B.C.).
Diameters of the originals: 2.3 cm and 3 cm respectively. By permission of the Trustees of the British Museum, London (BMC Cnossus 24 and 41).

sical map—the maze—that was associated with several forms of emblematic representation.

The maze, or labyrinth, a feature well known from dynastic Egypt and elsewhere, was particularly associated in classical antiquity with the palace of Minos at Cnossos.[88] Sir Arthur Evans thought it was connected there with the double ax (*labrys*), and a maze pattern appears on one of the wall frescoes of the palace.[89] The city of Cnossos, proud of its Minoan heritage, minted coins in the classical period with rough plans of mazes, mostly having rectilinear paths, though the circular type also exists (fig. 14.12).[90] An early artifact showing one is a clay tablet from the palace of Nestor at Pylos in the Peloponnese.[91] On the reverse is a maze, thought to have been incised earlier than the obverse, that contains a list of ten men in the same Linear B script as occurs at Cnossos. Another example from Greece is a tile of the classical period from the acropolis at Athens, where there had much earlier been a Mycenaean stronghold.[92]

An Etruscan vase found at Tragliatella, near Rome,[93] shows among other features infantry performing a dance and cavalry next to a circular maze labeled TPVIA (Etruscan Truia = Troy). Here we have an allusion not merely to an early palace but obviously to the "Game of Troy" that, whether rightly or wrongly attributed to that city, is described by Virgil as a Roman game having labyrinthine movements.[94]

88. "Labirinto," in *Enciclopedia dell'arte antica, classica e orientale*, 7 vols. (Rome: Istituto della Enciclopedia Italiana, 1958–66), W. H. Matthews, *Mazes and Labyrinths: A General Account of Their History and Developments* (London: Longman, 1922); Janet Bord, *Mazes and Labyrinths of the World* (London: Latimer New Dimensions, 1976); Karl Kerényi, *Labyrinth-Studien: Labryrinthos als Linienreflex einer mythologischen Idee* (Zurich: Rhein-Verlag, 1950).

89. Arthur Evans, *The Palace of Minos*, 4 vols. (London: Macmillan, 1921–35); Matthews, *Mazes*, 32, fig. 8 (note 88).

90. Jean N. Svoronos (Ioannes N. Sbōrōnos), *Numismatique de la Crète ancienne accompagnée de l'histoire, la géographie et la mythologie de l'Île*, text and plates published separately (Macon: Imprimerie Protat Frères, 1890; reprinted Bonn: R. Habelt, 1972), 65 ff. and pls. IV–VIII; Warwick Wroth, *A Catalogue of the Greek Coins of Crete and the Aegean Islands*, ed. Reginald Stuart Poole (Bologna: A. Forni, 1963), 18–19, pls. 5 and 6; Matthews, *Mazes*, figs. 20–31 (note 88).

91. Mabel Lang, "The Palace of Nestor Excavations of 1957: Part II," *American Journal of Archaeology*, 2d ser., 62 (1958): 181–91. The clay tablet is Inventory Cn. 1287, discussed on p. 190 and pictured in pl. 46.

92. Ernst Buschor, *Die Tondächer der Akropolis* (Berlin and Leipzig: Walter de Gruyter, 1929–33), 1.45 ff., K 108; Hermann Kern, *Labirinti: Forme e interpretazione, 5000 anni di presenza di un archetipo manuale e file conduttore* (Milan: Feltrinelli, 1981), German ed., *Labyrinthe: Erscheinungsformen und Deutungen, 5000 Jahre Gegenwart eines Urbilds* (Munich: Prestel-Verlag, 1982).

93. Matthews, *Mazes*, figs. 133-35 and refs. (note 88).

94. Virgil *Aeneid* 5.580 ff.; see *The Aeneid of Virgil*, trans. Cecil Day-Lewis (London: Hogarth Press, 1952).

One of the best-known Roman specimens is a graffito at Pompeii showing a square mosaic with the wording LABYRINTHUS: HIC HABITAT MINOTAURUS (Labyrinth: the Minotaur lives here), referring to the Cnossos palace.[95] In fact the most common use of the motif in Roman times is in mosaics.[96] One such is also at Pompeii, in the Villa di Diomede. Another particularly well preserved mosaic at Salzburg shows, in the center of a square maze, Theseus about to kill the Minotaur. This same theme in the center of a circular maze, with recumbent towers at the corners, is found on a mosaic at Cormerod, Switzerland.[97] A partly preserved square mosaic, with a vase and scrolls surrounding the maze, was found in the churchyard of Caerleon (Isca).[98] A lost specimen from Sousse (Hadrumetum), Tunisia, included Theseus's ship and the words HIC INCLUSUS VITAM PERDIT (one enclosed here loses his life).[99]

The early Christian use of maze mosaics can be exemplified from the fourth-century basilica of Saint Reparatus at al-Aṣnam, Algeria, where the center of the square maze has a play on words on SANCTA ECCLESIA.[100] The classical tradition of maze plans continued into the Middle Ages; the connection of the labyrinth with Theseus and the Minotaur, sometimes then misinterpreted as a centaur, was not forgotten in medieval church ornamentation.

THE USE OF MAPS IN THE ROMAN PERIOD

Much has already been either said or implied about the widespread use of maps in Roman society. There is more evidence for the use of maps in the Roman period than in other periods of antiquity, but the evidence is still largely fragmentary. Such evidence is partly from literary or technical works and partly from inscriptions. It suggests that there was an expansion in the use of maps, even in periods that lacked marked technical or scientific advances in mapmaking. Thus, as a conclusion to the period as a whole, an attempt is made here to classify the uses of maps in Roman society, while accepting that there is clearly much overlap.

MAPS AS CADASTRAL AND LEGAL RECORDS

It was probably for land survey that Rome produced its earliest working maps, one lost specimen (pp. 209–10) dating from the second century B.C. The Roman land survey treatises mention maps available both in Rome and locally; these were used among other purposes by lawyers (or surveyors experienced in land law) to contest property disputes, by emperors to decide boundary disputes between local authorities, and for levying central or regional taxes. One object of depositing two copies of survey maps, one in Rome and the other in a local

office, was to save users unnecessary travel. The map came to be recognized as a legal document, both in land survey and for determining the use of aqueducts by property owners. Both writers on water supply, Vitruvius and Frontinus, show themselves familiar with maps: Vitruvius mentions river sources as painted and written about in world cosmographies;[101] for the use of aqueduct maps as mentioned by Frontinus, see page 232.

Another function of mapping is shared by religious and legal or surveying aspects: often a tomb and surrounding plot were given their measurements on an inscription. An actual plan of these was only very rarely attached. Such a plan could have helped lawyers in any dispute, such as about whether the land around a tomb belonged to the heirs. The users of the *Forma Urbis Romae* were on the one hand public organizations, on the other hand private individuals living in Rome. It seems unlikely that builders of new roads normally consulted maps, though there was a civil service maps department under the late empire.[102]

MAPS AS STRATEGIC DOCUMENTS

Contrary to Lloyd Brown's view, the story that a Phoenician captain was publicly rewarded for running his ship aground so as not to reveal to a following Roman ship his route from Gades (Cádiz) to the Cassiterides gives no indication of the use of maps.[103] It is thought

95. E. Pottier, "Labyrinthus," in *Dictionnaire des antiquités grecques et romaines*, 5 vols., ed. Charles Daremberg and Edmond Saglio (Paris: Hachette, 1877–1919), 3.2:882–83.

96. Pliny *Natural History* 36.19.85 (note 41) says one should not compare the Cretan labyrinth to "our mosaic pavements or to the mazes made in fields to amuse children"; author's translation.

97. C. Bursian, "Aventicum Helvetiorum," *Mittheilungen der Antiquarischen Gesellschaft zu Zürich* 16, no. 1 (1867–70), esp. pl. 29; Matthews, *Mazes*, fig. 36 (note 88).

98. George C. Boon, *Isca: The Roman Legionary Fortress at Caerleon, Mon.*, 3d ed. (Cardiff: National Museum of Wales, 1972); Matthews, *Mazes*, 42 (note 88).

99. Salomon Reinach, *Répertoire de peintures grecques et romaines* (Paris: E. Leroux, 1922), 214.1, fig. 1.

100. Henri Leclercq, "Labyrinthe," in *Dictionnaire d'archéologie chrétienne et de liturgie*, 8.1 (1928): cols. 973–82, esp. 974–75 and fig. 6549 (note 17).

101. Vitruvius *De architectura* 8.2.6 (note 33).

102. See p. 244 above and n. 68 of this chapter.

103. The story is in Strabo *Geography* 3.5.11 (note 83). The Cassiterides are called a group of islands, and their name is derived from the Greek word for tin; they are usually thought to have been the Scilly Islands, but Rivet and Smith, *Place-Names*, 42–43 (note 9), prefer to place them outside the British Isles. Lloyd A. Brown, *The Story of Maps* (Boston: Little, Brown, 1949; reprinted New York: Dover, 1979), 9, has a misleading account of this episode: in Strabo the sea captain is called Phoenician, not Carthaginian, and Strabo does not mention any ship's log or charts.

to refer to the period between the First and Second Punic Wars, 241–218 B.C. If Roman ships' captains wanted to consult a work that would help them navigate outside the Straits of Gibraltar, they could perhaps have turned to the periplus of pseudo-Scylax (p. 383), though in the form in which it has come down to us it contains nothing on sea areas north of Cádiz.

The only extant admonition to the soldier to use maps is in the late military writer Vegetius. Since, however, both Julius Caesar and Agrippa, the one a general, the other an admiral, were promoters of maps, there is good reason to think that the use of these, as well as of itineraries and periploi (see "Maps for Traveling," below) was established relatively early.

The Augustan period was one in which, perhaps for the first time in the Roman world, the use of maps by the man in the street was taken for granted. Rough ones could be traced on the spot. The first poem of Ovid's *Heroides* is supposedly a letter from Penelope to Ulysses, but typically the poet makes his Trojan heroes behave like contemporaries. One such, who has reached home after the sack of Troy, describes the course of events as he sketches the Troy area in wine on the table: "This is where the river Simois flowed; this is the land of Sigeum; here stood the high palace of old Priam; that is where Achilles encamped, that is where Ulysses encamped; this is the point where the mangling of Hector terrified the galloping horses." [104] Propertius (ca. 50 B.C. to ca. 16 B.C.) also uses the literary format of the letter, but from a Roman lady to her lover far away in the army. His Arethusa spends the winter nights studying maps: "I learn in what area the river Araxes, which is to be conquered, flows, how many miles the Parthian horse runs without water. I compel myself to learn painted worlds from the map [*tabula*] . . . ; which land freezes up, which crumbles from heat, which wind gives good sailing to Italy." [105] Editors mostly treat *tabula* here as map rather than picture, though the expression may include both a map and its commentary.

The extent to which Roman expeditions carried maps is open to dispute. If the word *forma* means "map" in several expedition accounts, both compilation and use are attested; but in any case there is at least one reference, and probably two, during the early empire to expedition maps sent back to Rome by commanders in the field. [106] The elder Pliny insists that the Caucasian Gates pass should be so called, not Caspian Gates. The mistake, he says, arose in expeditions to the East by Domitius Corbulo between A.D. 58 and 63; and "maps of the area [*situs*] painted and sent home have this name drawn on them." [107] The other reference is also in Pliny and concerns a party of Praetorian guards who explored the upper Nile south of Khartoum. [108] Unfortunately the word *forma* is ambiguous, since it can mean either

"shape" or "map." If it is taken to be the latter, the sentence may be rendered: "The map of Ethiopia became known, as mentioned, and when recently brought to the emperor Nero it showed that for 996 miles from Syene, the boundary of the empire, to Meroë there were few trees and that all of these were species of palm." [109] Since trees were often drawn on maps, though sometimes incorrect species, palms may have been drawn in as far as Syene and their dearth deduced from the appearance of very few upstream from there.

Such maps are likely to have been drawn by military surveyors but used with caution by strategists back in Rome. Certainly, by the second century A.D., educated Romans were becoming aware of the limitations of maps as a basis for action on the fringes of unexplored territory. Thus Plutarch (ca. A.D. 46 to after 120) dedicates his *Parallel Lives* to Sosius Senecio, consul four times between A.D. 98 and 107, with this simile: "Just as historians, Sosius Senecio, in their geographies squeeze onto the edges of their maps [*pinakes*] parts of the earth that escape their knowledge, with notes explaining 'Everything beyond is sandy desert with no water or full of wild animals' or 'unexplored marsh' or 'Scythian frost' or 'frozen sea,' so in writing my Lives . . . I might as well say of prehistory: 'Everything beyond is full of wonders.'" [110]

Such references suggest that the value of maps to the state and its generals was widely accepted. Together with coins and such monuments as Trajan's column, with its graphic representation of Rome's campaigns on the Danube, they could have had great propaganda value. But there were possible dangers also to an autocracy from knowledge that could be extracted from maps by movements hostile to the imperial system, so that in the wrong hands they could become a threat to security. Domitian, emperor from A.D. 81 to 96, was by nature autocratic, and knowing of previous conspiracies he was quick to

104. Ovid *Heroides* 1.33 ff., author's translation; see *Heroides [and] Amores*, trans. Grant Showerman, Loeb Classical Library (Cambridge: Harvard University Press; London: William Heinemann, 1958); cf. Tibullus *Elegies* 1.10.29–32; see *Elegies*, trans. Guy Lee, Liverpool Latin Texts 3, 2d ed. (Liverpool: Francis Cairns, 1982).

105. Propertius *Elegies* 4.3.33–40, author's translation; see *Propertius*, trans. H. E. Butler, Loeb Classical Library (Cambridge: Harvard University Press; London: William Heinemann, 1912).

106. Robert K. Sherk, "Roman Geographical Exploration and Military Maps," in *Aufstieg und Niedergang der römischen Welt*, ed. Hildegard Temporini (Berlin: Walter de Gruyter, 1972–), 2.1 (1974): 534–62, esp. 537–43.

107. Pliny *Natural History* 6.15.40 (note 41), author's translation.

108. Pliny *Natural History* 6.35.181 (note 41).

109. Pliny *Natural History* 12.8.19 (note 41), author's translation.

110. Plutarch *Theseus* 1.1; cf. Christian Jacob, "Lectures antiques de la carte," *Études Françaises* 21.2 (1985): 21–46, esp. 44–45.

suppress any hint of one against himself. Mettius Pompusianus was put to death by Domitian because, in addition to being born under an imperial constellation and giving Carthaginian generals' names to his slaves, he carried around a parchment map of the world, together with speeches of kings and generals from Livy.[111] This was a period when parchment was increasing in use as against papyrus;[112] it was certainly more transportable. Suspicious of spies, Domitian may have linked the map with fears of an uprising in North Africa; in an expanded empire there was plenty of scope for rebellion.

MAPS FOR TRAVELING

Ptolemy's *Geography*, with its tables of coordinates, was never intended for the traveler; but there is reason to think that the latitudes of Roman provinces on the reverse of portable sundials were indebted to Ptolemy, and these sundials were used by travelers and surveyors.

Itineraries for land journeys and simple periploi for sea journeys were also commonly used. A governor such as Arrian, second century A.D. (who was also a man of letters), chose to compile an expanded periplus for his sail around the Black Sea. This may have appeared in Latin for official purposes as well as in Greek for his reading public.[113] Whether it was based on a map we cannot be sure, but the Dura Europos shield proves that at least road maps of that area existed. It is also difficult to establish how far land itineraries were derived from maps; since road maps are called by Vegetius *itineraria picta*, painted itineraries, the priority may be the other way around. The most famous are the Antonine land and sea itineraries (pp. 235–36). These are not representative of the type carried around by the traveling public. They obviously detailed particular journeys made by people like the emperor Caracalla, whereas other travelers might want only a section of these. Among Christian itineraries, that of the journey from Bordeaux to Jerusalem (p. 237), by expanding on the basic list form, was designed to help the pilgrim traveling to the Holy Land.

Clearly anyone making a complicated road journey would have benefited from having an *itinerarium pictum* like the archetype of the Peutinger map. But ancient maps may not have stood up very well to travel, since they would normally be carried around loose, not in a *capsa* (cylindrical box), used for storing rolls of papyrus in the owner's house. Perhaps, however, the increasing use of parchment started to promote a greater mobility of maps in the late empire. Julian the Apostate, emperor 360–63, thanks his friend Alypius of Antioch thus: "I happened already to have recovered from my illness when you sent the geography; all the same I was glad to receive the map you sent. Not only does it have better

drawings [*diagrammata*] than previous ones, but you have made it more attractive by adding iambic verses."[114] This may have been a map of Britain, since Alypius was or had been *vicarius* of the British provinces. The Greek verses remind us of the Latin ones attached to the Theodosian map (see below, pp. 258–59).

RELIGIOUS AND PROPAGANDA FUNCTIONS FOR MAPS

One of the first Roman maps we hear of, the 174 B.C. map of Sardinia (p. 205), served both as a form of thanksgiving to the gods for victory and as a useful piece of propaganda showing Rome's expansion, just as Mussolini displayed maps of the Roman empire. Mosaics at Ostia (pp. 246–47) advertised a shipping corporation or the transport guild. A somewhat different element of public relations is visible in the chief Christian map of the early Byzantine period, the Madaba map (pp. 263–65), where Jerusalem is given specially enlarged and detailed treatment; the map was oriented so as to be easily seen by the congregation.

DIDACTIC AND SCHOLARLY USES FOR MAPS

It is clear from Varro that at least the more progressive landowners in Italy were regarded as being familiar with the use of maps. Both he and the elder Pliny were encyclopedic writers who absorbed vast quantities of Greek scholarly exposition. Varro may or may not have cited sources (his encyclopedic writing exists only in fragments). Pliny, in book 1, lists all his sources and thereafter frequently refers to them, but often in such a way that he does not tell us whether he is referring to a text or a map. Thus for the circumference of the Black Sea, apart from Agrippa's map and commentary (which he calls simply Agrippa), he cites four estimates, dating from about 50 B.C. to about A.D. 70, all of which may come from texts rather than maps. In the case of lengths of the shores of the Red Sea, his object seems to have

111. Suetonius *Domitian* 10.3, book 8 of *De vita Caesarum* (The lives of the Caesars), in *Suetonius*, 2 vols., trans. J. C. Rolfe, Loeb Classical Library (Cambridge: Harvard University Press; London: William Heinemann, 1913–14). Pascal Arnaud, "L'affaire Mettius Pompusianus, ou Le crime de cartographie," *Mélanges de l'Ecole Française de Rome: Antiquité* 95 (1983): 677–99, prefers to follow the account of Dio Cassius and Zonaras. According to this, Pompusianus had a world map painted on the walls of his bedroom. See Dio Cassius *Roman History* 67.12.4 (note 7).

112. C. H. Roberts, "The Codex," *Proceedings of the British Academy* 40 (1954): 169–204.

113. Arrian *Periplus Ponti Euxini*, in Müller, *Geographi Graeci minores*, 1:370–423 (note 20).

114. Julian *Epistles* 7, author's translation; see vol. 3 of *The Works of the Emperor Julian*, 3 vols., trans. Wilmer Cave Wright, Loeb Classical Library (Cambridge: Harvard University Press; London: William Heinemann, 1913–23).

been to correct the mapping by Eratosthenes with references to the much higher figures given by Artemidorus of Ephesus (fl. 104–101 B.C.) and by Agrippa. Varro had illustrations in some of his works, but none has survived; Pliny is not known to have had any. After Pliny we have no extant encyclopedias until the fifth century A.D. and later (Martianus Capella, Boethius, and Isidore); of these only Isidore, bishop of Seville A.D. 602–36, dealt specifically with geography.[115]

Of purely geographical Latin writings, the lost commentary of Agrippa was intended (as the many fragments show) to explain his map; it would appeal mainly to those, including Strabo and Pliny, who visited his colonnade. Pomponius Mela's *De chorographia* was designed for the ordinary reading public and had no maps. But the writer probably had one before his eyes when, for example, he writes of the Baltic: "The Codanus Gulf . . . is dotted with large and small islands."[116]

An interesting insight into the teaching of geography with the use of a terrestrial globe is found in a pupil's addendum to the *Cosmographia* of Julius Honorius: "So as to avoid errors, as the teacher has said, this book of extracts should not be separated from the globe."[117] The use of the map in schools continued after the fall of Rome: Cassiodorus (ca. A.D. 490 to ca. 583), who after many years as an important administrator under kings of the Goths in Italy devoted himself to Christianity and education, recommended for teaching purposes the *pinax* (map) accompanying the geographical poem of Dionysius Periegetes.[118] In addition to maps in schools, the words quoted from the rhetorician Eumenius show that, in Gaul of A.D. 298 at least, an idealist wanting to foster culture among the young after the ravages of war included a large map on a balcony wall as an important teaching aid.

Mapmaking clearly constituted an essential part of the training of land surveyors under the late Roman Empire. First and foremost was drawing up centuriation maps, which, although to some extent diagrammatic, had to be very accurate from the legal and administrative point of view. For this purpose apprentice surveyors would consult the official copies; yet in the *Corpus Agrimensorum* we find some maps that are unrealistic in topographical terms. This must represent a tradition of decades of specimen teaching maps, in which topographical accuracy and the representation of real landscapes were of secondary importance.

It must, however, be admitted that educated Romans, although they respected Greek scientific research, did not always understand the mathematical principles behind it. We do know that about 168 B.C. Crates of Mallos gave many lectures in Rome, illustrated by a globe (see description above, pp. 162–64). Even so, his globe is reported not in extant Latin writings of the

republic but in the Greek geography of Strabo. Astronomical and mathematical instruments were introduced into Rome as contacts with the Greek world increased. But it often took a long time for the proper use of an instrument to be appreciated, as in the case of the sundial. In 263 B.C. one such was brought from Catana to Rome, where it was set up near the speakers' platform in the forum. Nevertheless, for ninety-nine years it displayed the wrong time, since the necessary adjustment for a different latitude had not been made: only in 164 B.C. did Q. Marcius Philippus put up by its side a sundial correct for the latitude of Rome.[119] Lucretius was exceptional in expounding Greek physical theory in Latin verse. Cicero's contribution of this type consisted only of a translation of the *Phaenomena* of Aratus of Soli (b. ca. 315 B.C.), a meteorological work. He did promise his friend Atticus a work on geography, but though Atticus sent him, evidently for this purpose, a work of mathematical geography by Serapion, it never materialized;[120] Cicero, in thanking him, commented, "Between ourselves, I hardly understand one line in a thousand."[121] This is not, however, to say that Cicero never consulted maps; he probably did in at least one case. When Atticus criticizes him for having written that almost all the city-states of the Peloponnese were by the sea,[122] he replies that he got this from the *tabulae* of Dicaearchus.[123] Since he goes on to refer to Dicaearchus's work on the underground oracle of Trophonius,

115. Isidore of Seville *Etymologies*; see *Etymologiarum sive originum libri XX*, 2 vols., ed. W. M. Lindsay (Oxford: Clarendon Press, 1911); idem, *De natura rerum*, in *Traité de la nature*, ed. Jacques Fontaine, Bibliothèque de l'Ecole des Hautes Etudes Hispaniques, fasc. 28 (Bordeaux: Féret, 1960), 164–327.

116. Pomponius Mela *De chorographia* 3.31 (note 40), author's translation.

117. Julius Honorius *Excerpta* 50 (note 60), author's translation.

118. Cassiodorus *Institutiones divinarum et saecularium litterarum* 1.25.2; see *Institutiones*, ed. R. A. B. Mynors (Oxford: Clarendon Press, 1937), or, for an English translation, *An Introduction to Divine and Human Readings*, ed. and trans. Leslie Webber Jones (New York: Columbia University Press, 1946). For Dionysius Periegetes, see pp. 171–73.

119. Pliny *Natural History* 7.60.214; see corresponding notes in Schilling, *Histoire naturelle*, 261–62, esp. n. 214.3 (note 41). There was a dispute even in antiquity whether this was the earliest sundial erected in a public place in Rome: one of 293 B.C. is also mentioned by Pliny (*Natural History* 7.60.213); again, see Schilling, 261, n. 213.1. Cf. Censorinus *De die natali* 23.6-7 (cited by Schilling) in *De die natali liber*, ed. F. Hultsch (Leipzig: Teubner, 1867).

120. Cicero *Letters to Atticus* 2.4.3; see *Cicero's Letters to Atticus*, 7 vols., ed. and trans. D. R. Shackleton Bailey (Cambridge: Cambridge University Press, 1965–70).

121. Cicero *Letters to Atticus* 2.4.1 (note 120).

122. Cicero *The Republic* 2.4.8; see *La république*, 2 vols., trans. Esther Bréguet (Paris: Belles Lettres, 1980).

123. Cicero *Letters to Atticus* 6.2 (note 120).

the latest editor, Shackleton Bailey, doubts whether these *tabulae* were maps[124]; but Cicero is probably referring to Dicaearchus's lost *Gēs periodos*, a "geographical tour" of the world.

In general it may be concluded that, as far as we know at present, up to about 170 B.C. maps were relatively unfamiliar to most Romans. From that time onward their use steadily increased. By the time of Julius Caesar or Augustus they were used by a wide range of people, had become indispensable for the surveying of land, public works, and other engineering projects, and were important for legal, strategic, traveling, scholarly, and didactic purposes.

---

124. Shackleton Bailey, *Cicero's Letters*, 3:257 (note 120).

BIBLIOGRAPHY
CHAPTER 14 ITINERARIES AND GEOGRAPHICAL MAPS IN THE EARLY AND LATE ROMAN EMPIRE

Avienius. *Ora maritima*. Edited by André Berthelot. Paris: H. Champion, 1934.

Böker, Robert. "Windrosen." In *Paulys Realencyclopädie der classischen Altertumswissenschaft*, ed. August Pauly, Georg Wissowa, et al., 2d ser., 8.2 (1958): cols. 2325–81. Stuttgart: J. B. Metzler, 1894–.

Bosio, Luciano. *La Tabula Peutingeriana: Una descrizione pittorica del mondo antico*. I Monumenti dell'Arte Classica, vol. 2. Rimini: Maggioli, 1983.

Carmody, F. J. *La Gaule des itinéraires romains*. Berkeley: Carmody, 1977.

Castagnoli, Ferdinando. "L'orientamento nella cartografia greca e romana." *Rendiconti della Pontificia Accademia Romana di Archeologia* 48 (1975–76): 59–69.

Clemente, G. *La Notitia Dignitatum*. Saggi di Storia e Letteratura 4. Cagliari: Fossataro, 1968.

Cumont, Franz. "Fragment de bouclier portant une liste d'étapes." *Syria* 6 (1925): 1–15.

Cuntz, Otto, ed. *Itineraria Romana*. Vol. 1, *Itineraria Antonini Augusti et Burdigalense*. Leipzig: Teubner, 1929.

Dilke, O. A. W. *Greek and Roman Maps*. London: Thames and Hudson, 1985.

Elter, A. *Itinerärstudien*. Bonn: C. Georgi, 1908.

Finkelstein, I. "The Holy Land in the Tabula Peutingeriana." *Palestine Exploration Quarterly* 111 (1979): 27–34.

Frank, Johannes. "Beiträge zur geographischen Erklärung der *Ora maritima* Aviens." Dissertation, Würzburg. Sangerhausen: Schneider, 1913.

Gichon, Mordechai. "The Plan of a Roman Camp Depicted upon a Lamp from Samaria." *Palestine Exploration Quarterly* 104 (1972): 38–58.

Goodburn, R., and P. Bartholomew, eds. *Aspects of the Notitia Dignitatum*. British Archaeological Reports, Supplementary Series 15. Oxford: British Archaeological Reports, 1976.

Gross, H. *Zur Entstehungs-geschichte der Tabula Peutingeriana.* Bonn: H. Ludwig, 1913.

Itineraria et alia geographica. In *Corpus Christianorum*, Series Latina, vols. 175 and 176 (1965).

Janni, Pietro. *La mappa e il periplo: Cartografia antica e spazio odologico.* Università di Macerata, Pubblicazioni della Facoltà di Lettere e Filosofia 19. Rome: Bretschneider, 1984.

Kubitschek, Wilhelm. "Itinerarien." In *Paulys Realencyclopädie der classischen Altertumswissenschaft*, ed. August Pauly, Georg Wissowa, et al., vol. 9 (1916): cols. 2308–63. Stuttgart: J. B. Metzler, 1894–.

———. "Karten." In *Paulys Realencyclopädie der classischen Altertumswissenschaft*, ed. August Pauly, Georg Wissowa, et al., vol. 10 (1919): cols. 2022–2149. Stuttgart: J. B. Metzler, 1894–.

Levi, Annalina, and Mario Levi. *Itineraria picta: Contributo allo studio della Tabula Peutingeriana.* Rome: Erma di Bretschneider, 1967.

Maier, I. G. "The Giessen, Parma and Piacenza Codices of the 'Notitia Dignitatum' with Some Related Texts." *Latomus* 27 (1968): 96–141.

———. "The Barberinus and Munich Codices of the *Notitia Dignitatum Omnium*." *Latomus* 28 (1969): 960–1035.

Miller, Konrad. *Itineraria Romana.* Stuttgart: Strecker und Schröder, 1916.

———. *Die Peutingersche Tafel.* Stuttgart: F. A. Brockhaus, 1962.

Pliny. *Natural History.* 10 vols. Translated by H. Rackham et al. Loeb Classical Library. Cambridge: Harvard University Press; London: William Heinemann, 1940–63.

Polaschek, Erich. "Notitia Dignitatum." In *Paulys Realencyclopädie der classischen Altertumswissenschaft*, ed. August Pauly, Georg Wissowa, et al., 17.1 (1936): cols. 1077–116. Stuttgart: J. B. Metzler, 1894–.

Reed, N. "Pattern and Purpose in the Antonine Itinerary." *American Journal of Philology* 99 (1978): 228–54.

Riese, Alexander, ed. *Geographi Latini minores.* Heilbronn, 1878; reprinted Hildesheim: Georg Olms, 1964.

Rodwell, Warwick. "Milestones, Civic Territories and the Antonine Itinerary." *Britannia* 6 (1975): 76–101.

Seeck, Otto, ed. *Notitia Dignitatum.* Berlin: Weidmann, 1876; reprinted Frankfort: Minerva, 1962.

Stewart, Aubrey, trans. *Itinerary from Bordeaux to Jerusalem: "The Bordeaux Pilgrim."* Palestine Pilgrims Text Society, vol. 1, no. 2. London: Palestine Exploration Fund, 1896.

Temporini, Hildegard, and Wolfgang Haase, eds. *Aufstieg und Niedergang der römischen Welt.* Berlin: Walter de Gruyter, 1972–.

Uhden, Richard. "Bemerkungen zu dem römischen Kartenfragment von Dura Europos." *Hermes* 67 (1932): 117–25.

Van der Poel, Halsted B., et al., eds. *Corpus topographicum pompeianum.* Part 5, Cartography. Rome: Edizione dell'Elefante, 1981.

Ward, J. H. "The British Sections of the Notitia Dignitatum: An Alternative Interpretation." *Britannia* 4 (1973): 253–63.

Weber, Ekkehard, ed. *Tabula Peutingeriana: Codex Vindobonensis 324.* Graz: Akademische Druck- und Verlagsanstalt, 1976.

Zicàri, Italo. "L'anemoscopio Boscovich del Museo Oliveriano di Pesaro." *Studia Oliveriana* 2 (1954): 69–75.

15 · Cartography in the Byzantine Empire

O. A. W. DILKE
WITH ADDITIONAL MATERIAL SUPPLIED BY THE EDITORS

Of all the civilizations of the classical world, the Byzantine is probably the least known from the cartographic point of view. The Byzantine state was the richest, the most powerful, and the most civilized in Europe and the Middle East at that time.[1] Although the territorial boundaries of its empire fluctuated,[2] there was a continuity in political organization, in cultural influences, and in religion for over a thousand years from A.D. 330, when Constantinople was founded, to the fall of Trebizond in 1461, eight years after the collapse of the capital.

It is paradoxical, however, that a literate society, heir to Greek and Roman learning, should have left so few traces of an interest in mapping. At least some of the necessary conditions for the development of such an interest were present. In late Roman times the eastern empire, from its base in Constantinople, had access to the practical skills of the Roman land surveyors, including mapmaking. The cartographic needs of Byzantine emperors in connection with administration, military conquest and subjugation, propaganda, land management, and public works were apparently similar to those of Rome itself. Moreover, the revival of classical culture, consequent on the restoration of literary Greek, gave the educated classes a reading knowledge of classical Greek and Latin. Finally, it is known that astronomical and geographical texts, both containing maps, were in circulation even before the so-called Renaissance of the tenth century A.D. Certainly these were available during the later Byzantine Empire when Maximus Planudes (ca. 1260–1310) was able to initiate a successful search for the manuscripts of Ptolemy's *Geography*.[3]

At the same time, there were other factors that weakened this continuity of classical learning. These included the decline of the Byzantine Empire in the seventh and eighth centuries; the religious movement known as iconoclasm, which may have resulted in the destruction of some images relevant to cartography; the capture of Constantinople by the Crusaders in 1204; and the removal of other manuscripts to western Europe by refugee scholars. Thus the transmission of original Greek and Latin manuscripts through the centuries was far from being a simple process. Literary and artistic currents that mingled in maritime and commercial centers such as Alexandria, Antioch, Ephesus, and Thessalonica—and above all in Constantinople itself—were complex. They were not only the contacts with the heartland of the old classical world but also links with the Islamic and other societies to the east.[4] Byzantine cities became entrepôts through which astronomical and geographical learning (including a knowledge of maps) was handed on in many directions.

Notwithstanding their complexity, some of these conditions should have been favorable to the survival of classical cartographic knowledge. It is disappointing, therefore, that so few maps have come down to us from the whole of the Byzantine millennium. Moreover, it is quite clear that these few are representative neither of the theoretical cartography developed by the Greeks nor of the applied mapping practiced by the Romans. In addition, there are fewer literary allusions to maps from the Byzantine period than from the Roman period, so that once again our expectations cannot be matched by actual evidence.[5]

ROMAN INFLUENCES: THE THEODOSIAN MAP AND THE RAVENNA COSMOGRAPHY

Despite the gaps in our knowledge, there are no grounds for believing in a hiatus, in the fifth and sixth centuries, between mapping in the late Roman Empire and map-

---

1. Robert Browning, *The Byzantine Empire* (New York: Charles Scribner's Sons, 1980), 7. Not until the western empire was overthrown in A.D. 476 can one consider Byzantium to have been acting entirely on its own.

2. At its greatest extent Byzantium not only retained the eastern provinces of the Roman Empire but also, as under Justinian (emperor 527–565), took over Italy, North Africa, and parts of Spain.

3. For the Greek and Latin manuscripts containing celestial maps of the ninth and tenth centuries, see volume 3 of the *History* and Paul Lemerle, *Le premier humanisme byzantin* (Paris: Presses Universitaires de France, 1971).

4. For Arab links with Byzantium see volume 2 of the present *History*.

5. In view of the large extant literature from the Byzantine period, including much of a philosophical and technological nature, we may hope that a detailed search for cartographic material (which has not hitherto been undertaken) would yield further references to the existence or use of maps.

ping as it would develop in the eastern empire. On the contrary, there was a conscious preservation of all things Roman. The Byzantines called themselves not Byzantines but *Romaioi* (Romans),[6] and they liked to see themselves as heirs of the Roman Empire. In cartography there were deliberate imitations of some of the maps of the earlier era, especially where these were perceived as fulfilling imperial purposes, such as the glorification of the greatness of Byzantium at a date when it was still possible to believe in the reconstitution of the Roman Empire as a whole.[7]

The map of the Byzantine Empire that was issued on the orders of Theodosius II (emperor of the East from A.D. 408 to 450) can be interpreted in this light. The map itself has not survived, but we know about it from the poem that was attached.[8] Although Greek was the common language of the eastern empire, this poem is in Latin hexameters, Latin being at the time the official language of both parts of the empire. In the original text, the date mentioned is the fifteenth *fasces* of Theodosius. This does not mean, as the Irish geographer Dicuil (fl. A.D. 814–25) thought, the fifteenth year of Theodosius's reign[9] but refers to his fifteenth consulship, A.D. 435. The poem may thus be rendered:

> This famous work—including all the world,
> Seas, mountains, rivers, harbors, straits and towns,
> Uncharted areas—so that all might know,
> Our famous, noble, pious Theodosius
> Most venerably ordered when the year
> Was opened by his fifteenth consulship.
> We servants of the emperor (as one wrote,
> The other painted), following the work
> Of ancient mappers, in not many months
> Revised and bettered theirs, within short space
> Embracing all the world. Your wisdom, sire,
> It was that taught us to achieve this task.[10]

Dicuil took these lines to indicate that two members of the imperial staff were instructed to travel around the empire. A more appropriate interpretation would be that the instructions were to edit and update a map and, perhaps, a commentary. The latter would have been almost certainly derived from Agrippa, always recognized during Byzantine times as the official source, rather than the works of Marinus or of Ptolemy. The use of Latin would point to it, as do the measurements in Roman miles. The map itself was still extant when Dicuil was writing in the early ninth century. Dicuil also noted that the map's authors calculated the length of Syria, from the borders of Asia Minor to Arabia and Lower Egypt, as 470 Roman miles.[11]

It is possible that Theodosius's map was exceptional and that other manuscripts that are linked with his name were not necessarily accompanied by maps even when they dealt with subjects where maps would have been appropriate. This was certainly the case with the anon-

ymous treatise *Urbs Constantinopolitana nova Roma*, dedicated to Theodosius II.[12] Nor is the more detailed description of Constantinople by one Marcellinus—which lists the fourteen districts of the city and its most important buildings—linked to any large-scale plan comparable to the *Forma Urbis Romae*.[13] Moreover, even in matters where the emperor dealt with the legal organization and codification of lands within the empire, cadastral surveys and mapping do not seem to have been undertaken, and certainly not in the manner recommended in the *Corpus Agrimensorum*. There is a written survey of property law, said to have been instituted by Theodosius II and to have dealt not only with a resurvey of the Nile valley but also with conditions in other provinces of the Roman Empire,[14] but again the surviving part of the text makes no mention of maps.

To extrapolate from such scraps of evidence, it is possible to suggest that while the Byzantine emperors retained maps for propaganda and (as will be seen) religious purposes, the many practical uses for mapping so characteristic of the western empire steadily declined. Such an interpretation is borne out by the periploi, books of sailing directions, which continued to lack accom-

6. Browning, *Byzantine Empire*, 8 (note 1).

7. A representation of an orb or globe in an imperial context is associated with the colossal statue of an emperor erected in Barletta and found in the sea off the town. Symbolizing Byzantine power in the West, it is two and a half to three times life size, with an orb in the emperor's hand, but there are no markings on the orb. It has traditionally been considered to be a likeness of Heraclius (emperor 610–41); but according to an alternative theory it is of Valentinian I (emperor 364–75). See *Enciclopedia italiana di scienze, lettere ed arti*, originally 36 vols. ([Rome]: Istituto Giovanni Treccani, 1929–39), 6:197, col. 2 and photo 196.

8. Emil Baehrens, ed., *Poetae Latini minores*, 5 vols. (Leipzig: Teubner, 1879–83; reprinted New York: Garland, 1979), 5:84; Wanda Wolska-Conus, "Deux contributions à l'histoire de la géographie: I. La diagnôsis Ptolémeenne; II. La 'Carte de Théodose II,'" in *Travaux et mémoires*, Centre de Recherche d'Histoire et Civilisation Byzantines, 5 (Paris: Editions E. de Baccard, 1973), 259–79.

9. Dicuil *De mensura orbis terrae* (On the measurement of the earth) 5.4; see *Liber de mensura orbis terrae*, ed. and trans. James J. Tierney, Scriptores Latini Hiberniae, no. 6 (Dublin: Dublin Institute for Advanced Studies, 1967); Tierney discusses Dicuil's errors of interpretation in his introduction, pp. 23–24.

10. Dicuil *De mensura* 5.4 (note 9), translation by O. A. W. Dilke.

11. Dicuil *De mensura* 2.4 (note 9).

12. In *Geographi Latini minores*, ed. Alexander Riese (Heilbronn, 1878; reprinted Hildesheim: Georg Olms, 1964), 133–39.

13. Marcellinus's cosmography was recommended by Cassiodorus *Institutiones divinarum et saecularium litterarum* 1.25.1; see *Institutiones*, ed. R. A. B. Mynors (Oxford: Clarendon Press, 1937), or, for an English translation, *An Introduction to Divine and Human Readings*, ed. and trans. Leslie Webber Jones (New York: Columbia University Press, 1946).

14. Text "Imperator Theodosius et Valentinianus," in *Die Schriften der römischen Feldmesser*, ed. Friedrich Blume et al., 2 vols. (Berlin: Georg Reimer, 1848–52; reprinted Hildesheim: G. Olms, 1967), 1:273–74.

panying maps as earlier in the Roman period. The term "portolan" or "portulan" has been somewhat misleadingly used for these by some modern scholars, but though eight periploi are known, none that are extant have maps.[15] An anonymous periplus of the Black Sea—again without maps—has also survived, and it is preceded by a summary measurement of the whole *oikoumene*.[16]

In the case of Byzantine land itineraries, too, there is no known graphic version, no *itinerarium pictum* comparable to the Peutinger map of the Roman period. The principal geographical listing of places to have survived is known as the Ravenna cosmography.[17] While this is clearly indebted to the earlier Roman models, the work perpetuates the written rather than the graphic form of such documents for travelers or other interested readers. It takes its name from what was the center of Byzantine power in Italy from A.D. 540 to 751. It is a list, in Latin, of some five thousand geographical names arranged in approximate topographical order, gathered from maps of most of the known world, the compiler proceeding roughly from west to east. It was not an official document of the Byzantine bureaucracy but was worked up by an unknown cleric (referred to now as the Ravenna cosmographer) for a fellow cleric called Odo, perhaps soon after the year 700. He gave his sources as Castorius (frequently quoted), Christian historians such as Orosius (fl. A.D. 414–17) and Jordanes (sixth century A.D.), and various Gothic writers, but he seems to have varied his method of compilation from region to region.[18]

Rather unsystematically, the Ravenna cosmographer apparently set out to list all the main places (*civitates*) of each area, together with some rivers and islands. There is no sign of methodical selection, so if the text contains a reference to an unknown place near a known place, there is only a fair chance that the unknown locality was in fact close by, whereas mention of an unknown place between two known ones can fairly safely be taken as an indication of its true position. The cosmographer noted: "We could, with the help of Christ, have written up the harbors and promontories of the whole world and the mileages between individual towns,"[19] a comment that suggests one of his sources might have been similar to the Peutinger map. This has thus led some modern writers to attempt to see Castorius (thought to have lived in the fourth century A.D.), as the maker of the Peutinger map.[20] However, it seems unlikely that the Ravenna cosmographer would have been so erratic in his ordering of place-names had his principal source contained roads as the Peutinger map does. It has also been claimed that a very corrupt form of Ptolemy's *Geography* was used for some parts of Asia and for the islands. One may suggest, though, that what looks like a corruption of Ptolemy's text may

have been a slightly less corrupt version of Marinus, on the assumption that Marinus's map was in fact available at Ravenna. Such questions are difficult to resolve, especially in view of the cosmographer's lack of method. This has often resulted either in omission of important places or in duplication of names, implying that the author was inexpert at reading map names in Greek. Sometimes he would give a contemporary regional name (e.g., Burgandia) in association with that of an ancient tribe (e.g., Allobroges).

However the sources of the Ravenna cosmography are interpreted, it is clear that a selection of Greek and Roman maps was available for consultation in Italy at this time in Byzantine history.[21] This listing, then, provides important evidence for the continuing use of maps, albeit in a very nontechnical way, even though the earlier impetus to produce new maps—or to revise older maps as new sources of information became available—no longer appears to have been given priority among scholars in early eighth-century Ravenna.

15. Armand Delatte, ed., *Les portulans grecs* (Liège: Bibliothèque de la Faculté de Philosophie et Lettres de l'Université de Liège, 1947); G. L. Huxley, "A Porphyrogenitan Portulan," *Greek, Roman and Byzantine Studies* 17 (1976): 295–300. See also pp. 237 and 383.

16. *De Ambitu Ponti Euxini*, in *Geographi Graeci minores*, 2 vols. and tabulae, ed. Karl Müller, (Paris: Firmin-Didot, 1855–56), 1:424–26. This defines the stade differently from the measurements that had been recognized in the western empire. On the matter of the modern value of the stade, see chap. 9 above, "The Growth of an Empirical Cartography in Hellenistic Greece," note 3.

17. Moritz Pinder and G. Parthey, eds., *Ravennatis anonymi Cosmographia et Guidonis Geographica* (Berlin: Fridericus Nicolaus, 1860; reprinted Aalen: Otto Zeller Verlagsbuchhandlung, 1962). The chief manuscripts are Urbinas Latinus 961 (Rome, Biblioteca Apostolica Vaticana), fourteenth century; MS. Lat. 4794 (Paris, Bibliothèque Nationale), thirteenth century; and Basiliensis F. V. 6 (Basel, Basel University Library), fourteenth to fifteenth century.

18. Louis Dillemann, "La carte routière de la *Cosmographie de Ravenne*," *Bonner Jahrbücher* 175 (1975): 165–70; Ute Schillinger-Häfele, "Beobachtungen zum Quellenproblem der *Kosmographie* von Ravenna," *Bonner Jahrbücher* 163 (1963): 238–51.

19. Translation by O. A. W. Dilke, *Ravenna cosmography* 1.18.10–15; see Pinder and Parthey, *Ravennatis anonymi Cosmographia*, 39 (note 17).

20. Konrad Miller, *Itineraria Romana* (Stuttgart: Strecker und Schröder, 1916), xxvii–xxix; J. Schnetz, *Untersuchungen über die Quellen der Kosmographie des anonymen Geographen von Ravenna*, Sitzungsberichte der Akademie der Wissenschaften, Philosophisch-historische Abteilung 6 (Munich: Verlag der Bayerischen Akademie der Wissenschaften, 1942).

21. For example, in Britain an otherwise unrecorded map of the Severan period seems to have been used. See Louis Dillemann, "Observations on Chapter V, 31, Britannia, in the Ravenna Cosmography," *Archaeologia* (1979): 61–73; A. L. F. Rivet and Colin Smith, *The Place-Names of Roman Britain* (Princeton: Princeton University Press, 1979), 185–215.

Religious Cartography: Cosmas Indicopleustes and the Map Mosaics

Christianity clearly distinguishes the Byzantine Empire from the preceding Roman Empire. This was to become both the state religion and that of the majority of its citizens. By the sixth century, Christian modes of thought and Christian imagery permeated the political, intellectual, and artistic life of the society and indeed gave it many of its characteristic qualities. The prominence of the church throughout the Byzantine world similarly imparted a religious tone to much of the cartography of the period. It is no accident that the principal surviving maps—those of Cosmas Indicopleustes and the mosaic maps at Nicopolis and Madaba—as much as the later *mappaemundi* of western Europe, reflected the superimposition of these new ideas on a classical foundation. The Nicopolis mosaic and some of the maps of Cosmas Indicopleustes are intelligible only in a religious context. Cosmas had traveled widely around the Red Sea and in adjacent areas, but he was much more interested in theology than in geography or cartography. The Madaba map also reveals its religious function: it was laid out in a church, and it gives great importance to Jerusalem, together with biblical quotations.

Cosmas Indicopleustes

During the sixth century, traditional teaching still flourished in the Byzantine world, but signs of decline were already appearing. On the one hand there was, at the end of the fifth century, the copying of Strabo's *Geography*; remains of this copy have been preserved in a palimpsest. Later, in southern Italy, the Christian Cassiodorus (ca. A.D. 487–583) advised young monks to learn geography and cosmography through Dionysius's map and Ptolemy's *Geography*.[22] In Alexandria, still functioning as the greatest intellectual center of the late Roman world, another Christian, Johannes Philoponus (ca. A.D. 490–570), commented on Aristotle's works and taught, like him, that the earth was spherical and lay in the center of the celestial sphere.

There were some, however, who considered Aristotelian and Hellenistic teaching about the universe to be in contradiction with the Scriptures. A polemic developed in which Cosmas (called Indicopleustes meaning, literally, "Indian sea traveler"; fl. A.D. 540), took an active part. Cosmas, a merchant in Alexandria, was a self-taught man; he had traveled much, though he probably did not go as far as India, as his nickname suggests. In Persia he had attended the lectures of Patrikios, a Christian teacher, and had been converted to Nestorian Christianity. Coming back to Alexandria, wishing to propagate what he considered the true Christian teaching, he composed a *Geography*, an *Astronomy* (both

now lost), and a *Christian Topography* in twelve books, of which three manuscripts have been preserved.[23]

Cosmas thought that the earth was flat and that the cosmos was shaped like a huge rectangular vaulted box. He sharply attacked "people from outside" (that is to say, infidels) who believed the world to be spherical and mocked their representations of the sky and earth. The *Christian Topography* is illustrated throughout with diagrams and paintings that are part of his demonstration of these beliefs. The manuscripts that have come down to us are thought to be fairly faithful to the original, so that we can take their illustrations as similar to the ones actually drawn by Cosmas himself; in general, however, his importance to the history of cosmography has been greatly exaggerated.[24]

In the *Christian Topography* the cosmos is represented schematically as a rectangular box, vaulted along its length. It is divided into two parts, an upper and a lower, by the firmament, which serves as a screen separating the two. The lower part represents the visible world, in which men and the angels live. The upper part represents the invisible world, the realm of God. Two diagrams illustrate this conception: on one only the narrow end of the box is drawn, with its semicircular top; on the other, the whole box is shown, in oblique perspective (fig. 15.1).

Another of the text's pictures presents Cosmas's concept of a flat earth. It is a map of the inhabited world drawn as a rectangle surrounded by an ocean with a rectangular frame (fig. 15.2). Four gulfs of the ocean break the regular outline of the inhabited world: the Caspian on the northern side; the Arabian (Red Sea) and Persian gulfs on the southern side; and the Mediterranean (called the Romaic Gulf) on the western side, the only major sea to be shown.[25] Beyond the narrow Asian

22. Cassiodorus *Institutiones* 1.25.2 (note 13).

23. Cosmas Indicopleustes *Topographie chrétienne*, ed. Wanda Wolska-Conus, in *Sources Chrétiennes*, nos. 141 (1968), 159 (1970), and 197 (1973); Wanda Wolska, *La Topographie chrétienne de Cosmas Indicopleustès: Théologie et science au VIᵉ siècle*, Bibliothèque Byzantine, Etudes 3 (Paris: Presses Universitaires de France, 1962). Cosmas's text is preserved in the Vat. Gr. 699, Rome, Biblioteca Apostolica Vaticana (ninth century), copied in Constantinople; the Sinaïticus Gr. 1186, Mount Sinai, Monastery of Saint Catherine (eleventh century), written in Cappadocia; and Plut. 9.28, Florence, Biblioteca Medicea Laurenziana (eleventh century), copied on Mount Athos.

24. See below, p. 348.

25. As in Eratosthenes' and Strabo's geographies, the Caspian is represented as a gulf of the ocean. Herodotus before them and Ptolemy afterward thought the Caspian was an enclosed sea. The same kind of representation of a rectangular earth, surrounded by the ocean and partly divided by a deeply penetrating gulf, was still used by twelfth-century illuminators. It illustrated the creation of the cosmos, Gen. 1:1–24, in Octateuch manuscripts (Seragliensis 8, fol. 32v, and Smyrnaeus A1, fol. 7v, both in the Sultan's Library, Istanbul). But the

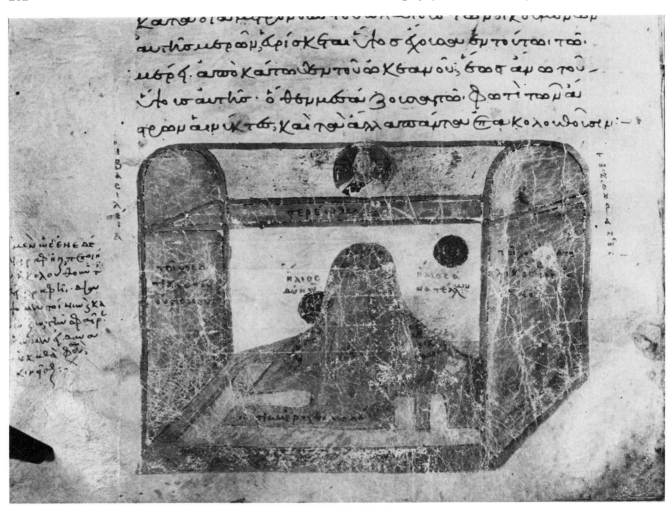

FIG. 15.1. THE UNIVERSE OF COSMAS INDICO-
PLEUSTES. The arched vault of heaven is represented above
a flat, rectangular earth, where the sun is shown both rising
and setting around the great mountain in the north. The firmament is at the meeting of the vault and the lower region.

Size of the original: 10.2 × 13 cm. By permission of the
Biblioteca Medicea Laurenziana, Florence (Plut. 9.28, fol.
95v).

side of the rectangular world, and beyond the ocean, a
small rectangle is drawn figuring paradise, blooming
with flowers and trees. Four rivers flow from paradise
into the inhabited world, passing under the ocean. One
of them, the Gihon (Nile), flows into the Romaic Gulf;
the others, the Tigris, the Euphrates, and the Pishon
(Indus), flow into the Persian Gulf. Surrounding the rectangular ocean is "the earth beyond the ocean." In the
upper part of this, men lived before the Flood. Taken
as a whole, this particular map of Cosmas's is a strange
mixture of classical and Hellenistic knowledge (the four
gulfs of the ocean, the length double the breadth, the
rectangular frame) and of biblical teaching (paradise and
its four rivers, the four corners of the earth, the earth
beyond the ocean). Cosmas attributes such a map to the

historian Ephorus (ca. 405–330 B.C.), and he may in
fact be adapting Ephorus's work.

In other illustrations, Cosmas presents his own concepts of the world, wishing to prove their validity. One
diagram shows four large men standing at right angles

drawing of the inhabited world (inhabited only by animals, since it
illustrates the creation on the sixth day), is far less accurate, with only
one gulf on the ocean. Wolska, *Topographie chrétienne de Cosmas*,
137–38 (note 23), thinks that both kinds of illustrations, in Cosmas
and in the Octateuch, are derived from an early prototype. Cynthia
Hahn, "The Creation of the Cosmos: Genesis Illustration in the Octateuch," *Cahiers Archéologiques* 28 (1979): 29–40, advances the opinion that "the geographical configuration of the Octateuch miniature
was derived from the *Topography*" (p. 35), the illustrator adding
animals and foliage to the map structure of the *Topography* miniature.
This would explain the degradation in the drawing of the map itself.

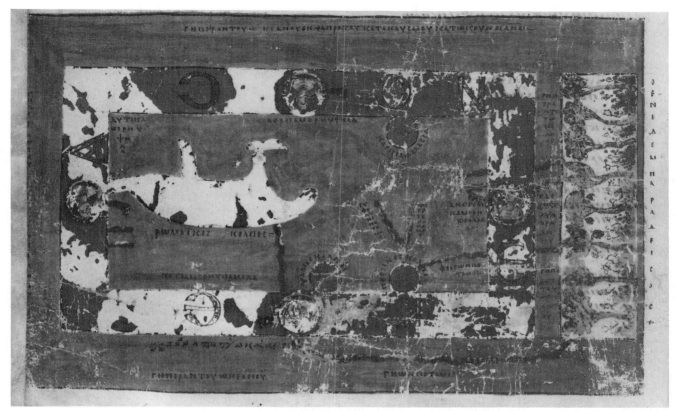

FIG. 15.2. THE WORLD ACCORDING TO COSMAS IN-
DICOPLEUSTES. In a mixture of Hellenistic and biblical geography, Cosmas envisaged a rectangular inhabited world surrounded by an ocean. To the east, beyond the ocean, is
paradise. Beyond the ocean at the top (south) is the uninhabited
world.

Size of the original: 23.3 × 31.5 cm. Photograph from the
Biblioteca Apostolica Vaticana, Rome (Vat. Gr. 699, fol. 40v).

to each other on a small round earth; this was enough,
in Cosmas's eyes, to demonstrate the stupidity of those
who judged the existence of the Antipodes possible, even
probable. In another diagram, the orbits of the planets
were drawn as circles with the earth as the central one;
on the upper circle, the twelve zodiacal signs are represented in a rough diagram.

For Cosmas, the drawing was part of the teaching.
But his drawing was of necessity sketchy; the mind is
left to transpose the diagram into the complex reality.
It seems as though Cosmas never thought of the transposition: a flat map meant a flat earth; biblical teaching
should be taken literally. Despite the assumed importance of Cosmas's teaching by modern writers of historical texts, however, the work appears to have had little
influence on medieval thought. The spherical world of
the Greeks (and the Romans, for that matter) was never
forgotten.[26]

THE MAP MOSAICS AT NICOPOLIS AND MADABA

Byzantine art was predominantly religious, and just as
maps furnished the theme of many Christian wall paintings in western Europe, so too in the Byzantine Empire
they were incorporated into mosaics. Together with frescoes, mosaics were the most magnificent expression of
Byzantine art, and this is clearly reflected in the cartography of the period.[27] Besides the primary examples of
Nicopolis and Madaba, a number of zodiac mosaics are
known from other locations. The function of the larger
of these maps—as with the *mappaemundi*—was no
doubt to instruct the faithful by presenting the allegories
of biblical lore.

26. See chapter 18 in this volume, "Medieval *Mappaemundi*," pp.
286–370.
27. On Byzantine mosaics see Otto Demus, *Byzantine Mosaic Decoration: Aspects of Monumental Art in Byzantium* (London: Routledge and Kegan Paul, 1948).

FIG. 15.3. THE NICOPOLIS MOSAIC. From room X, Basilica α (basilica of Saint Demetrius), Nicopolis, Greece, sixth century A.D.

Size of the original: 2.35 × 3.01 m. Photograph courtesy of Marie Spiro, University of Maryland, College Park.

An unusual early Byzantine mosaic may be seen in situ in the remains of the Byzantine church at Nicopolis, near Preveza in Epirus (fig. 15.3).[28] Nicopolis ("victory city"), which has the largest and most extensive Roman remains in Greece, was founded to commemorate the victory of Augustus over Antony and Cleopatra at the battle of Actium in 31 B.C. The mosaic at Nicopolis was evidently set up by an Archbishop Dometios (there were two of that name, both of whom seem to have flourished in the sixth century). The Greek verses incorporated in the mosaic may be rendered:

> Here you can see the boundless ocean run
> Carrying in its midst the earth, wherein
> All that can breathe and creep is here portrayed
> Using the skillful images of art.
> Noble archpriest Dometios founded this.[29]

The mosaic itself is rectangular (and thus reminiscent of Cosmas's concept of the world) and has around its sides an ocean with fish swimming in it. In the middle, in a rectangular centerpiece somewhat similar to that of the Mount Nebo mosaic,[30] are trees and birds. To say that without the inscription one would have thought it represented paradise is perhaps to prejudice the interpretation: the mosaic could represent the earth as it is, but it could also represent the earth before the creation

of man, when it was paradise.[31] In either case, biblical and pagan Greek concepts are mingled in this mosaic, the outer edges reflecting either the Homeric or subsequent representations of the ocean that surrounds the earth.[32]

The Madaba map—probably the best known example of Byzantine cartography—is a mosaic of Palestine and Lower Egypt with legends in Greek. The remaining parts are well preserved. It was rediscovered in 1884 in the old church at Madaba, Jordan (plate 7).[33] In 1896 Father

28. Ernst Kitzinger, "Studies on Late Antiquity and Early Byzantine Floor Mosaics. I. Mosaics at Nikopolis," *Dumbarton Oaks Papers* 6 (1951): 81–122, figs. 18–19.

29. Translation by O. A. W. Dilke; see Kitzinger, "Nikopolis," fig. 18 (note 28).

30. Sylvester J. Saller, *The Memorial of Moses on Mount Nebo*, 3 vols., Publications of the Studium Biblicum Franciscanum, no. 1 (Jerusalem: Franciscan Press, 1941–50), vol. 1, 230 ff., and vol. 2, pls. 103.1, 106, 107.

31. Kitzinger, "Nikopolis," 100 (note 28).

32. This concept was sometimes given expression in art. Thus it is thought by R. Hinks, *Myth and Allegory in Ancient Art* (London: Warburg Institute, 1939), 30 and pl. 1b, that a bronze disk in the British Museum represents Oceanus embracing the three continents.

33. Herbert Donner and Heinz Cüppers, *Die Mosaikkarte von Madeba*, Abhandlungen des Deutschen Palästinavereins (Wiesbaden: O.

Kleopas Koikylides, librarian to the Greek Orthodox patriarch, found that work on the new church was causing damage to the mosaic, which was then preserved, and he and others made drawings of it. In 1965 it was restored by Heinrich Brandt and handed over to the Greek Orthodox patriarch.

The mosaic is datable to between A.D. 542 and 565 (the year of Justinian's death). The main surviving fragment measures about 10.5 by 5 meters, but originally the entire mosaic could have accommodated a map as large as 24 by 6 meters. The mosaic squares that have been preserved from the map are in five or six shades of green and blue and four shades of red, with others of black, white, brown, violet, yellow, and gray. The map, thought to have shown the area from Byblos and Damascus to Mount Sinai and Thebes in Egypt, would have required over two million mosaic squares.

The Madaba map was clearly intended to instruct the faithful. It faced the part of the church that the laity would frequent, though they would have been looking at it through a screen. The portion that has been preserved extends from Aenon in the Jordan valley to the Canopic branch of the Nile. The scale varies from about 1:15,000 for central Judaea to about 1:1,600 for Jerusalem, which was given a prominent position and an exaggerated size. The map seems to have had a fairly straight Mediterranean coast so that, whereas it has east at the top in Palestine, in Egypt it has southeast or south at the top. Place-names are numerous considering the coarse nature of the medium. They are mostly based on the *Onomasticon* of Eusebius of Caesarea (ca. A.D. 260–340), but partly also, no doubt, on a road map, since road stations are given four and nine miles out of Jerusalem. In Egypt some villages are called by the names of Greek or Roman owners, such as "the [village] of Nicias."[34] Cartographic signs for towns or churches are somewhat similar to those in the Peutinger map.

Evidently the compiler was eager to fill all empty spaces either with pictures of boats or trees or with historical explanations. Some of these explanations come from the Old Testament, some from the New: examples are "Floor of Atath, now Bethagla";[35] "Salt Lake or Pitch Lake or Dead Sea"; "Selo, where the ark once was";[36] "Ephron or Ephraia, where the Lord went";[37] "Ailamon, where the moon stood still one day in the time of Joshua son of Nun";[38] "Rama: a voice was heard in Rama";[39] "the church of Saint Philip where they say the eunuch Candaces was baptized";[40] "borders of Egypt and Palestine"; "Zabulon shall dwell by the sea and its border shall be unto Sidon";[41] and "Sareptha, the long village where a child was brought back to life that day."[42]

Parts of the map reproduce architectural details. The church of Saint Zacharias near Bethzachar is promi-

nently outlined. It has a porch with sloping roof, surmounted by a facade with three windows. A semicircular court with columns, presumably at the rear, is shown in oblique perspective as if it were above the porch and facade. Jerusalem is portrayed in great detail as an oval walled city with its principal gateway in the north (plate 8). The central street is shown with its east colonnade the right way up but its west colonnade upside down. The western area is viewed from outside and the eastern area from inside. In the center foreground is the Church of the Holy Sepulcher. Like the west colonnade, it is shown upside down, with a staircase at the top and a dome at the base. Numerous other buildings of Byzantine Jerusalem have been identified from the mosaic. The latest to be so recognized, at the south end of the east colonnade, is thought to be the New Church (Nea) of the Theotokos, consecrated in 542. To the north of Jerusalem is the Damascus Gate. Its Arabic name, Bābil ʿamūd, commemorated a Byzantine column in the nearby square. This column, on which the statue of an emperor had been replaced by a cross, is depicted on the mosaic. Another city that is shown in some detail, though the mosaic is only partly preserved, is Gaza, with what looks like a Greek theater in the background.

The coloring in mountainous areas includes red, pink, green, and dark brown, with dark brown surrounds in roughly parallel curves. It is possible that some sort of form lines were intended, an effect that looked most realistic in a fragment of the mosaic now thought to be lost. A drawing of this fragment made in 1897 shows Agbaron (Akhbara), with a village to the left between two mountains, one of which has form lines or terracing.[43]

Harassowitz, 1977); Michael Avi-Yonah, *The Madaba Mosaic Map* (Jerusalem: Israel Exploration Society, 1954); R. T. O'Callaghan, "Madaba (Carte de)," in *Dictionnaire de la Bible: Supplement*, ed. L. Pirot and A. Robert (Paris: Letouzey et Ané, 1928–), vol. 5 (1957), 627–704.

34. This is the correct form, not Nicius as given in Avi-Yonah, *Madaba Mosaic* 76, no. 133 (note 33).

35. Accad in Gen. 10:10.

36. Shiloh in Josh. 18:1.

37. John 11:54: "Accordingly Jesus no longer went about publicly in Judaea, but left that region for the country bordering on the desert, and came to a town called Ephraim."

38. Avi-Yonah, *Madaba Mosaic* (note 33), points out that Aijalon, in Josh. 10:12, was actually at Yalu, three kilometers from Nicopolis in Palestine, as indicated by Jerome's translation of the *Onomasticon*; but the mapmaker follows Eusebius.

39. Ramah in Jer. 31:15.

40. Not so, he was a eunuch of Candace (Kandake), queen of Ethiopia: Acts 8:26–27.

41. Gen. 49:13: "Zebulun dwells by the seashore."

42. Elijah's miracle at Zarethan, now Sarafand, 1 Kings 7:8 ff.

43. Avi-Yonah, *Madaba Mosaic*, 76, fragment A (note 33).

The third type of mosaic map dating from the Byzantine period depicted the zodiac, perpetuating a subject already noted as popular during the Roman Empire (p. 248).[44] In the period after the Hammath mosaic, for example, there developed a Byzantine tradition of mosaics with Hebrew inscriptions in the interiors of synagogues (exteriors were not allowed to be decorated). At Beth-Alpha, Hefzibah, a well-preserved mosaic from the early sixth century A.D. portrays the sacrifice of Isaac and also the signs of the zodiac (fig. 15.4).[45] These are set out counterclockwise in a circle around the chariot of the sun. In the corners are personifications of the four seasons: spring (top left), summer, fall, and winter (counterclockwise). The artistic style is vigorous but naive. The sun-god (a Greco-Roman concept) appears as a head with *corona radiata*; the head and front legs of each of his four horses are shown, but not the rest of their bodies. Around him are a crescent moon and twenty-three stars, irregularly spaced. Or again, the same Byzantine style is manifest in the representation of the signs of the zodiac: Virgo, for example, is shown seated frontally, but her feet are turned sideways.

At Susiya, near Hebron, there was originally a similar zodiac mosaic alongside a picture showing Daniel in the lions' den. But since mosaics were disapproved of by the Jews as graven images, they were both removed. In other mosaics of the Byzantine period from the Holy Land, the zodiac is represented only by the names of its signs rather than by their graphic representations.

THE GREEK REVIVAL AND PTOLEMY'S *GEOGRAPHY*

Byzantine cartography is not easy to classify. Such maps as are known from the earlier part of the period—as is clear from the mixture of Greek, Roman, and Christian ideas in Cosmas and the Madaba map—may be regarded as a synthesis of all three traditions. Later, however, there developed a strand of cartographic representation in which Greek influence eventually came to predominate. While the Byzantines did not for most of their long history call themselves Hellenes, the dominant language and the dominant culture of the empire were always Greek.[46] By the ninth century, however, the conscious reactivation of Hellenic traditions—by restoring, for instance, the use of Atticizing literary Greek, virtually forgotten during the seventh and eighth centuries—was more actively taken up by scholars in the ruling class. In the tenth century and later, when this fashion had become an even greater preoccupation of literary and linguistic scholarship, many Greek texts were recovered and transcribed into new books, themselves an important factor in the revival of this aspect of classical culture.[47]

The short-term significance of this renaissance of Hellenic traditions by Byzantine scholars for mapping is not easy to assess. Though we may assume that Greek astronomical and geographical works were being recopied by the Byzantine scribes, or in some cases digested into encyclopedias,[48] there is little evidence that these activities quickly gave rise to a renewed interest in mapmaking per se. If anything, before Planudes' search for Ptolemy, to be described below, there seems to have been a continuing decline of practical knowledge about mapmaking. Apart from the scrap of information (even if it can be accepted as of Byzantine origin), about Agathodaimon's drawing a world map according to Ptolemy's instructions, also discussed below (pp. 271–72), the few Byzantine writers of geographical texts known to us exhibit scant interest in compiling maps. This was evidently the case with the ninth-century scholar Epiphanius of Jerusalem. Epiphanius wrote a guide to Syria, Jerusalem, and the holy places that has survived.[49] It is true also of Eustathius of Constantinople (died ca. 1194), who had already written a large number of scholarly works when he became metropolitan of Thessalonica in 1174–75.[50] These works include a commentary on the description of the world by Dionysius Periegetes. In this, quotations from works of earlier geographers, some no longer extant, are given together with citations from the complete *Ethnika* of Stephanus. Eustathius attempted to reconcile the two differing accounts of the Caspian. He agreed that one could indeed walk around it but suggested that perhaps it discharged into the northern ocean by means of an underground stream. India he

44. Michael Avi-Yonah, *Ancient Mosaics* (London: Cassell, 1975); Michael Avi-Yonah and Meyer Schapiro, eds., *Israel Ancient Mosaics*, UNESCO World Art Series, no. 14 (Greenwich, Conn.: New York Graphic Society, 1960). For signs of the zodiac in Hebrew manuscripts see Bezalel Narkiss, *Hebrew Illuminated Manuscripts* (Jerusalem: Encyclopaedia Judaica, 1969), 32–33.

45. Avi-Yonah, *Ancient Mosaics*, 56–59 (note 44); Avi-Yonah and Schapiro, *Israel Ancient Mosaics*, 7, 14, 25, pls. XI, XII (note 44); Eleazar L. Sukenik, *The Ancient Synagogue of Beth Alpha* (Jerusalem: University Press, 1932).

46. Browning, *Byzantine Empire*, 8 (note 1).

47. See L. D. Reynolds and N. G. Wilson, *Scribes and Scholars: A Guide to the Transmission of Greek and Latin Literature*, 2d ed. (Oxford: Clarendon Press, 1974), esp. 50–53. On the development of Byzantine scholarship, see John Edwin Sandys, *A Short History of Classical Scholarship from the Sixth Century B.C. to the Present Day* (Cambridge: Cambridge University Press, 1915), 92–110.

48. Fritz Saxl, "Illustrated Mediaeval Encyclopaedias. 1. The Classical Heritage; 2. The Christian Transformation," in his *Lectures*, 2 vols. (London: Warburg Institute, 1957), 1:228–54 (lectures delivered at the Warburg Institute, University of London, February 1939).

49. Herbert Hunger, *Die hochsprachliche profane Literatur der Byzantiner* (Munich: Beck, 1978–), 1:517, n. 47.

50. Eustathius *Commentary* (on Dionysius's *Periegesis*), in Müller, *Geographi Graeci minores*, 2:201–407, esp. 401 on the shape of India (note 16).

FIG. 15.4. THE BETH-ALPHA MOSAIC. An example of a tradition of Byzantine mosaics depicting the Zodiac and seasons with Hebrew inscriptions.
From Michael Avi-Yonah and Meyer Schapiro, eds., *Israel Ancient Mosaics*, UNESCO World Art Series, no. 14 (Greenwich, Conn.: New York Graphic Society, 1960), 7. Copyright UNESCO 1960; reproduced by permission of UNESCO.

described as having the shape of a rhombus, which he illustrated as in figure 15.5. In fact, this is merely a variation on the rhombus of Dionysius, where the Indus River is in the west rather than the southwest and the other places are adjusted accordingly. But no diagram exists of Dionysius's scheme, and however much or little Eustathius may have consulted maps, his coastline, albeit schematic, is more correct than Ptolemy's.

THE GREEK MANUSCRIPTS OF PTOLEMY'S *GEOGRAPHY*

Given such antecedents, the subsequent recognition by Byzantine scholars of the importance of the manuscripts of Ptolemy's *Geography* is both remarkable and of great significance in the development of mapping. No Greek manuscript of Ptolemy's *Geography* earlier than the thir-

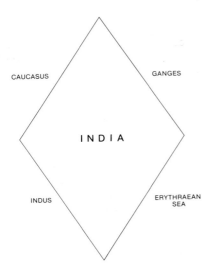

FIG. 15.5. INDIA ACCORDING TO EUSTATHIUS OF BY-
ZANTIUM. The representation of India as a rhombus owes
much to the ideas of Dionysius and, perhaps, Eratosthenes (see
fig. 9.6).

teenth century has been preserved (appendix 15.1).[51] In
that century Maximus Planudes (ca. 1260–1310) was
reviving scientific studies, collecting many precious manuscripts for the library in his monastery at Chora. Planudes was an active scholar:[52] he revised Aratus's poem
to relate it to Ptolemy's *Almagest*; he restored an old
copy of Strabo's *Geography* (the ninth-century MS. Gr.
1397, Paris, Bibliothèque Nationale) that had been
gnawed by rats; and he acquired another copy of Strabo
(the thirteenth-century MS. Gr. 1393, Paris, Bibliothèque Nationale) and gathered extracts from it. These
are preserved in the manuscript completed before his
death known as Plut. 69.30, Florence, Biblioteca Medicea Laurenziana.

In the summer of 1295—he tells us in a letter—Planudes was in search of a copy of Ptolemy's *Geography*, a
work long neglected. Certainly it was not until long after
the Arabs had been working on Ptolemy's text that Byzantine scholars gave serious attention to it. That it was
the sight of Arab maps from Ptolemy that stimulated
Planudes' interest can only be speculation.[53] It seems
unlikely that Planudes' thorough examination of Greek
mathematics and science—including his work on various
Greek texts and his advocacy of the introduction of
Arabic numerals, as well as his resurrection of Ptolemy—
was undertaken in isolation and without at least a small
circle of contemporary scholars interested in such studies.[54] What is perhaps most remarkable, anticipating the
teaching of Manuel Chrysoloras and the Italian scholars
in the fifteenth century, is Planudes' identification of the
Geography as a key text in geographical science.

In any event, Planudes seems to have found a copy of
the *Geography* rapidly. This may be surmised from the

fact that a Greek copy (Vat. Gr. 177, Rome, Biblioteca
Apostolica Vaticana), dated to the end of the thirteenth
century includes a note that points to Planudes, in Chora
monastery, as its former owner. "Heroic verses" added
to the manuscript describe Planudes' efforts to find this
work and his disappointment at encountering a copy in
which the maps were lacking.[55] This particular copy has
no maps, although the intended number is indicated in
the text: there should have been twenty-six. So Planudes
had maps drawn to accompany the text, and both text
and maps were admired by the emperor Andronicos II
Palaeologus (1282–1328), who had a copy made for
himself by Athanasius (formerly patriarch in Alexandria,
then retired in Constantinople, 1293–1308).

This in itself is a pointer to the level of cartographic
understanding in thirteenth-century Constantinople.
Planudes was not alone in being able to have a set of
maps drawn to order. Evidently there were draftsmen
in Constantinople sufficiently skilled to follow Ptolemy's
instructions for mapmaking, possessed of a technical
understanding of the relevant map projections, able to
translate coordinates into map images, and able to employ cartographic signs in the fashion of the maps in the
Greek manuscripts (described below).

A precious thirteenth- or fourteenth-century manuscript, the Vat. Gr. 191, in the Biblioteca Apostolica
Vaticana, Rome, is generally held to be evidence for
scholarly activity in Constantinople during Andronicos's
reign. It contains various astronomical, astrological, and
geographical texts. Among these are the collection

51. The reception of Ptolemy's manuscripts in western Europe, their
impact on cartography, and the development of celestial maps from
his materials will be dealt with in volume 3 of the *History*. Maps were
also sometimes drawn to accompany the copies of the *Geography*
transmitted to the Arab world. We know, for example, through the
Arab historian al-Masūdī, who lived in the tenth century, that at least
one copy of Ptolemy's *Geography* existed and was richly illustrated
with colored maps showing towns, rivers, seas, and mountains: see
Marcel Destombes, ed., *Mappemondes A.D. 1200–1500: Catalogue
préparé par la Commission des Cartes Anciennes de l'Union Géographique Internationale* (Amsterdam: N. Israel, 1964); and al-Masūdī, *Les prairies d'or*, 9 vols., trans. C. Barbier de Meynard and Pavet
de Courteille, Société Asiatique Collection d'Ouvrages Orientaux
(Paris: Imprimerie Impériale, 1861–1917), 1:76–77.

52. On Planudes' activity see Carl Wendel, "Planudes, Maximos,"
in *Paulys Realencyclopädie der classischen Altertumswissenschaft*, ed.
August Pauly, Georg Wissowa, et al. (Stuttgart: J. B. Metzler,
1894–), 20.2 (1950): cols. 2202–53. See also note 54 below.

53. Armando Cortesão, *History of Portuguese Cartography*, 2 vols.
(Coimbra: Junta de Investigações do Ultramar-Lisboa, 1969–71),
1:179–85.

54. "The reign of Andronicus II Palaeologus (1282–1328) has the
paradoxical distinction of being the period in which the signs of in-
curable political and economical weakness in the empire became unmistakable and yet the level of cultural life rose to a height as great
as had ever been seen." N. G. Wilson, *Scholars of Byzantium* (London:
Duckworth, 1983), 229–41, quotation on 229.

55. See MS. Gr. 43 (Milan, Biblioteca Ambrosiana).

known as the *Little Astronomy*, the *Commentary on Eudoxus's and Aratus's Phaenomena* by Hipparchus, and also Ptolemy's *Geography*. Once again the *Geography* lacks maps, but a note following the description of the last map reads: "There are twenty-seven maps instead of twenty-six in this copy, since the tenth map of Europe has been divided into two parts representing (1) Macedonia, (2) Epirus, Achaea, the Peloponnese, Crete, and Euboea."[56] We may infer from this that the archetype of the Vat. Gr. 191 was accompanied by maps.

THE MANUSCRIPTS WITH TWENTY-SEVEN MAPS: THE A RECENSION[57]

From allusions described above, it is clear that by the late thirteenth and early fourteenth centuries in Constantinople there were circulating, and being actively copied, a number of versions of Ptolemy's *Geography*. Some were without maps.[58] But the three oldest surviving manuscripts with maps of Ptolemy's *Geography* also date from the end of the thirteenth century. These three are: the Urbinas Graecus 82 in the Biblioteca Apostolica Vaticana, Rome; the Seragliensis 57 in the Sultan's Library, Istanbul; and the Fragmentum Fabricianum Graecum 23 in the Universitetsbiblioteket, Copenhagen (of which only a bifolium remains). Paul Schnabel pointed out that these three manuscripts are similar in size, number of lines to a page, and arrangement of the maps.[59] Diller has added that of the two hands in the Seragliensis 57, one is identical to that of the Fragmentum Fabricianum Graecum 23 and the other to that of the Paris MS. Gr. 1393 (the copy of Strabo owned by Planudes).[60] Diller also noted that certain special features common to the three manuscripts are very clear in the Urbinas Graecus 82, less so in the Seragliensis 57, and rather vague in the Fragmentum Fabricianum Graecum 23. On this basis he concluded that the three manuscripts had been copied in that order, as if, moreover, by one (perhaps Planudes') impulse. Diller speculated, plausibly, that the Urbinas Graecus 82 was perhaps copied for the emperor and the Seragliensis 57 for Planudes' personal use. A supplementary argument in favor of Diller's hypothesis could be that the Urbinas Graecus 82 and the Seragliensis 57, in spite of their similarity, have at least one major difference. In the case of the Urbinas Graecus 82, the map of the inhabited world is drawn on Ptolemy's first projection with straight converging meridians; in the Seragliensis 57 it is drawn on his second projection, whose curved meridians were much more difficult to execute. This could be taken to imply that the former was drawn first and that Planudes had the more advanced projection executed for his personal use.

The Urbinas Graecus 82 is a beautiful copy of Ptolemy's text. It is large (570 × 415 mm), written in two

columns (fifty-seven lines to a page), and has frequent marginal notes, many of which refer to Strabo. At the end of book 7, the comprehensive map of the inhabited world is drawn on two pages (fols. 60v–61r); this is followed (from book 8, fols. 3–28) by the twenty-six regional maps. In the first projection, the inhabited world is surrounded by winds, drawn as human faces blowing a kind of horn, and by the signs of the zodiac within red circles. The network of curved parallels and straight meridians is carefully drawn. Mountains are represented as colored straight features that, in the Asian part, indicate the limits between neighboring countries. Rivers are drawn as undulating blue lines; oceans are also colored blue (plate 9).

The regional maps in Urbinas Graecus 82 are drawn on Marinus's rectangular projection. Meridians and parallels form a red network. The meridians are drawn every five degrees (each degree is divided into halves in the external frame). Parallels are spaced at intervals of a quarter of an hour of maximum daylight (but degrees of latitude are numbered in the external frame). Names of towns are inscribed inside red vignettes, which are represented with or without towers according to the importance of the town. Outlines of coasts and courses of rivers are rather approximate and only schematically drawn. As in the other manuscripts of that type and as in Ptolemy's original specification, there are thus twenty-seven maps in all in the Urbinas Graecus 82: the world

56. Translation by O. A. W. Dilke from the Vat. Gr. 191 (Rome, Biblioteca Apostolica Vaticana). In fact, from book 5 to the end, coordinates are lacking for all towns and geographical places.

57. Ptolemy, *Claudii Ptolemaei Geographiae Codex Urbinas Graecus 82*, 2 vols. in 4, ed. Joseph Fischer, Codices e Vaticanis Selecti Quam Simillime Expressi, vol. 19 (Leiden: E. J. Brill; Leipzig: O. Harrassowitz, 1932), 1:208–415; Erich Polaschek, "Ptolemaios als Geograph," in *Realencyclopädie*, suppl. 10 (1965): cols. 680–833, esp. 734–53 (note 52); Leo Bagrow, "The Origin of Ptolemy's Geographia," *Geografiska Annaler* 27 (1945): 318–87.

58. In addition, the fourteenth-century Vat. Gr. 176 (Rome, Biblioteca Apostolica Vaticana) contains a *diorthosis* of Ptolemy's *Geography* made by Nicephoros Gregoras (ca. 1295–1360), who was an active scholar in Constantinople. In fact this revision concerns only book 1, the first chapter of book 2, and from book 7, chapter 5, to book 8, chapter 2, the only ones copied in this manuscript. Rodolphe J. Guilland, *Essai sur Nicéphore Grégoras* (Paris: P. Geuthner, 1926), 27b, argues that this revision was not carried out by Gregoras, but the manuscript note says it was. There is also a scholium written by the monk Isaac Argyros about the first system of projection proposed by Ptolemy. The manuscript ends with Ptolemy's *Harmonics*, which Nicephoros Gregoras had undertaken to revise, but death intervened.

59. Paul Schnabel, *Text und Karten des Ptolemäus*, Quellen und Forschungen zur Geschichte der Geographie und Völkerkunde 2 (Leipzig: K. F. Koehlers Antiquarium, 1938).

60. Aubrey Diller, "The Oldest Manuscripts of Ptolemaic Maps," *Transactions of the American Philological Association* 71 (1940): 62–67. See also Fischer, *Urbinas Graecus 82* (note 57).

map, ten maps for Europe, four for Libya (Africa), and twelve for Asia.[61]

THE MANUSCRIPTS WITH SIXTY-FIVE MAPS: THE B RECENSION

A second group of the manuscripts of Ptolemy's *Geography*, known to Fischer as the "B Redaction," are distinguished by containing sixty-five maps, distributed throughout the text, instead of the twenty-seven maps of manuscripts of the A recension. The interest of the B manuscripts, in which the format and arrangement of the maps depart from Ptolemy's original scheme for an atlas, lies in the fact that they point to different stemmata for the descent of Ptolemy's cartographic ideas and raw materials from the fourth century A.D. into the Middle Ages. Indeed, so different are the A and B recensions that their divergences can hardly be accounted for by textual corruption. Leaving aside minor variations between the individual Greek manuscripts, some scholars have been led to ask whether Ptolemy might not at some point have modified his original instructions; whether the extant manuscripts result from additions made cumulatively or sporadically after his death; and most radically, whether Ptolemy may properly even be regarded as the sole author of these variant manuscripts surviving from the late Byzantine period.[62] Such questions remain largely unresolved through lack of evidence. Yet there is no doubt that by the early fourteenth century, if not earlier, the separate stemmata epitomized in the A and B recensions were well established.

Of the manuscripts of the B recensions, the Plut. 28.49 in the Biblioteca Medicea Laurenziana, Florence, dating from the beginning of the fourteenth century, is the earliest. It is a parchment manuscript, smaller (340 × 260 mm) than the first three of the A recension.[63] It was carefully written in two columns, with forty-nine lines to a page. Maps are provided from book 2 to book 7. There are thirteen in book 2 and twelve in book 3, making twenty-five all together for Europe; eight in book 4, dealing with Africa; thirteen in book 5, fifteen in book 6, three in book 7 (making a total of thirty-one for Asia). Following these sixty-four regional maps, a comprehensive map of the inhabited world is drawn on Ptolemy's first projection, similar to the one in the Urbinas Graecus 82. Curiously at odds with actual numbers of maps included, the summary relates to four of the books of the *Geography* and refers to the traditional number of maps in the A recension (not the B recension).

The regional maps in the Plut. 28.49 are rectangular, of various sizes, and each is included in a graduated frame. Towns are indicated by vignettes, sometimes with a cross on top (for Iouernis and Rhaiba, for instance, in Ierne island—Ireland), and sometimes with one or several towers (three for London). Marginal signs in the text (indicating tribes) are reproduced as part of the settlement sign on the relevant maps. Coasts are drawn as a single blue line; mountains are long brown rectangles. When the maps are out of their regular order (which happens a few times), the copyist has indicated where to find them. The map of the inhabited world (fols. 98v–99r) follows the last regional one, the map of Taprobane (Sri Lanka). The end of book 7 is lacking. Book 8 begins with the next folio, but apparently is in a different hand and less carefully copied.

This Florentine manuscript is the oldest of a series dating from the fourteenth and fifteenth centuries, which can display significant variations in their maps. One of this family, the Burney Gr. 111, now in the British Library, London, and dated to the end of the fourteenth century, has the same sixty-five maps.[64] But in the Biblioteca Ambrosiana's MS. Gr. 997, Milan (dated by Schnabel to the fourteenth century) the sixty-five usual maps are followed by four modern maps, one showing Europe, one Africa, and two Asia (northern and southern Asia). The Seragliensis 27 in the Sultan's Library, Istanbul (late fourteenth-early fifteenth centuries) is a copy of the MS. Gr. 997, except that two maps are lacking, the Peloponnese map and the map of the inhabited world. Yet another copy, the Urbinas Gr. 83 in

61. For example, the *Vatopedi 655*, preserved in Mount Athos and elsewhere, was copied from the Urbinas Graecus 82 at the beginning of the fourteenth century, in Constantinople. Ptolemy's *Geography* (with its maps, as in the Urbinas Graecus 82) is followed by the corpus of minor geographers, copied from the tenth-century Pal. 398, in the Universitätsbibliothek, Heidelberg, and by Strabo's *Geography*. See *Géographie de Ptolémée* (reproduction of the Vatopedi manuscript), ed. Victor Langlois (Paris: Firmin-Didot, 1867); Fischer, *Urbinas Graecus 82*, 1:234–43 (note 57); Aubrey Diller, "The Vatopedi Manuscript of Ptolemy and Strabo," *American Journal of Philology* 58 (1937): 174–84. The Ptolemy map of the inhabited world (from the Vatopedi manuscript), drawn on his first projection, is incorporated in a London manuscript, British Library Add. MS. 19391. A penciled note by Leo Bagrow in the British Library Catalogue states that leaves of this manuscript are also found in Leningrad: "Einige Blätter von diese[m] MS. auch in Petersburg im Museum d. Gesellschaft Lyubetelei Drevney Pismennosti i Pechati."

62. The line was most strongly argued by Bagrow, "Origin" (note 57).

63. The need to accommodate the maps to this smaller format may have been one reason for their fragmentation in the B recension.

64. Joseph Fischer, "Der Codex Burneyanus Graecus 111," in *75 Jahre Stella Matutina*, 3 vols., Festschrift (Feldkirch: Selbstverlag Stella Matutina, 1931), 1:151–59. In fact the Burney Gr. 111 contains sixty-six maps. These are listed in British Museum, *Catalogue of the Manuscript Maps, Charts, and Plans, and of the Topographical Drawings in the British Museum*, 3 vols. (London: Trustees of the British Museum, 1844–61), vol. 1, though some pages are given incorrectly. Burney Gr. 111 contains two maps of Taprobane Insula. The first forms the frontispiece, and it is obvious from differences not only of the paint itself but also of presentation (the sea, for example, is fully colored whereas on all other maps only the coastline is outlined in a brighter blue) that this is not one of the original series.

the Biblioteca Apostolica Vaticana, Rome (dated to the middle of the fifteenth century), has the sixty-four regional maps and the four modern ones, but the space left vacant for the map of the inhabited world has not been filled. Such variations taken all together, between manuscripts that have the same basic genealogy, serve to confirm the inconsistencies and errors inherent in the transmission process of a manuscript age. If such difficulties could emerge in the century or so after the compilation of the Plut. 28.49, then the scope for similar changes in the much longer period between the death of Ptolemy and late Byzantine times was potentially much greater.

THE AGATHODAIMON ENIGMA

The difficulty of reconstructing the precise path of Ptolemy's texts as transmitted through the early Middle Ages is further increased by the relatively few surviving fragments of evidence that point to the modification of the manuscripts in this period. In particular, there is tantalizingly little evidence that will illuminate the crucial process of the descent and modification of the maps—and their addition or otherwise to the texts—in this long period.[65] Thus the proper interpretation of what at first sight might seem to be a relatively minor scrap of evidence—the so-called Agathodaimon endorsement on some Ptolemy manuscripts—becomes an important part of our reconstruction.

The endorsement takes the form of a note appended to book 8 of the *Geography* in some manuscripts. It may be rendered: "Agathos Daimon" (or "I, Agathos Daimon"), "a technician [*mechanikos*] of Alexandria, drew [have drawn] the whole world from Ptolemy's *Geography*."[66] This statement is found at the end of a number of copies of Ptolemy's *Geography*. It occurs, for instance, in the thirteenth-century Vat. Gr. 177 (which has no maps); in the thirteenth-century Urbinas Gr. 82 (which has twenty-seven maps); in the fifteenth-century Conv. Sopp. 626 [Abbatiae] 2380 in the Biblioteca Medicea Laurenziana, Florence (twenty-seven maps); and in the fifteenth-century MS. Gr. 1402 in the Bibliothèque Nationale, Paris (twenty-seven maps). In the fifteenth-century Paris manuscript, MS. Gr. 1401, for which Urbinas Graecus 82 was used as a model, a variant slips into the note: the wording is "has drawn" instead of "I have drawn." The main manuscript of the B recension, Plut. 28.49, has the same note at the end of book 8 but uses another form of the verb, though also in the third-person singular. It is generally held that Agathos Daimon ("good spirit") should be taken as a variant on Agathodaimon (the name of some real technician) and not as a humorous appellation.[67]

That being the case, three questions may be asked: Are the maps that we now see in the surviving manu-

scripts copies of maps drawn previously by Agathodaimon? When did Agathodaimon live? Did he make all the regional maps or only the map of the inhabited world? The answer to the third question seems the most straightforward. It is that Agathodaimon, described in the note on the manuscripts as *mechanikos*, drew only the world map. It was only this map, drawn on a projection like Ptolemy's first, that required the geometric knowledge and drafting skill that may have been possessed by such a "technician," albeit Agathodaimon may in fact have contented himself with drawing the network of parallels and meridians just according to Ptolemy's instructions.[68]

The other two questions can be answered only by conjecture. As regards the second, Agathodaimon could have lived at any time between Ptolemy's lifetime and the thirteenth century. It was not uncommon for scribes of the postclassical period to continue copying out notes of this nature as if they were part of the original manuscript. Some guidance may be inferred from the fact that he was an Alexandrian and that Alexandria did not have Greek technicians after the sixth century. Thus it would be reasonable to speculate that Agathodaimon might have been alive in the fifth or sixth century.[69] If so, he could have been, perhaps, a contemporary of Cosmas Indicopleustes, author of a number of small maps of the inhabited world, and of Johannes Philoponus, a good graphic technician as well as a commentator on Aristotle's work. The suggestion of a fairly early date for Agathodaimon is reinforced by the fact that the note in question appears in manuscripts belonging to different families, indicating that it had been inserted into an early copy of the *Geography*, before the splitting

65. Erich Polaschek, "Ptolemy's *Geography* in a New Light," *Imago Mundi* 14 (1959): 17–37, citing evidence from the Greek geographer Marcianus of Heraclea (ca. A.D. 400), who has already been noted as revising the eastern half of the map, suggests that "before the second half of the 4th century alterations were being made in the geographical work of Ptolemy" (p. 37).

66. Translation by O. A. W. Dilke. Two of several manuscripts containing Agathodaimon's note are the Paris MSS. Gr. 1401 and 1402 (see appendix 15.1).

67. Joseph Fischer, "Ptolemaeus und Agathodämon," *Kaiserliche Akademie der Wissenschaften in Wien*, Philosophisch-Historische Klasse, 59 (1916): 3–25; Polaschek, "Ptolemaios" (note 57).

68. There is a striking example of the plotting of the network of parallels and meridians before the drawing of the map, in MS. Gr. 1402 (Paris, Bibliothèque Nationale): the network is complete, but not even half of the map is drawn. It is clear that the general map of the inhabited world required a particularly skillful draftsman; it was often executed after the regional maps and by a different specialist. Likewise, in the Urbinas Graecus 82 the map of the inhabited world is the only one lacking, though it was situated between two sets of existing maps.

69. Bagrow, "Origin," 350 ff. (note 57), seems to ignore the fact that the whole of Greek scientific writing in Alexandria came to an end in the sixth century. For a more positive approach, see Polaschek, "Ptolemy's *Geography*," 17–37 (note 65).

of the stemma, and one that had been provided with a comprehensive map by this Agathodaimon.

The first question, whether the maps that have come down to us were copied from Agathodaimon's world map (if it existed), raises different problems. It is doubtful, in fact, that such a direct link can be accepted, even were the existing world maps to be, as they probably are, very similar to that planned by Ptolemy and drawn by Agathodaimon or some other "technician." It can be argued that Ptolemy's intention was to provide anyone with a means of drawing a special map or, rather, two types of maps. He knew that one projection was easier to execute than the other and that mapmakers would choose the easier one. Planudes, who proved to be such an enthusiastic scholar and so interested in geography, seems exactly the kind of man to order the maps remade according to Ptolemy's various instructions. Hence the different mapping for the inhabited world in the Urbinas Graecus 82 and in the Seragliensis 57; hence too, perhaps, the different arrangement for the regional maps in recension A and recension B. If such an interpretation is acceptable, and without prejudicing the suggestion that Ptolemy may also have made maps in his own day, it reinforces the conclusion that the Byzantine maps as a whole group are not so much the lineal descendants of some Ptolemaic prototype as the creation of scholars in a society also interested in maps, stimulated by the availability of Ptolemy's instructions for their manufacture.

By the thirteenth century, and by way of the Byzantine Empire, Ptolemy's *Geography* had thus rejoined the mainstream of cartographic history in Europe. A century later it was to be translated into Latin in Italy, where its importance was appreciated by scholars and where the many copies so characteristic of the Italian Renaissance would start to proliferate. The rediscovery of Ptolemy may be regarded as the seminal contribution of Byzantine scholarship to the long-term development of mapmaking.

APPENDIX 15.1 GREEK MANUSCRIPTS OF PTOLEMY'S *GEOGRAPHY*

Repository and Collection Number	Date	Description (number of leaves refers to whole manuscript)
Bologna, Ecclesiae S. Salvatoris 305 (present whereabouts unknown)	1528	(C) 33 × 23 cm, no maps
Chicago, Newberry Library, Ayer Collection, Ayer MS. 743	Fifteenth century	40 × 28 cm, 144 leaves
Copenhagen, Universitetsbiblioteket, Fragmentum Fabricianum Graecum 23	Thirteenth century	(F, C) Fragment with 3 maps, 56.5 × 42.5 cm
El Escorial, Real Biblioteca, Gr. Ω I.1	Sixteenth century	(M) 212 leaves
Florence, Biblioteca Medicea Laurenziana, Plut. 28.49	Fourteenth century	33.5 × 26.0 cm, 113 leaves, 64 maps and a world map; (C) 34 × 27 cm, 64 maps
Florence, Biblioteca Medicea Laurenziana, Plut. 28.42	1445	29 × 21 cm, 147 leaves
Florence, Biblioteca Medicea Laurenziana, Plut. 28.9	Fifteenth century	(C) 29 × 22 cm, 132 leaves, no maps
Florence, Biblioteca Medicea Laurenziana, Plut. 28.38	Fifteenth century	(M) 24 × 16 cm, 177 leaves
Florence, Biblioteca Medicea Laurenziana, Conv. Sopp. 626 (Abbatiae 2380)	Fifteenth century	(F) 59.5 × 44.0 cm, 104 leaves, 27 maps
Istanbul, Sultan's Library, Seragliensis 57	Thirteenth century	(F) Reference A10
Istanbul, Sultan's Library, Seragliensis 27	Late fourteenth/ early fifteenth century	(F) Reference B5; (C) 41 × 29 cm, 88 leaves, books 2.3–8
Leiden, Bibliotheek der Rijksuniversiteit, Voss. Gr. F. 1	Sixteenth century	54.5 × 42.5 cm, maps only
London, British Library, Burney Gr. 111	Late fourteenth/ early fifteenth century (F); fifteenth century (C)	43.1 × 30.4 cm, 115 leaves, 66 maps
London, British Library, Add. MS. 19391. Originally in Mount Athos, Vatopedi (see below)	Fifteenth century; fourteenth century (M)	(F) 34.5 × 25.5 cm, 8 leaves; (M) books 7–8, chap. 4

Repository and Collection Number	Date	Description (number of leaves refers to whole manuscript)
Milan, Biblioteca Ambrosiana, Codex D 527 inf.	Fourteenth or fifteenth century (F); thirteenth century (M)	(F) 40.3 × 28.5 cm, 101 leaves, 69 maps
Milan, Biblioteca Ambrosiana, Codex N 289	Fifteenth century (M); fifteenth and sixteenth centuries (C)	(M) 27 × 16 cm, books 1–7, chaps. 1 and 2
Mount Athos, Vatopedi 9 (655)	Fifteenth century (F); late thirteenth/early fourteenth century (C)	(F) Reference A2, 58 leaves, see London, British Library, Add. MS. 19391; some leaves also in Leningrad. Facsimile by V. Langlois, *Géographie de Ptolémée* (Paris, 1867). Probably a copy of Rome, Vatican, Urbinas Graecus 82
Oxford, Bodleian Library, 3376(46)-Qu. Catal. i(Greek), Cod. Seld. 41	Fourteenth century	(M) 274 leaves; 25 × 16 cm, books 1–8, chap. 28
Oxford, Bodleian Library, 3375(45)-Qu. Catal. i(Greek), Cod. Seld. 40	1482	176 leaves; (M) 29 × 18 cm
Oxford, Bodleian Library, Laud. 52	1568	(M) 32 × 21 cm, 83 leaves, books 1–7, chap. 5, and book 8, chaps. 1 and 2
Paris, Bibliothèque Nationale, MS. Gr. 2423	Thirteenth century	(M) 158 leaves; 25 × 16 cm, book 1, chaps. 7–24, book 2, chap. 6, beginning of book 3
Paris, Bibliothèque Nationale, MS. Gr. 2399	Thirteenth/fourteenth centuries	(M) 122 leaves; extracts from book 8
Paris, Bibliothèque Nationale, Gr. Suppl. 119	Fourteenth century	(M) 232 leaves; 20 × 14 cm
Paris, Bibliothèque Nationale, Coislin 337	Fourteenth/fifteenth centuries	(M) 278 leaves; 23 × 14 cm, books 1–8, chap. 28
Paris, Bibliothèque Nationale, MS. Gr. 1407	1438	(M) 215 leaves; 27 × 19 cm, parts of book 8
Paris, Bibliothèque Nationale, MS. Gr. 2027	1449	(M) 235 leaves; 20 × 14 cm, book 1, chap. 22 and part of chap. 23
Paris, Bibliothèque Nationale, MS. Gr. 1401	Fifteenth century	(F) 101 leaves; 58.8 × 43.7 cm, 27 maps
Paris, Bibliothèque Nationale, MS. Gr. 1402	Fifteenth century	(F) 72 leaves, 59.3 × 44.5 cm, 5 unfinished maps; (M) books 1–7 and chapters of book 8
Paris, Bibliothèque Nationale, MS. Gr. 1403	Fifteenth century	(M) 225 leaves; 28 × 19 cm, books 1–8, chap. 27
Paris, Bibliothèque Nationale, MS. Gr. 1404	Fifteenth century	175 leaves; 30 × 22 cm, books 1–7, chap. 5
Paris, Bibliothèque Nationale Coislin 173	Fifteenth century	311 leaves, books 1, 2, and 8
Paris, Bibliothèque Nationale, Gr. Suppl. 673	Fifteenth century	28 leaves; fragments of books 1 and 7
Paris, Bibliothèque Nationale, MS. Gr. 1411	Fifteenth or sixteenth century	(M) 585 leaves, 20 × 13 cm, copy of MS. Gr. 1407
Rome, Biblioteca Apostolica Vaticana, Vat. Gr. 191	Late twelfth or early thirteenth century	23.7 × 36.5 cm, 397 leaves
Rome, Biblioteca Apostolica Vaticana, Urbinas Graecus 82	Thirteenth century; ca. twelfth century (F); eleventh century (C)	(F) 57.5 × 41.8 cm, 111 leaves, 27 maps; facsimile by J. Fischer
Rome, Biblioteca Apostolica Vaticana, Vat. Gr. 177	Thirteenth century; fourteenth century (C)	24 × 16 cm, 240 leaves; (M) books 1–7, chap. 4, and parts of book 8
Rome, Biblioteca Apostolica Vaticana, Vat. Gr. 178	Fourteenth century	18.3 × 12.5 cm, 216 leaves; (M) books 1–7, chap. 4, and parts of book 8
Rome, Biblioteca Apostolica Vaticana, Vat. Gr. 176	Fourteenth century	27.0 × 17.3 cm, 193 leaves; (M) books 1–7, chap. 5, book 8, chaps. 1 and 2

APPENDIX 15.1—*continued*

Repository and Collection Number	Date	Description (number of leaves refers to whole manuscript)
Rome, Biblioteca Apostolica Vaticana, Vat. Gr. 193	Early fifteenth century	29.2 × 20.0 cm, 181 leaves; (M) (M) books 1–2.1, chap. 5, book 8, chaps. 1 and 2
Rome, Biblioteca Apostolica Vaticana, Pal. Gr. 388	Early fifteenth century	(M) 36 × 24 cm, 280 leaves
Rome, Biblioteca Apostolica Vaticana, Barberinianus 163	Fifteenth century	24.7 × 15.6 cm, 233 pages
Rome, Biblioteca Apostolica Vaticana, Barberinianus 128	Sixteenth century	(M) 19 × 15 cm
Rome, Biblioteca Apostolica Vaticana, Pal. Gr. 261	Fifteenth century	[Excerpta ex initio Ptolemaei Geographiae]
Rome, Biblioteca Apostolica Vaticana, Urbinas Graecus 83	Fifteenth century	42 × 29 cm, 118 leaves (C) 65 maps; (F) 69 maps; copy of Milan, Biblioteca Ambrosiana, Codex D 527 inf.
Rome, Biblioteca Apostolica Vaticana, Pal. Gr. 314	Late fifteenth century	28 × 19 cm, 224 leaves, books 1–8, chap. 29; no maps
Rome, Biblioteca Apostolica Vaticana, Christinae Reginae 82	Sixteenth century	(M) 22 × 22 cm, 166 leaves, no maps
Rome, San Gregorio Magno al Celio, 15 (present whereabouts unknown)	Fifteenth century	(M) 40 × 28 cm
Venice, Biblioteca Nazionale Marciana, Gr. 516	Fifteenth century (F); fourteenth century (C)	(F) 33 × 22 cm, 138 leaves, 24 maps; (C) 30 × 21 cm, 208 leaves, 22 maps and 2 half maps
Venice, Biblioteca Nazionale Marciana, Gr. 103	Fifteenth century	(M) 20 × 15 cm, book 3, chap. 14
Venice, Biblioteca Nazionale Marciana, Gr. 388	Fifteenth century	(F) 59.0 × 44.2 cm, 101 leaves, 27 maps
Vienna, Österreichische Nationalbibliothek, Vindobonensis historicus Graecus	1454	(F) 59.5 × 44.0 cm, 99 leaves, 27 maps

Based on Douglas W. Marshall, "A List of Manuscript Editions of Ptolemy's *Geography*," *Bulletin of the Geography and Map Division, Special Libraries Association* 87 (1972): 17–38.
Additional sources:

(C) = Otto Cuntz, ed., *Die Geographie des Ptolemaeus: Galliae Germania, Raetia, Noricum, Pannoniae, Illyricum, Italia* (Berlin: Weidmann, 1923).

(F) = Joseph Fischer, ed., *Claudii Ptolemaei Geographiae Codex Urbinas Graecus 82*, 2 vols. in 4, Codices e Vaticanus Selecti Quam Simillime Expressi, vol. 19 (Leiden: E. J. Brill; Leipzig: O. Harrassowitz, 1932).

(M) = Karl Müller, "Rapports sur les manuscrits de la géographie de Ptolémée," *Archives des Missions Scientifiques et littéraires*, 2d ser., 4 (1867): 279–98.

BIBLIOGRAPHY
CHAPTER 15 CARTOGRAPHY IN THE BYZANTINE EMPIRE

Avi-Yonah, Michael. *The Madaba Mosaic Map*. Jerusalem: Israel Exploration Society, 1954.

Bagrow, Leo. "The Origin of Ptolemy's Geographia." *Geografiska Annaler* 27 (1945): 318–87.

Browning, Robert. *The Byzantine Empire*. New York: Charles Scribner's Sons, 1980.

Cosmas Indicopleustes. *Topographie chrétienne*. Edited by Wanda Wolska-Conus. In *Sources Chrétiennes*, nos. 141 (1968), 159 (1970), and 197 (1973).

Delatte, Armand, ed. *Les portulans grecs*. Liège: Bibliothèque de la Faculté de Philosophie et Lettres de l'Université de Liège, 1947.

Dilke, O. A. W. *Greek and Roman Maps*. London: Thames and Hudson, 1985.

Dillemann, Louis. "La carte routière de la *Cosmographie de Ravenne*." *Bonner Jahrbücher* 175 (1975): 165–70.

Diller, Aubrey. "The Oldest Manuscripts of Ptolemaic Maps." *Transactions of the American Philological Association* 71 (1940): 62–67.

———. *The Tradition of the Minor Greek Geographers*. Philological Monographs, no. 14. Lancaster, Penn: American Philological Association, 1952.

———. "The Greek Codices of Palla Strozzi and Guarino Veronese." *Journal of the Warburg and Courtauld Institutes* 24 (1961): 313–21.

Donner, Herbert, and Heinz Cüppers. *Die Mosaikkarte von Madeba*. Abhandlungen des Deutschen Palästinavereins. Wiesbaden: O. Harrassowitz, 1977.

Huxley, G. L. "A Porphyrogenitan Portulan." *Greek, Roman and Byzantine Studies* 17 (1976): 295–300.

Kitzinger, Ernst. "Studies on Late Antique and Early Byzantine Floor Mosaics: I. Mosaics at Nikopolis." *Dumbarton Oaks Papers* 6 (1951): 81–122.

Kubitschek, Wilhelm. "Die sogenannte B-Redaktion der ptolemäischen Geographie." *Klio* 28 (1935): 108–32.

O'Callaghan, R. T. "Madaba (Carte de)." In *Dictionnaire de la Bible: Supplement*, ed. L. Pirot and A. Robert, vol. 5 (1957): 627–704. Paris: Letouzey et Ané, 1928–.

Pekkanen, T. "The Pontiac Civitates in the Periplus of the Anon. Ravennas." *Arctos* 13 (1979): 111–28.

Pinder, Moritz, and Gustav Parthey, eds. *Ravennatis anonymi Cosmographia et Guidonis Geographica*. Berlin: Fridericus Nicolaus, 1860; reprinted, Aalen: Otto Zeller Verlagsbuchhandlung, 1962.

Polaschek, Erich. "Ptolemy's *Geography* in a New Light." *Imago Mundi* 14 (1959): 17–37.

———. "Ptolemaios als Geograph." In *Paulys Realencyclopädie der classischen Altertumswissenschaft*, ed. August Pauly, Georg Wissowa, et al., suppl. 10 (1965): cols. 680–833. Stuttgart: J. B. Metzler, 1894–.

Richmond, Ian A., and Osbert G. S. Crawford. "The British Section of the Ravenna Cosmography." *Archaeologia* 93 (1949): 1–50.

Schillinger-Häfele, Ute. "Beobachtungen zum Quellenproblem der *Kosmographie* von Ravenna." *Bonner Jahrbücher* 163 (1963): 238–51.

Schnabel, Paul. *Text und Karten des Ptolemäus*. Quellen und Forschungen zur Geschichte der Geographie und Völkerkunde 2. Leipzig: K. F. Koehlers Antiquarium, 1938.

Schnetz, Joseph. *Untersuchungen über die Quellen der Kosmographie des anonymen Geographen von Ravenna*. Sitzungsberichte der Akademie der Wissenschaften, Philosophisch-historische Abteilung 6. Munich: Verlag der Bayerischen Akademie der Wissenschaften, 1942.

Stolte, B. H. *De Cosmographie van den anonymus Ravennas: Een studie over de bronnen van Boek II–V*. Zundert, n.d.

Wolska, Wanda. *La topographie chrétienne de Cosmas Indicopleustès: Théologie et science au VIᵉ siècle*. Bibliothèque Byzantine, Etudes 3. Paris: Presses Universitaires de France, 1962.

Wolska-Conus, Wanda. "Deux contributions à l'histoire de la géographie: I. La diagnôsis Ptoléméenne; II. La 'Carte de Théodose II.'" In *Travaux et mémoires*, 259-79. Centre de Recherche d'Histoire et Civilisation Byzantines, 5. Paris: Editions E. de Baccard, 1973.

16 · Cartography in the Ancient World: A Conclusion

O. A. W. Dilke

In reaching a concluding assessment of the conceptual and practical status of cartography in the ancient world, several themes emerge. Even if allowance is made for the severe lack of map artifacts from the period, it is possible to conclude from literary evidence that no one civilization had a monopoly on a particular variety or function of map and that the number of map functions was considerable. In Mesopotamia and in Egypt, as well as in the Greek and Roman centers, both celestial and terrestrial maps existed. Large-scale maps, fulfilling a multitude of functions, were also found in all these societies, although it must be said that there is more evidence for the use of maps in the Roman period than in other periods of antiquity. These functions included the use of maps as cadastral and legal records, as aids to the traveler, to commemorate military and religious events, as strategic documents, as political propaganda, and for academic and educational purposes. Whereas up to about 170 B.C. maps were apparently unfamiliar to most Romans, after that date their use increased steadily. But while the evidence for the use of maps in Roman society is more plentiful, it should not be forgotten that similar uses are likely to have been present in civilizations normally regarded as having a less practical bent—such as classical Greece.

Maps varied considerably in scale, from depictions of the cosmos and the universe at one end of the continuum to large-scale plans of rooms or tombs at the other end. The extent to which the makers of maps in the ancient world were aware of the concept of a metrical scale is still not settled. We have apparently accurate Babylonian plans of properties, houses, temples, cities, and fields from about 2300 to 500 B.C., and there is evidence of the use of some sort of graphic scale on the plan on the statue of Gudea (ca. 2100 B.C.). But it is not until later in the period that a clear concept of ratio is explicit, when an instruction in the *Corpus Agrimensorum* is thought to refer apprentice surveyors to a scale of 1:5,000, corresponding to one Roman foot to a Roman mile, and the *Forma Urbis Romae* may have been consciously planned at a general scale of 1:240 or 1:250. We now know that at the Temple of Apollo at Didyma architects worked at a scale of 1:16.

The orientation of these early maps varied. Unlike one Babylonian map (the clay tablet of Nuzi), classical maps do not contain an explicit indication of the cardinal points, but north must have been at the top in the archetypes of Ptolemy's maps and in the Peutinger map. The widespread use of globes in Hellenistic Greece, with the inhabited world occupying an upper quadrant and the *climata* in parallel zones perpendicular to the earth's axis, may also have encouraged the early use of north as a primary orienting direction. South and east may also have been favored in the Middle East long before their established use by Arabic and Christian mapmakers.

The accuracy of maps in this early period varied appreciably. The Greeks were great sailors and astronomers whereas the Romans were above all road makers, soldiers, and farmers. Perhaps had more Egyptian maps been preserved, we should find in at least some of them the degree of accuracy manifest in the pyramid measurements. Since calculation of distances on sea routes was always more difficult and astronomical bearings were used rather sporadically, we may expect greater accuracy, where this mattered, in Roman than in Greek maps. Distances given in texts or on maps usually indicated the maximum length and width of a province, region, or island. Marinus, for instance, included some land distances as well as coordinates. His coordinates may have been based on a longitude running east of the Canaries, like Ptolemy's, and a latitude that either was similar to Ptolemy's or was based on Rhodes, though he was never consistent in giving both latitude and longitude. The idea of the use of coordinates was developed first in celestial cartography, itself a Greek rather than a Roman concern, and was later adapted for terrestrial use. It must be pointed out, however, that the precision of Ptolemy's coordinates of places, estuaries, and promontories was largely illusory, since few scientific measurements of longitude or perhaps even latitude had been made. Most of the figures were based on estimates of land or sea distances derived from sources of varying reliability.

It is not until the classical period of Greek cartography that we can start to trace a continuous tradition of theoretical concepts about the size and shape of the earth. To appreciate how this period laid the foundations for the developments of the ensuing Hellenistic period, it is necessary, as we have seen, to draw on a wide range of Greek writings containing references to maps. In some cases the authors of these texts are not normally thought of in the context of geographic or cartographic science, but nevertheless they reflect a widespread and often critical interest in such questions. Aristotle's writings, for example, provide a summary of the theoretical knowledge that underlay the construction of world maps by the end of the Greek classical era. At the time when Alexander the Great set off to conquer and explore Asia and when Pytheas of Massalia was exploring northern Europe, therefore, the sum of geographic and cartographic knowledge in the Greek world was already considerable and was demonstrated in a variety of graphic and three-dimensional representations of sky and earth.

Terrestrial maps and celestial globes were widely used as instruments of teaching and research. It has been shown how these could have appealed to the imagination not only of an educated minority—for whom they sometimes became the subject of careful scholarly commentary—but also of a wider Greek public that was already learning to think about the world in a physical and social sense through the medium of maps. If a literal interpretation was followed, the cartographic image of the inhabited world, like that of the universe as a whole, was often misleading; it could create confusion or it could help establish and perpetuate false ideas. The celestial globe had reinforced the belief in a spherical and finite universe such as Aristotle had described; the drawing of a circular horizon, however, from a point of observation, might have perpetuated the idea that the inhabited world was circular, as might also the drawing of a sphere on a flat surface. There was evidently no consensus between cartographic theorists, and there was a gap in particular between the acceptance of the most advanced scientific theories and their translation into map form. In spite of the assertions of Democritus, Eudoxus, and Aristotle, maps of the inhabited world remained circular, with their outer limits very vague. Knowledge even of the Mediterranean was incompletely established. Although just before the invasion of Sicily (415 B.C.) average Athenians may have been able to sketch the outline of the island and indicate Libya and Carthage in relation to it, they generally knew little about its size. It can be said, with hindsight, that by the end of the classical Greek era the need to find a means of drawing maps to scale, and of making a systematic study of the inhabited world, was urgent.

The importance of the Hellenistic period in the history of cartography in the ancient world has thus been clearly established. Its outstanding characteristic was the fruit-ful marriage of theoretical and empirical knowledge. It has been demonstrated beyond doubt that the geometric study of the sphere, as expressed in theorems and physical models, had important practical applications and that its principles underlay the development both of mathematical geography and of scientific cartography as applied to celestial and terrestrial phenomena. With respect to celestial mapping, the poem about the stellar globe by Aratus (though removed in time from Eudoxus) had encouraged the more systematic study of real globes such as that on the archetype of the Farnese Atlas or those constructed by Archimedes. The main constellations on these artifacts were equated with religious beliefs or legends, mainly in human or animal form. This practice in turn had stimulated a closer study of the sky and its groups of stars. By the end of the Hellenistic period, the celestial globes, although they were artistically decorated, were regarded as credible scientific representations of the sky that in turn could be given astrological uses, as in the compilation of horoscopes, in Greek society at large.

In the history of geographical (or terrestrial) mapping, the great practical step forward was to locate the inhabited world exactly on the terrestrial globe. Eratosthenes was apparently the first to accomplish this, and his map was the earliest scientific attempt to give the different parts of the world represented on a plane surface approximately their true proportions. On his map, moreover, one could have distinguished the geometric shapes of the countries, and one could have used the map as a tool to estimate the distances between places.

Thus it was at various scales of mapping—from the purely local to the representation of the cosmos—that the Greeks of the Hellenistic era enhanced and then disseminated a knowledge of maps. By so improving the *mimesis* or imitation of the world, founded on sound theoretical premises, they made other intellectual advances possible and helped to extend the Greek vision far beyond the Aegean. To Rome, Hellenistic Greece left a seminal cartographic heritage—one that, in the first instance at least, was barely challenged in the intellectual centers of Roman society.

The culmination of Greek cartographic thought is seen in the work of Claudius Ptolemy, who worked within the framework of the early Roman Empire. Our review of Ptolemaic scholarship offers nothing to revise the long-held consensus that he is a key figure in the long-term development of scientific mapping. The present *History* has not set out to identify the "cartographic geniuses" who "revolutionize" mapping. Yet Ptolemy—as much through the accidental survival and transmission of his texts when so many others perished as through his comprehensive approach to mapping—does nevertheless stride like a colossus over the cartographic knowledge of the later Greco-Roman world and the Renaissance. This is perhaps more remarkable in that his work was primarily instructional and theoretical, and it remains debatable if he bequeathed a set of images that could be automatically copied by an uninterrupted succession of manuscript illuminators. Ptolemy's principal legacy was thus to cartographic method, and both the *Almagest* and the *Geography* may be regarded as among the most influential works in cartographic history. It would be wrong to overemphasize, as so much of the topographical literature has tended to do, a catalog of Ptolemy's "errors": what is vital for the cartographic historian is that his texts were the carriers of the idea of celestial and terrestrial mapping long after the factual content of the coordinates had been made obsolete through new discoveries and exploration. Finally, our interpretation has progressively come down on the side of the opinion that Ptolemy or a contemporary probably did make at least some of the maps so clearly specified in his texts.

When we turn to Roman cartography, it has been shown that by the end of the Augustan era many of its essential characteristics were already in existence. Drawing on the theoretical knowledge of Greek scholars and technicians, both geographical maps at a small scale and large-scale cadastral maps were brought into more regular use. The primary stimulus to the former seems to have been the recognition by the Roman rulers not only that maps were of practical assistance in the military, political, and commercial integration of the empire, but also that a publicly displayed map of its extent could serve for the people as a symbol of its reality and territorial power. Similarly, the cadastral maps, given the force of law by the end of the period, were designed to record and to help uphold a system of property rights and agrarian production in which the state had a vested interest. Maps had thus become the tools of statecraft at a number of territorial scales. It was these motives, rather than disinterested intellectual curiosity, that led to an extension and diversification of mapping as the empire was further consolidated in the period from Tiberius to Caracalla.

In the course of the early empire large-scale maps were harnessed to a number of clearly defined aspects of everyday life. Roman surveyors were capable of constructing complex maps to a consistent scale. These were used particularly in connection with the land attached to colonies, settlements often set up to provide veterans with smallholdings. In the countryside, although only a few fragments of stone cadaster have survived—and none of the bronze maps that recorded land ownership—many thousands of such maps must originally have been made for centuriation and other schemes. Similarly, in the towns, although only the *Forma Urbis Romae* is

known to us in detail, large-scale maps were recognized as practical tools recording the lines of public utilities such as aqueducts, displaying the size and shape of imperial and religious buildings, and indicating the layout of streets and private property. Some types of Roman maps had come to possess standard formats as well as regular scales and established conventions for depicting ground detail. Yet it is perhaps in the importance accorded the map as a permanent record of ownership or rights over property—whether held by the state or by individuals—that Roman large-scale mapping most clearly anticipated the modern world. In this respect, Rome had provided a model for the use of maps that was not to be fully exploited in many parts of the world until the eighteenth and nineteenth centuries.

Maps in the period of the decline of the empire and its sequel in the Byzantine civilization were of course greatly influenced by Christianity. In its most obvious aspect, the exaggerated size of Jerusalem on the Madaba mosaic map was no doubt an attempt to make the Holy City not only dominant but also more accurately depicted in this difficult medium. Pilgrims from distant lands obviously needed itineraries like that starting at Bordeaux, giving fairly simple instructions. But more realistic geographical maps were not entirely lacking: the choice in the fifth century A.D. for a depiction of the Roman world would perhaps lie between the map commissioned by Theodosius II, which may have revised that of Agrippa, and one based on the ancestor of the Peutinger map.

Continuity between the classical period and succeeding ages was interrupted, and there was disruption of the old way of life with its technological achievements, which also involved mapmaking. Some aspects of a partial cartographic heritage, however, may be suggested. When we come to consider the mapping of small areas in medieval western Europe, it will be shown that the Saint Gall monastery map (pp. 466–68) is very reminiscent of the best Roman large-scale plans. Similarly, it will be made clear to what extent the *mappaemundi* were indebted to a number of classical sources, including Greek maps showing *climata* and the simple tripartite T-O maps (which may have arisen in Roman works involving Africa in the first century B.C.), together with, probably, the map of Agrippa as a common archetype. However, the maps of Marinus and Ptolemy, one of the latter containing thousands of place-names, were at least partly known to Arabic geographers of the ninth to the tenth century. But the transmission of Ptolemy's *Geography* to the West came about first through reconstruction by Byzantine scholars and only second through its translation into Latin (1406) and its diffusion in Florence and elsewhere. In the case of the sea charts of the Mediterranean, it is still unresolved whether the earliest portolan charts of the thirteenth century had a classical antecedent. If they had, one would suppose it to be a map connected with the periploi (sea itineraries). But none of these either has a map or, in the present state of our knowledge, can be shown to have ever had one.

The Byzantine Empire—though providing essential links in the chain—remains something of an enigma for the history of the long-term transmission of cartographic knowledge from the ancient to the modern world. In both western Europe and Byzantium relatively little that was new in cartography developed during the Dark Ages and early Middle Ages, although monks were assiduously copying out and preserving the written work of many past centuries available to them. Some maps, along with other illustrations, were transmitted by this process, but too few have survived to indicate the overall level of cartographic awareness in Byzantine society. While almost certainly fewer maps were made than in the Greco-Roman period, nevertheless the key concepts of mapping that had been developed in the classical world were preserved in the Byzantine Empire. The most ac-

complished Byzantine map to survive, the mosaic at Madaba, is clearly closer to the classical tradition than to maps of any subsequent period. But as the dichotomy increased between the use of Greek in the East and Latin in the West, the particular role of Byzantine scholars in perpetuating Greek texts of cartographic interest becomes clearer. Byzantine institutions, particularly as they developed in Constantinople, facilitated the flow of cartographic knowledge both to and from western Europe and to the Arab world and beyond. Our sources point to only a few late glimpses of these transfers—as when Planudes took the lead in Ptolemaic research, for example. But in order to reach an understanding of the historical processes involved in the period, we must examine the broader channels for Christian, humanistic, and scientific ideas rather than a single map, or even the whole corpus of Byzantine cartography. Viewed in this context, some of the essential cartographic impulses of the fifteenth-century Renaissance in Italy are seen to have been already active in late Byzantine society.

PART THREE

CARTOGRAPHY IN MEDIEVAL EUROPE AND THE MEDITERRANEAN

17 · Medieval Maps: An Introduction

P. D. A. HARVEY

Few maps were drawn in medieval Europe. Certainly those we have today are merely the survivors of the far larger number that must have been produced. Almost certainly too, not all the survivors have yet come to light; we probably have much more to learn about local mapping in medieval Spain, Portugal, and Italy, for instance. But the pattern of the medieval maps known to us makes it clear that maps then were simply not drawn for the multitude of everyday purposes for which they might have been used. Rather, they were confined to particular areas and to particular occasions for which their use had become established by custom. It is rare to find cartography applied outside these limits; when we do we should probably see it as a bold conceptual initiative by some particularly imaginative individual.

What in fact we have are several quite distinct traditions of medieval maps, notably the *mappaemundi*, the portolan charts, the regional and local maps, and a relatively small corpus of celestial maps. The first three form the basis of the division between the next three chapters; the treatment of the celestial maps is deferred until volume 3, where they will be described in a single essay, along with the Renaissance material. The three-fold division for the chapters in this section presents no problems; the three traditions form mainly discrete groups. It is seldom that we find contact between them: the possible use of local maps from Reichenau in the Ebstorf world map, the construction of regional maps of Italy on the basis of portolan charts, and the links between the portolan charts and the books of islands are all exceptional instances of cross-fertilization. Even within each of the three main divisions—world maps, portolan charts, and regional and local maps—we are often dealing with distinct subgroups of maps that developed in isolation from the rest. The classification of the various types of world maps points in this direction. So too does the development of portolan charts, which from the mid-fourteenth century ceased to form a single tradition, a single body of pooled information, and split up into regional schools. Among the regional and local maps, where there is any discoverable connection at all between one map and another we find isolated pockets of particular sorts of map: plans of Italian cities, regional maps from northern Italy, local maps from coastal areas of the Low Countries, and a few others. Given that none of these sorts of maps were in widespread everyday use, it is arguable that scholars in the Middle Ages would not have recognized the products of these varying traditions, these groups and subgroups, as constituting a single class of object—that they would not have seen them, as we do, as maps distinct from diagrams on the one hand and from pictures on the other.

But if few maps were being drawn in the fifteenth century, there were even fewer in the twelfth. World maps seem to have been produced at a fairly steady rate from the eighth century onward, but most other medieval maps date from the fourteenth and fifteenth centuries. None of the surviving portolan charts is older than about 1300, and the earliest known reference to them is in 1270. Of the regional and local maps, few of those from Italy or England are earlier than the fourteenth century and all the maps known to us from France, the Low Countries, and Germany date from later. This cannot be explained simply by the later maps being more likely to survive; the fourteenth and fifteenth centuries clearly saw at least a modest acceptance in western Europe of the value and use of maps in some particular circumstances. How this came about is one of the many unsolved problems of medieval cartography. Another, which also stems from our lack of early medieval maps, is its relationship to the cartography of antiquity. Only in the world maps do surviving examples link medieval maps with the culture of Greece and Rome. Elsewhere, if we are to see any connection at all, we must postulate either the rediscovery of ancient maps, certainly the case with Ptolemy's *Geography* in the thirteenth century and more controversially with the coastal outline used in portolan charts in the thirteenth, or else a tenuous continuity maintained by maps mostly now lost, as may have happened in plans of Italian cities and in maps of the Holy Land. Nor, of course, did classical antiquity provide the only possible external influence on medieval mapping: there may have been some connections with Arab cartography, though this is harder to maintain now than it was once, and the possibility of even remoter links with the cartography of China cannot be entirely ruled out.

Not only were the world maps, the portolan charts, and the regional and local maps of medieval Europe mostly produced in quite separate traditions, but each served a distinctive purpose. There was no such thing in the Middle Ages as a general map designed to be put to a wide variety of uses. Any one map was drawn with one particular purpose, even one particular occasion, in mind. It follows that if we are to evaluate a medieval map, or even understand it at all, we must first know just why it was made. The purpose of the *mappaemundi* was philosophical and didactic: a schematic representation of the earth that in the more detailed examples was extended to give a great deal of information about its inhabitants and their relationship to the deity. Many sources were drawn on in compiling these more elaborate world maps; if these sources included maps that produced on the *mappamundi* more than an approximate, diagrammatic outline of some part of the earth's surface, this was quite incidental, and it certainly does not follow that a detailed geographic outline was an aim of the compilers. This would be quite irrelevant to the map's purpose. The same point comes out again in comparing the early portolan charts with another fourteenth-century map, the Gough map of Great Britain. This, like the portolan charts, was drawn as a guide for travel, and like them it was probably based simply on measuring directions and distances over many journeys. But whereas the Gough map was for travel by land along principal roads, of which there was a fair network but still a finite number, the portolan charts were for travel by sea, on which the ship could move freely over an infinite number of routes. It follows that all that was needed for the Gough map to serve its purpose was a diagram of the road network, showing how routes interconnected, what places they passed through and, by notes of mileage, how far one place was from the next. To show directions and distances between places off these routes, to provide, that is, a complete scale map of the country, would be irrelevant to what it was trying to do. The portolan charts, on the other hand, if they were to be of any use in navigating a freely sailing ship, had to show directions and distances between an unlimited number of points, and the only way to do this was by a scale map with coastal outlines drawn as accurately as possible. The portolan chart looks like a map as we know it, the Gough map fails to meet our modern expectation of what a map should be. But each served its particular purpose with equal efficiency. In assessing medieval maps we should thus always try to discover the intention of their makers and judge how far this was achieved; to compare them with the maps of later centuries is to apply a quite inappropriate set of standards.

Some of the hardest problems in medieval cartography concern those maps that seem most like the general maps of the sixteenth century and later: maps drawn with attention to accuracy of outline and consistency of scale. The problems are those of concept and of execution: how the idea for such maps arose or was transmitted, and how the measurements underlying them were made. We have seen already that the coastal outlines of the portolan charts may or may not have descended from classical antiquity; the origins of Pietro Vesconte's maps of Italy and the Holy Land in the early fourteenth century are no less obscure. The source of inspiration of the early fifteenth-century scale map of Vienna is equally mysterious, and there is a good deal of doubt over how—and even when or by whom—the maps attributed to Nicolas of Cusa were compiled. Behind at least the last two of these problems lies the possible role of the medieval interest in scientific geography, in precisely locating places, an interest we see in compilations of geographical coordinates from the eleventh-century Toledo tables by al-Zarkali to the work of the fifteenth-century geographers of Vienna and Klosterneuburg.

Recent writers have tended to be skeptical of the role of geographical theory in medieval mapping; the criticisms of the view that Klosterneuburg produced a series of maps, now lost, are symptomatic of this. In fact very few scholars have dealt in any detail with medieval mapping in all its aspects; the general work of Joachim Lelewel, *Géographie du Moyen Age* (four volumes and epilogue, 1850–57), and the chronologically narrower study of John K. Wright, *The Geographical Lore of the Time of the Crusades* (1925), are both exceptional. Partly this reflects the natural division of medieval maps into their several quite separate groups, but the work on the individual groups has still been less than adequate. Even the systematic work under the editorship of Marcel Destombes, *Mappemondes A.D. 1200–1500* (1964), has been found in need of revision; much work on portolan charts is vitiated by a curious preoccupation with conflicting national claims to particular innovations; and there is no single monograph at all covering the whole field of medieval regional and local mapping. It is a pity that medieval maps have not been more thoroughly studied, for they probably have much to tell us about the history of cartography in general and about the development of its concepts and techniques. Many features of medieval Europe's cartography can be paralleled elsewhere: the use of maps only for particular purposes or by restricted groups of people, the growth of the idea of precise scale, and so on. Medieval Europe is a relatively well-documented society. We can, however cautiously, argue from the silences in our evidence as well as from what it actually tells us. At least some of the processes and developments it reveals to us may well have occurred in other societies from which our evidence is scantier. Potentially the three chapters that follow have a significance in the history of cartography beyond the bounds of medieval Europe.

18 · Medieval *Mappaemundi*

DAVID WOODWARD

In the millennium that links the ancient and modern worlds, from about the fifth to the fifteenth century after Christ, there developed a genre of world maps or map-paintings originating in the classical tradition but adopted by the Christian church. The primary purpose of these *mappaemundi*, as they are called, was to instruct the faithful about the significant events in Christian history rather than to record their precise locations. They rarely had a graticule or an expressed scale, and they were often schematic in character and geometric—usually circular or oval—in shape. Although several maps fitting this description are also found in the medieval Arabic culture or the cosmographies of South and East Asia during this period (as described in volume 2), the western *mappaemundi* form a well-defined group. They provide a body of documents whose form, content, and meaning reflect many aspects of medieval life.

THE CONTEXT AND STUDY OF *MAPPAEMUNDI*

MAP AND TEXT

In the Middle Ages, the word (especially the oral word) was predominant over the image and was prescribed as such by the nature of the biblical narrative and the views of the early church fathers. Saint Gregory the Great stated that pictures were for the illiterate what the Scriptures were for those who could read.[1] What then was the role of the *mappaemundi*, and at what audience were they aimed? Were they merely illustrations, subservient to the text and adding little in the way of information, or were they independently valuable?

The answers to these questions depend greatly on the type of *mappamundi* under discussion. The making of world maps was not an identifiably separate activity in the medieval period. Their makers were not called cartographers and did not form a characteristic group as, for example, the portolan chartmakers seem to have done by the fourteenth century. Some 900 of the 1,100 surviving *mappaemundi* are found in manuscript books. Moreover, they seem not to have required the services of a specialized scribe: the lettering on the maps and the adjacent text, for example, can usually be identified as being in the same hand. The vast majority of the maps that survive were produced as ipso facto book illustrations. In the late Middle Ages of the fourteenth and fifteenth centuries, there was a tendency to place maps on the first or second page of a codex, which may reflect the growing importance of maps in giving the reader an overview of the text.[2]

The relation between map and text is also seen in the frequent reliance on early texts as sources for the compilation of *mappaemundi*. This raises the general question of how efficiently a map could be drawn from verbal directions, particularly without benefit of a list of coordinates from which places could be plotted. Modern reconstructions from textual sources of the lost maps of Herodotus, Eratosthenes, Strabo, Agrippa, the Ravenna cosmographer, Marco Polo, and others, attempted by geographers and historians in the nineteenth and early twentieth centuries, illustrate the potential difficulties of such exercises.

However, there were large and detailed *mappaemundi*, particularly in the later Middle Ages, that were conceived and drawn as independent documents, though only a handful survives. Since these contained extensive text or rubrics, they can hardly have been designed only for the illiterate. There is other evidence that such maps appealed strongly to a learned audience. Jacques de Vitry, the thirteenth-century bishop of Acre, specifically

The author gratefully acknowledges the assistance of Peter Arvedson, Tony Campbell, William Courtenay, O. A. W. Dilke, P. D. A. Harvey, Frank Horlbeck, George Kish, Mark Monmonier, and Juergen Schulz in the preparation of this chapter.

1. Sixten Ringbom, "Some Pictorial Conventions for the Recounting of Thoughts and Experiences in Late Medieval Art," in *Medieval Iconography and Narrative: A Symposium*, ed. Flemming G. Andersen et al. (Odense: Odense University Press, 1980), 38–69, esp. 38.

2. Uwe Ruberg, "Mappae Mundi des Mittelalters im Zusammenwirken von Text und Bild," in *Text und Bild: Aspekte des Zusammenwirkens zweier Künste in Mittelalter und früher Neuzeit*, ed. Christel Meier and Uwe Ruberg (Wiesbaden: Ludwig Reichert, 1980), 550–92, esp. 558–60. Some hold that the texts are more interesting than the maps, but this would clearly depend on the individual circumstances. See Neil Ker, review of *Mappemondes A.D. 1200–1500: Catalogue préparé par la Commission des Cartes Anciennes de l'Union Géographique Internationale*, ed. Marcel Destombes, in *Book Collector* 14 (1965): 369–73, esp. 370.

mentioned that he found a *mappamundi* to be a useful source of information.[3] Fra Paolino Veneto, an early fourteenth-century Minorite friar, was also explicit in endorsing their value:

> I think that it is not just difficult but impossible without a world map to make [oneself] an image of, or even for the mind to grasp, what is said of the children and grandchildren of Noah and of the Four Kingdoms and other nations and regions, both in divine and human writings. There is needed moreover a twofold map, [composed] of painting and writing. Nor wilt thou deem one sufficient without the other, because painting without writing indicates regions or nations unclearly, [and] writing without the aid of painting truly does not mark the boundaries of the provinces of a region in their various parts sufficiently [clearly] for them to be descried almost at a glance.[4]

TERMS

The term *mappamundi* (plural *mappaemundi*) is from the Latin *mappa* (a tablecloth or napkin) and *mundus* (world).[5] Since their geometric construction was by no means consistent, *mappaemundi* can thus be distinguished from the planisphere (Italian *planisfero*), which usually refers to a world map that has been consciously constructed according to the principles of transformation from a spherical to a flat surface and whose primary purpose is locational. The early use of the planisphere was in astronomical charts employing a stereographic projection, as in Ptolemy's *Planisphaerium*.

It should be stressed that this rather restrictive meaning of the term *mappamundi* was not the contemporaneous use. In the thirteenth and fourteenth centuries, for example, the term was used generically to mean any map of the world, whether in the style of the portolan chart or not. Thus in a contract for world maps at Barcelona in 1399–1400, the terms *mapamundi* or *mappamondi* and *carta da navigare* or *charte da navichare* were all used interchangeably.[6] In modern Italian, the term *mappamondo* is of broad significance and even specifically includes globes.

Nor was the term used in classical Latin of the late Roman era, where the preference was for *forma*, *figura*, *orbis pictus*, or *orbis terrarum descriptio*. *Figura* was usually reserved for the small diagrams in manuscripts that functioned as scientific illustrations. The eighth-century Beatus of Liebana used *formula picturarum*.[7] For medieval Latin, Du Cange defines *mappa mundi* as an "expository chart or map, in which a description of the earth or the world is contained."[8] In the late Middle Ages other terms were also used, such as *imagines mundi*, *pictura*, *descriptio*, *tabula*, or even the *estoire* of the Hereford map, although *mappamundi* was by far the most common. On the Ebstorf map we find a rubric

that may be rendered: "A map is called a figure, whence a mappa mundi is a figure of the world."[9] *Imago mundi* usually indicated a theoretical treatment of cosmography rather than a graphic description.[10]

It is unwise to assume that *mappamundi* necessarily meant a graphic depiction of the world.[11] It is common to find the term used to mean a verbal description in a metaphorical sense, much as we talk today of "mapping a strategy." For example, when Ranulf Higden wrote of a *mappamundi* in the *Polychronicon*, he was referring not to the world map that frequently accompanies it, but to a verbal description of the world.[12] A manuscript in the British Library entitled "Mappa mundi sive orbis descriptio" is also purely a textual account.[13] Peter of Beauvais was the author of a French verse "*mappemonde*" for Philip of Dreux, bishop of Beauvais (fl. 1175–1217).[14] This use of the term was still common

3. Jacobus de Vitriaco, *Libri duo, quorum prior orientalis, sive Hierosolymitanae: Alter, occidentalis historiae nomine inscribitur* (Douai, 1597; republished Farnborough: Gregg, 1971), 215; John Block Friedman, *The Monstrous Races in Medieval Art and Thought* (Cambridge: Harvard University Press, 1981), 42.

4. Paolino Veneto, Vat. Lat. 1960, fol. 13, Biblioteca Apostolica Vaticana. The translation is from Juergen Schulz, "Jacopo de' Barbari's View of Venice: Map Making, City Views, and Moralized Geography before the Year 1500," *Art Bulletin* 60 (1978): 425–74, quotation on 452.

5. According to Thomas Phillipps, "Mappae Clavicula: A Treatise on the Preparation of Pigments during the Middle Ages," *Archaeologia* 32 (1847): 183–244, the word *mappa*, as in *Mappae clavicula*, the late twelfth-century technical treatise, could also mean drawing or painting. In classical Latin the term could also mean a starting cloth for chariot races.

6. R. A. Skelton, "A Contract for World Maps at Barcelona, 1399–1400," *Imago Mundi* 22 (1968): 107–13.

7. Richard Uhden, "Zur Herkunft und Systematik der mittelalterlichen Weltkarten," *Geographische Zeitschrift* 37 (1931): 321–40, esp. 322.

8. "Charta vel mappa explicata, in qua orbis seu mundi descriptio continetur." Charles Du Fresne Du Cange, "Mappa mundi," in *Glossarium mediae et infimae latinatis conditum a Carolo Du Fresne, domino Du Cange, cum supplementis integris D. P. Carpenterii*, 7 vols. (Paris: Firmin Didot, 1840–50), author's translation.

9. "Mappa dicitur forma. Inde mappa mundi id est forma mundi." Konrad Miller, *Mappaemundi: Die ältesten Weltkarten*, 6 vols. (Stuttgart: J. Roth, 1895–98), 5:8. For volume titles, see bibliography, p. 369.

10. *Imago mundi* (or its translated equivalent) appears as the title of several medieval cosmographical works, including those by Honorius, Gautier de Metz, and Pierre d'Ailly.

11. This theme is well developed in Ruberg, "Mappae Mundi," 552–55 (note 2).

12. Churchill Babington and J. R. Lumby, eds., *Polychronicon Ranulphi Higden, Together with the English Translation of John Trevisa and of an Unknown Writer of the Fifteenth Century* (London: Longman, 1865–86).

13. London, British Library, Harl. MS. 3373.

14. The Peter of Beauvais poem is described by Charles Victor Langlois, *La vie en France au Moyen Age, de la fin du XIIᵉ au milieu du XIVᵉ siècle*, 4 vols. (Paris: Hachette, 1926–28), vol. 3, *La connaissance de la nature et du monde*, 122–34.

into the eighteenth century: thus an eighteenth-century manuscript version of the thirteenth-century Spanish geography, the *Semeiança del mundo*, was entitled *Mapa mundi*.[15] The late twelfth- to early thirteenth-century chronicler Gervase of Canterbury described a gazetteer of religious houses in England, Wales, and part of Scotland as a *mappa mundi*.[16]

REALISM VERSUS SYMBOLISM

Two themes relating to the geographical utility of medieval world maps can be identified in the literature since the late nineteenth century. On the one hand, Beazley's desire to view the *mappaemundi* as a static phase in the gradually improving representation of the earth's features resulted from an assumption, shared by many other authors, that the sole function of maps was to provide correct locations of geographical features. In his basic work on medieval geography, he was to dismiss two of the most celebrated *mappaemundi* with the following words: "the non-scientific maps of the later Middle Ages . . . are of such complete futility . . . that a bare allusion to the monstrosities of *Hereford* and *Ebstorf* should suffice."[17] This view was challenged by John K. Wright who pointed out that since geometric accuracy in the *mappaemundi* was not a primary aim, the lack of it could hardly be criticized.[18] We are now accustomed to the notion that Euclidean geometry is by no means the only effective graphic structure for ordering our thoughts about space: distance-decay maps, in which logarithmic or other scalars modify conventional latitude and longitude, were among the first products of the digital mapping age, but the concept is far from new. The twelfth-century map of Asia known as one of the two "Jerome" maps exaggerates Asia Minor—its main point of interest—to the point that it is almost as large as the representation of the rest of Asia (fig. 18.1).[19] A legend on the Matthew Paris map of Britain also demonstrates how map scale could be adjusted to fit the circumstances: "if the page had allowed it, this whole island would have been longer."[20]

The geographical content of the *mappaemundi* was not always solely symbolic and fanciful, however. Crone has demonstrated that, in the case of the Hereford map, its content was expanded from time to time using available resources, providing a more or less continuous cartographic tradition from the Roman Empire to the thirteenth century. The scribe of the Hereford map seems to have systematically plotted lists of place-names on the map from various written itineraries, in an attempt to fulfill a secular as well as a spiritual need. Far from being a mere anthology of mythical lore, the map was thus also a repository of contemporary geographical information of use for planning pilgrimages and stimulating the intended traveler.[21]

The second theme, which Bevan and Phillott introduced as early as 1873, draws attention to the historical or narrative function of the medieval world maps.[22] This theme has recently been developed in detail by Anna-Dorothee von den Brincken in a series of articles where the *mappaemundi* are seen as pictorial analogies to the medieval historical textual chronicles.[23] Von den

15. William E. Bull and Harry F. Williams, *Semeiança del Mundo: A Medieval Description of the World* (Berkeley and Los Angeles: University of California Press, 1959), 1.

16. Gervase of Canterbury, *The Historical Works of Gervase of Canterbury*, ed. William Stubbs, Rolls Series 21 (London: Longman, 1879–80), 417–18.

17. Charles Raymond Beazley, *The Dawn of Modern Geography: A History of Exploration and Geographical Science from the Conversion of the Roman Empire to A.D. 900*, 3 vols. (London: J. Murray, 1897–1906), 3:528. This view had also been expressed long before. For example, the Abbé Lebeuf in 1743 included in his description of a fourteenth-century world map in a manuscript of the Chronicles of Saint-Denis (published in 1751): "it had such inexact proportions that it could only show how imperfect geography had been in fourteenth-century France." See "Notice d'un manuscrit des Chroniques de Saint Denys, le plus ancien que l'on connoisse," *Histoire de l'Académie Royale des Inscriptions et Belles-Lettres* 16 (1751): 175–85, quotation on 185, author's translation. For a development of this theme, see David Woodward, "Reality, Symbolism, Time, and Space in Medieval World Maps," *Annals of the Association of American Geographers* 75 (1985): 510–21.

18. John Kirtland Wright, *The Geographical Lore of the Time of the Crusades: A Study in the History of Medieval Science and Tradition in Western Europe*, American Geographical Society Research Series no. 15 (New York: American Geographical Society, 1925; republished with additions, New York: Dover Publications, 1965), 248. The study of *mappaemundi* had not normally been the province of cartographers up to this time. For example, Arthur Hinks was able to express in 1925 "his indebtedness to Mr. Andrews for much instruction in a subject he had hitherto regarded as outside his own province, which was limited to maps based on latitude and longitude." See Michael Corbet Andrews, "The Study and Classification of Medieval Mappae Mundi," *Archaeologia* 75 (1925–26): 61–76, quotation on 75.

19. London, British Library, Add. MS. 10049, fol. 64r.

20. "Si pagina pateretur, hec totalis insula longior esse deberet." London, British Library, Royal MS. 14.C.VII(a), fol. 5v. See Richard Vaughan, *Matthew Paris* (Cambridge: University Press, 1958), 243.

21. Gerald R. Crone, "New Light on the Hereford Map," *Geographical Journal* 131 (1965): 447–62; Woodward, "Reality, Symbolism, Time, and Space," 513–14 (note 17).

22. W. L. Bevan and H. W. Phillott, *Medieval Geography: An Essay in Illustration of the Hereford Mappa Mundi* (London: E. Stanford, 1873).

23. Anna-Dorothee von den Brincken, "Mappa mundi und Chronographia," *Deutsches Archiv für die Erforschung des Mittelalters* 24 (1968): 118–86; and her more general summary "Zur Universalkartographie des Mittelalters," in *Methoden in Wissenschaft und Kunst des Mittelalters*, ed. Albert Zimmermann, Miscellanea Mediaevalia 7 (Berlin: Walter de Gruyter, 1970), 249–78. See also her "Europa in der Kartographie des Mittelalters," *Archiv für Kulturgeschichte* 55 (1973): 289–304. Juergen Schulz has adapted this idea to city views of the Renaissance in his "Jacopo de' Barbari's View of Venice: Map Making, City Views, and Moralized Geography before the Year 1500," *Art Bulletin* 60 (1978): 425–74. This article is far more than an essay on de' Barbari's view, and it contains important general material on the *mappaemundi*, medieval surveying, and cartography.

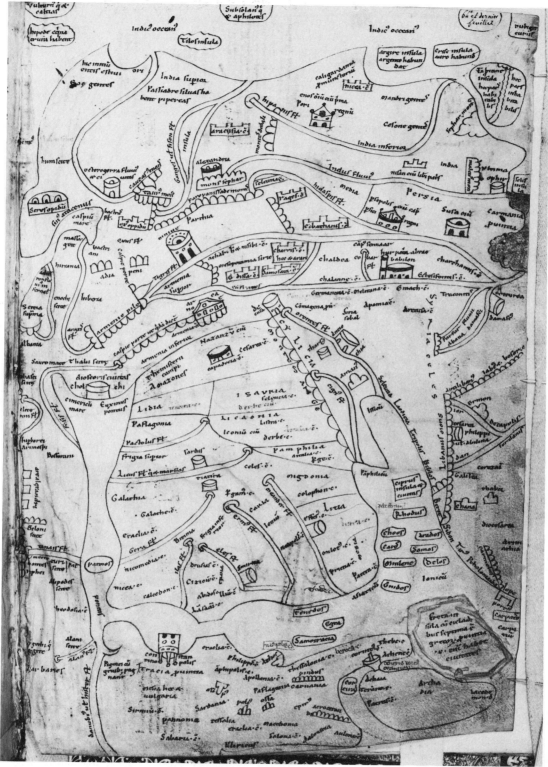

FIG. 18.1. "JEROME" MAP OF ASIA. This twelfth-century manuscript map illustrating the writings of Saint Jerome (fourth/fifth century) exaggerates its principal area of interest, Asia Minor, to be almost as large as the rest of Asia. Such variations in metrical scale were common in *mappaemundi*.

Size of the original: 35.6 × 22.9 cm. By permission of the British Library, London (Add. MS. 10049, fol. 64r).

Brincken illustrates this historical function by listing, in a series of tables, the place-names appearing on twenty-one selected maps. In addition to the expected frequent occurrence of the centers of Christianity (Jerusalem, Rome, Constantinople, Antioch, and Patmos), a surprising number of secular places of historical interest are found—such as Olympus, Taprobane, and Pergamon—together with several secular places of particular interest at the time, such as Kiev, Novgorod, Samarkand, and Georgia.[24] More specialized studies on the early appearance of place-names on medieval maps confirm this view. For example, the tenth-century Cotton map contains an early reference to Bulgaria.[25]

The *mappaemundi* may thus be seen as analogous to the narrative medieval pictures that portray several events separated by time and included within the same scene. Instead of being presented in sequence as in a frieze or cartoon, they are placed in their logical positions in the picture. For the *mappaemundi*, this meant the approximate geographical or topological location of the event.[26]

The origins of the didactic world map can be traced to late antiquity. It seems that maps had a place in everyday life. Eumenius, a teacher and orator of distinction, delivered a discourse in A.D. 297 on the subject of restoring the well-known Roman school at Augustodunum in Gaul, present-day Autun. Among other admonitions, he advised that schoolboys should be made to study geography, using furthermore the *mappamundi* found in the portico of the Autun school.[27]

The medieval view of the *mappaemundi* seems to have been expressed by Hugh of Saint Victor about 1126: "We must collect a brief summary of all things . . . which the mind may grasp and the memory retain with ease. The mind chiefly esteems events by three things: the persons by whom deeds were done, the places in which they were done, and the times when they were done."[28]

There was more than a mnemonic function, however. The monumental size and method of display of some of these world maps suggest that there was also a public iconographic role: thus the Agrippa map (see pp. 207–9) and the one referred to by Eumenius above may have stood for the dominance of the Roman Empire over most of the world. Medieval literature and the *mappaemundi* both mirrored this classical symbolism and adapted this function to religious ends. The medieval romances, particularly those describing the exploits of the classical heroes, frequently use a *mappamundi* as a symbol of military dominance. In medieval religious life, a *mappamundi* might stand as a representation of the world, for the transitoriness of earthly life, the divine wisdom of God, the body of Christ, or even God himself. The Godlike image is best seen in the Ebstorf map, where the head, hands, and feet of Christ are represented at

the four cardinal directions, with the map itself standing for the body of Christ (figs. 18.2 and 18.3).[29]

Another illustration of a similar metaphor is seen in the many diagrammatic views of the tripartite globe represented as an orb held in the left hand of a sovereign, Christ (as *Salvator mundi*), or God the Father. Usually the threefold division is drawn in perspective so as to conform to the shape of the globe, as in plate 10. The representation of the orb as a symbol of imperial or royal power was derived from Roman times where it appears on many coins of the late Roman period.[30] A simple version of the globe also sometimes appears under Christ's feet in representations of the Last Judgment, as in plate 11. Less schematic but still decorative and symbolic representations are found in the much-reproduced world map in Jean Mansel, *La fleur des histoires*, which clearly represents a spherical earth divided among the three sons of Noah (plate 12).[31]

24. Von den Brincken, "Mappa mundi und Chronographia," 160–67 (note 23).

25. For examples of studies of individual regions on the *mappaemundi*, see Peter St. Koledarov, "Nai-Ranni Spomenavanniya na Bilgaritye virkhu Starinnitye Karty" (The earliest reference to the Bulgarians on ancient maps), *Izvestija na Instituta za Istorija* 20 (1968): 219–54; Kyösti Julku, "Suomen tulo maailmankartalle" (Appearance of Finland on medieval world maps), *Faravid* 1 (1977): 7–41.

26. On narrative painting, to which the *mappaemundi* seem to be analogous, see Ringbom, "Pictorial Conventions" (note 1); Otto Pächt, *The Rise of Pictorial Narrative in Twelfth-Century England* (Oxford: Clarendon Press, 1962); and Henrietta Antonia Groenewegen-Frankfort, *Arrest and Movement: An Essay on Space and Time in the Representational Art of the Ancient Near East* (Chicago: University of Chicago Press, 1951). The theme is developed in Woodward, "Reality, Symbolism, Time, and Space" (note 17).

27. Eumenius *Oratio pro instaurandis scholis* 20, 21; see 9(4) in *XII* [*Duodecim*] *Panegyrici Latini*, ed. R. A. B. Mynors (Oxford: Clarendon Press, 1964). Crone, "New Light," 453 (note 21) reports that the map referred to by Eumenius may have survived as late as the seventeenth century, when it was possibly described by C. M. Grivaud, "Sur les antiquités d'Autun (I)," *Annales des Voyages, de la Géographie et de l'Histoire* 12 (1810): 129–66. See also Gaston Boissier, "Les rhéteurs gaulois du IVe siècle," *Journal des Savants* (1884): 1–18; Beazley, *Dawn of Modern Geography*, 2:379 (note 17), and Dilke (above, p. 209) for this and other classical references to maps in public places.

28. Hugh of Saint Victor, *De tribus maximis circumstantiis gestorum*. See transcription by von den Brincken, "Mappa mundi und Chronographia," 124 (note 23), or her "Universalkartographie," 253 (note 23), and the translation by Schulz, "Moralized Geography," 447 (note 23).

29. For a thorough summary of the symbolism of the *mappaemundi* in the French medieval romances, see Jill Tattersall, "Sphere or Disc? Allusions to the Shape of the Earth in Some Twelfth-Century and Thirteenth-Century Vernacular French Works," *Modern Language Review* 76 (1981): 31–46, esp. 41–44. On the symbolism of the Ebstorf map, see Ruberg, "Mappae Mundi," 563–85 (note 2); Schulz, "Moralized Geography," 449 (note 23).

30. Miller, *Mappaemundi*, 3:129–31 (note 9).

31. Brussels, Bibliothèque Royale, MS. 9231. See Marcel Destombes, ed., *Mappemondes A.D. 1200–1500: Catalogue prepare par*

FIG. 18.2. CHRIST'S HEAD IN THE EBSTORF MAP. The thirteenth-century *mappamundi* known as the Ebstorf map represents the world as the body of Christ. In this detail, Christ's head is shown at the top of the map—the east—next to Paradise. For an illustration of the entire map, see figure 18.19.
From Walter Rosien, *Die Ebstorfer Weltkarte* (Hanover: Niedersächsisches Amt für Landesplanung und Statistik, 1952). By permission of the Niedersächsisches Institut für Landeskunde und Landesentwicklung an der Universität Göttingen.

FIG. 18.3. CHRIST'S LEFT HAND IN THE EBSTORF MAP. In the Ebstorf map, Christ's arms span north and south. In this detail, his left hand gathers in even the monstrous races. For an illustration of the entire map, see figure 18.19.
From Walter Rosien, *Die Ebstorfer Weltkarte* (Hanover: Niedersächsisches Amt für Landesplanung und Statistik, 1952). By permission of the Niedersächsisches Institut für Landeskunde und Landesentwicklung an der Universität Göttingen.

RELATIONSHIP OF *MAPPAEMUNDI* TO OTHER MEDIEVAL MAPS

With an obvious exception in the curious maps of Opicinus de Canistris and the transitional maps discussed later,[32] most medieval *mappaemundi* share no obvious formal or functional similarities with other maps of the period such as the portolan charts and the regional, topographical, or cadastral maps. The geographical content of the first portolan charts in the late thirteenth century bears no apparent relationship to that of the *mappaemundi* of the time.[33] It is difficult to agree with Beazley that "the absurdities of Dark Age map-making are precursors of the first accurate charts and modern atlases"[34] unless the term "precursor" is simply used chronologically. Indeed, the fact that the Carte Pisane (to which Beazley was referring) and the Hereford map are products of the same age exemplifies how two cartographic genres can exist side by side. These two maps appear to have been compiled in quite different environments, assuming entirely different functions and structured in different ways. The former is of mercantile origin, the second monastic.

la Commission des Cartes Anciennes de l'Union Géographique Internationale (Amsterdam: N. Israel, 1964), 179.

32. The maps of Opicinus de Canistris (1296–ca. 1350) are an unusual hybrid between *mappaemundi* and the portolan charts of the Mediterranean, sharing the symbolic attributes of the former with the coastal accuracy of the latter. They were introduced by Richard Georg Salomon in his *Opicinus de Canistris: Weltbild und Bekenntnisse eines Avignonesischen Klerikers des 14. Jahrhunderts*, Studies of the Warburg Institute, vols. 1A and 1B (text and plates) (London: Warburg Institute, 1936), and followed up in his "A Newly Discovered Manuscript of Opicinus de Canistris," *Journal of the Warburg and Courtauld Institutes* 16 (1953): 45–57, pls. 12 and 13, and his "Aftermath to Opicinus de Canistris," *Journal of the Warburg and Courtauld Institutes* 25 (1962): 137–46 and pls. 26d and 27. More recently, Jörg-Geerd Arentzen has focused attention on the maps in his *Imago Mundi Cartographica: Studien zur Bildlichkeit mittelalterlicher Welt- und Ökumenekarten unter besonderer Berücksichtigung des Zusammenwirkens von Text und Bild*, Münstersche Mittelalter-Schriften 53 (Munich: Wilhelm Fink, 1984).

33. The portolan charts do not appear to have had any visible influence on other maps before the thirteenth century, thus joining other strong evidence that controverts the hypothesis of Charles H. Hapgood, *Maps of the Ancient Sea Kings: Evidence of Advanced Civilization in the Ice Age*, rev. ed. (New York: E. P. Dutton, 1979), and other writers that the origin of portolan charts extends back to preclassical times.

34. Beazley, *Dawn of Modern Geography*, 1:18 (note 17). Armando Cortesão, *History of Portuguese Cartography*, 2 vols. (Coimbra: Junta de Investigações do Ultramar-Lisboa, 1969–71), 1:151, also follows Beazley on this.

In the later Middle Ages, three distinct methods of compiling maps existed side by side. The portolan chart seems to have been constructed incrementally (from the inside out, as it were), relying on the natural closures provided by the basins of the Mediterranean Sea and being bounded only by the natural shape of the vellum on which it was drawn. The *mappaemundi* appear to have been compiled with the assumption that there was a finite amount of information to be fitted into a predetermined bounding shape, be it a rectangle, circle, oval, or other geometrically definable figure. This space is often partitioned schematically into segments. A third system assumed a regular net of parallels and meridians into which geographical information could be placed. Although described in an astronomical, astrological, and geometric context in the Middle Ages long before the reception of Ptolemy's *Geography* into the West, rectangular and spherical coordinate systems for terrestrial mapping were not fully accepted until the fifteenth century. These three cartographic systems existed in largely separate traditions until the portolan charts began to influence the later *mappaemundi* in the early fourteenth century and the Ptolemaic manuscripts of the *Geography* overturned Western notions of mapmaking in the fifteenth.

Such was the practical value of the portolan charts, however, that by the fourteenth century their influence was being revealed in the *mappaemundi*. Although the usually circular form of the map was retained, therefore, accurate outlines of the Mediterranean Sea and other areas traditionally found on the portolan charts, together with their characteristic rhumb lines, are frequently found on *mappaemundi* from the fourteenth century. In the fifteenth, even graphic scales were sometimes added.

There was a closer and earlier affinity between the *mappaemundi* and the regional maps and itineraries. Regional maps were also compiled by authors in the monastic tradition, and the larger-scale maps were no doubt used as source material for the smaller, their style and content often being similar. In some cases the extent of the regional maps was so large, as in the "Jerome" map of Asia, that they have been mistaken for fragments of world maps.[35] The use of pilgrim and trade-route itineraries, some of which dated from Roman times, was also a common practice in compiling the *mappaemundi*. For example, Crone has made a careful analysis of the use of these sources in the Hereford map.[36]

PROBLEMS IN THE STUDY OF *MAPPAEMUNDI*

As in other aspects of the history of cartography, scholars wishing to study medieval *mappaemundi* have found major difficulties. These include the incompleteness of the record, the difficulty of compiling general works summarizing the widely scattered literature from many fields, and the large capital cost of preparing published catalogs and facsimile atlases from which comparisons could be made. The need for these tools was recognized as the value of *mappaemundi* as cartographic, historical, and artistic documents came to be fully realized. This was not until the middle of the nineteenth century, but since then there have been several landmark texts that have improved the situation.

Skelton believed that the wastage or loss of maps up to the sixteenth century was more severe than that for any other type of historical document.[37] Although we may prefer to be less categorical, there is direct literary evidence that many medieval maps have not come down to us. Some of these would have been large and considered important at the time. The list of major *mappaemundi* in appendix 18.2 provides many examples. We also know by inference from later versions of such world maps as those of Orosius, Isidore, and Macrobius that the key prototypes in the early medieval period are missing. For the later period, the few inventories of monastic libraries that have been published are excellent sources of references to *mappaemundi* that apparently existed as separate items.[38] The frequency of these allusions suggests that many more large *mappaemundi* were lost than have come down to us. This underlines the need to admit to the imperfect or provisional nature of the conclusions drawn from such an incomplete sample.[39]

The first general study of *mappaemundi* was that of Manuel Francisco de Barros e Sousa, second viscount of Santarém (1791–1856). Although Santarém had drawn attention to those of his predecessors who had shown more than passing interest in the subject, such as William Playfair (1759–1823) and Placido Zurla (1769–1834), it was Santarém himself who first attempted a general synthesis of the subject. His work, accompanied by a magnificent facsimile atlas of 117 *mappaemundi*, of which only 21 had previously been published, is still a useful summary.[40] Major contemporaries of Santarém

---

35. Crone, "New Light," 453 (note 21).
36. Crone, "New Light," 451–55 (note 21).
37. R. A. Skelton, *Maps: A Historical Survey of Their Study and Collecting* (Chicago: University of Chicago Press, 1972), 26.
38. [Leo Bagrow], "Old Inventories of Maps," *Imago Mundi* 5 (1948): 18–20, and additions in later volumes of Akademie der Wissenschaften, Vienna, *Mittelalterliche Bibliothekscataloge Österreichs* (Vienna, 1915–71), and Akademie der Wissenschaften, Munich, *Mittelalterliche Bibliothekscataloge Deutschlands und der Schweiz* (Munich, 1918–62); see also Schulz, "Moralized Geography," 449–50 (note 23).
39. Skelton, *Maps*, 26 (note 37).
40. Manuel Francisco de Barros e Sousa, Viscount of Santarém, *Essai sur l'histoire de la cosmographie et de la cartographie pendant*

(in some cases his rivals) who made significant contributions to the general history of medieval world maps included Edme-François Jomard (1777–1862), Joachim Lelewel (1786–1861), and Marie Armand Pascal d'Avezac-Macaya (1799–1875). The contribution of Jomard, the head of the map department of the Bibliothèque Impériale, was a rival facsimile atlas that contained thirty medieval world maps.[41] Lelewel's work, again accompanied by a small facsimile atlas, stressed the Arabic and not the Western contribution to the genre, clearly an unusual slant for the period. It was the subject of a detailed review by Santarém.[42] D'Avezac-Macaya, although he helped Santarém with his facsimile atlas, for which he is acknowledged, is better known for his work on individual maps and the history of projections.[43] However, nothing rivaling the importance of Santarém's study and atlas appeared until the six-volume survey of *mappaemundi* by Konrad Miller (1866–1944).[44] This thorough and careful work was extremely well received and was rapidly accepted as the standard text, as is shown by reviews.[45]

While Miller's volumes were being published, Charles Raymond Beazley (1858–1951) was producing his three-volume history of geographical travel and exploration in the Middle Ages.[46] Beazley did not always appreciate the full meaning of the *mappaemundi*, but he was well aware of the importance of maps in revealing the geographical spirit of the age. He thoroughly described almost all the major world maps of the period in a series of chapters and appendixes, arranged chronologically, and his work—along with Miller's—still provides a wealth of detail not available elsewhere.

On balance, Beazley's three-volume work was more a contribution to the history of geographical exploration than to the history of geographical thought. It was the historians of science who developed the framework for the history of medieval cosmographical concepts. Pierre Duhem's multivolume survey still remains a standard source for the subject,[47] despite more recent claims that his approach suffers from "precursorism."[48] Other historians of science and technology, including the founder of the modern field of that study in Europe and America, George Sarton, made detailed if scattered contributions to the subject in *Introduction to the History of Science*, as did the team of historians working for the seven-volume *History of Technology* under the leadership of Charles Singer.[49] The influence of the Harvard historian of science Charles Haskins must also be specifically mentioned:[50] his student John K. Wright's doctoral dissertation led to his *Geographical Lore*, a masterly work with several chapters on the cartography of the period and an excellent bibliography.[51] Among the most original contributions to the study of the late period of medieval cartography, however, was Dana

Bennett Durand's monograph on the Vienna-Klosterneuburg map corpus, based on his doctoral dissertation submitted to Harvard's history department under the supervision of Sarton. Durand demonstrated the previously unrecognized existence of a group of maps in the fifteenth century that was partly independent of both the Ptolemaic and the medieval traditions of regional and world maps and that appeared to form a transitional link between medieval and Renaissance cartography. He also provided a useful summary on the cultural context of these maps.[52]

le Moyen-Age et sur les progrès de la géographie après les grandes découvertes du XVe siècle, 3 vols. (Paris: Maulde et Renou, 1849–52), and *Atlas composé de mappemondes, de portulans et de cartes hydrographiques et historiques depuis le VIe jusqu'au XVIIe siècle* (Paris, 1849; Facsimile reprint, Amsterdam: R. Muller, 1985).

41. Edme-François Jomard, *Les monuments de la géographie; ou, Recueil d'anciennes cartes européennes et orientales* (Paris: Duprat, etc., 1842–62).

42. Joachim Lelewel, *Géographie du Moyen Age*, 4 vols. and epilogue (Brussels: Pilliet, 1852–57; reprinted Amsterdam: Meridian, 1966). Santarém's review was not published until 1914. See Cortesão, *History of Portuguese Cartography*, 1:38 (note 34).

43. Marie Armand Pascal d'Avezac-Macaya, "Note sur la mappemonde historiée de la cathédrale de Héréford, détermination de sa date et de ses sources," *Bulletin de la Société de Géographie*, 5th ser., 2 (1861): 321–34; idem, "La mappemonde du VIIIe siècle de St. Béat de Liébana: Une digression géographique à propos d'un beau manuscrit à figures de la Bibliothèque d'Altamira," *Annales des Voyages, de la Géographie, de l'Histoire et de l'Archéologie* 2 (1870): 193–210; and idem, *Coup d'oeil historique sur la projection des cartes de géographie* (Paris: E. Martinet, 1863), first published as "Coup d'oeil historique sur la projection des cartes de géographie," *Bulletin de la Société de Géographie*, 5th ser., 5 (1863): 257–361, 438–85.

44. Miller, *Mappaemundi* (note 9). Miller's sequel to this work, *Mappae Arabicae*, 6 vols. (Stuttgart, 1926–31), was less successful, as will be discussed in the Asian volume of this *History* (volume 2).

45. For example, those by Charles Raymond Beazley, "New Light on Some Mediæval Maps," *Geographical Journal* 14 (1899): 620–29; 15 (1900): 130–41, 378–89; 16 (1900): 319–29.

46. Beazley, *Dawn of Modern Geography* (note 17).

47. Pierre Duhem, *Le système du monde: Histoire des doctrines cosmologiques de Platon à Copernic*, 10 vols. (Paris: Hermann, 1913–59).

48. David C. Lindberg, ed., *Science in the Middle Ages* (Chicago: University of Chicago Press, 1978), vii.

49. George Sarton, *Introduction to the History of Science*, 3 vols. (Baltimore: Williams and Wilkins, 1927–48), contains many notes relating to the history of medieval cartography. There is also a relevant section in Charles Singer et al., eds. *A History of Technology*, 7 vols. (Oxford: Clarendon Press, 1954–78), vol. 2, *The Mediterranean Civilizations and the Middle Ages, c. 700 B.C. to c. A.D. 1500.*

50. See Charles Homer Haskins, *Studies in the History of Mediaeval Science* (Cambridge: Harvard University Press, 1927), and his *Renaissance of the Twelfth Century* (New York: Meridian, 1957).

51. Wright, *Geographical Lore* (note 18).

52. Dana Bennett Durand, *The Vienna-Klosterneuburg Map Corpus of the Fifteenth Century: A Study in the Transition from Medieval to Modern Science* (Leiden: E. J. Brill, 1952).

By far the most useful reference work for the comparison of medieval *mappaemundi* yet to appear is the sixteen-volume facsimile atlas initiated and financed by Prince Youssouf Kamal but compiled by Frederik Caspar Wieder (1874–1943).[53] Although confined to maps illustrating the exploration and discovery of Africa, it contains almost all major medieval maps that include Africa, reproduced photographically, making it the single most valuable source of illustrations of these maps. The work has two main drawbacks: first, it lacks specific descriptions of the maps reproduced, except where they relate to the discovery of Africa, and second, the distribution of the work was limited to one hundred copies.[54]

In addition to the many accounts and chapters in general works on the history of maps of varying completeness and accuracy, there have also been some outstanding encyclopedia articles on the subject.[55] The most valuable recent general book-length treatment of the historiography, context, form, and allegorical content of *mappaemundi* is the doctoral dissertation of Jörg-Geerd Arentzen. This work also has a particularly valuable general bibliography.[56]

Systematic comparative work on *mappaemundi* depends on a general census. Some catalogs of maps in national libraries and listings of maps (including *mappaemundi*) held in particular countries, such as the one for Italy by Uzielli and Amat di San Filippo and the one for Germany by Ruge, had been published by 1916,[57] but the idea for a general listing of medieval maps was not proposed until 1949, by Marcel Destombes at the Sixteenth International Geographical Congress in Lisbon, and a Commission on Early Maps was formed to prepare a four-volume catalog of medieval maps, as follows: 1. *mappaemundi*; 2. nautical charts; 3. regional maps, including Ptolemy; and 4. printed maps. Volume 4 appeared in preliminary form in 1952,[58] and the revised and enlarged version awaits publication. Volume 1, covering the manuscript *mappaemundi*, appeared in 1964. Work for the other volumes has not yet been undertaken.[59]

CLASSIFICATION SYSTEMS

These works contain several attempts at the classification of *mappaemundi*, summarized in table 18.1. A satisfactory classification would be useful to the scholar wishing to bring order to the diverse images of the *mappaemundi* by drawing attention to differences in form and origin and by providing a satisfactory vocabulary for describing the maps. For *mappaemundi* the availability of the International Geographical Union's census makes this task much easier. The utility of the classification can be tested by checking the number of entries in the catalog failing to fit the categories provided. Previously developed systems of classification are now reviewed, and the system proposed is presented in table 18.2.

It requires considerable care to classify a large number of scattered artifacts into empirically satisfactory categories. While Santarém may be credited with the idea of publishing a large facsimile atlas of medieval *mappaemundi*, making comparison possible for the first time, he settled for a simple chronological ordering rather than a classification of the maps according to their sources.[60] Nor did Konrad Miller propose a systematic classification, his book being subdivided rather by the emphasis he wished to place on certain single maps or maps by a single author. Thus, for example, Beatus, the

53. Youssouf Kamal, *Monumenta cartographica Africae et Aegypti*, 5 vols. in 16 pts. (Cairo, 1926–51). For full contents, see p. 40.

54. Norman J. W. Thrower, "Monumenta Cartographica Africae et Aegypti," *UCLA Librarian*, suppl. to vol. 16, no. 15 (31 May 1963): 121–26.

55. Although the following is only a small sample, such general accounts include W. W. Jervis, *The World in Maps: A Study in Map Evolution* (London: George Philip, 1936), 68–86; George H. T. Kimble, *Geography in the Middle Ages* (London: Methuen, 1938), 181–204; Lloyd A. Brown, *The Story of Maps* (Boston: Little, Brown, 1949; reprinted New York: Dover, 1979), 81–112; Gerald R. Crone, *Maps and Their Makers: An Introduction to the History of Cartography*, 5th ed. (Folkestone, Kent: Dawson; Hamden, Conn.: Archon Books, 1978), 5–9, 19–33; Joachim G. Leithäuser, *Mappae mundi: Die geistige Eroberung der Welt* (Berlin: Safari-Verlag, 1958), chaps. 2 and 3; Leo Bagrow, *History of Cartography*, rev. and enl. R. A. Skelton, trans. D. L. Paisey (Cambridge: Harvard University Press; London: C. A. Watts, 1964), 41–73; S. M. Ziauddin Alavi, *Geography in the Middle Ages* (Delhi: Sterling, 1966); Cortesão, *History of Portuguese Cartography*, 1:150–215 (note 34). Encyclopedia articles include those by various authors in *Paulys Realencyclopädie der classischen Altertumswissenschaft*, ed. August Pauly, Georg Wissowa, et al. (Stuttgart: J. B. Metzler, 1894–); Otto Hartig, "Geography in the Church," in *The Catholic Encyclopedia*, 15 vols. (New York: Robert Appleton, [1907–12]), 6:447–53; Giuseppe Caraci, "Cartografia," in *Enciclopedia italiana di scienze, lettere ed arti*, originally 36 vols. ([Rome]: Istituto Giovanni Treccari, 1929–39), 9:232; Ernest George Ravenstein, "Map," in *Encyclopaedia Britannica*, 11th ed., 32 vols. (New York: Encyclopaedia Britannica, 1910–11), 17:629–63, esp. 633–46; and Vincent Cassidy, "Geography and Cartography, Western European," in *Dictionary of the Middle Ages*, ed. Joseph R. Strayer (New York: Charles Scribner's Sons, 1982–), 5: 395–99.

56. Arentzen, *Imago Mundi Cartographica* (note 32).

57. Gustavo Uzielli and Pietro Amat di San Filippo, *Mappamondi, carte nautiche, portolani ed altri monumenti cartografici specialmente italiani dei secoli XIII–XVII*, 2d ed., 2 vols., Studi Biografici e Bibliografici sulla Storia della Geografia in Italia (Rome: Società Geografica Italiana, 1882; reprinted Amsterdam: Meridian, 1967); Sophus Ruge, "Älteres kartographisches Material in deutschen Bibliotheken," *Nachrichten von der Königlichen Gesellschaft der Wissenschaften zu Göttingen*, Philologisch-Historische Klasse (1904): 1–69; (1906): 1–39; (1911): 35–166; suppl. (1916).

58. Marcel Destombes, ed., *Catalogue des cartes gravées au XVᵉ siècle* (Paris: International Geographical Union, 1952). The enlarged version is by Tony Campbell of the British Library.

59. Destombes, *Mappemondes* (note 31).

60. Santarém, *Essai* (note 40).

TABLE 18.1 Comparison of the Main Features of Classifications of *Mappaemundi*

Simar[a] (1912)	Andrews[b] (1926)	Uhden[c] (1931)	Destombes[d] (1964)	Arentzen[e] (1984)	Woodward[f] (1987)
A[g] Roman	I Oecumenical Tripartite Simple	I Roman	Oecumenical A Schematic D Geographical	Oecumenical maps	Tripartite Schematic Nonschematic
B Greek	III Hemispheric	II Crates	C Greek	World maps	Zonal
AB Combination	II Intermediate	III Combination	B Fourth continent		Quadripartite
					Transitional

[a]Théophile Simar, "La géographie de l'Afrique Centrale dans l'antiquité et au Moyen-Age," *Revue Congolaise* 3 (1912–13): 1–23, 81–102, 145–69, 225–52, 289–310, 440–41.

[b]Michael Corbet Andrews, "The Study and Classification of Medieval Mappae Mundi," *Archaeologia* 75 (1925–26): 61–76.

[c]Richard Uhden, "Zur Herkunft und Systematik der mittelalterlichen Weltkarten," *Geographische Zeitschrift* 37 (1931): 321–40.

[d]Marcel Destombes, ed., *Mappemondes A.D. 1200–1500: Catalogue préparé par la Commission des Cartes Anciennes de l'Union Géographique Internationale* (Amsterdam: N. Israel, 1964).

[e]Jörg-Geerd Arentzen, *Imago Mundi Cartographica: Studien zur Bildlichkeit mittelalterlicher Welt- und Ökumenekarten unter besonderer Berücksichtigung des Zusammenwirkens von Text und Bild,* Münstersche Mittelalter-Schriften 53 (Munich: Wilhelm Fink, 1984), esp. 63–66.

[f]Present work.

[g]Designations follow those of the original authors.

TABLE 18.2 Proposed Classification of *Mappaemundi*

Tripartite	Zonal	Quadripartite	Transitional
Schematic T-O	Macrobius Martianus Capella Alphonse and d'Ailly	Tripartite/zonal Beatus	Portolan chart influence Ptolemaic influence
Isidore			
Sallust			
Gautier de Metz			
Miscellaneous and unknown authors			
T-O reverse			
Y-O with Sea of Azov			
V-in-square and T-in-square			
Nonschematic			
Orosius			
Orosius-Isidore			
Cosmas			
Higden			

Note: See also appendix 18.1.

small T-O maps, the Hereford map, and the Ebstorf map all have volumes to themselves.[61] Beazley made no attempt at classification, writing a straightforward chronological narrative.

The first rational attempt was made by Théophile Simar in 1912. He proposed a simple threefold classification based on the main sources of the maps, as was later fully explained by John K. Wright.[62] Simar distinguished two main types, Roman and late Greek, and a third intermediate category containing characteristic features of both. Michael Andrews offered his classification in 1926,[63] the result of systematic examination of some six hundred *mappaemundi*. In its general lines it was based on that of Simar, but Andrews subdivided the three main families into divisions, genera, and species. Another classification, by Richard Uhden, used the same main categories but divided them into subgroups based on key examples. However, Uhden made no reference

61. Miller, *Mappaemundi* (note 9).

62. Théophile Simar, "La géographie de l'Afrique Centrale dans l'antiquité et au Moyen-Age," *Revue Congolaise* 3 (1912–13): 1–23, 81–102, 145–69, 225–52, 289–310, 440–41. Wright, *Geographical Lore*, 389–90, n. 114 (note 18).

63. Andrews, "Classification of Mappae Mundi," 61–76 (note 18).

to the earlier work of Simar or Andrews and does not seem to have been acquainted with either.[64]

The Andrews classification was adopted by the International Geographical Union's Commission on Early Maps (Destombes) with one important modification: a fourth category, D, was made from Andrews's "oecumenical simple" division. The basis for this change was that these maps, which exhibited far more geographical information than the schematic tripartite variety (T-O maps) needed a category to themselves. While this was an understandable modification, the system of numbering and lettering the subgroups in the four main categories is unclear and not fully explained in the volume.[65]

Jörg-Geerd Arentzen has pointed out that there are really only *two* fundamentally different types of *mappamundi*: those based on the Greek view of the entire terrestrial hemisphere (the world maps) and those depicting a smaller cultural area, the inhabited tripartite world. These two types of images from different cultural origins exist side by side and are not viewed as being opposed to each other.[66] He believes that the intermediate type traditionally formed by the Beatus maps and the zonal maps integrating a T-O pattern in the Northern Hemisphere should be included in the "world map" category. For the early Middle Ages, there is no question that this simplification has merit. But as the medieval period wore on these two traditional types of map became less distinct, and the profound later modifications to them should be recognized in any classification.

The system developed for this chapter, summarized at the beginning of appendix 18.1, thus identifies four main categories: tripartite, zonal, quadripartite, and transitional (figs. 18.4–18.7). Figures 18.8 and 18.9 show the absolute and relative numbers of maps in each category from the eighth to the fifteenth centuries.

Tripartite

Since all the maps in Andrews's oecumenical category are broadly tripartite, this term has been adopted. Within this category, schematic and nonschematic types are recognized. The latter are by far the more complicated and carry a greater density of geographical information, and accordingly they have been renamed "nonschematic" rather than using Andrews's misleading term "simple."[67] Several subgroups within this category have also been recognized based on their predominant source, whether Orosius, Cosmas, or Higden, for example.[68] The tripartite category presented here thus includes those maps that represent the inhabited world of late Roman times with three continents. Each category can be divided further into classes according to whether they are clearly diagrammatic or whether, while preserving the general positions of the three continents, they are nonschematic in nature.

In these T-O *mappaemundi*, the parts of the T are represented by the three major waterways believed by medieval scholars to divide the three parts of the earth: Tanais (the river Don) dividing Europe and Asia; the Nile dividing Africa and Asia; and the Mediterranean Sea dividing Europe and Africa.[69] In most cases the four cardinal directions are provided in Latin: Septentrio (*septemtriones*—the seven plow-oxen from the stars of the Great Bear or Little Bear); Meridies (for the position of the sun at midday); Oriens (from the direction of the rising sun); and Occidens (from the direction of the setting sun).

Zonal

This category of maps corresponds broadly to Andrews's term "hemispheric." The grounds for changing his terminology are that some tripartite maps belonging to the first category also represent part of the Southern Hemisphere. This general class of maps is characterized by orientation to the north or south and the representation of latitudinal zones or *climata*.

Quadripartite

Intermediate between the tripartite and the zonal categories of *mappaemundi* is a third category, here named "quadripartite" (corresponding broadly to Andrews's "intermediate"), which contains maps bearing the characteristics of each. Although these are not numerous, they are sufficiently distinctive to warrant a separate category.

Transitional

One shortcoming in all previous classifications is their inability to accommodate what is recognized here as a profound change in *mappaemundi* that took place in the fourteenth and fifteenth centuries. The late maps included in this category differ fundamentally from the Macrobian or Sallustian models of the late Roman world

---

64. Uhden, "Herkunft und Systematik" (note 7).

65. Destombes, *Mappemondes* (note 31). For example, the annotation "AZ" on pages 30, 31, and 48 is not explained, nor is the meaning of A1, A2, and A3 in the explanation of symbols on page 29.

66. Arentzen, *Imago Mundi Cartographica*, 321 (note 32).

67. Andrews, "Classification of Mappae Mundi," 69 (note 18).

68. See John B. Conroy, "A Classification of Andrews' Oecumenical Simple Medieval World Map Species into Genera" (M.S. thesis, University of Wisconsin, Madison, 1975), who also strongly questioned Andrews's terminology (pp. 209–16) but deferred to it in his thesis.

69. The biblical division of the world among the three sons of Noah is described in Genesis 10. Gervase of Tilbury points out that, as the firstborn, it is appropriate that Shem have the most land; see Gervase of Tilbury, *Otia imperialia* 2.2; one edition is *Otia imperialia*, ed. Felix Liebrecht (Hanover: C. Rümpler, 1856).

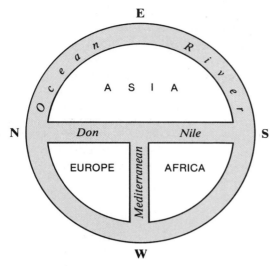

FIG. 18.4. THE TRIPARTITE TYPE OF *MAPPAMUNDI*. Also known as the T-O category, this type can be further subdivided into schematic (as shown in this diagram) or non-schematic, in which the general tripartite pattern is preserved but considerable embellishments of content are added.

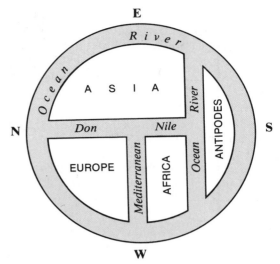

FIG. 18.6. THE QUADRIPARTITE TYPE OF *MAPPA-MUNDI*. This category includes characteristics of both the tripartite and zonal types, consisting of a tripartite model in the Northern Hemisphere and, in the Southern Hemisphere, a fourth continent, either uninhabited or inhabited by the Antipodeans.

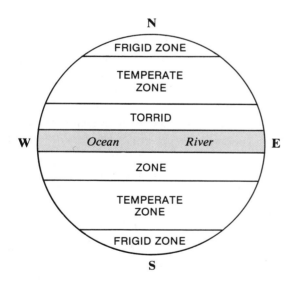

FIG. 18.5. THE ZONAL TYPE OF *MAPPAMUNDI*. The north- or south-oriented maps showing the parallel zones of the Greek *climata* form the second main category of *mappae-mundi*. They consist of a central uninhabited hot equatorial zone flanked by two inhabited temperate zones, and cold uninhabited zones in the polar areas.

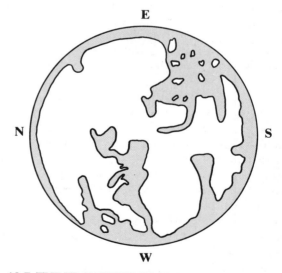

FIG. 18.7. THE TRANSITIONAL TYPE OF *MAPPAMUNDI*. These maps differ sufficiently from the previous three main types to warrant their own category. Dating from the fourteenth and fifteenth centuries, they show the influence of the portolan charts, particularly in the Mediterranean, and later, the world views of Ptolemy as the *Geography* was integrated into Western cartographic thought.

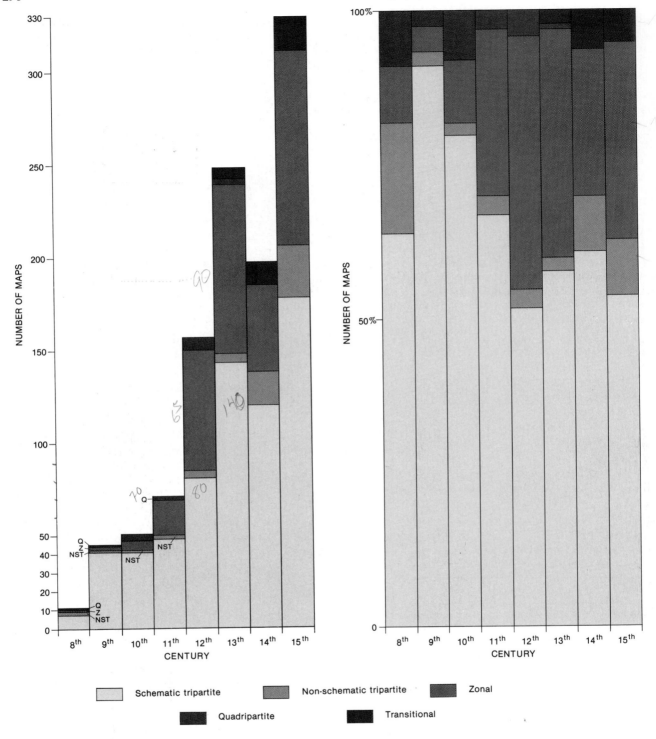

CENTURY

NUMBER OF MAPS

CENTURY

NUMBER OF MAPS

☐ Schematic tripartite   ☐ Non-schematic tripartite   ■ Zonal

■ Quadripartite   ■ Transitional

FIGS. 18.8. and 18.9. EXTANT *MAPPAEMUNDI*: ABSO-
LUTE AND RELATIVE NUMBERS BY CATEGORY FROM
THE EIGHTH TO THE FIFTEENTH CENTURY. Based on
the tables in Marcel Destombes, ed., *Mappemondes A.D.*

*1200–1500: Catalogue préparé par la Commission des Cartes
Anciennes de l'Union Géographique Internationale* (Amsterdam: N. Israel, 1964), 21–23, with modifications.

and anticipate in many ways the Renaissance. They have as their basis the configuration of the Mediterranean commonly found in portolan charts and rely in some degree on contemporary exploration, especially the Portuguese voyages to the Atlantic islands and along the west coast of Africa. The various regional or historical traditions, such as that of the Catalan chartmakers or the Ptolemaic influence, can provide the basis for further subdivision.

MAIN PERIODS OF *MAPPAEMUNDI*

It has long been recognized that the Middle Ages were not an undifferentiated millennium of ignorance and disorder between two periods of enlightened civilization. Not only were engineering, architecture, and mechanics greatly advanced during this period, but the humanistic legacy of Greek and Roman civilization was in evidence in every century. There are, however, fundamental differences among the maps that make it possible for the historian of cartography to recognize four major subperiods for *mappaemundi*, the last three of which were marked by their own renaissances. As with all historical periods, these interlock and overlap rather than meeting neatly at common boundaries. The first period—from the beginning of the fifth century to the end of the seventh, corresponding approximately to the patristic period of the fathers of the church from Lactantius (ca. 240–320) to Gregory the Great (ca. 540–604)—saw three fundamental cartographic traditions, here named after the authors who popularized them: Macrobius (ca. 395–436), Orosius (ca. 383–post 417), and Isidore (ca. 560–636). These three types of maps were to continue to have great influence in the rest of the Middle Ages, coexisting in derived forms until the Renaissance. In the second period, from the beginning of the eighth century to the beginning of the twelfth, the accelerated production of books and manuscripts for schools in cathedrals and monasteries in the "Carolingian renaissance" of the eighth century heralded what Bagrow called "the golden age of Church cartography."[70] The third period, from the beginning of the twelfth century to the end of the thirteenth, saw the influx of dozens of Arabic and Greek classics into western Europe, especially the *Almagest* of Ptolemy: Haskins called it "the renaissance of the twelfth century."[71] Finally, from the beginning of the fourteenth century to the middle of the fifteenth, we can identify a transitional period between the Middle Ages and the Renaissance, with world maps that have the characteristics of both.

MACROBIUS TO ISIDORE: THE LATE GRECO-ROMAN AND PATRISTIC PERIOD (CA. 400 TO CA. 700)

After the administrative division of the Roman Empire in the fourth century and following the establishment of Constantinople as the eastern capital, the secular influence of the Greco-Roman civilization went into decline, and the church enjoyed formal recognition and a steady growth of its power as a state authority. The maps of this first period were shaped by two opposing streams of thought: the Greco-Roman philosophical tradition and the teachings of the church fathers. The pagan geographical writings of the late Latin authors Macrobius, Martianus Capella, and Solinus were among the most influential in the Western world and had their basis in the works of Pliny and Pomponius Mela as well as in a theoretical Greek tradition handed down from Pythagorean times to Posidonius. Both Macrobius and Martianus Capella were to transmit parts of this tradition to the later Middle Ages. Solinus, on the other hand, copied Pliny and Pomponius Mela (without the slightest acknowledgment) in his *Collection of Remarkable Facts* and earned for himself the nickname "Pliny's ape." Nevertheless, the *Collection* provided a ready compendium of much of the geographical mythology that is found on *mappaemundi* up to the fifteenth century and was the subject of numerous printed editions.[72] Although it was immensely popular, the *Collection* provides a striking example of how classical science deteriorated in the Middle Ages through constant borrowing and plagiarism.

The attitude of the early church fathers to the pagan desire for knowledge was mixed. The church had no specific rulings on matters geographic or cosmographic and at worst regarded them as irrelevant to the Christian life. Lactantius (early fourth century) declared that scientific pursuits were unprofitable, and Saint Damian asked, "What can Christians gain from science?"[73] On the other hand, Saint Jerome (340–420), who is known to have been fascinated by and devoted to pagan learning, is traditionally considered to have compiled maps of Palestine and Asia, yet these are known only from late (twelfth-century) recensions. Certainly he was aware of the way in which maps could express information concisely, because he refers to "those who draw a region of the world on a small tablet."[74]

70. Bagrow, *History of Cartography*, 42 (note 55).

71. Haskins, *Renaissance of the Twelfth Century* (note 50).

72. The standard edition of Solinus is Gaius Julius Solinus, *Collectanea rerum memorabilium*, ed. Theodor Mommsen (Berlin: Weidmann, 1895). See also William Harris Stahl, *Roman Science: Origins, Development, and Influence to the later Middle Ages* (Madison: University of Wisconsin Press, 1962), and Beazley, *Dawn of Modern Geography*, 1:248–73 (note 17).

73. Bagrow, *History of Cartography*, 41 (note 55).

74. "Sicut ii qui in brevi tabella terrarum situs pingunt. . ." St. Jerome, *Epistola* 60, pt. 7 (336).

Macrobius

The type of *mappamundi* known as the Macrobian, or zonal map, is derived from the cosmographic section of Macrobius's early fifth-century commentary on Cicero's *Dream of Scipio* (51 B.C.). This in turn derived its cosmography from Posidonius (ca. 135 to ca. 51–50 B.C.), Serapion of Antiocheia (second or first century B.C.), Crates of Mallos (ca. 168 B.C.), Eratosthenes (ca. 275–194 B.C.), and—ultimately—from a Pythagorean concept. The earliest stage in this sequence about which anything is known starts with Crates, who made a large globe with four inhabited quarters separated by two belts of ocean that divided the hemispheres into north, south, east, and west.[75] Two of these continents constituted the known hemisphere, separated by an ocean river, Alveus Oceani, thought to flow just below the surface of the sea. This hemisphere was divided into five climatic zones (six if the central zone is considered to be divided by the ocean river) following parallels of latitude. The width of each zone conforms to precise measurements prescribed by Macrobius (fig. 18.10).[76] The two polar zones were held to be frigid and uninhabitable, and the equatorial zone, *zona perusta*, uncrossable because of its heat. It was the temperate zones between these two extremes that were habitable. The southern temperate zone, according to the original Greek concept, was inhabited by the Antipodeans. Over 150 *mappaemundi* drawn according to the Macrobian schema are found in manuscripts of the *Commentary on the Dream of Scipio* from the ninth century to the fifteenth, and throughout several other works such as the *Liber floridus* of Lambert of Saint-Omer (ca. 1120) and the *De philosophia* of William of Conches (ca. 1130).[77]

Often associated with Macrobius is the fifth-century encyclopedist Martianus Capella (fl. 410–39), who continued to popularize the zonal map in his *Marriage of Philology and Mercury*. This was an allegorical treatise on the seven liberal arts, the trivium of grammar, dialectic, and rhetoric and the quadrivium of geometry, arithmetic, astronomy, and music. Martianus's cosmographical writings were to be directly used in the *Liber floridus*.[78]

Orosius

The second major source used for *mappaemundi* of this period is the text of Paulus Orosius's *History against the Pagans*. The outstanding difference between Orosius's text and those of Macrobius and Martianus Capella is that it was directed against pagan writings. Orosius's initial encouragement seems to have come from Saint Augustine (354–430), to whom the book was dedicated.[79]

FIG. 18.10. THE MACROBIAN MODEL OF *MAPPA-MUNDI*. Based on a variety of Greek authors, this world map shows the five zones or *climata* of the earth, in which the tropical zone is divided by the "ocean river." The widths of the zones conform to precise measurements stated by Macrobius: 36, 30, 24 + 24, 30, 36 (in degrees from pole to pole). Diameter of the original: 14.3 cm. From a printed edition of Macrobius's *In somnium Scipionis expositio* (Brescia, 1485). By permission of The Huntington Library, San Marino, California (HEH 91528).

75. See above, pp. 162–63 and fig. 10.2.

76. Jacques Flamand, *Macrobe et le néo-Platonisme latin, à la fin du IVᵉ siècle* (Leiden: E. J. Brill, 1977), 464–82.

77. William Harris Stahl, "Astronomy and Geography in Macrobius," *Transactions and Proceedings of the American Philological Society* 35 (1942): 232–38, and his edition of Macrobius, *Commentary on the Dream of Scipio*, ed. and trans. William Harris Stahl (New York: Columbia University Press, 1952; second printing with supplementary bibliography, 1966).

78. William Harris Stahl, *The Quadrivium of Martianus Capella: Latin Traditions in the Mathematical Sciences, 50 B.C.–A.D. 1250* (New York: Columbia University Press, 1971), and Martianus Capella, *The Marriage of Philology and Mercury*, trans. William Harris Stahl and Richard Johnson with E. L. Burge (New York: Columbia University Press, 1977), vols. 1 and 2, respectively, of the series Martianus Capella and the Seven Liberal Arts. On Lambert of Saint-Omer, see *Lamberti S. Audomari Canonici liber floridus*, ed. Albert Derolez (Ghent: Story-Scientia, 1968), and Albert Derolez, ed., *Liber floridus colloquium* (Ghent: Story-Scientia, 1973).

79. Cortesão, *History of Portuguese Cartography*, 1:151–64 with a translation of certain geographical passages in appendix C, 241–42 (note 34). A modern translation of Paulus Orosius is found in *The Seven Books of History against the Pagans*, trans. Roy J. Deferrari (Washington, D.C.: Catholic University of America Press, 1964).

Orosius nowhere mentions a map in his text, but Bately reports a theory that in compiling his history he may have used a *mappamundi* in addition to the more expected textual sources.[80] She goes on to show, however, that there is no evidence (in his text) to support the idea that the use of a *mappamundi* was inevitably required.[81]

Orosius's text was widely used during all of the Middle Ages. In Cortesão's words, "practically every author after Orosius who wrote on geography and history, from St. Isidore to Roger Bacon and Dante, based his work on that of Orosius, drew more or less freely on it, or borrowed entirely from it."[82] Maps that are thought to bear at least some influence of the Orosian writings include the Albi map (eighth century), the Cotton "Anglo-Saxon map" (tenth century), the world map of Henry of Mainz (twelfth century), two Matthew Paris maps (thirteenth century), and the Hereford *mappamundi* (thirteenth century). However, the ambiguous nature of the primary evidence should always be borne in mind. In the absence of any map known to have been drawn by Orosius himself, it is not possible to decide whether maps bearing the influence of the Orosian writings were based on a single map tradition from the time of Orosius or whether several independent map traditions were based on later versions of the text. In addition, many other maps can be said to owe part of their origin to Orosius, though also modified by other authors, notably Isidore of Seville. In view of such problems, the stemma of the sources of the Orosian tradition clearly needs a detailed separate study.

Isidore

The third and best-known group of *mappaemundi* deriving from this period are the schematic tripartite diagrams of the world known as the T-O maps. Their name derives from the insertion of a capital T within an O, and the name was apparently coined in *La sfera*: "The drawing shows a "T" within an "O" as the earth was divided in three parts."[83]

Two major works of Isidore provide most of the maps in the schematic T-O category. Isidore of Seville (ca. 560–636) was one of the foremost encyclopedists and historians in the early Middle Ages. In about 600 he succeeded his brother as bishop of Seville, and through wide reading in both Roman and Christian sources he amassed an unparalleled fund of knowledge. This he distilled into some thirty titles, although his *Etymologiarum sive originum libri XX* (between 622 and 633) and the *De natura rerum* (between 612 and 615) are probably the most important. For his geographical and cosmographical knowledge, Isidore relied heavily on the popular writings of Roman authors and the early Chris-

tian fathers, particularly Ambrose, Augustine, Boethius, Cassiodorus, Lucretius, Lucan, Macrobius, Orosius, Pliny the Elder, Sallust, Servius, and Solinus. Isidore apparently knew no Greek, but this did not prevent his continuing the tradition of inserting Greek words and phrases in Latin texts that had been handed down through generations of compilers.[84]

In its broad sense, and in its derivations in later centuries, the Isidore schema is found in over 660 examples listed by Destombes. Its popularity in the Middle Ages is further illustrated by its appearance in several printed editions of the *Etymologies*. The original seventh-century Isidorian T-O no longer survives, but we may assume that it would have been a simple tripartite diagram. A second type of Isidorian map appears in the eighth century in which the Meotides Paludes (classical Palus Maeotis), or Sea of Azov, has been added. Since both versions are found in fifteenth-century printed editions of Isidore's *Etymologies*, we may assume that both continued as parallel traditions in the intervening period

80. Janet M. Bately, "The Relationship between Geographical Information in the Old English Orosius and Latin Texts Other Than Orosius," in *Anglo-Saxon England*, ed. Peter Clemoes (Cambridge: Cambridge University Press, 1972–), 1:45–62, esp. 45–46.

81. Bately, "Orosius," 62 (note 80).

82. Cortesão, *History of Portuguese Cartography*, 1:156 (note 34).

83. "Un T dentro ad un O mostra il disegno—Chome in tre parte fu diviso il mondo." Leonardo di Stagio Dati, trans. Goro (Gregorio) Dati, *La sfera* 3.11 (ca. 1425). See *La sfera: Libri quattro in ottava rima*, ed. Enrico Narducci (Milan: G. Daelli, 1865; reprinted [Bologna]: A. Forni, 1975), where it is pointed out (p. vi) that a manuscript in the Biblioteca Nazionale, Florence attests to the translation from Latin into Italian by Leonardo's brother Goro. The transcription given here is according to Roberto Almagià, *Monumenta cartographica Vaticana*, 4 vols. (Rome: Biblioteca Apostolica Vaticana, 1944–55), vol. 1, *Planisferi, carte nautiche e affini dal secolo XIV al XVII esistenti nella Biblioteca Apostolica Vaticana*, 118.

84. Isidore of Seville, *Traité de la nature*, see *Traité de la nature*, ed. Jacques Fontaine, Bibliothèque de l'Ecole des Hautes Etudes Hispaniques, fasc. 28 (Bordeaux: Féret, 1960). Ernest Brehaut, *An Encyclopedist of the Dark Ages: Isidore of Seville*, Studies in History, Economics and Public Law, vol. 48, no. 1 (New York: Columbia University Press, 1912). Clara LeGear, in *Mappemondes*, ed. Destombes, 54 (note 31), says that Isidore was well versed in Greek, Latin, and Hebrew, but Haskins and Stahl disagree. See Haskins, *Mediaeval Science*, 279 (note 50), and Stahl, *Roman Science*, 216 (note 72). See also Jacques Fontaine, *Isidore de Séville et la culture classique dans l'Espagne visigothique*, 2 vols. (Paris: Etudes Augustiniennes, 1959); Fritz Saxl, "Illustrated Mediaeval Encyclopaedias: 2. The Christian Transformation," in his *Lectures*, 2 vols. (London: Warburg Institute, 1957), 1:242–54; Wesley M. Stevens, "The Figure of the Earth in Isidore's 'De Natura Rerum,'" *Isis* 71 (1980): 268–77; and Ingeborg Stolzenberg, "Weltkarten in mittelalterlichen Handschriften der Staatsbibliothek Preußischer Kulturbesitz," in *Karten in Bibliotheken: Festgabe für Heinrich Kramm zur Vollendung seines 65. Lebensjahres*, ed. Lothar Zögner, Kartensammlung und Kartendokumentation 9 (Bonn-Bad Godesberg: Bundesforschungsanstalt für Landeskunde und Raumordnung, Selbstverlag, 1971), 17–32, esp. 20–21.

(figs. 18.11 and 18.12).[85] A further development, dating from at least the thirteenth century, was the addition of a representation of paradise and its four rivers in a rectangle.

An intermediate type between the Isidorian tripartite world with the representation of the Sea of Azov and the Beatus maps (discussed below) contains the tripartite diagram joined by a fourth continent that is sometimes shown as inhabited, sometimes not. This fourth continent is either in the Southern Hemisphere or tacked on strangely at a tangent to the circle representing the traditional known world, without apparent regard for geographical position (fig. 18.13). The earliest known form of this type of *mappamundi* (fig. 18.14) was thought by Miller to be a late seventh- or early eighth-century vestige in a palimpsest with, otherwise, ninth-century contents. Miller based his inference on the differences in lettering on the map, claiming that the rustic capitals are much earlier than the other hands. If this is correct, this is the earliest medieval *mappamundi* known and would occupy a key transitional place between the Roman and medieval traditions.[86]

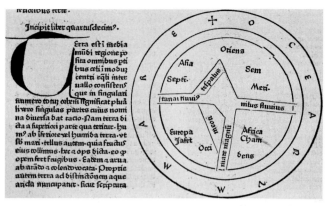

FIG. 18.12. ISIDORIAN T-O MAP WITH THE SEA OF AZOV. A more elaborate version of Isidore's original schema (fig. 18.11.), this includes the addition of the Meotides Paludes, the Sea of Azov.
Diameter of the original: 11.1 cm. From Isidore of Seville, *Etymologiae* (Cologne, 1478). By permission of The Huntington Library, San Marino, California (HEH 89025).

BEDE TO LAMBERT OF SAINT-OMER (CA. 700 TO CA. 1100)

Despite the renewed interest in natural science in this second subperiod, the *mappaemundi* of this time tend to be secondary versions of the Greco-Roman sources transmitted largely through the works of Macrobius, Orosius, and Isidore. It is, however, the first period in the entire history of European cartography to yield a reasonable sample of artifacts. Over 175 *mappaemundi* dating from the eighth through the eleventh century are known to have survived. They are largely in historical and geographical texts, copies of the Psalter, and the *Commentary* by Beatus of Liebana. Moreover, library catalogs of the period contain frequent mentions of *mappaemundi* as apparently separate items. Among texts by individual authors, three works by the Venerable Bede

FIG. 18.11. ISIDORIAN T-O MAP. The simplest version of Isidore of Seville's original type of schematic *mappamundi* is here reproduced from the fifteenth-century printed version. Diameter of the original: 6.4 cm. From Isidore of Seville, *Etymologiarum sive originum libri XX* (Augsburg: Günther Zainer, 1472). Courtesy of The Newberry Library, Chicago.

85. Isidore of Seville, *Etymologies*; see *Etymologiarum sive originum libri XX* (Augsburg: Günther Zainer, 1472) and several other incunable editions. See Rodney W. Shirley, *The Mapping of the World: Early Printed World Maps 1472–1700* (London: Holland Press, 1983), 1, who does not, however, describe or reproduce the second version with the Sea of Azov.

86. Stiftsbibliothek St. Gallen, Codex 237. Miller, *Mappaemundi*, 6:57–58 (note 9), dates it to the end of the seventh century, Destombes, *Mappemondes*, 30, map 1.6 (note 31), dates it to the eighth century, and Kamal, *Monumenta cartographica* (note 53), dates it to the ninth century. But as Ker states on page 370 of his review of *Mappemondes* (note 2), "MSS are not dated in the 8th century without reference to *Codices Latini Antiquiores*: if they are not in *CLA* it is best to think again." This map is not in *CLA*. The map clearly needs further detailed study, but following Miller, we have tentatively placed it on the stemma in figure 18.15 at the end of the seventh century.

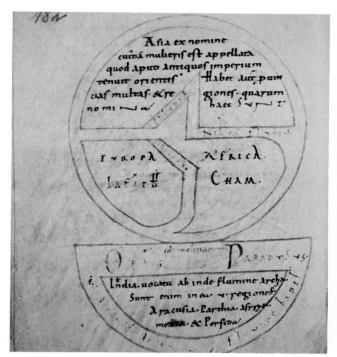

FIG. 18.13. ISIDORIAN T-O MAP WITH THE FOURTH CONTINENT. The distinguishing characteristic of this type is that a fourth continent in the style of the Beatus maps (see plate 13) is added to either variant of Isidore's schematic T-O map. In this example, India is shown in an unusual western location.
Diameter of the original: 11 cm. By permission of the Stiftsbibliothek, Einsiedeln (Codex Eins. 263 [973], fol. 182r).

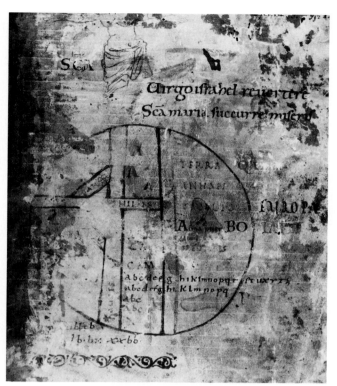

FIG. 18.14. *MAPPAMUNDI* FROM ISIDORE. According to Miller, the main part of this map dates from the late seventh or early eighth century, with additions in later hands. He therefore believes it represents the oldest known extant *mappamundi*.
Diameter of the original: 13.5 cm. From a ninth-century codex of Isidore's *Etymologies*. By permission of the Stiftsbibliothek, Saint Gall (Codex 237, fol. 1r).

(672/73–735) contain *mappaemundi* (of which fifteen examples are extant).[87] Nor is the period devoid of references to monumental maps that perhaps demonstrate a degree of infusion of geographical interest into everyday life: for example, Pope Zacharias (pope 741–52) is known to have had a world map painted on the wall of the Lateran palace,[88] and Charlemagne possessed three silver tables described in the *Vita Karoli Magni*: one of Constantinople, one of Rome, and a third, a "description of the whole world," which has been reconstructed and interpreted by Estey and others as a celestial map.[89]

The maps in the surviving manuscripts of the *Commentary on the Apocalypse of Saint John* by the Benedictine abbot Beatus of Liebana (fl. 776–86) provide perhaps the only spark of innovation.[90] Two main types of maps are found in the Beatus *Commentary*. Best known are the large, usually rectangular, maps (of which fourteen survive) that can be traced back to the now lost prototype of 776–86. They form a well-defined group

87. Destombes, *Mappemondes*, 35–36 (note 31). See also T. R. Eckenrode, "Venerable Bede as a Scientist," *American Benedictine Review* 21 (1971): 486–507.

88. Schulz, "Moralized Geography," 448 (note 24).

89. Eckenrode, "Venerable Bede as a Scientist," 486–507 (note 87). F. N. Estey, "Charlemagne's Silver Celestial Table," *Speculum* 18 (1943): 112–17. Estey's interpretation is based on several authors, including Georg Thiele, *Antike Himmelsbilder, mit Forschungen zu Hipparchos, Aratos und seinen Fortsetzern und Beiträgen zur Kunstgeschichte des Sternhimmels* (Berlin: Weidmann, 1898), 141 n. 1, and a passage in the *Annales Bertiniani* that describes both the table and its destruction in 842: Georg Waitz, ed., *Annales Bertiniani, Scriptores rerum Germanicorum: Monumenta Germaniae historica* (Hanover: Impensis Bibliopolii Hahniani, 1883), 4, 27.

90. The most recent work is Peter K. Klein, *Der ältere Beatus-Kodex Vitr. 14-1 der Biblioteca Nacional zu Madrid: Studien zur Beatus-Illustration und der spanischen Buchmalerei des 10. Jahrhunderts* (Hildesheim: Georg Olms, 1976), but the maps are not heavily emphasized. In some works, such as Georgiana Goddard King, "Divagations on the Beatus," in *Art Studies: Medieval, Renaissance and Modern*, 8 vols., ed. members of Departments of Fine Arts at Harvard and Princeton universities (Cambridge: Harvard University Press, 1923–30), 8:3–58, descriptions of the maps (as distinct from the other illustra-

and fall into their own transitional category in appendix 18.1. Their graphic style can be described as "Mozarabic," that is to say, showing the Arabic influence in Spain, with bright, opaque colors and arabesque illumination. They are all oriented to the east, with paradise enclosed in a square vignette with the four rivers flowing from it (plate 13). Around the edge is the ocean sea containing decorative representations of fishes in an unmistakable Islamic style. Their main characteristic however, is the representation of a fourth continent in addition to the traditional tripartite world. The context of the map is evangelistic, following the subject of the work in which they appear: the apostles were to go into every corner of the earth, including the fourth continent, which Beatus considered to be inhabited. Various legends are found written on the representations of this continent, to inform the viewer that: "outside the three parts of the world there is a fourth part, the farthest from the world, beyond the ocean, which is unknown to us on account of the heat of the sun. We are told that the Antipodeans, around whom revolve many fables, live within its confines."[91]

The second type of map found in the manuscripts of the *Commentary* are small Isidorian maps that also show a fourth continent. Their occurrence in the same manuscripts as the large Beatus maps has led Menéndez-Pidal to postulate that the latter were derived from the former.[92] A general stemma for the large maps is provided in figure 18.15.

The *Liber floridus* by Lambert, canon of Saint-Omer, marks the end of this second period of *mappaemundi*. The original illustrated manuscript of 1120 is still preserved in Ghent and is a text in the Isidorian tradition of great encyclopedias. Despite the breadth of knowledge it contains, there is nothing startlingly new. Lambert's sources are as might be expected. He usually cites them by name: Pliny, Macrobius, Martianus Capella, the Latin fathers, Isidore, and Bede.[93]

HENRY OF MAINZ TO RICHARD OF HALDINGHAM (CA. 1100–1300)

Whatever effect the Crusades (1096–1270) may have had on medieval Europe in general, they had little direct effect on the content of the *mappaemundi*. Lach was able to write "the Crusades themselves changed almost nothing in Europe's pictorial image of Asia," and the same can be said for other continents.[94] There was a great dissemination of knowledge about the Holy Land and the routes of pilgrimage to it, however, that reached most segments of the population and was reflected in cartography, such as in the regional and itinerary maps of Matthew Paris.

At about the same time, the influx of new knowledge

into western Europe between 1100 and 1200, some through Italy and Sicily but most through the Muslim authors in Spain, was facilitated by dozens of translations of Arabic and Greek classics, particularly in philosophy, mathematics, astronomy, and the physical and natural sciences.[95] Ignorance of the Greek language by most scholars in western Europe, with some important exceptions, effectively closed the early and High Middle Ages to the best classical work. For cartography, this meant that Ptolemy's *Almagest* was unavailable to the non-Greek reader between the second and twelfth centuries, and his *Geography* between the second and the fifteenth.

Table 18.3 summarizes the dates of the main translations of texts of interest to cosmography and cartography. Although the early translations were literal and the choice of works to be translated hardly systematic, they were convenient and popular and eventually stimulated original thinking. The main legacy of the "renaissance of the twelfth century" lay in the expounding of the principles of empirical science. Such scholars in the following century as Roger Bacon (ca. 1214–94), John Duns Scotus (ca. 1265–1306), and William of Occam (ca. 1290 to ca. 1349) were all Franciscans, and their work was a natural outgrowth of the philosophy of this movement with its intense curiosity about the natural world. Founded in 1209, the Franciscan order nurtured many distinguished experimental scientists and travelers whose interests frequently turned to cosmography and geography, including the compilation of *mappaemundi*, in a way that was mirrored several centuries later by the Jesuits. John of Plano Carpini, a companion and disciple of Saint Francis of Assisi, undertook the first of the missionary journeys to Asia (1245–47) as an envoy of Pope Innocent IV. The stated aims of the journeys were to discover the history and customs of the Mongols, convert the Grand Khan, and seek an alliance with him against their common enemy, the Muslims.

tions) are deliberately avoided. See also Destombes, *Mappemondes*, 40–42 and 79–84 (note 31), and Jesús Domínguez Bordona, *Die spanische Buchmalerei vom siebten bis siebzehnten Jahrhundert*, 2 vols. (Florence, 1930).

91. Turin, Biblioteca Nazionale Universitaria, MS. I.II.1 (old D.V.39), author's translation.

92. G. Menéndez-Pidal, "Mozárabes y asturianos en la cultura de la alta edad media en relación especial con la historia de los conocimientos geográficos," *Boletin de la Real Academia de la Historia* (Madrid) 134 (1954): 137–291.

93. Derolez, *Liber floridus colloquium*, 20 (note 78).

94. Donald F. Lach, *Asia in the Making of Europe*, 2 vols. in 5 (Chicago: University of Chicago Press, 1965–77), 1:24.

95. On the transmission of Arabic science, see Haskins, *Renaissance of the Twelfth Century* (note 50), and Richard Walzer, *Arabic Transmission of Greek Thought to Medieval Europe* (Manchester: Manchester University Press, 1945).

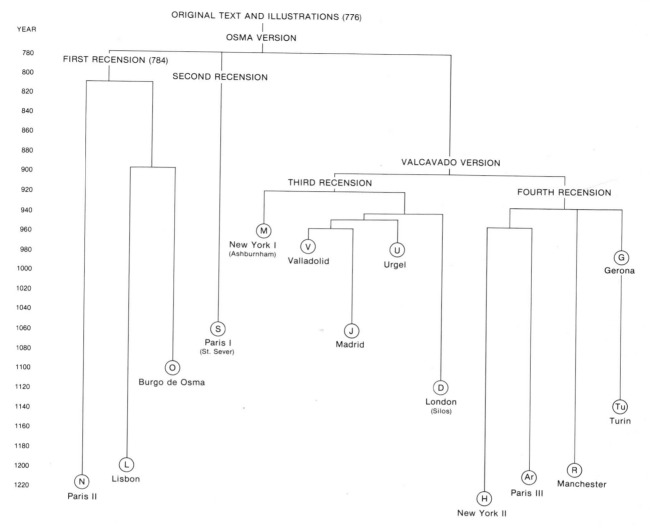

YEAR

ORIGINAL TEXT AND ILLUSTRATIONS (776)

OSMA VERSION

FIRST RECENSION (784)

SECOND RECENSION

VALCAVADO VERSION

THIRD RECENSION

FOURTH RECENSION

FIG. 18.15. GENERAL STEMMA FOR THE LARGE BEA-
TUS MAPS. This genealogy shows the lineage of the extant
Beatus manuscripts containing full-page maps. Two main versions or drafts are shown, from the eighth and ninth/tenth
centuries respectively, each with two recensions. For full references to the extant manuscripts, designated by letters in circles, see appendix 18.2. The circles are not intended to imply

the shape of the maps. Names in parentheses are those commonly used for identification. Adapted from Peter K. Klein,
*Der ältere Beatus-Kodex Vitr. 14–1 der Biblioteca Nacional
zu Madrid: Studien zur Beatus-Illustration und der spanischen
Buchmalerei des 10. Jahrhunderts* (Hildesheim: Georg Olms,
1976).

Another Franciscan friar, William of Rubruck (ca. 1200
to after 1256) was sent on the same mission by Louis
IX, and the detailed report of his observations was later
used by Roger Bacon. Unfortunately, no maps survive
in these reports.[96]

For practical navigation, the major achievement was
by Ramón Lull (ca. 1233–1315), a Majorcan Franciscan, who wrote about the science of navigation based
on his own direct experience at sea. He was also the first
to describe the nautical chart. But it was in the contribution of Roger Bacon that the Franciscan aptitude for

cartography was most evident, as we see in his discussion
of projections and coordinate systems in the *Opus majus*
(1268).[97]

The burgeoning of knowledge and cosmopolitan
awareness in eleventh-century Europe led naturally to
the development of several kinds of permanent institu-

96. Cortesão, *History of Portuguese Cartography*, 1:191 (note 34).
97. Roger Bacon, *The Opus Majus of Roger Bacon*, 2 vols., trans.
Robert Belle Burke (1928; reprinted New York: Russell and Russell,
1962). Cortesão, *History of Portuguese Cartography*, 1:193–98 (note
34). See also the section on projections and coordinate systems below.

TABLE 18.3 Dates of Translation of the Main Greek and Arabic Manuscripts of Cartographic Interest into Latin

Author	Dates	Work	Latin Translator	Place and Date of Translation
Al-Khwarizmi	9th century	Astronomical tables	Adelard of Bath	ca. 1126
Aristotle	384–322 B.C.	*De caelo et mundo*	Gerard of Cremona	Toledo, 12th century
Averroes	1126–98	*De caelo et mundo*	Michael Scot	Early 13th century
Euclid	ca. 330–260 B.C.	*Elements*	Adelard of Bath	Early 12th century
Ptolemy	ca. 90–168	*Almagest*	Gerard of Cremona	Toledo, 1175
Ptolemy	ca. 90–168	*Geography*	Jacobus Angelus	Florence, 1406–7
Ptolemy	ca. 90–168	*Planisphaerium*	Hermann of Carinthia	Toulouse, 1143

Sources: Adapted from Jean Gimpel, *The Medieval Machine: The Industrial Revolution of the Middle Ages* (New York: Penguin Books, 1977), 176–77; G. J. Toomer, "Ptolemy," in *Dictionary of Scientific Biography*, 16 vols., ed. Charles Coulston Gillispie (New York: Charles Scribner's Sons, 1970–80), 11:186–206; and George Sarton, *Introduction to the History of Science*, 3 vols. (Baltimore: Williams and Wilkins, 1927–48), 2:173.

tions of higher learning, both practical and theoretical.[98] The University of Salerno (tenth century) is the earliest such institution: it specialized in medicine. Bologna, Paris, and Oxford were the twelfth-century ancestors of the modern universities in the sense of academic guilds, deriving their name not from the idea of universal knowledge but from the banding together of a universal group of professors and students. Of the four subjects taught in the quadrivium in medieval universities—arithmetic, astronomy, geometry, and music—the activity of cartography related directly to three.[99] The place of man in the terrestrial, celestial, and spiritual world was a central concern for medieval philosophers, and such geographical issues as the nature, shape, and size of the earth were of perennial interest.[100]

The universities of Oxford and Paris were particularly strong centers of a cosmographical and geographical culture that reached its climax in Europe in the thirteenth century. Sacrobosco (also known as John of Holywood or Halifax; d. 1256), though born in England and possibly educated at Oxford, was admitted as a member of the University of Paris in 1221. He is best known for his work *De sphaera*, which probably appeared in the 1220s or 1230s. It was a textbook for beginners in cosmography, fully illustrated with world maps and diagrams, and thanks to its clarity and brevity it enjoyed widespread use in multiple versions and printed editions until the seventeenth century, continuing to be used long after the Copernican theory had been accepted (fig. 18.16). It almost certainly predated the *De sphaera* of Robert Grosseteste (ca. 1175–1253), first chancellor of Oxford University and bishop of Lincoln.[101]

The English geographical culture in the thirteenth century is also revealed in the unusual circumstance that four important thirteenth-century *mappaemundi*—the Vercelli, "Duchy of Cornwall," Ebstorf, and Hereford

maps—either are English or appear to have strong English connections.[102] The Vercelli map (84 × 70–72 cm) (fig. 18.17), is the smallest of the three. It now resides in the Archivio Capitolare in Vercelli and has been dated by Carlo Capello to between 1191 and 1218. Its inspiration may well have been English. Capello believes that the map was carried to Vercelli by cardinal Guala-Bicchieri on his return from England about 1218–19 as papal legate to Henry III.[103] He also argues that the figure on the map of a king in Mauretania named "Philip" is intended to represent Philip II of France (1180–1223) and not Philip III (1270–85) (fig. 18.18).

98. Hastings Rashdall, *The Universities of Europe in the Middle Ages*, ed. F. M. Powicke and A. B. Emden (Oxford: Oxford University Press, 1936), and Charles Homer Haskins, *The Rise of Universities* (New York: Henry Holt, 1923).

99. The concept of the seven liberal arts gained its popularity largely from the *De nuptiis Philologiae et Mercurii* of Martianus Capella (see note 78), and the division into the trivium (grammar, rhetoric, and dialectic) and the more advanced quadrivium (music, arithmetic, geometry, and astronomy) dates from the time of Alcuin (735–804). The quadrivium provided the outline for the natural sciences that was filled out by the experimental studies of the twelfth-century Renaissance. See Rashdall, *Universities*, 34–36 (note 98).

100. See Wright, *Geographical Lore* (note 18).

101. A good summary of Sacrobosco's life is that by John F. Daly, "Sacrobosco," in *Dictionary of Scientific Biography*, 16 vols., ed. Charles Coulston Gillispie (New York: Charles Scribner's Sons, 1970–80), 12:60–63. See also Lynn Thorndike, ed. and trans., *The Sphere of Sacrobosco and Its Commentators* (Chicago: University of Chicago Press, 1949).

102. The major English figures of this period are listed in Charles Singer, "Daniel of Morley: An English Philosopher of the XIIth Century," *Isis* 3 (1920): 263–69. Along with the lack of important surviving *mappaemundi* from the European continent, there is a parallel lack of regional maps to compare with the Matthew Paris and Gough maps of Great Britain.

103. Carlo F. Capello, *Il mappamondo medioevale di Vercelli (1191–1218?)*, Università di Torino, Memorie e Studi Geografici, 10 (Turin: C. Fanton, 1976).

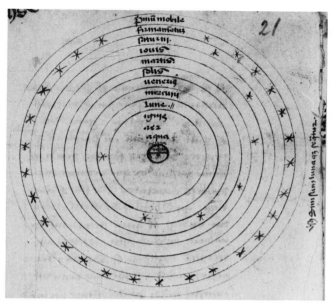

FIG. 18.16. THE TREATISE OF SACROBOSCO. Originally appearing in the early thirteenth century, Sacrobosco's *De sphaera* became a popular textbook of cosmography, being copied in many versions—both manuscript and printed—until the seventeenth century. His earth-centered diagrams of the solar system were thus in use long after they had been superseded by the Copernican theory. From a fifteenth-century manuscript of the work.
Diameter of the original detail: 10.3 cm. By permission of the Biblioteca Nacional, Lisbon (Codex ALC. 285, fol. 21).

On stylistic grounds, he similarly places the map earlier rather than later in the thirteenth century and draws particular attention to the fact that it is not centered on Jerusalem as were maps later in the century.

A parchment fragment of a *mappamundi* has recently been discovered among the records of the Duchy of Cornwall. Carbon dating at Oxford University has established that its most likely date is between 1150 and 1220. The fragment measures 61 centimeters high by 53 centimeters wide. The original circular map measured approximately 1.57 meters in diameter. The surviving segment of the map, depicting part of Africa, suggests an original form similar to the Vercelli, Hereford, and Ebstorf maps. Some details strongly resemble elements in these maps. For example, the fragment contains a gazetteer that alludes to the traditional classical surveyors, information that is incorporated into the Hereford map. In addition, the marginal text is similar to that found on the Hereford map. Several of the monstrous races are clearly shown in their traditional location. Forming a border along the bottom edge of the fragment is a series of finely executed line drawings of figures apparently depicting stages of life; each figure delivers

a cautionary message. They include a woman at vespers, an old man bent with age, a figure in purgatory holding a bowl of fire, and an angel (plate 14).[104]

The Ebstorf map—while its English connections are tenuous at best—has been linked to Gervase of Tilbury (ca. 1160–1235?) (fig. 18.19). Gervase was a teacher of canon law in Bologna who may possibly be identified with the provost of the abbey of Ebstorf who died in 1235. In his historical work *Otia imperialia* (1211), he refers to a "world map," and his text has been recognized as the latest known source of information from which the author of the Ebstorf map might have drawn.[105]

Discovered in the Benedictine abbey of Ebstorf in 1830 and made public in an article in a Hanover newspaper in 1832, the Ebstorf map was moved in 1834 to the Museum of the Historical Society of Lower Saxony in Hanover, where it remained until 1888. It was then taken to Berlin for restoration, at which point it was separated into thirty vellum sheets and photographed for the edition by Sommerbrodt. This remains the only full-sized photographic reproduction (unfortunately not in color).[106] It was returned to Hanover, where it was destroyed in an air raid in 1943. Since the original no longer exists, the accuracy of the existing facsimiles is crucial. Even as early as 1896, Miller had pointed out the problems associated with the Sommerbrodt photographic edition, which was touched up in the faded areas.[107] Miller's own edition was a hand-drawn copy reproduced in color and thus was also subjective in its interpretation.

The controversy surrounding the authorship and dating of the map has been well summarized by Arentzen.[108] The date of 1284 in arabic numerals on the map appears to have been added in a later hand, and the earliest date for its appearance is probably 1234, after the death of

104. I am indebted to Graham Haslam, archivist of the Duchy of Cornwall, for providing this paragraph and the transparency for plate 14. We look forward to the full study of the map that Dr. Haslam is planning to publish.

105. Miller, *Mappaemundi*, 5:75 (note 9). The ensuing controversy concerning Gervase's connection with the map is summarized by Arentzen, *Imago Mundi Cartographica*, 140 (note 32). As a supporter of Gervase's authorship, along with Richard Uhden, "Gervasius von Tilbury und die Ebstorfer Weltkarte," *Jahrbuch der Geographischen Gesellschaft zu Hannover* (1930): 185–200, may be added Jerzy Strzelczyk, *Gerwazy z Tilbury: Studium z dziejów uczoności geograficznej w średniowieczu*, monograph 46 (Warsaw: Zakład Narodowy im. Ossolińskich, 1970).

106. On the discovery, see Arentzen, *Imago Mundi Cartographica*, 138 (note 32). On its restoration and reproduction, see Ernst Sommerbrodt, *Afrika auf der Ebstorfer Weltkarte*, Festschrift zum 50-Jährigen Jubiläum des Historischen Vereins für Niedersachsen (Hanover, 1885).

107. Miller, *Mappaemundi*, 5:3 (note 9).

108. Arentzen, *Imago Mundi Cartographica*, 138–47 (note 32).

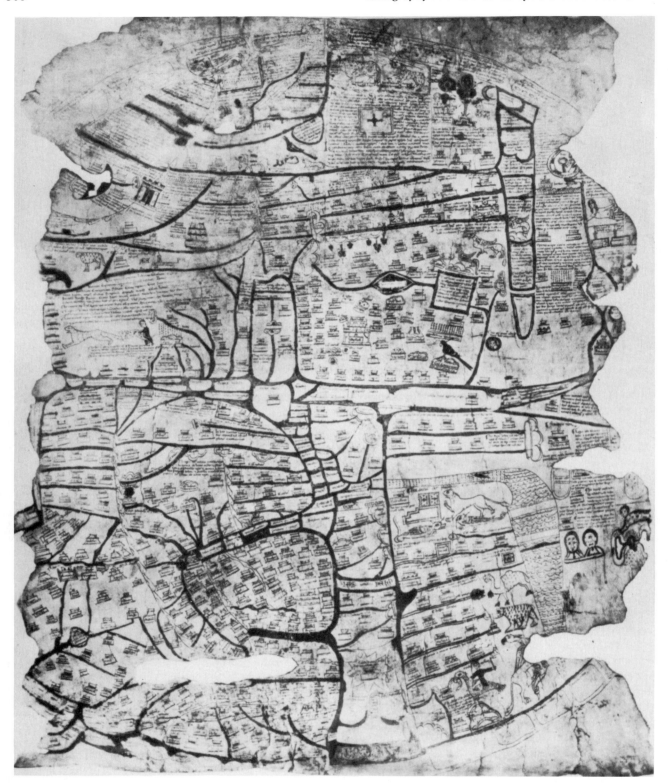

FIG. 18.17. THE VERCELLI MAP. One of two large thir-teenth-century *mappaemundi* to have survived (the other being the Hereford map), this too is probably of English origin. It is thought to have been brought to Italy about 1219 by a papal legate to Henry III.

Size of the original: 84 × 70–72 cm. From Marcel Destombes, ed. *Mappemondes A.D. 1200: Catalogue préparé par la Commission des Cartes Anciennes de l'Union Géographique Internationale* (Amsterdam: N. Israel, 1964), pl. XXIII. By permission of the Archivio Capitolare del Duomo di Vercelli.

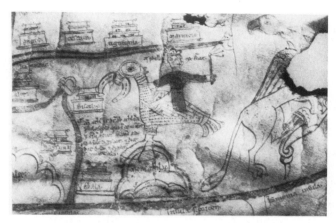

FIG. 18.18. THE VERCELLI MAP: DETAIL OF PHILIP. This figure, placed in Mauretania, is either Philip II of France (1180–1223) or Philip III (1270–85), probably the former. This provides a clue to its early thirteenth-century date. From Carlo F. Capello, *Il mappamondo medioevale di Vercelli (1191–1218?)*, Università di Torino, Memorie e Studi Geografici, 10 (Turin: C. Fanton, 1976), pl. 23. By permission of the Archivio Capitolare del Duomo di Vercelli.

Johannes Marcus, dean of the cathedral of Hildesheim, who may have ordered the map to be drawn.[109]

With dimensions of 3.58 by 3.56 meters, the Ebstorf map is the largest *mappamundi* to have been recorded. Although its main intended use was no doubt to demonstrate the historical events in the Christian life—for example, the burial places of Mark, Bartholomew, Philip, and Thomas are shown—the author also had some more directly practical use in mind, as he himself made clear. In the upper right-hand corner of the map, he writes: "it can be seen that [this work] is of no small utility to its readers, giving directions for travelers, and the things on the way that most pleasantly delight the eye".[110] We also find an allusion to the traditional cartographic proclamation of Julius Caesar: "How Julius Caesar first constructed [a *mappamundi*], for the breadth of the whole earth, legates having been sent, collecting the regions, provinces, islands, cities, quicksands, marshes, plains, mountains, and rivers as if to be seen on one page" (author's translations).[111]

It is the well-known Hereford map that represents the culmination of the Orosian type (fig. 18.20).[112] The map contains a clear and direct reference to its origin: "Orosius's description of the *ornesta* of the world, as displayed within."[113] Partly as a result of its sheer size (1.65 × 1.35 m), it contains more information than any other surviving pre-fifteenth-century *mappamundi*. In addition to Orosius and the Bible, its sources definitely include Isidore, Augustine, Jerome, Pliny, Strabo, and the Antonine itinerary. There is also a reference—unusual indeed for any medieval world map—to its authorship, in a note in the bottom left-hand corner:

Let all who have this history,
Or shall hear or read or see it,
Pray to Jesus in His Divinity,
To have pity on Richard of Haldingham and Lafford,
Who has made and planned it,
To whom joy in heaven be granted.[114]

Around the map's border on the left side, we read that the world began to be measured by Julius Caesar. In the lower left corner we find a drawing of the emperor Augustus handing out his edict (see fig. 12.4).[115] Pliny alludes to a large world map of Vipsanius Agrippa displayed in Rome at the time of the emperor Augustus (ca. A.D. 14), which may have resulted from the survey of the provinces ascribed by tradition to Julius Caesar.[116]

109. Richard Drögereit, "Die Ebstorfer Weltkarte und Hildesheim," *Zeitschrift des Vereins für Heimatkunde im Bistum Hildesheim* 44 (1976): 9–44, esp. 43.

110. "Que scilicet non parvam prestat legentibus utilitatem, viantibus directionem rerumque viarum gratissime speculationis directionem." Transcription from Walter Rosien, *Die Ebstorfer Weltkarte* (Hanover: Niedersächsisches Amt für Landesplanung und Statistik, 1952), 80.

111. Also in the lower right corner, before the quotation in note 105: "Quam Julius Cesar missis legatis per totius orbis amplitudinem primus instituit: regiones, provincias, insulas, civitates, syrtes, paludes, equora, montes, flumina quasi sub unius pagine visione coadunavit." Rosien, *Ebstorfer Weltkarte*, 80 (note 110).

112. The standard facsimile and description of the Hereford map is Gerald R. Crone, *The World Map by Richard of Haldingham in Hereford Cathedral*, Reproductions of Early Manuscript Maps 3 (London: Royal Geographical Society, 1954). Further research is found in his "New Light," 447–62 (note 21), and his " 'Is leigen fünff perg in welschen landt' and the Hereford Map," *Erdkunde* 21 (1967): 67–68. See also Destombes, *Mappemondes*, 197 (note 31).

113. "Descriptio Orosii de ornesta mundi sicut interius ostenditur." Transcription from the Royal Geographical Society facsimile (note 112), author's translation. The word *ornesta* is thought to refer generically to medieval world maps, derived from a contraction of *orosii mundi historia*. See Crone, "New Light," 448 (note 21).

114.   Tuz ki cest estoire ont,
            Ou oyront ou lirront ou veront,
            Prient a ihesu en deyte,
            De Richard de Haldingham e de Lafford eyt pite,
            Ki lat fet e compasse,
            Ki ioie en cel li seit done.
Transcription from the Royal Geographical Society facsimile (note 112), author's translation. See also Arthur L. Moir, *The World Map in Hereford Cathedral*, 8th ed. (Hereford: Friends of the Hereford Cathedral, 1977).

115. The text of the edict, "Exiit edictum ab Augusto Cesare ut describeretur huniversus orbis" (Luke 2:1) is above the Caesar's head and not completely shown in figure 12.4. Transcription from Royal Geographical Society facsimile (note 112). This follows the Vulgate: *Biblia sacra juxta vulgatam Clementinam* (Rome: Typis Societatis S. Joannis Evang., 1956). The modern translation reads: "In those days a decree was issued by the emperor Augustus for a registration to be made throughout the Roman world." The usual meaning of the word *describeretur* involves not simply registration but a *survey*, leading perhaps to a confusion by the author of the Hereford map between the two events (and the two Caesars).

116. See above, p. 205, and Beazley, *Dawn of Modern Geography*, 1:382 (note 17).

FIG. 18.19. THE EBSTORF MAP. This thirteenth-century *mappamundi* (destroyed in World War II) represents the world as the body of Christ. Christ's head is situated next to Paradise (for a detail, see fig. 18.2), the feet in the west, and the hands gathering in the north and south (for a detail, see fig. 18.3). Jerusalem, the navel of the world, is at the center.

Size of the original: 3.56 × 3.58 m. From Walter Rosien, *Die Ebstorfer Weltkarte* (Hanover: Niedersächsisches Amt für Landesplanung und Statistik, 1952). By permission of the Niedersächsisches Institut für Landeskunde und Landesentwicklung an der Universität Göttingen.

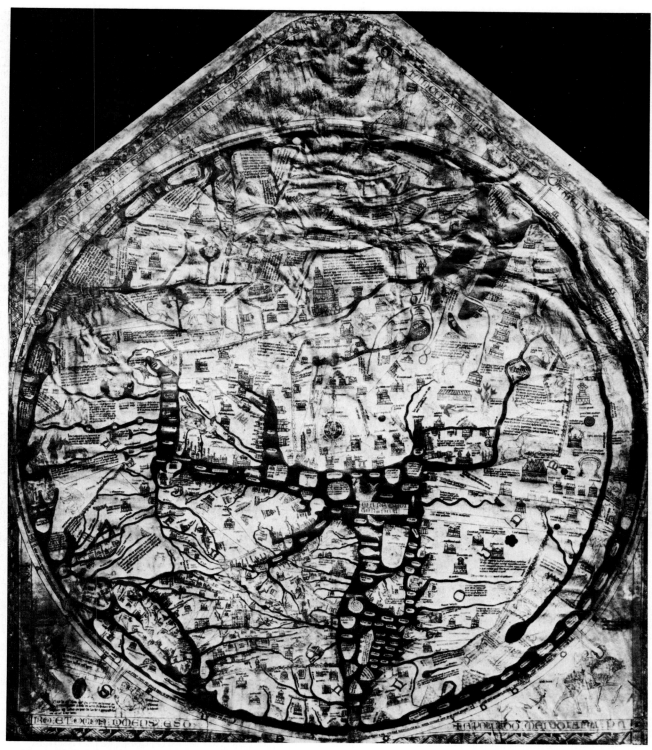

FIG. 18.20. THE HEREFORD MAP, CA. 1290. This famous map in Hereford Cathedral represents the culmination of the type based on the history of Paulus Orosius (fourth century). Its compiler, Richard de Bello, also drew on the works of Strabo, Pliny, Augustine, Jerome, the Antonine itinerary, and Isidore. See also figure 11.2.

Size of the original: 1.65 × 1.35 m.; the diameter is 1.32 m. From a negative of the original by permission of the Royal Geographical Society, London.

One of the issues surrounding the map relates to its authorship. Most authorities agree that these verses point to Richard of Haldingham, who has been identified as one Richard de Bello, prebend of Sleaford (Lafford) in the diocese of Lincoln in 1277. However, a Richard de Bello was also apparently prebend of Norton in the diocese of Hereford in 1305 and did not die until 1326. Some scholars, among them Denholm-Young and Emden, have thus argued that to span such a long career, there may have been *two* Richard de Bellos.[117] Yates has recently summarized the issue, leaning to the speculative conclusion that there was only one Richard de Bello and that he made the Hereford map. Yates also points out that further physical analysis of the map, especially of its pigments and calligraphy, might well reveal the period of time over which additions were made to it and thus increase the precision with which its contents might be dated.[118]

The close connection in content between the Hereford map and the early twelfth-century map by Henry of Mainz has been described by several authors.[119] The authorship of this map and of the manuscript of the *Imago Mundi* in which it appears has been the source of some confusion. While there is now general agreement that the basic text of the *Imago Mundi* is by Honorius of Autun,[120] the identity of the editor and the dedicatee, both named "Henry," is more in question. The editor is identified as Henry of Mainz from the list of contents: "this Henry who edited this book was a canon in the Church of Saint Mary in Mainz,"[121] and the compiler of the map is assumed to be this same Henry of Mainz. The Henry to whom the book is dedicated is a source of more controversy, but recent research points to English connections.[122]

English *mappaemundi* of the fourteenth century are represented by the maps in the *Polychronicon*, one of the most popular Latin histories of the fourteenth and fifteenth centuries, edited by the Benedictine monk Ranulf Higden (ca. 1299–1363). The world map, derived as usual from a variety of Roman sources, is found in the first book, and some twenty-one extant examples can be traced to the 1342 London manuscript of the *Polychronicon* for which stemmata have been provided by Miller with modifications by Skelton (fig. 18.21).[123]

Higden's maps, while having similar geographical content, differ widely in the shape of their frameworks. Three categories can be recognized: oval, circular, and mandorla.[124] Jerusalem and Rome are always prominent, but rarely in the center.

The large oval map in the British Library is thought to be closest to the original lost prototype, despite the claims of Galbraith, who believes that the Huntington (San Marino) copy of the *Polychronicon* is the author's working copy in his own handwriting (plate 15).[125] The circular Higden maps are thought to be simplifications of the earlier oval maps, and they certainly appear in generally later manuscripts, as can be seen from the chronological table in appendix 18.2. Finally, the almond-shaped mandorla maps (also known as *vesica piscis*, fish bladder) form the third group of Higden maps (fig. 18.22). They are generally later simplifications, and Skelton believes that the example in the National Library of Scotland (see appendix 18.2), with its truncated top and rounded point, represents a transition to the true almond shape.[126]

The oval shape of Higden's maps and its simplification, the *vesica piscis*, is a particular characteristic of his, but not original. Skelton implies that a lost prototype (which may have been a large world map such as that referred to by Matthew Paris a century earlier) was probably circular and that the oval shape was an adaptation to the shape of the codex leaf. It is more likely, however,

117. Noël Denholm-Young, "The *Mappa Mundi* of Richard of Haldingham at Hereford," *Speculum* 32 (1957): 307–14. A. B. Emden, *A Biographical Register of Oxford University to A.D. 1500* (Oxford: Clarendon Press, 1957–59).

118. W. N. Yates, "The Authorship of the Hereford Mappa Mundi and the Career of Richard de Bello," *Transactions of the Woolhope Naturalist's Field Club* 41 (1974): 165–72. But Yates's interpretation of the "authorship" rubric on the map is puzzling. While he draws attention to the possible specific meanings of the words *estoire* and *compasse*, maintaining correctly that *estoire* can refer to either a design of a picture or a history and *compasse* can refer equally to actually doing or to abstract planning, he neglects to translate the words *ki lat fet (qui l'a fait)*, which seem to mean "who [i.e., Richard of Haldingham] has made it."

119. See Crone, *World Map*, 15 (note 112).

120. Valerie I. J. Flint, "Honorius Augustodunensis Imago Mundi," *Archives d'Histoire Doctrinale et Littéraire du Moyen Age* 57 (1982): 7–153. Miller, *Mappaemundi*, 3:22 (note 9), regarded this attribution as "without doubt," but Sarton, *Introduction to the History of Science*, 2:201 (note 49), disagrees.

121. See Montague Rhodes James, *A Descriptive Catalogue of the Manuscripts in the Library of Corpus Christi College Cambridge*, 2 vols. (Cambridge: Cambridge University Press, 1912), 1:138–39.

122. Flint, "Honorius," 10–13 (note 120).

123. Miller, *Mappaemundi*, 3:95 (note 9); R. A. Skelton, in *Mappemondes*, ed. Destombes, 149–60, esp. 152–53 (note 31). See also John Taylor, *The "Universal Chronicle" of Ranulf Higden* (Oxford: Clarendon, 1966). In 1985, a small parchment fragment about 25 × 15 centimeters covering the Mediterranean from the Canaries to the Holy Land was acquired by the British Library. It is apparently part of a large late fourteenth-century *mappamundi* used as a wall map, and it is speculated that the map became extremely faded before the fragment was used as a binding for a Norfolk rental book from 1483, in which it was found. The original outline was apparently shaped as a mandorla, and its information bears some kinship to the Higden maps found in the *Polychronicon*, perhaps suggesting that it was akin to one of the wall maps available in the fourteenth century on which Higden based his reduction. I owe this note to Peter Barber of the Department of Manuscripts, British Library.

124. Skelton, in *Mappemondes*, ed. Destombes, 150–51 (note 31).

125. V. H. Galbraith, "An Autograph MS of Ranulph Higden's *Polychronicon*," *Huntington Library Quarterly* 34 (1959): 1–18.

126. Skelton, in *Mappemondes*, ed. Destombes, 153 (note 31).

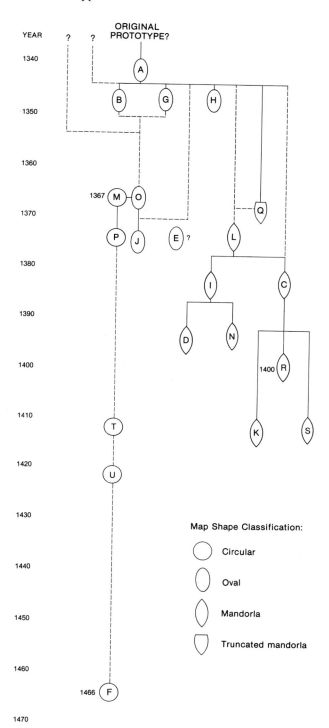

FIG. 18.21. GENERAL STEMMA FOR THE EXTANT WORLD MAPS OF RANULF HIGDEN. A provisional genealogy based on Konrad Miller, *Mappaemundi: Die ältesten Weltkarten*, 6 vols. (Stuttgart: J. Roth, 1895–98), 3:95, and Skelton in Marcel Destombes, ed., *Mappemondes A.D. 1200–1500: Catalogue préparé par la Commission des Cartes Anciennes de l'Union Géographique Internationale* (Amsterdam: N. Israel, 1964), 151–53.

that the oval shape was derived from the practice—described by Hugh of Saint Victor—of drawing maps in the supposed shape of Noah's ark.[127]

In this period of the twelfth and thirteenth centuries, then, though an attempt was made at a more exact understanding of the natural world, this tendency toward realism was only barely seen in the *mappaemundi*. For the most part the maps continued to reflect a mixture of much earlier Roman sources as well as the stock-in-trade of Macrobius and Isidore. At the same time, there are glimpses of the new concepts and techniques that

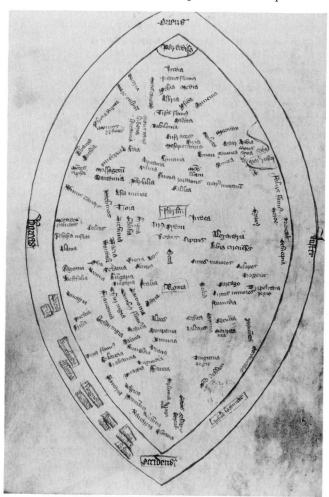

FIG. 18.22. A HIGDEN WORLD MAP:MANDORLA TYPE, MID-FOURTEENTH CENTURY. The mandorla or almond-shaped maps of Higden—perhaps representing a common Christian symbol of the aureole surrounding Christ—are generally later simplifications of the oval (see plate 15) and circular maps.

Size of the original: 35.5 × 21 cm. By permission of the British Library, London (Royal MS. 14.C.xii, fol. 9v).

127. Skelton, in *Mappemondes*, ed. Destombes, 150–51 (note 31). On Hugh of Saint Victor, see p. 334 below.

were to transform the *mappaemundi* of the fourteenth and fifteenth centuries into a cross between the maps of the medieval and modern worlds.

PIETRO VESCONTE TO FRA MAURO: THE TRANSITIONAL PERIOD FROM 1300 TO 1460

In the late Middle Ages, several trends can be noticed in the *mappaemundi* that lend a Renaissance character to maps of the period, though the basic frameworks of representation are medieval. The nature of the *mappaemundi* made between the times of Vesconte and of Fra Mauro is so different from that of the earlier maps that they warrant separate treatment, a view also reflected in the classification offered here in appendix 18.1. The transition was of course not abrupt: we have already seen that the experimental philosophers of the twelfth and thirteenth centuries heralded the thinking of the fifteenth and sixteenth centuries, and the tenacity of such characteristically medieval authors as Isidore and Sacrobosco is shown by their popularity well into the Renaissance.[128] In some parts of Europe, the medieval period seems (in cartographic terms) to have extended far beyond its normal chronological bounds: for example, a Russian broadside world map, drawn in the first half of the seventeenth century directly from medieval sources, continued to appear in print until the nineteenth century.[129]

The transition is marked by a convergence of three conceptual frameworks of world maps represented by the traditional, confined *mappamundi*, the expanding portolan chart, and the Ptolemaic coordinate system. There was also a trend toward the emergence of the map as an independent artifact rather than as a mere addition to a text. This had originated in earlier periods but gathered momentum during the Renaissance, as maps and atlases were published for profit in their own right. This emerging identity of the map is reflected in the work of Pietro Vesconte, the first professional cartographer in western Europe to routinely sign and date his works. Although he was chiefly a compiler of nautical charts, world maps made by him are found in the manuscripts of Marino Sanudo's *Liber secretorum fidelium crucis super Terrae Sanctae recuperatione et conservatione*, 1306–21, a book written as a means of arousing interest in a crusade.[130]

The Vesconte *mappaemundi*, together with the contemporary examples found in the chronicle compiled by Franciscan Minorite friar Fra Paolino, clearly show the influence of the portolan charts in three major characteristics (plate 16). First, the outline of the Mediterranean Sea is derived directly from such charts. Second, a network of rhumb lines is also provided, even across wide expanses of land where they could be of little use.

Such lines were later to become a hallmark of precision on world maps ostensibly made for navigation, but that in fact had no practical use. Third, graphic scales were inserted on later *mappaemundi*, as in the Genoese world map of 1457 (fig. 18.23) or the map of Andreas Walsperger, 1448.

The geographical expansion of the traditional bounds of the portolan chart is well illustrated by the famous Catalan atlas [1375], which is perhaps the finest example of a *mappamundi* in its final transitional state (plate 17).

FIG. 18.23. A SCALE ON A *MAPPAMUNDI*. On the later world maps, the influence of the portolan charts and Ptolemy's *Geography* is seen in the addition of the apparatus of the navigator, such as the two scales on the 1457 Genoese world map, of which one is shown.
Length of the original scale: ca. 10 cm. By permission of the Biblioteca Nazionale Centrale, Florence (Port. 1).

128. Elizabeth L. Eisenstein, *The Printing Press as an Agent of Change: Communications and Cultural Transformations in Early Modern Europe*, 2 vols. (Cambridge: Cambridge University Press, 1979), 2:510.

129. Leo Bagrow, "An Old Russian World Map," *Imago Mundi* 11 (1954): 169–74.

130. For the best general study, see Bernhard Degenhart and Annegrit Schmitt, "Marino Sanudo und Paolino Veneto," *Römisches Jahrbuch für Kunstgeschichte* 14 (1973): 1–137. See also Schulz, "Moralized Geography," 445 and 452 (note 24).

More is known about the manufacture of this atlas than most, thanks to the request in 1381 of an envoy of Charles VI of France (Guillaume de Courcy) to Pedro IV of Aragon (1336–87) for a copy of the latest available world map. The completed work has remained in the French royal library (now the Bibliothèque Nationale in Paris) ever since. Charles's request may also illustrate the high regard in which the Catalan cartographers were held at the time, particularly Cresques Abraham (1325–1387) and his son Jefuda Cresques.[131]

The Catalan atlas is actually a multisheet "*mapamundi*" and is so titled. It consists of twelve leaves mounted on boards to fold like a screen. Although the eastern section shows a circular edge, indicating its medieval roots, the compiler's main interest is evidently in the eastern and western extension of the Mediterranean. This forms a rectangular piece taken out of the traditional circular medieval world map. Other medieval vestiges include the approximately central position of Jerusalem and the west-east river in North Africa. The Mediterranean and Black seas reflect a standard portolan chart configuration.

But it is the wealth of information on central Asia, gleaned from the travel narratives of the thirteenth century, that makes the Catalan atlas the object of particular interest. It is the first map that bears the unequivocal influence of the travels of Nicolò, Maffeo, and Marco Polo (1260–69, 1271–95), although it appeared more than three-quarters of a century after their return to Venice. Although Marco Polo was a keen observer and recorder, and the first to give Europe a reasonably accurate description of East Asia, there is no evidence—if we ignore the probably apocryphal maps relating to northeastern Asia—that he drew any maps recording his experience.[132] Although Marco Polo did not allude specifically to maps in his narrative, there are three passages that merely mention the charts of mariners in the Indian Ocean without providing further detail. However, he did provide some compass bearings in the text that, along with other geographical information in the narrative, have been used by later scholars to reconstruct a map.[133] It is difficult to find his influence on the maps of Vesconte and Sanudo, although the earlier travels of Carpini and William of Rubruck were well known to the latter.[134]

Apart from its influence on the Catalan atlas, it appears that the narrative of Marco Polo had very little effect on world cartography of the time—certainly much less than the novelty of its geographical information would lead us to expect. There is some evidence that a map illustrating Marco Polo's discoveries was drawn on the wall of the Sala dello Scudo (now the Sala delle Due Mappe) in the ducal palace in Venice. In 1426 Don Pedro of Portugal received a map from the Signoria that may

have been a copy of such a map; it, or one like it, certainly existed in the mid-fifteenth century, for in 1459 the Senate ordered that such a map be repainted on the wall. Unfortunately, it was destroyed by fire in 1483.[135]

It is in the map made by Fra Mauro in 1459 that the greatest influence of the Marco Polo narratives is seen before the printed editions of them began to be disseminated (plate 18). This map stands at the culmination of the age of medieval cartography, although Bagrow may have exaggerated when he called it "the summit of Church cartography"[136] for it is far more secular in nature than, for example, the Ebstorf map. It is transitional in the sense that it included information derived from portolan charts, from Ptolemy's *Geography*, and from the new discoveries in Asia. Fra Mauro, working from the Camaldulian monastery on the island of Murano, was already an experienced cartographer. Detailed records relating to his mapmaking activities show that he made a map of a district in Istria as early as 1443, and in 1448–48 he was apparently at work on a *mappamundi*. Neither map has survived. The world map, now preserved in the Biblioteca Marciana in Venice, is a copy of a map commissioned by Afonso V, king of Portugal, and finished in April 1459 with the help of his assistant Andrea Bianco. The extant copy was made at the request of the Signoria, it is assumed in the same year, perhaps from notes that Fra Mauro and Bianco

131. Gonçal (Gonzalo) de Reparaz, "Essai sur l'histoire de la géographie de l'Espagne de l'antiquité au XVᵉ siècle," *Annales du Midi* 52 (1940): 137–89, 280–341, esp. 296, 307–9; and see also *El atlas catalán de Cresques Abraham: Primera edición con su traducción al castellano en el sexto centenario de su realización* (Barcelona: Diáfora, 1975), also published in Catalan; Georges Grosjean, ed., *The Catalan Atlas of the Year 1375* (Dietikon-Zurich: Urs Graf, 1978); Pinhas Yoeli, "Abraham and Yehuda Cresques and the Catalan Atlas," *Cartographic Journal* 7 (1970): 17–27, and Campbell below, chapter 19.

132. Leo Bagrow, "The Maps from the Home Archives of the Descendants of a Friend of Marco Polo," *Imago Mundi* 5 (1948): 3–13.

133. Marco Polo, *The Book of Ser Marco Polo*, 3d ed., 2 vols. ed. and trans. Sir Henry Yule, rev. in accordance with discoveries by Henri Cordier (New York: Charles Scribner's Sons, 1903), 2:245 n. 7, 312, 424; Yule's reconstructed map is in vol. 1, facing *108*.

134. Cortesão, *History of Portuguese Cartography*, 1:279, 290 (note 34).

135. Juergen Schulz, "Maps as Metaphors: Mural Map Cycles of the Italian Renaissance," in *Art and Cartography: Six Historical Essays*, ed. David Woodward (Chicago: University of Chicago Press, 1987). See also Rodolfo Gallo, "Le mappe geografiche del palazzo ducale di Venezia," *Archivio Veneto*, 5th ser., 32 (1943): 47–89; Jacopo Morelli, *Operette di Iacopo Morelli*, 2 vols. (Venice: Tipografia di Alvisopoli, 1820), 1:299; and Polo, *Book of Marco Polo*, 1:*111* (note 133). It was again painted over by Giacomo Gastaldi in the mid-sixteenth century with four new maps, and once again by Francesco Griselini at the direction of the doge Marco Foscarini in 1762, each repainting obliterating the previous version, so that all that remain visible are the eighteenth-century murals.

136. Bagrow, *History of Cartography*, 72 (note 55).

had made. In its circular framework it is clearly medieval, and the southern orientation shows some Arabic influence, but the Mediterranean coasts are modeled on portolan charts and there is an allusion to its debt to the Ptolemaic tradition.[137]

The influence of Marco Polo's travels on the content of the later Renaissance maps was profound. Information about the Indian Ocean gleaned from his voyage from Zaiton to Hormuz via Java, Sumatra, Ceylon, and India was incorporated into the maps of Henricus Martellus Germanus, the globe of Martin Behaim, and early sixteenth-century printed maps such as the Ruysch map of 1507. Madagascar too appears on these maps much as Marco Polo reported it: about one thousand miles south of Socotra and four thousand miles in circuit.

Another major influence on the *mappaemundi* of this transitional period was from the *Geography* of Claudius Ptolemy. After its translation into Latin by Jacobus Angelus about 1406–7, the popularity of this work increased steadily throughout the fifteenth century, as reflected in the frequency of printed editions from 1475 onward. An early world map showing such influence—displaying, for example, the closed Indian Ocean of Ptolemy—is the Pirrus de Noha map accompanying a manuscript of Pomponius Mela about 1414 (see plate 19 and fig. 18.79).[138]

To understand the Ptolemaic influence, it is necessary first to be aware of a school of science under the leadership of the mathematician and astronomer Johannes de Gmunden at the University of Vienna and the prelate Georg Müstinger at the Augustinian monastery of Klosterneuburg, now in suburban Vienna.[139] The school flourished from the early 1420s until 1442, when both scholars died. Its contributions to cartography were but a fraction of its legacy of scientific manuscripts, including astronomical treatises, star catalogs, and tables of planetary motions, eclipses, and conjunctions, as well as general works on mathematics, including trigonometry. Most of these were recopied versions of earlier medieval works, but nevertheless Klosterneuburg constituted a seedbed of scientific innovation. In particular, the maps and coordinate tables associated with this school help to fill in a period of relative cartographic obscurity between the Claudius Clavus map of about 1425 and the *tabulae modernae* of the later Ptolemaic manuscripts about 1450. The earliest maps, two rough plots of coordinates in the Vatican Library probably prepared by Conrad of Dyffenbach in 1426, were based on versions of the Toledo tables (a detail from the first of these maps is illustrated in fig. 18.24).[140] Between 1425 and 1430, Müstinger and his collaborators were working on a map genre that assimilated the Jerusalem-centered medieval world map with elements from Ptolemy and the portolan charts, which when reconstructed are similar in their general geographical configuration to the circular Vesconte-Sanudo maps which have already been described.

Although only coordinate tables survive for the earliest versions of these circular world maps of the Vienna-Klosterneuburg school, Durand reconstructed maps from the tables, most of which are to be found in a 522-page codex in the Bayerische Staatsbibliothek.[141] There are, however, two surviving original maps that Durand believes are based on this genre: the Walsperger map of 1448 and the Zeitz map of about 1470.[142] To these may be added the fragment of the world map acquired by the James Ford Bell Collection in 1960.[143]

This evidence suggests that fifteenth-century cartographers were clearly impressed with the Ptolemaic model and took pains to demonstrate that, although they did not agree with all of Ptolemy's information or method of using coordinates, the tradition was to be revered. Fra Mauro felt it necessary to apologize for not following the parallels, meridians, and degrees of the *Geography* on his world map of 1459, because he found them too confining to show discoveries (presumably in Asia) unknown to Ptolemy. Andreas Walsperger, in his *mappamundi* of 1448, stated that it was "made from the cosmography of Ptolemy proportionally according to longitude, latitude, and the divisions of climate." He exiled the monstrous races found in Africa on earlier maps to Antarctica.[144] Later in the century, Henricus Martellus Germanus developed the second Ptolemaic projection for his world maps and fitted the new discoveries into it, but his efforts belong to the Renaissance, along with the globe of Martin Behaim.[145]

The maps of Giovanni Leardo, a Venetian cosmographer of the mid-fifteenth century, provide useful examples of a genre of late medieval *mappaemundi* cen-

137. Tullia Gasparrini Leporace, *Il mappamondo di Fra Mauro* (Rome: Istituto Poligrafico dello Stato, 1956). Bagrow, *History of Cartography*, 72–73 (note 55). See also Campbell below, chapter 19.

138. Rome, Biblioteca Apostolica Vaticana, Archivio di San Pietro, H. 31. The original author of this map is unknown; Pirrus de Noha was simply a copyist. See Destombes, *Mappemondes*, 187–88 (note 31).

139. Durand, *Vienna-Klosterneuburg*, 52–60 (note 52).

140. Durand, *Vienna-Klosterneuburg*, 106–13 (note 52).

141. Durand, *Vienna-Klosterneuburg*, 174–208 (note 52). Munich, Bayerische Staatsbibliothek, Clm. 14583.

142. Durand, *Vienna-Klosterneuburg*, 209–15 (note 52). Rome, Biblioteca Apostolica Vaticana, Pal. Lat. 1362, and Zeitz, Stiftsbibliothek, MS. Lat. Hist., fol. 497.

143. John Parker, "A Fragment of a Fifteenth-Century Planisphere in the James Ford Bell Collection," *Imago Mundi* 19 (1965): 106–7.

144. Friedman, *Monstrous Races*, 56–57 (note 3). See also Paul Gallez, "Walsperger and His Knowledge of the Patagonian Giants, 1448," *Imago Mundi* 33 (1981): 91–93.

145. Although Behaim's globe, the Laon globe, and the Martellus planispheres are included in Destombes as pre-1500 world maps, they belong in the Renaissance period and will be dealt with in volume 3.

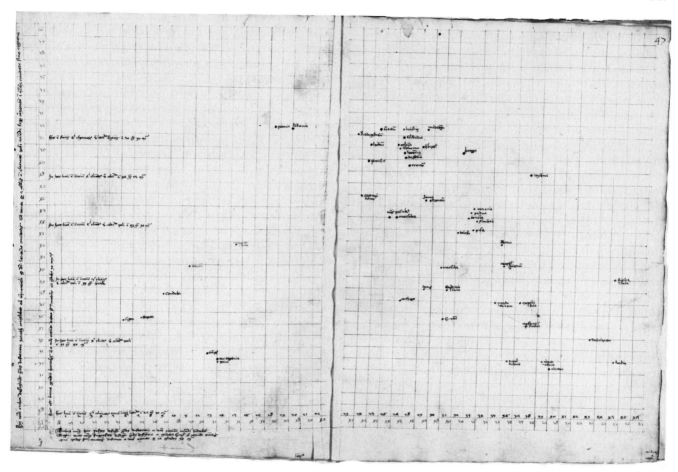

FIG. 18.24. SKETCH FROM THE VIENNA-KLOSTERNEU-BURG CORPUS. Derived from versions of the Toledo tables, this sketch map of places in Europe and North Africa was plotted on a framework of latitude and longitude coordinates. It was probably prepared by Conrad of Dyffenbach in 1426.

Size of the original: 39.4 × 58.6 cm. Photograph from the Biblioteca Apostolica Vaticana, Rome (Pal. Lat. 1368, fols. 46v–47r).

tered on Jerusalem on the eve of the age of exploration of the Western Hemisphere. Almost nothing is known of Leardo other than from his three surviving world maps (all signed and dated respectively 1442, 1448, and 1452), and the description of one now lost (also signed by Leardo and dated 1447). The three existing maps all have calendars that show the zodiac, the dates of Easter, and, for the largest two (1448 and 1452), the phases of the moon. The maps use similar signs despite their varied size, and the torrid and frigid zones are displayed prominently and colored appropriately (plate 20).[146]

Such maps were to give way to those that assimilated the new discoveries into the Ptolemaic framework, abandoning the convention of placing Jerusalem at the center of a circular map. Since the traditional frame no longer held the new discoveries in the fifteenth century (Andrea Bianco's world map of 1436 literally breaches its circular border in East Asia), it became a practical impossibility

to center the maps on Jerusalem. Several mid-fifteenth-century *mappaemundi* reflect this, including Andrea Bianco's map, the Catalan world (Estense) map of about 1450, Walsperger's map of 1448, the Borgia map on metal, the Genoese map of 1457, and Fra Mauro's map of 1459.[147]

---

146. On Leardo and his maps, see P. Durazzo, *Il planisfero di Giovanni Leardo* (Mantua: Eredi Segna, 1885), and John Kirtland Wright, *The Leardo Map of the World, 1452 or 1453, in the Collections of the American Geographical Society*, American Geographical Society Library Series, no. 4 (New York, 1928), who does not, however, mention Durazzo and his clear transcription of the date "1452." On the 1448 map preserved in Vicenza, see *Teatro del cielo e della terra: Mappamondi, carte nautiche e atlanti della Biblioteca Civica Bertoliana dal XV al XVIII secolo: Catalogo della mostra* (Vicenza: Biblioteca Civica Bertoliana, 1984), 16–17 and unnumbered plate.

147. Fra Mauro, on his [map of the world], 1459, Venice, Biblioteca Nazionale Marciana, rationalizes his placing of Jerusalem away from the center of the map by stating that he has used the center of population. See the transcription of the legend by Gasparrini Leporace,

In this section we have tried to show that there were clear differences in the character of medieval world maps depending on the subperiod in which they were created; it is thus not possible to generalize accurately for the *mappamundi* of this thousand-year period. In the patristic period, from about 400 to 700, three basic cartographic traditions—the Macrobian, Orosian, and Isidorian—were established, and these do recur throughout the Middle Ages. In the second period, from about 700 to 1100, in which a reasonable sample of *mappaemundi* first appears, little innovation is seen except in the maps of Beatus, despite the renewed interest in natural science. It is not until the third period, from about 1100 to 1300, with the influx and translation of numerous Arabic and Greek manuscripts, especially the *Almagest*, that scientific interest reawakens. The last period, from about 1300 to 1460, stands apart from the earlier tradition of *mappaemundi* and acts as a transitional stage between the medieval and modern worlds of mapping. The three frameworks of maps—monastic, nautical, and Ptolemaic—which had for a while each enjoyed a separate and parallel development, came together in the fifteenth century and set the stage for the technical advances of the Renaissance.

Themes in the Study of *Mappaemundi*

form

Contemporary evidence on the methods of construction of *mappaemundi* is extremely scanty, the brief description by Hugh of Saint Victor being unusual.[148] The artifacts themselves often speak eloquently about how they were made, but much more intensive scrutiny of the original artifacts needs to be done. The aim of this section is to treat thematically some of the points arising from the previous chronological survey. This will include a discussion of the framework, concepts of the shape of the earth, projections and coordinate systems, the production of *mappaemundi* (inks and pigments, lettering, signs, and color), and the content and meaning of the maps as revealed in the factual aspects of their geography, the more fanciful legendary traditions, and their complex symbolism.

The Frame of the *Mappaemundi*

We have already suggested that the medieval world maps were conceived within a preestablished frame of a limited selection of geometric shapes: circular, oval, rectangular, or mandorla, each shape having its own symbolic connotation.[149] This is borne out by Hugh of Saint Victor's description of how to draw a *mappamundi* in the shape of an ark, his instructions clearly being more related to the mystical functions of the map than to any geographical use. In the absence of a firsthand descrip-

tion of the compilation methods of maps of the size and complexity of the Hereford or Ebstorf maps it is difficult to imagine how places could be fitted into the outline. Since no graticule was apparently drawn, one must assume that once the border, the center, and the tripartite division were established the countries and other details were broadly sketched in and adjusted until they fitted the designer's intentions. This assumption is supported by the unfinished state of paradise on the pair of oval world maps by Higden in the British Library. The uncolored part reveals a faint underlying sketch (fig. 18.25).[150]

Close physical scrutiny of a large sample of the original documents might well yield further evidence about these frameworks in the same way that calligraphers are now finding detailed clues to the history of their craft by examining medieval manuscripts with such technical questions in mind.[151] A parallel study for maps has yet to be systematically undertaken, although it must be admitted that the lack of large *mappaemundi* is a major barrier to this approach. Had the Ebstorf map been examined with this in mind and the results properly documented before its destruction in 1943, some further important clues might have been revealed.

Concepts of the Shape of the Earth

In geography and cartography, the persistent influence of classical Greek learning in medieval times is shown

---

Mappamondo di Fra Mauro, 38 (note 137). The attribution to Andrea Bianco in Woodward, "Reality, Symbolism, Time, and Space," 517 (note 17) is incorrect. It should also be pointed out, however, that several fifteenth-century maps *were* centered on Jerusalem, such as the world map in the *Rudimentum novitiorum* (1475), the world maps of Hanns Rüst and Hanns Sporer, or the three extant maps of Giovanni Leardo (1442, 1448, and 1452). But these were, without exception, based on much earlier models that had appeared when the practice of centering the map on Jerusalem was more usual.

148. See the description below by Hugh of Saint Victor, p. 334.

149. This topic is discussed in detail by Arentzen, *Imago Mundi Cartographica*, 29–37 (note 32). The structural shape of *mappaemundi* has also been the subject of three articles by Osvaldo Baldacci: "Ecumene ed emisferi circolari," *Bollettino della Società Geografica Italiana* 102 (1965): 1–16; "Geoecumeni quadrangolari," *Geografia* 6 (1983): 80–86; and "L'ecumene a mandorla," *Geografia* 6 (1983): 132–38. In the first article, Baldacci stresses the fundamental difference between the circular shape of the oikoumene and the implied sphericity of the zonal hemispheric system. In the second and third articles, he argues for the influence of Strabo and Marinus of Tyre on both the rectangular and mandorla shapes of medieval *mappaemundi*, but since the ideas of Marinus were transmitted through Ptolemy's *Geography*, which was not available to the West until the fifteenth century, the influence of Marinus at least is difficult to accept.

150. British Library, Royal MS. 14.C.ix, fols. 2v, 3r, and 3v.

151. The work of Michael Gullick, as reflected in Donald Jackson, *The Story of Writing* (New York: Taplinger, 1981), provides a particular example of this.

FIG. 18.25. PARADISE ON A HIGDEN WORLD MAP. As with several examples of the Higden maps, a sketch representing paradise is unfinished in this example.
Size of the original detail: 4.75 × 5 cm. By permission of the British Library, London (Royal MS. 14.C.ix, fol. 2v).

partly by the tenacity of the notion of the earth's sphericity, despite modern popular writers who have assumed that medieval (and even early Renaissance) man believed the earth was flat.[152] This myth may have been perpetuated by some historians who have tended to emphasize the unusual beliefs of the period and even to accept these as the norm. For example, many general histories devote undue consideration to the concept of a flat, rectangular, four-cornered earth with a vaulted heaven from the sixth-century *Christian Topography* of Cosmas Indicopleustes.[153] It is important to realize that Cosmas's text, now preserved only in two manuscripts, was not thought worthy of mention by medieval commentators, with the exception of Photius of Constantinople, who said not only that "the style is poor, and the arrangement hardly up to the ordinary standard" but also that "he may fairly be regarded as a fabulist rather than a trustworthy authority."[154]

The relationship of the concept of the Antipodes to that of the earth's sphericity has been a source of confusion. The fathers of the church were embarrassed by a doctrine that implied the existence of a race not descended from the sons of Adam. But it was intellectually possible to believe that the earth was a sphere without subscribing to the idea of the Antipodes. It was about the latter that Virgil of Salzburg and Pope Zacharias confronted each other in the ninth century, not about the sphericity of the earth. The shape of the earth seems to have been much less a subject of debate.[155]

A further confusion resulting from literal interpretation of biblical sources arose from the apparent incompatibility of the circular form of the earth and the four corners referred to in the Bible. The German encyclopedist Rabanus Maurus (ca. 776–856), for example,

asked how circular and quadrate shapes could agree and went on to relate this problem to that of Euclid's squaring the circle.[156] The medieval cartographer's solution was either to portray the circular earth within a square, leaving convenient spaces in the corners for iconographically suitable images, such as the symbols for the four evangelists, or to place the square within the circle so that the four cardinal directions and the circular earth could be combined.

Despite the difficulties of literal biblical interpretation, most early fathers of the church agreed that the earth was a sphere. Augustine specifically mentioned it at least twice.[157] The works of the popular secular writers such as Pliny, Macrobius, and Martianus Capella also contain many references to its sphericity.[158] Perhaps in overre-

152. For a summary of the complexities of the question, see Woodward, "Reality, Symbolism, Time, and Space," 517–19 (note 17). Recent sources used in this study include W. G. L. Randles, *De la terre plate au globe terrestre: Une mutation épistémologique rapide (1480–1520)*, Cahiers des Annales 38 (Paris: Armand Colin, 1980), and Tattersall, "Sphere or Disc?" (note 29).

153. Cosmas's concepts were derived from the following biblical passages: Isa. 40:22, "God sits throned on the vaulted roof of earth"; Matt. 24:31, "With a trumpet blast he will send out his angels, and they will gather his chosen from the four winds, from the farthest bounds of heaven on every side"; and Rev. 7:1, "After this I saw four angels stationed at the four corners of the earth, holding back the four winds." Charles W. Jones, "The Flat Earth," *Thought: A Quarterly of the Sciences and Letters* 9 (1934): 296–307, esp. 305, places Cosmas in true perspective.

154. Photius of Constantinople *Bibliotheca* 36; see *The Library of Photius*, trans. J. H. Freese (London: Macmillan, 1920), 1:31–32. For example, Randall wrote that Cosmas "had great popularity among even the educated till the twelfth century." See John Herman Randall, Jr., *The Making of the Modern Mind: A Survey of the Intellectual Background of the Present Age* (Boston: Houghton Mifflin, 1926), 23.

155. This problem has been summarized by F. S. Betten, "St. Boniface and the Doctrine of the Antipodes," *American Catholic Quarterly Review* 43 (1918): 644–63. The letter from Pope Zacharias threatening Virgil with excommunication is in *Monumenta Germaniae historica: Epistolarum*, 8 vols. (Berlin: Wiedmann, 1887–1939), 3:356–61, esp. 360.

156. Cortesão, *History of Portuguese Cartography*, 1:172 (note 34). G. L. Bertolini, "I quattro angoli del mondo e la forma della terra nel passo di Rabano Mauro," *Bollettino della Società Geografica Italiana* 47 (1910): 1433–41.

157. Saint Augustine, *De civitate Dei* 16.9: "They fail to observe that even if the world is held to be global or rounded in shape . . . it would still not necessarily follow that the land on the opposite side is not covered by masses of water," trans. Eva Matthews Sanford and William McAllen Green, vol. 5 of Saint Augustine, *The City of God against the Pagans*, 7 vols., Loeb Classical Library (Cambridge: Harvard University Press, 1965), 5:51. Also see Saint Augustine *De genesi ad litteram libri duodecim* 1.10, in *Corpus scriptorum ecclesiasticorum Latinorum* 28 (1894): 15, l. 6, and Saint Augustine *Quæstionum evangelicarum libri* 2.14, in *Patrologiæ cursus completus*, 221 vols. and suppls., ed. J. P. Migne (Paris, 1844–64; suppls., 1958–), 35:1339.

158. The classical sources for the idea of the spherical earth have already been discussed above, p. 145. Less well known is Ovid's description in the *Metamorphoses* 1.32–36:

action to these "pagan" works, Severianus and Lactantius were to take the opposite view, but the importance of their works, which have interested historians perhaps because of their controversial nature, has probably been exaggerated.[159]

The case of Isidore of Seville perhaps merits particular attention in view of the widespread influence of his writings, especially the *Etymologies* and *De natura rerum*. Isidore is clear about the sphericity of the universe: "The sphere of the heavens is rounded and its center is the earth, equally shut in from every side. This sphere, they say, has neither beginning nor end, for the reason that being rounded like a circle it is not easily perceived where it begins and where it ends."[160] While he uses the word *globus* several times in *De natura rerum* in connection with the moon or the planets,[161] he neglects to comment directly on the sphericity of the earth itself except in the following passage: "The ocean, spread out on the peripheral regions of the globe, bathes almost all the confines of its orb".[162] What appears to be Isidore's leaning toward a belief in a spherical earth is supported by the *Epistula Sisebuti*, an astronomical poem written as a letter to Isidore by Sisebut, king of the Goths, to whom Isidore had dedicated *De natura rerum*.[163] In explaining an eclipse, Sisebut uses the word *globus* for the earth coming between the sun and the moon.[164]

Other passages in his texts have been used to support the idea that Isidore thought the world was flat. In one place, he described the earth as a wheel: "The circle of lands [orbis] is so called from its roundness, which is like that of a wheel, whence a small wheel is called *orbiculus*".[165] In another passage, he seems to have misunderstood the Greek concept of parallel zones from his reading of the *Poeticon Astronomicon* of Hyginus. He took too literally the statement that the lines separating the zones should be drawn as circles on a globe, and disregarded the possibility that these might look different when drawn on a flat surface. The zones thus appeared as five circles mechanically placed on a disk (fig. 18.26): "In describing the universe the philosophers mention five circles, which the Greeks call parallels, that is, zones, into which the circle of lands is divided. . . . Now let us imagine them after the manner of our right hand, so that the thumb may be called the Arctic Circle, uninhabitable because of cold; . . . the northern and southern circles, being adjacent to each other, are not inhabited, for the reason that they are situated far from the sun's course."[166] Such an interpretation can hardly be taken as evidence of Isidore's belief in a flat earth, however, when it reflects his inability to grasp the basic geometry of the Greek concept of the *climata*.

In another passage, Isidore seems to say that, when it rises, the sun is visible at the same time to people in both east and west: "The sun is similar for the Indians and the Bretons in the same moment that both see it rising. It does not seem smaller for the Orientals when it is setting; and the Occidentals, when it rises, do not find it any smaller than the Orientals."[167] Two interpretations are possible of the phrase "in the same moment that both see it rising." It could mean that the rising sun is visible at the same time to people in both east and west, thus implying a flat earth. It could also be interpreted to mean that the size of the sun appears the same to those in the east and west at the time of its rising.

Despite Isidore's apparent confusion about the shape of the earth revealed in these passages, the evidence appears to confirm that he thought the earth, like the universe, was a sphere. He was joined in this view by other influential Christian writers, some of whom explained the reasons thoroughly. For example, the Venerable Bede (672/73–735) was careful in his explanation: "The cause of the unequal length of the days is the globular shape of the earth, for it is not without reason that the

---

Whatever god it was, who out of chaos
Brought order to the universe, and gave it
Division, subdivision, he molded earth,
In the beginning, into a great globe,
Even on every side.

See Ovid, *Metamorphoses*, trans. Rolfe Humphries (Bloomington: University of Indiana Press, 1957), 4.

159. Jones, "Flat Earth" (note 153). Anna-Dorothee von den Brincken, "Die Kugelgestalt der Erde in der Kartographie des Mittelalters," *Archiv für Kulturgeschichte* 58 (1976): 77–95, summarizes the history of the concepts of the spherical shape of the earth but avoids the controversy over Isidore's views.

160. Brehaut, *Isidore of Seville* (note 84).

161. Isidore, *Traité de la nature*, ed. Fontaine, 223 (planets) and 231, 239, and 277 (moon) (note 84).

162. "Oceanus autem regione circumductionis sphaerae profusus, prope totius orbis adluit fines." Isidore, *Traité de la nature*, ed. Fontaine, 325 (note 84).

163. Isidore, *Traité de la nature*, ed. Fontaine, 151 (note 84).

164. Isidore *Epistula Sisebuti*, in *Traité de la nature*, ed. Fontaine, 333 line 40 (note 84).

165. Isidore *Etymologies* 14.2.1: "*Orbis* a *rotunditate* circuli dictus, quia sicut rota est; unde brevis etiam rotella *orbiculus* appellatur," in *Patrologiæ cursus completus*, ed. Migne, 82:495 (note 157), author's translation.

166. "In definitione autem mundi circulos aiunt philosophi quinque, quos Graeci parallelois, id est zonas uocant, in quibus diuiditur orbis terrae. . . . Sed fingamus eas in modum dexterae nostrae, ut pollex sit circulus arcticos, frigore inhabitabilis; . . . At contra septentrionalis et australis circuli sibi coniuncti idcirco non habitantur quia a cursu solis longe positi sunt." Isidore, *Traité de la nature*, ed. Fontaine, 209–11 (note 84), author's translation. This degeneration of the original *climata* concept was transmitted to the Muslim world, but with *seven* circles. See George Sarton, review of Ahmed Zeki Valīdī Togan, "Bīrūnī's Picture of the World," *Memoirs of the Archaeological Survey of India* 53 [1941] in *Isis* 34 (1942): 31–32.

167. "Similis sol est et Indis et Brittanis; eodem momento ab utrisque uidetur cum oritur, nec cum uergit in occasu minor apparet Orientalibus, nec Occidentalibus, cum oritur, inferior quam Orientalibus extimatur." Isidore, *Traité de la nature*, ed. Fontaine, 231 (note 84), author's translation.

Sacred Scriptures and secular letters speak of the shape of the earth as an orb, for it is a fact that the earth is placed in the center of the universe not only in latitude, as it were round like a shield, but also in every direction, like a playground ball, no matter which way it is turned."[168] Saint Thomas Aquinas (ca. 1227–74) argued that the earth must be spherical because changes in the position of constellations occur as one moves over the earth's surface.[169]

a score of others.[171] Dante used the idea of a spherical earth to set his *Divine Comedy*, probably the most widely disseminated vernacular work of its type. Moreover, he apparently felt not the slightest need to justify his view.[172] Even John Mandeville, whose *Travels* (ca. 1370) were immensely popular (albeit later ridiculed), explained that the earth was spherical and that the Antipodes could indeed exist.[173]

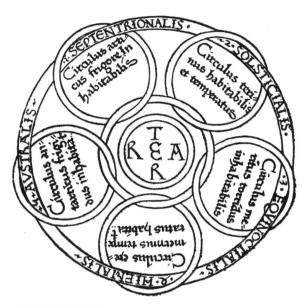

FIG. 18.26. ISIDORE'S VIEW OF THE EARTH'S FIVE ZONES. Isidore applied the Greek concept of zones not to a sphere but to a flat, circular earth, but this probably results from his misunderstanding of the nature of the concept and does not imply his ignorance of the earth's sphericity.
Diameter of the original: 13.5 cm. After George H. T. Kimble, *Geography in the Middle Ages* (London: Methuen, 1938).

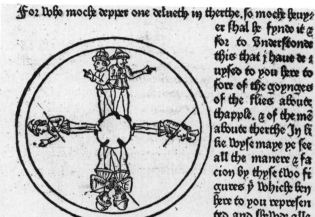

FIG. 18.27. DEMONSTRATION OF THE EARTH'S SPHERICITY IN THE THIRTEENTH CENTURY. Gautier de Metz explained that if two travelers left from the same place in opposite directions they would meet at the other side of the earth.
Diameter of the original detail: 8.2 cm. From a printed edition of *Image du monde* (London: Caxton, 1481). By permission of The Huntington Library, San Marino, California.

Late medieval commentators generally agreed that the earth was a sphere. Aristotle's elegant three-part demonstration of the sphericity of the earth and the astronomical works of Ptolemy—for which the concept was essential—were well known to the West after the twelfth century. The text of the Catalan atlas [1375] clearly states that the world is a sphere 180,000 stades in circumference. With the exception of a few polemical works against the idea—such as Zachariah Lilio's *Contra Antipodes*—the medieval scholar would have agreed with Gautier de Metz that "a man could go around the world as a fly makes the tour of an apple"[170] (fig. 18.27). The same theme is echoed in the writings of William of Conches, Hildegard of Bingen, Adam of Bremen, Lambert of Saint-Omer, Vincent of Beauvais, Albertus Magnus, Robert Grosseteste, Sacrobosco, Roger Bacon, and

168. Bede *De temporum ratione* 32, author's translation; see *The Complete Works of Venerable Bede*, 12 vols., ed. John Allen Giles (London: Whittaker, 1843–44), 6:210.

169. Saint Thomas Aquinas *Summa theologica* 1.47.3.3; see *Summa theologica/St. Thomas Aquinas*, 5 vols., trans. Fathers of the English Dominican Province (New York: Benziger Brothers, 1947–48).

170. For a translation of the relevant text in the Catalan atlas, see Grosjean, *Catalan Atlas*, 40 (note 131). For the Gautier de Metz quotation, see William Caxton (after Gautier de Metz), *Mirrour of the World*, Early English Text Society Extra Series 110 (London: Kegan, Paul, Trench, Trübner, 1913), 52.

171. A summary of the views of these scholars is found in Sarton, *Introduction to the History of Science*, vols. 2 and 3 (note 49). For a discussion of Hildegard's concepts, see Charles Singer, *Studies in the History and Method of Science*, 2 ed., 2 vols. (London: W. Dawson, 1955), 1:1–55.

172. See Mary Acworth Orr, *Dante and the Early Astronomers* (New York: A. Wingate, 1956). Arthur Percival Newton, *Travel and Travellers of the Middle Ages* (New York: Alfred A. Knopf, 1926), 9.

173. John Mandeville, *Mandeville's Travels*, 2 vols., ed. Paul Hamelius (London: Published for the Early English Text Society by K. Paul et al., 1919–23), 1:120–24. Newton, *Travel*, 12–13 (note 172).

Projections and Coordinate Systems

In the broadest sense, any transformation from one surface to another, and thus from a sphere to a plane, involves the process we call projection.[174] It could be argued, for example, that even the simple Macrobian diagrams with their parallel *climata* drawn on a circle were drawn on a projection crudely approximating an orthographic (equatorial aspect). The circular *climata* on the globe were thus portrayed with straight parallel boundaries on the flat map. It is possible to extend this argument to all *mappaemundi* and to point out, for example, that the world map of Matthew Paris and the "Jerome" map of Asia seem to have been constructed on "projections" approaching the azimuthal logarithmic, where the central part of the map—of most interest—is enlarged in scale.[175] Tobler has drawn our attention to a similar pattern of deformation on the Hereford map.[176]

Interest in this aspect was also shown by d'Avezac-Macaya, who described the projection system apparently used by the seventh-century writer known as the "Ravenna cosmographer" as the basis for his map. It is difficult to visualize this system, since it can be reconstructed only from the verbal description of the author, but d'Avezac-Macaya assumed that it was an oval map with twelve zones radiating from Ravenna. Each zone corresponded to the position of the sun overhead at hourly intervals during the day, from India in the morning to France (Brittany) in the evening, rather like a sundial superimposed on a world map.[177] Implied in this system is an azimuthal projection, although the center of the projection is still a point of discussion.[178]

Deliberate systems of projection, however, that reveal a conscious knowledge on the part of their compilers of a transformation of coordinate positions, are not found in the Middle Ages until the time of Roger Bacon. In his *Opus majus* (1268), Bacon describes a map, which has not survived, that he appended to the work, which seems to demonstrate that he had a clear idea of the value of using a systematic coordinate system to transform and inventory the positions of places: "Since these climates and the famous cities in them cannot be clearly understood by means of mere words, our sense must be aided by a figure. In the first place, then, I shall give a drawing of this quarter with its climates, and I shall mark the famous cities in their localities by their distance from the equinoctial circle, which is called the latitude of the city or region; and by the distance from the west or east, which is called the longitude of the region."[179] Then he goes on to describe a system of projection (which he calls a "device") in which the positions of places may be known by their distance from the equator and central meridian. The parallels are equally spaced on the meridian quadrant 90° east or west of the central meridian

(not on the central meridian itself; figure 18.28). This implies that the spacing of the parallels on the central meridian would decrease toward the pole. The meridians are equally spaced on the equator. From such a description it is clear that Bacon's "device" was certainly not the orthographic projection that Cortesão reports.[180]

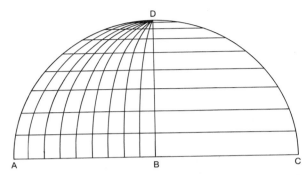

FIG. 18.28. RECONSTRUCTION OF ROGER BACON'S MAP PROJECTION. In this thirteenth-century map projection, Bacon fixes the position of a point by its distance from the equator and a central meridian. The parallel of latitude is drawn straight and parallel to the equator through the place's latitude on the colure (AD and DC). The meridians are represented as arcs of circles through the pole and the longitude of the place on the equator, except the central meridian, which is straight.
Author's reconstruction from the text of Bacon's *Opus majus*.

Most modern maps are based not only on a specific projection but also on a system of mathematically constructed coordinates. However, since the primary function of *mappaemundi* was not locational (other than in the crudest topological sense), sophisticated coordinate systems are not to be expected. They were not, anyway, widely available in medieval Europe until the translation of Ptolemy's *Almagest* into Latin in the twelfth century and the *Geography* in the fifteenth. These two texts may have provided medieval mapmakers with the crucial idea of an ordered space by the use of a pair of unique coordinates. On such a graph, information about the sky and the earth could be systematically inventoried. One of the earliest of these, dating from the first quarter of

174. Waldo R. Tobler, "Medieval Distortions: The Projections of Ancient Maps," *Annals of the Association of American Geographers* 56 (1966): 351–60, esp. 351.

175. London, British Library, Cotton Nero MS. D.V., fol. 1v, and British Library, Add. MS. 10049.

176. Tobler, "Medieval Distortions," 360 (note 174).

177. D'Avezac-Macaya, "Projection des cartes," 289–91 (note 43).

178. Other authorities center the map on other places, such as Constantinople, Rhodes, or Jerusalem. See Beazley, *Dawn of Modern Geography*, 1:390 (note 17).

179. Bacon, *Opus Majus*, 1:315 (note 97).

180. The description of Bacon's projection has been partly reconstructed from the translation by Cortesão in his *History of Portuguese Cartography*, 1:194–98 (note 34).

the eleventh century, is a curious graph showing the passage of the sun and the planets through the zodiac (fig. 18.29). Here there is evidence of a clear notion of celestial longitude and latitude that would probably have been derived from Pliny's encyclopedia. It includes thirty parts of longitude and twelve parts of latitude within the zodiac.[181]

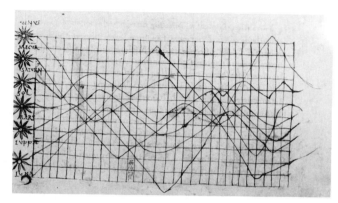

FIG. 18.29. AN ELEVENTH-CENTURY GRAPH. One of several graphs from this period, the diagram shows the passage of the sun and planets around the zodiac.
Size of the original: 13 × 22 cm. By permission of the Bayerische Staatsbibliothek, Munich (Clm. 14436, fol. 58r).

The implications of graphing went beyond a mere inventory function. The graphic representations of time, speed, distance, and instantaneous velocity by Nicole Oresme (ca. 1320–1382), Giovanni de Casali, and other mathematicians were essential to the understanding of these concepts. In the words of a modern historian of mathematics: "The development of graphical representation forged a link between the intuitive concepts of continuously varying quantities arising from physical phenomena and the geometry of the Greeks."[182] The connection between the graphing concept and cartography is seen in Oresme's use of the terms "longitude" and "latitude" for the independent and dependent variables plotted on a graph. Once these ideas were associated with algebraic symbolism, the seventeenth-century mathematicians René Descartes and Pierre de Fermat were able to formulate analytical geometry in the form familiar today.[183]

Although both the *Almagest* and the *Geography* remained unknown to the Western medieval world before the twelfth century, the concept of longitude and latitude had nevertheless filtered into northwestern Europe by the early eleventh century, largely through contacts with Islamic scientists in Spain. For example, al-Zarkali (ca. 1029 to ca. 1087), a Spanish Muslim from Cordova, was the principal composer of the Toledo tables. These tables contain a long list of geographical coordinates based on the prime meridian of the Canaries. For the

first time, the length of the Mediterranean was given correctly as 42° of longitude.[184]

There had also been attempts to measure longitude in the eleventh and twelfth centuries. Petrus Alphonsus (1062–1110) gave an explanation of the relation between time and longitude in his *Dialogi cum Judaeo*. Walcher's observation of lunar eclipses on 19 October 1091, on 18 October 1092, and in 1107–12 demonstrated a clear understanding that longitude could be expressed as a difference in time between two places: a lunar eclipse in Italy was seen shortly before dawn, whereas in England it had been observed in the middle of the night.[185] Later in the same century, Roger of Hereford reported that the eclipse of 12 September 1178 was observed simultaneously in Hereford, Marseilles, and Toledo and calculated the longitude of these places in relation to the meridian of Arin, the mythical center of the Islamic world.[186]

As Durand has shown, neither the early techniques of graphic representation of coordinates in Oresme nor the ability to measure longitude as the difference in time between two places can be shown to have had a direct influence on medieval cartography. Coordinates, for example, were used exclusively to calculate the relative time differences of places required in astrology rather than to aid in locating them on a map or globe. But although there is no clear testimony of the use of geographical coordinates in Europe between Roger Bacon and the first Vienna-Klosterneuburg maps of about 1425, the principles must have remained latent. The lack of maps drawn on this principle in this period, therefore, may have had more to do with the availability of reliable positional data than with the existence of a method of plotting it.[187]

181. Harriet Pratt Lattin, "The Eleventh Century MS Munich 14436: Its Contribution to the History of Coordinates, of Logic, of German Studies in France," *Isis* 38 (1947): 205–25. She discusses the contributions of Siegmund Günther, "Die Anfänge und Entwickelungsstadien des Coordinatenprincipes," *Abhandlungen der Naturhistorischen Gesellschaft zu Nürnberg* 6 (1877): 1–50, esp. 19, and H. Gray Funkhouser, "Notes on a Tenth-Century Graph," *Osiris* 1 (1936): 260–62, among others.

182. Margaret E. Baron, *The Origins of the Infinitesimal Calculus* (Oxford: Pergamon Press, 1969), 5.

183. Howard Eves, *An Introduction to the History of Mathematics* (New York: Holt, Rinehart and Winston, 1969), 281.

184. The Toledo Tables were adjusted for the location of Marseilles in the twelfth century by Raymond of Marseilles (Paris, Bibliothèque Nationale, MS. Lat. 14704, fol. 119v). See also Haskins, *Mediaeval Science*, 96–98 (note 50).

185. John Kirtland Wright, "Notes on the Knowledge of Latitudes and Longitudes in the Middle Ages," *Isis* 5 (1922): 75–98; Wright, *Geographical Lore*, 244–56 (note 18); Cortesão, *History of Portuguese Cartography*, 1:182–83 (note 34).

186. See below, appendix 18.1.

187. Durand, *Vienna-Klosterneuburg*, 94–105 (note 52). Wright, *Geographical Lore*, 246 (note 18), may have been too categorical in

The Production of *Mappaemundi*: Parchment, Inks, Pigments, Color, and Lettering

Mappaemundi were regarded as paintings in the early Middle Ages. Since their makers were map painters rather than cartographers in the modern sense of the word, the methods, tools, and materials used for these maps were those of the medieval artist in general. In particular, since the vast majority of these maps were produced for manuscript books, the techniques involved are indistinguishable from those used in manuscript illumination. Although yielding its place as a major art to architecture and sculpture in the course of the twelfth century, illumination was the focus for many major medieval artists and arguably constituted the greatest of the early medieval arts.[188]

The manuscript book was not the only vehicle for *mappaemundi*. The images appear in a variety of forms and materials. They are seen in stained-glass windows, frescoes, and floor mosaics, in reredos and tympana decoration, as sculpture, and even carved in benches.[189] Most commonly, however, they are found in manuscript encyclopedias, Bibles, and psalters. Thus, the vast majority were drawn and painted on parchment with a variety of inks and pigments.[190]

Records relating to the cost of *mappaemundi* or of the materials on which they they are drawn are scanty. There is a mention in the account books of the monastery at Klosterneuburg of a series of payments for a "*mappa*." Durand believes that the high cost of this map (thirty florins)—and the probable reference to making a case for it (payment of six talers for a locksmith)—suggests that it was large and elaborate.[191] Other sources of information, unfortunately now lost, were the account books of the monastery of San Michele di Murano, in which was found a notice concerning the copying and transmittal of the *mappamundi* (presumably for King Afonso V of Portugal) in the workshop of Fra Mauro, but without the details of the expenses.[192]

The attitude of medieval artists toward imperfections seems to have often been casual, as regards either the parchment or the drawing of the maps. For example, on the "Jerome" map of Asia, a hole in the vellum (about 3 × 5 cm) had been patched and sewn with another small piece before the map was drawn. The patch itself was then used to represent Crete, its shape preordained by the defect in the material (fig. 18.30). On the verso of the same leaf, on which a map of Palestine is drawn, the edge of the patch becomes the Caucasus Mountains from which the Ganges, Indus, and Tigris rivers are shown to spring.[193]

Several treatises on the materials and pigments used by medieval illuminators can help in reconstructing the methods used in the technical creation of the *mappae-*

mundi. Three are outstanding for their detail: *Mappae clavicula* (late twelfth century), *De arte illuminandi* (late fourteenth century), and the *Libro dell'arte* of Cennino Cennini (late fourteenth century).[194] These treatises are the recipe books of painting; they describe natural elements, minerals, and vegetable extracts as well as the artificial, manufactured salts used in preparing pigments.

Two types of ink were known and used in the Middle Ages. One was a suspension of carbon and the other a suspension of black organic salts of iron. Those *mappaemundi* drawn and lettered in ink used the same materials as any other manuscript, and the iron inks became the more common writing medium. They were sometimes mixed with gallic and tannic acid obtained from oak galls, providing an intense purple-black ink that darkened with age.[195]

The complex systems of map signs employed in modern cartography were less developed in the classical period and the Middle Ages. Instead, map features were often described with rubrics or legends, some of which

his statement that the influence of geographical coordinates on the cartography of the twelfth and thirteenth centuries "was absolutely *nil*."

188. Daniel V. Thompson, *The Materials and Techniques of Medieval Painting* (New York: Dover, 1956), 24.

189. [John K. Wright?], "Three Early Fifteenth Century World Maps in Siena," *Geographical Review* 11 (1921): 306–7; Giuseppe Caraci, "Tre piccoli mappamondi intarsiati del sec. XV nel Palazzo Pubblico di Siena," *Rivista Geografica Italiana* 28 (1921): 163–65; Bernhard Brandt, *Mittelalterliche Weltkarten aus Toscana*, Geographisches Institut der Deutschen Universität in Prag (Prague: Staatsdruckerei, 1929).

190. Parchment is any kind of animal skin prepared for writing or drawing. It is a general word for such material and does not specify the animal, whether sheep, calf, goat, or whatever. Vellum was sometimes used to refer to calfskin and fine parchment, but the distinction has become less clear in recent times. Uterine vellum—from an aborted animal, was extremely rare. The distinction between parchment and vellum is also discussed by W. Lee Ustick, "Parchment and Vellum," *Library*, 4th ser., 16 (1935): 439–43. See also Daniel V. Thompson, "Medieval Parchment-Making," *Library*, 4th ser., 16 (1935): 113–17.

191. Durand, *Vienna-Klosterneuburg*, 123–24 (note 52), also refers to the original work on these account books by Berthold Černik, "Das Schrift- und Buchwesen im Stifte Klosterneuburg während des 15. Jahrhunderts," *Jahrbuch des Stiftes Klosterneuburg* 5 (1913): 97–176. For another description of technical details—from a contract for world maps at the end of the fourteenth century—see Skelton, "Contract," 107–13 (note 6), and Campbell, chapter 19 below.

192. Gasparrini Leporace, *Mappamondo di Fra Mauro*, 15 n. 2 (note 137). A transcription of the note is found in Antonio Bertolotti, *Artisti veneti in Roma nei secoli XV, XVI e XVII: Studi e ricerche negli archivi romani* (Venice: Miscellanea Pubblicata dalla Reale Deputazione di Storia Patria, 1884, reprinted Bologna: Arnaldo Forni, 1965), 8.

193. London, British Library, Add. MS. 10049, fols. 64r–64v.

194. Franco Brunello, *"De arte illuminandi" e altri trattati sulla tecnica della miniatura medievale* (Vicenza: Neri Pozza Editore, 1975), contains a valuable bibliography of recent literature on the topic.

195. Thompson, *Medieval Painting*, 81–82 (note 188).

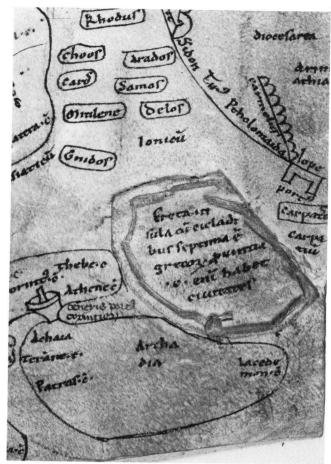

FIG. 18.30. MAP ON VELLUM SHOWING REPAIR. The vellum of this twelfth-century map has been ingeniously repaired with a patch that then represents Crete.
Length of the patch: 5.3 cm. From the "Jerome" map. By permission of the British Library, London (Add. MS. 10049, fol. 64r).

been routinely ruled up for text before the map was drawn and there is sometimes an attempt to follow the lines. This can be seen, for example, in one of the Ranulf Higden maps in the British Library, or in the Cotton "Anglo-Saxon" map, where the map was drawn on the verso of a page on which lines had been scored.[197] The scored lines show through the page, and the artist obviously made a conscious attempt either to line up the lettering or to avoid them. In some reproductions of the Cotton map, such as those in Beazley or Miller, the lines can be seen, but it is important to realize what they are and that they have no substantive meaning.[198] Such a point is a reminder of the importance of examining originals in order to avoid unfounded conclusions.

The inclusion of explanatory matter on the face of the map obviated the need for separate keys of signs. This topic has recently been explored by von den Brincken and by Delano Smith.[199] The occurrence of what have been called "silent maps" lacking any lettering was exceptional among *mappaemundi*. Von den Brincken cites only one example, which she discusses at length, the fourteenth-century world map in the *Livre dou trésor* of Brunetto Latini. She suggests, intriguingly and plausibly, that it lacks lettering because Latini could have used an Arabic model on which the legends were in Arabic, a language he could not transcribe.[200] In the later Middle Ages, explanations of the map painter's intentions are sometimes found on the map itself, as in the case of the world map of Andreas Walsperger (1448). Walsperger explains his system of distinguishing between Christian and Islamic cities: "The earth is indeed white, the seas of a green color, the rivers blue, the mountains variegated, likewise the red spots are cities of the Christians, the black ones in truth are the cities of the infidels on land and sea"[201] (plate 21).

could be extremely long. *Mappaemundi* were thus quite as much written as drawn. Calligraphic styles follow those prevailing in the texts of the time and thus can provide at least a very rough guide to the origin and chronology of the maps. For instance, there are the national hands of the sixth to the eighth century (although very few maps survive from this period), the Carolingian minuscules of the eighth to the twelfth century, and the Gothic or black letter in its various forms of the twelfth to the fifteenth century. Also common on *mappaemundi* are the semiformal crossbred current styles known as *littera bastarda*, combinations of the cursive everyday secretarial hand and the more formal black letter.[196]

Lettering was not usually laid out on the *mappaemundi* in a systematic manner, nor was there usually an attempt to rule guidelines. In some cases the vellum had

196. David Woodward, "The Manuscript, Engraved, and Typographic Traditions of Map Lettering," in *Art and Cartography* (note 135).

197. The Higden map is British Library, Add. MS. 10104, fol. 8r. The Cotton map is British Library Cotton MS. Tiberius B.V., fol. 56v.

198. Beazley, *Dawn of Modern Geography*, 2:560 (note 17). Miller, *Mappaemundi*, 1, pl. 10 (note 9).

199. Anna-Dorothee von den Brincken, "Die Ausbildung konventioneller Zeichen und Farbgebungen in der Universalkartographie des Mittelalters," *Archiv für Diplomatik: Schriftgeschichte Siegel- und Wappenkunde* 16 (1970): 325–49; Catherine Delano Smith, "Cartographic Signs on European Maps and Their Explanation Before 1700," *Imago Mundi* 37 (1985): 9–29.

200. Von den Brincken, "Zeichen und Farbgebungen" (note 199). It should be pointed out that the date for the Latini map is thought on stylistic grounds to be later than the dates for the compilation of the work while Latini was in exile (1260–66), but it probably predates 1320.

201. Von den Brincken, "Zeichen und Farbgebungen," 345 (note 199). See also below, Campbell, chapter 19. The use of color in medieval cartography is also treated in Ulla Ehrensvärd, "Color in Cartography: An Historical Survey," in *Art and Cartography* (note 135).

The use of color is widely varied on the *mappaemundi*, but certain deep-seated conventions, such as blue or green for water and red for the Red Sea, are usually followed. Occasionally, unusual coloring is seen, like the bright Mozarabic colors of the Beatus maps or the gray sea and orange rivers of the Cotton "Anglo-Saxon" map (plate 22). A list based on von den Brincken, of thirty selected maps on which the color has remained intact and unfaded, is provided as table 18.4. Considerable variation is noted, except in such conventions as the use of red for the Red Sea.

Signs for towns and mountains on *mappaemundi* had to be designed to overcome the problem of representing something in plan. Mountains were shown by chains of curves or spikes, teeth, heaps, lobes, or plaited ornamentation (guilloche). Towns were differentiated by stylistic pictures of groups of buildings seen from the side. Their realism varied depending on the mapmaker's familiarity with the place. The Arabic world maps are generally more abstract in their use of signs, using circles for cities.[202]

CONTENT AND MEANING

The content of *mappaemundi* may be conveniently discussed under three headings: the historical and geographical facts; the marvels, legends, and traditions; and the symbolic content. Of these, as has already been pointed out in this chapter, the greatest emphasis in the literature has traditionally been on the first two categories, particularly—it often seems—in order to demonstrate the shortcomings of medieval learning, such as the errors in the location of places and features on the earth and the curiosities associated with medieval fable and legend. The third category, symbolic content, has received little attention until recently, but its importance in understanding the meaning and historical significance of the *mappaemundi* will be demonstrated here.

Historical and Geographical Information

The factual information on medieval world maps is a blending of historical events and geographical places, a projection of history onto a geographical framework.[203] As with the medieval popular illustrations, in which a story is told by the simultaneous portrayal of various stages of the narrative within a single frame, a *mappamundi* not only represents static geography but is also an aggregation of historical information the mapmaker considered important with regard to his audience, no attempt being made to separate or identify the two types of information. This dual problem of man's status in the world and the universe—which Bertrand Russell has called "chronogeography"—was a prime question of the medieval philosophers.[204]

The sources of historical and geographical information available to the makers of *mappaemundi* were both classical and biblical. The emphasis on the latter increased toward the end of the Middle Ages. Both traditions were rich in historical and geographical lore—the commemoration of famous events and places being sometimes inseparable. The biblical tradition in the *mappaemundi* is usually derived from the Old rather than the New Testament. In early Judaism the importance of the location of events was emphasized, but early Christianity showed little interest in such things, with certain important exceptions such as the journeys of Saint Paul.[205] The teachings of Christ emphasized the spiritual and not the physical world.[206] In addition, although the Bible is full of references to places of local interest, there are few allusions indeed to cosmography: the words sphere, globe, or hemisphere in the geographical sense are nowhere found in its pages.[207]

In reaction to the classical geographers, the early fathers of the church were also anxious to stress that knowledge of the earth was of strictly secondary importance to the Christian, whose eyes should be on a higher spiritual plane. In outlining the characteristics of a true believer, Saint Augustine commented that "a man who has faith in you . . . though he may not know the track of the Great Bear, is altogether better than another who measures the sky and counts the stars and weighs the elements."[208]

202. Von den Brincken, "Zeichen und Farbgebungen," 336 (note 199).

203. Von den Brincken, "Mappa mundi und Chronographia," 118 ff. (note 23).

204. Bertrand Russell, *Philosophy* (New York: W. W. Norton, 1927), 283.

205. Robert North, *A History of Biblical Map Making*, Beihefte zum Tübinger Atlas des Vorderen Orients, B32 (Wiesbaden: Reichert, 1979), 76: "The earliest Christians showed no sentimental interest in the exact location of their own most sacred events." *BS·*

206. John 4:19–24. In response to the question whether to build a shrine at Gerizim or Jerusalem, Christ's answer was that one should be less concerned with location than with motivation.

207. The phrase in Ps. 83:11, "surface of the globe," is now considered to have no geographical significance. See *The Anchor Bible* (New York: Doubleday, 1964–), vol. 17, Mitchell Dahood, trans., *Psalms II: 51–100*, 275 n. 11. In addition, the frequent references to "the round world" in the original (sixteenth-century) *Book of Common Prayer*, as in Ps. 89:12, 96:10, 98:8 (Psalms 88, 95, and 97 in the Bible) express circularity rather than sphericity, from the Latin of the Vulgate, *orbis terrae*. The only specific mention of a "map" (or at least a town view) that I have been able to find in the Bible is in Ezek. 4:1: "Man, take a tile [Vulgate: *laterem*] and set it before you. Draw a city on it, the city of Jerusalem." It is also possible, as Menashe Har-El believes, that maps were in use for the extensive survey (register) dividing the tribes of Israel, found in Joshua 13–19, especially Josh. 18:5. See Menashe Har-El, "Orientation in Biblical Lands," *Biblical Archaeologist* 44, no. 1 (1981): 19–20.

208. Saint Augustine, *Confessions*, trans. R. S. Pine-Coffin (London: Penguin Books, 1961), 95.

TABLE 18.4 Survey of Representational Styles of Selected Medieval Maps

Date	Author	Seas	Red Sea	Rivers	Relief Representation	Settlements
8th–9th century	Cosmas	Blue	Blue	Green	None	None
ca. 775	Isidore	Blue/green	Red	Blue/green	Red jagged chains	Six eight-pointed stars
10th century	Anglo-Saxon	Gray	Red	Orange	Green chains of "teeth"	Double towers, rotundas
ca. 1050	Beatus-Saint Sever	Blue	Red	Blue	Jagged and arched chains	Yellow battlemented buildings
1055	Theodulf	Blue/green	Red	Blue/green	Brown jagged lines	Square stone buildings
ca. 1109	Beatus-Silos	Blue	Red	Blue	Green and red arch clusters	Only legends
ca. 1110	Henry of Mainz	Green	Red	Violet	Red lobed chains	Double towers, ramparts
1119	Guido of Pisa	Blue	Red	Green	Double leaves, green inside	Legends only
13th	Psalter map	Green	Red	Blue	Natural-colored lobed chains	Ocher triangles
After 1342	Higden	Green	Red	Green	Green/red, green/black mountain chains	Large vignettes
ca. 1430	Borgia map	--------------No coloring--------------			Rows of "teeth"	Triple towers
1448	Andreas Walsperger	Green	Red	Blue	Brown or green shapes	Red or black circles; individual buildings
1452	Giovanni Leardo	Blue	Red	Gray	Red/green three-tiered mountains	Building clusters
1457	Genoese map	Blue	Red	Gray	Green patches, gray/white hill drawings	Red, pink, and white tower clusters
1459	Fra Mauro	Blue	Blue	Blue	Green/blue hill drawings	Red, green, and blue tower clusters

Source: After Anna-Dorothee von den Brincken, "Die Ausbildung konventioneller Zeichen und Farbgebungen in der Universalkartographie Mittelalters," Archiv für Diplomatik: Schriftgeschichte Siegel- und Wappenkunde 16 (1970):325–49.

In the absence of a grid of latitude and longitude, the main locational structure of the *mappaemundi* was provided by prominent hydrographic features. Three of these, the river Don, the Nile or the Red Sea, and the Mediterranean provided the boundaries within the tripartite world. Around the entire world was the encircling ocean, an enduring tradition since the time of Homer. Indenting the edge of the circular world are the prominent gulfs of the Red Sea and the Mediterranean; the Caspian Sea is also often shown as a small gulf in the northeast. The Gulf of Azov—the Palus Maeotis of classical times which becomes Meotides Paludes on the *mappaemundi*—also sometimes appears as a small gulf of the surrounding ocean, as on the Corpus Christi College, Oxford, version of Higden's map or the world map of Guido of Pisa (1119).[209] This idea appears to have been derived from the passage in 2 Esdras prescribing that all the earth's hydrography had to be connected in some way, a point taken up by Saint Basil.[210]

Although the four rivers of paradise—Tigris, Euphrates, Pishon, and Gihon—are usually shown on *mappaemundi* as fanning out from the location of paradise in a simple, stylized fashion, they were also represented as real rivers: the Tigris, Euphrates, Ganges, and Indus, as on the "Jerome" map of Palestine (fig. 18.31).[211] The Nile is sometimes equated with the Gihon and shown as an extension of this river, as in a map found in a tenth-century manuscript of one of Isidore's works (fig. 18.32).[212] The persistence of the confusion over the correct location of the rivers of paradise is shown by Columbus, who, on hearing a report that his men in the caravel *Correo* had seen four rivers at the head of the Gulf of Paria on the third voyage in 1498, thought they were the rivers of paradise.[213]

Many fourteenth- and fifteenth-century *mappaemundi* contain a representation of the River of Gold, Strabo's Pactolus and the Rio del Oro of the Middle Ages.[214] The River of Gold was thought to be the flood reaches of the Niger above Timbuktu, and there were several attempts during the fourteenth century to develop a route to it from the coast of West Africa. It appears on the Catalan atlas, the Borgia map, the Catalan (Estense) map, and Fra Mauro's map of 1459 (to cite only the better-known world maps), usually in the form of a bulging lake in the course of the river, into which four or five rivers flow from the western Mountains of the Moon.[215]

Information regarding human settlements on *mappaemundi* was also derived from a mixture of classical and biblical sources.[216] The names of classical peoples, tribes, regions, and cities took their place with the names of the newly formed bordering nations in eastern and northern Europe. For example, the regions of the Slavs, Bulgaria, Norway, and Iceland all appear on the Cotton

"Anglo-Saxon" map of the tenth century. The Henry of Mainz map includes Denmark and Russia. The Psalter map shows Hungary and Russia, and Bohemia, Poland, and Prussia appear first on the Ebstorf map and then on the Hereford map and on maps by Higden and Fra Paolino. Sweden first appears on the maps of Lambert of Saint-Omer, and Finland is found on the Vesconte and Fra Paolino world maps and on the printed world map in the *Rudimentum novitiorum*. Despite its publication date of 1475, this last work was derived from a much earlier source.

Similarly, together with such classic regions as Gallia, Germania, Achaea, and Macedonia, the names of more recently organized provinces and states of commercial importance came to be inserted, as with the appearance of Genoa, Venice, and Bologna in Italy or Barcelona and Cádiz in Spain. Some cities had ceased to exist long before the maps were drawn but their historical importance merited their mention, such as Troy in Asia Minor and Leptis Magna and Carthage in North Africa. Other cities were included in the maps because of their contemporary political importance, Rome and Constantinople among them.

As the influence of the classical tradition declined, biblical sources became more prominent. Although originally Roman, the basic structure of the tripartite diagrams now owed their form to the tradition of the peopling of the earth by the descendants of Noah. The families of Shem, Ham, and Japheth are sometimes listed on the maps in full, taken from the passage in Genesis

209. Miller, *Mappaemundi*, 3:97–98 (note 9). Beazley, *Dawn of Modern Geography*, 2:632 (note 17).

210. Saint Basil *Homily* 4.2–4; see *Exegetic Homilies*, trans. Sister Agnes Clare Way, The Fathers of the Church, vol. 46 (Washington, D. C.: Catholic University of America Press, 1963). Kimble, *Middle Ages*, 33–34 (note 55), and 2 Esd. 6:42: "On the third day you ordered the waters to collect in a seventh part of the earth; the other six parts you made into dry land."

211. Miller, *Mappaemundi*, 2, pl. 12 (note 9).

212. Madrid, Biblioteca de la Real Academia de la Historia, Codex 25, fol. 204. Kamal, *Monumenta cartographica*, 3.2:667 (note 53).

213. Samuel E. Morison, *Admiral of the Ocean Sea: A Life of Christopher Columbus*, 2 vols. (Boston: Little, Brown, 1942), 2:283.

214. Strabo: "The Pactolus . . . anciently brought down a large quantity of gold-dust, whence it is said, the proverbial wealth of Croesus and his ancestors obtained renown." See Eva G. R. Taylor, "Pactolus: River of Gold," *Scottish Geographical Magazine* 44 (1928): 129–44.

215. Kimble, *Middle Ages*, 107–8 (note 55). Charles de La Roncière, *La découverte de l'Afrique au Moyen Age: Cartographes et explorateurs*, Mémoires de la Société Royal de Géographie d'Egypte, vols. 5, 6, 13 (Cairo: Institut Français d'Archéologie Orientale, 1924–27), is the fullest account, with an excellent bibliography.

216. I would like to acknowledge the help of George Kish with this section. See also von den Brincken, "Mappa mundi und Chronographia," 169 (note 23), who makes a very useful systematic survey of place-names on selected *mappaemundi*, providing seven tables of place-names as found on twenty-one maps.

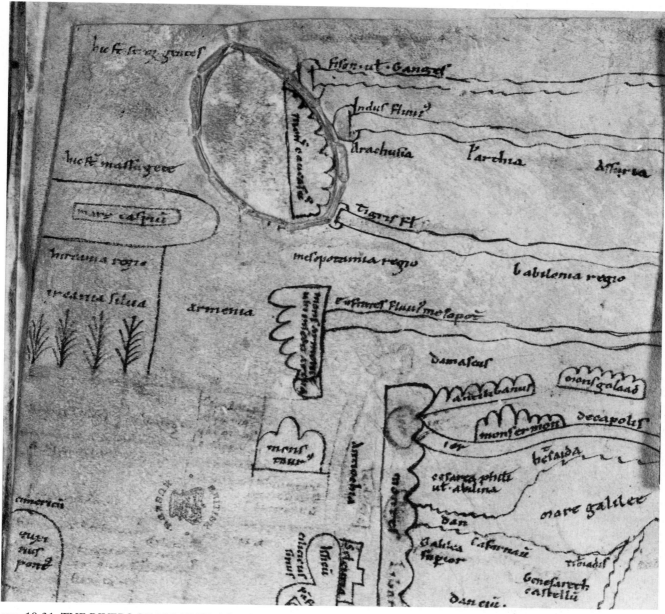

FIG. 18.31. THE RIVERS OF PARADISE. This detail from the twelfth-century "Jerome" map of Palestine shows the four rivers of paradise represented as real watercourses. The rivers are (*top to bottom*) the Ganges, Indus, Tigris, and Euphrates.

Size of the original detail: 15.7 × 15.7 cm. By permission of the British Library, London (Add. MS. 10049, fol. 64v).

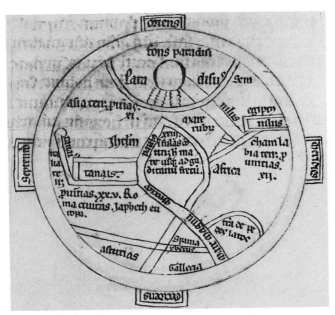

FIG. 18.32. THE NILE AS AN EXTENSION OF THE FOUR RIVERS OF PARADISE. This schematic T-O map, from a tenth-century manuscript of Isidore of Seville, shows the Nile with two sources: one in paradise and another in Africa. Diameter of the original: 11.5 cm. By permission of the Biblioteca Medicea Laurenziana, Florence (Plut. 27 sin. 8, fol. 64v).

(fig. 18.33).[217] Noah's ark, Mount Sinai, the Tower of Babel, Babylon, the Dead Sea, the river Jordan, Samaria, and the twelve tribes of Israel are also shown. Although the New Testament provided much less of the content of the *mappaemundi* by the later Middle Ages, those places that evoke the life of Christ and the apostles tended to be marked. In addition to Jerusalem, we find Bethlehem, Nazareth, the Sea of Galilee, Damascus, Ephesus, Antioch, Nicaea, Tarsus, and even the tombs of Saint Thomas, Saint Philip, and Saint Bartholomew identified on the Beatus (Saint Sever) and Ebstorf maps.[218]

Pilgrimage goals were often emphasized on *mappaemundi*, and the associated itineraries provided the source for many of the place-names, as Crone has demonstrated for the Hereford map.[219] Santiago de Compostela in Spain and Mont Saint-Michel in Brittany were commonly shown. Not surprisingly, Rome appears on almost every map, reflecting its multiple role as the old imperial capital of the West, the seat of the papacy, and the city of many churches where indulgence was offered to pilgrims. Jerusalem's importance as the greatest of all Christian pilgrims' goals is underlined not only by its appearance on most *mappaemundi*, but also by the popularity in the same period of detailed maps of the Holy Land, and plans of the Holy City as well, as Harvey describes in chapter 20 of this volume.

Marvels and Legends

Representations of monstrous races and historical legends on *mappaemundi* reflected the medieval craving for the bizarre and fantastic.[220] In classical times, especially in Greece, such a demand had been expressed in the invention of mythical creatures with religious associations, such as centaurs, sirens, and satyrs. Nonreligious images were formed of monstrous races of men who inhabited progressively more remote areas as more of the earth became known. Many of these ideas derived from empirical observation—for example, the Amyctyrae with protruding lower lips could well have been based on remote contact with the Ubangi tribe.[221] Table 18.5 summarizes the main groups of semimythical races that appear on *mappaemundi*.

The sources of the monstrous races go back at least to the fifth century B.C. to writers such as Herodotus, Ctesias of Cnidos (fl. 398 B.C.), and Megasthenes (ca. 303 B.C.). The last two had apparently traveled to India, where most of the marvels were assumed to be found.[222] With Alexander the Great's invasion of India in 326 B.C., a body of legend grew out of his travels that was revived in the Middle Ages in the form of the Alexander romances. Although the Greek geographer Strabo (64/63 B.C. to A.D. 21) disdained the reports of these marvels and monstrous races, being "seized with disgust for such worthless writings that contribute neither to adorn nor to improve life," Pliny the Elder was less critical, and his writings had considerably more influence on medieval thought. His *Historia naturalis* (ca. A.D. 77) contained a vast collection of geographical lore culled from hundreds of sources. Much of Pliny's encyclopedic work is of great descriptive value, but it was largely the bizarre that was transmitted to the Middle Ages. The *Collectanea rerum memorabilium* of Gaius Julius Solinus (third century A.D.), for example, emphasized the marvels and little else. Popular writers like Macrobius and Martianus Capella, although enlightened in several matters such as the zonal concept and the sphericity of the earth, also perpetuated the monster legends in later medieval times. All the great encyclopedias of the later Middle Ages contain references to

217. Gen. 10.

218. Von den Brincken's tables in "Mappa mundi und Chronographia" (note 23) allow the reader to trace all these place-names and others to the respective maps.

219. Crone, "New Light," 451–53 (note 21).

220. Rudolf Wittkower, "Marvels of the East: A Study in the History of Monsters," *Journal of the Warburg and Courtauld Institutes* 5 (1942): 159–97, esp. 159.

221. Friedman, *Monstrous Races*, 24 (note 3).

222. Wittkower, "Marvels," 160 (note 220). See also Jean Céard, *La nature et les prodiges: L'insolie au 16ᵉ siècle*, Travaux d'Humanisme et Renaissance, no. 158 (Geneva: Droz, 1977).

FIG. 18.33. DESCENDANTS OF NOAH. The families of Noah's three sons, "Sem," "Cham," and "Sefet" (i.e., Shem, Ham, and Japheth), are shown in this genealogical diagram from an eleventh-century manuscript of the *Commentary on the Apocalypse of Saint John* of Beatus of Liebana. The *mappamundi*, top right (detail, fig. 18.52), is used to illustrate the division of the world between the three sons.

Size of the original: 37 × 55.3 cm. Photograph from the Bibliothèque Nationale, Paris (MS. Lat. 8878, fols. 6v–7r).

TABLE 18.5 List of the Main Semimythical Races Found on *Mappaemundi*

Name	Characteristics	Location	Maps
Amyctyrae	Protruding lower or upper lip		Ebstorf
Anthropophagi	Man-eaters; drink from skulls	Scythia; Africa	Walsperger; Ebstorf
Antipodeans	Opposite-footed	Antipodes	Beatus
Artibatirae	Walk on all fours		Psalter
Astomi	Mouthless; apple smellers	Ganges	Walsperger
Blemmyae	Faces on chests; no necks; also known as Acephali	Libya (Africa)	Walsperger
Cyclopes	One-eyed; also known as Monoculi	Sicily; India	Walsperger
Cynocephali	Dog-headed	India	Borgia; Hereford; Ebstorf; Walsperger; Psalter
Epiphagi	Eyes on shoulders (similar to Blemmyae)		Psalter
Hippopodes	Horses' hooves		Ebstorf
Maritimi	Hold bow and arrow (four eyes)	Africa	Psalter; Ebstorf
Martikhora	Four-legged beasts with men's heads		Hereford; Ebstorf
Sciopods	Shadow-footed (sometimes also called Monocoli, from the Greek, causing confusion with the Monoculi above)	India	Beatus; Hereford; Psalter
Troglodytes	Cave dwellers	Ethiopia	Walsperger; Psalter

monsters: Isidore, Rabanus Maurus, Honorius, Gautier de Metz, Gervase of Tilbury, Bartolomeus Anglicus, Brunetto Latini, Vincent of Beauvais, and Pierre d'Ailly. Inevitably, maps incorporated into these works also featured them, right into the fifteenth century.[223] However, there was also the skeptical and perhaps nostalgic view, such as that expressed by François Rabelais (ca. 1495–1553): "I saw an incredible number of attentive men and women . . . they held a *mappamundi* and spoke eloquently of prodigies . . . of the pyramids and the Nile . . . and of Troglodytes, of Himantopodes, of Blemmyae. . . of Cannibals. . . . There I saw Herodotus, Pliny, Solinus . . . and many other ancients . . . all writing beautiful lies."[224]

The monstrous races posed a number of problems for the fathers of the church. If they existed—and there was general agreement that they did—were they human? And if they were human, were they descended from Adam and Noah, possessing souls that could be saved? Several biblical passages stated that the gospel must be preached to all nations of men, which was taken to include the monsters. Hence, the main target of the medieval missionaries were the Cynocephali—the dog-headed peoples sometimes associated with Islam—whose conversion would have created a dramatic demonstration of the power of the gospel.[225] These creatures are thus found on the didactic *mappaemundi*. The Borgia map, for instance, contains a representation of the dog-headed Saracen, under the rubric: "Ebinichibel is a Saracen Ethiopian king with his dog-headed people" (fig. 18.34).[226]

The placing of the monstrous races on the world map varied according to the three main types of *mappaemundi*. In tripartite maps, the races were usually crowded into a band in the southernmost part of Africa, no particular attempt being made to link the position of these peoples with climatic or other physical factors. This represents a location derived from Pliny. An additional advantage, in the eyes of medieval ecclesiastics, would have been their being shown as far as possible from the civilized center of the Earth—Jerusalem—but, as the Ebstorf map so vividly shows, still within the reach of the left arm of Christ.

In the zonal maps, the Antipodes have to be taken as the guide to the location of the monstrous races, since usually very few descriptions—verbal or graphic—are given on the maps themselves. In contrast, those maps that show a fourth continent, especially the Beatus type of map, which aimed to illustrate the mission of the church in the conversion of all peoples of the world, contain the earliest extant representations of the monstrous races, together with detailed rubrics.[227]

The fear of races and spiritual forces outside Christianity gave rise to two other legends that appear so frequently on *mappaemundi* that they merit individual

FIG. 18.34. CYNOCEPHALI ON THE BORGIA MAP. A dog-headed people, associated with Islam and supposed to exist in Ethiopia, the Cynocephali were prime candidates for conversion by medieval missionaries and were therefore frequently shown on the didactic *mappaemundi* as in the upper right here. Size of the original detail: 12 × 7.2 cm. Photograph from the Biblioteca Apostolica Vaticana, Rome (Borgiano XVI).

explanation. These are the legends of the mythical Christian king Prester John and the suggested existence of nations associated with the names Gog and Magog. Gog, and his subjects Magog, appear in Ezekiel and in Revelation, where they are described as the forces of the Antichrist who will be loosed at the Day of Judgment to overrun the civilized earth.[228] Alexander the Great is said to have built a wall, with a great brass gate in the Caucasus Mountains, in order to contain them. On the *mappaemundi*, Gog and Magog were personified as two

223. The marvels continued to appear in printed illustrated histories and cosmographies, such as those by Hartmann Schedel, Sebastian Münster, André Thevet, Sebastian Franck, and other Renaissance figures.

224. François Rabelais, *Pantagruel*, 5.31, in *Oeuvres complètes*, ed. Jacques Boulenger (Paris: Gallimard, 1955), 844.

225. Friedman, *Monstrous Races*, 59–86 (note 3).

226. "Ebinichibel rex est sarracenos ethiopicos cum populo suo habiens caninam." Almagià, *Vaticana*, 1:27–29 and pl. XI (note 83).

227. Friedman, *Monstrous Races*, 50 (note 3).

228. Ezek. 38:1–9, Rev. 20:7–8.

giants situated somewhere in the northern or north-eastern part of Asia. Sometimes they were shown contained by Alexander's wall, often mistaken for a representation of the Great Wall of China (fig. 18.35).[229]

The Prester John legend, which Cortesão has called "the greatest hoax in the history of geography,"[230] concerns the existence of a mythical Christian king. Prester John, it was hoped, would act as a rearguard ally of the Christians in their struggles with the Islamic empire.[231]

with a cross. As successive expeditions failed to find him, the choice of possible locations was progressively narrowed, and his image appeared to migrate accordingly (fig. 18.36).

FIG. 18.36. PRESTER JOHN. A mythical Christian king, Prester John first appeared in Asia. But as successive expeditions found no trace of this possible ally in the West's struggles with Islam, he was moved to various parts of Africa.
From Diogo Homem's atlas of about 1565. By permission of the British Library, London (Add. MS. 5415a).

FIG. 18.35. THE WALL ABOUT THE KINGDOM OF MA-GOG. The purpose of this wall—which derived from the Alexander legend—was to contain Gog, whose hordes in the kingdom of Magog were supposed to overrun the world at the Day of Judgment. Since it was situated in Asia, such depictions have been mistaken for the Great Wall of China. The detail is from the thirteenth-century Psalter map.
Size of the original detail: 3.3 × 2.5 cm. By permission of the British Library, London (Add. MS. 28681, fol. 9r).

According to Cortesão the story did not appear on any map until Carignano's chart of about 1307, where the king is found in Ethiopia, albeit rather indistinctly. In the Vesconte and Sanudo world maps of about 1320, Prester John is shown in India. On several maps thereafter until well into the sixteenth century, the king is featured in India, China, and several parts of Africa, usually as a throned monarch holding a staff surmounted

229. Andrew R. Anderson, *Alexander's Gate, Gog and Magog, and the Inclosed Nations* (Cambridge, Mass.: Medieval Academy of America, 1932). An enormous body of legend was generated by the travels of Alexander the Great and found its way into medieval thought—and hence onto the *mappaemundi*—by way of the Alexander romances. See also W. J. Aerts et al., eds., *Alexander the Great in the Middle Ages: Ten Studies on the Last Days of Alexander in Literary and Historical Writing*, Symposium Interfacultaire Werkgroep Mediaevistiek, Groningen, 12–15 October, 1977 (Nijmegen: Alfa Nijmegen, 1978). Friedman, *Monstrous Races*, 33 (note 3).

230. Cortesão, *History of Portuguese Cartography*, 1:255–75 (note 34).

231. The story of Prester John started in Rome in the early twelfth century. It was given credence by a forged letter of 1163 purporting to be from the mysterious priest-king John in India to Emmanuel of Constantinople and Frederick Barbarossa, describing the wealth and power of his kingdom. Pope Alexander III replied to this letter in 1177, asking if Prester John would pledge his support to reconquer Jerusalem for Christendom. The original letter (which is known to us in a hundred manuscripts and many fifteenth- and sixteenth-century printed editions) was to influence several attempts to find and make political contact with this mythical king. The efforts of Prince Henry the Navigator were particularly noteworthy in this regard; he sent his chamberlain, Antão Gonçalves to explore the coast of West Africa in 1441 with the instruction that "he not only desired to have knowledge of that land, but also of the Indies, and of the land of Prester John, if he could." See Cortesão, *History of Portuguese Cartography*, 1:264 (note 34).

Symbolism: History, Power, and Orientation

The function of medieval *mappaemundi* was largely exegetic, with symbolism and allegory playing major roles in their conception. This was acknowledged at the time. Hugh of Saint Victor (ca. 1097–1141) defined a symbol as "a collecting of visible forms for the demonstration of invisible things."[232] It can be inferred from this that Hugh was assuming symbols to have graphic form, whereas modern writers of medieval history and literature tend to refer to symbolic imagery in a strictly verbal rather than a graphic sense. The modern medieval historian is also more concerned with the abstract, mystical meaning of symbolism—the cross as a symbol of the Passion, for example—than with the spatial symbolism relating to the shape of the cross as representing the four directions of the universe in which the influence of God is found: height, depth, length, and breadth.[233] There is, however, support for the notion that medieval man thought in concrete and literal ways in addition to the mystical and allegorical. Ladner has pointed out that Saint Gregory of Nyssa (fourth century) even extended the spatial imagery of the cross to the two-dimensional view: the four quarters of the world and the four cardinal directions, and even to the four-part division of Christ's clothing after the Crucifixion.[234]

Many such visible forms representing spiritual concepts of the Christian church are evident in the *mappaemundi*. Sometimes the whole map is presented as a symbol of Christian truths. The central theme is the earth as a stage for a sequence of divinely planned historical events from the creation of the world, through its salvation by Jesus Christ in the Passion, to the Last Judgment. Such an interpretation bears out von den Brincken's view that the maps are as much historical chronicles as geographical inventories.[235]

In such maps, the creation of the world is symbolized by the way the tripartite schema is used to divide the earth into the three continents as peopled by the sons of Noah. The three-part structure is thus a symbol of the historical beginning of man's life on earth. With varying amounts of detail, the families of Shem, Ham, and Japheth are depicted on individual maps according to their biblical listing in Genesis, Shem's family having the largest share (Asia) to reflect his primogeniture. The Semitic, Hamitic, and Japhetic peoples derive from this division.

But the T-O map can also be seen as a symbol of the Passion of Christ. It is probable, as Lanman suggests, that the T in the T-O schemata represented a cross, but of the tau variety (the *crux commissa*). This is particularly noticeable when the ends of the crossbar are angled or truncated, as in figure 18.37.[236] When the body of Christ is superimposed on the map of the earth in an all-embracing dying gesture, as in the Ebstorf map, the map itself becomes a clear symbol of the salvation of the world. Even the twenty-four monstrous races are embraced by the arms of Christ, although symbolically they are by his left hand at the very extremity of the world.

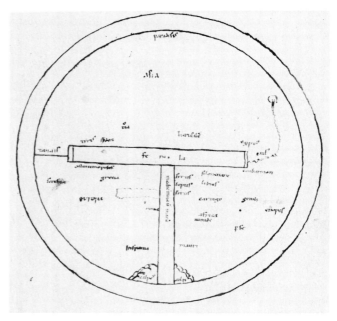

FIG. 18.37. T-O MAP WITH TAU CROSS. Such images reinforce the symbolization of the Passion of Christ that is inherent within the T-O schema, the T-O representing the tau cross (*crux commissa*). The map shown here is dated to the eleventh century.
Diameter of the original: 16.2 cm. From a manuscript of Sallust, *De bello Jugurthino*. By permission of the Universitätsbibliothek Rostock, GDR (Codex Philol. 27, fol. 1v).

232. Gerhart B. Ladner, "Medieval and Modern Understanding of Symbolism: A Comparison," *Speculum* 54 (1979): 223–56, quotation on 225. Hugh of Saint Victor's graphic bent is also shown by his indication that, when seeking the meaning of a passage of Scripture, he used to help himself by drawing diagrams. See Beryl Smalley, *The Study of the Bible in the Middle Ages*, 2d ed. (Oxford: Blackwell, 1952), 96.

233. Eph. 3:18.

234. Gerhart B. Ladner, "St. Gregory of Nyssa and St. Augustine on the Symbolism of the Cross," in *Late Classical and Mediaeval Studies in Honor of Albert Mathias Friend, Jr.*, ed. Kurt Weitzmann (Princeton: Princeton University Press, 1955), 88–95, esp. 92–93. John 19:23.

235. See above, pp. 288–90.

236. Jonathan T. Lanman, "The Religious Symbolism of the T in T-O Maps," *Cartographica* 18, no. 4 (1981): 18–22. This is strongly confirmed by the small *mappamundi* functioning as an initial T with a crucified figure. Florence, Biblioteca Medicea Laurenziana, Conventus soppressus 319, fol. 90v. See Arentzen, *Imago Mundi Cartographica*, 235–36 and pl. 79 (note 32).

The inclusion of "Christ in Glory" at the head of several *mappaemundi* demonstrates the third symbolic stage of Christian history, that of the Last Judgment. The figures of Christ or of God the Father may be surrounded by a mandorla, an aura of light used to symbolize holiness and common in Christian art from the fifth century until the Renaissance. The almond shape of Higden's maps and of the Genoese world map of 1457 is thus probably no accident. It reflects the use of this widespread symbol to denote the entire world as the domain of Christ.[237]

A *mappamundi* could thus represent simultaneously the complete history of the Christian world: its creation, salvation, and final judgment. Such a powerful message would not have gone unnoticed by those who saw either the small maps in monastic texts or the great wall maps—no longer surviving but to which we possess many allusions—that hung in churches and palaces.[238]

A special example of the spatial significance of a religious symbol lies in the association of the cross with the four cardinal directions, most commonly seen in the cruciform plan of churches, with the apse and altar in the east. The symbolism of the number four in Christian literature has its roots in classical times, as is illustrated by a diagram from Bede's *De natura rerum*. The relationship between the four cardinal directions, the four seasons, and the four climates demonstrated by Bede goes back to Aristotelian thought (fig. 18.38).[239] In this, the close relation between man and the heavens, the root of astrology, was also shown by the correspondence of the four peripatetic elements—fire, water, air, and earth—with the four humors of the human body, itself a microcosm of the universe.[240] Isidore presented a similar diagram of the elements and their relation to the cardinal directions and climates (fig. 18.39).[241] Table 18.6 summarizes the relations between the cardinal directions and various classical and medieval attributes.

The most specific allusion to the importance of such symbolism is given by Hugh of Saint Victor. Hugh's *On the Mystical Noah's Ark* not only provides us with an all too rare account of the making of a *mappamundi*, but also shows how the symbolic meanings were deliberately incorporated:

> the perfect ark is circumscribed with an oblong circle, which touches each of its corners, and the space the circumference includes represents the earth. In this space, a world map is depicted in this fashion: the front of the ark faces the east, and the rear faces the west. . . . In the apex to the east formed between the circle and the head of the ark is paradise. . . . In the other apex, which juts out to the west, is the Last Judgment, with the chosen to the right and the reprobates to the left. In the northern corner of this apex is hell, where the damned are thrown with the apostate spirits. Around this above-mentioned circle is

drawn one a little wider so that the zones may be effectively seen; the atmosphere is in this space. In this second space, the four parts of the earth and the four seasons are represented: spring to the east, summer to the south, autumn to the west, and winter to the north.[242]

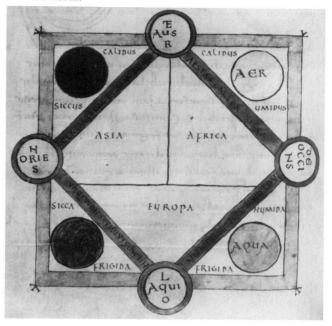

FIG. 18.38. THE SYMBOLISM OF THE NUMBER FOUR. From a ninth-century manuscript of Bede's *De natura rerum*, this diagram indicates the relationships perceived in the Middle Ages to have existed between the four cardinal directions (and the three continents), the four seasons, the four elements, and the four material properties (hot, cold, wet, dry).
Size of the original detail: 24 cm square. By permission of the Bayerische Staatsbibliothek, Munich (Clm. 210, fol. 132v).

237. For illustrations of the various forms of the aureole or nimbus, including the mandorla, see F. R. Webber, *Church Symbolism* (Cleveland: J. H. Jansen, 1927), 154.

238. For example, the allusion by Matthew Paris to three such maps on his "world" map in the *Chronica majora* (British Library, Cotton MS. Nero D.V., fol. 1v).

239. Karl A. Nowotny, *Beiträge zur Geschichte des Weltbildes* (Vienna: Ferdinand Berger, 1970), 26. For a summary of element diagrams, see John Emery Murdoch, *Antiquity and the Middle Ages*, Album of Science (New York: Charles Scribner's Sons, 1984), 346–59. As a magnificent iconographical statement of the interconnectedness of the seasons, climates, elements, rivers of paradise, and other cosmographical subjects, the stained glass rose window in Lausanne cathedral has few equals. See Ellen Judith Beer, *Die Glasmalereien der Schweiz vom 12. bis zum Beginn des 14. Jahrhunderts*, Corpus Vitrearum Medii Aevi, Schweiz, vol. 1 (Basel: Birkhäuser, 1956).

240. Ernst Cassirer, *The Individual and the Cosmos in Renaissance Philosophy* (Oxford: Clarendon Press, 1963). Leonard Barkan, *Nature's Work of Art: The Human Body as Image of the World* (New Haven: Yale University Press, 1975).

241. Isidore, *Traité de la nature*, ed. Fontaine, fig. 2, 202 ff. (note 84).

242. Hugh of Saint Victor *De arca Noe mystica* XIV, in *Patrologiæ cursus completus*, ed. Migne, 176:700 (note 157), author's translation.

TABLE 18.6 The Number Four and Its Symbolic Association with Medieval *Mappaemundi*

	Cardinal Directions			
	North	East	South	West
Major world divisions				
Continents	Europe	Asia	Africa	Fourth continent
Peoples	Japheth	Shem	Ham	Antipodes
Astronomy and astrology				
Winds (Isidore)	Septentrio	Subsolanus	Auster	Favonius
Winds (Aristotle)	Boreas	Apeliotes	Notos	Zephyros
Seasons	Winter	Spring	Summer	Fall
Times of day	Midnight	Morning	Midday	Evening
Elements	Earth	Air	Fire	Water
Climates	Cold humid	Hot humid	Hot dry	Cold dry
Humors	Black bile	Blood	Choler	Phlegm
Classical mythology				
Personifications	Vulcan	Flora/Venus	Ceres	Bacchus
Planets	Jupiter	Venus	Mars	Neptune
Bible				
Evangelists	Matthew	Mark	Luke	John
How Gospel begins	Angel	Voice in wilderness	Sacrifice	Winged word
Form	Man	Lion	Ox	Eagle
Attribute	Wisdom	Strength	Patience	Freedom
Chartres south transept window	Isaiah	Daniel	Jeremiah	Ezekiel
Fathers of church	Ambrose	Jerome	Augustine	Gregory
Rivers of paradise	Pishon	Tigris	Gihon	Euphrates
Representation of rivers	Ganges	Tigris	Nile (Indus)	Euphrates
Colors				
Persia	Black	Red	White	Yellow
Israel	Black	Red	Green	White
Greece	Black	Blue green	Red	White
Roman chariot races	White	Green	Red	Blue
Aristotle	Black	Red	Yellow	White

The significance of the number four went beyond the physical characteristics of the earth and heaven. By their evangelistic association with the four corners of the earth in the Book of Revelation, the authors of the four Gospels are often shown in the northeast, southeast, southwest, and northwest corners of the world as in the Leardo map of 1452 (fig. 18.40). Since the time of Irenaeus, bishop of Lyons (ca. 180 A.D.), each was associated with one of the four winged creatures in Revelation.[243]

Even the orientation of the *mappamundi* has a symbolic meaning. The term "orientation" itself comes from primitive societies' preoccupation with the east as a primary means of ordering space. The prominence of the four cardinal directions on the *mappaemundi*, together with appropriate symbolic wind heads, thus undoubtedly has a far deeper significance than simply showing the reader which way the map is to be read. The maps

"Adjecimus tamen quaedam, quae breviter commemorabimus. Hoc modo arca perfecta, circumducitur et circulus oblongus, qui ad singula cornua eam contingat, et spatium quod circumferentia ejus includit, est orbis terrae. In hoc spatio mappa mundi depingitur ita ut caput arcae ad orientem convertatur, et finis ejus occidentem contingat, ut mirabili dispositione ab eodem principe decurrat situs locorum cum ordine temporum, et idem sit finis mundi, qui est finis saeculi. Conus autem ille circuli, qui in capite arcae prominet ad orientem, Paradisus est, quasi sinus Abrahae, ut postea apparebit majestate depicta. Conus alter, qui prominet ad occidentem, habet universalis resurrectionis judicium in dextra electos, in sinistra reprobos. In cujus coni angulo Aquilonari est infernus, quo damnandi cum apostatis spiritibus detrudentur. Post haec supradicto circulo alter paulo laxior circumducitur, ut quasi zonam videatur efficere, et hoc spatium aer est. In quo spatio secundum quatuor partes mundi quatuor anni tempora disponuntur, ita ut ver sit ad orientem, ad austrum aestas, ad occidentem autumnus, ad aquilonem hiems."

243. Irenaeus, *Five Books of S. Irenaeus, Bishop of Lyons, against Heresies*, trans. John Keble (Oxford: J. Parker, 1872), 125. See also James Strachan, *Early Bible Illustrations: A Short Study Based on Some Fifteenth and Early Sixteenth Century Printed Texts* (Cam-

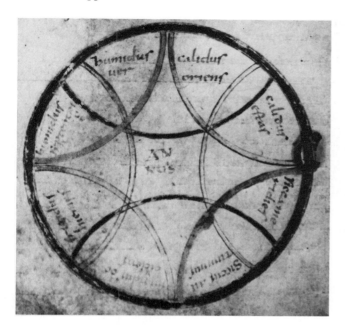

FIG. 18.39. DIAGRAM OF THE ELEMENTS FROM ISIDORE. Similar to Bede's representation (see fig. 18.38), this again indicates the combination of properties that are held by the four elements. Thus, fire is hot and dry, air is hot and wet, water is cold and wet, and earth is cold and dry. Isidore's use of circles reflects his misunderstanding of the earth's climatic zones (see fig. 18.26). From a manuscript of various treatises on astronomy and chronology dated to about 850.
Diameter of the original: 14 cm. By permission of the Bibliothèque Municipale, Rouen (MS. 524, fol. 74v).

of the sunrise, followed by that of the sunset, is the object of a deeply rooted human curiosity. It has been observed that the directions east and west tend to be named in early languages before north and south. Of the two, words for east commonly precede those for west.[246] The importance of east in social and religious practice is also shown in the origin of the words of many languages for the cardinal directions. For example, "north" was described by its position to the "left" of east, and it thus became associated with sinister behavior, left-handedness, and evil. In Celtic languages, the words for "north" and "left" are closely related.[247]

Mappaemundi also became symbols of royal and imperial power, thus reflecting the secular influences behind their creation. The orb and scepter were accepted regalia in representations of royalty, not only in ceremony and art but also on coins of the realm. Some of the earliest extant pictures of the tripartite and spherical earth are found on coins.[248] The tradition of representing the earth as a sphere on Roman coins started in the first century B.C. with a gold medal bearing on its reverse three circles representing the tripartite world.[249] This is particularly significant considering the paucity of other references to the knowledge of the sphericity of the earth in Roman times. This symbolic incorporation of the world map or globe as an item of regalia continued throughout the Middle Ages. It was extended to paintings of God reigning in glory, depicted holding an orb surmounted by a cross in, usually, the left hand.

Another symbolic theme in the *mappaemundi* is the representation of the earth as a scene of vain pursuits. The *vanitas* symbol, as art historians call it, has been

are found oriented in all four directions, but east, north, and south are the most common, in that order. An eastern orientation is usually, but by no means exclusively, found on the tripartite *mappaemundi*, and it follows the late Roman Sallustian tradition adopted by the Christian world. The northern orientation is found on the other large group of *mappaemundi* that can be traced back to earlier classical Greek sources and whose geometry was centered on the earth's axis and the *climata*. The southern orientation is probably derived from Arabic influence, since world maps of the Arabic culture were characteristically oriented to the south. There may be two reasons for this. First, the early people conquered by the Arabs were the Zoroastrians, for whom south was sacred. Second, since the early cultural centers were in this newly conquered territory, north of Mecca, the holy direction toward which all Muslims prayed became south.[244]

The cardinal directions thus not only were an abstract means of orientation, but became mythical entities in their own right.[245] As is well documented, the position

bridge: Cambridge University Press, 1957). Occasionally each evangelist is shown holding a vessel, the symbolic source of the four rivers emanating from paradise.

244. B. L. Gordon, "Sacred Directions, Orientation, and the Top of the Map," *History of Religions* 10 (1971): 211–27, esp. 218. Later, of course, the habitual outlook of the Arabic culture was on the Indian Ocean, thus confirming south as a favored direction. See also Salvatore Cusa, "Sulla denominazione dei venti e dei punti cardinali, e specialmente di nord, est, sud, ovest," *Terzo Congresso Geografico Internazionale, Venice, 1881*, 2 vols. (Rome: Società Geografica Italiana, 1884), 2:375–415.

245. Cassirer, *Individual and the Cosmos*, 98 (note 240).

246. Cecil H. Brown, "Where Do Cardinal Direction Terms Come From?" *Anthropological Linguistics* 25 (1983): 121–61. But see Gordon, "Sacred Directions," 211 (note 244), who points out the variability of the position of sunrise and sunset according to season and latitude.

247. Brown, "Cardinal Direction Terms," 124 (note 246). The Cornish word for both left and north is *cleth*; the Welsh for left is *cledd* and for north is *gogledd*.

248. Miller, *Mappaemundi*, 3:129–31, fig. 66 (note 9).

249. Miller, *Mappaemundi*, 3:131 (note 9) dates it 22 B.C. If so, the name on the medal—M. Cocceius Nerva—clearly does not refer to the emperor of the same name who lived A.D. 35–98.

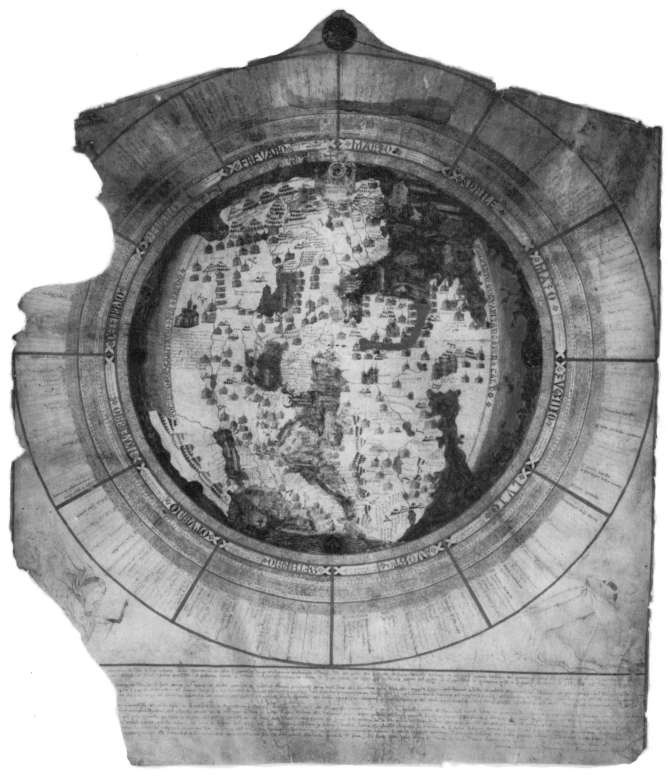

FIG. 18.40. LEARDO WORLD MAP, 1452. The authors of the four Gospels are shown in the four corners of this map following their evangelical association with the four corners of the earth in the Book of Revelation.

Size of the original: 73 × 60 cm. By permission of the American Geographical Society Collection, University of Wisconsin, Milwaukee.

FIG. 18.41. THE MOSAIC OF THE GODDESS FORTUNA. This twelfth-century floor mosaic combines a world map with a depiction of Fortuna's wheel as the central circle, a common theme in medieval art representing the earth as a place of vain pursuits.

Size of the original: 6.5 × 7 m. By permission of the Museo Civico, Turin.

well documented for the post-Renaissance period, but its sources date from much earlier.[250] The allegory of the goddess Fortuna with a wheel or standing on a globe is found on Roman coins. In the medieval period, Fortuna's wheel was combined with a world map on the twelfth-century floor mosaic now in the Museo Civico, Turin (fig. 18.41).[251] Here the central circle is an allegory of Fortuna's wheel, while the imagery around the edge is clearly intended to be cartographic. Henry III's decoration of the hall in Winchester castle included both a world map (1236) and a wheel of fortune (1239).[252] A poem by Baudri de Bourgueil (ca. 1100) refers to a *mappamundi* on the floor of the chamber of Adela, countess of Blois, probably also intended as a *vanitas* symbol. Nothing of this has survived, but the map was described in such detail by Baudri that it is unlikely it was a product of mere imagination: he even refers to a glass top placed on it to seal out the dust.[253]

250. The standard late baroque source was Cesare Ripa, *Iconologia* (3d ed., 1603; facsimile reprint, Hildesheim and New York: Georg Olms, 1970). See also the work of James A. Welu, most recently "The Sources of Cartographic Ornamentation in the Netherlands," in *Art and Cartography* (note 135).

251. Ernst Kitzinger, "World Map and Fortune's Wheel: A Medieval Mosaic Floor in Turin," *Proceedings of the American Philosophical Society* 117 (1973): 344–73.

252. See H. M. Colvin, ed., *History of the King's Works,* 6 vols. (London: Her Majesty's Stationery Office, 1963–82), 1:127, 497, 2:859, 861.

253. Baudri de Bourgueil, *Les oeuvres poétiques de Baudri de Bourgueil (1046–1130),* ed. Phyllis Abrahams (Paris: Honoré Champion, 1926), lines 719–948 (pp. 215–21). Professor O. A. W. Dilke, from his reading of Baudri's ambiguous Latin, believes that the map was not a mosaic, as has been reported, but probably either a painting on marble or a painting or embroidery on silk over marble (personal communication). The glass top is described in lines 727–28: "Ne vero pulvis picturam laederet ullus, Tota fuit vitrea tecta superficie" (So that the dust would not damage the picture, its surface was completely covered with glass—Author's translation).

Mappaemundi could also reflect the two main ways the individual was considered in the Middle Ages to be related to the universe. Both the microcosmic and the anthropocentric concepts were pervasive themes in medieval cosmological thought. According to the microcosmic theme, the human body was viewed as an epitome of the universe, in which the elements, humors, and organs of the body (the microcosm) were directly related to and controlled by the universe (the macrocosm). It was the central purpose of astrology to explain these connections.[254]

Mappaemundi were themselves graphic epitomes of the earth, and the physical relationships between the earth and the universe are well illustrated, for example, in the Isidorian diagrams. Other diagrams show the human body in a mandorla-shaped framework surrounded by graduations of the zodiac or the earth as one of four concentric circles representing the elements. *Mappaemundi* thus belong to a much wider family of spatial representations and ideas found in architecture as well as in cartography. Byzantine churches were often laid out with their main doors facing east, and later in the Middle Ages, particularly in northern Europe, the buildings were so oriented that the congregation faced the altar in the east. A dome, representing the heavens above the four directions of the earth, was often built above the intersection of the transepts and nave. In this way the building expressed the same symbolic spatial concepts as the *mappamundi*, a microcosm of earth and heaven.[255]

The second concept, the anthropocentric, placed the individual in the center of an abstract geometric system of cardinal directions or in relation to some prominent feature in the landscape, such as a river. This was a natural world view in those societies where livelihood depended largely on the immediate visible environment and in which the daily and seasonal positions of the sun, moon, and stars were strong orienting influences. During the Middle Ages, however, man was not at the center of the world. The idea of the City of Man, at least in medieval Europe, as opposed to Augustine's City of God, was to await the European Renaissance. If anything was depicted at the center of the *mappaemundi*, it was not the monastic centers where the maps were made but the symbolic biblical centers, such as Jerusalem or Mount Sinai, or classical centers such as the sacred isle of Delos or Rome. For the Christian, there was clear biblical justification for centering maps on Jerusalem.[256] There was also a sensitive awareness of space in the Old Testament that gave location an integral role in the events of Jewish history.[257] Adamnan, abbot of Iona, in his *De locis sanctis*, speaks of: "a very high column which stands in the center of the city. . . . It is remarkable how this column . . . fails to cast a shadow at midday during the summer solstice, when the sun reaches the center of the heavens.

. . . And so this column . . . proves Jerusalem to be at the center of the world . . . and its navel."[258]

Such an observation of the sun is impossible astronomically (unless the column was leaning ten degrees toward the south), Jerusalem being some ten degrees north of the Tropic of Cancer. However, the attempt to prove a traditional concept with a scientific observation reflects the newfound respectability of science. Such scientific precision was also found in the description of Bernard the Wise (ca. 870), who reported that the walls of the four main churches in Jerusalem enclosed an unroofed porch, over which four chains were strung from each church to join in a point over the center of the world.[259]

Despite such beliefs, Jerusalem was not shown as the center of most medieval *mappaemundi*.[260] This is apparent in those maps not in the diagrammatic T-O category, such as the Beatus, Orosius, or Higden or important twelfth-century maps like that by Henry of Mainz (fig. 18.42). It is true that three particularly well known *mappaemundi*—the Ebstorf, Hereford, and Psalter maps—are all precisely centered on Jerusalem, and it is this that has perhaps led historians and geographers to overgeneralize. It has also been wrongly assumed that, since the T in the schematic T-O maps represents the meeting of the Mediterranean with the Don–Black Sea–Aegean–Nile axis, the Holy Land is near enough that intersection for Jerusalem to be at the center of the map. But there are not only many examples of where intersection of the stem and the crossbar of the T is far above the center, but also many where Jerusalem is placed at some distance from this intersection.[261]

254. Yi-Fu Tuan, *Topophilia: A Study of Environmental Perception, Attitudes, and Values* (Englewood Cliffs, N. J.: Prentice-Hall, 1974), and idem, *Space and Place: The Perspective of Experience* (Minneapolis: University of Minnesota Press, 1977). Barkan, *Nature's Work of Art* (note 240). Cassirer, *Individual and the Cosmos* (note 240).

255. Mieczyslaw Wallis, "Semantic and Symbolic Elements in Architecture: Iconology as a First Step towards an Architectural Semiotic," *Semiotica* 8 (1973): 220–38, esp. 224–28; Mircea Eliade, *The Sacred and the Profane: The Nature of Religion* (New York: Harcourt, Brace and World, 1959).

256. "This city of Jerusalem I have set among the nations, with the other countries round about her" (Ezek. 5:5).

257. See also Eliade, *Sacred and Profane*, 42–47 (note 255). Robert L. Cohn, *The Shape of Sacred Space: Four Biblical Studies* (Chico, Calif.: Scholars Press, 1981), 2. For a general discussion of the concept of space and place in Jewish philosophy, see Israel Isaac Efros, *The Problem of Space in Jewish Mediaeval Philosophy* (New York: Columbia University Press, 1917).

258. Quoted in Friedman, *Monstrous Races*, 219 n. 23 (note 3).

259. J. H. Bernard, trans., *The Itinerary of Bernard the Wise*, Palestine Pilgrims Text Society 3 (London, 1893; reprinted New York: AMS Press, 1971), 8.

260. Beazley, *Dawn of Modern Geography*, 1:339 (note 17).

261. For example, Paris, Bibliothèque Nationale, MS. Lat. 7676 (Reg. 6067), fol. 161. Destombes, *Mappemondes*, 63 (28.13) and fig. IIIb (note 31).

FIG. 18.42. THE TWELFTH-CENTURY WORLD MAP OF HENRY OF MAINZ. Representing a class of *mappaemundi* which did not place Jerusalem at the center, this map is thought to have influenced later maps of the same type. Derived from an ancient Greek tradition, the center is the Cyclades, the islands circling the sacred isle of Delos, shown in this detail.

Size of the original detail: ca. 8 × 11 cm. By permission of the Master and Fellows of Corpus Christi College, Cambridge (MS. 66, p. 2).

Thus, while there is a clear biblical justification for centering these maps on Jerusalem and an empirical reason for doing so (it did occur roughly in the middle of the then known world), the idea does not seem to have been taken as literally as was previously thought. One reason for not centering maps on Jerusalem derives from the original use not of a Christian model, but of a Greco-Roman one in the *mappaemundi*, which was perpetuated through the Orosian tradition. The strengthening of the idea of Jerusalem as the spiritual center, a natural outcome of the Crusades, may have been responsible for a noticeable shift in the structure of *mappaemundi* from 1100 to 1300, toward centering the maps on Jerusalem.

Although many pilgrimages to the Holy Land had taken place in early medieval times—owing to the efforts of Saint Helena, a number took place in the fourth century—it was only after the Crusades that widespread popular attention was focused on the central position of Jerusalem. The trend toward centralization is seen when we compare the world map of Henry of Mainz (ca. 1110) with the Hereford map (ca. 1290), at either end of this period. This characteristic has been used to date the Vercelli map (which is not centered on Jerusalem) early in the thirteenth century, in contrast to the later Hereford and Ebstorf maps (both of which are so centered). By the fourteenth and fifteenth centuries, the practice of placing Jerusalem at the center became common, but this was by no means true for the entire medieval period, or even the most of it.[262]

CONCLUSIONS

Traditional histories of cartography contain a number of misconceptions concerning the *mappaemundi*. The three most important of these are the assumption that geographical accuracy was the prime function of the *mappaemundi* (and hence that their goal was poorly achieved); the assumption that Jerusalem was almost invariably placed at the center of the maps; and the notion that the *mappaemundi* illustrated and confirmed the popularly held view of the earth as a flat disk in the Middle Ages.

Although Crone drew attention to what he considered to be the route-planning function of some world maps, such as the representation of pilgrimage routes on the Hereford map, no amount of twentieth-century historiographic ingenuity can counteract the overwhelming evidence that the function of the *mappaemundi* was primarily didactic and moralizing and lay not in the communication of geographical facts. The history of cartography, like the history of science, is moving away from being primarily a search for precursors and is attempting to understand the developments in various periods on their own terms. In the light of this interpretive shift, it now seems strange to read the views of the older historians of geography, such as Charles Beazley, who simply refused to describe such unambiguously cartographic manifestations of medieval culture as the Hereford and Ebstorf maps on the grounds that they appeared as retrogressions to an ever improving literal geographical picture of the world. In Beazley's view, the only purpose of maps was precisely that of providing an accurate representation of the distribution of places and events in an increasingly "correct" continental outline.

The importance of the symbolic content of the *mappaemundi* has thus now been established. This symbolism is a blend of the historical and the geographical. The maps consist of historical aggregations or cumulative inventories of the major events in both the Christian and the secular legendary history of the world, particularly the former. The three major events in the Christian history of the world—its creation, salvation by Christ, and the Last Judgment—commonly are symbolically portrayed on the maps or by the maps themselves, as in the Ebstorf map, which is a clear representation of the world as the body of Christ. There are also many examples where details in religious and secular history that span a thousand years appear on a single map without any differentiation between historical and geographical information. They are projections of history on a geographical base.

It has also been shown that the practice of placing Jerusalem at the center of the *mappaemundi* was by no means a universal convention throughout the Middle Ages but was largely confined to the post-Crusade period in the thirteenth and fourteenth centuries. Once interest was focused particularly on Jerusalem after the main period of the Crusades, there does appear to have been a trend in this direction until the end of the Middle Ages, when the assimilation of new geographical information and frameworks from Ptolemy's *Geography*, the development of the portolan charts, and the Renaissance discoveries led to a redefinition of the outer borders of the world map and a displacement of the traditional center.

It is also commonly assumed that the best-known form of *mappamundi*, the T-O map, with its tripartite division of the inhabited world and the surrounding ocean river, was prima facie evidence for universal medieval belief in a flat earth, a misconception still perpetuated in some school history texts in the context of Columbus's discovery of the New World. On the contrary, it has been shown that the influential Isidore of Seville, despite the ambiguity in his writings, was probably quite aware of the earth's sphericity, and a score of medieval church fathers, scholars, and philosophers in almost every century from the fifth to the fifteenth stated this categorically. Furthermore, by the fourteenth century, thinkers such as Roger Bacon not only knew the earth was spherical but described the need for map projections to satisfactorily transform the curvature of the earth to a flat plane.

The study of *mappaemundi* is well served—in comparison with other types of medieval maps—by general checklists and facsimile atlases. Sadly lacking are the detailed studies of individual maps and groups of maps in their cultural context along the lines of the work done by Durand for the fifteenth-century Vienna-Klosterneuburg map corpus. Obvious priorities would include regional studies on the *mappaemundi* associated with the geographical culture in thirteenth-century England or on the general role of the medieval Franciscans in the development of systematic cartography. There also is a need to develop the construction of stemmata to show the pedigree of maps of the eighth century and later. Stemmata for selected map types such as those included in this chapter (Beatus, Higden) may help to clarify influences and lines of descent, but much more detailed work needs to be undertaken in order to date and place the artifacts more accurately.

It is perhaps ironic that one of the most thorough studies of a single medieval world map—the Vinland

---

262. This point is also made by Wright, *Geographical Lore*, 259 (note 18). The concept of placing Jerusalem at the center of the world seems to have been introduced in the seventh century but was not generally established until the twelfth or even the thirteenth.

map—dealt with an alleged forgery. The importance of the use of modern techniques of physical analysis of parchment, pigment, and ink on the medieval *mappaemundi* cannot be overstressed. Such analysis would provide some much-needed benchmarks in dating and locating the place of manufacture of key artifacts. A case in point is the obscure origin of T-O maps representing both a fourth continent and the Meotides Paludes, a type that may include the earliest surviving world map, variously dated from the seventh to the ninth century (Saint Gall Stiftsbibliothek Codex 237). A study of the relation between this map and the T-O diagrams found in many manuscripts of the Beatus *Apocalypse of Saint John* may offer important insights into the transmission of cartographic ideas in the *mapppaemundi* of the seventh and eighth centuries. This topic, along with others suggested in this chapter, calls for an unusual blend of historical and geographical scholarship combined with an awareness of the importance of graphic artifacts in the study of medieval culture.

APPENDIX 18.1
REFERENCE GUIDE TO TYPES OF *MAPPAEMUNDI*

This appendix is a graphic reference guide to the main types of *mappaemundi* based on the classification outlined in table 18.2 above. It provides an illustration of each type and briefly describes its characteristics and context.

Schematic Tripartite
ISIDORE T-O TYPE

Over two hundred examples of this type are listed by Destombes.[1] They are found in two major works of Isidore of Seville (Isidorus Hispalensis; ca. 560–636): *Etymologiarum sive originum libri XX* (between 622 and 633) and *De natura rerum* (between 612 and 615).

The maps may be purely diagrammatic, bearing few or no names. In other cases the names of the sons of Noah are added or there is text describing the number of countries in each of the three major zones. Other maps include geographical features, such as place-names or bodies of water. For example, some of the maps in Isidore's *De natura rerum* represent the Gulf of Tunis (fig. 18.43).

SALLUST T-O TYPE

The versions of this map are found in approximately sixty manuscripts of the *De bello Jugurthino* of Gaius Sallustius Crispus (Sallust; 86–34 B.C.), of various dates from the ninth to the fourteenth century.[2] Its popularity in the fifteenth century is attested by the appearance of some fifty-five printed editions between 1470 and 1500.

FIG. 18.43. ISIDORE T-O MAP. From a late ninth-century manuscript of Isidore's *De natura rerum*.
Diameter of the original: 12.5 cm. By permission of the Burgerbibliothek, Bern (Codex 417, fol. 88v).

The Sallust maps usually are less diagrammatic than their counterparts in Isidore's works. The Don and the Nile rivers are frequently curved at the ends to reflect more closely the supposed courses of these rivers, and the maps usually include pictures of fortified towns or churches symbolizing major cities. Orientation is usually to the east, but it may also be to the south or west, as in figures 18.44 and 18.45. In cases with southern orientation, Africa may take up half the circle, with Asia and Europe sharing the other half (fig. 18.46), a configuration also alluded to in some medieval romances, for example *Aspremont* (late twelfth century) or *Sone de Nansay* (late thirteenth century).[3] The ends of the crossbar of the T may be truncated at an angle as in figure 18.47.

1. Marcel Destombes, ed., *Mappemondes A.D. 1200–1500: Catalogue préparé par la Commission des Cartes Anciennes de l'Union Géographique Internationale* (Amsterdam: N. Israel, 1964), 29–34 and 54–64.

2. Destombes, *Mappemondes*, 37–38 and 65–73 (note 1). See also A. D. Leeman, *A Systematical Bibliography of Sallust, 1879–1950* (Leiden: E. J. Brill, 1952); Bernhard Brandt, "Eine neue Sallustkarte aus Prag," *Mitteilungen des Vereins der Geographen an der Universität Leipzig* 14–15 (1936): 9–13; Johannes Keuning, "XVIth Century Cartography in the Netherlands (Mainly in the Northern Provinces)," *Imago Mundi* 9 (1952): 35–64; and Ingeborg Stolzenberg, "Weltkarten in mittelalterlichen Handschriften der Staatsbibliothek Preußischer Kulturbesitz," in *Karten in Bibliotheken: Festgabe für Heinrich Kramm zur Vollendung seines 65. Lebensjahre*, ed. Lothar Zögner, Kartensammlung und Kartendokumentation 9 (Bonn-Bad Godesberg: Bundesforschungsanstalt für Landeskunde und Raumordnung, Selbstverlag, 1971), 17–32, esp. 21–22.

3. Jill Tattersall, "Sphere or Disc? Allusions to the Shape of the Earth in Some Twelfth-Century and Thirteenth-Century Vernacular French Works," *Modern Language Review* 76 (1981): 31–46.

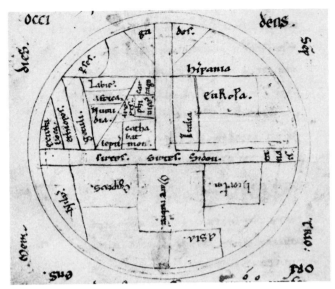

FIG. 18.44. SALLUST T-O MAP, WEST ORIENTATION. From a manuscript of the *De bello Jugurthino* of Sallust. Diameter of the original: 6.8 cm. Photograph from the Bibliothèque Nationale, Paris (MS. Lat. 6253, fol. 52v).

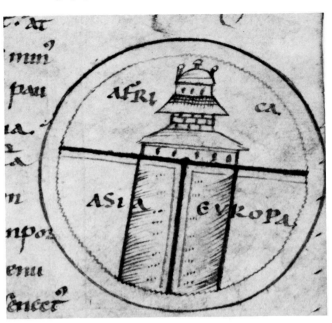

FIG. 18.46. SALLUST T-O MAP, AFRICA AS THE LARGEST CONTINENT. From a twelfth-century manuscript of the *De bello Jugurthino* of Sallust. Diameter of the original: 4 cm. Photograph from the Bibliothèque Nationale, Paris (MS. Lat. 5751, fol. 18r).

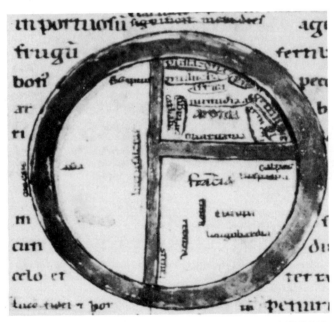

FIG. 18.45. SALLUST T-O MAP, SOUTH ORIENTATION. From a thirteenth-century manuscript of the *De bello Jugurthino* of Sallust. Diameter of the original: 4.3 cm. Photograph from the Bibliothèque Nationale, Paris (MS Lat. 6088 [Reg. 5974], fol. 33v).

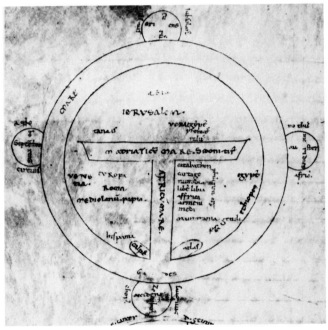

FIG. 18.47. SALLUST T-O MAP WITH TRUNCATED RIVERS. From a thirteenth-century manuscript of the *De bello Jugurthino* of Sallust. Diameter of the original: 16.5 cm. By permission of the Biblioteca Medicea Laurenziana, Florence (Plut. 16.18, fol. 63v).

GAUTIER DE METZ T-O TYPE

Gautier de Metz, about whom very little is known, is the supposed author of an encyclopedic poem called *L'image du monde* in more than six thousand verses in Lorraine dialect, dating about 1245. More than one hundred manuscripts survive of two recensions in verse and two in prose.[4]

The two types of *mappaemundi* found in these manuscripts are derived from book 14 of the *Etymologies* of Isidore. The first is in the form of a circle oriented to the east with a simple north-south line dividing the circle into equal parts. The four cardinal directions are shown, with the words "Aise la grant" (Asia major). The second is a more complete T-O map similar to the Isidore version but in French and occasionally surrounded with the names of the winds (fig. 18.48).

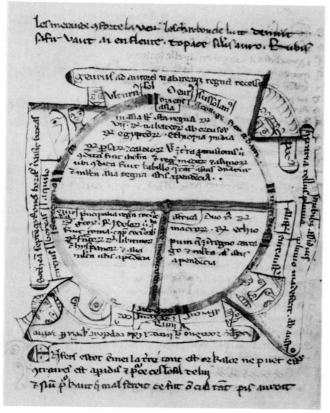

FIG. 18.48. GAUTIER DE METZ T-O MAP. From a thirteenth-century manuscript of *L'image du monde* of Gautier de Metz. The original, previously in the Bibliothèque Municipale, Verdun, is lost.
Diameter of the original: 6.6 cm. From Marcel Destombes, ed., *Mappemondes A.D. 1200–1500: Catalogue préparé par la Commission des Cartes Anciennes de l'Union Géographique Internationale* (Amsterdam: N. Israel, 1964), pl. Va.

MISCELLANEOUS AND UNKNOWN AUTHORS

Several modifications of the standard Isidore T-O characterize the maps in this category, which include the T-O maps of several authors, such as Lucan, Macrobius (excluding the zone maps, which form their own category), the Venerable Bede,

Guido of Pisa, and William of Tripoli, whose works are not numerous enough to warrant separate categories.[5] Modifications include the use of "Libya" for Africa, Y-shaped rivers (fig. 18.49), a truncated and notched T, the addition of two symmetrical rivers (fig. 18.50), a modified representation of the Nile (fig. 18.51), and the crossbar of the T a little higher than usual so that the areas of the three parts of the earth are approximately the same (fig. 18.52).

REVERSE T-O MAP

The names of Africa and Europe are here reversed on the traditional T-O diagram, interpreted by Destombes as scribal error.[6] However, Stevens has shown that this class of maps forms a well-defined subgroup based on an intentional reversal.[7] The rationale for the reverse T-O lies in the different viewpoints of the tripartite geometry. The traditional form certainly represents the three continents in their correct topological positions when viewed from above and oriented to the east. But if the tripartite division is projected onto the heavens, with the observer facing *west* and looking from the earth *out*, Asia will remain at the top but Africa and Europe will be reversed. In the case of the Hereford map, however, the transposition of the names Africa and Europe is clearly a scribal error: the remaining names and geographical details of these continents are not reversed (fig. 18.53).

Y-O MAP WITH SEA OF AZOV

These maps, which are usually found along with the conventional T-O map in manuscripts of the *Etymologies* of Isidore from the ninth century on, contain a representation of the Sea or Gulf of Azov and its surrounding marshes and lakes (or Meotides Paludes) as two arms of the river Don at an angle. In antiquity, the Sea of Azov was thought to have been much larger in extent than its present size of approximately 150 miles by 200 miles.[8] Its appearance on many maps in this category underlines its importance as a geographical feature in dividing the three main continents. Modifications include the addition of the four rivers of paradise, one of which is sometimes connected with the Nile.[9] Menéndez-Pidal believes both these versions are directly linked to the development of the Beatus maps in the ninth and tenth centuries (fig. 18.54).[10]

4. Destombes, *Mappemondes*, 117–48 (note 1).
5. Destombes, *Mappemondes*, 39, 46–49, 74–78, and 164–90 (note 1).
6. Destombes, *Mappemondes*, 67 (note 1).
7. Wesley M. Stevens, "The Figure of the Earth in Isidore's 'De Natura Rerum,'" *Isis* 71 (1980): 268–77, esp. 275 n. 24.
8. Roger Bacon, *The Opus Majus of Roger Bacon*, 2 vols., trans. Robert Belle Burke (1928; reprinted New York: Russell and Russell, 1962), 375.
9. Destombes, *Mappemondes*, map 26.9 (note 1).
10. G. Menéndez-Pidal, "Mozárabes y asturianos en la cultura de la alta edad media en relación especial con la historia de los conocimientos geográficos," *Boletín de la Real Academia de la Historia* (Madrid) 134 (1954): 137–291.

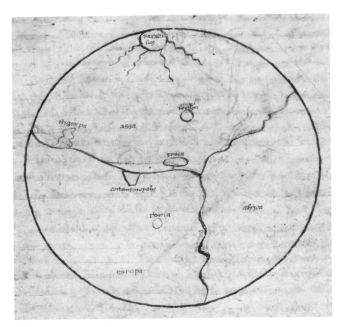

FIG. 18.49. MISCELLANEOUS T-O MAP, Y-O VARIANT. From a twelfth-century manuscript of Macrobius's *Commentarium in somnium Scipionis*.
Diameter of the original: 8.7 cm. Photograph from the Bibliothèque Nationale, Paris (MS. Lat. 16679, fol. 33v).

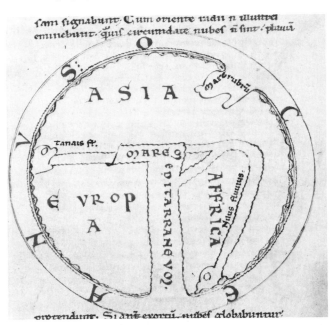

FIG. 18.51. MISCELLANEOUS T-O MAP, MODIFIED RIVER NILE. From a twelfth-century manuscript of Bede's *De natura rerum*.
Diameter of the original: 8.1 cm. Photograph from the Bibliothèque Nationale, Paris (MS. Lat. 11130, fol. 82r).

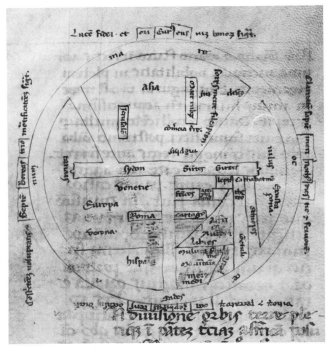

FIG. 18.50. MISCELLANEOUS T-O MAP, SYMMETRICAL RIVERS. From a thirteenth-century manuscript of Sallust.
Diameter of the original: 10.5 cm. By permission of the Master and Fellows of Gonville and Caius College, Cambridge (MS. 719/748, fol. 37v).

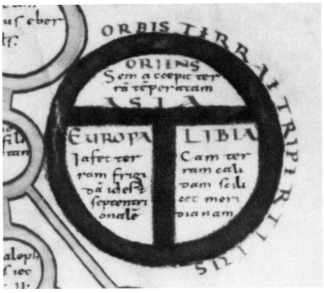

FIG. 18.52. MISCELLANEOUS T-O MAP, HIGH CROSS-BAR. From an eleventh-century manuscript of the *Commentary on the Apocalypse of Saint John* of Beatus of Liebana. See also figure 18.33.
Diameter of the original: 4.7 cm. Photograph from the Bibliothèque Nationale, Paris (MS. Lat. 8878, fol. 7r).

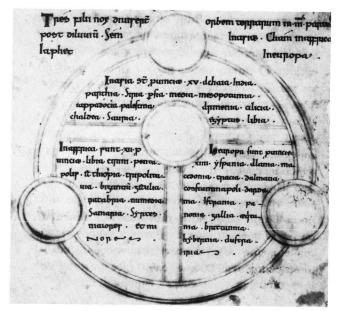

FIG. 18.53. REVERSE T-O MAP. From a twelfth-century manuscript of Isidore's *De natura rerum*.
Diameter of the original: 19 cm. By permission of the Dean and Chapter of the Cathedral Church of Exeter (MS. 3507, fol. 67r).

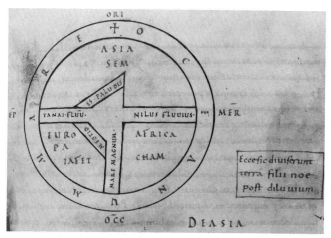

FIG. 18.54. Y-O MAP WITH SEA OF AZOV. From a tenth-century manuscript of the *Etymologies* of Isidore.
Diameter of the original: 9.7 cm. By permission of the Stiftsbibliothek, Saint Gall (Codex 236, fol. 89r).

V-IN-SQUARE AND T-IN-SQUARE MAPS

These variations on the T-O schema appear in various works; they may include a V or a T in a square oriented either to the east or to the south (fig. 18.55). An excellent example of the T-in-square map is also found in a manuscript of *De natura rerum* of Bede, in which the elements and seasons are related to the four cardinal directions from Ptolemy's astrological work, the *Quadripartitum* or *Tetrabiblos* (fig. 18.38).[11]

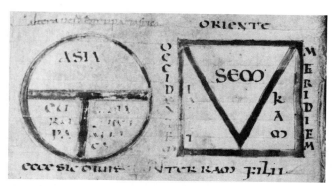

FIG. 18.55. V-IN-SQUARE MAP. From a ninth-century manuscript of various treatises (see also fig. 18.38).
Size of the original: 4.8 cm square. By permission of the Bibliothèque Municipale, Rouen (MS. 524, fol. 74v).

Nonschematic Tripartite

This group contains those maps that retain the general tripartite distribution of the three inhabited continents but are not rigidly drawn to the T-O schema. They are subdivided here according to the historical origin of their content.

OROSIAN

These maps are based directly on the *Historia adversum paganos* of Paulus Orosius.[12] They usually emphasize the Mediterranean basin, and their coastlines are almost always generalized in undulating style. Maps belonging to this group include the Albi map (fig. 18.56),[13] the Cotton "Anglo-Saxon" map (fig. 18.57),[14] two Matthew Paris maps (fig. 18.58),[15] the world map of Henry of Mainz (fig. 18.59),[16] and the Hereford *mappamundi* (figs. 12.4, 18.20, and 18.60).[17]

OROSIAN-ISIDORIAN

These maps, although owing their ultimate origin to Orosius, have been modified by the influence of Isidore of Seville. When comparing, for example, the oval Isidore map in figure 18.61

11. Destombes, *Mappemondes*, map 6.1, p. 36 (note 1). See also Karl A. Nowotny, *Beiträge zur Geschichte des Weltbildes* (Vienna: Ferdinand Berger, 1970), 26.

12. For a modern translation of Paulus Orosius, see *The Seven Books of History against the Pagans*, trans. Roy J. Deferrari (Washington, D.C.: Catholic University of America Press, 1964).

13. Destombes, *Mappemondes*, 22.1 (note 1). See also Charles Raymond Beazley, *The Dawn of Modern Geography: A History of Exploration and Geographical Science from the Conversion of the Roman Empire to A.D. 900*, 3 vols. (London: J. Murray, 1897–1906), 2:586; and Y. Janvier, *La géographie d'Orose* (Paris: Belles Lettres, 1982).

14. Destombes, *Mappemondes*, map 24.6 (note 1).

15. Richard Vaughan, *Matthew Paris* (Cambridge: Cambridge University Press, 1958).

16. Destombes, *Mappemondes*, map 25.3 (note 1).

17. Destombes, *Mappemondes*, 197–202 (note 1).

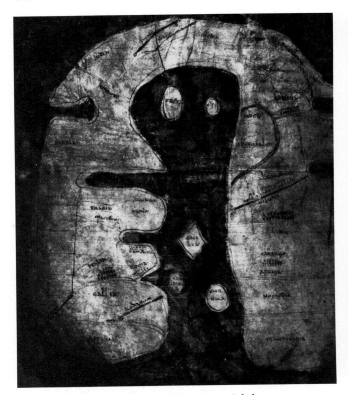

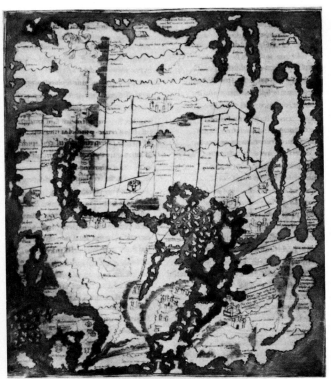

FIG. 18.56. THE ALBI MAP. From an eighth-century manuscript of miscellanea.
Size of the original: 29 × 23 cm. By permission of the Bibliothèque Municipale d'Albi (MS. 29 Albi, fol. 57v).

FIG. 18.57. THE "ANGLO-SAXON" MAP. From a tenth-century manuscript of the *Periegesis* of Priscian, included in a codex of divers authors. See also plate 22.
Size of the original: 21 × 17 cm. By permission of the British Library, London (Cotton MS. Tiberius, BV, fol. 56v).

with the Hereford map (figs. 18.20 and 18.60), one can see a difference in lineage: Jerusalem is not at the center, paradise is not situated due east, and the graphic generalization is far more angular and schematic. The Taurus-Caucasus Mountains are heavily emphasized, forming a chain containing the Gog-Magog region of northeastern Asia. Similar in general conception is the 1119 *mappamundi* by Guido of Pisa (fig. 18.62), which, in addition to its Orosian and Isidorian heritage, contains information derived from the Antonine itinerary, the Ravenna cosmography, and the *Notitia Urbis*.[18]

Also belonging to this group are the so-called Psalter map (fig. 18.63),[19] the Wiesbaden fragment (fig. 18.64),[20] the Vercelli map (fig. 18.17),[21] and the Ebstorf map (fig. 18.65).[22]

COSMAS INDICOPLEUSTES

The illustrations to the *Christian Topography* of Cosmas Indicopleustes form a small, well-defined separate group of medieval world maps, the importance of which has tended to be exaggerated because of their curiously fundamentalist flavor, once thought to characterize the medieval period. They are Christian, exegetic, and didactic in nature, but are regarded

as an extension of the Greco-Roman Byzantine tradition discussed in chapter 15 of this volume (figs. 15.2 and 18.66).

RANULF HIGDEN

The *mappaemundi* of Ranulf Higden are found in the first book of the *Polychronicon*.[23] The large oval map in the British Library (fig. 18.67) is believed to be closest to the original lost prototype. The circular maps, which form the smallest group (fig. 18.68), are thought to be later simplifications of the oval maps. The almond-shaped mandorla maps (also later variants) form a third group (fig. 18.69).

18. Beazley, *Dawn of Modern Geography*, 2:632–33 (note 13).
19. London, British Library, Add. MS. 28681, fol. 9.
20. Destombes, *Mappemondes*, 202–03 (note 1).
21. Destombes, *Mappemondes*, 193–94 (note 1), and Carlo F. Capello, *Il mappamondo medioevale di Vercelli (1191–1218?)*, Università di Torino, Memorie e Studi Geografici, 10 (Turin: C. Fanton, 1976).
22. Destombes, *Mappemondes*, 194–97 (note 1).
23. R. A. Skelton, in Destombes, *Mappemondes*, 149–60 (note 1).

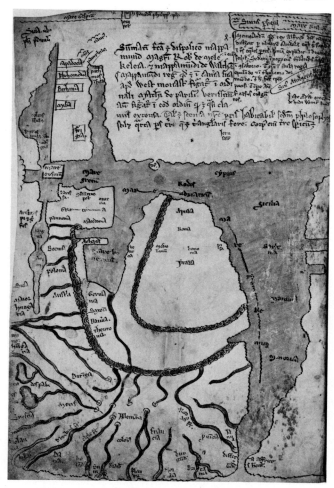

FIG. 18.58. *MAPPAMUNDI* BY MATTHEW PARIS. From the first part of a thirteenth-century manuscript of Matthew Paris's *Chronica majora*.

Size of the original: 35.4 × 23.2 cm. By permission of the Master and Fellows of Corpus Christi College, Cambridge (MS. 26, p. 284).

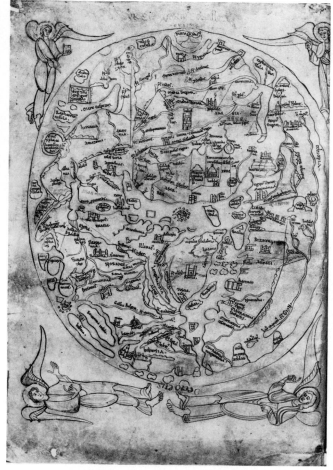

FIG. 18.59. THE HENRY OF MAINZ WORLD MAP. From a twelfth-century manuscript of the *Imago Mundi* by Honorius of Autun. See also figure 18.42 for detail.

Size of the original: 29.5 × 20.5 cm. By permission of the Master and Fellows of Corpus Christi College, Cambridge (MS. 66, p. 2).

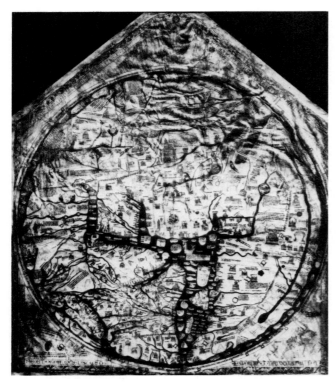

FIG. 18.60. THE HEREFORD MAP, CA. 1290. See also figures 12.4 and 18.20.
Diameter of the original: 1.32 m. By permission of the Royal Geographical Society, London.

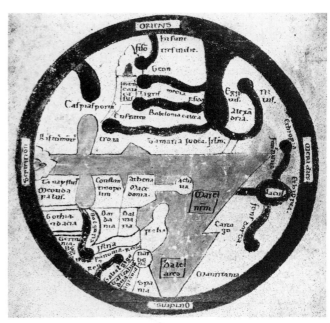

FIG. 18.62. *MAPPAMUNDI* BY GUIDO OF PISA, 1119. Diameter of the original: 13 cm. Copyright Bibliothèque Royale Albert Ier, Brussels (MS. 3897–3919 [cat. 3095], fol. 53v).

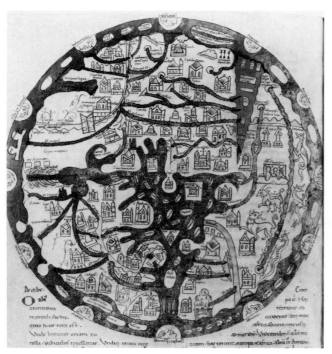

FIG. 18.61. OROSIAN-ISIDORIAN *MAPPAMUNDI*. From an eleventh-century copy of Isidore's *Etymologies*.
Diameter of the original: 26.5 cm. By permission of the Bayerische Staatsbibliothek, Munich (Clm. 10058, fol. 154v).

FIG. 18.63. THE PSALTER MAP, THIRTEENTH CENTURY. See also figure 18.35 for detail.
Size of the original: 14.3 × 9.5 cm. By permission of the British Library, London (Add. MS. 28681, fol. 9r).

FIG. 18.65. THE EBSTORF MAP. See also figs. 18.2, 18.3, and 18.19.

Size of the original: 3.56 × 3.58 m. From Walter Rosien, *Die Ebstorfer Weltkarte* (Hanover: Niedersächsisches Amt für Landesplanung und Statistik, 1952). By permission of the Niedersächsisches Institut für Landeskunde und Landesentwicklung an der Universität Göttingen.

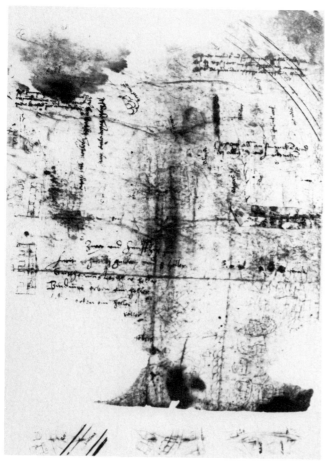

FIG. 18.64. THE WIESBADEN FRAGMENT.

Size of the original: 75 × 59 cm. By permission of the Hessisches Hauptstaatsarchiv, Wiesbaden (MS. A.60).

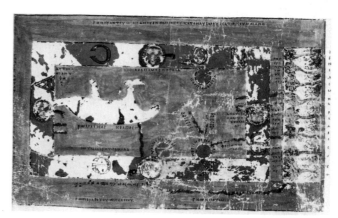

FIG. 18.66. COSMAS INDICOPLEUSTES' MAP FROM THE *CHRISTIAN TOPOGRAPHY*. See also figure 15.2.

Size of the original: 23.3 × 31.5 cm. Photograph from the Biblioteca Apostolica Vaticana, Rome (Vat. Gr. 699, fol. 40v).

FIG. 18.67. HIGDEN'S OVAL *MAPPAMUNDI*. From a four-teenth-century manuscript of Higden's *Polychronicon*.
Size of the original: 46.5 × 34.2 cm. By permission of the British Library, London (Royal MS. 14.C.ix, fols. 1v–2r).

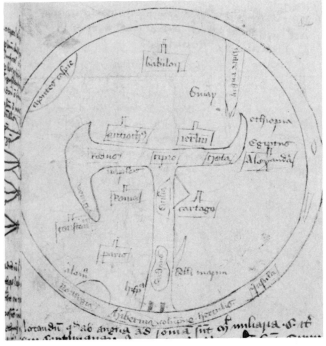

FIG. 18.68. HIGDEN'S CIRCULAR *MAPPAMUNDI*. From a manuscript of miscellanea dated 1466.
Diameter of the original: 14 cm. By permission of the British Library, London (Harl. MS. 3673, fol. 84r).

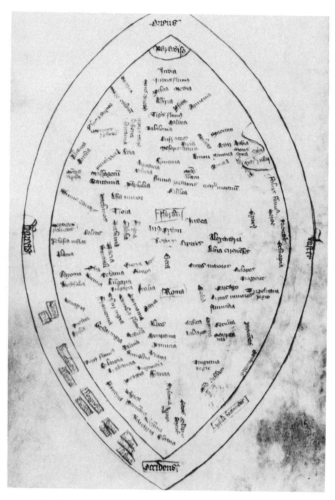

FIG. 18.69. HIGDEN'S MANDORLA-SHAPED *MAPPA-MUNDI*. From a fourteenth-century manuscript of Higden's *Polychronicon*.
Size of the original: 35.5 × 21 cm. By permission of the British Library, London (Royal MS. 14.C.xii, fol. 9v).

Zonal

The maps in this category are circular representations of the known hemisphere, usually oriented to the north, containing five or seven climatic zones that follow parallels of latitude. There are three main types: one early medieval type derived directly from Macrobian sources; a second from the work of Martianus Capella; and a third, arising later in the medieval period, that shows the influence of the zonal concept transmitted through Ptolemy and the Arab world.

MACROBIAN

The Macrobian map is derived from the cosmographical section (chaps. 5–8 of book 2) in Macrobius's early fifth-century commentary on Cicero's *Dream of Scipio* (51 B.C.).[24] Over

150 maps drawn according to the Macrobian schema are found in manuscripts of the *Commentary on the Dream of Scipio* from the ninth century to the fifteenth.[25]

These maps have five climatic zones. The Alveus Oceani (ocean river) divides the hemisphere into two equal parts, surrounded by the Mare Oceanum (ocean sea) (fig. 18.70). The Orcades (Orkney) islands are sometimes represented to the west. Reference is usually made to the circumference of the earth as measured by Eratosthenes (252,000 stades) and to the impossibility of crossing the central zone.

TYPE DERIVED FROM MARTIANUS CAPELLA

The maps in this category are primarily found in the several versions of the *Liber floridus* of Lambert of Saint-Omer (ca. 1050–1125?), beginning with the Ghent manuscript of 1120 (fig. 18.71).[26] They are derived from the work of Martianus Capella (fl. 410–439), *The Marriage of Philology and Mercury*, a fifth-century encyclopedia of the seven liberal arts.[27] Similar maps are also found in the *De philosophia mundi* (ca. 1130?) of William of Conches (ca. 1080 to ca. 1154) (fig. 18.72).[28]

The Martianus Capella maps also contain the equatorial ocean but are of a quite different style than the Macrobian maps. The ecliptic is usually shown, with the twelve signs of the zodiac, and the generalization of the coastlines is rounded in nature. The maps are characteristically oriented to the east (although some are oriented to the north) and have a large amount of text in the southern continent. The zones may or may not be explicitly shown. Regularly shaped islands are usually found in the ocean surrounding the northern continent.

LATER MAPS BY ALPHONSUS AND D'AILLY

Petrus Alphonsus (1062–1110) was a learned Spanish astronomer and geographer whose map appears in the *Dialogi cum Judaeo*. Pierre D'Ailly (1350–1420) was a French cardinal whose *Imago Mundi* (ca. 1410) appeared in several manuscript versions and a printed edition of 1480 or 1483. In some ways, the book forms a bridge between the medieval and Renaissance

24. Macrobius, *Commentary on the Dream of Scipio*, ed. and trans. William Harris Stahl (New York: Columbia University Press, 1952; second printing with supplementary bibliography, 1966). See also Carlos Sanz, "El primer mapa del mundo con la representacíon de los dos hemisferios," *Boletín de la Real Sociedad Geográfica* 102 (1966): 119–217, who provides a list of printed editions of Macrobius's *Commentary* with an index to maps in these editions.

25. Destombes, *Mappemondes*, 85–95 (note 1).

26. Destombes, *Mappemondes*, 96–116 (note 1).

27. William Harris Stahl, *The Quadrivium of Martianus Capella: Latin Traditions in the Mathematical Sciences, 50 B.C.–A.D. 1250* (New York: Columbia University Press, 1971), and Martianus Capella *The Marriage of Philology and Mercury*, trans. William Harris Stahl and Richard Johnson with E. L. Burge (New York: Columbia University Press, 1977), vols. 1 and 2, respectively, of the series Martianus Capella and the Seven Liberal Arts.

28. Lambert of Saint-Omer, *Liber floridus*; see *Liber floridus colloquium*, ed. Albert Derolez (Ghent: Story-Scientia, 1973); and *Lamberti S. Audomari Canonici liber floridus*, ed. Albert Derolez (Ghent: Story-Scientia, 1968).

periods in that it transmitted directly to Columbus Roger Bacon's idea that the sailing distance westward from Portugal to India was only half the corresponding land distance eastward from Portugal to India.[29]

These maps show the influence of the zonal concept transmitted through Ptolemy and modified by the Arabic geographers. Prominent is the mythical town of Aryn (Arin, Arym, etc.), the Islamic center of the earth, lying on the central meridian bisecting the inhabited world. No central ocean is portrayed. Two versions, by Petrus Alphonsus and Pierre D'Ailly, may be identified. The Alphonse version is oriented south and contains three town symbols representing "Aren civitas" in the southern part (fig. 18.73). In the maps by d'Ailly, the three continents are named in the northern part, the meridian of Aryn is prominently marked, and the map is oriented to the north.

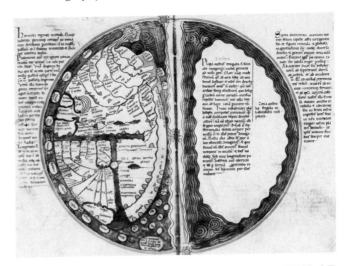

FIG. 18.71. ZONAL *MAPPAMUNDI* BY LAMBERT OF SAINT-OMER. From a twelfth-century manuscript of the *Liber floridus* of Lambert of Saint-Omer.
Diameter of the original: 41.3 cm. By permission of the Herzog August Bibliothek, Wolfenbüttel (Codex Guelf. 1 Gud. Lat. [cat. 4305], fols. 69v–70r).

---

29. Destombes, *Mappemondes*, 161–63 (note 1). See also Armando Cortesão, *History of Portuguese Cartography*, 2 vols. (Coimbra: Junta de Investigações do Ultramar-Lisboa, 1969–71), 1:195–98. Among other works, Columbus copiously annotated a copy of D'Ailly's *Imago Mundi* [Louvain: 1480 or 1483] now in the Biblioteca Colombina, Seville. For a discussion of this link, and other references to the geographical conceptions of Columbus, see Pauline Moffitt Watts, "Prophecy and Discovery: On the Spiritual Origins of Christopher Columbus's 'Enterprise of the Indies,'" *American Historical Review* 90 (1985): 73–102, esp. 82.

FIG. 18.70. MACROBIAN ZONAL *MAPPAMUNDI*. From a fifteenth-century manuscript of Macrobius's *Commentarium in somnium Scipionis*.
Diameter of the original: 12.5 cm. Photograph from the Biblioteca Apostolica Vaticana, Rome (Ottob. Lat. 1137, fol. 54v).

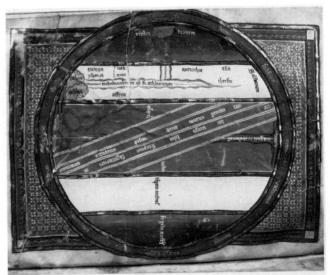

FIG. 18.72. ZONAL *MAPPAMUNDI* BY WILLIAM OF CONCHES. From a twelfth-century manuscript of the *De philosophia mundi* of William of Conches.
Diameter of the original: 12.8 cm. By permission of the Bibliothèque Sainte-Geneviève, Paris (MS. 2200, fol. 34v).

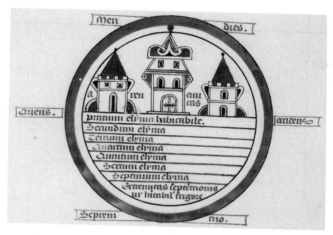

FIG. 18.73. ZONAL *MAPPAMUNDI* BY PETRUS ALPHON-SUS. From an early fifteenth-century manuscript of his *Dialogi duodecim cum Moyse Judaeo.*
Diameter of the original: 9 cm. By permission of the Bodleian Library, Oxford (Laud. Misc. 356, fol. 120r).

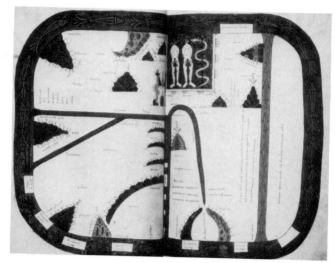

FIG. 18.75. QUADRIPARTITE *MAPPAMUNDI*: BEATUS TYPE. From a twelfth-century manuscript of the *Commentary on the Apocalypse of Saint John.* See also plate 13.
Size of the original: 32 × 43 cm. By permission of the British Library, London (Add. MS. 11695, fols. 39v–40r).

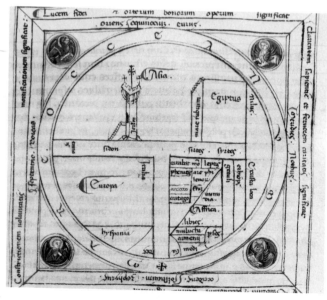

FIG. 18.74. TRIPARTITE/ZONAL *MAPPAMUNDI.* From a fourteenth-century manuscript of the *Opera* of Sallust.
Size of the original: 13 cm square. By permission of the Biblioteca Nazionale Marciana, Venice (Lat. Z.432, [MS. 1656], fol. 40r).

FIG. 18.76. VESCONTE'S *MAPPAMUNDI.* From a fourteenth-century manuscript of the *Liber secretorum fidelium crucis* of Marino Sanudo. See also plate 16.
Diameter of the original: 35 cm. By permission of the British Library, London (Add. MS. 27376*, fols. 187v–188r).

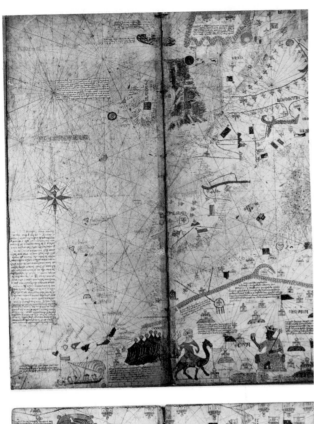

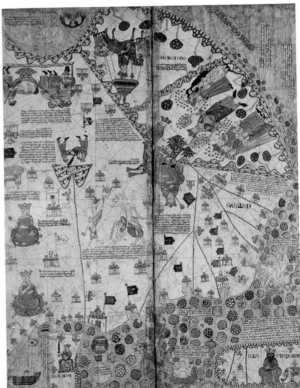

FIG. 18.77. THE CATALAN ATLAS, [1375]. See also plate 17.

Size of the originals: 65 × 50 cm. Photographs from the Bibliothèque Nationale, Paris (MS. Esp. 30).

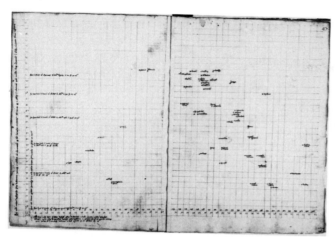

FIG. 18.78. *MAPPAMUNDI* FROM THE VIENNA-KLOS-TERNEUBURG SCHOOL. See also figure 18.24.
Size of the original: 39.4 × 58.6 cm. Photograph from the Biblioteca Apostolica Vaticana, Rome (Pal. Lat. 1368, fols. 46v–47r).

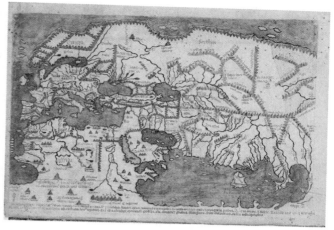

FIG. 18.79. *MAPPAMUNDI* SHOWING PTOLEMAIC IN-FLUENCE. From an early fifteenth-century incipit by Pirrus de Noha of the *De cosmographia* of Pomponius Mela. See also plate 19.
Size of the original: 18 × 27 cm. Photograph from the Biblioteca Apostolica Vaticana, Rome (Archivio di San Pietro H. 31, fol. 8r).

Quadripartite

These maps share characteristics of both the tripartite and the zonal categories. The first type includes maps that are based on a clear T-O schema in the inhabited world with zones or a fourth continent added in the southern part. A second category may also be recognized in which the maps are derived from the now lost eighth-century *mappamundi* of Beatus.

TRIPARTITE/ZONAL TYPE

These maps follow the tripartite structure in the northern half, but the southern half either is left blank or contains *climata*. A central ocean river may or may not be present, and some portray the signs of the zodiac around the circumference (fig. 18.74).

These maps are usually found in manuscripts of the works of Sallust or Isidore. The Saint Gall map (fig. 18.14) is thought by Miller to be the earliest *mappamundi* known.[30] Von den Brincken describes a curious hybrid in the Chronicle of John of Wallingford that contains a Y-shaped division of the continents in the Northern Hemisphere superimposed over seven zones, and a textual description in the Southern Hemisphere.[31]

BEATUS TYPE

The fourteen extant large Beatus maps are all thought to stem from one lost eighth-century prototype of Beatus of Liebana in his *Commentary on the Apocalypse of Saint John*.[32] The map illustrates the mandate of the apostles to travel in all parts of the earth to preach the gospel (fig. 18.75). A stemma indicating the relationship of the illustrated manuscripts is provided as figure 18.15. The smaller maps found in the Beatus codices can be traced to Isidorian models.

The circular, oval, and rectangular maps of Beatus share several characteristics. They are oriented to the east and are usually richly drawn and illuminated in a Spanish-Arabic style. The surrounding ocean sea contains representations of fish, and paradise is represented in a square vignette, occasionally including its four rivers. There is also shown an ocean river, sometimes associated with the Red Sea (Mare Rubrum). Beyond this barrier, a rubric informs the reader that outside the three known parts of the world there is a fourth part that is unknown on account of the sun's heat, but inhabited by the Antipodeans.

Transitional Type

Many later *mappaemundi* clearly show the influence first of the portolan chart in the fourteenth century and then of the Ptolemaic world map in the fifteenth, forming a separate transitional category between the medieval and Renaissance periods.

PORTOLAN CHART INFLUENCE

The first influence of the portolan chart is seen in the *mappaemundi* of Fra Paolino and Pietro Vesconte in the 1320s

30. Konrad Miller, *Mappaemundi: Die ältesten Weltkarten*, 6 vols. (Stuttgart: J. Roth, 1895–98), vol. 6, *Rekonstruierte Karten* (1898), 57.

31. Anna-Dorothee von den Brincken, "Die Klimatenkarte in der Chronik des Johann von Wallingford—ein Werk des Matthaeus Parisiensis?" *Westfalen* 51 (1973): 47–57.

32. Destombes, *Mappemondes*, 40–42 and 79–84 (note 1).

(fig. 18.76).[33] Both Catalan and Italian traditions of chart-makers are reflected in maps of this category, which include the world map in the Medici atlas, the map of Albertin de Virga,[34] the Topkapi Library fragment, the Catalan atlas of 1375 (fig. 18.77), the maps of Buondelmonti 1420, and Andrea Bianco 1436, the maps of Giovanni Leardo (pl. 20 and fig. 18.40), the Catalan (Estense) map, the Borgia map, the Genoese map of 1457 (fig. 18.23), and the map by Fra Mauro (plate 18).

The maps of this type are often circular, with a well-delineated Mediterranean and Black Sea area directly derived from the portolan charts. The accuracy falls off dramatically outside the Mediterranean basin. The cartographic signs and generalization are similar in style to those of the portolan charts, as is the network of rhumb lines radiating from the center of the map. Biblical sources predominate, especially for the land areas toward the edges of the map. The explorations in Asia of the thirteenth century and the Portuguese expansion down the coast of West Africa of the fifteenth century are reflected in many of the later maps.

PTOLEMAIC INFLUENCE

The maps are either circular or rectangular and reflect the influence of Ptolemy's *Geography* (closed Indian Ocean, Mediterranean Sea twenty degrees too long, Mountains of the Moon, etc.), which appeared after the introduction and translation of this work to western Europe in the early fifteenth century. Some belong to a subgroup of maps called the Vienna-Klosterneuburg map corpus, the world maps of which were compiled with the help of coordinates (fig. 18.78). Other examples include the Pirrus de Noha map of about 1414 (plate 19 and fig. 18.79), the fragment in the James Ford Bell Library, and the world map by Andreas Walsperger of 1448.

33. Bernhard Degenhart and Annegrit Schmitt, "Marino Sanudo und Paolino Veneto," *Römisches Jahrbuch für Kunstgeschichte* 14 (1973): 1–137.

34. Destombes, *Mappemondes*, 205–7 (note 1).

APPENDIX 18.2 CHRONOLOGICAL LIST OF MAJOR MEDIEVAL *MAPPAEMUNDI*, A.D. 300–1460

Content Date[a]	Date of Artifact	Description	Location	References			
				Miller[b]	Uhden[c]	Kamal[d]	Destombes[e]
THIRD CENTURY							
	297	Eumenius reference	†		3		
	ca. 1150	Jerome[f]	London, British Library, Add. MS. 10049	2, pl. 12			
	4th century	Julius Honorius	†	6:69	5		
FIFTH CENTURY							
	418	Paulus Orosius	†	6:61	6		
	12th century	Martianus Capella (Lambert of Saint-Omer)	Ghent, Rijksuniversiteit, MS. 92	3:45	4		43.1
	12th century	Lambert of Saint-Omer	Wolfenbüttel, Herzog August Bibliothek, Codex Guelf. 1 Gud. Lat. (cat. 4305), fols. 69v–70r				43.2
	13th century	Lambert of Saint-Omer	Paris, Bibliothèque Nationale, MS. Lat. 8865 (Suppl. 10–2)	3:46			43.3
	14th century	Lambert of Saint-Omer	Leiden, Rijksuniversiteit, Codex Voss. Lat., fol. 31	3:45 and pl. 4			43.4
	14th century	Lambert of Saint-Omer	Genoa, Biblioteca Durazzo				43.5
	14th century	Lambert of Saint-Omer	Chantilly, Musée Condé				43.6
	15th century	Lambert of Saint-Omer	The Hague, Koninklijke Bibliotheek MS. 72.A.23 (old Y392)	3:45			43.7
SIXTH CENTURY							
	8th–9th century	Cosmas	Rome, Biblioteca Apostolica Vaticana, Vat. Gr. 699, fol. 40v	3:60		2.3:370	22.5
	11th century	Cosmas	Florence, Biblioteca Medicea Laurenziana, Plut. 9.28, fol. 92v	3:60		2.3:371	24.3
SEVENTH CENTURY							
	7th century	Isidore	Saint Gall, Stiftsbibliothek, Codex 237, fol. 1r	6:57	7	3.1:512	1.6

APPENDIX 18.2—*continued*

Content Date[a]	Date of Artifact	Description	Location	References			
				Miller[b]	Uhden[c]	Kamal[d]	Destombes[e]
	ca. 775	Isidore	Rome, Biblioteca Apostolica Vaticana, Vat. Lat. 6018, fols. 64v–65				
	11th century	Isidore	Munich, Bayerische Staatsbibliothek, Clm. 10058, fol. 154v			3.3:694	4.6
	ca. 650	Ravenna cosmographer	†	6:5	8		
EIGHTH CENTURY							
	ca. 750	Pope Zacharias	†	3:151	9		
	ca. 730	Merovingian map	Albi, Bibliothèque Municipale, MS. 29 (old 23), fol. 57v	3:57	10	3.1:500	22.1
	ca. 950–60	Beatus (M)[g]	New York, Pierpont Morgan Library, MS. M644, fols. 33v–34	1:12		3.1:563	17.1
	10th century	Beatus (U)	Urgel, Archivo Diocesano, Codex 4	1:18			17.2
	970	Beatus (V)	Valladolid, Biblioteca Universitaria, MS. 1789, fols. 36v–37	1:14		3.2:640	17.3
	970	Beatus[h]	†				
	975	Beatus (G)	Gerona, Museo de la Catedral, MS. 10	1:16		3.2:641	17.5
	ca. 1047	Beatus (J)	Madrid, Biblioteca Nacional, Vitr. 14.2 (old B.31), fols. 63v–64	1:15			17.6
	ca. 1050	Beatus (S)	Paris, Bibliothèque Nationale, MS. Lat. 8878 (S. Lat. 1075), fol. 45	1:11	11	3.3:709	17.7
	ca. 1086	Beatus (O)	Burgo de Osma, Archivo de la Catedral, MS. 1, fols. 35v–36	1:12		3.3:744	17.8
	ca. 1109	Beatus (D)	London, British Library, Add. MS. 11695, fols. 39v–40	1:15		3.3:766	17.9
	ca. 1100–1150	Beatus (Tu)	Turin, Biblioteca Nazionale Universitaria, MS. I.II.1 (old D.V.39) fols. 38v–39	1:17		3.3:752	17.10
	1189	Beatus (L)	Lisbon, Arquivo Nacional da Torre do Tombo, Codex 160			3.3:745	17.11

Content Date[a]	Date of Artifact	Description	Location	References			
				Miller[b]	Uhden[c]	Kamal[d]	Destombes[e]
	12th–13th century	Beatus (N)	Paris, Bibliothèque Nationale, NAL 1366, fols. 24v–25	1:11		3.4:918	17.12
	12th–13th century	Beatus (R)	Manchester, John Rylands Library, MS. Lat. 8, fols. 43v–44			3.4:871	17.13
	1220	Beatus (H)	New York, Pierpont Morgan Library, MS. 429, fols. 31v–32			3.5:947	35.1
	13th century	Beatus (Ar)	Paris, Bibliothèque Nationale, NAL 2290, fols. 13v–14	1:17		3.4:919	35.2
NINTH CENTURY							
	ca. 800	Charlemagne[i]	†	3:151	12		
	1055	Theodulf	Vatican		13		24.11
	ca. 840	Author unknown	Saint Riquier†	3:151	14		
	842	Author unknown	Reichenau†	3:151	15		
	ca. 870	Author unknown	Saint Gall†§	3:151	16		
TENTH CENTURY							
	983	Gerbert (Sylvester II)	†	3:151	17		
	10th century	Anglo-Saxon	London, British Library, Cotton MS. Tiberius B.V., fol. 56v	3:29	18	3.1:545	24.6
ELEVENTH CENTURY							
	11th century	Authors unknown	Tegernsee (2 maps)†	3:151			
	ca. 1050	Asaph Judaeus	Paris, Bibliothèque Nationale, MS. Lat. 6556 (Reg. 4764)	3:150	19	3.3:820	50.17
TWELFTH CENTURY							
	ca. 1100	Adela, countess of Blois	[In poem of Baudri de Bourgueil: see above, p. 339]†		20		
	ca. 1110	Henry of Mainz	Cambridge, Corpus Christi College, MS. 66, p. 2	3:21	21	3.3:785	25.3
	1112–23	Authors unknown	Bamberg (3 maps)†				

APPENDIX 18.2—*continued*

Content Date[a]	Date of Artifact	Description	Location	References			
				Miller[b]	Uhden[c]	Kamal[d]	Destombes[e]
	1119	Guido of Pisa	Brussels, Bibliothèque Royale Albert Ier, MS. 3897–3919 (cat. 3095), fol. 53v		22	3.3:774	25.2
	1120	Henry of Huntingdon	Oxford, Bodleian Library, MS e Musaeo 223 (S.C. 3538)				49.13
	ca. 1150	Author unknown	Northeim†		23		
	1195	Author unknown	Durham†				
	12th century	Author unknown	Lincoln Cathedral†				
	12th century	Author unknown	Muri†				
	12th century	Authors unknown	Göttwieg (2 maps)†				
	12th century	Author unknown	Elno, Saint Amand†§				
	12th century	Author unknown	Weihenstephan†§	3:151	24		
THIRTEENTH CENTURY	ca. 1214	Gervase of Tilbury	†		25		
	ca. 1200	Vercelli map	Vercelli, Archivio Capitolare			3.5:997	52.1
	13th century	Psalter map	London, British Library, Add. MS. 28681, fol. 9r	3:37	26	3.5:998	49.8
	ca. 1235	Ebstorf map	Hanover†	5: whole vol.	27	4.1:1117	52.2
	1236	Henry III[i]	†				
	ca. 1250	Matthew Paris	Cambridge, Corpus Christi College, MS. 26, p. 284	3:71		3.5:1000	54.2
	ca. 1250	Matthew Paris	London, British Library, Cotton MS. Nero D.V., fol. 1v	3:70	28		54.1
	ca. 1250	Robert of Melkeley	†	3:72	29		
	ca. 1250	Robert of Melkeley	Waltham†	3:72	30		
	ca. 1250	Matthew Paris	†	3:72	31		
	1265	Conrad of Basle	†	3:151	32		
	1268	Roger Bacon	†				

Content Date[a]	Date of Artifact	Description	Location	References			
				Miller[b]	Uhden[c]	Kamal[d]	Destombes[e]
	ca. 1290	Richard de Bello	Hereford Cathedral	4: whole vol.	33	4.1:1077	52.3
	13th century	Albertus Magnus	†	3:151			
	1299	Edward I inventory[k]	†				
	13th century	Wiesbaden fragment	Wiesbaden, Hessisches Hauptstaats-archiv, MS. A.60				52.4
FOURTEENTH CENTURY							
	ca. 1337	Romance map	†				
	ca. 1320	Fra Paolino	Rome, Biblioteca Apostolica Vaticana, Vat. Lat. 1960, fol. 264v	3:132			54.3
	ca. 1320	Fra Paolino[l]	Paris, Bibliothèque Nationale, MS. Lat. 4939, fol. 9	3:132			54.10
	ca. 1321	Pietro Vesconte	Brussels, Bibliothèque Royale Albert Ier, MS. 9347–48, fols. 162v–163	3:132			54.4
	ca. 1321	Pietro Vesconte	Brussels, Bibliothèque Royale Albert Ier, MS. 9404–5, fols. 173v–174	3:132			54.5
	ca. 1321	Pietro Vesconte	Florence, Biblioteca Medicea Laurenziana, Plut. 21.23, fols. 138v–139				54.6
	ca. 1321	Pietro Vesconte	London, British Library, Egerton MS. 1500, fol. 3				54.7
	ca. 1321	Pietro Vesconte	London, British Library, Add. MS. 27376*, fols. 8v–9				54.8
	ca. 1321	Pietro Vesconte	Oxford, Bodleian Library, Tanner 190, fols. 203v–204	3:132			54.9
	ca. 1321	Pietro Vesconte	Rome, Biblioteca Apostolica Vaticana, Pal. Lat. 1362A, fol. 2	3:132			54.11

APPENDIX 18.2—*continued*

Content Date[a]	Date of Artifact	Description	Location	References			
				Miller[b]	Uhden[c]	Kamal[d]	Destombes[e]
	ca. 1321	Pietro Vesconte	Rome, Biblioteca Apostolica Vaticana, Reg. Lat. 548, fols. 138v–139				54.12
	ca. 1321	Pietro Vesconte	Rome, Biblioteca Apostolica Vaticana, Vat. Lat. 2972, fols. 112v–113				54.13
	1329–39	Author unknown	Venice, Palazzo Ducale†				
	1335–36	Opicinus de Canistris	Rome, Biblioteca Apostolica Vaticana, Vat. Lat. 1993				
	1341	Opicinus de Canistris	Rome, Biblioteca Apostolica Vaticana, Vat. Lat. 6435				
	After 1342	Higden (A)[m]	London, British Library, Royal MS. 14.C.ix, fols. 1v–2r	3:95		4.2:1265	47.1
	After 1342	Higden (B)	London, British Library, Royal MS. 14.C.ix, fol. 2v	3:96		4.2:1266	47.2
	After 1377	Higden (C)	London, British Library, Royal MS. 14.C.xii, fol. 9v	3:97		4.2:1269	47.9
	After 1377	Higden (D)	London, British Library, Add. MS. 10104, fol. 8				47.10
	14th century	Higden (E)	London, Lambeth Palace, MS. 112, fol. 2v				47.15
	1466	Higden (F)	London, British Library, Harl. MS. 3673, fol. 84r			4.3:1382	47.21
	ca. 1350	Higden (G)	San Marino, Huntington Library, HM 132, fol. 4v				47.3
	After 1347	Higden (H)	Oxford, Bodleian Library, Tanner 170 (S.C. 9996), fol. 15v				47.4
	14th century	Higden (I)	Oxford, Magdalen College, MS. 190, fol. 1v				47.8
	14th century	Higden (J)	Oxford, Corpus Christi College, MS. 89, fol. 13v	3:97		4.2:1267	47.13

Content Date[a]	Date of Artifact	Description	Location	References			
				Miller[b]	Uhden[c]	Kamal[d]	Destombes[e]
early 15th century	Higden (K)	Oxford, Bodleian Library, Digby 196 (S.C. 1797), fol. 195v					47.17
early 15th	Higden (L)	Warminster, Library of the Marquess of Bath, Longleat House, MS. 50, fol. 7v					47.5
1367	Higden (M)	Cambridge, University Library, Add. MS. 3077, fol. 11		3:98			47.6
14th century	Higden (N)	Cambridge, Corpus Christi College, MS. 21, fol. 9v		3:97			47.12
After 1367	Higden (O)	Paris, Bibliothèque Nationale, MS. Lat. 4922, fol. 2		3:96		4.2:1268	47.7
14th century	Higden (P)	Paris, Bibliothèque Nationale, MS. Lat. Lat. 4126, fol. 1v				4.3:1381	47.14
14th century	Higden (Q)	Edinburgh, National Library of Scotland Adv. MS. 33.4.12, fol. 13v		3:97			47.11
1400	Higden (R)	Winchester College, MS. 15, fol. 13r		3:99			47.16
15th century	Higden (S)	Rome, Biblioteca Apostolica Vaticana, Reg. Lat. 731					47.18
15th century	Higden (T)	Glasgow, University Library, MS. T 3.10, fol. 15r					47.19
15th century	Higden (U)	Lincoln, Cathedral Library, MS. A.4.17					47.20
1344	Ambrogio Lorenzetti[n]	Siena†					
ca. 1350	Johannes Utinensis	Stuttgart, Württembergische Landesbibliothek, Theol. Fol. 100 fol. 3v		3:146	36		51.29
ca. 1350	Johannes Utinensis	Munich, Bayerische Staatsbibliothek, Clm. 721, fol. 3v					51.14
ca. 1370	Saint-Denis Chronicles	Paris, Bibliothèque Sainte Geneviève, MS. 782, fol.374v		3:136	37	4.2:1270	50.19

APPENDIX 18.2—*continued*

Content Date[a]	Date of Artifact	Description	Location	References			
				Miller[b]	Uhden[c]	Kamal[d]	Destombes[e]
	1363–81	Author unknown	Heiligenkreuz†§				
	[1375]	Catalan atlas	Paris, Bibliothèque Nationale, MS. Esp. 30. See above, pp. 314–15.				
	ca. 1380	Catalan fragment	Istanbul, Topkapi Sarayi, Kutuphané no. 1828 (old 2758, 49361)				52.5
FIFTEENTH CENTURY							
	1405	Author unknown	Bourges†§				
	15th century	Medici atlas	Florence, Biblioteca Medicea Laurenziana, Gad. Rel. 9				
	1410–12	Author unknown	Library of Amplionus von Rotinck†§				
	ca. 1411–15	Albertin de Virga	Location unknown			4.3:1377	52.6
	1414?	Pirrus de Noha	Rome, Biblioteca Apostolica Vaticana, Archivio di San Pietro H.31, fol. 8r				
	1416	Authors unknown	Duc de Berry (3 maps)†				
	1417	Pomponius Mela	Reims, Bibliothèque de la Ville, MS. 1321, fol. 13	3:138	38		51.27
	ca. 1430	Borgia map	Rome, Biblioteca Apostolica Vaticana, Borgiano XVI (galerie)	3:148		5:1493	53.1
	15th century	Anonymous Venetian	Rome, Biblioteca Apostolica Vaticana, Borgiano V				52.15
	15th century	Catalan (Estense)	Modena, Biblioteca Estense, C.G.A. 1				52.12
	15th century	Author unknown	Minneapolis, University of Minnesota, James Ford Bell Collection				52.11
	15th century	Bartholomaeus Anglicus	Wolfenbüttel, Herzog August Bibliothek, Codex Helmstedt 422 (cat. 477)		39		51.39

Content Date[a]	Date of Artifact	Description	Location	References			
				Miller[b]	Uhden[c]	Kamal[d]	Destombes[e]
	15th century	Circular Ptolemy	Zeitz, Stiftsbibliothek, MS. Lat. Hist., fol. 497, fol. 48				54.17
	15th century	Author unknown	Venice, Palazzo Ducale, Sala dello Scudo†				
	15th century	Jan van Eyck[o]	†				
	1436	Andrea Bianco	Venice, Biblioteca Nazionale Marciana, MS. Fondo Ant. It. Z.76	3:143			54.16
	1440?	Vinland map[p]	New Haven, Beinecke Library, Yale University				
	1448	Andreas Walsperger	Rome, Biblioteca Apostolica Vaticana, Pal. Lat. 1362b	3:147			52.10
	1442	Giovanni Leardo	Verona, Biblioteca Comunale				52.7
	1443	Fra Mauro	†				
	1447	Giovanni Leardo	†				
	1448	Giovanni Leardo	Vicenza, Biblioteca Civica Bertoliana	3:145			52.8
	1448–49	Fra Mauro	†				
	1452	Giovanni Leardo	Milwaukee, University of Wisconsin, American Geographical Society Collection				52.9
	1457	Genoese map	Florence, Biblioteca Nazionale Centrale, Port. 1			5:1494	52.13
	1459	Fra Mauro	[To Portugal]†				
	1459	Fra Mauro	Venice, Biblioteca Nazionale Marciana			5:1495	52.14

APPENDIX 18.2—*continued*

Note: Small T-O maps by Macrobius, Sallust, and others have been omitted from this list.

†Not extant.

§Maps referred to by Leo Bagrow, "Old Inventories of Maps," *Imago Mundi* 5 (1948): 18–20.

[a]This column is ordered by the century to which the content of the map refers.

[b]Konrad Miller, *Mappaemundi: Die ältesten Weltkarten,* 6 vols. (Stuttgart: J. Roth, 1895–98).

[c]Richard Uhden, "Zur Herkunft und Systematik der mittelalterlichen Weltkarten," *Geographische Zeitschrift* 37 (1931): 321–40.

[d]Youssouf Kamal, *Monumenta cartographica Africae et Aegypti* 5 vols. in 16 parts (Cairo, 1926–51).

[e]Marcel Destombes, ed., *Mappemondes A.D. 1200–1500: Catalogue préparé par la Commission des Cartes Anciennes de l'Union Géographique Internationale* (Amsterdam: N. Israel, 1964).

[f]This map of Asia is strictly speaking a regional map but contains so much of the world that it is included here. The other "Jerome" map on the verso is a regional map of Palestine and is not included here.

[g]Letters refer to manuscript designations in the Beatus stemma in figure 18.17. Only the large Beatus maps are included here.

[h]The large map in Madrid, Archivo Histórico Nacional, MS. 1240 is missing.

[i]This map is interpreted by some to be a celestial map. See above, p. 303.

[j]Ernest William Tristram, *English Medieval Wall Painting,* 2 vols. (London: Oxford University Press, 1944–50), vol. 2, *The Thirteenth Century,* 180, 610.

[k]Otto Lehmann-Brockhaus, *Lateinische Schriftquellen zur Kunst in England, Wales und Schottland, vom Jahre 901 bis zum Jahre 1307,* 5 vols. (Munich: Prestel, 1955–60), 3:no. 6261.

[l]Bernhard Degenhart and Annegrit Schmitt, "Marino Sanudo und Paolino Veneto," *Römisches Jahrbuch für Kunstgeschichte* 14 (1973), 1–137, esp. 107, pl. 145.

[m]Letter designation refers to copies in the stemma in figure 18.21.

[n]Aldo Cairola and Enzo Carli, *Il Palazzo Pubblico di Siena* (Rome: Editalia, 1963), 139–40.

[o]Charles Sterling, "Le mappemonde de Jan van Eyck," *Revue de l'Art* 33 (1976): 69–82.

[p]The authenticity of the Vinland map has been the source of much controversy, which still continues. The content of the map was thoroughly studied by Skelton in R. A. Skelton, Thomas E. Marston, and George D. Painter, *The Vinland Map and the Tartar Relation* (New Haven: Yale University Press, 1965), 107–239; he concluded that the map was drawn in the second quarter of the fifteenth century (p. 230) and was "the oldest surviving map of American lands" (p. 232). The publication stimulated several studies of both its content and physical form, summarized in Helen Wallis et al., "The Strange Case of the Vinland Map: A Symposium," *Geographical Journal* 140 (1974): 183–214. The testing of the ink by Walter McCrone Associates suggested a date of about 1920 and appeared to close the issue, but recent proton beam analysis by Thomas A. Cahill and his colleagues at the Crocker Nuclear Laboratory, University of California—Davis casts doubt on the McCrone analysis and has revived the controversy. Their findings will be published in a forthcoming issue of *Analytical Chemistry.*

BIBLIOGRAPHY
CHAPTER 18 MEDIEVAL *MAPPAEMUNDI*

Almagià, Roberto. *Monumenta cartographica Vaticana*. Vol. 1, *Planisferi, carte nautiche e affini dal secolo XIV al XVII esistenti nella Biblioteca Apostolica Vaticana*. Rome: Biblioteca Apostolica Vaticana, 1944.

Andrews, Michael Corbet. "The Study and Classification of Medieval Mappae Mundi." *Archaeologica* 75 (1925–26): 61–76.

Arentzen, Jörg-Geerd. *Imago Mundi Cartographica: Studien zur Bildlichkeit mittelalterlicher Welt- und Ökumenekarten unter besonderer Berücksichtigung des Zusammenwirkens von Text und Bild*. Münstersche Mittelalter-Schriften 53. Munich: Wilhelm Fink, 1984.

Bagrow, Leo. *History of Cartography*. Revised and enlarged by R. A. Skelton. Translated by D. L. Paisey. Cambridge: Harvard University Press; London: C. A. Watts, 1964.

Barkan, Leonard. *Nature's Work of Art: The Human Body as Image of the World*. New Haven: Yale University Press, 1975.

Beazley, Charles Raymond. *The Dawn of Modern Geography: A History of Exploration and Geographical Science from the Conversion of the Roman Empire to A.D. 900*. 3 vols. London: J. Murray, 1897–1906.

————. "New Light on Some Mediæval Maps." *Geographical Journal* 14 (1899): 620–29; 15 (1900): 130–41, 378–89; 16 (1900): 319–29.

Bevan, W. L., and H. W. Phillott. *Medieval Geography: An Essay in Illustration of the Hereford Mappa Mundi*. London: E. Stanford, 1873.

Brehaut, Ernest. *An Encyclopedist of the Dark Ages: Isidore of Seville*. Studies in History, Economics and Public Law, vol. 48, no. 1. New York: Columbia University Press, 1912.

Brincken, Anna-Dorothee von den. "Mappa mundi und Chronographia." *Deutsches Archiv für die Erforschung des Mittelalters* 24 (1968): 118–86.

————. "Die Ausbildung konventioneller Zeichen und Farbgebungen in der Universalkartographie des Mittelalters." *Archiv für Diplomatik: Schriftgeschichte Siegel- und Wappenkunde* 16 (1970): 325–49.

————. "Zur Universalkartographie des Mittelalters." In *Methoden in Wissenschaft und Kunst des Mittelalters*, edited by Albert Zimmermann, 249–78. Miscellanea Mediaevalia 7. Berlin: Walter de Gruyter, 1970.

————. "Europa in der Kartographie des Mittelalters." *Archiv für Kulturgeschichte* 55 (1973): 289–304.

————. "Weltbild der Lateinischen Universalhistoriker und -Kartographen." In *Popoli e paesi nella cultura altomedievale*, 2 vols., Settimane di Studio del Centro Italiano di Studi sull'alto Medioevo, 19, 1:377–408 and plates. Spoleto: Presso la Sede del Centro, 1983.

Cortesão, Armando. *History of Portuguese Cartography*. 2 vols. Coimbra: Junta de Investigações do Ultramar-Lisboa, 1969–71.

Crone, Gerald R. *The World Map by Richard of Haldingham in Hereford Cathedral*. Reproductions of Early Manuscript Maps 3. London: Royal Geographical Society, 1954.

————. "New Light on the Hereford Map." *Geographical Journal* 131 (1965): 447–62.

————. *Maps and Their Makers: An Introduction to the History of Cartography*. 5th ed. Folkestone, Kent: Dawson; Hamden, Conn.: Archon Books, 1978.

Destombes, Marcel, ed. *Catalogue des cartes gravées au XVᵉ siècle*. Paris: International Geographical Union, 1952.

————. *Mappemondes A.D. 1200–1500: Catalogue préparé par la Commission des Cartes Anciennes de l'Union Géographique Internationale*. Amsterdam: N. Israel, 1964.

Duhem, Pierre. *Le système du monde: Histoire des doctrines cosmologiques de Platon à Copernic*. 10 vols. Paris: Hermann, 1913–59.

Durand, Dana Bennett. *The Vienna-Klosterneuburg Map Corpus of the Fifteenth Century: A Study in the Transition from Medieval to Modern Science*. Leiden: E. J. Brill, 1952.

Friedman, John Block. *The Monstrous Races in Medieval Art and Thought*. Cambridge: Harvard University Press, 1981.

Glacken, Clarence J. *Traces on the Rhodian Shore*. Berkeley and Los Angeles: University of California Press, 1967.

Gordon, B. L. "Sacred Directions, Orientation, and the Top of the Map." *History of Religions* 10 (1971): 211–27.

Isidore of Seville. *Traité de la nature*. Edited by Jacques Fontaine. Bibliothèque de l'Ecole des Hautes Etudes Hispaniques, fasc. 28. Bordeaux: Féret, 1960.

Jomard, Edme-François. *Les monuments de la géographie; ou, Recueil d'anciennes cartes européennes et orientales*. Paris: Duprat, 1842–62.

Jones, Charles W. "The Flat Earth." *Thought: A Quarterly of the Sciences and Letters* 9 (1934): 296–307.

Kamal, Youssouf. *Monumenta cartographica Africae et Aegypti*. 5 vols. in 16 pts. Cairo, 1926–51.

Kimble, George H. T. *Geography in the Middle Ages*. London: Methuen, 1938.

Klein, Peter K. *Der ältere Beatus-Kodex Vitr. 14–1 der Biblioteca Nacional zu Madrid: Studien zur Beatus-Illustration und der spanischen Buchmalerei des 10. Jahrhunderts*. Hildesheim: Georg Olms, 1976.

Leithäuser, Joachim G. *Mappae mundi: Die geistige Eroberung der Welt*. Berlin: Safari-Verlag, 1958.

Lelewel, Joachim. *Géographie du Moyen Age*. 4 vols. and epilogue. Brussels: J. Pilliet, 1852–57; reprinted Amsterdam: Meridian, 1966.

Menéndez-Pidal, G. "Mozárabes y asturianos en la cultura de la alta edad media en relación especial con la historia de los conocimientos geográficos." *Boletin de la Real Academia de la Historia* (Madrid) 134 (1954): 137–291.

Migne, J. P., ed. *Patrologiæ cursus completus*. 221 vols. and suppls. Paris, 1844–64; suppls. 1958–.

Miller, Konrad. *Mappaemundi: Die ältesten Weltkarten*. 6 vols. Stuttgart: J. Roth, 1895–98. Vol. 1, *Die Weltkarte des Beatus* (1895). Vol. 2, *Atlas von 16 Lichtdruck-Tafeln* (1895). Vol. 3, *Die kleineren Weltkarten* (1895). Vol. 4, *Die Herefordkarte* (1896). Vol. 5, *Die Ebstorfkarte* (1896). Vol. 6, *Rekonstruierte Karten* (1898).

Murdoch, John Emery. *Antiquity and the Middle Ages*. Album of Science. New York: Charles Scribner's Sons, 1984.

North, Robert. *A History of Biblical Map Making.* Beihefte zum Tübinger Atlas des Vorderen Orients, B32. Wiesbaden: Reichert, 1979.

Santarém, Manuel Francisco de Barros e Sousa, Viscount of. *Atlas composé de mappemondes, de portulans et de cartes hydrographiques et historiques depuis le VIe jusqu'au XVIIe siècle.* Paris, 1849. Reprint, Amsterdam: R. Muller, 1985.

———. *Essai sur l'histoire de la cosmographie et de la cartographie pendant le Moyen-Age et sur les progrès de la géographie après les grandes découvertes du XVe siècle.* 3 vols. Paris: Maulde et Renou, 1849–52.

Sanz, Carlos. "El primer mapa del mundo con la representacíon de los dos hemisferios." *Boletín de la Real Sociedad Geográfica* 102 (1966): 119–217.

Sarton, George. *Introduction to the History of Science.* 3 vols. Baltimore: Williams and Wilkins, 1927–48.

Schulz, Juergen. "Jacopo de' Barbari's View of Venice: Map Making, City Views, and Moralized Geography before the Year 1500." *Art Bulletin* 60 (1978): 425–74.

Shirley, Rodney W. *The Mapping of the World: Early Printed World Maps 1472–1700.* London: Holland Press, 1983.

Simar, Théophile. "La géographie de l'Afrique Centrale dans l'antiquité et au Moyen-Age." *Revue Congolaise* 3 (1912–13): 1–23, 81–102, 145–69, 225–52, 289–310, 440–41.

Skelton, R. A. "A Contract for World Maps at Barcelona, 1399–1400." *Imago Mundi* 22 (1968): 107–13.

Skelton, R. A., Thomas E. Marston, and George D. Painter. *The Vinland Map and the Tartar Relation.* New Haven: Yale University Press, 1965.

Stahl, William Harris. *Roman Science: Origins, Development, and Influence to the later Middle Ages.* Madison: University of Wisconsin Press, 1962.

Tattersall, Jill. "Sphere or Disc? Allusions to the Shape of the Earth in Some Twelfth-Century and Thirteenth-Century Vernacular French Works." *Modern Language Review* 76 (1981): 31–46.

Thompson, Daniel V. *The Materials and Techniques of Medieval Painting.* New York: Dover, 1956. Unabridged and unaltered republication of the first edition, 1936, titled *The Materials of Medieval Painting.*

Tobler, Waldo R. "Medieval Distortions: The Projections of Ancient Maps." *Annals of the Association of American Geographers* 56 (1966): 351–60.

Uhden, Richard. "Zur Herkunft und Systematik der mittelalterlichen Weltkarten." *Geographische Zeitschrift* 37 (1931): 321–40.

Wittkower, Rudolf. "Marvels of the East: A Study in the History of Monsters." *Journal of the Warburg and Courtauld Institutes* 5 (1942): 159–97.

Woodward, David. "Reality, Symbolism, Time, and Space in Medieval World Maps." *Annals of the Association of American Geographers* 75 (1985): 510–21.

Wright, John Kirtland. "Notes on the Knowledge of Latitudes and Longitudes in the Middle Ages." *Isis* 5 (1922): 75–98.

———. *The Geographical Lore of the Time of the Crusades: A Study in the History of Medieval Science and Tradition in Western Europe.* American Geographical Society Research Series no. 15. New York: American Geographical Society, 1925; republished with additions, New York: Dover, 1965.

19 · Portolan Charts from the Late Thirteenth Century to 1500

TONY CAMPBELL

INTRODUCTION

To the historian of late medieval and early modern European cartography[1] the portolan charts are fundamental documents, if mysterious in their origin and precocious in their precision. Their importance has long been acknowledged, and "The First True Maps" was the enthusiastic title of an article by Charles Raymond Beazley in 1904.[2] More recently, Armando Cortesão considered the "advent of the portolan chart . . . one of the most important turning points in the whole history of cartography."[3] Alberto Magnaghi went further, describing them as a unique achievement not only in the history of navigation but in the history of civilization itself.[4] For Monique de La Roncière the work of the first named practitioner, Pietro Vesconte, was so exact that the Mediterranean outlines would not be improved until the eighteenth century.[5] In terms of the economic history of cartography, Vesconte and his contemporaries may have been the first, in the plausible opinion of a recent writer, "to pursue mapmaking as a full-time commercial craft."[6]

From the earliest extant copies, probably a little before 1300, the outline they gave for the Mediterranean was amazingly accurate. In addition, their wealth of place-names constitutes a major historical source. Their improvement over the Ptolemaic maps relating to the same area is obvious at a glance, and the North African coast with its clearly defined Syrtes is the most striking advance. Moreover, the Ptolemaic maps began to circulate widely through Europe only in the fifteenth century, by which time the portolan charts were well established. Though a linear scale was implied on Ptolemy's maps by their grid of longitude and latitude, the medieval sea charts were the first cartographic documents to regularly display one.[7] This should be contrasted with the history of European topographical mapping, which shows that the first local map since Roman times to be drawn explicitly to scale was a plan of Vienna dating from about 1422.[8] As P. D. A. Harvey further points out, virtually

I wish to acknowledge the generous assistance of the following persons who read a preliminary draft for this essay, suggesting many improvements and identifying a number of errors—though not, of course, sharing any responsibility for those that remain: Janet Backhouse, Peter Barber, and Sarah Tyacke (all of the British Library), P. D. A. Harvey (University of Durham), Thomas R. Smith (University of Kansas), and David W. Waters (formerly of the National Maritime Museum). Thanks are also due to others not specifically mentioned in these notes: Dudley Barnes (Paris) for the loan of enlarged photographs of the Vesconte charts; William Crampton of the Flag Institute, Chester, for lending unobtainable copies of flag articles; O. A. W. Dilke (University of Leeds) for comments on the classical aspect of portolan chart origins; J. P. Hudson (British Library) for assistance on paleographic points; Georges Pasch (Paris) for generously sharing the fruits of many years' research into portolan chart flags; and Vladimiro Valerio (University of Naples) for help with obscure Italian references. In addition, I received invaluable assistance from librarians and curators, unfortunately too numerous to mention individually.

1. A terminal date of 1500 has been adopted for two main reasons: first, the extension southward and eastward to include the Cape of Good Hope and the route to the Indies occurs close to that date, as does the first cartographic representation of Columbus's discoveries; second, the earliest surviving charts to incorporate a latitude scale—and thus in some opinions to have outgrown the term "portolan chart"—also date from the very first years of the sixteenth century, see below, p. 386.

2. Charles Raymond Beazley, "The First True Maps," *Nature* 71 (1904): 159–61, esp. 159.

3. Armando Cortesão, *History of Portuguese Cartography*, 2 vols. (Coimbra: Junta de Investigações do Ultramar-Lisboa, 1969–71), 1:215–16.

4. Alberto Magnaghi, "Nautiche, carte," in *Enciclopedia italiana di scienze, lettere ed arti*, originally 36 vols. ([Rome]: Istituto Giovanni Treccani, 1929–39), 24:323–31, esp. 330b.

5. Monique de La Roncière, "Les cartes marines de l'époque des grandes découvertes," *Revue d'Histoire Economique et Sociale* 45 (1967): 5–28, esp. 18. Other writers have made the same point even more strongly. Bojan Beševliev, "Basic Trends in Representing the Bulgarian Lands in Old Cartographic Documents up to 1878," *Etudes Balkaniques* 2 (1980): 94–123, esp. 100, found that "even up to the middle of the last century there existed no more accurate representations of the Black Sea than those on the portolans" (i.e., charts); and Luigi Piloni, *Carte geografiche della Sardinia* (Cagliari: Fossataro, 1974), unpaginated caption to pl. v, considered the Carte Pisane's outline of Sardinia so exact that it differed little from a modern map of equivalent scale.

6. John Noble Wilford, *The Mapmakers* (New York: Alfred A. Knopf; London: Junction Books, 1981), 50.

7. Eva G. R. Taylor, *The Haven-Finding Art: A History of Navigation from Odysseus to Captain Cook* (London: Hollis and Carter, 1956), 111.

8. P. D. A. Harvey, *The History of Topographical Maps: Symbols, Pictures and Surveys* (London: Thames and Hudson, 1980), 80.

no local maps produced during the period under discussion, that is up to 1500, made "the slightest attempt at consistency of scale."[9]

An even greater gulf divided the portolan charts from the medieval *mappaemundi*, the cartographic content of which was largely shaped by their theological message. It is worth recalling that the earliest known portolan chart is thought to be almost exactly contemporary with the Hereford world map. It cannot be claimed, of course, that the portolan charts were totally free from what today we call superstition, but neither were medieval sailors. Yet Prester John, the four rivers of paradise, the mythical Atlantic islands, and other legendary features found on some charts are all placed in the little-known interior or around the periphery. The continental coastlines that constitute the charts' primary purpose are in no way affected. The unidentified author of the Genoese world map of 1457,[10] whose depiction of the Mediterranean is based on the portolan charts, neatly sums up the chartmakers' attitude: "This is the true description of the world of the cosmographers, accommodated to the marine [chart], from which frivolous tales have been removed."[11]

The medieval sea chart is the clearest statement of the geographic and cartographic knowledge available in the Mediterranean. Occasionally the coverage was extended to the East, as in the case of the Catalan atlas. Contact with China, however, ceased after the mid-fourteenth century with the collapse of that Tatar empire at which the Polos had marveled. But in the West the portolan charts of the fourteenth and fifteenth centuries provide the best, and at times the only, documentation of the first chapter of Renaissance discovery—the exploration of the Atlantic islands and the charting of Africa's entire west coast. The Spanish and Portuguese seaborne empires whose foundations were to be laid by Christopher Columbus and Vasco da Gama were the fruits of these preparatory voyages.

The medieval *mappaemundi* are the cosmographies of thinking landsmen. By contrast, the portolan charts preserve the Mediterranean sailors' firsthand experience of their own sea, as well as their expanding knowledge of the Atlantic Ocean. They are strikingly original, signaling, as Gerald R. Crone pointed out, a "complete break with tradition."[12] Whatever their antecedents might have been, these cannot be identified with any confidence today; but this is only one of the many unanswered questions these documents pose. How was the prototype constructed and when? How were copies manufactured for some four hundred years without steadily increasing distortion? Did the Catalans influence the Italians, or vice versa? And most fundamental of all, what was their function?

A general study has already been devoted to the *map-*

paemundi,[13] but no broad survey of the pre-Columbian portolan charts has been attempted in English since Nordenskiöld's in 1897.[14] This lack of recent reassessment has meant that portolan charts have benefited only slightly from the more rigorous analytical methods of contemporary scholarship. Past discussions of portolan charts have tended toward one of two extremes: either sweeping generalizations based on a priori reasoning, or myopic studies of individual works. Where the first approach tended to stretch the limited available evidence beyond the breaking point, the second missed most of the opportunities for comparative analysis, giving as much weight to extraordinary features as to typical ones. This essay attempts to steer a middle course by drawing together the strands scattered among numerous detailed studies and spinning into a single thread—tenuous though it often is—the little that is known of the history of these charts. Various aspects will be considered in turn: the question of their origin, the way they were drawn, their changing content, the social standing of their creators, the likely identity of their first owners, and the purposes for which they were made. But the charts themselves are more important and reliable witnesses than any secondary authorities. New and compelling evidence about the relation of one chartmaker to another and about their response to changing external realities has emerged from a close comparative examination of surviving charts.

The feature subjected to particularly close examination in this way is the toponymy of the early charts. Contrary to the belief in the essential conservatism of the portolan charts through the centuries, strongly voiced by Nordenskiöld and others, a recent survey of their place-names has revealed extensive toponymic change up to at least the middle of the fifteenth century.[15] Less marked in areas little frequented by trading vessels but strongly evident along the northern Mediterranean littoral (and most particularly in the Adriatic), these con-

9. Harvey, *Topographical Maps*, 103 (note 8).

10. Florence, Biblioteca Nazionale Centrale, Port. 1; see Marcel Destombes, ed., *Mappemondes A.D. 1200–1500: Catalogue préparé par la Commission des Cartes Anciennes de l'Union Géographique Internationale* (Amsterdam: N. Israel, 1964), 222.

11. "Hec est vera cosmographorum cum marino accordata des[crip]cio quorundam frivolis naracionibus rejectis." The translation is from Gerald R. Crone, *Maps and Their Makers: An Introduction to the History of Cartography*, 5th ed. (Folkestone, Kent: Dawson; Hamden, Conn.: Archon Books, 1978), 28.

12. Crone, *Maps and Their Makers*, 11 (note 11).

13. Destombes, *Mappemondes* (note 10). For the historiography of the *mappaemundi*, see pp. 292–99.

14. A. E. Nordenskiöld, *Periplus: An Essay on the Early History of Charts and Sailing-Directions*, trans. Francis A. Bather (Stockholm: P. A. Norstedt, 1897).

15. Nordenskiöld, *Periplus*, 45, 56 (note 14). See also pp. 415–28.

stantly changing patterns of place-names proclaim a hitherto unsuspected vitality in the early charts. To Nordenskiöld and his disciples "the most perfect map of the Middle Ages, the Iliad of Cartography," was the result of a single act of creation, that of the first chart (his "normal-portolano").[16] Now that we can point to a process of continual toponymic revitalization—marked in the fourteenth century even if diminishing in the fifteenth—the portolan charts must be reinterpreted as a living record of Mediterranean self-knowledge, undergoing constant modification. This is the single most important discovery to have emerged from this investigation.

Through the place-name analysis it has also proved possible to suggest more reliable dates for a number of unsigned works. Approximately half of the atlases and charts assigned to the period up to 1500 lack both signature and date. Agreement about the dating of these documents is an essential precondition for introducing them into a history of portolan charts. Unfortunately, this requirement has not been met. The dates proposed for some important charts have fluctuated widely, while the arbitrary use of unreliable and conflicting dating criteria has led many researchers to adopt untenable positions. Once a conclusion on dating had been reached, it has often been repeated without explanation by subsequent commentators. The student of today thus inherits a legacy in which a number of unfounded dating estimates, and the conclusions unwisely based on them, have come to be treated as received wisdom.[17]

Though certainly not a perfect method, the place-name lists provide a far better system of dating than any previously devised, and they enable the undated works to be integrated with some confidence into the general historical account.[18] On the basis of an extensively amended chronological list of fourteenth- and fifteenth-century charts, fresh conclusions can be drawn about the stages of their development, and about the interrelationships between Catalan, Genoese, and Venetian practitioners.

SURVIVAL

The approximately 180 charts and atlases that can now be assigned to the fourteenth and fifteenth centuries must be a minute fraction of what was originally produced, and they are not necessarily representative.[19] Any conclusions based on the body of extant charts must acknowledge this incompleteness. Few charts can have been as highly prized as Gabriel de Valseca's ornamental production of 1439[20] (see plate 24), for which Amerigo Vespucci (1454–1512) was prepared to pay the handsome sum of 130 "ducati di ora di marco."[21] Nevertheless, the dozen references to portolan charts in inventories accompanying Genoese probate documents of the period 1384–1404 are evidence that they were then considered to be of some significance, even when, as in one instance, described as being already old.[22] These documents also give a clue to the extent of wastage over the centuries.

The factors leading to the destruction of those used at sea are obvious, but the survival of charts in landsmen's hands was by no means guaranteed. The bookseller in Anatole France's *La rôtisserie de la reine Pédauque*, who admitted having "nailed old Venetian maps on the doors,"[23] might have been fictional, but obsolete sea charts often fared no better in the real world. When Sir Thomas Phillipps was assiduously collecting manuscripts of all kinds in the nineteenth century, he often found that his rivals were not other bibliophiles, but goldbeaters, glue makers and tailors, all of whom derived some advantage from the destruction of vellum manuscripts.[24] A number of unadorned charts suffered the indignity of being dismembered for use in bookbinding. Several fragments, sometimes displaying the needle holes, testify to this. One chart was even chopped up into small pieces by a lawyer for bookmarks.[25] Sadly,

16. Nordenskiöld, *Periplus*, 45 (note 14).

17. Giuseppe Caraci was one of the few to question these assumptions; see "A proposito di alcune carte nautiche di Grazioso Benincasa," *Memorie Geografiche dall'Istituto di Scienze, Geografiche e Cartografiche* 1 (1954): 283–90.

18. There are too many important works among the undated atlases and charts for it to be feasible to restrict such an account to securely dated examples.

19. This figure of 180 charts and atlases derives from a census of early portolan charts in which this contributor is currently engaged and which is to be published in *Imago Mundi*. This total must remain approximate because there are as yet no reliable diagnostic criteria for distinguishing the unadorned works of the fifteenth century from those of the sixteenth.

20. For the location of Valseca's chart, and of all the signed, dated, or named charts subsequently mentioned in this essay, see appendixes 19.2, 19.3, and 19.4. Locations of pre-1430 unsigned works are given in table 19.3 (pp. 416–20). For references to reproductions of the atlases and charts referred to in this essay, see appendixes 19.2 and 19.3.

21. Nordenskiöld, *Periplus*, 62 (note 14). Cortesão, *History of Portuguese Cartography*, 2:148–49 (note 3), goes in detail into the George Sand spilled inkwell incident, which left its disfiguring mark on this chart.

22. Paolo Revelli, *Cristoforo Colombo e la scuola cartografica genovese* (Genoa: Consiglio Nazionale delle Ricerche, 1937), 452–58, esp. 453 (no. xviii).

23. Ruthardt Oehme, "A Cartographical Certificate by the Cologne Painter Franz Kessler," *Imago Mundi* 11 (1954): 55–56, quotation on 56.

24. Armando Cortesão, *The Nautical Chart of 1424 and the Early Discovery and Cartographical Representation of America: A Study on the History of Early Navigation and Cartography* (Coimbra: University of Coimbra, 1954), 4.

25. Paris, Bibliothèque Nationale, Rés. Ge. D 3005; see Ernest Théodore Hamy, "Note sur des fragments d'une carte marine catalane du XVe siècle, ayant servi de signets dans les notules d'un notaire de

loss and destruction continue. The 1463 Grazioso Benincasa atlas was stolen from the Royal Army Medical Corps Library, London, in 1930, and an eastern Mediterranean fragment vanished after the war from the Archivio di Stato, Venice.[26] World War II took its toll as well: the Giovanni da Carignano map was totally destroyed and the 1490 Andrea Benincasa chart was partially burned in the 1944 bombing of Ancona—as was Grazioso Benincasa's written portolano.[27]

There are, besides, a considerable number of references to charts that vanished long ago.[28] This applies particularly to Portugal's contribution. "Although it is beyond dispute," wrote Cortesão and Teixeira da Mota in 1960, "that many Portuguese charts were drawn in the time of the Infante [Prince Henry the Navigator] and soon after, possibly some as early as the fourteenth century, it is very odd indeed that only one such chart and a fragment of another have survived."[29] Despite this statement, Cortesão could find no reference to charts in Portuguese records earlier than 1443, and the oldest known Portuguese chart is either one by Reinel, dated to about 1483, or an unsigned work in Modena (Biblioteca Estense, C.G.A.5c), assigned to the last quarter of the fifteenth century.[30] The Reinel chart's suggested date of 1483 might, however, have to be modified if the fleur-de-lis on the flag of Marseilles is taken as a reference to the town's transfer from Provence to France in 1486. This unusual flag form, replacing the normal blue cross, is also found on some charts of a century or more later.[31] The 1492 Jorge de Aguiar chart and a fragment in Lisbon, datable "after 1493, perhaps before the end of the century," bring to four the number of known Portuguese works supposedly produced before 1500. The Aguiar chart's existence was first made known at a meeting in Portugal in 1968, when it was described by Alexander O. Vietor.[32]

Insofar as the lost charts referred to Portuguese discoveries in western Africa and the Atlantic, their substance did nevertheless find its way onto portolan charts, evidently with Italians formerly in Portuguese employ acting as intermediaries.[33] As a sign of early French activity, de La Roncière drew attention to documents of 1476 commissioning two painters, Jehan Robert and Jehan Morel, to "portray" the coast around the Seine estuary.[34] This would presumably have been a panoramic view and as such no exception to the rule that portolan charts (besides Andrea Bianco's London production of 1448) are not known to have been produced outside southern Europe and the Muslim world before the sixteenth century.[35] Besides the few original Arab works that survive, one early Western chart, Bertran's of 1482, displays annotations in Arabic lettering.[36] On the other hand, the supposed Arabic lettering on the undated chart in Dijon has proved to be imaginary.[37] In addition to

Perpignan (1531–1556)," *Bulletin du Comité des Travaux Historiques et Scientifiques: Section de Géographie Historique et Descriptive* (1897): 23–31, esp. 24; reprinted in *Acta Cartographica* 4 (1969): 219–27.

26. On the Benincasa atlas see the unsigned note "Der gestohlene Gratiosus Benincasa," *Imago Mundi* 1 (1935): 20; information on the fragment formerly in the Archivio di Stato, Venice, comes in a personal communication from the director, Maria Francesca Tiepolo.

27. Cortesão, *History of Portuguese Cartography*, 1:219, 2:193 (note 3). Paolo Revelli (who sent questionnaires to Italian institutions after World War II) was relieved to find less damage than he expected; see "Cimeli geografici di biblioteche italiane distrutti o danneggiati dalla guerra," *Atti della XIV Congresso Geografico Italiano, Bologna, 1947* (1949): 526–28; and idem, "Cimeli geografici di archivi italiani distrutti o danneggiati dalla guerra," *Atti della XV Congresso Geografico Italiano, Torino, 1950*, 2 vols. (1952), 2:879.

28. See, for example, Julio Rey Pastor and Ernesto García Camarero, *La cartografía mallorquina* (Madrid: Departamento de Historia y Filosofía de la Ciencia, 1960), 59–60, 63, 65–66, 84–86.

29. Armando Cortesão and Avelino Teixeira da Mota, *Portugaliae monumenta cartographica*, 6 vols. (Lisbon, 1960), 1:xxxiv.

30. See Cortesão, *History of Portuguese Cartography*, 2:118 and 211 (note 3), and Cortesão and Teixeira da Mota, *Portugaliae monumenta cartographica*, 1:3–4 (note 29).

31. I owe this point to Georges Pasch.

32. The quotation is from Cortesão, *History of Portuguese Cartography*, 2:218 (note 3). See Alexander O. Vietor, "A Portuguese Chart of 1492 by Jorge Aguiar," *Revista da Universidade de Coimbra* 24 (1971): 515–16, and also Cortesão, *History of Portuguese Cartography*, 2:212–16 (note 3). The Dijon chart, described as Portuguese in Cortesão and Teixeira da Mota, *Portugaliae monumenta cartographica*, 5:187 (note 29), is not mentioned in Cortesão's later *History*; the so-called Columbus chart in Paris, Bibliothèque Nationale, Rés. Ge. AA 562, is not considered Portuguese by Cortesão, *History of Portuguese Cartography*, 2:220; and the Munich chart has been reassigned in this essay to the sixteenth century (see p. 386).

33. Alvise da Cadamosto, whose discovery of the Cape Verde Islands was recorded by Grazioso Benincasa in 1468, was a Venetian in Portuguese service, and Fra Mauro on his world map of 1459 claimed to have Portuguese charts in his possession; see Cortesão, *History of Portuguese Cartography*, 2:85, 176 (note 3).

34. Charles de La Roncière, *Les portulans de la Bibliothèque de Lyon*, fasc. 8 of *Les Portulans Italiens* in Lyon, Bibliothèque de la Ville, *Documents paléographiques, typographiques, iconographiques* (Lyons, 1929), 793.

35. A chart, supposedly fifteenth-century French but actually sixteenth-century Italian, is described in Gustavo Uzielli and Pietro Amat di San Filippo, *Mappamondi, carte nautiche, portolani ed altri monumenti cartografici specialmente italiani dei secoli XIII–XVII*, 2d ed., 2 vols., Studi Biografici e Bibliografici sulla Storia della Geografia in Italia (Rome: Società Geografica Italiana, 1882; reprinted Amsterdam: Meridian, 1967), vol. 2, no. 403. See also Roberto Almagià, *Monumenta cartographica Vaticana*, 4 vols. (Rome: Biblioteca Apostolica Vaticana, 1944–55), vol. 1, *Planisferi, carte nautiche e affini dal secolo XIV al XVII esistenti nella Biblioteca Apostolica Vaticana*, 84.

36. See Theobald Fischer, *Sammlung mittelalterlicher Welt- und Seekarten italienischen Ursprungs und aus italienischen Bibliotheken und Archiven* (Venice: F. Ongania, 1886; reprinted Amsterdam: Meridian, 1961), 95 (where, following Uzielli, it was wrongly described as a chart of 1491).

37. Paul Gaffarel, "Etude sur un portulan inédit de la Bibliothèque de Dijon," *Mémoires de la Commission des Antiquités de la Côte-d'Or* 9 (1877): 149–99, esp. 160. For corrective, see Roberto Almagià, "Una carta nautica di presunta origine genovese," *Rivista Geografica Italiana* 64 (1957): 58–60, esp. 59, and Isabelle Raynaud-Nguyen,

Muslim works, three charts were drawn by Jehuda ben Zara in Alexandria and Galilee (see appendix 19.2). Lost charts that Bartolomeo Colombo made in England, one of which he presented to Henry VII on 13 February 1488, would need to be set alongside Bianco's 1448 chart as evidence of early chartmaking in England.[38]

TERMINOLOGY

Ideally, the terminology of a subject should provide a common platform for those working in it. With portolan charts, however, the basic nomenclature continues to divide; it is itself part of the controversy. Most English-speaking writers use the term "portolan chart" (or sometimes the variant "portulan chart"). Derived from the Italian word *portolano*, for a collection of written sailing directions, this stresses (whether intentionally or not) the way the charts are assumed to complement the written account. The term "portolan chart" has been traced back no further than the 1890s.[39] The earlier, and incorrect, shorthand form "portolan" continues to cause unnecessary confusion between the charts and the written directions. The British Museum in its 1844 printed catalog of manuscript maps referred to a "portolano or collection of sea charts."[40] This ambiguous usage has recurred regularly since.

Cortesão and Teixeira da Mota sum up a general feeling that, while far from ideal, the designation "portolan chart" is now too well established to be altered.[41] There have, however, been dissenters. In 1925 Max Eckert suggested they be called "rhumb line charts."[42] It was apparently Arthur Breusing who first proposed in 1881 the charged term "loxodromic charts."[43] A loxodrome is a line of constant compass bearing; its employment here thus begs a number of questions about the part played by the magnetic compass in both the construction and the use of the charts. "Loxodromic charts" has found few champions since. A similar term, "compass charts,"[44] has certainly been in existence for more than a century but has the same drawbacks. To avoid all these overtones, French, Italian, Portuguese (when not writing in English), and Spanish scholars often refer to them simply as "nautical charts" or some variant thereof.[45] While free from unwanted connotations, the term is too broad to distinguish portolan charts from any other type of marine chart, including those produced today.

Contemporary usage is of little assistance. Eva G. R. Taylor cataloged the following terms employed at the time: *carta de Navegar, carta pro Navigando, mappamundi, mappae maris,* even the confusing *compasso,* which could equally well mean a *portolano.*[46] Pietro Vesconte used the Latin words *carta* and *tabula* for his own charts; a Catalan ordinance of 1354 and an official Portuguese document of 1443 mentioned *carta de marear;* and Antonio Pelechan in 1459 turned the subject of his specialized chart of the Adriatic into a description of the sheet it was drawn upon, terming it *cholfo* (i.e., gulf).[47] Despite all these designations, the term "portolan chart" seems to be the most convenient for present-day use, and it will accordingly be employed throughout this essay.

"L'hydrographie et l'événement historique: Deux exemples" (paper prepared for the Fourth International Reunion for the History of Nautical Science and Hydrography, Sagres-Lagos, 4–7 July 1983).

38. Fernando Colombo, *Historie del Signor Don Fernando Colombo: Nelle quali s'hà particolare, & vera relatione della vita, e de' fatti dell'Ammiraglio Don Christoforo Colombo, suo padre* (Venice, 1571), fol. 31v.

39. "Portolankarte." Franz R. von Wieser, "A. E. v. Nordenskiöld's Periplus," *Petermanns Mitteilungen* 45 (1899): 188–94, used the term over thirty times in one review. Although Nordenskiöld habitually referred to the charts as "portolani," he did use the term "portolano maps" at least once in an earlier work—see A. E. Nordenskiöld, "Résumé of an Essay on the Early History of Charts and Sailing Directions," *Report of the Sixth International Geographical Congress, London, 1895* (1896): 685–94, esp. 694; reprinted in *Acta Cartographica* 14 (1972): 185–94, esp. 194 (I am grateful to Francis Herbert for the second reference). The statement that "the term 'portolan chart' first occurs in Italy in the thirteenth century," made in a translated article by Hans-Christian Freiesleben, is apparently without foundation; see his "The Still Undiscovered Origin of the Portolan Charts," *Journal of Navigation* (formerly *Navigation: Journal of the Institute of Navigation*) 36 (1983): 124–29, esp. 124.

40. British Museum, *Catalogue of the Manuscript Maps, Charts, and Plans, and of the Topographical Drawings in the British Museum,* 3 vols. (London, 1844–61), 1:16, referring to the 1467 Benincasa atlas. The original title of a recent general study of portolan charts was *Les portulans,* followed by the subtitle, *Cartes marines du XIIIᵉ au XVIIᵉ siècle* (Fribourg: Office du Livre, 1984). It is the English version that is cited in this essay: Michel Mollat du Jourdin and Monique de La Roncière with Marie-Madeleine Azard, Isabelle Raynaud-Nguyễn, and Marie-Antoinette Vannereau, *Sea Charts of the Early Explorers: 13th to 17th Century,* trans. L. le R. Dethan (New York: Thames and Hudson, 1984).

41. Cortesão and Teixeira da Mota, *Portugaliae monumenta cartographica,* 1:xxvi (note 29).

42. "Rhumbenkarten." See Max Eckert, *Die Kartenwissenschaft: Forschungen und Grundlagen zu einer Kartographie als Wissenschaft,* 2 vols. (Berlin and Leipzig: Walter de Gruyter, 1921–25), 2:59.

43. Arthur A. Breusing, "Zur Geschichte der Kartographie: La Toleta de Marteloio und die loxodromischen Karten," *Kettlers Zeitschrift für Wissenschaft: Geographie* 2 (1881): 129–33, 180–95; reprinted in *Acta Cartographica* 6 (1969): 51–70.

44. Sophus Ruge, *Ueber Compas und Compaskarten,* Separat Abdruck aus dem Programm der Handels-Lehranstalt (Dresden, 1868).

45. *Cartes nautiques, carte nautiche,* and *cartas náuticas.*

46. Taylor, *Haven-Finding Art,* 115–17 (note 7).

47. On Vesconte's *carta* and *tabula* see Lelio Pagani, *Pietro Vesconte: Carte nautiche* (Bergamo: Grafica Gutenberg, 1977), 7; on the Catalan ordinance see Ernest Théodore Hamy, "Les origines de la cartographie de l'Europe septentrionale," *Bulletin du Comité des Travaux Historiques et Scientifiques: Section de Géographie Historique et Descriptive* 3 (1888): 333–432, esp. 416; on the Portuguese document see Cortesão, *History of Portuguese Cartography,* 2:118 (note 3); and on Pelechan's *cholfo* see pp. 433–34 and note 433.

CHARACTERISTICS AND DEFINITION

A number of features, albeit not necessarily present in all cases, set the portolan charts apart from sea charts in general. The charts of the two centuries we are considering (that is, up to 1500) are almost always drawn in inks on vellum.[48] Though the larger charts might require more than one piece of vellum, most use a single animal skin. The "neck," which has sometimes been shaped, is often clearly visible at one side.[49] Charts were normally rolled[50]—although many have since been straightened out—and a few are still attached to what may well be their original wooden rollers. A leather thong would have fastened the chart, sometimes being passed through paired incisions visible on the necks of some surviving examples—among them Pietro Vesconte's of 1311.[51] Atlases, which were usually the equivalent of a loose chart spread over several sheets, were necessarily treated differently. Although the separate vellum sheets might be handled like a book and provided with a typical binding, Pietro Vesconte had from the outset appreciated the advantages of pasting the vellum sheets onto wooden boards—a procedure that would have obviated distortion or shrinkage in salt water. Though the boards no longer survive from his 1313 atlas, they are still in evidence on the two he produced

in 1318.[52] Thick cardboard was an adequate substitute for wood, as Grazioso Benincasa found in the 1460s (fig. 19.1).

Turning to the content of both single charts and atlases, the most obvious of the common denominators that link the earliest survivor (the Carte Pisane) to those of several centuries later is the network of interconnecting rhumb lines.[53] At first glance an apparent jumble, on closer examination these will prove to be arranged in a coherent pattern. Around the circumference of one or sometimes a tangential pair of "hidden" circles (usually occupying the maximum available area) are sixteen equidistant intersection points or "secondary centers."[54] Each is joined to most or all of the others to provide thirty-two directions, which are thus repeated

FIG. 19.1. PHYSICAL CHARACTERISTICS OF A PORTOLAN ATLAS. Vellum charts could be mounted on wood, but Grazioso Benincasa used cardboard for this 1469 atlas. As usual, the charts are backed onto one another.
Height of the original: 32.7 cm. By permission of the British Library, London (Add. MS. 31315).

48. Konrad Kretschmer predicted that any drawn on paper would be exceptional; see *Die italienischen Portolane des Mittelalters: Ein Beitrag zur Geschichte der Kartographie und Nautik,* Veröffentlichungen des Instituts für Meereskunde und des Geographischen Instituts an der Universität Berlin, vol. 13 (Berlin, 1909; reprinted Hildesheim: Georg Olms, 1962), 35. Only two have so far been noted: Rome, Biblioteca Apostolica Vaticana, Rossi. 676 and the so-called Lesina chart of the Caspian (which probably dates from the first quarter of the sixteenth century); see E. P. Goldschmidt, "The Lesina Portolan Chart of the Caspian Sea" (with a commentary by Gerald R. Crone), *Geographical Journal* 103 (1944): 272–78. Paper was not produced in Europe in large quantities until the early fifteenth century; see Janet Backhouse, *The Illuminated Manuscript* (Oxford: Phaidon, 1979), 7–8. Nevertheless, Luca del Biondo, writing from Bruges in 1398 to a Florentine correspondent in Majorca, requested a chart on paper; see Charles de La Roncière, "Une nouvelle carte de l'école cartographique des Juifs de Majorque," *Bulletin du Comité des Travaux Historiques et Scientifiques: Section de Géographie* 47 (1932): 113–18, esp. 118.

49. For a discussion of alternative western and eastern necks, see below, p. 444.

50. A commentator writing in 1404 described how the sailors "opened their charts"; see Gutierre Díaz de Gámez, *The Unconquered Knight: A Chronicle of the Deeds of Don Pero Niño, Count of Buelna,* trans. and selected by Joan Evans from *El Vitorial* (London: Routledge, 1928), 97.

51. Pagani, *Vesconte,* 20 (note 47).

52. On the 1313 atlas boards see Myriem Foncin, Marcel Destombes, and Monique de La Roncière, *Catalogue des cartes nautiques sur vélin conservées au Département des Cartes et Plans* (Paris: Bibliothèque Nationale, 1963), 10; on the 1318 atlases' boards see Pagani, *Vesconte,* 20, 27 (note 47).

53. The time-hallowed term "rhumb line" is retained for convenience throughout this essay. This should not be taken to imply acceptance of the idea that these are rhumb lines in the true sense of the word. (See below, p. 385, for discussion on this point.)

54. The unsatisfactory twenty-four intersection point network on the general chart of the British Library's Cornaro atlas (Egerton MS. 73; see plate 23) is an exception to the general rule, as are the simplified networks devised, for reasons of limited space, on the small Luxoro and Pizigano atlases. See Thomas R. Smith, "Rhumb-Line Networks on Early Portolan Charts: Speculations Regarding Construction and Function" (paper prepared for the Tenth International Conference on the History of Cartography, Dublin, 1983), for the various arrangements that early chartmakers devised for the junction of the two rhumb systems on twin-circle charts.

a number of times across the chart.[55] This network shares a common orientation with the coastal outlines but is otherwise unrelated to them. The standard practice was for the eight "winds" (i.e., north, northeast, east, etc.) to be drawn in black or brown, the next eight half-winds (north-northeast, east-northeast, etc.) to be in green, and the sixteen quarter-winds (north by east, northeast by north, northeast by east, etc.) to be in red. This consistent convention allowed the navigator to pick his wind or direction without having to count around from one of the recognizable primary directions.[56]

In terms of their geographical scope, portolan charts would usually cover at least the area of Nordenskiöld's "normal-portolano"—the Mediterranean and Black seas—sometimes adding to this the Atlantic coasts from Denmark to Morocco and the British Isles (plate 23). The scale varies considerably from one chart to another. At a rough estimate, a typical chart might measure about 65 by 100 centimeters and be drawn to an approximate scale of 1:6 million.[57] The early charts are not provided with a graticule, latitude being first indicated at the beginning of the sixteenth century. A scale bar was usually provided, and one or more of its varying number of larger divisions would be subdivided into five sections, each representing ten *miglia*.[58] Unfortunately, no key to the unit of measurement was supplied, and this has led to much discussion of the scale(s) involved.[59]

From the Carte Pisane onward there are clear indications of simplification and exaggeration in the coastal drawing. Because of their greater navigational significance, islands and capes tend to be enlarged.[60] In some sections the stretch between headlands has been formalized into regular arcs, owing more to geometry and aesthetics than to hydrographic reality.[61] The headlands themselves frequently conform to one of a number of repeated types: pointed, rounded, or wedge shaped. River estuaries are regularly conventionalized as short parallel lines leading inward. The tendency toward simplification becomes more noticeable in regions outside the Mediterranean for which there was little or no first-hand experience, such as the Atlantic, Baltic, and inland areas. While the artificiality of these coastal conventions reduces our confidence in the accuracy of the very small hydrographic details, it suggests that the draftsman's main concern was to locate headlands (which had to be rounded) and estuaries (which provided both fresh water and access to the interior). With these features as fixed points, a remarkably accurate overall picture of the Mediterranean was achieved—at least after improvements had been made to the very earliest attempts. These constantly repeated coastlines and their steadily evolving array of place-names provide the portolan charts with their two most significant features.

Certain conventions were standard. So that there

should be no interference with the detailing of the coast or its offshore hazards, the place-names were written inland, at right angles to the shore. This practice meant that the names have no constant orientation but follow one another in a neat unbroken sequence around the entire continental coastlines. To avoid ambiguity, the names of nearby islands run in the opposite direction from names on the mainland. On the basis of north orientation, the west coasts of Italy and Dalmatia, for example, are the "right way" around, whereas Italy's Adriatic coastline is "upside down." The quotation

55. That the rhumb lines terminate at the intersection points on some of the earliest charts makes these appear noticeably different from later examples whose rhumb lines continue to the edges of the chart. No particular significance should be attached to this. Most later draftsmen carried the rhumb lines through the intersection points so as to fill all the sea area with wind directions, but the particular pattern adopted in each case was probably determined as much by a desire for mathematical balance as by any navigational considerations.

56. These and other conventions are described by Bartolomeo Crescenzio in his *Nautica Mediterranea* (Rome, 1602). Nordenskiöld, *Periplus*, 18 (note 14), gives a transcription; an English translation appears in Peter T. Pelham, "The Portolan Charts: Their Construction and Use in the Light of Contemporary Techniques of Marine Survey and Navigation" (master's thesis, Victoria University of Manchester, 1980), 8–9. Silvanus P. Thompson, "The Rose of the Winds: The Origin and Development of the Compass-Card," *Proceedings of the British Academy* 6 (1913–14): 179–209, cited a number of works, published between 1561 and 1671, that specified the rhumb line colors to be used (p. 197).

57. Hans-Christian Freiesleben, "Map of the World or Sea Chart? The Catalan Mappamundi of 1375," *Navigation: Journal of the Institute of Navigation* 26 (1979): 85–89, esp. 87. See also Nordenskiöld, *Periplus*, 24 (note 14).

58. James E. Kelley, Jr., "The Oldest Portolan Chart in the New World," *Terrae Incognitae: Annals of the Society for the History of Discoveries* 9 (1977): 22–48, esp. 32. The Carte Pisane is the exception to this, with ten subdivisions, each worth five *miglia*. A note on the Carignano map explains how the scale worked; for variant transcriptions of its wording, see Bacchisio R. Motzo, "Note di cartografia nautica medioevale," *Studi Sardi* 19 (1964–65): 349–63, esp. 357–58.

59. Pagani, *Vesconte*, 14 n. 32 (note 47); see also Kelley, "Oldest Portolan Chart," 36–39, 46–48 (note 58), and below, p. 388–89.

60. Magnaghi, "Nautiche, carte," 324b (note 4). The same kinds of features that are emphasized in this way on the charts serve as the intended destinations for the direct voyages described in the mid-thirteenth-century *Lo compasso da navigare*; see Massimo Quaini, "Catalogna e Liguria nella cartografia nautica e nei portolani medievali," in *Atti del 1° Congresso Storico Liguria-Catalogna: Ventimiglia-Bordighera-Albenga-Finale-Genova, 14–19 ottobre 1969* (Bordighera: Istituto Internazionale di Studi Liguri, 1974), 549–71, esp. 558–59.

61. Avelino Teixeira da Mota detected a new style of realism in the coastal drawing of the Portuguese chart in Modena, Biblioteca Estense e Universitaria, C.G.A. 5c, which he dated to between 1471 and 1485; see "Influence de la cartographie portugaise sur la cartographie européenne à l'époque des découvertes," in *Les aspects internationaux de la découverte océanique aux XVe et XVIe siècles: Actes du Vème Colloque Internationale d'Histoire Maritime*, ed. Michel Mollat and Paul Adam (Paris: SEVPEN, 1966), 223–48, esp. 227.

marks warn against twentieth-century attitudes. Intended to be rotated, portolan charts have no top or bottom. It is only when there are nonhydrographic details designed to be viewed from one particular direction that we can ascribe any definite orientation to the chart concerned. Examples would be the corner portraits of saints on some of Vesconte's atlases (both those of 1318, the undated Lyons atlas, and the 1321 Perrino Vesconte atlas in Zurich), which establish them as oriented to the south, and the smaller of the two signatures on his chart of 1311. A similar conclusion can be drawn from the majuscules denoting the continents on the Carignano map (in contrast to the majuscules and notes on the Dalorto and Dulcert charts, which are conveniently arranged to face the nearest outside edge). For most of the early charts, however—and this includes the Carte Pisane—there is no way of telling which, if any, of the four main directions they were primarily intended to be viewed from. Nor can it readily be determined which is the front of an atlas, and hence which way its charts are oriented.[62]

Further indication that the portolan charts belong to one self-conscious family comes from a number of consistent color conventions (plate 24). The three different inks habitually used for the rhumb line network have already been mentioned. To those can be added the use of red to pick out the more significant places. These are not necessarily ports, as has often been assumed:[63] for example, among the red names are found cities like Bilbao, Pisa, and Rome, which had their own named outlets (fig. 19.2).

Islands would often be picked out in different colors to distinguish them from one another and from the adjacent mainland, and important river deltas (particularly the Rhône, Danube, and Nile) tended to be treated in the same way.[64] This attractive device also served the more practical function of emphasis. A few islands were singled out for special treatment. Lanzarote in the Canaries was covered with a red cross, possibly on a silver ground, from the time of its first appearance on Angelino Dulcert's chart of 1339. Although Dulcert attributed its discovery to the Genoese Lanzarotto Malocello, and despite the red cross of Saint George, it appears that Genoa never laid claim to the island.[65] Khios, too, was occasionally overlaid with the Genoese cross of Saint George. The earliest instances of this are on two charts produced in Majorca by Valseca in 1439 and 1447 and on three undated, or controversially dated, atlases that probably belong to the first half of the fifteenth century.[66] Rhodes, home of the Knights Hospitalers from 1309 onward, was often identified by a white or silver cross on a red ground.[67] Despite the fact that the Knights were forced to leave Rhodes in 1523, this custom was continued long afterward. It was later applied to their new home, Malta, as well.

Other conventions found on the more ornate Catalan-style charts will be referred to later. They share with Italian work, however, a consistent approach to navigational symbols. A cross or a series of black dots meant rocks, while red dots indicated sandy shallows.[68] These oblique references provide the only information on the depth of water.[69]

It must be admitted that this description of a typical portolan chart falls short of a watertight definition; it is in a sense a list of superficial characteristics. What links

62. Since, for religious reasons, medieval *mappaemundi* were usually oriented to the east, there was no well-established tradition in 1300 that north should be at the top. Many later maps, Fra Mauro's of 1459 and Erhard Etzlaub's of 1500, for example, were oriented to the south, as were the sheets in Bianco's 1436 atlas and a chart in Rome, Biblioteca Apostolica Vaticana, Borgiano V. The separate sheets in a portolan atlas would sometimes be oriented to different points of the compass so as better to accommodate the shapes involved. On this see also Cornelio Desimoni, "Elenco di carte ed atlanti nautici di autore genovese oppure in Genova fatti o conservati," *Giornale Ligustico* 2 (1875): 47–285, esp. 283–85.

63. For example, Nordenskiöld, *Periplus*, 18 (note 14); Crone, *Maps and Their Makers*, 12 (note 11); Derek Howse and Michael Sanderson, *The Sea Chart* (Newton Abbot: David and Charles, 1973), 19.

64. On the islands see Magnaghi, "Nautiche, carte," 324b (note 4). Georges Pasch pointed out that the islands around Aigues-Mortes (Rhône delta) were habitually colored yellow and blue; see his "Drapeau des Canariens: Témoignage des portulans," *Vexillologia: Bulletin de l'Association Française d'Etudes Internationales de Vexillologie* 3, no. 2 (1973): 51.

65. Cortesão and Teixeira da Mota, *Portugaliae monumenta cartographica*, 1:xxix (note 29). Aguiar also placed a cross over Flores (Azores) in 1492.

66. London, British Library, Add. MS. 19510 ("Pinelli-Walckenaer atlas"); Lyons, Bibliothèque Municipale, MS. 179; Venice, Biblioteca Nazionale Marciana, It. VI, 213 ("Combitis atlas"). On their dating see table 19.3, pp. 416–20. The convention continued until at least the end of the sixteenth century, although the Genoese were expelled in 1566 after having controlled Khios for two centuries.

67. The chart acquired by Nico Israel of Amsterdam at Sotheby's in 1980 and tentatively dated 1325 gave the island an unusual green backing; Sotheby's *Catalogue of Highly Important Maps and Atlases*, 15 April 1980, Lot A; Nico Israel, Antiquarian Booksellers, *Interesting Books and Manuscripts on Various Subjects: A Selection from Our Stock . . .*, catalog 22 (Amsterdam: N. Israel, 1980), no. 1.

68. Although the Carte Pisane has many instances of the cross symbol, it is not until the 1311 Vesconte chart that the use of stippling for shoals is encountered. Magnaghi made the unconvincing suggestion that the simple isolated crosses, found from the time of the Carte Pisane onward and in deep water, were intended to indicate localized and up-to-date magnetic declination; see Magnaghi, "Nautiche, carte," 328a (note 4). Yet the cross off the southern coast of Italy on the Carte Pisane has the word *Guardate* (Beware) written twice beside it and was clearly intended for a rock. On the hydrographic symbols of early charts, see Mary G. Clawson, "Evolution of Symbols on Nautical Charts prior to 1800" (master's thesis, University of Maryland, 1979).

69. References to depths stated in *parmi* (palms) in Magnaghi, "Nautiche, carte," 325a (note 4), seem to apply more properly to the sixteenth century or later. On early soundings, see Marcel Destombes, "Les plus anciens sondages portés sur les cartes nautiques aux XVIe et XVIIe siècles," *Bulletin de l'Institut Océanographique*, special no. 2 (1968): 199–222.

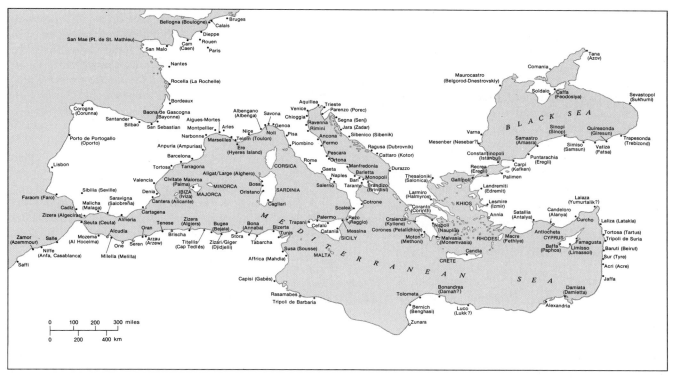

FIG. 19.2. MAJOR PLACE-NAMES ON MEDIEVAL POR-
TOLAN CHARTS. This map shows place-names that were
habitually picked out in red on fourteenth- and fifteenth-century portolan charts as being of greater importance. Modern
equivalents are given in parentheses, and questionable locations are indicated with open circles.

the charts is imitation; yet, as will be demonstrated later,
this continuous copying failed to curb a constant and
wide-ranging development. Thus the charts could
change in certain essential respects by barely perceptible
stages. The addition of latitude scales in the sixteenth
century does not necessarily mark the advent of a new
type of chart. Indeed, drafts that have strong claims to
be termed portolan charts were still being produced
throughout the seventeenth century.[70]

An additional complication concerns the overlap of
the charts and a number of contemporary world maps,
but though the latter's authors frequently incorporated
the portolan chart outlines, the scale was rarely sufficient
for more than a sprinkling of names.[71] These maps
would have lacked any possible navigational application. So too would the simplified, distorted extracts from
the portolan charts that illustrated the margins of fifteenth-century manuscripts of Leonardo Dati's *La
sfera*.[72] Of more questionable status are those manuscripts that borrowed the portolan chart outlines and
place-names but not their rhumb lines. Examples can be
found in the fifteenth-century island books (*isolarii*) of
Cristoforo Buondelmonti and Henricus Martellus Ger-

70. See the comment by the French pilot Dechales in 1677 that
charts without latitude graduations were still in use in the Mediterranean, quoted by Avelino Teixeira da Mota, "L'art de naviguer en
Méditerranée du XIIIe au XVIIe siècle et la création de la navigation
astronomique dans les océans," in *Le navire et l'économie maritime
du Moyen-Age au XVIIIe siècle principalement en Méditerranée: Travaux du IIème Colloque Internationale d'Histoire Maritime*, ed. Michel
Mollat (Paris: SEVPEN, 1958), 127–54, esp. 139.

71. For instance, the Catalan world map at Modena, the Genoese
world map in Florence, and the acknowledged works of Giovanni
Leardo, Fra Mauro, Pirrus de Noha, and Albertin de Virga all incorporated portolan chart outlines. On these see Destombes, *Mappemondes* (note 10 and above, chap. 18). Opicinus de Canistris (1296
to ca. 1350), a Pavian who worked at the papal court in Avignon,
drew a series of imaginative maps, while acknowledging in a text
written between 1334 and 1338 his use of nautical charts; see Roberto
Almagià, "Intorno alla più antica cartografia nautica catalana," *Bollettino della Reale Società Geografica Italiana*, 7th ser., 10 (1945):
20–27, esp. 23–25; and Motzo, "Cartografia nautica medioevale,"
349–59 (note 58).

72. Almagià, *Vaticana*, 1:128–29 (note 35). Almagià also disposes
of Nordenskiöld's belief that the Dati designs were the direct descendants of the detailed skipper charts, to which the latter attributed the
origin of the portolan charts; see Nordenskiöld, *Periplus*, 45 (note
14); idem, "Dei disegni marginali negli antichi manoscritti della *Sfera*
del Dati," *Bibliofilia* 3 (1901–2): 49–55.

manus (plate 25).[73] These have sometimes been treated as nautical charts, even though they clearly had no navigational function.[74] The same may apply to "charts" reported in Ptolemaic manuscripts.[75]

A definition that insisted on at least potential marine use would also exclude the Giovanni da Carignano map (Florence, Archivio di Stato, CN 2, destroyed in 1943). Produced at some point in the early fourteenth century, this has been seen by a number of commentators as the most important portolan chart after the Carte Pisane. Despite this, the Carignano map's few place-names are mostly written in the sea and in the opposite direction from those on all other surviving charts. It has, for example, almost exactly half the Carte Pisane's total for Italy. Islands and coastal features thus become confused, and its priestly author can hardly have had sailors in mind.[76]

THE ORIGIN AND COMPILATION OF THE PORTOLAN CHARTS

Among the research problems connected with the portolan charts, the question of their origin is perhaps the most intractable.[77] Although a number of the conflicting theories have had their committed champions, the skeptics are probably in the majority, particularly among modern writers. The title of a very recent pronouncement, "The Still Undiscovered Origin of the Portolan Charts," is a case in point.[78] Despite the thousands of scholarly words expended on the subject, most of the hypotheses about portolan chart origins have remained just that. In the absence of corroborating data they often appear to be less explanations than creation myths. Cortesão's comment on portolan chart origins, made fifteen years ago, that "no satisfactory solution has yet been reached," remains a valid judgment.[79] Instead of simply endorsing any single existing theory, however venerable, it seems preferable to summarize briefly the principal lines of earlier arguments. Theories of ancient and medieval origin will be contrasted, and the supposed involvement of the magnetic compass in the charts' compilation will be reviewed, as will other related issues: the nature of any discernible projection, the various ways the initial regional charts might have been constructed, and the portolan charts' most likely place of origin.

The earliest reliably documented references to the portolan charts date from the late thirteenth century, the first of them to 1270 (see below, p. 439). Regardless of the fact that this date almost coincides with that often assigned to the oldest surviving chart, the Carte Pisane, many attempts have been made to justify an older beginning.[80] Cortesão, for example, proposed an early thirteenth-century date, and Richard Oldham was for pressing still further back to the twelfth or even the eleventh

century.[81] For all their differences of detail, however, these theories remained compatible with the idea of a medieval invention. On the other hand, a sizable body of scholarly opinion over the past century or so has speculated instead that the portolan charts were the resurrected masterpieces of the ancient world.

ANCIENT ORIGIN

Even among what might be termed the "ancient" rather than the "medievalist" school, there has been great divergence of opinion. Most extreme, in terms of both age and plausibility, is Hapgood's contention that the inception of the portolan charts should be traced back to Neolithic times.[82] Less controversial, but still little sup-

73. Aegean and Black Sea sheets are reproduced in *The Netherlands—Bulgaria: Traces of Relations through the Centuries—Material from Dutch Archives and Libraries on Bulgarian History and on Dutch Contacts with Bulgaria*, ed. P. Kolev et al. (Sofia: State Publishing House "Septemvri," 1981), pls. 4 and 5. An Aegean sheet is illustrated in Pietro Frabetti, *Carte nautiche italiane dal XIV al XVII secolo conservate in Emilia-Romagna* (Florence: Leo S. Olschki, 1978), pl. VII. For a discussion on Buondelmonti's *isolario*, the earliest known, see below, chap. 20, pp. 482–84.

74. They were treated as nautical charts, for example, by Frabetti, *Carte nautiche italiane*, 33 (note 73).

75. For example, Paris, Bibliothèque Nationale, Département des Manuscrits, MS. Lat. 4801, and Rome, Biblioteca Angelica, MS. 2384.

76. This is somewhat surprising considering that Carignano was rector of a church on the waterfront (San Marco al Molo) and that in 1314 he fell foul of his archbishop for storing sails and other nautical paraphernalia in and around the church and in the clergy house; see Arturo Ferretto, "Giovanni Mauro di Carignano Rettore di S. Marco, cartografo e scrittore (1291–1329)," *Atti della Società Ligure di Storia Patria* 52 (1924): 33–52, esp. 43. Arthur R. Hinks pointed out that the Carignano map's color conventions are also atypical; see his *Portolan Chart of Angellino de Dalorto 1325 in the Collection of Prince Corsini at Florence, with a Note on the Surviving Charts and Atlases of the Fourteenth Century* (London: Royal Geographical Society, 1929), 8. Nevertheless, it would be unnecessarily pedantic to omit the Carignano map altogether from a history of the portolan charts.

77. Though we do not necessarily have to be as pessimistic about the chances of solving it as is Youssouf Kamal, *Hallucinations scientifiques (les portulans)* (Leiden: E. J. Brill, 1937), 2.

78. Freiesleben, "Still Undiscovered Origin," 124–29 (note 39).

79. Cortesão, *History of Portuguese Cartography*, 1:223 (note 3).

80. There is no validity for the 1260–69 date suggested for the chart-inspired Brunetto Latini world map, see above, p. 325 n. 200.

81. Cortesão, *History of Portuguese Cartography*, 1:229 (note 3). Richard D. Oldham, "The Portolan Maps of the Rhône Delta: A Contribution to the History of the Sea Charts of the Middle Ages," *Geographical Journal* 65 (1925): 403–28. Another who supported an eleventh-century date for the portolan chart origin was George Sarton, *Introduction to the History of Science*, 3 vols. (Baltimore: Williams and Wilkins, 1927–48), vol. 2, *From Rabbi ben Ezra to Roger Bacon*, 1047, and he added a strange suggestion of possible Scandinavian origin—on the basis of the hardly relevant Adam of Bremen periplus.

82. Charles H. Hapgood, *Maps of the Ancient Sea Kings: Evidence of Advanced Civilization in the Ice Age*, rev. ed. (New York: E. P. Dutton, 1979).

ported, have been Cortesão's further suggestions that the Phoenicians or Egyptians were responsible for developing the charts, notwithstanding the conflict with his support for a medieval origin.[83]

It is, however, to the worlds of ancient Greece and Rome that we have most often been directed in the search for a solution to this mystery. Strabo, Agathemerus, and Pliny have all been cited as sources for the contention that sea charts were used in ancient times,[84] and one writer has detected traces of the work of Eratosthenes in the medieval charts.[85] Yet the name most frequently mentioned is Marinus of Tyre, known to us through the writings of his near contemporary, Ptolemy. It was Marinus who introduced projections into mapmaking about A.D. 100; according to some accounts of a disputed text in the *Geography*, Marinus's work has been interpreted as a sea chart.[86] On this single reference hangs the repeated assertion that the medieval charts were little more than revivals of his work.[87] Laguarda Trías made the specific claim that the supposedly fifteenth-century chart of the Mediterranean in Istanbul (Topkapi Sarayi, Deissmann 47) was nothing less than a reproduction of the lost Marinus chart.[88] If, as he suggests, the rhumb line system replaced the original square-grid network, the former, being therefore astronomically determined, would point to true, not magnetic north; but there are weighty arguments for considering the portolan charts to be compass inspired (as discussed below, pp. 384–85). Nor is there any justification for Laguarda Trías's further claim that the atypical and not especially early Istanbul sheet reflects the appearance of the prototype chart.[89]

Attempts have been made to bridge the gap of more than a thousand years between Ptolemy's comment and the medieval charts, but with little conviction, since the tenth-century Arab reference to Marinus was apparently concerned with a world map and not a sea chart.[90] Nor is it easy to assign to the Arabs the role of intermediaries between the ancient and medieval worlds in this context.[91] The few early Arab charts that survive are lacking in originality, and there are many points of dissimilarity between the best Arab work, that of al-Idrīsī, and the earliest Western charts.[92] Nor is any influence traceable to the imprecisely described Indian Ocean charts, of the type shown to Marco Polo at the end of the thirteenth century.[93]

A theory of Roman origin has, however, recently been revived by Georges Grosjean.[94] His contention is not that the Romans produced sea charts as such but that a dependable scaled map of the Mediterranean would have been the indirect result of Roman centuriation. This hypothesis has two major weaknesses besides the absence of irrefutable evidence. First, current archaeological findings indicate that no more than sections of the

Roman Empire were centuriated,[95] and second, even Grosjean admitted there was virtually no trace of Roman influence in the portolan chart toponymy.[96]

MEDIEVAL ORIGIN

Passing briefly over two further suggestions—that a lost map used by the Ravenna cosmographer (soon after A.D. 700) might have supplied the missing link,[97] and that the credit for inventing the portolan charts be accorded to the Byzantines shortly after the year 1000[98]—the argument moves on to the medieval period. The "medi-

83. Cortesão, *History of Portuguese Cartography*, 1:223 and 229 (note 3).

84. Richard Uhden, "Die antiken Grundlagen der mittelalterlichen Seekarten," *Imago Mundi* 1 (1935): 1–19, esp. 2–4.

85. Rolando A. Laguarda Trías, *Estudios de cartología* (Madrid, 1981), 29–41.

86. The word *pinax* used by Ptolemy in connection with Marinus and the mapmakers who followed him (*Geography* 1.17.1) simply means "map" rather than "chart." I owe this comment to Professor O. A. W. Dilke.

87. Repeated, for example, by Nordenskiöld, *Periplus*, 48 (note 14); see also Laguarda Trías, *Estudios de cartología*, 22–28 (note 85). For the opposing view, denying discernible links between the ancient periploi and medieval portolan charts, see O. A. W. Dilke, *Greek and Roman Maps* (London: Thames and Hudson, 1985), 143.

88. Laguarda Trías, *Estudios de cartología*, 24 (note 85).

89. Laguarda Trías, *Estudios de cartología*, 24–25 (note 85).

90. On the reference to Marinus see Manuel Francisco de Barros e Sousa, Viscount of Santarém, *Essai sur l'histoire de la cosmographie et de la cartographie pendant le Moyen-Age et sur les progrès de la géographie après les grandes découvertes du XVe siècle*, 3 vols. (Paris: Maulde et Renou, 1849–52), 1:337.

91. Cortesão, *History of Portuguese Cartography*, 1:224 (note 3).

92. Kamal, *Hallucinations*, 15–16 (note 77).

93. Marco Polo, *The Travels of Marco Polo*, trans. Ronald Latham (London: Folio Society, 1968; reprinted Penguin Books, 1972), 240, 259, 303. Even more speculative is the unsubstantiated hypothesis that the portolan chart developed in South China in the twelfth century through Japanese intermediacy, subsequently reaching Europe via Persia; see an editorial note in *Imago Mundi* 12 (1955): 160.

94. Georges Grosjean, ed., *The Catalan Atlas of the Year 1375* (Dietikon-Zurich: Urs Graf, 1978), 17–18 (also an edition in German). A similar thesis had been proposed earlier in a work not mentioned in Grosjean's bibliography: Attilio Mori, "Osservazioni sulla cartografia romana in relazione colla cartografia tolemaica e colle carte nautiche medioevali," in *Atti del III Congresso Nazionale di Studi Romani*, 5 vols. (Bologna: Cappelli, 1934), 1:565–75. This was discussed in the Monthly Record section of the *Geographical Journal* 87 (1936): 90–91.

95. O. A. W. Dilke, *The Roman Land Surveyors: An Introduction to the Agrimensores* (Newton Abbot: David and Charles, 1971), 134–58.

96. Grosjean, *Catalan Atlas*, 18 (note 94).

97. Giovanni Marinelli, "Venezia nella storia della geografia cartografica ed esploratrice," *Atti del Reale Istituto Veneto di Scienze, Lettere ed Arti*, 6th ser., 7 (1888–89): 933–1000, esp. 946–47; Uhden, "Die antiken Grundlagen," 10–12 (note 84).

98. Matteo Fiorini, *Le projezioni delle carte geografiche* (Bologna: Zanichelli, 1881), 648.

evalist" school is generally agreed that the portolan charts originated in the period leading up to their first documented existence in the late thirteenth century. Notwithstanding this broad consensus on the "when" of the problem, there have been widely differing answers to the "how" and "where" components of the same question. It is convenient, first of all, to divide this further group of conflicting opinions into two sections: those positing a single master copy and those offering theories of gradual or collaborative origin.

Among single-origin hypotheses, undoubtedly the most intriguing is Destombes's hint that the Knights Templars might have been involved.[99] The members of this powerful order would certainly have had wide experience of the Near East before it was suppressed in 1312. Although the similarity between their red cross and that used to indicate east on Angelino de Dalorto's chart of 1325/30 has been mentioned, this suggestion, like so many others, must languish for want of evidence.

Attempts have even been made to identify by name the supposed originator of these early sea charts. Noting that the Genoese admiral Benedetto Zaccaria had, from 1261 onward, served under different masters throughout the Mediterranean and Black seas, in commissions that ranged as far north as Scotland and France, de La Roncière wondered if he might have been the person responsible.[100] This theory, however, assumes but does not demonstrate the vital step from navigational experience to hydrographic innovation. Nor does de La Roncière's claim that Zaccaria should be credited with improvements to the Atlantic toponymy between the time of the Carte Pisane and that of Vesconte provide proof of the admiral's hydrographic abilities.[101]

Another suggestion was made by Nordenskiöld, that Ramón Lull was "if not the author at least the guiding spirit in the compilation of this master-piece" (the prototype chart).[102] This hypothesis flowed naturally from Nordenskiöld's conviction that the portolan charts had a Majorcan origin; yet this cannot be substantiated. On the other hand, the claim made by Motzo in 1947 that he had identified, in general terms, the author of a single prototype chart has attracted favorable comment.[103] In his commentary on the mid-thirteenth-century *Lo compasso da navigare*, the oldest systematic portolano, or collection of sailing directions, that survives for the Mediterranean,[104] Motzo concluded that *Lo compasso da navigare* and the prototype chart (not necessarily the Carte Pisane) formed part of the same work. In his opinion they were composed by the same person and based on the same data.[105] He proposed that the chart's author might be looked for in the mathematical school of Leonardo Pisano (Fibonacci) or of his pupil Campano da Novara.[106]

Examination of place-names in *Lo compasso da na-*

vigare pointed to its compilation between 1232, the date of the reestablishment of Agosta (Augusta) in Sicily (or possibly 1248, about which time a port was constructed in Aigues-Mortes), and the creation of Manfredonia in 1258.[107] Thorough comparison, however, of the toponymic lists extracted from *Lo compasso da navigare* and the Carte Pisane exposes notable discrepancies, and these make Motzo's thesis less appealing. When the Black Sea is excluded, we find that roughly 40 percent of the Carte Pisane's continental names are not to be found in *Lo compasso da navigare*.[108] This is unexpected, since *Lo compasso da navigare*'s author was free from the space restrictions imposed on the compiler of the Carte Pisane and frequently goes into far greater detail than was possible on a chart.

99. Marcel Destombes, "Cartes catalanes du XIVᵉ siècle," in *Rapport de la Commission pour la Bibliographie des Cartes Anciennes*, 2 vols., International Geographical Union (Paris: Publié avec le concours financier de l'UNESCO, 1952), vol. 1, *Rapport au XVIIᵉ Congrès International, Washington, 1952 par R. Almagià: Contributions pour un catalogue des cartes manuscrites, 1200–1500*, ed. Marcel Destombes, 38–63, esp. 38–39.

100. Charles de La Roncière, *La découverte de l'Afrique au Moyen Age: Cartographes et explorateurs*, Mémoires de la Société Royale de Géographie d'Egypte, vols. 5, 6, 13 (Cairo: Institut Français d'Archéologie Orientale, 1924–27), 1:40. However, Roberto Lopez in a special study on Zaccaria was unable to find any actual evidence in support of de La Roncière's theory; see Roberto Lopez, *Genova marinara nel duecento: Benedetto Zaccaria ammiraglio e mercante* (Messina-Milan: Principato, 1933), 202–3, 212 n. 106.

101. De La Roncière, *Afrique*, 1:41–42 (note 100). See also Pagani, *Vesconte*, 17 (note 47). The improved toponymy found on Vesconte's earliest charts affects all parts of the Mediterranean, not just the French coasts about which Zaccaria supposedly had special knowledge.

102. Nordenskiöld, *Periplus*, 34 (note 14).

103. Bacchisio R. Motzo, "Il Compasso da navigare, opera italiana della metà del secolo XIII," *Annali della Facoltà di Lettere e Filosofia della Università di Cagliari* 8 (1947): I–137.

104. An earlier fragment, covering a journey between Acre and Venice, survives in Venice, Biblioteca Nazionale Marciana, It. XI, 87. Kretschmer, *Die italienischen Portolane*, 200 (note 48), thought it belonged to the thirteenth century; see pp. 235–37 for his transcription.

105. Motzo, "Compasso da navigare," XLVIII (note 103).

106. Motzo, "Compasso da navigare," LI–LIV (note 103).

107. Motzo, "Compasso da navigare," XXVII, XXX (note 103). See elsewhere for other estimates of *Lo compasso*'s date: between about 1250 and 1265 (p. v) and about 1245 and 1255 (p. XLVIII). When Louis IX acquired Aigues-Mortes, the area was uninhabited and there was no artificial harbor. However, the natural refuge Louis developed was already referred to as a port in a document of 1226; hence the mention of Aigues-Mortes in *Lo compasso* does not necessarily provide a terminus post quem of 1248; see Jules Papezy, *Mémoires sur le port d'Aiguesmortes* (Paris: Hachette, 1879), 36, 84–90.

108. The Black Sea is considered to have been added later to *Lo compasso*, though certainly before 1296, and the area is largely obliterated on the Carte Pisane.

Lo compasso da navigare's strange lacunae, involving whole stretches of coastline—Manfredonia to Fermo, Venice to Trieste (and with scattered exceptions, to Durazzo), Valona to Moton, Landrimiti to Bodrum, and the Libyan coast between Tolometa and Tripoli—provide only a partial explanation for the disagreement between the two works. Even if *Lo compasso da navigare*'s omitted coastlines are excluded from the comparison, on the grounds that these sections might have been present in the original manuscript and then been lost during subsequent copying, some 30 percent of the Carte Pisane's names still derive from a source independent of *Lo compasso da navigare*. Given the importance of Genoa at the time, it is also surprising that *Lo compasso da navigare* should omit the ports between Savona and Genoa, which were invariably named on the charts. Among the noteworthy individual omissions on *Lo compasso da navigare* are Arles, Amalfi, Rimini, and Sousse (Tunisia). All these are picked out in red on the Carte Pisane. Simonetta Conti came to similar conclusions through analysis of the coastal names between Spain and Venice.[109] She also contrasted the pure Italian of *Lo compasso da navigare* with the Carte Pisane's varied dialects. In short, it is hard not to conclude that, since only one generation, or at most two, divides the two works, their dissimilarities point to separate origins rather than to progressive stages of a single development.

It has often been assumed that charts were designed to accompany a written portolano, or even that a chart was derived from a series of sailing directions. The acid test, of course, is to construct a chart using nothing more than *Lo compasso da navigare*. This has recently been attempted. To overcome the manuscript's shortcomings, however, occasional adjustments were allowed, and these introduced the risk that, albeit inadvertently, hindsight was leading the researcher's hand toward the familiar Mediterranean shape.[110] The creator of the first chart would have had no mental map against which to test the portolano's deficiencies, and he would unknowingly have introduced inevitable errors. These would have led to cumulative distortion as he worked around the coast. Given further that *Lo compasso da navigare* does no more than relate one place to the next by bare statements of distance[111] and direction, the predictably angular and simplified outline that resulted from this attempted reconstruction is noticeably dissimilar to the sophisticated coastal patterns found on the earliest surviving charts.[112]

Any skepticism about *Lo compasso da navigare*'s potential in this respect must be far greater in face of the claims made for earlier sailing directions. Though properly belonging with the previous comments on the ancient world, they involve an extrapolation similar to that made by Motzo and are more conveniently considered

here. The earliest of these periploi, as the ancient sailing directions were termed, was a Mediterranean pilot-book, allegedly compiled by Scylax of Caryanda, an admiral of the late sixth century B.C. but actually written in the fourth century B.C.. This, it has been argued, "possibly served as explanatory text to a map or chart of the Mediterranean and the Black Sea, which, however, is no longer extant, and is not expressly referred to in the text."[113] The existence of this hypothetical chart was assumed by some scholars, even though no wind directions were supplied and most distances were loosely conveyed in terms of the sailing days required.[114] Another Greek periplus, the *Stadiasmus*, or measurement in stades of the great sea (Mediterranean), was compiled about the third or fourth century A.D.[115] The same unsubstantiated claim for an accompanying map was repeated.[116]

From such theories of a single or portolano-inspired beginning, we can move to the last main group of claimants: arguments that see the earliest portolan chart as a piecemeal creation. A belief in multiple origins unites many scholars of past and present. The details, however, are disputed. Nordenskiöld saw the prototype chart of the Mediterranean and Black seas (his "normal-portolano") as cobbled together from a number of separately compiled "sketch-maps and reports by skippers."[117] Unfortunately, he went on to identify the fifteenth-century Dati illustrations as revivals of these same skipper

109. Simonetta Conti, "Portolano e carta nautica: Confronto toponomastico," in *Imago et mensura mundi: Atti del IX Congresso Internazionale di Storia della Cartografia*, 2 vols., ed. Carla Clivio Marzoli (Rome: Enciclopedia Italiana, 1985), 1:55–60.

110. The exercise was discussed theoretically by Eva G. R. Taylor, "Early Charts and the Origin of the Compass Rose," *Navigation: Journal of the Institute of Navigation* (now *Journal of Navigation*) 4 (1951): 351–56, esp. 355. The actual attempt was made by Jonathan T. Lanman, "On the Origin of Portolan Charts" (paper prepared for the Eleventh International Conference on the History of Cartography, Ottawa, 1985). Lanman noted, for example, that twelve of the instructions omitted direction and two left out distance; he also identified gaps at the straits of Otranto and Gibraltar.

111. The distances were rounded off, normally to the nearest ten miles, and usually understated; see Taylor, "Early Charts," 355 (note 110).

112. Greece and Asia Minor are good examples of this; see Lanman, "Origin of Portolan Charts," figs. 3a and 3b (note 110). Expressing a contrary view that inverted the supposed progression from portolano to portolan charts, Kelley, "Oldest Portolan Chart," 47 (note 58), thought that some of the information in the pilot books was derived from the charts, not vice versa.

113. Nordenskiöld, *Periplus*, 5 (note 14). See also Dilke, *Greek and Roman Maps* 130–44, esp. 133–37 (note 87).

114. For example, Lloyd A. Brown, *The Story of Maps* (Boston: Little, Brown, 1949; reprinted New York: Dover, 1979), 120.

115. Nordenskiöld, *Periplus*, 10–14 (note 14).

116. For example, Brown, *Story of Maps*, 120 (note 114).

117. Nordenskiöld, *Periplus*, 45 (note 14).

charts—an assertion that has been roundly refuted.[118] On a broader geographical front than that involved in the small Dati sections, Nordenskiöld noted regional discrepancies. It seemed to him "as though a map of the East Mediterranean and one of the West had been joined to coast-maps of the Black Sea, England, the countries around Gibraltar etc."[119] Working from the more reliable evidence of scale variation, rather than differences in rhumb line grids (or "loxodrome nets") to which Nordenskiöld called attention, Kelley amended the areas concerned and demonstrated the apparently independent origin of the Atlantic, Mediterranean, and Black Sea surveys by reference to their conflicting scales.[120]

THE COMPASS AND MAGNETIC VARIATION

Central to much of the argument about a medieval origin has been the part that the compass played, or did not play, in this process. Its involvement has been both assumed and denied at a theoretical level and supposedly detected in cartometric analysis. A brief résumé of the controversy surrounding the magnetic compass is also an essential preface to any discussion of the related issue of magnetic variation.

The mariner's compass is considered to have come into use in the Mediterranean in the thirteenth century, but a simplified lodestone, consisting of a magnetized needle pushed through a floating piece of wood, can be traced back to the preceding century.[121] Though Amalfi's claim to have invented the compass by the beginning of the twelfth century rests on slender evidence, the writings of the English monk Alexander Neckham provide confirmation of its existence by the last two decades of that century.[122] Besides Neckham's account, several other references from the period leading up to that of the earliest surviving charts show that the use of the magnetic compass at sea was unexceptional. One commentator, Jacques de Vitry, a bishop of Acre, even wrote in 1218 of its necessity for navigation.[123] The rhumb line network found on the charts enabled them to be used with a compass, but it remains a matter of debate whether the instrument was actually concerned in the initial construction of the portolan charts.[124] There is, however, some measure of agreement that the compass card, divided into sixteen points or multiples thereof, came into being about the year 1300.[125] It has even been suggested that this compass card imitated the system on the charts instead of inspiring it.[126]

In support of the contention that the early charts were "compass charts," some scholars have claimed that their overall distortion is consistent with the magnetic variation supposedly in force at the time. This phenomenon seems not to have been appreciated until the fifteenth century,[127] and any bearings taken during the charts'

initial compilation would certainly have been from magnetic north.[128] Unfortunately, there is no agreement either about the extent of distortion on the earliest charts or about the degree of magnetic variation in the later thirteenth century. Working from the charts' own repeated north-south rhumb line, several commentators have detected an easterly variation on the charts, but estimates of its extent have ranged from four to eleven degrees.[129]

Reconstructing the true medieval variation is fraught with difficulties, although recent calculations have been made of historical variation of declination, for example, at Mount Etna.[130] If further research could identify with more certainty the degree of variation—on the one hand that embodied in the charts, on the other that in force

118. See note 72.

119. Nordenskiöld, *Periplus*, 56 (note 14).

120. Kelley, "Oldest Portolan Chart," 48 (note 58). See Magnaghi, "Nautiche, carte," 326a (note 4), who points out that the fifteenth-century Pietro de Versi portolano gives Atlantic distances in leagues and Mediterranean figures in *miglia*; and also Laguarda Trías, *Estudios de cartología*, 3–11 (note 85).

121. Joseph Needham, *Science and Civilisation in China* (Cambridge: Cambridge University Press, 1954–), vol. 4, *Physics and Physical Technology (Part 1: Physics)*, 245–49.

122. Taylor, *Haven-Finding Art*, 92, 95 (note 7).

123. On de Vitry see Taylor, *Haven-Finding Art*, 94–95 (note 7).

124. Among those who considered the compass to be relevant to the charts' construction were Fischer, *Sammlung*, 56 (note 36); Cortesão, *History of Portuguese Cartography*, 1:230 (note 3); and Crone, *Maps and Their Makers*, 16 (note 11). Among those who denied the connection were Nordenskiöld, *Periplus*, 47 (note 14); Eckert, *Kartenwissenschaft*, 2:59 (note 42); Uhden, "Die antiken Grundlagen," 5 (note 84); Kamal, *Hallucinations*, 15 (note 77); and Frederic C. Lane, "The Economic Meaning of the Invention of the Compass," *American Historical Review* 68, no. 3 (1963): 605–17, esp. 615–17.

125. Kretschmer, *Die italienischen Portolane*, 74 (note 48).

126. Pelham, "Portolan Charts," 110 (note 56).

127. David W. Waters, *Science and the Techniques of Navigation in the Renaissance*, 2d ed., Maritime Monographs and Reports no. 19 (Greenwich: National Maritime Museum, 1980), 4.

128. Taylor, *Haven-Finding Art*, 102 (note 7).

129. A. Clos-Arceduc, "L'énigme des portulans: Etude sur le projection et le mode de construction des cartes à rhumbs du XIVᵉ et du XVᵉ siècle," *Bulletin du Comité des Travaux Historiques et Scientifiques: Section de Géographie* 69 (1956): 215–31, esp. 225, found a range from 4° to 9°; Hapgood, *Sea Kings*, 98 (note 82), estimated the error of the 1339 Dulcert chart at 6°; Magnaghi, "Nautiche, carte," 327b (note 4), discerned a 6°–7° variation in the Tyrrhenian Sea and 11° in the eastern Mediterranean; and Heinrich Winter, "A Late Portolan Chart at Madrid and Late Portolan Charts in General," *Imago Mundi* 7 (1950): 37–46, esp. 40, cited a 10° variation.

130. Pelham, "Portolan Charts," 84 (note 56), citing J. C. Tanguy, "An Archaeometric Study of Mt. Etna: The Magnetic Direction Recorded in Lava Flows Subsequent to the Twelfth Century," *Archaeometry* 12 (1970): 115–128; idem, "L'Etna: Etude pétrologique et paléomagnetique, implications volcanologiques" (Ph.D. diss., Université Pierre et Marie Curie, Paris, 1980). On palaeomagnetism see also Robert W. Bremner, "An Analysis of a Portolan Chart by Freduci d'Ancone" (paper prepared for the Eleventh International Conference on the History of Cartography, Ottawa, 1985).

in the Mediterranean at different dates in the thirteenth century—it might prove possible to place a firmer date on the initial compilation of the portolan charts. Viewed in this light, secular magnetic variation is potentially as valuable in the history of cartography as the radiocarbon method in archaeology, though the calibrations have yet to be worked out.[131] In addition, researchers have exploited the fact that there are differences in variation between the separate Mediterranean basins and that these differences remain in a constant relationship to one another.[132] If a full-scale cartometric analysis were to confirm that localized distortions on the earliest charts coincide with the pattern of regional magnetic variation,[133] then the part played by the compass in the compilation of the portolan charts would be definitely established.[134]

THE PROJECTION

Relevant to many of the arguments about portolan chart origins is the nature of the projection on which the charts were constructed. Here again, a controversy that has remained alive for a century and a half shows no sign of producing any single theory able to command general acceptance. Those who attributed authorship of the portolan charts to Marinus of Tyre maintained that it was his cylindrical projection that was involved, even if the lines of longitude and latitude had been discarded.[135] Fiorini, for his part, believed he could detect the equidistant azimuthal projection; and a more recent Portuguese historian, António Barbosa, asserted, without clarifying the point, that the lines of longitude and latitude were both curved.[136]

The foregoing theories presuppose the existence of an intentional projection according to which the outlines of the earliest charts were laid down. Majority opinion, though, has rejected that view, considering instead that the portolan charts were projectionless or that any projection was accidental. A projection on a terrestrial map will normally be recognized by the way the meridians and parallels have been treated. But in the absence of any evidence beyond what can be derived from the charts themselves, scholars have had to impose a grid of longitude and latitude on examples that survive; yet these charts show no awareness of astronomically determined information of that kind. If, however, a distortion grid is laid over an early chart, by linking places that are known to lie on the same meridian or parallel, this may disclose the underlying projection; but it will also reveal error in the coastal outlines, both of a localized and of a general kind. Unscrambling these two elements presents a circular problem, since the nature and extent of the distortion depend on the projection deemed to be involved.

Most commentators have concluded that both the meridians and the parallels inherent in the early charts were straight lines. Scholars differed, though, on whether the longitude and latitude grid created a pattern of squares or rectangles. Those who favored the former held that the portolan charts were akin to plane charts, although, as a conscious projection, the plane chart (variously termed *plate carrée* and *carta plana quadrada*) is seen as a Portuguese invention of the second half of the fifteenth century.[137] Seeing the charts as constructed on a basis of observed distance, with directions obtained by means of a magnetic compass, this interpretation denied that any compensation had been made for the sphericity of the earth.

The only way that the habitual network of rhumb lines on the portolan charts could be treated as genuine loxodromes (i.e., lines of constant compass bearing) would be if the charts were drawn on the Mercator projection.[138] First demonstrated to sailors on Mercator's world map of 1569, this allowed any compass course to appear as a straight line; but the widening gap between latitudes as they moved toward the poles caused distance to be increasingly exaggerated. Following the logic of their contention that the straight rhumb lines on the charts represented compass bearings, some writers have claimed that the portolan charts were drawn, albeit unconsciously, on the Mercator projection, or something closely akin to it. Nordenskiöld had already come to this conclusion by 1897, but it was Clos-Arceduc who explored the theory more fully, superimposing a modern outline drawn on the Mercator projection

---

131. Clos-Arceduc, "Enigme des portulans," 226 (note 129), supposed that the process of compiling the prototype chart would have lasted no more than twenty years and that the magnetic variation incorporated into the finished product, and not changed thereafter, would thus date, on average, from ten years before.

132. Clos-Arceduc, "Enigme des portulans," 222 (note 129).

133. As suggested by Magnaghi, "Nautiche, carte," 327b (note 4).

134. Heinrich Winter, "Scotland on the Compass Charts," *Imago Mundi* 5 (1948): 74–77, esp. 74 n. 3, found that the European Atlantic coasts did not demonstrate the effects of magnetic variation as did those of the Mediterranean.

135. Laguarda Trías, *Estudios de cartología*, 25 (note 85).

136. Fiorini, *Projezioni*, 689–96 (note 98); António Barbosa, *Novos subsídios para a história da ciência náutica portuguesa da época dos descobrimentos* (Oporto, 1948), 179 ff. I owe this reference and several other points in this discussion of projections to Luís de Albuquerque. On projections see also Johannes Keuning, "The History of Geographical Map Projections until 1600," *Imago Mundi* 12 (1955): 1–24, esp. 4 and 15–17.

137. Charles Cotter, "Early Tabular, Graphical and Instrumental Methods for Solving Problems of Plane Sailing," *Revista da Universidade de Coimbra* 26 (1978): 105–22.

138. For a definition of loxodromes see David W. Waters, *The Art of Navigation in England in Elizabethan and Early Stuart Times* (London: Hollis and Carter, 1958), 71–72.

over the coastal shapes of the earliest charts.[139] The surprisingly close match between the two convinced him that the Mediterranean and Black Sea sections of the portolan charts were drawn on the Mercator projection, while the part covering the Atlantic was a plane chart.[140] Clos-Arceduc even found that the 1339 Dulcert chart, when treated as if drawn on the Mercator projection, gave more accurate outlines for the Mediterranean than Mercator's own world map of 1569.[141]

Clos-Arceduc does not, however, attempt to explain the mechanism by which the supposed thirteenth-century compilers of the original portolan chart (or of its individual sections) managed to overcome the technical cartographic problems caused by the conflicting demands of straight-line compass directions and converging meridians. A full mathematical solution to this would have to wait to the very end of the sixteenth century. In essence, the Mercator projection hypothesis attempts to embrace two contradictory principles: on the one hand that of intentional and remarkably sophisticated cartographic manipulation, on the other that of pure observation. It is possible that what Clos-Arceduc interpreted as a system of unequal parallels reflects instead cartographic error. Certainly, the distortion grid devised for the Catalan atlas by Grosjean[142] shows no obvious signs of the Mercator projection, although the diagram's small scale and the two-degree intervals involved may have led to oversimplification. The element of distortion for the Mediterranean and Black seas on the early charts may, alternatively, be attributable to the inevitable compromise forced on their compilers when attempting to reconcile discrepancies arising from the unsuspected convergence of the meridians.

Although the Portuguese apparently developed a new type of chart about 1485, one graduated with latitudes, no surviving latitude scale has been convincingly dated before the sixteenth century.[143] There have, however, been a number of claims. Cortesão and Teixeira da Mota dated to about 1500 the unsigned Portuguese chart in Munich that displays a latitude scale. Working from their reproduction, however, it appears that the Spanish flag is placed over Oran and Bougie (Bejaia).[144] Yet these fell to Spain only in 1509 and January 1510, respectively, while Algiers, which was captured in July 1510, still has its Arab flag. Similar comments can be made about another work preserved in Munich (Universitätsbibliothek, 8° Codex MS. 185, sheets 2 and 3), which again displays the Spanish flag over Oran. The latitude scales on its first two sheets appear to have been added later, anyway, as seems to be the case with the supposedly late fifteenth-century chart in the Henry E. Huntington Library (HM 1548) and the 1403 Francesco Beccari chart at Yale.

The earliest chart with a latitude scale is probably one of those undated works assigned to the first decade of

the sixteenth century, for example, the King-Hamy chart of about 1504 (Henry E. Huntington Library, HM 45) or the Caverio chart of about 1505 (Bibliothèque Nationale, Département des Cartes et Plans, S. H. Archives no. 1). A latitude scale also occurs on the unreproduced Gap chart, but this must raise doubts about the pre-1453 date suggested by Charles de La Roncière on the unreliable grounds that Constantinople still flies the Byzantine flag.[145] It is significant that when latitude scales came to be added in the sixteenth century to charts whose Mediterranean outlines were essentially copies of earlier models, the parallels were equidistant.[146]

MEDITERRANEAN NAVIGATION BEFORE THE PORTOLAN CHART

Any attempt to assess the likelihood that a chart like the Carte Pisane might have been built up as a composite jigsaw puzzle must consider the navigational techniques of the time.[147] Unfortunately, documentation on the methods practiced in the thirteenth century is as elusive as that for the charts themselves. It is firmly established, however, that the science of astronomical navigation was introduced by the Portuguese in the fifteenth century, specifically in response to problems encountered outside the Mediterranean.[148] Hence it must be assumed that in the period leading up to the first portolan charts navigation depended almost entirely on the pilot's stored experience. Dead reckoning—the estimate of the distance and direction run—would owe little to any instrument, except perhaps to a primitive precursor of the log for gauging speed, a sandglass for reckoning time, a lodestone, and a lead and line.[149] Like other sailors at

---

139. Nordenskiöld, *Periplus*, 16–17 (note 14); Clos-Arceduc, "Enigme des portulans," 217–228 (note 129).

140. Clos-Arceduc, "Enigme des portulans," 223 (note 129).

141. Clos-Arceduc, "Enigme des portulans," 225 (note 129).

142. Grosjean, *Catalan Atlas*, 16–17 (note 94).

143. See Teixeira da Mota, "Art de naviguer," 134 (note 70).

144. See Cortesão and Teixeira da Mota, *Portugaliae monumenta cartographica*, 1:23–24, pl. 7 (note 29). Munich, Bayerische Staatsbibliothek, Codex Icon. 138/40, fol. 82.

145. See Charles de La Roncière, "Le portulan du XVᵉ siècle découvert à Gap," *Bulletin du Comité des Travaux Historiques et Scientifiques: Section de Géographie Historique et Descriptive* 26 (1911): 314–18. For the canard that Benincasa introduced latitude scales on his charts of the second half of the fifteenth century, see Alexander von Humboldt, *Examen critique de l'histoire de la géographie du nouveau continent et des progrès de l'astronomie nautique au XVᵉ et XVIᵉ siècles*, 5 vols. (Paris: Gide, 1836–39), 1:291.

146. For examples, see Nordenskiöld, *Periplus*, pls. xxviii–xxxi (note 14).

147. For a discussion of the navigational techniques relating to the use rather than origin of the charts, see below, pp. 441–44.

148. Cortesão, *History of Portuguese Cartography*, 2:221 ff. (note 3).

149. Whereas some commentators contend that an experienced seaman could have estimated distances at sea with considerable accuracy,

different times and in different places, these early thirteenth-century Mediterranean pilots presumably carried with them a mental chart of the regions they frequented. This was no doubt adequate for their purposes, just as Geoffrey Chaucer's fourteenth-century "shipman" knew all the havens from the island of Gotland to Finisterre without recourse to a chart.[150]

It must be obvious that the navigational abilities of at least some Mediterranean sailors in the mid-thirteenth century would have to have been as sophisticated as the cartographic accuracy of the earliest portolan charts— if a theory of medieval origin is accepted for them. A pilot who was unable to navigate with any confidence, say from Palma to Acre, could have made no contribution to the compilation of the earliest charts, nor would he have had much use for the finished product, at least when the ship was out of sight of land. If early thirteenth-century French statutes could prescribe the death penalty for a pilot whose negligence led to the loss of his ship,[151] a considerable degree of navigational expertise must have existed among pilots operating along France's Atlantic coasts. It is only fair to suppose that equivalent skills were available in the Mediterranean, where conditions were easier than in the Atlantic, the North Sea, or the Baltic. Many of the Mediterranean ships, for example, were galleys, and the straight courses they could pursue made it far easier to estimate position.[152] The limited tidal range in the Mediterranean further simplified matters. So too did the normally clear air in the summer months, when almost all voyages took place until the compass-inspired revolution of the late thirteenth century.[153] All these factors, combined with the frequency with which high land rises up from the coast,[154] increased the chance of a good landfall. Nor can it still be maintained that medieval ships routinely hugged the shore.[155] It is precisely inshore waters, the most likely to conceal rocks and shoals, that are the greatest navigational hazards. Moreover, just as the periplus of Scylax can be cited as evidence that the ancients regularly made direct open-sea sailings, even by night,[156] so descriptions of long passages (*peleggi*) given in *Lo compasso da navigare* indicate that such voyages were actually undertaken, and with considerable navigational accuracy.[157]

THE METHOD OF COMPILATION

It does not necessarily follow from an ability to navigate successfully around and across the Mediterranean and Black seas that there was either the inclination or the skill to record the experience in chart form. Nevertheless, a recent writer has proposed that a simple form of triangulation ("resection and intersection") could have been used to control a running survey.[158] The problem with

this interpretation is that however elementary the geometry, and however unsophisticated the "plane table, sextant or similar instrument" involved,[159] there is no evidence that any of these were available until long after the thirteenth century.

The particular nature of the Mediterranean Sea may lead to another theory—yet to be tested—about the origin of the portolan charts.[160] As Braudel has shown in detail, the physical configuration of the Mediterranean is better understood not as one continuous sea, but as a series of basins separated by peninsulas.[161] The most clear-cut division—between the western and the eastern portions—runs from Tunisia to Sicily, crossing a strategically important channel (the Strait of Sicily) only ninety miles wide, guarded by the islands of Pantelleria and Malta. The two divisions tended to become distinct geopolitical entities. Within the western portion, there are three basins: the Alboran Sea between Spain and North Africa; the Balearic Sea; and the Tyrrhenian Sea. The eastern portion consists of the Adriatic, Ionian, and Aegean seas, as well as the most easterly basin surrounded by Asia Minor, Syria, the Holy Land, and Egypt. The Black Sea—which routinely appears on the portolan charts—provides yet another self-contained body of water with its own subbasin in the Sea of Azov.

for example, Taylor, *Haven-Finding Art*, 121 (note 7), and Freiesleben, "Catalan Mappamundi," 87 (note 57), others deny this—for instance, Grosjean, *Catalan Atlas*, 17 (note 94).

150. Geoffrey Chaucer, *Prologue to the Canterbury Tales*, lines 401–9; see *The General Prologue to the Canterbury Tales*, ed. James Winny (Cambridge: Cambridge University Press, 1965).

151. David W. Waters, *The Rutters of the Sea: The Sailing Directions of Pierre Garcie—A Study of the First English and French Printed Sailing Directions* (New Haven: Yale University Press, 1967), 36–39, 122, 385. The statutes of Oléron were codified in the thirteenth and fourteenth centuries but are undoubtedly of much earlier date.

152. Pagani, *Vesconte*, 9 (note 47). Since it was known how far the galley traveled with each oar stroke, measurement of the distance run was obtained simply by counting the strokes. Bartolomeo Crescenzio described this method in 1602; see his *Nautica Mediterranea*, 245–53 (note 56).

153. Lane, "Invention of the Compass," 608 (note 124); on atmospheric conditions see Fernand Braudel, *The Mediterranean and the Mediterranean World in the Age of Philip II*, 2 vols., trans. Siân Reynolds (London: Collins, 1972–73), 1: 232–34.

154. Vincenzo Coronelli, *Specchio del mare* (Venice, 1693), chap. 3.

155. Lane, "Invention of the Compass," 607 (note 124).

156. Nordenskiöld, *Periplus*, 8 n. 3 (note 14).

157. Pagani, *Vesconte*, 14 (note 47). Several of the voyages described in *Lo compasso* were 500 *miglia* long (about 600 km), and some were as much as 700 *miglia* (roughly 900 km); the hoped-for landfall was often a cape or a small island.

158. Pelham, "Portolan Charts," 104 (note 56).

159. Pelham, "Portolan Charts," 109 (note 56).

160. The rest of this section on the traverse-trilateration theory (to p. 388) has been written by David Woodward.

161. Braudel, *Mediterranean*, 1:103–38 (note 153).

The cumulative experience of several centuries of coastal and other shipping in each of these basins could have led to the independent recording of traditionally known distances. It is probable that navigators in the Mediterranean during the late medieval and Renaissance periods used both coastal traverses and cross-basin routes. The average distances derived from these sailings between pairs of ports—both along the coast and across the sea—could then have been used in the construction of a series of separate charts of the individual basins. If these routes were plotted to form networks in each of the basins listed above, each network might have assumed the form of a self-correcting closed traverse approximating the shape of each basin. The rigidity of this structure would, however, have depended on the availability of the cross-basin distances, acting as braces to the framework. It is thus postulated that some system of empirical or stepwise graphic method of correcting these frameworks was used to achieve a "least-squares" result. These discrete compilations could then have been amalgamated into charts of the entire Mediterranean.

It must be stressed, however, that this theory does not require that modern methods or instruments of trilateration or triangulation be available in the thirteenth century. Indeed, there is no evidence that such techniques were available until the fifteenth century. The terms are simply used as an analogy for the natural structure that may have underlain the charts.

By using distances only, such a system could have worked independently of the magnetic compass. It might also have given rise to the approximately four- to eleven-degree shift in the axis of the early charts from the parallel of Rhodes. The longitudinal extent of the Mediterranean is such that the curvature of the earth must be taken into account in any accurate cartographic representation. If the charts were constructed piecemeal from different self-correcting compilations of the basins, the natural tendency would thus have been for the whole framework to be skewed farther north as the accretion continued.

Furthermore, this theory might also explain the rapid deterioration of these charts' structural accuracy in those areas where empirical information was lacking for one or more sides of the basins, as with the Bay of Biscay or the North Sea. In the course of constructing charts by this method, we would also expect to find that some of the individual basins (for example, the Black Sea) could have been tacked on at varying angles to adjacent basins while still maintaining their individual integrity or that land areas between basins could have become constricted. At first sight this appears to have been the case, but empirical studies to test these deformations on selected charts still have to be undertaken.[162] Finally, it is likely that—if this method was used—it was employed

in combination with a number of other techniques. It does not exclude the use of the compass, for example, which would simply add more directional stability, however crude, to the orientation of the individual basins of the Mediterranean Sea.

THE CHARTS' PLACE OF ORIGIN

Even assuming a thirteenth-century origin, the question remains, Which Mediterranean center should be seen as the cradle of the portolan charts? This is probably the thorniest issue of all. There are not many forces more potent than nationalism, and the upheavals of this century have done little to dampen its effect on the historiography of the subject. Few Italian writers have failed to insist on an Italian origin, and Spanish scholars have tended to react as predictably. Among those neutral by birth, Nordenskiöld and Winter were committed to the Catalan cause, while Fischer and Kretschmer, for example, supported a theory of Italian origin.[163] The possible involvement of the Portuguese was not ruled out by Kamal;[164] nevertheless, this did not lead to any serious claims for a Lusitanian invention by Cortesão or other Portuguese historians.

Despite its length and intensity, the nationalist debate has thrown up little of lasting value. The writings of the late thirteenth-century Majorcan polymath Ramón Lull are often cited in support of early Catalan chartmaking. Yet there are no indications that the sea charts to which he refers were actually of Catalan workmanship. Nor, on the Italian side, have demonstrations that the Norman kingdom of Sicily acted as a catalyst between the Arab and Christian worlds necessarily brought us any nearer to a solution of the problem.[165] That a climate sympathetic to the creation of portolan charts existed in thirteenth-century Sicily is of itself no proof that they were actually produced.

Two particular features have been harnessed for use in this argument: first, the value of the scale unit; second, the language or dialect thought to predominate in the

162. Such studies are currently the subject of a doctoral dissertation by Scott Loomer, University of Wisconsin–Madison.

163. Nordenskiöld, *Periplus*, 47 (note 14); Heinrich Winter maintained the Catalan position in his numerous writings on the subject; Fischer, *Sammlung*, 81–97 (note 36); Kretschmer, *Die italienischen Portolane*, 103–4 (note 48).

164. Youssouf Kamal, *Quelques éclaircissements épars sur mes Monumenta cartographica Africae et Aegypti* (Leiden: E. J. Brill, 1935), 188.

165. For example, Grosjean, *Catalan Atlas*, 15 (note 94). A recent article makes the claim that the origin of the portolan charts should be looked for in Sicily during the time of the emperor Frederick II of Hohenstaufen (1194–1250), but no convincing evidence is adduced; see Hans-Christian Freiesleben, "The Origin of Portolan Charts," *Journal of Navigation* 37 (1984): 194–99.

toponymy of the earliest charts. If the length of the "portolan mile" could be determined, and if it were shown to approximate a known unit of measure, this could be a useful pointer to the charts' place of origin. Nordenskiöld, for example, believed the charts were drawn according to a scale of Catalan leagues (*legua*), and this fitted conveniently into his general thesis of a Catalan invention.[166] Yet neither his explanation nor Kelley's has been successful in imposing any kind of order on the numerous different interpretations.[167] While it is known that each of the smaller scale divisions on the portolan charts represented ten *miglia* (sometimes termed *mia* or *milliaria* instead),[168] that is the extent of unanimity. Taking the value of a Catalan league as 5.83 kilometers, Nordenskiöld, having suggested that each ten-*miglia* unit contained two portolan miles, arrived at a figure of 1.16 kilometers for a single *miglio* ($5.83 \div 5$). He then proposed that this was an error resulting from an attempt by Italian draftsmen to "fit the Italian mile-measure with the portolan-scale," and the more correct value of a *miglio* was 1.457 ($5.83 \div 4$).[169] Nordenskiöld's successors have variously supported the five-*miglia* and the four-*miglia* portolan mile.[170]

With disagreement about fundamentals, and given a general rejection of the supposed relevance of the Catalan league, there has been wide scope for alternative estimates of the length of the portolan mile. Even though majority opinion has settled for an approximate value of 1.25 kilometers, the issue is far from settled. Leaving aside the significant differences of scale involved in the Mediterranean, Black Sea, and Atlantic sections of the charts (see below, p. 414), there remain discrepancies in the various calculations,[171] the significance of which may have been overlooked through averaging out. Alternatively, the variation in the figures might be interpreted as evidence of regional differences in the value of the *miglio*, leading to fluctuations in the overall scale. It is hardly possible, anyway, to arrive at a precise estimate of the value of an unstated unit of measurement on charts whose method of construction remains unknown and whose accuracy is clearly uneven. If internal inconsistency was confirmed, this would make irrelevant the task of matching the stated scale divisions to a single portolan mile unit, and it would also confirm the "mosaic" theory of portolan chart origins.

Language also played a significant part in the particular Italian and Catalan controversy that is most associated with the names of Caraci and Winter.[172] If, the argument runs, the earliest charts revealed in their toponymy clear traces of one particular language, this would identify their place of origin. Again, the same data have been interpreted differently.[173] Winter discerned Catalan name forms, which "speak for a Catalan origin," whereas Guillén y Tato supposed the model was Cantabrio-Castilian.[174] Although himself a Spaniard, Reparaz detected Italianisms on Catalan charts, even along the Spanish coasts, and Caraci was also convinced that the name forms were Italian.[175] The problem goes beyond the simple Catalan/Italian distinction to embrace the contradictory claims of different Italian cities. Although the Carte Pisane is conventionally considered to be of Genoese construction, at least one writer was more inclined to see it as Venetian.[176] A strong argument in favor of Genoese origin is the fact that the earliest reference to a portolan chart, in 1270, occurred on board a Genoese ship.[177] Conti's contention that the Carte Pisane's toponymy embodies a range of languages and dialects may explain why Italian historians have not always been convincing in their interpretations of the dialect showing through in the place-names of the early charts.[178] This very uncertainty adds substance to the view that portolan chart origin should be looked for in terms of separate regional sources.

166. Nordenskiöld, *Periplus*, 24 (note 14). For criticism of this see Salvador García Franco, "The 'Portolan Mile' of Nordenskiöld," *Imago Mundi* 12 (1955): 89–91.

167. Kelley, "Oldest Portolan Chart," 46–48 (note 58).

168. This is stated in a legend on the Carignano map; see Nordenskiöld, *Periplus*, 22 (note 14), and see note 58.

169. Nordenskiöld, *Periplus*, 22–23 (note 14).

170. Magnaghi, "Nautiche, carte," 325a (note 4), the five-*miglia*; Kelley, "Oldest Portolan Chart," 47 (note 58), the four-*miglia*. Armando Cortesão, "The North Atlantic Nautical Chart of 1424," *Imago Mundi* 10 (1953): 1–13, esp. 2, referred to a league of four miles— a "league" and a "portolan mile" being apparently interchangeable in this context.

171. For example, Nordenskiöld, *Periplus*, 20 (note 14).

172. See note 201.

173. See, for example, the inconclusive discussion about the nationality of a fragment in Rome, Biblioteca Apostolica Vaticana, Vat. Lat. 14207 in Destombes, "Cartes catalanes," 63 (note 99).

174. Heinrich Winter, "Catalan Portolan Maps and Their Place in the Total View of Cartographic Development," *Imago Mundi* 11 (1954): 1–12, esp. 5, 7; Julio F. Guillén y Tato, "A propos de l'existence d'une cartographie castillane," in *Les aspects internationaux de la découverte océanique aux XVᵉ et XVIᵉ siècles: Actes du Vᵉᵐᵉ Colloque Internationale d'Histoire Maritime*, ed. Michel Mollat and Paul Adam (Paris: SEVPEN, 1966), 251–53, esp. 253.

175. Gonçal (Gonzalo) de Reparaz, "Essai sur l'histoire de la géographie de l'Espagne de l'antiquité au XVᵉ siècle," *Annales du Midi* 52 (1940): 137–89, 280–341, esp. 303; Giuseppe Caraci, *Italiani e Catalani nella primitiva cartografia nautica medievale* (Rome: Istituto di Scienze Geografiche e Cartografiche, 1959), 83–187, 302–7, English summary on 351–53. See also note 329.

176. Motzo, "Compasso da navigare," LXIII (note 103). For the Genoese case, see O. Pastine, "Se la più antica carta nautica medioevale sia di autore genovese," *Bollettino Ligustico* 1 (1949): 79–82.

177. De La Roncière, *Afrique*, 1:39 (note 100).

178. Conti, "Portolano e carta nautica" (note 109); see below, p. 424 and note 364, where Venetian origin is proposed for the Luxoro atlas—a work that has until now been treated as Genoese.

THE ARGUMENTS SUMMARIZED

In the light of this sometimes ambiguous and often conflicting evidence, it is to be doubted if many of the confident claims of the past have much advanced the cause of truth. Lacking any one compelling theory, we must deal instead with a balance of probabilities. Attempts to locate the charts' origin elsewhere than in the thirteenth century run counter to the information available. Following the earliest identified literary reference to a portolan chart and the oldest surviving example of the genre—both roughly contemporary at the end of that century—there are more than 30 charts and atlases attributable to the fourteenth century and almost 150 that probably derive from the fifteenth century. Working backward along this pattern of steady expansion to its beginning, it is hard to see the justification for extending the vanishing point much further. If portolan charts existed before the thirteenth century, they have failed to leave any discernible trace. Those who suggest a great leap backward over a thousand years to the world of ancient Rome must find an explanation for the fact that the most striking error on the Mediterranean portion of the Carte Pisane, when compared with its immediate successors, concerns no less a region than Italy. Similarly, since the earliest Catalan chart is preceded by the work of four or possibly five Italian practitioners, there seems to be little substance to claims for a Catalan origin.[179]

Even the most plausible of the many proffered explanations—that the Carte Pisane's Mediterranean and Black seas represent a mosaic in which are preserved the separate navigational experiences in the various different basins—leaves a number of questions unanswered. Who could have provided the necessary cartographic skill to produce this confident patchwork without leaving any visible joins? How, even if the trilateration thesis is accepted as the charts' constructional basis, could large errors have been avoided in the relationship of one distant shore to another, given that a voyage, say, from Sicily to the Holy Land could take several weeks?[180] Is there evidence of localized distortion and variation of scale between one basin and another, and does this remain hidden behind measurements that have been presented as averages? What can be learned in this context from distortion grids, like those compiled by Grosjean and Romano?[181] A general awareness of the vital importance of such questions as these and the development of systematic diagnostic techniques to test these cartometric points remain long overdue.

One day, no doubt, these issues will be conclusively resolved. Until that happens, and as an oblique comment on the whole question of origin, it is worth drawing particular attention to the notable developments that can be discerned among the earliest surviving charts. These will be discussed in due course. Whether of changes to the shape of Italy, of additions or corrections to the Atlantic coastlines, or of a constant toponymic updating, they all point to a remarkable vitality. If the moment of portolan chart birth cannot be precisely pinpointed, the creative process can be clearly seen at work in the early fourteenth century.

DRAFTING

There are no signs that the basic techniques involved in producing a chart were much different in 1300 and 1500—though no study of portolan chart draftsmanship has yet been made. Unfortunately, since no contemporary account has survived of how a chart was drawn, we are left to infer the procedures from the charts themselves. The vellum on which the chart was to be drawn would have been purchased, fully prepared, from a specialist parchment maker.[182] Scholarly opinion has been almost equally divided about what happened next: whether the rhumb line network or the coastal outline was laid down first.[183] To test this point, four of the British Library charts (two from the fourteenth century and two from the fifteenth) were therefore examined through a microscope. In three cases the order of superimposition showed definitely that the rhumb lines were under both coastal outlines and place-names, while the fourth instance was ambiguous but pointed the same way.[184]

179. The earliest Catalan chart is Dalorto's of 1325/30, when judged by its language and leaving aside the controversy surrounding the author's possible Italian origin. Preceding Italian practitioners include the unknown authors of the Carte Pisane and the Cortona chart, Pietro and Perrino Vesconte, and possibly Giovanni da Carignano—see below, pp. 406–7, for comments on the suggestion that only one Vesconte was involved and above, p. 380, for the judgment that Giovanni da Carignano's "map" was not a true portolan chart.

180. Taylor, *Haven-Finding Art*, 109 (note 7).

181. Grosjean, *Catalan Atlas*, 16–17 (note 94); Virginia Romano, "Sulla validità della *Carte Pisana*," *Atti dell'Accademia Pontaniana* 32 (1983): 89–99, esp. 96–97.

182. Daniel V. Thompson, *The Materials of Medieval Painting* (London: G. Allen and Unwin, 1936), 24.

183. Kretschmer, *Die italienischen Portolane*, 39 (note 48), cited sheets that had rhumb lines but no outlines. An example, from the Lyons Vesconte atlas, is illustrated in Mollat du Jourdin and de La Roncière, *Sea Charts*, 12 (note 40). Several commentators, however, were convinced that the outlines were drawn first. Thompson, "Rose of the Winds," 194 (note 56), made the surprising suggestion that while the black and green rhumb lines were laid down first, the red ones were often added after the outlines.

184. London, British Library, Add. MS. 27376* (Vesconte), Add. MS. 25691 (Dulcert-type), and Egerton MS. 2855 (Benincasa, 1473). The doubtful instance was Add. MS. 18665 (attributed to Giroldi). The maps were examined under a low-angled light from a fiber-optic lamp. I owe these comments to the assistance and expertise of A. E. Parker, senior conservation officer of the Department of Manuscripts, British Library.

Speculation remains as to the precise way this rhumb line network was constructed. Most fourteenth- and fifteenth-century charts—but significantly not all—reveal a "hidden circle" scraped into the vellum with a compass point. The sole function of this circle was to define the sixteen intersection points whose subsequent connection created the rhumb line network. Martin Cortés, writing in his *Arte de navegar* of 1551, advocated drawing in the circle with a piece of lead whose marks could later be rubbed out.[185] This may correctly reflect sixteenth-century usage, but there are no signs that chartmakers of the two previous centuries had access to an erasable pencil. There are certainly no traces of it on Benincasa's work, which dates from the second half of the fifteenth century. A ruler, a pair of dividers, a pen, and various inks seem to have been the standard equipment of the early chartmakers.[186] It was vital that the sixteen intersection points (or double that number when two networks were involved) be precisely located if a symmetrical pattern was to be achieved. From the holes usually visible at each of the intersection points, it is likely that the circle, which was presumably first quartered by single vertical and horizontal lines, was then further subdivided into sixteenths using a pair of dividers.[187] The holes thus formed would afterward be joined to one another by ruled lines.[188]

COPYING

That the Mediterranean outlines on portolan charts were copied from one another for four centuries is obvious. But the statement begs the question: How did the early chartmakers actually copy from an existing chart?[189] The Spaniard Cortés, in another passage from the 1551 *Arte de navegar*, described how to use oiled tracing paper in conjunction with smoked carbon paper.[190] Yet the Italian Bartolomeo Crescenzio, writing half a century later, prescribed two quite different alternatives.[191] The first involved perforating a sheet of paper with a succession of closely spaced pinholes to define the coastlines. A fine powder (pounce) was then rubbed over the perforated sheet, leaving small deposits on the underlying vellum. Using these as a guide, the coastlines could then be inked in. Crescenzio's second method was to stretch the model and the fresh vellum together over a frame with a light source behind, then make a freehand copy.

It might reasonably be supposed that one of these three methods had been employed from the outset; yet there is a major drawback to this interpretation. All the described procedures would have produced direct facsimiles, identical in coverage and scale to their model. Yet this is patently not the case. Though Nordenskiöld and Kelley, for example, found that a number of the charts they considered had approximately the same scale,[192]

they manifestly failed to identify any one single scale in general use. In a recent catalog, Pietro Frabetti estimated the representative fraction for eight fourteenth- and fifteenth-century works and found that these ranged from 1:4.5 million to 1:8 million.[193] Variations of scale are also encountered within an atlas and even on a single sheet.[194]

The propensity of vellum to distort and cockle when wet[195] must place a question mark beside the many portolan chart measurements that have been obtained in the face of these difficulties. Nevertheless, this evident fluctuation of scale seems perverse, since it must always

185. Thomas R. Smith, "Manuscript and Printed Sea Charts in Seventeenth-Century London: The Case of the Thames School," in *The Compleat Plattmaker: Essays on Chart, Map, and Globe Making in the Seventeenth and Eighteenth Centuries*, ed. Norman J. W. Thrower (Berkeley: University of California Press, 1978), 45–100, esp. 90, quoting from Martin Cortés, *The Arte of Navigation*, trans. Richard Eden (London: Richard Jugge, 1561).

186. On this see H. W. Dickinson, "A Brief History of Draughtsmen's Instruments," *Transactions of the Newcomen Society* 27 (1949–50 and 1950–51): 73–84; republished in the *Bulletin of the Society of University Cartographers* 2, no. 2 (1968): 37–52.

187. Cortés, *Arte of Navigation*, fol. lvi verso (note 185), gives instructions for the quadrants to be divided "in the middest with a pricke or puncte."

188. The charts of the British Library's Cornaro atlas (Egerton MS. 73), which are in many respects unusual for portolan chart work, were constructed by pricking through the intersection points from a master copy. The holes can be made out on one of its sheets that is otherwise cartographically blank.

189. It is as well to dispose here of Nordenskiöld's unfounded speculation about the 1467 Grazioso Benincasa atlas, "that the inscriptions were partly produced by mechanical means through printing or stamping, a method of production which naturally was ready to the hand of so prolific a portolan manufacturer as Benincasa," *Periplus*, 126 (note 14). Nordenskiöld may have been misled by the later use of hand stamps for some decorative features. I owe to Christopher Terrell the information that a 1548 Vesconte Maggiolo chart used stamps for tents, animals, monarchs, towns, and a ship (National Maritime Museum, G. 230/10 MS). Stamped town symbols have also been seen on a 1520 Juan Vespucci chart (in private hands), but no fifteenth-century instances have yet been recorded. Alternatively, Nordenskiöld was led down this blind alley by Uzielli and Amat di San Filippo, *Mappamondi*, 92 (note 35), who inexplicably used the word "impressa" (printed) in connection with the Andrea Benincasa chart of 1490. No credence should be given either to Nordenskiöld's further suggestion that "mechanical means were probably used for the reproduction of the land-outlines on the charts that pass under the names of Sanudo and Vesconte," *Periplus*, 126 (note 14). On this see also Cortesão, *History of Portuguese Cartography*, 2:94–95 (note 3).

190. See Smith, "Thames School," 90 (note 185).

191. Crescenzio, *Nautica Mediterranea* (note 56). A translation is provided in Pelham, "Portolan Charts," 27–28 (note 56).

192. Nordenskiöld, *Periplus*, 24 (note 14); Kelley, "Oldest Portolan Chart," 38 (note 58).

193. Frabetti, *Carte nautiche italiane*, 1–40 (note 73).

194. Examples of the latter would be the Aegean sheet in the 1313 Vesconte atlas and the Italy and Adriatic sheet in the 1373 Pizigano atlas.

195. Ronald Reed, *Ancient Skins, Parchments and Leathers* (London: Seminar Press, 1972), 123.

be more complicated to enlarge or reduce than to make a copy to the same scale. Perhaps the Grazioso Benincasa corpus provides us with a partial explanation. Comparison of the British Library's holding of his five atlases and a single loose chart, when considered in conjunction with measurements cited for his two atlases in the Bibliothèque Nationale (see the Biographical Index, appendix 19.2, pp. 449–56), reveals that all eight works were apparently drawn at one or the other of two distinct scales. Five of the major scale divisions measure approximately either 52 millimeters or 64 millimeters. While these differences would have enabled Benincasa to offer his customers a choice of atlas size, this apparent use of two specific scales—which extended to the separate chart as well—probably reflects a desire to make 1:1 copies where possible. Further measurements are needed to test this hypothesis and to explain the great variety of scales encountered elsewhere.

A standard way that cartographic scale was altered in the past was the square grid method. Once a grid had been placed over the original, the contents of each square could be copied in turn onto the equivalent square (now enlarged or contracted) on the new vellum. Later the grid lines would be rubbed out. It remains to be explained, though, how a temporary grid could have been drawn before the availability of the graphite pencil.[196] The Carte Pisane is sometimes cited in this connection, but all its small squared sections fall outside the hidden circle—thus having nothing to do with the construction of the chart as a whole—and both grid and circles are indelibly inked in.[197]

Another early instance of an underlying grid is to be found in the 1320 Sanudo-Vesconte atlas (Biblioteca Apostolica Vaticana, Pal. Lat. 1362A), but its use on the map of Palestine merely accentuates the non-portolan-chart characteristics of that particular sheet. A more relevant, though equivocal, example is that of the undated and anonymous atlas in the Topkapi Sarayi, Istanbul.[198] Its general chart is covered with a network of small squares but has no rhumb lines. There are too many uncertainties about this work, however, for any conclusions to be drawn.[199] A final possibility, that the rhumb line network itself might have served as a substitute copying grid, is ruled out by inconsistencies in the rhumb line placings on otherwise identical charts.[200] No commentator has yet managed to detect traces of the copying method on any early portolan chart. Again, this remains a challenge for future research.

STYLISTIC CONTENT
CATALAN AND ITALIAN DIFFERENCES

Once rhumb lines, coastal outlines, and names had been completed, the chart could then be embellished with

inland detail and decoration. This point is significant both as defining the stage at which geographic fact gave way to artistic expression and also as marking the divide between Italian austerity and Catalan flamboyance. Yet caution is necessary here. Of the many arguments that have surrounded the portolan charts for a century or more, the fiercest have concerned nationality. With full patriotic fervor, Spaniards (or at times their foreign champions) have claimed certain practitioners or innovations as Catalan, only to have Italian scholars lodge a counterclaim.[201] Nor have the Portuguese stood idly by. The arguments have often been more emotional than substantial; the gain for history has been minimal. The terms "Catalan style" and "Italian style" will therefore be used. This is not simply an evasive compromise; it reflects the degree of overlap between the work of Catalans and Italians—a factor not sufficiently acknowledged by most earlier writers.

Italian-style charts might show part of the Danube; beyond that, their interiors are usually empty. They tend

196. Cortés, *Arte of Navigation* (note 185), described the square grid method, and Smith, "Thames School," 90 (note 185), detected places on the seventeenth-century English charts "Where a careless hand has left a bit of guideline [for the coast] uncovered." But all this relates to a later period, when graphite was generally available. Very recently, a commentator detected "an initial dry-point sketch" on the 1409 Virga chart, which the final coastal outlines did not always follow. This intriguing observation invites further study; see Mollat du Jourdin and de La Roncière (i.e., Isabelle Raynaud-Nguyen), *Sea Charts*, 204 (note 40).

197. The circles and grid squares are red, the diagonal lines across the latter are green; see Mollat du Jourdin and de La Roncière, *Sea Charts*, 198 (note 40). This conflicts with the interpretation given by Motzo, "Compasso da navigare," LXXI (note 103).

198. G. Adolf Deissmann, *Forschungen und Funde im Serai, mit einem Verzeichnis der nichtislamischen Handschriften im Topkapu Serai zu Istanbul* (Berlin and Leipzig: Walter de Gruyter, 1933), no. 47, where the atlas is attributed to Grazioso Benincasa.

199. The chart is reproduced in Marcel Destombes, "A Venetian Nautical Atlas of the Late Fifteenth Century," *Imago Mundi* 12 (1955): 30. Destombes interpreted the grid on the Istanbul chart as "simply a scale of miles extended throughout the surface of the map," but see also above, p. 381. A further instance of an underlying square grid is to be found on one sheet of the British Library's Cornaro atlas, but this belongs to a group of sheets on which the draftsman was evidently experimenting with alternatives to the normal rhumb line network.

200. Nordenskiöld, *Periplus* 17 (note 14), and Motzo, "Compasso da navigare," LXXV (note 103). Yet the claim that the rhumb line network "served as a framework for the plotting of coastlines" has recently been revived in Mollat du Jourdin and de La Roncière, *Sea Charts*, 12 (note 40).

201. Giuseppe Caraci, for example, devoted most of his *Italiani e Catalani* (note 175) to the demolition of Heinrich Winter's claims in favor of the Catalans. See also Alberto Magnaghi, "Alcune osservazioni intorno ad uno studio recente sul mappamondo di Angelino Dalorto (1325)," *Rivista Geografica Italiana* 41 (1934): 1–27, esp. 6–14 on the Dalorto controversy. A rare corrective to the nationalistic arguments was provided by Quaini, "Catalogna e Liguria," 551 n. 3 and 563–66 (note 60).

to be virtually, if not completely, devoid of everything for which there was no functional necessity (plate 26). By contrast, rivers, mountains, and a host of ornamental features make the standard Catalan chart immediately recognizable (plate 27). Most of these elements appear on the oldest surviving Catalan-style charts: Dalorto's of 1325/30 (on its dating, see below, p. 409), Dulcert's of 1339, and the unsigned British Library example, which is closely related to them.[202]

Already stylized, many of these Catalan-style conventions continue throughout our period and beyond, with only minor modifications. Rivers cross the interior, sometimes drawn as elongated corkscrews emerging from almond-shaped lakes. Mountain ranges, picked out in green, are also given distinctive forms. The largest of these, the Atlas chain, seems like a bird's leg, with two, and later three claws at the eastern end[203] and a spur halfway along, while Bohemia is typically enclosed within a horseshoe of green mountains. The Red Sea, appropriately colored, is cut into at its northwest end to mark the miraculous crossing of the Israelites; the sea itself might be covered in parallel wavy lines. Important shrines are represented by a simplified drawing of a church, and the more significant towns are accorded a distinctive sign formed of a circular castle with a red interior, shown in a bird's-eye view. Majorca, often picked out in solid gold, is sometimes striped in the colors of Aragon,[204] and Tenerife (Inferno) occasionally displays a white disk in its center, probably a reference to the snow-covered Pico de Teide.

Discursive notes are another hallmark of the Catalan-style charts,[205] as are the names of provinces and kingdoms. The many other devices scattered over such charts not only add immeasurably to their beauty but also convey a wide range of further information. Flags flying above a tent or town sign identify, if not always accurately, the ruling dynasty, just as crowned figures represent real kings. The occasional ships and fishes are doubtless intended to convey a specific message too, like the North Atlantic whaling scene on the 1413 Mecia de Viladestes chart and the beautiful vignettes on the 1482 Grazioso Benincasa chart (though the latter appear to have been added later). Around the periphery of these Catalan charts there will usually be disks to locate the eight main wind directions.

It should always be borne in mind that these decorative elements might have been the work of specialist artists. Such was certainly the case with a series of world maps that the chartmaker Francesco Beccari contracted to produce in collaboration with Jefuda Cresques for a Florentine merchant in 1399.[206] Beccari, who was responsible for the map's ornamentation, even charged individually for the figures and animals, ships and fishes, flags and trees.

For the most part, the division between charts drawn by Catalans and Italians matches the stylistic differences just described. But there are enough exceptions to demonstrate the ease with which the alternative style could be adopted. Chartmakers give frequent proof of both versatility and a desire for variety. Guillermo Soler, who evidently worked in Majorca, signed two surviving charts. One is undated and typically Catalan; the other of 1385 is in the Italian manner with virtually no inland detail or decoration. Other unsigned and unadorned Catalan charts survive from the late fourteenth century, and only their characteristic town signs distinguish them at a glance from Italian work.[207] The Venetian Pizigani brothers ornamented their 1367 chart in the Catalan style; yet the atlas signed by Francesco Pizigano alone six years later is subdued and typically Italian. The prolific Grazioso Benincasa produced at least seventeen atlases in the austere Italian manner; but his latest known production, a chart of 1482, is thoroughly Catalan in style. If these examples demolish the hard-and-fast divisions between the work of Catalans and Italians, others—the 1447 Valseca chart, for example (fig. 19.3)—illustrate a style midway between the two extremes. Thus there was a regular stylistic interchange between the Catalan-speaking chartmakers of Majorca and their Italian counterparts. This is most noticeable in the way Genoese practitioners imitated the decorative devices of the Catalans. It is also corroborated by the borrowing of place-names to be discussed below.

Nevertheless, there was one important way Catalan draftsmen consistently differed from their Italian counterparts. Catalan chartmakers—or their clients—apparently had no use for bound volumes of charts.[208] Conversely, the Italians of the fourteenth and fifteenth centuries seem never to have decorated their atlases in the Catalan style (though they may have added cornerpieces to otherwise unadorned charts), reserving these

202. London, British Library, Add. MS. 25691; Heinrich Winter, "Das katalanische Problem in der älteren Kartographie," *Ibero-Amerikanisches Archiv* 14 (1940/41): 89–126, esp. 89.

203. Youssouf Kamal, *Monumenta cartographica Africae et Aegypti*, 5 vols. in 16 pts. (Cairo, 1926–51), 4.4:1469.

204. This convention is found for the first time on the Catalan atlas of 1375.

205. Those relating to Africa have been transcribed, translated, and analyzed by Kamal, *Monumenta cartographica*, 4.4:1472–77 (note 203).

206. R. A. Skelton, "A Contract for World Maps at Barcelona, 1399–1400," *Imago Mundi* 22 (1968): 107–13, esp. 107–9.

207. Florence, Biblioteca Nazionale Centrale, Port. 22; Venice, Biblioteca Nazionale Marciana, It. IV, 1912.

208. The so-called Catalan atlas is not an exception to this, since it was originally mounted on six wooden panels; see Grosjean, *Catalan Atlas*, 10 (note 94). However, two lost fourteenth-century productions seem to have been Catalan atlases; see Rey Pastor and García Camarero, *Cartografía mallorquina*, 66 (note 28).

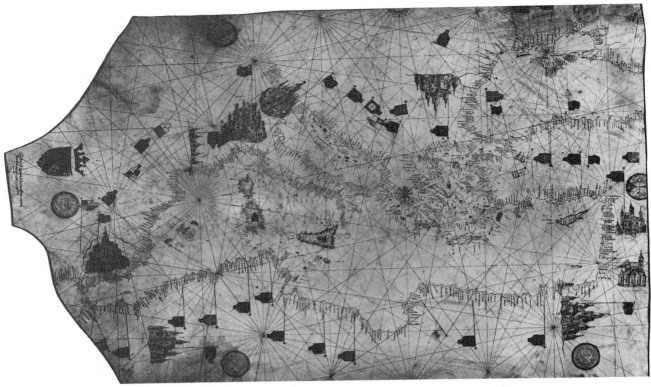

FIG. 19.3. A CHART IN AN INTERMEDIATE STYLE. Drawn by the Majorcan Gabriel de Valseca in 1447, this style is midway between the two extremes of Catalan flamboyance and Italian austerity. Flags, town vignettes, and wind disks are typically Catalan, the lack of inland detail typically Italian. Note the compass roses at the center and top of the rhumb line network. Valseca apparently revived this convention, first found in the Catalan atlas (see fig. 19.5). The coat of arms at the neck of the chart belongs to the Lauria family, perhaps denoting the Francesco de Lauria whose name occurs in the Aragonese archives in the 1440s.

Size of the original: 59 × 94 cm. Photograph from the Bibliothèque Nationale, Paris (Rés. Ge. C 4607).

flourishes for those drawn on a single skin. Here, at least, the distinctions are valid.

With their concern to reveal the nature of the interior, Catalan-style charts are simultaneously terrestrial maps. It is not surprising, therefore, that some Catalan draftsmen should have continued their work eastward to take in regions whose coastlines and hinterland were both little known. This effectively meant the countries beyond the Caspian Sea and Persian Gulf. The former, realistically treated in the Medici atlas, was well known to the Italians; as Marco Polo tells us, they had even launched their own ships there. Some of those Catalan productions that ventured east of the Caspian blend into *mappaemundi*.[209] The farther they go, the more unreliable the coastal information becomes, as greater weight is given in Asia to interior details and less to navigational information. It is hard to see how an Italian-style portolan chart, with its bold statement of factual coastal outlines, could have expanded in this speculative way.

Italian world maps might borrow their Mediterranean outlines from the charts, but in other respects they belong to a separate tradition.[210] The overall stylistic con-

209. It was strongly argued by Reparaz, "Essai," 300 (note 175), that the two-sheet Dulcert chart of 1339 had originally had a further eastern section. Although this possibility was denied by Destombes, "Cartes catalanes," 51 (note 99), Cortesão, *History of Portuguese Cartography*, 2:40 (note 3), more recently proposed that, when constructed, the Dulcert map would have had *two* further sheets to the east, providing a total coverage similar to that of the Catalan atlas. Hinks, *Dalorto*, 9 (note 76), supposed there had been a similar eastern section for the Catalan chart in Florence, Biblioteca Nazionale Centrale, Port. 16.

210. Rome, Biblioteca Apostolica Vaticana, Borgiano V (once attributed to Fra Mauro), does extend some way east of the Caspian, but it is far from being a simplified Italian chart, having its interior packed with notes and details. So too is the Italian fragment in Istanbul; see Marcel Destombes, "Fragments of Two Medieval World Maps at the Topkapu Saray Library," *Imago Mundi* 12 (1955): 150–52, esp. plate facing p. 150, where it is incorrectly captioned as "Catalan"; see *Imago Mundi* 13 (1956): 193.

sistency of the Catalan examples may serve to disguise the joins between those regions based on portolan chart information and those derived from travelers' accounts or theoretical cartography, but it must be emphasized that the eastern extension has little to do with the portolan charts. Catalan world maps of this sort are properly discussed in the chapter on medieval *mappaemundi* and belong rather to an account of European discovery and knowledge of the Orient.

STYLISTIC DEVELOPMENT

It is to be regretted that no attempt has yet been made to provide a generalized index—preferably a visual one—of the common and distinguishing characteristics of portolan charts. Style obviously has great potential for assigning an anonymous work to its correct author or, failing that, to its most logical "school." Yet the limited use so far made of this method has been highly selective. Heinrich Winter, who, more than anyone else, concerned himself with the ornamentation and design of portolan charts, appreciated that perceived changes in stylistic content could be a useful dating aid. Certain characteristics, for instance, are peculiar to the earliest charts. The scale is placed inside a circle on the Carte Pisane, the Cortona chart, Pietro Vesconte's chart of 1311, and the latter's 1313 atlas (fig. 19.4). These are probably the four earliest survivors, and this circle device is not found later. Then again, as Kelley appreciated, borders made up of a recurring chevron pattern are found only on charts that can be ascribed to the first half of the fourteenth century.[211]

It was the compass rose that Winter considered "one of the most important elements which make it possible, in the case of anonymous charts, to determine their national origin and to some extent also their date."[212] Unfortunately, his investigation was marred by its reliance on undated charts, whose ascription to a particular period can be challenged. That compass rose design is indicative of both the place and the time of construction seems clear, but we can only hint here at what seem to be the major developments.

The terminology first needs unscrambling. Many writers confuse the terms "wind rose" and "compass rose."[213] If "wind rose" is to be used at all it should apply only to the unembellished intersection point at which the rhumb lines meet, while "compass rose" should denote the circular compass design into which the intersection point was sometimes elaborated (fig. 19.5). All portolan charts have wind roses, though not necessarily complete with the full thirty-two points; the compass rose, however, seems to have been a Catalan innovation. Whereas the Italian charts from the Carte Pisane onward often indicated the eight main wind di-

rections,[214] the Catalan chartmakers preferred to place disks around the perimeter. Among the symbols they used was an eight-pointed Pole Star indicating north. It needed only a small additional step to remove the star from its enclosing disk, enlarge it, and place it in a prominent part of the chart. The earliest instance of this occurs on the Catalan atlas of 1375 (see fig. 19.5). In this case, and this alone, the compass rose has been imposed on the rhumb line system rather than growing naturally out of a preexisting intersection point. As a result, only four of the rhumb lines, those for the cardinal points, pass through the heart of its compass rose. This awkwardness strongly suggests that the Catalan atlas instance was a first, or at least a very early, attempt. Thereafter the compass roses would be fully integrated with the rhumb

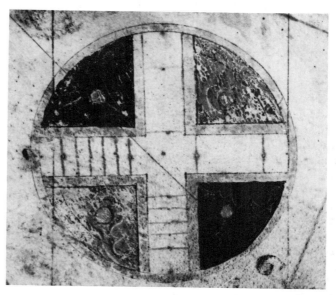

FIG. 19.4. SCALE BAR ON AN EARLY PORTOLAN CHART. A scale found within a circle occurs only on the earliest datable charts. This example from the 1313 atlas of Pietro Vesconte is the latest of four known examples of this convention.
Diameter of the original detail: 6.3 cm. Photograph from the Bibliothèque Nationale, Paris (Rés. Ge. DD 687, pl. 5).

211. The Carignano map, some of the Vesconte work, the Dalorto and Dulcert charts, and three incomplete anonymous charts: Rome, Biblioteca Apostolica Vaticana, Vat. Lat. 14207; Washington, D.C., Library of Congress; and that acquired by Nico Israel of Amsterdam in 1980 (see note 67). See Kelley, "Oldest Portolan Chart," 32 (note 58).

212. Winter, "Late Portolan Charts," 37 (note 129); page 38 illustrates different compass rose types, as does Thompson, "Rose of the Winds," pls. I–V (note 56).

213. See, for example, Nordenskiöld, *Periplus*, 47 (note 14) and the footnote.

214. By name, initial, or in the case of the 1327 Perrino Vesconte chart, by wind heads.

line system, although it was not until the second half of the fifteenth century that compass roses became commonplace, subsequently increasing in both number and intricacy.[215] A further development, the addition of a north-pointing fleur-de-lis outside the compass disk itself, is first found on Portuguese work, specifically on the Jorge de Aguiar chart of 1492 (fig. 19.6).[216]

The later part of our period demonstrates a second way the rhumb line network could be elaborated. On the earlier charts there are just sixteen lines running out from the center of the hidden circle (the core of the system), connecting it with each of the secondary centers on its circumference. It seems to have been Petrus Roselli who decided to double the number of spokes radiating from the main center by adding a further sixteen red lines between the existing ones. These run out to the edge of the chart, avoiding all the secondary centers. Roselli's three earliest charts, two of 1447 and one of 1449, have the basic form; his six later works (1456–68) display the expanded network (fig. 19.7).[217] This feature is considered here under "stylistic development" since, in duplicating existing compass directions, the added lines serve no obvious practical function.[218] Later, however, as the anonymous chart of 1487 and Aguiar's chart of 1492 demonstrate, this enlarged center could

FIG. 19.6. COMPASS ROSE FROM THE 1492 CHART OF JORGE DE AGUIAR. This compass rose is apparently the earliest to include a north-pointing fleur-de-lis outside the compass disk itself.
Diameter of the original detail: 6.5 cm. By permission of the Beinecke Rare Book and Manuscript Library, Yale University, New Haven.

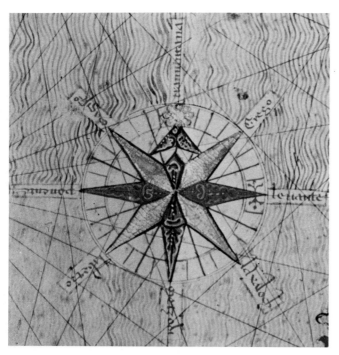

FIG. 19.5. COMPASS ROSE FROM THE CATALAN ATLAS. The Catalan atlas [1375] provides the earliest example of a compass rose, although it is unique in being positioned away from any of the intersection points.
Diameter of the original detail: 5.6 cm. Photograph from the Bibliothèque Nationale, Paris (MS. Esp. 30).

215. The Combitis and Pinelli-Walckenaer atlases, which have in the past been treated as roughly contemporary with the Catalan atlas, display simplified compass roses, but they have been reassigned in this chapter to the early fifteenth century (see table 19.3, pp. 416–20). The second dated appearance of a compass rose occurs on the unlocatable Sentuzo Pongeto chart of 1404. See the reproduction in Weiss und Co., Antiquariat, *Codices manuscripti incunabula typographica, catalogus primus* (Munich: Weiss, 1926), no. 55. (I am grateful to Peter Barber for bringing to my attention the copy of this now scarce catalog in the reference library of the Department of Manuscripts, British Library.) It should be pointed out, however, that Giuseppe Caraci, working only from the Weiss description and partial reproduction, expressed doubt about the Pongeto chart's authenticity; see Giuseppe Caraci, "Carte nautiche in vendita all'estero," *Rivista Geografica Italiana* 34 (1927): 135–36, esp. 135. The Albertin de Virga chart of 1409 introduced a cloverleaf device at a number of the intersection points, but the compass rose proper does not reappear until the Gabriel de Valseca charts of 1439 and later.

216. Cortesão, *History of Portuguese Cartography*, vol. 1, frontispiece (note 3).

217. See appendix 19.2. This point could not be determined for the 1469 chart, known to have been in the possession of Otto H. F. Vollbehr in 1935.

218. N. H. de Vaudrey Heathcote, "Early Nautical Charts," *Annals*

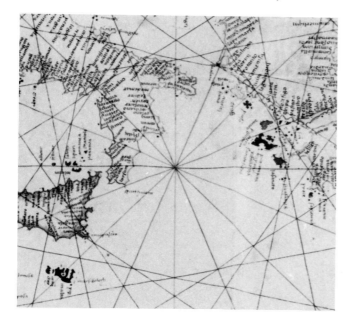

(a)

(b)

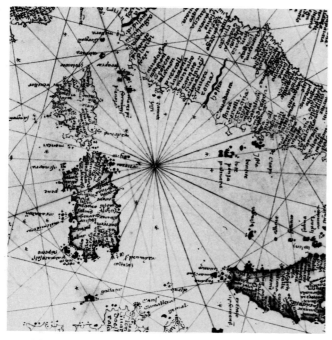

FIG. 19.7. RHUMB LINE CENTERS. In the mid-fifteenth century, the number of lines radiating from the main center of the portolan chart was increased from sixteen to thirty-two. Petrus Roselli appears to have been the first chartmaker to do this, as is shown by these details from his charts of 1449 (*a*) and 1456 (*b*).

The 1449 chart by permission of the Badische Landesbibliothek, Karlsruhe (MS. S6); the 1456 chart courtesy of the Edward E. Ayer Collection, The Newberry Library, Chicago (MS. Map 3).

be more readily elaborated into a thirty-two-point compass rose. Roselli's innovation was repeated on other fifteenth-century charts but has not been observed on Italian work before 1500. This can thus contribute to the determination of date.[219]

Realistic town views appear on portolan charts at about the same time as the compass rose. Though rudimentary urban symbols had already been a feature of the 1327 Perrino Vesconte chart and of the earliest Catalan productions, the first drawings to betray actual observation are Italian towns on Italian charts.[220] The 1367 chart by the Venetian Pizigani brothers includes an unmistakable Campanile in its Venice vignette, as does the little-known Sentuzo Pongeto chart of 1404. What may be the first attempt in a portolan atlas to depict Genoa is found in the calendar of the Francesco Pizigano atlas of 1373. Like the Venice scene with which it is paired, this is a mixture of the true and the fanciful (fig. 19.8).[221] The 1403 chart drawn by the Genoese Francesco Beccari includes a vignette that unquestionably depicts the crescent-shaped harbor of his native city, with the lighthouse at the end of the western arm.[222] The models for these simplified sailor's-eye views have yet to be identified, and it is possible they may be the work of the chartmakers themselves.[223] Notwithstand-

of Science 1 (1936): 1–28, esp. 21, was under the impression that all mid-fifteenth-century charts had thirty-two point centers. He also made the strange assertion that this expanded network was essential to the charts' use.

219. It adds, for example, further doubts to the dates suggested by Nordenskiöld for his own chart and that in Uppsala; see *Periplus*, pls. XIX and XXIII (note 14), see note 357. It must also place a question mark beside the Bibliothèque Nationale's fragments (Rés. Ge. D 3005); see Foncin, Destombes, and de La Roncière, *Catalogue des cartes nautiques*, 18 (note 52), even though their suggested early fifteenth-century date is corroborated by the toponymic evidence; see table 19.3 (pp. 416–20).

220. The urban sign was already used on the Dalorto chart to distinguish Christian towns with a cross, and the Catalan atlas added domes to the Muslim centers. Early Catalan charts emphasized shrines rather than commercial centers. Simplified town signs had already occurred on Roman maps and early medieval *mappaemundi*.

221. Reproduced in Nordenskiöld, *Periplus*, 51 (note 14). Paolo Revelli, *La partecipazione italiana alla Mostra Oceanografica Internazionale di Siviglia (1929)* (Genoa: Stabilimenti Italiani Arti Grafiche, 1937), 183, identified the cathedral of San Lorenzo in the view. Carlo Errera, "Atlanti e carte nautiche dal secolo XIV al XVII conservate nelle biblioteche pubbliche e private di Milano," *Rivista Geografica Italiana* 3 (1896): 91–96, reprinted in *Acta Cartographica* 8 (1970): 225–52, pointed out that the outlines in question were drawn on the back of the original charts no earlier than 1381. Examination of the toponymic content of the charts that belong to this same intercalated supplement has not, however, detected any later names. The two elements of this atlas may thus be very close in date.

222. The lighthouse had already been referred to in the mid-thirteenth-century *Lo compasso da navigare*; see Taylor, *Haven-Finding Art*, 107 (note 7).

223. Although comparable views of Genoa were included about 1400 in Giovanni Sercembi's "Chronicles," see the reproductions in

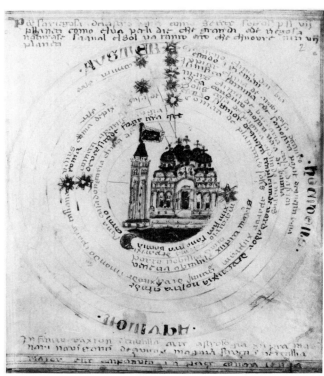

FIG. 19.8. GENOA AND VENICE IN A FOURTEENTH-CENTURY CALENDAR DIAGRAM. Two vignettes from the 1373 Francesco Pizigano atlas. On the left are the lighthouse and cathedral at Genoa, and on the right are the Campanile and church of Saint Mark in Venice.

By permission of the Biblioteca Ambrosiana, Milan (S.P., 10, 29).

ing this increasing tendency for Genoa and Venice to be realistically portrayed, a number of fifteenth- and sixteenth-century chartmakers persisted with fictitious town views.

The 1426 Batista Beccari chart was, in Winter's opinion, the first to emphasize the coasts by color.[224] The same article drew attention to another datable development: the placing of a Madonna and Child vignette on the neck of the chart.[225] The earliest instance of this known to Winter was by Roselli in 1464 (plate 28), although the Nicolo Florino chart of two years earlier had included the IHS monogram. Neither of these chartmakers, however, was the first to make a public gesture of piety. On four of their early fourteenth-century atlases, Pietro and Perrino Vesconte had filled the corners of the sheets with portraits of various saints.[226] These cornerpieces were to recur in the other (anonymous) Lyons atlas (dated by its place-names to the beginning of the fifteenth century), in atlases of the Giacomo Giroldi type, and in the Cornaro atlas. This practice of adding corner portraits to portolan atlases seems to have been a Venetian hallmark. Certainly Vesconte worked

there, Giroldi was a native of Venice, and the Cornaro atlas was produced in that city.

❧

FLAGS

If a systematic study of portolan chart style and embellishment would pay dividends, present indications are that a comparable investigation of the flags arrayed on

Luigi Volpicella, "Genova nel secolo XV: Note d'iconografia panoramica," *Atti della Società Ligure di Storia Patria* 52 (1924): 255–58. Volpicella identifies the oldest recorded view of Genoa as one dating from 1365 (p. 254).

224. Heinrich Winter, "Petrus Roselli," *Imago Mundi* 9 (1952): 1–11, esp. 4.

225. Winter, "Roselli," 6 (note 224).

226. Both Pietro's 1318 atlases and the undated volume in Lyons, as well as Perrino's 1321 Zurich atlas. For an art-historical interpretation of these miniatures see Bernhard Degenhart and Annegrit Schmitt, "Marino Sanudo und Paolino Veneto," *Römisches Jahrbuch für Kunstgeschichte* 14 (1973): 1–137.

about a third of the surviving fourteenth- and fifteenth-century charts might prove disappointing.[227] This paradox needs explaining, because the political statement embodied in a city banner or a nation's arms is precisely the kind to raise expectations of accuracy and contemporaneity. There are countless instances in the extensive literature where an unsigned chart has been given a definite terminus ad quem on the grounds that it omits any reference to a change that took place in that particular year or because it displays a flag that was superseded at that time. Yet no methodical check was made to see whether dated charts were accurate in that respect. When considering the Dalorto chart, whose date has been variously read as 1325 or 1330 (see below, p. 409), Gerola argued in favor of the latter year on the grounds that Cagliari was flying the Aragonese colors—a reference to the subjugation of Sardinia in 1326. This verdict rested, however, on the further, and questionable, assumption that chartmakers recognized only de facto and not de jure rule, since Sardinia had been nominally Aragonese since 1297.[228]

Suppositions such as these need to be tested against the overall pattern of portolan chart flags. Unfortunately, though a highly visible element of a chart, the boldly colored flags tend to blur to unrecognizability in reproduction. A further problem is the extent to which the development of distinctive flags for particular places roughly parallels that of the charts themselves.[229] Standardization in the design of flags had not yet taken place; hence it is not surprising that a great variety of forms were adopted by the different early chartmakers. Designs would be simplified or altered at will, and colors switched.[230] The crescent, for example, was frequently used as a convenient shorthand for the Muslim cities of North Africa, even though several fourteenth-century charts already accorded these their own distinctive banners. It is also difficult for the researcher to be sure which place was intended, because the flag-bearing standard, pointing to the relevant place-name, was often located imprecisely. Where banners flew over unnamed town vignettes, this problem increased.

Flags are found on four of the Vesconte works (plate 29), on the Carignano map (though in the form of atypical, simplified disks), on the three Dalorto/Dulcert charts, on the mid-fourteenth-century *Libro del conoscimiento*, and on a number of later works.[231] Of those atlases and charts attributable to the fourteenth and fifteenth centuries, probably upward of seventy (or about 40 percent) are involved, with more than sixty flags found at times on a single work. A comprehensive study of these is awaited; meanwhile some provisional observations can be made. In the first place, although certain flags recurred regularly, there were numerous variations in the selection of featured places, even within the work

of a single practitioner. In some cases an apparently superseded form reappeared much later.[232] Such inconsistency limits the generalizations that can usefully be made at present. Nevertheless, the selected instances presented in table 19.1, p. 400, are probably the most instructive.

What emerges most clearly from these examples is the difference in the chartmakers' attitudes toward victory and defeat. Whatever their nationality, none begrudged the Portuguese their success at Ceuta, for this was a Christian triumph over Islam. Yet the Ottoman expansion through Asia Minor (culminating in the capture of Constantinople) and around the Black Sea into Europe was ignored. Instead, the chartmakers used their flags to deny an unpleasant reality. This tells us about psy-

227. The main studies devoted to this subject are as follows. First, two by Giuseppe Gerola: "L'elemento araldico nel portolano di Angelino dall'Orto," *Atti del Reale Istituto Veneto di Scienze, Lettere ed Arti* 93, pt. 1 (1933–34): 407–43; and "Le carte nautiche di Pietro Vesconte dal punto di vista araldico," in *Atti del Secondo Congresso di Studi Coloniali, Napoli, 1–3 October 1934*, 7 vols. (Florence: Leo S. Olschki, 1935), 2:102–23. Second, three by Georges Pasch in *Vexillologia: Bulletin de l'Association Française d'Etudes Internationales de Vexillologie*: "Les drapeaux des cartes-portulans: L'atlas dit de Charles V (1375)," 1, nos. 2–3 (1967): 38–60; "Les drapeaux des cartes-portulans: Drapeaux du 'Libro del Conoscimiento,'" 2, nos. 1–2 (1969): 8–32; "Les drapeaux des cartes-portulans [portulans du groupe Vesconte]," 3, no. 2 (1973): 52–62. Finally, Anna-Dorothee von den Brincken, "Portolane als Quellen der Vexillologie," *Archiv für Diplomatik: Schriftgeschichte Siegel- und Wappenkunde* 24 (1978): 408–26.

228. Gerola, "Dall'Orto," 423 (note 227). But see Magnaghi, "Alcune osservazioni," 23–27 (note 201), where many objections are raised to Gerola's theory and it is argued, in support of the 1325 date, that the flag was intended to refer to the entire island, not just to Cagliari.

229. Flags developed out of the gonfanon (a war pennant carried on a lance) and the streamers flown at a ship's masthead. The general adoption of heraldic symbols has been dated to the mid-thirteenth century; see von den Brincken, "Vexillologie," 409–10 (note 227). See also E. M. C. Barraclough and W. G. Crampton, *Flags of the World* (London: Warne, 1978), 12–14.

230. Pasch, "Libro del Conoscimiento," 10 (note 227), and Pasch, "Atlas de Charles V," 44 (note 227), where he talks of the "psychophysiology of color."

231. The Vesconte works are the Vatican atlases of [1320] and [1321], the British Library's atlas of about 1325, and the Perrino Vesconte chart of 1327; the Dalorto/Dulcert charts are Dalorto (1325/30), Dulcert (1339), and British Library, Add. MS. 25691; the *Libro del conoscimiento* was compiled by a Spanish Franciscan friar (born 1304). His written survey was apparently composed with the aid of a Catalan world map. Between them, the three surviving manuscript recensions illustrate about 110 different flags. The text refers to a battle that took place in 1348, but attempts to narrow further the date of its construction are not convincing. See Clements R. Markham, ed. and trans., *Libro del Conoscimiento: Book of the Knowledge of All Kingdoms* (London: Hakluyt Society, 1912), 2d ser., 29, and Pasch, "Libro del Conoscimiento" (note 227), both of whom reproduce the flags.

232. For example, the 1421 Francesco de Cesanis chart has several forms associated with the Vescontes a century earlier.

TABLE 19.1 Flags and Chartmakers' Response to Political Change

Place	Historical Event	Chartmakers' Response
England	From 1340 the fleur-de-lis was quartered with the three lions when Edward III claimed the French throne.	This form appeared in the *Libro del conoscimiento* (mid-fourteenth century), on the Soler chart (ca. 1385) in Paris, and on most charts thereafter. Portuguese charts at the end of the fifteenth century replaced the quarterings with the cross of Saint George.
Seville (or Spain)	United under Ferdinand and Isabella in 1479.	Where previously the arms of Spain had been the quartered lion and castle, most late fifteenth-century charts seem to have incorporated the alternating bands of Aragon.[a]
Granada	Taken from the Moors in 1492.	Probably recognized immediately.
Valencia	Acquired by Aragon in 1238; toward the end of the fifteenth century transferred to Castile.	Fourteenth-century charts have the alternating bands of Aragon; fifteenth-century charts tend to include a crown as well.[b]
Montpellier	Passed from Aragon to France in 1349.	The bands of Aragon, forming half the flag, are replaced by fleur-de-lis. The 1426 Beccari chart was the first instance noted of the later form.
Marseilles	Transferred from Provence to France in 1486.	Instead of the normal blue cross there is a fleur-de-lis on the [1483?] Reinel chart—see p. 374.
Rome	Republican at various times, e.g., under Cola di Rienzo (1347 and 1353–54).	SPQR is found in place of the usual crossed keys on the anonymous Corsini chart[c] and in the *Libro del conoscimiento*. The 1473 Benincasa atlas in Bologna includes both forms.[d]
Salonika (Thessalonica)	Changed hands several times between the Byzantine Empire, Venice, and the Ottoman Turks, until finally passing to the Turks in 1430.	The Paleologue arms of the Byzantine emperors (4 Bs on their sides) were apparently retained throughout the fifteenth century.
Constantinople	Fell to the Ottoman Turks in 1453.	Subsequent charts either retained the Paleologue form or omitted a flag altogether.
Curcho (in Armenia Minor)	Overrun by the Ottomans in 1375.	Its flag—white crosses on a blue ground—continued for at least a century.
Ceuta	Taken by Portugal in 1415.	No post-1415 instance has been found showing the preconquest device (twin keys) apart from the 1421 Cesanis chart.

[a]It is likely that the rulers depicted on some charts will also indicate changing political realities. However, the attempt by one commentator to use in this way the Spanish sovereigns on two Bertran charts is unconvincing, since neither the unification of Spain nor the conquest of Granada—both of which were supposed to be reflected—occurred during the period in question (1482–89). See Niccolò Rodolico, "Di una carta nautica di Giacomo Bertran, maiorchino," in *Atti del III Congresso Geografico Italiano, Florence, 1898*, 2 vols.(1899), 2:544–50, esp. 545.

[b]Georges Pasch noted that the bands were horizontal in the fifteenth century and then vertical in the sixteenth and seventeenth centuries;

see *Vexillologia: Bulletin de l'Association Française d'Etudes Internationales de Vexillologie* 2, no. 6 (1972): 19. Confusingly, late fifteenth-century Portuguese charts reverted to the earlier form.

[c]For which reason a mid-fourteenth-century date was suggested for it; see Pietro Amat di San Filippo, "Recenti ritrovamenti di carte nautiche in Parigi, in Londra ed in Firenze," *Bollettino della Società Geografica Italiana*, 3d ser., 1 (1888): 268–78, esp. 273; reprinted in *Acta Cartographica* 9 (1970): 1–11.

[d]Marina Emiliani (later Marina Salinari), "Le carte nautiche dei Benincasa, cartografi anconetani," *Bollettino della Reale Società Geografica Italiana* 73 (1936): 485–510, esp. 498–99.

chological attitudes, not about historical events. Hence bizarre results are produced when attempts are made to draw dating inferences from flags flying over towns that were to be overrun by the Ottoman Turks.[233] Errors of at least a century can be produced in this way. On the other hand, the rapidity with which the capture of Ceuta was generally acknowledged gives some justification for treating the earlier Arab insignia as a terminus ad quem of 1415. In the same way, the Spanish flag seems to have been speedily inserted over Granada after 1492 and beside Oran, Bougie (Bejaia), and Tripoli in Libya after those towns were subdued in 1509–10.[234] If the presence of these flags has caused some charts to be moved out of the fifteenth century,[235] their absence provides reasonable (though inconclusive) grounds for a date before 1510.[236]

Once an element of selective truth is admitted, it is no longer possible to see the flags as giving "valuable information to a visiting ship."[237] Many a Christian sailor would have ended up a galley slave had he relied on his chart to distinguish friend from foe. It is surely fair to assume anyway that medieval mariners would have been aware of those political developments taking place in Mediterranean or Black Sea ports that carried implications for them.

Even if we agree with Georges Pasch, the most assiduous modern commentator on this subject, that the portolan chart vexillology was no better or worse than that of other documents of the period,[238] it is as well to underline its limited value to historians of cartography. The flags have often been interpreted as a straightforward account of shifting political reality. Yet they made a very limited response to the turmoil of the centuries concerned as the fortunes of the Genoese and Venetian colonies fluctuated in the face of the military expansion of the Ottoman Turks. Many towns continued to display an unaltered flag despite changes of overlord, presumably because the design belonged exclusively to the town rather than to the imperial power that dominated it. Other factors may sometimes have overridden considerations of political accuracy: aesthetic or practical judgments must have played their part in the choice of flags. If the place-names left limited space, eastern Adriatic flags and those for Rome and Florence might be omitted to avoid a cramped appearance.[239] Sometimes more positive reasons were involved, as when a chartmaker turned the flags to his own political purpose. There is probably no better example of this than the dozen Genoese flags the chartmaker of that city, Albino da Canepa, placed around the Black Sea on his chart of 1489. This exercise in medieval flag-waving turned the Black Sea into a Genoese lake as much as fifteen years after a process of contraction had culminated in the fall of its final stronghold, Caffa (Feodosiya).[240]

PALEOGRAPHIC EVIDENCE

Calligraphy is less obvious as an indicator of style, but this ever-present element must represent one of the most important pointers to the place and date of a chart's production. A faithful copy reproducing an archaic hand could be made a long time afterward, though, with the result that on occasion there may be considerable disparity between the dates of a chart's compilation and its execution. One well-known instance—the British Library's Cornaro atlas (Egerton MS. 73)—makes no attempt to disguise its use of obsolete charts. Though its thirty-four sheets of charts are apparently drawn in the same hand throughout[241] (and one of them is dated 1489), most are acknowledged copies of the work of earlier chartmakers. Among the authors cited are Francesco Beccari and Nicolo de Pasqualini, for whom single signed works are known, dated respectively 1403 and 1408. The same intentional reproduction of out-of-date charts might help explain the conflicting drafts of the

233. Despite the warning given in 1882 by Amat di San Filippo in the preface to Uzielli and Amat di San Filippo, *Mappamondi*, iv (note 35). For example, Gaffarel, "Dijon," 166 (note 37), argued from the absence of a crescent over Salonika that the Dijon chart must have been drawn before 1429; yet the Spanish flags attached to Granada (taken from the Moors in 1492) and the Canaries (conquered 1493–96) point to the very end of the fifteenth century at the earliest. On this, see also Raynaud-Nguyen, "Hydrographie" (note 37). Rhodes, occupied by the Knights Hospitalers until 1523, retained its red cross on some charts drawn more than a century afterward, for instance, the Caloiro chart of 1665; see Frabetti, *Carte nautiche italiane*, pl. xxxvii (note 73).

234. For example, on the 1512 Maggiolo atlas, see Georges Grosjean, ed., *Vesconte Maggiolo, "Atlante nautico del 1512": Seeatlas vom Jahre 1512* (Dietikon-Zurich: Urs Graf, 1979).

235. See p. 386.

236. For example, the chart reproduced in Edward Luther Stevenson, *Facsimiles of Portolan Charts Belonging to the Hispanic Society of America*, Publications of the Hispanic Society of America, no. 104 (New York, 1916), pl. V, and that described successively in E. P. Goldschmidt, Booksellers, *Manuscripts and Early Printed Books (1463–1600)*, catalog 4 (London: E. P. Goldschmidt, [1924–25]), no. 125, and Ludwig Rosenthal, Antiquariat, catalog 163, no. 1237. On the latter, see also Caraci, "Estero," 136 (note 215).

237. Taylor, *Haven-Finding Art*, 113 (note 7). Nor should the opposite view, that the portolan chart flags are usually of no more than secondary and ornamental significance, be accepted; see Magnaghi, "Alcune osservazioni," 19 (note 201).

238. Pasch, "Vesconte," 56 (note 227).

239. Aesthetic values must have been involved when Francesco Beccari inserted 340 flags on his large world map. Since this was three times the total available on the model from which he worked, the additions must simply have been repeats of existing flags or imaginary flags; see Skelton, "Contract," 108 (note 206).

240. Magnaghi, "Alcune osservazioni," 3 (note 201), cites several mid-sixteenth-century examples of the Genoese flag being retained for Black Sea ports.

241. But see Cortesão, *History of Portuguese Cartography*, 2:196 (note 3), for a note on contrary opinions.

Adriatic in the Medici atlas (see appendix 19.1 and fig. 19.19).

Nordenskiöld was too dismissive of the use of paleography in dating charts when he declared that "it is perfectly impossible to decide the age of portolanos [i.e., charts] by their style, for even the hand-writing is slavishly copied."[242] Yet it must be accepted that dating based on script can only be approximate. Moreover, the style of the compressed calligraphy forced on the chartmakers by limitations of space was probably slow to change. The combined outlines, toponymy, and style of the Cortona chart all consistently suggest that it antedates the 1311 Pietro Vesconte chart. Should we then follow Armignacco in assigning the chart to about 1350, purely on paleographic grounds?[243] A different kind of conflict between writing and content was reported by Cortesão and Teixeira da Mota. They dated a Portuguese chart in the Arquivo Nacional, Lisbon, to the very end of the fifteenth century from internal evidence but admitted that "the hand-writing of the place-names might suggest an earlier date."[244]

A systematic examination by appropriate specialists of the script on all supposed fourteenth- and fifteenth-century portolan charts is clearly long overdue. If the peculiarities of known chartmakers were codified, it might prove possible to test the numerous, and frequently conflicting, attributions of unsigned work that have been casually put forward on supposed paleographic grounds alone. Until a thorough survey has been made of this type of evidence it must be handled with considerable caution (figs. 19.9 and 19.10 show how paleography can corroborate the toponymic analysis— see also p. 424).

HYDROGRAPHIC DEVELOPMENT

It has been natural to emphasize the very obvious visual differences between (usually Italian) austerity and (usually Catalan) exuberance. Indeed, the dissimilarity between the two types might raise doubts that they really belong to the same family. Yet if the Catalan ornamentation is disregarded, it is apparent that the underlying hydrographic content is essentially the same on charts of equivalent age, wherever they were made. While regional variations might lead to differences in the form in which the charts were presented, these differences were softened (in the early stages at least) by the unifying effect of a shared hydrographic development.

It has always been assumed, no doubt correctly, that portolan chartmakers were accomplished copyists. What is not justified, however, is the commonly expressed extension of this: that they slavishly imitated unchanging models. As we shall show, the portolan charts of the fourteenth and fifteenth centuries display a continuous and wide-ranging development. It is convenient to discuss this under three headings: changes in the outlines of existing coastlines; the addition of new information beyond the Mediterranean; and the evolving toponymy of the areas that lie at the heart of the portolan charts.

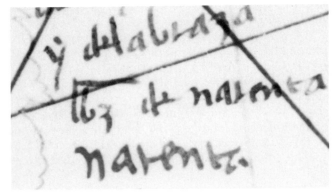

(a)

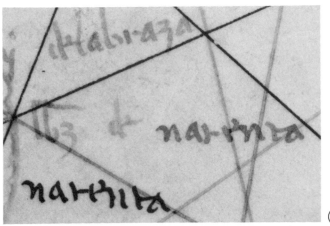

(b)

FIG. 19.9. PALEOGRAPHIC COMPARISON (1). The use of the otherwise rare abbreviation *lbz* for "gulf" on both the Combitis (*a*) and the Pinelli-Walckenaer (*b*) atlases is one of the pieces of evidence pointing to their common authorship. The portion of the two atlases illustrated is the *lbz de Narenta*, on the Dalmatian coast.
By permission of (*a*) the Biblioteca Nazionale Marciana, Venice (It. VI, 213 [MS. 5982]), and (*b*), the British Library, London (Add. MS. 19510, fol. 5r).

242. Nordenskiöld, *Periplus*, 58 (note 14).

243. Vera Armignacco, "Una carta nautica della Biblioteca dell'Accademia Etrusca di Cortona," *Rivista Geografica Italiana* 64 (1957): 185–223, esp. 192. Armignacco and others have argued that the Cortona chart might be a copy of one that antedated the Carte Pisane. The absence of Manfredonia, founded in 1258 and present on the Carte Pisane, has been cited in support of this. The dating of this important chart was also discussed, though not resolved, in Caraci, *Italiani e Catalani*, 275–79, esp. 278 (note 175). Caraci suggested that remarks about Palestine, written in an old hand on the reverse of the chart, were reminiscent of proposals for a crusade made at the Council of Lyons in 1274. If verified, this could have important implications for the dating of the Cortona chart and, by extension, of the Carte Pisane.

244. Cortesão and Teixeira da Mota, *Portugaliae monumenta cartographica*, 1:5 (note 29).

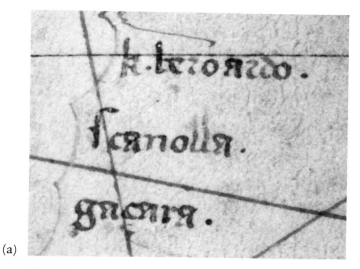

(a)

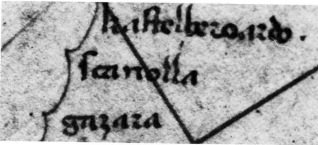

(b)

FIG. 19.10. PALEOGRAPHIC COMPARISON (2). As a draftsman, Francesco de Cesanis was distinctive in using a period after most place-names as well as a wavy *l*. This illustration compares the names *scanolla* and *kastelberoardo* along the Palestine coast taken from the 1421 chart of Cesanis (*a*) with their forms in the Luxoro atlas (*b*). Other shared characteristics, particularly patterns of place-names, make it likely that both are by the same hand. The Luxoro atlas would thus be a Venetian work of the early fifteenth century, rather than a Genoese production of the early fourteenth century, as generally supposed.
By permission of (*a*) the Civico Museo Correr, Venice (Collezione Correr, Port. 13), and (*b*) the Biblioteca Civica Berio, Genoa.

CHANGING COASTAL OUTLINES

It has been generally accepted that alterations to the presented shape of coastal features on portolan charts demonstrate hydrographic progress. However, the unwarranted evolutionary assumptions on which this depends, and the part presumably played by the chance effects of copying in molding particular features, have been insufficiently considered. The more rigorous methods of plotting shape—whether with the aid of a computer, with a camera, or by means of multiple measurements[245]—are of recent development and have yet to be systematically applied to the early charts. Instead, we have been offered selected tracings of particular features for comparison and have been invited to

detect supposed similarities and stages of development. Yet this method is innately subjective and has proved of only limited value. The differences are evident; their significance is a matter of conjecture. R. A. Skelton warned of the dangers of basing a "heavy load of theory . . . on a visual impression."[246]

A few of the many specific coastal outlines that have been considered in this way can be mentioned: the Rhône and Danube deltas, and the Syrtes (by Kelley); the Crimean peninsula, mouth of the Nile, Majorca, and the Straits of Gibraltar (by Nordenskiöld and Caraci), Rathlin Island, off the coast of Northern Ireland (by M. C. Andrews), and France (by Numa Broc).[247] Besides these, Rey Pastor and García Camarero offered comparative sketches of various mountain ranges and rivers as they occurred on Catalan charts.[248] The most thorough investigation of this kind was carried out by Andrews, who analyzed the shape of the British Isles in three important articles.[249] He classified the charts of the three centuries according to which "type" they exhibited and was able to identify thirteen variant shapes for Scotland, twelve for England, and seven for Ireland.[250] Andrews taught us much about the influence of one chartmaker on another, but the method suffers from a basic weakness: in assigning charts to one or another of the identified types, the countless (and possibly significant) mi-

245. For computer-aided methods see Joan M. Murphy, "Measures of Map Accuracy Assessment and Some Early Ulster Maps," *Irish Geography* 11 (1978): 88–101, and Jeffrey C. Stone, "Techniques of Scale Assessment on Historical Maps," in *International Geography 1972*, ed. W. P. Adams and F. M. Helleiner (Toronto: University of Toronto Press, 1972), 452–54; on camera method see Harry Margary, "A Proposed Photographic Method of Assessing the Accuracy of Old Maps," *Imago Mundi* 29 (1977): 78–79; and for multiple measures see Elizabeth Clutton, "Some Seventeenth Century Images of Crete: A Comparative Analysis of the Manuscript Maps by Francesco Basilicata and the Printed Maps by Marco Boschini," *Imago Mundi* 34 (1982): 51–57.

246. R. A. Skelton, *Looking at an Early Map* (Lawrence: University of Kansas Libraries, 1965), 6.

247. Kelley, "Oldest Portolan Chart," figs. 2–4 (note 58); see also Oldham, "Rhône Delta," 403–28 (note 81); Nordenskiöld, *Periplus*, 23 (note 14); Giuseppe Caraci, "The Italian Cartographers of the Benincasa and Freducci Families and the So-Called Borgiana Map of the Vatican Library," *Imago Mundi* 10 (1953): 23–49, esp. 41; Michael Corbet Andrews, "Rathlin Island in the Portolan Charts," *Journal of the Royal Society of Antiquaries of Ireland* 55 (1925): 30–35, esp. 30; Numa Broc, "Visions Médiévales de la France," *Imago Mundi* 36 (1984): 32–47, esp. 40, 42.

248. Rey Pastor and García Camarero, *Cartografía mallorquina*, 26–28 (note 28).

249. Michael Corbet Andrews, "The Boundary between Scotland and England in the Portolan Charts," *Proceedings of the Society of Antiquaries of Scotland* 60 (1925–26): 36–66; idem, "The British Isles in the Nautical Charts of the XIVth and XVth Centuries," *Geographical Journal* 68 (1926): 474–81; idem, "Scotland in the Portolan Charts," *Scottish Geographical Magazine* 42 (1926): 129–53, 193–213, 293–306.

250. Andrews, "Scotland in the Portolan Charts," 132 (note 249).

nor variations have to be ignored. Although Andrews was able to document frequent change, he stressed that it seldom involved improvement in content. Some of his types were in use for a century or more, and several were found to coexist. His analysis was thus more a guide to lineage than to date. Yet despite the imperfections of this method, some infusions of new information can nevertheless be discerned for the earliest period. The areas involved are large enough and the changes sufficiently marked for there to be little danger of subjectivity in their interpretation.

Ideally, the next step would be to parade the earliest charts in their chronological sequence, examining them for these changes in content. Before this is attempted, however, it is necessary to clarify some basic issues. In the first place, we must establish which charts belong to this crucial formative period of the early fourteenth century. Second, we have to clarify which works can be attributed to the period's most prolific practitioners (or practitioner), Pietro and Perrino Vesconte. Unfortunately, many of the supposedly earliest charts are the objects of controversy precisely about their date, and since a number remain unpublished, it is difficult to test earlier judgments on this point. Furthermore, the 1311 Vesconte chart and that in the Library of Congress portray only the eastern half of the Mediterranean,[251] the Cortona chart is trimmed at its western edge, and some of the others are even more incomplete.[252] All of this makes it more difficult to set within a chronological framework the first major improvements to be found on the portolan charts.

The Carte Pisane (plate 30) is accepted as being the oldest surviving portolan chart. It owes its title to Jomard, who believed it came from a Pisan family. Nevertheless, the chart is often considered to be Genoese in workmanship and to date from the late thirteenth century.[253] The first step is to compare its outlines with those of its presumed closest successors: the little-known Cortona chart (fig. 19.11), the Carignano map, and the plentiful productions of Pietro Vesconte from 1311 onward. None of these repeats the Carte Pisane's very apparent broadening of Italy, nor do they imitate its distortion of Spain's Mediterranean outline. The thoroughly unreal outline of the Atlantic coasts of southwestern Europe was possibly repeated on the Cortona chart. Although its Spain has been trimmed away, the small extant section of the Belgian and French coasts imitates very closely that found on the Carte Pisane. As apparently the second earliest surviving chart (in terms of compilation, though not necessarily in terms of drafting), it deserves further study.[254]

A work that has featured prominently in almost all previous discussions on the earliest charts is the Giovanni da Carignano map. Since it was destroyed in 1943,

it must now be approached through reproductions, none of which is particularly clear.[255] All that can be said with certainty is what is stated in its author's legend, namely that it was drawn by Giovanni the priest, rector of Saint Mark's in the port of Genoa.[256] Beyond that we move off firm ground. A variety of dates have been proposed for it, ranging from 1291 to 1400,[257] although Ferretto reduced the possibilities by showing that Carignano died between September 1329 and May 1330.[258] The forty-year tolerance thus allowed still left open the possibilities that the Carignano map might antedate the Carte Pisane or belong to a period later than the Dalorto chart and the entire Vesconte output.

Although Carignano's name features fairly frequently in the Genoese archives, there are no mentions of his mapmaking interests.[259] We learn of these instead from a later chronicler, Jacopo Filippo Foresti of Bergamo (known as Bergomensis). In 1306 an Ethiopian embassy on its way back from Avignon and Rome stopped off in Genoa.[260] There they met and were interviewed by

251. For the 1311 Vesconte chart see Nordenskiöld, *Periplus*, pl. V (note 14). The Library of Congress chart is reproduced in Kelley, "Oldest Portolan Chart," 22 (note 58), and in color in Charles A. Goodrum, *Treasures of the Library of Congress* (New York: H. N. Abrams, 1980), 93.

252. Like Rome, Biblioteca Apostolica Vaticana, Vat. Lat. 14207, and the chart acquired by Nico Israel in 1980 (see note 67).

253. The approximations "ca. 1275" and "ca. 1300" are often given to it. The more precise "ca. 1290" makes the unwarranted assumption that a cross would not have been placed beside Acre after that town fell to the Turks in 1291. The even more exact "ca. 1284" (applied to the Carte Pisane during an exhibition in Paris in 1931) seems to derive from a careless reading of a passage in de La Roncière, *Afrique*, 1:41 (note 100). I owe this last suggestion to Mireille Pastoureau. One recent commentator misinterpreted the toponymic evidence to reach the unwarranted conclusion that the Carte Pisane was drawn about 1265; see Romano, "Carte Pisana," 89–90 (note 181). Another concluded that it could not have been drawn before 1275, a statement not supported by the cited evidence; see Mollat du Jourdin and de La Roncière, *Sea Charts*, 16 (note 40). For a bibliographical note on the various opinions about the Carte Pisane's date see Revelli, *Partecipazione italiana*, lxix–lxx (note 221). Not all Italian scholars are convinced about the Carte Pisane's supposed Genoese authorship; see above p. 389 and note 176.

254. See above, p. 402 and note 243.

255. Ferdinando Ongania, *Raccolta di mappamondi e carte nautiche del XIII al XVI secolo* (Venice, 1875–81), pt. 3; Nordenskiöld, *Periplus*, pl. V (note 14); Kamal, *Monumenta cartographica*, 4.1:1138 (note 203).

256. "Johannes presbyter, rector Sancti Marci de portu Janue me fecit."

257. Revelli, *Partecipazione italiana*, lxx, lxxviii–lxxx (note 221).

258. Ferretto, "Carignano," 45 (note 76). This corrected an earlier belief that he had died in 1344.

259. Ferretto, "Carignano," 36–45 (note 76).

260. R. A. Skelton, "An Ethiopian Embassy to Western Europe in 1306," in *Ethiopian Itineraries circa 1400–1524, Including Those Collected by Alessandro Zorzi at Venice in the Years 1519–24*, ed. Osbert G. S. Crawford, Hakluyt Society, ser. 2, 109 (Cambridge:

FIG. 19.11. THE CORTONA CHART. It is likely that this little-known portolan chart is in fact the second oldest extant example. Unfortunately, much of the western portion has been trimmed away, but what remains of the Belgian and French coasts shows close similarities with the outlines of the Carte Pisane.

Size of the original: 47 × 60 cm. By permission of the Pubblica Biblioteca Comunale e dell'Accademia Etrusca, Cortona (Portolano).

the unnamed rector of Saint Mark's, who made both a written record of what he had heard and a *mappamundi*.[261] Since it is not generally believed that the recently destroyed Florence map should be identified with this *mappamundi*, which was apparently drawn close to 1306,[262] the Bergomensis account does not help with the dating of the former. Authorities can be cited to narrow the date of construction of the Florence map to the period 1321–26, but these do not inspire total confidence. Baldacci's assertion that the map could not have been drawn before 1321 because of the presence of Aragonese stripes over Sardinia relies on a type of evidence

Cambridge University Press for the Hakluyt Society, 1958), 212–16, esp. 212.

261. Skelton, "Ethiopian Embassy," 214 (note 260), points out that the first edition of Forestri's text included the strange statement (translated from Latin) that the rector of Saint Mark's had "published a treatise, which he also called a map."

262. Though some commentators have expressed this view and Cortesão, *History of Portuguese Cartography*, 1:220 (note 3), dated the surviving map ca. 1307. Paolo Revelli, "Cimeli cartografici di Archivi di Stato italiani distrutti dalla guerra," *Notizie degli Archivi di Stato* 9 (1949): 1–3, esp. 1, was one of those who argued that the map's construction seemed to antedate the papal move to Avignon in 1307 because the flag of Rome still displayed the papal keys. This argument is invalid, however, since the papal standard remained over Rome on portolan charts throughout the fourteenth century.

that has already been questioned.[263] Similarly, Fischer's view that the map must have been drawn before 1326, since Bursa is not tinted black like other towns in northwestern Turkey that had fallen to the Ottomans, depends on absent features—a notoriously suspect line of argument.[264]

It must be concluded that the dates so far suggested for the Carignano map's construction, on biographical or historical grounds, carry little conviction. Instead, therefore, of making an arbitrary assumption about its date and then weaving an account of portolan chart evolution around that, we will attempt to compare the maturity of the Carignano map's geographical content with whatever development can be discerned in the reliably dated charts of the early fourteenth century.

Those who have attempted to date the Carignano map to the very beginning of the fourteenth century[265] would have to explain how a priest, using secondhand information from nautical charts, could present outlines that were not achieved on datable charts for a further two or three decades. Although the earliest of the visible changes through which the charts passed included, inter alia, improvements to the shape of Italy as it had been portrayed on the Carte Pisane, all subsequent developments of this kind—some of which are already reflected on the Carignano map—occurred beyond the Pillars of Hercules. The first region to be affected was the British Isles, and we can actually observe the gradual emergence of outlines as mature as Carignano's in the successive productions of the Vescontes.

Without knowing the mechanism by which a steadily more plausible picture of the British Isles reached Mediterranean chartmakers, we cannot be sure of the relationship, if any, between the Vesconte charts and the Carignano map. Nevertheless, the latter's inclusion of Ireland, albeit sketchily, and the Bristol Channel is broadly equivalent to the stage reached on the steadily evolving Vesconte charts about 1325. The early dates suggested for the Carignano map, like Nordenskiöld's "about 1300,"[266] should surely be abandoned. On the evidence of the British Isles, the Carignano map needs to be set in the shadow of the Vescontes' work and placed in the closing stages of the priest's life, which ended in 1329 or 1330.[267]

The first medieval chartmaker whose name is known to us, Pietro Vesconte, declared himself to be Genoese, like Carignano, but was working in Venice on the two occasions when his author's legend mentioned the place of production.[268] A measure of scholarly disagreement notwithstanding, it seems clear which works can be reliably attributed to him. The chart of 1311, the atlas of 1313, both those of 1318, the 1320 atlas in the Vatican (Pal. Lat. 1362A), and that at Lyons all bear his signature. The other Vatican atlas (Vat. Lat. 2972) is un-

signed but has enough shared features to substantiate earlier opinions that it too is Vesconte's work—at least as far as its strictly nautical charts are concerned.[269] In both the Vatican works the Mediterranean charts form part of a cartographic supplement to a manuscript by Marino Sanudo entitled *Liber secretorum fidelium crucis*. Vat. Lat. 2972 is accepted as being one of the two copies of Sanudo's treatise urging a further crusade that the author presented to the pope in September 1321.[270] This date makes it more probable that the slightly less developed Pal. Lat. 1362A really belongs to 1320, and not some later year—a possibility left open by the closely trimmed roman date, MCCCXX.[271] Also unsigned are the nautical charts attached to the British Library's copy of Sanudo's *Liber* (Add. MS. 27376*), but these can be confidently ascribed to Vesconte.[272]

Pietro Vesconte's work evidently consists, therefore, of one chart of the eastern half of the Mediterranean and seven atlases.[273] It has been suggested that Perrino

263. Osvaldo Baldacci, "La cartonautica medioevale precolombiana," in *Atti del Convegno Internazionale di Studi Colombiani 13 e 14 ottobre 1973* (Genoa: Civico Istituto Colombiano, 1974), 123–36, esp. 125. On the early fourteenth-century Sardinian arms, see Gerola, "Dall'Orto," 423 (note 227), and above, p. 399.

264. Fischer, *Sammlung*, 119 (note 36). Kamal, *Monumenta cartographica*, 4.1:1139 (note 203), transcribed and translated into French Carignano's Latin note about the coloring of non-Christian towns.

265. For example, Nordenskiöld, *Periplus*, 20 (note 14).

266. Nordenskiöld, *Periplus*, 20 (note 14).

267. See note 258. It is also worth noting that Carignano cites four extra features below Cape Bojador, the Vescontes' invariable western African terminus.

268. The 1318 atlas in Venice and the undated atlas in Lyons. He invariably included the words *de janua* in his author's legends. The two autograph productions of Perrino Vesconte, who might be the same as Pietro, were also signed from Venice.

269. Almagià, *Vaticana*, 1:18b (note 35); Degenhart and Schmitt, "Sanudo und Veneto," 66–67 (note 226).

270. Almagià, *Vaticana*, 1:17a (note 35).

271. Almagià, *Vaticana*, 1:14b (note 35). See the reproduction in Kamal, *Monumenta cartographica*, 4.1:1161 (note 203). Nevertheless, Revelli, *Partecipazione italiana*, lxxxi–lxxxii (note 221), detected retouching to the world map, which he considered to be in Vesconte's hand and to date from 1321.

272. Degenhart and Schmitt, "Sanudo und Veneto," 24 (note 226). Andrews, "Scotland in the Portolan Charts," 136 (note 249), proposed, equally plausibly, that its charts might be the work of Perrino Vesconte.

273. The attribution to Vesconte of the Luxoro atlas (Genoa, Biblioteca Civica Berio) by Revelli, *Colombo*, 80, 235–36, 250–51 (note 22), is clearly unfounded. A quite different hand is involved, and in any case the Luxoro atlas should be assigned to the first half of the fifteenth century on the basis of its names (see table 19.3, pp. 416–20), and probably to Francesco de Cesanis; see p. 424. Almagià, *Vaticana*, 1:20b (note 35), disputes the further attribution to Vesconte by Revelli, *Colombo*, 241 (note 22), of the maps in another Sanudo manuscript (Rome, Biblioteca Apostolica Vaticana, Reg. Lat. 548). Nor should any credence be given to the suggestion that the Combitis atlas (Venice, Biblioteca Nazionale Marciana, It. VI, 213) is similar

Vesconte, who signed the Zurich atlas of 1321 and the 1327 chart in Venice, was the same person as Pietro—Perrino being a diminutive form of Pietro.[274] Comparison of the writing on all the signed or attributed Vesconte works reveals strong similarities, but it is dangerous to draw conclusions from this, since a formal book hand of that kind would have been in common use at the time. Weighing against the theory that Pietro latterly signed himself Perrino is the fact that the illegibly dated Pietro atlas in Lyons can be placed on developmental grounds between the two signed Perrino works. Nevertheless, whether one man or two collaborating members of the same family are involved, the charts and atlases concerned exhibit improvements over their predecessors and a steady growth in geographical knowledge thereafter. This can be best demonstrated in the context of the British Isles.

THE BRITISH ISLES

Neither the Cortona chart nor Vesconte's chart of 1311 extends far enough west to take in the British Isles; hence Vesconte's atlas of 1313 invites direct comparison with the Carte Pisane. The latter had presented Britain as a misshapen rectangle lying on an east-west axis, with London (one of only six names) set into the middle of the south coast (fig. 19.12). Vesconte, in 1313, was able to align the British Isles more correctly north-south. Certain features are clearly recognizable: for example, the Cornish peninsula and South Wales (fig. 19.13).[275] There are now thirty-six names. By 1320, the date of Pal. Lat. 1362A, this figure has been increased to forty-six, and Ireland has made its first appearance on a dated portolan chart (fig. 19.14).[276] This presumably indicates either information received by Vesconte after 1318 or a new appreciation of the island's commercial importance.[277]

By the following year, as can be made out on Vat. Lat. 2972 and the Zurich atlas signed by Perrino Vesconte, Ireland has been radically redrawn. At the same time, it has been fitted out with up to forty names, where none had been included before.[278] Further refinements, this time to the east coast of Ireland, can be seen on the British Library's undated Sanudo atlas and on the 1327 Perrino Vesconte chart (fig. 19.15). The hollow of Dundalk Bay, for example, and the southeastern promontory of Carnsore Point are now clearly defined.

The increasing sophistication of the British Isles in the work of the Vescontes thus allows us to arrange their dated and undated productions in one assured chronological sequence. Besides other innovations, the 1321 Vatican and Zurich atlases introduce an unnamed Isle of Man, absent from the atlas of 1320. This and the 1321 form of Ireland are repeated on the illegibly dated

Lyons atlas, which can thus be assigned to 1321 at the earliest.[279] The final Vesconte contribution to the cartography of the British Isles occurs in almost identical fashion on the British Library atlas and the 1327 Perrino chart. The Isle of Man is now named, as are several Irish islands, and the Bristol Channel is shown for the first time. The named features around the coasts of Britain have also increased from the forty-nine found on the Lyons atlas to more than sixty. Ireland shows a similar increase from forty to fifty-four.

It is easy to document successive improvements to the outline and toponymy of the British Isles made by the Vescontes in 1320, 1321, and 1327; but it is another matter to identify the mechanism by which this information reached them, presumably in Venice. Since this process seems to owe nothing to earlier maps of Britain

to Vesconte's of 1318; see Guglielmo Berchet, "Portolani esistenti nelle principali biblioteche di Venezia," *Giornale Militare per la Marina* 10 (1865): 1–11, esp. 5. The Combitis atlas is probably fifteenth century; see table 19.3, pp. 416–20.

274. Nordenskiöld, *Periplus*, 58 (note 14). However, Revelli, *Colombo*, 268 (note 22), surmised that Perrino was probably a son or nephew of Pietro. On this reading Perrino might be a way of saying "Pietro the younger." The partial chart acquired by Nico Israel in 1980 (see note 67) has also been attributed to Perrino, but it has some un-Vesconte features. The white cross over a green Rhodes, a scale bar whose alternate divisions are subdivided, and the additional name *cauo ferro* beneath Mogador in western Africa are examples of these. Its names are also not typical of Vesconte's work, as can be seen in table 19.3, pp. 416–20.

275. Crone, *Maps and Their Makers*, 17 (note 11), pointed out, in the context of the 1327 Perrino Vesconte chart, that "as southern England is too small compared with the rest of the country, it is plain that a piece of relatively accurate survey has been fitted to an older, highly generalised outline of the whole island."

276. It has been argued that there was insufficient space on the relevant sheets in the earlier atlases for Ireland to have been accommodated. It seems safe to assume, though, that the cartographer would have planned his sheet arrangement carefully to leave room for what he wished to show.

277. See two articles by Thomas Johnson Westropp in *Proceedings of the Royal Irish Academy*, vol. 30, sect. C (1912–13): "Brasil and the Legendary Islands of the North Atlantic: Their History and Fable. A Contribution to the 'Atlantis' Problem," 223–60; idem, "Early Italian Maps of Ireland from 1300 to 1600 with Notes on Foreign Settlers and Trade," 361–428. Both are reprinted in *Acta Cartographica* 19 (1974): 405–45, 446–513.

278. The most pronounced feature of this second type is the large island-crowded bay set into the west coast. Plausibly identified as Clew Bay, this has above it a westward-projecting island, presumably Achill; see Michael Corbet Andrews, "The Map of Ireland: A.D. 1300–1700," *Proceedings and Reports of the Belfast Natural History and Philosophical Society for the Session 1922–23* (1924): 9–33, esp. 17 and pl. II, 1. Andrews's analysis of Ireland's shape on the portolan charts emphasizes Vesconte's importance in this respect, since the outline he introduced in 1321 was still being repeated by Grazioso Benincasa a century and a half later.

279. Almagià, *Vaticana*, 1:15 (note 35), and Mollat du Jourdin and de La Roncière (i.e., Isabelle Raynaud-Nguyen), *Sea Charts*, 199 (note 40), support this reasoning. However, de La Roncière, *Lyon*, 8 (note 34) had dated it 1319(?).

(though the final Vesconte form with the Bristol Channel reveals similarities to the Gough map of Great Britain, ca. 1350), it might reasonably be attributed to crew members of trading vessels returning to Venice. True, the state-controlled Venetian fleets, the so-called Flanders galleys, made regular visits to the Flemish ports from 1314 onward, but this proves to be a disappointing lead.[280] The earliest occasion on which this fleet is known to have been directed to England was in 1319, and a violent affray at Southampton led to the postponement of further visits by the Venetians for another twenty years.[281] Thus the Flanders galleys could have contributed neither to the improvement of the 1313 Vesconte atlas over the Carte Pisane nor to further updating in the 1320s.[282]

The quantity of surviving Vesconte works, the confidence with which they can be dated, and the narrow time band involved (ten works spread over a mere fifteen years) permit a fairly full account of the Vescontes' increasing knowledge about the British Isles. While there

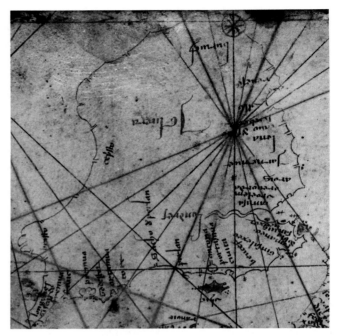

FIG. 19.13. THE CHANGING CONFIGURATION OF THE BRITISH ISLES (2). By the time of the 1313 Pietro Vesconte chart, Great Britain has developed into a more correctly aligned island, with some identifiable features, such as the Cornish peninsula. Instead of the Carte Pisane's mere six names, this chart lists thirty-six.
Photograph from the Bibliothèque Nationale, Paris (Rés. Ge. DD 687, pl. 5).

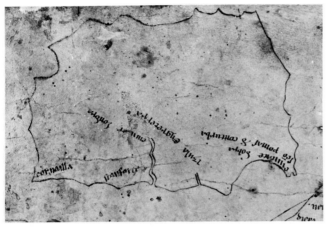

FIG. 19.12. THE CHANGING CONFIGURATION OF THE BRITISH ISLES (1). On the Carte Pisane, Great Britain is represented by a single misshapen rectangle with only six places marked, including *civitate londra* (London), which is placed in the middle of the south coast.
Photograph from the Bibliothèque Nationale, Paris (Rés. Ge. B 1118).

280. Alwyn A. Ruddock, *Italian Merchants and Shipping in South-ampton, 1270–1600* (Southampton: University College, 1951), 22.
281. Ruddock, *Italian Merchants*, 25–27 (note 280).
282. Nor does the fact that Vesconte acted as adviser to the first Flanders fleet appear to be relevant. If the channel of communication by which geographical information about the British Isles reached Mediterranean chartmakers was not Venetian, it was presumably Genoese or Majorcan. Genoese galleys had been visiting England since at least 1278, and a Majorcan galley is recorded at London three years later; see Ruddock, *Italian Merchants*, 19, 21 (note 280). The cartographic contribution of such ventures must remain a matter for speculation, given the individually organized and haphazardly documented nature of Genoese and Catalan operations.

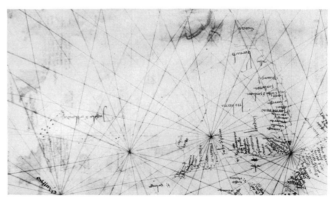

FIG. 19.14. THE CHANGING CONFIGURATION OF THE BRITISH ISLES (3). The first appearance of Ireland on a dated portolan chart is on the 1320 chart by Pietro Vesconte. Although Ireland has no place-names, the total for Great Britain has increased to forty-six.
Photograph from the Biblioteca Apostolica Vaticana, Rome (Pal. Lat. 1362A, fol. 7r).

is insufficient space in this essay to analyze every other change to the content of fourteenth- and fifteenth-century portolan charts, it is possible to sketch in the broad lines of subsequent expansion beyond the Mediterranean. The areas affected were the Baltic, the island groups of the central Atlantic, the western coastline of Africa, and the islands of the North Atlantic.

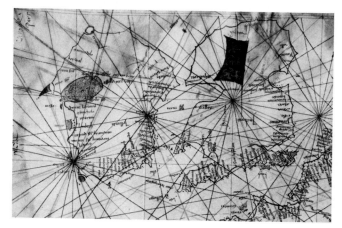

FIG. 19.15. THE CHANGING CONFIGURATION OF THE BRITISH ISLES (4). The chart in the atlas attributed to Pietro Vesconte and datable to about 1325 reveals greater knowledge of the Irish coastline. Dundalk Bay, for example, is now clearly defined.
By permission of the British Library, London (Add. MS. 27376*, fol. 181r).

THE BALTIC

The extension of the portolan charts to include the North Sea and Baltic marks the next major step in their development. The Carignano map is certainly an important witness in this respect, but it is impossible to state with any confidence whether it or a chart signed by Angelino de Dalorto should be considered the first to include Scandinavia (fig. 19.16). To the controversy about the dating of the Carignano map should now be added variant readings of the date on Dalorto's chart. Preserved in the private library of Prince Corsini in Florence, this has an indistinct roman date that has been read as MCCCXXII, MCCCXXV and MCCCXXX. The earliest possibility, 1322, has found few supporters, but opinions have been strongly divided between the other two.[283] To emphasize this continuing uncertainty, the Dalorto chart will be described hereafter as "1325/30."

By 1330 at the latest, therefore, portolan charts were being extended northward beyond the west coast of Jutland to recognize, if imperfectly, the existence of the Baltic and the lands to the north of it. A version of the Marino Sanudo world map of about 1320 has been shown to be a precursor of the portolan chart outlines, as has its immediate antecedent, the world map of Fra

Paolino (Paolino Veneto).[284] These are clearly from a quite different source, though, than the outline displayed on the Carignano and Dalorto versions, which rotate the Baltic away from Sanudo's north-south orientation to give it a more correct alignment. Considering that charts did not form part of the standard navigating equipment in northern waters,[285] this represents a significant advance. The Jutland peninsula, with an attempt

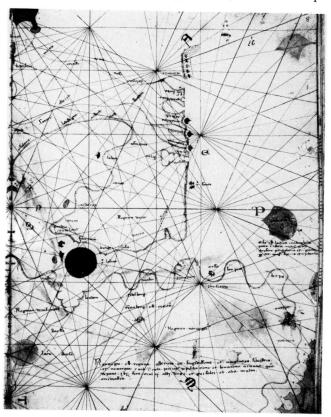

FIG. 19.16. SCANDINAVIA BY BIANCO. Denied access to the Baltic after the 1320s, Mediterranean sailors obtained their limited information about it secondhand. Many chartmakers, particularly Italian ones, omitted northern Europe altogether. One notable exception was Andrea Bianco, whose atlas of 1436 included this detailed separate chart of the region.
Size of the original: 37.3 × 26.5 cm. By permission of the Biblioteca Nazionale Marciana, Venice (It. Z.76, c. 6).

283. See particularly Magnaghi, "Alcune osservazioni," 20–23 (note 201).

284. Almagià, *Vaticana*, 1:15b (note 35); Heinrich Winter, "The Changing Face of Scandinavia and the Baltic in Cartography up to 1532," *Imago Mundi* 12 (1955): 45–54, esp. 45–46.

285. Commenting on the practice of northern mariners, William Bourne, in the mid-sixteenth century, had cause to complain of "auncient masters of shippes" who mocked the use of sea charts—quoted in Waters, *Navigation in England*, 15 (note 138). Nevertheless, the Barcelona archives record that in 1390 the merchant Domenech Pujol sent a consignment of eight *cartes de navegar* to Flanders; see Claude

at the Danish islands to its east, guards the entrance to a Baltic whose shape gives no hint of the Gulf of Bothnia. The Baltic's most prominent feature is a much exaggerated Gotland, whose capital, Visby, was one of the major trading centers of northern Europe. Though Norway is not present on the Carignano map, the Dalorto chart already gives it the heavily shaded outline (like a slanting, upside-down capital A) that would typically be found on many later charts.

That the early fourteenth-century picture of Scandinavia should have persisted on portolan charts until well into the following century has been explained in terms of Hanse influence.[286] Having no chartmaking tradition of its own, the Hanseatic League was in no position to supply hydrographic details about the Baltic, even had it wanted to.[287] In addition, the establishment of a staple at Bruges in 1323[288] denied southern ships direct access to the Baltic and hence opportunities for their pilots to make firsthand observations. It follows from this that the Baltic outlines given by Carignano and Dalorto were presumably gathered before 1323. Not only was there to be little improvement to these outlines, but many later charts omitted northern Europe entirely.[289] Generally speaking, Italian chartmakers from Vesconte onward more frequently omitted Scandinavia, whereas the Catalans tended to follow Dalorto's example and include it.

THE ATLANTIC ISLANDS

While information about northern Europe remained static on the portolan charts for large parts of the fourteenth and fifteenth centuries, awareness of what lay to the west and south was to grow steadily over the same period. Indeed, the charts themselves were to play an important part in broadcasting knowledge, or theories, about the Atlantic archipelagoes and the western coast of Africa. Because the islands depicted on the charts were stepping-stones for later voyages to America or have been treated as evidence of pre-Columbian discoveries of the new continent itself, this aspect of the subject has attracted more comment than any other. It would require an entire volume to summarize the complex and contradictory arguments about the apparently imaginary islands of Man, Brazil, Antilia, and others.[290]

The increasing array of names attached to Atlantic islands on fourteenth- and fifteenth-century charts has been conveniently tabulated by Armando Cortesão, the most tireless worker in this particular field.[291] The identification of these names with the islands of today is, however, a matter of interpretation, and some of Cortesão's conclusions have been challenged. Even when both modern and medieval islands bear the same names, this is no proof of identity. As Admiral Morison ob-

served, "it would have been natural enough for Prince Henry to have used the names of legendary islands for the actual islands that his men discovered."[292] It is thus with considerable caution that we should approach apparent representations on the early charts of the four main archipelagoes of the central Atlantic: the Canaries, the Madeiras, the Azores, and the Cape Verde Islands.

The Canaries present the least problem in this respect. Because of their shape and position, there can be no doubt about the authenticity of Lanzarote and Fuerteventura, which both appear (thus named) on Dulcert's chart three years after their documented discovery in 1336.[293] Cortesão's claim that the other three groups were depicted on charts a considerable time before their first mentions in the archival record has not been so readily accepted. He maintained that the Madeira islands are represented on the Dulcert chart, even though the islands are only known to have been discovered in 1418–19.[294] For the Azores, Cortesão considered that the 1367 Pizigani chart's *insula de bracir* indicated Terceira, and he interpreted several further instances in the Medici atlas (which he therefore redated to ca. 1370) as

Carrère, *Barcelone: Centre économique à l'époque des difficultés, 1380–1462*, Civilisations et Sociétés 5, 2 vols. (Paris: Moulton, 1967), 1:201 n. 4.

286. Anna-Dorothee von den Brincken, "Die kartographische Darstellung Nordeuropas durch italienische und mallorquinische Portolanzeichner im 14. und in der ersten Hälfte des 15. Jahrhunderts," *Hansische Geschichtsblätter* 92 (1974): 45–58, esp. 54.

287. Von den Brincken, "Nordeuropas," 46 (note 286).

288. Von den Brincken, "Nordeuropas," 54 (note 286).

289. Though the separate sheet devoted to this region in Bianco's atlas of 1436 deserves particular mention; see R. A. Skelton, Thomas E. Marston, and George D. Painter, *The Vinland Map and the Tartar Relation* (New Haven: Yale University Press, 1965), 116, 164.

290. See, for example, Cortesão, *Nautical Chart of 1424* (note 24); idem, *History of Portuguese Cartography*, 2:52–73 (note 3); James E. Kelley, Jr., "Non-Mediterranean Influences That Shaped the Atlantic in the Early Portolan Charts," *Imago Mundi* 31 (1979): 18–35, esp. 27–33; Samuel Eliot Morison, *The European Discovery of America: The Northern Voyages* (New York: Oxford University Press, 1971), 81–111; idem, *Portuguese Voyages to America in the Fifteenth Century* (Cambridge: Harvard University Press, 1940).

291. Cortesão, *History of Portuguese Cartography*, 2:58–59, table I (note 3). This is a significantly reworked version of similar tables published earlier by Cortesão.

292. Morison, *Portuguese Voyages*, 13 (note 290).

293. Cortesão, *History of Portuguese Cartography*, 2:72 (note 3).

294. Cortesão, *Nautical Chart of 1424*, 47 (note 24); Cortesão, *History of Portuguese Cartography*, 2:55, 58 (note 3). That the first archival mention of Madeira dates from 1418–19 is not sufficient reason for assuming that sheet 2 of the Pasqualini atlas must be later than its stated date of 1408 because it names Madeira thus, rather than using its earlier name *do legname*; see Petar Matković, "Alte handschriftliche Schifferkarten in der Kaiserlichen Hof-Bibliothek in Wien," *Programm des königlichen kaiserlichen Gymnasiums zu Wrasdin* (Agram: L. Gaj, 1860), 9. The Pasqualini toponymy is generally ahead of its time.

references to other islands in the group.[295] The first definite knowledge of the Azores, however, dates from 1427.[296] Lying much too close to Portugal, these fourteenth-century instances have been dubbed the "false Azores" by Cortesão's detractors.[297] Last of all were the Cape Verde Islands, discovered in 1455–56 but already, in Cortesão's view, partially represented on the 1413 Mecia de Viladestes chart.[298] Their earliest unequivocal appearance is in the two 1468 Grazioso Benincasa atlases.

More controversial still is the island of Antilia, first found on the 1424 Zuane Pizzigano chart, to which Cortesão devoted a separate study.[299] His assertion that Antilia and the islands close to it (*satanazes, ymana,* and *saya*) "are intended to represent the easternmost part of the American hemisphere"[300] has won little acceptance outside Portugal. It is perhaps more profitable in this context to consider the attitude of other fifteenth-century chartmakers. Where many twentieth-century historians have become mesmerized by Antilia and its immediate neighbors, contemporary chartmakers often ignored them. Grazioso Benincasa, for example, would show these legendary islands when space allowed (as on his 1470 and 1482 charts), but having inserted them on an early atlas in 1463, he failed to make room for them in his later volumes.[301] Yet Benincasa was one of the main conduits through whom the details of Portuguese discoveries reached the portolan charts.

WESTERN AFRICA

Compared with that of the offshore islands, the charting of Africa's west coast was more straightforward. The process by which successive Portuguese captains worked their way patiently down the coast, in a series of planned leapfrogs, has been frequently told. So too has the part played after 1415 by Prince Henry the Navigator, particularly in inspiring his men to pass the dreaded Cape Bojador.

The first documented expedition to round Bojador was that of Gil Eanes in 1434; yet information beyond this Mauritanian cape had already been appearing on portolan charts for a century. As with the Atlantic islands, it is hard to tell whether this derived from actual voyages, secondhand information, or pure conjecture. What is clear is that the coverage of the northwestern African coastlines increased steadily on the earliest charts.[302]

Table 19.2, p. 412, shows the farthest points reached by successive expeditions. Its right-hand column indicates the first charts to take note of them, thereby supplying an index, albeit an imperfect one, to the dissemination of geographical knowledge.

The Carte Pisane terminated at roughly 33° north,

Vesconte's atlas of 1318 showed a continuation southward for a further two degrees,[303] and the 1325/30 Dalorto chart added another two to that.[304] In its turn, the 1339 Dulcert chart included a few extra names, although these cannot be identified today. Dulcert refers to as *caput de non* what the Catalan atlas and later charts would call *cabo de buyetder.* Although somewhat misplaced, this was evidently intended for Cape Bojador; it clearly represented the limit of fourteenth-century coastal knowledge.[305] Beyond that, the early charts offer an ill-defined southeasterly or southerly outline, where

295. Cortesão, *History of Portuguese Cartography,* 2:58 (note 3).

296. Cortesão, *Nautical Chart of 1424,* 47 (note 24). The evidence comes from a note on the 1439 Valseca chart. However, F. F. R. Fernández-Armesto, "Atlantic Exploration before Columbus: The Evidence of Maps," *Renaissance and Modern Studies* (forthcoming), finds in favor of the fourteenth-century Azores.

297. Heinrich Winter, "The Fra Mauro Portolan Chart in the Vatican," *Imago Mundi* 16 (1962): 17–28, esp. 18 n. 5.

298. Cortesão, *Nautical Chart of 1424,* 47–48 (note 24).

299. Cortesão, *Nautical Chart of 1424* (note 24). Reactions to its claims were discussed by Cortesão in two later works: *History of Portuguese Cartography,* 2:134–39 (note 3); and "Pizzigano's Chart of 1424," *Revista da Universidade de Coimbra* 24 (1970): 477–91. Although this is the first reference to Antilia on a surviving chart, Pedro de Medina mentioned its presence on a Ptolemy manuscript presented to Pope Urban, evidently Urban VI (1378–89); Cortesão, "North Atlantic Nautical Chart," 8 (note 170). The 1424 chart in the James Ford Bell Library, Minneapolis, is not to be confused with that in the Zentralbibliothek der Deutschen Klassik, Weimar. The latter, often attributed to Conte Hectomano Freducci, was at one time thought to bear the date 1424 and was, on that basis, introduced into discussions about Antilia.

300. Cortesão, *Nautical Chart of 1424,* 3 (note 24). For a more recent interpretation that also sees Antilia and the other Atlantic islands as references to parts of the American continent see Kelley, "Non-Mediterranean Influences," 27–33 (note 290).

301. Benincasa's cavalier treatment of Antilia should be contrasted with the invariable appearance of the Cape Verde group in the atlases he drew after 1468. The occasional reappearance of Antilia on sixteenth-century charts—for instance, Georgio Calapoda's of 1560 in the National Library of Scotland—highlights the danger of overliteral interpretation by historians of cartography, since Calapoda's atlas of 1552 had contained an adequate representation of America; Nordenskiöld, *Periplus,* pl. XXVI (note 14). To Calapoda, Antilia was presumably just one of the features that belonged on a traditional chart.

302. See the analysis of African names in Kamal, *Monumenta cartographica,* 4.4:1468 (note 203).

303. The 1313 atlas, the earliest of his works to take in northwestern Africa, is incomplete at this point.

304. The extent of the Carignano map cannot now be determined from the available reproductions.

305. Notwithstanding the claim by Paolo Revelli, "Una nuova carta di Batista Beccari ('Batista Becharius')?" *Bollettino della Società Geografica Italiana* 88 (1951): 156–66, esp. 156–57 n. 1, that the Genoese rounded Bojador in the fourteenth century. He cites as evidence the Combitis atlas (described merely as Biblioteca Nazionale Marciana MS VI.213), whose supposed 1368 date derives from a misreading of the Irish reference to the 368 islands; see Fiorini, *Projezioni,* 676 (note 98). Hinks, *Dalorto,* 12 (note 76), corrected this error. On the dating of the Combitis atlas see table 19.3, pp. 416–20.

TABLE 19.2 The Cartographic Record of the Western Coast of Africa

Latitude	Longitude	Modern Name	Name on the Charts	First Recorded Sighting		First Chart or Map to Include It
				Date	Discovered by	
33.17N[a]		Azemmour	Zamor			Carte Pisane
31.31N		Essaouira	Mogador			1318 Vesconte
29.05N		Oued Noun River	[Aluet Nul][b]			[1325/30 Dalorto]
28.44N		Cap Drâa (Cabo Noun)				
27.57N		Cap Juby	[Cauo de Sabium]			[1375 Catalan atlas]
26.08N		Capo Bojador	Buyetder	1434	Eanes	1375 Catalan atlas (although the 1339 Dulcert chart, which names this *caput de non*, includes four names beyond this point)
24.40N		Angra de los Ruivos (Gurnet Bay)		(1435)	Eanes and Baldaia	
[23.36N]		Rio de Oro		1436	Baldaia	Referred to in a note on the 1339 Dulcert chart
22.11N		Ilha Piedra Galha		1436	Baldaia	
20.46N		Cap Blanc/ Capo Blanco		1441	Tristão	
20.40N		Baie d'Arguin (beyond Baie de Saint-Jean, 19.27N)	Terra dos Negros	1443 1444	Tristão Tristão	
14.43N		Cap Vert		1444	Dias	1448 Bianco
14.32N		Cap de Naze	Cabo dos Mastos	1445	Fernandes	
12.20N		Cabo Roxo		1446	Fernandes	[1448 Bianco], 1463 Benincasa, 1459 Fra Mauro[c] world map, Catalan world map in Modena (Estense C.G.A. 1)
12.16N		Ponta Varela	Cabo Vela	1446	Fernandes	
11.40N		Canal do Gêba	Rio Grande	1446	Tristão	
9.31N		Conakry		1447	Fernandes	
8.30N		Cape Sierra Leone	Capo Liedo	1460	Sintra	
6.19N		Cape Mesurado		1461	Sintra	
6.06N		Bassa Point	Cauo de Sancta Maria	1461	Sintra	1468 Benincasa (Great Britain, private collection)
4.22N	7.44W	Cape Palmas		1470	Costa	
	3.44W	Komóe River	Rio de Suero	1470	Costa	
	1.38W	Shama		[1470–71]	Santarém and Escobar	1492 Aguiar

Latitude	Longitude	Modern Name	Name on the Charts	First Recorded Sighting		First Chart or Map to Include It
				Date	Discovered by	
	3.23E	Lagos River	Rio de Lago	1471	Santarém and Escobar	Portuguese chart in Modena (Estense C.G.A. 5c)
Equator				1473	Gonçalves	
1.52S		Pointe Sainte-Catherine		1474	Gonçalves and Sequeira	
6.05S		Mouth of the Congo River		1483	Cão	Reinel chart in Bordeaux
13.25S		Cabo de Santa Maria	Cabo Lobo, and Pradro	1483	Cão	[1489] Cornaro atlas
21.47S		Cape Cross		1485	Cão	
34.21S	18.28E	Cape of Good Hope		1488	Dias	1492 Behaim globe, Martellus world maps
33.29S	27.08E	Great Fish River				

ᵃThe coordinates and modern place-name forms in this table have been taken from the continuing series of gazetteers issued by the United States Board on Geographic Names, Office of Geography, Department of the Interior, Washington, D.C.

ᵇThere is still doubt about a number of the western African discoveries and of the voyages that made them. This table does not pretend to be authoritative in this respect. Nor is it clear what genuine features are shown between Mogador and Bojador on fourteenth-century charts. On the voyages, much use has been made of Armando Cortesão, *History of Portuguese Cartography*, 2 vols. (Coimbra: Junta de Investigações do Ultramar-Lisboa, 1969–71), and of Boies Penrose,

Travel and Discovery in the Renaissance, 1420–1620 (Cambridge: Harvard University Press, 1952). On the names, see Avelino Teixeira da Mota, *Topónimos de origem Portuguesa na costa ocidental de Africa desde o Cabo Bojador ao Cabo de Santa Caterina*, Centro de Estudos da Guiné Portuguesa no. 14 (Bissau: Centro de Estudos da Guiné Portuguesa, 1950).

ᶜThe claim in Paolo Revelli, ed., *La partecipazione italiana alla Mostra Oceanografica Internazionale di Siviglia (1929)* (Genoa: Stabilimenti Italiani Arti Grafiche, 1937), xciii, that the Fra Mauro map took note of Portuguese discoveries dating from 1458 appears to lack foundation.

the genuine coast trends southwesterly. The various fourteenth-century names (up to four on any one chart) that occur beyond this cape carry little conviction. Among them is Cape Gozola,[306] often cited as marking the end of Africa.

Another legendary element found at the southern extremity of some charts was the supposed western branch of the Nile. Shown as flowing into the sea below Bojador, it was sometimes (on the Carignano map, for example) dubbed the River of Gold. One Jaime Ferrer set out in 1346 to search for it, a fact recorded on the Catalan atlas.[307] When, ninety years later, Afonso Gonçalves Baldaia discovered what is now the Rio de Oro he bestowed the name on a bay that has no river flowing into it at all.

Despite this modest evidence from the portolan charts, certain broad claims have been made: that the Gulf of Guinea was featured on maps before its documented discovery in 1470–72, and even that the rounding of the Cape of Good Hope in 1488 had been anticipated.[308] The Medici atlas (see appendix 19.1) and various fifteenth-century world maps have been cited in support

of these theories, but reliably documented western African discovery and the systematic colonization of the offshore islands began only with the capture of Ceuta in 1415.

Though the Portuguese are known to have recorded their discoveries on charts, the earliest surviving Portuguese work dates from the end of the fifteenth century.[309] We must turn instead to charts made by foreigners such as Andrea Bianco and Grazioso Benincasa for the cartographic record of these voyages. In only two cases are the charts in question dated. Indeed, most have been assigned a date—which is really no more than a *terminus post quem*—precisely because of the discoveries they include. Little can be learned, therefore, about the time it took for new information to become available

306. Revelli, *Colombo*, 371–75 (note 22).

307. Kamal, *Monumenta cartographica*, 4.2:1235 (note 203).

308. On the Gulf of Guinea see Destombes, *Mappemondes*, 220 (note 10), citing Eva G. R. Taylor, "Pactolus: River of Gold," *Scottish Geographical Magazine* 44 (1928): 129–44, and on the Cape of Good Hope see Nordenskiöld, *Periplus*, 122 (note 14).

309. See above, p. 374.

outside Portugal. Yet if Bianco in 1448 and Grazioso Benincasa in 1468 were in any way typical, news seems to have traveled fairly fast, since Bianco incorporated discoveries dating from four (or possibly only two) years previously and Benincasa those made seven years before.[310]

It has been claimed that an official policy of secrecy (*sigilo*) prohibited chartmakers from showing the later African discoveries. King Manuel's edict of 1504 demanding silence about the trend of the coast below the river Congo has often been cited. This theory has been challenged,[311] however, and if *sigilo* was operative in the fifteenth century it does not seem to have caused any obvious cartographic delays. Diogo Cão's discoveries south of the Congo in 1483 appear on a Venetian collection of copied charts, one of which is dated 1489 (the Cornaro atlas). Similarly, the results of Bartolomeo Dias's voyage found their way soon afterward onto the world maps of Henricus Martellus Germanus and Martin Behaim's globe of 1492.

THE NORTH ATLANTIC

During the fifteenth century, while Africa's west coast was being plotted, developments were also taking place in the North Atlantic. Arguably the most significant adjustment to have been made to the portolan charts after the early fourteenth century was the correction of the long-standing mismatch of scale between the Mediterranean and Atlantic sections. This was announced by Francesco Beccari on his chart of 1403. Atlantic distances had previously been understated by between 16 and 30 percent.[312] It remains to be tested which of those chartmakers who followed Beccari incorporated the amended Atlantic scale.[313] This improvement notwithstanding, the main trend of the coastline from Flanders to the northern tip of Jutland would not be properly understood until the mid-fifteenth century. A. W. Lang pinpointed Roselli's chart of 1462 as the first to introduce the fresh outline.[314]

In the same way that the portolan charts are an important, if ambiguous, witness to developing knowledge of the Atlantic archipelagoes, so they are frequently cited as evidence of partial Mediterranean understanding of the islands in the North Atlantic. In the process they have inspired further controversy. Winter considered that Iceland was first represented on Jaime Bertran's chart of 1482, though Revelli believed this island was intended by the earlier *archania* of Bartolomeo de Pareto's 1455 chart.[315] Others saw this innovation as having been foreshadowed by the oval *stillanda* found to the north of Scotland from the time of the earliest Catalan charts onward.[316] A larger body of opinion prefers to identify Iceland with the mysterious island of Frisland

(*frixlanda*).[317] Referred to in the account of the questionable Nicolò Zeno voyage to the North Atlantic in the late fourteenth century, Frisland seems first to appear on a dated chart in 1500.[318] This must raise justified doubts about the proposed fifteenth-century dates for anonymous charts that include it.[319]

CONSERVATISM

In isolating specific elements of the stylistic and hydrographic content of fourteenth- and fifteenth-century

---

310. It needs emphasizing that these discoveries clearly meant less to Benincasa than they do to twentieth-century historians. Having recorded the west coast of Africa as far as *cauo de sancta maria* on one of the 1468 atlases, he usually terminated his later atlases at *rio de palmeri* (river Sherbro), 1½° farther north.

311. Bailey W. Diffie, "Foreigners in Portugal and the 'Policy of Silence,'" *Terrae Incognitae* 1 (1969): 23–34.

312. Kelley, "Non-Mediterranean Influences," 22 (note 290), suggested an understatement of about 16 percent; Clos-Arceduc, "Enigme des portulans," 228 (note 129), proposed a figure of almost 30 percent.

313. In a brilliant piece of cartometric analysis, and long before the discovery of the 1403 Francesco Beccari chart with its explanatory note, Hermann Wagner noticed the improvement on (among others) the Batista Beccari chart of 1436 [i.e., 1435]; see Herman Wagner, "The Origin of the Mediaeval Italian Nautical Charts," in *Report of the Sixth International Geographical Congress, London, 1895* (London: Royal Geographical Society, 1896), 695–702, esp. 702; reprinted in *Acta Cartographica* 5 (1969): 476–83. See also Kelley, "Non-Mediterranean Influences," (note 290).

314. Arend Wilhelm Lang, "Traces of Lost North European Sea Charts of the Fifteenth Century," *Imago Mundi* 12 (1955): 31–44, esp. 36–37.

315. Winter, "Catalan Portolan Maps," 4 (note 174); Paolo Revelli, ed., *Elenco illustrativo della Mostra Colombiana Internazionale* (Genoa: Comitato Cittadino per le Celebrazioni Colombiane, 1950), 151.

316. Harald Sigurðsson, *Kortasaga Islands frá öndverðu til loka 16. aldar* (Reykjavik: Bókaútgáfa Menningarsjóðs og Þjóðvinafélagsins, 1971), 258 (English summary on pp. 257–67). Iceland's first cartographic appearance was on the Anglo-Saxon map of ca. 1000 (p. 257).

317. Skelton, Marston, and Painter, *Vinland Map*, 166 (note 289). See also Oswald Dreyer-Eimbcke, "The Mythical Island of Frisland," *Map Collector* 26 (1984): 48–49.

318. Ernesto García Camarero, "Deformidades y alucinaciones en la cartografía ptolemeica y medieval," *Boletin de la Real Sociedad Geográfica* 92 (1956): 257–310, esp. 289. Revelli, *Colombo*, 339 (note 22), discerned Frisland on the 1480 Canepa chart (his plate 80), but the island in question is Stililant.

319. For example, Paris, Bibliothèque Nationale, Département des Cartes et Plans, Rés. Ge. AA 562, and Département des Manuscrits, Ital. 1704. Several writers, among them Sigurðson, *Kortasaga Islands*, 258 (English summary), n. 241 (note 316), dated to the second half of the fifteenth century the undated Catalan chart in the Biblioteca Ambrosiana, Milan, which includes Fixlanda. The various reproductions made of this, for example by Sigurðsson, p. 61, are restricted to details of northwestern Europe and omit the chart's manuscript number. Nevertheless, it seems clear that S.P. 2, 36 (formerly S.P. II, 5) is involved; yet this cannot be earlier than 1530 because of the red cross over Malta.

charts and seeking to identify the moment at which these elements made their first appearance, there is danger that a false impression of constant evolution might have been conveyed in the previous sections.[320] It must be stressed that many charts ignored new information, and it would be quite wrong to see the history of portolan charts purely in terms of successive innovations and unrelenting progress. A number of sixteenth-century chartmakers continued to produce work that is indistinguishable, as far as the outlines are concerned, from that of their predecessors.[321] Scholars, whose currency is knowledge, have tended to explain these lapses in terms of cartographic ignorance. The truth is probably more mundane: that chartmakers often omitted fresh discoveries because they lacked relevance for themselves and their clients or because of practical limitations imposed by the material on which they worked.

To many historians of cartography it has been the geographical innovations that have seemed significant. The charts' contemporary users would probably have considered irrelevant the inclusion of any areas they were unlikely to visit if the correlation with trading activity (to be discussed below, pp. 444–45) is accepted. From the commercial viewpoint of most Mediterranean seafarers, the Atlantic coasts would have been of interest only between Morocco and Flanders; indeed, a number of charts stop at the exit from the Mediterranean. The frequent omission of Scandinavia from fourteenth- and fifteenth-century charts can be explained in this way; so, surely, can the refusal of some chartmakers to accommodate either distant Atlantic islands, or western African coastlines where after 1481 the controlling power, Portugal, threatened interlopers with death.[322]

Even if it must remain a mere hypothesis that the apparent shortcomings of some charts reflect the interests of their original purchasers, there is little doubt about the limitations imposed by the size of a single skin. Here the form of the chart must frequently have dictated its content. Though sections of vellum were sometimes joined to offer a larger surface, most charts used only one skin. For a fifteenth-century chartmaker to have incorporated the steadily growing western African coastline would have meant reducing the scale of the Mediterranean, the traditional heart and purpose of his chart.[323] This he was obviously not prepared to do. Grazioso Benincasa's atlases regularly included a special sheet for western Africa; his separate charts stopped at Bojador. Since they made no atlases, the Catalan draftsmen thereby denied themselves the best opportunity for showing the Portuguese discoveries.

Those were simple omissions. As several writers have pointed out, many of the charts were in no sense up to date.[324] They sometimes continued with outmoded forms or even, as in the case of Petrus Roselli, reverted

to an earlier design.[325] Nordenskiöld considered the 1339 Dulcert chart superior in some respects to Bianco's work of a century later.[326] At best, there was often no visible improvement in the Mediterranean outlines. This led one authority, Heinrich Winter, to conclude that the portolan charts reached their peak with the Pizigani chart of 1367, after which deterioration set in.[327] This harsh judgment has not gone unchallenged. Cortesão and Teixeira da Mota, for example, believed that the charts exhibited "successive improvements," not decline.[328] The best way to test these contradictory opinions is by examining the Mediterranean place-names.

TOPONYMIC DEVELOPMENT

Toponymy is the lifeblood of the portolan charts, providing an unrivaled diagnostic source, and one that can readily be quantified. The names' density and their spread to every part of the chart allow conclusions in which the accidental and local elements are neutralized. If the potential value of the analysis of place-names has been appreciated in the past, the daunting size of the task has presumably inhibited its systematic use.[329]

---

320. Michael J. Blakemore and J. B. Harley, *Concepts in the History of Cartography: A Review and Perspective*, Monograph 26, *Cartographica* 17, no. 4 (1980): 17–23, warn against making evolutionary assumptions.

321. For example, the Freducci family; see Caraci, "Benincasa and Freducci," 42 (note 247).

322. John Horace Parry, *Europe and a Wider World, 1415–1715*, 3d ed. (London: Hutchinson, 1966), 30.

323. The general chart in the British Library's Cornaro atlas (Egerton MS. 73) demonstrates the quite different format required if western Africa was to be included.

324. Andrews noted the lack of development in the representation of the British Isles in "Boundary," 46 (note 249), and "Map of Ireland," 14–15 (note 278).

325. Noted independently by Andrews, "Scotland in the Portolan Charts," 142 (note 249), and Lang, "Lost North European Sea Charts," 40 n. 1 (note 314).

326. Nordenskiöld, *Periplus*, 58 (note 14).

327. Winter, "Catalan Portolan Maps," 7 (note 174).

328. Cortesão and Teixeira da Mota, *Portugaliae monumenta cartographica*, 1:xxvi (note 29).

329. For example, Cornelio Desimoni appreciated the dating potential of toponymy, even if his interpretations about a handful of Ligurian names cannot now be supported. See the "Rendiconto" section of *Atti della Società Ligure di Storia Patria* 3 (1864): CVII. Caraci, too, was well aware of the fundamental importance of toponymy and the need for it to be subjected to minute analysis. His concern, though, was less with the incidence of place-names than with their form, and he used observed linguistic differences to argue in favor of the priority of Italian over Catalan chartmakers; see Giuseppe Caraci, "A proposito di una nuova carta di Gabriel Vallsecha e dei rapporti fra la cartografia nautica italiana e quella maiorchina," *Bollettino della Società Geografica Italiana* 89 (1952): 388–418, esp. 393–99. This theme was expanded in Caraci's later *Italiani e Catalani* (note 175).

TABLE 19.3 Significant Place-Name Additions from Dated Works Applied to Undated Atlases and Charts:

The twenty-one works listed across the top represent dated or datable charts and atlases produced between 1313 and 1426.[a] They are arranged in chronological order, with a note on their origin (Catalan, Genoese, Italian, or Venetian). Each is further supplied with a figure representing the number of "significant" new names it introduced (see appendix 19.5 for explanation of this term).

The first column on the left lists by number twenty-four undated works supposedly produced during this same period.[b] An examination was made of each undated example to see how many of the toponymic innovations assignable to the successive dated works it contained. The columns of figures indicate the number of new place-names present out of the totals listed at the top. Where charts are very incomplete,

DATED OR DATABLE WORKS

	Vesconte	Vesconte (Venice and Vienna atlases combined)[e]	Vesconte (Vatican, Pal. Lat. 1362A)	Vesconte (Vatican, Vat. Lat. 2972 and Zurich atlases combined)	Vesconte (Lyons)	[Vesconte—London]	Vesconte	Dalorto (treated as later than 1327)	Dulcert	Pizigani	Pizigano
	1313 It. (24)	1318 It. 39	(1320) It. 8	1321 It. 5	ca. 1322 It. 4	ca. 1325 It. 36	1327 It. 3	1325/30 Cat. 79	1339 Cat. 20	1367 Ven. 25	1373 Ven. 23
ARABIC											
1		3/10				1/11		4/17	1/10		
CATALAN											
2	12	21	3	2	1	15	1	72	19		
3	4	16	1	1	1	14	2	40/41	17		
4	14	20	3	1	1	12	1	75	17	1	1
5	15	23	3	2	1	15	2	76	20		
6	15	22	3	2	1	14	1	68	19	1	
7	12	22	3	1	1	13	1	76	19	1	
8	3/4	1/4	1/1	1/1				13/15	3/3		
9	11	21	2	2		16	1	62	20	4	3
ITALIAN											
10	5	2	1			1		1		1	1
11	4/6	2/6	1/2			1/14		10/14			
12	2/11	13/27	1/7			5/30	1/1	11/44	1/15		1/19
13	11	11	3		1	3		26			2
14	14	21	3	2	1	15	1	74	19		
15	4/8	7/10	3/5	1/2	1/3	8/10	1/1	7/12	3/4	16/18	10/11
16	3/3	1/3				7/11		9/13	1/2	3/3	2/6
17	16	19	2	2	3	20	2	41	4	16	14
18	15	19	3	2	1	17	2	49	5	5	5
19	1/3	7/9	4/5	2/2	3/3	5/7	1/1	4/6	3/5	10/16	7/11
20	18	24	5	4	2	24	3	57	9	21	15
21	17	26	5	4	3	29	3	54	6	19	17
22	9	24	2	2	2	16		38	4	3	7
23	10	22	2	2	3	18		38	4	5	8
24	13/15	11/12	4/5	2/2	3/3	9/10	1/1	9/12	3/4	17/18	13/15

[a]Because this study begins with Vesconte and is concerned with documenting the addition of names, the 1313 atlas was considered only for its toponymic innovations in the areas previously covered by his earliest chart, dated 1311 (i.e., the Black Sea and the eastern half of the Mediterranean). Two dated charts were omitted from the analysis: the untraceable 1404 Pongeto chart and the Arabic Kâtibî chart of 1413. Three others were found to contain no significant additions: the Viladestes charts of 1413 and 1428 and the Briaticho chart of 1430.

[b]Undated charts that could not be included in this analysis were as follows: the Carignano map; the chart formerly in the Prince Youssouf Kamal collection, Cairo (reproduced in Kamal, *Monumenta cartographica Africae et Aegypti*, 5 vols. in 16 pts. (Cairo, 1926–51), 4.2:1206; and the fragment in the Public Record Office, London, MPB 38 (all three on the grounds of legibility); the Catalan fragment stolen from the Archivio di Stato, Venice; and the unsigned chart in the Corsini collection, Florence. (I am grateful to Geraldine Beech for drawing my attention to the Public Record Office chart.) The chart in the Hispanic

A Guide to Dating and Interrelationship (Considering the Continuous Coastline between Dunkirk and Mogador)

a second figure has been added to show the number of that year's additions that fall within the chart's reduced area.[c]

If allowance is made for marked differences between Catalan and Italian works, these figures point to the most likely time slot for each undated work, by reference to the 415 toponymic innovations found on the dated examples. In the list of charts that follows, the dates proposed in the past are contrasted with those suggested by this analysis. It must be stressed that these cannot on their own provide reliable dates of construction, merely more plausible dates for their toponymic content.[d]

Catalan Atlas	Soler	Beccari	Pasqualini	Virga	Cesanis	Viladestes	[Pizzigano]	Beccari	Giroldi	Post-1430 Additions	TOTAL
(1375) Cat. 11	1385 Cat. 10	1403 Gen. 50	1408 Ven. 17	1409 Ven. 1	1421 Ven. 1	1423 Cat. 1	1424 It. 9	1426 Gen. 17	1426 Ven. 1	31	415
4	1										
7	1										
9	4										
8	4										
9	9					1					
1/1		3/10									
8	5	7				1	3	4		2	
1		1	1[f]								
		3/28									
			1								
2		1									
		3/10									
	1	4	4	1							
	1	7	3								
			2/7		1						
2	1	7	15		1						
		2	10	1	1						
1	2	3	2	1				1	1		
1	2	5	2	1				1	1		
		1/6	7/9		1						

Society, New York (chart 1), attributed to Giroldi with a date about 1425, was only partially legible from Stevenson's reproduction. Its names, however, conformed to the pattern of the 1426 Giroldi chart. A number of other works have been implausibly assigned by previous researchers to the period before 1430. The suggestion that the chart in Venice, Museo Civico Collezione Correr, Port. 40 (Morosini Gatterburg 469) should be dated to about 1400—on the grounds of its supposed similarities to the Pizigani and Virga charts of 1367 and 1409—was noticed too late for the chart to be included in this exercise; see Marcel Destombes, "La cartographie florentine de la Renaissance et Verrazano," in *Giornate commemorative di Giovanni da Verrazzano*, Istituto e Museo di Storia della Scienza, Biblioteca 7 (Florence: Olschki, 1970), 19–43, esp. 23–24 (where referred to under its former number "Correr 38"). The earlier catalog of the collection, to which Destombes did not refer, assigned the chart, without stated reason, to the sixteenth century; see Lucia Casanova, "Inventario dei portolani e delle carte nautiche del Museo Correr," *Bollettino dei Musei Civici Veneziani* 3–4 (1957): 17–36, esp. 32, 34.

TABLE 19.3—*continued*

ᶜFor example, where the Arabic Maghreb chart has only four of the seventy-nine Dalorto names, the fraction 4/17 shows that it still contains roughly a quarter of those that fall within its limits. So that the overall patterns should not be completely smothered, this second figure has normally been omitted. Where charts are incomplete or partially illegible, the figures inevitably give a distorted picture.

ᵈNontoponymic factors must be taken into account as well. Because of the lack of Catalan toponymic innovations between 1385 and 1430, the dates proposed for Majorcan work might be far too early. For example, the first of the Catalan charts assigned to the late fourteenth century has a toponymic profile similar to that of the 1423 Viladestes chart.

ᵉConflicting opinions about the priority of one 1318 Vesconte atlas over another are discussed by Paolo Revelli, *La partecipazione italiana alla Mostra Oceanografica Internazionale di Siviglia (1929)* (Genoa: Stabilimenti Italiani Arti Grafiche, 1937), lxxvi. Since the place-name analysis cannot separate the two, it is logical to treat them as a single entity.

ᶠApparent "erratics" on this and other undated charts should probably be interpreted as unrecognized innovations. These would demonstrate that the significant additions were not necessarily new names; they might be the traditional names for less important natural features. Every chart has its unique features, some names occur erratically over a long period, and there are numerous inconsistencies within the work of a single chartmaker.

THE UNDATED WORKS AND THEIR NEWLY PROPOSED DATES

ARABIC

1. Milan, Biblioteca Ambrosiana, S.P. 2, 259 (Maghreb chart).

    Previously proposed date: late 13th to mid-16th century.
    Earliest date from table 19.3: first half of 14th century.

For reproductions see Youssouf Kamal, *Monumenta cartographica Africae et Aegypti*, 5 vols. in 16 pts. (Cairo, 1926–51), 4.3:1336, and Juan Vernet-Ginés, "The Maghreb Chart in the Biblioteca Ambrosiana," *Imago Mundi* 16 (1962): 1–16, esp. 1.

CATALAN

2. London, British Library, Add. MS. 25691.
    Previously proposed date: ca. 1320–50.
    Earliest date from table 19.3: ca. 1339.

For a reproduction see Kamal, *Africae et Aegypti*, 4.3:1334 (1 above).

3. Florence, Biblioteca Nazionale Centrale, Port. 22.
    Previously proposed date: 14th to 16th century.
    Earliest date from table 19.3: late 14th century.

4. Venice, Biblioteca Nazionale Marciana, It. IV, 1912.
    Previously proposed date: ca. 1330 to 15th century.
    Earliest date from table 19.3: late 14th century.

5. Naples, Biblioteca Nazionale Vittorio Emanuele III, Sala dei MSS. 8.2.
    Previously proposed date: ca. 1390–1429.
    Earliest date from table 19.3: late 14th century.

For a reproduction see Kamal, *Africae et Aegypti*, 4.3:1331 (1 above).

6. Paris, Bibliothèque Nationale, Rés. Ge. AA 751.
    Previously proposed date: end 14th century to ca. 1416.
    Earliest date from table 19.3: late 14th century.

For reproductions see Kamal, *Africae et Aegypti*, 4.4:1396 (1 above) and Armando Cortesão, *History of Portuguese Cartography*, 2 vols. (Coimbra: Junta de Investigações do Ultramar-Lisboa, 1969–71), 2:51.

7. Paris, Bibliothèque Nationale, Rés. Ge. B 1131 (Soler).
    Previously proposed date: ca. 1380–90.
    Earliest date from table 19.3: late 14th century.

For a reproduction see Kamal, *Africae et Aegypti*, 4.3:1322 (1 above).

8. Paris, Bibliothèque Nationale, Rés. Ge. D 3005 (fragments).
    Previously proposed date: beginning to mid-15th century.
    Earliest date from table 19.3: early 15th century, but see note 219.

9. Florence, Biblioteca Nazionale Centrale, Port. 16.
    Previously proposed date: mid-14th century to ca. 1439.
    Earliest date from table 19.3: first half of 15th century.

The Portuguese flag over Ceuta provides a terminus post quem of 1415 if the unconvincing argument that the flag was a later addition is disregarded; see Alberto Magnaghi, "Alcune osservazioni intorno ad uno studio recente sul mappamondo di Angelino Dalorto (1325)," *Rivista Geografica Italiana* 41 (1934): 1–27, esp. 5. For reproductions see Kamal, *Africae et Aegypti*, 4.4:1463–64 (1 above), and Charles de La Roncière, *La découverte de l'Afrique au Moyen Age: Cartographes et explorateurs*, Mémoires de la Société Royale de Géographie d'Egypte, vols. 5, 6, 13 (Cairo: Institut Français d'Archéologie Orientale, 1924–27), pl. XVIII.

ITALIAN

10. Cortona, Biblioteca Comunale e dell'Accademia Etrusca (Cortona chart).
    Previously proposed date: mid-14th century.
    Earliest date from table 19.3: early 14th century.

For a reproduction see Vera Armignacco, "Una carta nautica della biblioteca dell'Accademia Etrusca di Cortona," *Rivista Geografica Italiana* 64 (1957): pls. I–III. See fig. 19.11.

11. Rome, Biblioteca Apostolica Vaticana, Vat. Lat. 14207 (fragment).
    Previously proposed date: 14th century.
    Earliest date from table 19.3: first half of 14th century.

Roberto Almagià discussed its place-names, noting that all were written in black ink and that several significant towns were absent; see *Monumenta cartographica Vaticana*, 4 vols. (Rome: Biblioteca Apostolica Vaticana, 1944–55), vol. 1, *Planisferi, carte nautiche e affini dal secolo XIV al XVII esistenti nella Biblioteca Apostolica Vaticana*, 24–26. Neither he nor Giuseppe Caraci, *Italiani e Catalani nella primitiva cartografia nautica medievale* (Rome: Istituto di Scienze Geografiche e Cartografiche, 1959), 312–13, mentions the strong likelihood—confirmed by my personal examination—that the red pigment used for the important names on this badly damaged chart has proved to be fugitive. There are several other instances where this has definitely occurred: for example, on the 1413 and 1423 Mecia de Viladestes charts and on the 1469 Benincasa atlas in the British Library. Because of the Vatican chart's incompleteness and semilegibility, no comment is possible about the presence or absence of Dulcert names and hence about Almagià's claim (p. 24) that this fragment is one of the earliest surviving examples of Catalan work. Almagià's firm belief in the chart's Catalan origin was later modified in favor of possible Genoese authorship; see Marcel Destombes, "Cartes catalanes du XIVᵉ siècle," in *Rapport de la Commission pour la Bibliographie des Cartes Anciennes*, 2 vols., International Geographical Union (Paris: Publié avec le concours financier de l'UNESCO, 1952), 1: 38–63, esp. 38–39. This point has yet to be resolved.

12. Amsterdam, Nico Israel.
    Previously proposed date: ca. 1320–25.
    Earliest date from table 19.3: there are insufficient data for a judgment.
The chart was sold to Nico Israel, antiquarian booksellers, and appears in their Fall 1980 catalog, *Interesting Books and Manuscripts on Various Subjects: A Selection from Our Stock . . .*, catalog 22 (Amsterdam: N. Israel, 1980), no.1, and Sotheby's *Catalogue of Highly Important Maps and Atlases*, 15 April 1980, lot A. Both catalogs reproduce the chart. See also note 274.

13. Washington, D.C., Library of Congress, vellum chart 3.
    Previously proposed date: ca. 1320–50.
    Earliest date from table 19.3: mid-14th century.
For a reproduction see Charles A. Goodrum, *Treasures of the Library of Congress* (New York: H. N. Abrams, 1980), 93.

14. Florence, Biblioteca Medicea Laurenziana, Gaddi 9 (Medici atlas—except larger-scale Adriatic and Aegean).
    Previously proposed date: 1351 to post-1415.
    Earliest date from table 19.3: ca. 1351.
For reproductions see Kamal, *Africae et Aegypti*, 4.2:1246–48 (1 above), and A. E. Nordenskiöld, *Periplus: An Essay on the Early History of Charts and Sailing-Directions*, trans. Francis A. Bather (Stockholm: P. A. Norstedt, 1897), 115, pl. X.

15. Milan, Biblioteca Ambrosiana, S.P. 10, 29 (Pizigano atlas 1373—larger-scale Adriatic and Aegean).
    Previously proposed date: post-1381.
    Earliest date from table 19.3: ca. 1373.

16. Venice, Museo Storico Navale (fragments).
    Previously proposed date: 13th century to ca. 1400.
    Earliest date from table 19.3: beginning 15th century.

17. Lyons, Bibliothèque de la Ville, MS. 179.
    Previously proposed date: 14th century to ca. 1400.
    Earliest date from table 19.3: beginning 15th century.
For a reproduction see Charles de La Roncière, *Les portulans de la Bibliothèque de Lyon*, fasc. 8 of *Les portulans Italiens*, in Lyon, Bibliothèque de la Ville, *Documents paléographiques, typographiques, iconographiques* (Lyons, 1929), pls. X–XIV.

18. Barcelona, Archivo de la Corona de Aragón, Caja II.
    Previously proposed date: 14th century to ca. 1550.
    Earliest date from table 19.3: early 15th century.
Though this is considered to be Catalan (e.g., by Julio Rey Pastor and Ernesto García Camarero, *La cartografía mallorquina* [Madrid: Departamento de Historia y Filosofía de la Ciencia, 1960], 51), the name pattern is that found on Italian charts. Among its strange (and probably misleading) features is the absence of the Canaries, included wherever there was room on all charts after Dalorto's of 1325/30.

19. Florence, Biblioteca Medicea Laurenziana, Gaddi 9 (Medici atlas—larger-scale Adriatic and Aegean).
    Previously proposed date: 1351 to post-1415.
    Earliest date from table 19.3: first half of 15th century.

20. London, British Library, Add. MS. 18665 (Giroldi?).
    Previously proposed date: ca. 1425.
    Earliest date from table 19.3: first half of 15th century.

21. Genoa, Biblioteca Civica Berio (Luxoro atlas).
    Previously proposed date: late 13th century to ca. 1350.
    Earliest date from table 19.3: first half of 15th century.
For reproductions see Kamal, *Africae et Aegypti*, 4.2:1245 (1 above); Cornelio Desimoni and Luigi Tommaso Belgrano, "Atlante idrografico del medio evo," *Atti della Società Ligure di Storia Patria* 5 (1867): 5–168, pls. X–VIII (reengraved copies); and Nordenskiöld, *Periplus*, pl. XVIII (14 above).

22. Venice, Biblioteca Nazionale Marciana, It. VI, 213 (Combitis atlas).
    Previously proposed date: 1368 to ca. 1400.
    Earliest date from table 19.3: first half of 15th century.
For a reproduction see Kamal, *Africae et Aegypti*, 4.3:1333 (1 above). See also note 305.

There are, however, inevitable dangers inherent in this type of analysis, as Cortesão and Caraci have warned.[330] Despite these limitations, place-names provide the best evidence of the relation of one chart to another. They also have a major role to play in revealing less obvious patterns of progress or retrogression, and in helping to date charts that lack an author's legend.

A mistake commonly made in the past was the failure to distinguish between reliably dated charts and those whose dating represented no more than an estimate. Any conclusions drawn from data that incorporated charts of the second category are inherently fallible, and the so-called Tammar Luxoro atlas (named after a former owner and now preserved in the Biblioteca Civica Berio, Genoa) provides a good example of this. Nordenskiöld's parallel transcriptions from four works[331] supposedly offered a useful spread of three centuries, because he dated the first of these, the Luxoro atlas, to the beginning of the fourteenth century. A recent comparison by Kelley of names at the head of the Adriatic[332] also included the Luxoro atlas, now assigned to about 1350. As can be demonstrated from an overall examination of its names, the work almost certainly belongs to the fifteenth century (see table 19.3, pp. 416–20). It is obviously dangerous to base arguments on charts that have been frequently redated in the past, since they may be moved again in the light of fresh information. Nearly all the undated (but supposedly early) charts and atlases have been misused in this way.

The conclusions that follow are based on a new analysis of the mainland names from Dunkirk southward to Gibraltar and then around the Mediterranean and Black seas to Mogador (Essaouira) in Morocco. Its findings (summarized in table 19.3 and explained in appendix 19.5), some expected and others surprising, invite a radical reassessment of the nature and evolution of the portolan charts. Altogether, forty-seven out of fifty-seven charts and atlases definitely or supposedly produced in the period up to 1430 were studied in detail. In this way the arbitrary selection of data usually employed in studies of this kind was avoided. In the past, toponymic analysis of portolan charts usually consisted either of columns of names extracted from, at most, a handful of works or of comparisons covering only limited stretches of coastline. Whether because of insufficient charts or limited areas, previous attempts have always left two issues unresolved: whether the same conclusions hold good for the work of other chartmakers, and whether they could be applied to other areas.

This fresh survey, by uncovering different patterns of toponymic development for the various parts of the region under consideration, demonstrated that past extrapolations from small samples were unrepresentative and frequently misleading.[333] Unwitting emphasis was also given in some earlier studies to the occasional omission of standard names, probably the result of carelessness by the chartmaker. The analysis further revealed that inconsistencies within the output of a single chartmaker can even be matched on overlapping sections of a single atlas.

It can be confirmed, though, that after the early fourteenth century there was little increase in the total number of names listed between Dunkirk and Mogador. One of the latest Vesconte works, the British Library atlas of about 1325, includes more names, for instance, than the Catalan atlas of half a century later, or even the 1593 atlas by Vincenzo di Demetrio Volcio.[334] Except for the very earliest period when regular additions were being incorporated, therefore, a simple name total is no easy pointer to the date of compilation.

Such general assertions still leave several basic questions unanswered. Might not a work drawn to a much larger scale display a corresponding increase in its toponymy? If a chart of 1330 and another of 1530 proved to be indistinguishable in terms of simple name totals, could anything of value be learned from examining the

---

330. Cortesão, "North Atlantic Nautical Chart," 6 (note 170); Caraci, "Benincasa and Freducci," 34–35 and 39 n. 5 (note 247).

331. Nordenskiöld, *Periplus*, 25–44 (note 14).

332. Kelley, "Oldest Portolan Chart," 42 (note 58).

333. William C. Brice, "Early Muslim Sea-Charts," *Journal of the Royal Asiatic Society of Great Britain and Ireland* 1 (1977): 53–61, esp. 60, considered one of the more static sections of coastline, that between Bejaia (Bougie) and Annaba (Bône). This led him to the false general conclusion that "the Italian-Catalan repertory of names remained remarkably consistent."

334. Nordenskiöld, *Periplus*, 25–44 (note 14), giving the Latinized form of the author's name. The respective totals for the continuous coastline between Dunkirk and Mogador are 1,191 (late Vesconte), 1,121 (Catalan Atlas), and 1,076 (Volcio).

incidence and form of individual names? From the answers to these questions come new and important insights into the development, dating, and interrelations of the early charts.

RELATION OF SCALE TO TOPONYMIC DENSITY

It might reasonably have been expected that the larger the chart, the denser the toponymy. Not only has this assumption proved wide of the mark, it almost certainly inverts the correct order of priority between scale and place-names. No instance has yet been encountered of a chartmaker's reducing the number of names to fit a smaller format.[335] Indeed, it is highly unlikely that most draftsmen would have possessed this kind of editorial skill.[336] The more likely inference is that the overriding need to accommodate the full complement of place-names imposed its own minimum limits on the scale. These restrictions were not absolute, of course, but were relative to the size of handwriting. In its turn, the size of the writing on a particular chart fluctuated according to the amount of space available.

Areas of special name density sometimes provide an exception to this rule, when large writing or small scale necessitated a number of omissions. Peninsulas and sharp turns in the coast, like the southern extremities of Italy and Greece, are instances of this. But where the names were spread out along an uncomplicated coastline—as for Turkey's Black Sea shore or North Africa—it can safely be said that the scale was limited by the place-names and not vice versa. Even the smallest atlases, like the Tammar Luxoro and the 1318 (Vienna) Vesconte, conform to this pattern despite containing charts that measure a mere 11 by 15 centimeters and 19 by 20 centimeters respectively. Broadly speaking, the smaller the atlas format, the more leaves required. Thus the Luxoro atlas is spread over eight sheets and the Vienna Vesconte over nine.

Although the place-names' indispensability set minimum restrictions of scale, there were no upper limits. The relation of scale to toponymic density can be tested in another way by considering examples drawn at increased scales. Unlike a chart, an atlas could be constructed at varying scales. Vesconte did this regularly, and although his earliest surviving atlas of 1313 has the Aegean drawn at twice the scale of the other sheets, no region was consistently favored in his atlases. Later draftsmen began to use scale to emphasize the Adriatic, Aegean, and Black seas. In some cases this was probably inadvertent—the Black Sea, for instance, naturally commanded a sheet to itself regardless of how the Mediterranean was divided up. From the time of the 1373 Francesco Pizigano atlas onward, though, there are indications of a conscious manipulation of scale. By

these means the Pizigano atlas enlarged both the Adriatic and the Aegean, as did, for example, the Medici, Pinelli-Walckenaer, and 1426 Giroldi atlases. These two seas were of great importance for the eastward-looking trading and colonizing ventures of Genoa and Venice.

There is little evidence, however, that the larger-scale sheets in portolan atlases contained more names,[337] although, as with so many generalizations about portolan charts, this too may need qualifying. A handful of surviving fifteenth-century examples indicates that more specialized charts were sometimes produced, limited to a small section of the regions normally covered.[338] The 1424 chart seems to be the first of these. Renowned for its Antilia, it needs to be valued also for its contribution to the toponymy of France and the Iberian peninsula. A number of unusual names are to be found also on the neglected chart of the Adriatic, drawn by Antonio Pelechan in 1459.[339] Similarly, the 1470 Nicolo chart of the Adriatic and Aegean was described by Almagià as being "rich in names."[340] From these examples it is clear that the extra space permitted by the larger scale of detailed separate charts was occasionally used to increase the number of place-names.

This did not happen, however, with Grazioso Benincasa's separate chart of the Adriatic, drawn in 1472.[341] Its total of 166 names includes only 16 that are not to be found on the relevant sheet in the atlas of 1473; and 9 of the names in the atlas are absent from the chart.

335. Unless the small and sparsely lettered Arabic chart in the Biblioteca Ambrosiana, Milan (the Maghreb chart), should be interpreted in this way.

336. It is significant that the marginal illustrations in *La sfera* (attributed to Leonardo Dati) merely extracted from a portolan chart the red names (for the more important places), ignoring the more numerous black ones; see Almagià, *Vaticana*, 1:128 (note 35), and also note 72.

337. The Medici atlas's larger Adriatic sheet has far more names but can be shown to be later; see appendix 19.1 and table 19.3, pp. 446–48 and 416–20, respectively.

338. The small-format Maghreb chart should not be seen as an example of this (see p. 445 and note 533), nor should a sheet like the one in the Museo Storico Navale, Venice, which is clearly the sole survivor of an atlas.

339. Although noted by Uzielli and Amat di San Filippo, *Mappamondi*, 75 (note 35), this important chart has been consistently ignored by commentators since.

340. Roberto Almagià, "Intorno ad alcune carte nautiche italiane conservate negli Stati Uniti," *Atti della Accademia Nazionale dei Lincei: Rendiconti, Classe di Scienze Morali, Storiche e Filologiche*, 8th ser., 7 (1952): 356–66, esp. 360. Since the Nicolo chart is evidently unreproduced, this point could not be checked.

341. Preserved in the Museo Correr, Venice, this was discussed by Marina Salinari (formerly Marina Emiliani), "Notizie su di alcune carte nautiche di Grazioso Benincasa," *Rivista Geografica Italiana* 59 (1952): 36–42. She transcribed the names, comparing them with the atlas of 1473. Caraci, "Grazioso Benincasa," 287 (note 17), arrived at name totals slightly different from ours. He also suggested that this sheet had formed part of an atlas (p. 286), but all the indications are to the contrary.

Yet the larger scale certainly allowed a far more realistic outline of the intricate Dalmatian coastline[342]—an indication of the distinction that needs to be made between hydrographic and toponymic development.[343]

SIGNIFICANT PLACE-NAME ADDITIONS

The total number of place-names on charts after about 1325 remains relatively static, therefore, increasing neither with the passage of time nor (generally) with enlargement of scale. But once the names are examined individually, there emerges a changing pattern that has remained unsuspected by most previous commentators. Winter's phrase "the agreed conservatism of the chartmakers" sums up a common attitude.[344] Nordenskiöld found complete correspondence between the work of Vesconte and "all portolanos of the normal type from the 14th–17th centuries."[345] More recently, R. A. Skelton concluded that the prototype chart "was reproduced with no structural alterations for nearly four centuries."[346] Steadily changing patterns of place-names might not normally be considered to form part of the "structure" of a class of maps. But given that the Mediterranean and Black Sea names are applied on the portolan charts to broadly unchanging coastlines, they demonstrate the importance that contemporaries attached to this particular element. Nordenskiöld, and the many others who followed him, based their generalizations about toponymic conservatism on the admittedly static majority. But what mattered was the sizable changing minority now revealed: more than five hundred place-names for coastlines where an average chart would have fewer than three times that number in all.[347] The extent and frequency of these toponymic changes invite a new respect for the early portolan charts, as well as a new awareness of the living, and not static, tradition they represent.[348]

It might have been anticipated that fresh names would have been added infrequently, marking out as significant prototypes the occasional portolan charts that conveyed the new information. True, some are more important in this respect than others. But what is particularly surprising, about the early charts at least, is that all the dated or datable ones produced up to 1408 are innovators. Each one injects into the communal bloodstream at least a few names that show up in later work.[349]

The third row in table 19.3 can be read off on its own to provide the total of significant place-name additions found on successive dated or datable charts. Nowhere is this development more marked than in the earliest stages, particularly in the work of Pietro and Perrino Vesconte (identified in table 19.3 as Italian because, although Genoese, they worked partially or entirely in Venice). Pietro's earliest production, the eastern Medi-

terranean chart of 1311, brought in a wealth of names not found in *Lo compasso da navigare*, the earliest surviving *portolano* of the Mediterranean, or on the Carte Pisane.[350] But when attention is focused exclusively on improvements the two Vescontes made to their own work—in other words, starting with the 1313 atlas for the areas covered in 1311 and the 1318 atlases for the remainder—we find they contributed in total no fewer

342. Salinari, "Notizie," 38–39 (note 341). Her belief that this hydrographic improvement was not entirely due to the increase of scale is hard to accept and was challenged by Caraci, "Grazioso Benincasa," 286 (note 17). The 1472 chart provides confirmation of Salinari's earlier conclusion that the expected connection does not exist between Grazioso Benincasa's charts and the sailing directions for the Adriatic and points east that he produced between 1435 and 1445; see Marina Emiliani (later Marina Salinari), "L'Arcipelago Dalmata nel portolano di Grazioso Benincasa," *Archivio Storico per la Dalmazia* 22 (1937): 402–22, esp. 407.

343. Two other undated charts should be mentioned here: a supposedly fifteenth-century chart of the archipelago that was in private hands in Tortona a century ago (see Uzielli and Amat di San Filippo, *Mappamondi*, 101 [note 35]), and the Lesina chart of the Caspian Sea, which, though probably dating from the first quarter of the sixteenth century, might well be based on an earlier prototype; see note 48 and also Renato Biasutti, "Un'antica carta nautica italiana del Mar Caspio," *Rivista Geografica Italiana* 54 (1947): 39–42.

344. Heinrich Winter, "The True Position of Hermann Wagner in the Controversy of the Compass Chart," *Imago Mundi* 5 (1948): 21–26, esp. 22.

345. Nordenskiöld, *Periplus*, 56 (note 14). Page 45 contains a very clear statement of the views of the "conservative" school. In the opinion of Taylor, *Haven-Finding Art*, 113 (note 7), "a single master-copy appears to have been available from the outset, from which all later ones show merely deviations in detail."

346. R. A. Skelton, *Maps: A Historical Survey of Their Study and Collecting* (Chicago: University of Chicago Press, 1972), 8.

347. See note 334.

348. Ironically, an earlier study that purported to demonstrate this very point by highlighting toponymic development on sixteenth- and seventeenth-century charts proves, on closer examination, to require reinterpretation. Giuseppe Caraci, "Inedita cartografica—1, Un gruppo di carte e atlanti conservati a Genova," *Bibliofilia* 38 (1936): 170–78, considered 301 mainland names between Marseilles and Cattaro (Kotor) on six works dated between 1563 and 1676. Comparing these with the listing of pre-1500 names given by Kretschmer, *Die italienischen Portolane* (note 48), Caraci found that some 14 percent of the names he was considering were missing from Kretschmer's catalog, and he concluded that most of these referred to new or expanded settlements (p. 170). Unfortunately, Kretschmer had missed a number of early names, and a check of Caraci's supposedly innovative mainland names found that only 6 percent were actually absent from charts of the fourteenth and fifteenth centuries and from the 1512 Vesconte Maggiolo atlas. In contrast, therefore, to what will be shown for the early period, this indicates that sixteenth- and seventeenth-century charts were toponymically static.

349. The addition of fresh names seems to conflict with the static overall totals. The explanation is that some of the new names replaced existing ones and others were exclusive to one particular chartmaking center.

350. The wide variations in the names of *Lo compasso* (ca. 1250—but see note 107), the Carte Pisane (late thirteenth century), and the 1311 Vesconte chart made it impractical to include the first two in the general place-name analysis.

than 119 significant additions. These are spread out among the different productions, serving, incidentally, to confirm the chronological sequence of Vesconte works earlier indicated by the evolving cartography of the British Isles.[351] There is nothing here to support Hinks's judgment that the Vescontes' work showed little originality.[352]

The insertion of new names has no significance for the present analysis until these innovations are imitated. In developmental terms, unique names can be ignored. Two important and complementary insights flow from an investigation of the way fresh names were introduced and then later repeated. In the first place, when viewed diachronically this analysis identifies the first dated appearance of names that were destined to be regularly repeated thereafter. This is, pure and simple, a dating aid. On the other hand, when approached synchronically the data demonstrate the interdependence or mutual isolation of Catalan, Genoese, and Venetian chartmakers by emphasizing those names whose inclusion or omission is indicative of a particular "school." These two aspects will be considered in turn.

TOPONYMY AS A DATING TOOL

Once the toponymic development on datable portolan charts from the Carte Pisane to 1430 has been systematically analyzed, it is possible to approach the undated charts with more confidence. Past attempts supply instructive warnings, as James E. Kelley pointed out.[353] He noted a two-and-a-half-century disparity in the case of the Arabic chart in the Biblioteca Ambrosiana, Milan (the Maghreb chart). This was dated to the late thirteenth century by Uzielli in 1882, to perhaps the mid-sixteenth century by Nordenskiöld in 1897, and to various points in between by other researchers.[354] Uzielli even suggested dates two hundred years apart within ten pages of the same work.[355]

Extraordinary though it may seem, this difference is not without parallel. To previous generations, no chart could be left undated (or usually unauthored). When no evidence was available, guesswork took over. The anonymous Catalan fragment in the Archivio di Stato, Venice, which was dated to the period 1490–1502 in a 1907 exhibition, had to be reassigned by Marcel Destombes to the first half of the fourteenth century.[356] Unfortunately it has since been stolen and has never, apparently, been reproduced in full. The chart in Uppsala and the so-called Richelieu atlas in the Bibliothèque Nationale (Département des Manuscrits, Français 24909) have both been ascribed at different times to the fourteenth, fifteenth, and sixteenth centuries.[357] The most extreme case is that of the Matteo Prunes chart in the Biblioteca Monumento Nazionale, Cava de' Tirreni.

Visibly signed by this sixteenth-century chartmaker, the work has nevertheless been assigned to the fifteenth and fourteenth centuries and even to the beginning of the thirteenth.[358] Clearly, a more scientific approach than this had to be found.

The place-name data provide for the first time an objective yardstick for dating the early Mediterranean and Black Sea charts—works that can rarely be fixed in relation to documented discoveries. The way this information can be applied to otherwise undatable charts offers approximation only, not precise or foolproof answers. Like all other exercises of this type, it can only work from the assumption that the document in question is typical of its period. Each undated chart is therefore assigned to its most logical chronological position in the documented evolution of the toponymy found on dated charts.

This method cannot, of course, distinguish between a later copy and its model, nor can it readily give credit for any innovations that might be present on an undated chart.[359] If used uncritically, without consideration of any other factors, this approach might provide misleading results, suggesting too early a date for a slavish copy and too late a slot for one that was ahead of its time. To obviate errors in the new dates proposed, paleo-

---

351. See above, pp. 407–9. For the totals of significant additions introduced on successive Vesconte productions see the top line of table 19.3.

352. Hinks, *Dalorto*, 3 (note 76).

353. Kelley, "Oldest Portolan Chart," 25 (note 58).

354. Uzielli and Amat di San Filippo, *Mappamondi*, 229 (note 35); Nordenskiöld, *Periplus*, 46–47 (note 14).

355. Uzielli and Amat di San Filippo, *Mappamondi*, 229, 237 (note 35).

356. *Catalogo delle Mostre Ordinate in Occasione del VI Congresso Geografico Italiano* (Venice, 1907), 73; Destombes, "Cartes catalanes," 53–54 (note 99).

357. Both have been reassigned to the sixteenth century: for Uppsala see Kretschmer, *Die italienischen Portolane*, 148 (note 48); for the Richelieu atlas see Winter, "Late Portolan Charts," 39 (note 129).

358. This chart was assigned to the beginning of the thirteenth century by Manuel Francisco de Barros e Sousa, Viscount of Santarém, *Estudos de cartographia antiga*, 2 vols. (Lisbon: Lamas, 1919–20), 1:52. The fourteenth-century suggestion came from Giuseppe de Luca, "Carte nautiche del medio evo disegnate in Italia," *Atti dell'Accademia Pontaniana* (1866): 3–35, esp. 11; reprinted in *Acta Cartographica* 4 (1969): 314–48. Rey Pastor and García Camarero, *Cartografía mallorquina*, (note 28), provide three distinct entries (pp. 86–87, 96, 97) for charts at the Biblioteca Monumento Nazionale, located in the monastery of La Trinità della Cava, near Cava de' Tirreni (between Naples and Salerno). Nevertheless, these sketchy and clearly second-hand descriptions, two of which mention the second half of the fifteenth century, can all apply to the faded Prunes chart reproduced in "Mostra cartografica" in *Atti del XI Congresso Geografico Italiano, Naples, 1930* (1930), 4:32–42, reproduction facing 326. The director of the Biblioteca confirms that only one chart is involved.

359. The Cornaro atlas is the best example of an archaic work; the 1403 Francesco Beccari chart is the most striking instance of innovation.

graphic expertise might also have to be invoked (though we must remember that scribes would sometimes even reproduce the handwriting of the manuscript they were copying).

The application of the place-name analysis to undated charts or atlases involved noting in each case the number of significant additions they included. Since every significant new name had been bracketed with the year of its first observed appearance, it was possible to prepare separate "toponymic profiles" for the undated works (see appendix 19.5).[360] The results for the works that have traditionally been assigned to the period up to 1430 are found in table 19.3 where the dates proposed in the past for each chart or atlas are compared with those indicated by its toponymic makeup.

In a number of cases earlier suggestions (or perhaps one from a range of alternatives) are reinforced. Several instances, however, challenge long-held assumptions. The Dalorto/Dulcert type chart in the British Library (Add. MS. 25691), considered by Heinrich Winter to be earlier than the Dalorto chart, actually contains as many as possible of the names first found on the definitely later Dulcert chart.[361] The place-name analysis also contradicts received opinions about five of the early Italian atlases. It indicates, for example, that the supposedly homogeneous Medici atlas in reality contains two sheets incorporating names not in circulation at its presumed date of execution, 1351 (see appendix 19.1). Conversely, the same data, when applied to the 1373 Francesco Pizigano atlas, reveal remarkable similarities between the core of the work and the larger-scale sheets, which are usually considered to be later. More drastic still is the clear indication that three "fourteenth-century" atlases—the Combitis, Luxoro, and Pinelli-Walckenaer volumes—should be removed altogether from that century.

In this context, the information summarized in table 19.3, pp. 416–20, has a further role to play. Common patterns of significant additions spotlight links between particular charts and can reveal a chartmaker's hidden signature. There is, for example, a close correlation in these respects between the Pinelli-Walckenaer and Combitis atlases. This substantiates their close match of style. Indeed, idiosyncrasies in the handwriting leave little doubt that both are in the same hand (see fig. 19.9).[362] In the same way, the remarkable consistency of the toponymic profiles extracted from the Luxoro atlas and the 1421 Francesco de Cesanis chart prompted their closer comparison through reproductions (see appendixes 19.2 and 19.3). Despite the early fourteenth-century date often claimed for the former (see above, p. 420), its writing shares a number of peculiarities with the reliably dated Cesanis chart (see fig. 19.10).[363] Although the divergence of scale between the two, and

hence the size of handwriting, results in inevitable differences, it is permissible to make a fairly confident attribution of the Luxoro atlas to Cesanis.[364]

TOPONYMY AS A GUIDE TO THE INTERRELATIONS OF CHARTS

The data displayed in table 19.3, pp. 416–20, can be used in yet another way to reveal the influence exerted by one chartmaker, or one center of production, over another. The most striking conclusion to emerge from this reading of the evidence is that only up to perhaps 1350 was there free interchange of information between different chartmakers. For this first period, the practitioners can be characterized, however loosely, as members of a single school, sharing at least partially the same constantly renewed sources of toponymic information. Several of the Italian charts that seem to belong to this period were influenced at least as much by Catalan place-

---

360. For explanation of the particular use of the term "significant" see appendix 19.5.

361. Winter read the Dalorto date as 1330 as in Winter, "Catalan Portolan Maps," 7 (note 174), and considered the British Library chart to be "prior to 1324 because in it the Aragon colours do not yet appear on Sardinia," Winter, "Fra Mauro," 17n. (note 297). The British Library chart incorporates nineteen of the twenty Dulcert innovations; the place where the twentieth (Gux) would fall on the west coast of Morocco is illegible. Giuseppe Caraci reached a similar conclusion about the relation of Add. MS. 25691 to the Dalorto and Dulcert charts; see Giuseppe Caraci, "The First Nautical Cartography and the Relationship between Italian and Majorcan Cartographers," *Seventeenth International Geographical Congress, Washington D.C., 1952—Abstracts of Papers*, International Geographical Union (1952): 12–13. His assertion that Add. MS. 25691 is Italian cannot, however, be accepted.

362. Both are datable to the first half of the fifteenth century on the strength of their names (see table 19.3, pp. 416–20), despite the 1384 commencement date of the Pinelli-Walckenaer atlas's calendar (see appendix 19.1) and the fictitious 1368 date proposed for the Combitis atlas (see note 305). That the two atlases have the same author seems not to have been remarked before, although Cortesão, *History of Portuguese Cartography*, 2:50 (note 3), came close to it.

363. For instance, a period after each word and an intermittent wavy *l*. Previous writers had anticipated the relegation of the Luxoro atlas to the fifteenth century. Cesare Paoli, "Una carta nautica genovese del 1311," *Archivio Storico Italiano*, 4th ser., 7 (1882): 381–84, esp. 382, thought the handwriting later than commonly supposed; more recently Quaini, "Catalogna e Liguria," 554 (note 60) assigned it to the mid-fifteenth century; and Revelli, *Colombo*, 251–52 (note 22), pointed out that the work's binding was reminiscent of one from the fifteenth century.

364. Since Cesanis was a Venetian, this attribution overturns the commonly held assumption that the Luxoro atlas is a Genoese work. See, for example, Giuseppe Piersantelli, "L'Atlante Luxoro," in *Miscellanea di geografia storica e di storia della geografia nel primo centenario della nascita di Paolo Revelli* (Genoa: Bozzi, 1971), 115–41, esp. 127–29. Nevertheless, Venetian authorship had already been suggested in 1864; see the "Rendiconto," CV–CVI (note 329), and Marinelli, "Venezia," 954 (note 97), who made a definite claim for Venice.

name innovations as by those of the Vescontes. Thereafter the regionalism, of which there are already hints at the outset, grows steadily more evident, until by about 1375 it has become the dominant tendency.

This shift can be demonstrated as follows. The names introduced by the Vescontes and by Dalorto win a lasting place on the portolan charts. Even if a higher proportion of Vesconte names will appear on Italian charts and of Dalorto names on Majorcan productions, almost all charts drawn before 1430 include a clear majority of those available from both sources. Yet the twenty Dulcert innovations of 1339 found little favor with Italian chartmakers, who often incorporated fewer than a third of them.[365] There then follows a thirty-year hiatus before dated charts resume in 1367. Now the development of separate regional toponymic "vocabularies" becomes more marked, with the forty-eight Pizigani innovations being almost entirely ignored by later Catalan chartmakers and the twenty-one names introduced on the Catalan atlas and Soler chart making an equally small impact on subsequent Italian work. Indeed, the local influences are so strong that "Italian" is a misnomer. The innovative 1426 chart of the Genoese Batista Beccari adopts almost all the 1375–85 Catalan additions but has fewer than half of those first disseminated by the Venetian Pizigani (1367–73). For names, as for decorative embellishment, the Genoese looked to Majorca instead of their more distant rival Venice. The largely unoriginal chart by the Venetian Giroldi (also of 1426) shows the pattern in reverse, being strong on Pizigani names and weak on Catalan ones.

By the late fourteenth century, and on at least to 1430, the presence or absence of new names provides a pointer, therefore, to the likely place of production. That chartmakers remained largely ignorant of (or unimpressed by) the names introduced in other centers allows us with some confidence to designate the Dulcert innovations as Catalan, the Pizigani additions as Venetian, the Beccari ones as Genoese, and so on. With so few original charts extant, there must remain the strong possibility that even earlier lost works deserve the credit for introducing some of the names onto the portolan charts. In other words, our dates for a number of new names might be too late. But the consistency of the regional patterns leaves little doubt that most of the innovations have been ascribed to the correct chartmaking center. If "Venetian" names, for example, had actually been borrowed from lost Catalan charts, they would show up in later Catalan work. Yet this does not happen. It thus becomes clear that the impetus from Majorca, so strong with Dalorto/Dulcert, fades thereafter. Only a further 21 names (1375–85) are attributable to that source over the next hundred years, compared with 135 additions by Italian chartmakers.[366] As a result, the Catalan chart produced by Mecia de

Viladestes in 1423 includes no fifteenth-century innovations at all, since these had all been Italian.[367]

Had that particular chart lost its author's legend, it would have been assigned, using toponymic evidence alone, to about 1380. It is essential, therefore, that the charts of the three main producing centers, Majorca, Genoa, and Venice, be related to the distinctive toponymic patterns of their places of origin before dating can be attempted. No unsigned Genoese charts have been identified with certainty for the period up to 1430. Were any to emerge, it seems likely that their names would be closer to those on the Beccari charts than are those of any surviving works.

It is possible, though this remains to be investigated, that the tendency to regional isolation may later have been reversed. The 1468 productions by the Majorcan Roselli and the Anconitan Grazioso Benincasa were specifically chosen for the pilot place-name study to reveal the extent of mid-fifteenth-century toponymic divergence between Majorcan and Italian work (see appendix 19.5). In fact, they proved to be remarkably similar.

Whereas many names found their way, if belatedly, onto the charts produced in other centers, some remained the almost exclusive hallmark of one particular "school." A few examples, covering the period up to 1430, must suffice. Near Sibenik, the names *port l'ospital*, *artadur*, and *zarona* appear only on Catalan charts (exceptions being provided by the Medici atlas and, in the first two cases, by the 1403 and 1426 Beccari charts). The Greek port Lepanto (Naupaktos), which is rendered *nepanto* on the earliest Italian charts, is given additionally as *lepanto* by Catalan draftsmen (with the 1403 Beccari chart the only Italian example here). On the other hand, the Adriatic offers several instances of names entirely ignored by the early Catalan chartmakers. Between Manfredonia and Rimini the following examples can be cited: *lesna, fortor, salline, cerano,* and *fumissino* (with the almost exclusively Catalan *potencia* nearby).[368] *Novegradi* (the Croatian port of Novi)[369] is similarly omitted from Catalan charts.

---

365. Except the Genoese Beccari family and the anonymous compiler of the Medici atlas.

366. The loss of the Catalan duchy of Athens and Thebes about 1380, marking as it did the beginning of Catalan withdrawal from the eastern Mediterranean, can plausibly be linked to this decline in cartographic inspiration.

367. This is the opposite conclusion from that reached by de La Roncière, *Afrique,* 1:139 (note 100), who considered the toponymy of the 1426 Batista Beccari chart to be a mere echo of the steadily developing Catalan work of the period. This is one of many instances where misleading results have been produced by superficial toponymic comparisons.

368. The non-Catalan instances of *potencia* are found on the Library of Congress and Nico Israel charts and on the Medici atlas.

369. Kretschmer, *Die italienischen Portolane,* 627 (note 48). This essay's conclusions about the injection of new names and the regional

THE SIGNIFICANCE OF PLACE-NAME CHANGES

So far the names have been treated collectively to construct a developmental framework against which undated charts could be assessed and regional interrelationships demonstrated. When considered individually, each added or abandoned name makes its own small contribution to the history of the Mediterranean. Cumulatively, this represents a vital and little-used source.[370] Because of the scarcity of regional or local maps in the Middle Ages,[371] the portolan charts are the earliest surviving cartographic documents to name a great many settlements, as well as numerous natural features. We have already shown how, contrary to expectation, the toponymy of the Mediterranean was under constant review by the portolan chartmakers. Enough names can be recognized today or related to contemporary documents to make it clear that most derived from sound authority.[372] With their toponymic credentials established, the portolan charts can be consulted, with all necessary caution, as primary documents. What they show and what they omit should be considered an important comment on aspects of the medieval world. Had the influx of new names or the purging of obsolete ones occurred only at widely separated intervals, the portolan charts would be a much less useful and convincing record of the growth and decay of coastal Mediterranean settlements. Additional significance attaches to a name first found, say, on a dated chart of the fifteenth century because of the existence of successive earlier charts that had failed to show it.[373]

Before the portolan chart toponymy can offer any insights into the changing political, commercial, navigational, or even religious importance of different places, we need to have some idea of the time lag involved—the gap between the historical event and its recognition in cartographic form. Unfortunately, few coastal villages were the result of a conscious creation (though the date of their first church might sometimes be a useful indicator), and it is impossible to identify the moment they passed from obscurity to significance. There are nevertheless a few towns whose origins are documented, and the labors of local historians may yet add to this number.

The Apulian port of Manfredonia was founded in 1258 by King Manfred, who transferred the population of Siponto to a fresh and healthy site nearby. Although the old name survived on charts for centuries, the new form was already recorded on the Carte Pisane (though not, significantly, on *Lo compasso da navigare* or the Cortona chart).[374] Belforte (at the head of the Adriatic) owes its origin, as its name suggests, to the Venetian fort constructed there in 1274.[375] Its first dated appearance on a portolan chart is not, as might have been anticipated, on one of Vesconte's Venetian productions,

but on the chart drawn by Dulcert in Majorca in 1339. The castle at Mola di Bari dates from 1278, but it was a century before the 1373 Francesco Pizigano atlas named it. Novi, the Croatian village that Kretschmer identified with *noue gradi*, was founded in 1288 and added to the charts for the first time by Vesconte about 1325.[376] Though built on the site of Olivule in 1295,[377] Villefranche did not supplant its predecessor on the portolan charts until the mid-fifteenth century. Vico Equense (south of Naples), which was reconstructed at the end of the thirteenth century, was also ignored by the charts for a century and a half. Like Novi, Bilbao illustrates the rapidity with which new information could be absorbed by the chartmakers. Founded about the year 1300 on a site some eight miles inland, it and its harbor Portugalete (*galleto*), were already noted by Dulcert in 1339.

Accentuated by the Black Death, which swept across western Europe after 1348, the fourteenth and fifteenth centuries were a period of drastic demographic and economic decline. Overall, it took until the early sixteenth century for population numbers in Europe to return to their 1300 levels.[378] It is not surprising, therefore, that for our next examples we should have to turn to the late fifteenth century, when population was expanding once again. Giulianova (south of Ancona) was founded in 1470, and the castle at Pizzo (near the toe of Italy) was built in 1486 by Ferdinand of Aragon. Of these only the latter is included in Vesconte Maggiolo's atlas of 1512. Unfortunately, the unevenness of this admittedly patchy sample allows us to draw only one conclusion

component of toponymic lists find general corroboration in a detailed study of one small area, Cyprus, although the island's unique history, particularly its capture by Venice in 1489, had special cartographic implications. See Tony Campbell, "Cyprus and the Medieval Portolan Charts," *Kupriakai Spoudai: Deltion tēs Etaireias Kupriakōn Spoudōn, Brabeuthen upo tēs Akadēmias Athēnōn* 48 (1984): 47–66, esp. 52–58 and the tables.

370. But see Quaini, "Catalogna e Liguria," 551–53 (note 60).

371. Harvey, *Topographical Maps*, 88 (note 8).

372. See Kretschmer, *Die italienischen Portolane*, 559–687 (note 48).

373. Obsolete names, however, are likely to have survived longer through inertia. See the glossary at the end of Kretschmer's *Die italienischen Portolane* (note 48), for instances of this.

374. Motzo, "Compasso da navigare," XXX (note 103); Armignacco, "Carta nautica," 197 (note 243).

375. Kretschmer, *Die italienischen Portolane*, 625 (note 48).

376. Kretschmer, *Die italienischen Portolane*, 627 (note 48). 1325 is the most likely date for the Sanudo-Vesconte atlas in the British Library (Add. MS. 27376*).

377. Motzo, "Compasso da navigare," XXVIII (note 103).

378. M. M. Postan, E. E. Rich, and E. Miller, eds., *The Cambridge Economic History of Europe*, 3 vols. (Cambridge: Cambridge University Press, 1963), vol. 3, *Economic Organization and Policies in the Middle Ages*, 37. See also J. C. Russell, "Late Ancient and Medieval Population," *Transactions of the American Philosophical Society*, n.s. 48, pt. 3 (1958): 5–152.

about the pace of toponymic absorption on the portolan charts: that it was unpredictable and erratic.[379]

Besides those places whose origin can be pinpointed, there were others, ultimately of more importance, that grew sufficiently during the fourteenth and fifteenth centuries, either in physical size or in perceived significance, for the chartmakers to take note of them. A number of these were even picked out in red on their first recorded appearance.[380] Several of the best-known places are listed below. The form in which they are rendered on the charts and the date of their first identified mention are given in parentheses.

Harfleur (*arefloe*, 1385)
Cherbourg (*ceriborg*, 1318)
Gijón (*gigon*, 1426)
Viareggio (*viaregio*, by 1512)
Livorno (*ligorna*, 1426)
Pozzuoli (*poçuol*, 1403)
Monfalcone (*montfarcom*, 1339)

Livorno was already recorded under that name at the beginning of the tenth century, but it was not until the brief period of Genoese control (1407–21) that it came to supplant its close neighbor Porto Pisano as the chief port at the mouth of the Arno. It is no coincidence that Livorno is first named on surviving charts in 1426, five years after its new overlord, Florence, acknowledged the town's superiority over Porto Pisano.[381] What is given cartographic recognition, therefore, is not Livorno's foundation but its coming of age. Many similar instances could be cited. The Basque harbor Pasajes had been a whaling center since the tenth century, yet it is a late addition to the charts. Gijón and Taggia have early medieval buildings and are (at least in the latter case) of certain Roman origin, but they are both missing from the early charts.

In attempting to match the cartographical and historical realities, it needs to be emphasized that, while a noncartographic approach to portolan chart toponymy may unearth valuable evidence, this must always remain secondary to that derived from the charts themselves. Whether the inclusion or omission of a particular place is a true reflection of its importance at the time in question is a matter for the local historian. To the historian of cartography, the toponymic information has intrinsic value. It is a primary record of the knowledge available to the seafaring peoples of the Mediterranean at different periods. Distorted or fictitious names are as much a part of the pattern as those that are clearly recognizable. Indeed, unfamiliarity with an area's place-names may reflect the infrequent contact of those who supplied the charts' raw data and hence provide pointers to the latter's identity.

THE CHARTMAKERS' INFORMANTS

The travels of Marco Polo left a very obvious mark on the Asian section of some medieval world maps, but changes to the Mediterranean area were subtle and anonymous. Among those who might have acted as informants for the chartmakers would have been travelers, some of them on pilgrimage. For example, Marino Sanudo, the Venetian for whom Pietro Vesconte drew several atlases, returned from Palestine in 1306.[382] Yet if his firsthand knowledge was made available to Vesconte, it did not result in a single addition to the toponymy of the eastern Mediterranean.[383] Had Francesco Beccari not decided to "make public for the removal from all persons of any matter of doubt" the improvements that he had recently made to the standard portolan chart, we should have been left to guess how most of the new information must have reached the chartmakers.[384]

The Francesco Beccari chart in question (now at Yale) is dated 1403. In his long "address to the reader" the author explains how he had lengthened the Atlantic distances on all the charts he himself had drawn since 1400 (fig. 19.17, and see above, p. 414): "the marrow of the truth having been discovered concerning these [things] aforesaid through the efficacious experience and most

379. That a number of names present in the mid-thirteenth-century *Lo compasso da navigare* (note 103) are not found on charts until the fifteenth century or later warns of divergence between the written and the cartographic records. Examples are: *capo de lardiero* and nearby *san trope*, and *capo de sancta maria* (at Italy's heel). Another would be *fanaro* at the entrance to the Bosphorus; this was repeated by Vesconte but then lapsed for a century; see Elisaveta Todorova, "More about 'Vicina' and the West Black Sea Coast," *Etudes Balkaniques* 2 (1978): 124–38, esp. 129, where Fanaro was thought to have first appeared on charts in the fifteenth century.

380. The assertion in Nordenskiöld, *Periplus*, 18 (note 14), that the list of red names remained static through the centuries is not correct. Working from monochrome reproductions, however, it was impossible to make a systematic check on this important barometer of contemporary significance.

381. Giuseppe Gino Guarnieri, *Il porto di Livorno e la sua funzione economica dalle origini ai tempi nostri* (Pisa: Cesari, 1931), 32. The first dated chart to show Livorno is Batista Beccari's of 1426. Its inclusion on the Pinelli-Walckenaer atlas is another reason for doubting the latter's traditional date of 1384 (see table 19.3, pp. 416–20). Porto Pisano remained active as a harbor up to the fifteenth century; thus the attempt by Nordenskiöld, *Periplus*, 46 (note 14), to date the construction of a prototype portolan chart before its supposed destruction in 1290 is unconvincing.

382. De La Roncière, *Lyon*, 11 (note 34).

383. The coast between Alexandretta (Iskenderun) and Alexandria is toponymically one of the most static stretches on the portolan charts. There is no difference here between the Vesconte charts of 1311 and 1327. Nor does the Sanudo *portolano* add any names for this section; see Kretschmer, *Die italienischen Portolane*, 237–46 (note 48), for a transcription.

384. Francesco Beccari, "Address to the Reader." This and the following two quotations are taken with kind permission from the translation given in H. P. Kraus, Booksellers, *Twenty-five Manuscripts*, catalog 95 (New York: H. P. Kraus, [1961]), 63–64.

FIG. 19.17. THE ADDRESS TO THE READER ON THE 1403 BECCARI CHART. Francesco Beccari puts forward, among other things, his reasons for lengthening the Atlantic distances and for adjusting the position of Sardinia. In both cases, as he explains, modifications were made after complaints and advice were received from the seafarers themselves.

By permission of the Beinecke Rare Book and Manuscript Library, Yale University, New Haven.

sure report of many, i.e. masters, ship-owners, skippers and pilots of the seas of Spain and those parts and also of many of those who are experienced in sea duty, who frequently and over a long period of time sailed those regions and seas."[385] There was another point on which "the forms and traces of old masters" had led him astray. He continues, "It was several times reported to me . . . by many owners, skippers and sailors proficient in the navigational art, that the island of Sardinia which is in the Sea, was not placed on the charts in its proper place by the above mentioned masters. Therefore, in Christ's name, having listened to the aforesaid persons, I placed the said island in the present chart in its proper place where it ought to be."[386]

From this invaluable statement we learn of frequent contact between the chartmaker Beccari and assorted masters, shipowners, skippers, pilots, and sailors to whose criticisms he had responded. His meaning is unequivocal: improvements were the result of comments from those who had used the charts at sea and found them wanting. This passage is as crucial as it is unique, for it explains both the charts' practical use and the mechanism by which their content was changed.[387] Beccari refers specifically to two cartographic features, the understated Atlantic distances and the location of Sardinia. Both of these were sufficiently major to be worth mentioning. He is mute about lesser changes, such as those affecting place-names. But this is hardly surprising; as we have shown, the toponymy was under almost continuous review. It is safe to assume, though, that Beccari and his fellow chartmakers derived this type of information in the same way and from the same seafaring sources.[388] Indeed, there is no plausible alternative, since it is anachronistic fancy to suppose that any medieval expedition would have been sent out to verify hydrographic or toponymic details about the Mediterranean. If chartmakers acquired the new names by word

of mouth from returning sailors, it might be expected that the toponymic input would reflect the prevailing patterns of trade. The way a number of the additional names occur next to important commercial centers—for example, Seville, Valencia, Genoa, Rome, Sibenik, Sevastopol, Izmir, Alexandria, and Algiers—supports that view.

THE BUSINESS OF CHARTMAKING
SCRIBAL TRADITIONS

By the fourteenth century, the monastic scriptoria had lost their monopoly to workshops run on strictly business lines. The Venetian artists' guild that had been set up by 1271 was essentially an organization of independent masters with their attendant journeymen and apprentices.[389] Yet if there are no documented links between the production of the earliest charts and the monastic scriptoria, there are signs that some of the chartmaking conventions represented old habits put to a new use. Considering that the earliest practitioners must presumably have been trained as scribes, this is hardly surprising. The use of red, for example, to emphasize important words, or "red-letter days," was an

---

385. Beccari, "Address to the Reader," 63–64 (note 384).

386. Beccari, "Address to the Reader," 64 (note 384). Wagner, "Italian Nautical Charts," 479 (note 313), observed, before the discovery of the Beccari chart, that Sardinia was placed too close to Africa on the early charts.

387. Kelley, "Non-Mediterranean Influences," 18 (note 290), approaching from the cartometric angle, arrived at a very similar interpretation of the mechanism of change on the portolan charts.

388. Though piracy could well have played an important part in distributing information; see below, pp. 439–40 and note 485 for the *San Nicola* incident.

389. Elena Favaro, *L'arte dei pittori in Venezia e i suoi statuti*, Università di Padova, Pubblicazione della Facoltà di Lettere e Filosofia, vol. 55 (Florence: Leo S. Olschki, 1975), 25.

established medieval tradition, perpetuated by the chart-makers in their special treatment of the more significant place-names. Then again, many liturgical works, and almost all books of hours, included a calendar for use in calculating the paschal moon. Just as the prediction of future full moons was essential for fixing the date of Easter, and hence the greater part of the church's year, so the same lunar information was vital to the mariner. Forearmed with knowledge of the moon's age, he could calculate the time of high water, essential for safe pilotage outside the Mediterranean.[390] Thus portolan atlases, from the earliest survivor of 1313 onward, often start with a lunar calendar (see also appendix 19.1).

As Kelley observed, the oldest charts contain more name contractions than the later ones (though the fifteenth-century author of the Combitis and Pinelli-Walckenaer atlases was an exception to that rule), and he convincingly interprets this as "a carryover from the highly abbreviated shorthand of Latin texts."[391] This is additional evidence of a continuing tradition rather than a fresh start. The mixing of colored inks and their application to prepared sheets of vellum, the careful copying of words in a neat and uniform hand—these were the skills already required of those who produced books. No doubt the chart imposed some special demands: the precise and repeated reproduction of coastal outlines, sometimes at altered scales, for example. But as far as the tools of the trade and their application were concerned, the portolan chart draftsmen obviously belonged to an existing tradition. Nor should we assume that chartmaking became totally separate from other related activity. Cresques Abraham was described as being both a "master of maps of the world" and a compass maker, just as Mecia de Viladestes was classified as a compass maker in 1401, twelve years before the date of his earliest known chart.[392] We may reasonably anticipate, therefore, the future discovery of other noncartographic documents by chartmakers, like Arnaldo Domenech's table of weights and measures.[393]

WORKSHOPS OR SINGLE INDIVIDUALS?

"Little or nothing," as Eva G. R. Taylor admitted, "can be said about the way a professional chartmaker organized his business and ordered his workroom."[394] Yet it is commonly assumed that portolan charts were constructed in workshops, even though no documentary evidence has been adduced in support.[395] Nor has any detailed paleographic examination been made of specific charts or atlases to test if more than one hand was involved. Moreover, the involvement of a workshop rather than a solitary chartmaker removes one of the main justifications for any attribution—namely the unique handwriting of a specific individual.[396] Whether single

or multiple hands are involved and whether one chartmaker's signed work has consistent and distinctive characteristics are points that could be resolved. No doubt they will be one day. At this stage we would merely caution against the automatic assumption that no chartmaker ever worked alone.

Occasional instances exist where some form of collaboration is openly acknowledged. This is made explicit on a pair of charts where the author's legends bracket the names of two chartmakers: first, that produced by the Pizigani brothers in 1367, and second, Bertran and Ripoll's chart of 1456. In neither case is the nature of the cooperation specified. Indeed, both author's legends are strange in that the singular form *composuit* is employed. To deepen the confusion, the variant readings proposed for the wording on the 1367 chart leave unresolved the name of Francesco Pizigano's collaborator and even whether there might have been more than two brothers involved.[397]

Although an illuminated manuscript might be produced by a single individual, the labor would often be divided between a scribe, a rubricator, and one or more painters. It is therefore likely that the corner miniatures in Vesconte's atlases, for example, were the work of another man. It is also possible that Francesco Pizigano's collaborator(s) was responsible for the artistic flourishes of the 1367 chart, since the 1373 atlas he signs alone is perfectly plain. A similar interpretation is proposed by

390. The Catalan atlas of 1375 contains the earliest surviving diagram showing, for various named harbors in Brittany and along the English Channel, the "establishment of the port": in other words, the bearing of the new moon at the time of high and low water on the day concerned. For explanation see Taylor, *Haven-Finding Art*, 137–38 (note 7), and Grosjean, *Catalan Atlas*, 38 (note 94).

391. Kelley, "Oldest Portolan Chart," 43 (note 58).

392. Grosjean, *Catalan Atlas*, 13 (note 94). As Grosjean points out, in this context "compass maker" implies not a precision-instrument maker but an artist who painted the decorative compass disk (pp. 13–14). On Viladestes see de La Roncière, *Afrique*, 1:126–27 (note 100). In sixteenth-century Portugal, it was common to combine the roles of chartmaker and manufacturer of nautical instruments; see Teixeira da Mota, "Influence," 228 (note 61). Another later example is provided by the sixteenth-century English compass maker and chartmaker Robert Norman; see Eva G. R. Taylor, *The Mathematical Practitioners of Tudor and Stuart England* (Cambridge: Cambridge University Press, 1954), 173–74.

393. Walter W. Ristow and R. A. Skelton, *Nautical Charts on Vellum in the Library of Congress* (Washington, D.C.: Library of Congress, 1977), 3–4.

394. Taylor, *Haven-Finding Art*, 113 (note 7).

395. For example, see Almagià, *Vaticana*, 1:43b (note 35); Cortesão, *History of Portuguese Cartography*, 2:216 (note 3).

396. Pietro Vesconte has been credited (without justification) with workshops in both Genoa and Venice. On Genoa, see Revelli, *Colombo*, 237 (note 22); on Venice, see Degenhart and Schmitt, "Sanudo und Veneto," 6, 67 (note 226).

397. Cortesão, *Nautical Chart of 1424*, 20 n. 1 (note 24).

Almagià to explain absent or incorrect initial letters for some of the inscriptions on the Vatican's unsigned chart, Borgiano V.[398] It is unfortunate that the only clear account so far unearthed of how functions were divided in practice should describe an exceptional group of world maps rather than a typical portolan chart. Nevertheless, it allows too many insights into contemporary working methods to be ignored.

From surviving legal documents we know that in 1399 a Florentine merchant, Baldassare degli Ubriachi, commissioned from Jefuda Cresques and Francesco Beccari, both then in Barcelona, four large and elaborate world maps for presentation to various European monarchs. Jefuda Cresques was the son of Cresques Abraham, the supposed author of the Catalan atlas. These 1399 world maps would probably have been similar to that work, although considerably larger, and like the Catalan atlas would have been built up around a portolan chart core. The documents, which were interpreted by R. A. Skelton,[399] clearly differentiated the contributions to be made by the two men. The Majorcan Jew Jefuda Cresques (here given his postconversion name, Jacme Ribes) was dubbed *maestro di charta da navichare* and was to draw the basic maps; the *dipintore* Beccari would then embellish them. Since Ubriachi's agent was required to collect the unfinished maps from Cresques and deliver them to Beccari, it is clear that the two men worked independently. Another passage anticipated that Beccari might require an assistant artist, the decision being left up to him. This incident provides, as Skelton pointed out, "a record of a Genoese mapmaker established, if temporarily, in Catalonia and collaborating with a Mallorcan, thus exemplifying the cultural continuum of the Western Mediterranean area."[400]

It is when we pass from these well-attested cases of temporary cooperation between mature chartmakers to consider the possiblity of permanent workshops that we move off firm ground. A workshop implies both a unit containing several individuals and a system of apprenticeship to convey skills from one generation to another. To assess the likelihood that chartmaking was carried out in ateliers we are forced into an oblique approach, because no early chartmaker has left any account of how he operated.

The thirteenth-century Venetian artists' guild, already referred to, embraced a wide range of craftsmen. Its statutes relating to apprenticeship and the ordering of workshops offer a relevant analogy.[401] Indeed, Pietro Vesconte and the other early chartmakers active in Venice might well have been members of the organization, since, from the beginning of the fourteenth century, all who practiced as artists were obliged to belong to it.[402]

The guild's statutes, the oldest of their kind in Italy,[403] indicate that the small workshop was the normal unit of production. The owner was officially limited to two qualified assistants and a single apprentice, though he could apply for a special license to exceed those numbers.[404] If that is the background against which we should set Vesconte, an example from fifteenth-century Genoa indicates an even smaller scale of operations. Although his name is not found on any surviving chart, Agostino Noli is known to us from a petition he addressed to the doge and Council of Elders in 1438, in which he claimed to have been the only chartmaker then active in Genoa.[405] His plea for remission of taxes was granted, with the proviso that he instruct his brother in the mysteries of chartmaking. The Genoese authorities would hardly have made that stipulation if Noli had belonged to a workshop or if he had already taken on an apprentice. A mere fifteen years later we encounter a second, similar instance. The Genoese priest Bartolomeo de Pareto, in a document dated 1453, was described as the city's most experienced chartmaker.[406] A single chart in his hand, dated 1455, has come down to us. Nevertheless, his documented ecclesiastical appointments, which included a spell as a papal acolyte in Rome, are hard to reconcile with the idea of a permanent cartographic workshop.[407]

---

398. Almagià, *Vaticana*, 1:32–33 (note 35).

399. Skelton, "Contract" (note 206). The forms Cresques Abraham and Jefuda Cresques are used because Jews during this period used patronymics rather than surnames.

400. Skelton, "Contract," 109 (note 206).

401. The statutes are discussed in Favaro, *Arte dei pittori in Venezia* (note 389).

402. Favaro, *Arte dei pittori in Venezia*, 26 (note 389). Unfortunately, records relating to practitioners survive only for the period after 1530.

403. Favaro, *Arte dei pittori in Venezia*, 15, 27 (note 389).

404. Favaro, *Arte dei pittori in Venezia*, 26 (note 389).

405. Marcello Staglieno, "Sopra Agostino Noli e Visconte Maggiolo cartografi," *Giornale Ligustico* 2 (1875): 71–79; and more accessibly, Revelli, *Colombo*, 460–61 (note 22). Revelli, *Mostra Colombiana*, 39 (note 315), proposed, without supporting evidence, that Noli might have drawn the Genoese world map (Florence, Biblioteca Nazionale, Port. 1).

406. "Pro Bartolomeo Pareto," *Atti della Società Ligure di Storia Patria* 4 (1866): 494–96, esp. 495. The relevant sentence reads: "Hac itaque animadversione commoti erga egregium presbiterum Bartolomeum de pareto peritum in arte ipsa conficiendarum cartarum navigabilium et quod alius nullus sit in hac urbe huius ministerii edoctus quodque predictum hoc eius ingenium ars et ministerium non modo utile verum etiam necessarium sit Januensibus navigantibus" (With this thought in mind they turned their attention to the distinguished priest Bartolomeo de Pareto, a man experienced in the art of constructing sailing charts, both because there was no one else in this city [Genoa] who was so skilled in this craft and because his specialized talents, already mentioned, were not only useful but of genuine necessity for Genoese sailors. Author's translation).

407. On his ecclesiastical appointments see Michele G. Canale, *Storia del commercio dei viaggi, delle scoperte e carte nautiche degl'Italiani* (Genoa: Tipografia Sociale, 1866), 456–57.

APPRENTICESHIP

Writing about 1400, Cennino Cennini described the awesome variety of accomplishments expected of an artist.[408] In his case these had taken twelve years to acquire.[409] A chartmaker's initiation would presumably have been less rigorous, even when we remember that several of those working in the austere Italian manner also showed themselves capable of artistic flourishes in the Catalan style when the occasion demanded (unless the ornamentation was done by someone else). Yet there are indications that some form of apprenticeship for chartmakers might well have been involved.

What has been seen as direct evidence of this occurs in the author's legend of Roselli's chart of 1447 (preserved in Volterra). Roselli declared that he had drawn this, "de arte Baptista Beccarii"—a reference to the Genoese chartmaker, whose charts of 1426 and 1435 survive. This vital phrase has been the subject of much argument: Winter, for example, interpreted it as an "expression of esteem," and Revelli as an acknowledgment to a teacher.[410] This dispute formed part of the broader controversy about nationality. Roselli was claimed for both Spain and Italy, although all his extant charts are thoroughly Catalan in style and are signed from Majorca.[411] It is not known where Beccari practiced his trade, but his 1426 chart demonstrates his ability to work in the Catalan manner that Roselli would later repeat. What might be a third link in the same apprenticeship chain—if that is what it is—occurs in the inscription found on Arnaldo Domenech's chart of 148– (the final digit is unclear), where he signs himself "dizipolus petri Rossel."[412]

Apprenticeship is sometimes indicated by the charts themselves. Occasional clumsiness—for example, several attempts at scraping the hidden circles on sheets of one of the British Library's Grazioso Benincasa atlases (Add. MS. 6390) or an abandoned circle on the 1424 chart—suggests the inexperienced hand of an apprentice. Kelley also noticed sloppy work, "almost as if the job was left to a junior member of the staff."[413] In general, however, the portolan charts display the competence of their creators. The insertion of hundreds of names in a neat and consistent hand was probably the most difficult part of the training. Although it was possible to remove a mistake by scraping the vellum surface, in practice this was rarely attempted.[414] Once a wrong name was started it would be crossed out, abandoned, or merged with the correct one. Lapses of concentration of this kind are found on most charts, even if infrequently. Since any blemishes would be permanent ones, accuracy must have been one of the most essential skills for a novice draftsman.

Hard evidence from the fourteenth and fifteenth centuries is so sparse on the topic of apprenticeship that it is justifiable to invoke briefly the analogy of the well-documented seventeenth-century Drapers' (or Thames) School in London.[415] To produce by hand charts that in complexity fall roughly midway between the Italian and Catalan-style productions of the earlier period, English apprentices had to serve a minimum of seven years. While this example reveals how apprenticeship within a single organization could create a "school" of chartmakers, it would be unwise to assume that a similar mechanism operated in Italy before 1500.

In the first place, the shared style of the English chartmakers had led them to be designated a "school" before their interrelationship via the Drapers' Company of the City of London was discovered.[416] There are occasional signs of a common style in fourteenth- and fifteenth-century Mediterranean work. The treatment of inland lakes is strikingly similar on the 1408 Pasqualini atlas and the productions of Giroldi some decades later, just as a third Venetian, Francesco de Cesanis, signs his chart of 1421 across the neck in exactly the manner adopted by Giroldi in the following year. If this points to the existence of a Venetian school, similar shared features of style hint at what may well turn out to be a comparable organization in Majorca (or more strictly Palma).

408. Cennino d'Andrea Cennini da Colle di Val d'Elsa, *Il libro dell'arte: The Craftsman's Handbook*, trans. Daniel V. Thompson, Jr. (New Haven: Yale University Press, 1933).

409. Cennini, *Libro dell'arte*, 2 (note 408).

410. Winter, "Roselli," 4 (note 224); Revelli, *Colombo*, 312 (note 22).

411. And a recently discovered chart of 1447 has the author's legend in Catalan; see Kenneth Nebenzahl, *Rare Americana*, catalog 20 (Chicago: Kenneth Nebenzahl, 1968), no. 164.

412. The unsubstantiated claim that Berenguer Ripoll, who jointly signs with Jaime Bertran the chart of 1456, might have been the latter's apprentice was made by Rey Pastor and García Camarero, *Cartografía mallorquina*, 82 (note 28).

413. Kelley, "Oldest Portolan Chart," 38 (note 58).

414. The Dijon chart is the only instance so far noted in which names have been scraped off and rewritten; see Raynaud-Nguyen, "Hydrographie" (note 37).

415. Tony Campbell, "The Drapers' Company and Its School of Seventeenth Century Chart-Makers," in *My Head Is a Map: Essays and Memoirs in Honour of R. V. Tooley*, ed. Helen Wallis and Sarah Tyacke (London: Francis Edwards and Carta Press, 1973), 81–106; Smith, "Thames School," 45–100 (note 185). While this essay was in press, an apprenticeship document was published; see Giovanna Petti Balbi, "Nel mondo dei cartografi: Battista Beccari maestro a Genova nel 1427," in Università di Genova, Facoltà di Lettere, *Columbeis I* (Genoa: Istituto di Filologia Classica e Medievale, 1986), 125–32. In the agreement, dated 17 August 1427, the nine-year-old boy Raffaelino Sarzana, son of a sailor ("navigator"), was apprenticed for eight years to Batista Beccari to learn the art of making charts ("artem faciendi cartas et signa pro navigando"). I owe this note to Corradino Astengo.

416. Ernesto García Camarero, "La escuela cartográfica inglesa 'At the Signe of the Platt,'" *Boletín de la Real Sociedad Geográfica* 95 (1959): 65–68.

There are certainly affinities between the two signed Soler charts, the Catalan atlas, and other Catalan work of that general period. De La Roncière pointed out that these practitioners were mostly Jews, who had a virtual monopoly of chartmaking in Majorca.[417] What he termed "L'école cartographique des Juifs de Majorque" included Cresques Abraham and his son Jefuda Cresques as well as Jaime Bertran and the converts Mecia de Viladestes and Gabriel de Valseca. Reparaz hinted also at a possible Jewish origin for Petrus Roselli.[418] He interpreted the reference in 1387 to "the" Christian master as an indication that there was only one non-Jewish cartographer working in Majorca at the time, perhaps Guilermo Soler. In the choice of "Jerusalem" rather than "Santo Sepulcro" as the label for the Holy City vignette—for example, in the Catalan atlas—de La Roncière detected the hidden signature of a Jewish cartographer.[419]

The signed Italian output from this period, on the other hand, is more notable for its stylistic dissimilarity. A further problem is raised by the peripatetic career of Grazioso Benincasa. The author's legends of his surviving productions, which range from 1461 to 1482, chronicle his movements: Genoa (1461),[420] Venice (1463–66), Rome (1467), Venice again (1468–69), Ancona (his hometown, 1470), Venice once more (1471–74), and Ancona again (1480–82).[421] How could even one apprentice have followed in this hectic wake?

Whatever doubts there might be about apprenticeship, it is fair to assume that chartmaking skills were often passed on within a family.[422] This might have been the case with all those who shared a surname; it must certainly have applied to Pietro and Perrino Vesconte (unless only one individual was involved), to Grazioso Benincasa and his son Andrea, and to Cresques Abraham and his son Jefuda. This pattern would be continued in the sixteenth century by the Caloiro y Oliva, Freducci, Maggiolo, Oliva, Olives, and Prunes families. Conte Hectomano Freducci imitated the Benincasa style so closely (he was, like them, from Ancona) that it is highly probable he learned his skills from one or other of the Benincasas, presumably transmitting them in turn to his own son Angelo.

THE PRACTITIONERS

At present there are some forty-six men known to us by name as active chartmakers during the fourteenth and fifteenth centuries.[423] These are listed, with their productions, in appendix 19.2. Unfortunately, no details are available about most of these individuals beyond what can be gleaned from the author's legends of their charts. Five of them would have been entirely forgotten

were it not for acknowledged copies of their work in the Cornaro atlas (British Library, Egerton MS. 73): Alvixe Cesanis, Zuane di Napoli, Cristoforo and Zuane Soligo, and Domenico de Zuane. Agostino Noli is another for whom we have faint echoes but no substantial legacy, while Nicolo de Pasqualini described himself as the "son of Nicolo"—presumably a reference to a chartmaking father.[424] That the names of fifteen others have come down to us only through mention on a unique portolan atlas or chart demonstrates the narrow line dividing recognition from oblivion. And then, of course, there are all those who decided to remain anonymous or whose handiwork has failed to survive (at least as far as the author's legend is concerned).

The fragmentary nature of the biographical information so far available about the known chartmakers makes a composite social picture all the more difficult to sketch in. Not surprisingly, some of their number were, or had been, sailors. For example, Andrea Bianco specifically described himself on his chart of 1448 as *comito di galia* (a senior officer on a galley), and official documents survive that link him with almost annual galley sailings throughout the period 1437–51.[425]

417. De La Roncière, *Afrique*, 1:121–41, esp. 126–28 (note 100).

418. Reparaz, "Essai," 322 (note 175).

419. De La Roncière, "Une nouvelle carte," 117 (note 48). See also Oton Haim Oren, "Jews in Cartography and Navigation (from the XIth to the Beginning of the XVth Century)," *Communication du Premier Congrès International d'Histoire de l'Océanographie* 1 (1966): 189–97; reprinted in *Bulletin de l'Institut Océanographique* 1, special no. 2 (1968): 189–97.

420. According to a legal document of 1460, he was already then domiciled in Genoa; see Marina Emiliani (later Marina Salinari), "Le carte nautiche dei Benincasa, cartografi anconetani," *Bollettino della Reale Società Geografica Italiana* 73 (1936): 485–510, esp. 486.

421. The other recorded instances of early chartmakers on the move are these: the Genoese Vescontes worked in Venice; the Genoese Francesco Beccari was in Barcelona in 1399–1400; and the Majorcan Domenech was in Naples at some point in the 1480s (the last digit of his chart's date is illegible). Carignano and Mecia de Viladestes made journeys to Sicily—the former in 1316 (Ferretto, "Carignano," 44 [note 76]), the latter in 1401 (Reparaz, "Essai," 325 [note 175]). Others were sailors at some stage in their careers.

422. This natural tendency was actively encouraged in the Venetian statutes: a master who took on a relative was exempt from the usual dues, whereas a son setting up on his own had to pay a fine; see Favaro, *Arte dei pittori in Venezia*, 25 (note 389).

423. This figure includes Giovanni da Carignano, whose known production is more properly a land map than a chart. It also makes the assumption that Dalorto and Dulcert were a single individual and that there were two Vescontes.

424. Almagià, "Stati Uniti," 360 (note 340), surmised that Nicolo de Pasqualini might be the same as Nicolo de Nicolo, though the dates of their respective charts (1408 and 1470) make this highly improbable.

425. Cornelio Desimoni, "Le carte nautiche italiane del Medio Evo—a proposito di un libro del Prof. Fischer," *Atti della Società Ligure di Storia Patria* 19 (1888–89): 225–66, esp. 260, interpreted

Bianco signed his 1448 chart from London. That was the only year in the period 1445–51 for which his destination is not independently documented. No doubt, as in 1446, 1449, and 1450, he was an officer on one of the Flanders galleys. Three ships were certainly fitted out by the Venetian Senate in February 1448, two of them intending to call at London. Presumably Bianco drew the chart ashore during the three and a half months allotted for cargo loading and customs clearance.[426] Bianco is also recorded as having collaborated with Fra Mauro on his celebrated world map, as payments made to him between 1448 and 1459 testify.[427] Although his will survives—one was made on 15 September 1435, the year before the earlier of his two surviving works—it is, unfortunately, silent about his professional activities.[428]

Another chartmaker, Grazioso Benincasa, though giving no hint of this in the signatures on his charts, had been a shipowner or captain (*padrone*) in the period leading up to the first of his many surviving charts and atlases, dated 1461. It was the loss of his ship to a Genoese corsair (as revealed in legal documents of 1460–61) that apparently ended his career afloat.[429] That Benincasa could call on at least a quarter of a century's experience of practical sailing is clear from a collection of notes in his hand that survived (until World War II) in his native Ancona. Dated from 1435 to 1445, these comprised sailing directions for the Adriatic, Aegean, and Black seas, "ascertained and seen with my own eyes."[430] Grazioso Benincasa's eldest son Andrea seems to have followed closely in his father's footsteps, being active both as a chartmaker and as a galley commander.[431] A fourth chartmaker with practical seafaring experience may possibly be added to this brief list. Cortesão has suggested a tentative identification between Jorge de Aguiar (the author of a 1492 chart that is the first extant to be signed and dated by a Portuguese) and a nobleman navigator of that name who disappeared in 1508 on a voyage to India.[432]

The author's legend to a little-discussed fifteenth-century chart informs us that its compiler, Antonio Pelechan, was also connected with the sea, though in a land-based administrative post. Pelechan described himself as

comito as second-in-command; Revelli, *Mostra Colombiana*, 174 (note 315), described Bianco as a galley commander, adding the unsupported statement that he had sailed along the west coast of Africa. Pompeo Gherardo Molmenti, *Venice: Its Individual Growth from the Earliest Beginnings to the Fall of the Republic*, 6 vols. in 3 pts., trans. Horatio F. Brown (London: J. Murray, 1906–8), pt. 1, *The Middle Ages*, 134, noted that galley commanders were known as *comiti* from the thirteenth century, their title changing to *sopra-comiti* during the fifteenth. The regulations drawn up in 1428 by Andrea Mocenigo and copied out in the Cornaro atlas of ca. 1489 are addressed to the *patronj e sora chomiti de galie*. See British Museum, *Catalogue of the Man-*

uscript *Maps*, 1:20 (note 40). If this leaves Bianco's status in some doubt (see also note 433), another source indicates that he had indeed risen to the rank of commander, at least by 1460. Freddy Thiriet, ed., *Délibérations des assemblées vénitiennes concernant la Romanie*, 2 vols. (Paris: Mouton, 1966–71), 2:221, records that in June of that year Bianco was one of nine candidates for the post of admiral of the Cyprus galleys. Though he was not successful, his qualities were recognized in his appointment as counselor to the new admiral. On Bianco's galley sailings see Biblioteca Nazionale Marciana and Archivio di Stato, *Mostra dei navigatori veneti del quattrocento e del cinquecento*, exhibition catalog (Venice, 1957), nos. 180–89.

426. See Rawdon Brown et al., eds., *Calendar of State Papers and Manuscripts, Relating to English Affairs, Existing in the Archives and Collections of Venice, and in Other Libraries of Northern Italy*, 38 vols., Great Britain Public Record Office (London: Her Majesty's Stationery Office, 1864–1947), 1:67–71.

427. See Tullia Gasparrini Leporace, *Il Mappamondo di Fra Mauro* (Rome: Istituto Poligrafico dello Stato, 1956), 5 ("Presentazione" by Robert Almagià) and 11.

428. Venice, Archivio di Stato, *Notarile Testamenti*, folder 1000, testament 303. I owe this reference to David Woodward. I am grateful to the director of the Archivio di Stato, Maria Francesca Tiepolo, for transcribing the difficult Venetian hand involved, and to Timothy Burnett, Department of Manuscripts, British Library, for helping with the translation.

429. Emiliani, "Carte nautiche," 486 (note 420). Revelli, *Mostra Colombiana*, 92 (note 315), pointed out that Benincasa's earliest dated chart was issued from Genoa because the unfinished litigation forced him to remain in that city.

430. "Tochate chon mano, et vegiute cholli occhi." See Ernesto Spadolini, "Il portolano di Grazioso Benincasa," *Bibliofilia* 9 (1907–8): 58–62, 103–9, 205–34, 294–99, 420–34, 460–63, esp. 104; reprinted in *Acta Cartographica* 11 (1971): 384–450. Spadolini also disposes of the often repeated fiction that the Benincasa portolano contained charts (p. 61).

431. Spadolini, "Benincasa," 60 (note 430), cites an unverifiable source for the assertion that Andrea Benincasa was given command by the Venetian authorities of "una galera per andare in corso." Revelli, *Mostra Colombiana*, 70 (note 315), adds the unsupported statement that this occurred during the war between Venice and Turkey (presumably the first war of 1463–79 or the second of 1499–1503).

432. Cortesão, *History of Portuguese Cartography*, 2:212 (note 3). An even more tentative identification is that suggested by an unsigned entry in the *Diccionari biogràfic*, 4 vols. (Barcelona: Albertí, 1966–70), vol. 1, s.v. "Jaume Bertran." This Bertran was a mariner who captured a pirate off Majorca in 1453. Another scholar records a Jacme Bertran as *patron* of the Majorca galley, both in that same year and in 1455; see Carrère, *Barcelone*, 2:638, 926 n. 1 (note 285). About the middle of the fifteenth century, Jacme, who belonged to a family of converted Jews settled in Majorca and Valencia, went to live in Genoa (Carrère, *Barcelone*, 2:584). In 1456, "Jachobus" Bertran signed the first of his three known charts. Although the surname was a fairly common one at the time, the possibility that the chartmaker was a seaman finds an echo in the suggestion by Rodolico that the prefix "Mestra" on his 1489 chart might denote (as with Juan de la Cosa in 1500) the status of a pilot. See Niccolò Rodolico, "Di una carta nautica di Giacomo Bertran, maiorchino," *Atti del III Congresso Geografico Italiano, Florence, 1898*, 2 vols. (1899) 2:544–50, esp. 545. A similar suggestion is made in an unsigned entry in *Enciclopedia universal ilustrada Europeo-Americana*, 70 vols. and annual supplements (Madrid and Barcelona: Espasa-Calpe, 1907–83), 68:1187, that the Matias Viladestes who commanded a galley belonging to Francés Burgés in 1415 should probably be identified with the Mecia de Viladestes who signed the 1413 chart. In the same work (66:838–39) Gabriel Valseca is also termed a *navegante*.

armiraio de Rutemo.[433] At that time, 1459, Retimo (Rethymnon) in Crete formed part of the Venetian dominions, and its *armiraio* was entrusted with the practical administration of the port. This office would have demanded experience in both seamanship and navigation. An unadorned chart of the Adriatic is all that survives in his hand. A close parallel exists between Pelechan's position and that occupied in 1496 by Andrea Benincasa. As *capitano del porto* of Ancona, he was responsible for the harbor's fortifications.[434] Taken in conjunction with other official posts he is known to have held, Andrea's atlas of 1476 and charts of 1490 and 1508 must represent the fruits of a less than full-time occupation.[435]

Pelechan and Andrea Benincasa were not the only chartmakers to enjoy privileged status. Vesconte was consulted by the Venetian authorities when the first Flanders fleet was being organized in the early fourteenth century.[436] Cresques Abraham, the supposed author of the Catalan atlas, was accorded special rights by King Pedro of Aragon—a reflection of his ability, since like other Jews he had suffered initially from discrimination.[437] A more remarkable example comes just after our period when the Genoese authorities used a hundred-lire annuity to lure Vesconte Maggiolo back from Naples in 1519.[438]

These instances are of rewards for skill; they tell us little about the social origins of the individuals concerned. Grazioso Benincasa, on the other hand, is known to have been of noble birth, and Pietro Vesconte evidently belonged to one of Genoa's ruling families.[439] It is possibly no coincidence, therefore, that the only recorded portraits of early chartmakers should have concerned these two. One of the cornerpieces on the Black Sea sheet of Pietro Vesconte's 1318 (Venice) atlas features a man seated at an angled table and working on a chart (plate 31).[440] It is only natural to speculate that Vesconte himself was probably the subject. The second instance, consisting of paired portraits of Grazioso and Andrea Benincasa, was set into a world map. Mentioned in 1536, this has unfortunately failed to survive.[441]

It would be quite wrong to suppose from this brief catalog that we could sketch in a similar profile of high birth and exalted social status for all the other named and nameless chartmakers of the period. It is precisely patricians like Benincasa whom history remembers; his humbler colleagues have no memorial but their charts.[442] A fairer picture of a chartmaker's true social position is probably the one that emerges from Agostino Noli's petition of 1438. Describing himself as "very poor," Noli managed to persuade the Genoese authorities to grant him ten years' tax exemption—among other reasons because they accepted that his work, though time-consuming, was not very lucrative.

433. Venice, Archivio di Stato. Uzielli and Amat di San Filippo, *Mappamondi*, 75 (note 35), misread the author's name as Antonio Pelegan e Miraro of Resina. The correct transcription of the author's legend—for which the assistance of Maria Francesca Tiepolo is gratefully acknowledged—would be: "antonio pelechan armiraio / de rutemo o fato questo chol/fo 1459 adi 4 luio." A similar instance is provided by the sixteenth-century chartmaker Antonio Millo, who signed his "Arte del navegar" as "Armiraglio del Zante"; see Uzielli and Amat di San Filippo, *Mappamondi*, 216 (note 35). For an interpretation of *armiraio* as a chief navigating officer on a ship or for a fleet, in which sense the term was applied to Andrea Bianco, see Frederic C. Lane, *Venice: A Maritime Republic* (Baltimore: Johns Hopkins University Press, 1973), 169, 277, 343–44.

434. Emiliani, "Carte nautiche," 488 (note 420).

435. As must be the work of other "amateurs," like the priest Pareto and the Tripoli physician Ibrāhīm al-Mursī—on the latter see Ettore Rossi, "Una carta nautica araba inedita di Ibrāhīm al-Mursī datata 865 Egira = 1461 Dopo Christo," in *Compte rendu du Congrès Internationale de Géographie* 5 (1926): 90–95 (11th International Congress, Cairo, 1925).

436. It is not clear whether this refers to Pietro or Perrino Vesconte; see de La Roncière, *Afrique*, 1:43 (note 100), and Crone, *Maps and Their Makers*, 17 (note 11).

437. Grosjean, *Catalan Atlas*, 13 (note 94).

438. Revelli, *Colombo*, 472–78 (note 22). This sum was still being paid to Vesconte's successors in 1650.

439. On Benincasa see Emiliani, "Carte nautiche," 485 (note 420); on Vesconte see Revelli, *Colombo*, 237, 418 (note 22). Matkovič, "Wien," 7 (note 294), described his family as holding important posts in Venice during the period 1270 to 1339. Recently a document was published purporting to show that Vesconte was a surgeon; see Piersantelli, "Atlante Luxoro," 135–38 (note 364). Given the proficiency of Vesconte's charts, it is surprising to learn from a Venetian legal dispute of 1326 or 1327 that one "Petrus Visconte" from Genoa was well thought of as a surgeon. He was able to command a large fee when called out to a Treviso lawyer, Pietro Flor, who was thought to be on the verge of death from acute dropsy. However, the document, whose Latin text is transcribed in full by Piersantelli, does not provide a totally convincing identification between the surgeon and the chartmaker. In the first place, Pietro and Perrino Vesconte signed between them eight surviving works, stating their last name as Vesconte or Vesssconte, never Visconte. Nor is any reference made in the document to chartmaking, unless the name of one of those who testified on his behalf, Rigo da le Carte, is construed in that light. The last mention of Pietro Vesconte, according to another document cited by Piersantelli, was in Genoa in 1347 (pp. 137–38).

440. Pagani, *Vesconte*, 20 (note 47), and Mollat du Jourdin and de La Roncière, *Sea Charts*, 14 (note 40).

441. Emiliani, "Carte nautiche," 489 (note 420). A slightly later example of what has been tentatively identified as a self-portrait set into a sea atlas is that of Jean Rotz; see Helen Wallis, ed., *The Maps and Text of the Boke of Idrography Presented by Jean Rotz to Henry VIII, Now in the British Library* (Oxford: Viscount Eccles for the Roxburghe Club, 1981), 38; idem, "The Rotz Atlas: A Royal Presentation," *Map Collector* 20 (1982): 40–42, esp. 42.

442. Though Desimoni, "Elenco," 48 (note 62), refers to the Beccari family sepulcher in Genoa. Exceptions are also provided by the Genoese priests Giovanni da Carignano and Bartolomeo de Pareto. Carignano featured in contemporary records a dozen times between 1291 and 1329; see Ferretto, "Carignano," 36–45 (note 76). For documents concerning Pareto, see Canale, *Storia del commercio*, 88, 457 (note 407). The only chartmakers featured so far in the *Dizionario biografico degli Italiani* (Rome: Istituto della Enciclopedia Italiana), whose publication started in 1960, are Batista Beccari, the two Benincasas, and

THE CHART TRADE

Our general ignorance about the individual chartmakers is matched by the very limited information available about their customers. That many charts must have been drawn for navigational use, and hence acquired by sailors, is discussed in a later section. But specific documentary evidence tends to relate to the more flamboyant productions, especially world maps in the Catalan style. Those that were commissioned from Jefuda Cresques and Francesco Beccari in 1399 for presentation to royalty have already been mentioned. The Catalan atlas of 1375 was probably made for the king of France; and other examples, since vanished, were ordered by members of the Aragonese royal family.[443] Another possible royal commission has been identified among the less obviously regal productions. The prominent arms of Castile and Leon on the 1426 Batista Beccari chart led Winter to suppose that it might have been made for the "Spanish crown" (more correctly, the king of Castile).[444] There is no doubt, however, that the unsigned Vatican atlas that Pietro Vesconte drew for Marino Sanudo was one of two presented to Pope John XXII in 1321.[445] Pareto's chart of 1455, although not now in the Vatican, was probably made for Pope Nicholas V, who died that same year.[446] Both men were Genoese, and Pareto served as one of the papal acolytes. Another prince of the church, Cardinal Raffaello Riario, was the recipient of Benincasa's unusually ornate chart of 1482.[447]

It is from the author's legend itself that we learn the identity of the individual who commissioned one of Benincasa's 1468 atlases (British Library, Add. MS. 6390), the Genoese doctor and diplomat Prospero da Camogli.[448] It is possible that the author's legend on the 1426 Batista Beccari chart would have resolved the doubts about its original owner, since the final words before it becomes completely illegible have been read as "mense novembris ad requisicionem et nomine."[449]

Coats of arms on contemporary bindings are another pointer, if an oblique one, to the identity of the original recipient of the atlas concerned. The arms of the Venetian Cornaro family are featured on two anonymous atlases: on the outer covers of one in Lyons and on the bookplate of the Cornaro atlas in the British Library.[450] Another undated atlas, recorded last century in Ventimiglia, bore the arms of the celebrated Usodimare family.[451] A similar personal mark, this time at the edge of the 1447 Valseca chart, was identified by Hamy as that of Francesco de Lauria.[452] A coat of arms (since overpainted) identifies Borso d'Este (d. 1471) as the recipient—probably in 1466—of a manuscript Ptolemy, whose final double folio contains a portolan chart that is considered to form an integral part of the work.[453]

It is also likely that some of the Italian families who are recorded as the earliest known owners of charts and atlases now in public collections may have been those who actually commissioned the works in question.[454] Unfortunately no proof of this, with the invaluable commercial documentation that might accompany it, has yet come to light. Among indications of a more general kind we can cite the instance of the Cortona chart, whose prominent naming of that town led Armignacco to suspect a Cortonese commission.[455]

Bianco (in notes by Angela Codazzi). The four-volume Catalan equivalent, *Diccionari biogràfic* (note 432), has brief unsigned notes on Dulcert and Macià Viladestes only. The *Enciclopedia italiana di scienze, lettere ed arti* (note 4) has separate headings for Grazioso Benincasa and Pietro Vesconte only (both by Roberto Almagià); the *Enciclopedia universal ilustrada* (note 432) has unsigned notes on Valseca and Matias Viladestes only.

443. Reparaz, "Essai," 293–97 (note 175), citing, for the full transcriptions, Antoni Rubió y Lluch, *Documents per l'historia de la cultura catalana mig-eval* (Barcelona: Institut d'Estudis Catalans, 1908–21). See also Rey Pastor and García Camarero, *Cartografía mallorquina*, 66 (note 28).

444. Winter, "Roselli," 2 (note 224). The Italian atlas in the Bibliothèque Nationale, Département des Manuscrits, MS. Lat. 4850, belonged to Louis XII (1499–1515); see Georges Deulin, *Répertoire des portulans et pièces assimilables conservés au Département des Manuscrits de la Bibliothèque Nationale* (typescript, Paris, 1936), 20.

445. Rome, Biblioteca Apostolica Vaticana, Vat. Lat. 2972, see Almagià, *Vaticana*, 1:17a (note 35).

446. Uzielli and Amat di San Filippo, *Mappamondi*, 74 (note 35).

447. Riario is not specifically mentioned by name, but the family's arms occur three times on the chart, surmounted by a cardinal's hat; see Emiliani, "Carte nautiche," 501 (note 420). Manuel Francisco de Barros e Sousa, Viscount of Santarém stated that the 1321 Perrino Vesconte atlas in Zurich was made for the doge of Venice; see his "Notice sur plusieurs monuments géographiques inédits du Moyen Age et du XVIe siècle qui se trouvent dans quelques bibliothèques de l'Italie, accompagné de notes critiques," *Bulletin de la Société de Géographie*, 3d ser., 7 (1847): 289–317, esp. 295 n. 1; reprinted in *Acta Cartographica* 14 (1972): 318–46. However, this stems from a misreading of the later note added beneath the author's legend.

448. This is the only pre-1500 chart or atlas that actually spells out the name of the dedicatee, "Prospero Camulio Medico." On Camogli see Revelli, *Colombo*, 354, 469 (note 22).

449. Desimoni, "Elenco," 48 (note 62).

450. On the Lyons atlas see de La Roncière, *Lyon*, 15 (note 34); on the Cornaro atlas (Egerton MS. 73) see British Museum, *Catalogue of the Manuscript Maps*, 1:17 (note 40).

451. Uzielli and Amat di San Filippo, *Mappamondi*, 101 (note 35).

452. Foncin, Destombes, and de La Roncière, *Catalogue des cartes nautiques*, 23 (note 52).

453. Paris, Bibliothèque Nationale, Département des Manuscrits, MS. Lat. 4801. See Elisabeth Pellegrin, "Les manuscrits de Geoffroy Carles, président du Parlement de Dauphiné et du Sénat de Milan," in *Studi di bibliografia e di storia in onore di Tammaro de Marinis*, 4 vols. (Verona: Stamperia Valdonega, 1964), 3:313–17. I am grateful to the conservateur en chef, Denise Bloch, for this reference and for other observations.

454. This suggestion has been made in connection with one of the 1447 Roselli charts, formerly owned by the Martelli family of Florence; see Nebenzahl, *Rare Americana*, no. 164 (note 411).

455. Armignacco, "Una carta nautica," 192 (note 243).

How these early chartmakers operated—whether they always worked on commission or sometimes drew charts for stock—remains largely a matter of speculation.[456] Under the terms of the 1399 agreement, Cresques and Beccari were contracted to work exclusively for Ubriachi, not leaving Barcelona until the task was completed.[457] Besides fees, Beccari, at least, received living expenses. This type of all-embracing patronage, however, was probably atypical.

There seem to be a number of parallels between the production of charts and that of books of hours, "the late medieval bestseller."[458] Indeed, their histories are roughly contemporary, since the earliest surviving book of hours dates from the middle of the thirteenth century.[459] Just as no two charts are alike, so books of hours tend to reflect, in the regional or even local variations of their text and decoration, the fact that most were tailor-made for particular clients.[460] Since those who commissioned hours included shopkeepers and kings,[461] the quality ranges from the workaday to the sumptuous. With the charts, this breadth of choice might not have been available if a particular practitioner worked only in one style, whether plain or ornate. But it would be wrong to rule out the possibility that the degree of ornamentation, or total lack of it, sometimes reflected the size of the customer's purse.

With books of hours as well as charts, the more extravagant productions, while representing the pinnacle of the pyramid, were valued as art objects from the outset and must reasonably be assumed to have come down to us in disproportionate numbers. The unpretentious hours for actual devotional use, the functional chart for navigational use—these have hardly survived at all, even though they represent the vast majority of those produced. A history of portolan charts ought to be largely concerned with these everyday charts, which were destined to be casually discarded once salt water and constant unrolling had obliterated their outlines. With survival, with glamour, and occasionally with documentation on their side, it is inevitable, though, that the deluxe productions should have received the lion's share of comment.[462]

A contractual link between an artist and his patron was the standard procedure in medieval Italy. Unfortunately, verbal agreements were considered sufficient for all but major projects.[463] Notarized contracts usually laid down fees and a completion date, but this potential source of valuable information was often sidestepped. It is probably for this reason that up to now only one contract involving chartmakers has come to light. Because of the unusually large scale of the world maps commissioned from Cresques and Beccari in 1399, the eleven months that the latter claimed to have expended on the two productions[464] is of little use in computing

the time required to draw even the most elaborate portolan chart. More relevant would be the three weeks quoted by a mid-seventeenth-century English chartmaker.[465]

Despite the weeks or months that each chart would have entailed, a number of signed fourteenth- and fifteenth-century works are dated to a particular day.[466] The topicality this can give is admirably illustrated by the chart drawn at Alexandria in 1497 by the Jewish chartmaker Jehuda ben Zara. Dated 8 February, its author's legend refers to the deposition of the sultan of Cairo a few days before (fig. 19.18).[467] This precision in dating could have proved embarrassing had the charts normally been made for stock and then failed to sell. It would have been a simple matter, however, to have left them unsigned until a purchaser was found.

FIG. 19.18. A TIMELY POSTSCRIPT ON A PORTOLAN CHART. Written by Jehuda ben Zara, this legend dated 8 February 1497 refers to the deposition of the sultan of Cairo a few days earlier. Such precision is, however, rare on surviving charts, which, if dated at all, usually refer only to the year. Photograph from the Biblioteca Apostolica Vaticana, Rome (Borgiano VII).

456. When the future Juan I of Aragon requested a world map from Majorca in 1379, he expressed a readiness either to buy an existing one or to commission a fresh example; see Rubió y Lluch, *Documents*, 2:202 (note 443).

457. Skelton, "Contract," 108 (note 206).

458. L. M. J. Delaissé's phrase, quoted in John Harthan, *Books of Hours and Their Owners* (London: Thames and Hudson, 1977), 9.

459. Harthan, *Books of Hours*, 13 (note 458).

460. Harthan, *Books of Hours*, 12 (note 458).

461. Harthan, *Books of Hours*, 31 (note 458).

462. And, one can add, of reproduction: most of the purely functional charts remain unpublished. Harthan makes a similar point about books of hours; see his *Books of Hours*, 31 (note 458).

463. Hannelore Glasser, *Artists' Contracts of the Early Renaissance* (New York: Garland Publishers, 1977), 1.

464. Skelton, "Contract," 108 (note 206).

465. Nicholas Comberford; see Smith, "Thames School," 91–92 (note 185).

466. As is frequently found in contemporary manuscript books of the period; see Andrew G. Watson, *Catalogue of Dated and Datable Manuscripts c. 700–1600 in the Department of Manuscripts, the British Library* (London: British Library, 1979).

467. Almagià, *Vaticana*, 1:47 (note 35).

An invaluable glimpse of the chart trade in action is provided by a series of late fourteenth-century documents in Barcelona. Relating in each case to the merchant Domenech Pujol, they describe how, on several occasions in 1390 and 1392, he entrusted *cartes de navegar* in batches of four to the mariners Pere Folch or Pere Jalbert for them to sell in various parts of the Mediterranean.[468] Alexandria, Genoa, Naples, Pisa, and Sicily were all mentioned. These documents are surprising in two ways. In the first place, Barcelona is not known to have been a chartmaking center of note in the fourteenth or fifteenth century, and the only reference to that city in an author's legend occurs over sixty years later on the Bertran and Ripoll chart of 1456. Second, these records, which are unique in demonstrating the activities of a medieval dealer in sea charts, suggest that the practice was widespread. It is improbable that Pujol drew the charts himself, and he seems to have handled them like any other merchandise. Jalbert, for example, was instructed to barter his consignment in Alexandria for pepper, and the recipient would presumably have sold the charts in his turn. Between chartmaker and chart user, therefore, there might have been three or more intermediaries.

The relevance and usefulness of portolan charts to all the mariners of the Mediterranean and Black seas presumably explains the risk Pujol took in shipping them without having definite orders in advance. Folch, for instance, in the final commission of 23 October 1392 was asked to sell his charts in either Sicily or Pisa. The Pujol documents also state the values of the consigned charts.[469] The four groups of four charts ranged in price from 8 *libra*, 16 *sueldos* to 7 *libra*, 17 *sueldos*. Each single chart would thus have been valued at approximately 2 *libra*. This figure can be compared with the sums agreed upon by Francesco Beccari for work on large world maps he prepared about ten years later (also in Barcelona) in collaboration with Jefuda Cresques. The smaller size (150 × 310 cm) commanded 60 Aragonese florins, the larger (368 cm square) 100 florins.

Data derived from the cathedral records of Burgos allow us to relate those three figures to contemporary wages and prices. A laborer, for example, would have had to work some 26 days to earn the price of one of the Pujol charts, about 480 days for the smaller of the Beccari-Cresques world maps, and 800 days for the larger. The world maps were respectively some seven and twenty-one times the size of an average chart (65 × 100 cm). Though no more than a single skin would presumably have been involved, the price of each Pujol chart was equivalent to about twenty large sheets of unused parchment. If this indicates that the cost of raw materials formed only a small part of the chart's selling price, Beccari's claim for expenses of roughly three times

the daily laboring wage confirms that chartmakers were reasonably well paid.[470]

That twenty-four charts should have been shipped out by one man in the space of two years from a city not otherwise known to have supported a single resident chartmaker[471] shows how unwise it would be to rely exclusively on the author's legends of surviving charts for information about the various centers of production. Nevertheless, this evidence is of value if approached with caution. Of sixty fourteenth- and fifteenth-century works whose place of construction is stated, roughly a third were made in Majorca and another third in Venice.[472] The strength and importance of Venice need no substantiating, but the leading role of Majorca might seem surprising. However, a modern map of the Mediterranean, compiled on the basis of the routes described in the mid-thirteenth-century *Lo compasso da navigare*, demonstrates the extent to which the Balearics functioned as a crossroads in terms of maritime communications.[473] This would have made Palma the natural center for chartmaking activities in the Catalan-speaking world. Alexandria, Ancona, Genoa, and Rome are all mentioned more than once in the author's legends of known charts, but—like Barcelona—Lisbon, London, Naples, Rethymnon, Savona, and Tripoli are each named only on a single document.[474] These figures are distorted because people like Roselli invariably included a full author's legend while others equally consistently

468. Carrère, *Barcelone*, 1:201 n. 4 (note 285). The earliest reference is to a group of eight charts, sent via an unnamed agent to Flanders.

469. See Carrère, *Barcelone*, 1:201 n. 4 (note 285).

470. On the Beccari-Cresques contract, see above p. 430. The information on currencies and values derives from the following, to which I was kindly directed by Angus MacKay; see his *Money, Prices and Politics in Fifteenth-Century Castile* (London: Royal Historical Society, 1981), 141, 144, 150; also Peter Spufford and Wendy Wilkinson, *Interim Listing of the Exchange Rates of Medieval Europe* (North Staffordshire: Department of History, University of Keele, 1977), 189. The figures were based on the following currency equivalents: for 1390, 1 florin = 0.55 *libra* = 23 *maravedíes*; for 1400, 1 florin = 48 new *maravedíes*.

471. Though it is not impossible that Pujol himself imported the charts into Barcelona from Majorca.

472. References to the author's birthplace (e.g., "de Janua") have sometimes been misinterpreted as the place of construction (e.g., "in Janua"). For the individual instances see appendixes 19.2 and 19.4. For Majorca: "Civitate maioricarum," that is, Palma, whenever the town was stated.

473. Quaini, "Catalogna e Liguria," 569; see also 560 (note 60).

474. In connection with Naples, mention should be made of Zuane di Napoli, one of those known only from the Cornaro atlas. Tripoli is cited in the author's legend of the 1461 al-Mursī chart, without specifying whether the Libyan or Lebanese town was involved. Rossi, "Carta nautica araba," 91 (note 435), detected a Maghreb hand and seems to have assumed that the chart was drawn in North Africa. It is significant that Francesco Beccari, in the note to his 1403 chart, distinguished Catalan, Venetian, and Genoese practitioners "as well as others who made navigational charts in past times" (see note 384).

omitted it. A third of the total is contributed by the peripatetic Benincasa alone.

Though there has been a tendency to see the first two centuries of portolan chartmaking as spread evenly between Majorca, Genoa, and Venice, Genoa's contribution has sometimes been overstated.[475] This is largely because Genoese historians have been more active in this field than their Venetian counterparts. Virtually no information about chartmaking has emerged from the Venetian archives (yet it is hard to believe that none exists). Indeed, it is from documents unearthed in Genoa's archives that we see how nearly chartmaking died out completely in that city, as exemplified by the fifteenth-century Noli and Pareto documents already mentioned. Nor does the dispatch by Pujol of four charts from Barcelona to Genoa in 1392 suggest that the native industry was flourishing at the end of the preceding century. Of the five other Genoese involved (leaving aside the possibly relevant Dalorto/Dulcert), Giovanni da Carignano belongs strictly outside the portolan chart tradition, neither Pietro or Perrino Vesconte nor Batista or Francesco Beccari ever definitely worked in Genoa, and only Albino da Canepa seems to have displayed wholehearted loyalty to his native city. The willingness of Genoese chartmakers to emigrate is also illustrated by the case of Vesconte Maggiolo, referred to earlier.

The appearance of the few unquestionably Genoese charts is additional testimony against the existence of a distinctive and continuous tradition of chartmaking in that city. The strong influence of Majorcan models is immediately apparent. If it is objected that the ornamented Catalan style is first found on the charts of Dalorto/Dulcert, for whom Genoese origin is claimed—and hence that it was the Majorcans who were the imitators—there are strong counterarguments. First, no surviving chart earlier than Pareto's of 1455 actually states that it was drawn in Genoa (if we except the Carignano map), and second, the work of known Genoese chartmakers incorporates only some of the Catalan characteristics. As far as the decorative elements are concerned, it is most reasonable to see Genoese work as an imitation or précis of the Catalan style.[476]

Genoa's slender contribution (at least after the earliest period) must be contrasted with the preeminent role Venice played in portolan chart construction before 1500. To the twenty-one works that were signed from there must be added a further handful that have been plausibly attributed to Venetian practitioners on stylistic grounds.[477] Marked in most cases by the absence of any details beyond those essential for navigation, Venetian work has as its hallmark a distinctive austerity. If future research succeeds in tying down the numerous unsigned works to a particular place of origin, on stylistic or orthographic grounds, it seems likely that Majorca (pre-

sumably Palma) and Venice will be seen as the only centers capable of supporting established schools of chartmaking during our period.

THE FUNCTION OF THE PORTOLAN CHARTS

It is not unreasonable to assert that a portolan chart becomes of greater interest once it has reached its first owner. It is undoubtedly true that a chart, whether considered as an artifact or as a cartographic record, derives added significance from the way it was used. Indeed, this question of function is arguably the most crucial of all.

ARCHIVAL PURPOSE

Two dimensions need to be separated, the instructional and the practical, though inevitably they tend to overlap. The best example of work intended for instructional rather than shipboard use is the British Library's Cornaro atlas. Its thirty-four sheets include eight versions of the standard portolan chart (variously on one, two, or three sheets).[478] Among the charts devoted to particular areas, the Aegean is treated five times and the Black Sea four times. This duplication extends to variant outlines for the Black Sea and Adriatic, paired for comparison on single sheets.

The most likely explanation for this strange Venetian collection is that it comprises archival copies of what was considered to be the best Italian, Catalan, and Portuguese work. There are strong indications that the atlas dates from about 1489.[479] Its juxtaposition of Portuguese discoveries in Angola, made just six years earlier, and charts of the Mediterranean, some of whose models

475. The claim has even been made that early fourteenth-century Genoa was the center of chartmaking with its own official municipal workshop; see Laura Secchi, *Navigazione e carte nautiche nei secoli XIII–XVI*, catalog of an exhibition held at the Palazzo Rosso, Genoa, May to October 1978, 38, citing Giuseppe Piersantelli, *L'atlante di carte marine di Francesco Ghisolfi e la storia della pittura in Genova nel Cinquecento* (Genoa, 1947), 1–7. In this context a distinction needs to be made between "Genoa" and the "Genoese."

476. Though this reverses the judgment of Andrews, "Scotland in the Portolan Charts," 138 (note 249). The toponymic originality of Batista and Francesco Beccari should, however, be emphasized here (see table 19.3, pp. 416–20). The presence of Francesco Beccari in Barcelona in 1399 (see p. 430) also stresses the cartographic links between Genoa and Catalonia, as do both the title and theme of Quaini, "Catalogna e Liguria" (note 60), and the 1392 shipment of charts from Barcelona to Genoa, see above, p. 437.

477. For instance, attributions to Giroldi by Andrews, "Rathlin Island," 33–34 n. 1 (note 247).

478. British Museum, *Catalogue of the Manuscript Maps*, 1:17–20 (note 40).

479. Cortesão gives a full collation of this work in his *History of Portuguese Cartography*, 2:195–200 (note 3), and discusses earlier descriptions of it.

had been drawn at the beginning of the century, suggests a continuing respect for work that we might have supposed would be considered out of date.[480] P. D. A. Harvey has pointed out that there was a unique appreciation in late fifteenth-century Venice of the value of maps to an administration.[481] Might we be seeing here the nautical aspect of this same cartographic consciousness?

A similar archival interpretation can perhaps be placed on some of the other surviving atlases. The alternative drafts of the Adriatic found in the Medici atlas (see appendix 19.1) probably have the same explanation as those in the Cornaro volume (fig. 19.19). Besides copies made for the record, we can also point to the insertion of additional material into the Pinelli-Walckenaer and Pizigano atlases as possibly reflecting a similar documentary purpose.[482] A note about a pilgrimage to the Holy Land on the reverse of the Cortona chart[483] draws attention to another function that the charts could have performed: namely, to plan or record a voyage.

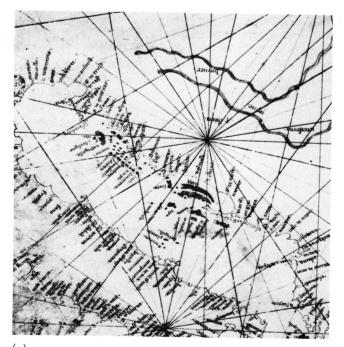

(a)

(b)

SHIPBOARD USE

These examples can only give us a small part of the picture, however, because the evidence that portolan charts were used on board ship is overwhelming. The earliest discovered reference to a medieval sea chart already makes this clear. When the French king Louis IX (Saint Louis) set out from Aigues-Mortes for Tunis in 1270 he was forced to make for Cagliari in a storm. To reassure the king that they were close to land, the captain showed him what must have been a chart, although the account of this voyage was written in Latin and used the words *mappâ mundi*.[484] The same ambiguous term recurred in 1294 when the prince of Aragon demanded restitution for a ship, the *San Nicola* of Messina, which had been seized by Italian pirates.[485] The inventory listed

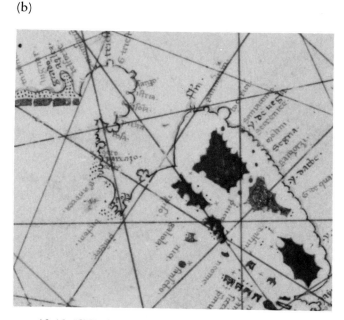

FIG. 19.19. ISTRIA IN THE MEDICI ATLAS. The occurrence of markedly different drafts of the Adriatic in this apparently homogeneous atlas suggests that they were intended as archival copies. The detail of the Istrian section from one of two smaller-scale charts (*a*) displays the range of names typical of the mid-fourteenth century. By contrast, larger-scale sheet (*b*) includes names such as *san iacomo* (first found on a dated chart in 1367) and *setrenice* (first found in 1408), while the earlier *cauo de osero* has been dropped.
By permission of the Biblioteca Medicea Laurenziana, Florence (Gaddi 9, charts 6 and 3 respectively).

---

480. The accompanying text also includes fourteenth- and early fifteenth-century material; see Revelli, *Colombo*, 351 (note 22).

481. Harvey, *Topographical Maps*, 60–61 (note 8). For a similar appreciation in mid-fifteenth-century Genoa, see the Noli petition referred to on p. 430.

482. On the Pinelli-Walckenaer atlas see Marie Armand Pascal d'Avezac-Macaya, "Fragments d'une notice sur un atlas manuscrit de la Bibliothèque Walckenaer: Fixation des dates des diverses parties dont il se compose," *Bulletin de la Société de Géographie*, 3d ser., 8 (1847): 142–71, esp. 171. On the Pizigano see Errera, "Atlanti," 91–96 (note 221).

483. Armignacco, "Una carta nautica," 186 (note 243), and Caraci, *Italiani e Catalani*, 278 (note 175).

484. Guillaume de Nangis, "Gesta Sanctæ Memoriæ Ludovici" (Life of Saint Louis), in *Recueil des historiens des Gaules et de la France*, 24 vols., ed. J. Naudet and P. Daunou (Paris: Imprimerie Royale, 1738–1904), 20 (1840): 309–465, esp. 444.

485. Charles de La Roncière, "Un inventaire de bord en 1294 et les origines de la navigation hauturière," *Bibliothèque de l'Ecole des Chartes* 58 (1897): 394–409.

no fewer than three charts—if the text has been correctly interpreted. More demonstrative still is the Aragonese ordinance of 1354 decreeing that each galley should carry a pair of sea charts.[486] If that seems excessive, a Genoese ship seized three years later had no fewer than four charts on board.[487] Further evidence of shipboard use can be found in the fifteenth century. Francesco Beccari specifically addressed the important note on his 1403 chart to "all those who do or shall sail the ocean sea,"[488] and the submissions of Noli and Pareto to the Genoese authorities stressed the navigational value—indeed, necessity—of sea charts, bracketing them firmly with the magnetic compass.

Thus, documentary evidence that early portolan charts formed a recognized part of marine equipment has to be reconciled with the equally clear indications, set out in the previous section, that other examples were constructed with the pleasure and enlightenment of landsmen in mind.[489] There is no ready agreement, however, about the line of demarcation between the two types and about how surviving charts should be distinguished according to the supposed alternatives of navigational or ornamental function. In the most radical interpretation, all the practical wayfinding charts have failed to survive (except for the occasional fragment). The lack of navigational markings on those charts available for study today (a point to be discussed in the following section) has been seen as further proof that almost all extant examples should be considered a different species from those intended for shipboard use.[490] Systematic chemical analysis of those charts that betray evident water staining, to test for traces of salt, would inject some needed data into what has been a largely theoretical discussion.

It is commonly assumed, for example, that none of the more elaborate productions would have been taken to sea, and that it was the plain Italian-style charts that were used, and normally worn out, on board ship. Though there is no firm evidence to support this thesis, and though the distinction functional versus artistic has little relevance in a medieval context, it does seem a plausible interpretation. Charts shipped to meet the requirements of the Aragonese decree or just for operating needs would presumably have been inexpensive, and hence unadorned, in marked contrast to lavish productions like the celebrated Valseca chart of 1439. But the earlier warning against oversimplification along nationalist lines must be repeated. Each of the main chart-making centers, and probably most of the individual practitioners, was presumably capable of manufacturing navigational or luxury charts as required.

If there has been disagreement about the function of the separate charts, it has been generally supposed that volumes of portolan charts were intended from the outset for the library shelf.[491] This theory, however, has

several drawbacks. In the first place, an atlas offered the mariner a number of practical advantages over a chart. Portuguese discoveries in the Atlantic islands and along the west coast of Africa, for instance, could not be properly accommodated on a normal chart.[492] Then again, a succession of overlapping charts, sometimes fixed to rigid boards and protected by a leather binding, would have been both easier to use and more durable, since they would have been less subject to crinkling and distortion and would have presented a flat surface to a parallel ruler. The paschal calendar found in some atlases might have a purely religious significance, but the lunar calendar that more frequently prefaces such volumes was an essential part of a navigator's equipment if the Atlantic tides were to be accurately predicted (for calendars see appendix 19.1).

It has to be admitted, however, that the earliest recorded explanation of the "establishment of the port" is found in the Catalan atlas, and the most extensive set of navigational rules and sailing directions is in the Cornaro atlas[493]—both of which have survived precisely because they were not risked at sea. The first can be interpreted as a cosmological compendium, the second as a collection of file copies; but there is every reason to suppose that the essential navigational information contained in those works found its way to sea in less ostentatious forms. The most likely vehicle would be a portolan atlas, and the example drawn in 1436 by the galley officer Andrea Bianco may well be a case in point, since it is prefaced by a sheet that includes mathematical

486. Hamy, "Origines," 416 (note 47). Some authorities have cited other dates for this ordinance: 1352—for example, Reparaz, "Essai," 286 (note 175); and 1359—Clements R. Markham, *The Story of Majorca and Minorca* (London: Smith Elder, 1908), 172. It is worth noting that a comparable ordinance of 1331 makes no reference to sea charts; see José Maria Madurell y Marimon, "Ordenanzas marítimas de 1331 y 1333," *Anuario de Historia del Derecho Español* 31 (1961): 611–28.

487. De La Roncière, *Afrique*, 1:123 n. 2 (note 100).

488. See note 384.

489. Giuseppe Caraci, "Un'altra carta di Albertin da Virga," *Bollettino della Reale Società Geografica Italiana* 63 (1926): 781–86, esp. 783, employed the term *carta d'uso* for those in the first category; Mollat du Jourdin and de La Roncière (i.e., Isabelle Raynaud-Nguyen), *Sea Charts*, 200 (note 40), described an example of the second category as "a work of art, a collector's piece." It needs to be emphasized that there are no differences of hydrographic or toponymic content between the two types.

490. For example, Giuseppe Caraci, "An Unknown Nautical Chart of Grazioso Benincasa, 1468," *Imago Mundi* 7 (1950): 18–31, esp. 20, talked of "the very rare examples known up to this date of nautical maps destined to be really used."

491. Almagià, *Vaticana*, 1:viii (note 35).

492. The Catalan solution was to produce a world map, but these could never have been taken to sea.

493. Derek Howse, "Some Early Tidal Diagrams," *Revista da Universidade de Coimbra* 33 (1985): 365–85, esp. 366–68. British Museum, *Catalogue of the Manuscript Maps*, 1:20–21 (note 40).

tables, known as the *Toleta* (to be discussed below). It can only be speculated that Bianco's atlas was taken to sea; but in another case there are strong indications that this was definitely the case.

Among the Vatican's collections is an anonymous atlas dated 1452.[494] Almagià considered it to be Venetian in origin, probably from the workshop of Giroldi. He also detected several significant additions in a number of different later hands. The lunar table (whose nineteen-year cycle could be endlessly repeated) had been annotated to identify the relevant letters for the years 1561, 1571, and 1618; a comment, probably in a sixteenth-century hand, had been placed beside Paxos; and four notes had been added, in three different sixteenth-century hands, describing the approaches to various harbors. Thus, for more than a century and a half this atlas (which came to the Vatican only in the nineteenth century) was being updated with practical navigational information.[495] When taken with the other arguments, the Vatican atlas demonstrates the possibility, if not the probability, that, like portolan charts, some portolan atlases were taken to sea, even if their extra cost might have restricted them to the larger ships.

NAVIGATIONAL PRACTICE

If the foregoing points firmly to a practical seafaring role for at least some portolan charts and atlases, it throws no light on their precise function. For this it is necessary to take a brief look at contemporary navigational practice. Faced with their modern-looking coastal outlines, it is tempting to take for granted supposed analogies between the use of medieval charts and the function of their modern counterparts. The little that can be learned of early navigation advises caution in this respect.[496] Instruments such as the cross-staff and quadrant were available from at least the early fourteenth century,[497] but more than three hundred years afterward they had still failed to make any discernible impact on Mediterranean sailing. Countering various claims that the Portuguese had merely taken over for their own use techniques of astronomical navigation already being practiced in the Mediterranean, Teixeira da Mota convincingly demonstrated that Mediterranean sailors adopted scientific techniques only in the eighteenth century.[498] Until then, charts with rhumb lines had proved adequate for the traditional methods of dead reckoning (estimates of position in terms of the direction and distance traveled).

There were two main reasons for this. First, the relatively small distances involved in the Mediterranean meant that it was most unusual for a ship to be more than a week out of sight of land;[499] indeed, in the separate Mediterranean basins the coast would be seen every day and errors would never be allowed to accu-

mulate.[500] Second, the early astrolabes—accurate, at best, to the nearest one-sixth of a degree of latitude (or about 18 km)—while acceptable for oceanic sailing, were too inexact for the smaller Mediterranean distances.[501] The authorities cited by Teixeira da Mota were writing in the sixteenth and seventeenth centuries, but there is no reason to doubt that their consistent descriptions of Mediterranean navigational methods are equally applicable to the medieval period. Of particular significance is António de Naiera's comment, made in 1628, that Mediterranean pilots took no note of compass variation, since this discloses an unbroken link with the fourteenth and fifteenth centuries when the users of the earlier charts were apparently unaware of the phenomenon.

The earliest extant account of Mediterranean navigational practice, though written by a poet and thus less reliable than the later descriptions of experienced mariners, confirms this continuity. In his "Documenti d'amore," which was composed at the very beginning of the fourteenth century, close to the time of the earliest charts, Francesco da Barberino referred to just three navigational aids: chart, lodestone (magnetized needle), and *larlogio* (sandglass).[502] Though Barberino provides no corroboration for this, it is conceivable that by this early date sailors were already aware of a mathematical device designed to calculate both the effective distance gained at sea when the ship had been forced off its direct course, for example by headwinds, and the new optimum direction involved.

The method was first described, as far as is known, in "Arbre de Sciencia," a work composed between 1295 and 1296 by Ramón Lull.[503] Termed by subsequent writers the *Raxon de marteloio*, this method was accompanied by mathematical tables (*Toleta*) enabling its

494. Rome, Biblioteca Apostolica Vaticana, Vat. Lat. 9015. Discussed in Almagià, *Vaticana*, 1:43–44 (note 35).

495. The latitude scales, inserted later on a number of early charts, should be viewed in the same light; see above, p. 386.

496. For an account of Mediterranean navigation in the period before the development of the portolan charts, see above, pp. 386–87.

497. George Sarton, *Introduction to the History of Science*, 3 vols. (Baltimore: Williams and Wilkins, 1927–48), vol. 3, *Science and Learning in the Fourteenth Century*, 600–601, 696.

498. Teixeira da Mota, "Art de naviguer," 140 (note 70).

499. Teixeira da Mota, "Art de naviguer," 137 (note 70), quoting Alonso de Santa Cruz, ca. 1555.

500. Teixeira da Mota, "Art de naviguer," 138 (note 70), quoting António de Naiera, 1628.

501. Teixeira da Mota, "Art de naviguer," 130 (note 70).

502. Francesco Egidi, ed., *I Documenti d'amore di Francesco da Barberino secondo i MSS originali*, 4 vols., Società Filologica Romana: Documenti di Storia Letteraria 3 (Rome: Presso la Società, 1905–27), 3:125–26.

503. Cortesão, *History of Portuguese Cartography*, 1:205 (note 3). However, it has been suggested that Italian sailors were using the method in the second half of the thirteenth century, and the *Toleta's*

actual use at sea.[504] The two oldest surviving examples of the *Toleta* are both Venetian and considered to date from the early fifteenth century: one was copied out in the late fifteenth-century Cornaro atlas (fig. 19.20), and another inserted into Bianco's atlas of 1436.[505] A *martilogium* is mentioned, however, in a 1390 Genoese inventory,[506] and in 1382 Cresques Abraham had supplied the king of Aragon with "certain tables"[507]—a reference plausibly interpreted as denoting the *Toleta*, particularly since "navigating tables" and a world map were linked in a single request nine years later.[508] It seems likely, though, that the *Toleta*'s origin should be pushed back a further century because the passage in Lull's "Arbre de Sciencia," though ambiguous, contains the word "instrument," and this is considered to refer to the explanatory tables.[509]

Armed with the *Toleta* alone, the navigator would have been involved in the multiplication calculations referred to by both Lull and Bianco.[510] To avoid this, a "circle and square" diagram was devised. Appropriately, an example of this is included in Bianco's atlas.[511] Once he had absorbed the *Raxon* and equipped himself with

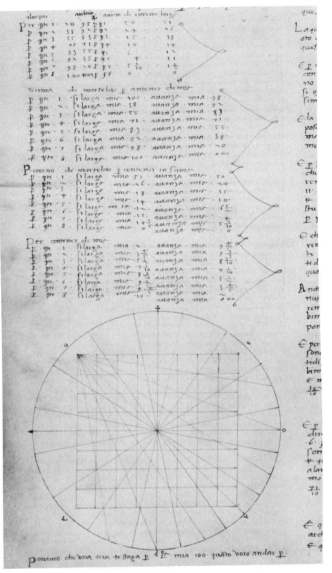

FIG. 19.20. A *TOLETA*. Such mathematical tables were sometimes included in portolan atlases to help the mariner calculate the effective distance gained when his ship had been forced off course and the best new direction to follow. This is one of the oldest extant examples, copied into the late fifteenth-century Cornaro atlas. It seems likely, however, that the *Toleta* was already in existence by the end of the fourteenth century.
Size of the original: 41.8 × 14.5 cm. By permission of the British Library, London (Egerton MS. 73, fol. 47v).

(mathematical tables) creation was ascribed to the mathematical school associated with Leonardo Pisano and his pupil Campano da Novara, see Motzo, "Compasso da navigare," LI (note 103).

504. Opinions differ as to the correct translation of *Raxon de marteloio*. Nordenskiöld, *Periplus*, 53 (note 14), seeking evidence of Catalan inspiration, looked for an explanation in the Spanish words for counting and hammer—a reference, he suggested, to the striking of the ship's bell to mark the watches; others saw it as Italian, the first word being interpreted as rule, the second remaining a mystery. For detailed explanations of how the *Toleta* worked, see Cotter, "Problems of Plane Sailing," 5–11 (note 137); Nordenskiöld, *Periplus*, 53 (note 14); and Taylor, *Haven-Finding Art*, 118–21 (note 7). The navigator, having estimated the distance run and the angle between the actual and desired courses (expressed as a number of quarter-winds), could then read off from the tables the distance the ship was off her course as well as the length of the course made good. Confusingly, a recent book uses the term *marteloio* to describe the underlying rhumb-line network of the standard portolan chart instead of the "circle and square" diagram; compare Mollat du Jourdin and de La Roncière, *Sea Charts* 12 (illustration), 276–77 (glossary) (note 40), and Taylor, *Haven-Finding Art*, pl. VIII (note 7).

505. On the Cornaro atlas (Egerton MS. 73) see British Museum, *Catalogue of the Manuscript Maps*, 1:20 (note 40). The section entitled "La Raxom del Marteloio" is undated, and there is no apparent justification for coupling it with the next passage, containing "Hordeni e chomandamenti" issued by Andrea Mocenigo in 1428, as does Revelli, *Colombo*, 351 (note 22). On Bianco's atlas see Nordenskiöld, *Periplus*, 53 (note 14).

506. Revelli, *Colombo*, 453 (note 22).

507. "Quasdam tabulas in quibus est figura mundi"; see Rubió y Lluch, *Documents*, 2:253 (note 443). Taylor, *Haven-Finding Art*, 117 (note 7) interpreted this as a reference to navigating tables.

508. "Nostre mapa mundi e les taules de navegar"; see Rubió y Lluch, *Documents*, 1:364 (note 443).

509. Cortesão, *History of Portuguese Cartography*, 1:206–7 (note 3); Taylor, *Haven-Finding Art*, 118 (note 7).

510. On Lull see Cortesão, *History of Portuguese Cartography*, 1:206 (note 3). On Bianco see Cotter, "Problems of Plane Sailing," 11 (note 137); Taylor, *Haven-Finding Art*, 116 (note 7). Lane, *Venice*, 169 (note 433) quotes Eva G. R. Taylor that "sailors were the first professional group to use mathematics in their everyday work."

511. Vincenzio Formaleoni, *Saggio sulla nautica antica de' Veneziani con una illustrazione d'alcune carte idrografiche antiche della Biblioteca di San Marco, che dimostrano l'isole Antille prima della scoperta di Cristoforo Colombo* (Venice: Author, 1783), 30.

both the *Toleta* and its graphic solution, any pilot should have been able to "resolve a traverse": in other words, make the necessary adjustments when tacking or if blown off course. But the *Toleta*'s effectiveness depended on a correct appreciation of the initial position and an accurate estimate of the course sailed. Once again, the essential element was the navigator's skill and experience, and for this the *Toleta* was no substitute.

If the *Toleta* was employed in Mediterranean navigation—certainly from the early fifteenth and probably from the late thirteenth century—it would have coexisted with the majority of the charts that survive from the period up to 1500, if not with all of them. But when an attempt is made to confirm that these tables were actually used in conjunction with the portolan charts, there is conflicting evidence. In particular, there is confusion about the use of dividers. Bianco specifically stated that the *Toleta* was used without either ruler or dividers.[512] Nevertheless, the equivalent Lull passage included a phrase that has been interpreted as a reference to dividers;[513] several of the charts itemized in the late fourteenth-century Genoese inventories were accompanied by one or two pairs of dividers;[514] and the leather case enclosing the small 1321 Perrino Vesconte atlas in Zurich has loops through which a pair of dividers would have been slotted.[515] This juxtaposition of chart and dividers can hardly have been accidental. That dividers did indeed play their part in medieval navigation is confirmed by other kinds of documents.

Two fifteenth-century landsmen left a record of their shipboard experiences, and while allowance needs to be made for the men's technical ignorance, both state clearly that the chart was marked in some way during navigational calculations. Pedro Niño's biographer, writing of events in 1404, described how the sailors "opened their charts and began to prick and measure with the Compass [i.e., dividers]."[516] The account by the German cleric Felix Fabri, written in 1483, differs in its details but confirms that the chart would have been annotated. The seamen, he relates, can "see where they are, even when they can see no land, and when the stars themselves are hid by clouds. This they find out on the chart by drawing a curve from one line to another, and from one point to another with wondrous pains."[517] However these two passages might be interpreted, they both characterize the chart as a working document. Similarly, Jean Rotz's navigational treatise, presented to Henry VIII in 1542, prescribed the use of two pairs of dividers to plot a position directly onto the chart.[518]

From these written descriptions it would be natural to expect extant charts to display traces of navigational use, either Fabri's scored circles or the divider holes of Niño and Rotz. Yet no modern scholar has apparently been able to identify marks of this kind on any of the surviving charts produced before the sixteenth century.[519] Magnaghi's assertion, on the basis of sixteenth- and seventeenth-century accounts, that any course plotting would have been done in lead pencil and then erased so as to extend the usefulness of an expensive piece of equipment, is unconvincing for the earlier period.[520] As discussed above, there is no evidence that an erasable pencil formed part of chart draftsmanship before the sixteenth century; hence there is no justification for assuming its presence on board ship. None of the three, Niño, Fabri, or Rotz, mentions a lead pencil, and a pair of dividers presumably substituted for a drawing instrument.

It remains difficult to reconcile statements implying that the portolan charts would have been pierced by divider holes with the absence of any disfigurement of that kind on those charts that survive. One possible interpretation is that careful use of sharp dividers would have avoided making holes large enough to be visible. It is also likely that new vellum would have retained sufficient suppleness to close up in cases where the dividers did pass right through. Alternatively, the holes

512. "Senca mexura e senca sesto"; see Formaleoni, *Saggio*, 30 (note 511).

513. "Carta e compàs"; see Cortesão, *History of Portuguese Cartography*, 1:207 (note 3), although Taylor, *Haven-Finding Art*, 118 (note 7), translated this as "chart [and] compass (*da navigare*)."

514. Revelli, *Mostra Colombiana*, 36–37 (note 315).

515. Leo Cunibert Mohlberg, *Katalog der Handschriften der Zentralbibliothek Zürich*, 2 vols. (Zurich: Zentralbibliothek Zürich, 1951), 1:89.

516. Díaz de Gámez, *Unconquered Knight*, 97 (note 50). The confusion between the magnetic compass and compasses (i.e., a pair of dividers) exists in other languages besides English.

517. Felix Fabri, *The Wanderings of Felix Fabri, circa 1480–1483 A.D.*, trans. Aubrey Stewart, Palestine Pilgrims' Text Society, vols. 7–10 (London: Palestine Exploration Fund, 1897), 1:135. What may be a more reliable account of fifteenth-century navigational practices is that given in the recently discovered but still unpublished Arte del navigare in the Biblioteca Medicea Laurenziana, Florence. This was compiled in 1464 by an unidentified naval officer who had been at sea since 1434; his background and experiences were thus parallel and contemporary with Grazioso Benincasa's. For a summary of the navigational content of the manuscript see Claudio de Polo, "*Arte del navigare*: Manuscrit inédit daté de 1464–1465," *Bulletin du Bibliophile* 4 (1981): 453–61.

518. Wallis, *Jean Rotz*, 81 (note 441). A similar procedure was described by Martin Cortés in 1551; see Waters, *Navigation in England*, 75–76 (note 138), for extracts from the 1561 English edition of his manual.

519. If we discount as irrelevant the divider holes in the scale or through the intersection points, both of which have been cited as evidence of navigational use; see Cortesão and Teixeira da Mota, *Portugaliae monumenta cartographica*, 5:3 (note 29), and the description of the chart now with Nico Israel of Amsterdam (see note 67). Pelham, "Portolan Charts," 26 (note 56), failed to find navigational marks after having specifically searched for evidence of this kind in various European libraries.

520. Magnaghi, "Nautiche, carte," 324a (note 4).

might have disappeared during the subsequent aging of the skin.[521] A microscopic examination of extant charts with this in mind is just one of the many tasks awaiting future researchers. If the findings were positive, it might be possible to reconstruct actual voyages from the pattern of prick marks. Many medieval voyages, of course, were made without losing sight of land. On an inshore voyage, the chart could have supplied valuable information on the sequence of coastal features, the location of offshore islands, the relationship of one Aegean island to another, and so on. But use in this way would leave no trace.

CONNECTION WITH TRADE

Besides mariners, the only groups for whom a portolan atlas or chart could have represented meaningful information would have been those engaged in maritime trade[522] or in the administration of a colonizing power like Venice. Whether the shipping they manipulated was engaged in cargo carrying or war, all ultimately served the same master, commerce. Genoese and Venetian colonies were essentially trading entrepôts, protected by warships in the same way as Venetian convoys to the North Sea. The use of a portolan chart as an aide-mémoire in a mercantile house has yet to be documented, but it seems likely enough.[523] The inevitable links between navigation and trade are illustrated, first by the inclusion of a table of tariffs for the port of Alexandria[524] in the Cornaro atlas's navigational appendix, and second by the guide to weights and measures produced by the chartmaker Arnaldo Domenech.[525] Some writers have even seen the creation of the portolan charts as a response to commercial prompting.[526] Cortesão and Teixeira da Mota were more specific, noting that in area the early portolan charts "corresponded roughly to the regions to which the Genoese and Venetians extended their trade by sea."[527] Those who followed Nordenskiöld in giving the portolan charts a Catalan origin naturally disagreed. What kind of a match, then, can be discerned between the trading patterns of the late Middle Ages and contemporary portolan charts?

To test in detail the hypothesis that the portolan charts were commercial as well as navigational instruments would require a special study of its own. At present we can only offer a few general comments. If the major changes affecting the trading patterns of the Mediterranean and Black Sea littorals were to be briefly summarized, a contrast would be drawn between the inexorable advance of the Ottoman Turk in the East and the partially compensatory Christian victories in the West. Notable among the latter were the Portuguese capture of the Moroccan stronghold of Ceuta in 1415 and the

expulsion of the Moors from Granada in 1492. With Muslims in control of North Africa, Palestine, Asia Minor, the Black Sea, and a large part of the Balkans, it is not surprising that the western Mediterranean and Atlantic took on increasing importance for the Christian trading states.

Kelley suggested that the neck on the portolan charts might have been switched from east to west in response to this commercial change.[528] Most of the eastern necks are certainly found on fourteenth-century charts; but this is better explained in terms of southward orientation.[529] There is a further objection to Kelley's theory. Writing of "the beginning of a shift from the Mediterranean as a focal point of business to the Atlantic seaboard," the historian Denys Hay concluded that "this had certainly not happened by 1500."[530] Yet a listing of restricted area charts—that is, those that, while complete, do not embrace the entire Mediterranean and Black seas—does seem to document for the fifteenth century the swing from east to west that Kelley proposed for the fourteenth and Hay for the sixteenth.

The 1311 Vesconte and (mid-fourteenth century?) Library of Congress charts are both confined to the eastern half of the Mediterranean.[531] Like most Italian charts, they take the eastern side of the Black Sea as their right-hand limit and extend west as far as required—or to the edge of the vellum. To save space, some chartmakers (and Albertin de Virga in 1409 was evidently the first) even detached the Black Sea and moved it sufficiently westward to align it with the Levant coast.[532] To clarify

521. Experiments made at my behest by a modern-day binder, John Llewellin, showed that if the vellum was kept slightly damp, as would presumably have been the case on board ship, no marks could be seen two years later, whereas holes pierced through dry vellum remained clearly visible.
522. Sea charts occur in the inventories of Barcelona merchants in 1457 and 1472; see Carrère, *Barcelone*, 1:201 n. 2 (note 285).
523. Though the 1432 instance of a Genoese firm's sending its Milanese representative a "beautiful sea chart" comes close to it; recounted in Revelli, *Colombo*, 459–60 (note 22).
524. British Museum, *Catalogue of the Manuscript Maps*, 1:20 (note 40).
525. Ristow and Skelton, *Nautical Charts*, 3–4 (note 393).
526. For instance, Kamal, *Eclaircissements*, 186 (note 164).
527. Cortesão and Teixeira da Mota, *Portugaliae monumenta cartographica*, 1:xxvi (note 29).
528. Kelley, "Oldest Portolan Chart," 24 (note 58).
529. See note 62.
530. Denys Hay, *Europe in the Fourteenth and Fifteenth Centuries* (London: Longmans, 1966), 388.
531. Cortesão, *History of Portuguese Cartography*, 1:220 (note 3), believed that Vesconte would have provided a matching sheet to the west, but no examples of multisection Italian charts have survived from this period.
532. Caraci, "Virga," 784 (note 489), was evidently the first to notice this. He cited also a late sixteenth-century English example. Another instance is the 1421 Francesco de Cesanis chart.

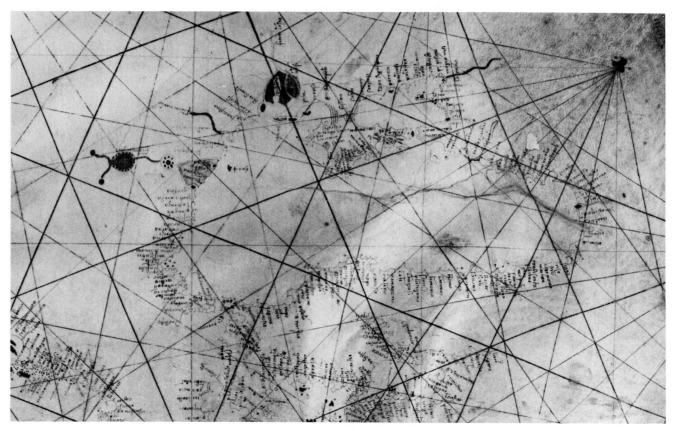

FIG. 19.21. THE BLACK SEA ON THE 1409 CHART OF ALBERTIN DE VIRGA. By detaching the Black Sea from the rest of the outlines (with the Bosphorus repeated), the chartmaker was able to push the Levant coast closer to the right-hand side of the vellum, thereby making space for more details of the Atlantic. It appears that Virga was the first to do this. Size of the original detail: 15 × 24 cm. Photograph from the Bibliothèque Nationale, Paris (Rés. Ge. D 7900).

what they had done, they showed the Bosphorus twice (fig. 19.21).

On the other hand, the Maghreb chart, if correctly ascribed to the early fourteenth century, is the first to concentrate exclusively on the western Mediterranean.[533] It seems probable, however, that this is either a sheet from an atlas or a copy of one. The 1424 chart by Zuane Pizzigano is thus the earliest of the dated survivors to concentrate on the Atlantic coasts. The undated chart in Florence (Biblioteca Nazionale Centrale, Port. 22), which was probably produced at some point after 1380, is perhaps the first complete chart to exclude the eastern Mediterranean and Black Sea. This example would be followed by several other chartmakers during the course of the fifteenth century. Though not of this type, the 1413 Mecia de Viladestes chart has a similar bias, reflecting in its whaling scene the commercial interests of the Atlantic Europeans (fig. 19.22).

CONCLUSIONS

The importance of the early portolan charts can be summed up as follows. First, it can be stated incontrovertibly that they were the most geographically realistic maps of their time. In many ways the early portolan charts are very modern. The Mediterranean's true shape can be recognized immediately on the earliest charts, yet

533. The early fourteenth century was proposed by Juan Vernet-Ginés, "The Maghreb Chart in the Biblioteca Ambrosiana," *Imago Mundi* 16 (1962): 1–16, esp. 4, and corroborated by the place-name analysis (see table 19.3, pp. 416–20, and also above, p. 423). There is little modern support for the view that the Maghreb chart is the oldest survivor of all, as suggested by Ernest Théodore Hamy, *Etudes historiques et géographiques* (Paris: Leroux, 1896), 31. Since the rhumb line network is incomplete on two sides, the chart could not have been constructed in the normal way on the basis of a hidden circle but must have been directly reproduced, rather than freely copied, from a slightly larger work—see the illustration to Vernet-Ginés's article.

the same could hardly be said of their near contemporary, the Hereford *mappamundi*. Such a close approximation to the true outlines of the Mediterranean and Black seas on the oldest known charts cannot be accidental.

Second, an almost exclusive interest in the real world sets the portolan charts aside from other mapmaking activities of the Middle Ages. Usually made for (and sometimes by) sailors, most of the charts must have been drawn to satisfy a commercial demand. To mariners especially, distance and direction were vital. Since these components were less in demand by scholars, rulers, or administrators, little use was made of compass or scale on maps uninfluenced by the portolan charts. Already impressively accurate at the time of the oldest survivor, these charts reveal considerable hydrographic improvement in the early stages and continuing adjustments to the coastal place-names thereafter. The deterioration of both outlines and toponymy after the mid-sixteenth century confirms the extent to which their earlier vitality had depended on their practical function. By the later period they had ceased to be indispensable. While the knowledge of existing coastlines was being refined, the charts were steadily extending their range to accommodate fresh information. That the charts and their sixteenth-century successors were almost alone in providing a cartographic record of the pre-Renaissance and Renaissance discoveries constitutes their third main claim to historical importance.

Accurate, responsive to change, and essential for navigation, the charts were a necessary if specialized element of medieval life. Like the pen, the breastplate, or the stirrup, they were unremarkable and hence usually passed unremarked. In a sense, this essay is dedicated to the thousands of ordinary charts that served their purpose and then perished. The obvious beauty of a richly illuminated chart is less noteworthy than the perpetuation of the initial coastal accuracy on unadorned charts for more than two centuries.

Appendix 19.1
Calendars as a Guide to Dating: The Case of the Medici and Pinelli-Walckenaer Atlases

A number of atlases are preceded by astronomical tables. Whether designed to show the precise time of each new moon or to facilitate calculation of the date of Easter and its allied festivals, these are constructed around the nineteen-year metonic cycle. This related the lunar month to the solar year. Successive years were numbered from 1 to 19 in an endlessly repeated series of golden numbers.[1] One cycle, for instance, began in 1368; hence 1375 was year 8, as the author of the Catalan atlas explains (plate 32).[2] Since some calendars stipulated the golden numbers for a period of a century or more, chartmakers could make a literal copy of an existing calendar

FIG. 19.22. A WHALING SCENE. From the fifteenth century onward, occasional charts emphasized the Atlantic at the expense of the eastern Mediterranean. The shift in commercial interests from the Mediterranean to the Atlantic is illustrated in this vignette from the 1413 chart by Mecia de Viladestes depicting whalers off the coast of Iceland.
Size of the original detail: ca. 12 × 13 cm. Photograph from the Bibliothèque Nationale, Paris (Rés. Ge. AA 566).

instead of bringing the golden numbers up to date, without the document's becoming obsolete. Thus, when a calendar's commencement year is indicated, this may differ significantly from the date given for the construction of the atlas to which it belongs.

Grazioso Benincasa, for example, included a lunar calendar and an Easter table in four of his atlases (fig. 19.23).[3] Yet the twin tables sometimes have different starting points, and none of the eight would have supplied the correct date for the atlas had the author's legend been missing. With its signature gone, the 1468 atlas would presumably have been assigned to 1451 on the strength of its lunar calendar. Though not a portolan chart, the Albertin de Virga world map of 141– provides a more startling example, with an Easter table starting more than a hundred years before its stated date of construction.[4] In some cases, however, the dates derived from calendars do

1. On calendars see Christopher R. Cheney, *Handbook of Dates for Students of English History* (London: Royal Historical Society, 1945), and also extensive handwritten notes in the M. C. Andrews Collection at the Royal Geographical Society, London.
2. Georges Grosjean, ed., *The Catalan Atlas of the Year 1375* (Dietikon-Zurich: Urs Graf, 1978), 38.
3. Those of 1468, 1469, 1473, and 1474—respectively London, British Library (Add. MS. 6390); Milan, Ambrosiana; London, British Library (Egerton MS. 2855); and Budapest.
4. Marcel Destombes, ed., *Mappemondes A.D. 1200–1500: Catalogue préparé par la Commission des Cartes Anciennes de l'Union Géographique Internationale* (Amsterdam: N. Israel, 1964), 205.

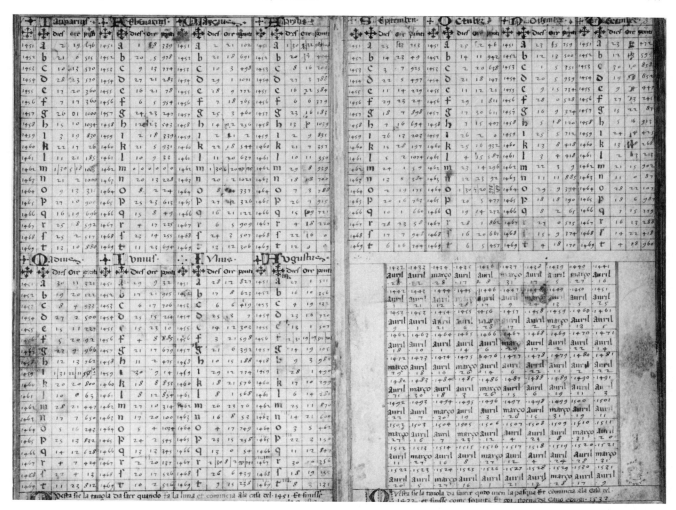

FIG. 19.23. A COMBINED LUNAR CALENDAR AND EASTER TABLE FROM A PORTOLAN ATLAS. Grazioso Benincasa included tables of this kind in four of his atlases. Three-quarters of this sheet is occupied by a calendar from which the precise time of any new moon could be calculated, and from that the time of high tide. Each list comprises the nineteen golden numbers (here lettered a–t) followed by the day, hour, and number of points (each worth 3.3 seconds) for the years

between 1451 and 1469. The table for calculating the date of Easter for the entire century after 1432 fills the lower right quarter of the sheet. Such calendars and tables must be carefully used when found in undated atlases. This example might have suggested a date of 1451 for its atlas had Benincasa not clearly dated the work 1468.

Size of the original: 27.4 × 33.9 cm. By permission of the British Library, London (Add. MS. 6390, fol. 3r).

confirm those on the main document, and occasionally the calendar date can be confidently extended to the whole work. The explanatory note in the Catalan atlas, for instance, refers to the year 1375 in the present tense.[5] Likewise, an unsigned atlas in the Vatican carries the reference in its lunar table to "el presente ano" 1452.[6]

might have been completed some years after 1377; see his "Viaggi fra Venezia e il levante fino al XIV secolo e relativa produzione cartografica," in *Venezia e il levante fino al secolo XV*, 2 vols., ed. Agostino Pertusi (Florence: L. S. Olschki, 1973–74), 1:147–84, esp. 178 (footnote). These find no favor with Grosjean or with the various authors of a recent collection of essays: *L'atlas Catalá de Cresques Abraham: Primera edició completa en el sis-cents aniversari de la seva realització* (Barcelona: Diáfora, 1975), also published in Spanish. The 1376 in the center of the wheel diagram was interpreted by Grosjean, *Catalan Atlas* 9 (note 2), as either the year of the atlas's completion or that of the next leap year.

5. Grosjean, *Catalan Atlas*, 38 (note 2). However, Gonçal (Gonzalo) de Reparaz, *Catalunya a les mars: Navegants, mercaders i cartògrafs catalans de l'Edat Mitjana i del Renaixement (Contribucío a l'estudi de la història del comerç i de la navegació de la Mediterrània)* (Barcelona: Mentova, 1930), 83, and Giuseppe Caraci, *Italiani e Catalani nella primitiva cartografia nautica medievale* (Rome: Istituto di Scienze Geografiche e Cartografiche, 1959), 315, 333, cited arguments in support of a 1377 date. More recently Caraci speculated that the atlas

6. Roberto Almagià, *Monumenta cartographica Vaticana*, 4 vols. (Rome: Biblioteca Apostolica Vaticana, 1944–55), vol. 1, *Planisferi, carte nautiche e affini dal secolo XIV al XVII esistenti nella Biblioteca Apostolica Vaticana*, 43a.

Faced with tables and calendars offering, in almost equal measure, dates that either confirm or contradict those of the works they accompany, we should obviously approach with caution those atlases whose dating has depended entirely on their calendars. The Medici and Pinelli-Walckenaer atlases are the most important of these. Their calendars start in 1351 and 1384 respectively, and many past commentators have, with insufficient justification, assumed those to be the dates of construction.

The world map at the beginning of the Medici atlas has attracted considerable interest because of its possible hint of South Africa. Many writers believed this map was reworked later, but opinion as to the date of its initial construction was strongly divided, with some supporting the calendar's date of 1351. However, as has been remarked, calendars often started in the first year of the century or half-century.[7] Anyway, in Wieser's opinion, 1351 was referred to as a year that had already passed.[8] Other scholars followed Wieser in proposing an early fifteenth-century date for the atlas. Since no one has suggested that the work is in more than one hand (except for the possible redrawing of southern Africa) it follows that these dates for the world map would have to be applied also to the portolan atlas that accompanies it. One way to test the validity of the 1351 date derived from the calendar would be to look for dating indications in the regional charts.

Cortesão—one of the few to have considered the charts in the Medici atlas—proposed a date about 1370, purely because of the relative sophistication of the Canary and Madeira archipelagoes.[9] Using the place-name analysis, it can now be shown that Santarém was nearest the mark when he asserted in 1852 that the Medici atlas comprised a "collection of marine charts of different periods."[10]

Besides the calendar and world map, the atlas contains three small-scale overlapping charts covering the Mediterranean and larger-scale sheets for the Adriatic, Aegean, and Black seas. There is also an important half-sheet devoted to the Caspian. Given that there are no surviving dated charts between the 1339 Dulcert and the 1367 Pizigani, it cannot be predicted what place-names a chart of 1351 might be expected to contain. Nevertheless, the three general sheets of the Mediterranean include none of the "significant" additional names first encountered in 1367 or later.[11] A date about 1351 is therefore quite plausible for these on strictly toponymic grounds. It is probably applicable as well to the enlarged drafts of the Aegean and Black seas. Despite having a handful of extra names, these include between them only three significant additions from 1367 onward—not an indication of any notable input of new names. But when the detailed Adriatic sheet is compared with the two similar, but smaller-scale versions, it is clear that quite different toponymic generations are involved.

Separate entries for the two elements of the Medici atlas in table 19.3, pp. 416–20, show how the toponymic profile of the larger-scale Adriatic sheet suggests a date in the first half of the fifteenth century. The strictly numerical relationship of the three Medici atlas treatments of the Adriatic is considered in table 19.4. The large-scale sheet has some 50 percent more names than the other two. Lest this be interpreted as merely a result of its increased size, it should be noted that 20 percent of the names on one of the smaller versions were ignored by

the larger. When these sheets are considered as part of the broader analysis of Adriatic names, they are found to reflect the widespread differences between the patterns current in the mid-fourteenth and the early fifteenth centuries.

If the sheets of the Medici atlas were all drawn at the same time (though some would be copies of earlier work), it must follow from the toponymic evidence that the often accepted date of 1351 has to be abandoned (for its charts and world map alike) in favor of one from the first half of the fifteenth century. Similarly, the analysis of the Pinelli-Walckenaer atlas's names in table 19.3 indicates that its calendar-inspired date of 1384 should be replaced by one from that same general period.

TABLE 19.4 Adriatic Names between Otranto and Vlorë (Valona) on the Three Relevant Sheets of the Medici Atlas (see fig. 19.19)

	Larger-Scale Adriatic (1)	Smaller-Scale Northwestern Europe (2)	Smaller-Scale Eastern Mediterranean (3)
Totals	151	106	91[a]
Not on (1)		22	15
Not on (2)	66		4
Not on (3)	74[a]	14	
Not on the other two sheets	63	6	0

[a]A small section around Venice is excluded, with the omission of perhaps six names.

7. Biblioteca Nazionale Marciana and Archivio di Stato, Venice, *Mostra dei navigatori veneti del quattrocento e del cinquecento,* exhibition catalog (Venice, 1957), 91–92.

8. Franz R. von Wieser, *Die Weltkarte des Albertin de Virga aus dem Anfange des XV. Jahrhunderts in der Sammlung Figdor in Wien* (Innsbruck: Schurich, 1912), 12; reprinted in *Acta Cartographica* 24 (1976): 427–40.

9. Armando Cortesão, *History of Portuguese Cartography,* 2 vols. (Coimbra: Junta de Investigações do Ultramar-Lisboa, 1969–71), 2:58.

10. Manuel Francisco de Barros e Sousa, Viscount of Santarém, *Essai sur l'histoire de la cosmographie et de la cartographie pendant le Moyen-Age et sur les progrès de la géographie après les grandes découvertes du XV⁰ siècle,* 3 vols. (Paris: Maulde et Renou, 1849–52), 3:LXIX (author's translation).

11. For explanation of the toponymic analysis, see appendix 19.5.

APPENDIX 19.2 BIOGRAPHICAL INDEX TO THE ATLASES AND CHARTS PRODUCED UP TO 1500

	Date	Atlas (A) or Chart (C)	Where Made	Where Preserved	Kamal[a]	Selected Accessible Works
ABENZARA. See under Jehuda ben Zara						
AGUIAR, Jorge de (Portuguese)	1492	C	Lisbon	New Haven, Beinecke Library, Yale University		Cortesão, *History of Portuguese Cartography,* vol. 1, frontispiece; vol. 2, fig. 90[b]
BECCARI, Batista[c] (misread as Beclario, Bedrazius, Bescario) (of Genoa)	1426	C		Munich, Bayerische Staatsbibliothek, Codex Icon. 130	4.4:1453	De La Roncière, *Afrique,* vol. 6, pl. XXII[d]
	1435	C[e]		Parma, Biblioteca Palatina, II, 21, 1613		Frabetti, *Carte nautiche italiane,* pl. 2 (note e)
Attributed[f]		C		[London, Christie's, 2 December 1964, lot 77, ex Rex Beaumont]		
		C		[Genoa, Amedeo Dallai—in 1951]		Revelli, "Beccari," 162–63[g]
BECCARI, Francesco (Bechaa, Ircharius) (of Genoa) working in Barcelona, 1399–1400	1403	C	Savona	New Haven, Beinecke Library, Yale University		
Acknowledged copy	[1489]	C		London, British Library, Egerton MS. 73		
BENINCASA, Andrea (of Ancona), son of Grazioso	1476	A5[h]		Geneva, Bibliothèque Publique et Universitaire, MS. Lat. 81		Atlantic: Cortesão, *Nautical Chart of 1424,* pl. XV; Lelewel, *Géographie,* pls. 34–35[i]
	1490	C		Ancona, Museo Nazionale, 253		Revelli, *Partecipazione italiana,* no. 35[j]
	1508	C	Ancona	Rome, Biblioteca Apostolica Vaticana, Borgiano VIII		Almagià, *Vaticana,* vol. 1, pl. XX[k]
BENINCASA, Grazioso (of Ancona)	[1461?]	C	Genoa	Florence, Archivio di Stato, CN6		
	1461	C	Genoa	Florence, Archivio di Stato, CN5		
	1463	A5	Venice	London, British Library, Add. MS. 18454		Atlantic: Cortesão, *Nautical Chart of 1424,* pl. XII (note i)
	1463	A4	Venice	London (Royal Army Medical Corps—stolen 1930)		
	1465	A5	Venice	Vicenza, Biblioteca Civica Bertoliana, 598b		
	1466	A5	Venice	Paris, Bibliothèque Nationale, Rés. Ge. DD 2779		
	1467	A5	Rome	London, British Library, Add. MS. 11547		Eastern Mediterranean: Stylianou and Stylianou, *Cyprus,* 177–78[l]

APPENDIX 19.2—*continued*

Date	Atlas (A) or Chart (C)	Where Made	Where Preserved	Kamal[a]	Selected Accessible Works (Reproductions)
1467	A5	Rome	Nogent-sur-Marne, Bibliothèque Nationale Annexe		Nordenskiöld, *Periplus*, pls. XXXIII, XL[m]
1467	A5	Rome	Paris, Bibliothèque Nationale, Rés. Ge. DD 1988	5.1:1497 [Atlantic]	
1468	A7	Venice	Great Britain, private collection—ex Lanza di Trabia and Kraus		
1468	A6	[Genoa]	London, British Library, Add. MS. 6390		Atlantic: Cortesão, *History of Portuguese Cartography*, vol. 2, fig. 83 (note b)
1468	C	Venice	Palma de Mallorca, Fundación Bartolomé March Servera		Caraci, "Grazioso Benincasa," 18[n]
1469	A6	Venice	London, British Library, Add. MS. 31315		
1469	A6	Venice	Milan, Biblioteca Ambrosiana, S.P., 2, 35		Atlantic: Cortesão, *History of Portuguese Cartography*, vol. 1, fig. 84 (note b)
1470	C	Ancona	London, British Library, Add. MS. 31318A		Atlantic: Cortesão, *Nautical Chart of 1424*, pl. XIII (note i)
1471	A6	Venice	Rome, Biblioteca Apostolica Vaticana, Vat. Lat. 9016		Almagià, *Vaticana*, vol. 1, pl. XVII–XVIII (note k); Santarém, *Atlas*, pls. VIII–IX[o]
1472	A8		[Milan, Luigi Bossi—before 1882]		
1472	C	Venice	Venice, Museo Civico Collezione Correr, Port. 5		Salinari, "Notizie," 40[p]
1473	A6	Venice	Bologna, Biblioteca Universitaria, MS. 280		North Atlantic: Frabetti, *Carte nautiche italiane*, pl. III (note e)
1473	A6	Venice	London, British Library, Egerton MS. 2855		
1474	A6	Venice	Budapest, Magyar Nemzeti Muzeum, Codex Lat. M.A. 353		
1480	A6	Ancona	Vienna, Österreichische Nationalbibliothek, MS. 355		
1482	C	Ancona	Bologna, Biblioteca Universitaria, Rot. 3		Frabetti, *Carte nautiche italiane*, pl. IV (note e)
Acknowledged copies [1489]	C2		London, British Library, Egerton MS. 73		
Attributed	C	[Venice?]	Florence, Archivio di Stato, CN9		Western section: Cortesão, *Nautical Chart of 1424*, pl. XIV (note i)

	Date	Atlas (A) or Chart (C)	Where	Where Preserved	Kamal[a]	Selected Accessible Works
		A5		Milan, Biblioteca Trivulziana, Codex 2295		
BERTRAN, Jaime (of Barcelona), Jewish	1456	C	Barcelona	Greenwich, National Maritime Museum, G230:1/7 MS—co-signed with Ripoll		Howse and Sanderson, *Sea Chart,* 18[q]
	1482	C	Majorca	Florence, Archivio di Stato, CN7	5.1:1503	Western section: Cortesão, *Nautical Chart of 1424,* pl. XVII (note i); Winter, "Catalan Portolan Maps," 4[r]
	1489	C	Majorca	Florence, Biblioteca Marucelliana		
BIANCO, Andrea (of Venice), galley officer	1436	A7		Venice, Biblioteca Nazionale Marciana, It. Z.76		Nordenskiöld, *Periplus,* 19, pls. XX–XXI (note m)
	1448	C	London	Milan, Biblioteca Ambrosiana, F.260, inf. (1)	5.1:1492	Cortesão, *History of Portuguese Cartography,* 2:143 (note b)
BOSCAINO, Ponent[s]						
BRIATICHO, Cola de	1430	A3		Siena, Biblioteca Comunale, SV2	4.4:1460–61	Atlantic: Cortesão, *Nautical Chart of 1424,* pl. IV (note i)
BUONDELMONTI (see pp. 379–80)						
CANEPA, Albino da (of Genoa)	1480	C	Genoa	Rome, Società Geografica Italiana		Revelli, *Colombo,* pl. 80[t]
	1489	C	Genoa	Minneapolis, University of Minnesota, James Ford Bell Collection, B1489mCa		Cortesão, *Nautical Chart of 1424,* pl. XVI—wrongly captioned as 1480 (note i)
CARIGNANO, Giovanni da (of Genoa), priest	Early 14th century	Map	Genoa	Florence, Archivio di Stato, CN2—destroyed 1943	4.1:1138–39	Nordenskiöld, *Periplus,* pl. V. (note m)
CESANIS, Alvixe (Aloyse, Luigi) (Cexano)						
Acknowledged copy	[1489]	C		London, British Library, Egerton MS. 73		
CESANIS, Francesco de (Cexano) (of Venice)	1421	C		Venice, Museo Civico, Collezione Correr, Port. 13	4.4:1417	
Acknowledged copy	[1489]	C		London, British Library, Egerton, MS. 73		
Attributed		A8		Genoa, Biblioteca Civica Berio—the Luxoro atlas	4.2:1245	(See p. 420 and notes 273, 363, and 364) and under "Luxoro atlas" in appendix 19.3

APPENDIX 19.2—*continued*

	Date	Atlas (A) or Chart (C)	Where Made	Where Preserved	Kamal[a]	Selected Accessible Works
					Reproductions	
COLUMBUS, Christopher						
Attributed (but not generally accepted)	Late 15th or early 16th century	C		Paris, Bibliothèque Nationale, Rés. Ge. AA 562		Bagrow, *History of Cartography*, pl. LIV;[u] de La Roncière, *Afrique*, vol. 13, pl. XXXVIII (note d)
CONPIMENTO DELCEXANO[v]						
CRESQUES ABRAHAM (of Majorca), Jewish						
Attributed[w]	[1375]	6 panels		Paris, Bibliothèque Nationale, MS. Esp. 30–the Catalan atlas	4.3:1301–3	(See under "Catalan atlas" in appendix 19.3)
DALORTO, Angelino de—possibly the same as Dulcert	1325/30	C		Florence, Prince Filippo Corsini	4.2:1197–98	Hinks, *Dalorto*[x]
DOMENECH, Arnaldo	[148–]	C	Naples	Greenwich, National Maritime Museum, G230:1/9 MS		
(Guide to weights and measures, not a chart)	1484			Washington, D.C., Library of Congress, vellum chart 4		Ristow and Skelton, *Nautical Charts on Vellum in the Library of Congress*, 3[y]
DULCERT, Angelino (misread as Dolcedo)—possibly the same as Dalorto	1339	C	Palma	Paris, Bibliothèque Nationale, Rés. Ge. B 696	4.2:1222	Nordenskiöld, *Periplus*, pls. VIII–IX (note m); Putman, *Early Sea Charts*, pl. 1[z]
Attributed		C		London, British Library, Add. MS. 25691	4.3:1334	Winter, "Das katalanische Problem"[aa]
FLORINO, Nicolo (of Venice)	1462	A2		Vienna, Österreichische Nationalbibliothek, K.II. 100.725		
Acknowledged copy	[1489]	C		London, British Library, Egerton MS. 73		
FREDUCCI, Conte Hectomano[bb] (of Ancona)	1497	C		Wolfenbüttel, Herzog August Bibliothek, Codex Guelf 99		Nordenskiöld, *Periplus*, pl. XXII (note m)
Attributed		C		Weimar, Zentralbibliothek der Deutschen Klassik		Bagrow, *History of Cartography*, pl. XLV (note u); Cortesão, *Nautical Chart of 1424*, pl. VIII (note i)
GIOVANNI, Giorgio (of Venice)	1494	C	Venice	Parma, Biblioteca Palatina, II, 30, 1622		Frabetti, *Carte nautiche italiane*, pl. 5 (note e)
GIROLDI, Giacomo (Zeroldi) (of Venice)	1422	C		Paris, Bibliothèque Nationale, Rés. Ge. C 5088	4.4:1420	

	Date	Atlas (A) or Chart (C)	Where Made	Where Preserved	Kamal[a]	Reproductions
						Selected Accessible Works
	1426[cc]	A6		Venice, Biblioteca Nazionale Marciana, It. VI, 212	4.4:1452	Nordenskiöld, *Periplus*, pl. IV (note m)
	1443	A6		Milan, Biblioteca Ambrosiana, S.P., 2, 38		
	1446	A6		Florence, Accademia Toscana di Scienze e Lettere "La Colombaria," 229		
Attributed		A6		Chicago, Newberry Library, Ayer Collection, MS. Map 2		
		A5		London, British Library, Add. MS. 18665		
		A6		Milan, Biblioteca Ambrosiana, S.P., 2, 39		
		C		New York, Hispanic Society of America, K4		Stevenson, *Facsimiles*, pl. 1[dd]
		A5		Rome, Biblioteca Apostolica Vaticana, Rossiano 676		Almagià, *Vaticana*, vol. 1, pl. XVI (note k)
	1452	A3		Rome, Biblioteca Apostolica Vaticana, Vat. Lat. 9015		
JEHUDA BEN ZARA (Abenzara, Ichidabruzara)	1497	C	Alexandria	Rome, Biblioteca Apostolica Vaticana, Borgiano VII		Almagià, *Vaticana*, vol. 1, pl. XIX (note k)
	1500	C	Alexandria	Cincinnati, Klau Library, Hebrew Union College		Dürst, *Iehuda ben Zara*, 11; Kraus, *Remarkable Manuscripts*, 83; Roth, "Judah Abenzara"[ee]
	1505	C	Safed, Galilee	New Haven, Beinecke Library, Yale University		
KĀTIBĪ, Tunuslu Ibrāhīm[ff]	[1413]	C		Istanbul, Topkapi Sarayi		
MARCH/MARE, Nicolo[gg]						
MARTELLUS GERMANUS, Henricus See pp. 379–80						
MIRARO[hh]						
AL-MURSĪ, Ibrāhīm (of Tripoli), physician	[1461]	C	Tripoli (Libya?)	Istanbul, Deniz Muzesi		Rossi, "Carta nautica araba"[ii]
NICOLO, Nicolo de	1470	C		New York, Hispanic Society of America, K6		
NOLI, Agostino (of Genoa), see pp. 430 and 434						
PARETO, Bartolomeo de (of Genoa), priest	1455	C	Genoa	Rome, Biblioteca Nazionale, CN1		Revelli, *Colombo*, pl. 79 (note t)

APPENDIX 19.2—*continued*

	Date	Atlas (A) or Chart (C)	Where Made	Where Preserved	Kamal[a]	Selected Accessible Works
					Reproductions	
PASQUALINI, C.[ji] Father of Nicolo de Pasqualini						
PASQUALINI, Nicolo de (Pasqualin, G.—mistake for Nicolo Pasqualini)	1408	A6		Vienna, Österreichische Nationalbibliothek, Codex 410*	4.3:1349	
Acknowledged copy	[1489]	C		London, British Library, Egerton MS. 73		
PELECHAN, Antonio (*Armiralo* of Rethymnon, Crete)	1459	C	Rethymnon	Venice, Archivio di Stato, LXXXV no. 1		
PESINA, Benedetto	1489	C	Venice	London, British Library, Egerton MS. 73		
PIZIGANO, Domenico or Marco	1367	C	Venice	Parma, Biblioteca Palatina, Parm. 1612—signed with Francesco	4.2:1285–86; 4.4:1483	
PIZIGANO, Francesco (Piçegany) (of Venice)	1367	C	Venice	Parma, Biblioteca Palatina, Parm. 1612—signed with Domenico or Marco	4.4:1483	
	1373	A5	Venice	Milan, Biblioteca Ambrosiana, S.P., 10, 29	4.2:1289	Nordenskiöld, *Periplus*, 29, 31, 51, 55 (note m)
[PIZZIGANO, Zuane]	1424	C		Minneapolis, University of Minnesota, James Ford Bell Collection, B1424mPi		Cortesão, *Nautical Chart of 1424* (note i); Cortesão, *History of Portuguese Cartography*, 2:126–27 (note b)
PONGETO, Sentuzo	1404	C		[Munich, Weiss und Co., in 1926—see p. 396 and note 215]		(Partial) in Weiss und Co., *Codices*[kk]
REINEL, Pedro (Portuguese)	[1483?]	C		Bordeaux, Archives Départementales de Gironde, 2 Z 1582 bis.		Cortesão, *History of Portuguese Cartography*, vol. 2, frontispiece (note b); Cortesão and Teixeira de Mota, *Portugaliae monumenta cartographica*, 5:521[ll]
RIPOLL, Berenguer	1456	C	Barcelona	Greenwich, National Maritime Museum, G230:1/7 MS—co-signed with Bertran		Howse and Sanderson, *Sea Chart*, 18 (note q)
ROSELLI, Petrus (Pere Rossell)	1447	C	Majorca	North America, private collection—ex Nebenzahl, Chicago		Nebenzahl, *Rare Americana*[mm]
	1447	C	Palma	Volterra, Museo e Biblioteca Guarnacciana, MS. C.N. 1BG		Revelli, *Colombo*, pl. 35 (note t)
	1449	C	Palma	Karlsruhe, Badische Landesbibliothek, S6		
	1456	C	Palma	Chicago, Newberry Library, Ayer Collection, MS. Map 3		
	1462	C	Palma	Paris, Bibliothèque Nationale, Rés. Ge. C 5090		

	Date	Atlas (A) or Chart (C)	Where Made	Where Preserved	Kamal[a]	Reproductions — Selected Accessible Works
	1464	C	Palma	Nuremberg, Germanisches Nationalmuseum, Codex La. 4017		Cortesão, *Nautical Chart of 1424*, pl. X (note i); Winter, "Roselli," 1[nn]
	1465	C	Palma	London, British Library, Egerton MS. 2712		
	1466	C	Palma	Minneapolis, University of Minnesota, James Ford Bell Collection		Cortesão, *Nautical Chart of 1424*, pl. XI (note i)
	1468	C	Palma	New York, Hispanic Society of America, K35		Stevenson, *Facsimiles*, pl. 2 (note dd)
	1469	C	Majorca	[Washington, D.C., Otto H. F. Vollbehr—in 1935]		
Acknowledged copy	[1489]	C		London, British Library, Egerton MS. 73		
Attributed		C		Modena, Biblioteca Estense e Universitaria, C.G.A. 5b		
		C		[Munich, Ludwig Rosenthal, catalog 167]		
		C		Paris, Bibliothèque Nationale, Rés. Ge. C 5096		
RUSSO, Pietro[oo]						
SOLER, Guilermo	1385	C		Florence, Archivio di Stato, CN3	4.3:1320	Nordenskiöld, *Periplus*, pl. XVIII (note m)
		C		Paris, Bibliothèque Nationale, Rés. Ge. B 1131	4.3:1322	
SOLIGO, Cristoforo						
Acknowledged copy	[1489]	C		London, British Library, Egerton MS. 73	5.1:1510	
SOLIGO, Zuane						
Acknowledged copy	[1489]	C		London, British Library, Egerton MS. 73		
VALSECA, Gabriel de (of Majorca), converted Jew	1439	C	Majorca	Barcelona, Museo Marítimo, inv. no. 3236	5.1:1491	De La Roncière, *Afrique*, vol. 5, pl. XII (note d)
	1447	C	Majorca	Paris, Bibliothèque Nationale, Rés. Ge. C 4607		Kish, *Carte*, pl. 43[pp]; Putman, *Early Sea Charts*, pl. 2 (note z)
	1449	C	Majorca	Florence, Archivio di Stato, CN22		
VESCONTE, Perrino (of Genoa), possibly the same as Pietro	1321	A4	Venice	Zurich, Zentralbibliothek, R.P.4		
	1327	C	Venice	Florence, Biblioteca Medicea Laurenziana, Med. Pal. 248	4.2:1205	Nordenskiöld, *Periplus*, pl. VIIA (note m)
Attributed		C		Amsterdam, Nico Israel— but see note 274		Sotheby's *Highly Important Maps and Atlases*; N. Israel, *Interesting Books*[qq]

APPENDIX 19.2—*continued*

	Date	Atlas (A) or Chart (C)	Where Made	Where Preserved	Reproductions Kamal[a]	Reproductions Selected Accessible Works
VESCONTE, Pietro (of Genoa), possibly the same as Perrino	1311	C		Florence, Archivio di Stato, CN1	4.1:1140	Nordenskiöld, *Periplus*, pl. V (note m)
	1313	A6		Paris, Bibliothèque Nationale, Rés. Ge. DD 687	4.1:1147–49	
	1318	A6	Venice	Venice, Museo Civico, Collezione Correr, Port. 28	4.1:1154	
	1318	A10		Vienna, Österreichische Nationalbibliothek MS. 594		Nordenskiöld, *Periplus*, pl. VI (note m); Pagani, *Vesconte*[rr]
	[1320?]	A5		Rome, Biblioteca Apostolica Vaticana, Pal. Lat. 1362A	4.1:1160–61	Almagià, *Vaticana*, vol. 1, pls. IV–VIII (note k)
	ca. 1322	A7	Venice	Lyons, Bibliothèque de la Ville, MS. 175		De La Roncière, *Lyon*, pls. I–IX[ss]
Attributed (possibly by Perrino)	[1321]	A5		Rome, Biblioteca Apostolica Vaticana, Vat. Lat. 2972	4.1:1170	Almagià, *Vaticana*, vol. 1, pls. VIII–IX (note k)
	ca. 1325	A5		London, British Library, Add. MS. 27376*		
VILADESTES, Johanes de	1428	C	Palma	Istanbul, Topkapi Sarayi, 1826	4.4:1457	Winter, "Catalan Portolan Maps," 1 (note r)
VILADESTES, Mecia de (Matias) (of Majorca), converted Jew	1413	C		Paris, Bibliothèque Nationale, Rés. Ge. AA 566	4.3:1368	Western section: Cortesão, *Nautical Chart of 1424*, pl. III (note i)
	1423	C	Palma	Florence, Biblioteca Medicea Laurenziana, Ashb. 1802		
VIRGA, Albertin de (of Genoa), author of world map dated 141–. See p. 357 and note 34	1409	C	Venice	Paris, Bibliothèque Nationale, Rés. Ge. D 7900	4.3:1350	
ZUANE, Domenico de (Zane, misread as Dezane)						
Acknowledged copy	[1489]	C		London, British Library, Egerton MS. 73		
ZUANE DI NAPOLI						
Acknowledged copy	[1489]	C		London, British Library, Egerton MS. 73		

[a]Youssouf Kamal, *Monumenta cartographica Africae et Aegypti*, 5 vols. in 16 pts. (Cairo, 1926–51). For a list of contents by volume, see p. 40.

[b]Armando Cortesão, *History of Portuguese Cartography*, 2 vols. (Coimbra: Junta de Investigações do Ultramar-Lisboa, 1969–71).

[c]There is still no consensus on how this and other names should be spelled. The forms used here follow the best recent authority in each case.

[d]Charles de La Roncière, *La découverte de l'Afrique au Moyen Age: Cartographes et explorateurs*, Mémoires de la Société Royale de Géographie d'Egypte, vols. 5, 6, 13 (Cairo: Institut Français d'Archéologie Orientale, 1924–27).

[e]In this instance the chartmaker's first name has disappeared, but its final *a* can be read, and the chart is invariably assigned to Batista; see Pietro Frabetti, *Carte nautiche italiane dal XIV al XVII secolo conservate in Emilia-Romagna* (Florence: Leo S. Olschki, 1978), 8.

[f]There have been far more attributions of authorship than are mentioned in this list. Only plausible claims are included here.

[g]Paolo Revelli, "Una nuova carta di Batista Beccari ('Batista Becharius')?" *Bollettino della Società Geografica Italiana* 88 (1951): 156–66.

[h]Throughout appendixes 19.2, 19.3, and 19.4 the figure denotes the number of sheets containing charts.

[i]Armando Cortesão, *The Nautical Chart of 1424 and the Early Discovery and Cartographical Representation of America: A Study on the History of Early Navigation and Cartography* (Coimbra: University of Coimbra, 1954); Joachim Lelewel, *Géographie du Moyen Age*, 4 vols. and epilogue (Brussels: J. Pilliet, 1852–57; reprinted Amsterdam: Meridian, 1966).

[j]Paolo Revelli, *La partecipazione italiana alla Mostra Oceanografica Internazionale di Siviglia (1929)* (Genoa: Stabilimenti Italiani Arti Grafiche, 1937).

[k]Roberto Almagià, *Monumenta cartographica Vaticana*, 4 vols. (Rome: Biblioteca Apostolica Vaticana, 1944–55), vol. 1, *Planisferi, carte nautiche e affini dal secolo XIV al XVII esistenti nella Biblioteca Apostolica Vaticana* (1944).

[l]Andreas Stylianou and Judith A. Stylianou, *The History of the Cartography of Cyprus*, Publications of the Cyprus Research Centre, 8 (Nicosia, 1980).

[m]A. E. Nordenskiöld, *Periplus: An Essay on the Early History of Charts and Sailing-Directions*, trans. Francis A. Bather (Stockholm: P. A. Norstedt, 1897).

[n]Giuseppe Caraci, "An Unknown Nautical Chart of Grazioso Benincasa, 1468," *Imago Mundi* 7 (1950): 18–31.

[o]Manuel Francisco de Barros e Sousa, Viscount of Santarém, *Atlas composé de mappemondes, de portulans et de cartes hydrographiques et historiques depuis le VIe jusqu'au XVIIe siècle* (Paris, 1849; facsimile reprint Amsterdam: R. Muller, 1985).

[p]Marina Salinari (formerly Marina Emiliani), "Notizie su di alcune carte nautiche di Grazioso Benincasa," *Rivista Geografica Italiana* 59 (1952): 36–42.

[q]Derek Howse and Michael Sanderson, *The Sea Chart* (Newton Abbot: David and Charles, 1973).

[r]Heinrich Winter, "Catalan Portolan Maps and Their Place in the Total View of Cartographic Development," *Imago Mundi* 11 (1954): 1–12.

[s]The title of one sheet in the British Library's Cornaro atlas, interpreted as a personal name.

[t]Paolo Revelli, *Cristoforo Colombo e la scuola cartografica genovese* (Genoa: Consiglio Nazionale delle Ricerche, 1937).

[u]Leo Bagrow, *History of Cartography*, rev. and enl. R. A. Skelton, trans. D. L. Paisey (Cambridge: Harvard University Press; London: C. A. Watts, 1964).

[v]The title of one sheet in the British Library's Cornaro atlas, this presumably means "complement of Cexano's chart" rather than being a personal name.

[w]Though the Catalan atlas is habitually attributed to Cresques Abraham, doubts have been raised. See this author's review of Georges Grosjean, *The Catalan Atlas of the Year 1375*, in *Imago Mundi* 33 (1981): 115–16, esp. 116.

[x]Arthur R. Hinks, *Portolan Chart of Angellino de Dalorto 1325 in the Collection of Prince Corsini at Florence, with a Note on the Surviving Charts and Atlases of the Fourteenth Century* (London: Royal Geographical Society, 1929).

[y]Walter W. Ristow and R. A. Skelton, *Nautical Charts on Vellum in the Library of Congress* (Washington, D.C.: Library of Congress, 1977), 3.

[z]Robert Putman, *Early Sea Charts* (New York: Abbeville Press, 1983).

[aa]Heinrich Winter, "Das katalanische Problem in der älteren Kartographie," *Ibero-Amerikanisches Archiv* 14 (1940/41): 89–126.

[bb]Most of Freducci's work dates from the sixteenth century. On this see Giuseppe Caraci, "The Italian Cartographers of the Benincasa and Freducci Families and the So-Called Borgiana Map of the Vatican Library," *Imago Mundi* 10 (1953): 23–49, and the further references given there.

[cc]Though the date is normally read as 1426, Armando Cortesão thought 1427 or 1432 more likely; see *History of Portuguese Cartography*, 2:124, n. 87 (note b).

[dd]Edward Luther Stevenson, *Facsimiles of Portolan Charts Belonging to the Hispanic Society of America*, Publications of the Hispanic Society of America, no. 104 (New York: Hispanic Society of America, 1916).

[ee]The contention by Arthur Dürst that there were actually two 1497 charts is unconvincing; see *Seekarte des Iehuda ben Zara: (Borgiano VII) 1497*, pamphlet accompanying a facsimile edition of the chart (Zurich: Belser Verlag, 1983), p. 2 and n. 7. The now untraceable chart found by Santarém (actually Hommaire de Hell) in 1847 in the Collegio di Propaganda Fide is clearly identifiable with Borgiano VII, as was noted by earlier authorities, among them Roberto Almagià, *Vaticana*, 1:47 (note k). It is known that the Borgia material reached the Vatican via the Propaganda Fide. Dürst's other claim—that there were also two 1500 charts—can be swiftly dismissed. Photographs of the "Kraus" and "Cincinnati" charts show them to be one and the same; see H. P. Kraus, Booksellers, *Remarkable Manuscripts, Books and Maps from the IXth to the XVIIIth Century*, catalog 80 (New York: H. P. Kraus, 1956), and Cecil Roth, "Judah Abenzara's Map of the Mediterranean World, 1500," *Studies in Bibliography and Booklore* 9 (1970): 116–20.

[ff]On Arab chartmakers, see also pp. 374–75, the entry under al-Mursī in this appendix, and the entry for the Maghreb chart in appendix 19.3.

[gg]Attempted decipherment of the illegible author's legend on the 1487 chart in Florence (Archivio di Stato, CN8).

[hh]Misreading of *aermiralo* in the author's legend to Pelechan's chart; see entry under Pelechan and note 433.

[ii]Ettore Rossi, "Una carta nautica araba inedita di Ibrāhīm al-Mursī datata 865 Egira = 1461 Dopo Cristo," in *Compte rendu du Congrès Internationale de Géographie* 5 (1926): 90–95 (Eleventh International Congress, Cairo, 1925).

[jj]The initial C is a misreading of the word *de* in the inscription "Nicollaus fillius de Pasqualini Nicollai . . ." on the 1408 atlas—an indication, perhaps, that there were two chartmakers with the name Nicolo de Pasqualini.

[kk]Weiss und Co., Antiquariat, *Codices manuscripti incunabula typographica catalogus primus* (Munich: Weiss und Co., 1926), no. 55.

[ll]Armando Cortesão and Avelino Teixeira da Mota, *Portugaliae monumenta cartographica*, 6 vols. (Lisbon, 1960).

[mm]Kenneth Nebenzahl, *Rare Americana*, catalog 20 (Chicago: Kenneth Nebenzahl, 1968), no. 1.

[nn]Heinrich Winter, "Petrus Roselli," *Imago Mundi* 9 (1952): 1–11.

APPENDIX 19.2—*continued*

[oo]Pietro Russo of Messina, who, it has been suggested, might have produced charts in the fifteenth century, has been omitted from this census. See Julio Rey Pastor and Ernest García Camarero, *La cartografía mallorquina* (Madrid: Departamento de Historia y Filosofía de la Ciencia, 1960), 92. None of his surviving work is dated before 1508. On his output see Roberto Almagià, "I lavori cartografici di Pietro e Jacopo Russo," *Atti della Accademia Nazionale dei Lincei* 12 (1957): 301–19.

[pp]George Kish, *La carte: Image des civilisations* (Paris: Seuil, 1980).

[qq]Sotheby's, *Catalogue of Highly Important Maps and Atlases*, 15 April 1980, lot A. The chart was sold to Nico Israel, antiquarian booksellers, and also appears in their Fall 1980 catalog, *Interesting Books and Manuscripts on Various Subjects: A Selection from Our Stock*, catalog 22 (Amsterdam: N. Israel, 1980), no. 1.

[rr]Lelio Pagani, *Pietro Vesconte: Carte nautiche* (Bergamo: Grafica Gutenberg, 1977).

[ss]Charles de La Roncière, *Les portulans de la Bibliothèque de Lyon*, fasc. 8 of *Les Portulans Italiens* in Lyon, Bibliothèque de la Ville, *Documents paléographiques, typographiques, iconographiques* (Lyons, 1929), 793.

Appendix 19.3 Atlases and Charts Known by Name

Name	Atlas (A) or Chart (C)	Where Preserved	Kamal[a]	Selected Accessible Works
				Reproductions
Atlante Mediceo *See* Medici atlas				
Carte Pisane/Carta Pisana	C	Paris, Bibliothèque Nationale, Rés. Ge. B 1118	4.1:1137	Bagrow, *History of Cartography*, pl. XXXII[b]
Catalan atlas	6 panels	Paris, Bibliothèque Nationale, MS. Esp. 30	4.3:1301–3	*Atlas Catalá*; Grosjean, *Catalan Atlas*; Nordenskiöld, *Periplus*, pls. XI–XIV;[c] Bagrow, *History of Cartography*, pls. XXXVII–XL, (note b)
Combitis atlas	A4	Venice, Biblioteca Nazionale Marciana, It. VI 213 [MS. 5982]	4.3:1333	
Cornaro atlas	(A34)	London, British Library, Egerton MS. 73	5.1:1508–12	
Cortona chart	C	Cortona, Biblioteca Comunale e dell'Accademia Etrusca		Armignacco, "Una carta nautica," pls. I–III[d]
Laurentian portolano/Portolano Laurenziano-Gaddiano *See* Medici atlas				
Lesina chart	C	Lost?		Goldschmidt, "Lesina Portolan Chart"[e]
(Tammar) Luxoro atlas	A8	Genoa, Biblioteca Civica Berio	4.2:1245	Desimoni and Belgrano, "Atlante idrografico," pls. I–VIII (reengraved copies);[f] Nordenskiöld, *Periplus*, pl. XVIII (note c)
Maghreb chart/Carta Mogrebina	C	Milan, Biblioteca Ambrosiana, S.P. II, 259	4.3:1336	Vernet-Ginés, "The Maghreb Chart," 1[g]
Medici/Medicean atlas	A6	Florence, Biblioteca Medicea Laurenziana, Gaddi 9	4.2:1246–48	Nordenskiöld, *Periplus*, 21, 115, pl. X (note c); Bagrow, *History of Cartography*, pl. XXXVI (world map) (note b)
Pinelli-Walckenaer atlas/Portolano Pinelli	A6	London, British Library, Add. MS. 19510	4.3:1316–19	Nordenskiöld, *Periplus*, pls. XV–XVII (note c)

[a]Youssouf Kamal, *Monumenta cartographica Africae et Aegypti*, 5 vols. in 16 pts. (Cairo, 1926–51).

[b]Leo Bagrow, *History of Cartography*, rev. and enl. R. A. Skelton, trans. D. L. Paisey (Cambridge: Harvard University Press; London: C. A. Watts, 1964).

[c]*L'atlas catalá de Cresques Abraham: Primera edició completa en el sis-cents aniversari de la seva realització* (Barcelona: Diáfora, 1975), also published in Spanish; Georges Grosjean, ed., *The Catalan Atlas of the Year 1375* (Dietikon-Zurich: Urs Graf, 1978); A. E. Nordenskiöld, *Periplus: An Essay on the Early History of Charts and Sailing-Directions*, trans. Francis A. Bather (Stockholm: P. A. Norstedt, 1897).

[d]Vera Armignacco, "Una carta nautica della Biblioteca dell'Accademia Etrusca di Cortona," *Rivista Geografica Italiana* 64 (1957): 185–223.

[e]E. P. Goldschmidt, "The Lesina Portolan Chart of the Caspian Sea" (with a commentary by Gerald R. Crone), *Geographical Journal* 103 (1944): 272–78, reproduction between 274 and 275.

[f]Cornelio Desimoni and Luigi Tommaso Belgrano, "Atlante idrografico del medio evo," *Atti della Società Ligure di Storia Patria* 5 (1867): 5–168.

[g]Juan Vernet-Ginés, "The Maghreb Chart in the Biblioteca Ambrosiana," *Imago Mundi* 16 (1962): 1–16.

APPENDIX 19.4
CHRONOLOGICAL INDEX OF DATED AND
DATABLE ATLASES AND CHARTS PRODUCED
UP TO 1500

Date		Atlas (A) or Chart (C)	Where Made	Date		Atlas (A) or Chart (C)	Where Made
1311	Pietro Vesconte	C		1459	Pelechan	C	Rethymnon, Crete
1313	Pietro Vesconte	A6					
1318	Pietro Vesconte	A6	Venice[a]	1461	G. Benincasa	C	Genoa
1318	Pietro Vesconte	A10		[1461]	al-Mursī	C	Tripoli, [Libya?]
[1320?]	Pietro Vesconte	A5		1462	Florino	A2	
[1321]	[Pietro Vesconte]	A5		1462	Roselli	C	Palma
1321	Perrino Vesconte	A4	Venice	1463	G. Benincasa	A5	Venice
ca. 1322	Pietro Vesconte	A7	Venice	1463	G. Benincasa	A4	Venice
ca. 1325	[Vesconte]	A5		1464	Roselli	C	Palma
1327	Perrino Vesconte	C	Venice	1465	G. Benincasa	A5	Venice
1325/30	Dalorto	C		1465	Roselli	C	Palma
1339	Dulcert	C	Palma	1466	G. Benincasa	A5	Venice
1367	F. Pizigano (with D. or M. Pizigano)	C	Venice	1466	Roselli	C	Palma
				1467	G. Benincasa	A5	Rome
				1467	G. Benincasa	A5	Rome
1373	F. Pizigano	A5	Venice	1467	G. Benincasa	A5	Rome
[1375]	Catalan atlas	6 panels		1468	G. Benincasa	A6	
1385	Soler	C		1468	G. Benincasa	A7	Venice
1403	F. Beccari	C	Savona	1468	G. Benincasa	C	Venice
1404	Pongeto	C		1468	Roselli	C	Palma
1408	Pasqualini	A6		1469	G. Benincasa	A6	Venice
1409	Virga	C	Venice	1469	G. Benincasa	A6	Venice
1413	M. de Viladestes	C		1469	Roselli	C	Majorca
[1413]	Kâtibî	C		1470	G. Benincasa	C	Ancona
1421	F. de Cesanis	C		1470	Nicolo	C	
1422	Giroldi	C		1471	G. Benincasa	A6	Venice
1423	M. de Viladestes	C	Palma	1472	G. Benincasa	A8	
1424	[Z. Pizzigano]	C		1472	G. Benincasa	C	Venice
1426	B. Beccari	C		1473	G. Benincasa	A6	Venice
1426[b]	Giroldi	A6		1473	G. Benincasa	A6	Venice
1428	J. de Viladestes	C	Palma	1474	G. Benincasa	A6	Venice
1430	Briaticho	A3		1476	A. Benincasa	A5	
1435	[B.] Beccari	C		1480	G. Benincasa	A6	Ancona
1436	Bianco	A7		1480	Canepa	C	Genoa
1439	Valseca	C	Majorca	1482	G. Benincasa	C	Ancona
1443	Giroldi	A6		1482	Bertran	C	Majorca
1446	Giroldi	A6		1487[c]		C	Majorca
1447	Roselli	C	Majorca	1489	Bertran	C	Majorca
1447	Roselli	C	Palma	1489	Canepa	C	Genoa
1447	Valseca	C	Majorca	1489	Pesina	C	Venice
1448	Bianco	C	London	148–	Domenech	C	Naples
1449	Roselli	C	Palma	1490	A. Benincasa	C	
1449	Valseca	C	Majorca	1492	Aguiar	C	Lisbon
1452	[Giroldi?]	A3		1494	Giovanni	C	Venice
1455	Pareto	C	Genoa	1497	Freducci	C	
1456	Bertran and Ripoll	C	Barcelona	1497	Jehuda ben Zara	C	Alexandria
1456	Roselli	C	Palma	1500	Jehuda ben Zara	C	Alexandria

Note: For present-day locations and a note of reproductions, see appendixes 19.2 and 19.3.

[a]The Venetian year began on 1 March; some of the charts signed from Venice could therefore actually belong to the following year.

[b]Though the date is normally read as 1426, Armando Cortesão thought 1427 or 1432 more likely; see his *History of Portuguese Cartography*, 2 vols. (Coimbra: Junta de Investigações do Ultramar-Lisboa, 1969–71), 2:124 n. 87.

[c]Florence, Archivio di Stato, CN8. Reproduced in Youssouf Kamal, *Monumenta cartographica Africae et Aegypti*, 5 vols. in 16 pts. (Cairo, 1926–51), 5.1:1506, and the western portion in Armando Cortesão, *The Nautical Chart of 1424 and the Early Discovery and Cartographical Representation of America: A Study on the History of Early Navigation and Cartography* (Coimbra: University of Coimbra, 1954), pl. XVIII.

APPENDIX 19.5
METHODOLOGY OF THE TOPONYMIC ANALYSIS

The toponymic analysis set out in table 19.3 (pp. 416–20) has not been published before, nor has its methodology been explained elsewhere. Since its findings underpin many of the new conclusions of this essay, it is necessary that its procedures be briefly explained. Despite the exclusion of all names north of Dunkirk and south of Mogador (areas frequently left off the charts themselves) as well as the omission of islands,[1] there were still twelve hundred names (of varying degrees of legibility) to be considered on a typical chart. If a comparative analysis was to be made of more than a handful of charts, there was a clear need for a research technique that would reduce this mass of place-names to manageable proportions. The method adopted comprised four stages, as follows.

1. Seven works were selected for a pilot study, as representatives of different periods and various centers of production. A complete transcription was made of their mainland names between Dunkirk and Mogador:

		Published transcriptions
a.	1311 Pietro Vesconte chart (substituting the 1313 atlas for the omitted western section)	
b.	ca. 1325 Vesconte atlas, British Library, Add. MS. 27376*	
c.	1375 Catalan atlas	Grosjean, *Catalan Atlas*, 53–77[2]
d.	1426 Giacomo Giroldi atlas	Nordenskiöld, *Periplus*, 25–44[3]
e.	1468 Grazioso Benincasa atlas, British Library, Add. MS. 6390	
f.	1468 Petrus Roselli chart[4]	
g.	1512 Vesconte Maggiolo atlas	Grosjean, *Maggiolo*,[5]

2. The parallel columns of the pilot study lists were compared and a note was made when names were added or removed or when the name form underwent drastic change. The names first recorded on works (b) to (f) and repeated at least once thereafter were termed "significant." Including 18 names first discerned on the 1512 Maggiolo atlas, these came to the surprisingly large total of 415. Names that were unique to any work in the pilot study except the last were excluded from this analysis; the concern was with the transmission of information, not with individuality. Variations in spelling, which might point to corruption or local dialect, were not considered in this context, and clearly distinct forms were treated as separate names even if they seemed to refer to the same place. Over the same period, about 100 names apparently became obsolete, but a larger time span would be needed to check that they were not revived later, as sometimes happened (for an example of this, see text footnote 379).

3. The significant additions were listed and their presence or absence was checked on the dated and datable charts and atlases produced up to 1430—the approximate point at which the portolan chart reproductions end in Kamal, *Monumenta cartographica*, 4.1–4 and 5.1.[6] Once this exercise had been completed, each significant name could be paired with the dated work on which it had made its first certain appearance on the portolan charts.

4. The results of (3) were then applied to those undated works that had in the past been realistically assigned to the period up to 1430.[7] In this way each chart was provided with a "toponymic profile" (as set out in tabulated form in appendix 19.3), which indicated the number of names present out of each group of datable innovations. The toponymic profile, a generalized and quantified record, could then indicate the most likely chronological slot for the chart concerned by means of a comparison with equivalent profiles on dated works.

1. Although islands were not considered in the general toponymic analysis, a special study of Cyprus names considered forty-three charts and atlases from the late thirteenth century to 1497; see Tony Campbell, "Cyprus and the Medieval Portolan Charts," *Kupriakai Spoudai: Deltion tēs Etaireias Kupriakōn Spoudōn, Brabeuthen upo tēs Akadēmias Athēnōn* 48 (1984): 47–66, esp. 52–58 and the tables.
2. Georges Grosjean, ed., *The Catalan Atlas of the Year 1375* (Dietikon-Zurich: Urs Graf, 1978).
3. A. E. Nordenskiöld, *Periplus: An Essay on the Early History of Charts and Sailing-Directions*, trans. Francis A. Bather (Stockholm: P. A. Norstedt, 1897).
4. Edward Luther Stevenson, *Facsimiles of Portolan Charts Belonging to the Hispanic Society of America*, Publications of the Hispanic Society of America, no. 104 (New York: Hispanic Society of America, 1916), pl. 2.
5. Georges Grosjean, ed., *Vesconte Maggiolo "Atlante Nautico del 1512": Seeatlas vom Jahre 1512* (Dietikon-Zurich: Urs Graf, 1979).
6. Youssouf Kamal, *Monumenta cartographica Africae et Aegypti*, 5 vols. in 16 pts. (Cairo, 1926–51). All but two of the twenty-eight works involved were examined directly or through reproductions—note b to table 19.3, pp. 416–20, explains the omissions.
7. It was possible to consider twenty-one out of the twenty-seven works concerned, see table 19.3, note a, for the details of those that could not be examined.

BIBLIOGRAPHY
CHAPTER 19 PORTOLAN CHARTS FROM THE LATE THIRTEENTH CENTURY TO 1500

Almagià, Roberto. *Monumenta cartographica Vaticana.* 4 vols. Rome: Biblioteca Apostolica Vaticana, 1944–55.

Andrews, Michael Corbet. "The Boundary between Scotland and England in the Portolan Charts." *Proceedings of the Society of Antiquaries of Scotland* 60 (1925–26): 36–66.

———. "The British Isles in the Nautical Charts of the XIVth and XVth Centuries." *Geographical Journal* 68 (1926): 474–81.

———. "Scotland in the Portolan Charts." *Scottish Geographical Magazine* 42 (1926): 129–53, 193–213, 293–306.

Armignacco, Vera. "Una carta nautica della Biblioteca dell'Accademia Etrusca di Cortona." *Rivista Geografica Italiana* 64 (1957): 185–223.

L'atlas català de Cresques Abraham: Primera edició completa en el sis-cents aniversari de la seva realització. Barcelona: Diáfora, 1975. Also published in Spanish.

Brincken, Anna-Dorothee von den. "Die kartographische Darstellung Nordeuropas durch italienische und mallorquinische Portolanzeichner im 14. und in der ersten Hälfte des 15. Jahrhunderts." *Hansische Geschichtsblätter* 92 (1974): 45–58.

Canale, Michele G. *Storia del commercio dei viaggi, delle scoperte e carte nautiche degl'Italiani.* Genoa: Tipografia Sociale, 1866.

Caraci, Giuseppe. *Italiani e Catalani nella primitiva cartografia nautica medievale.* Rome: Istituto di Scienze Geografiche e Cartografiche, 1959.

Clos-Arceduc, A. "L'énigme des portulans: Etude sur le projection et le mode de construction des cartes à rhumbs du XIVe et du XVe siècle." *Bulletin du Comité des Travaux Historiques et Scientifiques: Section de Géographie* 69 (1956): 215–31.

Cortesão, Armando. *The Nautical Chart of 1424 and the Early Discovery and Cartographical Representation of America: A Study on the History of Early Navigation and Cartography.* Coimbra: University of Coimbra, 1954.

———. *History of Portuguese Cartography.* 2 vols. Coimbra: Junta de Investigações do Ultramar-Lisboa, 1969–71.

Crone, Gerald R. *Maps and Their Makers: An Introduction to the History of Cartography.* 1st ed., London: Hutchinson University Library, 1953; 5th ed., Folkestone, Kent: Dawson; Hamden, Conn.: Archon Books, 1978.

Desimoni, Cornelio. "Elenco di carte ed atlanti nautici di autore genovese oppure in Genova fatti o conservati." *Giornale Ligustico* 2 (1875): 47–285.

———. "Le carte nautiche italiane del Medio Evo—a proposito di un libro del Prof. Fischer." *Atti della Società Ligure di Storia Patria* 19 (1888–89): 225–66.

Destombes, Marcel. "Cartes catalanes du XIVe siècle." In *Rapport de la Commission pour la Bibliographie des Cartes Anciennes.* 2 vols. International Geographical Union. Paris: Publié avec le concours financier de l'UNESCO, 1952. Vol. 1, *Rapport au XVIIe Congrès International, Washington, 1952 par R. Almagià: Contributions pour un catalogue des cartes manuscrites, 1200–1500,* ed. Marcel Destombes, 38–63.

Emiliani, Marina (later Marina Salinari). "Le carte nautiche dei Benincasa, cartografi anconetani." *Bollettino della Reale Società Geografica Italiana* 73 (1936): 485–510.

Fiorini, Matteo. *Le projezioni delle carte geografiche.* Bologna: Zanichelli, 1881.

Fischer, Theobald. *Sammlung mittelalterlicher Welt- und Seekarten italienischen Ursprungs und aus italienischen Bibliotheken und Archiven.* Venice: F. Ongania, 1886; reprinted Amsterdam: Meridian, 1961.

Foncin, Myriem, Marcel Destombes, and Monique de La Roncière. *Catalogue des cartes nautiques sur vélin conservées au Département des Cartes et Plans.* Paris: Bibliothèque Nationale, 1963.

Frabetti, Pietro. *Carte nautiche italiane dal XIV al XVII secolo conservate in Emilia-Romagna.* Florence: Leo S. Olschki, 1978.

García Camarero, Ernesto. "Deformidades y alucinaciones en la cartografia ptolemeica y medieval." *Boletin de la Real Sociedad Geográfica* 92 (1956): 257–310.

Grosjean, Georges, ed. *The Catalan Atlas of the Year 1375.* Dietikon-Zurich: Urs Graf, 1978.

Guarnieri, Giuseppe Gino. *Il Mediterraneo nella storia della cartografia nautica medioevale.* Leghorn: STET, 1933.

———. *Geografia e cartografia nautica nella loro evoluzione storia e scientifica.* Genoa, 1956.

Hinks, Arthur R. *Portolan Chart of Angellino de Dalorto 1325 in the Collection of Prince Corsini at Florence, with a Note on the Surviving Charts and Atlases of the Fourteenth Century.* London: Royal Geographical Society, 1929.

Kamal, Youssouf. *Monumenta cartographica Africae et Aegypti.* 5 vols. in 16 pts. Cairo, 1926–51.

———. *Hallucinations scientifiques (les portulans).* Leiden: E. J. Brill, 1937.

Kelley, James E., Jr. "The Oldest Portolan Chart in the New World." *Terrae Incognitae: Annals of the Society for the History of Discoveries* 9 (1977): 22–48.

———. "Non-Mediterranean Influences That Shaped the Atlantic in the Early Portolan Charts." *Imago Mundi* 31 (1979): 18–35.

Kretschmer, Konrad. *Die italienischen Portolane des Mittelalters: Ein Beitrag zur Geschichte der Kartographie und Nautik.* Veröffentlichungen des Instituts für Meereskunde und des Geographischen Instituts an der Universität Berlin, vol. 13. Berlin, 1909; reprinted Hildesheim: Georg Olms, 1962.

Laguarda Trías, Rolando A. *Estudios de cartologia.* Madrid, 1981.

Lane, Frederic C. "The Economic Meaning of the Invention of the Compass." *American Historical Review* 68, no. 3 (1963): 605–17.

La Roncière, Charles de. *La découverte de l'Afrique au Moyen Age: Cartographes et explorateurs.* Mémoires de la Société Royale de Géographie d'Egypte, vols. 5, 6, 13. Cairo: Institut Français d'Archéologie Orientale, 1924–27.

Longhena, Mario. "La carte dei Pizigano del 1367 (posseduta dalla Biblioteca Palatina di Parma)." *Archivio Storico per le Province Parmensi,* 4th ser., 5 (1953): 25–130.

Magnaghi, Alberto. "Il mappamondo del Genovese Angellinus de Dalorto (1325)." *Atti del III Congresso Geografico Italiano, Florence, 1898.* 2 vols. (1899) 2:506–43.

———. "Sulle origine del portolano normale del Medio Evo e della cartografia dell'Europa occidentale." *Memorie geografiche pubblicate come supplemento alla Rivista Geografica Italiana* 8 (1909): 115–87.

———. "Nautiche, carte." In *Enciclopedia italiana di scienze, lettere ed arti*, originally 36 vols., 24:323–31. [Rome]: Istituto Giovanni Treccani, 1929–39.

Marinelli, Giovanni. "Venezia nella storia della geografia cartografica ed esploratrice." *Atti del Reale Istituto Veneto di Scienze, Lettere ed Arti*, 6th ser., 7 (1888–89): 933–1000.

Mollat du Jourdin, Michel, and Monique de La Roncière, with Marie-Madeleine Azard, Isabelle Raynaud-Nguyen, and Marie-Antoinette Vannereau. *Les portulans: Cartes marines du XIIIᵉ au XVIIᵉ siècle.* English edition, *Sea Charts of the Early Explorers: 13th to 17th Century.* Translated by L. le R. Dethan. New York: Thames and Hudson, 1984.

Motzo, Bacchisio R. "Il Compasso da navigare, opera italiana della metà del secolo XIII." *Annali della Facoltà di Lettere e Filosofia della Università di Cagliari* 8 (1947): I–137.

Nordenskiöld, A. E. *Periplus: An Essay on the Early History of Charts and Sailing-Directions.* Translated by Francis A. Bather. Stockholm: P. A. Norstedt, 1897.

Oldham, Richard D. "The Portolan Maps of the Rhône Delta: A Contribution to the History of the Sea Charts of the Middle Ages." *Geographical Journal* 65 (1925): 403–28.

Pagani, Lelio. *Pietro Vesconte: Carte nautiche.* Bergamo: Grafica Gutenberg, 1977.

Pelham, Peter T. "The Portolan Charts: Their Construction and Use in the Light of Contemporary Techniques of Marine Survey and Navigation." Master's thesis, Victoria University of Manchester, 1980.

Piersantelli, Giuseppe. "L'Atlante Luxoro." In *Miscellanea di geografia storica e di storia della geografia nel primo centenario della nascita di Paolo Revelli*, 115–41. Genoa: Bozzi, 1971.

Quaini, Massimo. "Catalogna e Liguria nella cartografia nautica e nei portolani medievali." In *Atti del 1° Congresso Storico Liguria-Catalogna: Ventimiglia-Bordighera-Albenga-Finale-Genova, 14–19 ottobre 1969*, 549–71. Bordighera: Istituto Internazionale di Studi Liguri, 1974.

Reparaz, Gonçal (Gonzalo) de. *Catalunya a les mars: Navegants, mercaders i cartògrafs catalans de l'Edat Mitjana i del Renaixement (Contribucío a l'estudi de la història del comerç i de la navegació de la Mediterrània).* Barcelona: Mentova, 1930.

———. "Essai sur l'histoire de la géographie de l'Espagne de l'antiquité au XVᵉ siècle." *Annales du Midi* 52 (1940): 137–89, 280–341.

Revelli, Paolo. *Cristoforo Colombo e la scuola cartografica genovese.* Genoa: Consiglio Nazionale delle Ricerche, 1937.

———. *La partecipazione italiana alla Mostra Oceanografica Internazionale di Siviglia (1929).* Genoa: Stabilimenti Italiani Arti Grafiche, 1937.

———. *Elenco illustrativo della Mostra Colombiana Internazionale.* Genoa: Comitato Cittadino per le Celebrazioni Colombiane, 1950.

Rey Pastor, Julio, and Ernesto García Camarero. *La cartografía mallorquina.* Madrid: Departamento de Historia y Filosofía de la Ciencia, 1960.

Santarém, Manuel Francisco de Barros e Sousa, Viscount of. "Notice sur plusieurs monuments géographiques inédits du Moyen Age et du XVIᵉ siècle qui se trouvent dans quelques bibliothèques de l'Italie, accompagné de notes critiques." *Bulletin de la Société de Géographie*, 3d ser., 7 (1847): 289–317; reprinted in *Acta Cartographica* 14 (1972): 318–46.

———. *Essai sur l'histoire de la cosmographie et de la cartographie pendant le Moyen-Age et sur les progrès de la géographie après les grandes découvertes du XVᵉ siècle.* 3 vols. Paris: Maulde et Renou, 1849–52.

———. *Estudos de cartographia antiga.* 2 vols. Lisbon: Lamas, 1919–20.

Skelton, R. A. "A Contract for World Maps at Barcelona, 1399–1400." *Imago Mundi* 22 (1968): 107–13.

Stevenson, Edward Luther. *Portolan Charts: Their Origin and Characteristics with a Descriptive List of Those Belonging to the Hispanic Society of New York.* New York: [Knickerbocker Press], 1911.

Taylor, Eva G. R. *The Haven-Finding Art: A History of Navigation from Odysseus to Captain Cook.* London: Hollis and Carter, 1956.

Teixeira da Mota, Avelino. "L'art de naviguer en Méditerranée du XIIIᵉ au XVIIᵉ siècle et la création de la navigation astronomique dans les océans." In *Le navire et l'économie maritime du Moyen-Age au XVIIIᵉ siècle principalement en Méditerranée: Travaux du IIᵉᵐᵉ Colloque Internationale d'Histoire Maritime*, ed. Michel Mollat, 127–54. Paris: SEVPEN, 1958.

Uhden, Richard. "Die antiken Grundlagen der mittelalterlichen Seekarten." *Imago Mundi* 1 (1935): 1–19.

Uzielli, Gustavo, and Pietro Amat di San Filippo. *Mappamondi, carte nautiche, portolani ed altri monumenti cartografici specialmente italiani dei secoli XIII–XVII.* 2d ed., 2 vols. Studi Biografici e Bibliografici sulla Storia della Geografia in Italia. Rome: Società Geografica Italiana, 1882; reprinted Amsterdam: Meridian, 1967.

Wagner, Hermann. "Historische Ausstellung, betreffend die Entwickelung der Seekarten vom XIII.–XVIII. Jahrhundert oder bis zur allgemeinen Einführung der Mercator-Projektion und der Breitenminute als Seemeile." *Katalog der Ausstellung des XI. Deutschen Geographentages*, part of the *Verhandlungen.* Bremen, 1895.

———. "The Origin of the Mediaeval Italian Nautical Charts." In *Report of the Sixth International Geographical Congress, London, 1895*, 695–702. London: Royal Geographical Society, 1896; reprinted *Acta Cartographica* 5 (1969): 476–83.

Winter, Heinrich. "Das katalanische Problem in der älteren Kartographie." *Ibero-Amerikanisches Archiv* 14 (1940/41): 89–126.

———. "Petrus Roselli." *Imago Mundi* 9 (1952): 1–11.

———. "Catalan Portolan Maps and Their Place in the Total View of Cartographic Development." *Imago Mundi* 11 (1954): 1–12.

Wuttke, Heinrich. *Zur Geschichte der Erdkunde im letzten Drittel des Mittelalters: Die Karten der seefahrenden Völker Südeuropas bis zum ersten Druck der Erdbeschreibung des Ptolemäus.* Dresden, 1871.

Zurla, Placido. *Di Marco Polo e degli altri viaggiatori veneziani più illustri.* 2 vols. Venice, 1818.

20 · Local and Regional Cartography in Medieval Europe

P. D. A. HARVEY*

SCOPE AND CHARACTERISTICS

This chapter covers all terrestrial maps from medieval Christendom that are neither world maps nor portolan charts nor the rediscovered maps of Ptolemy.[1] They are relatively few in number but highly varied in character. They range from maps of the whole of Palestine carefully constructed on a measured grid to a painted picture of three villages and their surrounds on the border of Burgundy; from Francesco Rosselli's detailed view of Florence to a few sketched lines representing strips in a field in East Anglia. Despite their variety, they have features in common. Whether they show a large area or a small one, they are all conceived as showing it from above, either vertically or obliquely, viewed from a position often unattainable in reality; this has been taken as defining a map for our purpose. These are the products of medieval Europe that are typologically, historically, or conceptually the precursors of the large-scale and topographic maps of the sixteenth century and later. A very few are maps of entire countries: the maps of Palestine, the Matthew Paris and Gough maps of Britain, the maps of Germany and central Europe by Nicolas of Cusa and Erhard Etzlaub. But most are maps of small areas: local maps covering an area, whether a single field or half a province, that would lie within the normal experience of an individual. Where these maps show features above ground level they nearly always take pictorial form; the most elaborate, like Rosselli's map of Florence, are straightforward bird's-eye views, realistic, fully detailed pictures of landscape as seen from above, and indeed medieval local maps are related as much to the bird's-eye views as to the large-scale maps of later centuries (fig. 20.1). In bringing together these very diverse representations of landscape and calling them all maps, we are acting with the hindsight of later cartographic development; no one in the Middle Ages would have seen them as a single class of objects, and no language of medieval Europe had a word corresponding exactly to our *map*. Representations of this kind were in any case very unusual.

Medieval Europe was in fact a society that knew little of maps. It was not just that the regular use of maps and plans for government or business was confined to a very few particular areas and crafts. The idea of drawing a casual sketch map to show some topographical relationship—the way from one place to another, the layout of fields, the sequence of houses in a street—was one that seldom occurred to people in the Middle Ages. That this was so, and that in the sixteenth century people suddenly became aware of the value of maps, can be shown by the numbers of local maps (of every kind, down to the very roughest sketches) that survive from medieval England: from the mid-twelfth to the mid-fourteenth century we have only three in all, from each half-century between 1350 and 1500 about ten, and from the half-century 1500–1550 about two hundred.[2] These figures point to changes in the actual production and use of maps, not merely to their better chance of survival as time went on: the survival of other sorts of documents from medieval England follows quite a different pattern.[3] Medieval maps of small areas or regions were in their own time quite abnormal productions, often displaying great originality and imaginative ingenuity on the part of the draftsman, who may never have seen a map drawn by anyone else. In the Middle Ages, the normal way of setting out and recording topographical relationships was in writing, so in place of maps we have written descriptions: itineraries, urban surveys, field terriers, and so on. These might be of great complexity; a terrier might list hundreds, even thousands, of individual plots of land in a set of fields, giving the exact location

*With a contribution by ELIZABETH CLUTTON on the *Isolarii*, pp. 482–84.

1. A fuller account of the topographical maps from medieval Europe, with many illustrations, is in P. D. A. Harvey, *The History of Topographical Maps: Symbols, Pictures and Surveys* (London: Thames and Hudson, 1980), esp. chaps. 3–5 and 9. I am most grateful to Messrs. Thames and Hudson for permission to repeat here the evidence and the conclusions on medieval maps that are presented in that book.

2. P. D. A. Harvey, "The Portsmouth Map of 1545 and the Introduction of Scale Maps into England," in *Hampshire Studies*, ed. John Webb, Nigel Yates, and Sarah Peacock (Portsmouth: Portsmouth City Records Office, 1981), 33–49, esp. 35.

3. R. A. Skelton and P. D. A. Harvey, eds., *Local Maps and Plans from Medieval England* (Oxford: Clarendon Press, 1986), 4, 34–35.

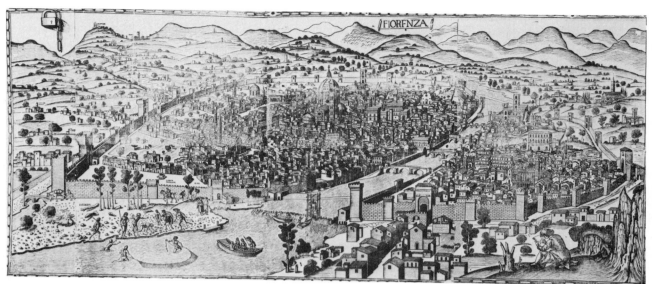

FIG. 20.1. ROSSELLI'S MAP OF FLORENCE, CA. 1485. One of the finest oblique pictorial views surviving from the fifteenth century, this "map with the chain" by Francesco Rosselli shows the city from a southwesterly vantage point. Buildings are shown in elevation, as on many maps during the medieval period.

Size of the original: 58.5 × 131.5 cm. By permission of the Staatliche Museen zu Berlin, DDR, Kupferstichkabinett.

of each,[4] so that it was a veritable "carte parlante" as de Dainville put it.[5] But that this should be preferred to putting the information graphically on a map points to a way of thought quite different from our own. It is with this in mind that we should view those regional and local maps and plans that we have from the Middle Ages.

Nor was it a matter of a gradual but steady increase in mapmaking, with growing awareness of the value of maps. With hindsight we can see in the fifteenth century some small pointers to the change that was to come in the sixteenth. But no more than that; the growth of mapmaking after 1500 was as sudden as it was rapid. The numbers of local maps surviving from England are significant here: there seems to have been no increase in production between 1350 and 1500. But it is significant too that there are hardly any English local maps earlier than 1350. The fact is that if we have few maps of small areas from medieval Europe, we have very few indeed earlier than the mid-fourteenth century: nearly all date from the last 150 years before 1500. Almost certainly this is a change not just in the pattern of survival but in the pattern of production, reflecting some sort of modest and restricted spread of the idea of drawing maps. But we can do little more than guess how or why this happened. Here, as in other questions about medieval maps of small areas, we are hampered by the patchiness of the research so far done on them. Some groups of maps,

such as those from the Low Countries and Britain or the regional maps from northern Italy, have been recorded and studied fairly systematically and can thus be viewed comprehensively. Others, such as those from France and Germany, have been brought to light more or less casually, so that we cannot tell how far those known are typical of the entire surviving corpus. From some areas, such as Spain and Portugal, no medieval maps of small areas have been reported, but this need not mean that none exist. There and elsewhere many such maps may have escaped notice simply because it has not been generally appreciated how unusual maps of any sort were before the sixteenth century, how the merest sketch map, of no cartographic importance in later periods, may be of great significance if it dates from the Middle Ages. Because our knowledge is so patchy, it is difficult to reach general conclusions about the mapping of small areas in medieval Europe; future discoveries may radically change the picture.

4. For example, Catherine P. Hall and J. R. Ravensdale, eds., *The West Fields of Cambridge* (Cambridge: Cambridge Antiquarian Records Society, 1976), a mid-fourteenth-century terrier.

5. François de Dainville, "Rapports sur les conférences: Cartographie Historique Occidentale," in *Ecole Pratique des Hautes Etudes, IVe section: Sciences historiques et philologiques. Annuaire 1968–1969* (Paris, 1969), 401–2.

However, one conclusion can be safely drawn from the evidence now available. Scale, the observance of a fixed proportion between distances on the map and distances on the ground, played practically no part in medieval maps of small areas; they were scarcely ever based on measured surveys. Sometimes, indeed, we find a broad correspondence in shape and distance between the map and the actual terrain. But often the mapmaker is concerned only to show the sequence of features along a very few routes, or some other topographical relationships that can be set out diagrammatically without any regard to scale at all. And sometimes consistency of scale is deliberately ignored, as in some northern Italian regional maps where the city at the center is drawn to a far larger scale than the rest of the map, to emphasize its importance. Apart from, just possibly, some Italian city plans, scale maps of small areas were unknown for most of our period, perhaps making a hesitant appearance only at its start in the ninth century with the Saint Gall plan, and its finish in the fifteenth. Conceptually related to the lack of scale is the fact that these maps are essentially picture maps: any detail above ground level is shown pictorially, not in plan. The form and artistry of this pictorial element vary enormously. At one extreme it may consist of no more than a roughly sketched outline or perspective view of a single feature, shown conventionally or even by means of a sign: a church for a village, or walls for a town. At the other extreme the entire map may be a realistic and accurate bird's-eye view, drawn in strict perspective throughout. The picture maps of the Middle Ages were the ancestors not only of the large-scale maps of the sixteenth century and later, but of the bird's-eye views as well, and they are all probably best thought of as a particular sort of topographical picture, drawn as though from a viewpoint (or often, indeed, more than one viewpoint) above ground level. Certainly many of the changes we see in the medieval picture maps are simply developments in artistic styles and techniques.[6]

ORIGINS AND DEVELOPMENT

It may seem surprising that scale maps were virtually unknown in medieval Europe in view of the extremely competent and complex scale maps produced by the Roman surveyors from the first to the third century A.D.[7] What debt, if any, medieval mapping of small areas owed to classical Roman precedent is very difficult to determine. If some medieval Italian city plans were based on measured surveys (and this is far from certain), we should not rule out the possibility of a direct line of tradition from the Roman surveyors. But it seems more likely that the Roman tradition of scale maps died out altogether, leaving as its latest surviving relics Arculf's

plans of the holy places and the monastic plan in Saint Gall Abbey library. Both of these are like the Roman surveyors' maps, and unlike any other medieval maps, in showing buildings simply in outline ground plan and possibly also in being drawn to a fixed scale.[8] The plans of Arculf, a Frankish bishop, are of four buildings in the Holy Land: the Churches of the Sepulcher (fig. 20.2), of Mount Zion, and of the Ascension in Jerusalem, and the Church of Jacob's Well at Nablus. They illustrate the account of his pilgrimage there in 670, an account written out for posterity by Adamnan, abbot of Iona, where Arculf is said to have stayed after his ship had been carried far off course when returning to Gaul. The plans are described as having been drawn by Arculf on wax tablets, but they were not copied in all the surviving medieval manuscripts of the text, and the earliest in which they are found dates from the ninth century. Imprecise copying makes it impossible to tell whether the original plans were drawn to scale; we can only say that their style is entirely that of the Roman surveyors' scale plans.[9] Of the plan at Saint Gall (fig. 20.3), on the other hand, we have only the single original that was sent, as its dedicatory inscription tells us, to Gozbert, who was abbot there from 816 to 837. There has been much scholarly research on the plan, culminating in the recent work of Horn and Born, who have shown how the buildings and plots of ground it depicts conform to a single unit of length, so that all their alignments coincide with the grid that underlies the entire plan.[10] These buildings

6. Harvey, *Topographical Maps*, 48 (note 1).

7. Gianfilippo Carettoni et al., *La pianta marmorea di Roma antica: Forma Urbis Romae*, 2 vols. (Rome: Comune di Roma, 1960); besides the plan of Rome that is the subject of this work, other Roman scale plans are discussed and reproduced on pages 207–10 and pl. Q. André Piganiol, *Les documents cadastraux de la colonie romaine d'Orange*, *Gallia* suppl. 16 (Paris: Centre Nationale de la Recherche Scientifique, 1962).

8. Harvey, *Topographical Maps*, 131–32 (note 1).

9. Titus Tobler and Augustus Molinier, eds., *Itinera Hierosolymitana et descriptiones Terrae Sanctae* (Paris: Société de l'Orient Latin, 1879), xxx–xxxiii, 149, 160, 165, 181; related plans are in manuscripts of Bede's *De locis sanctis*, for example, London, British Library, Add. MS. 22653, fols. 44–46v; see Reinhold Röhricht, "Karten und Pläne zur Palästinakunde aus dem 7. bis 16. Jahrhundert, II," *Zeitschrift des Deutschen Palästina-Vereins* 14 (1891): 87–92, esp. 91–92.

10. Walter Horn and Ernest Born, *The Plan of St. Gall: A Study of the Architecture and Economy of, and Life in, a Paradigmatic Carolingian Monastery*, 3 vols. (Berkeley: University of California Press, 1979). A different interpretation of the underlying grid is given by Eric Fernie, "The Proportions of the St. Gall Plan," *Art Bulletin* 60 (1978): 583–89. Of earlier writings on the plan, Walter Horn and Ernest Born, "New Theses about the Plan of St. Gall," in *Die Abtei Reichenau: Neue Beiträge zur Geschichte und Kultur des Inselklosters*, ed. Helmut Maurer (Sigmaringen: Thorbecke, 1974), 407–76, is a guide to recent discussion, while Hans Reinhardt, *Der St. Galler Klosterplan* (Saint Gall: Historischer Verein des Kantons St. Gallen, 1952), is a useful short introduction to the plan.

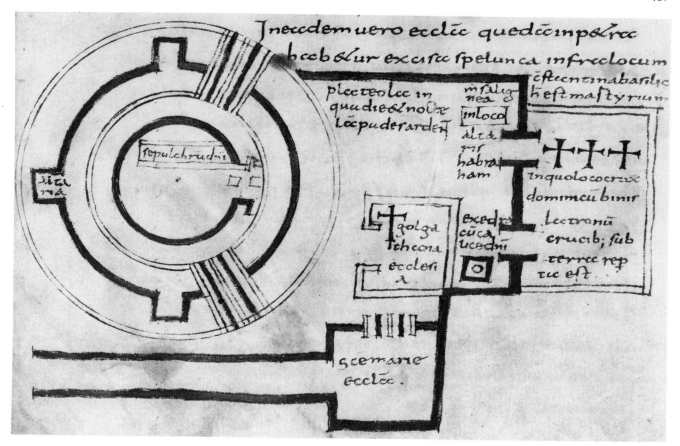

FIG. 20.2. PLAN OF THE HOLY SEPULCHER. This and other plans of holy places were drawn by Arculf, a Frankish bishop, to illustrate his pilgrimage to Jerusalem in 670. Like the monastery plan from Saint Gall, it belongs to the Roman tradition of surveyors' plans.

Size of the original: 11.5 × 17.5 cm. Photograph from the Bibliothèque Nationale, Paris (MS. Lat. 13048, fol. 4v).

and grounds are an abbey church and its precincts, with accommodation for the monks and their servants, cloisters, gardens, and houses for the estate workers and livestock. It has been much debated whether this was an existing monastery, a plan for one intended to be built, or simply an ideal schema. Nor is it certain that it was drawn as a scale map; the measurements entered on it for certain features are not entirely consistent with the proportions of the map itself.[11] It is in any case an impressive monument to the art and thought of the Carolingian age, and one of the outstanding cartographic productions of medieval Europe. But it had no direct successor. It is possible, even likely, that knowledge of the Saint Gall plan lies behind the plan of Canterbury cathedral and its priory that was drawn in the mid-twelfth century—the style of the inscriptions particularly suggests a connection—but if so it was only the idea of a plan that was transmitted, not its cartographic concept

or method, for the Canterbury plan is a picture map, showing the buildings in elevation, not an outline ground plan (fig. 20.4).[12] The Saint Gall plan was linked to the past, not the future. It is the last known surviving map in the tradition of the Roman surveyors.

If the idea of drawing maps to scale failed to pass from ancient Rome to medieval Europe, did any other aspects of cartography meet with better success? Did any traditions of mapping survive from classical times? Einhard, Charlemagne's courtier and biographer, tells us that when Charlemagne died he had among his pos-

---

11. See, for instance, the doubts expressed by David Parsons, "Consistency and the St. Gallen Plan: A Review Article," *Archaeological Journal* 138 (1981): 259–65.

12. Cambridge, Trinity College, MS. R.17.1, fols. 284v–285r; William Urry, "Canterbury, Kent, *circa* 1153 × 1161," in Skelton and Harvey, *Local Maps and Plans*, 43–58 (note 3).

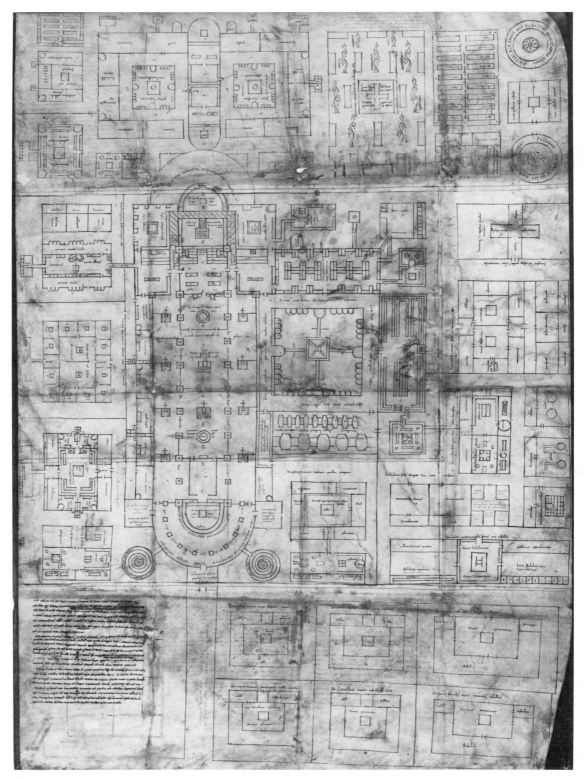

FIG. 20.3. THE PLAN OF SAINT GALL. This architectural plan of an idealized monastery and its associated buildings, dated 816–837, is remarkable in its consistent unit of measurement and planimetric style, reminiscent more of Roman than of medieval cartography.

Size of the original: 114 × 74.5 cm. By permission of the Stiftsbibliothek, Saint Gall (Codex 1092).

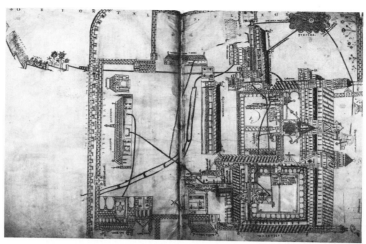

FIG. 20.4. PLAN OF CANTERBURY CATHEDRAL AND ITS PRIORY. Drawn in the mid-twelfth century, this is a picture map depicting features above the ground in elevation. Only the water supply, which consists of pipes from springs not on the map, is shown in plan.

Size of the original: 45.7 × 66 cm. By permission of the Master and Fellows of Trinity College, Cambridge (MS. R.17.1, fols. 284v–285).

sessions tables of gold and silver engraved with representations of the world and of Rome and Constantinople.[13] There may have been a continuing tradition of plans of both these cities, but those surviving to us from the Middle Ages give no hint of classical antecedents. It is more likely that we have a continuous tradition of maps of Palestine. From the twelfth century onward we have various maps of the Holy Land, drawn in western Europe; since we also have from the earlier period both written itineraries and maps (the Peutinger map, the Madaba mosaic) covering the same area, it should be possible to discover whether they owe anything to precedent.[14] Certainly there are at least superficial similarities between the sixth-century Madaba mosaic and the medieval tradition that includes the early fourteenth-century maps by Pietro Vesconte of Venice; but the exact relationships between even the medieval maps have not been fully worked out—the question is not a simple one—and this would be an essential preliminary to detailed comparison with the earlier material. Medieval plans of Jerusalem, which also date from the twelfth century onward, may well also owe something to classical tradition, though this would be harder to determine. There is a diagram of Jerusalem with Arculf's seventh-century text, but otherwise the only earlier plan is the representation of the city on the Madaba mosaic.

It is thus possible, but very far from certain, that medieval maps of Palestine and perhaps of other areas derived from Roman models. But it is only when we look at the detailed features of medieval maps that we find an element that can convincingly be shown to lie in a direct tradition from classical antiquity. This is the way of showing a town by a stylized picture of city walls, seen in bird's-eye view, usually with towers and some-

times with one or more buildings inside. It is a very natural convention to adopt; indeed, it occurs in Chinese cartography as well as European.[15] But it can be traced in all sorts of contexts from ancient Greece down to the High Middle Ages, and there seems no doubt that it reached medieval maps through a continuing artistic tradition of great antiquity.[16] Lavedan has well named it the city ideogram ("l'idéogramme urbain"), and we shall see that it formed the basis for very much more elaborate plans of some of the towns of medieval Europe.

If the maps of small areas that were drawn in medieval Europe owed little to the classical past, where did they draw their inspiration? One answer lies in the diagrams that medieval writers used to illustrate all kinds of relationships—philosophical, scientific, administrative, and so on. Many of the maps, especially the earliest ones, were simply applications of the same kind of drawing to topographical relationships. An early thirteenth-century English plan showing the source of Waltham Abbey's water supply at Wormley (Hertfordshire) (fig. 20.5) is very like the diagrams that Matthew Paris was drawing at Saint Albans Abbey at about the same time to illustrate astrological and other works, and a direct

13. Einhard, *Early Lives of Charlemagne*, ed. and trans. A. J. Grant (London: Moring, 1905), 54; but see F. N. Estey, "Charlemagne's Silver Celestial Table," *Speculum* 18 (1943): 112–17, who interprets the world map as being of the heavens.

14. Otto Cuntz, ed., *Itineraria Romana* (Leipzig: Teubner, 1929–), vol. 1, *Itineraria Antonini Augusti et Burdigalense*; Ekkehard Weber, ed., *Tabula Peutingeriana: Codex Vindobonensis 324* (Graz: Akademische Druck- und Verlagsanstalt, 1976); Michael Avi-Yonah, *The Madaba Mosaic Map* (Jerusalem: Israel Exploration Society, 1954).

15. For example, Harvey, *Topographical Maps*, pls. V, VI (note 1).

16. Pierre Lavedan, *Représentation des villes dans l'art du Moyen Age* (Paris: Vanoest, 1954), 33–35, pl. XVII; Carl H. Kraeling, ed., *Gerasa, City of the Decapolis* (New Haven: American Schools of Oriental Research, 1938), 341–51.

connection is not impossible.[17] One early map of Palestine shows Jerusalem as a series of large concentric circles, surrounded by tiny circles linked with straight lines to represent other towns, roads, the river Jordan, and the coastline.[18] The oldest known map from the Netherlands, dated 1307, consists simply of place-names and other notes, written in an arrangement corresponding to their positions on the ground, with drawn outlines of gable ends representing two churches (fig. 20.6).[19] Nor is it difficult to find later maps that show just as clearly the techniques of the diagram applied topography. One example from 1441 shows an estate on both sides of the Rhine at Wantzenau, north of Strasbourg: drawn in red and black ink, it shows the estate as a square, divided by wavy lines representing the river and subdivided into rectangles to mark the individual farms.[20] One English topographical diagram is a list of all the houses in Gloucester in 1455, arranged in two columns to correspond with the two sides of each street and with thumbnail sketches of principal buildings and other landmarks drawn in at the appropriate places.[21] Here the line between map and noncartographic diagram becomes difficult to draw, and we can see how the use of diagrams for quite different purposes could easily lead to the topographical diagram and the map.

Building plans too could well have led to the idea of drawing maps. We have abundant evidence that drawing plans on parchment or paper was a normal technique of the late medieval architect or builder—indeed from the fifteenth century we have substantial numbers of plans surviving. One notable collection is from Saint Stephen's, Vienna, but it includes plans of other buildings, even as far away as the Rhine, that were probably brought to Vienna by itinerant masons.[22] Earlier, in the thirteenth century, the notebook of Villard de Honnecourt confirms that architects were accustomed to think in terms of drawn outline ground plans.[23] From England only a single medieval building plan survives, showing part of Winchester College and dating from about 1390 (fig. 20.7), but we have references to plans in written

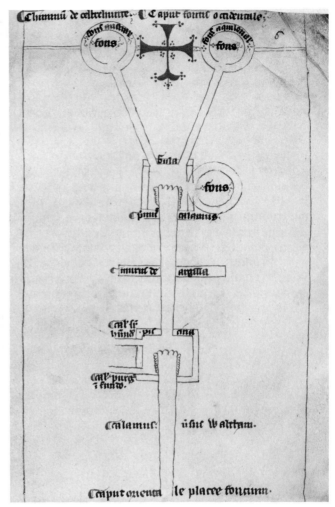

FIG. 20.5. PLAN OF A WATER SUPPLY. This schematic early thirteenth-century plan shows the source, at Wormley, Hertfordshire, of the water supply for Waltham Abbey. The cross flanked by the northern and southern springs has a splayed foot representing the east, perhaps the earliest direction pointer known on any map. It is possible that the style of this plan was inspired by the work of Matthew Paris, of nearby Saint Albans Abbey, who drew many schematic diagrams to illustrate various philosophical relationships at the time.
Size of the original: 21.8 × 14.7 cm. By permission of the British Library, London (Harl. MS. 391, fol. 6r).

17. London, British Library, Harl. MS. 391, fol. 6r; P. D. A. Harvey, "Wormley, Hertfordshire, 1220 × 1230," in Skelton and Harvey, *Local Maps and Plans*, 59–70 (note 3); Richard Vaughan, *Matthew Paris* (Cambridge: Cambridge University Press, 1958), 254–55, 257–58, pls. XX, XXIb.

18. London, British Library, Harl. MS. 658, fol. 37v; Reinhold Röhricht, "Karten und Pläne zur Palästinakunde aus dem 7. bis 16. Jahrhundert, III," *Zeitschrift des Deutschen Palästina-Vereins* 14 (1891): 137–141, esp. 140–41, pl. 5.

19. Lille, Archives Départementales du Nord, B 1388/1282 bis; M. K. Elisabeth Gottschalk, *Historische geografie van Westelijk Zeeuws-Vlaanderen*, 2 vols. (Assen: Van Gorcum, 1955–58), vol. 1, *Tot de St-Elizabethsvloed van 1404*, 148–49; Harvey, *Topographical Maps*, 89 (note 1).

20. Strasbourg, Archives Départementales du Bas-Rhin, G 4227 (8);

Franz Grenacher, "Current Knowledge of Alsatian Cartography," *Imago Mundi* 18 (1964): 60–61.

21. Gloucester, Gloucestershire Records Office, GDR 1311; W. H. Stevenson, ed., and Robert Cole, comp., *Rental of All the Houses in Gloucester A.D. 1455* (Gloucester: Bellows, 1890); Harvey, *Topographical Maps*, 90–91 (note 1).

22. Hans Koepf, *Die gotischen Planrisse der Wiener Sammlungen* (Vienna: Böhlau, 1969).

23. Paris, Bibliothèque Nationale, fr. 19093; H. R. Hahnloser, ed., *Villard de Honnecourt: Kritische Gesamtausgabe des Bauhüttenbuches*, 2d ed. (Graz: Akademische Druck- und Verlagsanstalt, 1972).

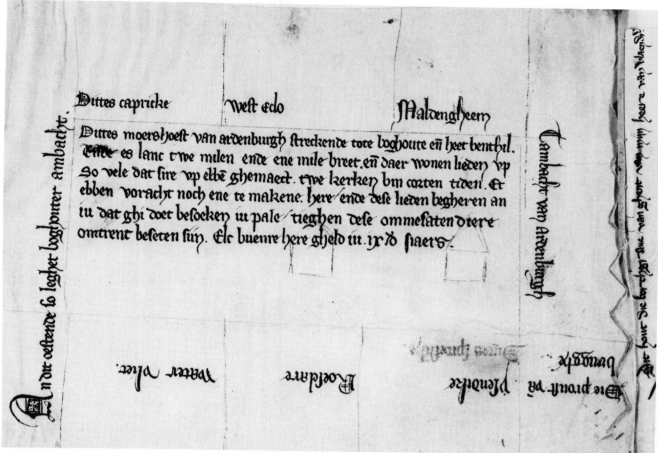

FIG. 20.6. MAP OF AN AREA NEAR SLUIS, ZEELAND. The oldest extant map from the Low Countries, dated 1307, uses place-names and notes arranged geographically. Two gable ends, representing churches, are faintly visible.

By permission of the Archives Départementales du Nord, Lille (B 1388/1282 bis).

building contracts from 1380 onward.[24] The modest increase in mapmaking that we find in England from the mid-fourteenth century may well reflect, if not the introduction of drawing plans as a technique of the builder's craft, at least a growing custom of showing these plans to their clients so that more people became familiar with this sort of representation. Whether consciously or not, building plans may well have inspired the mapping of small areas in the Middle Ages.

It is natural to ask whether these maps of small areas owed anything to other types of medieval map—to the world maps, to portolan charts, or, in the fifteenth century, to the reintroduction of Ptolemy's maps to the West. Nearly all the world maps of every sort, from the simplest to the most complex, belong to the diagram-drawing traditions of the Middle Ages, just like many of the maps of smaller areas. A few of the most elaborate represent Jerusalem at the center by a plan recognizably in the tradition of local maps of the city.[25] But beyond this there seems to be no connection between the two types of map; the world maps were so different in origin and concept that it is difficult to see what influence they can have had on the mapping of smaller areas. On the other hand, personal links suggest that at least some medieval draftsmen did not see these two types of map as entirely unconnected. In the thirteenth century Matthew Paris of Saint Albans drew a world map and maps

24. Winchester, Winchester College Muniments, 22820, inside front and back covers; John H. Harvey, "Winchester, Hampshire, *circa* 1390," in Skelton and Harvey, *Local Maps and Plans*, 141–46 (note 3); Louis Francis Salzman, *Building in England down to 1540* (Oxford: Clarendon Press, 1952), 14–22.

25. Even if only in its circular battlemented walls, as on the Ebstorf map.

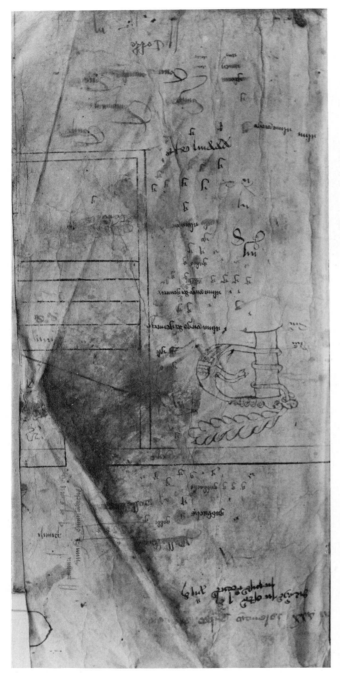
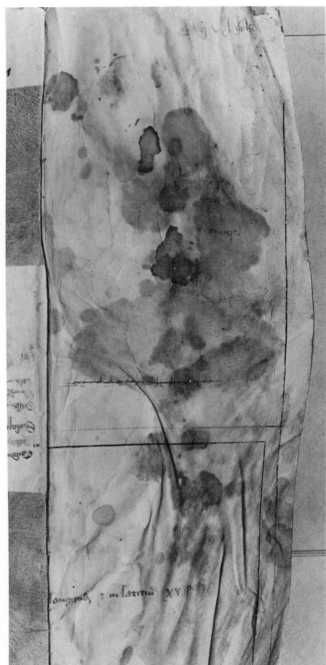

FIG. 20.7. PLAN FROM WINCHESTER COLLEGE. Although we have references to plans in building contracts from 1380 onward, this plan of about 1390 seems to be the sole survivor of the English tradition.

Size of the original: 31.2 × 26.0 cm. By permission of the Warden and Fellows of Winchester College (Winchester College Muniments, 22820, inside front and back covers).

of Britain, the Holy Land, and the route from England to southern Italy and, as we have seen, may have inspired the production of an English local map.[26] In the early fourteenth century Pietro Vesconte, who came from Genoa but worked in Venice, not only drew portolan charts but also provided a world map, a map of the Holy Land, and plans of Acre and Jerusalem for Marino Sanudo's *Liber secretorum fidelium crucis*; a little later the similar collection of maps illustrating Paolino Veneto's *Chronologia magna* included a world map, a portolan chart, maps of Italy and the Holy Land, a regional map of the lower Po, and plans of towns in both Italy and Palestine.[27] Certainly the *isolarii* (island books) of the fifteenth century suggest a clear link between the portolan charts and the mapping of small areas, but an earlier one can be found in the maps of peninsular Italy from the early fourteenth century onward, though the exact relationship has still to be demonstrated in both cases. In Italy at least, the portolan charts probably served as a model or inspiration for other maps of smaller areas; some of the plans of cities and the northern Italian regional maps may well have owed something to them, even though there was no direct borrowing of information. Outside Italy there seems to be no trace of a connection between portolan charts and the mapping of small areas, and it is unlikely that they helped to disseminate even the idea of drawing maps elsewhere in Europe. In the mid-fifteenth century English local maps suddenly begin to show a decided preference for northern orientation; it is conceivable that this shows the influence of Ptolemy's maps, but a more likely explanation would be the growing use of the land compass with north-pointing needle.[28] One further point of contact between local maps and maps of large areas is late fifteenth-century Germany. Here we see both types of map in the work of Erhard Etzlaub of Nuremberg, physician and instrument maker and author not only of general maps of central Europe from the Baltic to Rome, but also of a regional map of the area around Nuremberg and probably also of local estate maps (though none has survived).[29] Earlier there may have been a connection between the scale plan of Vienna and Bratislava that was drawn about 1422 (fig. 20.8) and geographical work at the monastery of Klosterneuburg that may have included mapmaking;[30] probably, however, the plan derived from the city plans that were being drawn in Italy.

All this might seem to point to strong connections between the maps of small areas and other types of medieval map and even to suggest that local and regional mapping in the Middle Ages was an offshoot of small-scale geographic mapping. This would be misleading. The clearest connections between the two sorts of map are in the work of a few outstanding individuals: Matthew Paris, Pietro Vesconte, Erhard Etzlaub. The tra-

ditions and concepts of geographic maps underlie only a few of the maps of small areas, and it is neither arbitrary nor artificial to draw a clear distinction between the two. Most of the maps of small areas took their inspiration from other sources; we may reasonably suppose that among these were diagrams and building plans, but others may still come to light.

MAPS OF PALESTINE AND ITS CITIES

Many medieval maps of small areas seem to have been individual productions, drawn for a particular purpose from first principles, owing nothing to precedent or example. We can, however, distinguish several clear traditions of mapmaking. Of these, maps of the Holy Land and of Jerusalem were the earliest to appear. Indeed, as we have seen, they may well owe something to classical models, and we have an early precursor of the medieval tradition in the plans of Jerusalem and the holy places that accompany the account of Arculf's pilgrimage in 670. However, a regular sequence begins only in the twelfth century after the capture of Jerusalem in 1099 and the establishment of the Crusader states. One of the earliest plans of the city to survive from this period, dating from the 1140s, is also one of the most impressive, for it is clearly based on direct knowledge. It shows the city walls in rhomboid outline, naming the gates and some of the towers and other features; inside it shows a few of the main streets, but only the principal buildings and churches are marked, so that much of the area is simply left blank. The walls, buildings, and hills are all shown pictorially, in elevation.[31] Another outstanding plan of Jerusalem, with the irregular outline of the medieval walls drawn fairly accurately and showing more

26. Vaughan, *Matthew Paris*, 235–50, pls. XII–XVII (note 17).

27. Bernhard Degenhart and Annegrit Schmitt, "Marino Sanudo und Paolino Veneto," *Römisches Jahrbuch für Kunstgeschichte* 14 (1973): 1–137, esp. 60–87, 105–30.

28. P. D. A. Harvey's Introduction in Skelton and Harvey, *Local Maps and Plans*, 36–37 (note 3).

29. Fritz Schnelbögl, "Life and Work of the Nuremberg Cartographer Erhard Etzlaub (†1532)," *Imago Mundi* 20 (1966): 11–26.

30. Vienna, Historisches Museum, I.N.31.018; S. Wellisch, "Der älteste Plan von Wien," *Zeitschrift des Oesterreichischen Ingenieur- und Architekten-Vereines* 50 (1898): 757–61; Max Kratochwill, "Zur Frage der Echtheit des 'Albertinischen Planes' von Wien," *Jahrbuch des Vereins für Geschichte der Stadt Wien* 29 (1973): 7–36; Harvey, *Topographical Maps*, 80–81 (note 1); Dana Bennett Durand, *The Vienna-Klosterneuburg Map Corpus of the Fifteenth Century: A Study in the Transition from Medieval to Modern Science* (Leiden: E. J. Brill, 1952); Ernst Bernleithner, "Die Klosterneuburger Fridericuskarte von etwa 1421," in *Kartengeschichte und Kartenbearbeitung*, ed. Karl-Heinz Meine (Bad Godesberg: Kirschbaum Verlag, 1968), 41–44; Fritz Bönisch, "Bemerkungen zu den Wien-Klosterneuburg-Karten des 15. Jahrhunderts," in *Kartengeschichte und Kartenbearbeitung* (above), 45–48.

31. Cambrai, Bibliothèque Municipale, MS. 466, fol. 1r; Ludwig

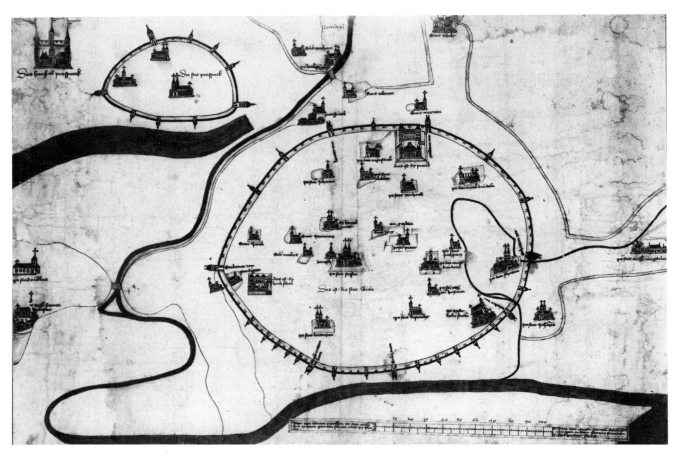

FIG. 20.8. PLAN OF VIENNA AND BRATISLAVA. The earliest European local map to be explicitly drawn to scale, the "Albertinischer plan," as it is called, is a mid-fifteenth-century copy of a 1421–22 original. The graphic scale (bottom right) is graduated in paces. Bratislava is shown in the upper left, with its castle beside it in a style reminiscent of Italian city plans.

Size of the original: 39.7 × 57.6 cm. By permission of the Historisches Museum der Stadt Wien (I.N. 31.018).

streets but fewer buildings than the plan of the 1140s, must date from before the Crusaders' final loss of the city in 1244 but is known only in early fourteenth-century copies in the works of Marino Sanudo and Paolino Veneto (it is accompanied there by a similarly impressive plan of Acre and, in Paolino Veneto's work, by one of Antioch too).[32] But most medieval maps of Jerusalem—and many dozens survive—give it a circular form and though likewise showing its walls and a selection of its chief monuments are mostly much more stylized and diagrammatic in appearance (fig. 20.9).[33] The Holy Land had long been lost to the Christians by the time more detailed and realistic plans of Italian towns were being drawn in the form of bird's-eye views, but we have one such view of Jerusalem, owing nothing to earlier plans, in the account of Bernard von Breydenbach's jour-

ney to the Holy Land, published in l486; it was drawn by Erhard Reuwich of Utrecht, who accompanied him.[34]

H. Heydenreich, "Ein Jerusalem-Plan aus der Zeit der Kreuzfahrer," in *Miscellanea pro arte*, ed. Joseph Hoster and Peter Bloch (Cologne: Freunde des Schnütgen-Museums, 1965), 83–90, pls. LXII–LXV; Harvey, *Topographical Maps*, 70–71 (note 1).

32. Reinhold Röhricht, "Marino Sanudo sen. als Kartograph Palästinas," *Zeitschrift des Deutschen Palästina-Vereins* 21 (1898): 84–126, pls. 2–11; Degenhart and Schmitt, "Sanudo und Veneto," 78–80, 105, 120–22 (note 27).

33. For example, those reproduced in Reinhold Röhricht, "Karten und Pläne zur Palästinakunde aus dem 7. bis 16. Jahrhundert, IV," *Zeitschrift des Deutschen Palästina-Vereins* 15 (1892): 34–39, and esp. pls. 1–5.

34. Bernard von Breydenbach, *Peregrinatio in Terram Sanctam* (Mainz: Reuwich, 1486); Reinhold Röhricht, "Die Palästinakarte Bernhard von Breitenbach's," *Zeitschrift des Deutschen Palästina-Ver-*

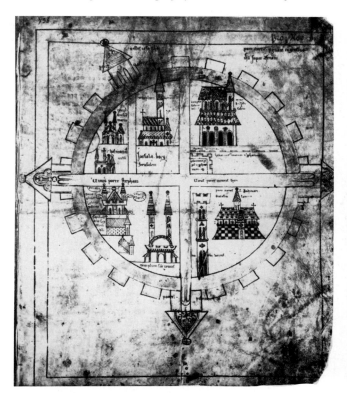

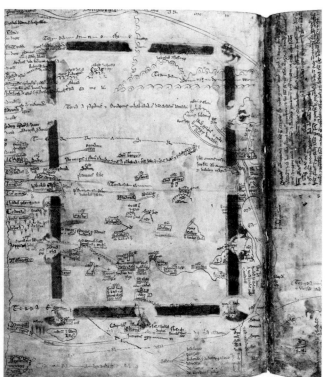

FIG. 20.9. JERUSALEM. This stylized map of Jerusalem—a fourteenth-century copy of a ca. 1180 original—is typical of the many that survive from the Middle Ages: within the circular diagrammatic wall are shown two main thoroughfares in the form of a cross and a selection of landmarks.
Size of the original: 26.2 × 21.7 cm. By permission of the Arnamagnæan Commission, Copenhagen (MS. 736 I, 4to, fol. 2r).

FIG. 20.10. PALESTINE. Unlike most general maps of the Holy Land, which are oriented to the east, this thirteenth-century example by Matthew Paris has north at the top. Along the coast are measurements in days of travel or leagues.
Size of the original: 47.5 × 35.0 cm. By permission of the President and Fellows, Corpus Christi College, Oxford (MS. 2, fol. 2v).

The view of Jerusalem in Breydenbach's work is set, though vastly out of scale, in a general map of the Holy Land. One thirteenth-century diagram map of Jerusalem, showing other places from the Mediterranean to the river Jordan, has already been mentioned. Of the two maps of the Holy Land by Matthew Paris one is dominated by a plan of Acre (city walls with principal buildings), with the few other marked places shown conventionally and to a quite different scale.[35] But these are exceptional productions: the plans of towns in Palestine and maps of the Holy Land belonged to quite separate medieval traditions. Among the early general maps of Palestine the other map by Matthew Paris is of particular interest, though we have only what seems to be a draft (fig. 20.10); most maps of the Holy Land had east at the top, the direction of the pilgrim's approach by sea, but this one is oriented to the north, and along the coast it marks distances between towns as so many day's journeys or leagues.[36] However, the most detailed medieval map of the Holy Land comes from Italy and is known

in several versions. The first exists only in a late thirteenth-century copy, at Florence;[37] of the second we have some nine copies in the early fourteenth-century works of Marino Sanudo and Paolino Veneto, and it may well have originated in the workshop of Pietro Vesconte.[38] This second version marks many more places

eins 24 (1901): 129–35; Ruthardt Oehme, "Die Palästinakarte aus Bernhard von Breitenbachs Reise in das Heilige Land, 1486," *Beiheft zum Zentralblatt für Bibliothekswesen* 75 (1950): 70–83; Harvey, *Topographical Maps*, 82–83 (note 1).

35. For example, London, British Library, Royal MS. 14.C.vii, fols. 4v–5r (three other versions survive); Charles Raymond Beazley, "New Light on Some Mediæval Maps IV," *Geographical Journal* 16 (1900): 319–29, esp. 326; Vaughan, *Matthew Paris*, 241, 244–45, pl. XVI (note 17); Harvey, *Topographical Maps*, 56–57 (note 1).

36. Oxford, Corpus Christi College, MS. 2, fol. 2v; Vaughan, *Matthew Paris*, 245–47, pl. XVII (note 17).

37. Florence, Archivio di Stato; Reinhold Röhricht, "Karten und Pläne zur Palästinakunde aus dem 7. bis 16. Jahrhundert, I," *Zeitschrift des Deutschen Palästina-Vereins* 14 (1891): 8–11, pl. 1.

38. Röhricht, "Marino Sanudo," 84–126, pls. 2–11 (note 32); Degenhart and Schmitt, "Sanudo und Veneto," 76–78, 105, 116–19 (note

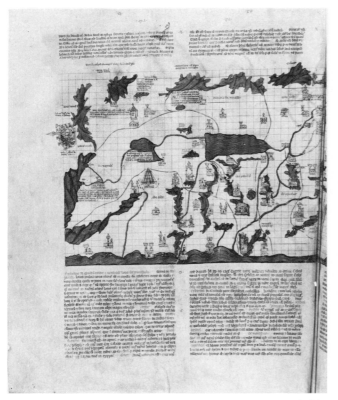

FIG. 20.11. PALESTINE BY VESCONTE. There are nine versions of this map found in the fourteenth-century works by Marino Sanudo and Fra Paolino Veneto, probably illustrated by Pietro Vesconte. One outstanding feature of this version is the use of an equidistant grid underpinning the whole map.

Size of the original: 51 × 81 cm. Photograph from the Bibliothèque Nationale, Paris (MS. Lat. 4939, fols. 10v–11).

than the first and (a feature of great interest, to which we shall return later) sets the whole map on a grid of equidistant lines (fig. 20.11). At the same time both versions are closely related: their patterns of hills and rivers are very similar, and the Twelve Tribes are named on both. Also related to them are the large maps that accompany William Wey's account of his journeys to the Holy Land in 1457–58 and 1462, and some other fifteenth-century maps.[39] Some inscriptions from these maps are echoed on the map illustrating Breydenbach's book of 1486, though this seems to be a quite new compilation by Reuwich, mostly independent of earlier models.[40] Certainly there was a clear tradition, perhaps more than one tradition, of medieval maps of the Holy Land, although their relationships to each other and to other sources have not yet been fully worked out.

Maps of Italian Cities

Besides these traditions of maps and plans of Palestine and its cities, which seem to have been known through-

out western Europe, there were several distinct traditions of small-area maps that were confined to Italy. The earliest was of plans of cities, and it is quite possible that the plans of Jerusalem served as models. Most fall within a single genre—the bird's-eye view. At its start are two of the three Italian town plans that survive from before the thirteenth century; they show tenth-century Verona and twelfth-century Rome as bird's-eye views of city walls—the city ideogram of ancient origin—with, crowded inside and around them, a recognizable selec-

27). For a description of a larger version of the Vesconte map, see Frederick B. Adams, *Seventh Annual Report to the Fellows of the Pierpont Morgan Library* (New York: Pierpont Morgan Library, 1957), 14–17.

39. Röhricht, "Marino Sanudo," 101–4, pl. 6 (note 32); idem, "Die Palästinakarte des William Wey," *Zeitschrift des Deutschen Palästina-Vereins* 27 (1904): 188–93; Adams, *Annual Report*, 16–17 (note 38); Sotheby and Company Sale Catalogue, 11 July 1978, lot 34 (in a manuscript of Gabriele Capodilista, *Itinerario di Terra Santa*, ca. 1475; I am grateful to M. Henri Schiller for telling me of this).

40. Röhricht, "Die Palästinakarte Bernhard von Breitenbach's," 131–32 (note 34).

FIG. 20.12. BIRD'S-EYE VIEW OF ROME. This twelfth-century bird's-eye view—the beginning of a tradition of maps of ancient Rome—shows the stylized city walls and a selection of seven principal monuments contained within them.
Size of the original: 20.8 × 13.9 cm. By permission of the Biblioteca Ambrosiana, Milan (MS. C246 Inf., fol. 3v).

tion of each city's principal monuments (seven at Rome, over twenty at Verona; fig. 20.12).[41] At the end of the sequence, in the late fifteenth century, we have realistic bird's-eye views showing entire cities in full and fairly accurate detail: of Rome and Florence by Francesco Rosselli in the 1480s (apart from one section of the Florence view, we know both only from derivative versions; those he produced of Pisa and Constantinople are entirely lost) and, an outstanding masterpiece, of Venice by Jacopo de' Barbari in 1500.[42] Between these extremes comes a range of representations, varying widely in artistry and accuracy, having in common only that they show cities as if from a height with at least a few of the most distinctive buildings visible within the walls. Among them are the view of Rome on a seal of the emperor Ludwig the Bavarian that was first used in 1328[43] and another, very much more elaborate, which appears at about the same time in Paolino Veneto's *Chronologia magna* but which is a far cruder production than the plans of Jerusalem and Acre in the same work.[44] Among them too are the first plan of Florence, a fresco of the 1350s in

the Loggia del Bigallo,[45] plans of Rome based on the account of its antiquities written by Flavio Biondo in 1444–46,[46] and a whole series of plans included in the various manuscripts of Ptolemy's *Geography* that were produced in the late fifteenth century in Pietro del Massaio's workshop at Florence: Rome (plate 33), Florence, Milan, Volterra, Jerusalem, Damascus, Constantinople, Adrianople, Alexandria, and Cairo.[47] Again, some maps of larger areas include miniature versions of town plans in this style; Genoa and Venice, for instance, appear on some portolan charts.[48] A turning point came in the fourteenth century, when the new realism in Italian painting came to be applied to topographical representations; a view in 1382 by Giusto de' Menabuoi, showing Padua as though from a small height, may mark its introduction to the tradition of city plans,[49] but a hundred years later plans were still being drawn in the

41. Verona, Biblioteca Capitolare, MS. CXIV, fol. 187 (eighteenth-century copy of the lost original); Vittorio Cavallari, Piero Gazzola, and Antonio Scolari, eds., *Verona e il suo territorio*, 2 vols. (Verona: Istituto per gli Studi Storici Veronesi, 1964), 2:39–42, 232–33, 481–85, pl. opp. 192; Vittorio Galliazzo, "Il ponte della pietra di Verona," *Atti e Memorie della Accademia di Agricoltura, Scienze e Lettere di Verona* 146 (1968–69): 533–70, esp. 564. Milan, Biblioteca Ambrosiana, MS. C246 Inf., fol. 3v; Annalina Levi and Mario Levi, "The Medieval Map of Rome in the Ambrosian Library's Manuscript of Solinus," *Proceedings of the American Philosophical Society* 118 (1974): 567–94.

42. L. D. Ettlinger, "A Fifteenth-Century View of Florence," *Burlington Magazine* 94 (1952): 160–67; Marcel Destombes, "A Panorama of the Sack of Rome by Pieter Bruegel the Elder," *Imago Mundi* 14 (1959): 64–73; Roberto Weiss, *The Renaissance Discovery of Classical Antiquity* (Oxford: Blackwell, 1969), 92–93; Juergen Schulz, "Jacopo de' Barbari's View of Venice: Map Making, City Views, and Moralized Geography before the Year 1500," *Art Bulletin* 60 (1978): 425–74.

43. Wilhelm Erben, *Rombilder auf kaiserlichen und päpstlichen Siegeln des Mittelalters*, Veröffentlichungen des Historischen Seminars der Universität Graz, 7 (Graz: Leuschner und Lubensky, 1931), 57–83, pl. III; Harvey, *Topographical Maps*, 74 (note 1).

44. G. B. de Rossi, *Piante iconografiche e prospettiche di Roma anteriori al secolo XVI* (Rome: Salviucci, 1879), 81–83, 139–41, pl. I; F. Ehrle and H. Egger, *Piante e vedute di Roma e del Vaticano dal 1300 al 1676*, illus. Amato Pietro Frutaz (Rome: Biblioteca Apostolica Vaticana, 1956), 9, pls. I, II; Degenhart and Schmitt, "Sanudo und Veneto," 86–87, 105, 125–27 (note 27); Harvey, *Topographical Maps*, 72–73 (note 1).

45. Rodolfo Ciullini, "Firenze nelle antiche rappresentazioni cartografiche," *Firenze* 2 (1933): 33–79, esp. 35; Schulz, "Moralized Geography," 462–63 (note 42).

46. Gustina Scaglia, "The Origin of an Archaeological Plan of Rome by Alessandro Strozzi," *Journal of the Warburg and Courtauld Institutes* 27 (1964): 137–63; the view that Biondi himself drew the prototype is, however, questioned by Weiss, *Renaissance Discovery*, 92 (note 42).

47. De Rossi, *Piante di Roma*, 90–92, 144–46, pls. II, III (note 44); Scaglia, "Origin of an Archaeological Plan," 137–40 (note 46).

48. Paolo Revelli, "Figurazioni di Genova ai tempi di Colombo," *Bollettino del Civico Istituto Colombiano* 3 (1955): 14–23, esp. 21–22.

49. Schulz, "Moralized Geography," 462–63 (note 42).

older stylized form. Of all the town plans in this tradition, we have far more of Rome than of any other city; but as we have seen, Italian draftsmen produced plans of this sort for quite a number of other cities in Italy and around the eastern Mediterranean.

But not all town plans from medieval Italy conformed to this pattern of conventional or realistic bird's-eye views. The third plan to survive from before the thirteenth century, of Venice, is quite different from those of Rome and Verona. We know it only in three fourteenth-century copies,[50] but as long ago as 1781 it was argued, from internal evidence, that its basic outline dates from the first half of the twelfth century and that the demonstrably later features that appear on it are copyists' additions.[51] This conclusion has been generally accepted, and it is interesting that a sixteenth-century note on one copy of the map says it was drawn up for the doge Ordelaffo Falier (1102–18). Cartographically it is a relatively advanced production; it marks churches (and some other buildings) by tiny sketches, but it sets these not on a supposed or actual bird's-eye view but on an outline plan of the city's major and minor waterways. This outline plan is startlingly accurate; its draftsman had clearly mastered the concept of consistent scale, and it raises the very interesting possibility that it derives from a measured survey. On the other hand it is not so accurate as to make this a certainty, and another possibility is that the outline was developed over a long period by gradually adjusting and correcting maps that were in official keeping.[52] That such a plan comes from Venice probably rules out a simple sequence of copying from scale plans of the classical Roman surveyors, for the city was not fully established until the seventh century; the measured survey, if any, was made in the Middle Ages. That maps of Italian cities might be drawn on the basis of measurement on the ground is suggested too by a letter written (250 years later, it should be noted) between 1377 and 1381 by Lapo di Castiglionchio, a lawyer at Florence, in which he describes how a young judge, Francesco da Barberino, had made a drawing of the whole city showing, among other things, "all the walls and their measurements" (*tutte le mura e la loro misura*).[53] Certainly the idea of a ground plan drawn to a uniform scale seems to underlie the way some of the fifteenth-century district maps from northern Italy show the cities on which they are centered; this is particularly clear in one of Verona (plate 34 and fig. 20.13).[54] In all these cases the measured survey need have consisted of no more than pacing out key distances on foot. Possibly we should see as part of this same Italian tradition certain plans of cities in the Holy Land (notably those of Jerusalem and Acre in the works of Marino Sanudo and Paolino Veneto) and also the scale plan of Vienna and Bratislava in about 1422. What gives all these city plans

such interest is not the techniques that were used; they can only have been rudimentary. Rather, it is their apparent acceptance of the basic cartographic concept of a map drawn to scale. In these plans we may have, as we have nowhere else in medieval Europe, a long tradition of large-scale maps. The earliest Italian town plan drawn specifically to scale is one of Imola by Leonardo da Vinci in 1502–3;[55] it in no way belittles his achievement to suggest that he may have been simply improving on an existing Italian tradition of outline city plans.

NORTHERN ITALIAN DISTRICT MAPS

Maps of particular districts in northern Italy, from Piedmont to Venice, form a tradition that we can trace back to the thirteenth century (they are listed below, appendix 20.1). A map of the area around Alba and Asti in a manuscript of 1291 is very badly damaged, but we have an intact copy from the fourteenth century; it marks over 160 settlements.[56] We have contemporary references to fourteenth-century maps of the area around Padua and of the whole of Lombardy,[57] but the only other to survive is of Lake Garda, showing in some detail the lakeside villages and fortifications.[58] The other surviving examples of this tradition of district maps—eleven in all—date from the fifteenth century (listed in appendix 20.1). One dated 1440 and another of about the same time both cover the whole of Lombardy (fig. 20.14),[59] while a third, dating from the end of the century, is a

50. Juergen Schulz, "The Printed Plans and Panoramic Views of Venice (1486–1797)," *Saggi e Memorie di Storia dell'Arte* 7 (1970): 9–182, esp. 16–17; Degenhart and Schmitt, "Sanudo und Veneto," 83, 86, 105, 124 (note 27); Harvey, *Topographical Maps*, 76–79 (note 1).

51. Tommaso Temanza, *Antica pianta dell'inclita città di Venezia delineata circa la metà del XII. secolo* (Venice: Palese, 1781).

52. Schulz, "Moralized Geography," 440–41, 445 (note 42).

53. Attilio Mori, "Firenze nelle sue rappresentazioni cartografiche," *Atti della Società Colombaria di Firenze* (1912): 25–42, esp. 30–31; Harvey, *Topographical Maps*, 78 (note 1).

54. Venice, Archivio di Stato; Harvey, *Topographical Maps*, pl. IV (note 1).

55. Windsor, Royal Library, MS. 12284; John A. Pinto, "Origins and Development of the Ichnographic City Plan," *Journal of the Society of Architectural Historians* 35, no. 1 (1976): 35–50, esp. 38–42, fig. 1; Harvey, *Topographical Maps*, 155 (note 1).

56. Roberto Almagià, "Un'antica carta del territorio di Asti," *Rivista Geografica Italiana* 58 (1951): 43–44.

57. Almagià, "Antica carta del territorio di Asti," 44 (note 56).

58. Verona, Biblioteca Civica; Roberto Almagià, *Monumenta Italiae cartographica* (Florence: Istituto Geografico Militare, 1929), 5, pl. VII; Harvey, *Topographical Maps*, 59 (note 1).

59. Paris, Bibliothèque Nationale, Rés. Ge.C.4990; Treviso, Museo Comunale; Almagià, *Monumenta*, 9, pl. VIII (note 58).

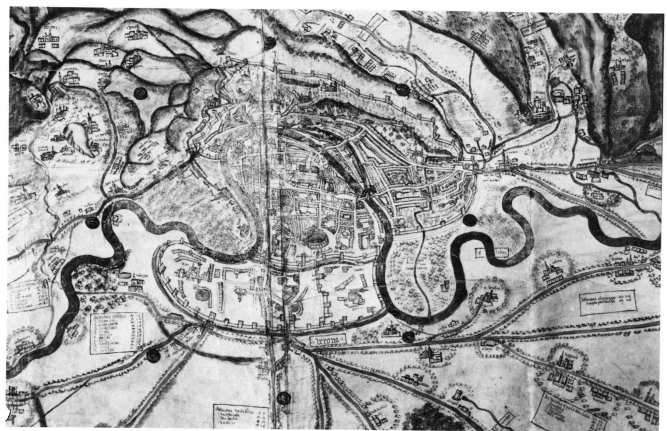

FIG. 20.13. VERONA. A detail of plate 34, it appears that a uniform scale was applied to the representation of the town itself, but not to the surrounding countryside.

Size of the original detail: 110 × 160 cm. By permission of the Archivio di Stato, Venice.

stylized map of the Venetian mainland territories, showing simply the main rivers, the lines of mountains, and towns with their fortifications, mostly as more or less elaborate city ideograms.[60] All the rest have a particular city at the center, showing it in its region; thus one of Brescia in 1469–70 includes Lakes Garda and Iseo and names some 280 places.[61] They are elaborate colored productions, showing a great deal of local topographic detail: woods, mountains, bridges, mills, and so on. Three of these fifteenth-century maps are signed by the author; one, of the area around Padua, is by Francesco Squarcione, a known Paduan painter (fig. 20.15).[62] On nearly all of them fortifications and roads are prominent, with distances between places marked either along the roads themselves or in tables beside the central city. Some show other features of military significance: the "Fossa Bergamasca" that marked the boundary between the lands of Bergamo and Milan, or a picture of an event in the war of 1437–41 between Milan and Venice, when

the Venetians carried six ships overland from the river Adige to Lake Garda.[63]

These northern Italian maps form a single tradition, but not in the sense that they were directly based on one another. Some minor links can be seen; there are, for instance, similarities between the two Lombardy maps

60. Istanbul, Topkapi Sarayi Muzesi; Rodolfo Gallo, "A Fifteenth Century Military Map of the Venetian Territory of *Terraferma*," *Imago Mundi* 12 (1955): 55–57; Harvey, *Topographical Maps*, 60–61 (note 1).

61. Modena, Biblioteca Estense; Mario Baratta, "Sopra un'antica carta del territorio bresciano," *Bollettino della Reale Società Geografica*, 5th ser., 2 (1913): 514–26, 1025–31, 1092.

62. The others are of Lombardy, 1440, by "Ioanes Pesato" and of the Padua area, 1449, by Annibale de Maggi of Bassano; Almagià, *Monumenta*, 9, 12 (note 58).

63. Almagià, *Monumenta*, 9 (note 58); Roberto Almagià, "Un'antica carta topografica del territorio veronese," *Rendiconti della Reale Accademia Nazionale dei Lincei: Classe di Scienze Morali, Storiche e Filologiche* 32 (1923): 63–83, esp. 74–75.

FIG. 20.14. MAP OF LOMBARDY. This is one of two maps, both produced about 1440, that cover the whole of Lombardy. Unlike most of the extant district maps of this period from northern Italy, neither has a particular city as its focus.
Size of the original: 38.5 × 55 cm. Photograph from the Bibliothèque Nationale, Paris (Rés. Ge. C 4990).

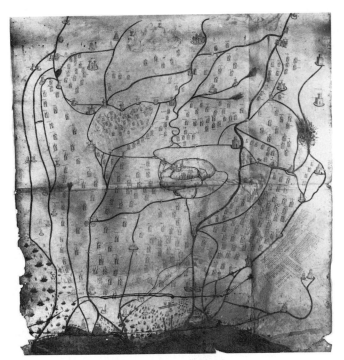

FIG. 20.15. THE DISTRICT OF PADUA. One of three fifteenth-century Italian district maps to have been signed, this was produced by Francesco Squarcione, a well-known Paduan painter.
Size of the original: 115 × 100 cm. By permission of the Museo Civico, Padua.

of about 1440 both in general style and in certain details including some (not all) of the distance figures. But it is only in the two early maps of the Alba and Asti area that we find one map copied from another. The tradition consists rather in the idea of maps of this kind, the idea that maps could serve a practical purpose in administration and government. The peculiarity (and, in the history of cartography, the importance) of this tradition is shown by the fact that we have no similar maps from other parts of late medieval Italy. The emphasis on fortifications and on other features of military significance suggests that it was the value of these maps in warfare that was particularly understood, and it seems that by the mid-fifteenth century this was especially appreciated at Venice. There is a clear Venetian emphasis in the later maps; they relate to Venetian wars or Venetian territories—apart from Parma all the cities on which these maps are centered were under Venetian rule. In 1460 the Venetian Council of Ten ordered local governors to have maps made of the areas under their control and to send them to Venice; none of the city-centered maps has been dated to just this time, but historians have associated several of them with this decree.[64] It seems that for official purposes Venice developed an existing tradition of mapmaking in northern Italy, thus becoming the only state in fifteenth-century Europe to make regular use of maps in the work of government.

Another map that might be thought to belong to the same northern Italian tradition is one from the early fourteenth century showing the Po delta and the lands behind; it is found in two manuscripts of Paolino Veneto's *Chronologia magna*, and in one it is entitled "Mapa Lombardie et Ferrarie."[65] In fact it belongs to a quite separate tradition of medieval Italian cartography—maps based on coastal outlines deriving from portolan charts but with often a great deal of added detail both on the coast and inland. One other regional map from medieval Italy belongs to the same tradition. This is of Tuscany; its earliest version, by Pietro del Massaio, is in a 1456 manuscript of Ptolemy's *Geography*, and others of 1469 and 1472 derive from this.[66] Otherwise the tradition consists of maps of the whole of Italy. The

64. With some uncertainty, however, over which are the relevant maps. Cf. Almagià, *Monumenta*, 11, 12 (note 58); Gallo, "Fifteenth Century Military Map," 55 (note 60); Roberto Almagià, *Scritti geografici* (Rome: Edizioni Cremonese, 1961), 613.

65. Venice, Biblioteca Marciana, Lat. Z 399, and Rome, Biblioteca Apostolica Vaticana, Vat. Lat. 1960; Almagià, *Monumenta*, 4–5 (note 58); Degenhart and Schmitt, "Sanudo und Veneto," 83–84, 105 (note 27).

66. Roberto Almagià, "Una carta della Toscana della metà del secolo XV," *Rivista Geografica Italiana* 28 (1921): 9–17; idem, *Monumenta*, 12, pl. XIII (note 58); Lina Genoviè, "La cartografia della Toscana," *L'Universo* 14 (1933): 779–85, esp. 780.

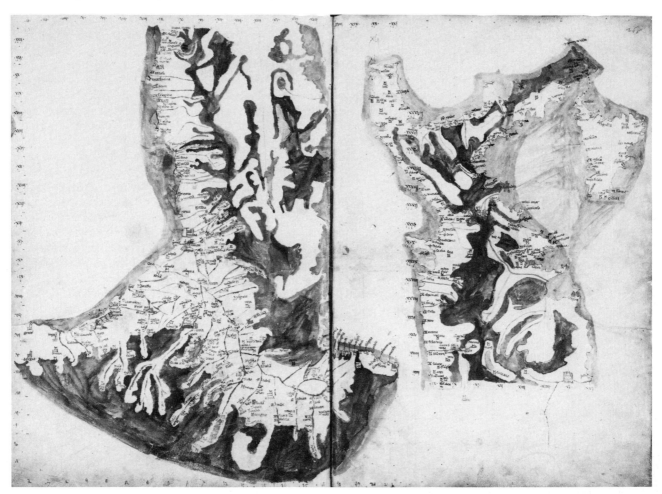

FIG. 20.16. MAP OF ITALY, CA. 1320. This is the larger and more detailed of two maps of Italy found in Paolino Veneto's work. The coastal information seems to have been drawn from the contemporary portolan charts, with added details of inland Italy. Oriented to the south, both maps use a notional grid.

Size of original: 34 × 74.5 cm. Photograph from the Biblioteca Apostolica Vaticana, Rome (Vat. Lat. 1960, fols. 267v–268r).

two earliest examples are both found in an early fourteenth-century manuscript of Paolino Veneto's work; one occupies a single page, the other shows the peninsula on a larger scale, divided between facing pages (fig. 20.16). The manuscript is thought to have been given by Paolino to King Robert of Naples, and it may be that the fifteenth-century writer Flavio Biondo had it in mind when he wrote of a map of Italy (*pictura Italiae*) drawn up by King Robert and his friend Francesco Petrarch.[67] Both maps are based on the same notional grid of equidistant lines (their positions are marked and numbered along the edge of each), and they are very alike in style and content, though the larger one gives more detail; it has been suggested even so that it is incomplete.[68] They show relief and rivers over the whole country, but settlements and names are given only along the coasts and in the northern Italian plain. This underlines their origin in nautical maps, but it suggests too that they drew on a northern Italian tradition of district maps that was already well established. We find some internal detail, notably the principal rivers, on some fourteenth-century portolan charts of the Italian coasts, but it is not until the fifteenth century that we find further detailed maps of the whole of Italy. Two seem to be unique productions, but five others form a single series of closely related maps from which further maps were indeed derived in

67. Rome, Biblioteca Apostolica Vaticana, Vat. Lat. 1960, fols. 266v, 267v, 268r; Almagià, *Monumenta*, 4–5 (note 58); Degenhart and Schmitt, "Sanudo und Veneto," 81–85, 105, 128–30 (note 27). Almagià considers it more likely that the map given to King Robert was a picture map drawn from classical Roman sources.

68. Almagià, *Monumenta*, 4–5 (note 58).

the early sixteenth century. Three of the five occur in copies of Ptolemy's *Geography*, one of them (in three versions) by Pietro del Massaio in the same manuscripts as the regional map of Tuscany.[69] All these maps are cartographically more advanced than any of the others discussed in this chapter; in concept as well as in origin they are related more closely to the portolan charts and to geographic maps than to the other medieval maps of small areas. That they come from Italy is significant, for Italy was by far the most map-conscious part of medieval Europe.

THE *ISOLARII*: BUONDELMONTI'S *LIBER INSULARUM ARCIPELAGI*

One distinctive cartographic phenomenon that has its roots in this period is the *isolario* (plural *isolarii*), or island book. As the name suggests, the *isolario* is essentially a collection of island materials, each island usually having its own map and associated text. Some *isolarii* also contain world or regional maps, showing the location of islands relative to each other and to the adjacent mainland. Some have extra illustrative material on individual islands—for example, perspective views, city and fortress plans, or drawings of inhabitants in local costume.

There are at least twenty different *isolarii* extant; the exact number depends on the precise definition used. The style, content, and accuracy of both the maps and the texts vary greatly from work to work. Some *isolarii* have the character of prototype atlases, others are more encyclopedic in nature, yet others have been put together by, and sometimes for, seamen and travelers, leading some authors to regard the maps from *isolarii* as sea charts. *Isolarii* exist dating from the early fifteenth century through the sixteenth and seventeenth centuries. Only one clearly belongs to the period before 1470. For this reason the main account of the *isolarii* will be found in volume 3. Problems of definition are there considered more fully; the main *isolarii* are listed, their contents are analyzed, and their possible relationships are discussed. This brief introductory account is restricted to a general description of the earliest *isolario* known to us: the *Liber insularum arcipelagi* by Cristoforo Buondelmonti, written about 1420.

The *Liber* was evidently a popular work in fifteenth-century Europe. Originally written by Buondelmonti and sent to his patron Cardinal Giordano Orsini in Rome, it was frequently copied and widely circulated. Many manuscript copies survive today, of varying dates and styles, and with some differences in content.[70] Buondelmonti also wrote a detailed account of Crete, the *Descriptio insule Crete*, and it is from this that his description of Crete in the *Liber* is derived.[71]

In the prologue to the *Liber*, Buondelmonti describes his work as: "an illustrated book of the islands of the Cyclades and of other islands scattered around them, together with [an account of] the actions that have taken place in them in ancient times [and] down to the present day." The seventy-nine islands and other localities included, and their sequence in the book, probably reflect the pattern of their author's travels over a period of several years. Written in debased Latin, the text is a disorderly mixture of fact, fiction, and fantasy, compiled from personal observation, hearsay, and a variety of historical and poetic sources whose authors are frequently named. In all the text there is no mention of a source or sources for the maps. If these are Buondelmonti's own work, it seems probable on stylistic grounds that the island outlines were derived at least in part from nautical charts carried on board the ships in which he traveled. Within the outlines, topographical details are shown in pictorial style: they may represent what was actually seen; they may be a visual interpretation of the information assembled in the text; or they may be a mixture of the two (figs. 20.17 and 20.18). Certainly there is an extremely close correlation between the details shown on individual maps and those described in their related texts.

When considering the content of the *Liber*, it is important to remember by whom and for whom it was compiled. Buondelmonti, "an ecclesiastic with a taste for antiquities,"[72] was traveling on behalf of his patron, a prince of the church and a scholar. His main purpose is said to have been to search for early Greek manuscripts.[73] Although this cannot be conclusively demonstrated, such an aim may have influenced the pattern of his travels within the Aegean, which included monastic retreats on remote rocky islets as well as the great urban repositories of Byzantine civilization. In his account of his travels, Buondelmonti is at pains to demonstrate his diligence, but his main aim is to interest and entertain his patron. The resulting descriptions of historical and natural wonders, interspersed with tales of storms at sea

---

69. Almagià, *Monumenta*, 7, 9–11 (note 58); Roberto Almagià, "Nota sulla cartografia dell'Italia nei secoli XV e XVI," *Atti della Accademia Nazionale dei Lincei: Rendiconti, Classe di Scienze Morali, Storiche e Filologiche*, 8th ser., 6 (1951): 3–8, esp. 4–5.

70. See discussion in volume 3 of the present *History*.

71. Cristoforo Buondelmonti, *"Descriptio insule Crete" et "Liber Insularum,"* cap. XI: Creta, ed. M.-A. van Spitael (Candia, Crete: Syllagos Politistikēs Anaptyxeōs Herakleiou, 1981).

72. Benedetto Bordone, *Libro . . . de tutte l'isole del mondo*, introduction by R. A. Skelton (Amsterdam: Theatrum Orbis Terrarum, 1966), V.

73. See, for example, A. E. Nordenskiöld in *Periplus: An Essay on the Early History of Charts and Sailing-Directions*, trans. Francis A. Bather (Stockholm: P. A. Norstedt, 1897), 59.

FIG. 20.17. CORFU BY CRISTOFORO BUONDELMONTI. Written in about 1420, Buondelmonti's *Liber insularum arcipelagi* is the only example of an extant *isolario* (island book) from before 1470. The book describes seventy-nine islands around the eastern Mediterranean and Aegean seas, illustrating the accounts with small vignettes. The style used to portray the coasts is derived from portolan charts.
Size of the original: 20.6 × 14 cm. Photograph from the Biblioteca Apostolica Vaticana, Rome (Rossiano 702, fol. 2r).

FIG. 20.18. COS BY CRISTOFORO BUONDELMONTI. A map of the island of Cos in the Aegean from the *Liber insularum arcipelagi*.
Size of the original: 25 × 14.6 cm. Photograph from the Biblioteca Apostolica Vaticana, Rome (Chigiano F.V.110, fol. 31v).

and dietary deficiencies and backed up by quotations from poets and historians, will be familiar to any reader of modern travelogues. The *Liber* is an early example of the genre, written with the interests of one particular armchair traveler in mind. As Buondelmonti puts it to Cardinal Orsini: "I have wished to send it to you . . . so that when you are tired you can, with this book, bring pleasure to your mind."

Although it was addressed to one man, the wider appeal of the *Liber* is evident from the relatively large number of manuscript copies of differing dates that have survived. There is also a manuscript of a translation into Greek.[74] The popularization of the work in its own time not only ensured its survival but led to its becoming a model for later *isolarii*, many of which are partially derived from it and exhibit close structural, textual, and

cartographic similarities. Widening horizons during the Renaissance resulted in the addition of much new material, but for some time this was little more than an accretion round the medieval core of Buondelmonti's *isolario*.

The question inevitably arises: Is Buondelmonti's the first *isolario* or merely the earliest example known to us of a tradition already well established by the early fifteenth century? The possibility, even the probability, of such a tradition must be acknowledged. Islands have a unique appeal that invites special treatment. More specifically, the islands of the Aegean and Adriatic seas had

74. Cristoforo Buondelmonti, *Description des îles de l'archipel*, ed. and trans. Emile Legrand, L'Ecole des Langues Orientales, ser. 4, 14 (Paris: Leroux, 1897).

been for many centuries the natural foci of a series of civilizations and the stepping stones of empires and kingdoms based on maritime trade. Periploi, portolani, and early portolan charts all bear witness to the detailed island knowledge that had been accumulated. Island shores were charted, island histories written: the data for an *isolario* as we know it had long been available. Only an assembly job was required, and the organization of the data into the characteristic *isolario* format, with its island-by-island, text-map sequence. Buondelmonti may have been the first to do this; alternatively, his *isolario* may have had its predecessors.

OTHER LOCAL MAPS

So far we have been looking at conscious traditions of mapmaking. The maps may have been copies or improvements of earlier maps or they may have been entirely original compilations, but in either case they were following well-established precedents. In looking at medieval maps of small areas outside these defined traditions, we must remember that not only were there normally no models for the map of any particular place, but there were no precedents for the idea of drawing maps at all.

Setting aside the vestiges of ancient cartography represented by Arculf's work and the Saint Gall plan, and setting aside too the traditions of maps from Palestine and Italy, it is only from England that we have maps of small areas earlier than the fourteenth century (fig. 20.19). The earliest English map is the plan, already mentioned, of Canterbury Cathedral with its attached priory, which probably dates from the 1150s (see fig. 20.4).[75] Strictly speaking it is a pair of plans; a continuation on a separate sheet shows the source of the water supply whose distribution around the cathedral precinct is one of the principal features of the main plan. It is an elaborate and detailed production, showing all buildings in elevation. The two thirteenth-century English plans are less impressive. One, already mentioned, is the plan of springs at Wormley (Hertfordshire) (see fig. 20.5); an intriguing detail is its inclusion of a direction pointer in the form of a cross with the foot splayed to mark the east, but this is probably evidence of ingenuity in diagram drawing rather than of the influence of any other sort of map. The other is another small, diagram-like plan showing the layout of pastures on Wildmore Fen in Lincolnshire.[76] All other local maps and plans from medieval England are later than 1350; in all there are thirty-two individual items or closely related groups (they are listed below, appendix 20.2).[77] They form a highly varied collection. The areas they cover are of every size and type: strips in arable fields, house plots in towns, rivers with mills and fisheries, whole tracts of country-

FIG. 20.19. PRINCIPAL PLACES IN ENGLAND ASSOCIATED WITH MEDIEVAL LOCAL MAPS.

side including towns and villages. There is no consistency in treatment. Some are in cartularies or other books, some are on separate sheets or rolls. Some are the merest rough sketches in ink, a few drawn lines with some added names, such as the plans of waterways at Cliffe (Kent) and of the village of Clenchwarton (Norfolk), which both date from the late fourteenth or early fifteenth century (fig. 20.20).[78] Others are substantial artistic works, carefully drawn and colored, such as the map of the Isle of Thanet (Kent) (plate 35) at the same period or the map of Dartmoor (Devon) about 1500.[79]

75. Cambridge, Trinity College, MS. R.17.1, fols. 284v–286r; William Urry, "Canterbury, Kent, *circa* 1153 × 1161," in Skelton and Harvey, *Local Maps and Plans*, 43–58 (note 3).

76. London, British Library, Harl. MS. 391, fol. 6r; P. D. A. Harvey, "Wormley, Hertfordshire, 1220 × 1230," in Skelton and Harvey, *Local Maps and Plans*, 59–70 (note 3); Loughlinstown (county Dublin), Library of Sir John Galvin, Kirkstead Psalter, fol. 4v; H. E. Hallam, "Wildmore Fen, Lincolnshire, 1224 × 1249," in Skelton and Harvey, *Local Maps and Plans*, 71–81 (note 3).

77. They are reproduced and fully discussed in Skelton and Harvey, *Local Maps and Plans* (note 3).

78. Canterbury, Archives of the Dean and Chapter, Charta Antiqua C.295; London, British Library, Egerton MS. 3137, fol. 1v; F. Hull, "Cliffe, Kent, Late 14th Century × 1408," 99–105, and Dorothy M. Owen, "Clenchwarton, Norfolk, Late 14th or Early 15th Century," 127–30, both in Skelton and Harvey, *Local Maps and Plans* (note 3).

79. Cambridge, Trinity Hall, MS. 1, fol. 42v; Exeter, Royal Albert Memorial Museum; F. Hull, "Isle of Thanet, Kent, Late 14th Century × 1414," 119–26, and J. V. Somers Cocks, "Dartmoor, Devonshire, Late 15th or Early 16th Century," 293–302, both in Skelton and Harvey, *Local Maps and Plans* (note 3).

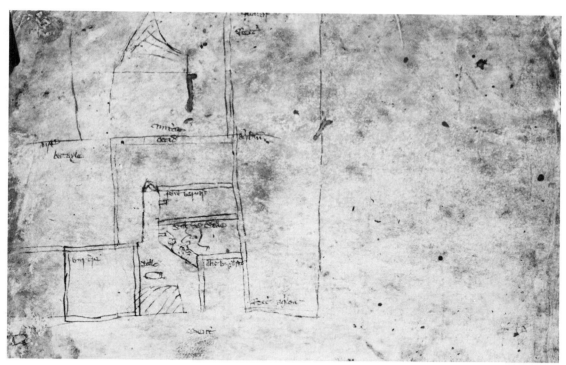

FIG. 20.20. PLAN OF CLENCHWARTON, NORFOLK. Produced ca. 1400, this map is an example of a simple sketch with few names that forms one stylistic extreme of medieval English maps.

Size of the original: 24.5 × 31.8 cm. By permission of the British Library, London (Egerton MS. 3137, fol. 1v).

Nor is there any consistency of style, except that features above ground level are nearly always shown pictorially; the extent and form of this pictorial element vary from the full (if stylized) bird's-eye view of Bristol about 1480 to the crude outlines of churches on the map accompanying the contemporary history of Barholm and Greatford (Lincolnshire).[80] The inscriptions on the maps also vary widely. Overall it is impossible to discover any traces of uniformity, any hint of a genuine tradition of mapmaking among these English local maps. Apart from the very few that have a common origin (two from Westminster Abbey, four from Durham cathedral priory), each seems a wholly individual production.

We find rather less variety among the medieval local maps from the Low Countries (listed below, appendix 20.3, see also fig. 20.21), though here too there is unlikely to have been any clear tradition of mapmaking. There are fewer surviving maps than from England: fifteen have so far been recorded. The earliest, already mentioned, is the diagram map of 1307 setting out the settlements of the area between Aardenburg (Zeeland) and Boechoute (East Flanders) (see fig. 20.6).[81] The next was drawn in 1357 in France, in a register of the University of Paris; dispute had arisen over which of the

university's *nationes* a student from Geertruidenberg belonged to, the Picard or the English (the latter included students from the Netherlands), and the map set out one party's view of the customary boundary between the two. It consists of no more than lines marking the river Meuse (Maas) and a few place-names (fig. 20.22).[82] Another map has been dated 1358 because it seems to be connected with an agreement made then between the bishop of Tournai and Saint Peter's Abbey at Ghent; it shows diagrammatically an area between Oostburg and IJzendijke (Zeeland), with lines marking the dikes and boundaries and with the names of the dikes and the

80. Bristol, Bristol Record Office, MS. 04720, fol. 5v; Lincoln, Lincolnshire Archives Office, Lindsey Deposit 32/2/5/1, fol. 17v; Elizabeth Ralph, "Bristol, *circa* 1480," 309–16, and Judith A. Cripps, "Barholm, Greatford, and Stowe, Lincolnshire, Late 15th Century," 263–88, both in Skelton and Harvey, *Local Maps and Plans* (note 3).

81. Lille, Archives Départementales du Nord, B 1388/1282 bis; Gottschalk, *Historische geografie*, 148–49 (note 19); Harvey, *Topographical Maps*, 89 (note 1).

82. Paris, Bibliothèque de la Sorbonne, Archives de l'Université de Paris, Reg. 2, vol. 2, fol. 35v; Gray C. Boyce, "The Controversy over the Boundary between the English and Picard Nations in the University of Paris (1356–1358)," in *Etudes d'histoire dédiées à la mémoire de Henri Pirenne* (Brussels: Nouvelle Société d'Editions, 1937), 55–66.

FIG. 20.21. PRINCIPAL PLACES IN THE LOW COUNTRIES ASSOCIATED WITH MEDIEVAL LOCAL MAPS.

assumed that there was a strong tradition of local cartography in the Low Countries long before the sixteenth century.[90]

In the Low Countries medieval maps, even of the sketchiest sort, have long been the subject of local antiquarian and scholarly interest. In Britain there has recently been a general search for this type of material. In both these areas, therefore, we can assume that, although more medieval maps may still come to light, enough have been found for us to view them comprehensively and to draw fairly confident conclusions on their numbers, their variety, and their chronology. This is not the case anywhere else in Europe (fig. 20.25). The number of English local maps previously unknown that were revealed by very straightforward search suggests that many may be waiting to be discovered in other countries. So too does the work of François de Dainville in France.[91] His systematic search in the departmental

landholders and some other notes.[83] These three are the only maps of small areas in the Low Countries earlier than the second half of the fifteenth century, though we have copies, made in 1565 from the original in the Gouda town archives, of a map dating from before 1421 of an area at the mouths of the Waal and Meuse.[84] Most of these fifteenth-century maps are no more than rough sketches in pen and ink. Thus one of 1471–72 shows the boundary, in a peat-cutting area, between lands of Elten Abbey and Gooiland (North Holland): a few lines with rough drawings of woods, a church, and a few other features.[85] Another map of part of Overflakkee (South Holland) in 1487 marks streams and boundaries with, again, a rough drawing of a windmill (fig. 20.23).[86] However, some are more elaborate, among them a panoramic painting of about 1480 showing Dordrecht and the surrounding country at the time of the so-called Saint Elizabeth's Day Flood in 1421,[87] and a map of the lower Scheldt in 1468, with its many pictures of ships and full perspective views of towns, villages, and castles (fig. 20.24).[88] Some confusion has been caused by sets of maps, rightly or wrongly associated with Egmond Abbey (Egmond aan Zee, North Holland) and often copied in the seventeenth and eighteenth centuries, showing areas of Zeeland and Flanders at various dates from 600 to 1288;[89] they are in fact historical reconstructions, but some scholars, notably F. C. Wieder, have taken them as stemming from genuine medieval maps and have thus

83. Ghent, Rijksarchief; M. K. Elisabeth Gottschalk, "De oudste kartografische weergave van een deel van Zeeuwsch-Vlaanderen," *Archief: Vroegere en Latere Mededelingen Voornamelijk in Betrekking tot Zeeland Uitgegeven door het Zeeuwsch Genootschap der Wetenschappen* (1948): 30–39.

84. Jan Henricus Hingman, *Inventaris der verzameling kaarten berustende in het Rijks-Archief* (The Hague: Nijhoff, 1867–71), 2:96 (nos. 811, 812); cf. A. J. H. Rozemond, *Inventaris der verzameling kaarten berustende in het Algemeen Rijksarchief zijnde het eerste en tweede supplement op de collectie Hingman* (The Hague: Algemeen Rijksarchief, 1969), suppl. 2, nos. 232, 476.

85. The Hague, Algemeen Rijksarchief, Grafelijkheid van Holland, Rekenkamer no. 755f; D. T. Enklaar, "De oudste kaarten van Gooiland en zijn grensgebieden," *Nederlandsch Archievenblad* 39 (1931–32): 185–205, esp. 188–92, pl. I.

86. Brussels, Archives Générales du Royaume, Grand Conseil de Malines, Appels de Hollande 188, sub G; A. H. Huussen, *Jurisprudentie en kartografie in de XVᵉ en XVIᵉ eeuw* (Brussels: Algemeen Rijksarchief, 1974), 7–8, pl. 1.

87. Amsterdam, Rijksmuseum, A 3147a,b; Rijksmuseum, *All the Paintings of the Rijksmuseum in Amsterdam: A Completely Illustrated Catalogue* (Amsterdam: Rijksmuseum, 1976), 633.

88. Brussels, Archives Générales du Royaume; M. K. Elisabeth Gottschalk and W. S. Unger, "De oudste kaarten der waterwegen tussen Brabant, Vlaanderen en Zeeland," *Tijdschrift van het Koninklijk Nederlandsch Aardrijkskundig Genootschap*, 2d ser., 67 (1950): 146–64; Johannes Keuning, "XVIth Century Cartography in the Netherlands (Mainly in the Northern Provinces)," *Imago Mundi* 9 (1952): 35–64, esp. 41.

89. For example, C. de Waard, *Rijksarchief in Zeeland: Inventaris van kaarten en teekeningen* (Middelburg: D. G. Kröber, Jr., 1916), XXIX, L, 1–6.

90. Frederik Caspar Wieder, *Nederlandsche historisch-geographische documenten in Spanje* (Leiden: E. J. Brill, 1915), 305–6; cf. S. J. Fockema Andreae and B. van 'tHoff, *Geschiedenis der kartografie van Nederland van den Romeinschen tijd tot het midden der 19de eeuw* (The Hague: Nijhoff, 1947), 12, and B. van 'tHoff, "The Oldest Maps of the Netherlands: Dutch Map Fragments of about 1524," *Imago Mundi* 16 (1962): 29–32, esp. 29.

91. His conclusions are given in de Dainville, "Rapports," 397–408 (note 5), and François de Dainville, "Cartes et contestations au XVᵉ siècle," *Imago Mundi* 24 (1970): 99–121.

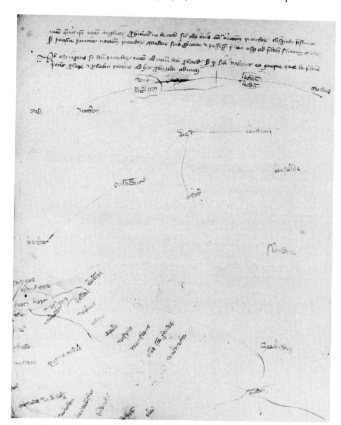

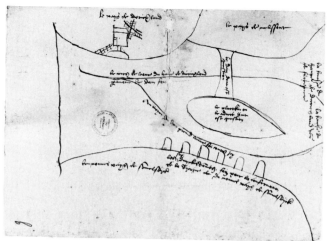

FIG. 20.23. MAP OF PART OF OVERFLAKKEE, SOUTH HOLLAND. Like most fifteenth-century maps from the Netherlands, this 1487 map is little more than a pen-and-ink sketch. It was drawn for a legal dispute concerning the ownership of waterways and embankments. Streams and boundaries are shown as well as a windmill.
Size of the original: 21.5 × 29 cm. By permission of the Archives Générales du Royaume, Brussels (Grand Conseil de Malines, Appels de Hollande 188, sub G).

FIG. 20.22. MAP OF THE PICARDY-NETHERLANDS BORDER, 1357. This plan was produced as part of a legal dispute concerning the residence of students attending the University of Paris. The map shows the river Maas (Meuse), the border between the *nationes* of Picardy and of the English.
Size of the original: 30 × 23 cm. By permission of the Bibliothèque de la Sorbonne, Paris (Archives de l'Université de Paris, Reg. 2, vol. 2, fol. 35v).

archives of southern and eastern France produced a most interesting collection of some ten medieval local maps, but this work has not been continued since his death, and we thus cannot tell whether other regions would produce similar material and can draw no general conclusions about local maps in medieval France. None yet found is older than the 1357 map of the boundary between *nationes* in the register of the University of Paris. Jehan Boutillier wrote in 1395 that maps ("figure et pourtraict") might be used in presenting cases to the Parlement, but the oldest in the group that de Dainville has described dates from 1422.[92] The medieval local maps from France are as varied in subject and style as those from England. They include plans of towns—one, now lost, of Avignon was commissioned in 1491, one survives of Rodez in 1495—and of large tracts of countryside, such as the mid-fifteenth-century map of the county of Gâpençais (Drôme, Basses Alpes, Hautes

Alpes), which covers an area some forty miles square (fig. 20.26).[93] But they also include some plans of very small areas: some pastures at Tillenay (Côte-d'Or) in 1467, for instance, or the pool of Scamandre in the Camargue (Bouches-du-Rhône) in 1479.[94] Some, as in England and the Low Countries, are no more than the roughest of sketches; one of the Valentinois and Diois region (Drôme, Isère) consists only of variously positioned notes with lines drawn between them.[95] Others are more elaborate. The plan of Rodez, an ink drawing, gives full elevations of street frontages throughout the town, showing architectural details, and is inscribed with many names and other notes. A map of 1460 showing the boundary of the duchy of Burgundy between Heuilley, Maxilly, and Talmay (Côte-d'Or) is painted with perspective views of the three villages (drawn to three horizons) with the river Saône and the stumps of

92. De Dainville, "Cartes et contestations," 99, 117 (note 91).

93. Rodez, Archives Départementales de l'Aveyron, FF 2; Grenoble, Archives Départementales de l'Isère, B 3751; Lavedan, *Représentation des villes*, 30, pl. XX (note 16); de Dainville, "Cartes et contestations," 107–9, 112–14 (note 91); Harvey, *Topographical Maps*, 81 (note 1).

94. Dijon, Archives Départementales de la Côte-d'Or, G 880; Nîmes, Archives Départementales du Gard, G 1181; de Dainville, "Cartes et contestations," 112, 113, 116 (note 91).

95. Grenoble, Archives Départementales de l'Isère, B 3495; de Dainville, "Cartes et contestations," 104–5 (note 91).

FIG. 20.24. THE LOWER SCHELDT, 1468. This is just one portion of an elaborate picture map which clearly stems from a well developed artistic tradition. The whole map is 5.2 meters in length.
Height of the original: 57 cm. By permission of the Archives Générales du Royaume, Brussels (inv. 1848, no. 351, secs. 5–7).

some cleared woodland in the background (plate 36).[96] Here again there is no consistent style that would suggest even a regional tradition of mapmaking.

From Italy, in view of the traditions of city plans and of northern Italian district maps, we might expect local maps of other kinds to appear in some numbers. In fact only five are recorded: a seal of the emperor Frederick II, used in 1226, bearing a map of the Straits of Messina; a thirteenth-century picture map of the Chiana valley (Tuscany); a plan of 1306 showing the harbor with projected building plots, at Talamone (Tuscany) (fig. 20.27); a fifteenth-century sketch that is part view, part map, of the lagoon near Venice; and, also from the fifteenth century, a small group of maps of areas around Ravenna.[97] It may well be that the relative profusion of medieval maps from Italy, proficiently drawn in well-defined traditions, has led historians to overlook lesser, apparently trivial productions. But little too has been recorded from Germany and central Europe. The plan of Vienna and Bratislava drawn to scale about 1422 (see fig. 20.8) and the map of 1441 showing estates at Wantzenau (Bas-Rhin and South Baden) have both been mentioned already.[98] The latter should perhaps be grouped with other maps from southwestern Germany that appeared at the end of the fifteenth century: a plan of Ulm has been dated 1480, and two picture maps of Beringsweiler (Württemberg-Baden) are from about 1500.[99] Certainly a superb and elaborate maplike engraving of Lake Constance must have been produced very soon after the war of 1499, since it includes scenes of its battles and other incidents.[100] As a more technical type of cartography (discussed below) we have, from the fifteenth century, not only the general maps of Nicolas of Cusa and Erhard Etzlaub, but also Etzlaub's map of the region around Nuremberg and Konrad Türst's map of Switzerland. From Poland come two sketch maps of 1464 showing the Pomeranian lands of the Teutonic order and marking simply the coastline, rivers, and settlements; and we have also an interesting reference to a map, painted on cloth, of the northern provinces of Poland, which royal envoys gave Pope Martin V in 1421 to further a claim against the Teutonic order.[101] For the Balkans nothing is recorded beyond an Italian map of 1453 showing the lower Danube with its delta.[102]

96. Dijon, Archives Départementales de la Côte-d'Or, B 263; de Dainville, "Cartes et contestations," 112, fig. 10 (note 91); Harvey, *Topographical Maps*, 97 (note 1).

97. Gustave Schlumberger, Ferdinand Chalandon, and Adrien Blanchet, *Sigillographie de l'Orient Latin* (Paris: Geuthner, 1943), 22, pl. I; Vittorio Fossombroni, "Illustrazione di un antico documento relativo all'originario rapporto tra le acque dell'Arno e quelle della Chiana," *Nuova raccolta d'autori italiani che trattano del moto dell' acque*, 6 vols., ed. F. Cardinali (Bologna: Marsigli, 1824), 3:333, 337, pl. 8; Siena, Archivio di Stato di Siena, Caleffo Nero, 4 April 1306; Wolfgang Braunfels, *Mittelalterliche Stadtbaukunst in der Toskana* (Berlin: Mann, 1979), 77–78; Modena, Biblioteca Estense; Almagià, "Antica carta topografica," 81 (note 63); Ravenna, Biblioteca Classense, Carte nn.520–24; Almagià, *Documenti cartografici dello stato pontificio* (Rome: Biblioteca Apostolica Vaticana, 1960), 10 n. 1.

98. Above, pp. 470, 473–74, 478.

99. Ruthardt Oehme, *Die Geschichte der Kartographie des deutschen Südwestens* (Constance: Thorbecke, 1961), 97; Karl Schumm, *Inventar der handschriftlichen Karten im Hohenlohe-Zentralarchiv Neuenstein* (Karlsruhe: Braun, 1961), 5.

100. Wilhelm Bonacker, "Die sogenannte Bodenseekarte des Meisters PW bzw. PPW vom Jahre 1505," *Die Erde* 6 (1954): 1–29.

101. Bolesław Olszewicz, *Dwie szkicowe mapy Pomorza z połowy XV wieku*, Biblioteka "Strażnicy Zachodniej" no. 1 (Warsaw: Nakładem Polskiego Związku Zachodniego, 1937); Karol Buczek, *The History of Polish Cartography from the 15th to the 18th century*, trans. Andrzej Potocki, 2d ed. (Amsterdam: Meridian, 1982), 22–24.

102. Marin Popescu-Spineni, *România în istoria cartografiei pâna la 1600* (Bucharest: Imprimeria Naţionala, 1938), 2:33.

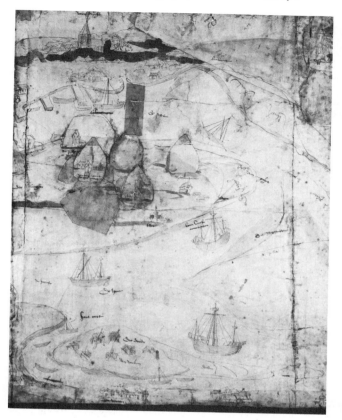

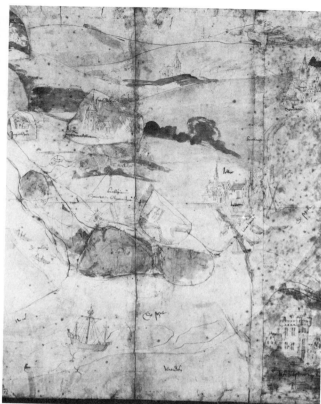

It seems likely that more—perhaps many more—medieval local and regional maps will come to light not only in Italy but in all these areas, and in others where so far we know of none at all; Spain and Portugal are obvious omissions. At the same time we should not take this for granted. It may well be that in many parts of medieval Europe no maps of small areas were ever drawn. The search for such maps in Britain substantially increased the number known from England but produced none at all from Wales, Scotland, or Ireland. The drawing of maps, even of the simplest sort, may have been confined to particular areas. We have seen how the northern Italian district maps cannot be paralleled in other parts of Italy, and there are clear signs of regional concentrations among the local maps elsewhere. Thus in England seven of the thirty-two fourteenth- and fifteenth-century maps come from an area around the Wash that can be defined as lying within twenty miles of Wisbech or King's Lynn. In the Low Countries the concentration is even more marked: nearly all the recorded maps come from within thirty miles of the coast in a long strip drawn from Haarlem and Hilversum in the north to Ostend and Ghent in the south. The few medieval maps recorded from Württemberg and Baden in southwestern Germany, taken with Martin Waldsee-müller's maps of 1507 and 1513, may point to another concentration in this area. We should not dismiss these concentrations as mere accidents of survival or discovery simply because the maps in any one of these groups do not look alike. We should remember that drawing any sort of map was a very unusual thing to do in the Middle Ages; it may have been that particular local circumstances especially favored the production of maps, not that there was any conscious tradition that would produce maps with traces of uniform style.

PURPOSE AND USE

What were the circumstances that led to the drawing of local and regional maps in the Middle Ages? Many seem to have been made to put to a court of law either as general evidence or in support of one side or the other in a particular case. The map shown to Pope Martin V by the Polish envoys in 1421 can be paralleled by others known to have been produced in lesser courts. The Rodez plan of 1495 was drawn for a case over fairs. The map of the lower Scheldt in 1468 was commissioned by a committee that was collecting evidence from both parties to a suit. The 1487 sketch map of part of Overflakkee comes from the ju-

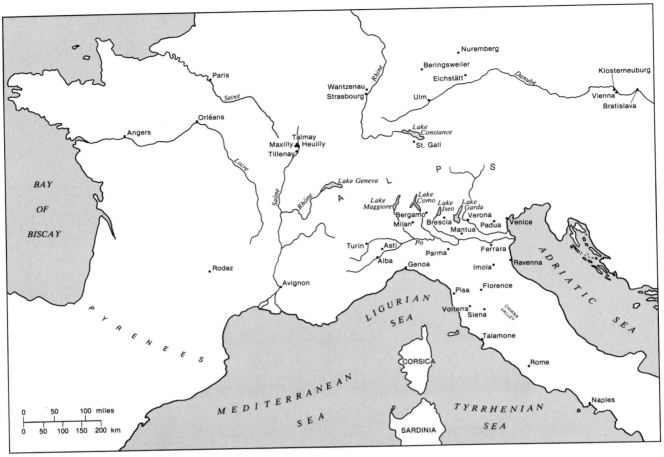

FIG. 20.25. PRINCIPAL PLACES IN CENTRAL EUROPE AND ITALY ASSOCIATED WITH MEDIEVAL LOCAL AND REGIONAL MAPS.

dicial archives of the Great Council at Mechelen. We have seen that Jehan Boutillier in his treatise of 1395 spoke of using maps in cases before the French Parlement; in fact we have earlier and weightier authority for drawing maps to put before courts of law. Local maps of this sort were known in Burgundy as *tibériades*, and de Dainville has shown that this refers to a treatise written in 1355 by the distinguished Italian lawyer Bartolo da Sassoferrato, *De fluminibus seu tiberiadis*; in it he discusses legal problems arising from rights in rivers and streams and shows how they can be solved by using plans.[103] The plans in the treatise itself are simply diagrams, and it is clear that even if the idea of using maps in law courts stemmed from da Sassoferrato's work, he cannot be given credit for the great variety of styles among the maps drawn for this purpose.

However, none of the maps from medieval England seems to have been intended for use in a court of law. This may well reflect the distinctive practice and traditions of the English legal system. Interestingly, the only

English medieval map known to have been drawn for legal purposes dates from the very end of the fifteenth century and forms part of a document by a public notary, working in the traditions of the continental civil law: a statement of title to properties in Exeter in 1499, with a plan of their bounds and elevations of the frontages at the top.[104] On the other hand, several of the English maps can be connected with particular disputes and lawsuits. One of Inclesmoor (near the confluence of the Trent and Ouse rivers, Yorkshire), for instance, seems to have been drawn between 1405 and 1408 when a case was in progress between the duchy of Lancaster and Saint Mary's Abbey, York, over rights to pasture and peat in the area; there are two versions of this map, both in the duchy's archives, and they will have served

103. De Dainville, "Cartes et contestations," 117–20 (note 91).
104. Exeter, Devon Record Office (East Devon Area), Exeter City Archives, ED/M/933; H. S. A. Fox, "Exeter, Devonshire, 1499," in Skelton and Harvey, *Local Maps and Plans*, 329–36 (note 3).

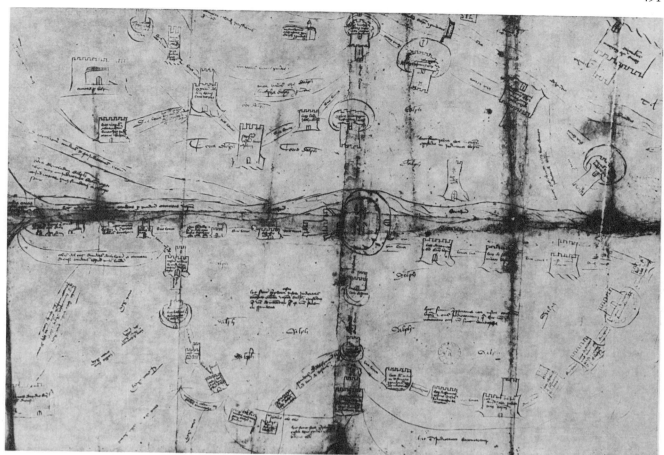

FIG. 20.26. MAP OF THE COUNTY OF GÂPENÇAIS. This mid-fifteenth-century map in four sheets covers an area of about forty by forty miles across the Drôme, Basses Alpes, and Hautes Alpes.

Size of the original: 59 × 86 cm. From F. de Dainville, "Cartes et contestations au XVᵉ siècle," *Imago Mundi* 24 (1970), fig. 7.

both as a guide to the officials conducting the case and as a permanent record of what was claimed (plate 37).[105] Outside England too some maps were drawn simply as a record for one party to a dispute rather than for formal submission to a court; thus in 1444 the duke of Burgundy, Philip the Good, disputing the duchy's boundary with the lands of the king of France, paid to have maps made "so as to see clearly the towns and villages that are included in the duchy, and also those that belong to the kingdom . . . so as to avoid and guard against the encroachments that are made every day by the people and officials of the king."[106]

Some maps served other administrative purposes. The Talamone plan of 1306 was drawn at a time when the city of Siena was concerned to develop and build up this recently acquired coastal site. But it is the English maps that met the most varied needs in administration. The two earliest (of Canterbury cathedral and of Waltham Abbey's springs at Wormley) both served as guides to

buried or concealed water pipes, and so too did a mid-fifteenth-century plan showing the water supply of the London Charterhouse.[107] Others were added to copies of documents simply to illustrate the relative positions of properties: of tenterframes at Exeter for instance, or of lands beside a stream at Staines (Middlesex).[108] Out-

105. London, Public Record Office, DL 42/12, fols. 29v–30r, and MPC 56, ex DL 31/61; M. W. Beresford, "Inclesmoor, West Riding of Yorkshire, *circa* 1407," in Skelton and Harvey, *Local Maps and Plans*, 147–61 (note 3).

106. De Dainville, "Cartes et contestations," 109 (note 91), author's translation.

107. London, Muniments of the Governors of Sutton's Hospital in Charterhouse, MP 1/13; W. H. St. John Hope, "The London Charterhouse and Its Old Water Supply," *Archaeologia* 58 (1902): 293–312; M. D. Knowles, "Clerkenwell and Islington, Middlesex, Mid-15th Century," in Skelton and Harvey, *Local Maps and Plans*, 221–28 (note 3).

108. London, Muniments of the Dean and Chapter of Westminster, 16805; Susan Reynolds, "Staines, Middlesex, 1469 × *circa* 1477," in Skelton and Harvey, *Local Maps and Plans*, 245–50 (note 3).

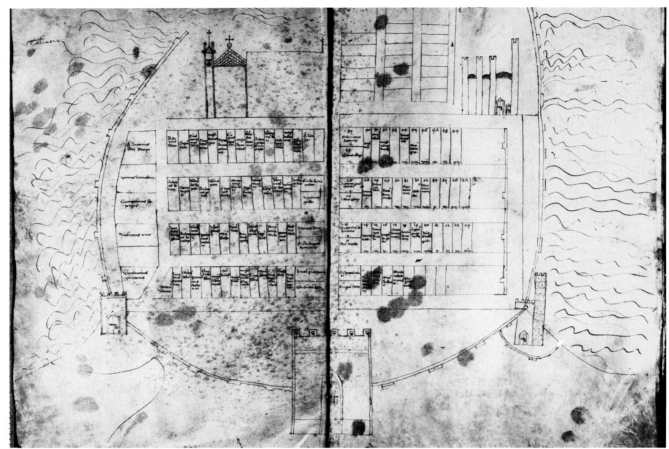

FIG. 20.27. TALAMONE HARBOR, 1306. This is one of few local maps known to have survived from Italy, despite the extensive traditions of city plans and northern Italian district maps. It shows existing (named) and projected (numbered) building plots within the Tuscan harbor and so might have been produced to fill some planning or administrative need.

Size of the original: 43.5 × 58.5 cm. By permission of the Archivio di Stato, Siena (Capitoli 3, fols. 25v–26).

line plans of four plots of land belonging to the Bridge House estates of London were entered in a register in the 1470s, apparently as a guide to their shape and dimensions.[109]

In the Italian tradition of city plans we find a certain antiquarian bias: where only a selection of buildings appears within the walls, they seem often to have been chosen not just because they are the city's most distinctive buildings, but because they are monuments of antiquity. This is particularly obvious in the late fifteenth-century maps of Rome by Alessandro Strozzi and Pietro del Massaio, which draw on Flavio Biondo's account of the city's antiquities, *Roma instaurata*, written in 1444–46.[110] There may be something of the same bias in the

medieval plans of Jerusalem, though here it is difficult to distinguish between the antiquarian and the pious interest in the city's ancient buildings. Several local maps from medieval England were drawn specifically to illustrate antiquarian works. Thomas of Elmham's history of Saint Augustine's Abbey, Canterbury, written at the

109. London, Corporation of London Records Office, Bridge House Deeds, Small Register, fols. 9r–11r; John H. Harvey, "Four Fifteenth-Century London Plans," *London Topographical Record* 20 (1952): 1–8, figs. 1–4; Philip E. Jones, "Deptford, Kent and Surrey; Lambeth, Surrey; London, 1470–1478," in Skelton and Harvey, *Local Maps and Plans*, 251–62 (note 3).

110. Scaglia, "Origin of an Archaeological Plan," 137–63 (note 46); Weiss, *Renaissance Discovery*, 5–6, 90–93 (note 42).

beginning of the fifteenth century, includes not only careful facsimiles of some of the abbey's early charters, but also a plan of the chancel of the abbey church, showing the layout of its altars, and a map of the Isle of Thanet (Kent) with the complicated line of the "run of the hind," the path which reputedly was followed by a pet deer of Queen Domneva of Mercia and which formed the boundary of the abbey's manor there.[111] The bird's-eye view of Bristol is another case in point; it was drawn in the chronicle written by Robert Ricart, the town clerk, and though it seems to show fifteenth-century Bristol, it was probably meant to represent the town at the time of its legendary foundation.[112] These antiquarian maps cannot be paralleled among the medieval maps so far recorded from France or the Low Countries. They anticipate the association between mapmaking and antiquarian interests that is so often found in the sixteenth and seventeenth centuries.

Another aspect of the Italian traditions of medieval mapping that was echoed elsewhere in Europe was the use of maps for permanent display. Clearly this was the purpose of the large, realistic bird's-eye views of cities in the late fifteenth century, such as de' Barbari's woodcut of Venice or the painting of Rome in the ducal palace at Mantua that derives from Rosselli's lost engraving.[113] But we find earlier examples too, such as the mid-fourteenth-century fresco showing Florence in the Loggia del Bigallo or the view of Rome that Taddeo di Bartolo painted in the Palazzo Comunale at Siena in 1413–14.[114] Outside Italy we have the large woodcut of Lake Constance showing scenes from the war of 1499 (it measures 50 by 110 cm) and, from the Netherlands, the painting of about 1480 of the Saint Elizabeth's Day Flood. None of the English maps falls in this category. But we have two references from France to maps of this sort that have failed to survive: one showing the course of the Loire, with its towns and bridges, that was on a cloth roll given to the duke of Orleans in 1440, and three maps painted on cloth that were recorded in 1472 among the contents of the castle at Angers.[115]

Behind these maps made for display lay very much more than the idea of providing simple decoration. They were products of the artistic tradition of their times and, despite the graphic realism of the later Middle Ages, the art of medieval and Renaissance Europe was not simply representational: it conveyed a wealth of connotation and implication, of symbolic, metaphorical meaning that attached to the visual image. In some maps this is obvious: the woodcut of Lake Constance with its battle scenes was a glorification of Swiss successes in the war of 1499. Usually it is subtler: Juergen Schulz has analyzed the complex of ideas presented by de' Barbari's view of Venice and other sixteenth-century panoramas of cities.[116] It would of course be absurd to read mean-

ings of this kind into sketch maps that were demonstrably drawn for practical and ephemeral purposes. But the more elaborate maps of even quite small areas may often have done more than just convey topographical information: such carefully drawn productions as Canterbury cathedral's plan of its water system or the duchy of Lancaster's maps of Inclesmoor should be seen, as was probably intended, as reflecting pride of ownership and assertion of rights. The Gough map of Britain may possibly have been compiled for functional official use; Matthew Paris's maps were not, and to appreciate their significance they must be seen in the context of the artistic and intellectual works of thirteenth-century England.[117] Beyond the factual information, the outwardly simple medieval map may have conveyed what contemporaries would see as a straightforward message but what to us can only be painfully worked out as a hidden inner meaning.

SURVEYING AND MAPPING

It may seem surprising that among these medieval maps of small areas there are no maps compiled as part of a survey, that is, a general description of a piece of property, whether a single field, an individual holding, a village community and its lands, or an entire estate. Where maps were drawn for administrative purposes it seems to have been always in order to make a single point: to show the arrangements of water pipes, for instance, or the line of a disputed boundary. The many surveys of landed property that survive from medieval Europe were drawn up as written descriptions, often extremely long and detailed ones, without mapping at any stage. Only two possible exceptions to this have come to light, both from England. One is a pair of plans showing small groups of strips in the fields of Shouldham (Norfolk) in

111. Cambridge, Trinity Hall, MS. 1, fols. 77r, 42v; William Urry, "Canterbury, Kent, Late 14th Century × 1414," 107–17, and F. Hull, "Isle of Thanet, Kent, Late 14th Century × 1414," 119–26, both in Skelton and Harvey, *Local Maps and Plans* (note 3).

112. Bristol, Bristol Record Office, MS. 04720, fol. 5v; Elizabeth Ralph, "Bristol, *circa* 1480," in Skelton and Harvey, *Local Maps and Plans*, 309–16 (note 3).

113. Schulz, "Moralized Geography" (note 42); de Rossi, *Piante di Roma*, 104–7, pls. VI–XII (note 44); C. Hülsen, "Di una nuova pianta prospettica di Roma del secolo XV," *Bullettino della Commissione Archeologica Comunale di Roma*, 4th ser., 20 (1892): 38–47; Ehrle and Egger, *Piante e vedute di Roma*, 14, pl. XI (note 44).

114. Ciullini, "Firenze rappresentazioni," 35 (note 45); Ehrle and Egger, *Piante e vedute di Roma*, 10–11, pl. IV (note 44).

115. De Dainville, "Rapports," 402–3 (note 5).

116. Schulz, "Moralized Geography," 441–72 (note 42); cf. J. B. Harley, "Meaning and Ambiguity in Tudor Cartography," in *English Map-Making, 1500–1650*, ed. Sarah Tyacke (London: British Library, 1983), 22–45.

117. Below, pp. 495–96.

1440–41.[118] The other is a pair of plans of a small area divided into closes at Tanworth in Arden (Warwickshire), drawn between 1497 and 1519.[119] In neither case are the plans part of the finished survey, which is solely a written description; they are found among the preliminary notes and drafts. It might be thought that sketch maps of this sort were a normal way of gathering information for a field survey, to be discarded once the final version had been set out in writing. This seems unlikely; the Shouldham plans are of only a tiny portion of the lands covered by the draft survey, and they are drawn with a lack of expertise, as though the draftsman was experimenting with a novel technique. In England there was no recognized profession of surveyors before the sixteenth century. In the Low Countries, on the other hand, we find people referred to as surveyors, suggesting that this was their full-time calling, from the thirteenth century on.[120] Yet despite this professionalism there is no evidence that they ever drew maps as a part of their craft, whether as a finished product or as a working technique. The earliest instructions on surveying from the Low Countries, in the early fifteenth-century lawbook from the town of Brielle (South Holland) by Jan Matthijssen, describe the surveyor's duties in some detail but give no hint that drawing maps or plans was part of his work.[121]

What the surveyor had to do was to produce a description of landed property, and from the ninth century onward we have more or less elaborate surveys of estates; the earliest come from the triangle between the mouths of the Seine and Rhine and the Bavarian mountains.[122] The Domesday Book, drawn up in England for William I in 1086, is the unique surviving survey of an entire kingdom, but from the early twelfth century onward surveys were an increasingly common record of estate management. These written surveys need not involve measurement of land: property might be described simply in terms of the rents and services it produced for the landlord or, if lands were described in more detail, they might be defined by referring to the number of selions, or plowing strips, they contained. In England precise measurement of land was known by the late ninth century, but it was only in the early thirteenth that surveys in general began to give the measured areas of the lands they described. It is unlikely that any more sophisticated methods were used to produce these figures than simple measurement by rod or cord followed by calculations of area by actual or assumed rectangles and, perhaps, right-angled triangles; this is what is described in Richard Benese's treatise of 1537, and we have many copies from the thirteenth century onward of tables that give the length of an acre of land for any given width. By the late fifteenth century the land compass was probably being used to define the alignment of lands; the

earliest printed English treatise on surveying, by John Fitzherbert in 1523, recommends its use.[123] Certainly we have much more to learn about the medieval surveyors of continental Europe, but if they in fact used more exact or more elaborate methods than their English counterparts, the evidence has still to come to light; the professional status of the surveyors of the medieval Netherlands probably reflected an enhanced legal standing rather than more skilled techniques.

But this is not to say that the Middle Ages had no knowledge of the methods of the ancient Roman surveyors or of geometric theory. The Roman surveying treatises were still being copied in the twelfth, thirteenth, and fourteenth centuries. A series of treatises on geometry gave medieval Europe access to both classical and Islamic methods, including the use of instruments: the *Geometria* of Gerbert (who was Pope Sylvester II in 999–1003) and the early twelfth-century *Liber embadorum* of Abraham bar Chiia are early examples that were heavily drawn on by later medieval writers. These treatises were practical in their approach: the early thirteenth-century *Practica geometriae* of Leonardo Pisano was the earliest to describe the plumb-bob level, used in finding the horizontal area of sloping ground, and the quadrant, applied to various exercises in surveying.[124] But despite their practical implications, these were works of scholarship and learning, and it need not follow that their precepts were known or used by those who were involved in the everyday business of surveying, of measuring and describing lands. Very much later in England Leonard Digges in 1556 and George Atwell in 1658 were to comment on the lack of contact between geometric theory and practical surveying in their own times, and

118. Norwich, Norfolk Record Office, Hare 2826, fols. 16v, 34v; P. D. A. Harvey, "Shouldham, Norfolk, 1440 × 1441," in Skelton and Harvey, *Local Maps and Plans*, 195–201 (note 3).

119. Stratford upon Avon, Shakespeare Birthplace Trust Records Office, DR37/box 74, B ii b,c; B. K. Roberts, "North-west Warwickshire; Tanworth in Arden, Warwickshire, 1497 × 1519," in Skelton and Harvey, *Local Maps and Plans*, 317–28 (note 3).

120. P. S. Teeling, "Oud-Nederlandse landmeters, III," *Orgaan der Vereniging van Technische Ambtenaren van het Kadaster* 7 (1949): 126–34, esp. 126–27; A. Viaene, "De landmeter in Vlaanderen, 1281–1800," *Biekorf* 67 (1966): 7; Cornelis Koeman, "Algemene inleiding over de historische kartografie, meer in het Bijzonder: Holland vóór 1600," *Holland* 7 (1975): 230.

121. Koeman, "Algemene inleiding," 231 (note 120).

122. Robert Fossier, *Polyptyques et censiers* (Turnhout: Brepols, 1978), 33.

123. Eva G. R. Taylor, "The Surveyor," *Economic History Review*, 1st ser., 17 (1947): 121–33; A. C. Jones, "Land Measurement in England, 1150–1350," *Agricultural History Review* 27 (1979): 10–18; P. D. A. Harvey's Introduction in Skelton and Harvey, *Local Maps and Plans*, 11–15 (note 3).

124. Edmond R. Kiely, *Surveying Instruments: Their History* (New York: Teachers College, Columbia University, 1947; reprinted Columbus, Ohio: Carben Surveying Reprints, 1979), 50–54.

in the eighteenth century some writers on surveying were still recommending methods that avoided measurement of angles.[125] Knowledge of sophisticated surveying methods among the learned in medieval Europe is no evidence of their actual use.

The surveyors of medieval Europe were not its mapmakers. Did the mapmakers—those who produced our few medieval local and regional maps—draw on the methods of either the practical surveyors or the theoretical geometricians? This is a harder question to answer. For most surviving maps it is clear that even the simplest measurements on the ground played no part in their construction. The fully detailed city views from late fifteenth-century Italy were more likely built up from mosaics of sketches from high buildings than based on actual measurement.[126] But we have seen that some form of measurement played a part in constructing probably some city plans from Italy and Palestine, and certainly the 1422 plan of Vienna and Bratislava. In the 1440s the Italian humanist scholar and architect Leon Battista Alberti wrote an account of Rome and its monuments, *Descriptio Urbis Romae*, in which he set out a method for placing the city's principal features on a scale map by means of measured radii from a central vantage point; in later works, *Ludi matematici* (1450) and *De re aedificatoria* (1452), he described more elaborate ways of constructing maps to scale, including triangulation. Although in his first work he provides such specific application of his method, including the coordinates of some features, it is not certain that he actually drew the map of Rome that he describes.[127] If he did it is now lost; it is no longer thought, as formerly, that Alberti's work underlies the map of Rome by Alessandro Strozzi in 1474.[128] Some other fifteenth-century maps may be based on measurement, even on a knowledge of geometric theory, but we should be cautious and not assume this without the strongest evidence.

ITINERARY MAPS AND THE DEVELOPMENT OF MAPS DRAWN TO SCALE

The remaining maps to be considered cannot be said to form a coherent, conscious tradition. On the other hand, there is a clear conceptual relationship between them, and in their particular approach to cartographic problems we see the gradual reemergence in later medieval Europe of the notion of strict scale applied to maps. What they have in common is that they are essentially maps of routes, itinerary maps, that take as their basis the simple sequence of places between one point and another. Several of the maps already discussed are of this sort: some of the northern Italian district maps with their emphasis on roads, or the water supply plans from Waltham Abbey and the London Charterhouse, where

the line of the conduit can be seen as the route. An outstanding English example is a map of Sherwood Forest (Nottinghamshire); it consists of little more than roughly radiating lists of places along about a dozen routes through the forest, an efficient guide to any one of these routes, but not meant as a general map of the area and quite inadequate as a means of finding one's way across from one of these routes to another.[129] It was of course exceptional for an itinerary to take the form of a map; normally it would be set out in writing, and we have a number of written itineraries listing routes often over long distances—an outstanding example is the fourteenth-century Bruges itinerary, which lists places and the distances between them along routes from Bruges to almost every part of Europe.[130] But the itinerary map seems to have been better understood in medieval Europe than some other kinds of cartography, and it is interesting that our knowledge of the Peutinger map, a fourth-century itinerary map of great elaboration, should stem from an apparently faithful copy made in the twelfth or early thirteenth century.[131]

It is interesting too, though it may be mere accident, that several of the most significant early itinerary maps come from England: the maps of Matthew Paris and the Gough map. We have four versions of the map of the route from London to Apulia (Italy) that Matthew Paris drew up in the mid-thirteenth century. It is a straightforward series of vertical strips showing the successive staging points by tiny thumbnail sketches, some based on the places' actual appearance; between the staging points are notes of the distances in terms of day's journeys, and for part of the way alternative routes are shown (plate 38).[132] This is the most obvious itinerary map by Matthew Paris, but it is by no means his only one. As we have already seen, of his two maps of the

125. H. C. Darby, "The Agrarian Contribution to Surveying in England," *Geographical Journal* 82 (1933): 529–35, esp. 532–33; A. W. Richeson, *English Land Measuring to 1800: Instruments and Practices* (Cambridge: Society for the History of Technology and MIT Press, 1966), 125–26, 152–53, 158.

126. Schulz, "Moralized Geography," 431–41 (note 42).

127. Joan Gadol, *Leon Battista Alberti: Universal Man of the Early Renaissance* (Chicago: University of Chicago Press, 1969), 167–78; Pinto, "Ichnographic City Plan," 36–38 (note 55).

128. Scaglia, "Origin of an Archaeological Plan," 137–41 (note 46); Weiss, *Renaissance Discovery*, 90–92 (note 42).

129. Belvoir, Archives of the Duke of Rutland, map 125; M. W. Barley, "Sherwood Forest, Nottinghamshire, Late 14th or Early 15th Century," in Skelton and Harvey, *Local Maps and Plans*, 131–39 (note 3).

130. Joachim Lelewel, *Géographie du Moyen Age*, 4 vols. and epilogue (Brussels: Pilliet, 1852–57; reprinted Amsterdam: Meridian, 1966), *Epilogue*, 281–308.

131. Above, pp. 238–42.

132. Vaughan, *Matthew Paris*, 242, 247–50 (note 17); Harvey, *Topographical Maps*, 67 (note 1).

Holy Land one, known only through what is probably a draft, gives distances between places along the coast and was probably based on a written itinerary. But his much better known map of Great Britain, which also survives in four versions, should also be seen as an itinerary map (plate 39). Its axis is the route from Newcastle upon Tyne to Dover, with some fifteen intermediate places, among them Matthew Paris's own monastery at Saint Albans; this route is drawn along a straight line, ignoring the fact that on the ground it turns through ninety degrees at London to run east instead of south. On either side of this route, and beyond it in Scotland, are many further details: towns, rivers, and coastlines, as well as a few topographical notes. The result bears some resemblance to a modern geographic map of Great Britain, but to view it in this light is to miss the crucial point of its construction on the basis of a single route; essentially it is simply an itinerary map with additions.[133] Probably this is how we should also see the far more elaborate Gough map of the mid- or late fourteenth century. This too is a map of the whole of Britain, but instead of having a single route as its basis it takes five main roads radiating from London, with some branches and crossroads and, in Lincolnshire and Yorkshire, some local roads too. Like Matthew Paris's map it marks towns, rivers, and coastlines beyond its basic routes as well as along them, and it does this with a thoroughness and accuracy that far surpass the achievement of Matthew Paris. But that its starting point and main center of interest are these primary routes is clear; in Cornwall, Wales, and Scotland, where it shows few roads or none, its general shape is least like what we would expect from a geographic map today (plate 40).[134]

Clearly both Matthew Paris and the author of the Gough map had some notion of consistent scale. This appears explicitly in a note on one version of Matthew Paris's map: "The whole island should have been longer if only the page had permitted" (*Si pagina pateretur hec totalis insula longior esse deberet*).[135] And since the places named on his basic route are more or less equidistant on the ground—he very likely viewed them as staging points—as well as on the map, where they are simply entered one below the other without much space between, they provide a very rough basis of scale as well as of relative position. On the Gough map the relative positions of some groups of places are so close to what we would expect from an accurate scale map that it has been argued, by E. J. S. Parsons, that the compiler made use of existing local maps drawn to scale, reproducing them on his general map at a scale of about 1:1,000,000.[136] From what we know of medieval mapping in general this seems unlikely, but it is a tribute to the map that the possibility can even be considered. In fact it cannot be claimed that the Gough map—let alone

Matthew Paris's map—was constructed, or even envisaged, as a scale map throughout. Along the roads are figures giving the distances from one place to another; these distances are in local (and very variable) customary miles, but the lengths of the roads on the map itself bear no fixed relation either to these figures or to the distances expressed by a standard measure. But even if these cannot be seen as scale maps, they show clearly that where a map, with some idea of scale behind it, is constructed around a route or routes with distances entered, it is conceptually only a very small step to set out the itinerary with its distances all in due proportion; and once that has been done the idea of scale has been introduced at the very heart of the map.

The map of the Holy Land found in the early fourteenth-century works of Marino Sanudo and Paolino Veneto can be placed in the same context, though it is conceptually further advanced than even the Gough map. Sanudo's text sets out the principle behind the map, how Palestine was imagined as divided by a grid into squares of one league and how each town could then be placed in the appropriate square; and besides entering them on the map itself he lists all the towns, defining their positions on the grid.[137] The result is, of course, a complete scale map, even if inaccuracies, particularly at the edges, mean that its execution falls short of its intention. The sudden appearance in medieval Europe of a land map drawn to scale on the basis of a grid is of extraordinary interest, and Needham has suggested that it may derive from the grid-based scale maps that we know were a little earlier being drawn in China; Arab maps consisting of place-names entered on a grid provide a possible connection and an immediate source for the method.[138] At first sight this may seem to have little to

133. British Museum, *Four Maps of Great Britain Designed by Matthew Paris about A.D. 1250* (London: British Museum, 1928); J. B. Mitchell, "Early Maps of Great Britain: I. The Matthew Paris Maps," *Geographical Journal* 81 (1933): 27–34; Vaughan, *Matthew Paris*, 241–44 (note 17); Harvey, *Topographical Maps*, 140–42 (note 1).

134. *The Map of Great Britain, circa A.D. 1360, Known as the Gough Map* (Oxford: Oxford University Press, 1958), published for the Royal Geographical Society and Bodleian Library with an accompanying memoir by E. J. S. Parsons, *Introduction to the Facsimile* (Oxford: Oxford University Press, 1958).

135. London, British Library, Royal MS. 14.C.vii, fol. 5v.

136. Parsons, *Introduction*, 9 (note 134).

137. Marino Sanudo, *Liber secretorum fidelium crucis*, vol. 2 in *Gesta Dei per Francos*, 2 vols., ed. Jacque Bongars (Hanover: Heirs of J. Aubrius, 1611), 246, for Sanudo's text; Röhricht, "Marino Sanudo," 84–126 (note 32); Degenhart and Schmitt, "Sanudo und Veneto," 76–78, 116–19 (note 27); Harvey, *Topographical Maps*, 144–46 (note 1).

138. Joseph Needham, *Science and Civilisation in China* (Cambridge: Cambridge University Press, 1954–), vol. 3, *Mathematics and the Sciences of the Heavens and the Earth*, 564, pls. LXXXVII, LXXXVIII.

do with the itinerary map. But defining the positions of the towns in the Holy Land so that they could be correctly placed on the grid can only have been done by measuring the routes between them; ultimately the map must be based on a whole series of measured itineraries, even though it does not show the roads or distances between one place and another. There is, of course, an obvious similarity between this map of the Holy Land and the large and small maps of Italy that appear in one manuscript of Paolino Veneto's *Chronologia magna*; these too, as we have seen, are based on a grid (the same for both maps), set out with dots and figures along each margin.[139] On the other hand, these maps of Italy were essentially extensions (to cover some inland features) of maps of the peninsula's coastline drawn as portolan charts. They may thus be a link between the portolan charts and the map of the Holy Land, being the first application of the ideas and techniques of the portolan charts to a map of a land mass viewed as something more than just a coastline.

Whatever the origin of the ideas and techniques behind this early fourteenth-century map of the Holy Land, they were not taken up by other mapmakers: it had no immediate successors. And when, in the mid-fifteenth century, the itinerary-based scale map reappeared in Germany it owed no direct debt to the sea-itinerary tradition of the portolan charts. Instead it almost certainly owed something, probably a great deal, to the interest in theoretical geography and in working out geographical coordinates that we find among scholars in late medieval Germany and Austria; significantly, degrees of latitude and longitude are now marked, suggesting that astronomical measurement may have been used to fix the positions of places north or south. But the circumstances of the scale map's reappearance are mysterious. What we have are two maps of Germany, both explicitly attributed to Nicolas of Cusa, philosopher and theologian, who died in 1464. But both maps in their surviving form date from long after his death. One is in manuscript, an addition to the maps of Ptolemy in an atlas drawn up by Henricus Martellus Germanus in 1490. The other is printed, a copper engraving which is dated at Eichstätt (Bavaria) in 1491 but which may well not have been completed and brought into use until the 1530s.[140] The two are quite dissimilar but may just possibly derive from the same original map, now lost. Whatever these maps owe to the work of the geographers in fixing key points on the maps, they are essentially itinerary maps in their detailed construction, based on measurements along many routes. These measurements will have been of angles as well as of length, and behind these maps lies the introduction of the magnetic compass for measuring direction on land; its use seems to have spread in the fifteenth century, particularly as a compo-

nent of pocket sundials.[141] But whatever the methods and authorship of these maps, they are of great importance: their accuracy argues a high degree of technical accomplishment, and they mark the beginning of a real tradition of topographical maps drawn to scale in Europe.

Although we know nothing of the origin of the maps attributed to Nicolas of Cusa, we know very much more about the maps that lie in direct succession to them in Germany. These were the work of Erhard Etzlaub of Nuremberg; significantly, he was a maker of pocket sundials. His most extensive map, already mentioned, covers Germany, central Europe and the Alps, and northern Italy. It exists in two versions, both woodcuts, and they are commonly distinguished by the opening words of the descriptive title above the map. One reads *Das ist der Rom-Weg . . .* ("This is the road to Rome"); it may have been intended to guide pilgrims to Rome in the holy year 1500, or it may have been produced rather earlier in the 1480s or 1490s. The other reads *Das sein dy lantstrassen durch das Römisch reych . . .* ("These are the roads through the Roman Empire") and bears the date 1501. The maps have two particular points of interest that reflect their origin as itinerary maps. One is that main land routes are shown prominently and are marked by series of dots, each representing one German mile, as is explained, with a scale, at the foot of each map. The other is that also at the foot of each map is a tiny compass rose with a note explaining how to orient the map with an actual compass. Etzlaub's maps, like Nicolas of Cusa's, were constructed from innumerable measurements of direction and distance. This applies equally to another map that Etzlaub must have compiled either as a by-product or as a preliminary sample of his larger maps; like them it has Nuremberg as its center, but it is simply a regional map, circular in shape, covering an area of sixteen German miles from the city. This too was published as a woodcut, dated 1492. But it was not only in producing these itinerary-based scale maps that Etzlaub is important in the history of cartography. In 1507 he was employed to survey an estate that had been bought by the city of Nuremberg; whether or not this included drawing a map, it is of great significance that we see here surveyor and mapmaker brought together in a single person, a union that foreshadowed important developments in large-scale cartography.[142]

139. Above, p. 481.

140. A. Wolkenhauer, "Über die ältesten Reisekarten von Deutschland aus dem Ende des 15. und dem Anfange des 16. Jahrhunderts," *Deutsche Geographische Blätter* 26 (1903): 120–38, esp. 124–28.

141. Cf. P. D. A. Harvey's Introduction in Skelton and Harvey, *Local Maps and Plans*, 37 (note 3).

142. Wolkenhauer, "Über die ältesten Reisekarten," 130–36 (note 140); Herbert Krüger, "Erhard Etzlaub's *Romweg* Map and Its Dating

So too, though in a different way, did the work of Konrad Türst, physician to the emperor Maximilian I, presage new developments. In 1495–97 he wrote an account of Switzerland, *De situ confoederatorum descriptio*, but he did not publish it, and it is known only from four manuscript copies. Two of these include a map of Switzerland, which again is drawn to scale on the basis of itineraries, an achievement the more remarkable in view of the difficulty of much of the Swiss terrain. But though this links the map with those of Nicolas of Cusa and Etzlaub, its general appearance is far closer to the district maps from northern Italy, for instead of the austere dots or circles that mark towns on the German maps it has a multiplicity of tiny thumb-nail sketches that are pictures of the places' actual appearance, drawn from life.[143] In bringing together these two strands of cartographic tradition from the Middle Ages, the itinerary map and the picture map, Türst seems to have anticipated developments that were to find their full expression in the Renaissance.

in the Holy Year of 1500," *Imago Mundi* 8 (1951): 17–26; Schnelbögl, "Life and Work of Etzlaub," 11–26 (note 29); Tony Campbell, "The Woodcut Map Considered as a Physical Object: A New Look at Erhard Etzlaub's *Rom Weg* Map of c. 1500," *Imago Mundi* 30 (1978): 79–91; Harvey, *Topographical Maps*, 147–49 (note 1).

143. Eduard Imhof, *Die ältesten Schweizerkarten* (Zurich: Füssli, 1939), 6–14, pls. 1, 2; Theophil Ischer, *Die ältesten Karten der Eidgenossenschaft* (Bern: Schweizer Bibliophile Gesellschaft, 1945).

APPENDIX 20.1 CHRONOLOGICAL LIST OF LOCAL MAPS FROM NORTHERN ITALY BEFORE 1500

1. 1291: Asti area. Turin, Biblioteca Nazionale, Codex Alfieri. See Roberto Almagià, "Un'antica carta del territorio di Asti," *Rivista Geografica Italiana* 58 (1951): 43–44.

2. First half of fourteenth century: Asti area. Asti, Codex de Malabaya. See Almagià, "Antica carta."

3. 1383–1400: Lake Garda. Verona, Biblioteca Civica. See Roberto Almagià, *Monumenta Italiae cartographica* (Florence: Istituto Geografico Militare, 1929), 5, pl. VII.

4. 1406–16: Brescia area. Brescia, Archivio Storico Civico, n. 434/3. See Giovanni Treccani degli Alfieri, *Storia di Brescia*, 4 vols. plus index (Brescia: Morcelliana, 1961), 1:870.

5. 1437–41: Lombardy. Paris, Bibliothèque Nationale, Rés. Ge. C. 4990. See Almagià, *Monumenta*, 9, pl. VIII (where is is cited as Ge.C.4090).

6. After 1439: Verona area. Venice, Archivio di Stato. See Roberto Almagià, "Un'antica carta topografica del territorio veronese," *Rendiconti della Reale Accademia Nazionale dei Lincei: Classe di Scienze Morali, Storiche e Filologiche* 32 (1923): 63–83; idem, *Monumenta*, 11, pl. XI.

7. 1440: Lombardy, by "Ioanes Pesato." Treviso, Museo Comunale. See Almagià, *Monumenta*, 9, pl. VIII.

8. 1449: Padua area, by Annibale di Maggi. Milan, Biblioteca Ambrosiana. See Almagià, *Monumenta*, 12.

9. Second half of fifteenth century: Veneto region. Istanbul, Topkapi Sarayi Muzesi. See Rodolfo Gallo, "A Fifteenth Century Military Map of the Venetian Territory of *Terraferma*," *Imago Mundi* 12 (1955): 55–57.

10. After 1460: Parma area. Parma, Archivio di Stato. See Almagià, *Monumenta*, 9.

11. 1465: Padua area, by Francesco Squarcione. Padua, Museo Civico. See Almagià, *Monumenta*, 12.

12. 1469–70: Brescia area. Modena, Biblioteca Estense. See Mario Baratta, "Sopra un'antica carta del territorio bresciano," *Bollettino della Reale Società Geografica*, 5th ser., 2 (1913): 514–26, 1025–31, 1092; Almagià, *Monumenta*, 12, pl. XII.

13. Before 1472: Brescia area. Brescia, Biblioteca Queriniana. See Almagià, *Monumenta*, 9, pl. VII.

14. 1479–83: Verona area. Venice, Archivio di Stato, Scuola di Carità, busta 36, n. 2530. See A. Bertoldi, "Topografia del Veronese (secolo XV)," *Archivio Veneto*, n.s., 18 (1888): 455–73; Almagià, *Monumenta*, 12; Gallo, "Fifteenth Century Military Map," 55.

APPENDIX 20.2 CHRONOLOGICAL LIST OF LOCAL MAPS AND PLANS FROM ENGLAND BEFORE 1500

Most of the following items are reproduced, and each is fully discussed in its local historical context, in *Local Maps and Plans from Medieval England*, ed. R. A. Skelton and P. D. A. Harvey (Oxford: Clarendon Press, 1986); I am most grateful to the Oxford University Press for permission to list the material here. The only items not included there are nos. 5, 6, 7, 17, and 23, which have recently come to light and have been brought to my attention by the kindness of Catherine Delano Smith, H. S. A. Fox, J. H. Harvey, and M. M. Condon. The date assigned to each map is that of its original compilation; in some cases the surviving maps are later copies.

1. About 1153–61: Canterbury (Kent). Cambridge, Trinity College, MS. R.17.1, fols. 284v–285r, 286r.

2. 1220–30: Wormley (Hertfordshire). London, British Library, Harl. MS. 391, fol. 6r.

3. 1224–49: Wildmore Fen (Lincolnshire). Loughlinstown (county Dublin), library of Sir John Galvin, Kirkstead Psalter, fol. 4v.

4. Mid- or late fourteenth century: Peterborough (North-amptonshire). Peterborough, Archives of the Dean and Chapter of Peterborough, MS. 1, fol. 368r.

5. Mid- or late fourteenth century: Fineshade (North-amptonshire). Lambeth Palace Library, Court of Arches, Ff.291, fol. 58v.

6. Late fourteenth century: Chute Forest (Hampshire and Wiltshire). Winchester, Winchester College Muniments, 2206.

7. Late fourteenth century: Clare (Suffolk). London, British Library, Harl. MS. 4835, fols. 66v–67r.

8. Late fourteenth century: Isle of Ely (Cambridgeshire) and Holland (Lincolnshire). London, Public Record Office, MPC 45.

9. Late fourteenth century–1408: Cliffe (Kent). Canterbury, Archives of the Dean and Chapter of Canterbury, Charta Antiqua C.295.

10. Late fourteenth century–1414: Canterbury (Kent). Cambridge, Trinity Hall, MS. 1, fol. 77r.

11. Late fourteenth century–1414: Isle of Thanet (Kent). Cambridge, Trinity Hall, MS. 1, fol. 42v.

12. Late fourteenth or early fifteenth century: Clenchwarton (Norfolk). London, British Library, Egerton MS. 3137, fol. 1v.

13. Late fourteenth or early fifteenth century: Sherwood Forest (Nottinghamshire). Belvoir (Leicestershire), Archives of the Duke of Rutland, map 125.

14. About 1390: Winchester (Hampshire). Winchester, Winchester College Muniments, 22820, inside front and back covers.

15. About 1407: Inclesmoor (Yorkshire). London, Public Record Office, DL 42/12, fols. 29v–30r, and MPC 56.

16. About 1420: Exeter (Devonshire). Exeter, Devon Record Office (East Devon Area), Exeter City Archives, book 53A, fol. 34r.

17. About 1420 to about 1430: Exeter (Devonshire). Exeter, Devon Record Office (East Devon Area), Exeter City Archives, Miscellaneous Roll 64, m.1d.

18. About 1430 to about 1442: Tursdale Beck (county Durham). Durham, Muniments of the Dean and Chapter of Durham, Miscellaneous Charter 6417.

19. 1439 to about 1442: Durham. Durham, Muniments of the Dean and Chapter of Durham, Miscellaneous Charter 5828/12.

20. 1440–41: Shouldham (Norfolk). Norwich, Norfolk Record Office, Hare 2826, fols. 16v, 34v.

21. 1440 to about 1445: Durham. Durham, Muniments of the Dean and Chapter of Durham, Miscellaneous Charter 7100.

22. 1444–46: Boarstall (Buckinghamshire). Aylesbury, Buckinghamshire Record Office, Fletcher Archives, Boarstall Cartulary, Boarstall section fol. 1r.

23. Mid-fifteenth century: Burnham Overy (Norfolk). London, Public Record Office, E.163.

24. Mid-fifteenth century: Clerkenwell and Islington (Middlesex). London, Muniments of the Governors of Sutton's Hospital in Charterhouse, MP 1/13.

25. Mid-fifteenth century: Witton Gilbert (county Durham). Durham, Muniments of the Dean and Chapter of Durham, Cartulary IV, fol. 301v.

26. Mid- or late fifteenth century: Chertsey (Surrey) and Laleham (Middlesex). London, Public Record Office, E.164/25, fol. 222r.

27. 1469 to about 1477: Staines (Middlesex). London, Muniments of the Dean and Chapter of Westminster, 16805.

28. 1470–78: Deptford (Kent and Surrey), Lambeth (Surrey), and London. London, Corporation of London Records Office, Bridge House Deeds, Small Register, fols. 8r–11r.

29. Late fifteenth century: Barholm, Greatford, and Stowe (Lincolnshire). Lincoln, Lincolnshire Archives Office, Lindsey deposit 32/2/5/1, fol. 17v.

30. Late fifteenth century: Deeping Fen (Lincolnshire). London, British Library, Cotton MS Otho B.xiii, fol. 1r.

31. Late fifteenth or early sixteenth century: Dartmoor (Devonshire). Exeter, Royal Albert Memorial Museum.

32. About 1478: Denham (Buckinghamshire) and Harefield (Middlesex). London, Muniments of the Dean and Chapter of Westminster, 432.

33. About 1480: Bristol. Bristol, Bristol Record Office MS. 04720, fol. 5v.

34. 1497–1519: Northwestern Warwickshire and Tanworth in Arden (Warwickshire). Stratford upon Avon, Shakespeare Birthplace Trust Records Office, DR 37/box 74, B ii a–c.

35. 1499: Exeter (Devonshire). Exeter, Devon Record Office (East Devon Area), Exeter City Archives, ED/M/933.

APPENDIX 20.3 CHRONOLOGICAL LIST OF LOCAL MAPS AND PLANS FROM THE LOW COUNTRIES BEFORE 1500

I am most grateful to Cornelis Koeman for his helpful comments and corrections to the first draft of this list. The date assigned to each map is that of its original compilation; in some cases the surviving maps are later copies.

1. 1307: Aardenburg (Zeeland) and Boechoute (East Flanders). Lille, Archives Départementales du Nord, B 1388/1282 bis. See M. K. Elisabeth Gottschalk, *Historische geografie van Westelijk Zeeuws-Vlaanderen*, 2 vols. (Assen: Van Gorcum, 1955–58), vol. 1, *Tot de St-Elizabethvloed van 1404* 148–49.

2. 1357: River Meuse. Paris, Bibliothèque de la Sorbonne, Archives de l'Université de Paris, Reg. 2, vol. 2, fol. 35v. See

Gray C. Boyce, "The Controversy over the Boundary between the English and Picard Nations in the University of Paris (1356–1358)," in *Etudes d'histoire dédiées à la mémoire de Henri Pirenne* (Brussels: Nouvelle Société d'Editions, 1937), 55–66.

3. 1358: Oostburg and IJzendijke (Zeeland). Ghent, Rijksarchief. See M. K. Elisabeth Gottschalk, "De oudste kartografische weergave van een deel van Zeeuwsch-Vlaanderen," *Archief: Vroegere en Latere Mededelingen Voornamelijk in Betrekking tot Zeeland Uitgegeven door het Zeeuwsch Genootschap der Wetenschappen* (1948): 30–39.

4. Fifteenth century: Dordrecht(?). Dordrecht, Gemeente-archief. See Cornelis Koeman, *Collections of Maps and Atlases in the Netherlands: Their History and Present State* (Leiden: E. J. Brill, 1961), 207.

5. Fifteenth century: Rivers Lek and Waal. The Hague, Algemeen Rijksarchief. See Jan Henricus Hingman, *Inventaris der verzameling kaarten berustende in het Rijks-Archief* (The Hague: Nijhoff, 1867–71), 2:35 (no. 236); B. van 'tHoff, "The Oldest Maps of the Netherlands: Dutch Map Fragments of about 1524," *Imago Mundi* 16 (1962): 29–32, esp. 30, fig. 2.

6. 1457: Houtrijk (North Holland). See S. J. Fockema Andreae and B. van 'tHoff, *Geschiedenis der kartografie van Nederland van den Romeinschen tijd tot het midden der 19de eeuw* (The Hague: Nijhoff, 1947), 11–12, pl. 12; Johannes Keuning, "XVIth Century Cartography in the Netherlands (Mainly in the Northern Provinces)," *Imago Mundi* 9 (1952): 35–64, esp. 41.

7. 1468: River Scheldt. Brussels, Archives Générales du Royaume; Antwerp, Gemeentearchief; Middelburg, Rijksarchief in Zeeland. See C. de Waard, *Rijksarchief in Zeeland: Inventaris van kaarten en teekeningen* (Middelburg: D. G. Kröber, Jr., 1916), 10–11 (no. 99); M. K. Elisabeth Gottschalk and W. S. Unger, "De oudste kaarten der waterwegen tussen Brabant, Vlaanderen en Zeeland," *Tijdschrift van het Koninklijk Nederlandsch Aardrijkskundig Genootschap*, 2d ser., 67 (1950): 146–64; Keuning, "XVIth Century Cartography," 41.

8. 1472: Gooiland (North Holland). The Hague, Algemeen Rijksarchief, Grafelijkheid van Holland, Rekenkamer no. 755f. See D. T. Enklaar, "De oudste kaarten van Gooiland en zijn grensgebieden," *Nederlandsch Archievenblad* 39 (1931–32): 185–205, esp. 188–92, pl. I.

9. 1480: Braakman (Zeeland). Ghent, Rijksarchief. See M. P. de Bruin, "Kaart van de Braakman van ca. 1480," *Tijdschrift van het Koninklijk Nederlandsch Aardrijkskundig Genootschap*, 2d ser., 70 (1953): 506–7.

10. About 1480: Dordrecht area, depicting the Saint Elizabeth's Day Flood, 1421. Amsterdam, Rijksmuseum, A 3147a,b. See Rijksmuseum, *All the Paintings of the Rijksmuseum in Amsterdam: A Completely Illustrated Catalogue* (Amsterdam: Rijksmuseum, 1976), 633.

11. 1487: Overflakkee (South Holland). Brussels, Archives Générales du Royaume, Grand Conseil de Malines, Appels de Holland 188, sub G. See A. H. Huussen, *Jurisprudentie en kartografie in de XVᵉ en XVIᵉ eeuw* (Brussels: Algemeen Rijksarchief, 1974), 7–8, pl. 1.

12. 1498: Gouda area. Gouda, Gemeentearchief. See Fockema Andreae and van 'tHoff, *Geschiedenis der kartographie*, 13.

13. About 1500: Hoeksche Waard (South Holland). The Hague, Algemeen Rijksarchief. See Hingman, *Inventaris*, 2:231 (no. 2081); Fockema Andreae and van 'tHoff, *Geschiedenis der kartografie*, 13; Maritiem Museum Prins Hendrik, *Eilanden en waarden in kaart en beeld: Tentoonstelling, 22 december, 1953–15 maart, 1954* (Rotterdam: Maritiem Museum Prins Hendrik, [1954]), no. 54.

14. Fifteenth century(?): South Holland. The Hague, Algemeen Rijksarchief. See Hingman, *Inventaris*, 2:212 (no. 1889).

15. Fifteenth century(?): Voorne (South Holland). The Hague, Algemeen Rijksarchief. See Hingman, *Inventaris*, 2:226 (no. 2028).

BIBLIOGRAPHY
CHAPTER 20 LOCAL AND REGIONAL CARTOGRAPHY IN MEDIEVAL EUROPE

Almagià, Roberto. *Monumenta Italiae cartographica.* Florence: Istituto Geografico Militare, 1929.

Bernleithner, Ernst. "Die Klosterneuburger Fridericuskarte von etwa 1421." In *Kartengeschichte und Kartenbearbeitung,* ed. Karl-Heinz Meine, 41–44. Bad Godesberg: Kirschbaum Verlag, 1968.

Bonacker, Wilhelm. "Die sogenannte Bodenseekarte des Meisters PW bzw. PPW vom Jahre 1505." *Die Erde* 6 (1954): 1–29.

Bönisch, Fritz. "Bemerkungen zu den Wien-Klosterneuburg-Karten des 15. Jahrhunderts." In *Kartengeschichte und Kartenbearbeitung,* ed. Karl-Heinz Meine, 45–48. Bad Godesberg: Kirschbaum Verlag, 1968.

Ciullini, Rodolfo. "Firenze nelle antiche rappresentazioni cartografiche." *Firenze* 2 (1933): 33–79.

Dainville, François de. "Rapports sur les conférences: Cartographie Historique Occidentale." In *Ecole Pratique des Hautes Etudes, IVᵉ section: Sciences historiques et philologiques. Annuaire 1968–1969,* 401–2. Paris, 1969.

———."Cartes et contestations au XVᵉ siècle." *Imago Mundi* 24 (1970): 99–121.

Degenhart, Bernhard, and Annegrit Schmitt. "Marino Sanudo und Paolino Veneto." *Römisches Jahrbuch für Kunstgeschichte* 14 (1973): 1–137.

Destombes, Marcel. "A Panorama of the Sack of Rome by Pieter Bruegel the Elder." *Imago Mundi* 14 (1959): 64–73.

Durand, Dana Bennett. *The Vienna-Klosterneuburg Map Corpus of the Fifteenth Century: A Study in the Transition from Medieval to Modern Science.* Leiden: E. J. Brill, 1952.

Ehrle, F., and H. Egger. *Piante e vedute di Roma e del Vaticano dal 1300 al 1676.* Illustrated by Amato Pietro Frutaz. Rome: Vaticana, 1956.

Erben, Wilhelm. *Rombilder auf kaiserlichen und päpstlichen Siegeln des Mittelalters.* Veröffentlichungen des Historischen Seminars der Universität Graz, 7. Graz: Leuschner und Lubensky, 1931.

Ettlinger, L. D. "A Fifteenth-Century View of Florence." *Burlington Magazine* 94 (1952): 160–67.

Fernie, Eric. "The Proportions of the St. Gall Plan." *Art Bulletin* 60 (1978): 583–89.

Fockema Andreae, S. J., and B. van 'tHoff. *Geschiedenis der kartografie van Nederland van den Romeinschen tijd tot het midden der 19de eeuw.* The Hague: Nijhoff, 1947.

Harvey, P. D. A. *The History of Topographical Maps: Symbols, Pictures and Surveys.* London: Thames and Hudson, 1980.

Heydenreich, Ludwig H. "Ein Jerusalem-Plan aus der Zeit der Kreuzfahrer." In *Miscellanea pro arte*, ed. Joseph Hoster and Peter Bloch, 83–90, pls. LXII–LXV. Cologne: Freunde des Schnütgen-Museums, 1965.

Horn, Walter, and Ernest Born. "New Theses about the Plan of St. Gall." In *Die Abtei Reichenau: Neue Beiträge zur Geschichte und Kultur des Inselklosters*, ed. Helmut Maurer, 407–76. Sigmaringen: Thorbecke, 1974.

————. *The Plan of St. Gall: A Study of the Architecture and Economy of, and Life in, a Paradigmatic Carolingian Monastery*, 3 vols. Berkeley: University of California Press, 1979.

Hülsen, C. "Di una nuova pianta prospettica di Roma del secolo XV." *Bullettino della Commissione Archeologica Comunale di Roma*, 4th ser., 20 (1892): 38–47.

Huussen, A. H. *Jurisprudentie en kartografie in de XVe en XVIe eeuw*. Brussels: Algemeen Rijksarchief, 1974.

Ischer, Theophil. *Die ältesten Karten der Eidgenossenschaft*. Bern: Schweizer Bibliophile Gesellschaft, 1945.

Koeman, Cornelis. "Algemene inleiding over de historische kartografie, meer in het Bijzonder: Holland vóór 1600." *Holland* 7 (1975): 230.

Kratochwill, Max. "Zur Frage der Echtheit des 'Albertinischen Planes' von Wien." *Jahrbuch des Vereins für Geschichte der Stadt Wien* 29 (1973): 7–36.

Lavedan, Pierre. *Représentation des villes dans l'art du Moyen Age*. Paris: Vanoest, 1954.

Lelewel, Joachim. *Géographie du Moyen Age*. 4 vols. and epilogue. Brussels: J. Pilliet, 1852–57; reprinted Amsterdam: Meridian, 1966.

Levi, Annalina, and Mario Levi. "The Medieval Map of Rome in the Ambrosian Library's Manuscript of Solinus." *Proceedings of the American Philosophical Society* 118 (1974): 567–94.

Oberhummer, E. "Der Stadtplan, seine Entwickelung und geographische Bedeutung." *Verhandlungen des Sechszehnten Deutschen Geographentages zu Nürnberg* (1907): 66–101.

Oehme, Ruthardt. "Die Palästinakarte aus Bernhard von Breitenbachs Reise in das Heilige Land, 1486." *Beiheft zum Zentralblatt für Bibliothekswesen* 75 (1950): 70–83.

Olszewicz, Bolesław. *Dwie szkicowe mapy Pomorza z połowy XV wieku*. Biblioteka "Strażnicy Zachodniej" no. 1. Warsaw: Nakładem Polskiego Związku Zachodniego, 1937.

Parsons, David. "Consistency and the St. Gallen Plan: A Review Article." *Archaeological Journal* 138 (1981): 259–65.

Parsons, E. J. S. *Introduction to the Facsimile*. Memoir accompanying *The Map of Great Britain, circa A.D. 1360, Known as the Gough Map*. Oxford: Oxford University Press, 1958.

Pinto, John A. "Origins and Development of the Ichnographic City Plan." *Journal of the Society of Architectural Historians* 35, no. 1 (1976): 35–50.

Röhricht, Reinhold. "Karten und Pläne zur Palästinakunde aus dem 7. bis 16. Jahrhundert." *Zeitschrift des Deutschen Palästina-Vereins* 14 (1891): 8–11, 87–92, 137–41, pls. 1, 3–5; 15 (1892): 34–39, 185–88, pls. 1–9; 18 (1895): 173–82, pls. 5–7.

————. "Marino Sanudo sen. als Kartograph Palästinas." *Zeitschrift des Deutschen Palästina-Vereins* 21 (1898): 84–126, pls. 2–11.

Rossi, G. B. de. *Piante iconografiche e prospettiche di Roma anteriori al secolo XVI*. Rome: Salviucci, 1879.

Scaglia, Gustina. "The Origin of an Archaeological Plan of Rome by Alessandro Strozzi." *Journal of the Warburg and Courtauld Institutes* 27 (1964): 137–63.

Schnelbögl, Fritz. "Life and Work of the Nuremberg Cartographer Erhard Etzlaub (†1532)." *Imago Mundi* 20 (1966): 11–26.

Schulz, Juergen. "Jacopo de' Barbari's View of Venice: Map Making, City Views, and Moralized Geography before the Year 1500." *Art Bulletin* 60 (1978): 425–74.

Skelton, R. A., and P. D. A. Harvey, eds. *Local Maps and Plans from Medieval England*. Oxford: Clarendon Press, 1986.

Teeling, P. S. "Oud-Nederlandse landmeters." *Orgaan der Vereniging van Technische Ambtenaren van het Kadaster* 7 (1949): 34–45, 90–98, 126–34, 158–70, 198–209; 8 (1950): 2–11.

Vaughan, Richard. *Matthew Paris*. Cambridge: Cambridge University Press, 1958.

Weiss, Roberto. *The Renaissance Discovery of Classical Antiquity*. Oxford: Blackwell, 1969.

21 · Concluding Remarks

J. B. HARLEY AND DAVID WOODWARD

This first volume of the *History of Cartography* provides an overview of the present state of knowledge of prehistoric, ancient, and medieval maps in Europe and the Mediterranean. It reflects research in progress and points the way to new avenues of research. It also represents a stage in the formulation of durable generalizations, based on the reconstruction of long-term trends and patterns, about the making, use, and historical significance of maps. Certain achievements can be noted. Our concept of what constitutes a map has been expanded. We have acquired a fuller sense of the antiquity and varieties of cartographic thought and expression. Our knowledge of the technical characteristics of the surviving map artifacts has also been greatly strengthened. Similarly, the meaning of these maps within their wider cultural and social context is starting to emerge. Nevertheless, the *History* offers only some tentative steps forward. Many questions remain in the study of early European maps. In these concluding remarks we highlight three problems that permeate the material presented in the individual chapters. They concern how far there has been a continuous history of the making and use of maps; the cognitive transformations involved in the emergence of early cartography; and the social contexts of mapping.

GAPS AND DISCONTINUITIES

The study of maps from the prehistoric, ancient, and medieval periods in Europe and the Mediterranean is fraught with difficulties arising from the nature of the evidence. In some respects these difficulties are insurmountable, the record being tantilizingly incomplete and often indirect. Lines of inquiry are frequently frustrated not only by a lack of original artifacts from the periods under discussion but also by the fact that the surviving maps are often descendents of earlier prototypes. It thus becomes doubly difficult to arrive at an origin for particular cartographic traditions. Sometimes the gaps in the temporal or geographical record are explained simply in terms of the nonsurvival of evidence; for instance, attention has been drawn in the present essays to the physical vulnerability of Babylonian clay tablet maps,

Egyptian plans on papyri, Roman bronze maps, and those portolan charts actually used on board ship. In other cases, however, the evidence is ambiguous rather than nonexistent. The lack of a specific or exclusive word denoting a map in ancient Greek and Latin, for example, makes the evidence in classical literary sources confusing, since it is often difficult to distinguish references to what may have been textual itineraries from references to graphic images.

There is yet another dimension to the incompleteness of the cartographic evidence. The textual allusions may be to maps that are themselves merely derivatives of a prototype of much earlier and distant origin. Moreover, it is not always clear that a map—rather than a verbal description—ever existed. The *Ora maritima* is a case in point. This is a marine itinerary poem of the fourth century A.D., thought to be based on a periplus of the same period, itself based on the second-century B.C. Pseudo-Scymnus, which in turn is thought to have been derived from a fifth-century B.C. model. The detail of the geographical information provided in the surviving text of the *Ora maritima* suggests that maps accompanied such itineraries, but there is inadequate evidence to be certain about this.

The same problem about the intangibility of each link in the chain of derivatives can apply even where the surviving evidence is a map, since its ancestors need not also have been maps in every case. Thus, while the Peutinger map, which dates from the twelfth or early thirteenth century A.D., may be taken to indicate the strong likelihood that graphic itineraries existed in the Roman period, as a cartographic artifact it is so distant from any demonstrated Roman original that the case for such a tradition must be regarded as not proved. Similarly, even with the Hereford *mappamundi*, whose stemma goes back to the fifth-century A.D. geographical treatise of Orosius, it is still far from clear whether the ultimate ancestor was in textual or graphic form. In short, the convoluted lineage of surviving map images, or references to such images, has so many discontinuities that in most cases it simply does not allow either the divination of the details of their original construction or confident extrapolation across the gaps in the record.

It is equally clear that these gaps cannot be interpreted solely as a result of physical destruction of evidence. In terms of the geographical distribution of recorded cartographic activity, all the periods discussed reveal extensive blank areas. For the later prehistoric period, for example, plan maps appear to be confined not only to those areas where rock art in general was particularly well developed but also to specific districts within those areas, notably Valcamonica and Mont Bégo in the Alps. In the classical world, a formal knowledge of maps was associated primarily with the urban centers of Greek and Roman learning and power, and in the Middle Ages the production of both portolan charts and local and regional maps was concentrated within relatively few areas.

Similar discontinuities can be observed on the time scale. Evidence of cartographic activity during the millennia covered in this volume occurs in relatively few periods. It is not possible to point to many uninterrupted sequences of continuous mapmaking. The exceptions are in the Greek and Roman period and, to a lesser extent, in the later Middle Ages. The Greco-Roman era is preeminent in this respect, and its influence was felt up to modern times: it saw developments as fundamental for the long-term growth of cartography as Eratosthenes' measurements and map, Roman large-scale surveys, and Ptolemy's coordinates (the use of which extended into the Renaissance).

It must, however, be stressed that much remains unknown and many questions of transmission remain unresolved. No direct links have yet been established between maps made in the prehistoric period and maps of the ancient Near East and Egypt, for instance. Nor has it yet been possible to tell how far the early Greeks may have been aware of, or influenced by, Babylonian or dynastic Egyptian mapping (indeed, later Egyptian cartography may have in fact been influenced more by Greek practice than vice versa). Similarly, in western Europe and in Byzantium there is a marked contrast between the number of maps known to us from the thirteenth century onward and the relatively few that survive from earlier centuries. As in the case of the mapless regions, such breaks in the chronological records cannot be attributed solely to loss of originals or even to the obscurity of contemporary references to maps. The likely conclusion is that for long periods—and in probably the greater part of the Mediterranean and European regions—very few maps were being made or used.

It is also evident that cartographic knowledge was sometimes developed and then forgotten. Even in the case of the *mappaemundi*—where the facts of a long-term continuity are much clearer—it cannot be assumed that such a continuity necessarily implied a changing cartography, adapting to new scientific, social, or religious circumstances. Until very late in the period under consideration, survival of a mapping tradition often meant little more than its fossilized preservation by copyists. The long-delayed rediscovery of Ptolemy should alone be sufficient to highlight the significance of the difference between archival preservation and the active and dynamic continuation of a cartographic tradition that entails the growth of mapmaking skills or an orderly modification of the content of existing maps.

There is probably a limit to how far this fragmented mosaic of cartographic activity will ever be filled out. Traditionally, classicists have sought to compensate for the lack of surviving maps by reconstructing them from the textual sources. While the general validity of these exercises is often accepted, they must be interpreted with caution. Such depictions of the world—as Eratosthenes or Strabo may have seen it—with precise representations of parallels and meridians are clearly no substitute for original map artifacts, and in some respects they must inevitably distort the image of the real classical cartography. A different approach lies in the continuing attempts to expand the corpus of known maps. There is undoubtedly still scope for filling in the factual picture, and individual authors, despite considerable achievements in this direction, have repeatedly made the point that some of the most basic tasks of listing their raw material still need to be done. In the case of prehistoric mapping, the corpus of cartographically relevant material needs to be extended by applying the new criteria to a much larger body of prehistoric art before analysis can proceed. For the history of the cartography of ancient Greece and Rome, systematic lists are still needed of all artifacts, such as coins, frescoes, and mosaics, that may contain maps or maplike representations. Even for the relatively well-worked medieval period, full lists of nautical charts still have to be published and comprehensive searches made for large-scale local maps in Germany, France, Italy, Spain, and Portugal. Such inventories are essential for the future enhancement of our knowledge about the directions, manner, and rate of the dissemination of maps and the concept of the map in the European and Mediterranean worlds before the end of the fifteenth century.

Though such inquiries are likely to yield the most important future discoveries, the significance of the gaps in time and space still needs to be kept under review. An understanding of the nature and cause of these gaps—which vary between different map traditions—would help to clarify, for example, why mapping originated and grew in some areas rather than in others; and more generally, it would also highlight the necessary and sufficient conditions for cartographic development in different cultures. Eventually, greater comprehension of the nature of these gaps may help establish whether mapmaking originated independently in a number of centers—as seems certain in some cases but remains unclear for other, crucial areas—or whether it was disseminated from ideas and practices developed in a single society. All that can be firmly stated at present is that throughout the period considered in this volume there were large areas of Europe and the Mediterranean that remained devoid of the knowledge and practice of mapmaking.

COGNITIVE TRANSFORMATIONS

What can also be identified in Europe and the Mediterranean is the apparently independent initiation of a number of sophisticated traditions of mapmaking. These increase our sense of the diverse origins of cartography within a number of cultures and—together with Asia—Europe and the Mediterranean rank as the major hearth for the subsequent development of cartography worldwide. What occurred can essentially be seen as a series of cognitive transformations leading toward an awareness of "the idea of the map" as a basic form of human communication and involving changes in modes of thinking about, and graphic representation of, the world at various scales. The history of cartography is thus a history, at least in part, of the means by which this developing picture of reality—what was actually perceived—was modified with the help of maps. It is a reciprocal process of cognition in which both perception and representation become increasingly structured by different map models. We regard its operation as essential to understanding the nature of the change cartography underwent.

There are two aspects to this cognitive transformation. The first is the recognition, by groups within a number of different societies from prehistoric times onward, that the particular type of image we call the map could record and structure human experience about space. Whether it was intuitive or conscious, a graphic "language of maps" (to use the modern simile) was being developed. Moreover, it was discovered that this language could be applied equally to the representation of cosmographic, celestial, or terrestrial space and that it could be articulated in two or three dimensions.

As a more concrete historical reality, this development can be defined in terms of a growing recognition that maps fulfilled particular functions within these early societies. The essays in this volume have revealed a considerable number of map functions in the prehistoric, ancient, and medieval periods. These can be grouped into four main categories of map purpose: geographical wayfinding and inventory of the real world; sacred and cosmological representation of the world of the religious mind; the promotion of secular ideology; and an aesthetic function or decoration. Within the time period surveyed, prehistoric maps seem to have been somewhat exceptional in that all those surviving appear to have served largely religious or ceremonial purposes. In the world of the ancients, however, and throughout medieval times, all four functions were represented, sometimes in a single map. The Thera fresco was probably designed for its decorative symbolism. Many of the portolan charts, however, managed to combine some locational accuracy with a highly decorative style. Likewise, whereas the didactic and symbolic *mappaemundi* served to present the faithful with moralized versions of Christian history from the Creation to the Last Judgment, Claudius Ptolemy's instructions on how to compile a map of the known world were strictly practical. Thus the history of cartography contains several different histories, each associated with these different functions and each relating to man's attempts to impose order on the external world.

The second cognitive transformation recognized in these essays is inextricable from the first. It involves the complex historical process by which maps have become deliberately designed graphic artifacts with distinctive geometrical structures and arrays of signs recognizable to the intended viewers. To use modern terminology, concepts such as the idea of the map's frame, its orientation, its centering, its reference lines, and its transformational relation to the earth or heavens in terms of scale and projection, together with the signs that codify its content, all can be seen to have been engaged in the service of cartography during our period. When and by what means such devices crept into the maps of western Europe and the Mediterranean does not lend itself to easy generalization. It cannot be understood in terms of a neat linear progression or seen as a slow accretion of cartographic knowledge and practice. Nor should we be misled by apparent "breakthroughs" in early cartography into thinking of a series of sudden conceptual and technical revolutions in which the mapmaker's art was abruptly transformed. Neither, for that matter, do any of the evolutionary or developmental models favored by some historians of art and others, whose trajectory is patterned on hypotheses concerning man's cognitive growth, easily fit the empirical record of mapping. For instance, in the case of map signs there is no evidence to support an evolutionary maturing of the different concepts, from pictorial to abstract signs or from oblique to planimetric angle of view, in any of the maps in the period under review. At one end of the time scale are the images dating from the Upper Paleolithic, in which the idea of plan representation was already present. At the other end, late medieval *mappaemundi* and local and regional maps tend to use pictorial signs shown in profile or from an oblique viewpoint. Likewise, the adoption of the various geometric structures that are found in maps of different function shows no simple line of unidirectional change. Topological relationships are found in medieval large-scale maps as well as in those of the prehistoric period. Even where the geometries involved formal calculations and instrumental measurement, there is little in this volume to support the notion that development in the mathematical or technological aspects of cartography was straightforward or cumulative in nature.

The distinctive geometrical structures of maps, whether topological or Euclidean, are so crucial in the history of cartography that it is worth elaborating in greater detail on this aspect of the cognitive transformation. All maps share a number of common elements, but the form of each of these not only can vary but is, moreover, often historically quite specific. For example, the space occupied by the map image itself has been bounded in quite fundamentally different ways in different cultures and through time. The apparently simple idea of putting a rectangular border around a map (the so-called neat line) does not routinely appear on maps until the Renaissance. To the modern map user, the bounding frame announces the completeness and consistency of what is within that line and separates the map space from the surrounding space. Thus the frame represents a fundamental concept. The depiction of one feature within that framed space implies that all like features will also be represented therein. No such rectangular frames, however, are found on most classes of maps in the period covered by this volume. Instead, in many cases the confines of the map image were dictated by the shape and dimensions of the medium on which the map was made. The simple rectangular format of the bound codex that was prevalent throughout the first millennium A.D., for example, placed constraints on the design and layout of maps. The well-known example of Matthew Paris's map of the British Isles, which "would have been longer had the page allowed it," is not exceptional. In the case of the Peutinger map, the world known to the Romans was compressed and stretched to fit the format of the scroll on which it was drawn. Distortion of the edges of portolan charts and some of their decorative elements can often be attributed to the draftsman's attempts to make the most economical and elegant use of the vellum. And prehistoric maps were drawn to fit the contours and extent of the rock.

Yet other factors were at work in bounding and shaping maps in the prehistoric, ancient, and medieval world.

To judge from later Greek writers such as Hipparchus and Strabo, the frame of ancient Greek maps was apparently modified over the course of Greek history. The earliest frame was the circular disk of Homer's world as depicted on the shield of Achilles; later there was the rectangular shape of the maps of the *pinaki* (wooden tablets). The latter was supposed to provide a truer depiction of the inhabited world, and for Ptolemy at least, the first step in constructing a map of the world was to draw a rectangle twice as long as it was broad. The idea of a circular map, however, persisted into the Middle Ages, together with other outlines of biblical significance, such as the oval (ark shaped), the mandorla (Christian aura), and the square (the four corners of the earth).

A second geometric feature of some of these maps is the way particular areas were given significantly different weight and map space on a single map, to the extent that any notion of a uniformly scaled image is absent. Two main types of representation are found: there are maps composed of heterogeneous space, and there are maps in which the entire space was treated homogeneously. The former includes maps with a strongly symbolic or didactic function, such as prehistoric maps or *mappaemundi*. Here, certain parts of the map may be endowed with particular meaning and importance. This is clear in the caricaturelike deformation of areas of specific interest on, for example, the "Jerome" map of Asia (in which Asia Minor consumes fully half the map space), while even the strikingly modern *Forma Urbis Romae* emphasizes some buildings at the expense of others. The centering of maps on a particular point of sacred or secular importance—Delos, Rome, Jerusalem—reflects the same sort of manipulation of the geometry of the map to fit a specific perception of the world.

Maps that use space homogeneously result in a more detached and abstract mode of mapping. The underlying concepts are quite different. Each point on the map is, in theory at least, accorded identical importance, thus reducing the power of the center. This was also lessened by the decision of classical geographers to move the reference meridian of their maps of the inhabited world from Alexandria, then the center of mathematical activity, to the Fortunate Isles, an arbitrary and convenient westerly point. In addition, the whole concept of unique pairs of coordinates to describe position—explained in a cartographic context by Ptolemy in the *Geography*—by its very nature implies homogeneous space. So does the development of rectangular reference grids. Such grids may be traced back at least to the division of the inhabited world into northern and southern sections by the parallel of Athens and Rhodes and by the subsequent addition of a perpendicular reference meridian passing through Rhodes. At a more practical level, the rectangular reference grid of most Roman land division systems also implied a homogeneous treatment of space. Spherical and rectangular coordinates, both transmitted from the classical period through medieval astronomy and mathematics, were to become the basis of the cartographic renaissance of the fifteenth century. The triumph of their elegant logic is still reflected in the structure of our present-day maps.

The final aspect of the geometric structure of maps, again of vital importance in influencing the cognition of the space they represent, is their projection and orientation. This too is already manifest in the ancient as well as the medieval period. The transformation of coordinates from a sphere onto a plane surface involved the development of formal map projections. In this, Hipparchus and Ptolemy were pioneers. By the second century A.D., what we would describe today as resembling rectangular, stereographic, simple conic, and conical equal-area projections were all used for astronomical or geographical purposes. Implicit in the use of parallels and meridians in these projections was the question of the alignment of each map frame with the earth's axis. It is known that the inhabited world was routinely drawn on large globes by the Greeks in the Northern Hemisphere, and this may be why these globes, and hence the flat maps derived from them, should have been oriented to the north. However, there were, once again, other variables at work apart from those of pure mathematical logic. In the medieval period, when maps such as the *mappaemundi* were not drawn on formal graticules, map orientation varied according to religious doctrine. On the other hand, as already noted, the main axis of the early portolan charts may have been fixed with a view to achieving the best fit of the shape of the Mediterranean (the axis of which is about ten degrees off the west-east line) to that of the vellum they were drawn on, with the result that they were not aligned to any specific cardinal direction.

When taken together, these geometric features found in early maps thus serve to define the sum of an important, albeit often tentative, cognitive development in ways of representing, structuring, and thinking about space. They point to the foundations of an awareness of quite sophisticated cartographic concepts. It is clear, from the various geometric manifestations on prehistoric, ancient, and medieval maps that the key elements of "the idea of the map" were already understood in these societies. They were, moreover, translated into practices that imply deliberate decisions about the design and layout of maps while also confirming the piecemeal emergence of appropriate technical skills for their execution and manufacture.

SOCIAL CONTEXTS

Maps are the product of decisions and actions taken by identifiable members of social groups in particular historical circumstances. More than a mirror of society, maps are a reciprocal part of cultural growth and influence the pattern of its development. In this final section we will suggest that the social contexts and processes relevant to cartography are rooted in the power structures, organization, and group perceptions of those societies. We will also suggest that an insight hitherto often missing from the literature of the history of maps is to be found in these social contexts. The influences that impinge on the making and use of maps have been shown to be very diverse. But if they are cultural, economic, intellectual, political, ideological, technological, ethical, and aesthetic, in the widest sense they are also social. They are integral elements of the fabric of each society. This volume has shown how an understanding of that fabric is essential for interpreting the long-term development of European cartography, just as a knowledge of maps would be central to any history of the social perception and use of space.

Throughout the period covered by this volume, mapmaking was one of the specialized intellectual weapons by which power could be gained, administered, given legitimacy, and codified. This was almost certainly the case in the oral communities of the prehistoric periods. In societies with writing, the making of maps was both initiated and nurtured by the few literati who were associated with the ruling classes, whether as priests, scholars, or bureaucrats. For example, it has been shown how closely maps were associated with the religious elite of dynastic Egypt and of Christian medieval Europe; with the intellectual elite of Greece and Rome; and with the mercantile elite of the city-states of the Mediterranean world during the later Middle Ages. Whether the focus is on the unidentified creators of prehistoric rock art, the crusading popes and kings of the medieval period who commissioned maps of the Holy Land, or the nobility of fifteenth-century Italy who, as humanists, helped create the conditions for the return to western Europe of Ptolemy's *Geography*, the critical factor in the transmission of cartography and its skills often seems to have been the patronage of mapmaking by such elites.

The links between the character of particular elites, the institutions through which their power in society was exercised, and the types of maps they produced can also be clearly seen throughout the period covered by this volume. These links are helpful in understanding the fundamental contrast between mapping that was practical, and usually geographical in nature, and that which was cosmological in content and motivation. Dealing first with the more practical or utilitarian aspect, it becomes evident that the impetus behind geographical mapping was usually the desire for territorial expansion and control, whether the context was colonial, commercial, military, or political and whether it involved cadastral mapping as a means not only of dividing land and levying taxes but also—and especially in the case of centuriation systems—of maintaining tighter political and administrative control. In this way a common thread links certain of the geographical plans of Babylonia with those of the Roman world. Similarly, in fourteenth- and fifteenth-century Europe the portolan charts and an incipient tradition of local and regional mapping are also so linked. Even where maps were deployed as emblems, as on some of the coins and public displays of Rome, they were addressing the same theme of imperial power, and today they can be seen as a reflection of that society's attachment to its territory. It is tempting to go on from such examples, to speculate on the possibility of a link between the development and practical use of geographical maps and the emergence of some of the territorially expanding states of pre-Renaissance Europe. Certainly, maps could be major agents in the geopolitical process of these societies.

Cosmological maps arise from man's endeavor to understand his universe. They represent a different but complementary tradition of cartography. It is likely that they would have been regarded as just as practical, and just as ideally accurate, as any geographical map. In the ancient and medieval periods, when the distinction between the material and the spiritual was not made as in the post-Reformation era, the two traditions of mapping were closely interrelated. Certainly the development of early European mapmaking described in this volume owes as much to cosmological as to scientific-geographical ideas: it was cosmological inquiry, we have been constantly reminded, that provided the underlying thrust of many important developments, as in the design of globes and maps to represent the celestial sphere. This duality of motivation led to two quite different concepts of map "accuracy." In some cases, as with the astronomical globes and maps of the classical period, the integration of the results of sustained empirical observations and their graphic representation by means of scientifically determined map projections should not blind us to the fact that these artifacts were often used primarily in connection with astrological practices and in a society where astrology was still fused with astronomy as one science. That Ptolemy would have been known to ancient and medieval scholars as the author of *Tetrabiblos* (an astrological treatise) as well as of the *Almagest* and the *Geography* reminds us of the social context of such mapping. Thus globes and maps, often illustrating mythological concepts within a world view, are related to the general belief systems and so-

cial values of these early societies as much as, and sometimes more than, to an arcane mathematical learning. In any case, cosmological maps often did not require the geographical and mathematical accuracy that has become the lodestar of so much modern mapping, and their styles of representation are consequently paradigmatic rather than factual. The form of the images used varied widely, ranging from the curious posture of the Egyptian goddess Nut, arching over the universe, and the division of the heavens into segments as in the Bronze Liver of Piacenza, to the epitomization of the Christian world in the Madaba mosaic or the images of the *mappaemundi*. Such maps were the emanations of the power of a clerical elite. They codify an entirely different way of seeing the world and record a different type of interpreted experience. They are also representations of a conception of universal order and of a socially constructed world view, albeit one not requiring the practical terrestrial mapping demanded by an administration, or needed for commerce, or useful in building and maintaining empires.

Social mechanisms were equally a part of the process of the transmission of cartographic knowledge. For the historian, the problem in studying transmission is not so much one of demonstrating the general cultural contacts of those societies that did make maps—and that are perhaps assumed to be linked cartographically in some way—as one of isolating the map from the wider contexts in which it invariably occurs. On the one hand, the volume clearly demonstrates that the pathway of cartography was also the pathway of art, literature, philosophy, science, religion, and much else. On the other hand, the record of explicit cartographic contacts is extremely meager. "Mapmakers," even if we accept the validity of such a generic noun in our period, were always embedded in much wider artisanal or social groups. Throughout the period covered by this volume, very few people who made maps did so either exclusively or to earn their livelihood, and this too complicates the study of transmission. There are exceptions such as the globe makers of ancient Greece, the artisans who made the thousands of maps related to landownership and public works in the Roman Empire, or the chartmakers of medieval Italy or Spain. Even so, it is fair to say that the processes of mapmaking were usually merged with others, such as those of the scholar, the theologian, the painter, the surveyor, the fortifications expert, or the compass maker. A constant and recurring question therefore is to what extent maps were regarded in these days as a special class of artifact. These essays have shown that cartography was often inseparable from didactic or religious art in the prehistoric, the ancient, and the medieval world, that it entered the discourse of drama, poetry, mathematics, or philosophy, that it was embedded in the Scholastic cultures of the Middle Ages in western Europe and in Byzantium, and that, even with documents as important as the *Corpus Agrimensorum*, diagrams in general were only a minor part of the manuscripts as a whole, and maps in particular constituted only a small fraction of these. Just as no branch of ancient or medieval knowledge existed in isolation, so too was mapmaking rarely regarded as a separate activity. In short, throughout our period the transmission of cartographic ideas—and of map models—can be made sense of as a historical process only when seen as part of the totality of a society's knowledge and when that knowledge is also seen as a manifestation of a socially constructed world.

Questions of social and cultural context thus go to the root of understanding in the history of cartography. The empirical record of the technical changes normally associated with mapmaking in the ancient and medieval periods offers many examples of the way technical innovation alone was insufficient either to initiate or to promote a spontaneous expansion of mapping activity. Such technical landmarks—the invention of a surveying instrument or the design of a new map projection, for example—must be included among conditions for the spread of mapping. In themselves, however, they were not sufficient to account for such changes. In some respects they were as much symptoms as causes of cartographic change, and they should be seen as only one strand in the wider process by which maps developed. Viewed in this way, some of the apparent anomalies in early European cartography as recorded in this volume are better understood. For instance, there is the question of the time lag between many key inventions and their adoption in mapmaking—the magnetic compass is but one example—that highlights the superficiality of explanations for increasing cartographic sophistication couched in terms of simple cause and effect relationships. Another example is the case of dynastic Egypt where, despite the existence of many of the necessary technical conditions for making cadastral maps, such as the means of measuring, calculating, and registering small areas of land, as well as of the skills appropriate to map drawing, there is still no evidence for the systematic development of this type of cartography. Similarly, in medieval Europe, for which it is possible to reconstruct in detail both instruments and techniques of land surveying, the technical developments did not in themselves give rise to a new, well-defined genre of local mapping, divorced from traditions of artistic representation. The decision to make use of available techniques of mapmaking rested within society, so cartographic history becomes a study of needs and wants rather than of just the ability to make maps in the technical sense. Throughout the whole period, cartographic advances were often due as much

to political and ideological factors as to the level of technological progress in surveying or the graphic arts.

The same conclusion holds about how far maps should be regarded as an agent of change or of continuity throughout early European history. Not only has the evidence for the making of maps been reviewed in this volume, but an attempt has also been made to gather complementary evidence for their use, potential impact, and meaning. These aspects are also socially determined, and they raise questions about the nature and role of map users in each period. It can be argued that though mapmaking was an elite activity, and though these elites manipulated maps for their own purposes, those who were ultimately reached and influenced by the knowledge symbolized in the maps must have constituted a very much larger group. Thus the potential impact of any individual map was probably far greater than its isolated occurrence might suggest. It follows that the capacity of cartography to influence human actions or to mold mental worlds must depend not only on the extent to which maps were actually seen but also on the way they, or their messages, were understood.

It is on such questions as these that our understanding of the context of the map user is at its most speculative. Evidence for the level of map consciousness in early societies, for example, is virtually nonexistent. We do not know how often, for how long, in what circumstances, and with what lasting effects (if any) the Bronze Age inhabitants of Valcamonica may have ascended the valley side and contemplated the images on the ice-polished surfaces above their fields. We do not even know how far a publicly displayed *mappamundi* such as the Hereford world map may have been actually used (as is often implied) to instruct the peasants and pilgrims who may have stood before it.

Only for a few periods and places, as in the Greek and Roman world and to a lesser extent in some parts of Europe in the later Middle Ages, do we stand on somewhat surer ground about the population of map users. It can be inferred that by the fifth century B.C. in Athens, not only were celestial globes and maps of the earth already widely used instruments of teaching and research for the educated minority, but ordinary Athenians would have become acquainted with some types of maps through the use of zodiacs and nativity charts in astrology or through allusions to maps in the dramas staged in their theaters. Evidence from the Roman period contains nothing to suggest that the general exposure to maps was any less. The practical, educational, and propagandist uses of maps, some of which were publicly displayed in Rome or depicted on coins, must have made them even more familiar to many ordinary citizens as well as to some of the more progressive landowners of the Italian peninsula or to the tenants of centuriated areas. Such inferences are derived from literary sources; but for the period after the fall of Rome it is reasonable to suppose that this acquaintance with maps dwindled significantly, that maps passed out of the popular view, and that their influence as graphic knowledge became confined to very limited ecclesiastical or courtly elites. Not until the later Middle Ages can the reemergence of maps as a potentially greater force in history again be traced. But just as the earlier isolated cartographic examples had been scattered, this was a piecemeal reappearance, most evident in the maritime cities of the Mediterranean and of Spain and Portugal, in northern Italy and in southern Germany, and in those parts of northern Europe where local maps were gaining a limited currency among lawyers and the literati. It seems, then, that in the Middle Ages a widespread use of maps and a universal understanding of their meanings was no more than embryonic. Indeed, it is the low level of what can be termed map literacy, as much as the relatively slight record of mapmaking itself, that marks off the early period dealt with in this volume from that of the European Renaissance and helps give it a distinctive place in the larger general history of cartography.

In the final analysis, therefore, maps have always been a social as well as a technical phenomenon. In these essays we have examined the complexity of maps as graphic artifacts, traced the mutability of their functions, viewed them as monuments to human skill and ingenuity, acknowledged them as sources for the reconstruction of past environments, recognized them as bearers of new mathematical and graphic concepts, and established them as the ideological tools of political, military, and religious power. But the underlying dynamic in the historical process of cartography was the map's ability to fulfill social as well as technical roles. It was this that gave rise to the birth of mapping in different places and in different contexts in the Western world from prehistoric times to the later Middle Ages. Important, if not paramount, has been the capacity of the map as an instrument of knowledge—a way of seeing and of structuring the external world of man—to expand human consciousness and to propel the mind away from its immediate environment into the intangible spaces beyond. Seeing maps in these ways has opened a window, allowing us to glimpse the history of the map through the eyes of those early societies in which it originated and developed.

Contributors

Germaine Aujac is professor at the University of Toulouse—Le Mirail.

Tony Campbell is deputy map librarian of the British Library.

Catherine Delano Smith is senior lecturer in geography at the University of Nottingham.

O. A. W. Dilke is professor emeritus of Latin at the University of Leeds.

J. B. Harley is professor of geography at the University of Wisconsin—Milwaukee, formerly Montefiore Reader in Geography at the University of Exeter.

P. D. A. Harvey is professor emeritus of medieval history at the University of Durham.

G. Malcolm Lewis is senior lecturer in geography at the University of Sheffield.

A. R. Millard is the Rankin Reader in Hebrew and Ancient Semitic Languages at the University of Liverpool.

A. F. Shore is professor of Egyptology at the University of Liverpool.

David Woodward is professor of geography at the University of Wisconsin—Madison.

Bibliographical Index

Bibliographical Access to This Volume

Three modes of access to bibliographical information are used in this volume: the footnotes; the bibliographies; and the Bibliographical Index.

The footnotes provide the full form of a reference the first time it is cited in each chapter with short-title versions in subsequent citations. In each of the short-title references, the note number of the fully cited work is given in parentheses.

The bibliographies following each chapter provide a selective list of major books and articles relevant to its subject matter.

The Bibliographical Index comprises a complete list, arranged alphabetically by author's name, of all works cited in the footnotes. Numbers in bold type indicate the pages on which references to these works can be found. This index is divided into two parts. The first part identifies the texts of classical and medieval authors. The second part lists the modern literature.

Texts of Classical and Medieval Authors

Aelianus, Claudius (Aelian). *Varia historia.* **139**

Agathemerus. *Geographiae informatio.* **134, 135, 137, 143, 152, 153, 243**

Ambrose, Saint. *Expositio in psalmum 118.* **237**

Apollonius Rhodius. *Argonautica.* **158**

Aratus. *Phaenomena.* **141, 142, 165**

Aristophanes. *The Clouds.* **139**

Aristotle. *De caelo.* **145, 146, 148**

———. *Meteorologica.* **135, 145, 153, 248, 249**

Arrian (Flavius Arrianus). *Anabasis.* **149**

———. *Indica.* **149**

———. *Periplus Ponti Euxini.* **254**

Augustine, Saint. *De civitate Dei.* **319**

———. *Confessions.* **326**

———. *De genesi ad litteram.* **319**

———. *Quæstionum evangelicarum libri.* **319**

Autolycus of Pitane. *On the Sphere in Motion.* **154**

Avienius. *Arati Phaenomena* (translation of Aratus's versification of Eudoxus's *Phaenomena*). **143**

———. *Carmina.* **243**

———. *Descriptio Orbis Terrae* (translation of Dionysius Periegesis). **171**

———. *Ora maritima.* **150**

Bacon, Roger. *Opus Majus.* **305, 322, 345**

Basil, Saint. *Homilies.* **328**

Bede, the Venerable. *De natura rerum.* **137**

———. *De temporum ratione.* **321**

Cassiodorus. *Institutiones divinarum et saecularium litterarum.* **172, 255, 259, 261**

Cato the Elder. *Origines.* **205**

Censorinus. *De die natali.* **255**

Chaucer, Geoffrey. *Prologue to the Canterbury Tales.* **387**

Cicero. *Arataea* (translation of Aratus's versification of Eudoxus's *Phaenomena*). **143**

———. *Letters to Atticus.* **255**

———. *De natura deorum.* **160, 168**

———. *The Republic.* **159, 160, 255**

———. *Tusculan Disputations.* **160**

Cleomedes. *De motu circulari.* **152, 154, 169**

Cosmas Indicopleustes. *Christian Topography.* **143, 144, 261**

Ctesias of Cnidus. *Indica.* **149**

———. *Persica.* **149**

Dicuil. *De mensura orbis terrae.* **208, 259**

Dio Cassius. *Roman History.* **207, 236, 254**

Diodorus Siculus. *Bibliotheca.* **125**

Diogenes Laertius. *Vitae philosophorum.* **134, 158**

Dionysius Periegetes. *Periegesis.* **171, 173**

Einhard. *Early Lives of Charlemagne.* **469.**

Eratosthenes. *Geographica* (fragments). **154**

Euclid. *Elements.* **154**

———. *Phaenomena.* **154**

Eudoxus. *Circuit of the Earth* (fragments). **143**

Eumenius. *Oratio pro instaurandis scholis.* **209, 290**

Eustathius. *Commentary* (on Dionysius's *Periegesis*). **171, 266**

Frontinus. *De agrorum qualitate.* **218**

———. *De aquis urbis Romae.* **232**

———. *De controversiis limitum.* **219**

———. *De limitibus.* **202**

Geminus. *Introduction to phenomena.* **135, 143, 159, 162, 170, 171**

Germanicus. *The Phenomena of Aratus* (translation of Aratus's versification of Eudoxus's *Phaenomena*). **143**

Modern Literature

Acanfora, Maria Ornella. "Lastra di pietra figurata da Triora." *Rivista di Studi Liguri* 21 (1955): 44–50. 90, 96
———. "Singolare figurazione su pietra scoperta a Triora (Liguria)." In *Studi in onore di Aristide Calderini e Roberto Paribeni*, 3 vols., 3:115–27. Milan: Casa Editrice Ceschina, 1956. 90
———. *Pittura dell'età preistorica.* Milan: Società Editrice Libraria, 1960. 68, 83, 97
Adams, Percy G. *Travelers and Travel Liars, 1660-1800.* Berkeley and Los Angeles: University of California Press, 1962. 10
Adams, Thomas R. "The Map Treasures of the John Carter Brown Library." *Map Collector* 16 (1981): 2–8. 16
Adler, Bruno F. "Karty pervobytnykh narodov" (Maps of primitive peoples). *Izvestiya Imperatorskogo Obshchestva Lyubiteley Yestestvoznaniya, Antropologii i Etnografii: Trudy Geograficheskogo Otdeleniya* 119, no. 2 (1910). 46, 47, 54, 65, 66
Aerts W. J., et al., eds. *Alexander the Great in the Middle Ages: Ten Studies on the Last Days of Alexander in Literary and Historical Writing.* Symposium Interfacultaire Werkgroep Mediaevistiek, Groningen, 12–15 October, 1977. Nijmegen: Alfa Nijmegen, 1978. 333

Agius, George, and Frank Ventura. "Investigation into the Possible Astronomical Alignments of the Copper Age Temples in Malta." *Archaeoastronomy* 4 (1981): 10–21. **83**

Alavi, S. M. Ziauddin. *Geography in the Middle Ages.* Delhi: Sterling, 1966. **294**

Allen, John L. "Lands of Myth, Waters of Wonder: The Place of the Imagination in the History of Geographical Exploration." In *Geographies of the Mind: Essays in Historical Geosophy in Honor of John Kirtland Wright,* ed. David Lowenthal and Martyn J. Bowden, 41–61. New York: Oxford University Press, 1976. **10**

Almagià, Roberto. "Una carta della Toscana della metà del secolo XV." *Rivista Geografica Italiana* 28 (1921): 9–17. **480**

———. "Un'antica carta topografica del territorio veronese." *Rendiconti della Reale Accademia Nazionale dei Lincei: Classe di Scienze Morali, Storiche e Filologiche* 32 (1923): 63–83. **479, 488, 498**

———. *Monumenta Italiae cartographica.* Florence: Istituto Geografico Militare, 1929. **478, 479, 480, 481, 482, 498**

———. *Monumenta cartographica Vaticana.* 4 vols. Rome: Biblioteca Apostolica Vaticana, 1944–55. **8, 301, 332, 374, 379, 406, 407, 409, 419, 421, 429, 430, 435, 436, 440, 441, 447, 457**

———. "Intorno alla più antica cartografia nautica catalana." *Bollettino della Reale Società Geografica Italiana,* 7th ser., 10 (1945): 20–27. **379**

———. "Un'antica carta del territorio di Asti." *Rivista Geografica Italiana* 58 (1951): 43–44. **478, 498**

———. "Nota sulla cartografia dell'Italia nei secoli XV e XVI." *Atti della Accademia Nazionale dei Lincei: Rendiconti, Classe di Scienze Morali, Storiche e Filologiche,* 8th ser., 6 (1951): 3–8. **482**

———. "Intorno ad alcune carte nautiche italiane conservate negli Stati Uniti." *Atti della Accademia Nazionale dei Lincei: Rendiconti, Classe di Scienze Morali, Storiche e Filologiche,* 8th ser., 7 (1952): 356–66. **421, 432**

———. *Rapport au XVIIe Congrès international: Contributions pour un catalogue des cartes manuscrites, 1200-1500.* Ed. Marcel Destombes, International Geographical Union, Commission on the Bibliography of Ancient Maps. [Paris], 1952. **18**

———. "Una carta nautica di presunta origine genovese." *Rivista Geografica Italiana* 64 (1957): 58–60. **374**

———. "I lavori cartografici di Pietro e Jacopo Russo." *Atti della Accademia Nazionale dei Lincei* 12 (1957): 301–19. **458**

———. *Documenti cartografici dello stato pontificio.* Rome: Biblioteca Apostolica Vaticana, 1960. **488**

———. *Scritti geografici.* Rome: Edizioni Cremonese, 1961. **480**

Amat di San Filippo, Pietro. "Recenti ritrovamenti di carte nautiche in Parigi, in Londra ed in Firenze." *Bollettino della Società Geografica Italiana,* 3d ser., 1 (1888): 268–78; reprinted in *Acta Cartographica* 9 (1970): 1–11. **400**

American Geographical Society. *Research Catalogue of the American Geographical Society.* 15 vols. and map supplement. Boston: G. K. Hall, 1962. Updated by *Current Geographical Publications: Additions to the Research Catalogue of the American Geographical Society.* New York: American Geographical Society, 1938–78; Milwaukee: American Geographical Society Collection, 1978–. **31**

Amodeo, Tony. *Mapline* 14 (July 1979). **203**

Anati, Emmanuel. "Rock Engravings in the Italian Alps." *Archaeology* 11 (1958): 30–39. **95**

———. "Les travaux et les jours aux Ages des Métaux du Val Camonica." *L'Anthropologie* 63 (1959): 248–68 and plates I–LIV. **95, 96**

———. *La civilisation du Val Camonica.* Paris: B. Arthaud, 1960. English edition, *Camonica Valley.* Trans. Linda Asher. New York: Alfred A. Knopf, 1961; reprinted, London: Jonathan Cape, 1964. **70, 71, 78, 95**

———. *La stele di Bagnolo presso Malegno.* 2d ed. Brescia: Camuna, 1965. **87, 88**

———. *Il masso di Borno.* Brescia: Camuna, 1966. **70, 75, 95**

———. "Magourata Cave." *Archaeology* 22 (1969): 92–100. **89, 93**

———. "Magourata Cave, Bulgaria." *Bollettino del Centro Camuno di Studi Preistorici* 6 (1971): 83–107. **89, 90, 93**

———. "La stele di Ossimo." *Bollettino del Centro Camuno di Studi Preistorici* 8 (1972): 51–119. **91**

———. "La stele di Triora (Liguria)." *Bollettino del Centro Camuno di Studi Preistorici* 10 (1973): 101–27. **90, 96**

———. *Capo di Ponte.* First English edition, trans. AFSAI, International Scholarships, Brescia Chapter. Capo di Ponte: Edizioni del Centro, 1975. **95**

———. *Evolution and Style in Camunian Rock Art.* Trans. Larryn Diamond. Capo di Ponte: Edizioni del Centro, 1976. **87, 95**

———. *L'arte rupestre del Negev e del Sinai.* Milan: Jaca Book, 1979. **61**

———. "Art with a Message That's Loud and Clear." *Times Higher Educational Supplement,* 12 August 1983, 9. **92**

Anderson, Andrew R. *Alexander's Gate, Gog and Magog, and the Inclosed Nations.* Cambridge, Mass.: Medieval Academy of America, 1932. **333**

Andree, Richard. "Die Anfänge der Kartographie." *Globus: Illustrierte Zeitschrift für Länder* 31 (1877): 24–27, 37–43. **45, 54**

———. *Ethnographische Parallelen und Vergleiche.* Stuttgart: Julius Maier, 1878. **45**

Andrés, Juan. *Dell'origine, progressi e stato attuale d'ogni letteratura di Giovanni Andrés.* New ed., 8 vols. Pisa: Presso Niccolò Capurro, 1829–30. **12**

Andrews, J. H. "Medium and Message in Early Six-Inch Irish Ordnance Maps: The Case of Dublin City." *Irish Geography* 6 (1969–73): 579–93. **36**

Andrews, Michael Corbet. "The Map of Ireland: A.D. 1300–1700." *Proceedings and Reports of the Belfast Natural History and Philosophical Society for the Session 1922–23* (1924): 9–33. **407, 415**

———. "Rathlin Island in the Portolan Charts." *Journal of*

the *Royal Society of Antiquaries of Ireland* 55 (1925): 30–35. **403, 438**

———. "The Boundary between Scotland and England in the Portolan Charts." *Proceedings of the Society of Antiquaries of Scotland* 60 (1925–26): 36–66. **403, 415**

———. "The Study and Classification of Medieval Mappae Mundi." *Archaeologia* 75 (1925–26): 61–76. **288, 295, 296**

———. "The British Isles in the Nautical Charts of the XIVth and XVth Centuries." *Geographical Journal* 68 (1926): 474–81. **403**

———. "Scotland in the Portolan Charts." *Scottish Geographical Magazine* 42 (1926): 129–53, 193–213, 293–306. **403, 406, 415, 438**

Anville, Jean-Baptiste Bourguignon d'. *Dissertation sur l'étendue de l'ancienne Jérusalem et de son temple, et sur les mesures hebraiques de longueur.* Paris: Prault Fils, 1747. **10**

———. *Traité des mesures itinéraires anciennes et modernes.* Paris: Imprimerie Royale, 1769. **10**

———. *Considérations générales sur l'étude et les connaissances que demande la composition des ouvrages de géographie.* Paris: Galeries du Louvre, 1777. **10**

———. *Géographie ancienne abrégée.* Paris: A. Delalain, 1782. **10**

Appleton, Jay. *The Experience of Landscape.* New York: John Wiley, 1975. **51**

Arentzen, Jörg-Geerd. *Imago Mundi Cartographica: Studien zur Bildlichkeit mittelalterlicher Welt- und Ökumenekarten unter besonderer Berücksichtigung des Zusammenwirkens von Text und Bild.* Münstersche Mittelalter-Schriften 53. Munich: Wilhelm Fink, 1984. **291, 294, 295, 296, 307, 318, 334**

Armignacco, Vera. "Una carta nautica della Biblioteca dell'Accademia Etrusca di Cortona." *Rivista Geografica Italiana* 64 (1957): 185–223. **402, 418, 435, 439, 459**

Arnaud, D. *Naissance de l'écriture.* Ed. Béatrice André-Leicknam and Christiane Ziegler. Paris: Editions de la Réunion des Musées Nationaux, 1982. **114**

Arnaud, Pascal. "L'affaire Mettius Pompusianus, ou Le crime de cartographie." *Mélanges de l'Ecole Française de Rome: Antiquité* 95 (1983): 677–99. **254**

Arnberger, Erik. "Die Kartographie als Wissenschaft und ihre Beziehungen zur Geographie und Geodäsie." In *Grundsatzfragen der Kartographie,* 1–28. Vienna: Österreichische Geographische Gesellschaft, 1970. **xvii**

"Art and Cartography: Two Exhibitions, October 1980-January 1981." *Mapline,* special no. 5 (October 1980). **22**

Arte e scienza per il disegno del mondo. Exhibition catalog, city of Turin. Milan: Electa Editrice, 1983. **21**

Ashby, Thomas. *The Aqueducts of Ancient Rome.* Oxford: Clarendon Press, 1935. **229**

Atkinson, Richard J. C. Review of *Megalithic Remains in Britain and Brittany,* by A. Thom and A. S. Thom. *Archaeoastronomy: Supplement to the Journal for the History of Astronomy* no. 1, suppl. to vol. 10 (1979): 99–102. **85**

El atlas catalán de Cresques Abraham: Primera edición con su traducción al castellano en el sexto centenario de su

realización. Barcelona: Diáfora, 1975. Catalan edition, *L'atlas catalá de Cresques Abraham: Primera edició completa en el sis-cents aniversari de la seva realització.* Barcelona: Diáfora, 1975. **315, 447, 459**

Aujac, Germaine. *Strabon et la science de son temps.* Paris: Belles Lettres, 1966. **173, 174**

———. "La sphéropée ou la mécanique au service de la découverte du monde." *Revue d'Histoire des Sciences* 23 (1970): 93–107. **170**

———. "Une illustration de la sphéropée: L'*Introduction aux phénomènes* de Géminos." *Der Globusfreund* 18–20 (1970): 21–26. **170**

———. "Le géocentrisme en Grèce ancienne?" In *Avant, avec, après Copernic: La représentation de l'univers et ses conséquences epistémologiques,* 19–28. Centre International de Synthèse, 31ᵉ semaine de synthèse, 1–7 June 1973. Paris: A. Blanchard, 1975. **146**

———. "Poseidonios et les zones terrestres: Les raisons d'un échec." *Bulletin de l'Association Guillaume Budé* (1976): 74–78. **169**

———. "De quelques représentations de l'espace géographique dans l'Antiquité." *Bulletin du Comité des Travaux Historiques et Scientifiques: Section de Géographie* 84 (1979): 27–38. **131**

Avezac-Macaya, Marie Armand Pascal d'. "Fragments d'une notice sur un atlas manuscrit de la Bibliothèque Walckenaer: Fixation des dates des diverses parties dont il se compose." *Bulletin de la Société de Géographie,* 3d ser., 8 (1847): 142–71. **439**

———. "Note sur la mappemonde historiée de la cathédrale de Héréford, détermination de sa date et de ses sources," *Bulletin de la Société de Géographie,* 5th ser., 2 (1861): 321–34. **293**

———. *Coup d'oeil historique sur la projection des cartes de géographie.* Paris: E. Martinet, 1863. Originally published as "Coup d'oeil historique sur la projection des cartes de géographie." *Bulletin de la Société de Géographie,* 5th ser., 5 (1863): 257–361, 438–85. **186, 293, 322**

———. "La mappemonde du VIIIᵉ siècle de St. Béat de Liébana: Une digression géographique à propos d'un beau manuscrit à figures de la Bibliothèque d'Altamira." *Annales des Voyages, de la Géographie, de l'Histoire et de l'Archéologie* 2 (1870): 193–210. **293**

Avi-Yonah, Michael. *The Madaba Mosaic Map.* Jerusalem: Israel Exploration Society, 1954. **265, 469**

———. *Ancient Mosaics.* London: Cassell, 1975. **248, 266**

Avi-Yonah, Michael, and Meyer Schapiro, eds. *Israel Ancient Mosaics.* UNESCO World Art Series, no. 14. Greenwich, Conn.: New York Graphic Society, 1960. **266, 267**

Babington, Churchill, and J. R. Lumby, eds. *Polychronicon Ranulphi Higden, Together with the English Translation of John Trevisa and of an Unknown Writer of the Fifteenth Century.* London: Longman, 1865–86. **287**

Backhouse, Janet. *The Illuminated Manuscript.* Oxford: Phaidon, 1979. **376**

Badawy, Alexander. *Le dessin architectural chez les anciens Egyptiens.* Cairo: Imprimerie Nationale, 1948. **118**

———. *A History of Egyptian Architecture: The Empire (the New Kingdom).* Berkeley: University of California Press, 1968. **118**

Baehr, U. *Tafeln zur Behandlung chronologischer Probleme.* Veröffentlichungen des Astronomischen Rechen-Instituts zu Heidelberg no. 3. Karlsruhe: G. Braun, 1955. **141**

Baehrens, Emil, ed. *Poetae Latini minores.* 5 vols. Leipzig: Teubner, 1879–83; reprinted New York: Garland, 1979. **259**

Bagrow, Leo. *Istoriya geograficheskoy karty: Ocherk i ukazatel' literatury* (The history of the geographical map: Review and survey of literature). *Vestnik arkheologii i istorii, izdavayemyy Arkheologicheskim Istitutom* (Archaeological and historical review, published by the Archaeological Institute). Petrograd, 1918. **24, 47, 54, 66, 73**

———. *A. Ortelii catalogus cartographorum.* Gotha: Justus Perthes, 1928; reprinted in *Acta Cartographica* 27 (1981): 65–357. **11**

———. Review of *The World of Maps: A Study in Map Evolution*, by W. W. Jervis. *Imago Mundi* 2 (1937): 98. **25**

———. "Sixteenth International Geographical Congress, 1938." *Imago Mundi* 3 (1939): 100–102. **18**

———. "The Origin of Ptolemy's Geographia." *Geografiska Annaler* 27 (1945): 318–87. **189, 190, 192, 269, 270, 271**

———. "The Maps from the Home Archives of the Descendants of a Friend of Marco Polo." *Imago Mundi* 5 (1948): 3–13. **315**

———. "Old Inventories of Maps." *Imago Mundi* 5 (1948): 18–20. **8, 292**

———. *Die Geschichte der Kartographie.* Berlin: Safari-Verlag, 1951. **25, 45, 47, 48, 54, 73**

———. "An Old Russian World Map." *Imago Mundi* 11 (1954): 169–74. **314**

———. *Meister der Kartographie.* Berlin: Safari-Verlag, 1963. **25, 47, 54, 73**

———. *History of Cartography.* Rev. and Enl. R. A. Skelton. Trans. D. L. Paisey. Cambridge: Harvard University Press; London: C. A. Watts, 1964; reprinted and enlarged, Chicago: Precedent Publishing, 1985. **xv, 25, 26, 45, 47, 48, 54, 73, 85, 96, 178, 234, 294, 299, 315, 316, 457, 459**

Baker, Alan R. H., ed. *Progress in Historical Geography.* Newton Abbot: David and Charles, 1972. **31**

Baldacci, Osvaldo. "Storia della cartografia." In *Un sessantennio di ricerca geografica italiana*, 507–52. Memorie della Società Geografica Italiana, vol. 26. Rome, 1964. **37**

———. "Ecumene ed emisferi circolari." *Bollettino della Società Geografica Italiana* 102 (1965): 1–16. **318**

———. "La cartonautica medioevale precolombiana." In *Atti del Convegno Internazionale di Studi Colombiani 13 e 14 ottobre 1973*, 123–36. Genoa: Civico Istituto Colombiano, 1974. **406**

———. "Geoecumeni quadrangolari." *Geografia* 6 (1983): 80–86. **318**

———. "L'ecumene a mandorla." *Geografia* 6 (1983): 132–38. **318**

Baldelli, Gabriele. *Novilara: Le necropoli dell'età del ferro.* Exhibition catalog. Pesaro: Museo Archelogico Oliveriano, Comune di Pesaro, IV Circoscrizione, n.d. **76**

Ball, John. *Egypt in the Classical Geographers.* Cairo: Government Press, Bulâq, 1942. **122**

Baratta, Mario. "Sopra un'antica carta del territorio bresciano." *Bollettino della Reale Società Geografica*, 5th ser., 2 (1913): 514–26, 1025–31, 1092. **479, 498**

Barbosa, António. *Novos subsídios para a história da ciência náutica portuguesa da época dos descobrimentos.* Oporto, 1948. **385**

Barguet, Paul. "Khnoum-Chou, patron des arpenteurs." *Chronique d'Egypte* 28 (1953): 223–27. **125**

———. *Le livre des morts des anciens Egyptiens.* Paris: Editions du Cerf, 1967. **119**

———. "Essai d'interprétation du Livre des deux chemins." *Revue d'Egyptologie* 21 (1969): 7–17. **120**

Barkan, Leonard. *Nature's Work of Art: The Human Body as Image of the World.* New Haven: Yale University Press, 1975. **335, 340**

Barley, M. W. "Sherwood Forest, Nottinghamshire, Late 14th or Early 15th Century." In *Local Maps and Plans from Medieval England*, ed. R. A. Skelton and P. D. A. Harvey, 131–39. Oxford: Clarendon Press, 1986. **495**

Baron, Margaret E. *The Origins of the Infinitesimal Calculus.* Oxford: Pergamon Press, 1969. **323**

Barraclough, E. M. C., and W. G. Crampton. *Flags of the World.* London: Warne, 1978. **399**

Barraclough, Geoffrey, ed. *The Times Atlas of World History.* Maplewood: Hammond, 1979. **49**

Barth, Fredrik. *Ritual and Knowledge among the Baktaman of New Guinea.* New Haven: Yale University Press, 1975. **59**

Barthel, W. "Römische Limitation in der Provinz Africa." *Bonner Jahrbücher* 120 (1911): 39–126. **198**

Barthes, Roland. *Elements of Semiology.* Trans. Annette Lavers and Colin Smith. New York: Hill and Wang, [1968]. **2**

Bately, Janet M. "The Relationship between Geographical Information in the Old English Orosius and Latin Texts Other Than Orosius." In *Anglo-Saxon England*, ed. Peter Clemoes, 1:45–62. Cambridge: Cambridge University Press, 1972–. **301**

Battaglia, Raffaello. "Incisioni rupestri di Valcamonica." In *Proceedings of the First International Congress of Prehistoric and Protohistoric Sciences, London, August 1-6, 1932*, 234–37. London: Oxford University Press, 1934. **78, 79, 95**

———. "Ricerche etnografiche sui petroglifi della Cerchia Alpina." *Studi Etruschi* 8 (1934): 11–48, and pls. I–XXII. **78, 79, 95, 96**

Battaglia, Raffaello, and Maria Ornella Acanfora. "Il masso inciso di Borno in Valcamonica." *Bollettino di Paletnologia Italiana* 64 (1954): 225–55. **75, 95**

Battisti, Eugenio. In "Astronomy and Astrology." In

Encyclopedia of World Art, 2:40. New York: McGraw-Hill, 1960. **81**

Baudouin, Marcel. *La préhistoire par les étoiles*. Paris: N. Maloine, 1926. **82**

Baudri de Bourgueil. *Les oeuvres poétiques de Baudri de Bourgueil (1046-1130)*. Ed. Phyllis Abrahams. Paris: Honoré Champion, 1926. **339**

Bausani, Alessandro. "Interpretazione paleo-astronomica della stele di Triora." *Bollettino del Centro Camuno di Studi Preistorici* 10 (1973): 127–34. **90, 96**

Beazley, Charles Raymond. *The Dawn of Modern Geography: A History of Exploration and Geographical Science from the Conversion of the Roman Empire to A.D. 900*. 3 vols. London: J. Murray, 1897–1906. **288, 290, 291, 293, 299, 309, 322, 325, 328, 340, 347, 348**

———. "New Light on Some Mediæval Maps." *Geographical Journal* 14 (1899): 620–29; 15 (1900): 130–41, 378–89; 16 (1900): 319–29. **293, 475**

———. "The First True Maps." *Nature* 71 (1904): 159–61. **371**

Becatti, Giovanni, ed. *Mosaici e pavimenti marmorei*. 2 pts. (1961). Both are vol. 4 of *Scavi di Ostia*. Rome: Istituto Poligrafico dello Stato, 1953–. **230, 231, 246, 247**

Beer, Ellen Judith. *Die Glasmalereien der Schweiz vom 12. bis zum Beginn des 14. Jahrhunderts*. Corpus Vitrearum Medii Aevi, Schweiz, vol. 1. Basel: Birkhäuser, 1956. **335**

Bellori, Giovanni Pietro. *Ichnographia veteris Romae*. Rome: Chalcographia R.C.A., 1764. **239**

Beloch, Julius. *Campanien*. 2d ed. Breslau, 1890; reprinted Rome: Erma di Bretschneider, 1964. **210**

Beltràn Lloris, Miguel. "Los grabados rupestres de Bedolina (Valcamonica)." *Bollettino del Centro Camuno di Studi Preistorici* 8 (1972): 121–58. **78, 79, 95**

Beltran-Martínez, Antonio, René Gailli, and Romain Robert. *La Cueva de Niaux*. Monografías Arqueologicas 16. Saragossa: Talleres Editoriales, 1973. **55**

Bentham, R. M. "The Fragments of Eratosthenes of Cyrene." Typescript for Ph.D. thesis, University of London, 1948. **154**

Berchem, D. van. "L'annone militaire dans l'empire romain au IIIᵉ siècle." *Bulletin de la Société Nationale des Antiquaires de France* 80 (1937): 117–202. **235**

Berchet, Guglielmo. "Portolani esistenti nelle principali biblioteche di Venezia." *Giornale Militare per la Marina* 10 (1865): 1–11. **407**

Berenguer, Magín. *Prehistoric Man and His Art: The Caves of Ribadesella*. Trans. Michael Heron. London: Souvenir Press, 1973. **55**

Beresford, M. W. "Inclesmoor, West Riding of Yorkshire, circa 1407." In *Local Maps and Plans from Medieval England*, ed. R. A. Skelton and P. D. A. Harvey, 147–61. Oxford: Clarendon Press, 1986. **491**

Berger, Hugo. "Entwicklung der Geographie der Erdkugel bei den Hellenen." *Die Grenzboten: Zeitschrift für Politik, Literatur und Kunst* 39.4 (1880): 403–17. **163**

Berger, Suzanne. "A Note on Some Scenes of Land-Measurement." *Journal of Egyptian Archaeology* 20 (1934): 54–56. **125**

Bergman, Jan. "Zum Zwei-Wege-Motiv:

Religionsgeschichtliche und exegetische Bemerkungen." *Svensk Exegetisk Årsbok* 41–42 (1976–77): 27–56. **120**

Berlin, Staatliche Museen, Preußischer Kulturbesitz. *Ägyptisches Museum, Berlin*. Berlin: Staatliche Museen, 1967. **125**

Bernard, J. H., trans. *The Itinerary of Bernard the Wise*. Palestine Pilgrims Text Society 3. London, 1893; reprinted New York: AMS Press, 1971. **340**

Bernardini, Enzo. *Le Alpi Marittime e le meraviglie del Monte Bego*. Genoa: SAGEP Editrice, 1979. **66, 67, 93, 95**

Bernleithner, Ernst. "Die Klosterneuburger Fridericuskarte von etwa 1421." In *Kartengeschichte und Kartenbearbeitung*, ed. Karl-Heinz Meine, 41–44. Bad Godesberg: Kirschbaum Verlag, 1968. **473**

Bernoulli, Carl Christoph. "Ein Karteninkunabelnband der öffentlichen Bibliothek der Universität Basel." *Verhandlungen der Naturforschenden Gesellschaft in Basel* 18 (1906): 58–82; reprinted in *Acta Cartographica* 27 (1981): 358–82. **6**

Berthelot, André. *L'Asie ancienne centrale et sud-orientale d'après Ptolémée*. Paris: Payot, 1930. **199**

Berthelot, René. *La pensée de l'Asie et l'astrobiologie*. Paris: Payot, 1949. **86**

Bertin, Jacques. *Sémiologie graphique: Les diagrammes, les réseaux, les cartes*. Paris: Gauthier-Villars, 1967. English edition, *Semiology of Graphics: Diagrams, Networks, Maps*. Ed. Howard Wainer. Trans. William J. Berg. Madison: University of Wisconsin Press, 1983. **34**

Bertoldi, A. "Topografia del Veronese (secolo XV)." *Archivio Veneto*, n.s., 18 (1888): 455–73. **498**

Bertolini, G. L. "I quattro angoli del mondo e la forma della terra nel passo di Rabano Mauro." *Bollettino della Società Geografica Italiana* 47 (1910): 1433–41. **319**

Bertolotti, Antonio. *Artisti veneti in Roma nei secoli XV, XVI e XVII: Studi e ricerche negli archivi romani*. Venice: Miscellanea Pubblicata dalla Reale Deputazione di Storia Patria, 1884; reprinted Bologna: Arnaldo Forni, 1965. **324**

Beševliev, Bojan. "Basic Trends in Representing the Bulgarian Lands in Old Cartographic Documents up to 1878." *Etudes Balkaniques* 2 (1980): 94–123. **371**

Betten, F. S. "St. Boniface and the Doctrine of the Antipodes." *American Catholic Quarterly Review* 43 (1918): 644–63. **319**

Bevan, W. L., and H. W. Phillott. *Medieval Geography: An Essay in Illustration of the Hereford Mappa Mundi*. London: E. Stanford, 1873. **288**

Biasutti, Renato. "Un'antica carta nautica italiana del Mar Caspio." *Rivista Geografica Italiana* 54 (1947): 39–42. **422**

Bicknell, Clarence M. "Le figure incise sulle rocce di Val Fontanalba." *Atti della Società Ligustica di Scienze Naturali e Geografiche* 8 (1897): 391–411, and pls. XI–XIII. **66, 93, 94**

———. *The Prehistoric Rock Engravings in the Italian Maritime Alps*. Bordighera: P. Gibelli, 1902. **66, 67, 75, 93, 94**

———. *Further Explorations in the Regions of the*

Prehistoric Rock Engravings in the Italian Maritime Alps. Bordighera: P. Gibelli, 1903. **66, 67, 93, 94**

——. *A Guide to the Prehistoric Rock Engravings in the Italian Maritime Alps.* Bordighera: G. Bessone, 1913. **66, 67, 74, 75, 77, 86, 94, 95**

Biggs, Robert D. "The Ebla Tablets: An Interim Perspective." *Biblical Archaeologist* 43, no. 2 (1980): 76–86. **107**

Birch, Samuel. "Upon an Historical Tablet of Ramses II., 19th Dynasty, relating to the Gold Mines of Æthiopia." *Archaeologia* 34 (1852): 357–91. **122**

Birley, Anthony R. *The Fasti of Roman Britain.* Oxford: Clarendon Press, 1981. **217**

Blacker, Carmen, and Michael Loewe, ed. *Ancient Cosmologies.* London: George Allen and Unwin, 1975. **86, 87, 112**

Blain, André, and Yves Paquier. "Les gravures rupestres de la Vallée des Merveilles." *Bollettino del Centro Camuno di Studi Preistorici* 13–14 (1976): 109–19. **67, 93**

Blakemore, Michael J. "Cartography." In *The Dictionary of Human Geography,* ed. R. J. Johnston, 29–33. Oxford: Blackwell Reference, 1981. **34**

——. "From Way-finding to Map-making: The Spatial Information Fields of Aboriginal Peoples." *Progress in Human Geography* 5, no. 1 (1981): 1–24. **47, 48, 58, 59, 60**

Blakemore, Michael J., and J. B. Harley. *Concepts in the History of Cartography: A Review and Perspective.* Monograph 26. *Cartographica* 17, no. 4 (1980). **xvii, 21, 28, 35, 38, 47, 48, 60, 61, 415**

Blatt, Sidney J. *Continuity and Change in Art: The Development of Modes of Representation.* Hillsdale, N.J.: Lawrence Erlbaum Associates, 1984. **2**

Blaut, James M., George S. McCleary, and America S. Blaut. "Environmental Mapping in Young Children." *Environment and Behavior* 2 (1970): 335–49. **52**

Bliss, Carey S. "The Map Treasures of the Huntington Library." *Map Collector* 15 (1981): 32–36. **16**

Blume, Friedrich, et al., eds. *Die Schriften der römischen Feldmesser.* 2 vols. Berlin: Georg Reimer, 1848–52; reprinted Hildesheim: G. Olms, 1967. **213, 217, 220, 226, 259**

Blumer, Walter. "The Oldest Known Plan of an Inhabited Site Dating from the Bronze Age, about the Middle of the Second Millennium B.C." *Imago Mundi* 18 (1964): 9–11. **54, 78, 95**

——. "Felsgravuren aus prähistorischer Zeit in einem oberitalienischen Alpental: Ältester bekannter Ortsplan, Mitte des zweiten Jahrtausends v. Chr." *Die Alpen,* 1967, no. 2. **54, 78, 95, 96**

——. "Ortsplan von Bedolina: Felsgravur um die Mitte des zweiten Jahrtausends v. Chr." *Kartographische Nachrichten* 18 (1968): 10–13. **95**

Board, Christopher. "Maps as Models." In *Models in Geography,* ed. Richard J. Chorley and Peter Haggett, 671–725. London: Methuen, 1967. **34**

——. "Maps and Mapping." *Progress in Human Geography* 1 (1977): 288–95. **34**

——. "Cartographic Communication." In *Maps in*

Modern Geography: Geographical Perspectives on the New Cartography, ed. Leonard Guelke, 42–78. Monograph 27, *Cartographica* 18, no. 2 (1981). **34**

Boardman, David. *Graphicacy and Geography Teaching.* London: Croom Helm, 1983. **30**

Boisse, Claude. *Le Tricastin des origines à la chute de l'Empire romain.* Valence: Sorepi, 1968. **222**

Boissier, Gaston. "Les rhéteurs gaulois du IVᵉ siècle." *Journal des Savants* (1884): 1–18. **290**

Böker, Robert. "Windrosen." In *Paulys Realencyclopädie der classischen Altertumswissenschaft,* ed. August Pauly, Georg Wissowa, et al., 2d ser., 8.2 (1958): cols. 2325–81. Stuttgart: J. B. Metzler, 1894–. **248**

Boll, F. "Die Sternkataloge des Hipparch und des Ptolemaios." *Biblioteca Mathematica,* 3d ser., 2 (1901): 185–95. **165**

Bonacker, Wilhelm. "The Egyptian *Book of the Two Ways.*" *Imago Mundi* 7 (1950): 5–17. **88, 120**

——. "Eine unvollendet gebliebene Geschichte der Kartographie von Konstantin Cebrian." *Die Erde* 3 (1951–52): 44–57. **25**

——. "Die sogenannte Bodenseekarte des Meisters PW bzw. PPW vom Jahre 1505." *Die Erde* 6 (1954): 1–29. **488**

——. "Lev Semenovič Bagrov (1888-1957): Ein Leben für die Geschichte alter Karten." *Petermanns Geographische Mitteilungen* 101 (1957): 308–9. **26**

——. "Kartographische Gesellschaften: Vorläufer und Wegbereiter der internationalen kartographischen Vereinigung." *Geographisches Taschenbuch* (1960–61) supp., 58–77. **32**

——. "Stellungsnahme zu dem Plan einer internationalen Zentralstelle für Geschichte der Kartographie." *Kartographische Nachrichten* 12 (1962): 147–50. **37**

——. "Über die Wertsteigerung von Einzelblättern aus zerfledderten alten Atlaswerken." *Kartographische Nachrichten* 13 (1963): 178–79. **6**

——. "The First International Symposium of the Coronelli Weltbund der Globusfreunde." *Imago Mundi* 18 (1964): 83–84. **22**

Bonfante, Giuliano, and Larissa Bonfante. *The Etruscan Language: An Introduction.* Manchester: Manchester University Press, 1984. **201**

Bönisch, Fritz. "Bemerkungen zu den Wien-Klosterneuburg-Karten des 15. Jahrhunderts." In *Kartengeschichte und Kartenbearbeitung,* ed. Karl-Heinz Meine, 45–48. Bad Godesberg: Kirschbaum Verlag, 1968. **473**

Bonner, John T. *The Evolution of Culture in Animals.* Princeton: Princeton University Press, 1980. **50**

Boon, George C. *Isca: The Roman Legionary Fortress at Caerleon, Mon.* 3d ed. Cardiff: National Museum of Wales, 1972. **252**

Borchardt, Ludwig. "Ein altägyptisches astronomisches Instrument." *Zeitschrift für Ägyptische Sprache und Altertumskunde* 37 (1899): 10–17. **125**

Bord, Janet. *Mazes and Labyrinths of the World.* London: Latimer New Dimensions, 1976. **88, 251**

Bordone, Benedetto. *Libro . . . de tutte l'isole del mondo.* Amsterdam: Theatrum Orbis Terrarum, 1966. **482**

Borgna, Cesare Giulio. "La mappa litica di rocio Clapier." *L'Universo* 49, no. 6 (1969): 1023–42. **80**

Bosinski, Gerhard. "Magdalenian Anthropomorphic Figures at Gönnersdorf (Western Germany)." *Bolletino del Centro Camuno di Studi Preistorici* 5 (1970): 57–97. **58**

Bosio, Luciano. *La "Tabula Peutingeriana": Una carta stradale romana del IV secolo*. Florence: 3M Italia and Nuova Italia, 1972. **241**

———. *La Tabula Peutingeriana: Una descrizione pittorica del mondo antico*. I Monumenti dell'Arte Classica, vol. 2. Rimini: Maggioli, 1983. **238**

Bottomore, Tom, ed. *A Dictionary of Marxist Thought*. Oxford: Blackwell Reference, 1983. **xviii**

Bouché-Leclercq, A. *L'astrologie grecque*. Paris: E. Leroux, 1899; reprinted Brussels: Culture et Civilisation, 1963. **166**

Boulding, Kenneth E. *The Image*. Ann Arbor: University of Michigan Press, 1956. **2**

Bourbaki, Nicolas. *Eléments d'histoire des mathématiques*. New ed. Paris: Hermann, 1974. **xvii**

Bowen, Margarita. *Empiricism and Geographical Thought from Francis Bacon to Alexander von Humboldt*. Cambridge: Cambridge University Press, 1981. **10**

Boxer, C. R. *The Portuguese Seaborne Empire, 1415-1825*. London: Hutchinson, 1969. **2**

Boyce, Gray C. "The Controversy over the Boundary between the English and Picard Nations in the University of Paris (1356–1358)." In *Etudes d'histoire dédiées à la mémoire de Henri Pirenne*, 55–66. Brussels: Nouvelle Société d'Editions, 1937. **485, 499**

Boyer, Carl B. *A History of Mathematics*. New York: John Wiley, 1968. **xvii**

Bradford, John. *Ancient Landscapes*. London: Bell, 1957. **219**

Brancati, Antonio. *La biblioteca e i musei Oliveriani di Pesaro*. Pesaro: Banca Popolare Pesarese, 1976. **248**

Brandt, Bernhard. *Mittelalterliche Weltkarten aus Toscana*. Geographisches Institut der Deutschen Universität in Prag. Prague: Staatsdruckerei, 1929. **324**

———. "Eine neue Sallustkarte aus Prag." *Mitteilungen des Vereins der Geographen an der Universität Leipzig* 14–15 (1936): 9–13. **343**

Braudel, Fernand. *The Mediterranean and the Mediterranean World in the Age of Philip II*. 2 vols. Trans. Siân Reynolds. London: Collins, 1972–73. **387**

Braunfels, Wolfgang. *Mittelalterliche Stadtbaukunst in der Toskana*. Berlin: Mann, 1979. **488**

Brehaut, Ernest. *An Encyclopedist of the Dark Ages: Isidore of Seville*. Studies in History, Economics and Public Law, vol. 48, no. 1. New York: Columbia University Press, 1912. **301, 320**

Bremner, Robert W. "An Analysis of a Portolan Chart by Freduci d'Ancone." Paper prepared for the Eleventh International Conference on the History of Cartography, Ottawa, 1985. **384**

Brendel, Otto J. *Symbolism of the Sphere: A Contribution to the History of Earlier Greek Philosophy*. Leiden: E. J. Brill, 1977. **171**

Brennan, Martin. *The Stars and the Stones: Ancient Art and Astronomy in Ireland*. London: Thames and Hudson, 1983. **84**

Breuil, Henri. *Les peintures rupestres schématiques de la Péninsule Ibérique*. 4 vols. Fondation Singer-Polignac. Paris: Imprimerie de Lagny, 1933. **68, 69, 97**

———. *Les roches peintes du Tassili-n-Ajjer*. Paris: Arts et Métiers Graphiques, 1954. **69, 70, 97**

———. "The Palaeolithic Age." In *Larousse Encyclopedia of Prehistoric and Ancient Art*, ed. René Huyghe, 30–39. London: Paul Hamlyn, 1962. **60**

Breuil, Henri, Hugo Obermaier, and W. Verner. *La Pileta a Benaojan (Malaga)*. Monaco: Impr. artistique Vve A. Chêne, 1915. **69**

Breusing, Arthur A. "Zur Geschichte der Kartographie: La Toleta de Marteloio und die loxodromischen Karten." *Kettlers Zeitschrift für Wissenschaft: Geographie* 2 (1881): 129–33, 180–95; reprinted in *Acta Cartographica* 6 (1969): 51–70. **375**

Breydenbach, Bernard von. *Peregrinatio in Terram Sanctam*. Mainz: Reuwich, 1486. **474**

Brice, William C. "Early Muslim Sea-Charts." *Journal of the Royal Asiatic Society of Great Britain and Ireland* 1 (1977): 53–61. **420**

Bricker, C., and R. V. Tooley. *Landmarks of Mapmaking: An Illustrated Survey of Maps and Mapmakers*. Brussels: Elsevier-Sequoia, 1968. **3, 22**

Brincken, Anna-Dorothee von den. "Mappa mundi und Chronographia." *Deutsches Archiv für die Erforschung des Mittelalters* 24 (1968): 118–86. **288, 290, 326, 328, 330**

———. "Die Ausbildung konventioneller Zeichen und Farbgebungen in der Universalkartographie des Mittelalters." *Archiv für Diplomatik: Schriftgeschichte Siegel- und Wappenkunde* 16 (1970): 325–49. **325, 326, 327**

———. "Zur Universalkartographie des Mittelalters." In *Methoden in Wissenschaft und Kunst des Mittelalters*, ed. Albert Zimmermann, 249–78. Miscellanea Mediaevalia 7. Berlin: Walter de Gruyter, 1970. **288, 290**

———. "Europa in der Kartographie des Mittelalters." *Archiv für Kulturgeschichte* 55 (1973): 289–304. **288**

———. "Die Klimatenkarte in der Chronik des Johann von Wallingford—ein Werk des Matthaeus Parisiensis?" *Westfalen* 51 (1973): 47–57. **357**

———. "Die kartographische Darstellung Nordeuropas durch italienische und mallorquinische Portolanzeichner im 14. und in der ersten Hälfte des 15. Jahrhunderts." *Hansische Geschichtsblätter* 92 (1974): 45–58. **410**

———. "Die Kugelgestalt der Erde in der Kartographie des Mittelalters." *Archiv für Kulturgeschichte* 58 (1976): 77–95. **320**

———. "Portolane als Quellen der Vexillologie." *Archiv für Diplomatik: Schriftgeschichte Siegel- und Wappenkunde* 24 (1978): 408–26. **399**

British Museum. *Catalogue of the Manuscript Maps, Charts, and Plans, and of the Topographical Drawings in the British Museum*. 3 vols. London: Trustees of the British

Museum, 1844–61. 19, 270, 375, 433, 435, 438, 440, 442, 444

————. *Catalogue of the Printed Maps, Plans and Charts in the British Museum.* 2 vols. London: W. Clowes by order of the Trustees of the British Museum, 1885. 19

————. *Cuneiform Texts from Babylonian Tablets, etc., in the British Museum.* London: British Museum, 1896–. 109, 110, 111, 112

————. *Four Maps of Great Britain Designed by Matthew Paris about A.D. 1250.* London: British Museum, 1928. 496

————. *Catalogue of Printed Maps, Charts and Plans.* 15 vols. and suppls. London: Trustees of the British Museum, 1967. 21

Broc, Numa. *La géographie de la Renaissance (1420-1620).* Paris: Bibliothèque Nationale, 1980. 9, 38

————. "Visions Médiévales de la France." *Imago Mundi* 36 (1984): 32–47. 403

Brody, Hugh. *Maps and Dreams.* New York: Pantheon Books, 1982. 59

Brown, Cecil H. "Where Do Cardinal Direction Terms Come From?" *Anthropological Linguistics* 25 (1983): 121–61. 337

Brown, Lloyd A. *The Story of Maps.* Boston: Little, Brown, 1949; reprinted New York: Dover, 1979. 3, 25, 26, 35, 45, 47, 168, 177, 252, 294, 383

Brown, Rawdon, et al., eds. *Calendar of State Papers and Manuscripts, Relating to English Affairs, Existing in the Archives and Collections of Venice, and in Other Libraries of Northern Italy.* 38 vols. Great Britain Public Record Office. London: Her Majesty's Stationery Office, 1864–1947. 433

Brown, Roger William. *Words and Things.* New York: Free Press, 1958. 60

Browne, George Forrest. *On Some Antiquities in the Neighbourhood of Dunecht House Aberdeenshire.* Cambridge: Cambridge University Press, 1921. 82

Browning, Robert. *The Byzantine Empire.* New York: Charles Scribner's Sons, 1980. 258, 259, 266

Bruin, M. P. de. "Kaart van de Braakman van ca. 1480." *Tijdschrift van het Koninklijk Nederlandsch Aardrijkskundig Genootschap,* 2d ser., 70 (1953): 506–7. 500

Brunello, Franco. *"De arte illuminandi" e altri trattati sulla tecnica della miniatura medievale.* Vicenza: Neri Pozza Editore, 1975. 324

Brunner, Hellmut. "Die Unterweltsbücher in den ägyptischen Königsgräbern." In *Leben und Tod in den Religionen: Symbol und Wirklichkeit,* ed. Gunter Stephenson, 215–28. Darmstadt: Wissenschaftliche Buchgesellschaft, 1980. 120

Bruns, C. G. *Fontes iuris Romani antiqui.* 7th ed. Tübingen: Mohr, 1909. 210

Buache, Jean Nicholas. *Mémoire sur la Géographie de Ptolémée et particulièrement sur la description de l'intérieur de l'Afrique.* Paris: Imprimerie Royale, 1789. 11

Buache, Philippe. "Dissertation sur l'île Antillia." In *Mémoires sur l'Amérique et sur l'Afrique donnés au mois d'avril 1752.* N.p., 1752. 10

Buchner, Edmund. "Römische Medaillons als Sonnenuhren." *Chiron* 6 (1976): 329–48. 214

————. "Horologium Solarium Augusti: Vorbericht über die Ausgrabungen 1979/80." *Mitteilungen des Deutschen Archäologischen Instituts, Römische Abteilung* 87 (1980): 355–73. 208

————. *Die Sonnenuhr des Augustus: Nachdruck aus RM 1976 und 1980 und Nachtrag über die Ausgrabung 1980/1981.* Mainz: von Zabern, 1982. 208

Buck, Adriaan de. *The Egyptian Coffin Texts VII.* Oriental Institute Publications, vol. 87. Chicago: University of Chicago Press, 1961. 120

Buczek, Karol. *History of Polish Cartography from the 15th to the 18th Century.* 2d ed. Trans. Andrzej Potocki. Amsterdam: Meridian, 1982. 37, 488

Buisseret, David. "Les ingénieurs du roi au temps de Henri IV." *Bulletin du Comité des Travaux Historiques et Scientifiques: Section de Géographie* 77 (1964): 13–84. 9

Bull, William E., and Harry F. Williams. *Semeiança del Mundo: A Medieval Description of the World.* Berkeley and Los Angeles: University of California Press, 1959. 288

Bulmer-Thomas, Ivor. "Conon of Samos." In *Dictionary of Scientific Biography,* 16 vols., ed. Charles Coulston Gillispie, 3:391. New York: Charles Scribner's Sons, 1970–80. 159

————. "Pappus of Alexandria." In *Dictionary of Scientific Biography,* 16 vols., ed. Charles Coulston Gillispie, 10:293–304. New York: Charles Scribner's Sons, 1970–80. 234

Bunbury, Edward Herbert. *A History of Ancient Geography among the Greeks and Romans from the Earliest Ages till the Fall of the Roman Empire.* 2d ed., 2 vols., 1883; New York: Dover, 1959. 130, 135, 152, 157, 172, 175, 184

Bunge, William. *Theoretical Geography.* Lund Studies in Geography, Ser. C, General and Mathematical Geography no. 1. Lund: C. W. K. Gleerup, 1962; 2d ed. 1966. 30

Buondelmonti, Cristoforo. *Description des îles de l'archipel.* Ed. and trans. Emile Legrand. L'Ecole des Langues Orientales, ser. 4, 14. Paris: Leroux, 1897. 483

————. *"Descriptio insule Crete" et "Liber Insularum,"* cap. XI: Creta. Ed. M.-A. van Spitael. Candia, Crete: Syllagos Politistikēs Anaptyxeōs Herakleiou, 1981. 482

Burgess, Colin. *The Age of Stonehenge.* London: J. M. Dent, 1981. 64

Burgstaller, Ernst. "Felsbilder in den Alpenländern Österreichs." In *Symposium International d'Art Préhistorique Valcamonica, 23–28 Septembre 1968,* 143–47. Union Internationale des Sciences Préhistoriques et Protohistoriques. Capo di Ponte: Edizioni del Centro, 1970. 87

————. *Felsbilder in Österreich.* Schriftenreihe des Institutes für Landeskunde von Oberösterreich 21. Linz, 1972. 87

————. "Zur Zeitstellung der Österreichischen Felsbilder." *Act of the International Symposium on Rock Art: Lectures at Hankø 6–12 August, 1972,* ed. Sverre

Marstrander, 238–46. Oslo: Universitetsforlaget, 1978.
87

Burkitt, M. C. "Rock Carvings in the Italian Alps."
Antiquity 3, no. 10 (1929): 155–64. 75

Bursian, C. "Aventicum Helvetiorum." *Mittheilungen der
Antiquarischen Gesellschaft zu Zürich* 16, no. 1 (1867–
70). 252

Buschor, Ernst. *Die Tondächer der Akropolis.* Berlin and
Leipzig: Walter de Gruyter, 1929–33. 251

Butterfield, Herbert. *The Origins of History.* New York:
Basic Books, 1981. 6

Butzmann, Hans, ed. *Corpus Agrimensorum: Codex
Arcerianus A der Herzog-August-Bibliothek zu
Wolfenbüttel.* Codices Graeci et Latini 22. Leiden: A. W.
Sijthoff, 1970. 217

Cairola, Aldo, and Enzo Carli. *Il Palazzo Pubblico di Siena.*
Rome: Editalia, 1963. 368

Calabrese, Omar, Renato Giovannoli, and Isabella Pezzini,
eds. *Hic sunt leones: Geografia fantastica e viaggi
straordinari.* Catalog of exhibition, Rome. Milan: Electa
Editrice, 1983. 22

Cameron, Alan. "Macrobius, Avienus, and Avianus."
Classical Quarterly, n.s., 17 (1967): 385–99. 243

Caminos, Ricardo A. *Late-Egyptian Miscellanies.* Brown
Egyptological Studies 1. London: Oxford University Press,
1954. 124

Campbell, Eila M. J. Review of *Maps and Their Makers: An
Introduction to the History of Cartography,* by Gerald R.
Crone. *Geographical Journal* 120 (1954): 107–8. 26

————. "The Beginnings of the Characteristic Sheet to
English Maps," pt. 2 of "Landmarks in British
Cartography." *Geographical Journal* 128 (1962): 411–15.
35

Campbell, Tony. "The Drapers' Company and Its School of
Seventeenth Century Chart-Makers." In *My Head Is a
Map: Essays and Memoirs in Honour of R. V. Tooley,* ed.
Helen Wallis and Sarah Tyacke, 81–106. London: Francis
Edwards and Carta Press, 1973. 431

————. "The Woodcut Map Considered as a Physical
Object: A New Look at Erhard Etzlaub's *Rom Weg* Map
of c. 1500." *Imago Mundi* 30 (1978): 79–91. 498

————. *Early Maps.* New York: Abbeville Press, 1981. 22

————. Review of *The Catalan Atlas of the Year 1375,* ed.
by Georges Grosjean. *Imago Mundi* 33 (1981): 115–16.
457

————. "Cyprus and the Medieval Portolan Charts."
*Kupriakai Spoudai: Deltion tēs Etaireias Kupriakōn
Spoudōn, Brabeuthen upo tēs Akadēmias Athēnōn* 48
(1984): 47–66. 426, 461

Camus, Pierre. *Le pas des légions.* Paris: Diffusion
Frankelve, 1974. 206

Canale, Michele G. *Storia del commercio dei viaggi, delle
scoperte e carte nautiche degl'Italiani.* Genoa: Tipografia
Sociale, 1866. 430, 434

Cantemir, Dimitrie, prince of Moldavia. *Operele principelui
Demetriu Cantemiru.* 8 vols. Bucharest, 1872–1901. 63

Capart, Jean. *Primitive Art in Egypt.* Trans. A. S. Griffith.
London: H. Grevel, 1905. 81

Capel, Horacio. "Institutionalization of Geography and
Strategies of Change." In *Geography, Ideology and Social
Concern,* ed. D. R. Stoddart, 37–69. Oxford: Basil
Blackwell; Totowa, N.J.: Barnes and Noble, 1981. 12, 14,
17

Capello, Carlo F. *Il mappamondo medioevale di Vercelli
(1191-1218?).* Università di Torino, Memorie e Studi
Geografici, 10. Turin: C. Fanton, 1976. 306, 309, 348

Caporiacco, Ludovico di, and Paolo Graziosi. *Le pitture
rupestri di Àin Dòua (el-Auenàt).* Florence: Istituto
Geografico Militare, 1934. 97

Caraci, Giuseppe. "Tre piccoli mappamondi intarsiati del
sec. XV nel Palazzo Pubblico di Siena." *Rivista
Geografica Italiana* 28 (1921): 163–65. 324

————. "Un'altra carta di Albertin da Virga." *Bollettino
della Reale Società Geografica Italiana* 63 (1926): 781–
86. 440, 444

————. "Carte nautiche in vendita all'estero." *Rivista
Geografica Italiana* 34 (1927): 135–36. 396

————. "Cartografia." In *Enciclopedia italiana di scienze,
lettere ed arti,* originally 36 vols., 9:232. [Rome]: Istituto
Giovanni Treccari, 1929–39. 294

————. "Inedita cartografica—1, Un gruppo di carte e
atlanti conservati a Genova." *Bibliofilia* 38 (1936): 170–
78. 422

————. "An Unknown Nautical Chart of Grazioso
Benincasa, 1468." *Imago Mundi* 7 (1950): 18–31. 440,
457

————. "The First Nautical Cartography and the
Relationship between Italian and Majorcan
Cartographers." *Seventeenth International
Geographical Congress, Washington D.C., 1952—
Abstracts of Papers,* International Geographical Union
(1952): 12–13. 424

————. "A proposito di una nuova carta di Gabriel
Vallsecha e dei rapporti fra la cartografia nautica italiana
e quella maiorchina." *Bollettino della Società Geografica
Italiana* 89 (1952): 388–418. 415

————. "The Italian Cartographers of the Benincasa and
Freducci Families and the So-Called Borgiana Map of the
Vatican Library." *Imago Mundi* 10 (1953): 23–49. 403,
415, 420, 457

————. "A proposito di alcune carte nautiche di Grazioso
Benincasa." *Memorie Geografiche dall'Istituto di Scienze,
Geografiche e Cartografiche* 1 (1954): 283–90. 373, 421,
422

————. *Italiani e Catalani nella primitiva cartografia
nautica medievale.* Rome: Istituto di Scienze Geografiche e
Cartografiche, 1959. 389, 392, 402, 415, 419, 439, 447

————. "La prima raccolta moderna di grandi carte murali
rappresentanti i 'quattro continenti.'" *Atti del XVII
Congresso Geografico Italiano, Trieste 1961,* 2 vols.
(1962), 2:49–60. 8

————. "Viaggi fra Venezia e il levante fino al XIV secolo e
relativa produzione cartografica." In *Venezia e il levante
fino al secolo XV,* 2 vols., ed. Agostino Pertusi, 1:147–84.
Florence: L. S. Olschki, 1973–74. 447

Carder, James Nelson. *Art Historical Problems of a Roman
Land Surveying Manuscript: The Codex Arcerianus A,*

Wolfenbüttel. New York: Garland, 1978. **7, 217, 218, 221**

Cardi, Luigi. *Indice degli Atti dei Congressi Geografici Italiani dall'undicesimo al ventesimo (1930–1967)*. Naples: Comitato dei Geografi Italiani, 1972. **21**

Carettoni, Gianfilippo, et al. *La pianta marmorea di Roma antica: Forma Urbis Romae*. 2 vols. Rome: Comune di Roma, 1960. **212, 225, 226, 227, 229, 466**

Carmody, Francis J. *L'Anatolie des géographes classiques: Etude philologique*. Berkeley: Carmody, 1976. **198**

———. *La Gaule des itinéraires romains*. Berkeley: Carmody, 1977. **236**

Carpenter, Kenneth E., ed. *Books and Society in History: Papers of the Association of College and Research Libraries Rare Books and Manuscripts Preconference, 24-28 June 1980, Boston, Massachusetts*. New York: R. R. Bowker, 1983. **5, 36**

Carrère, Claude. *Barcelone: Centre économique à l'époque des difficultés, 1380–1462*. 2 vols. Civilisations et Sociétés 5. Paris: Moulton, 1967. **410, 433, 437, 444**

Carter, Howard, and Alan H. Gardiner. "The Tomb of Ramesses IV and the Turin Plan of a Royal Tomb." *Journal of Egyptian Archaeology* 4 (1917): 130–58. **126**

Cartes et figures de la terre. Exhibition catalog. Paris: Centre Georges Pompidou, 1980. **21, 60, 142**

Carver, Jonathan. *Travels through the Interior Parts of North-America in the Years 1766, 1767, and 1768*. London, 1778. **11**

Casanova, Lucia. "Inventario dei portolani e delle carte nautiche del Museo Correr." *Bollettino dei Musei Civici Veneziani* 3–4 (1957): 17–36. **417**

Cassidy, Vincent. "Geography and Cartography, Western European." In *Dictionary of the Middle Ages*, ed. Joseph R. Strayer, 5:395–99. New York: Charles Scribner's Sons, 1982–. **294**

Cassirer, Ernst. *The Individual and the Cosmos in Renaissance Philosophy*. Oxford: Clarendon Press, 1963. **335, 337, 340**

Castagnoli, Ferdinando. *Le ricerche sui resti della centuriazione*. Rome: Edizioni di Storia e Letteratura, 1958. **210**

———. "L'orientamento nella cartografia greca e romana." *Rendiconti della Pontificia Accademia Romana di Archeologia* 48 (1975–76): 59–69. **208, 227**

Castner, Henry W. "Formation of the I.C.A. Working Group on the History of Cartography." *Proceedings of the Eighth Annual Conference of the Association of Canadian Map Libraries* (1974): 73–76. **33**

Catalogo delle Mostre Ordinate in Occasione del VI Congresso Geografico Italiano. Venice, 1907. **423**

Catling, H. W. "Archaeology in Greece, 1979–80." *Archaeological Reports* 1979–80, no. 26 (1980): 12, col. 2. **139**

Cavallari, Vittorio, Piero Gazzola, and Antonio Scolari, eds. *Verona e il suo territorio*. 2 vols. Verona: Istituto per gli Studi Storici Veronesi, 1964. **477**

Céard, Jean. *La nature et les prodiges: L'insolie au 16ᵉ siècle*. Travaux d'Humanisme et Renaissance, no. 158. Geneva: Droz, 1977. **330**

Cebrian, Konstantin. *Geschichte der Kartographie: Ein Beitrag zur Entwicklung des Kartenbildes und Kartenwesens*. Gotha: Perthes, 1922. **25**

Cennini da Colle di Val d'Elsa, Cennino d'Andrea. *Il libro dell'arte: The Craftsman's Handbook*. Trans. Daniel V. Thompson, Jr. New Haven: Yale University Press, 1933. **431**

Černik, Berthold. "Das Schrift- und Buchwesen im Stifte Klosterneuburg während des 15. Jahrhunderts." *Jahrbuch des Stiftes Klosterneuburg* 5 (1913): 97–176. **324**

Černý, Jaroslav. *A Community of Workmen at Thebes in the Ramesside Period*. Cairo: Institut Français d'Archéologie Orientale, 1973. **124, 127**

———. *The Valley of the Kings*. Cairo: Institut Français d'Archéologie Orientale, 1973. **126**

Chabas, François J. *Les inscriptions des mines d'or*. Chalon-sur-Saône: Dejussieu, 1862. Also published in *Bibliothèque Egyptologique* 10 (1902): 183–230. **122**

Chace, A. B. *The Rhind Mathematical Papyrus*. 2 vols. Oberlin, Ohio: Mathematical Association of America, 1927–29. **127**

Chapman, Hugh. "A Roman Mitre and Try Square from Canterbury." *Antiquaries Journal* 59 (1979): 403–7. **227**

Cheney, Christopher R. *Handbook of Dates for Students of English History*. London: Royal Historical Society, 1945. **446**

Chevallier, Raymond. "Sur les traces des arpenteurs romains." *Caesarodunum*, suppl. 2. Orléans-Tours, 1967. **196, 222**

———. *Les voies romaines*. Paris: Armand Colin, 1972. **235**

Chippindale, Christopher. *Stonehenge Complete*. London: Thames and Hudson, 1983. **81**

Chiron, Pierre. "Les Phénomènes d'Euclide." In *L'astronomie dans l'antiquité classique*, 83–89. Actes du Colloque tenu à l'Université de Toulouse-Le Mirail, 21–23 October 1977. Paris: Belles Lettres, 1979. **154**

Chouquer, Gérard, and Françoise Favory. *Contribution à la recherche des cadastres antiques*. Annales Littéraires de l'Université de Besançon 236. Paris, 1980. **219**

Chouquer, Gérard, et al., "Cadastres, occupation du sol et paysages agraires antiques." *Annales: Economies, Sociétés, Civilizations* 37, nos. 5–6 (1982): 847–82. **218**

Churchill, Awnsham, and John Churchill. *A Collection of Voyages and Travels*. 6 vols. London: J. Walthoe, 1732. **11**

Cirlot, Juan Eduardo. *A Dictionary of Symbols*. Trans. Jack Sage. London: Routledge and Kegan Paul, 1971. **86, 88**

Ciullini, Rodolfo. "Firenze nelle antiche rappresentazioni cartografiche." *Firenze* 2 (1933): 33–79. **477, 493**

Clapham, J. H., and Eileen Power. *The Cambridge Economic History of Europe from the Decline of the Roman Empire*. 7 vols. Cambridge: Cambridge University Press, 1941–78. **xviii**

Clark, George. "General Introduction: History and the Modern Historian." In *The Renaissance, 1493-1520*. Ed. G. R. Potter. Vol. 1 of *The New Cambridge Modern History*. Cambridge: Cambridge University Press, 1957–79. **xviii**

Clarke, Somers, and Reginald Engelbach. *Ancient Egyptian Masonry: The Building Craft.* London: Oxford University Press, 1930. **125, 126, 127**

Claval, Paul. *Essai sur l'évolution de la géographie humaine.* New ed. Paris: Belles Lettres, 1976. **31**

Clavel-Lévêque, Monique, ed. *Cadastres et espace rural: Approches et réalités antiques.* Paris: Centre National de la Recherche Scientifique, 1983. **219**

Clawson, Mary G. "Evolution of Symbols on Nautical Charts prior to 1800." Master's thesis, University of Maryland, 1979. **378**

Clay, Albert Tobias. "Topographical Map of Nippur." *Transactions of the Department of Archaeology, University of Pennsylvania Free Museum of Science and Art* 1, no. 3 (1905): 223–25. **111**

Clère, J. J. "Fragments d'une nouvelle représentation égyptienne du monde." *Mitteilungen des Deutschen Archäologischen Instituts, Abteilung Kairo* 16 (1958): 30–46. **121**

Clos-Arceduc, A. "L'énigme des portulans: Etude sur le projection et le mode de construction des cartes à rhumbs du XIVᵉ et du XVᵉ siècle." *Bulletin du Comité des Travaux Historiques et Scientifiques: Section de Géographie* 69 (1956): 215–31. **384, 385, 386, 414**

Clutton, Elizabeth. "Some Seventeenth Century Images of Crete: A Comparative Analysis of the Manuscript Maps by Francesco Basilicata and the Printed Maps by Marco Boschini." *Imago Mundi* 34 (1982): 51–57. **403**

————, ed. and comp. *International Directory of Current Research in the History of Cartography and in Carto-bibliography.* No. 5. Norwich: Geo Books, 1985. **37**

Codazzi, A. "With Fire and Sword." *Imago Mundi* 5 (1948): 37–38. **6**

Cohn, Robert L. *The Shape of Sacred Space: Four Biblical Studies.* Chico, Calif.: Scholars Press, 1981. **340**

Colini, A. M., and L. Cozza. *Ludus Magnus.* Rome: Comune di Roma, 1962. **230**

Collingwood, William Gershom. *Northumbrian Crosses of the Pre-Norman Age.* London: Faber and Gwyer, 1927. **91**

Collins, Desmond, and John Onians. "The Origins of Art." *Art History* 1 (1978): 1–25. **55**

Collins, Lydia. "The Private Tombs of Thebes: Excavations by Sir Robert Mond, 1905 and 1906." *Journal of Egyptian Archaeology* 62 (1976): 18–40. **127**

Collyer, Robert, and J. Horsfall Turner. *Ilkley: Ancient and Modern.* Otley: W. Walker, 1885. **86**

Colombo, Fernando. *Historie del Signor Don Fernando Colombo: Nelle quali s'hà particolare, & vera relatione della vita, e de' fatti dell'Ammiraglio Don Christoforo Colombo, suo padre.* Venice, 1571. **375**

Colvin, H. M., ed. *History of the King's Works.* 6 vols. London: Her Majesty's Stationery Office, 1963–82. **339**

Compte-rendu du Congrès des Sciences Géographiques, Cosmographiques et Commerciales. 2 vols. Antwerp: L. Gerrits and Guil. Van Merlen, 1872. **15**

Conroy, John B. "A Classification of Andrews' Oecumenical Simple Medieval World Map Species into Genera." M. S. thesis, University of Wisconsin, Madison, 1975. **296**

Conti, Carlo. *Corpus delle incisioni rupestri di Monte Bego: I.* Collezione di Monografie Preistoriche ed Archeologiche 6. Bordighera: Istituto Internazionale di Studi Liguri, 1972. **66, 67**

Conti, Simonetta. "Portolano e carta nautica: Confronto toponomastico." In *Imago et mensura mundi: Atti del IX Congresso Internazionale di Storia della Cartografia,* 2 vols., ed. Carla Clivio Marzoli, 1:55–60. Rome: Enciclopedia Italiana, 1985. **383, 389**

Cornell, James. *The First Stargazers: An Introduction to the Origins of Astronomy.* New York: Scribner, 1981. **81**

Coronelli, Vincenzo. *Specchio del mare.* Venice, 1693. **387**

————. *Cronologia universale.* Venice, 1707. **11**

Cortés, Martin. *The Arte of Navigation.* Trans. Richard Eden. London: Richard Jugge, 1561. **391, 392**

Cortesão, Armando. "The North Atlantic Nautical Chart of 1424." *Imago Mundi* 10 (1953): 1–13. **389, 411, 420**

————. *The Nautical Chart of 1424 and the Early Discovery and Cartographical Representation of America: A Study on the History of Early Navigation and Cartography.* Coimbra: University of Coimbra, 1954. **373, 410, 411, 429, 457, 461**

————. *History of Portuguese Cartography.* 2 vols. Coimbra: Junta de Investigações do Ultramar-Lisboa, 1969–71. **12, 13, 14, 18, 130, 140, 152, 153, 154, 155, 157, 166, 168, 169, 173, 179, 180, 185, 189, 207, 234, 268, 291, 293, 294, 300, 301, 305, 315, 319, 322, 323, 333, 354, 371, 373, 374, 375, 380, 381, 384, 386, 391, 394, 396, 401, 405, 410, 411, 413, 418, 424, 429, 433, 438, 441, 442, 443, 444, 448, 449, 457, 461**

————. "Pizzigano's Chart of 1424." *Revista da Universidade de Coimbra* 24 (1970): 477–91. **411**

Cortesão, Armando, and Avelino Teixeira da Mota. *Portugaliae monumenta cartographica.* 6 vols. Lisbon, 1960. **374, 375, 378, 386, 402, 415, 443, 444, 457**

Cotter, Charles. "Early Tabular, Graphical and Instrumental Methods for Solving Problems of Plane Sailing." *Revista da Universidade de Coimbra* 26 (1978): 105–22. **385, 442**

Coulton, J. J. *Ancient Greek Architects at Work: Problems of Structure and Design.* Ithaca: Cornell University Press, 1977. **139**

Cowling, E. T. "Cup and Ring Markings to the North of Otley." *Yorkshire Archaeological Journal* 33, pt. 131 (1937): 290–97. **87**

Coxhead, David, and Susan Hiller. *Dreams: Visions of the Night.* New York: Avon Books; London: Thames and Hudson, 1976. **86, 87**

Crane, Diana. *Invisible Colleges: Diffusion of Knowledge in Scientific Communities.* Chicago: University of Chicago Press, 1972. **37**

Crescenzio, Bartolomeo. *Nautica Mediterranea.* Rome, 1602. **377, 387, 391**

Cripps, Judith A. "Barholm, Greatford, and Stowe, Lincolnshire, Late 15th Century." In *Local Maps and Plans from Medieval England,* ed. R. A. Skelton and P. D. A. Harvey, 263–88. Oxford: Clarendon Press, 1986. **485**

Crone, Gerald R. "John Green: Notes on a Neglected

Eighteenth Century Geographer and Cartographer." *Imago Mundi* 6 (1949): 85–91. **10**

————. *Maps and Their Makers: An Introduction to the History of Cartography.* 1st ed., 1953. 5th ed. Folkestone, Kent: Dawson; Hamden, Conn.: Archon Books, 1978. xv, **3, 4, 25, 26, 47, 48, 154, 178, 207, 234, 294, 372, 378, 384, 407, 434**

————. *The World Map by Richard of Haldingham in Hereford Cathedral.* Reproductions of Early Manuscript Maps 3. London: Royal Geographical Society, 1954. **309, 312**

————. "Early Cartographic Activity in Britain," pt. 1 of "Landmarks in British Cartography." *Geographical Journal* 128 (1962): 406–10. **37**

————. "New Light on the Hereford Map." *Geographical Journal* 131 (1965): 447–62. **288, 290, 292, 309, 330**

————. " 'Is leigen fünff perg in welschen landt' and the Hereford Map." *Erdkunde* 21 (1967): 67–68. **309**

Crone, Gerald R., and R. A. Skelton, "English Collections of Voyages and Travels, 1625–1846." In *Richard Hakluyt and His Successors.* 2d ser., 93. London: Hakluyt Society, 1946. **10**

Crook, John Hurrell. *The Evolution of Human Consciousness.* Oxford: Clarendon Press, 1980. **51**

Cumont, Franz. "Fragment de bouclier portant une liste d'étapes." *Syria* 6 (1925): 1–15. **249**

————. *Fouilles de Doura Europos (1922–1923).* Text and atlas. Paris: P. Geuthner, 1926. **249**

Cuntz, Otto, ed. *Itineraria Romana.* Leipzig: Teubner, 1929–. **235, 237, 469**

Cusa, Salvatore. "Sulla denominazione dei venti e dei punti cardinali, e specialmente di nord, est, sud, ovest." *Terzo Congresso Geografico Internazionale, Venice, 1881,* 2 vols., 2:375–415. Rome: Società Geografica Italiana, 1884. **337**

Dainville, François de. *Le langage des géographes.* Paris: A. et J. Picard, 1964. **35, 60**

————. "Rapports sur les conférences: Cartographie Historique Occidentale." In *Ecole Pratique des Hautes Etudes, IVᵉ section: Sciences historiques et philologiques. Annuaire 1968–1969,* 401–2. Paris, 1969. **465, 486, 493**

————. "Cartes et contestations au XVᵉ siècle." *Imago Mundi* 24 (1970): 99–121. **486, 487, 488, 490, 491**

Daly, Charles P. "On the Early History of Cartography; or, What We Know of Maps and Map-Making, before the Time of Mercator." Annual Address. *Bulletin of the American Geographical Society* 11 (1879): 1–40. **13**

Daly, John F. "Sacrobosco." In *Dictionary of Scientific Biography,* 16 vols., ed. Charles Coulston Gillispie, 12:60–63. New York: Charles Scribner's Sons, 1970–80. **306**

Dams, Lya. *L'art paléolithique de la caverne de la Pileta.* Graz: Akademische Druck, 1978. **68, 69**

Daniel, Glyn. Review of *The Megalithic Art of Western Europe,* by Elizabeth Shee Twohig. *Antiquity* 55 (1981): 235. **58**

Darby, H. C. "The Agrarian Contribution to Surveying in England." *Geographical Journal* 82 (1933): 529–35. **495**

Daressy, Georges. *Ostraca.* Catalogue Général des Antiquités Egyptiennes du Musée du Caire, vol. 1. Cairo: Institut Français d'Archéologie Orientale, 1901. **126**

D'Arms, J. H., and E. C. Kopff, eds. *The Seaborne Commerce of Ancient Rome: Studies in Archaeology and History.* Memoirs of the American Academy in Rome, 36. Rome: American Academy in Rome, 1980. **230**

Darnton, Robert. "What Is the History of Books?" In *Books and Society in History: Papers of the Association of College and Research Libraries Rare Books and Manuscripts Preconference, 24–28 June 1980, Boston, Massachusetts,* ed. Kenneth E. Carpenter, 3–26. New York: R. R. Bowker, 1983. **36**

Dati, Leonardo di Stagio, trans. Goro (Gregorio) Dati. *La sfera: Libri quattro in ottava rima.* Ed. Enrico Narducci. Milan: G. Daelli, 1865; reprinted [Bologna]: A. Forni, 1975. **301, 421**

Davidson, H. R. Ellis. *Pagan Scandinavia.* London: Thames and Hudson, 1967. **53, 91**

————. "Scandinavian Cosmology." In *Ancient Cosmologies,* ed. Carmen Blacker and Michael Loewe, 175–97. London: George Allen and Unwin, 1975. **87**

Davies, Graham I. "The Wilderness Itineraries: A Comparative Study." *Tyndale Bulletin* 25 (1974): 46–81. **115**

————. *The Way of the Wilderness: A Geographical Study of the Wilderness Itineraries in the Old Testament.* Cambridge: Cambridge University Press, 1979. **115**

Davies, Norman de Garis. "An Architect's Plan from Thebes." *Journal of Egyptian Archaeology* 4 (1917): 194–99. **127**

Deacon, A. Bernard. "Geometrical Drawings from Malekula and the Other Islands of the New Hebrides." *Journal of the Royal Anthropological Institute of Great Britain and Ireland,* n.s., 64 (1934): 129–75. **88**

Degenhart, Bernhard, and Annegrit Schmitt. "Marino Sanudo und Paolino Veneto." *Römisches Jahrbuch für Kunstgeschichte* 14 (1973): 1–137. **314, 358, 368, 398, 406, 429, 473, 474, 475, 477, 478, 480, 481, 496**

Deissmann, G. Adolf. *Forschungen und Funde im Serai, mit einem Verzeichnis der nichtislamischen Handschriften im Topkapu Serai zu Istanbul.* Berlin and Leipzig: Walter de Gruyter, 1933. **392**

Delano Smith, Catherine. *Western Mediterranean Europe: A Historical Geography of Italy, Spain and Southern France since the Neolithic.* London: Academic Press, 1979. **59, 60**

————. "The Emergence of 'Maps' in European Rock Art: A Prehistoric Preoccupation with Place." *Imago Mundi* 34 (1982): 9–25. **48, 93, 94, 95, 96**

————. "Cartographic Signs on European Maps and Their Explanation Before 1700." *Imago Mundi* 37 (1985): 9–29. **325**

————. "The Origins of Cartography, an Archaeological Problem: Maps in Prehistoric Rock Art." In *Papers in Italian Archaeology IV.* Pt. 2, *Prehistory,* ed. Caroline Malone and Simon Stoddart, 205–19. British Archaeological Reports International Series no. 244. Oxford: British Archaeological Reports, 1985. **61**

———. "Archaeology and Maps in Prehistoric Art: The Way Forward?" *Bollettino del Centro Camuno di Studi Preistorici* 23 (1986): forthcoming. **61**

Delatte, Armand, ed. *Les portulans grecs.* Liège: Bibliothèque de la Faculté de Philosophie et Lettres de l'Université de Liège, 1947. **260**

Demus, Otto. *Byzantine Mosaic Decoration: Aspects of Monumental Art in Byzantium.* London: Routledge and Kegan Paul, 1948. **263**

Denholm-Young, Noël. "The *Mappa Mundi* of Richard of Haldingham at Hereford." *Speculum* 32 (1957): 307–14. **312**

Deregowski, Jan B. *Distortion in Art: The Eye and the Mind.* London: Routledge and Kegan Paul, 1984. **60**

Derolez, Albert, ed. *Liber floridus colloquium.* Ghent: Story-Scientia, 1973. **300, 304, 353**

Desimoni, Cornelio. "Elenco di carte ed atlanti nautici di autore genovese oppure in Genova fatti o conservati." *Giornale Ligustico* 2 (1875): 47–285. **378, 434, 435**

———. "Le carte nautiche italiane del Medio Evo—a proposito di un libro del Prof. Fischer." *Atti della Società Ligure di Storia Patria* 19 (1888–89): 225–66. **432**

Desimoni, Cornelio, and Luigi Tommaso Belgrano. "Atlante idrografico del medio evo." *Atti della Società Ligure di Storia Patria* 5 (1867): 5–168. **419, 459**

De Smet, Antoine. "Viglius ab Aytta Zuichemus, savant, bibliothécaire et collectionneur de cartes du XVIᵉ siècle." In *The Map Librarian in the Modern World: Essays in Honour of Walter W. Ristow,* ed. Helen Wallis and Lothar Zögner, 237–50. Munich: K. G. Saur, 1979. **9**

Destombes, Marcel. "Cartes catalanes du XIVᵉ siècle." In *Rapport de la Commission pour la Bibliographie des Cartes Anciennes,* 2 vols., International Geographical Union, 1:38–63. Paris: Publié avec le concours financier de l'UNESCO, 1952. **382, 389, 394, 419, 423**

———. "A Venetian Nautical Atlas of the Late Fifteenth Century." *Imago Mundi* 12 (1955): 30. **392**

———. "Fragments of Two Medieval World Maps at the Topkapu Saray Library." *Imago Mundi* 12 (1955): 150–52. **394**

———. "A Panorama of the Sack of Rome by Pieter Bruegel the Elder." *Imago Mundi* 14 (1959): 64–73. **477**

———. "Les plus anciens sondages portés sur les cartes nautiques aux XVIᵉ et XVIIᵉ siècles." *Bulletin de l'Institut Océanographique,* special no. 2 (1968): 199–222. **378**

———. "La cartographie florentine de la Renaissance et Verrazano." In *Giornate commemorative di Giovanni da Verrazzano,* 19–43. Istituto e Museo di Storia della Scienza, Biblioteca 7. Florence: Olschki, 1970. **417**

———, ed. *Catalogue des cartes gravées au XVᵉ siècle.* Paris: International Geographical Union, 1952. **294**

———, ed. *Mappemondes A.D. 1200–1500: Catalogue préparé par la Commission des Cartes Anciennes de l'Union Géographique Internationale.* Amsterdam: N. Israel, 1964. **8, 137, 268, 286, 290, 294, 295, 296, 298, 301, 302, 303, 304, 308, 309, 312, 313, 316, 340, 343, 345, 347, 348, 353, 354, 358, 368, 372, 379, 413, 446**

Detlefsen, D. (S. D. F.). *Die Entdeckung des germanischen Nordens im Altertum.* Quellen und Forschungen zur Alten Geschichte und Geographie 8. Berlin: Weidmann, 1904. **197**

———. *Ursprung, Einrichtung und Bedeutung der Erdkarte Agrippas.* Quellen und Forschungen zur Alten Geschichte und Geographie 13. Berlin: Weidmann, 1906. **208, 209**

Deulin, Georges. *Répertoire des portulans et pièces assimilables conservés au Département des Manuscrits de la Bibliothèque Nationale.* Typescript, Paris, 1936. **435**

Díaz de Gámez, Gutierre. *The Unconquered Knight: A Chronicle of the Deeds of Don Pero Niño, Count of Buelna.* Trans. and selected by Joan Evans from *El Vitorial.* London: Routledge, 1928. **376, 443**

Diaz, Noël L. "The California Map Society: First Years." *Bulletin of the Society of University Cartographers* 18, no. 2 (1984): 103–5. **23**

Dickinson, H. W. "A Brief History of Draughtsmen's Instruments." *Transactions of the Newcomen Society* 27 (1949–50 and 1950–51): 73–84; republished in the *Bulletin of the Society of University Cartographers* 2, no. 2 (1968): 37–52. **391**

Dickinson, Robert E. *The Makers of Modern Geography.* New York: Frederick A. Praeger, 1969. **14**

Dicks, D. R. "Ancient Astronomical Instruments." *Journal of the British Astronomical Association* 64 (1954): 77–85. **181**

———. "Solstices, Equinoxes, and the Presocratics." *Journal of Hellenic Studies* 86 (1966): 26–40. **134**

———. *Early Greek Astronomy to Aristotle.* Ithaca: Cornell University Press, 1970. **130, 134, 137, 142**

———. "Dositheus." In *Dictionary of Scientific Biography,* 16 vols., ed. Charles Coulston Gillispie, 4:171–72. New York: Charles Scribner's Sons, 1970–80. **159**

———. "Eratosthenes." In *Dictionary of Scientific Biography,* 16 vols., ed. Charles Coulston Gillispie, 4:388–93. New York: Charles Scribner's Sons, 1970–80. **154, 170**

———. "Hecataeus of Miletus." In *Dictionary of Scientific Biography,* 16 vols., ed. Charles Coulston Gillispie, 6:212–13. New York: Charles Scribner's Sons, 1970–80. **134**

———, ed. *The Geographical Fragments of Hipparchus.* London: Athlone Press, 1960. **136, 140, 141, 145, 147, 148, 150, 151, 157, 164, 166, 167**

Diels, H., and W. Kranz, eds. *Die Fragmente der Vorsokratiker.* 6th ed., 3 vols. Berlin: Weidmann, 1951–52. **131**

Diffie, Bailey W. "Foreigners in Portugal and the 'Policy of Silence.'" *Terrae Incognitae* 1 (1969): 23–34. **414**

Digges, Leonard. *A Geometrical Practise, Named Pantometria.* London: Henrie Bynneman, 1571. **35**

Dilke, O. A. W. "Maps in the Treatises of Roman Land Surveyors." *Geographical Journal* 127 (1961): 417–26. **218, 220**

———. "Illustrations from Roman Surveyors' Manuals." *Imago Mundi* 21 (1967): 9–29. **217**

———. *The Roman Land Surveyors: An Introduction to the Agrimensores.* Newton Abbot: David and Charles, 1971. **62, 125, 202, 210, 213, 214, 225, 381**

———. "Archaeological and Epigraphic Evidence of Roman

Land Surveys." In *Aufstieg und Niedergang der römischen Welt*, ed. Hildegard Temporini, 2.1 (1974): 564–92. Berlin: Walter de Gruyter, 1972–. **219**

———. "The Arausio Cadasters." In *Akten des VI. internationalen Kongresses für griechische und lateinische Epigraphik, München, 1972. Vestigia 17*, 455–57. Munich: C. H. Beck'sche Verlagsbuchhandlung, 1973. **224**

———. "Varro and the Origins of Centuriation." In *Atti del Congresso Internazionale di Studi Varroniani*, 353–58. Rieti: Centro di Studi Varroniani, 1976. **202**

———. *Roman Books and Their Impact*. Leeds: Elmete Press, 1977. **248**

———. "Mapping of the North African Coast in Classical Antiquity." In *Proceedings of the Second International Congress of Studies on Cultures of the Western Mediterranean*, 154–60. Paris: Association Internationale d'Etude des Civilisations Méditerranéennes, 1978. **198**

———. *Gli agrimensori di Roma antica*. Bologna: Edagricole, 1979. **224**

———. "Geographical Perceptions of the North in Pomponius Mela and Ptolemy." In *Exploring the Arctic*, ed. Louis Rey, 347–51. Fairbanks: University of Alaska Press, Comité Arctique International, and Arctic Institute of North America, 1984. **197**

———. *Greek and Roman Maps*. Ithaca: Cornell University Press; London: Thames and Hudson, 1985. **193, 205, 207, 235, 238, 248, 381, 383**

———. "Ground Survey and Measurement in Roman Towns." In *Roman Urban Topography in Britain and the Western Empire*, ed. Francis Grew and Brian Hobley, 6–13. Council for British Archaeology Research Report no. 59. London: Council for British Archaeology, 1985. **227**

Dilke, O. A. W., and Margaret S. Dilke. "Terracina and the Pomptine Marshes." *Greece and Rome*, n.s., 8 (1961): 172–78. **218**

———. "The Eternal City Surveyed." *Geographical Magazine* 47 (1975): 744–50. **229**

———. "Italy in Ptolemy's Manual of Geography." In *Imago et mensura mundi: Atti del IX Congresso Internazionale di Storia della Cartografia*, 2 vols., ed. Carla Clivio Marzoli, 2:353–60. Rome: Enciclopedia Italiana, 1985. **195**

Dillemann, Louis. "La carte routière de la *Cosmographie de Ravenne*." *Bonner Jahrbücher* 175 (1975): 165–70. **260**

———. "Observations on Chapter V, 31, Britannia, in the Ravenna Cosmography." *Archaeologia* (1979): 61–73. **260**

Diller, Aubrey. "The Vatopedi Manuscript of Ptolemy and Strabo." *American Journal of Philology* 58 (1937): 174–84. **270**

———. "The Oldest Manuscripts of Ptolemaic Maps." *Transactions of the American Philological Association* 71 (1940): 62–67. **192, 269**

———. "The Greek Codices of Palla Strozzi and Guarino Veronese." *Journal of the Warburg and Courtauld Institutes* 24 (1961): 313–21. **192**

———. "Dicaearchus of Messina." In *Dictionary of Scientific Biography*, 16 vols., ed. Charles Coulston Gillispie, 4:81–82. New York: Charles Scribner's Sons, 1970–80. **152**

Dion, Roger. "Où Pythéas voulait-il aller?" In *Mélanges d'archéologie et d'histoire offerts à André Piganiol*, 3 vols., ed. Raymond Chevallier, 3:1315–36. Paris: SEVPEN, 1966. **150**

Diringer, David. *The Alphabet: A Key to the History of Mankind*. 3d ed. rev. London: Hutchinson, 1968. **49**

Doig, Ronald P. "A Bibliographical Study of Gough's *British Topography*." *Edinburgh Bibliographical Society Transactions* 4 (1963): 103–36. **7**

Domínguez Bordona, Jesús. *Die spanische Buchmalerei vom siebten bis siebzehnten Jahrhundert*. 2 vols. Florence, 1930. **304**

Donner, Herbert, and Heinz Cüppers. *Die Mosaikkarte von Madeba*. Abhandlungen des Deutschen Palästinavereins. Wiesbaden: O. Harassowitz, 1977. **264**

Dorn, Ronald I., and David S. Whitley. "Chronometric and Relative Age Determination of Petroglyphs in the Western United States." *Annals of the Association of American Geographers* 74 (1984): 308–22. **58**

Downey, Glanville. *A History of Antioch in Syria: From Seleucus to the Arab Conquest*. Princeton: Princeton University Press, 1961. **239**

Drachmann, Aage Gerhardt. *The Mechanical Technology of Greek and Roman Antiquity: A Study of Literary Sources*. Acta Historic Scientiarum Naturalium et Medicinalium, 17. Copenhagen: Munksgaard, 1961. **159**

Dreyer-Eimbcke, Oswald. "The Mythical Island of Frisland." *Map Collector* 26 (1984): 48–49. **414**

Dröber, Wolfgang. *Geographie des Welthandels*. 2 vols. Stuttgart, 1857–72. **45**

———. "Kartographie bei den Naturvölkern" (Mapmaking among primitive peoples). Diss., Erlangen University, 1903; reprinted Amsterdam: Meridian, 1964. Summarized under the same title in *Deutsche Geographische Blätter* 27 (1904): 29–46. **45, 48, 54**

Drögereit, Richard. "Die Ebstorfer Weltkarte und Hildesheim." *Zeitschrift des Vereins für Heimatkunde im Bistum Hildesheim* 44 (1976): 9–44. **309**

Du Bus, Charles. "Les collections d'Anville à la Bibliothèque Nationale." *Bulletin du Comité Travaux Historiques et Scientifiques: Section de Géographie* 41 (1926): 93–145. **10**

———. "Edme-François Jomard et les origines du Cabinet des Cartes (1777–1862)." Union Géographique Internationale, *Comptes rendus du Congrès International de Géographie, Paris 1931*, 3 (1934): 638–42. **15**

Dufresnoy, Abbé Lenglet. *Catalogue des meilleures cartes géographiques générales et particulières*. Reprinted Amsterdam: Meridian, 1965. **10**

Duhem, Pierre. *Le système du monde: Histoire des doctrines cosmologiques de Platon à Copernic*. 10 vols. Paris: Hermann, 1913–59. **293**

Dunbabin, Katherine M. D. *The Mosaics of Roman North Africa: Studies in Iconography and Patronage*. Oxford: Clarendon Press, 1978. **248**

Duplessis, Georges. "Roger de Gaignières et ses collections

iconographiques." *Gazette des Beaux-Arts*, 2d ser., 3 (1870): 468–88. **9**

Durand, Dana Bennett. *The Vienna-Klosterneuburg Map Corpus of the Fifteenth Century: A Study in the Transition from Medieval to Modern Science.* Leiden: E. J. Brill, 1952. **293, 316, 323, 324, 473**

Durazzo, P. *Il planisfero di Giovanni Leardo.* Mantua: Eredi Segna, 1885. **317**

Durbin, Paul T., ed. *A Guide to the Culture of Science, Technology, and Medicine.* New York: Free Press, 1980. **30**

Dürst, Arthur. *Seekarte des Iehuda ben Zara: (Borgiano VII) 1497.* Pamphlet accompanying a facsimile edition of the chart. Zurich: Belser Verlag, 1983. **457**

Eastham, Anne, and Michael Eastham. "The Wall Art of the Franco-Cantabrian Deep Caves." *Art History* 2 (1979): 365–85. **57**

Ebert, Max. *Reallexikon der Vorgeschichte.* Berlin: Walter de Gruyter, 1928. **91**

Eckenrode, T. R. "Venerable Bede as a Scientist." *American Benedictine Review* 21 (1971): 486–507. **303**

Eckert, Max. "Die Kartographie als Wissenschaft." *Zeitschrift der Gesellschaft für Erdkunde zu Berlin* (1907): 539–55. **24**

——. "Die wissenschaftliche Kartographie im Universitäts-Unterricht." In *Verhandlungen des Sechszehnten Deutschen Geographentages zu Nürnberg,* ed. Georg Kollm, 213–27. Berlin: Reimer, 1907. **24**

——. "On the Nature of Maps and Map Logic." Trans. W. Joerg. *Bulletin of the American Geographical Society* 40 (1908): 344–51. **24**

——. *Die Kartenwissenschaft: Forschungen und Grundlagen zu einer Kartographie als Wissenschaft.* 2 vols. Berlin and Leipzig: Walter de Gruyter, 1921–25. **xix, 25, 375, 384**

Edgar, Campbell Cowan. *Zenon Papyri in the University of Michigan Collection.* Michigan Papyri vol. 1. Ann Arbor: University of Michigan Press, 1931. **128**

Edgerton, Samuel Y. "Florentine Interest in Ptolemaic Cartography as Background for Renaissance Painting, Architecture, and the Discovery of America." *Journal of the Society of Architectural Historians* 33 (1974): 275–92. **189**

Edwards, I. E. S. *The Pyramids of Egypt.* New and rev. ed. Harmondsworth: Viking, 1986. **126**

Edzard, Dietz Otto. "Itinerare." In *Reallexikon der Assyriologie und vorderasiatischen Archäologie,* ed. Erich Ebeling and Bruno Meissner, 5:216–20. Berlin: Walter de Gruyter, 1932–. **108**

Efros, Israel Isaac. *The Problem of Space in Jewish Mediaeval Philosophy.* New York: Columbia University Press, 1917. **340**

Egidi, Francesco, ed. *I Documenti d'amore di Francesco da Barberino secondo i MSS originali.* 4 vols. Società Filologica Romana: Documenti di Storia Letteraria 3. Rome: Presso la Società, 1905–27. **441**

Ehrensvärd, Ulla. "Color in Cartography: An Historical Survey." In *Art and Cartography: Six Historical Essays.*

Ed. David Woodward. Chicago: University of Chicago Press, 1987. **325**

Ehrle, F., and H. Egger. *Piante e vedute di Roma e del Vaticano dal 1300 al 1676.* Illus. Amato Pietro Frutaz. Rome: Biblioteca Apostolica Vaticana, 1956. **477, 493**

Eisenstein, Elizabeth L. *The Printing Press as an Agent of Change: Communications and Cultural Transformations in Early-Modern Europe.* 2 vols. Cambridge: Cambridge University Press, 1979. **314**

Eliade, Mircea. *The Sacred and the Profane: The Nature of Religion.* New York: Harcourt, Brace and World, 1959. **340**

——. *Images and Symbols: Studies in Religious Symbolism.* Trans. Philip Mairet. London: Harvill Press, 1961. **59**

——. *A History of Religious Ideas.* Trans. Willard R. Trask. Vol. 1, *From the Stone Age to the Eleusinian Mysteries.* Chicago: University of Chicago Press, 1978. **xvi, 4, 48, 55**

Elliott, Carolyn. "The Religious Beliefs of the Ghassulians, c. 4000–3100 B.C." *Palestine Exploration Quarterly,* January-June 1977, 3–25. **77**

Emden, A. B. *A Biographical Register of Oxford University to A.D. 1500.* Oxford: Clarendon Press, 1957–59. **312**

Emiliani, Marina (later Marina Salinari). "Le carte nautiche dei Benincasa, cartografi anconetani." *Bollettino della Reale Società Geografica Italiana* 73 (1936): 485–510. **400, 432, 433, 434, 435**

——. "L'Arcipelago Dalmata nel portolano di Grazioso Benincasa." *Archivio Storico per la Dalmazia* 22 (1937): 402–22. **422**

Enklaar, D. T. "De oudste kaarten van Gooiland en zijn grensgebieden." *Nederlandsch Archievenblad* 39 (1931–32): 185–205. **486, 500**

Erben, Wilhelm. *Rombilder auf kaiserlichen und päpstlichen Siegeln des Mittelalters.* Veröffentlichungen des Historischen Seminars der Universität Graz, 7. Graz: Leuschner und Lubensky, 1931. **477**

Ernst, A. "Petroglyphen aus Venezuela." *Zeitschrift für Ethnologie* 21 (1889): Verhandlungen 650–55. **66**

Erren, Manfred. *Die Phainomena des Aratos von Soloi: Untersuchungen zum Sach- und Sinnverständnis.* Wiesbaden: Franz Steiner, 1967. **142**

Errera, Carlo. "Atlanti e carte nautiche dal secolo XIV al XVII conservate nelle biblioteche pubbliche e private di Milano." *Rivista Geografica Italiana* 3 (1896): 91–96; reprinted in *Acta Cartographica* 8 (1970): 225–52. **397, 439**

Estey, F. N. "Charlemagne's Silver Celestial Table." *Speculum* 18 (1943): 112–17. **303, 469**

Ettlinger, L. D. "A Fifteenth-Century View of Florence." *Burlington Magazine* 94 (1952): 160–67. **477**

Evans, Arthur. *The Palace of Minos.* 4 vols. London: Macmillan, 1921–35. **251**

Eves, Howard. *An Introduction to the History of Mathematics.* New York: Holt, Rinehart and Winston, 1969. **323**

Fabri, Felix. *The Wanderings of Felix Fabri, circa 1480–*

1483 A.D. Trans. Aubrey Stewart. Palestine Pilgrims' Text Society, vols. 7–10. London: Palestine Exploration Fund, 1897. **443**

Falbe, C. T. *Recherches sur l'emplacement de Carthage.* Paris: Imprimerie Royale, 1833. **219**

Farmakovsky, Mstislav. "Arkhaicheskiy period v Rossii: Pamyatniki grecheskogo arkhaicheskogo i drevnego vostochnogo iskusstva, naidënnye v grecheskikh koloniyakh po severnomu beregu Chërnogo morya v kurganakh Skifii i na Kavkaze" (The archaic period in Russia: Relics of Greek archaic and ancient Eastern art found in the Greek colonies along the northern coast of the Black Sea in the barrows of Scythia and in the Caucasus). *Materialy po Arkheologii Rossii, Izdavayemye Imperatorskoy Arkheologicheskoy Komissiyey* 34 (1914): 15–78. **72**

Faulkner, Raymond O., ed. and trans. *The Ancient Egyptian Coffin Texts.* 3 vols. Warminster: Aris and Phillips, 1978. **120**

Favaro, Elena. *L'arte dei pittori in Venezia e i suoi statuti.* Università di Padova, Pubblicazione della Facoltà di Lettere e Filosofia, vol. 55. Florence: Leo S. Olschki, 1975. **428, 430, 432**

Febvre, Lucien, and Henri-Jean Martin. *L'apparition du livre.* Paris: Editions Albin, 1958. English edition, *The Coming of the Book: The Impact of Printing, 1450–1800.* New ed. Ed. Geoffrey Nowell-Smith and David Wootton. Trans. David Gerard. London: NLB, 1976. **5, 10**

Ferguson, John. "China and Rome." In *Aufstieg und Niedergang der römischen Welt*, ed. Hildegard Temporini and Wolfgang Haase, 2.9.2 (1978): 581–603. Berlin: Walter de Gruyter, 1972–. **178**

Fernández-Armesto, F. F. R. "Atlantic Exploration before Columbus: The Evidence of Maps." *Renaissance and Modern Studies* (forthcoming). **411**

Fernie, Eric. "The Proportions of the St. Gall Plan." *Art Bulletin* 60 (1978): 583–89. **466**

Ferretto, Arturo. "Giovanni Mauro di Carignano Rettore di S. Marco, cartografo e scrittore (1291–1329)." *Atti della Società Ligure di Storia Patria* 52 (1924): 33–52. **380, 404, 432, 434**

Ferro, Gaetano. "Geografia storica, storia delle esplorazioni e della cartografia" (Introduzione). In *Ricerca geografica in Italia, 1960–1980.* 317–18. Milan: Ask Edizioni, 1980. **38**

Field, N. H., trans. *Roman Roads.* London: Batsford, 1978. **235**

Fierro, Alfred. *La Société de Géographie, 1821–1946.* Geneva: Librairie Droz, 1983. **14**

Finkelstein, Jacob J. "Mesopotamia." *Journal of Near Eastern Studies* 21 (1962): 73–92. **111**

Fiorini, Matteo. *Le proiezioni delle carte geografiche.* Bologna: Zanichelli, 1881. **381, 385, 411**

———. "Le sfere cosmografiche e specialmente le sfere terrestri." *Bollettino della Società Geografica Italiana* 30 (1893): 862–88, 31 (1894): 121–32, 271–81, 331–49, 415–35. **141**

———. *Sfere terrestri e celesti di autore italiano oppure*

fatte o conservate in Italia. Rome: Società Geografica Italiana, 1899. **142, 143**

Fischer, Irene. "Another Look at Eratosthenes' and Posidonius' Determinations of the Earth's Circumference." *Quarterly Journal of the Royal Astronomical Society* 16 (1975): 152–67. **148**

Fischer, Joseph. "Ptolemaeus und Agathodämon." *Kaiserliche Akademie der Wissenschaften in Wien.* Philosophisch-Historische Klasse, 59 (1916): 3–25. **271**

———. "Der Codex Burneyanus Graecus 111." In *75 Jahre Stella Matutina.* 3 vols., 1:151–59. Festschrift. Feldkirch: Selbstverlag Stella Matutina, 1931. **270**

———, ed. *Claudii Ptolemaei Geographiae Codex Urbinas Graecus 82.* 2 vols. in 4. Codices e Vaticanis Selecti quam Simillime Expressi, vol. 19. Leiden: E. J. Brill; Leipzig: O. Harrassowitz, 1932. **177, 189, 192, 269, 270, 274**

Fischer, Norbert. "With Fire and Sword, III." *Imago Mundi* 10 (1953): 56. **6**

Fischer, Theobald. *Sammlung mittelalterlicher Welt- und Seekarten italienischen Ursprungs und aus italienischen Bibliotheken und Archiven.* Venice: F. Ongania, 1886; reprinted Amsterdam: Meridian, 1961. **374, 384, 388, 406**

Flamand, Jacques. *Macrobe et le néo-Platonisme latin, à la fin du IVᵉ siècle.* Leiden: E. J. Brill, 1977. **300**

Flint, Valerie I. J. "Honorius Augustodunensis Imago Mundi." *Archives d'Histoire Doctrinale et Littéraire du Moyen Age* 57 (1982): 7–153. **312**

Fockema Andreae, S. J., and B. van 'tHoff. *Geschiedenis der kartografie van Nederland van den Romeinschen tijd tot het midden der 19de eeuw.* The Hague: Nijhoff, 1947. **486, 500**

Foncin, Myriem, Marcel Destombes, and Monique de La Roncière. *Catalogue des cartes nautiques sur vélin conservées au Département des Cartes et Plans.* Paris: Bibliothèque Nationale, 1963. **376, 397, 435**

Fontaine, Jacques. *Isidore de Séville et la culture classique dans l'Espagne visigothique.* 2 vols. Paris: Etudes Augustiniennes, 1959. **301**

Forbes, Robert James. *Notes on the History of Ancient Roads and Their Construction.* Archaeologisch-Historische Bijdragen 3. Amsterdam: North-Holland, 1934. **135**

Fordham, Herbert George. *Studies in Carto-bibliography, British and French, and in the Bibliography of Itineraries and Road-Books.* Oxford: Clarendon Press, 1914; reprinted London: Dawsons, 1969. **20**

———. *Maps, Their History, Characteristics and Uses: A Handbook for Teachers.* 2d ed. Cambridge: Cambridge University Press, 1927. **25, 47, 48**

Formaleoni, Vincenzio. *Saggio sulla nautica antica de' Veneziani con una illustrazione d'alcune carte idrografiche antiche della Biblioteca di San Marco, che dimostrano l'isole Antille prima della scoperta di Cristoforo Colombo.* Venice: Author, 1783. **442, 443**

Fossier, Robert. *Polyptyques et censiers.* Turnhout: Brepols, 1978. **494**

Fossombroni, Vittorio. "Illustrazione di un antico documento relativo all'originario rapporto tra le acque dell'Arno e quelle della Chiana." *Nuova raccolta d'autori italiani che trattano del moto dell'acque.* 6 vols. Ed. F. Cardinali. Bologna: Marsigli, 1824. **488**

Fox, H. S. A. "Exeter, Devonshire, 1499." In *Local Maps and Plans from Medieval England,* ed. R. A. Skelton and P. D. A. Harvey, 329–36. Oxford: Clarendon Press, 1986. **490**

Frabetti, Pietro. *Carte nautiche italiane dal XIV al XVII secolo conservate in Emilia-Romagna.* Florence: Leo S. Olschki, 1978. **380, 391, 401, 457**

Franciscis, Alfonso de. "La villa romana di Oplontis." In *Neue Forschungen in Pompeji und den anderen vom Vesuvausbruch 79 n. Chr. verschütteten Städten,* ed. Bernard Andreae and Helmut Kyrieleis, 9–38. Recklinghausen: Aurel Bongers, 1975. **240**

Frankowski, Eugenjusz. *Hórreos y palafitos de la Península Ibérica.* Comisión de Investigaciones Paleontológicas y Prehistóricas, no. 18. Madrid: Museo Nacional de Ciencias Naturales, 1918. **97**

Freedman, Nadezhda. "The Nuzi Ebla." *Biblical Archaeologist* 40, no. 1 (1977): 32–33, 44. **113**

Freeman, Kathleen. *Ancilla to the Pre-Socratic Philosophers.* Cambridge: Harvard University Press, 1948. **131**

Freidel, Frank, ed. *Harvard Guide to American History.* Rev. ed., 2 vols. Cambridge: Belknap Press of Harvard University Press, 1954. **1**

Freiesleben, Hans-Christian. "Map of the World or Sea Chart? The Catalan Mappamundi of 1375." *Navigation: Journal of the Institute of Navigation* 26 (1979): 85–89. **377, 387**

——. "The Still Undiscovered Origin of the Portolan Charts." *Journal of Navigation* (formerly *Navigation: Journal of the Institute of Navigation*) 36 (1983): 124–29. **375, 380**

——. "The Origin of Portolan Charts." *Journal of Navigation* 37 (1984): 194–99. **388**

Freitag, Ulrich. "Semiotik und Kartographie: Über die Anwendung kybernetischer Disziplinen in der theoretischen Kartographie." *Kartographische Nachrichten* 21 (1971): 171–82. **34**

——. "Die Zeitalter und Epochen der Kartengeschichte." *Kartographische Nachrichten* 22 (1972): 184–91. **xviii, 36**

——. "Peuples sans cartes." In *Cartes et figures de la terre,* 61–63. Exhibition catalog. Paris: Georges Pompidou, 1980. **60**

——. "Zur Periodisierung der Geschichte der Kartographie Thailands." In *Kartenhistorisches Colloquium Bayreuth '82: Vorträge und Berichte,* 213–27. Berlin: Reimer, 1983. **xviii**

Friedman, John Block. *The Monstrous Races in Medieval Art and Thought.* Cambridge: Harvard University Press, 1981. **287, 316, 330, 332, 333, 340**

Frisch, Karl von. *The Dance Language and Orientation of Bees.* Trans. Leigh E. Chadwick. Cambridge: Belknap Press of Harvard University Press, 1967. **50**

Frobenius, Leo. *Ekade Ektab: Die Felsbilder Fezzans.* Leipzig: O. Harrassowitz, 1937. **69, 70, 89, 90, 96, 97**

Frobenius, Leo, and Douglas C. Fox. *Prehistoric Rock Pictures in Europe and Africa.* New York: Museum of Modern Art, 1937. **63**

Frobenius, Leo, and Hugo Obermaier. *Hádschra Máktuba: Urzeitliche Felsbilder Kleinafrikas.* Munich: K. Wolff, 1925. **97**

Frutaz, Amato Pietro. *Le piante di Roma.* 3 vols. Rome: Istituto di Studi Romani, 1962. **229**

Funkhouser, H. Gray. "Notes on a Tenth-Century Graph." *Osiris* 1 (1936): 260–62. **323**

Furse, P. "On the Prehistoric Monuments in the Islands of Malta and Gozo." *International Congress of Prehistoric Archaeology, Transactions of the Third Session, Norwich 1868* (1869): 407–16. **81**

Gaballa, G. A. *Narrative in Egyptian Art.* Mainz: Philipp von Zabern, 1976. **119**

Gadol, Joan. *Leon Battista Alberti: Universal Man of the Early Renaissance.* Chicago: University of Chicago Press, 1969. **495**

Gaerte, W. "Kosmische Vorstellungen im Bilde prähistorischer Zeit: Erdberg, Himmelsberg, Erdnabel und Weltenströme." *Anthropos* 9 (1914): 956–79. **89**

Gaffarel, Paul. "Etude sur un portulan inédit de la Bibliothèque de Dijon." *Mémoires de la Commission des Antiquités de la Côte-d'Or* 9 (1877): 149–99. **374, 401**

Galbraith, V. H. "An Autograph MS of Ranulph Higden's *Polychronicon.*" *Huntington Library Quarterly* 34 (1959): 1–18. **312**

Gallery, Leslie Mesnick. "The Garden of Ancient Egypt." In *Immortal Egypt,* ed. Denise Schmandt-Besserat, 43–49. Malibu: Undena Publications, 1978. **118**

Gallez, Paul. "Walsperger and His Knowledge of the Patagonian Giants, 1448." *Imago Mundi* 33 (1981): 91–93. **199, 316**

Galliazzo, Vittorio. "Il ponte della pietra di Verona." *Atti e Memorie della Accademia di Agricoltura, Scienze e Lettere di Verona* 146 (1968–69): 533–70. **477**

Gallo, Rodolfo. "Le mappe geografiche del palazzo ducale di Venezia." *Archivio Veneto,* 5th ser., 32 (1943): 47–89. **315**

——. "A Fifteenth Century Military Map of the Venetian Territory of *Terraferma.*" *Imago Mundi* 12 (1955): 55–57. **479, 480, 498**

García Camarero, Ernesto. "Deformidades y alucinaciones en la cartografía ptolemeica y medieval." *Boletín de la Real Sociedad Geográfica* 92 (1956): 257–310. **414**

——. "La escuela cartográfica inglesa 'At the Signe of the Platt.'" *Boletín de la Real Sociedad Geográfica* 95 (1959): 65–68. **431**

García Franco, Salvador. "The 'Portolan Mile' of Nordenskiöld." *Imago Mundi* 12 (1955): 89–91. **389**

Gardiner, Alan H. "The Map of the Gold Mines in a Ramesside Papyrus at Turin." *Cairo Scientific Journal* 8, no. 89 (1914): 41–46. **122**

——. *Late-Egyptian Miscellanies.* Bibliotheca Aegyptiaca 7. Brussels: Edition de la Fondation Egyptologique Reine Elisabeth, 1937. **124**

Gasparrini Leporace, Tullia. *Il mappamondo di Fra Mauro.* Rome: Istituto Poligrafico dello Stato, 1956. 316, 317, 324, 433

Gelb, Ignace J., et al., eds. *The Assyrian Dictionary.* Chicago: Oriental Institute, 1968. 109

Gennep, Arnold van. *Les rites de passage: Etude systématique des rites.* Paris: E. Nourry, 1909. English edition, *The Rites of Passage.* Trans. Monika B. Vizedom and Gabrielle L. Caffee. London: Routledge and Kegan Paul, 1960. 59

Genoviè, Lina. "La cartografia della Toscana." *L'Universo* 14 (1933): 779–85. 480

George, Frank. Review of *The Story of Maps,* by Lloyd A. Brown. *Geographical Journal* 116 (1950): 109. 26

Georgiev, Georgi Illiev. "Forschungsstand der alten Felskunst in Bulgarien." In *Acts of the International Symposium on Rock Art: Lectures at Hankø, 6-12 August, 1972,* ed. Sverre Marstrander, 68–84. Oslo: Universitetsforlaget, 1978. 93

Gerola, Giuseppe. "L'elemento araldico nel portolano di Angelino dall'Orto." *Atti del Reale Istituto Veneto di Scienze, Lettere ed Arti* 93, pt. 1 (1933–34): 407–43. 399, 406

———. "Le carte nautiche di Pietro Vesconte dal punto di vista araldico." In *Atti del Secondo Congresso di Studi Coloniali, Napoli 1–3 October 1934,* 7 vols., 2:102–23. Florence: Leo S. Olschki, 1935. 399

Gershenson, Daniel E., and Daniel A. Greenberg. "How Old Is Science?" *Columbia University Forum* (1964), 24–27. 30

Geyer, P., and Otto Cuntz. "Itinerarium Burdigalense." In *Itineraria et alia geographica,* in *Corpus Christianorum,* Series Latina, vols. 175 and 176 (1965). 237

Gialanella, Costanza, and Vladimiro Valerio. "Atlas Farnèse." In *Cartes et figures de la terre,* 84. Exhibition catalogue. Paris: Centre Georges Pompidou, 1980. 142

Gibson, Ackroyd. "Rock-Carvings Which Link Tintagel with Knossos: Bronze-Age Mazes Discovered in North Cornwall." *Illustrated London News* 224, pt. 1 (9 January 1954): 46–47. 88

Gibson, McGuire. "Nippur 1975: A Summary Report." *Sumer* 34 (1978): 114–21. 110

Gichon, Mordechai. "The Plan of a Roman Camp Depicted upon a Lamp from Samaria." *Palestine Exploration Quarterly* 104 (1972): 38–58. 250, 251

Giedion, Sigfried. *The Eternal Present: The Beginnings of Architecture.* Bollingen Series 35, vol. 6, 2 pts. New York: Bollingen Foundation, 1962. 82, 86, 89, 96

Gimpel, Jean. *The Medieval Machine: The Industrial Revolution of the Middle Ages.* New York: Penguin Books, 1977. 306

Ginsburg, Herbert, and Sylvia Opper. *Piaget's Theory of Intellectual Development.* 2d ed. Englewood Cliffs, N.J.: Prentice-Hall, 1979. 2

Gioffredo, Pietro. *Corografia delle Alpi Marittime.* 2 books (1824). Republished with his *Storia delle Alpi Marittime* in *Monumenta Historia Patriae,* vol. 3, *Scriptorum I.* Genoa, Augustae Taurinorum, 1840. 67

Gisinger, F. "Geographie." In *Paulys Realencyclopädie der classischen Altertumswissenschaft,* ed. August Pauly, Georg Wissowa, et al., suppl. 4 (1924): cols. 521–685. Stuttgart: J. B. Metzler, 1894–. 208

Glanville, S. R. K. "Working Plan for a Shrine." *Journal of Egyptian Archaeology* 16 (1930): 237–39. 127

Glasser, Hannelore. *Artists' Contracts of the Early Renaissance.* New York: Garland Publishers, 1977. 436

Glob, P. V. *Helleristninger i Danmark* (Rock carvings in Denmark). Jysk Arkaeologisk Selskabs Skrifter, vol. 7. Copenhagen: Gyldendal, 1969. 63, 87

Goff, Beatrice Laura. *Symbols of Prehistoric Mesopotamia.* New Haven: Yale University Press, 1963. 71, 86, 88, 96

Goldenberg, L. A., ed. *Ispol'zovaniye starykh kart v geograficheskikh i istoricheskikh issledovaniyakh* (The use of old maps in geographical and historical investigations). Moscow: Moskovskiy Filial Geograficheskogo Obschestva SSSR (Moscow Branch, Geographical Society of the USSR) 1980. 38

Goldschmidt, E. P. "The Lesina Portolan Chart of the Caspian Sea." *Geographical Journal* 103 (1944): 272–78. 376, 459

Goldschmidt, E. P., Booksellers. *Manuscripts and Early Printed Books (1463–1600).* Catalog 4. London: E. P. Goldschmidt, [1924–25]. 401

Goodburn, R., and P. Bartholomew, eds. *Aspects of the Notitia Dignitatum.* British Archaeological Reports, Supplementary Series 15. Oxford: British Archaeological Reports, 1976. 244

Goodrum, Charles A. *Treasures of the Library of Congress.* New York: H. N. Abrams, 1980. 404, 419

Goody, Jack, ed. *Literacy in Traditional Societies.* Cambridge: Cambridge University Press, 1968. 5

Gordon, B. L. "Sacred Directions, Orientation, and the Top of the Map." *History of Religions* 10 (1971): 211–27. 337

Gottschalk, H. B. "Notes on the Wills of the Peripatetic Scholarchs." *Hermes* 100 (1972): 314–42. 158

Gottschalk, M. K. Elisabeth. "De oudste kartografische weergave van een deel van Zeeuwsch-Vlaanderen." *Archief: Vroegere en Latere Mededelingen Voornamelijk in Betrekking tot Zeeland Uitgegeven door het Zeeuwsch Genootschap der Wetenschappen* (1948): 30–39. 486, 499

———. *Historische geografie van Westelijk Zeeuws-Vlaanderen.* 2 vols. Assen: Van Gorcum, 1955–58. 470, 485, 499

Gottschalk, M. K. Elisabeth, and W. S. Unger. "De oudste kaarten der waterwegen tussen Brabant, Vlaanderen en Zeeland." *Tijdschrift van het Koninklijk Nederlandsch Aardrijkskundig Genootschap,* 2d ser., 67 (1950): 146–64. 486, 500

Gough, Richard. *Anecdotes of British Topography . . .* London: W. Richardson and S. Clark, 1768. 11

———. *British Topography; or, An Historical Account of What Has Been Done for Illustrating the Topographical Antiquities of Great Britain and Ireland.* 2 vols. London: T. Payne and J. Nichols, 1780. 7, 11

Gould, Peter. Review of *The Mapmakers*, by John Noble Wilford. *Annals of the Association of American Geographers* 72 (1982): 433–34. **26, 31**

Gould, R. A. *Living Archaeology*. Cambridge: Cambridge University Press, 1980. **59**

Goyon, Georges. "Le papyrus de Turin dit 'Des mines d'or' et le Wadi Hammamat." *Annales du Service des Antiquités de l'Egypte* 49 (1949): 337–92. **122, 124**

Grand, Paule Marie. *Arte preistorica*. Milan: Parnaso, 1967. **71**

Grapow, Hermann. "Zweiwegebuch und Totenbuch." *Zeitschrift für Ägyptische Sprache und Altertumskunde* 46 (1909): 77–81. **120**

Graux, Charles, and Albert Martin. "Figures tirées d'un manuscrit des *Météorologiques* d'Aristote." *Revue de Philologie, de Littérature et d'Histoire Anciennes*, n.s., 24 (1900): 5–18. **146, 248**

Graves, Charles. "On a Previously Undescribed Class of Monuments." *Transactions of the Royal Irish Academy* 24, pt. 8 (1867): 421–31. **64**

Graziosi, Paolo. *L'Arte rupestre della Libia*. Naples: Edizioni della Mostra d'Oltremare, 1942. **97**

[Green, John]. *The Construction of Maps and Globes*. London: T. Horne, 1717. **10, 35**

Greenhood, David. "The First Graphic Art." *Newsletter of the American Institute of Graphic Arts* 78 (1944): 1. **45**

Gregorii, Johann Gottfried. *Curieuse Gedancken von den vornehmsten und accuratesten alt- und neuen Land-Charten*. Frankfurt and Leipzig: H. P. Ritscheln, 1713. **11**

Grenacher, Franz. "With Fire and Sword, VII." *Imago Mundi* 15 (1960): 120. **6**

———. "Current Knowledge of Alsatian Cartography." *Imago Mundi* 18 (1964): 60–61. **470**

———. Review of *Meister der Kartographie*, by Leo Bagrow. *Imago Mundi* 18 (1964): 100–101. **27**

Grivaud, C. M. "Sur les antiquités d'Autun (I)." *Annales des Voyages, de la Géographie et de l'Histoire* 12 (1810): 129–66. **290**

Groenewegen-Frankfort, Henrietta Antonia. *Arrest and Movement: An Essay on Space and Time in the Representational Art of the Ancient Near East*. Chicago: University of Chicago Press, 1951. **290**

Grose, S. W. *Fitzwilliam Museum: Catalogue of the McClean Collection of Greek Coins*. Cambridge: Cambridge University Press, 1929. **159**

Grosjean, Georges, ed. *The Catalan Atlas of the Year 1375*. Dietikon-Zurich: Urs Graf, 1978. **315, 321, 381, 386, 387, 388, 390, 393, 429, 434, 446, 447, 459, 461**

———, ed. *Vesconte Maggiolo, "Atlante nautico del 1512": Seeatlas vom Jahre 1512*. Dietikon-Zurich: Urs Graf, 1979. **401, 461**

Grosjean, Georges, and Rudolf Kinauer. *Kartenkunst und Kartentechnik vom Altertum bis zum Barock*. Bern and Stuttgart: Hallwag, 1970. **25**

Guarnieri, Giuseppe Gino. *Il porto di Livorno e la sua funzione economica dalle origini ai tempi nostri*. Pisa: Cesari, 1931. **427**

Guelke, Leonard, ed. *The Nature of Cartographic*

Communication. Monograph 19. *Cartographica* (1977). **34**

Guilland, Rodolphe J. *Essai sur Nicéphore Grégoras*. Paris: P. Geuthner, 1926. **269**

Guillén y Tato, Julio F. "A propos de l'existence d'une cartographie castillane." In *Les aspects internationaux de la découverte océanique aux XVᵉ et XVIᵉ siècles: Actes du Vᵉᵐᵉ Colloque Internationale d'Histoire Maritime*, ed. Michel Mollat and Paul Adam, 251–53. Paris: SEVPEN, 1966. **389**

Gullini, Giorgio. *I mosaici di Palestrina*. Supplemento di Archeologia Classica 1. Rome: Archeologia Classica, 1956. **118**

Gundlach, Rold. "Landkarte." In *Lexikon der Ägyptologie*, ed. Wolfgang Helck and Eberhard Otto, 3:cols. 922–23. Wiesbaden: O. Harrassowitz, 1975–. **117**

Günther, Siegmund. "Die Anfänge und Entwickelungsstadien des Coordinatenprincipes." *Abhandlungen der Naturhistorischen Gesellschaft zu Nürnberg* 6 (1877): 1–50. **323**

Guthrie, W. K. C. *A History of Greek Philosophy*. 6 vols. Cambridge: Cambridge University Press, 1962–81. **130**

Gutkind, E. A. *The International History of City Development*. 8 vols. New York: Free Press of Glencoe, 1964–72. **xix**

Gyula, Pápay. "A kartográfiatörténet korszakolásának módszertani kédései." *Geodézia és Kartografia* 35, no. 5 (1983): 344–48. **xix**

Hadas, Moses. *Imperial Rome*. Alexandria, Va.: Time-Life Books, 1979. **246**

Hadingham, Evan. *Ancient Carvings in Britain: A Mystery*. London: Garnstone Press, 1974. **64**

———. *Circles and Standing Stones: An Illustrated Exploration of Megalith Mysteries of Early Britain*. Garden City, N.Y.: Anchor Press/Doubleday, 1975. **64**

———. *Secrets of the Ice Age: The World of the Cave Artists*. New York: Walker, 1979. **84**

Hahn, Cynthia. "The Creation of the Cosmos: Genesis Illustration in the Octateuch." *Cahiers Archéologiques* 28 (1979): 29–40. **262**

Hahnloser, H. R., ed. *Villard de Honnecourt: Kritische Gesamtausgabe des Bauhüttenbuches*. 2d ed. Graz: Akademische Druck- und Verlagsanstalt, 1972. **470**

Hake, Günter. *Der wissenschaftliche Standort der Kartographie*. Wissenschaftliche Arbeiten der Fachrichtung Vermessungswesen der Universität Hannover, no. 100. Hanover, 1981. **xvii**

Hakluyt, Richard. *The Principall Navigations Voiages and Discoveries of the English Nation*. A photolithographic facsimile (original imprinted in London, 1589). Cambridge: For the Hakluyt Society and the Peabody Museum of Salem at the University Press, 1965. **12**

Hall, Catherine P., and J. R. Ravensdale, eds. *The West Fields of Cambridge*. Cambridge: Cambridge Antiquarian Records Society, 1976. **465**

Hallam, H. E. "Wildmore Fen, Lincolnshire, 1224 × 1249." In *Local Maps and Plans from Medieval England*, ed. R. A. Skelton and P. D. A. Harvey, 71–81. Oxford: Clarendon Press, 1986. **484**

Hallo, William W. "The Road to Emar." *Journal of Cuneiform Studies* 18 (1964): 57–88. **108**

Hallpike, Christopher R. *The Foundations of Primitive Thought.* New York: Oxford University Press; Oxford: Clarendon Press, 1979. **52, 59, 60, 85**

Hamy, Ernest Théodore. "Les origines de la cartographie de l'Europe septentrionale." *Bulletin du Comité des Travaux Historiques et Scientifiques: Section de Géographie Historique et Descriptive* 3 (1888): 333–432. **375, 440**

———. *Etudes historiques et géographiques.* Paris: Leroux, 1896. **445**

———. "Note sur des fragments d'une carte marine catalane du XVᵉ siècle, ayant servi de signets dans les notules d'un notaire de Perpignan (1531–1556)." *Bulletin du Comité des Travaux Historiques et Scientifiques: Section de Géographie Historique et Descriptive* (1897): 23–31; reprinted in *Acta Cartographica* 4 (1969): 219–27. **373**

Hapgood, Charles H. *Maps of the Ancient Sea Kings: Evidence of Advanced Civilization in the Ice Age.* Rev. ed. New York: E. P. Dutton, 1979. **197, 291, 380, 384**

Har-El, Menashe. "Orientation in Biblical Lands." *Biblical Archaeologist* 44, no. 1 (1981): 19–20. **326**

Harbison, Robert. *Eccentric Spaces.* New York: Alfred A. Knopf, 1977. **4**

Hardie, P. R. "Imago Mundi: Cosmological and Ideological Aspects of the Shield of Achilles." *Journal of Hellenic Studies* 105 (1985): 11–31. **131**

Harding, G. Lankester. "The Cairn of Hani'." *Annual of the Department of Antiquities of Jordan* 2 (1953): 8–56. **61**

Harley, J. B. Review of *History of Cartography,* by Leo Bagrow. *Geographical Review* 131 (1965): 147. **25**

———. "The Evaluation of Early Maps: Towards A Methodology." *Imago Mundi* 22 (1968): 62–74. **35**

———. "The Map User in Eighteenth-Century North America: Some Preliminary Observations." In *The Settlement of Canada: Origins and Transfer,* ed. Brian S. Osborne, 47–69. Proceedings of the 1975 British-Canadian Symposium on Historical Geography. Kingston, Ont.: Queen's University, 1976. **36**

———. "Meaning and Ambiguity in Tudor Cartography." In *English Map-making, 1500–1650,* ed. Sarah Tyacke, 22–45. London: British Library, 1983. **36, 60, 62, 493**

———. "The Iconology of Early Maps." In *Imago et mensura mundi: Atti del IX Congresso Internazionale di Storia della Cartografia,* 2 vols., ed. Carla Clivio Marzoli, 1:29–38. Rome: Enciclopedia Italiana, 1985. **36**

———. "*Imago Mundi*: The First Fifty Years and the Next Ten." Paper prepared for the Eleventh International Conference on the History of Cartography, Ottawa, 1985. **30**

Harley, J. B., and David Woodward. "The History of Cartography Project: A Note on Its Organization and Assumptions." *Technical Papers,* 43d Annual Meeting, American Congress on Surveying and Mapping, March 1982, 580–89. **xv**

———. "Why Cartography Needs Its History." Forthcoming. **9**

Harms, Hans. *Künstler des Kartenbildes: Biographien und Porträts.* Oldenburg: E. Völker, 1962. **25**

Harthan, John. *Books of Hours and Their Owners.* London: Thames and Hudson, 1977. **436**

Hartig, Otto. "Geography in the Church." In *The Catholic Encyclopedia,* 15 vols., 6:447–53. New York: Robert Appleton, [1907–12]. **294**

Hartshorne, Richard. *The Nature of Geography: A Critical Survey of Current Thought in the Light of the Past.* Lancaster, Pa.: Association of American Geographers, 1939. **30**

———. *Perspective on the Nature of Geography.* Chicago: Rand McNally for the Association of American Geographers, 1959. **30**

Harvey, David. *Explanation in Geography.* London: Edward Arnold, 1969; New York: St. Martin's Press, 1970. **3**

Harvey, John H. "Four Fifteenth-Century London Plans." *London Topographical Record* 20 (1952): 1–8. **492**

———. "Winchester, Hampshire, *circa* 1390." In *Local Maps and Plans from Medieval England,* ed. R. A. Skelton and P. D. A. Harvey, 141–46. Oxford: Clarendon Press, 1986. **471**

Harvey, P. D. A. *The History of Topographical Maps: Symbols, Pictures and Surveys.* London: Thames and Hudson, 1980. **xvi, xix, 21, 48, 54, 75, 95, 225, 226, 229, 234, 371, 372, 426, 439, 464, 466, 469, 470, 473, 474, 475, 477, 478, 479, 485, 487, 488, 495, 496, 497, 498**

———. "The Portsmouth Map of 1545 and the Introduction of Scale Maps into England." In *Hampshire Studies,* ed. John Webb, Nigel Yates, and Sarah Peacock, 33–49. Portsmouth: Portsmouth City Records Office, 1981. **464**

———. "Cartographic Commentary." *Cartographica* 19, no. 1 (1982): 67–69. **63, 84**

———. "Shouldham, Norfolk, 1440 × 1441." In *Local Maps and Plans from Medieval England,* ed. R. A. Skelton and P. D. A. Harvey, 195–201. Oxford: Clarendon Press, 1986. **494**

———. "Wormley, Hertfordshire, 1220 × 1230." In *Local Maps and Plans from Medieval England,* ed. R. A. Skelton and P. D. A. Harvey, 59–70. Oxford: Clarendon Press, 1986. **470, 484**

Haselberger, Lothar. "Werkzeichnungen am jüngeren Didymeion." *Mitteilungen des Deutschen Archäologischen Instituts, Abteilung Istanbul* 30 (1980): 191–215. **140**

———. "The Construction Plans for the Temple of Apollo at Didyma." *Scientific American* December 1985, 126–32. **140**

Haskins, Charles Homer. *The Rise of Universities.* New York: Henry Holt, 1923. **306**

———. *Studies in the History of Mediaeval Science.* Cambridge: Harvard University Press, 1927. **293, 301, 323**

―――. *Renaissance of the Twelfth Century.* New York: Meridian, 1957. **293, 299, 304**

Hauber, Eberhard David. *Versuch einer umständlichen Historie der Land-Charten.* Ulm: D. Bartholomäi, 1724. **10**

Haudricourt, A., and J. Needham. "Ancient Chinese Science." In *Ancient and Medieval Science from the Beginnings to 1450,* 161–77. Vol. 1 of *History of Science.* 4 vols. Ed. Rene Taton. Trans. A. J. Pomerans. London: Thames and Hudson, 1963–66. **84**

Hawkes, C. F. C. *Pytheas: Europe and the Greek Explorers.* Eighth J. L. Myres Memorial Lecture. Oxford: Blackwell, 1977. **150, 151**

Hawkins, Gerald S. *Mindsteps to the Cosmos.* New York: Harper and Row, 1983. **86**

Hay, Denys. *Europe in the Fourteenth and Fifteenth Centuries.* London: Longmans, 1966. **444**

Hayes, William C. *Ostraka and Name Stones from the Tomb of Sen-Mūt (no. 71) at Thebes.* Publications of the Metropolitan Museum of Art Egyptian Expedition, vol. 15. New York: Metropolitan Museum of Art, 1942. **126**

Head, C. Grant. "The Map as Natural Language: A Paradigm for Understanding." In *New Insights in Cartographic Communication.* Ed. Christopher Board. Monograph 31. *Cartographica* 21, no. 1 (1984): 1–32. **2, 34**

Heathcote, N. H. de Vaudrey. "Early Nautical Charts." *Annals of Science* 1 (1936): 1–28. **396**

Heggie, Douglas C. *Megalithic Science: Ancient Mathematics and Astronomy in Northwest Europe.* London: Thames and Hudson, 1981. **81**

―――, ed. *Archaeoastronomy in the Old World.* Cambridge: Cambridge University Press, 1982. **81**

Heidel, William Arthur. "Anaximander's Book: The Earliest Known Geographical Treatise." *Proceedings of the American Academy of Arts and Sciences* 56 (1921): 237–88. **134**

―――. *The Frame of the Ancient Greek Maps.* New York: American Geographical Society, 1937. **132, 140, 152**

Heinrich, Ernst, and Ursula Seidl. "Grundrißzeichnungen aus dem alten Orient." *Mitteilungen der Deutschen Orient-Gesellschaft zu Berlin* 98 (1967): 24–45. **109, 110**

Helck, Wolfgang. "Gartenanlage, -bau." In *Lexikon der Ägyptologie,* ed. Wolfgang Helck and Eberhard Otto, 2:cols. 378–80. Wiesbaden: O. Harrassowitz, 1975–. **118**

Hennessy, J. B. "Preliminary Report on a First Season of Excavations at Teleilat Ghassul." *Levant* 1 (1969): 1–24. **58**

Hessels, John Henry, ed. *Abrahami Ortelii (geographi antverpiensis) et virorum eruditorum ad eundem . . . Epistulae . . . (1524–1628).* Ecclesiae Londino-Batavae archivum, vol. 1. London: Nederlandsche Hervormde Gemeente, 1887. **9**

Heurgon, Jacques. "La date des gobelets de Vicarello." *Revue des Etudes Anciennes* 54 (1952): 39–50. **235**

Hewes, Gordon W. "Primate Communication and the Gestural Origin of Language." *Current Anthropology* 14, nos. 1–2 (1973): 5–24. **52**

Heydenreich, Ludwig H. "Ein Jerusalem-Plan aus der Zeit der Kreuzfahrer." In *Miscellanea pro arte,* ed. Joseph Hoster and Peter Bloch, 83–90. Cologne: Freunde des Schnütgen-Museums, 1965. **474**

Heywood, Nathan. "The Cup and Ring Stones on the Panorama Rocks, Near Rombald's Moor, Ilkley, Yorkshire." *Transactions of the Lancashire and Cheshire Antiquarian Society* 6 (1888): 127–28. **86**

Hill, George Francis. *Coins of Ancient Sicily.* Westminster: A. Constable, 1903. **158**

Hill, Gillian. *Cartographical Curiosities.* London: British Museum Publications, 1978. **22**

Hingman, Jan Henricus. *Inventaris der verzameling kaarten berustende in het Rijks-Archief.* The Hague: Nijhoff, 1867–71. **486, 500**

Hinks, Arthur R. *Portolan Chart of Angellino de Dalorto 1325 in the Collection of Prince Corsini at Florence, with a Note on the Surviving Charts and Atlases of the Fourteenth Century.* London: Royal Geographical Society, 1929. **380, 394, 411, 423, 457**

Hinks, R. *Myth and Allegory in Ancient Art.* London: Warburg Institute, 1939. **264**

Hinrichs, Focke Tannen. *Die Geschichte der gromatischen Institutionen.* Wiesbaden: Franz Steiner, 1974. **218**

Hodder, Ian. *Symbols in Action: Ethnoarchaeological Studies of Material Culture.* Cambridge: Cambridge University Press, 1982. **60**

Hodgkiss, Alan G. *Understanding Maps: A Systematic History of Their Use and Development.* Folkestone: Dawson, 1981. **25**

Hoehn, Philip. "The Cartographic Treasures of the Bancroft Library." *Map Collector* 23 (1983): 28–32. **16**

Hoff, B. van 't. "The Oldest Maps of the Netherlands: Dutch Map Fragments of about 1524." *Imago Mundi* 16 (1962): 29–32. **486, 500**

Honigmann, Ernst. *Die sieben Klimata und die πόλεις ἐπίσημοι.* Heidelberg: Winter, 1929. **182, 191**

Hooke, S. H. "Recording and Writing." In *From Early Times to Fall of Ancient Empires,* 744–73. Vol. 1 of *A History of Technology.* 7 vols. Ed. Charles Singer et al. Oxford: Clarendon Press, 1954–78. **60**

Hope, W. H. St. John. "The London Charterhouse and Its Old Water Supply." *Archaeologia* 58 (1902): 293-312. **491**

Horn, W. Review of *Die Geschichte der Kartographie,* by Leo Bagrow. *Petermanns Geographische Mitteilungen* 97 (1953): 222. **25**

Horn, Walter, and Ernest Born. "New Theses about the Plan of St. Gall." In *Die Abtei Reichenau: Neue Beiträge zur Geschichte und Kultur des Inselklosters,* ed. Helmut Maurer, 407–76. Sigmaringen: Thorbecke, 1974. **466**

―――. *The Plan of St. Gall: A Study of the Architecture and Economy of, and Life in, a Paradigmatic Carolingian Monastery.* 3 vols. Berkeley: University of California Press, 1979. **466**

Hough, Samuel J. *The Italians and the Creation of America: An Exhibition at the John Carter Brown Library.* Providence: John Carter Brown Library, 1980. **22**

Howse, Derek. "Some Early Tidal Diagrams." *Revista da Universidade de Coimbra* 33 (1985): 365–85. **440**

Howse, Derek, and Michael Sanderson. *The Sea Chart*. Newton Abbot: David and Charles, 1973. **378, 457**

Hull, F. "Cliffe, Kent, Late 14th Century × 1408." In *Local Maps and Plans from Medieval England*, ed. R. A. Skelton and P. D. A. Harvey, 99–105. Oxford: Clarendon Press, 1986. **484**

———. "Isle of Thanet, Kent, Late 14th Century × 1414." In *Local Maps and Plans from Medieval England*, ed. R. A. Skelton and P. D. A. Harvey, 119–26. Oxford: Clarendon Press, 1986. **484, 493**

Hülsen, C. "Di una nuova pianta prospettica di Roma del secolo XV." *Bullettino della Commissione Archeologica Comunale di Roma*, 4th ser., 20 (1892): 38–47. **493**

Humboldt, Alexander von. *Examen critique de l'histoire de la géographie du nouveau continent et des progrès de l'astronomie nautique au XVᵉ et XVIᵉ siècles*. 5 vols. Paris: Gide, 1836–39. **17, 386**

———. *Views of Nature*. Trans. E. C. Otté and H. G. Bohn. London: Bell and Daldy, 1872. **47**

Humphreys, Arthur L. *Old Decorative Maps and Charts*. London: Halton and Smith; New York: Minton, Balch, 1926. Revised by R. A. Skelton as *Decorative Printed Maps of the 15th to 18th Centuries*. London: Staples Press, 1952. **25**

Hunger, Herbert. *Die hochsprachliche profane Literatur der Byzantiner*. Munich: Beck, 1978–. **266**

Hutorowicz, H. de. "Maps of Primitive Peoples." *Bulletin of the American Geographical Society* 43, no. 9 (1911): 669–79. **46, 54**

Huussen, A. H. *Jurisprudentie en kartografie in de XVᵉ en XVIᵉ eeuw*. Brussells: Algemeen Rijksarchief, 1974. **486, 500**

Huxley, G. L. "Eudoxus of Cnidus." In *Dictionary of Scientific Biography*, 16 vols., ed. Charles Coulston Gillispie, 4:465–67. New York: Charles Scribner's Sons, 1970–80. **140**

———. "A Porphyrogenitan Portulan." *Greek, Roman and Byzantine Studies* 17 (1976): 295–300. **260**

Ibarra Grasso, Dick Edgar. *La representación de América en mapas romanos de tiempos de Cristo*. Buenos Aires: Ediciones Ibarra Grasso, 1970. **199**

Imhof, Eduard. *Die ältesten Schweizerkarten*. Zurich: Füssli, 1939. **498**

———. "Beiträge zur Geschichte der topographischen Kartographie." *International Yearbook of Cartography* 4 (1964): 129–53. **38**

Institut Géographique National. *Atlas des centuriations romaines de Tunisie*. Paris: Institut Géographique National, 1954. **198, 219**

International Cartographic Association. *Multilingual Dictionary of Technical Terms in Cartography*. Ed. E. Meynen. Wiesbaden: Franz Steiner Verlag, 1973. **xvi**

———. *Map-Making to 1900: An Historical Glossary of Cartographic Innovations and Their Diffusion*. Ed. Helen Wallis. London: Royal Society, 1976. **33**

———. *Cartographical Innovations: An International Handbook of Mapping Terms to 1900*. Ed. Helen Wallis and Arthur Robinson. Tring, Hertfordshire: Map Collector Publications, forthcoming. **33**

International Cartographic Association (British National Committee for Geography subcommittee). *Glossary of Technical Terms in Cartography*. London: Royal Society, 1966. **xvi**

Ischer, Theophil. *Die ältesten Karten der Eidgenossenschaft*. Bern: Schweizer Bibliophile Gesellschaft, 1945. **498**

Israel, Nico, Antiquarian Booksellers. *Interesting Books and Manuscripts on Various Subjects: A Selection from Our Stock . . . .* Catalog 22. Amsterdam: N. Israel, 1980. **378, 419, 443, 458**

Issel, A. "Le rupi scolpite nelle alte valli delle Alpi Marittime." *Bollettino di Paletnologia Italiana* 17 (1901): 217–59. **67**

Jackson, Donald. *The Story of Writing*. New York: Taplinger, 1981. **318**

Jacob, Christian. "Lectures antiques de la carte." *Etudes françaises* 21, no. 2 (1985): 21–46. **139, 253**

Jacoby, G. "Über die Gründung einer internationalen Zentralstelle für die Geschichte der Kartographie." *Kartographische Nachrichten* 12 (1962): 27–28. **37**

Jacquet, Jean. "Remarques sur l'architecture domestique à l'époque méroïtique: Documents recueillis sur les fouilles d'Ash-Shaukan." In *Aufsätze zum 70. Geburtstag von Herbert Ricke*, ed. Abdel Moneim Abubakr et al., 121–31. Beiträge zur Ägyptischen Bauforschung und Altertumskunde, no. 12. Wiesbaden: F. Steiner, 1971. **127**

James, Montague Rhodes. *A Descriptive Catalogue of the Manuscripts in the Library of Corpus Christi College Cambridge*. 2 vols. Cambridge: Cambridge University Press, 1912. **312**

James, Preston E., and Geoffrey J. Martin. *All Possible Worlds: A History of Geographical Ideas*. 2d ed. New York: John Wiley, 1981. **31**

Janni, Pietro. *La mappa e il periplo: Cartografia antica e spazio odologico*. Università di Macerata, Pubblicazioni della Facoltà di Lettere e Filosofia 19. Rome: Bretschneider, 1984. **237**

Janvier, Y. *La géographie d'Orose*. Paris: Belles Lettres, 1982. **347**

Jaynes, Julian. "The Evolution of Language in the Late Pleistocene." *Annals of the New York Academy of Sciences* 280 (1976): 322. **53**

———. *The Origins of Consciousness in the Breakdown of the Bicameral Mind*. Boston: Houghton Mifflin, 1976. **51**

Jensen, Hans. *Symbol and Script: An Account of Man's Efforts to Write*. 3d ed., rev. and enl. London: George Allen and Unwin, 1970. **49**

Jervis, W. W. *The World in Maps: A Study in Map Evolution*. London: George Philip, 1936. **25, 294**

Johnson, Elmer D. *A History of Libraries in the Western World*. New York and London: Scarecrow Press, 1965. **167**

Johnson, Samuel. *Rambler* 84, Sat., 29 Dec. 1750. **12**

Johnston, A. E. M. "The Earliest Preserved Greek Map: A New Ionian Coin Type." *Journal of Hellenic Studies* 87 (1967): 86–94. **158**

Johnston, R. J. *Geography and Geographers: Anglo-American Human Geography since 1945*. 2d ed. London: Edward Arnold, 1983. **31**

Jomard, Edme-François. *Considérations sur l'objet et les avantages d'une collection spéciale consacrée aux cartes géographiques et aux diverses branches de la géographie*. Paris: E. Duverger, 1831. **15**

———. *De l'utilité qu'on peut tirer de l'étude comparative des cartes géographiques*. Paris: Burgogne et Martinet, 1841; reprinted from *Bulletin de la Société de Géographie*, 2d ser., 15 [1841]: 184–94. **15**

———. *Les monuments de la géographie; ou, Recueil d'anciennes cartes européennes et orientales*. Paris: Duprat, 1842–62. **13, 293**

———. *Sur la publication des Monuments de la géographie*. Paris, 1847. **13**

———. *De la collection géographique créée à la Bibliothèque Royale*. Paris: E. Duverger, 1848. **16**

———. *Introduction à l'atlas des Monuments de la géographie*. Paris: Arthus Bertrand, 1879. **13, 18**

Jones, A. C. "Land Measurement in England, 1150–1350." *Agricultural History Review* 27 (1979): 10–18. **494**

Jones, Charles W. "The Flat Earth." *Thought: A Quarterly of the Sciences and Letters* 9 (1934): 296–307. **319, 320**

Jones, Philip E. "Deptford, Kent and Surrey; Lambeth, Surrey; London, 1470–1478." In *Local Maps and Plans from Medieval England*, ed. R. A. Skelton and P. D. A. Harvey, 251–62. Oxford: Clarendon Press, 1986. **492**

Josephson, Åke. *Casae litterarum: Studien zum Corpus Agrimensorum Romanorum*. Uppsala: Almqvist och Wiksell, 1950. **226**

Julku, Kyösti. "Suomen tulo maailmankartalle" (Appearance of Finland on medieval world maps). *Faravid* 1 (1977): 7–41. **290**

Kadmon, Naftali. "Cartograms and Topology." *Cartographica* 19, nos. 3–4 (1982): 1–17. **xvii**

Kamal, Youssouf. *Monumenta cartographica Africae et Aegypti*. 5 vols. in 16 pts. Cairo, 1926–51. **18, 294, 302, 328, 368, 393, 404, 406, 411, 413, 416, 418, 419, 420, 457, 459, 461**

———. *Quelques éclaircissements épars sur mes Monumenta cartographica Africae et Aegypti*. Leiden: E. J. Brill, 1935. **388, 444**

———. *Hallucinations scientifiques (les portulans)*. Leiden: E. J. Brill, 1937. **380, 381, 384**

Kandler, Pietro. *Indicazioni per riconoscere le cose storiche del Litorale*. Trieste, 1855. **219**

Kantor, Helene J. "Narrative in Egyptian Art." *American Journal of Archaeology* 61 (1957): 44–54. **119**

Karig, Joachim Selim. "Die Landschaftsdarstellung in den Privatgräbern des Alten Reiches." Ph.D. diss., University of Göttingen, 1962. **118**

Karrow, Robert W. "Cartobibliography." *AB Bookman's Yearbook*, pt. 1 (1976): 43–52. **19, 20**

———. "The Cartographic Collections of the Newberry Library." *Map Collector* 32 (1985): 10–15. **16**

Keates, J. S. *Understanding Maps*. New York: John Wiley, 1982. **2**

Kees, Hermann. *Totenglauben und Jenseitsvorstellungen der alten Ägypter*. Berlin: Akademie-Verlag, 1956. **120**

Kelley, James E., Jr. "The Oldest Portolan Chart in the New World." *Terrae Incognitae: Annals of the Society for the History of Discoveries* 9 (1977): 22–48. **377, 383, 384, 389, 391, 395, 403, 404, 420, 423, 429, 431, 444**

———. "Non-Mediterranean Influences That Shaped the Atlantic in the Early Portolan Charts." *Imago Mundi* 31 (1979): 18–35. **410, 411, 414, 428**

Ker, Neil. Review of *Mappemondes A.D. 1200–1500: Catalogue préparé par la Commission des Cartes Anciennes de l'Union Géographique Internationale*, ed. Marcel Destombes. *Book Collector* 14 (1965): 369–73. **286, 302**

Kerényi, Karl. *Labyrinth-Studien: Labyrinthos als Linienreflex einer mythologischen Idee*. 2d ed. Zurich: Rhein-Verlag, 1950. **88, 251**

Kerferd, G. B. "Democritus." In *Dictionary of Scientific Biography*, 16 vols., ed. Charles Coulston Gillispie, 4:30–35. New York: Charles Scribner's Sons, 1970–80. **137**

Kern, Hermann. *Labirinti: Forme e interpretazione, 5000 anni di presenza di un archetipo manuale e file conduttore*. Milan: Feltrinelli, 1981. German edition, *Labyrinthe: Erscheinungsformen und Deutungen, 5000 Jahre Gegenwart eines Urbilds*. Munich: Prestel-Verlag, 1982. **4, 88, 251**

Keuning, Johannes. "XVIth Century Cartography in the Netherlands (Mainly in the Northern Provinces)." *Imago Mundi* 9 (1952): 35–64. **343, 486, 500**

———. "The History of Geographical Map Projections until 1600." *Imago Mundi* 12 (1955): 1–24. **185, 385**

Kiely, Edmond R. *Surveying Instruments: Their History*. New York: Teachers College, Columbia University, 1947; reprinted Columbus: Carben Surveying Reprints, 1979. **213, 232, 494**

Kimble, George H. T. *Geography in the Middle Ages*. London: Methuen, 1938. **294, 321, 328**

King, Georgiana Goddard. "Divagations on the Beatus." In *Art Studies: Medieval, Renaissance and Modern*, 8 vols., ed. members of Departments of Fine Arts at Harvard and Princeton Universities, 8:3–58. Cambridge: Harvard University Press, 1923–30. **303**

Kirk, G. S., J. E. Raven, and M. Schofield. *The Presocratic Philosophers*. 2d ed. Cambridge: Cambridge University Press, 1983. **130, 136**

Kirkbride, Diane. "Umm Dabaghiyah 1974: A Fourth Preliminary Report." *Iraq* 37 (1975): 3–10. **58**

Kish, George. "The Japan on the 'Mural Atlas' of the Palazzo Vecchio, Florence." *Imago Mundi* 8 (1951): 52–54. **8**

———. Review of *Maps and Their Makers: An Introduction to the History of Cartography*, by Gerald R. Crone. *Geographical Review* 45 (1955): 448–49. **26**

———. Review of *History of Cartography*, by Leo Bagrow. *Geographical Review* 56 (1966): 312–13. **25**

———. *La carte: Image des civilisations*. Paris: Seuil, 1980. **22, 25, 88, 95, 96, 458**

———, ed. *A Source Book in Geography*. Cambridge: Harvard University Press, 1978. **237, 244**

———, ed. *Bibliography of International Geographical Congresses, 1871-1976.* Boston: G. K. Hall, 1979. **15**

Kitzinger, Ernst. "Studies on Late Antiquity and Early Byzantine Floor Mosaics: I. Mosaics at Nikopolis." *Dumbarton Oaks Papers* 6 (1951): 81–122. **264**

———. "World Map and Fortune's Wheel: A Medieval Mosaic Floor in Turin." *Proceedings of the American Philosophical Society* 117 (1973): 344–73. **339**

Klebs, Luise. *Die Reliefs und Malereien des Neuen Reiches.* Heidelberg: C. Winter, 1934. **118**

Klein, Peter K. *Der ältere Beatus-Kodex Vitr. 14-1 der Biblioteca Nacional zu Madrid: Studien zur Beatus-Illustration und der spanischen Buchmalerei des 10. Jahrhunderts.* Hildesheim: Georg Olms, 1976. **303, 305**

Klein, Robert. *Form and Meaning: Essays on the Renaissance and Modern Art.* Trans. Madeline Jay and Leon Wieseltier. New York: Viking Press, 1970. **89**

Klotz, A. "Die geographischen Commentarii des Agrippa und ihre Überreste." *Klio* 24 (1931): 38–58, 386–466. **208**

Knowles, M. D. "Clerkenwell and Islington, Middlesex, Mid-15th Century." In *Local Maps and Plans from Medieval England,* ed. R. A. Skelton and P. D. A. Harvey, 221–28. Oxford: Clarendon Press, 1986. **491**

Koeman, Cornelis. *Collections of Maps and Atlases in the Netherlands: Their History and Present State.* Leiden: E. J. Brill, 1961. **8, 16, 500**

———. "An Increase in Facsimile Reprints." *Imago Mundi* 18 (1964): 87–88. **18**

———. "Hoe oud is het woord kartografie?" *Geografisch Tijdschrift* 8 (1974): 230–31. **12**

———. "Algemene inleiding over de historische kartografie, meer in het Bijzonder: Holland vóór 1600." *Holland* 7 (1975): 230. **494**

———. "Moderne onderzoekingen op het gebied van de historische kartografie." *Bulletin van de Vakgroep Kartografie* 2 (1975): 3–24. **38**

———. "Sovremenniye issledovaniya v oblasti istoricheskoy kartografii i ikh znacheniye dlya istorii kul'tury i razvitiya kartograficheskikh nauk" (Modern investigations in the field of the history of cartography: Their contribution to cultural history and the development of the science of cartography). In *Puti razvitiya kartografii* (Paths to the evolution of cartography), 107–21. A collection of papers on the occasion of Professor K. A. Salishchev's seventieth birthday. Moscow: Izdatel'stvo Moskovskogo Universiteta, 1975. **23**

———. *Geschiedenis van de kartografie van Nederland: Zes eeuwen land- en zeekaarten en stadsplattegronden.* Alphen aan den Rijn: Canaletto, 1983. **37**

Koepf, Hans. *Die gotischen Planrisse der Wiener Sammlungen.* Vienna: Böhlau, 1969. **470**

Kohl, Johann Georg. "Substance of a Lecture Delivered at the Smithsonian Institution on a Collection of the Charts and Maps of America." *Annual Report of the Board of Regents of the Smithsonian Institution . . . 1856,* (1857), 93–146. **13, 16**

Kolev, P., et al., eds. *The Netherlands—Bulgaria: Traces of Relations through the Centuries—Material from Dutch Archives and Libraries on Bulgarian History and on Dutch Contacts with Bulgaria.* Sofia: State Publishing House "Septemvri," 1981. **380**

Körte, G. "Die Bronzeleber von Piacenza." *Mitteilungen des Kaiserlich Deutschen Archaeologischen Instituts, Römische Abteilung* 20 (1905): 348–77. **202**

Kosack, Hans-Peter, and Karl-Heinz Meine. *Die Kartographie, 1943-1954: Eine bibliographische Übersicht.* Kartographische Schriftenreihe, vol. 4. Lahr-Schwarzwald: Astra Verlag, 1955. **31**

Kraeling, Carl H., ed. *Gerasa, City of the Decapolis.* New Haven: American Schools of Oriental Research, 1938. **469**

Kramer, Samuel Noah. *From the Tablets of Sumer.* Indian Hills, Colo.: Falcon's Wing Press, 1956. **110**

———. *History Begins at Sumer.* 3d ed. Philadelphia: University of Pennsylvania Press, 1981. **110**

Kramer, Samuel Noah, and Inez Bernhardt. "Der Stadtplan von Nippur, der älteste Stadtplan der Welt." *Wissenschaftliche Zeitschrift: Gesellschafts- und Sprachwissenschaftliche Reihe* 19 (1970): 727–30. **110**

Kratochwill, Max. "Zur Frage der Echtheit des 'Albertinischen Planes' von Wien." *Jahrbuch des Vereins für Geschichte der Stadt Wien* 29 (1973): 7–36. **473**

Kraus, Fritz Rudolf. "Provinzen des neusumerischen Reiches von Ur." *Zeitschrift für Assyriologie und vorderasiatische Archäologie,* n.s., 17 (1955): 45–75. **108**

Kraus, H. P., Booksellers. *Remarkable Manuscripts, Books and Maps from the IXth to the XVIIIth Century.* Catalog 80. New York: H. P. Kraus, 1956. **457**

———. *Twenty-five Manuscripts.* Catalog 95. New York: H. P. Kraus, [1961]. **427, 437**

Kretschmer, Ingrid. "The Pressing Problems of Theoretical Cartography." *International Yearbook of Cartography* 13 (1978): 33–40. **30**

Kretschmer, I., J. Dörflinger, and F. Wawrik. *Lexikon zur Geschichte der Kartographie.* 2 vols. Vienna, 1986. **27**

Kretschmer, Konrad. *Die italienischen Portolane des Mittelalters: Ein Beitrag zur Geschichte der Kartographie und Nautik.* Veröffentlichungen des Instituts für Meereskunde und des Geographischen Instituts an der Universität Berlin, vol. 13. Berlin, 1909; reprinted Hildesheim: Georg Olms, 1962. **376, 382, 384, 388, 390, 422, 423, 425, 426, 427**

Krüger, Herbert. "Erhard Etzlaub's *Romweg* Map and Its Dating in the Holy Year of 1500." *Imago Mundi* 8 (1951): 17–26. **498**

Kubitschek, Wilhelm. "Itinerarien." In *Paulys Realencyclopädie der classischen Altertumswissenschaft,* ed. August Pauly, Georg Wissowa, et al., 9 (1916): cols. 2308–63. Stuttgart: J. B. Metzler, 1894–. **237**

———. "Karten." In *Paulys Realencyclopädie der classischen Altertumswissenschaft,* ed. August Pauly, Georg Wissowa, et al., 10 (1919): cols. 2022–2149. Stuttgart: J. B. Metzler, 1894–. **246**

Kühn, Herbert. *Wenn Steine reden: Die Sprache der Felsbilder.* Wiesbaden: F. A. Brockhaus, 1966. **87**

Kuhn, Thomas S. *The Structure of Scientific Revolutions.* Chicago: University of Chicago Press, 1962. **23**

Kupčík, Ivan. *Alte Landkarten: Von der Antike bis zum Ende des 19. Jahrhunderts.* Hanau am Main: Dausien, 1980. French edition, *Cartes géographiques anciennes: Evolution de la représentation cartographique du monde de l'antiquité à la fin du XIXe siècle.* Trans. Suzanne Bartošek. Paris: Edition Gründ, 1981. **25**

Lacey, Alan Robert. *A Dictionary of Philosophy.* London: Routledge and Kegan Paul, 1976. **51**

Lach, Donald F. *Asia in the Making of Europe.* 2 vols. in 5. Chicago: University of Chicago Press, 1965–77. **304**

Ladner, Gerhart B. "St. Gregory of Nyssa and St. Augustine on the Symbolism of the Cross." In *Late Classical and Mediaeval Studies in Honor of Albert Mathias Friend, Jr.,* ed. Kurt Weitzmann, 88–95. Princeton: Princeton University Press, 1955. **334**

———. "Medieval and Modern Understanding of Symbolism: A Comparison." *Speculum* 54 (1979): 223–56. **334**

Lagrange, J. L. "Sur la construction des cartes géographiques." *Nouveaux Mémoires de l'Académie Royale des Sciences et Belles-Lettres* (1779), 161–210. **xv**

Laguarda Trías, Rolando A. *Estudios de cartologia.* Madrid, 1981. **381, 384, 385**

Lajoux, Jean Dominique. *The Rock Paintings of Tassili.* Trans. G. D. Liversage. London: Thames and Hudson, 1963. **96, 97**

Lamberg-Karlovsky, C. C. "Trade Mechanisms in Indus-Mesopotamian Interrelations." *Journal of the American Oriental Society* 92 (1972): 222–29. **108**

Lambert, Wilfred G. "The Cosmology of Sumer and Babylon." In *Ancient Cosmologies,* ed. Carmen Blacker and Michael Loewe, 42–65. London: George Allen and Unwin, 1975. **86, 112**

Landels, J. G. *Engineering in the Ancient World.* London: Chatto and Windus, 1978. **210**

Landsberger, Benno. *Materialien zum Sumerischen Lexikon: Vokabulare und Formularbücher.* Rome: Pontifical Biblical Institute Press, 1937–. **107**

Lane, Frederic C. "The Economic Meaning of the Invention of the Compass." *American Historical Review* 68, no. 3 (1963): 605–17. **384, 387**

———. *Venice: A Maritime Republic.* Baltimore: Johns Hopkins University Press, 1973. **434, 442**

Lang, Arend Wilhelm. Review of *Die Geschichte der Kartographie,* by Leo Bagrow. *Erdkunde* 7 (1953): 311–12. **25**

———. "Traces of Lost North European Sea Charts of the Fifteenth Century." *Imago Mundi* 12 (1955): 31–44. **414, 415**

———. *Das Kartenbild der Renaissance.* Ausstellungskataloge der Herzog August Bibliothek, no. 20. Wolfenbüttel: Herzog August Bibliothek, 1977. **21**

Lang, Mabel. "The Palace of Nestor Excavations of 1957: Part II." *American Journal of Archaeology,* 2d ser., 62 (1958): 181–91. **251**

Langdon, Stephen H. "An Ancient Babylonian Map." *Museum Journal* 7 (1916): 263–68. **111**

Langer, Susanne K. *Philosophy in a New Key: A Study in the Symbolism of Reason, Rite, and Art.* 3d ed. Cambridge: Harvard University Press, 1957. **60**

Langlois, Charles Victor. *La vie en France au Moyen Age, de la fin du XIIe au milieu du XIVe siècle.* 4 vols. Paris: Hachette, 1926–28. **287**

Lanman, Jonathan T. "The Religious Symbolism of the T in T-O Maps." *Cartographica* 18, no. 4 (1981): 18–22. **334**

———. "On the Origin of Portolan Charts." Paper prepared for the Eleventh International Conference on the History of Cartography, Ottawa, 1985. **383**

La Roncière, Charles de. "Un inventaire de bord en 1294 et les origines de la navigation hauturière." *Bibliothèque de l'Ecole des Chartes* 58 (1897): 394–409. **439**

———. "Le portulan du XVe siècle découvert à Gap." *Bulletin du Comité des Travaux Historiques et Scientifiques: Section de Géographie Historique et Descriptive* 26 (1911): 314–18. **386**

———. *La découverte de l'Afrique au Moyen Age: Cartographes et explorateurs.* Mémoires de la Société Royale de Géographie d'Egypte, vols. 5, 6, 13. Cairo: Institut Français d'Archéologie Orientale, 1924–27. **328, 382, 389, 404, 418, 425, 429, 432, 434, 440, 457**

———. *Les portulans de la Bibliothèque de Lyon.* Fasc. 8 of *Les Portulans Italiens.* In Lyon, Bibliothèque de la Ville. *Documents paléographiques, typographiques, iconographiques.* Lyons, 1929. **374, 407, 419, 427, 435, 458**

———. "Une nouvelle carte de l'école cartographique des Juifs de Majorque." *Bulletin du Comité des Travaux Historiques et Scientifiques: Section de Géographie* 47 (1932): 113–18. **376, 432**

La Roncière, Monique de. "Les cartes marines de l'époque des grandes découvertes." *Revue d'Histoire Economique et Sociale* 45 (1967): 5–28. **371**

La Roquette, Jean Bernard Marie Alexander Dezos de. *Notice sur la vie et les travaux de M. Jomard.* Paris: L. Martinet, 1863. **19**

Latham, Robert, and William Matthews, eds. *The Diary of Samuel Pepys.* 11 vols. Berkeley: University of California Press, 1970–83. **9**

Lattin, Harriet Pratt. "The Eleventh Century MS Munich 14436: Its Contribution to the History of Coordinates, of Logic, of German Studies in France." *Isis* 38 (1947): 205–25. **323**

Lavedan, Pierre. *Représentation des villes dans l'art du Moyen Age.* Paris: Vanoest, 1954. **469, 487**

Layard, John W. *Stone Men of Malekula.* London: Chatto and Windus, 1942. **88**

Layton, Robert. "Naturalism and Cultural Relativity in Art." In *Form in Indigenous Art: Schematisation in the Art of Aboriginal Australia and Prehistoric Europe,* ed. Peter Ucko, 34–45. Australian Institute of Aboriginal Studies, Prehistory and Material Culture Series no. 13. London: Gerald Duckworth, 1977. **60**

Leach, Edmund. *Culture and Communication: The Logic by Which Symbols Are Connected: An Introduction to the Use of Structuralist Analysis in Social Anthropology.* Cambridge: Cambridge University Press, 1976. **2**

Lebeuf, Abbé. "Notice d'un manuscrit des Chroniques de

Saint Denys, le plus ancien que l'on connoisse." *Histoire de l'Académie Royale des Inscriptions et Belles-Lettres* 16 (1751): 175–85. **288**

Leclant, Jean. "Earu-Gefilde." In *Lexikon der Ägyptologie*, ed. Wolfgang Helck and Eberhard Otto, 1:cols. 1156–60. Wiesbaden: O. Harrassowitz, 1975–. **119**

Leclercq, Henri. "Itinéraires." In *Dictionnaire d'archéologie chrétienne et de liturgie*. 15 vols., ed. Fernand Cabrol and Henri Leclercq, 7.2 (1927): cols. 1841–1922. Paris: Letouzey et Ané, 1907–53. **237**

———. "Labyrinthe." In *Dictionnaire d'archéologie chrétienne et de liturgie*, 15 vols., ed. Fernand Cabrol and Henri Leclercq, 8.1 (1928): cols. 973–82. Paris: Letouzey et Ané, 1907–53. **252**

Lee, Ivan. "Polesini: Upper Palaeolithic Astronomy." *Archaeology 83: The Pro-Am Newsletter* 2 (1983). **84**

Leeman, A. D. *A Systematic Bibliography of Sallust, 1879–1950.* Leiden: E. J. Brill, 1952. **343**

Leff, Gordon. *History and Social Theory.* University: University of Alabama Press, 1969. **xviii**

Lehmann-Brockhaus, Otto. *Lateinische Schriftquellen zur Kunst in England, Wales und Schottland, vom Jahre 901 bis zum Jahre 1307.* 5 vols. Munich: Prestel, 1955–60. **368**

Leithäuser, Joachim G. *Mappae mundi: Die geistige Eroberung der Welt.* Berlin: Safari-Verlag, 1958. **294**

Lelewel, Joachim. *Géographie du Moyen Age.* 4 vols. and epilogue. Brussels: J. Pilliet, 1852–57; reprinted Amsterdam: Meridian, 1966. **11, 293, 457, 495**

Lemerle, Paul. *Le premier humanisme byzantin.* Paris: Presses Universitaires de France, 1971. **258**

Lenzen, H. J., Adam Falkenstein, and W. Ludwig, eds. *Vorläufiger Bericht über die von dem Deutschen Archäologischen Institut und der Deutschen Orient-Gesellschaft aus Mitteln der Deutschen Forschungsgemeinschaft unternommenen Ausgrabungen in Uruk-Warka.* Abhandlungen der Deutschen Orient-Gesellschaft, Winter 1953/54, Winter 1954/55. Berlin: Gebr. Mann, 1956. **110**

Leonardi, Piero. "Su alcuni petroglifi della Valcamonica e della Venezia Tridentina." In *Symposium International d'Art Préhistorique Valcamonica, 23–28 Septembre 1968*, 235–39. Union Internationale des Sciences Préhistoriques et Protohistoriques. Capo di Ponte: Edizioni del Centro, 1970. **95**

Lepsius, Richard. *Auswahl der wichtigsten Urkunden des ægyptischen Alterthums: Theils zum erstenmale, theils nach den Denkmälern berichtigt.* Leipzig: Wigand, 1842. **122**

Leroi-Gourhan, André. *Art of Prehistoric Man in Western Europe.* Trans. Norbert Guterman. London: Thames and Hudson, 1968. **57**

———. *The Dawn of European Art: An Introduction to Palaeolithic Cave Painting.* Trans. Sara Champion. Cambridge: Cambridge University Press, 1982. **57**

Lesko, Leonard H. "Some Observations on the Composition of the *Book of Two Ways.*" *Journal of the American Oriental Society* 91 (1971): 30–43. **120**

———. *The Ancient Egyptian Book of Two Ways.*

University of California Near Eastern Studies Publications, vol. 17. Berkeley: University of California Press, 1972. **120**

Levi, Annalina, and Mario Levi. *Itineraria picta: Contributo allo studio della Tabula Peutingeriana.* Rome: Erma di Bretschneider, 1967. **7, 237, 238, 239, 240, 246, 248**

———. "The Medieval Map of Rome in the Ambrosian Library's Manuscript of Solinus." *Proceedings of the American Philosophical Society* 118 (1974): 567–94. **477**

Lévi-Strauss, Claude. *Structural Anthropology.* Trans. Claire Jacobson and Brooke Grundfest Schoepf. New York: Anchor Books, 1967. **58**

Lewis, David. "Observations on Route Finding and Spatial Orientation among the Aboriginal Peoples of the Western Desert Region of Central Australia." *Oceania* 46, no. 4 (1976): 249–82. **59**

Lewis, G. Malcolm. "The Recognition and Delimitation of the Northern Interior Grasslands during the Eighteenth Century." In *Images of the Plains: The Role of Human Nature in Settlement*, ed. Brian W. Blouet and Merlin P. Lawson, 23–44. Lincoln: University of Nebraska Press, 1975. **36**

———. "Changing National Perspectives and the Mapping of the Great Lakes between 1775 and 1795." *Cartographica* 17, no. 3 (1980): 1–31. **36**

Lewis-Williams, J. David. "Ethnography and Iconography: Aspects of Southern San Thought and Art." *Man, The Journal of the Royal Anthropological Institute*, n.s. 15, no. 3 (1980): 467–82. **86**

———. *The Rock Art of Southern Africa.* Cambridge: Cambridge University Press, 1983. **57, 59, 61, 63**

———. "Testing the Trance Explanation of Southern African Rock Art: Depictions of Felines." *Bollettino del Centro Camuno di Studi Preistorici* 22 (1985): 47–62. **59**

Lhote, Henri. *The Search for the Tassili Frescoes.* Trans. Alan Houghton Brodrick. London: Hutchinson, 1959. **63**

———. *Les gravures rupestres du Sud-Oranais.* Mémoires du Centre de Recherches Anthropologiques Préhistoriques et Ethnographiques 16. Paris: Arts et Métiers Graphiques, 1970. **69**

Libault, A. *Histoire de la cartographie.* Paris: Chaix, 1959. **25**

Library of Congress. *A List of Geographical Atlases in the Library of Congress, with Bibliographical Notes.* 8 vols. Washington, D.C.: Government Printing Office, 1909–74. Vols. 1–4 (1909–20) ed. Philip Lee Phillips. Supp. vols. 5–8 (1958–74) ed. Clara Egli LeGear. **19**

Liddell, Henry George, and Robert Scott, comps. *A Greek-English Lexicon.* 2 vols., rev. and augmented Henry Stuart Jones. Oxford: Clarendon Press, 1940. **179**

Lindberg, David C., ed. *Science in the Middle Ages.* Chicago: University of Chicago Press, 1978. **293**

Ling, Roger. "Studius and the Beginnings of Roman Landscape Painting." *Journal of Roman Studies* 67 (1977): 1–16. **205, 246**

Linquist, Sverre. *Gotlands Bildsteine.* 2 vols. Stockholm: Wahlström och Widstrand, 1941–42. **91**

Lloyd, G. E. R. *Early Greek Science: Thales to Aristotle.* New York: W. W. Norton, 1970. **130**

———. "Greek Cosmologies." In *Ancient Cosmologies*, ed. Carmen Blacker and Michael Loewe, 198–224. London: George Allen and Unwin, 1975. **86**

Lloyd, Robert. "A Look at Images." *Annals of the Association of American Geographers* 72 (1982): 532–48. **31**

Lloyd, Seton. *Early Highland Peoples of Anatolia*. Library of the Early Civilizations. London: Thames and Hudson, 1967. **81**

Łodiński, Marian. "With Fire and Sword, VI." *Imago Mundi* 14 (1959): 117. **6**

Longrigg, James. "Thales." In *Dictionary of Scientific Biography*, 16 vols., ed. Charles Coulston Gillispie, 13:297. New York: Charles Scribner's Sons, 1970–80. **134**

Lopez, Roberto. *Genova marinara nel duecento: Benedetto Zaccaria ammiraglio e mercante*. Messina-Milan: Principato, 1933. **382**

Lorblanchet, M. "From Naturalism to Abstraction in European Prehistoric Rock Art." In *Form in Indigenous Art: Schematisation in the Art of Aboriginal Australia and Prehistoric Europe*, ed. Peter Ucko, 44–56. Australian Institute of Aboriginal Studies, Prehistory and Material Culture Series no. 13. London: Gerald Duckworth, 1977. **69**

Louis, Maurice, and Giuseppe Isetti. *Les gravures préhistoriques du Mont-Bego*. Bordighera: Institut International d'Etudes Ligures, 1964. **93, 94**

Luca, Giuseppe de. "Carte nautiche del medio evo disegnate in Italia." *Atti dell'Accademia Pontaniana* (1866): 3–35; reprinted in *Acta Cartographica* 4 (1969): 314–48. **423**

Lumley, Henry de, Marie-Elisabeth Fonvielle, and Jean Abelanet. "Les gravures rupestres de l'Âge du Bronze dans la région du Mont Bégo (Tende, Alpes-Maritimes)." In *Les civilisations néolithiques et protohistoriques de la France: La préhistoire française*, ed. Jean Guiliane, 2:222–36. Paris: Centre National de la Recherche Scientifique, 1976. **66, 67, 93, 94, 95**

———. "Vallée des Merveilles." *Union International des Sciences Préhistoriques et Protohistoriques, IXe Congrès, Nice 1976*. Livret-Guide de l'Excursion C1. Nice: University of Nice. **66, 67**

Lumsden, Charles J., and Edward O. Wilson. *Promethean Fire: Reflections on the Origin of Mind*. Cambridge: Harvard University Press, 1983. **50**

Luriya, A. R. *Cognitive Development: Its Cultural and Social Foundations*. Ed. Michael Cole. Trans. Martin Lopez-Morillas and Lynn Solotaroff. Cambridge: Harvard University Press, 1976. **58**

Lynam, Edward. Review of *The Story of Maps*, by Lloyd A. Brown. *Geographical Review* 40 (1950): 496–99. **26**

Lynch, John Patrick. *Aristotle's School: A Study of a Greek Educational Institution*. Berkeley: University of California Press, 1972. **158**

Lyons, Henry. "Two Notes on Land-Measurement in Egypt." *Journal of Egyptian Archaeology* 12 (1926): 242–44. **125**

Mabillon, Jean. *Traité des études monastiques*. Paris: Charles Robustel, 1691. **11**

Macalister, Robert Alexander Stewart. *The Excavation of Gezer, 1902–1905 and 1907–1909*. 3 vols. Palestine Exploration Fund. London: John Murray, 1911–12. **250**

McCorkle, Barbara. "Cartographic Treasures of the Yale University Library." *Map Collector* 27 (1984): 8–13. **16**

MacEachren, Alan M. Review of *The Mapmakers*, by John Noble Wilford. *American Cartographer* 9 (1982): 188–90. **26**

MacGregor, Arthur. "Collectors and Collections of Rarities in the Sixteenth and Seventeenth Centuries." In *Tradescant's Rarities: Essays on the Foundation of the Ashmolean Museum 1683, with a Catalogue of the Surviving Early Collections*, ed. Arthur MacGregor, 70–97. Oxford: Clarendon Press, 1983. **9**

MacKay, Angus. *Money, Prices and Politics in Fifteenth-Century Castile*. London: Royal Historical Society, 1981. **437**

McLuhan, Marshall. *Understanding Media: The Extensions of Man*. 2d ed. New York: New American Library, 1964. **36**

Madrid, Biblioteca Nacional. *La historia en los mapas manuscritos de la Biblioteca Nacional*. Exhibition catalog. Madrid: Ministerio de Cultura, Dirección General del Libro y Biblioteca, 1984. **21**

Madurell y Marimon, José Maria. "Ordenanzas marítimas de 1331 y 1333." *Anuario de Historia del Derecho Español* 31 (1961): 611–28. **440**

Magnaghi, Alberto. "Nautiche, carte." In *Enciclopedia italiana di scienze, lettere ed arti*, originally 36 vols., 24:323–31. [Rome]: Istituto Giovanni Treccani, 1929–39. **371, 377, 378, 384, 385, 389, 443**

———. "Alcune osservazioni intorno ad uno studio recente sul mappamondo di Angelino Dalorto (1325)." *Rivista Geografica Italiana* 41 (1934): 1–27. **392, 399, 401, 409, 418**

Maier, I. G. "The Giessen, Parma and Piacenza Codices of the 'Notitia Dignitatum' with Some Related Texts." *Latomus* 27 (1968): 96–141. **244**

———. "The Barberinus and Munich Codices of the *Notitia Dignitatum Omnium*." *Latomus* 28 (1969): 960–1035. **244**

Mainzer, Klaus. *Geschichte der Geometrie*. Mannheim: Bibliographisches Institut, 1980. **xvii**

Malhomme, Jean. *Corpus des gravures rupestres du Grand Atlas*. Fascs. 13 and 14. Rabat: Service des Antiquités du Maroc, 1959–61. **71, 73, 96**

Mallon, Alexis, Robert Koeppel, and René Neuville. *Teleilāt Ghassūl*. 2 vols. Rome: Institut Biblique Pontifical, 1934–40. **88**

Mann, Ludovic MacLellan. *Archaic Sculpturings: Notes on Art, Philosophy, and Religion in Britain 200 B.C. to 900 A.D.*. Edinburgh: William Hodge, 1915. **81**

———. *Earliest Glasgow: A Temple of the Moon*. Glasgow: Mann, 1938. **81**

Manzi, Elio. "La storia della cartografia." In *La ricerca geografica in Italia, 1960–1980*, 327–36. Milan: Ask Edizioni, 1980. **38**

Marcel, Gabriel A. *Reproductions de cartes et de globes relatifs à la découverte de l'Amérique du XVIᵉ au XVIIIᵉ siècle avec texte explicatif.* Paris: Ernest Leroux, 1893–94. **17**

Margary, Harry. "A Proposed Photographic Method of Assessing the Accuracy of Old Maps." *Imago Mundi* 29 (1977): 78–79. **403**

Marinatos, Spyridon. *Excavations at Thera VI (1972 Season).* Bibliothēkē tēs en Athēnais Archaiologikēs Hetaireias, 64. Athens: Archailogikē Hetaireia, 1974. **132**

Marinelli, Giovanni. "Venezia nella storia della geografia cartografica ed esploratrice." *Atti del Reale Istituto Veneto di Scienze, Lettere ed Arti*, 6th ser., 7 (1888–89): 933–1000. **381, 424**

Maringer, Johannes. *The Gods of Prehistoric Man.* 2d ed. Trans. Mary Ilford. London: Weidenfeld and Nicolson, 1960. **68, 69, 83**

Markham, Clements R. *The Story of Majorca and Minorca.* London: Smith Elder, 1908. **440**

——, ed. and trans. *Libro del Conoscimiento: Book of the Knowledge of All Kingdoms.* 2d ser., 29. London: Hakluyt Society, 1912. **399**

Marshack, Alexander. "Polesini: A Reexamination of the Engraved Upper Palaeolithic Mobiliary Materials of Italy by a New Methodology." *Rivista di Scienze Preistorici* 24 (1969): 219–81. **84**

——. *The Roots of Civilization: The Cognitive Beginnings of Man's First Art, Symbol and Notation.* London: Weidenfeld and Nicolson, 1972. **83, 86**

Marshall, Douglas W. "A List of Manuscript Editions of Ptolemy's *Geographia.*" *Bulletin of the Geography and Map Division, Special Libraries Association* 87 (1972): 17–38. **274**

——. "The Formation of a Nineteenth-Century Map Collection: A. E. Nordenskiöld of Helsinki." *Map Collector* 21 (1982): 14–19. **16**

Marshall, P. J., and Glyndwr Williams. *The Great Map of Mankind: British Perceptions of the World in the Age of Enlightenment.* London: J. M. Dent, 1982. **10, 11**

Marstrander, Sverre. "A Newly Discovered Rock-Carving of Bronze Age Type in Central Norway." In *Symposium International d'Art Préhistorique Valcamonica, 23-28 Septembre 1968*, 261–72. Union Internationale des Sciences Préhistoriques et Protohistoriques. Capo di Ponte: Edizioni del Centro, 1970. **76**

Martin, Robert Sidney. "Treasures of the Cartographic Library at the University of Texas at Arlington." *Map Collector* 25 (1983): 14–19. **16**

al-Masūdī. *Les prairies d'or.* 9 vols. Trans. C. Barbier de Meynard and Pavet de Courteille. Société Asiatique Collection d'Ouvrages Orientaux. Paris: Imprimerie Impériale, 1861–1917. **268**

Matkovič, Petar. "Alte handschriftliche Schifferkarten in der Kaiserlichen Hof-Bibliothek in Wien." *Programm des königlichen kaiserlichen Gymnasiums zu Wrasdin.* Agram: L. Gaj, 1860. **410, 434**

Matthews, W. H. *Mazes and Labyrinths: A General Account of Their History and Developments.* London: Longman, 1922. **251, 252**

Mayer, Dorothy. "Miller's Hypothesis: Some California and Nevada Evidence." *Archaeoastronomy: Supplement to the Journal for the History of Astronomy*, no. 1, suppl. to vol. 10 (1979): 51–74. **85**

Meek, Theophile James. *Old Akkadian, Sumerian, and Cappadocian Texts from Nuzi.* Vol. 3 of Harvard University, Semitic Museum. *Excavations at Nuzi.* 8 vols. Cambridge: Harvard University Press, 1929–62. **113**

——. "The Akkadian and Cappadocian Texts from Nuzi." *Bulletin of the American Schools of Oriental Research* 48 (December 1932): 2–5. **113**

Mees, Gregorius. *Historische atlas van Noord-Nederland van de XVI eeuw tot op heden.* Rotterdam: Verbruggen en Van Duym, 1865. **33**

Meiggs, Russell. *Roman Ostia.* Oxford: Clarendon Press, 1960. **245**

Melandrino, Carlo. *Oplontis.* Naples: Loffredo, 1977. **241**

Mellaart, James. "Excavations at Çatal Hüyük, 1963: Third Preliminary Report." *Anatolian Studies* 14 (1964): 39–119. **54, 58, 73, 74, 96**

——. *Çatal Hüyük: A Neolithic Town in Anatolia.* London: Thames and Hudson, 1967. **58, 73, 96**

Menéndez-Pidal, G. "Mozárabes y asturianos en la cultura de la alta edad media en relación especial con la historia de los conocimientos geográficos." *Boletin de la Real Academia de la Historia* (Madrid) 134 (1954): 137–291. **304, 345**

Merk, Conrad. *Excavations at the Kesslerloch Near Thayngen, Switzerland, a Cave of the Reindeer Period.* Trans. John Edward Lee. London: Longmans, Green, 1876. **65**

Mette, Hans Joachim. *Sphairopoiia: Untersuchungen zur Kosmologie des Krates von Pergamon.* Munich: Beck, 1936. **136, 162**

Mickwitz, Ann-Mari. "Dear Mr. Nordenskiöld, Your Offer Is Accepted!" In *The Map Librarian in the Modern World: Essays in Honour of Walter W. Ristow*, ed. Helen Wallis and Lothar Zögner, 221–35. Munich: K. G. Saur, 1979. **16, 17**

Migliorini, Elio. *Indice degli Atti dei Congressi Geografici Italiani dal primo al decimo (1892-1927).* Rome: Presso la Reale Società Geografica Italiana, 1934. **21**

Mikoś, Michael J. "Joachim Lelewel: Polish Scholar and Map Collector." *Map Collector* 26 (1984): 20–24. **18**

Milisauskas, Sarunas. *European Prehistory.* London: Academic Press, 1978. **65**

Millar, Fergus. "Emperors, Frontiers and Foreign Relations, 31 B.C. to A.D. 378." *Britannia* 13 (1982): 1–23. **239**

Millard, A. R. "Strays from a 'Nuzi' Archive." In *Studies on the Civilization and Culture of Nuzi and the Hurrians*, ed. Martha A. Morrison and David I. Owen, 433–41. Winona Lake, Ind.: Eisenbrauns, 1981. **113**

Miller, Konrad. *Mappaemundi: Die ältesten Weltkarten.* 6 vols. Stuttgart: J. Roth, 1895–98. **6, 208, 287, 290, 293, 295, 302, 307, 312, 313, 325, 328, 337, 357, 368**

——. *Itineraria Romana.* Stuttgart: Strecker und Schröder, 1916. **235, 260**

——. *Mappae Arabicae.* 6 vols. Stuttgart, 1926–31. **293**

———. *Die Peutingersche Tafel.* Stuttgart: F. A. Brockhaus, 1962. **238**

Mitchell, J. B. "Early Maps of Great Britain: I. The Matthew Paris Maps." *Geographical Journal* 81 (1933): 27–34. **496**

Moens, Marie-Francine. "The Ancient Egyptian Garden in the New Kingdom: A Study of Representations." *Orientalia Lovaniensia Periodica* 15 (1984): 11–53. **118**

Mohlberg, Leo Cunibert. *Katalog der Handschriften der Zentralbibliothek Zürich.* 2 vols. Zurich: Zentralbibliothek Zürich, 1951. **443**

Moir, Arthur L. *The World Map in Hereford Cathedral.* 8th ed. Hereford: Friends of the Hereford Cathedral, 1977. **309**

Moles, Abraham. *Information Theory and Esthetic Perception.* Trans. Joel E. Cohen. Urbana: University of Illinois Press, 1966. **52**

Mollat du Jourdin, Michel, and Monique de La Roncière. *Les portulans: Cartes marines du XIIIe au XVIIe siècle.* Fribourg: Office du Livre, 1984. English edition, *Sea Charts of the Early Explorers: 13th to 17th Century.* Trans. L. le R. Dethan. New York: Thames and Hudson, 1984. **375, 390, 392, 404, 407, 434, 440, 442**

Molmenti, Pompeo Gherardo. *Venice: Its Individual Growth from the Earliest Beginnings to the Fall of the Republic.* 6 vols. in 3 pts. Trans. Horatio F. Brown. London: J. Murray, 1906–8. **433**

Molt, Paul Volquart. *Die ersten Karten auf Stein und Fels vor 4000 Jahren in Schleswig-Holstein und Niedersachsen.* Lübeck: Weiland, 1979. **66, 80, 87**

Montelius, Oscar. "Sur les sculptures de rochers de la Suède." In *Congrès International d'Anthropologie et d'Archéologie Préhistoriques, compte rendu de la 7e Session, Stockholm 1874,* 453–74. Stockholm: P. A. Norstedt, 1876. **87**

Moore, George. *Ancient Pillar Stones of Scotland: Their Significance and Bearing on Ethnology.* Edinburgh: Edmonstone and Douglas, 1865. **81**

Morelli, Jacopo. *Operette di Iacopo Morelli.* 2 vols. Venice: Tipografia di Alvisopoli, 1820. **315**

Mori, Attilio. "Firenze nelle sue rappresentazioni cartografiche." *Atti della Società Colombaria di Firenze* (1912): 25–42. **478**

———. "Osservazioni sulla cartografia romana in relazione colla cartografia tolemaica e colle carte nautiche medioevali." In *Atti del III Congresso Nazionale di Studi Romani,* 5 vols., 1:565–75. Bologna: Cappelli, 1934. **381**

Morison, Samuel E. *Portuguese Voyages to America in the Fifteenth Century.* Cambridge: Harvard University Press, 1940. **14, 410**

———. *Admiral of the Ocean Sea: A Life of Christopher Columbus.* 2 vols. Boston: Little, Brown, 1942. **328**

———. *The European Discovery of America: The Northern Voyages.* New York: Oxford University Press, 1971. **410**

Morris, Ronald W. B. "The Prehistoric Petroglyphs of Scotland." *Bollettino del Centro Camuno di Studi Preistorici* 10 (1973): 159–68. **64**

———. *The Prehistoric Rock Art of Galloway and the Isle of Man.* Poole: Blandford Press, 1979. **68**

Morrison, Joel L. "Changing Philosophical-Technical Aspects of Thematic Cartography." *American Cartographer* 1 (1974): 5–14. **34**

———. "The Science of Cartography and Its Essential Processes." *International Yearbook of Cartography* 16 (1976): 84–97. **34**

Motzo, Bacchisio R. "Il Compasso da navigare, opera italiana della metà del secolo XIII." *Annali della Facoltà di Lettere e Filosofia della Università di Cagliari* 8 (1947): I–137. **382, 389, 392, 426, 442**

———. "Note di cartografia nautica medioevale." *Studi Sardi* 19 (1964–65): 349–63. **377, 379**

Muehrcke, Phillip C. *Thematic Cartography.* Commission on College Geography Resource Paper no. 19. Washington, D.C.: Association of American Geographers, 1972. **xvii**

———. "Maps in Geography." In *Maps in Modern Geography: Geographical Perspectives on the New Cartography,* ed. Leonard Guelke, Monograph 27, *Cartographica* vol. 18, no. 2 (1981): 1–41. **30**

Müllenhoff, Karl. *Deutsche Altertumskunde.* 5 vols. Berlin: Weidmann, 1890–1920. **163**

Müller, Karl, ed. *Geographi Graeci minores.* 2 vols. and tabulae. Paris: Firmin-Didot, 1855–56. **237, 260;** Agathemerus, **134, 135, 137, 143, 152, 153, 243;** Arrian, **254;** Avienius, **171;** Dionysius, **171, 173;** Eustathius, **171, 266;** Marcianus, **237**

———. "Rapports sur les manuscits de la géographie de Ptolémée." *Archives des Missions Scientifiques et Littéraires,* 2d ser., 4 (1867), 279–98. **274**

Munn, Nancy D. "Visual Categories: An Approach to the Study of Representational Systems." *American Anthropologist* 68, no. 4 (1966): 936–50. Reprinted in *Art and Aesthetics in Primitive Societies,* ed. Carol F. Jopling, 335–55. New York: E. P. Dutton, 1971. **58**

———. "The Spatial Presentation of Cosmic Order in Walbiri Iconography." In *Primitive Art and Society,* ed. Anthony Forge, 193–220. London: Oxford University Press, 1973. **87**

Munro, Robert. *Archaeology and False Antiquities.* London: Methuen, 1905. **65**

Munz, Peter. *When the Golden Bough Breaks: Structuralism or Typology?* London and Boston: Routledge and Kegan Paul, 1973. **58**

Murdoch, John Emery. *Antiquity and the Middle Ages.* Album of Science. New York: Charles Scribner's Sons, 1984. **334**

Murphy, Joan M. "Measures of Map Accuracy Assessment and Some Early Ulster Maps." *Irish Geography* 11 (1978): 88–101. **403**

Murray, G. W. "The Gold-Mine of the Turin Papyrus." *Bulletin de l'Institut d'Egypte* 24 (1941–42): 81–86. **122**

Mussche, H. F. *Thorikos: Eine Führung durch die Ausgrabungen.* Ghent and Nuremberg: Comité des Fouilles Belges en Grèce, 1978. **139**

Nangis, Guillaume de. "Gesta Sanctæ Memoriæ Ludovici" (Life of Saint Louis). In *Recueil des historiens des Gaules et de la France,* 24 vols., ed. J. Naudet and P. Daunou, vol. 20 (1840): 309–465. Paris: Imprimerie Royale, 1738–1904. **439**

Narkiss, Bezalel. *Hebrew Illuminated Manuscripts.* Jerusalem: Encyclopaedia Judaica, 1969. **266**

Nebenzahl, Kenneth. *Rare Americana.* Catalog 20. Chicago: Kenneth Nebenzahl, 1968. **431, 435, 457**

Needham, Joseph. *Science and Civilisation in China.* Cambridge: Cambridge University Press, 1954–. **xix, 7, 53, 60, 384, 496**

Nemet-Nejat, Karen Rhea. *Late Babylonian Field Plans in the British Museum.* Studia Pohl: Series Maior 11. Rome: Biblical Institute Press, 1982. **111**

Neugebauer, Otto. "Über eine Methode zur Distanzbestimmung Alexandria-Rom bei Heron." *Historisk-Filologiske Meddelelser udgivne af det Kongelige Danske Videnskabernes Selskab* 26 (1938–39), nos. 2 and 7. **230**

———. *The Exact Sciences in Antiquity.* 2d ed. Providence: Brown University Press, 1957. **xviii, 130, 134**

———. "Ptolemy's *Geography,* Book VII, Chapters 6 and 7," *Isis* 50 (1959): 22–29. **188, 189**

———. "Survival of Babylonian Methods in the Exact Sciences of Antiquity and the Middle Ages." *Proceedings of the American Philosophical Society* 107 (1963): 528–35. **130**

———. "A Greek World Map." In *Le monde grec: Hommages à Claire Préaux,* ed. Jean Bingen, Guy Cambier, and Georges Nachtergael, 312–17. Brussels: Université de Bruxelles, 1975. **248**

———. *A History of Ancient Mathematical Astronomy.* New York: Springer-Verlag, 1975. **130, 154, 164, 165, 167, 169, 180, 182, 185, 187, 188, 189, 232**

Neugebauer, Otto, and Richard A. Parker, eds. and trans. *Egyptian Astronomical Texts.* 3 vols. Providence and London: Lund Humphries for Brown University Press, 1960–69. **121**

Newbold, Douglas. "Rock-Pictures and Archaeology in the Libyan Desert." *Antiquity* 2, no. 7 (1928): 261–91. **63**

Newton, Arthur Percival. *Travel and Travellers of the Middle Ages.* New York: Alfred A. Knopf, 1926. **321**

Newton, R. R. *The Crime of Claudius Ptolemy.* Baltimore: Johns Hopkins University Press, 1977. **177, 182**

Nicodemi, Giorgio. *Catalogo delle raccolte numismatiche.* 2 vols. Milan: Bestetti, 1938–40. **214**

Nilsson, Martin Persson. *Primitive Time-Reckoning.* Lund: C. W. K. Gleerup, 1920. **85**

Nodelman, Sheldon Arthur. "A Preliminary History of Characene." *Berytus* 13 (1960): 83–121. **238**

Nordén, Arthur G. *Östergötlands Bronsålder.* Linköping: Henric Carlssons Bokhandels Förlag, 1925. **64**

Nordenfalk, Carl. *Die spätantiken Zierbuchstaben.* Stockholm: published by the author, 1970. **217**

Nordenskiöld, A. E. *Facsimile-Atlas to the Early History of Cartography.* Trans. Johan Adolf Ekelöf and Clements R. Markham. Stockholm, 1889. **18**

———. "Résumé of an Essay on the Early History of Charts and Sailing Directions." *Report of the Sixth International Geographical Congress, London, 1895*

(1896): 685–94; reprinted in *Acta Cartographica* 14 (1972): 185–94. **375**

———. *Periplus: An Essay on the Early History of Charts and Sailing-Directions.* Trans. Francis A. Bather. Stockholm: P. A. Norstedt, 1897. **6, 18, 372, 373, 377, 378, 379, 381, 382, 383, 384, 386, 387, 388, 389, 391, 392, 395, 397, 402, 403, 404, 406, 407, 411, 413, 415, 419, 420, 422, 423, 427, 442, 457, 459, 461, 482**

———. "Dei disegni marginali negli antichi manoscritti della *Sfera* del Dati." *Bibliofilia* 3 (1901–2): 49–55. **379**

North, Robert. *A History of Biblical Map Making.* Beihefte zum Tübinger Atlas des Vorderen Orients, B32. Wiesbaden: Reichert, 1979. **117, 326**

Nougayrol, Jean. *Le palais royal d'Ugarit, IV: Textes accadiens des Archives Sud (Archives Internationales).* Mission de Ras Shamra, 9. Paris: Imprimerie Nationale, 1956. **108**

Nowotny, Karl A. *Beiträge zur Geschichte des Weltbildes.* Vienna: Ferdinand Berger, 1970. **335, 347**

O'Callaghan, R. T. "Madaba (Carte de)." In *Dictionnaire de la Bible: Supplement,* ed. L. Pirot and A. Robert, vol. 5 (1957), 627–704. Paris: Letouzey et Ané, 1928–. **265**

Odell, C. B. Review of *Die Geschichte der Kartographie,* by Leo Bagrow. *Annals of the Association of American Geographers* 43 (1953): 69–70. **25**

Oehme, Ruthardt. "Die Palästinakarte aus Bernhard von Breitenbachs Reise in das Heilige Land, 1486." *Beiheft zum Zentralblatt für Bibliothekswesen* 75 (1950): 70–83. **475**

———. "A Cartographical Certificate by the Cologne Painter Franz Kessler." *Imago Mundi* 11 (1954): 55–56. **373**

———. *Die Geschichte der Kartographie des deutschen Südwestens.* Constance: Thorbecke, 1961. **488**

———. "German Federal Republic." In the Chronicle section. *Imago Mundi* 25 (1971): 93–95. **38**

———. *Eberhard David Hauber (1695–1765): Ein schwäbisches Gelehrtenleben.* Stuttgart: W. Kohlhammer, 1976. **10**

Ogrissek, Rudi. "Ein Strukturmodell der theoretischen Kartographie für Lehre und Forschung." *Wissenschaftliche Zeitschrift der Technischen Universität Dresden* 29, no. 5 (1980): 1121–26. **30**

Oldham, Richard D. "The Portolan Maps of the Rhône Delta: A Contribution to the History of the Sea Charts of the Middle Ages." *Geographical Journal* 65 (1925): 403–28. **380, 403**

Oliver, J. H. "North, South, East, West at Arausio and Elsewhere." In *Mélanges d'archéologie et d'histoire offerts à André Piganiol,* 3 vols., ed. Raymond Chevallier, 2:1075–79. Paris: SEVPEN, 1966. **222**

Olszewicz, Bolesław. *Dwie szkicowe mapy Pomorza z połowy XV wieku.* Biblioteka "Strażnicy Zachodniej" no. 1. Warsaw: Nakładem Polskiego Związku Zachodniego, 1937. **488**

Ong, Walter J. *Orality and Literacy: The Technologizing of the Word.* London and New York: Methuen, 1982. **58**

Ongania, Ferdinando. *Raccolta di mappamondi e carte nautiche del XIII al XVI secolo*. Venice, 1875–81. **404**

Oppenheim, A. Leo. "Man and Nature in Mesopotamian Civilization." In *Dictionary of Scientific Biography*, 16 vols., ed. Charles Coulston Gillispie, 15:634-666. New York: Charles Scribner's Sons, 1970–80. **112**

Ordnance Survey. *Map of Roman Britain*. 4th ed. Southampton: Ordnance Survey, 1978. **193**

Oren, Oton Haim. "Jews in Cartography and Navigation (from the XIth to the Beginning of the XVth Century)." *Communication du Premier Congrès International d'Histoire de l'Océanographie* 1 (1966): 189–97; reprinted in *Bulletin de l'Institut Océanographique* 1, special no. 2 (1968): 189–97. **432**

Ormeling, F. J. "Einige Aspekte und Tendenzen der modernen Kartographie." *Kartographische Nachrichten* 28 (1978): 90–95. **xvii**

Orr, Mary Acworth. *Dante and the Early Astronomers*. New York: A. Wingate, 1956. **321**

Owen, Dorothy M. "Clenchwarton, Norfolk, Late 14th or Early 15th Century." In *Local Maps and Plans from Medieval England*, ed. R. A. Skelton and P. D. A. Harvey, 127–30. Oxford: Clarendon Press, 1986. **484**

Pächt, Otto. *The Rise of Pictorial Narrative in Twelfth-Century England*. Oxford: Clarendon Press, 1962. **290**

Pagani, Lelio. *Pietro Vesconte: Carte nautiche*. Bergamo: Grafica Gutenberg, 1977. **375, 376, 377, 382, 387, 434, 458**

Pallottino, Massimo. *The Etruscans*. Ed. David Ridgway. Trans. J. Cremona. London: Allen Lane, 1975. **201, 203**

———. *Saggi di Antichità*. 3 vols. Rome: G. Bretschneider, 1979. **203, 204**

Panofsky, Erwin. *Studies in Iconology: Humanistic Themes in the Art of the Renaissance*. Oxford: Oxford University Press, 1939. **62**

Paoli, Cesare. "Una carta nautica genovese del 1311." *Archivio Storico Italiano*, 4th ser., 7 (1882): 381–84. **424**

Papezy, Jules. *Mémoires sur le port d'Aiguesmortes*. Paris: Hachette, 1879. **382**

Paris, Bibliothèque Nationale. *A la découverte de la terre, dix siècles de cartographie*. Trésors du Département des Cartes et Plans de la Bibliothèque Nationale, Paris, May to July 1979. **22**

Parker, John. "A Fragment of a Fifteenth-Century Planisphere in the James Ford Bell Collection." *Imago Mundi* 19 (1965): 106–7. **316**

———. "The Map Treasures of the James Ford Bell Library, Minnesota." *Map Collector* 20 (1982): 8–14. **16**

Parker, Richard A. *Demotic Mathematical Papyri*. Providence: Brown University Press, 1972. **127**

———. "Ancient Egyptian Astronomy." *Philosophical Transactions of the Royal Society of London*, ser. A, 276 (1974): 51–65. **121**

Parkington, John. "Symbolism in Palaeolithic Cave Art." *South African Archaeological Bulletin* 24, pt. 1, no. 93 (1969): 3–13. **57**

Parry, John Horace. *Europe and a Wider World, 1415–1715*. 3d ed. London: Hutchinson, 1966. **415**

———. "Old Maps Are Slippery Witnesses." *Harvard Magazine*. Alumni ed. April 1976, 32–41. **3, 18**

Parson, Edward Alexander. *The Alexandrian Library: Glory of the Hellenic World*. Amsterdam, London, New York: Elsevier Press, 1952. **149**

Parsons, David. "Consistency and the St. Gallen Plan: A Review Article." *Archaeological Journal* 138 (1981): 259–65. **467**

Parsons, E. J. S. *Introduction to the Facsimile*. Memoir accompanying *The Map of Great Britain, circa A.D. 1360, Known as the Gough Map*. Oxford: Oxford University Press, 1958. **496**

Pasch, Georges. "Les drapeaux des cartes-portulans: L'atlas dit de Charles V (1375)." *Vexillologia: Bulletin de l'Association Française d'Etudes Internationales de Vexillologie* 1, nos. 2–3 (1967): 38–60. **399**

———. "Les drapeaux des cartes-portulans: Drapeaux du 'Libro del Conoscimiento.'" *Vexillologia: Bulletin de l'Association Française d'Etudes Internationales de Vexillologie* 2, nos. 1–2 (1969): 8–32. **399**

———. "Drapeau des Canariens: Témoignage des portulans." *Vexillologia: Bulletin de l'Association Française d'Etudes Internationales de Vexillologie* 3, no. 2 (1973): 51. **378**

———. "Les drapeaux des cartes-portulans [portulans du groupe Vesconte]." *Vexillologia: Bulletin de l'Association Française d'Etudes Internationales de Vexillologie* 3, no. 2 (1973): 52–62. **399, 401**

Pastine, O. "Se la più antica carta nautica medioevale sia di autore genovese." *Bollettino Ligustico* 1 (1949): 79–82. **389**

Pastoureau, Mireille. "Collections et collectionneurs de cartes en France, sous l'ancien-régime." Paper prepared for the Tenth International Conference on the History of Cartography, Dublin 1983. **8**

———. *Les atlas français, XVIe–XVIIe siècles: Répertoire bibliographique et étude*. Paris: Bibliothèque Nationale, 1984. **10**

Peck, William H. *Drawings from Ancient Egypt*. London: Thames and Hudson, 1978. **126**

Peet, T. Eric. *The Rhind Mathematical Papyrus*. Liverpool: University Press of Liverpool, 1923. **127**

———. *The Great Tomb-Robberies of the Twentieth Egyptian Dynasty*. Oxford: Clarendon Press, 1930. **128**

Pekkanen, T. Review of *Scandinavien bei Plinius und Ptolemaios*, by Joseph Gusten Algot Svennung. *Gnomon* 49 (1977): 362–66. **197**

Pelham, Peter T. "The Portolan Charts: Their Construction and Use in the Light of Contemporary Techniques of Marine Survey and Navigation." Master's thesis, Victoria University of Manchester, 1980. **377, 384, 387, 391, 443**

Pellegrin, Elisabeth. "Les manuscrits de Geoffroy Carles, président du Parlement de Dauphiné et du Sénat de Milan." In *Studi di bibliografia e di storia in onore di Tammaro de Marinis*, 4 vols., 3:313–17. Verona: Stamperia Valdonega, 1964. **435**

Pelletier, Monique. "Jomard et le Département des Cartes et Plans." *Bulletin de la Bibliothèque Nationale* 4 (1979): 18–27. **15**

———. "L'accès aux collections cartographiques en France." In *Le patrimoine des bibliothèques: Rapport à Monsieur le directeur du livre et de la lecture par une commission de douze membres*, 2 vols., ed. Louis Desgraves and Jean-Luc Gautier, 2:253–59. Paris: Ministère de la Culture, 1982. **20**

Penrose, Boies. *Travel and Discovery in the Renaissance, 1420-1620*. Cambridge: Harvard University Press, 1952. **413**

Pestman, P. W., ed. *Greek and Demotic Texts from the Zenon Archive*. Papyrologica Lugduno-Batava 20. Leiden: E. J. Brill, 1980. **128**

Petchenik, Barbara Bartz. "Cognition in Cartography." In *Nature of Cartographic Communication*. Monograph 19, *Cartographica* (1977), 117–28. **34**

———. "A Map Maker's Perspective on Map Design Research, 1950–1980." In *Graphic Communication and Design in Contemporary Cartography*, ed. D. R. Fraser Taylor, 37–68. Progress in Contemporary Cartography, vol. 2. New York: John Wiley, 1983. **34**

Peters, Roger. "Communication, Cognitive Mapping, and Strategy in Wolves and Hominids." In *Wolf and Man: Evolution in Parallel*, ed. Roberta L. Hall and Henry S. Sharp, 95–107. New York and London: Academic Press, 1978. **52**

———. "Mental Maps in Wolf Territoriality." In *The Behavior and Ecology of Wolves: Proceedings of the Symposium on the Behavior and Ecology of Wolves Held on 23–24 May 1975 in Wilmington, N.C.*, ed. Erich Klinghammer, 122–25. New York and London: Garland STPM Press, 1979. **50**

Petrie, W. M. F. *Prehistoric Egypt, Illustrated by over 1,000 Objects in University College, London*. London: British School of Archaeology in Egypt, 1917. **118**

Petrikovits, H. von. "Vetera." In *Paulys Realencyclopädie der classischen Altertumswissenschaft*, ed. August Pauly, Georg Wissowa, et al., 2d ser., 8 (1958): cols. 1801–34. Stuttgart: J. B. Metzler, 1894–. **236**

Petti Balbi, Giovanna. "Nel mondo dei cartografi: Battista Beccari maestro a Genova nel 1427." In Università di Genova, Facoltà di Lettere, *Columbeis I*, 125–32. Genoa: Istituto di Filologia Classica e Medievale, 1986. **431**

Pettinato, Giovanni. "L'atlante geografico del Vicino Oriente antico attestato ad Ebla e ad Abū Salābīkh." *Orientalia*, n.s., 47 (1978): 50–73. **107**

Phillipps, Thomas. "Mappae Clavicula: A Treatise on the Preparation of Pigments during the Middle Ages." *Archaeologia* 32 (1847): 183–244. **287**

Phillips, Philip Lee. *A List of Maps of America in the Library of Congress*. Washington, D.C.: Government Printing Office, 1901; reprinted New York: Burt Franklin, [1967]. **19, 32**

Piaget, Jean, and Bärbel Inhelder. *The Child's Conception of Space*. Trans. F. J. Langdon and J. L. Lunzer. London: Routledge and Kegan Paul, 1956. **2, 52, 59, 69**

Piaget, Jean, Bärbel Inhelder, and Alina Szeminska. *The Child's Conception of Geometry*. New York: Basic Books, 1960. **2**

Piankoff, Alexandre. *The Wandering of the Soul*. Completed and prepared for publication by Helen Jacquet-Gordon. Egyptian Religious Texts and Representations, Bollingen Series 40, vol. 6. Princeton: Princeton University Press, 1974. **120**

Pidoplichko, Ivan Grigorévich. *Pozdnepaleoliticheskye zhilishcha iz kostey mamonta na Ukraine* (Late paleolithic dwellings of mammoth bone in the Ukraine). Kiev: Izdatelstvo "Naukova Dumka," 1969. **70, 71**

———. *Mezhiricheskye zhilishcha iz kostey mamonta* (Mezhirichi dwellings of mammoth bone). Kiev: Izdatelstvo "Naukova Dumka," 1976. **70**

Piersantelli, Giuseppe. *L'atlante di carte marine di Francesco Ghisolfi e la storia della pittura in Genova nel Cinquecento*. Genoa, 1947. **438**

———. "L'Atlante Luxoro." In *Miscellanea di geografia storica e di storia della geografia nel primo centenario della nascita di Paolo Revelli*, 115–41. Genoa: Bozzi, 1971. **424, 434**

Piganiol, André. *Les documents cadastraux de la colonie romaine d'Orange*. Gallia, suppl. 16. Paris: Centre National de la Recherche Scientifique, 1962. **220, 221, 466**

Piggott, Stuart. *Ancient Europe from the Beginnings of Agriculture to Classical Antiquity*. Edinburgh: Edinburgh University Press, 1965. **72**

Piloni, Luigi. *Carte geografiche della Sardinia*. Cagliari: Fossataro, 1974. **371**

Pinchemel, Philippe. "Géographie et cartographie, réflexions historiques et épistémologiques." *Bulletin de l'Association de Géographes Français* 463 (1979): 239–47. **38**

Pinder, Moritz, and Gustav Parthey, eds. *Ravennatis anonymi Cosmographia et Guidonis Geographica*. Berlin: Fridericus Nicolaus, 1860; reprinted Aalen: Otto Zeller Verlagsbuchhandlung, 1962. **260**

Pinto, John A. "Origins and Development of the Ichnographic City Plan." *Journal of the Society of Architectural Historians* 35, no. 1 (1976): 35–50. **478, 495**

Pitsch, Helmut. "Landschaft (-Beschreibung und -Darstellung)." In *Lexikon der Ägyptologie*, ed. Wolfgang Helck and Eberhard Otto, 3:923–28. Wiesbaden: O. Harrassowitz, 1975–. **118**

Platner, Samuel Ball. *A Topographical Dictionary of Ancient Rome*. Rev. Thomas Ashby. London: Oxford University Press, 1929. **210**

Pognon, Edmond. "Les collections du Département des Cartes et Plans de la Bibliothèque Nationale de Paris." In *The Map Librarian in the Modern World: Essays in Honour of Walter W. Ristow*, ed. Helen Wallis and Lothar Zögner, 195–204. Munich: K. G. Saur, 1979. **15**

Polanyi, Michael. *The Study of Man*. Lindsay Memorial Lectures. London: Routledge and Kegan Paul, 1959. **5**

Polaschek, Erich. "Notitia Dignitatum." In *Paulys Realencyclopädie der classischen Altertumswissenschaft*, ed. August Pauly, Georg Wissowa, et al., 17.1 (1936): cols. 1077–116. Stuttgart: J. B. Metzler, 1894–. **244**

———. "Ptolemy's *Geography* in a New Light." *Imago Mundi* 14 (1959): 17–37. **189, 192, 199, 271**

———. "Ptolemaios als Geograph." In *Paulys*

Realencyclopädie der classischen Altertumswissenschaft, ed. August Pauly, Georg Wissowa, et al., suppl. 10 (1965): cols. 680–833. Stuttgart: J. B. Metzler, 1894–. **187, 191, 192, 195, 197, 269**

Polo, Claudio de. "*Arte del navigare*: Manuscrit inédit daté de 1464–1465." *Bulletin du Bibliophile* 4 (1981): 453–61. **443**

Popescu-Spineni, Marin. *România în istoria cartografiei pâna la 1600.* Bucharest: Imprimeria Naţionala, 1938. **488**

Posener, Georges, et al. *A Dictionary of Egyptian Civilization.* Trans. Alix Macfarland. London: Methuen, 1962. **122**

Postan, M. M., E. E. Rich, and E. Miller, eds. *The Cambridge Economic History of Europe.* 3 vols. Cambridge: Cambridge University Press, 1963. **426**

Pottier, E. "Labyrinthus." In *Dictionnaire des antiquités grecques et romaines,* 5 vols., ed. Charles Daremberg and Edmond Saglio, 3.2:882–83. Paris: Hachette, 1877–1919. **252**

Preisigke, Friedrich. *Sammelbuch griechischer Urkunden aus Ägypten.* Strasburg: K. J. Trübner, 1915. **128**

Priuli, Ausilio. *Incisioni rupestri della Val Camonica.* Ivrea: Priuli and Verlucca, 1985. **75, 78, 79, 95, 96**

Puertos y Fortificaciones en América y Filipinas. Comision de Estudios Historicos de Obras Publicas y Urbanism, CEHOPU, 1985. **21**

Puglisi, Salvatore. In "Astronomy and Astrology." In *Encyclopedia of World Art,* 2:42–43. New York: McGraw-Hill, 1960. **83**

Purce, Jill. *The Mystic Spiral, Journey of the Soul.* London: Thames and Hudson, 1974. **88**

Putman, Robert. *Early Sea Charts.* New York: Abbeville Press, 1983. **22, 457**

Pyenson, Lewis. "'Who the Guys Were': Prosopography in the History of Science." *History of Science* 15 (1977): 155–88. **14**

Quaini, Massimo. "Catalogna e Liguria nella cartografia nautica e nei portolani medievali." In *Atti del 1° Congresso Storico Liguria-Catalogna: Ventimiglia-Bordighera-Albenga-Finale-Genova, 14–19 ottobre 1969,* 549–71. Bordighera: Istituto Internazionale di Studi Liguri, 1974. **377, 392, 424, 426, 437, 438**

Rabchevsky, George A., ed. *Multilingual Dictionary of Remote Sensing and Photogrammetry.* Falls Church, Va.: American Society of Photogrammetry, 1983. **xvii**

Rabelais, François. *Oeuvres complètes.* Ed. Jacques Boulenger. Paris: Gallimard, 1955. **332**

Radmilli, Arturo Mario. "The *Movable Art* of the Grotta Polesini." *Antiquity and Survival,* no. 6 (1956): 465–73. **84**

Raidel, Georg Martin. *Commentatio critico-literaria de Claudii Ptolemaei Geographia, eiusque codicibus tam manuscriptis quam typis expressis.* Nuremberg: Typis et sumptibus haeredum Felseckerianorum, 1737. **11**

Raisz, Erwin. "The Cartophile Society of New England." *Imago Mundi* 8 (1951): 44–45. **22**

Ralph, Elizabeth. "Bristol, *circa* 1480." In *Local Maps and Plans from Medieval England,* ed. R. A. Skelton and P. D. A. Harvey, 309–16. Oxford: Clarendon Press, 1986. **485, 493**

Ramilli, G. *Gli agri centuriati di Padova e di Pola nell'interpretazione di Pietro Kandler.* Trieste: Società Istriana di Archeologia e Storia Patria, 1973. **219**

Ramin, Jaques. *Le Périple d'Hannon/The Periplus of Hanno.* British Archaeological Reports, Supplementary Series 3. Oxford: British Archaeological Reports, 1976. **150**

Randall, John Herman, Jr. *The Making of the Modern Mind: A Survey of the Intellectual Background of the Present Age.* Boston: Houghton Mifflin, 1926. **319**

Randles, W. G. L. *De la terre plate au globe terrestre: Une mutation épistémologique rapide (1480–1520).* Cahiers des Annales 38. Paris: Armand Colin, 1980. **319**

Ranson, C. L. "A Late Egyptian Sarcophagus." *Bulletin of the Metropolitan Museum of Art* 9 (1914): 112–20. **121**

Raschke, Manfred G. "New Studies in Roman Commerce with the East." In *Aufstieg und Niedergang der römischen Welt,* ed. Hildegard Temporini and Wolfgang Haase, 2.9.2 (1978): 604–1361. Berlin: Walter de Gruyter, 1972–. **178**

Rashdall, Hastings. *The Universities of Europe in the Middle Ages.* Ed. F. M. Powicke and A. B. Emden. Oxford: Oxford University Press, 1936. **306**

Ratajski, Lech. "The Research Structure of Theoretical Cartography." *International Yearbook of Cartography* 13 (1973): 217–28. **35**

———. "The Main Characteristics of Cartographic Communication as a Part of Theoretical Cartography." *International Yearbook of Cartography* 18 (1978): 21–32. **34**

Ravenstein, Ernest George. "Map." In *Encyclopaedia Britannica,* 11th ed., 32 vols., 17:629–63. New York: Encyclopaedia Britannica, 1910–11. **294**

Rawlins, Denis. "The Eratosthenes-Strabo Nile Map." *Archive for History of Exact Sciences* 26 (1982): 211–19. **148**

Raynaud-Nguyen, Isabelle. "L'hydrographie et l'événement historique: Deux exemples." Paper prepared for the Fourth International Reunion for the History of Nautical Science and Hydrography, Sagres-Lagos, 4–7 July 1983. **374, 401, 431**

Reckert, Robert D. "A Message from the President of ACSM." *American Cartographer* 1 (1974): 4. **32**

Reed, Ronald. *Ancient Skins, Parchments and Leathers.* London: Seminar Press, 1972. **391**

Rees, Alwyn, and Brinley Rees. *Celtic Heritage: Ancient Tradition in Ireland and Wales.* London: Thames and Hudson, 1961. **92**

Reinach, Salomon. *Répertoire de peintures grecques et romaines.* Paris: E. Leroux, 1922. **252**

Reinhardt, Hans. *Der St. Galler Klosterplan.* Saint Gall: Historischer Verein des Kantons St. Gallen, 1952. **466**

Renfrew, Colin. *Towards an Archaeology of Mind.* Inaugural Lecture, University of Cambridge, 30 November 1982. Cambridge: Cambridge University Press, 1982. **5, 92**

Reparaz, Gonçal (Gonzalo) de. *Catalunya a les mars:*

Navegants, mercaders i cartògrafs catalans de l'Edat Mitjana i del Renaixement (Contribucío a l'estudi de la història del comerç i de la navegació de la Mediterrània). Barcelona: Mentova, 1930. **477**

———. "Essai sur l'histoire de la géographie de l'Espagne de l'antiquité au XVᵉ siècle," *Annales du Midi* 52 (1940): 137–89, 280–341. **315, 389, 394, 432, 435, 440**

Revelli, Paolo. *Cristoforo Colombo e la scuola cartografica genovese.* Genoa: Consiglio Nazionale delle Ricerche, 1937. **373, 406, 407, 413, 414, 424, 429, 430, 431, 434, 435, 439, 442, 444, 457**

———. *La partecipazione italiana alla Mostra Oceanografica Internazionale di Siviglia (1929).* Genoa: Stabilimenti Italiani Arti Grafiche, 1937. **397, 404, 406, 413, 418, 457**

———. "Cimeli cartografici di Archivi di Stato italiani distrutti dalla guerra." *Notizie degli Archivi di Stato* 9 (1949): 1–3. **405**

———. "Cimeli geografici di biblioteche italiane distrutti o danneggiati dalla guerra." *Atti della XIV Congresso Geografico Italiano, Bologna, 1947* (1949): 526–28. **374**

———. "Una nuova carta di Batista Beccari ('Batista Becharius')?" *Bollettino della Società Geografica Italiana* 88 (1951): 156–66. **411, 457**

———. "Cimeli geografici di archivi italiani distrutti o danneggiati dalla guerra." *Atti della XV Congresso Geografico Italiano, Torino, 1950* (1952). 2 vols., 2:879. **374**

———. "Figurazioni di Genova ai tempi di Colombo." *Bollettino del Civico Istituto Colombiano* 3 (1955): 14–23. **477**

———, ed. *Elenco illustrativo della Mostra Colombiana Internazionale.* Genoa: Comitato Cittadino per le Celebrazioni Colombiane, 1950. **414, 430, 433, 443**

Rey Pastor, Julio, and Ernesto García Camarero. *La cartografía mallorquina.* Madrid: Departamento de Historia y Filosofía de la Ciencia, 1960. **374, 393, 403, 419, 423, 431, 435, 458**

Reynolds, L. D., and N. G. Wilson. *Scribes and Scholars: A Guide to the Transmission of Greek and Latin Literature.* 2d ed. Oxford: Clarendon Press, 1974. **266**

Reynolds, Susan. "Staines, Middlesex, 1469 × circa 1477." In *Local Maps and Plans from Medieval England,* ed. R. A. Skelton and P. D. A. Harvey, 245–50. Oxford: Clarendon Press, 1986. **491**

Rhotert, Hans. *Libysche Felsbilder: Ergebnisse der XI. und XII. Deutschen Inner-Afrikanischen Forschungs-Expedition (DIAFE) 1933/1934/1935.* Darmstadt: L. C. Wittich, 1952. **97**

Richardson, Mervyn E. J. "Hebrew Toponyms." *Tyndale Bulletin* 20 (1969): 95–104. **108**

Richeson, A. W. *English Land Measuring to 1800: Instruments and Practices.* Cambridge: Society for the History of Technology and MIT Press, 1966. **495**

Ridley, Michael. *The Megalithic Art of the Maltese Islands.* Poole, Dorsetshire: Dolphin Press, 1976. **83, 84, 87**

Riese, Alexander, ed. *Geographi Latini minores.* Heilbronn, 1878; reprinted Hildesheim: Georg Olms, 1964. **205, 206, 208, 243, 259**; Julius Honorius, **244, 255**

Ringbom, Sixten. "Some Pictorial Conventions for the Recounting of Thoughts and Experiences in Late Medieval Art." In *Medieval Iconography and Narrative: A Symposium,* ed. Flemming G. Andersen et al., 38–69. Odense: Odense University Press, 1980. **286, 290**

Ripa, Cesare. *Iconologia.* 3d ed. 1603; facsimile reprint, Hildesheim and New York: Georg Olms, 1970. **339**

Ripinsky, Michael M. "The Camel in Ancient Arabia." *Antiquity* 49, no. 196 (1975): 295–98. **69**

Ristow, Walter W. *Facsimiles of Rare Historical Maps: A List of Reproductions for Sale by Various Publishers and Distributors.* Washington, D.C.: Library of Congress, 1960. **19**

———. "Chronicle" section. *Imago Mundi* 17 (1963): 106–14. **37**

———. "Chronicle" section. *Imago Mundi* 20 (1966): 90–94. **37**

———. "Recent Facsimile Maps and Atlases." *Quarterly Journal of the Library of Congress* 24 (1967): 213–29. **18**

Ristow, Walter W., and R. A. Skelton. *Nautical Charts on Vellum in the Library of Congress.* Washington, D.C.: Library of Congress, 1977. **429, 444, 457**

Ritter, Dale W., and Eric W. Ritter. "Medicine Men and Spirit Animals in Rock Art of Western North America." In *Acts of the International Symposium on Rock Art: Lectures at Hankø 6-12 August, 1972,* ed. Sverre Marstrander, 97–125. Oslo: Universitetsforlaget, 1978. **61**

Rivet, A. L. F. "Some Aspects of Ptolemy's Geography of Britain." In *Littérature gréco-romaine et géographie historique: Mélanges offerts à Roger Dion,* ed. Raymond Chevallier, 55–81. Caesarodunum 9 bis. Paris: A. et J. Picard, 1974. **192**

———. "Ptolemy's Geography and the Flavian Invasion of Scotland." In *Studien zu den Militärgrenzen Roms, II,* 45–64. Vorträge des 10. Internationalen Limeskongresses in der Germania Inferior. Cologne: Rheinland-Verlag in Kommission bei Rudolf Habelt, 1977. **194**

Rivet, A. L. F., and Colin Smith. *The Place-Names of Roman Britain.* Princeton: Princeton University Press, 1979. **192, 193, 194, 236, 239, 245, 252, 260**

Robert de Vaugondy, Didier. *Essai sur l'histoire de la géographie.* Paris: Antoine Boudet, 1755. **10, 11**

Roberts, B. K. "North-west Warwickshire; Tanworth in Arden, Warwickshire, 1497 × 1519." In *Local Maps and Plans from Medieval England,* ed. R. A. Skelton and P. D. A. Harvey, 317–28. Oxford: Clarendon Press, 1986. **494**

Roberts, C. H. "The Codex." *Proceedings of the British Academy* 40 (1954): 169–204. **254**

Robinson, Arthur H. *Elements of Cartography.* 1st ed., 1953. 5th ed. (Robinson et al.) New York: John Wiley, 1978. **34**

———. "The Uniqueness of the Map." *American Cartographer* 5 (1978): 5–7. **2**

———. *Early Thematic Mapping in the History of Cartography.* Chicago: University of Chicago Press, 1982. **5, 26, 35**

Robinson, Arthur H., and Barbara Bartz Petchenik. *The Nature of Maps: Essays toward Understanding Maps and Mapping.* Chicago: University of Chicago Press, 1976. **xvii, 2, 3, 4, 33**

Rochette, Désiré Raoul. *Peintures antiques inédites, précédées de recherches sur l'emploi de la peinture dans la décoration des édifices sacrés et publics, chez les Grecs et chez les Romains.* Paris: Imprimerie Royale, 1836. **158, 174**

Rödiger, Fritz. "Vorgeschichtliche Zeichensteine, als Marchsteine, Meilenzeiger (Leuksteine), Wegweiser (Waranden), Pläne und Landkarten." *Zeitschrift für Ethnologie* 22 (1890): Verhandlungen 504–16. **64**

——. "Vorgeschichtliche Kartenzeichnungen in der Schweiz." *Zeitschrift für Ethnologie* 23 (1891): Verhandlungen 237–42. **54, 64, 65**

——. "Erläuterungen und beweisende Vergleiche zur Steinkarten-Theorie." *Zeitschrift für Ethnologie* 23 (1891): Verhandlungen 719–24. **64**

Rodolico, Niccolò. "Di una carta nautica di Giacomo Bertran, maiorchino." *Atti del III Congresso Geografico Italiano, Florence, 1898,* 2 vols. (1899) 2:544–50. **400, 433**

Rodríguez Almeida, Emilio. *Forma Urbis Marmorea: Aggiornamento generale 1980.* Rome: Edizioni Quasar, 1981. **226**

Rodwell, Warwick. "Milestones, Civic Territories and the Antonine Itinerary." *Britannia* 6 (1975): 76–101. **236**

Roggero, Roberto. "Recenti scoperte di incisioni rupestri nelle Valli di Lanzo (Torino)." In *Symposium International d'Art Préhistorique Valcamonica, 23–28 Septembre 1968,* 125–32. Union Internationale des Sciences Préhistoriques et Protohistoriques. Capo di Ponte: Edizioni del Centro, 1970. **76**

Röhricht, Reinhold. "Karten und Pläne zur Palästinakunde aus dem 7. bis 16. Jahrhundert, I." *Zeitschrift des Deutschen Palästina-Vereins* 14 (1891): 8–11. **475**

——. "Karten und Pläne zur Palästinakunde aus dem 7. bis 16. Jahrhundert, II." *Zeitschrift des Deutschen Palästina-Vereins* 14 (1891): 87–92. **466**

——. "Karten und Pläne zur Palästinakunde aus dem 7. bis 16. Jahrhundert, III." *Zeitschrift des Deutschen Palästina-Vereins* 14 (1891): 137–141. **470**

——. "Karten und Pläne zur Palästinakunde aus dem 7. bis 16. Jahrhundert, IV." *Zeitschrift des Deutschen Palästina-Vereins* 15 (1892): 34–39. **474**

——. "Marino Sanudo sen. als Kartograph Palästinas." *Zeitschrift des Deutschen Palästina-Vereins* 21 (1898): 84–126. **474, 476, 496**

——. "Die Palästinakarte Bernhard von Breitenbach's." *Zeitschrift des Deutschen Palästina-Vereins* 24 (1901): 129–35. **474, 475, 476**

——. "Die Palästinakarte des William Wey." *Zeitschrift des Deutschen Palästina-Vereins* 27 (1904): 188–93. **476**

Röllig, Wolfgang. "Landkarten." In *Reallexikon der Assyriologie und vorderasiatischen Archäologie,* ed. Erich Ebeling and Bruno Meissner, 6:464–67. Berlin: Walter de Gruyter, 1932–. **111**

Romanelli, Giandomenico, and Susanna Biadene. *Venezia piante e vedute: Catalogo del fondo cartografico a stampa.* Venice: Museo Correr, 1982. **12**

Romano, Virginia. "Sulla validità della *Carte Pisana.*" *Atti dell'Accademia Pontaniana* 32 (1983): 89–99. **390, 404**

Rosien, Walter. *Die Ebstorfer Weltkarte.* Hanover: Niedersächsisches Amt für Landesplanung und Statistik, 1952. **291, 301, 310, 351**

Rossi, G. B. de. *Piante iconografiche e prospettiche di Roma anteriori al secolo XVI.* Rome: Salviucci, 1879. **477, 493**

Rossi, Ettore. "Una carta nautica araba inedita di Ibrāhīm al-Mursī datata 865 Egira = 1461 Dopo Christo." In *Compte rendu du Congrès Internationale de Géographie* 5 (1926): 90–95 (11th International Congress, Cairo, 1925). **434, 437, 457**

Rössler-Köhler, Ursula. "Jenseitsvorstellungen." In *Lexikon der Ägyptologie,* ed. Wolfgang Helck and Eberhard Otto, 3:cols. 252–67. Wiesbaden: O. Harrassowitz, 1975–. **120**

Rostovtzeff, Mikhail I. *Iranians and Greeks in South Russia.* Oxford: Clarendon Press, 1922. **72, 73, 96**

Roth, Cecil. "Judah Abenzara's Map of the Mediterranean World, 1500." *Studies in Bibliography and Booklore* 9 (1970): 116–20. **457**

Rotterdam, Maritiem Museum Prins Hendrik. *Eilanden en waarden in kaart en beeld: Tentoonstelling, 22 december, 1953-15 maart, 1954.* Rotterdam: Maritiem Museum Prins Hendrik, [1954]. **500**

Rozemond, A. J. H. *Inventaris der verzameling kaarten berustende in het Algemeen Rijksarchief zijnde het eerste en tweede supplement op de collectie Hingman.* The Hague: Algemeen Rijksarchief, 1969. **486**

Ruberg, Uwe. "Mappae Mundi des Mittelalters im Zusammenwirken von Text und Bild." In *Text und Bild: Aspekte des Zusammenwirkens zweier Künste in Mittelalter und früher Neuzeit,* ed. Christel Meier and Uwe Ruberg, 550–92. Wiesbaden: Ludwig Reichert, 1980. **286, 287, 290**

Rubió y Lluch, Antoni. *Documents per l'historia de la cultura catalana mig-eval.* Barcelona: Institut d'Estudis Catalans, 1908–21. **435, 436, 442**

Ruddock, Alwyn A. *Italian Merchants and Shipping in Southampton, 1270–1600.* Southampton: University College, 1951. **408**

Rudwick, Martin J. S. "The Emergence of a Visual Language for Geological Science, 1760–1840." *History of Science* 14 (1976): 149–95. **36**

Ruge, Sophus. *Ueber Compas und Compaskarten.* Separat Abdruck aus dem Programm der Handels-Lehranstalt. Dresden, 1868. **375**

——. "Älteres kartographisches Material in deutschen Bibliotheken." *Nachrichten von der Königlichen Gesellschaft der Wissenschaften zu Göttingen, Philologisch-Historische Klasse* (1904): 1–69; (1906): 1–39; (1911): 35–166; suppl. (1916). **294**

Ruggles, Richard I. "Research on the History of Cartography and Historical Cartography of Canada, Retrospect and Prospect." *Canadian Surveyor* 31 (1977): 25–33. **37**

Russell, Bertrand. *Philosophy.* New York: W. W. Norton, 1927. **326**

Russell, G. N. "Secrets of the Labyrinth." *Irish Times,* 16 December 1964, 10. **88**

Russell, J. C. "Late Ancient and Medieval Population."

Transactions of the American Philosophical Society, n.s. 48, pt. 3 (1958): 5–152. **426**

Rzepa, Zbigniew. "Stan i potrzeby badań nad historia Kartografii w Polsce (I Ogólnopolska Konferencja Historyków Kartografii)." *Kwartalnik Historii Nauki i Techniki* 21 (1976): 377–81. **38**

———. "Joachim Lelewel, 1786–1861." *Geographers: Biobibliographical Studies* 4 (1980): 103–12. **13, 18**

Sack, Robert David. *Conceptions of Space in Social Thought: A Geographic Perspective*. Minneapolis: University of Minnesota Press; London: Macmillan, 1980. **84**

St. Koledarov, Peter. "Nai-Ranni Spomenavanniya na Bilgarityе virkhu Starinnitye Karty" (The earliest reference to the Bulgarians on ancient maps). *Izvestija na Instituta za Istorija* 20 (1968): 219–54. **290**

Sakalis, Dimitrios. "Die Datierung Herons von Alexandrien." Inaugural dissertation. University of Cologne, 1972. **230**

Salinari, Marina (formerly Marina Emiliani). "Notizie su di alcune carte nautiche di Grazioso Benincasa." *Rivista Geografica Italiana* 59 (1952): 36–42. **421, 422, 457**

Salishchev, K. A. *Osnovy kartovedeniya: Chast' istoricheskaya i kartograficheskiye materialy* (Fundamentals of map science: Historical part and cartographic materials). Moscow: Geodezizdat, 1948. **73**.

Saller, Sylvester J. *The Memorial of Moses on Mount Nebo*. 3 vols. Publications of the Studium Biblicum Franciscanum, no. 1. Jerusalem: Franciscan Press, 1941–50. **264**

Sallmann, Nicolaus. "De Pomponio Mela et Plinio Maiore in Africa describenda discrepantibus." In *Africa et Roma: Acta omnium gentium ac nationum Conventus Latinis litteris linguaeque fovendis*, ed. G. Farenga Ussani, 164–73. Rome: Erma di Bretschneider, 1979. **242**

Salomon, Richard Georg. *Opicinus de Canistris: Weltbild und Bekenntnisse eines Avignonesischen Klerikers des 14. Jahrhunderts*. Studies of the Warburg Institute, vols. 1A and 1B (text and plates). London: Warburg Institute, 1936. **291**

———. "A Newly Discovered Manuscript of Opicinus de Canistris." *Journal of the Warburg and Courtauld Institutes* 16 (1953): 45–57. **291**

———. "Aftermath to Opicinus de Canistris." *Journal of the Warburg and Courtauld Institutes* 25 (1962): 137–46. **291**

Salviat, F. "Orientation, extension et chronologie des plans cadastraux d'Orange." *Revue Archéologique de Narbonnaise* 10 (1977): 107–18. **222**

Salzman, Louis Francis. *Building in England down to 1540*. Oxford: Clarendon Press, 1952. **471**

Sandys, John Edwin. *A Short History of Classical Scholarship from the Sixth Century B.C. to the Present Day*. Cambridge: Cambridge University Press, 1915. **266**

Sanson, Nicolas. *Introduction à la géographie*. Paris, 1682. **9**

Santarém, Manuel Francisco de Barros e Sousa, Viscount of. *Atlas composé de mappemondes, de portulans et de cartes hydrographiques et historiques depuis le VIᵉ jusqu'au XVIIᵉ siècle*. Paris, 1849. Facsimile reprint with explanatory texts by Helen Wallis and A. H. Sijmons, Amsterdam: R. Muller, 1985. **293, 457**

———. "Notice sur plusieurs monuments géographiques inédits du Moyen Age et du XVIᵉ siècle qui se trouvent dans quelques bibliothèques de l'Italie, accompagné de notes critiques." *Bulletin de la Société de Géographie*, 3d ser., 7 (1847): 289–317; reprinted in *Acta Cartographica* 14 (1972): 318–46. **13, 17, 435**

———. *Essai sur l'histoire de la cosmographie et de la cartographie pendant le Moyen-Age et sur les progrès de la géographie après les grandes découvertes du XVᵉ siècle*. 3 vols. Paris: Maulde et Renou, 1849–52. **12, 13, 17, 292, 294, 381, 448**

———. *Estudos de cartographia antiga*. 2 vols. Lisbon: Lamas, 1919–20. **423**

Santos Júnior, J. R. dos. "O abrigo pre-histórico da 'Pala Pinta.'" *Trabalhos da Sociedade Portuguesa de Antropologiae Etnologia* 6 (1933): 33–43. **83**

Sanz, Carlos. "El primer mapa del mundo con la representacíon de los dos hemisferios." *Boletín de la Real Sociedad Geográfica* 102 (1966): 119–217. **353**

Sartiaux, F. "Recherches sur le site de l'ancienne Phocée." *Comptes Rendus des Séances de l'Académie des Inscriptions et Belles-Lettres* (1914): 6–18. **159**

Sarton, George. *Introduction to the History of Science*. 3 vols. Baltimore: Williams and Wilkins, 1927–48. **3, 20, 173, 293, 306, 312, 321, 380, 441**

———. Review of "Bīrūnī's Picture of the World." *Memoirs of the Archaeological Survey of India* 53 [1941], by Ahmed Zeki Valīdī Togan. *Isis* 34 (1942): 31–32. **320**

Sauer, Carl O. "The Education of a Geographer." *Annals of the Association of American Geographers* 46 (1956): 287–99. **2**

Savvateyev, Yury A. *Risunki na skalakh* (Rock drawings). Petrozavodsk: Karelskoye Knizhnoye Izdelstvo, 1967. **76**

Saxl, Fritz. "Illustrated Mediaeval Encyclopaedias: 1. The Classical Heritage; 2. The Christian Transformation." In his *Lectures*. 2 vols., 1:228–54. London: Warburg Institute, 1957. **266, 301**

Scaglia, Gustina. "The Origin of an Archaeological Plan of Rome by Alessandro Strozzi." *Journal of the Warburg and Courtauld Institutes* 27 (1964): 137–63. **477, 492, 495**

Scamuzzi, Ernesto. *Museo Egizio di Torino*. Turin: Fratelli Pozzo, 1964. **122**

Schack-Schackenburg, Hans, ed. *Das Buch von den zwei Wegen des seligen Toten (Zweiwegebuch): Texte aus der Pyramidenzeit nach einem im Berliner Museum bewahrten Sargboden des mittleren Reiches*. Leipzig: J. C. Hinrich, 1903. **117**

Schäfer, Heinrich. *Ägyptische und heutige Kunst und Weltgebäude der alten Ägypter: Zwei Aufsätze*. Berlin: Walter de Gruyter, 1928. **121**

———. *Principles of Egyptian Art*. Ed. with epilogue by Emma Brunner-Traut. Ed. and Trans. with introduction by John Baines. Oxford: Clarendon Press, 1974. **117**

Schalk, Fritz. "Über Epoche und Historie." Part of "Studien

zur Periodisierung und zum Epochebegriff" by Hans Diller and Fritz Schalk. *Abhandlungen der Akademie der Wissenschaften und der Literatur, Mainz.* Geistes- und Sozialwissenschaftliche Klasse, (1972): 150–76. **xviii**

Scharfe, Wolfgang. "Geschichte der Kartographie—heute?" In *Festschrift für Georg Jensch aus Anlaß seines 65. Geburtstages,* ed. F. Bader et al., 383–98. Abhandlungen des 1. Geographischen Instituts der Freien Universität Berlin, 20. Berlin: Reimer, 1974. **37**

———. "Die Geschichte der Kartographie im Wandel." *International Yearbook of Cartography* 21 (1981): 168–76. **38**

———. "Max Eckert's 'Kartenwissenschaft'—The Turning-Point in German Cartography." Paper prepared for the Eleventh International Conference on the History of Cartography, Ottawa, 1985. **24**

Schilder, Günter. "Organization and Evolution of the Dutch East India Company's Hydrographic Office in the Seventeenth Century." *Imago Mundi* 28 (1976): 61–78. **9**

Schillinger-Häfele, Ute. "Beobachtungen zum Quellenproblem der *Kosmographie* von Ravenna." *Bonner Jahrbücher* 163 (1963): 238–51. **260**

Schlichtmann, Hansgeorg. "Codes in Map Communication." *Canadian Cartographer* 16 (1979): 81–97. **34**

———. "Characteristic Traits of the Semiotic System 'Map Symbolism.'" *Cartographic Journal* 22 (1985): 23–30. **34**

Schlott-Schwab, Adelheid. *Die Ausmaße Ägyptens nach altägyptischen Texten.* Wiesbaden: O. Harrassowitz, 1981. **125**

Schlumberger, Gustave, Ferdinand Chalandon, and Adrien Blanchet. *Sigillographie de l'Orient Latin.* Paris: Geuthner, 1943. **488**

Schnabel, Paul. *Text und Karten des Ptolemäus.* Quellen und Forschungen zur Geschichte der Geographie und Völkerkunde 2. Leipzig: K. F. Koehlers Antiquarium, 1938. **192, 269**

Schneider, Marius. *El origen musical de los animales-símbolos en la mitología y la escultura antiguas.* Monograph 1. Barcelona: Instituto Español de Musicología, 1946. **86**

Schnelbögl, Fritz. "Life and Work of the Nuremberg Cartographer Erhard Etzlaub (†1532)." *Imago Mundi* 20 (1966): 11–26. **473, 498**

Schnetz, Joseph. *Untersuchungen über die Quellen der Kosmographie des anonymen Geographen von Ravenna.* Sitzungsberichte der Akademie der Wissenschaften, Philosophisch-historische Abteilung 6. Munich: Verlag der Bayerischen Akademie der Wissenschaften, 1942. **260**

Schöne, Hermann. "Das Visirinstrument der Römischen Feldmesser." *Jahrbuch des Kaiserlich Deutschen Archäologischen Instituts* 16 (1901): 127–32. **213**

Schönfeld, M. "L'astronomie préhistorique en Scandinavie." *La Nature,* no. 2444, 5 February 1921, 81–83. **82, 93**

Schulten, Adolf. "Fundus." In *Dizionario epigrafico di antichità romane,* ed. E. de Ruggiero, 3:347. Rome, 1895–. **226**

———. "Römische Flurkarten." *Hermes* 33 (1898): 534–65. **218**

———. *Tartessos: Ein Beitrag zur ältesten Geschichte des Westens.* Hamburg: L. Friederichsen and Co., 1922. **150**

Schulz, Juergen. "The Printed Plans and Panoramic Views of Venice (1486–1797)." *Saggi e Memorie di Storia dell'Arte* 7 (1970), 9–182. **12, 478**

———. "Jacopo de' Barbari's View of Venice: Map Making, City Views, and Moralized Geography before the Year 1500." *Art Bulletin* 60 (1978): 425–74. **36, 287, 288, 290, 292, 303, 314, 477, 478, 493, 495**

———. "Maps as Metaphors: Mural Map Cycles of the Italian Renaissance." In *Art and Cartography: Six Historical Essays.* Ed. David Woodward. Chicago: University of Chicago Press, 1987. **8, 315**

Schumm, Karl. *Inventar der handschriftlichen Karten im Hohenlohe-Zentralarchiv Neuenstein.* Karlsruhe: Braun, 1961. **488**

Schurtz, Heinrich. *Istoriya pervobytnoy kul'tury* (History of primitive cultures). Moscow, 1923. Translated from the German *Urgeschichte der Kultur.* Leipzig and Vienna, 1900. **47**

Schütte, Gudmund. *Ptolemy's Maps of Northern Europe: A Reconstruction of the Prototypes.* Copenhagen: Royal Danish Geographical Society, 1917. **197, 198**

———. *Hjemligt Hedenskab: I Almenfattelig Fremstilling.* Copenhagen: Gyldendal, 1919. **82**

———. "Primaeval Astronomy in Scandinavia." *Scottish Geographical Magazine* 36, no. 4 (1920): 244–54. **82, 83, 93**

Secchi, Laura. *Navigazione e carte nautiche nei secoli XIII–XVI.* Catalog of an exhibition held at the Palazzo Rosso, Genoa, May to October 1978. **438**

Sedgley, Jeffrey P. *The Roman Milestones of Britain: Their Petrography and Probable Origin.* British Archaeological Reports no. 18. Oxford: British Archaeological Reports, 1975. **236**

Seeck, Otto, ed. *Notitia Dignitatum.* Berlin: Weidmann, 1876; reprinted Frankfort: Minerva, 1962. **244**

Seltman, Charles Theodore. *Greek Coins: A History of Metallic Currency and Coinage down to the Fall of the Hellenistic Kingdoms.* London: Methuen, 1933; revised 1955. **158**

Serejski, Marian Henryk. *Joachim Lelewel, 1786-1861: Sa vie et son oeuvre.* Warsaw: Zakład Narodowy imienia Ossolińskich, 1961. **13**

Sharpe, Kevin. *Sir Robert Cotton, 1586–1631: History and Politics in Early Modern England.* Oxford: Oxford University Press, 1979. **9**

Shee, Elizabeth. "Recent Work on Irish Passage Graves Art." *Bollettino del Centro Camuno di Studi Preistorici* 8 (1972): 199–224. **61**

Sherk, Robert K. "Roman Geographical Exploration and Military Maps." In *Aufstieg und Niedergang der römischen Welt,* ed. Hildegard Temporini, 2.1 (1974): 534–62. Berlin: Walter de Gruyter, 1972–. **210, 253**

Shibanov, F. A. "The Essence and Content of the History of Cartography and the Results of Fifty Years of Work by Soviet Scholars." In *Essays on the History of Russian Cartography, 16th to 19th Centuries,* ed. and trans. James R. Gibson, introduction by Henry W. Castner, 141–45. Monograph 13, *Cartographica* (1975). **35, 38**

Shirley, Rodney W. *The Mapping of the World: Early Printed World Maps 1472–1700*. London: Holland Press, 1983. **302**

Shotwell, James T. *The History of History*. New York: Columbia University Press, 1939. **7**

Siegel, Linda S., and Charles J. Brainerd. *Alternatives to Piaget: Critical Essays on the Theory*. New York: Academic Press, 1978. **2**

Sieveking, Ann. *The Cave Artists*. London: Thames and Hudson, 1979. **57, 86**

Sigurðsson, Harald. *Kortasaga Islands frá öndverðu til loka 16. aldar*. Reykjavik: Bókaútgáfa Menningarsjóðs og Þjóðvinafélagsins, 1971. **414**

Simar, Théophile. "La géographie de l'Afrique Centrale dans l'antiquité et au Moyen-Age." *Revue Congolaise* 3 (1912–13): 1–23, 81–102, 145–69, 225–52, 289–310, 440–41. **295**

Simmons, Leo W., ed. *Sun Chief: The Autobiography of a Hopi Indian*. New Haven: Yale University Press, 1942. **59**

Simpson, James Young. *Archaic Sculpturings of Cups, Circles, etc. upon Stones and Rocks in Scotland, England, and Other Countries*. Edinburgh: Edmonston and Douglas, 1867. **64**

Singer, Charles. "Daniel of Morley: An English Philosopher of the XIIth Century." *Isis* 3 (1920): 263–69. **306**

———. *Studies in the History and Method of Science*. 2 ed., 2 vols. London: W. Dawson, 1955. **321**

———. *A Short History of Scientific Ideas to 1900*. Oxford: Clarendon Press, 1959; reprinted 1966. **3, 10**

———, et al., eds. *A History of Technology*. 7 vols. Oxford: Clarendon Press, 1954–78. **xviii, xix, 3, 60, 293**

Skelton, R. A. "An Ethiopian Embassy to Western Europe in 1306." In *Ethiopian Itineraries circa 1400–1524, Including Those Collected by Alessandro Zorzi at Venice in the Years 1519–24*, ed. Osbert G. S. Crawford, 212–16. Hakluyt Society, ser. 2, 109. Cambridge: Cambridge University Press for the Hakluyt Society, 1958. **404, 405**

[———]. "Leo Bagrow: Historian of Cartography and Founder of *Imago Mundi*, 1881–1957." *Imago Mundi* 14 (1959): 4–12. **22, 26, 27**

———. *Looking at an Early Map*. Lawrence: University of Kansas Libraries, 1965. **33, 403**

———. "Historical Notes on *Imago Mundi*." *Imago Mundi* 21 (1967): 109–10. **27**

———. "A Contract for World Maps at Barcelona, 1399–1400." *Imago Mundi* 22 (1968): 107–13. **287, 324, 393, 401, 430, 436**

———. *Maps: A Historical Survey of Their Study and Collecting*. Chicago: University of Chicago Press, 1972. **xvii, 3, 6, 7, 8, 9, 10, 11, 12, 14, 15, 16, 17, 18, 19, 21, 23, 24, 28, 34, 35, 38, 292, 422**

Skelton, R. A., Thomas E. Marston, and George D. Painter. *The Vinland Map and the Tartar Relation*. New Haven: Yale University Press, 1965. **368, 410, 414**

Skelton, R. A., and John Summerson. *A Description of Maps and Architectural Drawings in the Collection Made by William Cecil, First Baron Burghley, Now at Hatfield House*. Oxford: Roxburghe Club, 1971. **9**

Skelton, R. A., and P. D. A. Harvey, eds. *Local Maps and Plans from Medieval England*. Oxford: Clarendon Press, 1986. **464, 467, 470, 471, 473, 484, 485, 490, 491, 492, 493, 494, 495, 498**

Skop, Jacob. "The Stade of the Ancient Greeks." *Surveying and Mapping* 10 (1950): 50–55. **148**

Slobbe, Annemieke van. *Kartobibliografieën in het Geografisch Instituut Utrecht*. Utrechtse Geografische Studies 10. Utrecht: Geografisch Instituut Rijksuniversiteit Utrecht, 1978. **20**

Smalley, Beryl. *The Study of the Bible in the Middle Ages*. 2d ed. Oxford: Blackwell, 1952. **334**

Smith, Mary Elizabeth. *Picture Writing from Ancient Southern Mexico: Mixtec Place Signs and Maps*. Norman: University of Oklahoma Press, 1973. **53**

Smith, R. W. "The Significance of Roman Glass." *Metropolitan Museum Bulletin* 8 (1949): 56. **239**

Smith, Thomas R. "Manuscript and Printed Sea Charts in Seventeenth-Century London: The Case of the Thames School." In *The Compleat Plattmaker: Essays on Chart, Map, and Globe Making in the Seventeenth and Eighteenth Centuries*, ed. Norman J. W. Thrower, 45–100. Berkeley: University of California Press, 1978. **391, 392, 431, 436**

———. "Rhumb-Line Networks on Early Portolan Charts: Speculations Regarding Construction and Function." Paper prepared for the Tenth International Conference on the History of Cartography, Dublin, 1983. **376**

Smith, William Stevenson. *Interconnections in the Ancient Near-East: A Study of the Relationships between the Arts of Egypt, the Aegean, and Western Asia*. New Haven: Yale University Press, 1965. **119**

Snyder, George Sergeant. *Maps of the Heavens*. New York: Abbeville Press, 1984. **22**

Somers Cocks, J. V. "Dartmoor, Devonshire, Late 15th or Early 16th Century." In *Local Maps and Plans from Medieval England*, ed. R. A. Skelton and P. D. A. Harvey, 293–302. Oxford: Clarendon Press, 1986. **484**

Sommerbrodt, Ernst. *Afrika auf der Ebstorfer Weltkarte*. Festschrift zum 50-Jährigen Jubiläum des Historischen Vereins für Niedersachsen. Hanover, 1885. **307**

Sotheby's. *Catalogue of Highly Important Maps and Atlases*, 15 April 1980. **378, 419, 458**

Spadolini, Ernesto. "Il portolano di Grazioso Benincasa." *Bibliofilia* 9 (1907–8): 58–62, 103–9, 205–34, 294–99, 420–34, 460–63; reprinted in *Acta Cartographica* 11 (1971): 384–450. **433**

Spiegelberg, Wilhelm. *Die demotischen Denkmäler*. 2 vols. Leipzig: W. Drugulin, 1904–8. **128**

Spufford, Peter, and Wendy Wilkinson. *Interim Listing of the Exchange Rates of Medieval Europe*. North Staffordshire: Department of History, University of Keele, 1977. **437**

Staglieno, Marcello. "Sopra Agostino Noli e Visconte Maggiolo cartografi." *Giornale Ligustico* 2 (1875): 71–79. **430**

Stahl, William Harris. "Astronomy and Geography in Macrobius." *Transactions and Proceedings of the American Philological Society* 35 (1942): 232–38. **300**

————. *Ptolemy's Geography: A Select Bibliography.* New York: New York Public Library, 1953. **190**

————. "By Their Maps You Shall Know Them." *Archaeology* 8 (1955): 146–55. **88**

————. "Cosmology and Cartography." Part of "Representation of the Earth's Surface as an Artistic Motif." In *Encyclopedia of World Art,* 3:cols. 851–54. New York: McGraw-Hill, 1960. **70, 96**

————. *Roman Science: Origins, Development, and Influence to the later Middle Ages.* Madison: University of Wisconsin Press, 1962. **299, 301**

————. *The Quadrivium of Martianus Capella: Latin Traditions in the Mathematical Sciences, 50 B.C.–A.D. 1250.* Martianus Capella and the Seven Liberal Arts, vol. 1. New York: Columbia University Press, 1971. **300, 353**

Stanchul, T. A. "Natsional'nye kartograficheskye obshchestva mira" (National cartographic societies of the world). *Doklady Otdeleniy i Komissiy* 10 (1969): 89–99 (Geograficheskogo obshchestva SSSR, Leningrad). **32**

Stechow, E. "Zur Entdeckung der Ostsee durch die Römer." *Forschungen und Fortschritte* 24 (1948): 240–41. **197**

Steers, J. A. Review of *The Mapmakers,* by John Noble Wilford. *Geographical Journal* 149 (1983): 102–3. **26**

Stegena, Lajos. "Minoische kartenähnliche Fresken bei Acrotiri, Insel Thera (Santorini)." *Kartographische Nachrichten* 34 (1984): 141–43. **132**

Steinmeyer-Schareika, Angela. *Das Nilmosaik von Palestrina und eine Ptolemäische Expedition nach Äthiopien.* Halbelts Dissertationsdrucke, Reihe Klassische Archäologie 10. Bonn: Halbelt, 1978. **118**

Stephens, John D. "Current Cartographic Serials: An Annotated International List." *American Cartographer* 7 (1980): 123–38. **32**

Stephenson, Richard W. "The Henry Harrisse Collection of Publications, Papers, and Maps Pertaining to the Early Exploration of America." Paper prepared for the Tenth International Conference on the History of Cartography, Dublin 1983. **20**

Sterling, Charles. "Le mappemonde de Jan van Eyck." *Revue de l'Art* 33 (1976): 69–82. **368**

Stevens, Henry N. *Ptolemy's Geography: A Brief Account of All the Printed Editions down to 1730.* 2d ed. London: Henry Stevens, Son and Stiles, 1908; reprinted, Amsterdam: Theatrum Orbis Terrarum, [1973]. **17**

————. *Recollections of James Lenox and the Formation of His Library.* Ed. Victor Hugo Paltsits. New York: New York Public Library, 1951. **17**

Stevens, Wesley M. "The Figure of the Earth in Isidore's 'De Natura Rerum.'" *Isis* 71 (1980): 268–77. **301, 345**

Stevenson, Edward Luther. *Facsimiles of Portolan Charts Belonging to the Hispanic Society of America.* Publications of the Hispanic Society of America, no. 104. New York, 1916. **401, 457, 461**

————. *Terrestrial and Celestial Globes: Their History and Construction, Including a Consideration of Their Value as Aids in the Study of Geography and Astronomy.* 2 vols. Publications of the Hispanic Society of America, no. 86. New Haven: Yale University Press, 1921; reprinted New York and London: Johnson Reprint Corporation, 1971. **141, 159, 163, 164, 171**

————, trans. *Geography of Claudius Ptolemy.* New York: New York Public Library, 1932. **177, 179, 186, 198**

Stevenson, W. H., ed., and Robert Cole, comp. *Rental of All the Houses in Gloucester A.D. 1455.* Gloucester: Bellows, 1890. **470**

Stewart, Aubrey, trans. *Itinerary from Bordeaux to Jerusalem: "The Bordeaux Pilgrim".* Palestine Pilgrims Text Society, vol. 1, no. 2. London: Palestine Exploration Fund, 1896. **237**

Stolzenberg, Ingeborg. "Weltkarten in mittelalterlichen Handschriften der Staatsbibliothek Preußischer Kulturbesitz." In *Karten in Bibliotheken: Festgabe für Heinrich Kramm zur Vollendung seines 65. Lebensjahres,* ed. Lothar Zögner, 17–32. Kartensammlung und Kartendokumentation 9. Bonn-Bad Godesberg: Bundesforschungsanstalt für Landeskunde und Raumordnung, Selbstverlag, 1971. **301, 343**

Stone, Jeffrey C. "Techniques of Scale Assessment on Historical Maps." In *International Geography 1972,* ed. W. P. Adams and F. M. Helleiner, 452–54. Toronto: University of Toronto Press, 1972. **403**

Strachan, James. *Early Bible Illustrations: A Short Study Based on Some Fifteenth and Early Sixteenth Century Printed Texts.* Cambridge: Cambridge University Press, 1957. **336**

Struve, W. W. *Mathematischer Papyrus des Staatlichen Museums der Schönen Künste in Moskau.* Berlin: J. Springer, 1930. **127**

Strzelczyk, Jerzy. *Gerwazy z Tilbury: Studium z dziejów uczoności geograficznej w średniowieczu.* Monograph 46. Warsaw: Zakład Narodowy im. Ossolińskich, 1970. **307**

Stylianou, Andreas, and Judith A. Stylianou. *The History of the Cartography of Cyprus.* Publications of the Cyprus Research Centre, 8. Nicosia, 1980. **457**

Suhm, Peter Frederik. *Samlinger til den Danske historie.* Copenhagen: A. H. Godishes, 1779–84. **63**

Sukenik, Eleazar L. *The Ancient Synagogue of Beth Alpha.* Jerusalem: University Press, 1932. **266**

Svennung, Joseph Gusten Algot. *Belt und Baltisch: Ostseeische Namenstudien mit besonderer Rücksicht auf Adam von Bremen.* Uppsala: Lundequistska Bokhandeln, 1953. **197**

————. *Scandinavien bei Plinius und Ptolemaios.* Uppsala: Almqvist och Wiksell, 1974. **197**

Svoronos, Jean N. (Ioannes N. Sborōnos). *Numismatique de la Crète ancienne accompagnée de l'histoire, la géographie et la mythologie de l'Âile.* Macon: Imprimerie Protat Frères, 1890; reprinted Bonn: R. Habelt, 1972. **251**

Taisbak, C. M. "Posidonius Vindicated at All Costs? Modern Scholarship versus the Stoic Earth Measurer." *Centaurus* 18 (1973–74): 253–69. **169**

Tanguy, J. C. "An Archaeometric Study of Mt. Etna: The Magnetic Direction Recorded in Lava Flows Subsequent to the Twelfth Century." *Archaeometry* 12 (1970): 115–128. **384**

————. "L'Etna: Etude pétrologique et paléomagnetique, implications volcanologiques." Ph.D. Diss., Université Pierre et Marie Curie, Paris, 1980. **384**

Tanselle, G. Thomas. "From Bibliography to *Histoire*

Totale: The History of Books as a Field of Study." *Times Literary Supplement*, 5 June 1981, 647–49. **21**

———. "The Description of Non-letterpress Material in Books." *Studies in Bibliography* 35 (1982): 1–42. **21**

Tarn, William Woodthorpe. *Alexander the Great*. 2 vols. Cambridge: Cambridge University Press; New York: Macmillan, 1948. **149, 151**

Tate, George. Address to members at the anniversary meeting held at Embleton, 7 September 1853. *Proceedings of the Berwickshire Naturalists' Club* 3, no. 4 (1854): 125–41. **64**

———. *The Ancient British Sculptured Rocks of Northumberland and the Eastern Borders, with Notices of the Remains Associated with These Sculptures*. Alnwick: H. H. Blair, 1865. **64, 65, 86**

Taton, Juliette. "Jean-Baptiste Bourguignon d'Anville." In *Dictionary of Scientific Biography*, 16 vols., ed. Charles Coulston Gillispie, 1:175–76. New York: Charles Scribner's Sons, 1970–80. **10**

Taton, René. *Histoire générale des sciences*. 3 vols. in 4 pts. Paris: Presses Universitaires de France, 1957–64. English edition, *History of Science*. 4 vols. Trans. A. J. Pomerans. London: Thames and Hudson, 1963–66. **xix, 84**

Tattersall, Jill. "Sphere or Disc? Allusions to the Shape of the Earth in Some Twelfth-Century and Thirteenth-Century Vernacular French Works." *Modern Language Review* 76 (1981): 31–46. **290, 319, 343**

Taubner, Kurt. "Zur Landkartenstein-Theorie." *Zeitschrift für Ethnologie* 23 (1891): Verhandlungen 251–57. **54, 63, 66**

Taylor, Eva G. R. "Pactolus: River of Gold." *Scottish Geographical Magazine* 44 (1928): 129–44. **328, 413**

———. "The Surveyor." *Economic History Review*, 1st ser., 17 (1947): 121–33. **494**

———. "Early Charts and the Origin of the Compass Rose." *Navigation: Journal of the Institute of Navigation* 4 (1951): 351–56. **383**

———. *The Mathematical Practitioners of Tudor and Stuart England*. Cambridge: Cambridge University Press, 1954. **429**

———. *The Haven-Finding Art: A History of Navigation from Odysseus to Captain Cook*. London: Hollis and Carter, 1956. **371, 375, 384, 387, 390, 397, 401, 422, 429, 442, 443**

Taylor, John. *The "Universal Chronicle" of Ranulf Higden*. Oxford: Clarendon, 1966. **312**

Teeling, P. S. "Oud-Nederlandse landmeters." *Orgaan der Vereniging van Technische Ambtenaren van het Kadaster* 7 (1949): 34–45, 90–98, 126–34, 158–70, 198–209; 8 (1950): 2–11. **494**

Teixeira da Mota, Avelino. *Topónimos de origem Portuguesa na costa ocidental de Africa desde o Cabo Bojador ao Cabo de Santa Caterina*, Centro de Estudos da Guiné Portuguesa no. 14. Bissau: Centro de Estudos da Guiné Portuguesa, 1950. **413**

———. "L'art de naviguer en Méditerranée du XIIIe au XVIIe siècle et la création de la navigation astronomique dans les océans." In *Le navire et l'économie maritime du Moyen-Age au XVIIIe siècle principalement en Méditerranée: Travaux du IIème Colloque Internationale d'Histoire Maritime*, ed. Michel Mollat, 127–54. Paris: SEVPEN, 1958. **379, 386, 441**

———. "Influence de la cartographie portugaise sur la cartographie européenne à l'époque des découvertes." In *Les aspects internationaux de la découverte océanique aux XVe et XVIe siècles: Actes du Vème Colloque Internationale d'Histoire Maritime*, ed. Michel Mollat and Paul Adam, 223–48. Paris: SEVPEN, 1966. **377, 429**

———. "Some Notes on the Organization of Hydrographical Services in Portugal before the Beginning of the Nineteenth Century." *Imago Mundi* 28 (1976): 1–60. **9**

Temanza, Tommaso. *Antica pianta dell'inclita città di Venezia delineata circa la metà del XII. secolo*. Venice: Palese, 1781. **478**

Thiele, Georg. *Antike Himmelsbilder, mit Forschungen zu Hipparchos, Aratos und seinen Fortsetzern und Beiträgen zur Kunstgeschichte des Sternhimmels*. Berlin: Weidmann, 1898. **166, 303**

Thiriet, Freddy, ed. *Délibérations des assemblées vénitiennes concernant la Romanie*. 2 vols. Paris: Mouton, 1966–71. **433**

Thom, Alexander. "Astronomical Significance of Prehistoric Monuments in Western Europe." In *The Place of Astronomy in the Ancient World*, ed. F. R. Hodson, 149–56. Joint symposium of the Royal Society and the British Academy. London: Oxford University Press, 1974. **81**

Thomas, Elizabeth. "Cairo Ostracon J. 72460." In *Studies in Honor of George R. Hughes*, 209–16. Studies in Ancient Oriental Civilization, no. 39. Chicago: Oriental Institute of the University of Chicago, 1976. **126**

Thompson, Daniel V. "Medieval Parchment-Making." *Library*, 4th ser., 16 (1935): 113–17. **324**

———. *The Materials of Medieval Painting*. London: G. Allen and Unwin, 1936. Republished as *The Materials and Techniques of Medieval Painting*. New York: Dover, 1956. **324, 390**

Thompson, Silvanus P. "The Rose of the Winds: The Origin and Development of the Compass-Card." *Proceedings of the British Academy* 6 (1913–14): 179–209. **377, 390, 395**

Thomson, J. Oliver. *History of Ancient Geography*. Cambridge: Cambridge University Press, 1948; reprinted New York: Biblo and Tannen, 1965. **130, 145, 150, 151, 152, 162, 163, 166, 179**

Thorndike, Lynn, ed. and trans. *The Sphere of Sacrobosco and Its Commentators*. Chicago: University of Chicago Press, 1949. **306**

Thornton, Robert. "Modelling of Spatial Relations in a Boundary-Marking Ritual of the Iraqw of Tanzania." *Man*, n.s., 17 (1982): 528–45. **48**

Thrower, Norman J. W. "Monumenta Cartographica Africae et Aegypti." *UCLA Librarian*, suppl. to vol. 16, no. 15 (31 May 1963): 121–26. **294**

———. *Maps and Man: An Examination of Cartography in Relation to Culture and Civilization*. Englewood Cliffs, N.J.: Prentice-Hall, 1972. **4, 25, 47, 54, 59**

———. "The Treasures of UCLA's Clark Library." *Map Collector* 14 (1981): 18–23. **16**

Thulin, Carl Olof. *Die etruskische Disciplin. . . .* 3 pts.

1906–9; reprinted Darmstadt: Wissenschaftliche Buchgesellschaft, 1968. **201, 203**

————. "Kritisches zu Iulius Frontinus." *Eranos* 11 (1911): 131–44. **217**

————, ed. *Corpus Agrimensorum Romanorum.* Leipzig, 1913; reprinted Stuttgart: Teubner, 1971. **213, 219, 220;** Frontinus, **202, 218, 219;** Hyginus Gromaticus, **210, 216, 217**

Tierney, James J. "The Map of Agrippa." *Proceedings of the Royal Irish Academy* 63, sec. C, no. 4 (1963): 151–66. **207**

Tindale, Norman B. *Aboriginal Tribes of Australia: Their Terrain, Environmental Controls, Distribution, Limits and Proper Names.* Berkeley, Los Angeles, and London: University of California Press, 1974. **53**

Tobler, Arthur J. *Excavations at Tepe Gawra: Joint Expedition of the Baghdad School and the University Museum to Mesopotamia.* 2 vols. Philadelphia: University of Pennsylvania Press, 1950. **70, 71, 72, 96**

Tobler, Titus, and Augustus Molinier, eds. *Itinera Hierosolymitana et descriptiones Terrae Sanctae.* Paris: Société de l'Orient Latin, 1879. **466**

Tobler, Waldo R. "Medieval Distortions: The Projections of Ancient Maps." *Annals of the Association of American Geographers* 56 (1966): 351–60. **322**

Todorova, Elisaveta. "More about 'Vicina' and the West Black Sea Coast." *Etudes Balkaniques* 2 (1978): 124–38. **427**

Tooley, Ronald V. *Maps and Map-makers.* 6th ed. London: B. T. Batsford, 1978. **85**

Toomer, G. J. "Hipparchus." In *Dictionary of Scientific Biography,* 16 vols., ed. Charles Coulston Gillispie, 15:220. New York: Charles Scribner's Sons, 1970–80. **164, 165, 167**

————. "Ptolemy." In *Dictionary of Scientific Biography,* 16 vols., ed. Charles Coulston Gillispie, 11:186–206. New York: Charles Scribner's Sons, 1970–80. **177, 180, 306**

————, trans. *Ptolemy's Almagest.* London: Duckworth, 1984. **164, 177, 181, 184**

Toulmin, Stephen. *The Philosophy of Science: An Introduction.* London: Hutchinson, 1953. **1**

Toulmin, Stephen, and June Goodfield. *The Discovery of Time.* London: Hutchinson, 1965. **58**

Toynbee, Arnold J. *A Study of History.* 12 vols. London: Oxford University Press, 1934–[61]. **xix**

Tozer, H. F. *A History of Ancient Geography.* 1897. 2d ed.; reprinted New York: Biblo and Tannen, 1964. **130, 134, 137**

Treccani degli Alfieri, Giovanni. *Storia di Brescia.* 4 vols. plus index. Brescia: Morcelliana, 1961. **498**

Treidler, Hans. "Ζάβαι." In *Paulys Realencyclopädie der classischen Altertumswissenschaft,* ed. August Pauly, Georg Wissowa, et al., 2d ser., 9 (1967): cols. 2197–220. Stuttgart: J. B. Metzler, 1894–. **199**

Tristram, Ernest William. *English Medieval Wall Painting.* 2 vols. London: Oxford University Press, 1944–50. **368**

Trump, David H. "I primi architetti: I costruttori dei templi Maltesi." Rome: Giorgio Bretschneider, 1979. Extract

from φιλίας χάριν, *Miscellanea in Onore di Eugenio Manni.* **81, 96**

Tschudi, Jolantha. *Pitture rupestri del Tasili degli Azger.* Florence: Sansoni, 1955. **97**

Tuan, Yi-Fu. *Topophilia: A Study of Environmental Perception, Attitudes, and Values.* Englewood Cliffs, N.J.: Prentice-Hall, 1974. **4, 340**

————. *Space and Place: The Perspective of Experience.* Minneapolis: University of Minnesota Press, 1977. **4, 340**

————. *Landscapes of Fear.* Oxford: Basil Blackwell, 1979. **86**

Tudeer, Lauri O. T. "On the Origin of the Maps Attached to Ptolemy's Geography." *Journal of Hellenic Studies* 37 (1917): 62–76. **192**

Turville-Petre, Edward O. G. *Myth and Religion of the North: The Religion of Ancient Scandinavia.* London: Weidenfeld and Nicolson, 1964. **91**

Turyn, Alexander. *Codices Graeci Vaticani saeculis XIII et XIV scripti.* Codices e Vaticanis Selecti quam Simillime Expressi, vol. 28. Rome: Bibliotheca Apostolica Vaticana, 1964. **192**

Twyman, Michael. "A Schema for the Study of Graphic Language." In *Processing of Visible Language,* ed. Paul A. Kolers, Merald E. Wrolstad, and Herman Bouma, 1:117–50. New York: Plenum Press, 1979. **36**

Tyacke, Sarah. *The Map of Rome 1625, Paul Maupin: A Companion to the Facsimile.* London: Nottingham Court Press with Magdalene College, Cambridge, 1982. **9**

Tyacke, Sarah, and John Huddy. *Christopher Saxton and Tudor Map-making.* London: British Library, 1980. **21**

Ucko, Peter J., and Andrée Rosenfeld. *Palaeolithic Cave Art.* New York: McGraw-Hill; London: Weidenfeld and Nicolson, 1967. **55**

————, ed. *Form in Indigenous Art: Schematisation in the Art of Aboriginal Australia and Prehistoric Europe.* Australian Institute of Aboriginal Studies, Prehistory and Material Culture Series no. 13. London: Gerald Duckworth, 1977. **60, 69**

Ugolini, Luigi M. *Malta: Origini della civiltà mediterranea.* Rome: Libreria dello Stato, 1934. **83**

Uhden, Richard. "Gervasius von Tilbury und die Ebstorfer Weltkarte." *Jahrbuch der Geographischen Gesellschaft zu Hannover* (1930): 185–200. **307**

————. "Zur Herkunft und Systematik der mittelalterlichen Weltkarten." *Geographische Zeitschrift* 37 (1931): 321–40. **287, 295, 296, 368**

————. "Bemerkungen zu dem römischen Kartenfragment von Dura Europos." *Hermes* 67 (1932): 117–25. **249**

————. "Die antiken Grundlagen der mittelalterlichen Seekarten." *Imago Mundi* 1 (1935): 1–19. **381, 384**

Unger, Eckhard. *Babylon, die heilige Stadt nach der Beschreibung der Babylonier.* Berlin: Walter de Gruyter, 1931. **110, 112**

————. "Ancient Babylonian Maps and Plans." *Antiquity* 9 (1935): 311–22. **88**

————. "From the Cosmos Picture to the World Map." *Imago Mundi* 2 (1937): 1–7. **88, 91, 95, 96, 112**

United Nations. *Modern Cartography: Base Maps for World*

Needs. Document no. 1949.I.19. New York: United Nations Department of Social Affairs, 1949. **xvii**

Urry, William. "Canterbury, Kent, *circa* 1153 × 1161." In *Local Maps and Plans from Medieval England,* ed. R. A. Skelton and P. D. A. Harvey, 43–58. Oxford: Clarendon Press, 1986. **467, 484**

———. "Canterbury, Kent, Late 14th Century × 1414." In *Local Maps and Plans from Medieval England,* ed. R. A. Skelton and P. D. A. Harvey, 107–17. Oxford: Clarendon Press, 1986. **493**

Ustick, W. Lee. "Parchment and Vellum." *Library,* 4th ser., 16 (1935): 439–43. **324**

Uzielli, Gustavo, and Pietro Amat di San Filippo. *Mappamondi, carte nautiche, portolani ed altri monumenti cartografici specialmente italiani dei secoli XIII–XVII.* 2d ed., 2 vols. Studi Biografici e Bibliografici sulla Storia della Geografia in Italia. Rome: Società Geografica Italiana, 1882; reprinted Amsterdam: Meridian, 1967. **294, 374, 391, 401, 421, 422, 423, 434, 435**

Vacano, Otto-Wilhelm von. *The Etruscans in the Ancient World.* Trans. Sheila Ann Ogilvie. London: Edward Arnold, 1960. **203**

Valerio, Vladimiro. "La cartografia Napoletana tra il secolo XVIII e il XIX: Questioni di storia e di metodo." *Napoli Nobilissima* 20 (1980): 171–79. **38**

———. "Per una diversa storia della cartografia." *Rassegna ANIAI* 3, no. 4 (1980): 16–19 (periodical of the Associazione Nazionale Ingegneri e Architetti d'Italia). **38**

———. "A Mathematical Contribution to the Study of Old Maps." In *Imago et mensura mundi: Atti del IX Congresso Internazionale di Storia della Cartografia,* 2 vols., ed. Carla Clivio Marzoli, 2:497–504. Rome: Enciclopedia Italiana, 1985. **38**

———. "Sulla struttura geometrica di alcune carte di Giovanni Antonio Rizzi Zannoni (1736–1814)." Published as offprint only. **38**

Vandier, Jacques. *Manuel d'archéologie égyptienne.* 6 vols. Paris: A. et J. Picard, 1952–78. **125**

Vaughan, Richard. *Matthew Paris.* Cambridge: Cambridge University Press, 1958. **288, 347, 470, 473, 475, 495, 496**

Venice, Biblioteca Nazionale Marciana and Archivio di Stato. *Mostra dei navigatori veneti del quattrocento e del cinquecento.* Exhibition catalog. Venice, 1957. **433, 448**

Verbrugghe, Gerald Sicilia, Ingemar König, and Gerold Walser, eds. *Itinera Romana.* 3 vols. Bern: Kümmerly und Frey, 1967–76. **236**

Verner, Coolie. "The Identification and Designation of Variants in the Study of Early Printed Maps." *Imago Mundi* 19 (1965): 100–105. **21**

———. "Carto-bibliographical Description: The Analysis of Variants in Maps Printed from Copper Plates." *American Cartographer* 1 (1974): 77–87. **21**

Vernet-Ginés, Juan. "The Maghreb Chart in the Biblioteca Ambrosiana." *Imago Mundi* 16 (1962): 1–16. **418, 445, 459**

Viaene, A. "De landmeter in Vlaanderen, 1281–1800." *Biekorf* 67 (1966): 7. **494**

Vicenza, Biblioteca Civica Bertoliana. *Teatro del cielo e della terra: Mappamondi, carte nautiche e atlanti della Biblioteca Civica Bertoliana dal XV al XVIII secolo: Catalogo della mostra.* Vicenza: Biblioteca Civica Bertoliana, 1984. **317**

Vietor, Alexander O. "A Portuguese Chart of 1492 by Jorge Aguiar." *Revista da Universidade de Coimbra* 24 (1971): 515–16. **374**

Virágh, Dénes. "A legrégibb térkép" (The oldest map). *Geodézia és Kartográfia* 18, no. 2 (1965): 143–45. **96**

Vittmann, Günther. "Orientierung (von Gebäuden)." In *Lexikon der Ägyptologie,* ed. Wolfgang Helck and Eberhard Otto, 4:cols. 607–9. Wiesbaden: O. Harrassowitz, 1975–. **126**

Vleeming, S. P. "Demotic Measures of Length and Surface, chiefly of the Ptolemaic Period." In *Textes et etudes de papyrologie grecque, démotique et copte,* P. W. Pestman et al., 208–29. Papyrologica Lugduno-Batava 23. Leiden: E. J. Brill, 1985. **125**

Volpicella, Luigi. "Genova nel secolo XV: Note d'iconografia panoramica." *Atti della Società Ligure di Storia Patria* 52 (1924): 255–58. **398**

Vries, Dirk de. "Atlases and Maps from the Library of Isaac Vossius (1618–1689)." *International Yearbook of Cartography* 21 (1981): 177–93. **9**

Vrij, Marijke de. *The World on Paper: A Descriptive Catalogue of Cartographical Material Published in Amsterdam during the Seventeenth Century.* Amsterdam: Theatrum Orbis Terrarum, 1967. **21**

Waard, C. de. *Rijksarchief in Zeeland: Inventaris van kaarten en teekeningen.* Middelburg: D. G. Kröber, Jr., 1916. **486, 500**

Wachsmuth, Carl. *De Crate Mallota.* Leipzig, 1860. **163**

Waerden, B. L. van der. "Mathematics and Astronomy in Mesopotamia." In *Dictionary of Scientific Biography,* 16 vols., ed. Charles Coulston Gillispie, 15:667–80. New York: Charles Scribner's Sons, 1970–80. **115**

Wagner, Hermann. "The Origin of the Mediaeval Italian Nautical Charts." In *Report of the Sixth International Geographical Congress, London, 1895,* 695–702. London: Royal Geographical Society, 1896; reprinted in *Acta Cartographica* 5 (1969): 476–83. **414, 428**

Waitz, Georg, ed. *Annales Bertiniani, Scriptores rerum Germanicorum: Monumenta Germanicae historica.* Hanover: Impensis Bibliopolii Hahniani, 1883. **303**

Walbank, Frank William. *A Historical Commentary on Polybius.* 3 vols. Oxford: Clarendon Press, 1957–79. **162**

Wallace-Hadrill, Andrew. Review of *Die Sonnenuhr des Augustus: Nachdruck aus RM 1976 und 1980 und Nachtrag über die Ausgrabung 1980/1981,* by Edmund Buchner. *Journal of Roman Studies* 75 (1985): 246–47. **208**

Wallis, Helen. "The Map Collections of the British Museum Library." In *My Head Is a Map: Essays and Memoirs in Honour of R. V. Tooley,* ed. Helen Wallis and Sarah Tyacke, 3–20. London: Francis Edwards and Carta Press, 1973. **8**

———. "Maps as a Medium of Scientific Communication."

In *Studia z dziejów geografii i kartografii: Etudes d'histoire de la géographie et de la cartographie,* ed. Józef Babicz, 251–62. Monografie z Dziejów Nauki i Techniki, vol. 87. Warsaw: Zakład Narodowy Imienia Ossolińskich Wydawnictwo Polskiej Akademii Nauk, 1973. **36**

———. "Working Group on the History of Cartography." *International Geographical Union Bulletin* 25, no. 2 (1974): 62–64. **33**

———. "The Royal Map Collections of England." *Publicaciónes do Centro de Estudos de Cartografia Antiga.* Série Separatas, 141. Coimbra, 1981. **8, 16**

———. "Cartographic Innovation: An Historical Pespective." In *Canadian Institute of Surveying Centennial Convention Proceedings,* 2 vols., 2:50–63. Ottawa: Canadian Institute of Surveying, 1982. **12**

———. "The Rotz Atlas: A Royal Presentation." *Map Collector* 20 (1982): 40–42. **434**

Wallis, Helen, et al. "The Strange Case of the Vinland Map: A Symposium." *Geographical Journal* 140 (1974): 183–214. **368**

Wallis, Helen, ed. *The Maps and Text of the Boke of Idrography Presented by Jean Rotz to Henry VIII, Now in the British Library.* Oxford: Viscount Eccles for the Roxburghe Club, 1981. **434, 443**

Wallis, Mieczyslaw. "Semantic and Symbolic Elements in Architecture: Iconology as a First Step towards an Architectural Semiotic." *Semiotica* 8 (1973): 220–38. **340**

Walters, Gwyn. "Richard Gough's Map Collecting for the British Topography 1780." *Map Collector* 2 (1978): 26–29. **11, 12**

Walzer, Richard. *Arabic Transmission of Greek Thought to Medieval Europe.* Manchester: Manchester University Press, 1945. **304**

Warmington, E. H. "Posidonius." In *Dictionary of Scientific Biography,* 16 vols., ed. Charles Coulston Gillispie, 11:104. New York: Charles Scribner's Sons, 1970–81. **168**

———. "Strabo." In *Dictionary of Scientific Biography,* 16 vols., ed. Charles Coulston Gillispie, 13:83–86. New York: Charles Scribner's Sons, 1970–80. **173**

Warren, Peter. "The Miniature Fresco from the West House at Akrotiri, Thera, and Its Aegean Setting." *Journal of Hellenic Studies* 99 (1979): 115–29. **132**

Waterbolk, E. H. "Viglius of Aytta, Sixteenth Century Map Collector." *Imago Mundi* 29 (1977): 45–48. **9**

Waters, David W. *The Art of Navigation in England in Elizabethan and Early Stuart Times.* London: Hollis and Carter, 1958. **385, 409, 443**

———. *The Rutters of the Sea: The Sailing Directions of Pierre Garcie—A Study of the First English and French Printed Sailing Directions.* New Haven: Yale University Press, 1967. **387**

———. *Science and the Techniques of Navigation in the Renaissance.* 2d ed. Maritime Monographs and Reports no. 19. Greenwich: National Maritime Museum, 1980. **384**

Watson, Andrew G. *Catalogue of Dated and Datable Manuscripts c. 700–1600 in the Department of Manuscripts, the British Library.* London: British Library, 1979. **436**

Watts, Pauline Moffitt. "Prophecy and Discovery: On the Spiritual Origins of Christopher Columbus's 'Enterprise of the Indies.' " *American Historical Review* 90 (1985): 73–102. **354**

Webber, F. R. *Church Symbolism.* Cleveland: J. H. Jansen, 1927. **335**

Weber, Ekkehard, ed. *Tabula Peutingeriana: Codex Vindobonensis 324.* Graz: Akademische Druck- und Verlagsanstalt, 1976. **7, 238, 469**

Weidner, Ernst F. *Handbuch der babylonischen Astronomie, der babylonische Fixsternhimmel.* Leipzig: Hinrichs, 1915; reprinted Leipzig: Zentralantiquariat, 1976. **115**

Weiss, Roberto. *The Renaissance Discovery of Classical Antiquity.* Oxford: Blackwell, 1969. **477, 492, 495**

Weiss und Co., Antiquariat. *Codices manuscripti incunabula typographica, catalogus primus.* Munich: Weiss, 1926. **396, 457**

Welland, James. *The Search for the Etruscans.* London: Nelson, 1973. **203**

Wellisch, S. "Der älteste Plan von Wien." *Zeitschrift des Oesterreichischen Ingenieur- und Architekten-Vereines* 50 (1898): 757–61. **473**

Wellmann, Klaus F. "Rock Art, Shamans, Phosphenes and Hallucinogens in North America." *Bollettino del Centro Camuno di Studi Preistorici* 18 (1981): 89–103. **87**

Welu, James A. "The Sources of Cartographic Ornamentation in the Netherlands." In *Art and Cartography: Six Historical Essays.* Ed. David Woodward. Chicago: University of Chicago Press, 1987. **339**

Wendel, Carl. "Planudes, Maximos." In *Paulys Realencyclopädie der classischen Altertumswissenschaft,* ed. August Pauly, Georg Wissowa, et al., 20.2 (1950): cols. 2202–53. Stuttgart: J. B. Metzler, 1894–. **268**

West, Martin Litchfield. *Hesiod, Works and Days: Edited with Prolegomena and Commentary.* Oxford: Clarendon Press, 1976. **85**

Westedt, Amtsgerichtsrath. "Steinkammer mit Näpfchenstein bei Bunsoh, Kirchspiel Albersdorf, Kreis Süderdithmarschen." *Zeitschrift für Ethnologie* 16 (1884): Verhandlungen 247–49. **54, 66**

Westropp, Thomas Johnson. "Brasil and the Legendary Islands of the North Atlantic: Their History and Fable. A Contribution to the 'Atlantis' Problem." *Proceedings of the Royal Irish Academy,* vol. 30, sect. C (1912–13): 223–60; reprinted in *Acta Cartographica* 19 (1974): 405–45. **407**

———. "Early Italian Maps of Ireland from 1300 to 1600 with Notes on Foreign Settlers and Trade." *Proceedings of the Royal Irish Academy,* vol. 30, sect. C (1912–13): 361–428; reprinted in *Acta Cartographica* 19 (1974): 446–513. **407**

Wheatley, Paul. *The Golden Khersonese: Studies in the Historical Geography of the Malay Peninsula before A.D. 1500.* Kuala Lumpur: University of Malaya Press, 1961. **198**

Whitehouse, Helen. *The Dal Pozzo Copies of the Palestrina Mosaic.* British Archaeological Reports, Supplementary Series 12. Oxford: British Archaeological Reports, 1976. **118**

Wichmann, H. "Geographische Gesellschaften, Zeitschriften, Kongresse und Ausstellungen." *Geographisches Jahrbuch* 10 (1884): 651–74. **14**

Wieder, Frederik Caspar. *Nederlandsche historisch-geographische documenten in Spanje.* Leiden: E. J. Brill, 1915. **486**

Wieser, Franz R. von. "A. E. v. Nordenskiöld's Periplus." *Petermanns Mitteilungen* 45 (1899): 188–94. **375**

———. *Die Weltkarte des Albertin de Virga aus dem Anfange des XV. Jahrhunderts in der Sammlung Figdor in Wien.* Innsbruck: Schurich, 1912; reprinted in *Acta Cartographica* 24 (1976): 427–40. **448**

Wildung, Dieter. "Garten." In *Lexikon der Agyptologie,* ed. Wolfgang Helck and Eberhard Otto, 2:cols. 367–78. Wiesbaden: O. Harrassowitz, 1975–. **118**

Wilford, John Noble. *The Mapmakers.* New York: Alfred A. Knopf; London: Junction Books, 1981. **4, 25, 371**

Wilkinson, J. Gardner. "The Rock-Basins of Dartmoor, and Some British Remains in England." *Journal of the British Archaeological Association* 16 (1860): 101–32. **64**

Willcock, Malcolm M. *A Companion to the Iliad.* Chicago: University of Chicago Press, 1976. **131**

Williamson, J. A. *The Voyages of John and Sebastian Cabot.* Historical Association Pamphlet no. 106. London: G. Bell, 1937. **3**

Wilson, David McKenzie, and Ole Klindt-Jensen. *Viking Art.* London: George Allen and Unwin, 1966. **91**

Wilson, N. G. *Scholars of Byzantium.* London: Duckworth, 1983. **268**

Winkler, Hans Alexander. *Rock Drawings of Southern Upper Egypt.* 2 vols. Egyptian Exploration Society. London: Oxford University Press, 1938. **63**

Winter, Heinrich. "Das katalanische Problem in der älteren Kartographie." *Ibero-Amerikanisches Archiv* 14 (1940/41): 89–126. **393, 457**

———. "Scotland on the Compass Charts." *Imago Mundi* 5 (1948): 74–77. **385**

———. "The True Position of Hermann Wagner in the Controversy of the Compass Chart." *Imago Mundi* 5 (1948): 21–26. **422**

———. "A Late Portolan Chart at Madrid and Late Portolan Charts in General." *Imago Mundi* 7 (1950): 37–46. **384, 395, 423**

———. "Petrus Roselli." *Imago Mundi* 9 (1952): 1–11. **398, 431, 435, 457**

———. "Catalan Portolan Maps and Their Place in the Total View of Cartographic Development." *Imago Mundi* 11 (1954): 1–12. **14, 389, 414, 415, 424, 457**

———. "The Changing Face of Scandinavia and the Baltic in Cartography up to 1532." *Imago Mundi* 12 (1955): 45–54. **409**

———. "The Fra Mauro Portolan Chart in the Vatican." *Imago Mundi* 16 (1962): 17–28. **411, 424**

Wiseman, Donald J., ed. *Peoples of Old Testament Times.* Oxford: Clarendon Press, 1973. **115**

Wittgenstein, Ludwig. *Tractatus Logico-Philosophicus.* Trans. D. F. Pears and B. F. McGuinness. London: Routledge and Kegan Paul, 1961. **51**

Wittkower, Rudolf. "Marvels of the East: A Study in the History of Monsters." *Journal of the Warburg and Courtauld Institutes* 5 (1942): 159–97. **330**

Wolkenhauer, A. "Über die ältesten Reisekarten von Deutschland aus dem Ende des 15. und dem Anfange des 16. Jahrhunderts." *Deutsche Geographische Blätter* 26 (1903): 120–38. **497**

Wolska, Wanda. *La topographie chrétienne de Cosmas Indicopleustès: Théologie et science au VIe siècle.* Bibliothèque Byzantine, Etudes 3. Paris: Presses Universitaires de France, 1962. **143, 261, 262**

Wolska-Conus, Wanda. "Deux contributions à l'histoire de la géographie: I. La diagnôsis Ptoléméenne; II. La 'Carte de Théodose II.'" In *Travaux et mémoires,* 259–79. Centre de Recherche d'Histoire et Civilisation Byzantines, 5. Paris: Editions E. de Baccard, 1973. **259**

Wolter, John A. "Geographical Libraries and Map Collections." In *Encyclopedia of Library and Information Science,* ed. Allen Kent, Harold Lancour, and Jay E. Daily, 9:236–66. New York: Marcel Dekker, 1968–. **15**

———. "The Emerging Discipline of Cartography." Ph.D. Diss. University of Minnesota, 1975. **15, 23, 30, 31, 32, 33**

———. "Research Tools and the Literature of Cartography." *AB Bookman's Yearbook,* pt. 1 (1976): 21–30. **19, 20**

Wolter, John A., et al. "A Brief History of the Library of Congress Geography and Map Division, 1897–1978." In *The Map Librarian in the Modern World: Essays in Honour of Walter W. Ristow,* ed. Helen Wallis and Lothar Zögner, 47–105. Munich: K. G. Saur, 1979. **16**

Wolter, John A., Ronald E. Grimm, and David K. Carrington, eds. *World Directory of Map Collections.* International Federation of Library Associations Publication Series no. 31. Munich: K. G. Saur, 1985. **8**

Wood, Denis. Review of *The History of Topographical Maps: Symbols, Pictures and Surveys,* by P. D. A. Harvey. *Cartographica* vol. 17, no. 3 (1980): 130–33. **38**

———. Review of *The Mapmakers,* by John Noble Wilford. *Cartographica* 19, nos. 3–4 (1982): 127–31. **4, 26**

Woodburn, James. "An Introduction to the Hadza Ecology." In *Man the Hunter,* ed. Richard B. Lee and Irven DeVore, 49–55. Chicago: Aldine, 1968. **86**

Woodward, David. "The Study of the History of Cartography: A Suggested Framework." *American Cartographer* 1, no. 2 (1974): 101–15. **xvii, 25, 35, 36, 38**

———. "The Form of Maps: An Introductory Framework." *AB Bookman's Yearbook,* pt. 1 (1976), 11–20. **35**

———. *The Hermon Dunlap Smith Center for the History of Cartography: The First Decade.* Chicago: Newberry Library, 1980. **37**

———. "Reality, Symbolism, Time, and Space in Medieval World Maps." *Annals of the Association of American Geographers* 75 (1985): 510–21. **288, 290, 318, 319**

———. "The Manuscript, Engraved, and Typographic Traditions of Map Lettering." In *Art and Cartography: Six Historical Essays.* ed. David Woodward. Chicago: University of Chicago Press, 1987. **325**

———, ed. *Five Centuries of Map Printing.* Chicago:

University of Chicago Press for the Newberry Library, 1975. **17**

Worringer, Wilhem. *Abstraction and Empathy: A Contribution to the Psychology of Style.* Trans. Michael Bullock. London: Routledge and Kegan Paul, 1953. **86**

Wosien, Maria-Gabriele. *Sacred Dance: Encounter with the Gods.* New York: Avon Books; London: Thames and Hudson, 1974. **87**

[Wright, John K.?]. "Three Early Fifteenth Century World Maps in Siena." *Geographical Review* 11 (1921): 306–7. **324**

Wright, John Kirtland. "Notes on the Knowledge of Latitudes and Longitudes in the Middle Ages." *Isis* 5 (1922): 75–98. **323**

————. *The Geographical Lore of the Time of the Crusades: A Study in the History of Medieval Science and Tradition in Western Europe.* American Geographical Society Research Series no. 15. New York: American Geographical Society, 1925; republished with additions, New York: Dover Publications, 1965. **288, 293, 295, 306, 323, 342**

————. *The Leardo Map of the World, 1452 or 1453, in the Collections of the American Geographical Society.* American Geographical Society Library Series, no. 4. New York, 1928. **317**

Wright, John Kirtland, and Elizabeth T. Platt. *Aids to Geographical Research: Bibliographies, Periodicals, Atlases, Gazetteers and Other Reference Books.* 2d ed. American Geographical Society Research Series no. 22. New York: Columbia University Press for the American Geographical Society, 1947. **31**

Wroth, Warwick. *A Catalogue of the Greek Coins of Crete and the Aegean Islands.* Ed. Reginald Stuart Poole. Bologna: A. Forni, 1963. **251**

Yates, Frances A. *The Art of Memory.* London: Routledge and Kegan Paul, 1966. **48**

Yates, W. N. "The Authorship of the Hereford Mappa Mundi and the Career of Richard de Bello." *Transactions of the Woolhope Naturalist's Field Club* 41 (1974): 165–72. **312**

Yoeli, Pinhas. "Abraham and Yehuda Cresques and the Catalan Atlas." *Cartographic Journal* 7 (1970): 17–27. **315**

Žába, Zbyněk. *L'orientation astronomique dans l'ancienne Egypte et la précession de l'axe du monde.* Archiv Orientálni, suppl. 2. Prague: Editions de l'Académie Tchécoslovaque des Sciences, 1953. **126**

Zammit, Themistocles. *The Neolithic Temples of Hal-Tarxien-Malta.* 3d ed. Valletta: Empire Press, 1929. **83**

————. *Prehistoric Malta: The Tarxien Temples.* London: Oxford University Press, 1930. **81, 96**

Zanetti, Girolamo Francesco. *Dell'origine di alcune arti principali appresso i Veneziani.* 2 vols. Venice: Stefano Orlandini, 1758. **11**

Zelinsky, Wilbur. "The First and Last Frontier of Communication: The Map as Mystery." *Bulletin of the Geography and Map Division, Special Libraries Association* 94 (1973): 2–8. **3**

Zicàri, Italo. "L'anemoscopio Boscovich del Museo Oliveriano di Pesaro." *Studia Oliveriana* 2 (1954): 69–75. **248**

Ziegler, Konrat, and Walther Sontheimer, eds. *Der kleine Pauly.* 5 vols. Stuttgart: Alfred Druckenmüller, 1964–75. **236**

Zögner, Lothar. "Die Kartenabteilung der Staatsbibliothek, Bestände und Aufgaben." *Jahrbuch Preußischer Kulturbesitz* 14 (1977): 121–32. **20**

————. "Die Carl-Ritter-Ausstellung in Berlin—eine Bestandsaufnahme." In *Carl Ritter—Geltung und Deutung,* ed. Karl Lenz, 213–23. Berlin: Dietrich Reimer Verlag, 1979. **16**

————. "25 Jahre 'Bibliographia Cartographica.' " *Zeitschrift für Bibliothekswesen und Bibliographie* 29 (1982): 153–56. **31, 32**

General Index

Italic page numbers indicate that the topic appears in an illustration or in its caption on the cited page; the topic may also appear in the text of that page.

Authors are listed in this index only when their ideas or works are discussed; full listings of their works as cited in this volume may be found in the Bibliographical Index.

Mola di Bari, 426

Monastir, 198

Money, Roman, 222 n.42

Monfalcone, 427

Mongols, 304

Monstrous races, *291*, 307, *330–32*, 334
  placement of, 316, 332

Mont-Saint-Michel, 330

Montélimar, 223

Montfarcom (Monfalcone), 427

Montpellier, on portolan charts, 400

Monumenta cartographica Africae et Aegypti (Kamal), 18, 294

Monumenta cartographica Europea, 18 n.141

Monuments, megalithic, 81

Moon
  on Achilles' shield, 131 n.6, 132
  in Archimedes' planetarium, 160
  in Greek cosmography, 138, 164 n.13
  paschal, 429
  phases of, on *mappaemundi*, 317
  on Piacenza liver, 203, 204

Moon, Mountains of the, 328, 358

Moordorf (Germany), *91*

Moore, George, 81

Moors, 444

Morel, Jehan, 374

Moriduno, 238

Morison, Samuel Eliot, 410

Morocco
  Great Disk from, 71–72, *73*
  on portolan charts, 415

Morris, Ronald W. B., 68

Morsynas River, 158

Mosaics, 171
  Byzantine, 263. *See also* Madaba mosaic
    at Nicopolis, 261, *264*
  Hebrew, 248, 266, *267*
  of labyrinths, 252
  losses and survivals, 106
  mappaemundi, 324, *339*
  Palestrina (Barberini) mosaic, 118 n.4, 246 n.75
  Roman, 226, 230, *231*, 246–48, 254, *339*

Moses, 11, 241

Motion, circular, 136

Motzo, Bacchisio R., 382, 383

Mount Zion, Church of, 466

Mountains
  Chinese character for, 60 n.33
  in Christian cosmography, 262
  Eastern (Egyptian), *89*
  on *mappaemundi*, 325, 326
  on portolan charts, 393
  in prehistoric maps, 71, 72, *73*
  on Ptolemaic maps, 269
  on Roman maps, 217, 239
  squares representing, 114
  triangles representing, 71, 72
  Western (Egyptian), *89*

"Mozarabic" style, 304, 326

Muller, Frederick, 16, 17, 22 n.177

Müller, Karl, 237

Mundus, 287

Munich, Portuguese chart in, 374 n.32, 386

Munich, Universitätsbibliothek, Codex MS. 185, 386

Münster, Sebastian, 332 n.223

Murano Island, 315

al-Mursī, Ibrāhīm, 434 n.435, 437 n.474

Musei Capitolini (Rome), 226

Museo Egizio (Turin), 121

Museo Storico Navale (Venice), portolan chart fragments, 419, 421 n.338

Muses
  in art, with globe, 171
  Athenian shrine of, 158

Museum of History of Science (Florence), 12 n.93

Music, 52

Muslims, 304, 307, 444

Mussolini, Benito, 254

Müstinger, Georg, 316

Mycenaeans, 251

Mythology
  Asian, 84 n.141
  Egyptian, 117, 120
  Greek, 138, 330
  Scandinavian, 91
  Sumerian, 86

Nablus, 466

Naiera, António de, 441

Naples, 195, 432 n.421, 434, 437

Naples, Bay of, 239 n.34, 240

Naqshah, xvi n.7

Naram-Sin, 107

Narbo, 162

Narbonne, 162

Narenta, Ibz de, *402*

Narona, 239

Nationalism
  in cartographic history, 28
  maps in shaping of, 14
  and portolan charts, 14, 388, 392

Nations
  on *mappaemundi*, 328
  Table of (Genesis 10), 115

Naupaktos (Lepanto), 425

Navigation
  astronomical, 85, 92, 276, 386, 441
  compass (magnetic) in, 384
  by dead reckoning, 386, 441
  history of, and histories of exploration, 17, 18
  by indigenous peoples, 47, 48, 59, 85
  instruments, 384, 386, 387, 429 n.392, 441. *See also individual instruments*
  Lull's contribution to, 305
  in Mediterranean, 85, 386–87, 388, 441, 443
  metaphor in Plato's *Republic*, 138
  and portolan charts, 284, 439–44
    "circle and square" diagram, *442*
    compilation, 386–87
    flags, 401

markings indicating use, 440, 443–44
    navigational signs, 378
    Roman, 253
    and winds, 146
    without charts, in northern waters, 409

Navigators, and Sinus Magnus, 199

Nazareth, 330

Near East, 106
  cartography, 107–15
    and prehistoric maps, 503
    sites, *108*

Nearchus, 149

Neat line, 505

Nebenzahl, Kenneth, 22 n.177

Nebo, Mount, 264

Necho, 136

Neckham, Alexander, 384

Necos, 136

Nederlandsche Vereniging voor Kartografie, 33 n.267

Needham, Joseph, 60 n.33, 496

Neolithic period, 55, 57. *See also* Borno stone; Çatal Hüyük
  cosmological beliefs, 86
  Hal Saflien temple, 87
  Mont Bégo figures dated to, 75
  Ossimo stela, 90
  plan maps, 62 n.49
  planimetric and vertical projections, 70 n.83
  portolan charts as deriving from, 380
  Saharan rock art, 69
  sculptured block from Tarxien, 81
  spirals, 88 n.176
  Tal Qadi "star stone," 83
  tombs, 72

Nepanto. See Lepanto

Nepos, Cornelius, 242

Le Neptune Français, 8 n.54

Neretva River, 239

Nero, 212, 240, 245, *246*, 253

Nerva, Marcus Cocceius, 232, 337 n.249

Nestor, palace of, 251

Netherlands. *See also* Low Countries
  boundary with Picardy, 485, *487*
  cartographic history in, 14 n.107, 36, 37, 38
  medieval maps, 493
    oldest extant, 470, *471*
  Rijksarchief, catalog of maps, 19
  survey in, 494

Neugebauer, Otto, xviii n.21

New Hebrides, 88

New World
  ethnographies, 47
  indigenous cartography, 49
  maps of, Humboldt's interest in, 17
  prehistoric period, 49

Newberry Library, Hermon Dunlap Smith Center for the History of Cartography, 37

Newcastle upon Tyne, 496

Niaux, 55 n.8

Nicaea, 239, 330